AF352059
PROPERTY OF THE U.S. ARMY
REDSTONE SCIENTIFIC INFORMATION CENTER
REDSTONE ARSENAL, ALABAMA

Conference Record
of the
1991 IEEE
Particle Accelerator
Conference

Accelerator Science and Technology

May 6–9, 1991
San Francisco, California

Volume 3 of 5

Organized by
Lawrence Berkeley Laboratory
Stanford Linear Accelerator Center
with Assistance of
Los Alamos National Laboratory

Under the Auspices of
Institute of Electrical and Electronics Engineers—
Nuclear and Plasma Sciences Society

Sponsored by
Department of Energy
  Offices of High Energy and Nuclear Physics
  Superconducting Super Collider
  Basic Energy Sciences
National Science Foundation
Office of Naval Research

IEEE Catalog Number 91CH3038-7
ISBN 0-7803-0135-8 (Softbound)
ISBN 0-7803-0136-6 (Casebound)
ISBN 0-7803-0137-4 (Microfiche)
Library of Congress Number 88-647453

Additional copies are available from

IEEE Service Center
445 Hoes Lane
P.O. Box 1331
Piscataway, NJ 08854-1331
1-800-678-IEEE

Cover Artwork by Sylvia MacBride, SLAC
Conference Photos by Tom Nakashima, SLAC

TABLE OF CONTENTS

Volume 1

Opening Plenary Session

Chairman: K. Berkner

Accelerator Technology III–Superconducting Components, Magnets

Chairman: P. Mantsch

Low- and Medium-Energy Accelerators and Rings

Chairman: R. Richardson

High Energy Accelerators and Colliders

Beam Dynamics II

Applications and New Methods of Acceleration

Applications Chairman: R. W. Hamm

New Methods

New Methods Chairman: D. Sutter

Linear Accelerators and Pulsed Power Devices

Chairman: B. Jameson

Accelerator Technology II–RF, Power Supplies, Operations

Volume 2

Beam Dynamics II

Chairman: R. Gluckstern

Synchrotron Radiation Sources/FELs

Chairman: H. Winick

Accelerator Technology I—Instrumentation, Control, Feedback

Volume 3

Beam Dynamics I

Ion Sources and Injectors

Accelerator Technology III—Superconducting Components, Magnets

Beam Dynamics I

Chairman: A. Chao

Accelerator Technology I—Instrumentation, Control, Feedback

Chairman: J. Hinkson

Applications and New Methods of Acceleration

Synchrotron Radiation Sources/FELs

Volume 5

Low- and Medium-Energy Accelerators and Rings

High-Energy Accelerators and Colliders

Chairman: W. Wallenmeyer

Accelerator Technology II, RF, Power Supplies, Operations

Chairman: R. L. Kustom

Linear Accelerators and Pulsed Power Devices

Linear Colliders

Closing Plenary Session

Chairman: J. O'Fallon

Author Index Follows Pages

Beam Diagnostics at the COSY Injection Beamline

F. Anton
Interatom GmbH
W-5060 Bergisch Gladbach 1
Germany

Abstract

The 100 m long injection beamline at KFA Jülich from the 45 MeV/n cyclotron JULIC to the 2.5 GeV cooler ring COSY is presently under construction. We present the beam diagnostic concept, which is related to the particle optics and to the control system. The diagnostic instrumentation comprises slits, wire grids, viewers, faraday cups and phase probes. Measurements of the cyclotron beam properties, like emittance and energy spread, yield a full knowledge of the injection parameters into COSY under variable conditions.

Introduction

The cyclotron JULIC [1], which is operating at KFA since 1969, will be used as an injector for the Cooler Synchrotron COSY [2]. The ions with a maximum energy of 45 MeV/n will be transported to COSY by a beam transport line with a length of approximately 100 m, the path beeing chosen in order to leave the maximum possible space for the experimental areas. Interatom in 1989 got the contract for the design, construction, and commissioning of the beamline, the diagnostic system and all other subsystems.

The beamline matches the phase space of the cyclotron beam to the acceptance of the storage ring, providing the needed flexibility for different tune values and injection modes of COSY. The beam diagnostic instrumentation system allows to measure the displacements and the phase space of the beam along the line and is the basis for the injection tuning.

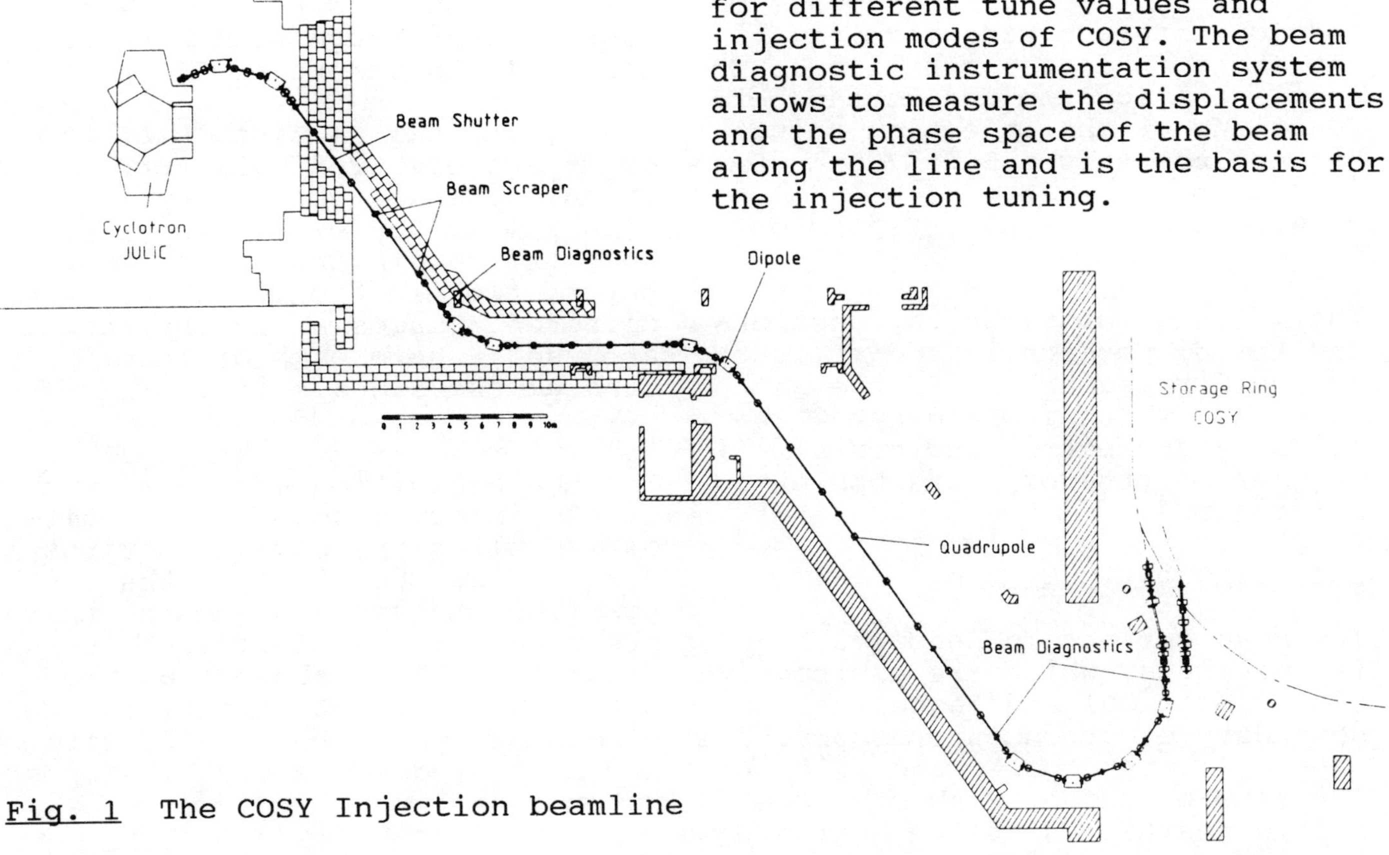

Fig. 1 The COSY Injection beamline

Layout of the Beamline

The layout of the beamline is shown in Fig. 1. It consists of the following subsections: The cyclotron achromat, a FODO transfer line, the transfer achromat, another FODO line, the injection achromat, and the injection section.

The beamline consists of six dipole magnets with a deflecting angle of $38.25°$ and four dipoles with an angle of $27.5°$. The 39 quadrupoles of the beamline have a magnetic length of 300 mm, the maximum gradient used is 14.5 T/m. For orbit correction, 12 x/y-steerer magnets are placed along the beam line.

The cyclotron beam has a rigidity between 0.9 and 2 Tm, an emittance of 3.2 π mrad mm horizontally resp. 6.4 π mrad mm vertically, a Dispersion of D_x = -0.12 m and $D_{x'}$ = -3.35 mrad/%, and a momentum spread of $\delta p/p$ = 0.15 %.

The different operational modes of the COSY storage ring require different injection conditions: The beamline allows to adjust the beta functions at the injection point between the values β_x = 7.0 m, β_y = 15.4 m and β_x = 15.4 m, β_y = -7.0 m, with a compensated dispersion function.

The matching of the beta functions and the dispersion is accomplished by the first achromat of the beamline, which is laid out slightly asymmetric and has six degrees of freedom (quadrupole parameters).

Beam Diagnostic System

The beam current is measured by faraday cups, which are equipped with electrical and magnetical secondary electron suppressors. The cups are water cooled and laid out for a beam load of 2 kW c.w. One cup is positioned at the beginning of the beamline, which at the same time serves as a security beam shutter. A second cup at this location, working as a security shutter only, gives the necessary redundancy. At the end of the first FODO line, and at the end of the injection achromat, respectively, further faraday cups are installed to control the beam transmission along the line.

The beamline is equipped with eigth x/y-beam profile monitors, located at the starting point of the beamline and at the end points of the seven subsections (achromats and FODO structures, respectively). Wire grid harps with 0.1 mm Tungsten-Rhenium wires are used. For each direction (x and y), a grid consists of a number of 39 wires, spaced by 1.5 mm each, thus giving a reasonable spatial resolution of the position and of the intensity profile of the beam. The wires are mounted with Duratherm springs on a ceramic frame with an aperture of 60 x 60 mm^2. Each harp is mounted on a linear feedthrough with a membrane bellow and a stroke of 100 mm.

The electrical nA signals of the grids are electronically processed in multichannel charge amplifiers with FET multiplexers. In order to keep the cabel lengths below 50 meter, two electronic cabinets at different places along the beamline are used, where each of them handles the signals of four x/y-grids.

For the definition of the emittance of the injected beam, three pairs of x/y-slits are used. The first is directly at the start of the beamline, the second and the third are placed at two locations in the first FODO line, about a quarter betatron wavelength apart. The two blades of each water cooled slit are moved independently by stepper motors.

Three phase detection probes allow to measure the longitudinal phase space of the cyclotron. One probe is at the beginning of the beamline. A probe at the beginning of the long FODO section and a probe at its end form a time of flight section.

The injection beamline has its own computer control system, which is based on a workstation and a VME front end system, connected to the workstation via an Ethernet branch. The menu oriented software allows to display and modify interactively the beamline parameters and handles all measured data of the beam diagnostic system. Application programs, such as an orbit correction program and an emittance measurement program, have direct access to the central database, which keeps all parameters of the beamline.

Control of electric drives for diagnostic components, like stepper motors for the slits, is done directly by VME interface cards. The multichannel amplifier electronics of the diagnostic grids are adressed via V24.

With its beam optical properties, the beamline provides an excellent tool for the diagnostics of the beam quality which is delivered by the cyclotron and injected into COSY. The beam emittance will be determined with the help of the grids in the downstream part of the beam line by variation of quadrupole focussing lengths.

Vacuum Considerations

The vacuum system of the beamline with all its diagnostic boxes is fully metal sealed. In the first two thirds of the beamline, the vacuum conditions will be similar to the vacuum in the cyclotron, i. e. between 10^{-6} and 10^{-7} mbar. In the downstream end, however, the pressure has to be brought down to the UHV of the COSY ring, and at the end of the beamline a vacuum of better than 10^{-10} mbar will be maintained. Therefore, all vacuum materials in the injection achromat and in the following section are subject to special specifications and will be baked out during fabrication. Also, the vacuum system in the last part of the beamline will be baked out in situ.

The two beam profile grids and the faraday cup in this section are correspondingly laid out for bake out.

Status of the Project

All diagnostic components for the beamline are delivered. The development of the control system is completed. Installation of the magnets and the vacuum system has started. The first beam will be transported in autumn 1991.

References

[1] W. Bräutigam et al., Upgrading JULIC as Injector for COSY-Jülich, 1st Eur. Part. Accel. Conf., Rome 1988

[2] U. Pfister et al., The COSY-Jülich Project, 2nd Eur. Part. Accel. Conf., Nice 1990

Design and Testing of the AGS Booster BPM Detector[*]

R. Thomas, D. J. Ciardullo, and W. Van Zwienen
AGS Department
Brookhaven National Laboratory
Upton, New York 11973-5000

Abstract

The AGS Booster beam position monitor system must accurately measure the position of beams and bunches over a wide range of intensity. The frequency of operation must also cover a wide range (600 kHz to 4.2 MHz) since the Booster accelerates both protons and heavy ions. Split-cylinder electrodes were chosen to monitor the position of the beam because of their good low frequency response and high linearity. The need to accelerate low-energy partially-stripped heavy ions requires the pick-up electrodes (PUEs) to operate in a 3×10^{-11} torr vacuum. The PUEs are to measure the beam position to an absolute accuracy of ±0.5 mm and must therefore be mechanically stable despite the requirement that they be vacuum fired at 950 °C and baked periodically to 300 °C. This presentation describes both the mechanical design of the PUEs and the automated test procedure used to ensure the stability, accuracy, and linearity of each unit.

I. DESIGN

Each capacitive detector is made from a diagonally split cylinder (Fig. 1). The position monitor also incorporates a calibration ring which couples equally to both electrodes. The ring provides a simulation of the electrical zero position that allows the monitor to be checked [1]. (See Fig. 1.)

The completely assembled position monitor must be vacuum-fired, i.e., heated in a vacuum, at 950 °C to remove dissolved hydrogen from the stainless steel components in order to achieve the desired 10^{-11} torr vacuum. Therefore a special design was required to support the components that permitted thermal expansion to occur without creating forces which could cause deformation or fracture. Stainless steel grows linearly by about 0.018 mm/mm when heated from room temperature to 950 °C. Hence, a stainless cylinder 127 mm in diameter and 127 mm long will grow approximately 2.3 mm in diameter and length. Spring material could not be used because at 950 °C, the stiffness coefficient of even high temperature spring material decreases by 90%.

In addition to accommodating the large thermal expansion, the various components must be electrically isolated from one another. The dielectric material needs to have the following properties: very low outgassing and very low porosity, ability to withstand vacuum firing to 950 °C, and good structural strength. High-density alumina, a ceramic containing more than 92% Al_2O_3, is suitable. The thermal growth of alumina, however, is about half that of stainless steel over the same temperature range, so in the design, the stainless always surrounds the alumina so that the stainless grows away from the alumina as the temperature increases.

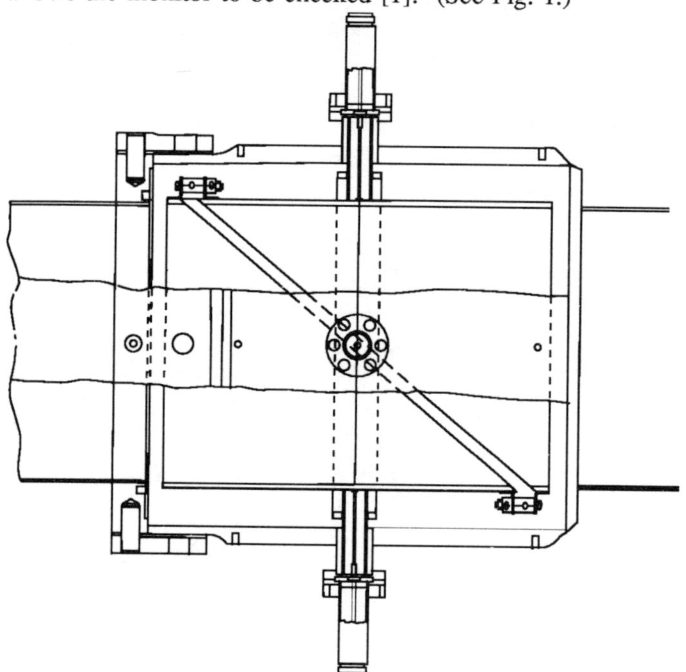

Figure 1. Beam position monitor (longitudinal view).

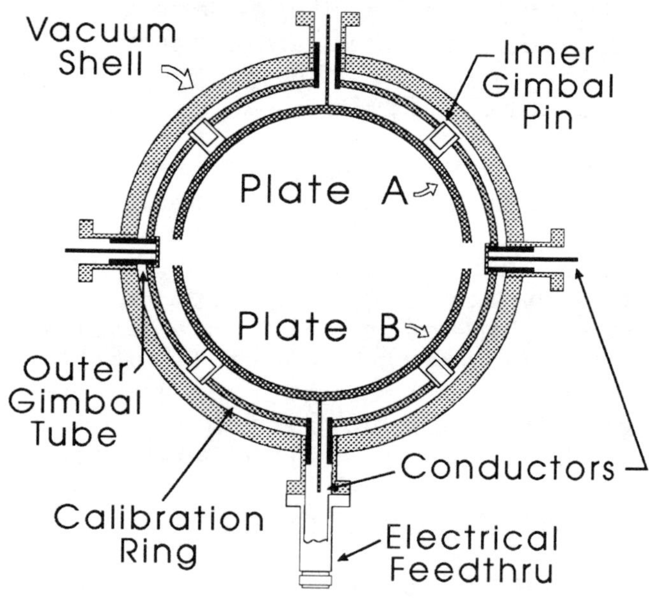

Figure 2. Double-gimballed beam position monitor.

Fig. 2 shows the arrangement of the four outer ceramic tubes which provide the gimbal support of the calibration ring from the outer shell. The four inner ceramic pins support the split-cylinder assembly from the calibration ring and thereby from the outer shell as well. Because all the support members are located in a plane perpendicular to the longitudinal (beam) centerline, expansion and contraction of the components can

[*]Work supported by the U. S. Department of Energy, Contract No. DOE 86-R-105.

occur freely in the longitudinal direction without inducing any loading on other components or on the ceramic supports. Radial movement is accommodated by the gimbal support between the outer shell and the calibration ring and between the calibration ring and the split-cylinder assembly. Differential radial expansion between any of the three cylindrical components, outer shell, calibration ring, or split-cylinder assembly, is accommodated by a radial sliding motion between either of the metal components and the ceramic supports without imposing loads on any of the components.

In addition to accommodating thermal movement, the design maintains the alignment of the components, including concentricity. The gimballed support insures that concentricity will be maintained even when thermal gradients produce relative dimensional changes.

To obtain the required accuracy of the location of the components, diametral clearances between the ceramic pins and tubes and their corresponding holes were limited to <0.03 mm. In order to permit assembly of the gimbals, the holes themselves had to be located with an similar accuracy. It was also necessary to fully anneal the stainless steel components before machining, and in some cases to stress relieve them prior to finish machining, in order to minimize distortion resulting from machining stresses or from the 950 °C vacuum firing. Any distortion during firing can cause binding at the gimbal connections which, in turn, imposes loads on the adjacent stainless and ceramic components that can result in damage to, or failure of, one or more components.

To maintain the required gap between the halves of the diagonally-cut cylinder while holding them together as a unit for gimbal mounting, it was necessary to fasten them together semi-rigidly while maintaining electrical isolation between the plates as shown in the Fig. 3. Two ceramic pins span the gap at opposite ends. Although little differential thermal growth was anticipated, the stainless steel angles to which the ceramic pins are attached flex to allow for any small changes in the gap during heating.

The plates of the detector are fabricated from a cylinder which was machined from fully annealed stainless tube stock. The cylinder was then stress relieved and the gimbal mounting holes added. The two pairs of holes are offset, so that once the halves are separated by the desired gap, the holes are co-planar. Even though it is a relatively slow process, electric discharge cutting was used so that no machining stresses were introduced into the part.

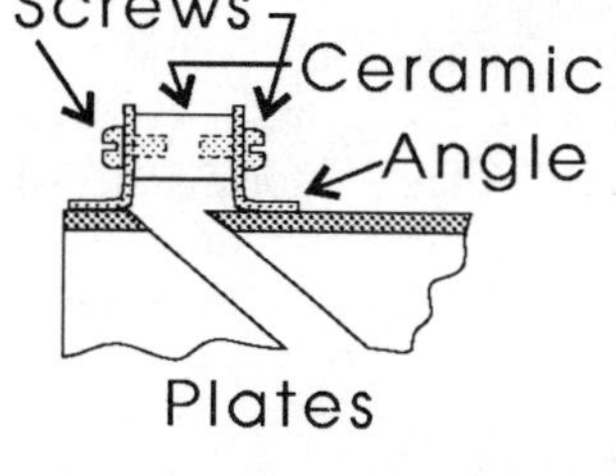

Figure 3. Split-cylinder assembly.

II. TEST

A. Description

After the split-cylinder pick-up electrodes and the associated calibration ring have been carefully assembled, the unit is inserted into the pick-up electrode body or shell. The pick-up electrode body is itself part of the vacuum chamber. It has openings for the four coaxial signal feed-throughs as well as mounting holes to allow the BPM to be precisely positioned in the Booster quadrupoles.

In order to allow the split cylinder to be inserted into the outer shell, one end is left open until it is welded to the flange that reduces the diameter of the chamber to that which passes through the quadrupoles. As a consequence, one of the halves of the split cylinder does not see the electrical environment it sees after being installed. Therefore, it is necessary to attempt to produce an electrically similar configuration. This is done by using a short extension.

After the extension has been attached, the detector is positioned vertically on a stand. Short rods are inserted into the same mounting holes which allow the BPM to be precisely mounted in the quadrupoles. These rods accurately position the detector on the mounting stand and prevent it from moving.

In order to obtain the scan measurement data, a signal wire is dropped through the inside of the pick-up electrode body. The wire (a 0.25-mm diameter piano wire) is connected to a 180 Ω terminating resistor and is then placed under tension produced by a strong spring. The signal wire is strung between the two arms of a large C-shaped support. The C-shaped support itself is mounted to the platform of an X/Y Cross Table (Techno Model HL32SBM2298).

A program running on a Hewlett Packard 9836S workstation/instrument controller automates the measurement process. The primary measuring instrument is the HP3577A Network Analyzer. The RF output port of the HP3577A provides the signal, and the A and B receiver ports measure the r.m.s. voltage on the each half of the split-cylinder of the pick-up electrode assembly. The equipment used is schematically indicated in Figure 4.

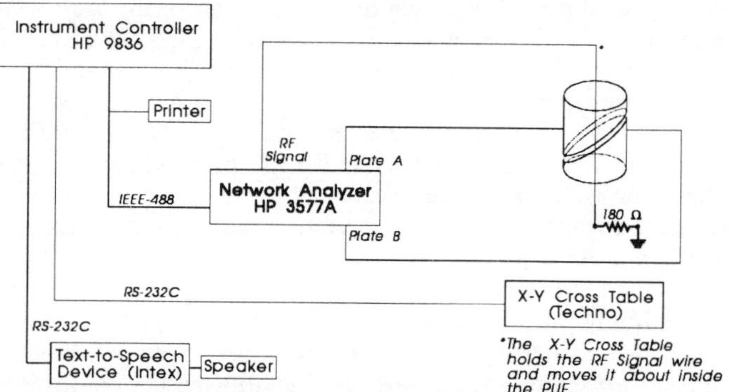

Figure 4. Set-up for the scanned wire measurements.

The X/Y Table Controller is attached to one of two RS-232C interfaces on the workstation. A complete scan consists of 261 measurements. By convention, the plane of highest sensitivity is designated y. The aperture is scanned at five x positions (-50, -30, 0, +30, +50). At each of the x positions, a measurement is taken every 2 mm from $y = -50$ mm to $y = +50$ mm (51 measurements). In addition to these 255 measurements, the wire is moved to (0,0) between each x position scan and at the beginning and the end of the entire measurement process, thus giving 6 additional measurements at position (0,0).

For each of the 261 measurements, the Network Analyzer takes 401 measurements each of the r.m.s. magnitude of the voltage of the signals at its A and B receivers. (These 401 measurements are called a "trace".) The program retrieves these values and computes the average, the difference between the maximum and minimum measurement, and the standard deviation for the two signals. These three values for each channel are stored on flexible discs along with the coordinates of the wire position.

B. X/Y Table Accuracy

The X/Y Cross Arm Table is an economical model and can not provide a continuous indication of its position. In fact, the table controller only knows when the platform has reached a single location on each of the axes, and this single location must be at one of the extremes of the arm movement. The controller knows that this position has been reached by the opening of a micro-switch when a protrusion mounted to the platform contacts it. All positions are then relative to the position of the platform when the switches on each axis opens (the "Home" position). The controller sends pulses to the two stepper motors to drive a lead screw which in turn moves the platform relative to the Home position. There are 100 half-steps per mm.

When the table is commanded to Home, the platform moves at the programmed acceleration until the switches are struck. The manufacturer specifies the accuracy of the homing operation to be ±0.01 mm. In addition to this error must be added the inaccuracy of determining the position of the mechanical center of the detector as mounted on the stand relative to the Home position.

The center of the stand (relative to the Home position) is determined by use of a special centering fixture. The centering fixture is mounted on the stand in place of the detector. The signal wire is dropped through a small hole in the fixture and made taut. Two micrometers are mounted over the center hole on a vertical bar. The deviation of the center of the wire—the wire has a radius of about 5 mils—from the true center is determined using the micrometers. The vertical mounting of two micrometers allows the C-shaped support to be adjusted to make the wire perpendicular to the stand. After the wire has been centered along one axis and made perpendicular in that plane, the centering fixture is rotated 90° and the procedure is repeated for the other axis. When the centering was repeatable to ±0.08 mm, it was deemed acceptable.

In addition to the difficulty of centering the wire, there was initially some jitter in the movement of the table platform itself. It was determined that this was due to either a bowing of the lead screw or to a loose fit that allowed the platform to tilt first to one side and then the other as it progressed along the lead screw. The effect was seen in the plots of the deviation of the data from a straight line where it manifested itself as a periodic variation instead of a smooth curve. This problem was solved by dismantling the entire X/Y Table and reassembling it with attention being given to making everything tight.

C. Results

Immediately after the data have been taken and stored, it is analyzed. All the results are plotted and a robust straight line fit is made to each of the five data sets (one for each of the five x positions) for the points which falls within -30 mm $< y < +30$ mm [2]. The fits are two-parameter fits (slope and offset) of $(V_A-V_B)/(V_A+V_B)$ as a function of the programmed y position, where V_A and V_B are the magnitudes of the voltages on plates A and B of the PUE, respectively. A final fit is performed for the same range of y values, but combining all the points in this range for the three central x scans (-30, 0, +30).

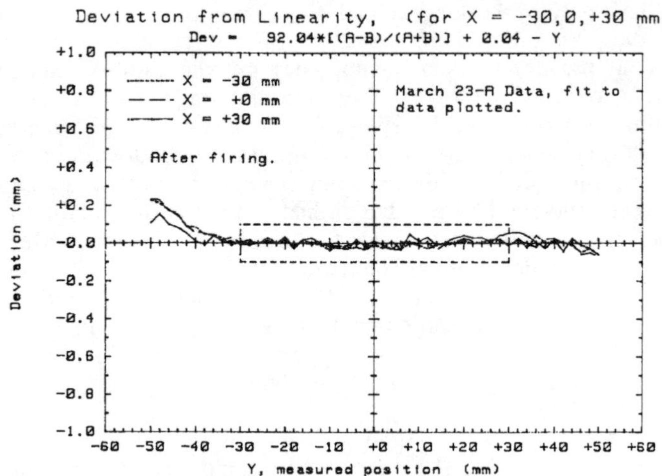

Figure 5. Typical scan wire measurement results.

The deviations from a straight line are plotted in all cases. Typical results are shown in Fig. 5.

Sixty detectors have been measured. (Forty-six are required for the Booster and six more are needed for the transfer line from the Booster to the AGS.) The average value for the slope, K, was 90.75 at 5 MHz. The standard deviation was 0.20, but the K value for each individual detector will be used by the Booster Beam Position Monitoring System. The offset, O, averaged -0.26 mm. All but three detectors had negative offsets. This is believed to result from the inability of the extender to fully simulate the electrical environment of the detector as it is after the shell has been welded to the next section of the vacuum chamber. Therefore, the offsets that are automatically acquired during operation by the Booster control software will be used instead [3].

III. ACKNOWLEDGMENTS

The authors wish to thank August Hoffmann, who carefully assembled and tested the position monitors, for his helpful suggestions, and James Savino for his work on the stand for the X/Y Cross Table.

IV. REFERENCES

[1] E. Beadle, *et. al.*, "Design of the AGS Booster Beam Position Monitor System," in *Proceedings of the 1989 IEEE Particle Accelerator Conference*, Vol. 3, Chicago, IL, March 1989, pp. 1536-1538.

[2] W. H. Press, *et. al.*, *Numerical Recipes*, Cambridge: Cambridge University Press, 1986, pp. 539-546.

[3] D. J. Ciardullo, *et. al.*, "The AGS Booster Beam Position Monitor System," in *Proceedings of the 1991 IEEE Particle Accelerator Conference*, San Francisco, CA, May 1991, to be published.

THE AGS BOOSTER BEAM POSITION MONITOR SYSTEM[*]

D.J. Ciardullo, A. Abola, E.R. Beadle, G.A. Smith, R. Thomas,
W. Van Zwienen, R. Warkentien, R.L. Witkover
Brookhaven National Laboratory
Upton, New York 11973

Abstract

To accelerate both protons and heavy ions, the AGS Booster requires a broadband (multi-octave) beam position monitoring system with a dynamic range spanning several orders of magnitude (2×10^{10} to 1.5×10^{13} particles per pulse). System requirements include the ability to acquire single turn trajectory and average orbit information with ± 0.1 mm resolution. The design goal of ± 0.5 mm corrected accuracy requires that the detectors have repeatable linear performance after periodic bakeout at 300°C. The system design and capabilities of the Booster Beam Position Monitor will be described, and initial results presented.

Introduction

The measurement parameters for this system are [1]:

Absolute accuracy:	± 0.5 mm
Precision:	± 0.1 mm
Circular aperture:	± 30 mm
Frequency range:	0.2 - 4.11 MHz
Intensity range:	2×10^{10} to 1.5×10^{13} ppp
Bunch width (at half max):	50 ns to 3750 ns

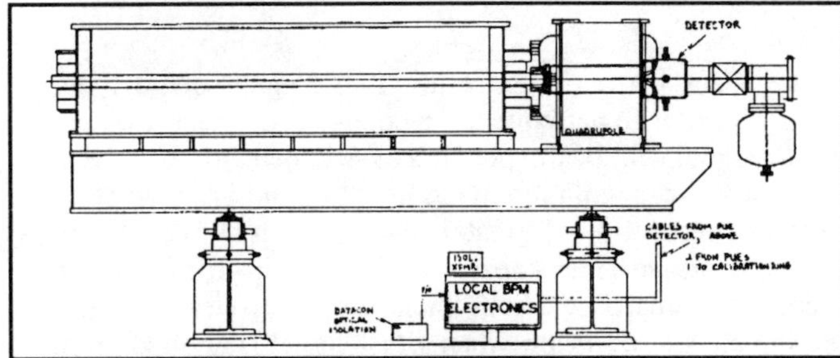

Fig. 2   A Typical Booster BPM Station.

Figure 1 is a block diagram of the AGS Booster BPM system. There are 46 BPM stations within the Booster ring, each responsible for average orbit measurement in a single plane

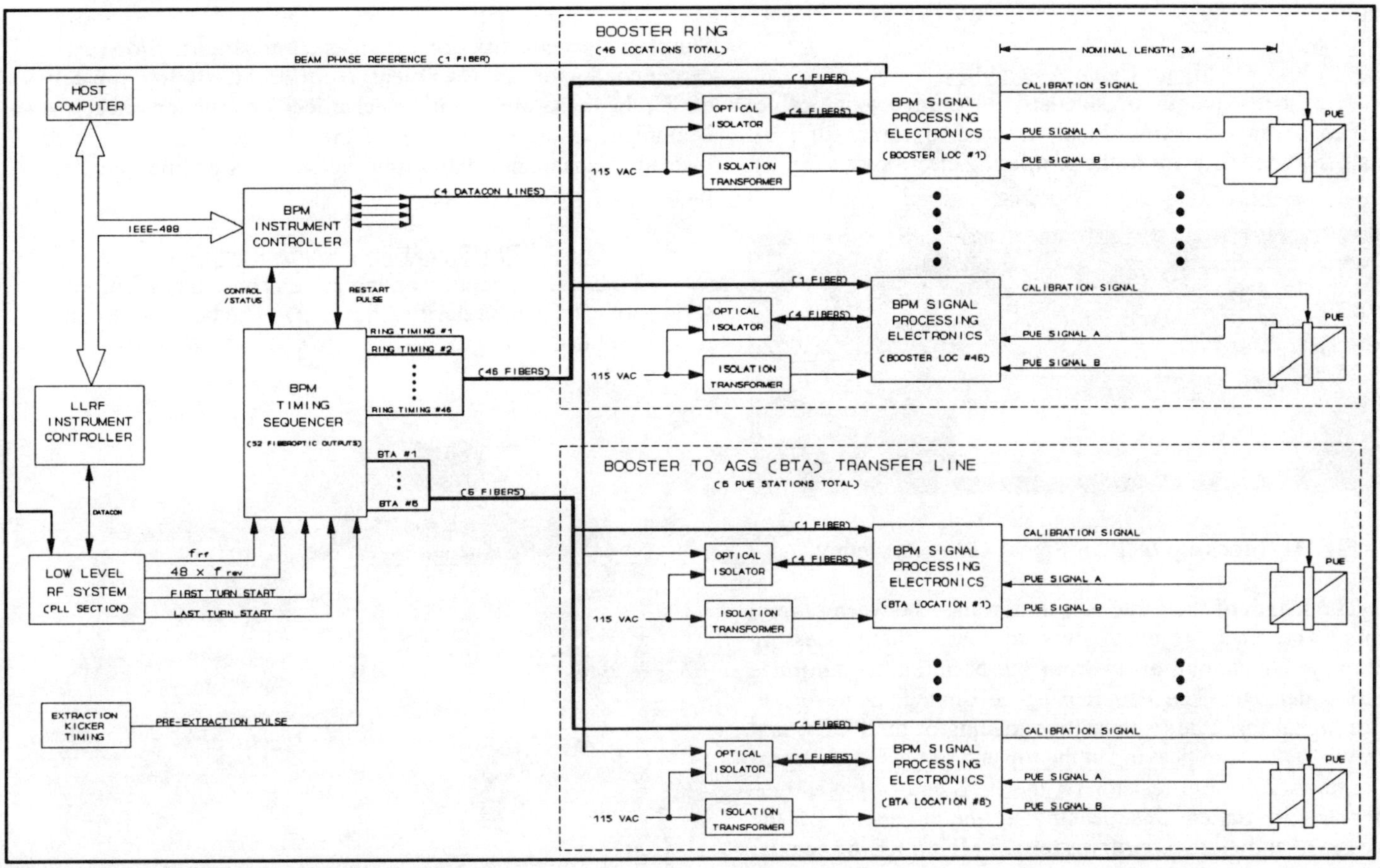

Fig. 1   Booster BPM System Block Diagram.

[*] Work performed under the auspices of the U.S. Department Of Energy.

(22 horizontal and 24 vertical). Six additional stations are located in the Booster To AGS (BTA) transfer line. At each station there is a beam sensor located at the appropriate betamax and locked to a quadrupole on the central orbit. Signals from the detector go to processing and acquisition hardware, located on the floor of the Booster just below each quadrupole (Figure 2). Digitized information is retrieved from each station up to six times per Booster cycle. The system provides average orbit information (over up to 80 turns) or first turn trajectory. In addition, analog real-time outputs of either the bunch signals or their sum and difference are available from any location for use by other systems.

The Detector

The beam sensor is the split cylinder electrostatic type, chosen for its flat low frequency response and linear induced electrode signal vs. beam position. Each detector has been tested prior to its installation in the Booster, and has a verified linearity and resolution of greater than $\pm$ 0.1 mm over the 60 mm circular measurement aperture [2]. A built-in calibration ring couples equally to both pick up electrodes (PUEs), simulating a centered beam when a test signal is applied from the electronics. This can be used to obtain a value for the total system offset, referenced to beampipe center. There are four rf N-type vacuum feedthrus which bring signals into and out of the detector; two for the calibration ring (one of which is terminated with 50 ohms), and one for each PUE of the split cylinder.

Signal Cable Assemblies

A 3 meter length of standard 93 ohm coaxial cable, modified to have a solid shield, is used to carry the PUE signals from the detector to the processing electronics (Fig. 3).

Fig. 3   Mock-up of PUE Signal Cable Assembly.

The length of the cable is a compromise between several factors. The main objective was to locate the processing electronics far enough away from the beamline to minimize radiation damage. The length must be limited, however, to reduce signal loss due to capacitive loading of the PUEs, and to prevent excessive peaking at the top end of the bunch signal response. A 200 ohm resistor (at the PUE end of the assembly) helps to flatten this peaking at the high end of the processing bandwidth (approximately 25 MHz). RG-62 coaxial cable was chosen for its low cost, moderate capacitance per unit length (46 pF/m) and reasonable radiation survivability (semi-solid polyethylene). The cable is enclosed by standard 3/8 inch copper tubing, which is electrically connected to the

shield at both ends of the assembly. Each assembly is measured on a network analyzer, and units are paired by matching their responses to within $\pm$ 0.01 dB (to 30 MHz).

The use of Thermal Isolators at each of the four vacuum feedthrus on the detector allows signal cables to remain connected during bakeout at 300°C. The isolators are effectively rf coaxial air lines, 14 cm in length, with male N type connectors on either end. This length is negligible as compared to the processing bandwidth of the electronics, and adds only 8 pF to the total capacitance load on each PUE.

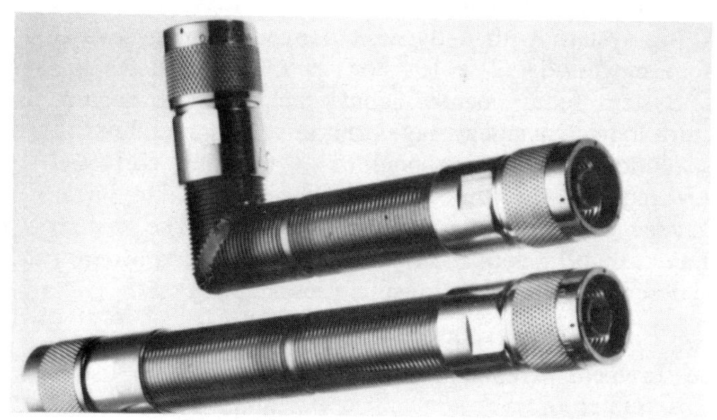

Fig. 4   Two versions of the Thermal Isolator.

Stainless steel construction provides a temperature drop (at the center conductor, 25°C ambient) from 200°C to approximately 60°C. In those areas with mechanical obstructions near the beampipe, an angled version of the Thermal Isolator is used instead. Capacitance variations between assemblies are less than $\pm$ 0.1 pF.

Local BPM Signal Processing Electronics

The Local Signal Processing crate is built into a Eurocard style subrack for modularity (Figure 5). The body of the subrack is isolated from ground and allowed to float at beampipe

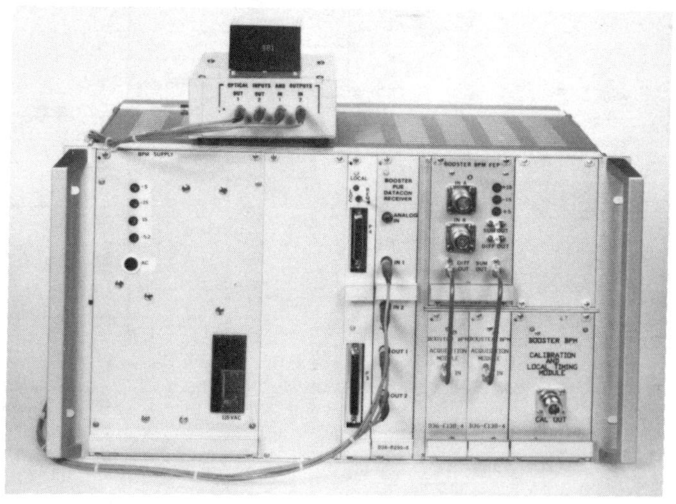

Fig. 5   Local BPM Processing Electronics.

potential for purposes of increased noise immunity. The top right slot is reserved for a dual analog fiberoptic transmitter module, used to isolate the electronics from any external systems requiring beam position information. A Test Access Panel (TAP) is included in each crate as a maintenance aide. This card allows accessibility to backplane signals, and has provision for switching the crate to local control.

The local processing electronics are divided into three major sections [3]; Front End Processing (FEP), Acquisition (ACQ) and Calibration and Local Timing (CLT). The FEP module provides appropriate PUE signal scaling and frequency compensation. Both channels of the FEP are matched to within 0.02 dB (to 20 MHz) in all three gain modes, resulting in a position accuracy for this module of better than 0.3 mm. A ± 1 dB flatness preserves the temporal response of the PUE signals for bunch widths down to 50 ns (half max) [4]. Each module has a secondary set of outputs which provides either buffered PUE signals or their sum and difference. These can be used for diagnostic purposes, or as real-time position information by other systems.

Two Acquisition (ACQ) modules and the Calibration and Local Timing (CLT) module form the next PUE signal processing step. Sum and difference signals from the FEP are baseline restored (the capacitively coupled PUE signals have zero average value), integrated then digitized by the ACQ. Local timing and control signals are generated by the CLT module. This module receives a properly phased pulse burst from the Sequencer and generates the BLR, integrate gating and reset pulses for each ACQ. The number of beam bunches over which to integrate can be selected from 1 to 255 (up to 85 turns), and the resulting integrals are digitized to 13 bits. In addition, the CLT module can generate pulses to drive the calibration ring of the detector at five selectable frequencies. This can be used to check system connections and to measure total system offset errors [5].

The Controller Interface Electronics (CIE) card is the local interface between the crate and the BPM Instrument Controller. This board passes digital data, control and status signals to and from the Controller via a fiberoptic version of DATACON which eliminates ground loops and allows the crate to float at beampipe potential. The CIE card also serves as the fiberoptic interface between the Sequencer and the CLT module.

Sequencer

The Sequencer is located remote from the Booster ring and connected to each of the 52 BPM stations via digital fiberoptic link [6]. The Booster Low Level RF (LLRF) system supplies the Sequencer with f_{rf} and $48xf_{rev}$, both of which are phase locked to the beam [7] (beam phase information is fed back to the LLRF system from the secondary sum output of an FEP module in the ring). These two signals are used to generate a burst (of specified duration) of BLR pulses for use by the Calibration and Local Timing module at each BPM station. The BLR pulses are phased such that they always occur between bunches, independent of harmonic number. The LLRF system also supplies the Sequencer with triggers which indicate the start of the first and last turns in the Booster cycle. Since the BPM stations in the extraction line (BTA) always see three bunches at the end of the acceleration cycle,

no baseline restoration of the sum and difference signals are necessary before integration. For the six BPM stations in the BTA line, the Sequencer always outputs a burst of three BLR pulses.

High Level Software Control

The BPM Instrument Controller interfaces the host computer (Apollo system) with the Sequencer and the local processing electronics, and provides reports to the software of PUE sum and difference data from one or more samples during the Booster cycle.

The "booster_orbit" program is the high level applications software for the Beam Position Monitor, running on the Apollo computer system. Its primary function is to graphically represent the equilibrium orbit of the Booster ring at any time during the cycle. The program allows for the acquisition of data with varying parameters such as number of turns (over which to integrate), number of BPM Booster groups to acquire information from, acquisition times and gain settings. The program also allows for a variety of displays using saved, current or difference orbits with varying modes of averaging and plotting parameters. Statistical display options are also available.

In addition to orbit acquisition and display, the scope of the program extends to diagnostic testing of the hardware, full calibration of each of the local BPM stations and display/update of static and dynamic BPM parameters (e.g. values of differential sensitivity, system offsets and operational status) used in the position calculations.

Acknowledgement

The authors wish to express their gratitude to J.M. Brennan for the considerable conceptual advice and expertise he has contributed throughout the course of this project.

References

[1] E. Beadle, et al, "Design of the AGS Booster Beam Position Monitor System," Proceedings of the 1989 IEEE Particle Accelerator Conference.

[2] R. Thomas, et al, "Design and Testing of the AGS Booster BPM Detector," this conference.

[3] D.J. Ciardullo, et al, "Design of the AGS Booster Beam Position Monitor Electronics," this conference.

[4] D.J. Ciardullo, private communication.

[5] D.J. Ciardullo, private communication.

[6] *op cit.*

[7] T. Hayes, A. Zaltsman "Indirect Phase Locking of RF Clock to the Beam for the BNL Booster BPM System," this conference.

Contemporary Approaches to Control System Specification and Design Applied to KAON

George A. Ludgate and Edwin A. Osberg, TRIUMF, Vancouver, B.C.
Don A. Dohan, SSC Laboratory, Dallas, Texas

Abstract

Large data acquisition and control systems have evolved from early centralized computer systems to become multi-processor, distributed systems. While the complexity of these systems has increased our ability to reliably manage their construction has not kept pace. Structured Analysis and Real-time Structured Analysis have been used successfully to specify systems but, from a project management viewpoint, both lead to different classes of problems during implementation and maintenance.

The KAON Factory central control system study employed a uniform approach to requirements analysis and architectural design. The methodology was based on well established object-oriented principles and was free of the problems inherent in the older methodologies. The methodology is presently being used to implement two systems at TRIUMF.

I. Introduction

Experiments data acquisition systems and accelerator control systems have evolved from early centralized systems to become distributed, multi-processor systems employing large, complex databases, artificial-intelligence software techniques and sophisticated direct manipulation user interfaces. While the complexity of these systems has increased several orders of magnitude over the last two decades our ability to reliably specify them and manage their construction has not kept pace with the other technological gains.

Text based requirements specifications are known to be too bulky and hide ambiguities and omissions. To address these concerns many graphical, model-based systems development methodologies were created during the late seventies, along with software tools to support their use on large projects. One of these popular techniques, called Structured Analysis (SA) [1], has been used successfully at CERN [2] and another, Real-time Structured Analysis (RT/SA) [3], has been used at TRIUMF [4]. Both suggest the use of Structured Design [5] and structured programming to complete the implementation.

II. Analyzing the KAON Factory Control System Requirements

The KAON Factory Project Definition Study [6] included an investigation whose goal was to produce a preliminary hardware and software design, and a cost estimate for the control system. A contemporary object-oriented requirements specification methodology [7] was chosen for this study to avoid the perceived problems with the use of SA or RT/SA on large system development projects. The methodology embodies a uniform approach to requirements analysis, architectural design, detailed design and coding based on well established object-oriented principles.

III. Criticisms of Structured Analysis

Structured Analysis followed by Structured Design has been used extensively in the business domain, and more recently in particle physics experiments and accelerator control systems. It is a "top-down" approach to system design that starts with the establishment of a Context Diagram. A Context Diagram shows the system to be built (denoted by a circle) communicating with other systems in its environment, called Terminators (denoted by boxes), by flows of data (denoted by directed lines). The new system is viewed as a single process capable of storing and transforming information in response to events in its environment.

An analysis of the system proceeds by developing a second diagram showing a decomposition, or refinement, of the Context Diagram process into several "lower level" processes, linked by data flows. The proposed lower level processes are provided the same external inputs as shown on the original Context Diagram and must produce the same external outputs. The cooperative working of these lower level processes must achieve the same effect as the single process shown on the Context Diagram for the proposed decomposition to be acceptable. Further analysis work proceeds by recursively "refining" each process until the analyst feels a process can be described textually and needs no further refinements.

Several major problems exist with using Structured Analysis on a large project:

- It is "top down". Detailed analysis of low level processes can only proceed after a "high-level" analysis of the system has been completed. Processes remaining to be refined obviously depend critically on refinement decisions made at higher levels. Any major changes to the high level process decomposition of a system may invalidate the analysis of all processes below that level.

- Analysts can only be added to the project as the number of refined processes increases.

- There is often no unique decompositions of a process into lower level processes. Analysts attempt to apply functional decomposition but there are as many understandings of the term "functional" as there are analysts. Thus, the "same" functionality observed by two different analysts may be decomposed differently, forever hiding the common features.

- At any point in the analysis of a new system it is not clear how many more levels of decomposition will be required.

- The *behavior* of a system is hidden "inside" process specifications. Changing the behavior of the system may involve modifying several processes that coordinate their actions through stored data.

Control system engineers employing SA tend to produce refinements of the Context Diagram that are naturally object-oriented; associating one or more (behavior determining) processes and (property specifying) stores with each Terminator [8].

IV. Criticisms of Real-time Structured Analysis

Real-time Structures Analysis [3] is an improvement over classical SA from the perspective of managing a large project. Like SA, the methodology advocates creating a Context Diagram for a new system first. The methodology then requires one or more analysts to model the environment of the new system by finding and recording all events in the environment which effect the system. The system's responses to each event must be planned and recorded along with the event, as must the mode of behavior before and after the occurrence of the event. These analysis tasks can be carried out by several analysts relatively independently of each other and, as long as only one master Event/Response List is maintained (which must also include all event synonyms) no duplication of requirements will occur.

The behavior and process structure of the new system is finally derived from the complete Event List by a process similar to the derivation of a classical control system transfer function from a *complete* specification of its inputs and outputs as continuous functions of time. The designer examines the Event List and constructs State Transition Diagrams (STDs), representing the behavior of the system at relative points in time. No heuristics were originally given to guide this "construction" process [3].

There are several major problems with the use of real-time systems analysis on a large project:

- Experience has shown that even with many analysts working concurrently, it is unlikely they will discover the *complete* Event List for a complex system - there being no recommended approach to partitioning the search for events nor for ensuring that all events are discovered.

- The internal process structure of the new system depends, *in principle*, on only the events discovered and the responses planned. Thus, while it is possible for several analysts to independently search for events and plan system's responses, it is not feasible for them to independently create models of the systems internals. Design work cannot start until the last event is discovered.

- Similarly, in principle, it is not possible to predict the amount of re-structuring of a system's internals required to accommodate the discovery of a *single* new event. In practise, partitioning the Event List by Terminators will limit the "propagation" of changes.

Well known approaches to using this methodology apply an object-oriented partitioning of the "event space" by factoring the search for events by Terminators and associating an STD with each terminator [7].

V. Benefits of the Object-Oriented Approach

A contemporary object-oriented approach to system analysis and design was selected to model requirements for the KAON Factory central control system [9-11]. This approach directs system analysts to initially study the application from the viewpoint of the users. For accelerator control systems the application domain includes operators (one kind of user) controlling accelerators composed of ion sources, magnets, rf cavities, beam diagnostics, production targets, vacuum equipment etc.; all working toward the goal of accelerating and transporting different kinds of beam from the ion source to one or more production targets according to a schedule. Each device type or conceptual entity (e.g. schedule) is considered to be an "object" for the purposes of this methodology and its role in the application domain is, therefore, subject to an object-oriented analysis (OOA).

Two analysis models are developed initially for each object identified to be part of the application; a dynamics model describing behavior and a statics model describing properties. The dynamics model, usually in the form of an STD, represents the object's modes of behavior and the allowed transitions between those modes. Conversely, the statics model of an object identifies both the properties that are needed to completely describe it at any instant in time and its relationships with other objects in the application domain.

Models of KAON Factory sub-systems were created in collaboration with sub-system engineers (experts) and validated for quality control. For example, a beam line magnet may only be in one of the modes {OFF, RAMPING UP, ON, RAMPING DOWN} at a time and no transitions are possible between ON and OFF states without going through one of the ramping behaviors. Similarly, values for the two properties {MAGNETIC FIELD, TEMPERATURE} may be required by the control system at all times to ensure the magnet is operated correctly.

In contrast to SA, the OOA approach is "bottom up". The analysis of each identified object can, in principle, proceed in parallel once the goals of the control system have been specified. In fact, the development of an object's dynamic and static models can also proceed in parallel, subject to the availability of experts familiar with the object being investigated. The two analyses of an object must be reconciled; discussions concerning behavior illuminate properties while discussions about the objects relationships to other objects will involve discussions of behavior modes.

Unlike RT/SA, the levelled internal structure of an object-oriented system is directly related to, and *derived* from, the statics model, while the detailed (process model) structure is *derived* from the dynamics models. Event responses appear explicitly in dynamics models, partitioned according to the object causing the event. The structural models are, therefore, more stable to change and to the addition or removal of objects from the application domain. Similarly, the list of events to which the control system must respond is *derived* from an examination of all of the dynamics models rather than searched for as in RT/SA.

Both SA and RT/SA use Structured Design to cast their "data-flow" specifications into a hierarchical form more easily adapted to the "subroutine call" form required by structured programming. In contrast, OOA is best followed by Object-Oriented Design (OOD). In OOD the static and dynamic models from OOA are *mapped* onto similar models of a design architecture (composed of processors, processes, inter-process flow and inter-processor flows, classes and messages etc.). The nature and form of the design models are unchanged as is the interpretation of the diagraming notation. No leap of faith is required to produce the design, unlike that required by practitioners of the "art" of Structured Design in going from data-flow diagrams to structure charts.

The OOA/OOD methodology, particularly when followed by coding in an object-oriented language, incorporates the best features of the earlier SA or RT/SA methodologies namely graphical models, simple notation, implementation independence, etc.. However, *logic*, the underlying basis for OOA and OOD, pre-dates all other system development methodologies and provides a firm, well known formal framework within which developers can work.

From a project management standpoint, relating the volume of analysis and design work to be completed to the number of objects discovered in the application domain is a useful metric for establishing the effort, and hence the manpower, needed to complete the two phases. This metric is determined early in the project, at the start of analysis. Similarly, the completion of analysis and design models for *each* object serves as milestones in charting the progress of the project.

VI. Conclusion

OOA and OOD have been presented as solutions to the system development problems associated with the use of the older technology SA or RT/SA methodologies. In addition, the approach leads to a better factoring of development effort and, therefore, to improved project management due to the occurrence of natural units of work.

The KAON Factory control system was designed using these two contemporary object-oriented methodologies and two projects are presently underway at TRIUMF to implement sub-system control systems using this same approach. Both systems will have object-oriented requirements specifications and be implemented using the C language running under the Vsystem product [12].

VII. References

[1] Tom DeMarco, *Structured Analysis and System Specification*, Yourdon Press, 1978, ISBN 0-13-854380-1.

[2] *Data Acquisition and Computing*, CERN/LEPC/85-9, March 1985.

[3] P.T. Ward and S.J. Mellor, *Structured Development for Real-time Systems*, Yourdon Press, 1985, ISBN 0-917072-51-0.

[4] G.A. Ludgate, B. Haley, L. Lee and Y.N. Miles, *The use of structured analysis and design in the engineering of the TRIUMF data acquisition and analysis system*, IEEE Trans. Nucl. Sci., NS-34, pp. 157-161, Feb. 1987.

[5] E. Yourdon and L.L. Constantine, *Structured Design*, Prentice Hall, 1979, ISBN 0-13-854471-9.

[6] *KAON Factory Study: Accelerator Design Report*, 1989, TRIUMF, 4004 Wesbrook Mall, Vancouver, B.C., Canada, V6T 2A3.

[7] C. Inwood, *Analysis of real-time requirements using dynamic object models*, Proprietary course notes, Inwood Real-time Systems, R.R. 2, Kinburn, Ontario, Canada K0A 2H0, April 1989.

[8] G. Morpurgo, *SASD and the CERN/SPS run-time co-ordinator*, Proc. ICALEPCS, Vancouver, Canada, 1989, Nucl. Instr. and Meth. A293 (1990) pp. 385-389.

[9] C. Inwood, G.A. Ludgate, D.A. Dohan, E.A. Osberg and S. Koscielniak, *Domain-driven specification techniques simplify the analysis of requirements for the KAON Factory central control system*, Proc. ICALEPCS, Vancouver, Canada, 1989, Nucl. Instr. and Meth. A293 (1990) pp. 390-393.

[10] E.A. Osberg, G.A. Ludgate, S. Koscielniak, D.A. Dohan and C. Inwood, *Dynamic object modelling as applied to the KAON control system*, Proc. ICALEPCS, Vancouver, Canada, 1989, Nucl. Instr. and Meth. A293 (1990) pp. 394-401.

[11] D.A. Dohan, G.A. Ludgate, E.A. Osberg, S. Koscielniak and C. Inwood, *Definition study of the TRIUMF KAON factory control system project*, Proc. ICALEPCS, Vancouver, Canada, 1989, Nucl. Instr. and Meth. A293 (1990) pp. 6-11.

[12] P. Clout, R. Rothrock, V. Martz, R. Westervelt, M. Geib, *Past, present and future of a commercial, graphically oriented control system*, Proc. ICALEPCS, Vancouver, Canada, 1989, Nucl. Instr. and Meth. A293 (1990) pp. 456-459.

Rejuvenation of the CPS Control System;
the First Slice

G. Benincasa, G. Daems, B. Frammery

P. Heymans, F. Perriollat, Ch. Serre, C.-H. Sicard

CERN 1211 Geneva 23 CH

Abstract

The consolidation of the CERN Accelerator Control Systems is, as a common official project of the CERN PS and SL Accelerator Divisions, expected to be completed in 1995. As a first slice for the CPS complex, the LEP Pre Injector (LPI) was chosen for its coherent wholeness as a particle source and as a system with a reasonable number of equipment to control. This machine is large enough to be a good validation testbed for the new control architecture, and modern enough to reduce the software porting problem. (Fig. 1)

The implementation of the new architecture, hardware as well as software, will be performed during the normal running time of the machines and step by step transferred into operation at the end of the annual long shut down of each set of accelerators. The report presents a general overview of this rejuvenation plan, and a detailed analysis of the first slice, including the estimation of the manpower, together with the selected options.

I. PROJECT OVERVIEW

The rejuvenation project is the result of a collaboration between the Accelerator Divisions which elaborate a common design performed by working groups, composed of people from the controls, operation and instrumentation groups. The first working group designed the common control system architecture and the front end processing. The second working group defined a common approach to the application software and the future layout of the work place in the control rooms. A third working group, mainly composed of users, elaborated a control protocol to connect, in a uniform way, equipment to the control system. The conclusions of these working groups were presented to the control system users, discussed and accepted by the different groups concerned. [1].

II. SOFTWARE AND HARDWARE ARCHITECTURES

The main aim of this project is to replace the obsolete 16 bit mini computers and microprocessors (ND100 family, PDP11, and TMS9900). They will be replaced by a distributed system including workstations, servers and equipment front end processors split in a clear two layer architecture linked by Ethernet network. (Fig. 2) Economical constraints enforce the reuse as far as possible of the hardware, mainly CAMAC, used to connect equipment to the control system, because of the enormous investment previously made.

- The Operation Layer in the control rooms which provides the operators with a reliable and user friendly set of tools to adjust the accelerator parameters, uses modern UNIX workstation platforms running commercial software packages (X-Windows and OSF Motif standards; DVDraw/DVTools based interface for synoptic representation). This layer includes also a number of Central Services executed in servers (more powerful machines of the same workstation family) using standard NFS service : coordination of various control tasks, central data and file storage, model computing and equipment alarm collection, RT relational Data Base (ORACLE and SQL interface).

- The Front End Computing Layer based on Device Stub controllers (DSC) made of VME crates with MC 68030-based controllers running a diskless version of the RT-UNIX (Posix 1003.4 compliant) multitasking system LynxOS. Equipment access through Field Buses (serial CAMAC, 1553 MIL and

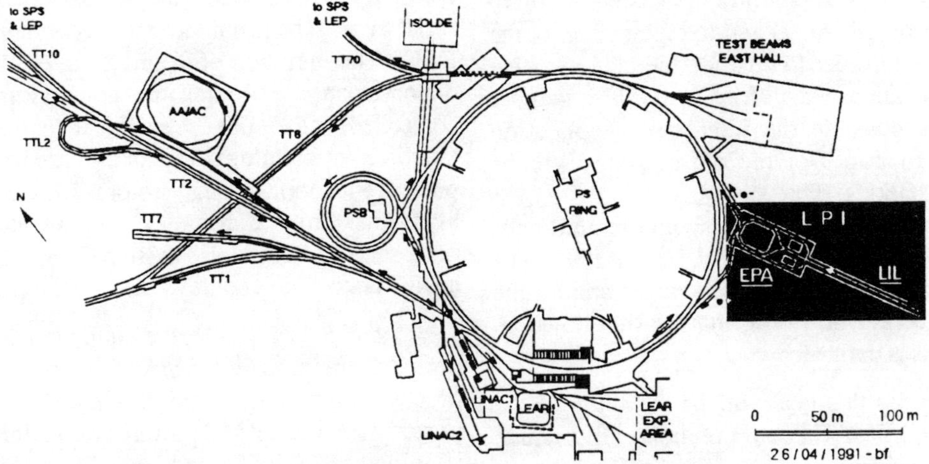

Figure 1. The CERN Proton Synchrotron (CPS) layout, with LPI accelerators highlighted

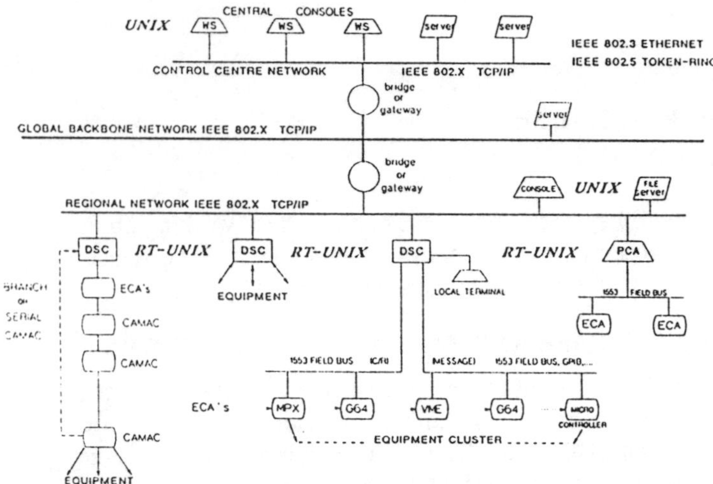

Figure 2. Control System Architecture

GPIB) is completed with local facilities where needed (display, touch-panel and Nodal interaction). The Operational Control Protocol will provide standard software interface with different types of equipment independent of the type of connection. [2].

The communications within and between the Control Room and Front End layers are based on modern networking standards using TCP/IP protocol suite. A remote Procedure Call with asynchronous call function, developed at CERN, completes the basic TCP/IP facility for easy implementation of distributed processing.

The Front End Layer will handle all the real time process constraints. At the level of workstations and servers no real time constraints must be treated. The only constraint is to react to the operator's requests in the shortest delay.

III. THE IMPLEMENTATION SCHEDULE

For the PS complex, the LEP Pre Injector has been chosen to be the first slice of the consolidation project. The main emphasis is on the level of Front End processing in order to remove completely two ND120 and the associated auxiliary CAMAC crate processor (SMACC) and to replace it with the new DSC architecture. In order to upgrade the man-machine interaction, DEC workstations were added to the slice; in order to reuse as much as possible the interactive application software, the Nodal functions of the old consoles will be emulated on the workstations. (Fig. 3)

The LEP Pre Injector [3] is continuously in use until mid-November 1991 and then turned off until March 1992. The upgrading should not disturb at all the operation and all the on-line tests have to be performed during the shut-down. A strict schedule of events is defined.

The first step to reach this goal will be the setup of the DSC basic system facilities. This environment will include remote boot (DSC are diskless machines for reliability), Nodal interpreter, Remote Procedure Call (RPC), device drivers for serial CAMAC and GPIB, external interrupts handler, and operation sequencer access routines. When needed a local video graphic facility will be added. DSC survey programs and the error reporting and logging package complete the basic software.

The second step is the transfer of the control application software to the new architecture from the ND120/SMACC processors. This action will partially starts in parallel with the previous one (basic system) at least for the initial phases. About 20% of the software, written in P+ (Pascal like language) and in NPL (Intermediate language proper to Norsk Data computers) in the ND120 will be rewritten; but the remaining 80%, written in C, will need only to be ported. As soon as the DSC basic implementation is setup for the Application programs, the new versions of the application programs on the DSCs will be generated and integrated, ready for the off-line tests.

The third step is the initial test period in November. When the LPI complex will be stopped, CAMAC crates will be connected to the newly installed DSC and on-line tests of the equipment control software modules in the DSC will be performed. Some interactive Nodal application programs will be integrated in the workstations and LPI will be run with beam as final test.

After these tests, the rest of the shut down between December and February will be spent to complete the transfer of the interactive programs in the workstations, prepare the modeling programs for the use with the workstations and with the network servers, and also to correct the bugs found during the preliminary tests. The last tests, and the complete system implementation, will be done at the end of the annual shutdown during the operational start-up period of the LPI machines (beginning of March 1992).

IV. DETAILED ANALYSIS OF THE LPI SLICE

Equipment control : The CAMAC embedded microcontrollers (SMACC) are removed. The 2 ND120 FEC minicomputers (LIL, EPA) are disconnected from the CAMAC serial loops. The serial loops are reconfigurated to be driven by DSC. The number of Camac crates driven by one DSC depends on the Real Time constraints of the equipment controlled, and on the geographical distribution of the crates in the different buildings. This is the case for power supplies (dc, pulsed, kickers), timing, motors, RF cavity, modulators and klystrons. For the complex instrumentation and when the necessary VME modules exist, a CAMAC crate is replaced by a DSC crate which controls directly the equipment.

On the software side, the equipment access package [4] is ported from the FEC+SMACC layout to the DSC. All the process equipment control modules from FEC+SMACC are ported to the VME microcontroller. These modules (Equipment Modules and Real Time tasks) are exclusively written in C.

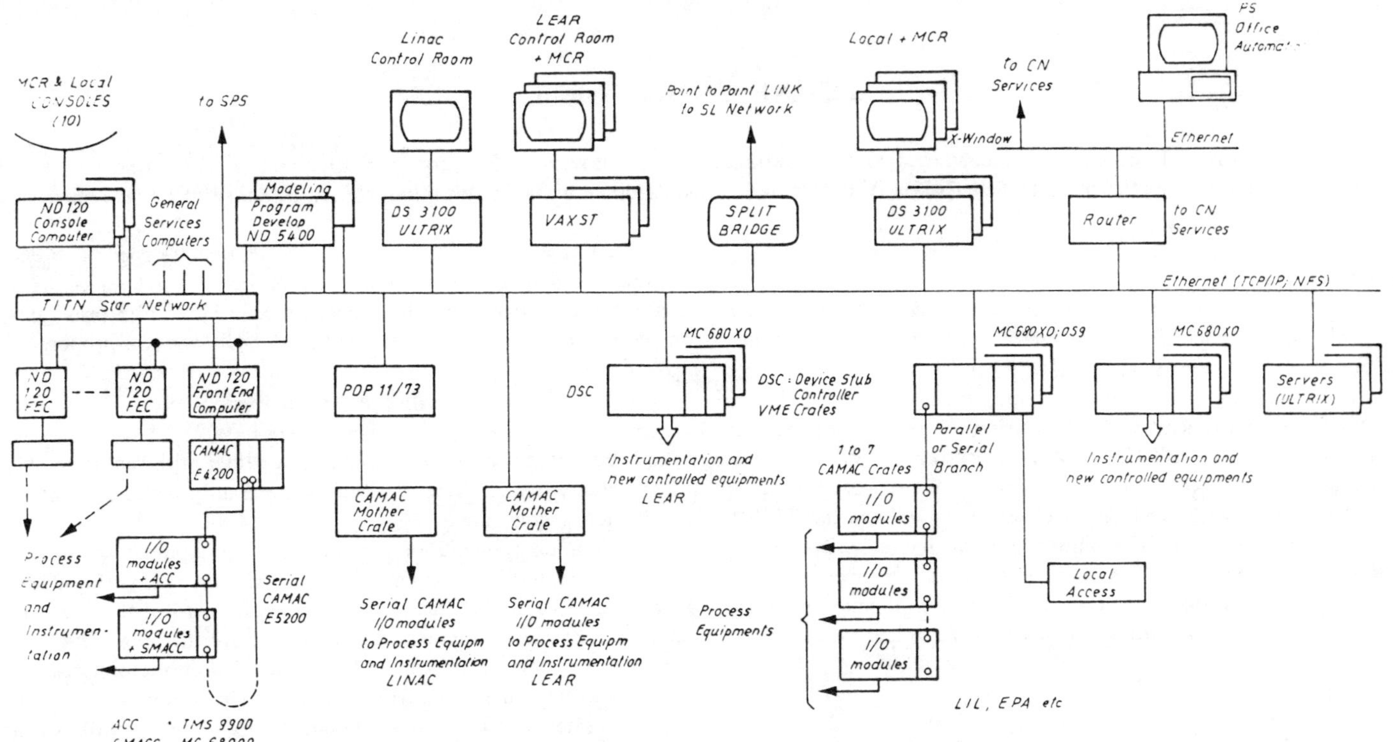

Figure 3. PS Control System during Transition Phase

The exploitation and test programs for the equipment interface remain in Nodal in the DSC.

Operation consoles : The equipment control part of the two consoles mainly used for LPI are replaced by two workplaces made of 3 DEC3100 workstations. The analog and video signal selection and observation systems will remain unchanged. Their future replacement will be done later during the control consolidation project. It will be based on developments underway using VME equipment and VXI bus capability.

The main part of the Nodal interactive application programs of the actual consoles are emulated on the workstations with a minimum of program editing. The generic programs developed in C for the workstations interaction prototype will be integrated; they include knobs parameter control, table and synoptic presentation and operation sequencer interaction. The LPI modeling programs (FORTRAN) will be ported in the workstations/servers network.

Manpower : the total effort requested to perform this first slice is estimated to 12.5 men-year which can be split into:

Hardware implementation of the DSCs -

Console modification :	~ 1.5	my
Basic software running in the DSCs :	~ 2.-	my
35 equipment modules+Real Time tasks :	~ 3.-	my
General application software :	~ 4.5	my

Nodal emulation+program transfers
generic C programs integration from
the prototype, transfer of more
specific programs :   ~ 1.5 my

V. CONCLUSIONS

Rejuvenating the control system of a running installation is not a trivial work, and it must be executed in coherent slices in order to preserve the integrity of the accelerator complex operation. The choice of the first slice to be renewed is one of the critical decisions to be taken in the management of such a project. This slice must be large enough, and well representative of the other parts of the complex to be a good validation testbed for the new architecture adopted and for the methodology used. Strange to say, the better choice is to take the newest part of the actual control system, the LPI complex in our case, to concentrate the effort only on the new architecture and technology and not to spend too much work on reengineering obsolete software modules.

With a unique annual shutdown to implement the major modifications, the planning of such a project is very constraining and any major event can result in a complete rescheduling. A good decoupling between interaction presentation and equipment access is very favorable because it gives the capability to validate step by step the control software architecture.

VI. REFERENCES

[1] The PS and SL Controls Groups, PS/SL Controls Consolidation Project, Technical Report, CERN PS/91-09 (CO), CERN SL/91-12 (CO), April 1991.

[2] Reported by G. Benincasa, Generalities on the operational Control Protocol, PS/CO/Note 91-01, April 1991.

[3] The LPI Beam Commissioning Team, J.H.B. Madsen et al., First experience with the LEP Pre-Injector (LPI) Proc. of the 1987 Particle Accelerator Conference, Washington, p.298.

[4] L. Casalegno et al., Distributed Application Software Architecture applied to the LPI Controls, IFAC Workshop, Bad Neuenahr, FRG, October 1986.

The UNK Control System

V.N. Alferov, V.L. Brook, A.F. Dunaitsev, S.G. Goloborodko, P.N.Kazakov,V.V. Komarov, A.F. Lukyantsev, M.S. Mikheev, N.N.Trofimov, V.P. Sakharov, E.D. Scherbakov, V.P. Voevodin, S.A. Zelepoukin, IHEP, Protvino, USSR, B. Kuiper, CERN,Geneva.

Abstract.

The design considerations and current status of the distributed control system for the UNK Accelerator and Storage Ring Complex is presented. All process devices will have intelligent equipment controllers using 8 and 16 bit processors, connected by a field bus to VME based frontend computers using 32 bit processors. The latter interconnect with UNIX based workstations and servers at the main control room. The network may have a Token Ring or FDDI backbone with Ethernet LAN segments and use the TCP/IP protocol. A maximum of commercially available hardware and software will be used. The user friendliness for operation and servicing is emphasized by early definition of application software and by providing facilities for local access and stand alone operation.

I. Introduction.

The UNK complex will combine - in one tunnel of 20.7 km circumference and 5.2 m diameter cross section - a 400 GeV conventional magnet synchrotron ("stage I") and a 3000 GeV superconducting synchrotron/storage ring ("stage II"). At a later stage a second superconducting ring ("stage III") may be added with the aim of doing proton-proton collider physics at 6 TeV.

The 400 GeV synchrotron is injected at 70 GeV from the existing proton synchrotron "U-70". For one filling up to 12 pulses from U-70 may be stacked in longitudinal phase space, accelerated to 400 GeV and transferred to the superconducting ring which in turn accelerates them up to 3 TeV.

Three main modes of operation are presently foreseen.
(i) Fixed Target at 3 TeV: fast or slow extraction will send the 3 TeV beam to the fixed target experimental area. During the acceleration of the superconducting ring, U-70 may produce beams for its own 70 GeV experimental area.
(ii) Colliding Beams at 3 + 0.6 TeV: the beams from, respectively, the superconducting ring and the conventional one are made to collide. For this the conventional ring is operated first as booster and, after field reversal, as a storage ring run at 600 GeV.
(iii) Colliding Beams at 3 + 3 TeV: the conventional ring will first inject into one superconducting ring and, after field reversal, into the second one.

The accelerator controls equipment will be distributed over near to 20 on-surface buildings situated mainly along the accelerator ring. In one of the these is the Main Control Room (MCR), the other ones house the remote nodes of the control system. The latter are totally controlled from the MCR and are in general not manned. The typical distance between any two adjacent buildings is about 1.8 km and the maximum is about 3.5 km.

The more than 2500 superconducting magnets require a cryogenic plant and elaborate distribution, recovery and safety installations   and their concomitant controls in the surface buildings around the ring tunnel.

The secondary beamlines and external experimental areas cover an area of roughly 12 km length. Controls for their equipment may follow closely the principles of the accelerator controls.

The upgrade of U-70, for meeting UNK injector specifications, requires intensive machine studies which in turn make a controls upgrade mandatory. since this must precede UNK, the principles and equipment may differ somewhat from UNK controls proper.

2. Hardware Architecture.

The size and complexity of UNK, together with real time and other requirements, dictate a multilevel multiprocessor controls architecture (see figure).

Various components of the accelerator equipment are driven by more than 4000 equipment controllers (EC) which perform low level control and data acquisition tasks in hard real time and provide a uniform equipment representation for the upper levels of the control system. The ECs are based on 8/16 bit microprocessors and vary in implementation from dedicated single board devices to modular crates with standard backplane buses, mainly Multibus I in Euromechanics.

A general timing system distributes reference events and clock trains to all ECs. A separate alarm and interlock network collects signals from all ECs monitoring vital accelerator subsystems. These signals may be used to trigger the beam abort system and inhibit beam injection.

The next higher level of control is represented by the front end computers (FEC) spread around the main UNK ring and beam transfer lines and interconnected with upper level computers by the

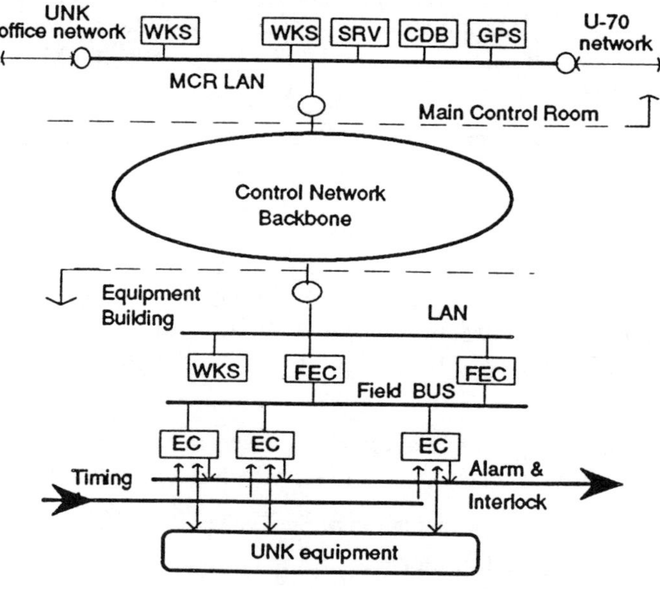

Figure 1. The UNK control system layout.

UNK controls network. The FECS will drive EC clusters through the 1 Mbit/s MIL-STD-1553 serial multidrop bus, which provides a cheap solution on a standard chip, noise immunity and galvanic insulation .

Physically, the FEC is a modular board assembly in a standard VMEbus crate. The basic FEC configuration will consist of a 32-bit processor board, a network interface and a number of MIL-1553 bus controllers. It may also include I/O modules for direct interfacing to equipment needing full network functionality or/and bandwidth to handle high data rates.

The FEC's main purpose is providing access for the upper level control software via network to the ECs. It may thus be considered as a gateway linking the MIL-1553 field bus to the UNK controls network and as a specific kind of network server, providing a set of equipment access services for application tasks running in the networked computers.

In addition to this general task, a number of FECs may be dedicated to certain functions, either through their own I/O modules (e.g. for sophisticated beam diagnostic equipment), or for functional segregating For example, one presently considers dedicating certain FECs to a more intense task in the real time control of various cryogenic processes. Finally, the FECs may run some application tasks, in particular for local equipment access, test and diagnostics. It will perform also ECs downloading and surveillance via the MIL-1553 field bus.

The network will be layered: a so-called backbone will interconnect the buildings and a number of LANs will interconnect equipment at the MCR and inside other buildings. The development infrastructure forms a sub-network which will be attached to the backbone. LANs will mostly be Ethernet. The backbone will be fiber optics FDDI or 16Mb fiber optics Token Ring. The LANs and backbone will be connected by bridges/routers (Internet standard). The high level network protocols will be TCP/IP suit. Most networking hardware and software will be standard commercial products.

UNK operator's consoles, 3-5 in total, will each be equipped with 3-5 graphics workstations and one server. The latter will provide common services (file, print, plot, etc.) and can also be used to execute certain run-time application software.

Each console configuration will include auxiliary equipment for supporting dedicated displays, analog signal observation and TV video links. The analog signal observation will be based on TDM channels, linking remote digitizers with digital to analog signal converters and scopes in the MCR. Signals are selected via the controls network, while the main data flow will be carried by dedicated, possibly very high bandwidth channels.

There will be a dedicated UNK control system's data base management server CDB and a general purpose server GPS for number crunching, modelling. The latter also caters for general user program development, thus supporting numerous workstations and terminals spread around the buildings.

3.Software architecture.

The run-time software is divided on two categories: Specific Applications to operate with the accelerator controls, and Application Environment, providing the applications with routine and unified services for communications, common data structures access and user interaction.

The diversity and multiplicity of process devices, requires some uniformisation, i.e. hiding the device specifics from the operational applications. For doing so, the device specifics are encapsulated in software envelopes having a standardized access protocol. The applications' vision is thus strictly limited to a small number of a well defined logical device types. The logical device abstraction has two closely related aspects: the device frame (a collection of data and procedural components to actually drive the device) and the device control protocol.

Off-line data preparation, including descriptions of machine and control system objects and relations, will be done using Oracle DBMS and relevant tools. One presently considers organizing all data inside of the control system in a specialized home-made real time DBMS. It should to contain both static read-only data derived from Oracle and dynamic data of the current machine state, some pre-defined number of UNK states for pulse-to-pulse modulation (PPM) and accelerator development, etc. This real time DBMS will support access to read/write data for fixed and off-line prepared data structures. It will serve a number of distributed data bases. Each DB is a standard file, containing data in the form of three-dimensional tables, some of which may be duplicated in the memory of specified computers. Physically the DB organization is optimized for real time usage.

The human interface is based on graphical workstations and windowing techniques (X-Windows, OSF/Motif, XUI). Much recent work in this field is concerned with specific extension of a commercial products (which are mostly some kinds of interactive graphics packages) to unify human interaction procedures as widely as possible.

Unix will be used in the operator workstations and servers while a Unix-like real-time system will be chosen for the time-critical fields (front end computers, equipment controllers). The TCP/IP protocol package, now a standard feature of most Unix and real-time systems, will be used in the general network and is also a basic communication mechanism in most X-Windows implementations.

4. Application software.

Experience in development of individual applications packages world wide shows that certain functionalities appear in the same or similar way in the majority of them. The trend, which we shall follow, is therefore to extract these parts from the specific applications and to supply them as more or less standardized common functionalities which may then be used by each specific application. This reduces multiple code and improves quality and maintainability. The software implementation of those common functionalities may be called the Applications Skeleton. But while systems software is computer or network oriented, the skeleton is controls oriented. It is supposed to provide a relatively stable environment and only evolve slowly.

A first systems analysis of the applications functionality was made using the SASD methodology. Thus the main procedures, data flows and data stores were identified. As a result, a first applications catalog was composed, containing the foreseen application procedures and data structures. This analysis, design and data structure development will be pursued using a modern integrated CASE tool package. The latter should also provide project management and documentation support, which are important when numerous programmers of different level must cooperate..

Standardization trends result in the appearance of commercial packages for industrial control which can be configured to realize a range of functionalities also in accelerators. Examples are DV-Draw /DV-Tool from Data Views for the user interface and ORACLE for off-line database management. Even certain parts of non-specific applications may be covered by commercial systems like V-System from VISTA or G2 from Gensym Corp. Some candidates will be evaluated and, where appropriate, be used while the market will be further monitored.

5. Cryogenics Aspects

The cryogenics and related equipment is a substantial part of the UNK project and falls into three broad groupings: (i) the helium liquifier plant with satellite refrigerators, distribution and recovery, consisting in turn of 4 subsystems: (a) compression, purification and storage; (b) the liquifiers proper, (c) satellite refrigetrators and distribution, and (d) nitrogen store and supply; (ii) the quench protection system and (iii) the superconducting magnet main power supplies with their ramping and dc programs.

By their nature these systems have a close internal binding and require only a weak coupling with the main UNK control system. A fair degree of autonomy and stand alone capability is therefore foreseen, which is helpful in commissioning and later servicing.

Cryogenics controls will, like the main UNK control system but with some special flavours, be based on Multibus-I with 16 bit processors for the ECs, connecting by MIL-STD-1553 to FECs. A specialized programmed logic controller module has been developed and will be used for controlling all valves and local feedback loops in the cryogenic system.

In collaboration with CEA, Saclay, a first version of a FNAL-like quench protection system will be tested on an experimental sector of 8 protection units, each consisting of 12 dipoles and 2 quadrupoles. The surveyance electronics and emergency heating power supplies are located in shielded cavities in the ring tunnel.

The main power supply controls will form a closed subsystem, losely coupled to the main controls and interlocked with the quench protection system.

6. External Beam Lines

The external beam zones will comprise a 12 km long neutrino channel, leading up to the neutrino experimental area, and three 6 km long hadron beamlines leading to their own experimental area each. An important part in the whole setup form the different target areas.

The technology of the equipment used in these beam lines and experimental areas is similar to the one of the accelerators proper, including the use of superconductivity hence cryogenics. A strongly different aspect, however, form the target areas with their radiation problems and remote handling requirements. Some advanced devices such as polarimeters and crystal bending and focusing may be used. One presently thinks of a controls architecture which is very close to the one described for the accelerators. There will be workstations with the modern windows and graphics oriented software, these will be interconnected with a TCP/IP based network, which in turn connects to the accelerator network, and at the frontend to VME based 32 bit microprocessors driving ECs over the MIL-STD-1553 fieldbus. In contrast with the accelerator system, the ECs of the external beam zones controls may still be largely CAMAC based.

Systems software will essentially be the same as for the accelerators. Applications will be strongly data based and model oriented and an expert system is being contemplated for operator support.

The operational patterns of the beamline zones are, by their nature, different from the accelerators. The essential aspect is the more frequent and rapid changes, following the experiment's requirements. Although routine operation may be done from a central MCR (combined with the accelerators), a strong local access and stand alone component remains essential.

7. U-70 Controls Upgrade

Contrary to the situation for the accelerator and beam zones controls, for which options were open, the U-70 injector group has a strong historical bias since both booster and U-70 proper are largely computer controlled. Moreover, the upgrade must be done virtually without interruption of the U-70 experimental programme. Finally, the urgency of this upgrade practically dictates using products now readily available in the USSR and using a stepwise implementation, converting small slices during the short planned shutdowns.

The FECs for the U-70 upgrade will be SM1810.30 computer crates with an Intel-like single board computer with the 16 bit 8086/8087 processor combination, using a suitable real time kernel. They will have appropriate RAM capacities, a LAN interface and a parallel branch highway CAMAC driver. Each of the about 12 FECs, spread over 4 buildings maximum 4 km apart, drives up to 3 CAMAC crates catering for I/O. Servers for files and data base, 4 to 6 in total, will be enhanced configurations of the FECs, featuring larger memory, hard disk and a more complete generation of the operating system. Interaction will be using PC-AT computers under DOS. A commercial LAN product is still being sought in the USSR.

8. Present Status

The Multibus-I based ECs have been largely defined, prototypes of most modules exist, preseries are expected in 6 months, industrial contracts are being negotiated. The MIL-STD-1553 fieldbus connection and remote terminal is under development. A conceptual design study of the upper part of the control system has been made and is accepted. Prototype partial integrations of the main control system and front end assemblies, as well as a quench protection test facility, are being prepared and should be available early 1992. An applications development environment with servers and workstations is being prepared. The external beam zones controls are in the conceptual design phase. The U-70 controls upgrade has been largely defined an frozen and implementation has started.

9. References

1. M. Mikheev, N. Trofimov, G. Sherbakov, S. Zelepukin, B. Kuiper, The UNK Control System, A conceptual Design Study, IHEP and CERN internal report, June 1990.
2. V. Alfiorov, S. Goloborodko, S. Logatchev, A. Skugarevski, The Control System for the UNK cryogenic Complex, IHEP internal report, Nov. 1990.
3. V. Sakharov et al., Preliminary Considerations on the Top Level Part of the Control System for UNK External Beam Channels, IHEP internal report, Nov. 1990.
4. S. Balakin, V. Brook, V. Voevodin, V. Tishin, The Control System for the Booster Synchrotron of IHEP, 19th School for Control Systems in Science Research, Novosibirsk, 1985, p.11.

Control System Specification for a Cyclotron and Neutron Therapy Facility

Jonathan Jacky, Ruedi Risler, Ira Kalet, Peter Wootton,
Alexandra Barke,* Stan Brossard, and Ralph Jackson

Department of Radiation Oncology†RC-08
University of Washington
Seattle, WA 98195

Abstract

It is usually considered an essential element of good practice in engineering to produce a specification for a system before building it. However, it has been found to be quite difficult to produce useful specifications of large software systems. We have nearly completed a comprehensive specification for the computer control system of a cyclotron and treatment facility that provides particle beams for cancer treatments with fast neutrons, production of medical isotopes, and physics experiments. We describe the control system as thoroughly as is practical using standard technical English, supplemented by tables, diagrams, and some algebraic equations. This specification comprises over 300 single-spaced pages. A more precise and compact specification might be achieved by making greater use of formal mathematical notations instead of English. We have begun work on a formal specification of our system, using the Z and Petri net notations.

1   Introduction

The Clinical Neutron Therapy System (CNTS) at the University of Washington is a cyclotron and treatment facility that provides particle beams for cancer treatments with fast neutrons, production of medical isotopes, and physics experiments [12,11]. The facility was installed in 1984, and includes a computer control system provided by the cyclotron vendor. Devices under computer control include a 900 amp electromagnet and a 30 ton rotating gantry, as well as four terminals at three operator consoles. The control system handles over one thousand input and output signals, and includes six programmable processors as well as some nonprogrammable (hard-wired) controls

The University is now developing a new, successor control system. This development project is motivated by requirements to make the system easier and quicker to use, easier to maintain, and able to accomodate future hardware and software modifications.

*Partially supported by National Institutes of Health grant number LM04174 from the National Library of Medicine

†Partially supported by NIH contract no. CM97282 from the National Cancer Institute

We have high reliability and safety requirements. There is growing recognition that the development of software which controls medical devices is not sufficiently systematic, that this is adversely affecting costs and safety, and there is much room for improvement [6].

We are attempting to achieve high reliability and safety by applying rigorous software development and quality assurance practices. We determined that our first step in this project should be the production of a comprehensive specification for the new control system, to serve as an authoritative and complete guide for software development, testing, and instruction of facility users.

2   Why write a specification?

It is usually considered an essential element of good practice in engineering to produce a specification for a system before building it. Modern recommendations for the development of safety-critical computer-based systems [2,8] emphasize the need for an explicit statement of functional and safety requirements, and a development process that can be shown to produce an implementation that meets the requirements. A central idea is that developers should provide a functional specification that allows system outputs to be predicted if system inputs are described.

However, it has been found to be quite difficult to create useful specifications for large software systems. A computer scientist has articulated the widely-held view that it is not practical to produce them:

> For many systems, it is simply not possible to construct a compact, concise specification. There are many different kinds of data and conditions, and a complete specification would have to detail the actions of the system for all of them. In practice such a specification is just a program. The working out of the many details only occurs as the system is being programmed. This has been observed to be a common feature of real-time and data processing systems [4].

Although we appreciate the difficulties, we do not agree with this view. Our experience reviewing the program code for the existing control system has made it clear that programs are no substitute for specifications.

3 Why programs are no substitute for specifications

We have been asked, "Why do you need to write a specification, when the program code that you write will tell you everything that the system will do?" We have found this attitude to be so widely held that it requires a detailed answer.

An obvious answer concerns project scheduling. We would like to have the acceptance test procedure and the instruction manuals for the users ready as soon as the programming is finished. To enable the acceptance tests to be designed and the manuals to be prepared while programming is underway, it is necessary to have a description of what the program will do, which is available before the program is complete. If this information has to be extracted from the program after it is written, testing and user instruction must be delayed.

Another answer is that there needs to be an independent standard of accuracy that the program can be tested against. This standard should be set by the users; in our case these are the physicists, engineers, and therapy technologists who use and maintain the facility. These are not the same people who will write the programs, and it is not reasonable for them to review program code to determine whether their needs will be met.

An answer that is less obvious to many software developers is that program code is not a suitable medium for expressing the information that needs to be in a specification. This is not merely a matter of the degree of expressivity provided by different notations. In fact, it is generally impossible to invert program code to determine the requirements that it is supposed to satisfy. When a program is written, requirements information is lost, but a great deal of information must be added that is not determined by the requirements, but is needed to code the program in the chosen programming language and get it to run on the available facilities.

4 Our specification

We have nearly completed a specification for the new control system. We believe it is as comprehensive as is practical to express using standard technical English, supplemented by tables, diagrams, and some algebraic equations.

The specification consists of four parts. Part I is an overview of the system that describes the facility, the hardware organization of controls, and introduces much of the vocabulary used in subsequent parts. It comprises almost 100 pages of single-spaced text. Part II is a detailed specification of operations which users perform at video terminals and control consoles. It comprises about 150 pages, including many illustrations of displays. Part III will be a detailed specifications of internal operations involving the cyclotron and therapy apparatus itself, which are only indirectly visible to users, and will comprise about 100 pages. the Part IV will be a largely tabular presentation of numerical operating parameters whose values can be changed independently of the rest of the specification. We plan to maintain these tables in a relational database.

Part I has been prepared for distribution outside our department as a technical report [7] and we anticipate releasing the other parts as they are completed in coming months.

5 How the specification was written

The specification is largely based on information gathered during meetings with facility users. The code and other documentation for the existing control system provide some useful information but have not been nearly as important as the interviews. The documentation for the existing system does not include a specification in the sense used here.

Most of the information was gathered in meetings that included the first two authors, a computer specialist (JJ) and the chief engineer of the facility (RR). Some meetings included other participants as well and there were meetings that did not include these authors. Most meetings lasted 2 to 3 hours. Computer specialist(s) asked questions; facility user(s) answered while a computer specialist took notes.

The first author drafted each chapter of the specification based on information gathered in a series of meetings, usually working from his own notes, sometimes from written material contributed by others. The chief engineer reviewed the chapter and the two met again. The review and ensuing discussion invariably revealed important errors, omissions and new information.

The first author then prepared a second draft. Second drafts were circulated among all authors for review. This review always resulted in some substantive revisions and many stylistic ones.

The first author then prepared the final version.

We believe that this lengthy process is the minimum necessary to achieve an acceptable specification because significant changes were always made after each cycle of review. In some cases more than two revisions were necessary.

The speed of the process was limited by the amount of time that facility users could spare from their other duties to meet with computer specialists and review drafts.

Most effort was devoted to creating and reviewing written material, but we also did some programming to experiment with the appearance of video screen displays and other aspects of the human-computer interface.

6 Formal specifications

Our specification is written in standard technical English, supplemented by tables, diagrams, and some algebraic equations. It is organized to help facility users read and understand it, so they can offer meaningful reviews. However, this organization is not necessarily the best for purposes of programming and analysis. An alternative is to make greater use of mathematical notations, annotated sparingly with English. Such specifications are called *formal specifications*.

We believe that there may be advantages to developing programs from formal specifications. They can be more precise and compact than English specifications. The can be checked by machine for certain kinds of errors. They can be used to prove or calculate whether the specificied behavior is consistent with certain intended properties, such as safety.

Formal specifications are distantly related to popular design notations based on data flow diagrams, which have been applied to acclerators [10]. We are investigating notations that provide more mathematical rigor, in the sense that they allow more properties to be inferred or calculated. Our preliminary experiments using the Petri net notation [9] are reported in [1], and some investigations into the Z notation [3,13] appear in [5].

7 Further work

We are preparing to implement the system we have specified. We will determine whether our specifications actually do provide sufficient information to build a useful system, and will assess the usefulness of formal specifications.

8 Acknowledgements

The authors thank Lee Hutter, Sandra Poirier, Astor Rask, and Jerry Sintay for sharing their knowledge of the facility.

References

[1] Alexandra Barke. *Use of Petri Nets to Model and Test a Control System.* Master's thesis, Department of Computer Science and Engineering, University of Washington, Seattle, Washington, 98195, 1990.

[2] Great Britain Health and Safety Executive. *Programmable Electronic Systems in Safety Related Applications. Volume 1: An Introductory Guide. Volume 2: General Technical Guidelines.* Her Majesty's Stationery Office, London, 1987.

[3] Ian Hayes, editor. *Specification Case Studies.* Prentice Hall International, Englewood Ciffs, NJ, 1987.

[4] William E. Howden. Validating programs without specifications. In Richard A. Kemmerer, editor, *Proceedings of the ACM SIGSOFT '89 Third Symposium on Software Testing, Analysis and Verification (TAV3)*, pages 2 – 9, ACM Press, 1989. (Also published as ACM SOFTWARE ENGINEERING NOTES, 14(8), Dec. 1989).

[5] Jonathan Jacky. Formal specifications for a clinical cyclotron control system. In Mark Moriconi, editor, *Proceedings of the ACM SIGSOFT International Workshop on Formal Methods in Software Development*, pages 45 – 54, Napa, California, USA, May 9 – 11 1990. (also in *ACM Software Engineering Notes*, 15(4), Sept. 1990).

[6] Jonathan Jacky. Programmed for disaster: software errors that imperil lives. *The Sciences*, 29(5):22–27, 1989. September/October.

[7] Jonathan Jacky, Ruedi Risler, Ira Kalet, and Peter Wootton. *Clinical Neutron Therapy System, Control System Specification, Part I: System Overview and Hardware Organization.* Technical Report 90-12-01, Radiation Oncology Department, University of Washington, Seattle, WA, December 1990.

[8] Nancy G. Leveson. Software safety: what, why and how. *ACM Computing Surveys*, 18(2):125–163, June 1986.

[9] Nancy G. Leveson and Janice L. Stolzy. Safety analysis using Petri nets. *IEEE Transactions on Software Engineering*, SE-13(3):386–397, 1987.

[10] G. A. Ludgate, B. Haley, L. Lee, and Y. N. Miles. The use of structured analysis and design in the engineering of the TRIUMF data acquisition and analysis system. *IEEE Transactions on Nuclear Science*, NS34(1):157 – 161, 1987.

[11] R. Risler, S. Brossard, J. Eenmaa, J. Jacky, I. Kalet, A. Rask, and P. Wootton. Routine operation of the Seattle cyclotron facility. In B. Martin and K. Ziegler, editors, *Cyclotrons and Their Applications: Proceedings of the Twelfth International Conference, Berlin, FR Germany, 8 – 12 May 1989*, World Scientific Publishing Co., Singapore, 1991.

[12] Ruedi Risler, Jüri Eenmaa, Jonathan P. Jacky, Ira J. Kalet, Peter Wootton, and S. Lindbaeck. Installation of the cyclotron based clinical neutron therapy system in Seattle. In *Proceedings of the Tenth International Conference on Cyclotrons and their Applications*, pages 428 – 430, IEEE, East Lansing, Michigan, May 1984.

[13] J. M. Spivey. *The Z Notation: A Reference Manual.* Prentice-Hall, New York, 1989.

The COSY Control System,
a Distributed Realtime Operating System:
First Practical Experience at the COSY-Injector

M. Stephan,* U. Hacker, K. Henn, A. Richert, K. Sobotta, A. Weinert
KFA Jülich, Germany

Abstract

The COSY control system is hierarchically organized with distributed intelligence and autonomous processing units for dedicated components. Data communication is performed via LAN and over a fieldbus. The hostsystems are UNIX-based, whereas the field-controllers are running a modular realtime operating-system RT/OS which has been developed at KFA. The computer-hardware consists of RISC mini computers, VME-computers in the field and G64 equipment-control-moduls in geographical expansion of the controller by a fieldbus based on the PDV-standard.

The man-machine interface consists of X-window based work stations. On top of X-window a graphical user interface based on object oriented methods is used. A distributed realtime data base allows access to the accelerator state from every workstation. A special highlevel language debugger hosted on the UNIX based workstation and connected over LAN to the VME targets will be used. Together with the software development system for UNIX applications an uniform view of the system appears to the programmer. First practical experience at the COSY injector is presented.

I. INTRODUCTION

The Cooler-Synchrotron COSY is under construction at the Research Center Juelich (KFA Jülich, Germany). This accelerator is intended to deliver high-intensity proton-beams with very low momentum-spread at an energy of 2.5 GeV. The existing variable-energy relativistic cyclotron JULIC is renovated to serve as COSY-injector delivering H_2^+–beams of 80 MeV. It is integrated into the newly desigend COSY-control to give a uniform operating-platform. Commissioning for the injector starts before commissioning of the COSY-ring, so it can be used to test the control hard- and software.

---

*Electronic mail address: martin@cecs.cc.kfa-juelich.de

II. COSY CONTROL HIERARCHY

The control-architecture is organized strongly hierarchically with distributed intelligence (Figure 1) and extensive use of standards.

At the top level of the computer-control-hardware workstations give the operators graphical access to the process. For these tasks Hewlett-Packard 9000/300 computers with Unix HP-UX and X Window System version 11 are currently in use. RISC-computers of the series 800 give computing power for model calculations and long-term data-bases.

These computers are interconnected using Ethernet (IEEE 802.3) and TCP/IP to the next layer of hardware: the workcells. Each workcell autonomously controls a subsection of the accelerator (e.g. workcell "woody" is dedicated to the COSY-injector and workcell "frieda" to the dynamic dipole-power-supplies). Workcells are HP9000/800 computers with Unix operating-system.

In each subsection a specific Ethernet-line connects all field-controllers to the corresponding workcell. Diskless VME-systems contain the field-controller and additional CPUs and I/O-cards. All CPUs are running the real-time operating-system RT/OS. For fast I/O the interface-cards are directly connected to the controlled device.

The VME-control-hierarchy for the COSY-injector has been distributed on 9 VME-systems. Each system is dedicated to a subtask in accelerator-control like control of the beamline-power-supplies or cyclotron-diagnostics. Typically these are multi-processor systems with all CPUs running RT/OS and communicating with each other over the VME-bus. In slot 1 of each system the field-controller handles the network-communication to the workcell and runs specific applications. Typically this is a CPU-board "E5" (Eltec company, Mainz, Germany) equipped with a 68020-CPU (16 MHz, 1 MByte triple-ported RAM) and a piggy-pack thin-wire Ethernet-interface. Additional CPUs are interfacing to the fieldbusses. CPU-boards "IBAM" (Eltec, Mainz, Germany) with 68010-CPUs (10 MHz, 512 kByte triple-ported RAM) carry up to two piggy-pack fieldbus-controllers.

For slow I/O a level of G64-systems [1] is introduced below the VME-level and accessed by PDV-bus [2] as fieldbus. This gives higher modularity and flexibility in inter-

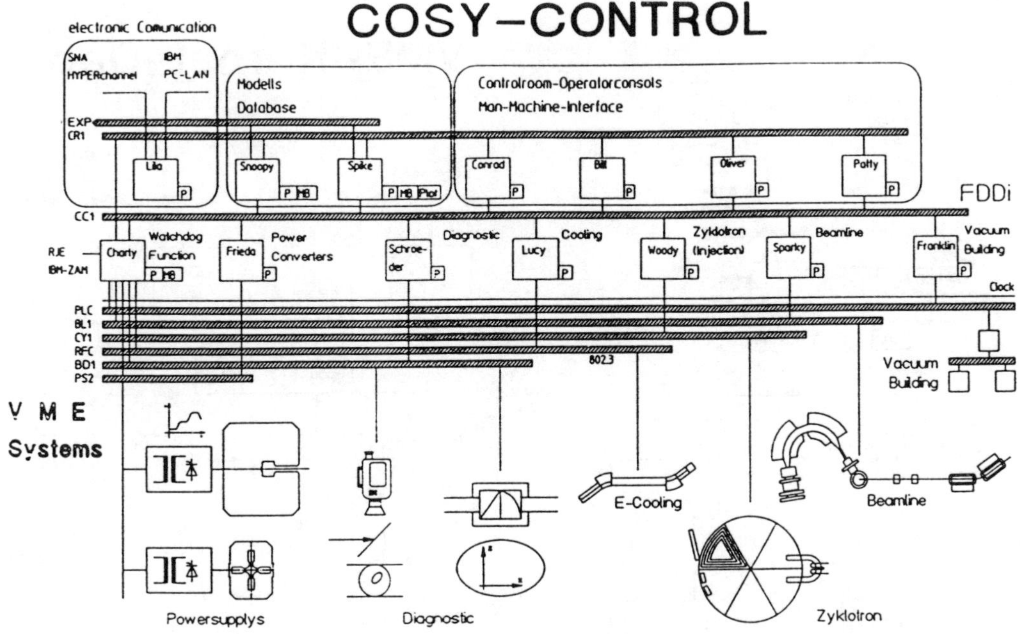

Figure 1: Cosy Control Hierarchy

facing to geographically distributed devices. Up to 4000 transactions per second have been reached on the PDV-bus. The PDV-controllers for VME and for G64 have been developed at the KFA. In G64/G96 typically 16 bit DACs are in use for magnet-power-supply control which are fully isolated against the bus. Additionally digital I/O is performed with optoisolated 16 bit parallel cards. The pulse-signals for stepper-motor power-stages are delivered by autonomous motor-controllers.

III. INJECTOR CONTROL SOFTWARE

A uniform computer access is given to accelerator operators and software developers by use of Unix and X Window System with fully integrated program development for the VME-targets. For the Software Development System (SDS, Figure 2) the GNU-software [3] has been extended to communicate with VME-targets via the network. The GNU-compiler gcc is taken as cross-compiler for the CPU-68000.

gdb gives dbx-like debugging-tools for programs running on VME handling also the communication with the rtdb on the target. Aside the standard language C also C++ is available on Unix-machines for object-oriented developments.

Operator access to process control is given by command-driven, menu-driven or graphic interaction. For rapid prototyping of visualisation three products have been used: DataViews, InterViews and ObjectWorks. The actual process picture is always contained in a "pmf" database and is modified by these operator interactions. The process picture can be accessed from different computers and also envents can be signaled. The "pmf" database has been developed at the KFA.

All VME-systems are equipped with a DataCom-CPU containing the remote-debugger rtdb in EPROM. Programs developed using SDS can be downloaded from the workcells to all CPUs in the VME-system. In particular the realtime operating-system RT/OS [4] has been developed at the KFA Jülich in this environment (Figure 3).

It is an implementation of the D-MOSI standard [5] and was designed with high modularity in all layers: Applications, kernel, device-drivers. A priority-controlled timeslice with preemption is provided for process control.

IV. PRACTICAL CONSIDERATIONS

AFTER INJECTOR COMMISIONING

For the instrumentation of the injector 9 VME-systems with 10 PDV-busses are installed serving 24 G64-crates. With the flexibility of the G64-systems existing local-control equipment has been made remote-controlled.

For substitution of missing hardware functions the special method of pseudo-device-drivers has been introduced. The tests of the process I/O is done integrated in the SDS of RT/OS. Rapid Prototyping of graphic user interfaces is performed by use of ObjectWorks for setup of operator control interaction.

Modularity of hard- and software gives sufficient opportunity in adaptation to existing hardware. The extense use of existing tools accelerates development times. Tools for

HP–PA÷VME

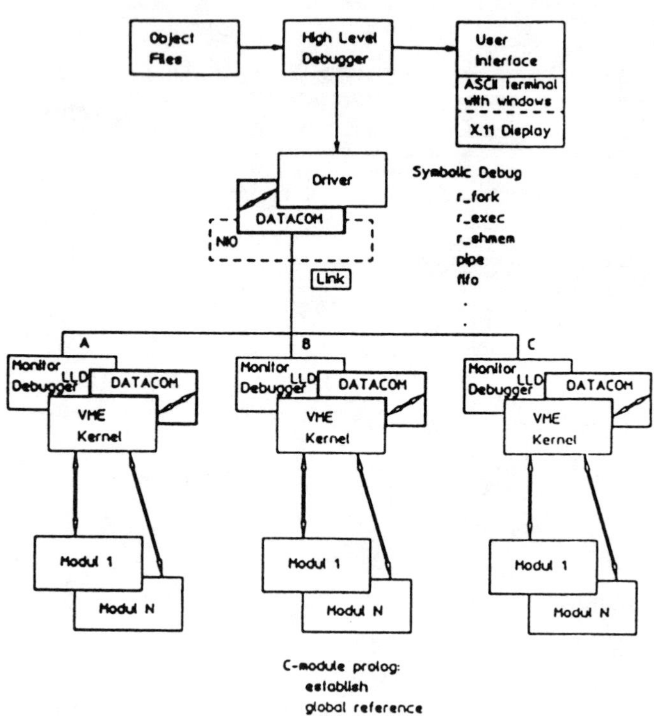

Figure 2: Software Development System

rapid integration of protocol-handlers are yet missing.

As a result a special language for process-description and control PBS has been developed. A cross-compiler has been implemented that produces C-source-code for applications running under RT/OS.

References

[1] Gespac G-64 Bus Specifications Manual, Version 3.1, January 1990

[2] DIN 19241, Bitserielles Prozeßbus-Schnittstellensystem, Juli 1985 (german standard)

[3] GNU-software, Free Software Foundation, Cambridge, MA 02139, USA

[4] anonymous ftp: cecs.cc.kfa-juelich.de

[5] IEEE std 855, IEEE Trial-Use Standard Specification for Microprocessor Operating Systems Interfaces, 1985

VME–Kernel–Debugger

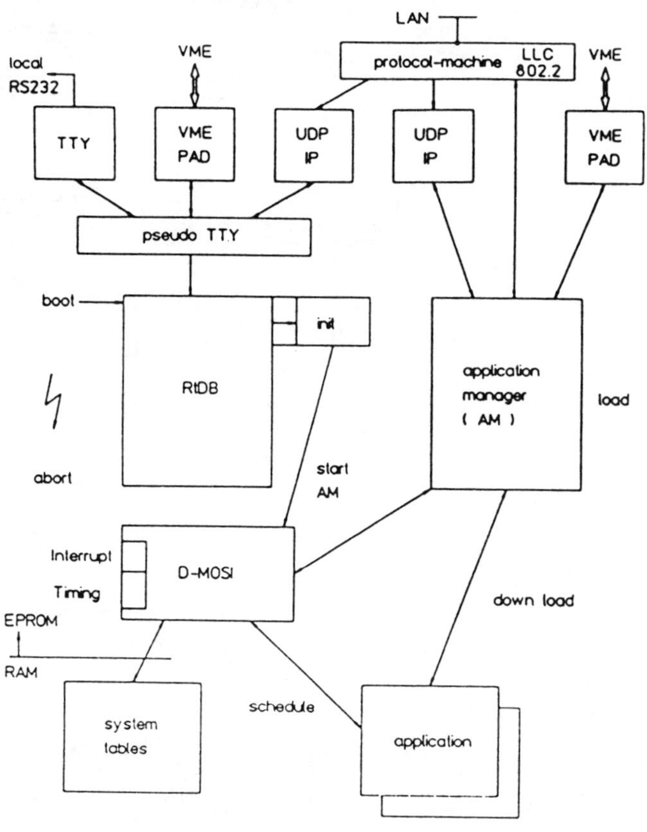

Figure 3: RT/OS

INITIAL CONTROL OF THE H⁻ ION SOURCE
AT THE SUPERCONDUCTING SUPER COLLIDER LABORATORY

G. Martinsen, S. Acharya, M. Allen, E. Faught, K. Low, J. Sage
SSC Laboratory *
2550 Beckleymeade Ave.
Dallas, Texas 75237

Abstract

The ion source produces a 30 mA beam of 35KeV H⁻ ions for SSC accelerators. The beam is chopped at 10 Hz into pulses of 7 to 100 μsec. The ion source presents an opportunity to implement and exercise software tools and techniques which will be useful in future SSC control systems. TACL, a software package from the Continuous Electron Beam Accelerator Facility, forms the core of the system. TACL controls several analog channels and monitors interlocks. Emittance measurement control is now under design. In addition to TACL, components of ISTK are also used to interactively acquire and display ion beam information. The integrated system, including VXI and CAMAC acquisition modules attached to a network of heterogenous computers, is described.

I. INTRODUCTION

The SSCL will have two H⁻ Ion Source types: a volume source to be built by Lawrence Berkeley Laboratory (LBL)[1] and a backup magnetron source built by the Texas Accelerator Center (TAC)[2]. The magnetron source, currently functioning as a prototype is discussed in this paper.

The beam current, specified at 30 mA and 35 kV, will be chopped to achieve a 7 to 100 μsec pulse width at 10 Hz. The design goal is an emittance of 0.2π mm-mrad

The Low Energy Beam Transport (LEBT) will be one of two types. A Helical Electrostatic Quadrupole (HESQ)[3] is the main design; with an Einzel Lens for backup. Due to low cost and simplicity of fabrication, the Einzel lens is now installed in the prototype and is discussed in this paper.

Diagnostics will consist of a phosphor screen monitored by television camera for remote viewing, charge collector, slit and collector in both transverse directions, two flying wires positioned 5 and 9 cm downstream from the LEBT, and a current toroid positioned at the exit of the LEBT.

II. CONTROL SYSTEM PROTOTYPE

A. Control Block Diagram

Figure 1 indicates that the present points available for monitoring and control are few, as scientists wish to have direct control while tuning the source.

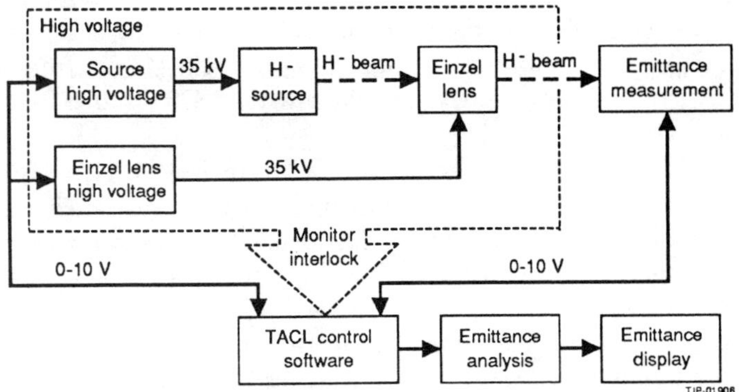

Figure 1: Control Block Diagram

The prototype control system monitors the high voltage power supply for the magnetron source, the safety interlock, and monitors and controls one of six high voltage power supplies for the einzel lens elements. Emittance measurement control is now under design.

The emittance measurement is done by independent slit and collector probes. The collector consists of 48 wires each separated by 0.254 mm, and is located 150 mm downstream from the slit. The slit width is adjustable but is nominally set at 0.1 mm. After each successful collection of data, the slit advances one slit width. Beam diameter is approximately 30 mm, which results in about 400 data points to fully traverse the beam and surrounding area. Ions not passing through the slit are collected from the slit plates and sampled. If the slit charge is not within tolerance from the expected value, the candidate data set from the collector is discarded and a new data set is taken, at the same position, on the next available pulse. Ions deposited on the collector yield intensity values for the 48 wires. It is expected that 6-10 wires will have significant values during each measurement. To ensure that all of the beam projection falls on active elements of the collector, data sets indicating excitation of wires too close to the ends of the detector are discarded and, after repositioning, a new data set is taken. An attempt is made to position the collector for each data collection so as to minimize the number of discarded data sets. Five analog-to-digital converters with 12 channels and 11 bit accuracy are used. All 60 channels are gated by the same NIM gate, which is triggered by the same master timer used to generate the beam pulse. Forty eight channels are assigned to the collector wires, while the slit, and each of the two flying wires, are assigned one channel. The width and delay of the NIM

*Operated by the Universities Research Association, Inc., for the U.S. Department of Energy under Contract No. DE-AC02-89ER40486.

gate is calibrated for the collector wires at midbeam. The current from the slit is attenuated by a resistive network before sampling. This allows the slit current which is much larger than the collector wire currents to be sampled in the same NIM gate interval.

B. TACL Control Software

Thaumaturgic Automated Control Logic (TACL) [4] is a hierarchical, distributed, and extensible control system. It is coded in C and operates in Unix on HP 9000 series computers. It interfaces mainly with CAMAC modules, but can be configured to interface with RS232 and GPIB instruments.

TACL has been extended at SSC to include VXI/VME interface capability. An oscilloscope in a VXI crate with an embedded processor running TACL has been used to monitor the ion source beam. Programming the oscilloscope interface was done in much the same manner as GPIB instruments have been done in the past, using ASCII string commands from within a custom user process. The beam trace data may be displayed by TACL and Kaspar, a plotting utility which will be described later. In addition, the VXI capability has been extended to register-based devices which allow memory mapping and fast backplane access to VXI devices. No custom code is required for these devices and when hardware becomes available, we expect to use this same technique to interface to VME cards in a VXI crate.

TACL uses two off-line editors to build the runtime system: a LOGIC editor and a DISPLAY editor. In the LOGIC editor, the user configures the hardware by selecting CAMAC modules from an extensible library and clicking on the desired slot on the screen icon graphically representing the CAMAC crate. Clicking on another icon displays the LOGIC screen which presents various menus and a two-dimensional grid.

In this grid, ladder logic is built which defines the control system algorithms. The elements making up the control logic may be selected from menus or custom-built "user-processes". The LOGIC array executes down the columns from left to right until the last column is executed and then starts at the beginning again. This constitutes a LOGIC cycle. CAMAC reads are automatic on each LOGIC cycle but writes can be either automatic or executed on demand.

Figure 2 shows the section of LOGIC array which controls the emittance measurement. The figure contains standard LOGIC elements plus two custom "user processes" written in C and executed by LOGIC when triggered.

Emittance measurement begins when the momentary switch labeled "start" is closed by the operator. This initiates the user process "qkck" which does four things: validates data when available, calculates motor movement for the next data collection, triggers motor movement by setting "trig", and signals completion of "qkck" by setting the "ckrdy flag". When the motors have stopped (busy3 is low), the CAMAC LAM signals are cleared and the NIM

gate is enabled in preparation for data collection. The second user process, "cklam", is initiated which: waits for LAMS to indicate signal digitization is complete, resets "ckrdy", sets "start" to repeat the above cycle.

The operator can enable/disable data gathering by toggling the enable button on the user screen.

The cluster of LOGIC in the lower left corner of the figure initiates motor movement.

The part of the LOGIC array for bit status monitoring and control is not shown in Figure 2 but consists of assigning the binary control points to a word to be sent to CAMAC and converting the output word from CAMAC to binary monitoring points. This allows the user screen to attach toggle symbols and/or pushbuttons to every binary point if desired.

Digital values reflecting the power supply settings are made available to the user screen before and after filtering by placing a "display output" symbol for each condition in the block diagram. Bar graphs are tied to the noisy signals while digital readouts are attached to the filtered values.

On the user screen (not shown), bit status is shown on a "byte" display which displays up to 16 bits. Each bit display changes color when the sytem response indicates that the bit was indeed set. Clicking on the bit display will set or reset the bit. Bits 9-16 are not settable by software but are reserved for monitoring. The interlock status bit is monitored at bit 16. Bits 1 and 2 are currently used for testing. All other bits are yet to be assigned. Analog control is done at read/write boxes by a point-click-enter sequence or by assigning a knob to the read/write box. Analog monitoring is done with bar graphs and digital readouts.

III. ISTK SOFTWARE

The ISTK collaboration [5] consists of a group of scientists, engineers, computer scientists, and industry representatives working on an informal basis to produce a collection of software products for use in engineering and science. The goal of this effort is to provide a tool kit that will contain a growing and evolving collection of scientific and engineering tools connected by a high-level integration architecture (a software bus). The following ISTK components are used during the operation of the ion beam source:

1. SDS (Self-describing Data Standard) - SDS is a data discipline for storing and retrieving binary data. It is portable across different architectures and supports different storage formats such as SYBASE, a commercial database, shared and process memory and disk.

2. Kaspar - Plotting utility for viewing self-describing data.

3. Camel - A control-panel designed for interfacing to and retrieving data from CAMAC crates. It stores the data it acquires as SDS datasets.

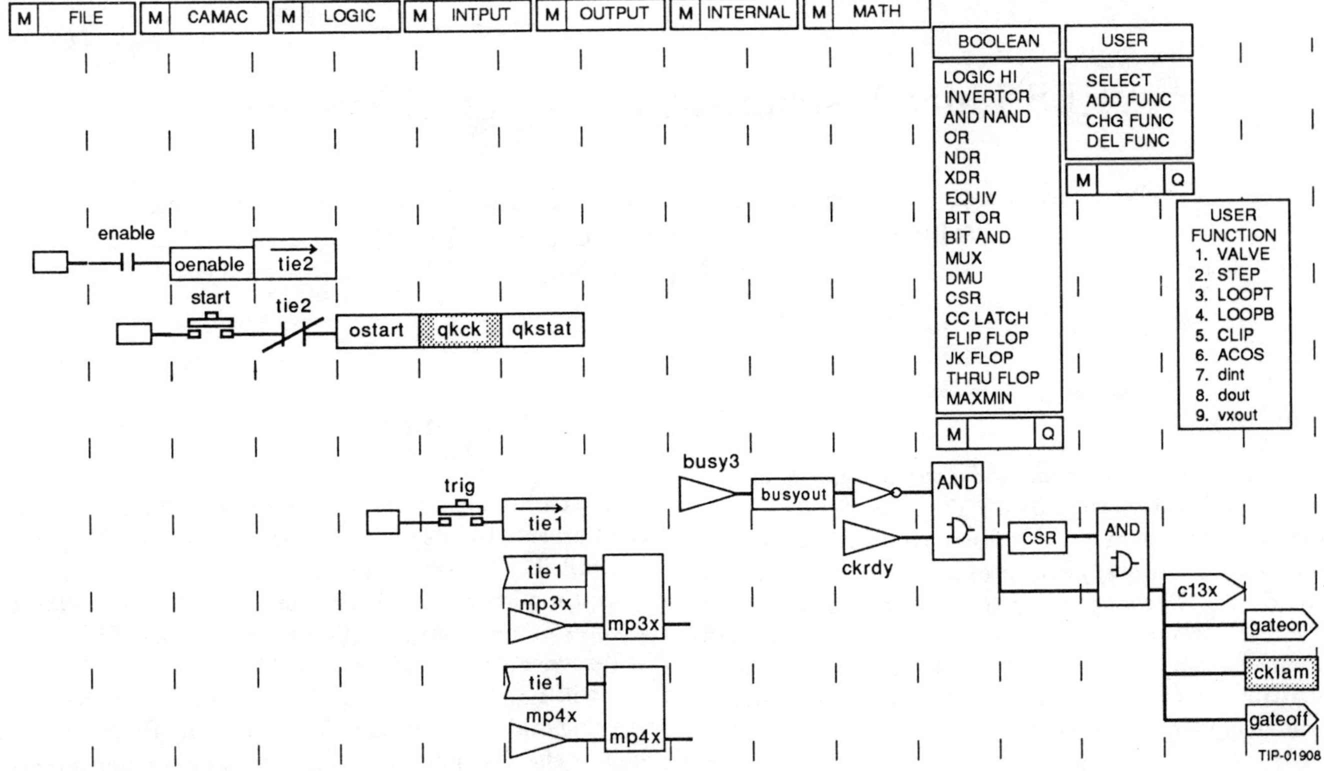

Figure 2: TACL LOGIC

The data acquisition and display sequence begins with a CAMAC data logger informing Camel that there is valid data available in its buffers. Camel retrieves this data, stores it as an SDS dataset in shared memory and informs kaspar. Kaspar is attached to this same SDS dataset and awaits a signal from Camel. When it arrives, Kaspar retrieves the data from the dataset, replots it on-screen and the entire sequence begins anew. During the preliminary operation of the ion source, Kaspar and Camel were connected via sockets. Ultimately, glish will be used to direct the sequence and oversee the process interconnections. Glish [5], also developed as part of the ISTK collaboration, is a system for accelerator control sequencing, providing a general mechanism for processes to exchange messages and to be sequenced according to both abstract and real (hardware) events. The sequencing is rule-based and processes can be controlled across the network.

IV. FUTURE PLANS

As the machine is tuned, more monitoring and control points will be made accessible. A current list identifies 117 monitoring and control points. In many instances, the additions are typical to tried and tested control LOGIC and DISPLAY and can be quickly duplicated. This allows the system designer to concentrate on designing LOGIC and DISPLAY of the new elements.

V. CONCLUSION

The above system has functioned well within the limited scope assigned. Physicists have been pleased with the ease of use and rapid development time. The addition of ISTK tools in parallel, allows browsing and plotting of data without interrupting operator control. The prototype has provided a valuable testbed to implement and test software tools and techniques to be used in future SSC control systems.

References

[1] K. N. Leung, C. A. Jauck, W. B. Kunkel, and S. R. Walther, Rev. Sci. Instrum., 61, 1110 (1990).

[2] An Optimized H^- Magnetron Ion Source/LEBT System, P. Tompkins, et.al Proceedings of the 1989 Particle Accelerator Conference, Chicago, Illinois, 1989

[3] Transport Properties of a Discrete Helical Electrostatic Quadrupole, C.R. Meitzler, P. Datte, F.R. Huson, L. Xiu, J. Ziegler, Texas Accelerator Center, and D. Raparia, SSC Laboratory, Paper MSC 7 at this conference.

[4] CEBAF CONTROL SYSTEM, *R. Bork, C. Grubb, G. Lahti, E. Navarro, J. Sage, Continuous Electron Beam Accelerator Facility* T. Moore, Lawrence Levermore National Laboratory, Proceedings of the 1988 Linear Accelerator Conference, Newport News, Virginia

[5] unpublished internal documents at SuperConducting Super Collider Laboratory, C. Saltmarsh, M. Allen, S. Acharya

PLS Linac Instrument and Control System*

C. Ryu, S.S. Chang, J.H. Kim, M.S. Kim, D.K. Liu,† S. Won, W. Namkung
Pohang Accelerator Laboratory/POSTECH
Pohang, Korea 790-600

Abstract

A plan for the instrument and control system of the PLS (Pohang Light Source) 2 GeV linear accelerator is described. Major beam diagnostic instruments consist of 2 *ns* beam current monitors, beam profile monitors, and beam loss monitors. The control system will adopt the VME bus. For the timing system, the basic timing pulses will be obtained from the storage ring's RF master oscillator and distributed to the gun pulser and klystrons through the time delay electronic modules.

1. Introduction

The PLS 2 GeV linac is a full energy injector for the PLS storage ring. This injector linac consists of a gun, 42 accelerating column, and 11 klystrons, with 10 energy doublers. The total length is about 150 *m*. In this PLS linac, electron beam will be accelerated to 2 GeV with the beam current of 200 *mA* in normal operation and of 500 *mA* in maximum. Beam pulse length will be 2 *ns* with the repetition rate of 10 *Hz* for the injection. The initial purpose of the PLS linac is to provide the storage ring with a 2 GeV electron beam. Later on, however, the linac will be used for other physics experiments, which will be carried out at facilities located at the beam analyzing stations.

Based on the above requirements, we plan to monitor the following beam characteristics as the essential physical quantities for the operation of the linac : beam current, beam position, beam emittance, beam profile, beam energy spread, beam bunch length, and beam loss location. For this purpose, we will employ the beam current monior, beam profile monitor, beam loss monitor, and beam bunch monitor.

For the control system, we plan to adopt a distributed control system based on VME bus. And for the timing system, we plan to take a scheme of distributing the timing pulse of the storage ring's R.F. master oscillator to the gun grid pulser and to high voltage modulators and klystrons.

---

*This project is suppoted by POSCO and MOST, KOREA.
†Permanent address : IHEP, Academia Sinica, PRC.

2. Beam Current Monitor

For accurate measurement of a 2 nanosecond short pulse beam current, 10 sets of beam current monitors(BCM) will be installed. These curresnt monitors are wall current type ones with cylindrical resistors surrounded by toroidal ferrite cores in metal shields. This type of monitor has a very fast response with excellent signal to noise ratio. Seven of them will be on the linac and three will be in the beam analyzing stations. During the operation, BCM's will be the most effective monitors. The operator can optimize the beam intensity by adjusting the RF phases, the quadrupole triplets, and the steering magnets based on the readings of these monitors. Fig.1 shows the schematic layout of the beam current monitors as well as other beam diagnostic instruments.

3. Beam Profile Monitor

In the commissioning period and in later operation, direct observation of the beam spot size and position is expected to provide very useful information. For this purpose, beam profile monitors using a fluorescent target are used. These monitors can be used to measure the beam energy, energy spread and emittance. Utilizing recent advances in image processing technology, video signals taken by a camera will be digitally processed. Digitally processed beam information will be much more accurate and useful than that observed by eye on a TV monitor. There will be a total of 9 profile monitors in the PLS 2 GeV linac. Three of them will be located on the linac, three of them will be in the transport line to measure the beam emittance, and three will be in the beam analyzing stations.

4. Beam Loss Monitor

The beam loss monitor is a long ion chamber which will be installed along the 2 GeV linac. It is a coaxial cable with a length of 6-8 *m*. When the electron beam hits the accelerating column, radiation will be emitted. The strength of the signal from the ion chamber is proportional

"

The considerations applied to this architecture are as follows:

- PLS linac control system can get greater benefits when it utilizes off-the-shelf items and already existing software. VME modules are getting very popular in the Factory Automation market. For this reason, they are adopted as control modules for the stroage ring. Using the common modules between the ring and the linac will reduce the overall development cost, and, later, maintenance efforts.

- As VME becomes more popular on the commercial market, more third vendors will produce compatible items, forming a more powerful development environment.

Based on the VME system, the architecture of the 2 GeV linac control system is configured as shown in Fig. 2.

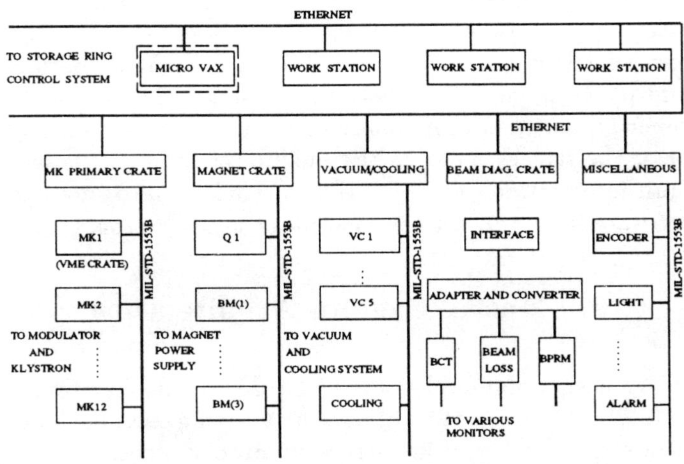

Fig.2 Control Architecture of 2 GeV Linac.

9. Timing system

A basic requirement of the timing system is that beams should fill up a single RF bucket of the storage ring in any pattern. For this purpose, the linac beam pulse needs to be synchronized with the storage ring's 500 MHz RF frequency. Thus timing signals are taken from the storage ring's RF master oscillator. The time interval between the RF buckets is 2 ns, and this 2 ns will be used as a basic time unit in the linac timing system. Thus, these precise 2 ns pulses synchronous with the ring RF oscillation will

be distributed to the gun, klystron and other devices on a pulse to pulse basis at up to 60 Hz.

Functions of the timing system will consist of pretriggering, acceleration and standby modes of operations.

For the triggering of klystrons, timing signals are generated by time delay modules. These triggers are sent to the MK and delayed by a programmable counter.

A schematic diagram of the timing system is shown in Fig. 3.

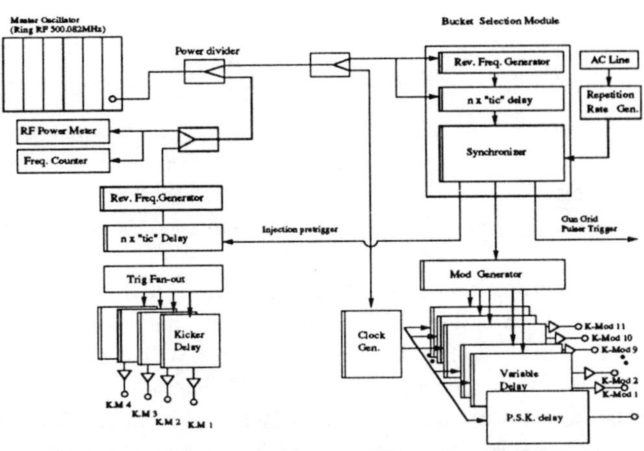

Fig.3 Schematic Diagram of the Timing System

References

[1] PLS Project, "Pohang Light Source Conceptual Design Report," POSTECH, 1990.

[2] R. B. Neal eds., "Stanford two mile accelerator," Benjamin, 1967.

[3] P. M. Lapostolle and A. L. Septier, eds. "Linear Accelerator," North Holland Amsterdam, 1970.

[4] E. Plouviez, "The automatic phasing system for the LEP Injector Linacs (LIL)," LAL/RT/84/03, April, 1984.

[5] BEPC, "Summary of the preliminary design of Beijing 2.2/2.8 GeV electron positron collider," BEPC design group, Dec., 1982.

[6] G. Stange, IEEE Trans. on Nucl. Sci., Vol. NS-26, No.3, June, 1979.

[7] M. C. Ross, "Beam diagnostics and control for SLC," 1987, PAC.

[8] T. Urano and K. Nakahara, "Trigger control system of the PF linac," NIM, A247, 226, 1986.

to the radiation. From this information, it is possible to identify the place where the beam is lost.

5. Bunch monitor

The bunch monitor consists of two RF cavities, one resonant at 2856 MHz and the other one at 14280 MHz. It is mounted on the beam line 10 m down from the gun. When a beam pulse passes through the bunch monitor, the monitor picks up signals whose frequencies are resonant with the cavities. The difference in power output between the two detected signals is

$$P_m = \frac{1}{2}R_m I_0\left(1 - \frac{1}{2}m^2\theta^2\right)$$

where $m = 5$ is a harmonic number, R is the shunt impedance of the cavity, I_0 is the fundamental component of the current, and θ is the phase width of the bunch stretch in radian. The smaller the θ is, the stronger is the output power P_m.

In principle, it is possible to use the bunch monitor to measure the absolute beam length, but there is some difficulty in reality. So the bunch monitor will be used as a qualitative measurement of the bunch length.

The cavities of the bunch monitor will have small apertures (9 mm in diameter), so it should be mounted on a special support to ensure that the beam passes through the cavities precisely.

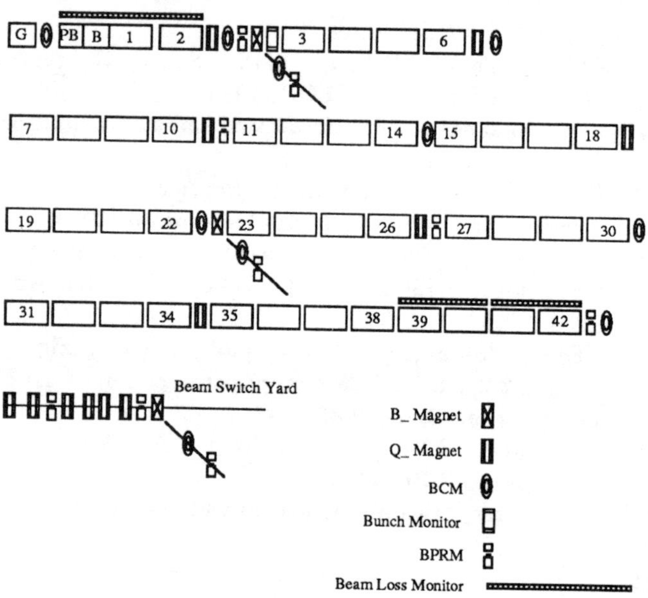

Fig.1 Beam Diagnostic Instrument Distribution in the Linac

For the same reason the bunch monitor will be located after the accelerating column number 6. At this position, the beam energy will be high enough to ensure the required small beam size.

6. Phase Amplitude Detector

The phase of the RF wave in the drive system will be measured for a comparison with the signal on the reference line. For this purpose, phase amplitude detectors(PAD) will be used. A PAD unit consists of a local phase shifter, nulling detector and amplitude monitor. The data obtained in the PAD unit will be sent to the MK control station and to the operator in the control room through a data link.

7. Modulator Klystron Support Unit

A klystron, modulator, and Energy Doubler(ED) cavities will be monitored and controlled by using a modulator klystron support unit(MKSU). The monitoring and control values taken care by an MKSU are; the amplitude and phase of the RF drive to the klystron, the ED cavity tuner position, current supply to the klystron electromagnets, the trigger pulse to the modulator, and status and fault information. The MKSU also provides hardware and software interlocks for the modulator and klystron.

8. Control System Architecture

The design philosophy of the PLS linac control system is to fulfil the requirements of linac by utilizing recent achievments in computer technology with minimum cost. Due to the rapid progress in VLSI technology, powerful intelligent controllers are readily available these days, and it has become possible to construct a powerful distributed control system at a reasonable cost.

In the PLS linac control system, a combination of front end intelligence controllers based on VME single board computers, microcomputers, and minicomputers are planned. A minicomputer (VAX 3100 size) will be used as a data acquisition computer and as a host computer. 32 bit workstations will be used as consoles and development stations. Thus, the system structure consists of computers, workstations, various intelligent controllers, networks and dedicated interfaces.

Between the linac and the storage ring control system, and between the linac host computer and the primary local VME intelligent crates, the Ethernet network standard (IEEE802.3) will be used. Between the primary VME crates and the secondary VME crates, MIL-STD-1553B will be used. The secondary VME crates will directly control the instruments, and their primitive database will be maintained in the primary VME crates.

Control System for the MLI Model 1.2-400 Synchrotron Light Source

B. Ng, R. Billing, R. Legg, K. Luchini, D. Meaney, S. Pugh, Y. Zhou
Maxwell Laboratories, Inc., Brobeck Division
4905 Central Avenue
Richmond, California 94804

Abstract

The control system for the MLI Model 1.2-400 Synchrotron Light Source is now being assembled and checked out. This system controls the injection of 200 MeV electrons from the linear accelerator into the storage ring, and the acceleration of the beam to 1.2 GeV, after accumulating current of 400 mA. The ring consists of 8 bending magnets, 20 quadrupoles, 16 sextupoles, an rf cavity driven at 500 MHz, and miscellaneous vacuum and utility support equipment. The control system also supports a transport line that consists of 2 dipoles, 11 quadrupoles, and 5 fast injection magnets. The storage ring is a synchrotron radiation source for research applications including development of x-ray lithography techniques. A detailed discussion of the approach, hardware and software implementation, and preliminary performance data are presented.

I. DESCRIPTION

The control system for the MLI Model 1.2-400 Synchrotron Light Source is designed for customers who plan to use the light source as a tool, rather than as an experiment. This emphasis on practicality implies an operator interface that is friendly to the operator. Many of the higher level operations are automated.

To make the control system economically attractive, development was minimized. Commercially available software has been used as much as possible. Hardware to support the chosen software was also selected to minimize development.

A. Software

MLI has selected a commercially available control system development platform (VISTA) that has a complete facility for user interface and a structure set up to handle the interface between hardware and user. It presents control and readback to the user via graphical screens; examples are shown in section II. The software has been described in an earlier paper [1].

The following have been developed inhouse by MLI:

- the database representing the real hardware connections has been defined.

- graphics interfaces connecting the user to the hardware control and readback have been drawn; these include:
  - magnet control and readback
  - vacuum control and readback
  - rf control and readback
  - injection timing control
  - diagnostic control and readback
  - diagnostic screen control, video digitizing, signal processing, and other beam monitoring systems

- the software drivers that the original system lacked have been written; these include:
  - individual CAMAC card handler
  - stepper motor drivers
  - PLC scanner handler
  - GPIB handler

- the higher level application software required to enable the operator to commission the machine has been written; this includes:
  - coordinated magnet ramping
  - automatic ramping data from accelerator physics parameters
  - orbit calculation correction
  - smooth restoring of settings for magnets that cannot tolerate any step change

B. Hardware

The hardware has been discussed in an earlier paper [2]; a short recap is presented here.

- Computing hardware:
  - Vaxstation 3100, as the user interface
  - MicroVax 3400, as the hardware connection to some of the interfacing electronics:
    - GPIB controller
    - video digitizing
    - industrial programmable logic controller (PLC) scanner interface
    - CAMAC controller

  Because it is the higher performance machine, the MicroVax is also used as the machine to maintain the realtime data base.

- Interfacing electronics:
  - CAMAC: because of the support already provided by the commercial package, this handles most of the magnet controls and diagnostics
  - GPIB interface: injection timer control
  - commercial PLC, for controlling the following:
    - vacuum interlock logic, control, and status (see Ref. [3])
    - dipole power supply protection logic, control, and status
    - injection magnet control and status
  - direct plug in card to the "Q bus" of the MicroVax: video digitizing

II. STATUS

A. Completed Development

The control system is now largely complete. First-cut menus are now available for all I/O points (approximately 600 points). The Video Monitoring menu (figure 1) is a typical graphics menu that makes use of the physical location of the unit to help the operator identify the device being controlled. The fluorescent screen at each location can be activated (inserted into the path of the electrons) by moving the cursor to the appropriate camera symbol and "clicking" on it with the mouse. The color of the camera indicates whether the screen reached its intended loca-

tion. For analog readback to the operator, both graphics and text are used in the display wherever possible. An example is the Beam Slits box in the Video Monitoring menu. Here the graphics display is also the control command; by "dragging" the end of the graphical slit with the mouse, the actual slit will be moved in the corresponding direction.

By activating the appropriate "button" on the menu, the Optical Monitoring Station menu will appear (figure 2). This is an example of a menu where the color and position of the symbol itself are part of the status indication. The video multiplexer will also be set up by the selection of the particular camera and will direct the signal to the TV monitor and the video-digitizer. By activating the appropriate "buttons" on the

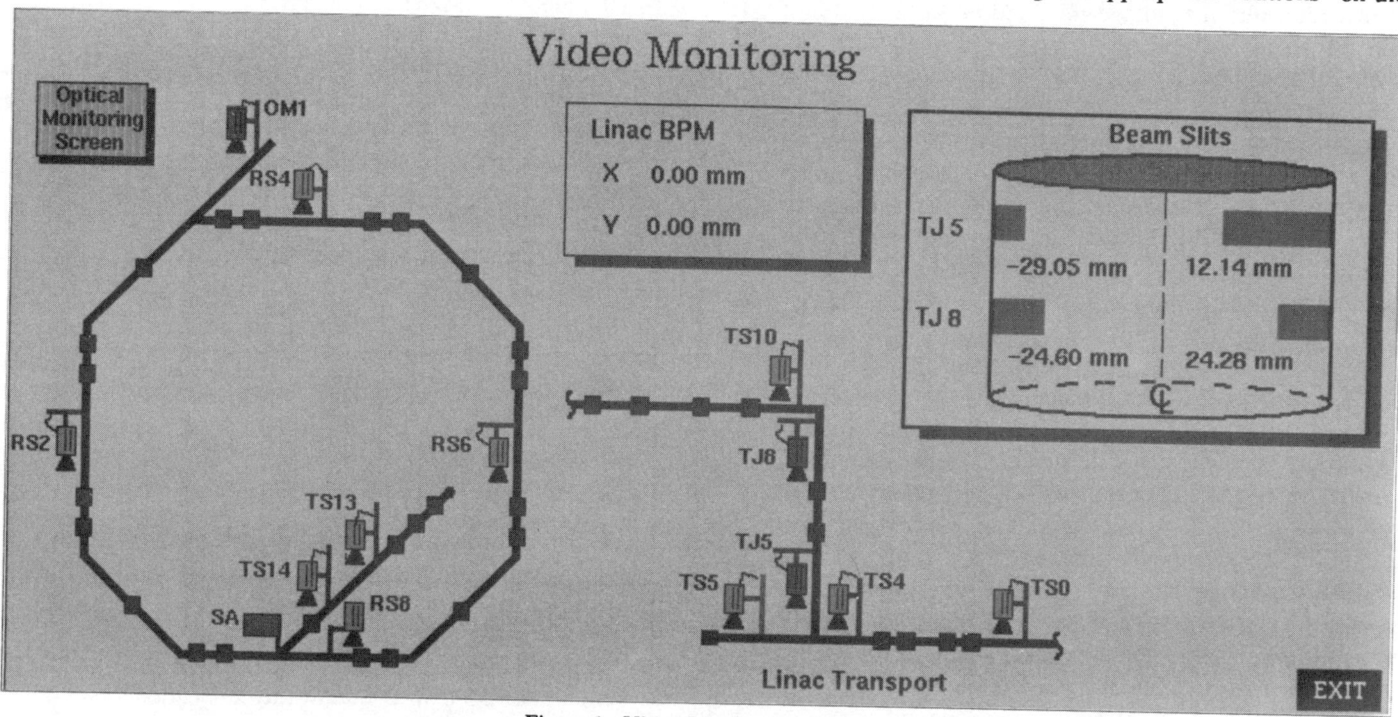

Figure 1. Video Monitoring menu.

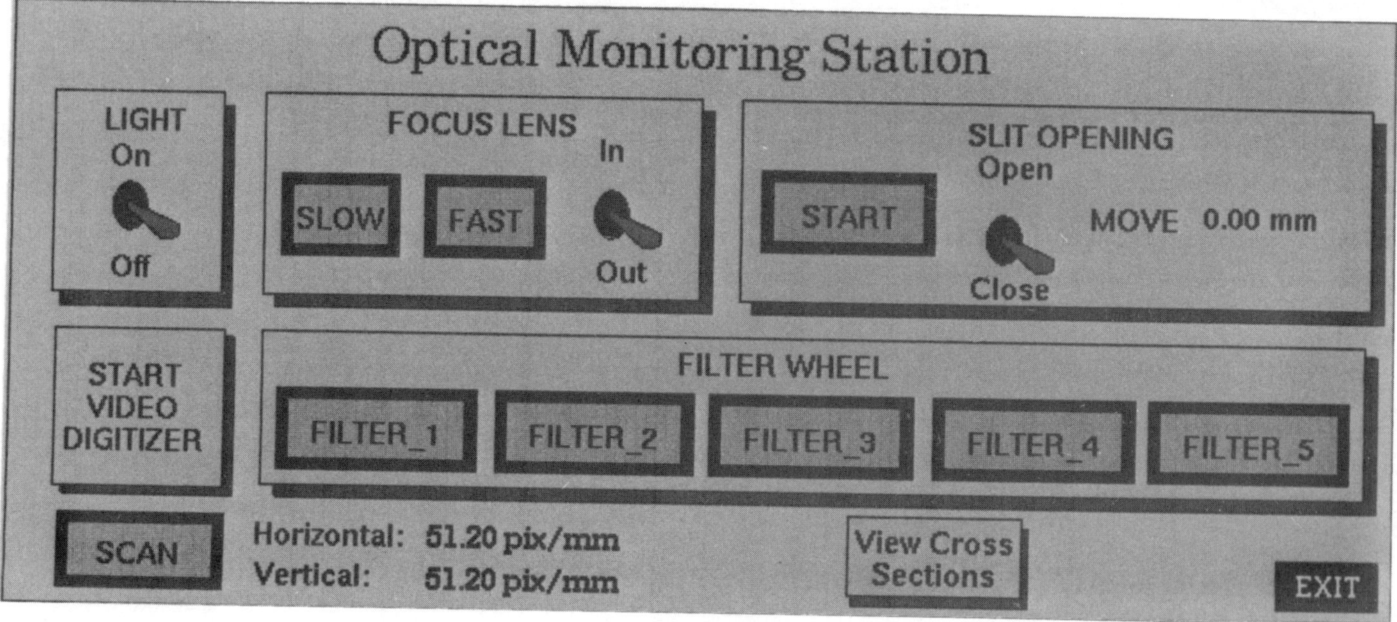

Figure 2. Optical Monitoring Station menu.

menu, the video digitizing process will be started and the video signal from the camera will be analyzed.

Significant mathematical calculation is required by some of the signals. Automatic beam monitor scanning, position calculation, and display is one example; the diagnostic screen TV signal is another. The digitizing of the TV signal is done by hardware. The image enhancement and the beam profile cross-section analysis (width, intensity, etc.) are done by software. An example of such signal processing is the pair of beam cross sections (horizontal, vertical) shown in figure 3. This kind of information will be used in the emittance calculation of the electron beam.

All interfaces to the outside world have been tested to the connector of the I/O cards. Software development is slightly ahead of the hardware development because not all the cable to connect the hardware to the computer interface is available yet. Although this has prevented us from moving on to the next step of testing, we do not expect any delay in the shipping schedule.

In terms of application software, the magnet power supplies can be ramped in a coordinated fashion from a list of predefined data points. A means to generate the magnet control ramping data from a list of estimated energy, horizontal and vertical tunes, and chromaticities stable points has been developed, but not yet fully tested.

B. Remaining Development

Testing to the final control devices via the intended cable will be completed before shipping, to check for noise and signal levels. The ramping data generation program will be completed. Work is underway on orbit correction calculation and an operator initiated automatic correction program.

III. PERFORMANCE

It is difficult to gauge the performance of a still unfinished control system. However, two statistics give some indication of what we may be able to expect. The first is the readback bandwidth of a typical set of display data. On the magnet control readback menu (the most important information for controlling a synchrotron), the maximum readback rate was tested to be 8 times per second. The second statistic is the time required to restore a complete set of control output (a total of 106 channels of data) to all the magnet power supplies; this is required after a shutdown. Complete restoration has been timed at 2.81 seconds. Both tests represent unoptimized results, and further improvement is still possible.

IV. REFERENCES

[1] B. Robinson and P. Clout, "Modularization and standardization of accelerator control systems," *Nucl. Inst. & Meth.,* Vol. A293, pp.316–320, 1990.

[2] B. Robinson et al., "The LSU electron storage ring control system," *Proc. 11th Int'l Conf. on the Application of Accelerators in Research & Industry*, Denton, Texas, Nov. 1990.

[3] S. Pugh and Brian Ng, "Vacuum control system for the MLI Model 1.2-400 Synchrotron Light Source," IEEE 1991 Particle Accelerator Conference, San Francisco, May 1991.

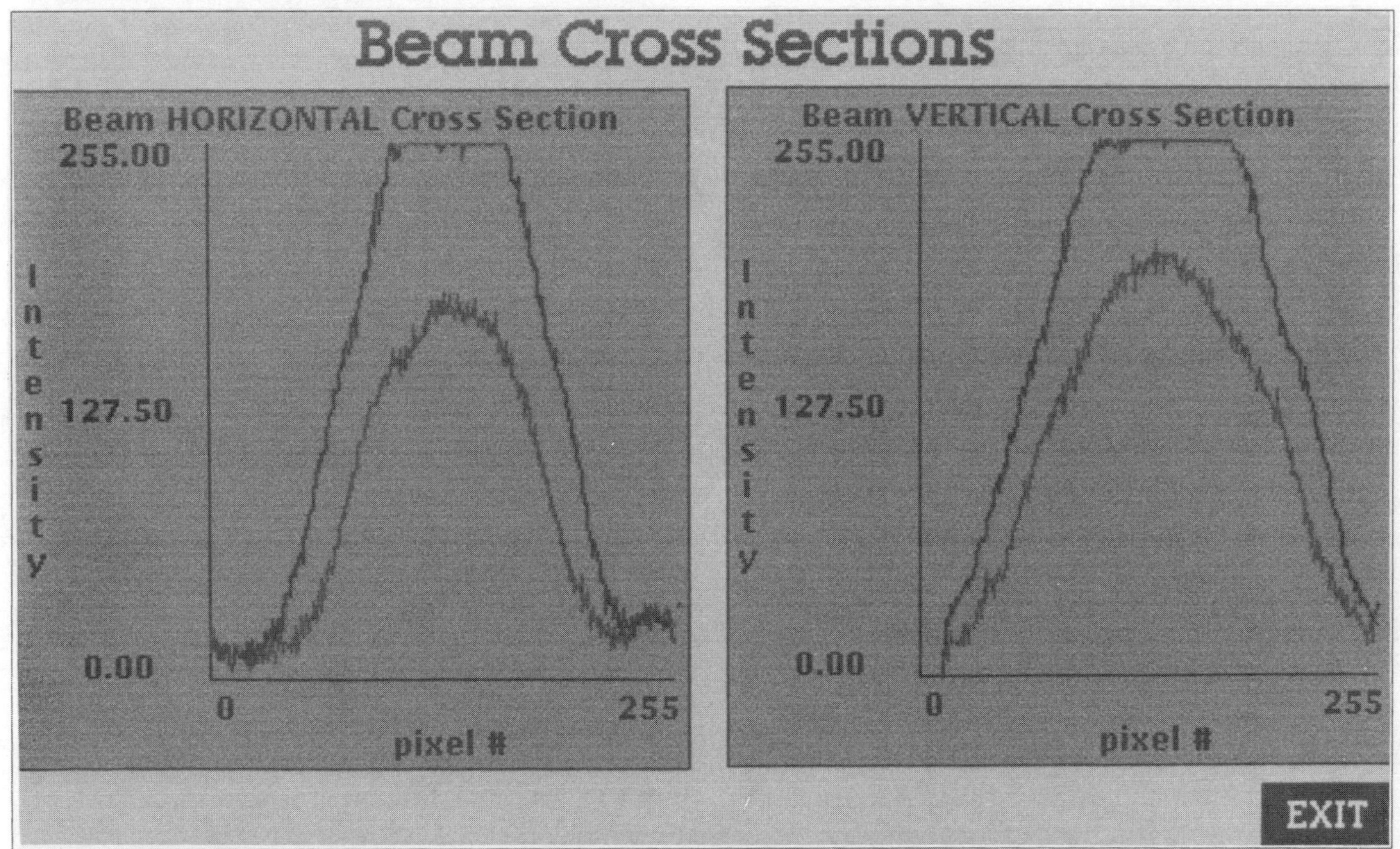

Figure 3. Beam cross-section data.

Control System of the Superconducting X-Ray Lithography (SXLS) at Brookhaven*

E. Desmond
National Synchrotron Light Source
Brookhaven National Laboratory
Upton, New York 11973

J. Galayda
Argonne National Laboratory
Argonne, IL 60439

W. Louie, B. Martin, R. Rose
Grumman Aerospace Corporation
Bethpage, NY 11714

Abstract

The design and implementation of a distributed real-time control system for a compact synchrotron will be discussed. Graphic generation of accelerator device control logic, CAMAC device interfaces and operator display screens is presented. Beam digitization techniques and results of beam position and profile measurements is presented. Methods for automation of routine operator procedures will be discussed.

I. INTRODUCTION

The SXLS machine is being developed as a prototype of an industrial X-Ray Lithography facility. As such the control system is required to provide for the flexible demands of a research machine and yet provide the foundation for an industrial control environment. Such a control system must easily adapt to the changing requirements of commissioning and the machine studies period. The industrialization requirements demands a system that is easy to learn and use and yet is sophisticated enough to provide automated sequencing of routine operator procedures, and meaningful diagnostics. In addition, the control system has to provide monitoring of system status, fault detection, alarm display, logging and plotting of data and a high degree of reliability. The decision was made early in the development cycle to adapt an existing control system rather than to build our own. This decision was based on our own manpower constraints while being mindful that the control system will become part of a commercial product. The CEBAF accelerator control system, developed at Newport News, Va was chosen as most closely meeting our needs [1].

II. MAJOR FEATURES

The control system features a set of graphic tools for configuring the computer system and hardware interface as well as for building control logic and operator display screens. Utilities are available for logging and plotting data, and defining state transitions for sequenced operation. The control system runs on Hewlett Packard workstations running the UNIX operating system. Machine hardware is interfaced to the control system through CAMAC interfaces. Diagnostic and measurement instrumentation is controlled via a GPIB interface. All code is written in C.

III. SYSTEM ARCHITECTURE

The control system is divided into a supervisory and a control layer.   The supervisory layer consists of HP workstations networked together via ethernet. This layer supports the operator display screens and is the main operator interface. This layer of workstation also maintains the central database. The supervisory level computers are connected via a separate ethernet port to diskless workstations which constitutes the local layer. These computers are connected to the CAMAC crates through which the accelerator hardware is controlled. These local computers repeatedly read all the CAMAC crate modules, execute the previously downloaded set of control logic functions, write out new setpoint values to the CAMAC crates and update the system database with new

*Work performed under the auspices of U.S. Department of Energy under contract DE-AC02-76CH00016 and funded by DoD/DARPA.

values. Our present configuration consists of 2 supervisory level computers and 2 local computers controlling six CAMAC crates. The CAMAC crates have been generally configured by function with one crate each controlling the magnets, RF, vacuum, frame grabber, diagnostics and motor systems. The present hardware configuration is shown in Figure 1.

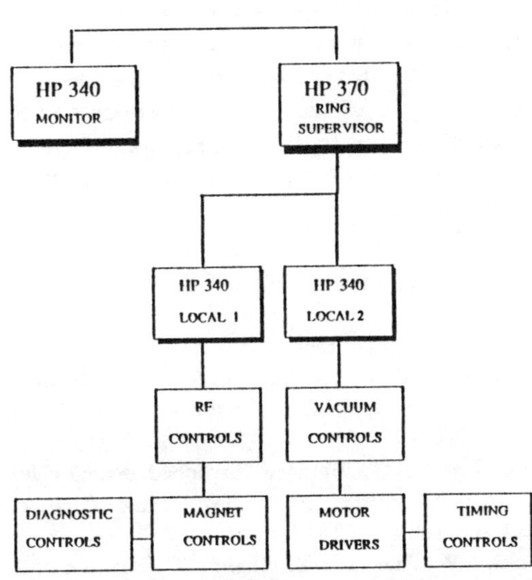

Figure 1

IV. OPERATOR INTERFACE

The operator interface consists of a workstation display screen along with a mouse and knobs for display screen control. A display screen typically depicts a subsystem of the accelerator and contains user defined icons which when clicked on with a mouse, can control individual accelerator components or can activate entire control sequences. At present we have built approximately 20 display screens for machine control. The display screens and icons are built with the aid of a graphic editor. This tool allows custom icons to be designed, built and assigned to machine devices in a few hours time. A standard set of display devices such as bar graphs and dial indicators is also available.

V. CONTROL LOGIC GENERATION

The control logic is executed through ladder logic. Ladder logic is implemented as an array of logic functions which are executed column by column. Control sequences are built out of a set of discrete logic functions which are graphically pasted onto the logic array to form the appropriate control sequences. The logic functions available include boolean, mathematical and user defined operators. A complete set of editing functions is available to modify existing logic arrays. Hardware devices are controlled by associating a name with individual hardware channels. A graphic utility is available to create the signal name to hardware connectivity.

VI. BEAM DIAGNOSTICS

The beam diagnostics on the SXLS machine is obtained through seven flags and six sets of pickup electrodes distributed about the ring and the transfer line. The flags are phosphorescent screens which are inserted into the beamline for beam position and profile measurements. Beam profile measurements are made through a synchrotron port located at the exit of the first dipole. The position and profile measurements are made by digitizing the video image of the electron beam. The video image is obtained through a set of Sony XC-77 CCD cameras. The cameras are positioned at each of the flags and at the synchrotron light port. The camera video signals are multiplexed to a Data Design Corporation AC100 frame grabber which digitizes the video image. The resulting 380 by 240 by 8 bit pixel array is read by the console computer where it is color mapped by pixel intensity before being displayed on the operator console. The digitized video image may be saved on disk along with the beam current and vacuum pressure for later analysis. Offline utilities have been written to calculate the average position and the rms width of the beam density distribution [2]. A sample display of data taken from the synchrotron light port is shown in Figure 2. The horizontal bar is the half maximum intensity. The two vertical bars show the rms beam width.

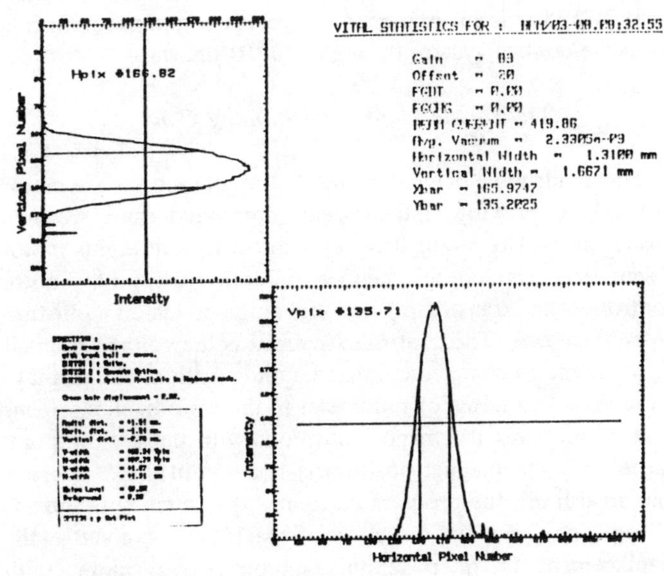

Figure 2

Beam Orbit Measurement

Beam orbit measurements are calculated from readings of six sets of pick-up electrodes (PUEs) within the synchrotron ring. Each set of PUEs in the two straight sections provides horizontal and vertical beam position information with the aid of RF Receivers [3]. Receivers in the two dipoles provide horizontal position information only. The PUE signals are read approximately four times per second. To avoid the display of jitter due to orbital variations, a time-averaged value is displayed along with the standard deviation. The beam position was found to depend on both the horizontal and vertical components in a non-linear way. A method of nonlinear reconstruction was used and a noniterative algorithm was developed to account for this [4]. Transport line and first turn beam position measurements were made by multiplexing processed PUE signals to a LeCroy 7200 oscilloscope. The processed PUE signals yield a difference and a sum signal of the individual PUE button. The LeCroy oscilloscope was programmed to calculate the beam position from a calibration constant and the processed PUE signals. This is done for the horizontal displacements according to the formula:

$$X = ((B + D) - (A + C))/Sum (A + B + C + D)$$

where A,C are pickup electrode buttons for the inside and B,D for the outside of the ring respectively [5].

Timing System and Beam Position Measurement

The septum and kicker magnets are synchronized to the LINAC electron bunches by triggering a Stanford Research Systems DG535 pulse delay generator with a synchronization pulse. The DG535 also provides a high precision pulse to the LeCroy 7200 oscilloscope for proper triggering on the electron beam bunches for the first turn beam position measurements. The DG535 was programmed to receive setpoint commands from the control system through a GPIB interface.

Quadrupole Magnet Positioning Control

The quadrupole magnets in the 2 straight sections are motor driven to provide independent horizontal and vertical positioning. The motor drive system consists of eights motor assemblies (including stepper motors, stepping motor controllers and translators), and eight high precision multi-turn potentiometers. The control system reads the voltage from all eight potentiometers, converts them into positions and displays these values in terms of millimeter to the user display screen. It also compares the magnet positions with the user setpoint values. If the magnet position is not within the 5 microns tolerance limit, the program calculates the new displacement, taking into account any gear backlash, converts the displacement to microstepping motor pulses, along with optimum speed and acceleration curve setpoints, sends the new value to the motor controller.

Special Features

Automated sequencing of routine operations is accomplished with state machine utility. This utility allows the definition of actions to be executed at given states and the transitions rules to pass from one state to another. The transition rules take the form of mathematical and logical operations on signal values in the system database. The state machine has been used to ramp the XLS ring magnets (Dipoles, Quadrupoles, Sextupoles and Trims) for low energy injection, beam orbit and beam lifetime studies.

Acknowledgements

The authors wish to thank Richard Heese and John Keane for their support and encouragement during this work. They also wish to thank the many people who contributed their technical expertise to this project including Denny Klein, Gloria Ramirez, Roy D'Alsace, Walter deBoer and Richard Biscardi. Special thanks to Rolf Bork, Joan Sage and the CEBAF Controls Group for their help and counsel with the CEBAF control system.

VI. REFERENCES

[1] R. Bork, private communications, 1990.

[2] S.L. Kramer and R. Rose, "XLS Video Beam Profile Measurement System," NSLS Technical Note SLK-4 (18 Feb. 1991).

[3] R. Biscardi and J.W. Bittner, "Switched detector for beam position monitor", *Proc. of the 1989 IEEE Particle Accel. Conf.*, March 1989; pp 1516-18.

[4] E. Bozoki, J.W. Bittner, J-Y. Huang; "Calculation of the orbit from PUE measurements using nonlinear reconstruction", NSLS Technical Note 355 (20 Sept. 1989).

[5] E. Bozoki, J.-Y. Huang, J.W. Bittner, "A noniterative method for calculating beam position from induced electric signals," *Proceedings of 1989 IEEE Particle Accelerator Conference*, Chicago, Illinois, Vol. 1, p. 470, 1989.

An Inexpensive PC-Based Ion Linac Control System

M. E. Hamm and J. M. Potter
AccSys Technology, Inc.
1177 Quarry Lane
Pleasanton, CA 94566

Abstract

A turn-key PC-based control system has been developed for the AccSys line of compact ion linear accelerators and rf power amplifiers. The control interface is based on the DZERO Rack Monitor Module[1], developed at Fermi National Accelerator Laboratory, communicating with a Ballard Technology MIL-STD-1553B controller board in a 286 or 386 personal computer. This cost effective and easy to operate control system features real-time control and monitoring of the linac/rf amplifier and can be customized for automatic start-up and unattended operation.

I. INTRODUCTION

As a manufacturer of rf linear ion accelerator systems in a variety of sizes and configurations, AccSys Technology needs a cost effective control system that can be readily tailored to the specific requirements of each accelerator. The control system must be compatible with the major points of our system philosophy. It must be simple, rugged, reliable and convenient to use. It is not our intention to invent "YACS" (Yet Another Control System), but to develop a commercially viable control system with a high performance/cost ratio. To that end we have adopted a data acquisition and control board developed by Fermi National Accelerator Laboratory (FNAL), used a commercially available board to interface it to a personal computer, and adapted an already developed, window-oriented data-base software utility to our needs. This system satisfies our basic needs and provides a versatile platform on which to build.

The computer control system, in its simplest form, is a replacement for the knobs, meters, switches, and indicator lamps of a manual control system. This approach is compatible with a basic tenet of our philosophy which requires that all interlocks and primary feedback circuits be implemented in hardware with no software dependence. The software features data screens with control, monitor and label fields which are defined in data files. The program is menu driven with on-line help. The screen update rate is fast (10 per second) for comfortable real-time control. The software has a log keeping function for periodically recording operating parameters and user configurable save and restore functions that allow resetting parameters to previous values. The system is versatile enough that automatic turn-on and optimization procedures may be easily added to the existing software.

A typical control system for a small accelerator is illustrated in Figure 1. Three data acquisition boards, one for the rf subsystem and two for the injector, are connected to the control computer using a MIL-1553B data bus. The control computer can be connected to the user's computer through a RS-232 serial data link, if desired.

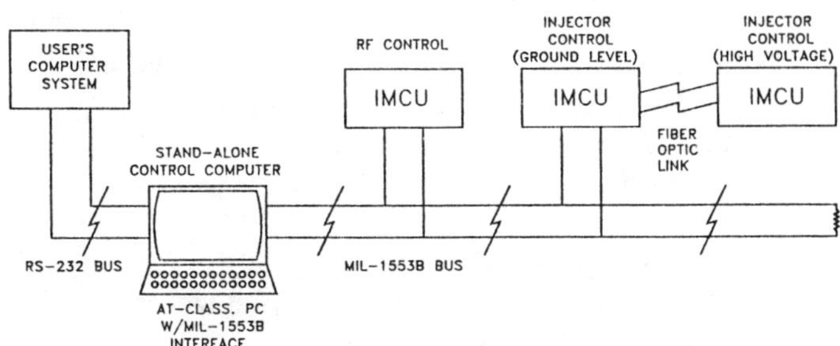

Figure 1. Typical control system configuration for a small accelerator system.

II. DESCRIPTION OF COMPONENTS

A. The Data Acquisition Board

The data acquisition and control board, the IMCU (Instrumentation Monitor and Control Unit), is a version of the DZERO Rack Monitor Module[1] developed at FNAL for their D0 experimental facility. The DZERO technology, transferred to AccSys under a formal technology transfer agreement, has been modified slightly to make it more suitable for accelerator control applications.

The IMCU is a single circuit board, housed in a 1.75" rack chassis, containing ADC, DAC, and binary I/O channels that interact with the control computer over the MIL-1553B bus. The IMCU does not have an on-board microprocessor; instead it uses programmed logic arrays to decode the incoming command and data stream and format the response data stream. Each IMCU has a switch-settable address, permitting multiple units to be connected to the MIL-1553B bus as shown in Figure 1. Typically, one IMCU is at ground potential and the other is at the ion source high voltage potential for the injector control. Both the ground level IMCU and the isolated IMCU have built-in fiber-optic transmitters and receivers to facilitate extending the MIL-1553B bus across the high voltage.

The IMCU has either 8 or 15 12-bit signed DAC channels, 64 12-bit signed differential ADC channels, and 64 bits of programmable binary I/O. Interface connections are made on the rear panel through 9 pin and 37 pin miniature "D" type connectors. The specifications of the IMCU are listed in Table I.

Table I.
IMCU specifications

<u>Analog to Digital Channels</u>
Number	64 differential
Input Range	-10 to +10 volts
Input Resolution	12 bits
Overvoltage Protect	35 volts to gnd max
Internal Impedance	$>10\ M\Omega$
Input Source Impedance	$2\ k\Omega$

<u>Digital to Analog Channels</u>
Number	up to 15, 8 standard
Output Voltage Range	-10 to +10 volts
Output Current	5 mA
Resolution	12 bits

<u>Digital I/O Channels</u>[#]
No. of bits	64 (4 16-bit words)
Input Levels [*]	TTL
Strobe Width	$1\ \mu sec$ nominal

<u>Data Bus</u> MIL-STD-1553B

<u>Power Input</u> 90-130 volts AC
30 watts, typical

[#] Data direction of each word is set by a front panel switch.
[*] Strobes indicate when data is read or written. Output data is valid on the trailing edge of the one microsecond positive strobe (high to low transition).

B. MIL-1553B Bus

The MIL-1553B bus is a two-wire, shielded-pair, 1 MHz serial data bus that can be extended to 300 m without repeaters. The bus was originally developed as a LAN for military aircraft control. The bus has a command/response protocol with one controller and multiple remote terminals. Up to 31 remote terminals may be connected to the bus. Each unit is connected to the bus through an isolation transformer for a high degree of noise immunity. The IMCUs function as remote terminals while the PC is the bus controller. Data transfers are initiated by the bus controller which sends a 16-bit command word consisting of the IMCU terminal address, the channel sub-address and the data type requested. The DAC channels are addressed one at a time. The binary channels are addressed as four 16-bit words. Multiple word transfers are supported, permitting data transfer to a multiplexed 16 bit binary channel. The D ADC channels are read in a block of 32, with all channels transmitted sequentially in response to a single read-ADCs command. Two transmit commands are necessary to read all 64 channels. With the high conversion rates of the IMCU ADCs and DACs, the system data rate is fast enough that a DAC channel can be incremented, all 64 ADC channels read back, and the screen display updated at over 10 times per second.

C. Software

The software package is written in Borland's Turbo Pascal which is convenient for interactive software development and can create a fast, low overhead run-time file. The program is menu driven with on-line help. It is based on a collection of fast window and screen routines developed by D. Shea of FNAL. Each operations window is configurable; any combination of control and response channels can be labeled as desired and displayed on the screen as specified by the user. The screen layouts and I/O database channel assignments are specified in ASCII text data files. The standard AccSys-supplied screen layouts can be swapped with custom windows for special applications or user convenience.

There are seven basic I/O channel types available: 12-bit signed analog-to-digital, 12-bit signed digital-to-analog, single bit read or write, 16-bit read or write word, and 4-bit external multiplexer address. The latter data type was created to permit multiplexing up to 16 pulsed diagnostic signals through an external sample-and-hold into a single ADC channel.

Currently, all control is through the computer keyboard. A function key accesses the on-line help and ESC brings up the menu. The menu can be bypassed and screens selected directly by entering the screen identification number. The keyboard arrow keys move the cursor on a user defined path through the output control fields. The selected field and its companion response field, if any, are highlighted. The channel is incremented or decremented in fine or coarse steps using the + and - keys, respectively, on the numeric keypad.

Several function keys are dedicated to operations that are useful for system troubleshooting. The analog input and output data can be displayed on the screen in raw counts (decimal and hex), in channel volts, or in engineering units. A function key toggles the display through the choices. Another function key toggles the DAC step size, which is displayed on the status line at the bottom of the display, between coarse and fine. All displayed DAC channels can be refreshed and a set of user-defined channel titles can be displayed or hidden, using other function keys.

For convenience and flexibility, there are several standard engineering unit conversions available, with up to three parameters specified in the channel assignment tables in the screen description files. The choices are: linear, inverse linear, quadratic, exponential, and logarithmic. Custom conversion formulae with up to three parameters are easily added.

The software has been tested at AccSys on inexpensive AT-type personal computers based on the 286 or 386SX microprocessor. A minimum of 2 Megabytes of RAM, a 101 style keyboard, and a hard disk are required. A math coprocessor is optional, but may enhance performance. A graphics display is not required and the monitor can be either color or monochrome. The update rate, defined as the rate at which a DAC channel can be stepped and the system response displayed on the screen, is 10 times per second, using a 386SX system with no math coprocessor.

III. Typical Application

Figure 2 shows a typical control screen for an AccSys Model PL-2 linac system. The screen is identified by a descriptive title and the screen number on the top line. The second line displays the date and time. This feature uniquely identifies the data when it is saved in a log file in screen-image form. The control (output) data is shown in the "SET" column along with the engineering units. If there is a corresponding response channel, it can be displayed on the same line, sharing the identifying text and making it evident that there is a response to the control change. Binary channels can be paired similarly, indicating, for example, a ready condition after a time delay. The binary data displayed in a control field is actually a read-back of the data sent out, verifying the current status of the bit.

For most applications, the screens are organized as a main system control screen where most normal operational adjustments are made, subsystem setup screens where less frequent adjustments are made and diagnostic screens where subsystem performance can be monitored. For a typical linac, there is one rf system setup and one rf diagnostic screen. There is one screen for the injector subsystem which controls the ion source, injector high voltage, system timing and beam transport elements. A separate screen controls and monitors the vacuum system.

IV. Planned System Improvements

Several improvements to the control system are planned. The IMCU will be enhanced by adding built-in EMI protection for use in hostile environments. This is currently done by conditioning the signals externally. The modification would simplify the interconnecting cables. The IMCU will be modularized to facilitate configuring the system by omitting unused I/O channels.

The software will be enhanced by augmenting the keyboard operation with control by a mouse or track ball. Histogram displays will be added for graphically monitoring the accelerator system performance. The data base will be modified to include user-configured operating limits and out-of-range alarms. Ultimately automatic turn-on sequences and optimization procedures will be added, minimizing the operator interaction required to run an accelerator.

V. Conclusion

Using straightforward hardware and software, AccSys has configured a cost-effective control system suitable for controlling small to medium size linear accelerator systems. The control system is easily customized for different applications and accelerator configurations. It features user-configurable displays for easy access to all control channels and diagnostic information, fast response for convenient adjustment of parameters, parameter initialization for simplified startup procedures and easy adaption to full turn-key operation with automatic start-up and optimization.

Acknowledgements

We would like to thank Mike Shea and Don Shea of FNAL for their assistance in adapting the DZERO unit and the data base software to our application. We also want to thank George Engeman of AccSys for his efforts in enhancing the performance of the IMCU.

VI. References

[1] Al Frank, Rich Mahler, and Mike Shea, *Dϕ Note No. 1501*, Fermi National Laboratory, November 1990.

```
┌─────────────────────────System Operation───────────────────────────1┐
   03/06/1991               --MONITOR--              --SET--      19:17:10
 ION SOURCE HV            28.907   kV              28.829   kV
 EXTRACTOR HV             17.706   kV              17.711   kV
 EINZEL LENS HV           26.817   kV              26.857   kV
 PLASMA ELECTRODE BIAS VOLTAGE                      0.547   V
 PLASMA RF PERIOD (FREQUENCY)                       0.546   usec
 PLASMA RF MICROPULSE (POWER)                       0.198   usec
 SPARE CHANNEL             1.279
 ION SOURCE AVERAGE CURRENT    3.662   mA
 EXTRACTOR P/S CURRENT    21.803   mA
 GAS FLOW                 -0.117   SCCM
 PLASMA CHAMBER PRESSURE  18.787   mT               0.100   msec
 ION GAUGE PRESSURE       9.82E-6  T
 RFQ RF AMPLIFIER POWER   -1.831   Vdet
                          --BINARY  1=SET/ON, 0=OK/OFF--
 INJECTOR INTERLOCK FAULT      0
 PLASMA RF WARM UP             1
 PLASMA RF HV UP               1
 PLASMA RF ENABLE                                   1
 INJECTOR HV ON/OFF                                 1
 RFQ AMPLIFIER ON/OFF                               1

└──────────────────────────────────────────────────────────────────────┘
 F1-Help F2-Units F3-DAC Inc= 50 F5-Refresh DACs F9-ADC Status  ON ESC-Menu RES
```

Figure 2. Typical linac control screen.

Control System at the Synchrotron Radiation Research Center

G. J. Jan*
Synchrotron Radiation Research Center
Hsinchu 30077, Taiwan, R. O. C.
* also Department of Electrical Engineering , National Taiwan University
Taipei 10764, Taiwan, R. O. C.

Abstract

A modern control system was designed for SRRC to control and monitor the facilities of storage ring, beam transport line and injection system. The SRRC control system is a distributed system which is divided into two logical levels. Several process computers and workstations at upper level provide the computing power for physics simulation, data storage and graphical user interfaces. VME-based Intelligent Local Controllers (ILC) are the backbone of the lower level system which handle the real time devices access and the closed loop control. Ethernet network provides the interconnection between these two layers using IEEE 802.3 and TCP/IP protocol. The software in upper level computers includes data base server, network server, simulation programs, various application codes and X windows based graphical user interfaces. Device drivers, application programs for devices control and communication programs are the major software components at the ILC level.

I. Introduction

The 1.3 GeV synchrotron radiation facility at SRRC has been designed as a third generation synchrotron light source with low emittance and high brilliance. The facility includes a turn-key full energy injector, a transport line and a storage ring with triple bend achromat lattice [1]. The turn-key injection system is composed of a 50 MeV linac and a 1.3 GeV booster synchrotron. The control system of SRRC is cost-effectively designed by using advanced technology. Two level hierarchical computer [2,3,4] systems were chosen to simplify the architecture of the control system. They are console level computers and intelligent local controllers (ILC). The ethernet network is used to link the console level computers and ILCs. The console level computers handle the system wide high level control function and provide a friendly operator interface. The ILC is a field level controller which performs data acquisition and local close loop control for the equipments of various subsystem.

II. Hardware Design Consideration and Configuration

Two level hierarchical control system of the synchrotron radiation facilities at SRRC provide good real time performance with update rate about 10Hz. The architecture is simple and easy to be expanded and maintained. The centralized database structure is implemented. The hardware configuration of the control system is shown in Figure 1. The open system configuration allows the centralized database to be changed

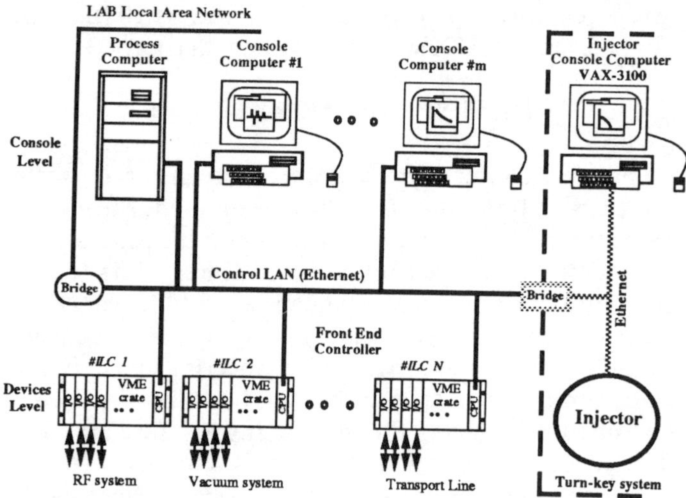

Figure 1. Hardware configuration of the control system at SRRC.

into a distributed database easily. Adding UNIX system into the control system is possible after modifying some software components.

The console level computers are composed of a process computer and several workstations. The process computer will be a VAX 6000 series model 510 which provides a large storage space, computing power and will maintain system-wide resources. The VAXstation 3100 series are chosen as workstations which provide a high performance platform to run graphical or other applications. The workstations are used as operator console. The graphical user interface standard X-Windows and OSF/Motif are used to design a friendly operator interface.

The ILC is a VME crate system which includes Motorola MVME-147 CPU board and a variety of interface cards. The MVME-147 CPU board consists of 68030 microprocessor, 68882 floating point coprocessor, 4 Mbyte on-board memory and ethernet interface. The ILCs are connected to the hardware devices via analog and digital input/output interfaces, IEEE-488 interface and serial communication interface. Data acquisition, closed loop control and monitoring of the equipments are handled by ILCs. Most of subsystems, such as magnet power supply, vacuum gauge controllers, general purpose measurement instruments, provide the IEEE-488 interface. The control and monitoring of equipments can be carried out with the workstation or personal computer through ethernet. The important feature is that the ILC which has a local console can serve as a stand-alone system to commission any sub-system and integrate it into the control system without modification. The RF controller, beam position acquisition

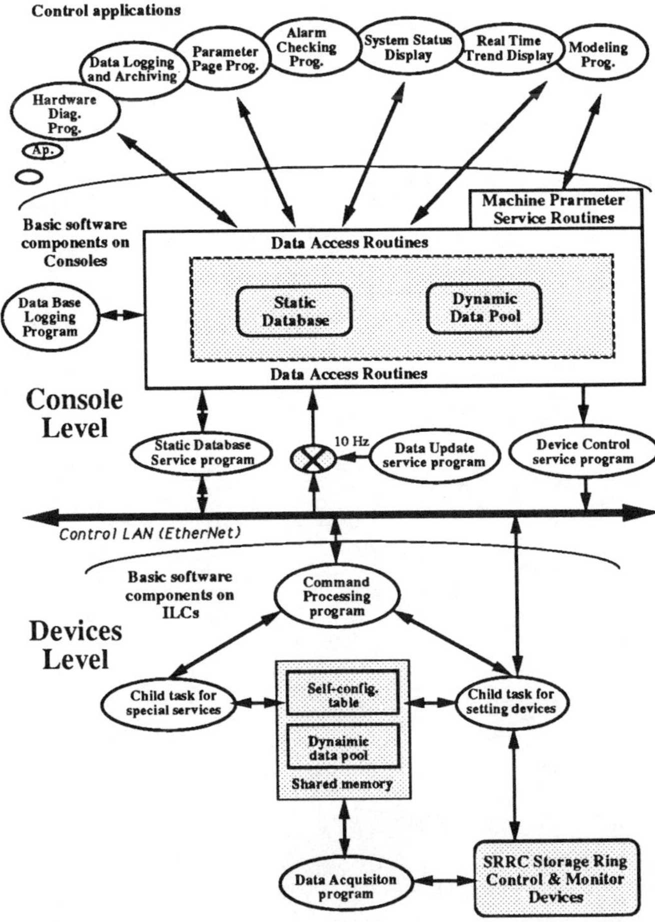

Figure 2. Block diagram of the software system.

system and power supply for correcting magnet are connected to ILCs through the general purpose input/output channel. All ILCs serving for different sub-systems of the beam transport line and the storage ring facility have the same priority level. The dynamic data can be updated into the database of console computers at the rate of about 10 Hz.

The ILC, Micro-VAX II and DECstation 5200 have been set up and linked successfully. The data acquisition system of the control and monitoring has been developed and tested with local console PC/AT-386. This stand-alone system is quite useful in testing the subsystem. The transfer rate of the ILC and DECstation 5200 through the ethernet with TCP/IP protocol was tested. The maximum number of ILC can be accessed from the workstations within 100 msec is about 15 sets. The performance of the ILC and VAX/VMS workstations is going to be tested and evaluated.

III. Software Design and Development

The software structure is divided into several logical layers as shown in Figure 2. There are device access, network access, database management, graphical user interface and applications. The goal to modularize the software into layers is to reduce the development time.

The devices access processes are run on ILC. The pSOS[+] real-time kernel provides the ILC with support for task scheduling, memory allocation, event handling and message queuing. The pNA[+] network support package provides socket network interface. The control tasks and various input and output tasks are also running on ILC. The hardware dependent device drivers are implemented. In order to test and to commission as well as for diagnosis of the subsystem, the PC/AT-386 is used as a local console to serve control and graphical monitoring system. The PC/AT-386 system is also used as program development station to develop the application programs using C language under MS-DOS environment. The control and monitoring programs for the magnet current power supply and simulator of the low level RF electronics as well as vacuum gauge controllers were implemented and tested successfully. The speed of the dynamic data uploading to console level computers is about 10 Hz. Downloading the database from the process computer into the ILC to form a local distributed database is underway.

The network access software is in charge of the data exchange between console level computers and ILCs. The protocol of the IEEE-802.3 is used to communicate with the turn-key injector system. Thus, the IEEE-802.3 protocol is still developed and coded until the completion of the machine commissioning. The TCP/IP protocol is using at present system and provides an open environment for further expansion. The IEEE-802.2 is also under intensive study, which is considered to be implemented in the control system to reduce the system overhead if needed in the future.

The console level computers are VAX/VMS system. The software package is developed using C language. The function of the process computer and workstations is slightly different. The process computer keeps the system-wide static database and maintains it. At system start-up, each workstation requests and receives a copy of the static database from the process computer. Each console computer then has all the database information necessary to process dynamic database frames received from the ILCs. The workstations are mainly for user interface. The upload sequence is requested by the process computer, the ILCs multicast the dynamic data sequentially. All of the console level computers receive dynamic data and updated into database at the same time. Hence, the console level computers can be expanded easily without increasing the network traffic.

The central database on console level computer is used as a buffer between the low level tasks at ILCs and the console level applications. The application programs access equipments parameters directly form database rather than from ILC. The application programs are devices transparent. The development of the application programs can be parallel with the development of the other programs at ILCs.

There are many applications run at console level computers. The data logging and archiving, alarm checking and machine modeling programs will run at process computer. The graphics-oriented applications such as real time trend display, machine parameters display can run at workstations. Since, the workstation has a powerful processor, some computation intensive tasks can run at workstation also.

IV. Summary

The control system of the synchrotron radiation facility at the SRRC has been designed and the implementation is under way. The two level computer system and ethernet data communication network are configured. The message exchange can be maintained at the 10 Hz transfer rate. The maximum number of the ILC is about 15 sets. The broadcasting mode is used in the ethernet currently.

Acknowledgements

The author would like to express his appreciation for many useful suggestion from Mr. R. W. Goodwin and Mr. M. F. Shea of FNAL, Dr. W. D. Klotz of ESRF as well as Prof. T. Katsura of KEK-PF.

References

[1] "SRRC design Handbook", Synchrotron Radiation Research Center, April 1991.

[2] A. J. Kozubal, D. M. Kerstiens, J. O. Hill and L. R. Dalesio, "Run-time environment and application tools for the ground test accelerator control system", Nucl. Instrum. and Meth. A293, (1990) 288.

[3] B. Robinson, and P. Clout, "Modularization and standardization of accelerator control systems", Nucl. Instrum. and Meth. A293 (1990) 316.

[4] J. P. Stott and D. E. Eisert, "The new ALADDIN control system", Nucl. Instrum. and Meth. A293 (1990) 107.

The SSRL Injector Control System*

C. Wermelskirchen[†], S. Brennan, T. Götz[‡], W. Lavender, R. Ortiz, M. Picard[‡], J. Yang
Stanford Synchrotron Radiation Laboratory, P.O. Box 4349, Bin 69, Stanford, CA 94309-0210

Abstract

The control system for the new SSRL injector for SPEAR is based on one central VAX/VMS workstation with additional workstations as consoles and standard CAMAC equipment for data transmission and hardware interfacing. The CAMAC crates are connected to the central workstation by using a serial highway.

The central computer has the data base that describes all the accelerator hardware characteristics and the actual accelerator status information. The application software is built on top of the central data base and is independent of the specific hardware characteristics.

The operator interface uses DECWindows (X-Windows) and provides menu driven access to all machine parameters and control applications. The application programs can react on parameter changes on an event driven basis.

I. INTRODUCTION

The new SSRL injector is a 10 Hz machine especially designed to accelerate electrons to energies up to 3.0 GeV to fill one or several buckets in SPEAR up to intensities of 100 mA for the synchrotron light experiments. The new accelerator includes a 150 MeV linac and a 3.0 GeV synchrotron [1] [2]. The control system for the injector was designed to make the operation of the accelerator user friendly, keeping in mind that the machine has to be operated from two different places (as stand–alone machine as well as from the SPEAR control room). The decision was made to interface all the digital and analog signals to a central computer by using standard CAMAC equipment without local intelligence. A VAX Workstation 3500 was chosen as the central control system computer, handling CAMAC as well as being the local operator console.

II. CONTROL SYSTEM KERNEL

The control system kernel is made out of several modules (see Figure 1) that access the data base, the communication with the front end system (CAMAC), the synchronization between different applications and the notification mechanism for parameter changes.

A. Data Base

The data base itself is kept in a shared memory section that is mapped from a disk file and written back to that file every minute to prevent loss of data in case of a system failure. A separate process, which is activated during

*Work supported by the Department of Energy, Office of Basic Energy Sciences, Division of Material Sciences.

[†]Now at Gesellschaft für Mathematik und Datenverarbeitung (GMD), Bonn, Germany.

[‡]Now at Physikalisches Institut der Universität Bonn, Germany.

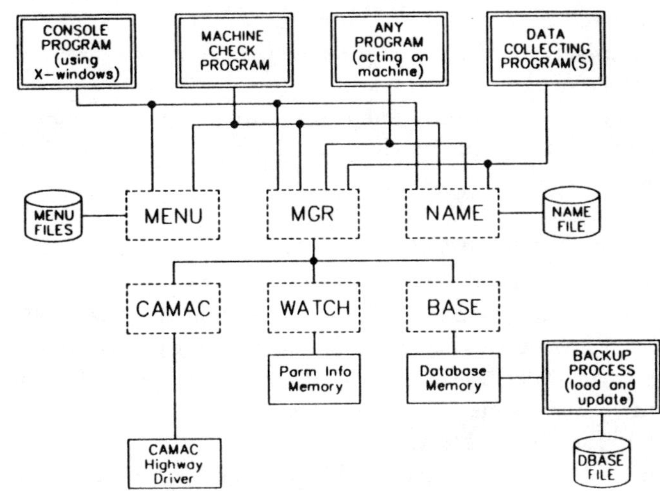

Figure 1
Control System Kernel Modules

system startup, is responsible for this task. Whenever it is activated it also reorganizes the data base to remove space from deleted devices. All other processes access the shared memory data base by using subroutines from the BASE package.

The data base keeps all information about hardware characteristics and actual parameter settings. It contains records that describe *devices* and each device contains all the *parameters* for that device. Each parameter has several fields, for example minimum/maximum allowed value, actual value, dimension, conversion data, addressing information, etc.

Parameters are handled in the following way: The direct physical parameter, for a power supply for example, is the output current. It is stored in the data base in exactly the form that is used by the hardware module (DAC, ADC). Data base routines transform the parameter to its floating point value, which represents the real current, for application programs and menu representation.

But parameters are not limited to direct physical values, they can also be linked to other parameters. In this case these parameters do not have a value by themselves, instead the value is taken from another parameter by using additional transformations. The data base access routines handle *simple* conversion of parameter values (factors, offsets) or they can call additional routines for special conversions or relations between different parameters. This way, for example, a parameter that defines the energy of a beam line can actually set the current of several magnets at the same time.

Currently there are some 300 devices defined in the data base with more than 1400 parameters total.

B. Maintenance Shell

All entries in the data base can be created, modified or deleted by using the maintenance shell, a program that can be run on any terminal. Devices and their parameters are described in normal text files and this information can be imported into the active data base. Devices can also be deleted at any time. The maintenance shell allows to interactively modify all characteristics of the devices and parameters.

There are tools included to generate different kinds of listings and reports to view the information in the data base.

C. Application Interface

The data base routines and the CAMAC interface software make the lower level of the central control system software. Between this layer and application programs there is the application interface layer. It completely separates the hardware dependencies of the machine parameters from the applications. For the programs on top of the application interface there exists a *virtual* accelerator, represented only by parameters and their values. These values may be *simple* values, like magnet currents, or they may represent more complex parameters, like beam line energies. For the application programs there is no difference between these parameters, they can be read and set, independent of the number of actually involved parameters that may be affected by setting a complex parameter.

The application interface also provides *real* names for all devices and parameters. Internally these are handled by using binary IDs. Externally there are names according to a defined scheme and there are modules that translate between these two representations. Each name is composed from four parts: The name of the accelerator or section to which the device belongs, the device name itself, the name of the parameter and the characteristic of the parameter. GUN_RF.PHASE_SHIFTER_AS, for example, is the analog setpoint (AS=characteristic) of the phase shifter (PHASE_SHIFTER=parameter name) for the RF (RF=device) of the gun (GUN=section).

Another important function that the application interface provides is a notification mechanism. Application programs can register themselves to be notified whenever a particular parameter is set by any program. This function provides the fastest way to react to changes of the value of any parameter and is used by all realtime applications, like the operator interface, the beam position readout program, or the waveform generating program.

III. CAMAC

The CAMAC crates are connected through a serial highway to a KineticSystems 2060 CAMAC driver module. The KineticSystems CAMAC library and a modified version of their VMS driver is used to execute the CAMAC commands. There is one CAMAC readout program that knows how to collect the data from the (standard) CA-

MAC modules (ADCs, digital inputs). It gets it's addressing information from the central data base and reads all the CAMAC modules every 200 msec. It detects changes in the parameter values and feeds the changed values into the data base. This also means that they are propagated to the applications by using the notification mechanism.

The readout program also monitors all the CAMAC crates and modules and invalidates values of parameters that are no longer available because of a CAMAC failure (powered down crates, removed modules). So the operator or application programs will get an identification about some CAMAC malfunction. The program tries to address lost modules/crates in certain time intervals and when it becomes available again, it will automatically be recovered by the program. The modules will be initialized to their previous settings and data will be collected again. This makes replacement of modules very easy, because after turning on a CAMAC crate again, normal operation is continued after only a short recovery time.

So far there are more than 40 CAMAC modules handled by the readout program. From these CAMAC modules about 600 words are read every 200 msec and more than 150 parameters are being set during a complete CAMAC initialization.

There are also some special programs that handle (non standard) modules like beam position monitors or waveform generators. These applications will be described in separate articles [3][4].

IV. Operator Interface

From the beginning the operator interface for the new SSRL injector was designed to be easy to use, be usable on workstations at different locations at the same time, be easy to adapt to the changing operator needs. A lot of final work could only be done during commissioning, when the entire control system had to be operational.

The hardware basis for the operator interface are VAX workstations that use X-Windows (DECWindows) as graphics standard. The operator interface was therefore made on top of DECWindows. Because X-Windows already has the complete networking functionality in it, all operator controls are accessible on all workstations in the network and no modifications have to be done to connect additional consoles.

The operator is controlling the injector by using *menus* which present the values of the machine parameters and allow the operator to manipulate these values. The menus are displayed in windows on the workstation screen and can be placed and arranged in any way in which the operator needs them. From one menu the operator can pop up additional menus, so it's possible to have additional parameters available to control areas or devices in greater detail.

In general a menu is made of graphic elements. These can either be static, like texts, lines or colored blocks, or be dynamic. Dynamic elements are connected to the parameters in the data base and can be used for input and output of parameter values. They are updated whenever the value of the parameter changes or is modified by any other program. The main dynamic elements are horizontal and vertical sliders and bars. The resolution of the horizontal input slider is switchable so that the operator can change the precision of the adjustments between a zero to hundred percent input and high resolution, where a pixel on the screen affects the least significant bit of the parameter.

The operator modifies parameter values and thus controls the machine by using the workstation's mouse as input device. He can either set parameters to fixed values by clicking at fields on the menu or he can drag sliders by using the mouse and so set parameters continuously to new values.

Several options to start/stop additional programs are available through pull down menus from the main menu. These are used to activate the standardization of beam lines, start the linac feedback program, etc. Another option allows to save or restore a complete machine configuration into separate disk files.

To create new menus or to modify existing menus there is a menu design program that also runs on the workstation. It is an interactive program that allows to manipulate all the graphic elements. They can be added, moved, changed, deleted, etc. The foreground and background colors of these elements can be set to 6*6*6=216 different colors, which is enough for the control system application and will allow to work with 8 color planes without problems.

V. Applications

Several application programs to handle complex settings or sequences of settings of parameters have been implemented by different people. Subroutines have been added to the data base routines that allow the setting of complete beam lines to different energies, affecting all magnets in these beam lines, using measured calibration data of the magnets. This is done whenever the operator uses the *energy knob* for such a beam line.

There are other programs running in a background mode, to perform tasks that take some time and can not be executed for each change of a certain parameter. One of these programs handles the standardization of magnets to reproduce the magnetic fields very precisely, once a beam line has been set to a new energy. This program is started by operator request from the main menu.

Another program stabilizes the linac for drifts in the electron energy. It gets the position of the electron beam after the first bend magnet downstream the linac from the data base and changes the linac parameters so that the position of the beam is kept, which means the electron energy is stabilized[5]. Once started, this feedback program performs its task without further interaction.

The program for the generation of the RF wave form is more interactive. It waits for notification of modification of some special control parameters, like phase, offset, amplitude by the operator and then recalculates a complete new table of values for the RF wave form table parameter. This table parameter is then loaded into the CAMAC module as part of the normal application interface function and is also displayed in a menu, so the operator will see the result of this calculation.

These programs are all coupled to the menus by either using the application notification function for watching parameters to perform their task or they can be started/stopped from the pull down menus of the main menu.

VI. Conclusion

The control system was operational when the first tests during the commissioning of the new injector had to be made. Not all devices were connected to it at that time and some devices were controlled locally to do special tests with machine components. During commissioning all devices have been connected to the control system and are monitored and controlled from the menus. Operators can use the local console as well as the remote consoles in the SPEAR control room.

The new SSRL injector is successfully being operated from the workstations, the menus are easily usable and have been adopted to the different operator needs. The injector reached its goal to produce an electron beam of designed intensity to fill SPEAR within several minutes.

VII. References

[1] H. Wiedemann, "3 GeV Booster Synchrotron for SPEAR," ACD-Note 45, SSRL, March 1987.

[2] H. Wiedemann, "3 GeV Injector Synchrotron for SPEAR." In these proceedings, 1991.

[3] S. Brennan, S. Baird, W. Lavender, H.-D. Nuhn, C. Wermelskirchen, and J. Yang, "The Control and Operation for the Programmable Waveform Generator of the SSRL Injector." In these proceedings, 1991.

[4] W. Lavender, S. Baird, M. Borland, S. Brennan, R. Hettel, H.-D. Nuhn, R. Ortiz, J. Safranek, J. Sebek, C. Wermelskirchen, and J. Yang, "The SSRL Injector Beam Position Monitoring Systems." In these proceedings, 1991.

[5] L. Emery, "Energy Feedback System for the SSRL Injector Linac." In these proceedings, 1991.

Longitudinal Damping System for the Fermilab Booster

I. Haberman   I. Rypshtein

Fermi National Accelarator Laboratory*

P.O Box 500 MS #341

Batavia, IL 60510

May 1991

Abstract

The paper describes the different building blocks of a new longitudinal damping system in the FERMILAB BOOSTER and the main design considerations.

INTRODUCTION

The general idea behind the damping system is to take from the beam-spectrum the instability frequency components (synchrotron sidebands), and to feed them back with the proper phase to create negative feedback.

As shown in figure 1, the filtering process is done at base-band and with tracking, digital periodic filters.

The frequency conversion to base-band is done by two identical channels, driven by quadrature local oscillator. The local oscillator is synthesized from the RF reference signal. By using a software controlled synthesizer the frequency coverage can be selected according to the following criteria:

- Instability frequencies.
- Form factor (maximum S/N).

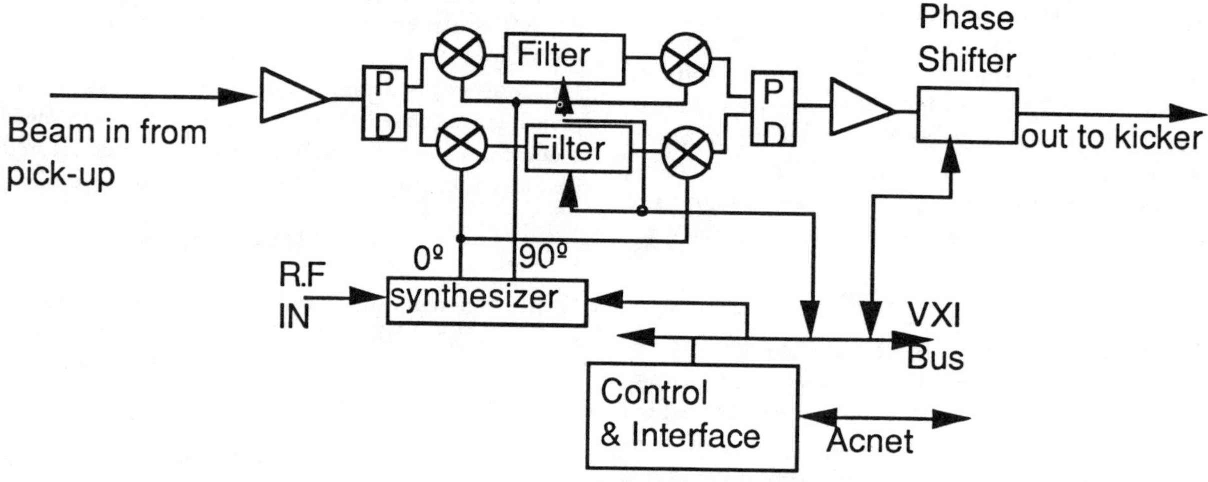

Figure 1 - Damper block diagram

PERIODIC FILTER

The synchrotron instabilities might appear in any of the 84 revolution frequencies. To minimize the hardware a "wide-band" approach has been taken.

The filter requirements (as shown in figure 2) are :
-180º phase difference between upper to lower synchrotron frequencies.

-Maximum attenuation of the revolution frequencies.
-Periodic in revolution frequency.(not shown)

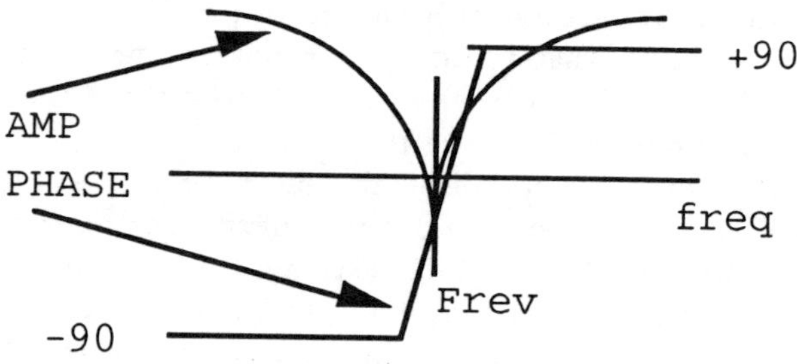

Figure 2 - Filter characteristics

----

* Operated by the Universities Research Association under contract with the U.S. Department of Energy

ACKNOWLEDGEMENTS: The authors wish to thank the members of the Booster and the P-bar Source departments who helped them define the general system concepts.

The filter response demands poles and zeros , which are implemented the by an IIR digital filter. The general transfer function of an IIR filter is (equation 1):

(1) $Y(t)= \sum X(t-n^{*}\Delta t)-\sum Y(t-n^{*}\Delta t)$

The digital filter is inherently periodic in 1/T, where T is the unit delay of the filter (in our case : the beam revolution time). The frequency coverage of the filter is determined by 1/2 the sampling clock frequency (Nyquist frequency). The system's selected clock is 10 Mhz near INJECTION.

The selected A/D & D/A are 12 bits components, so the dynamic range of the system is 72 db (theoretically). This high dynamic range is required in order to start the filtering process early in the booster cycle, when the synchrotron frequencies have low energy . Figure 3 shows the filter implementation.

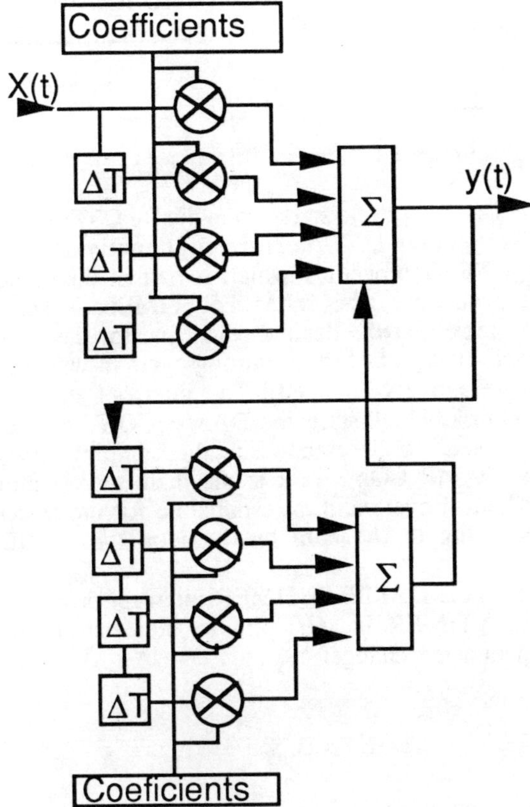

Figure 3 - IIR filter implementation

The filter hardware is designed to enable filter parameters change during the Booster cycle. The filter's clock is tracking the revolution frequency by using a D.D.S based synthesizer. The filter module is software controlled via VME bus. The controlled parameters are:

-filter coefficients (16 bit)
-filter period frequency
-filter delay (in 25 nS resolution)
-DC offset (.3 mV steps)

The filter section uses LSI LOGIC DSP series components, that can go up to 40 Mhz data rates, and allows a 4th order filter implementation.

Two-Path I.F section

The down and up frequency conversion is done by two L.O's with 90º phase difference. The two channels have identical IF sections.

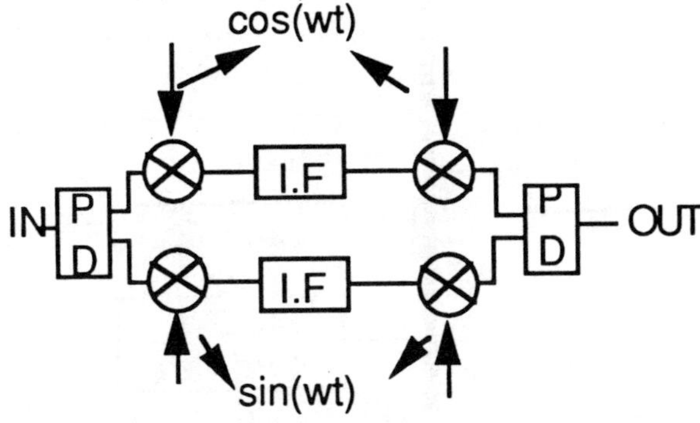

Figure 4 - Two path IF section

It can be shown that this type of implementation preserves the upper and lower frequency components of the beam signal and doubles the bandwidth of the system . For a sampling clock of 10 Mhz the theoretical B.W of the system shall be 10Mhz too. It means that 14 out of 84 revolution frequencies can be handled. The center frequency of the system is determined by the L.O.

R.F Front - end

The signal taken from pick-up is summed-up to get the beam profile in the longitudinal axis, amplified and divided to drive the two I.F sections.

R.F output stage

In the output R.F stage the signal is summed to reconstruct the information, amplified to drive the kicker and gets the right phase to suppress the instabilities.

Tracking synthesizer

The tracking synthesizer follows the R.F change to give a consistent coverage of the instabilities. The design is based on a D.D.S (Direct Digital Synthesizer) which is used as a reference to a P.L.L circuit.

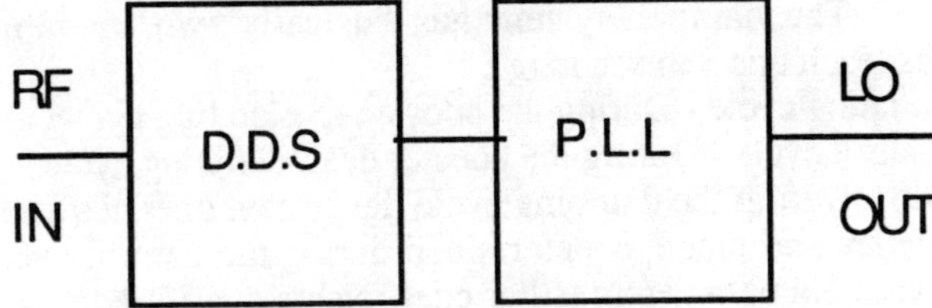

Figure 5 - Tracking synthesizer

The tracking error should be less then 500Hz , in order to maintain an effective filtering process.   The D.D.S tracking the R.F change (up to 7GHz/sec) and the P.L.L BW is the limiting factor.

Control section

The control section has the following tasks:
Hardware control, ACNET interface, BIT and diagnostics

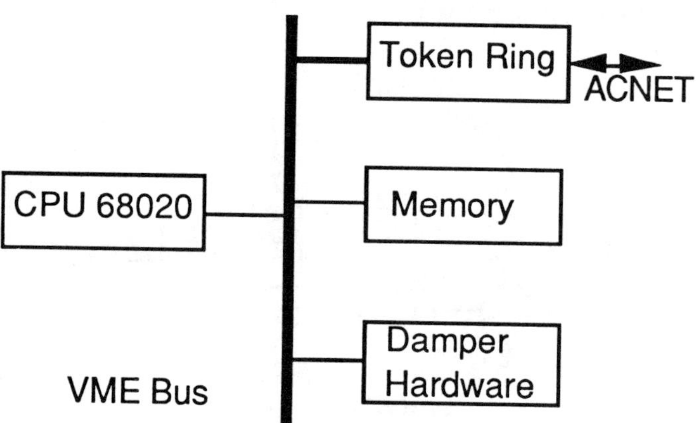

Figure 6 - Control section

The control section is based on Motorola's 68020 CPU card with VME interface. It includes a Token Ring card and general I/O and hardware interface. The VME cards are mounted on a VXI crate.

Damper software description

1) General

The damper software is based on the OOC speaking VMEbus standard system and is implemented by linking the OOC software with the damper application software.

The software is composed of the following parts:
a) MTOS - real time operating system.
b) OOC - Object Oriented Communication software.
c) ACNET Communication software.
d) Standard objects for:
    1) Local diagnostic terminal.
    2) System Service Module.
e) OOC drivers for damper application.
f) Damper application tasks.

2) Operation cycles

The damper system has basically two operation cycles (each one 33msec long):
a) Damping cycle - During the booster accelerating cycle.
b) Control cycle - During the booster deaccelerating cycle.

During the damping cycle the system operates at the frequency and mode as determined during the control cycle. The variable parameters (filter coefficients and IF delay) are The control cycle is used for ACNET communication and system parameters update.

3) Tasks Synchronization

Tasks synchronization is done by interrupts and task-level services of the MTOS operating system. Figure 7 shows the software flow chart.

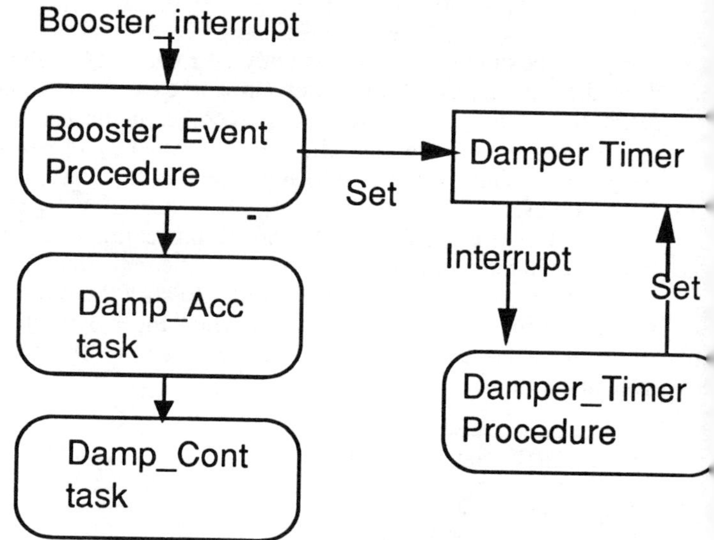

Figure 7 - Software flow chart

The Damping cycle starts when the BOOSTER_INT interrupt is received. This interrupt activates the BOOSTER_EVENT procedure which is part of the kernel. Within the procedure, the DAMPER_TIMER is set to activate an interrupt at the desired start time for the active Damping cycle (the part of the Damping cycle in which the parameters are actively changed). The interrupt starts the DAMP_ACC task, which starts the DAMP_CONT task and cause it to be paused for 35msec (during the Damping cycle).

The DAMP_CONT task is the main task running during the Control cycle and is responsible for the system control, according to the data received on the ACNET network.

Within the DAMPER_TIMER interrupt procedure, the DAMPER_TIMER is set again to the desired time interval for parameters change.

STATUS

The system design has been completed . The prototype is currently being built and tested. First system integration with the Booster is expected in August 1991.

REFERENCES

[1] Theory and performance of the longitudinal active damping system for the CERN PS Booster - 1977.
F. Pederson and F. Sacherer, CERN
[2] Electronics for the longitudinal active damping system for the CERN PS Booster - 1977.
B. Kriegbaum and F. Pederson CERN
[3] OOC - Object Oriented Communications L.J.Chapman, Fermilab.

DESIGN OF 4-8 GHz BUNCHED BEAM STOCHASTIC COOLING ARRAYS FOR THE FERMILAB TEVATRON

D. McGinnis, J. Budlong, G. Jackson, J. Marriner, and D. Poll
Fermi National Accelelerator Laboratory*

Abstract--Pickup and kicker electrodes that function in the 4-8 GHz frequency range were designed for bunched beam cooling applications in the Fermilab TEVATRON. The electrodes are planar and are fabricated with photolithographic techniques on woven Teflon board. The electrodes also have a variable transverse aperture ranging from 0.75" to 4" to accommodate the changing beam size in the TEVATRON during acceleration. To accommodate the plunging action of the electrodes, a unique flexible microwave transition piece was also incorporated into the design.

I. INTRODUCTION

A prototype bunched beam betatron stochastic cooling system was built for the Fermilab TEVATRON to increase the beam lifetime by combating emittance growth [1]. The 4-8 GHz frequency band was chosen as a compromise between a high cooling rate and low bad mixing between pickup and kicker. A cooling system is comprised of microwave pickup and kicker electrodes and an amplifier chain between the pickup and kicker. This paper will concentrate on describing the design of the pickup and kicker electrodes.

II. PLANAR LOOPS

A bunched beam produces a large longitudinal coherent signal that could saturate the electronics in the amplifier chain. The pickup and kicker electrodes are both comprised of two identical arrays that are positioned transversely to the beam axis. One way to reduce longitudinal signal is to build the two arrays to be as identical as possible so that the longitudinal signal induced on side of the pickup is cancelled by the signal on the other side of the pickup. This requirement can be met if the arrays consist of planar loops [2]. Planar loops, which are formed with photolithographic techniques, can be held to very small mechanical tolerances. Also, since planar loops require very little machining, the fabrication costs of planar arrays can be less than more conventional stripline arrays.

As shown in Figure 1, the planar loop is fabricated on a woven Teflon circuit board and is a coplanar waveguide line composed of two slots in a ground plane oriented parallel to the beam. Terminating the coplanar waveguide is another set of transverse slots, one upstream and one downstream. In the center of both transverse slots is a via-hole that is connected to a microstrip line located on the reverse side of the ground plane.

The behavior of the planar loop is similar to the behavior of the microstrip loop. If the planar loop is acting as a pickup, the image current flowing on the ground planes is intercepted

by the upstream and downstream slots forming a doublet response. The Fourier transform of a doublet response is:

$$Z(f) \propto \sin(2\pi f\tau) \tag{1}$$

where τ is the effective transit time for a signal to travel from one end of the electrode to the other end. The response has a maximum when the loop is a quarter wavelength long. The 3 dB bandwidth of the response is about one octave.

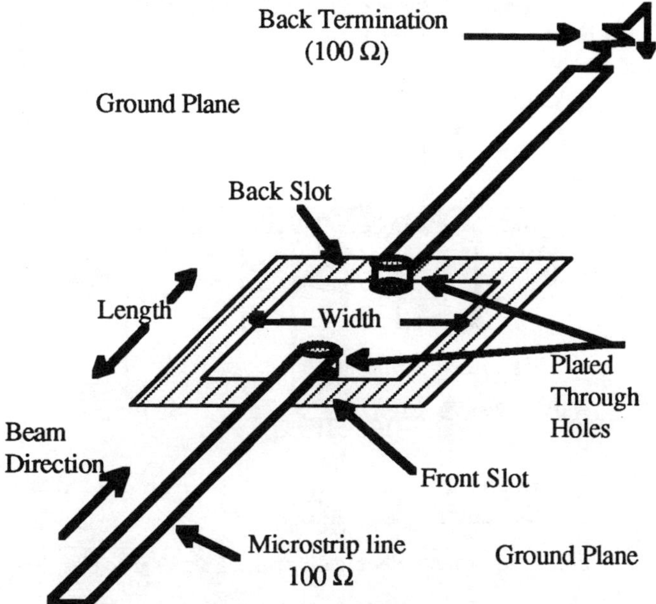

Fig. 1 Schematic view of a planar loop. The beam is on the other side of the ground plane.

The effect of pulse spreading [3] can be incorporated into the design by examining the image current density flowing on the ground planes surrounding the planar loop [2]. If the width of the slots in the ground plane is much smaller than a wavelength, then the slot lines may be modeled as non-dispersive transmission lines as shown in Figure 2. Since the field pattern in the slots is non TEM, the characteristic impedance of the slot lines is not uniquely defined. This paper will use the power impedance as the definition of the characteristic impedance of the slot lines. In the range from 4 to 8 GHz, a power impedance of 200 Ω and an effective dielectric constant of 1.2 can be obtained for a slot width of 1.9 mm patterned on a Teflon substrate thickness of 1.14 mm with a dielectric constant of 2.22.[4] Because the slot impedance changes very slowly with increasing slot width, impedances greater than 200 Ohms are difficult to obtain.

The width of the planar loop was made as large as possible to intercept the maximum amount of image current density flowing on the ground plane. The length of the planar loop was then adjusted so that the frequency response of the

---

* Operated by the Universities Research Association under contract wiht the United States Department of Energy.

electrode has a maximum at 6 GHz. These dimensions are shown in Figure 3. The input impedance of the loop is 100 Ω. The back termination of the loop was tapered from 100 Ω to 50 Ω and terminated with a commercial stripline termination. The input reflection coefficient of a single planar loop was less than -15 dB across the entire 4-8 GHz band.

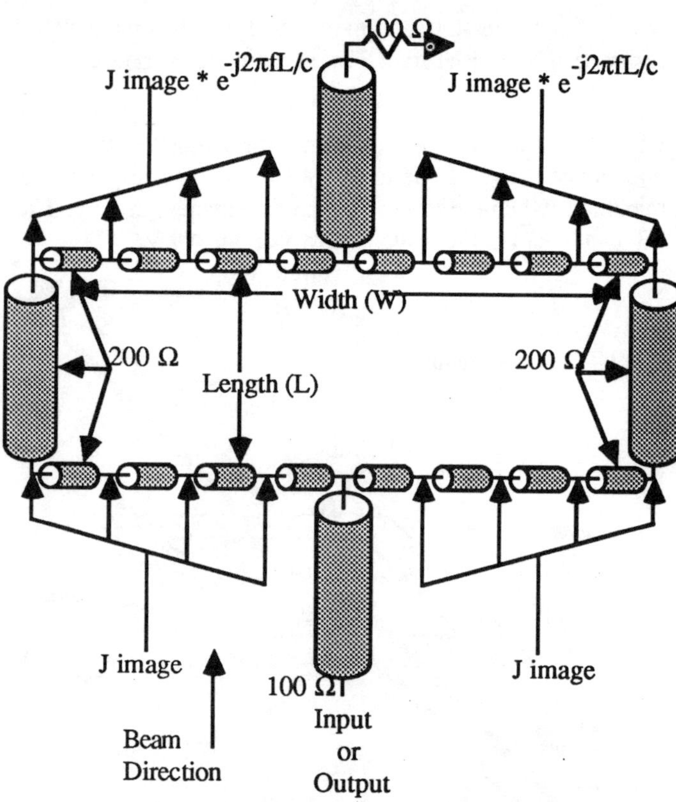

Fig. 2 Equivalent electrical circuit of a planar loop.

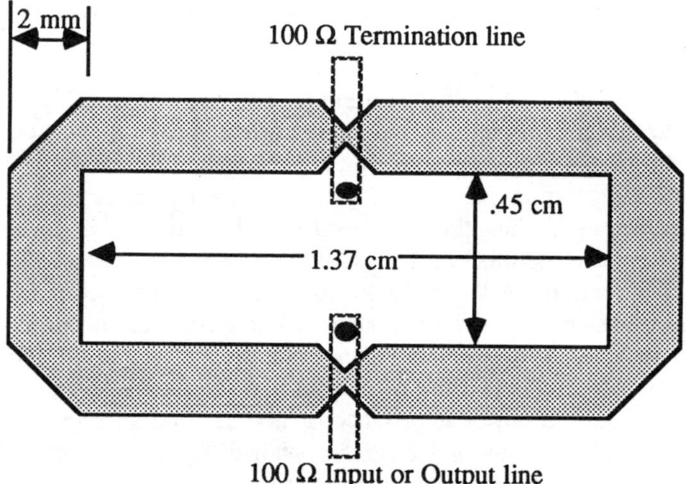

Fig. 3 Dimensions of planar loop used in the 4-8 GHz bunched beam cooling array.

III. ARRAY DESIGN

The array of planar loops is shown in Figure 4. The number of loops in an array is 16. On the reverse side of the array, the loops are combined in a binary tree as shown in Figure 5. The impedance transformation of 50-100Ω between different combiner levels are comprised of three 1/4 wave sections. Since a loop spacing greater than 1/2 wavelength will permit the combiner board to have a potential resonance in band, the spacing of the loops was selected to be 0.65 inches. This spacing is less than 1/2 wavelength for the entire 4-8 GHz band. The response of the array to a stretched wire measurement [3] is shown in Figure 6.

Fig. 4. Photo of 4-8 GHz planar loop array. This side faces towards the beam.

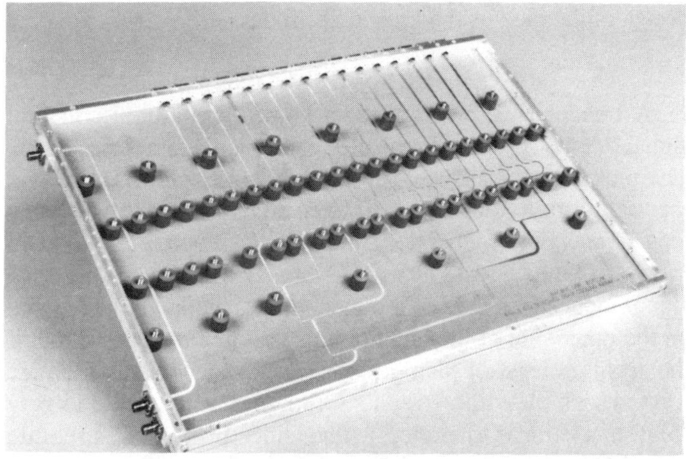

Fig. 5 Photo of the 4-8 GHz planar loop combiner board.

At the downstream end of each array are two beam position (BP) electrodes. The principle of operation of the BP electrodes is similar to the planar loops except that one of the slots at the end of the electrode is shorted. This makes the electrode sensitive to a beam travelling in either direction so that the position of both the proton and antiproton beams can be monitored simultaneously. The length of the BP electrodes was chosen so that the response has a maximum at about 200 MHz, which is close to a harmonic of the 53 MHz accelerating frequency of the TEVATRON.

The transverse separation between arrays in a tank can be varied from 0.75" to 4" so that the arrays would not scrape the beam during acceleration. The arrays were moved using a ball screw arrangement attached to a worm gear driven by a stepping motor. The transverse spacing resolution was well under 0.001".

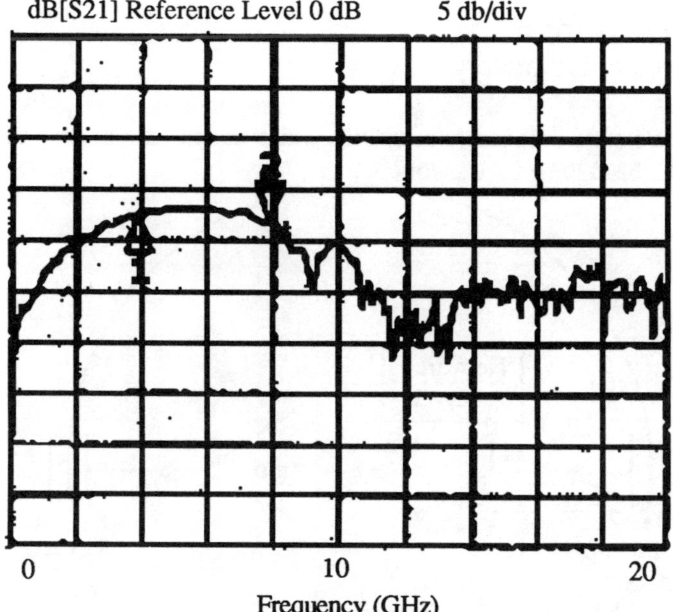

Fig. 6 Response of the 4-8 GHz array to a stretched wire measurement. Markers 1 and 2 border the 4-8 GHz band.

Fig. 7 End view of the stochastic cooling vacuum tank.

Microwave modes that could screen the Schottky signals will be excited if the image currents due to the beam encounter sharp discontinuities in the beam pipe. To minimize these effects, a flexible transition piece is placed between the ends of the array and the beam pipe as shown in Figure 7. The transition piece consists of a beryllium copper (BeCu) sleeve that fits over the bellows convolutions in the tank flange. Beryllium copper spring fingers line the edge of the sleeve to insure a conduction path for the image currents. Inside the sleeve is a pair of flexible BeCu flappers that transform the cylindrical geometry of the beam pipe to the parallel plate geometry of the array. The flexible flappers can follow the arrays throughout the entire range of motion.

To minimize the undesired microwave modes between the two arrays, a resistive material with an impedance of 300Ω/square was sprayed on a sheet of kapton and stretched between the two arrays as shown in Figure 7. Also, microwave modes inside the array are damped by placing ferrite washers around the screws as shown in Figure 5. Because these ferrite washers can be machined to tight tolerances, the washers also form the support for the teflon circuit board.

IV. RESULTS

An actual Schottky spectrum of a bunched beam in the TEVATRON is shown in Figure 8. More detail on the measurements can be found in Ref. 5.

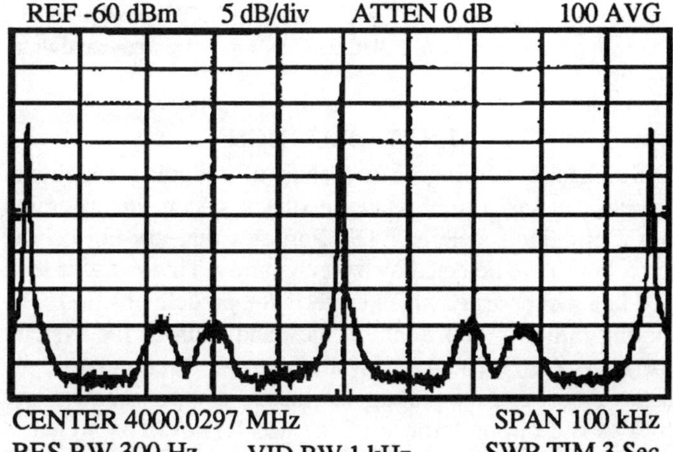

Fig. 8 Schottky signal response measured using the 4-8 GHz array. The lines at the center and edges of the graph are longitudinal lines. The other lines are the vertical betatron Schottky signals.

REFERENCES

1. G. Jackson "Bunched Beam Stochastic Cooling," Proceedings of this conference.
2. J. Petter, D. McGinnis, and J. Marriner,"Novel Stochastic Cooling Pickups and Kickers," Proceedings of the 1989 IEEE Particle Accelerator Conference, Vol. 1, 636, March 1989.
3. D. McGinnis, J. Petter, J. Marriner, and S. Y. Hsueh,"Frequency Response of 4-8 GHz Stochastic Cooling Electrodes," Proceedings of the 1989 IEEE Particle Accelerator Conference, Vol. 1, 639, March 1989.
4. R. Janaswamy, D. H. Schaubert, "Characteristic Impedance of a Wide Slotline on Low Permittivity Substrates," IEEE Transactions on Microwave Theory and Techniques, Vol. MTT-34, No.8., 901-902, August 1986.
5. G. Jackson, etal. "A Test of Bunched Beam Stochastic Cooling in the Fermilab TEVATRON Collider," Proceedings of this conference.

DESIGN OF 4-8 GHz STOCHASTIC COOLING EQUALIZERS FOR THE FERMILAB ACCUMULATOR

D. McGinnis and J. Marriner

Fermi National Accelelerator Laboratory*

Abstract--The 4-8 GHz core stochastic cooling system in the Fermilab Accumulator uses a coaxial cable transmission line to transmit signals from the pickup electrodes to the kicker electrodes. Because of the long length of the cable and the high frequency range of the cooling system, the system suffers a 40 dB gain slope and a 360° phase variation across the 4-8 GHz Band. To remedy this situation, a balanced stripline equalizer was designed featuring offset coupled lines and Schiffman phase shifters. The design was optimized by defining an effective bandwidth density and programming the OUTVAR and OUTEQN blocks of the microwave CAD program TOUCHSTONE to maximize bandwidth density.

I. INTRODUCTION

Stochastic cooling is a means of reducing the beam emittance using a microwave feedback system incorporated into a circular accelerator [1]. Particles that are not on the correct orbit are detected by a pickup array. This signal is then feed to a kicker array which corrects the particle position. The cooling rate is proportional to the bandwidth of the system. For the 4-8 GHz Betatron systems in the Accumulator, the distance between the pickup and kicker is approximately one third of the circumference of the accelerator. The signal is transmitted from the pickup to the kicker by means of a 1/2" diameter coaxial trunk line that spans the chord between pickup and kicker. The length of the coaxial line is approximately 350 ft. The insertion loss of the coaxial cable is 20 dB at 4 GHz and increases to 40 dB at 8 GHz. The signal level is boosted along the trunk line by placing amplifiers at the beginning and the end of the cable.

II. COOLING THEORY

As shown in Fig. 1, a network analyzer is inserted into the trunk line to measure the gain and phase of the cooling system at a number of Schottky bands throughout the bandwidth of the cooling system. The response is the product of the electronic gain of the cooling system and the beam transfer function from kicker to pickup. The response of the core horizontal betatron system before the equalizers were built is shown in Fig. 2. (The vertical betatron system response is similar to the horizontal response.) Because the stochastic cooling system is a feedback loop, the desired phase for all frequencies is 180°. Phase excursions greater than ±90° from 180° will cause emittance growth (heating). As shown in Fig. 2, there is a large region of the frequency band of the system that will cause the beam to heat. Also, the gain of the system

falls off significantly at 8 GHz which causes this portion of the band to be useless for cooling.

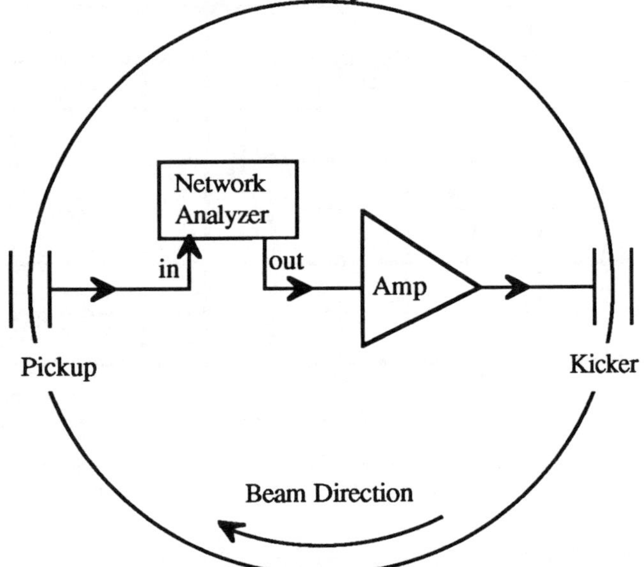

Fig. 1. Schematic of network analyzer measurement on a stochastic cooling system.

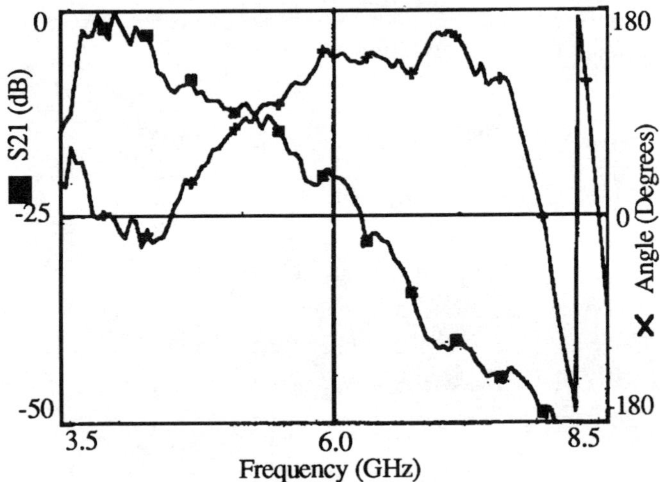

Fig. 2. Gain and phase response of the core horizontal stochastic cooling system without an equalizer.

The cooling rate for a betatron cooling system is [1]:

$$\frac{1}{\tau} = \frac{W}{M\,N_p}\left(2\,\mathrm{Re}[gM] - |gM|^2\right) \qquad (1)$$

were W is the bandwidth of the system, M is the mixing factor of the beam, N_p is the number of particles, and g is proportional to the electronic gain of the system and the number of particles. The product gM is proportional to the

---

* Operated by the Universities Research Association under contract from the United States Departement of Energy.

response measured by the network analyzer. The mixing factor is inversely proportional to frequency and can be written as:

$$M = \frac{M_c f_c}{f} \qquad (2)$$

where f_c is the center frequency of the system and M_c is the mixing factor at the center frequency. Since the gain and the mixing factor are a function of frequency, the cooling rate of Eqn. 1 should be a sum over all the Schottky bands. Since the spacing between the Schottky bands is much smaller than the bandwidth of the system, this sum can be replaced by an integral:

$$W_{eff} = \int_{bandwidth} \left(2\mathrm{Re}[\,gM\,] - |\,gM\,|^2 \right) \frac{f\,df}{f_c} \qquad (3)$$

where the cooling rate be rewritten as:

$$\frac{1}{\tau} = \frac{W_{eff}}{M_c N_p} \qquad (4)$$

The kernal of the integral of Eqn. 3 can be called the bandwidth density of the cooling system. The goal of the equalizer design is to maximize this bandwidth density.

III. EQUALIZER DESIGN

To maximize the bandwidth density, the equalizer should have high insertion loss at low frequencies and low loss at high frequencies. This response can be obtained by using the capacitive properties of electrically short coupled transmission lines. To obtain the proper gain slope a number of these are placed in series. To avoid resonances between other devices in the cooling system, the reflected power at low frequencies is terminated in a balanced arrangement of 90° hybrids. The mechanical tolerances of the 90° hybrid can be relaxed if the circuit is built in stripline.

The phase characteristic of the equalizer is shaped by Schiffman phase shifters [2]. These devices are all-pass tightly coupled transmission lines. The phase shape is determined by the length of the coupled lines. The magnitude of the phase excursion can be increased by adding a number of phase shifters in series. A schematic of the circuit is shown in Fig. 3.

IV. OPTIMIZATION OF EQUALIZER

The circuit was designed using a microwave CAD program called TOUCHSTONE™. To maximize the bandwidth density, the frequency domain response of the cooling system shown in Fig. 2 is multiplied by the equalizer response. The bandwidth density can be defined by using the OUTVAR and OUTEQN block. In the OUTVAR block, gM is defined as the product of the network analyzer response shown in Fig. 2 times the S21 response of the equalizer. The OUTEQN block then computes the square root of the kernal of the integral in Eqn. 3 at each frequency. Because the optimizer uses a least squares formulation and the square root of the bandwidth density is the optimized variable, the error function will be equal to the

effective bandwidth (W_{eff}). A random maximizer is used to maximize the error function.

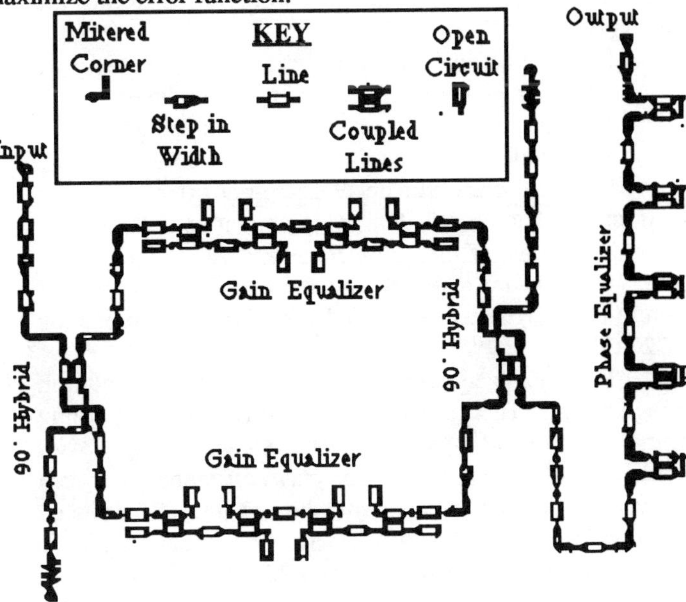

Fig. 3. Schematic of equalizer.

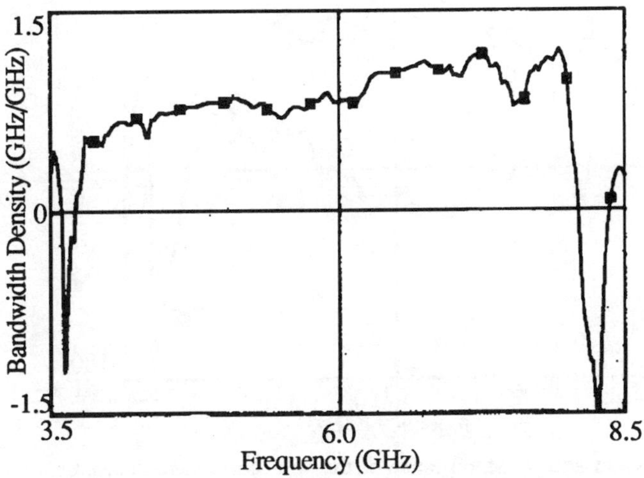

Fig. 4. Bandwidth density of cooling system with an equalizer.

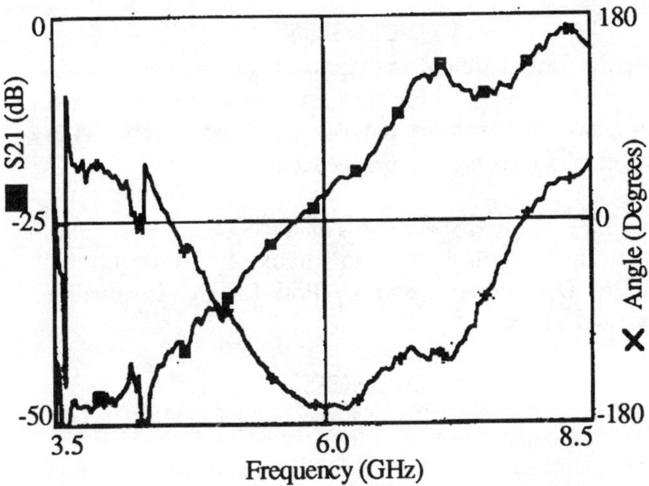

Fig. 5. Measured transfer response of the equalizer.

The following circuit parameters were varied during the optimization:

    1.) The electrical lengths of series coupled lines.
    2.) The impedances of the series coupled lines
    3.) The spacing between the series coupled lines.
    4.) The electrical lengths of the phase shifters.
    5.) An arbitrary gain,delay, and phase offset.

An optimized bandwidth density is shown in Fig. 4. The measured response of the equalizer is shown in Fig. 5. The product of the beam response and the equalizer response is shown in Fig. 6. The effective bandwidth for this combination is 3.720 GHz.

The equalizers were installed into the core betatron cooling systems. The gain and phase response of the cooling systems is very similar to the response shown in Fig. 6 with the exception of a phase offset. The phase differences between the measurement and the response showed in Fig. 6 reduced actual cooling bandwidth to 3.3 GHz. This situation can be corrected by flipping the polarity of a 90° hybrid in the trunk line of the cooling systems.

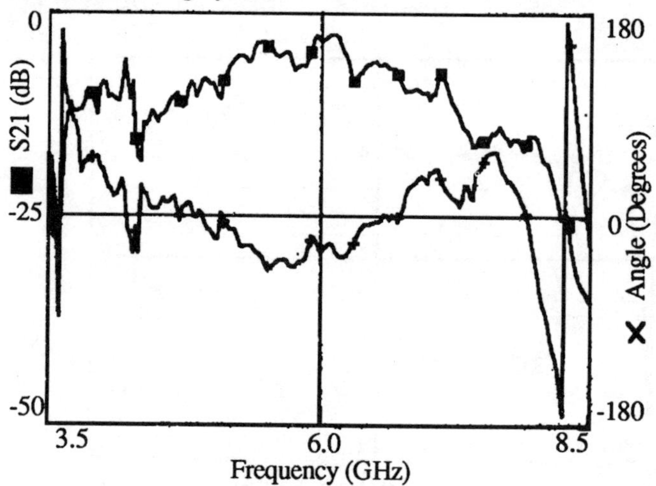

Fig. 6. Gain and phase response of the core horizontal stochastic cooling system with the equalizer.

V. CONCLUSIONS

Using the formulation of an effective bandwidth density, an equalizer can be designed for stochastic cooling systems that does not rely on the arbitrary judgements of the designer as to the gain and phase flatness across the band.

VI. ACKNOWLEDGEMENTS

The authors would like to thank J. Budlong, R. Pasquinelli, D. Peterson, and D. Poll for the invaluable assistance and advice.

REFERENCES

[1] F. Sacherer, "Stochastic Cooling Theory," CERN/ISR/ TH78-11 (1978), unpublished.
[2] B. M. Schiffman, "A New Class of Broadband 90° Phase Shifters,"MTT-6, No.2, April 1958, pp. 232-237

Bulk Acoustic Wave (BAW) Devices
for Stochastic Cooling Notch Filters

Ralph J. Pasquinelli

Fermi National Accelerator Laboratory*
Batavia , Illinois 60510

INTRODUCTION

Recursive notch filters have been in use for Stochastic Cooling in the Antiproton Source at Fermilab since its commissioning in 1985. They are important parts for providing the proper gain and phase shaping for the Stochastic Cooling feedback systems. Figure 1 is a simplified schematic of a typical "correlator" notch filter. It consists of a long and a short delay whose time of flight difference   sets the recurrence period of the filter's notch.

$$f_{notch} = \frac{1}{T_{long} - T_{short}}$$

In the past, two different types of filters have been successfully implemented using superconducting coaxial cable and single mode optical fiber. [1,2] Both have provided reliable performance. The reason for looking into a new technique of notch filters using BAW devices is one of long term stability and economics.

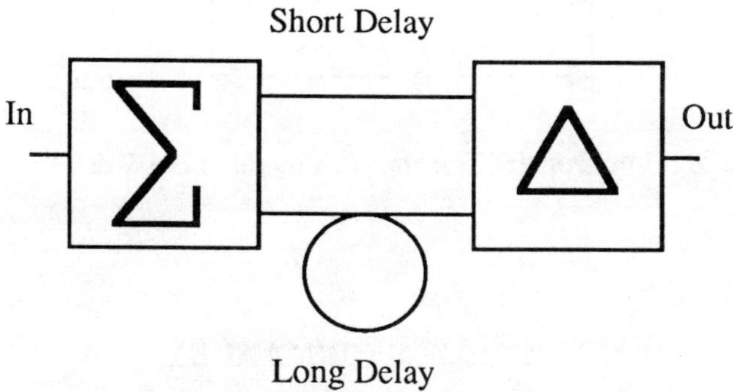

Figure 1. Simplified schematic of correlator notch filter.

Very low insertion loss and long term stability has been very good with the use of superconducting coax delay lines. [3] This is attributable to the fact that they are superconducting and passive devices. On the other hand, there is an economic burden with their usage. Liquid Helium is required to keep these delay lines in a superconducting state. Although liquid Helium is used extensively at Fermilab for the Tevatron, its only use in the Anti-proton Source are these filters.   There was an initial cost for installation of a special transfer line plus the added annual expense of maintaining three 100 liter Dewars full of liquid Helium. Elimination of these cryogenic filters and replacing them with BAW units will save an estimated $150,000 annually in operating costs.

Long delays using fiber optic links have been our other solution to making notch filters. Here the cost of cryogens has been eliminated but at the penalty of inserting several active devices in the filter, i.e. a laser and photodiode. Long term amplitude and phase stability contribute to overall filter performance. Although wideband optical transmitters have been improved over the years by the addition of peltier coolers on the laser, they do not attain the same level of amplitude and phase stability as a passive device. The present cost of a state of the art optical link is on the order of $20,000.

In recent years the reappearance of BAW delay lines has largely been spurred by advances and requirements of radar systems. Some of the features of BAW delays are :

Frequency of operation.........................500 MHz to 16 GHz

Bandwidth...up to one octave

Delay times.....................................200 nanoseconds
               to 30 microseconds

Delay tolerance................................+/- 0.001* delay,
               +/- 10 nanoseconds

Delay stability................................ 27 ppm/deg C

Attenuation stability...........................2 dB -54 to +71 C

Phase stability..................................+/- 10 degrees

* Operated by Universities Research Association Inc. under contract with the United States Department of Energy

HARDWARE IMPLEMENTATION OF FILTER

Work is proceeding on replacing all existing notch filters in the PBAR source with BAW notch filters. Three filters have already been designed and built for the Debuncher thus eliminating the Fiber Optic filters. These have been installed since the fall of 1990. New filters for replacing the super conducting filters in the Accumulator are now under construction.

The delay stability of the BAW device is on the order of 27 ppm per degree centigrade for a sapphire crystal. Maintaining notch frequency spacing requires a constant temperature environment hence the entire unit is mounted in a oven. The tolerance of the notch spacing is so tight that temperature alone is insufficient for keeping the delay time constant. The addition of a phase locked servo system is employed to achieve the required stability. A crystal stable microwave reference signal is split then injected on the long line and phase detected against the same reference signal. Any phase change is due solely to delay change in the BAW device. This phase error signal is used to drive a "trombone" (variable transmission line) in series with the BAW delay to keep the error voltage zero hence constant delay time. Figure 2 is a photo of the BAW delay and associated hardware installed in a temperature controlled oven.

Teledyne Microwave was awarded the bid to make specific units for Fermilab. In addition to some of the above mentioned parameters, the units would be octave bandwidths 1-2 GHz for the Stacktail systems, 2-4 GHz for the Debuncher systems. Delay times are 1.6 microseconds (one revolution period) and 3.2 microseconds (two revolution periods). The shorter time for momentum cooling and the longer for transverse cooling. Input/output matching required the addition of microwave circulators as well as a custom equalizer to meet overall amplitude and phase specifications. The insertion loss of the delay line can be considerable (40 to 50 dB) but is comparable to that of the Fiber Optic filters. The insertion loss stems from the fact that launching an acoustic wave onto a bulk medium is very inefficient. This inefficiency also produces an unwanted side effect in that the mismatch of the transducers leads to a triple travel feed through. This can be observed in the time domain as shown in Figure 3. If this undesirable response is not minimized the end result will be a notch filter that does not have uniform notch depth. Our units were specified to have a minimum of 20 dB suppression of this spurious response.

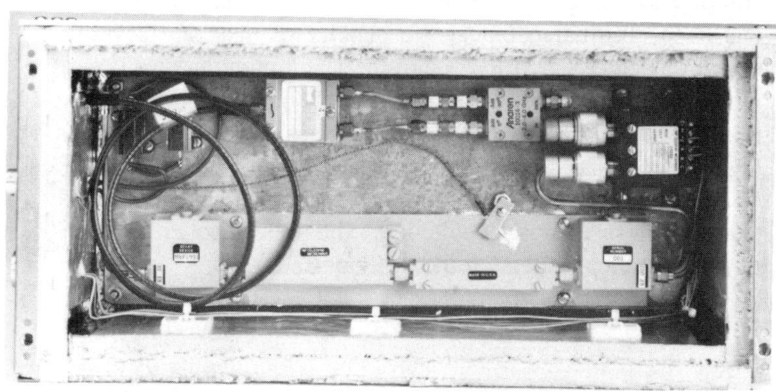

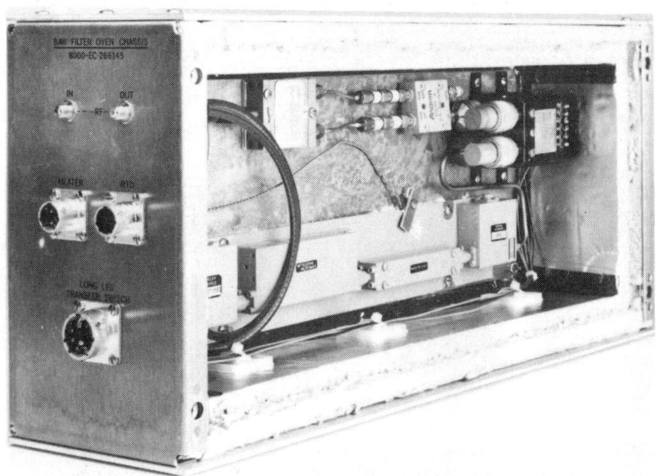

Figure 2. BAW delay and associated components installed in oven.

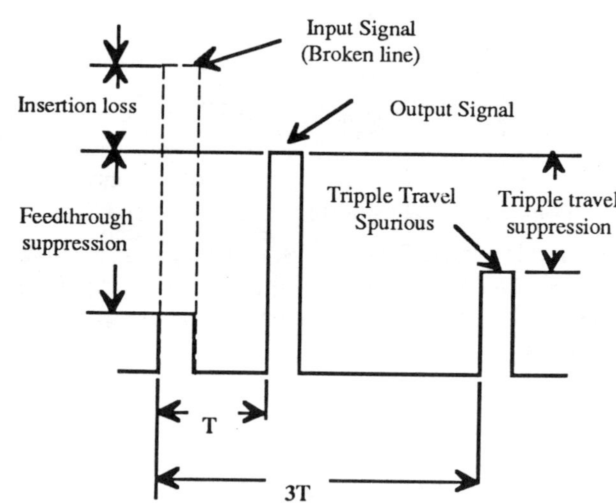

Figure 3. Time Domain Transmission mode of BAW delay line.

SYSTEM PERFORMANCE

The two main specifications for the notch filters are notch depth and notch frequency dispersion. Dispersion is defined as

$$\text{dispersion} = \frac{f_{notch} - N^* f_{rev}}{N^* f_{rev}}$$

where f_{notch} is the measured notch frequency, f_{rev} is the notch frequency spacing and N is the harmonic number of the notch. Matching of amplitude and phase characteristics over the octave bandwidths is critical. To achieve 25 dB deep notches with +/- 10 ppm notch dispersion requires an amplitude match of 1 dB and a phase match of 5 degrees across the frequency band of interest. A special equalizer was designed to compliment the amplitude and phase response of the long leg hence providing the required match. Figure 4 shows notch depth envelope and dispersion for one of the filters installed in the Debuncher Horizontal system. The fine grain ripple noticeable in the plots is a result of triple travel suppression (TTS). Because the suppression is not infinite, large excursions in notch depth can be seen. If the TTS was worse than 20 dB it would be impossible to have a minimum notch depth envelope of greater than 20 dB.

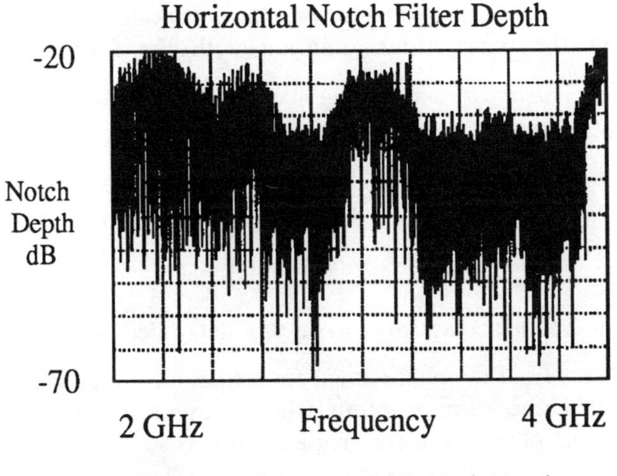

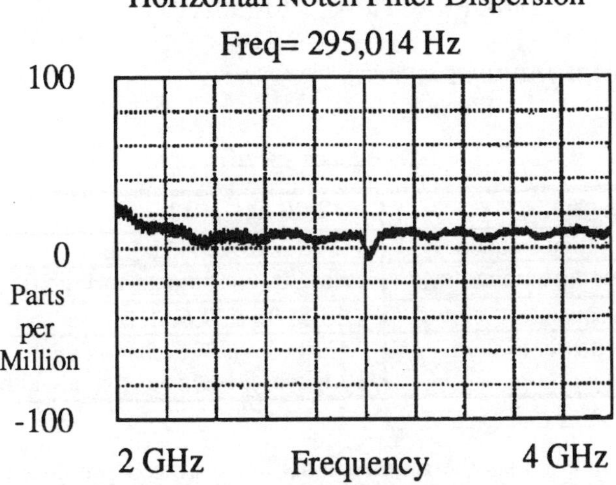

Figure 4. Notch depth (top) and dispersion (bottom) performance for filter installed in Debuncher cooling system.

CONCLUSIONS

A great deal of effort has been put into notch filter development at Fermilab over the years. There is no one technique that can be looked upon as the best choice for all applications. Cryogenic notch filters provided excellent results due to their very low loss, gain and phase flatness, and being totally passive. The cost and complication of liquid helium makes them impractical for new applications.

Fiber optic delays are also still viable solutions. A condition exists at Fermilab were a fiber optic link used in a notch filter is the only practical solution. For Bunched Beam Cooling in the Tevatron[5] a recursive notch filter with notch spacing of 47 kilohertz is required. This corresponds to a time delay of approximately 21 microseconds. To do this in a BAW delay would require a series of delay crystals almost 21 cm long. At this length the insertion loss and bandwidth limitation of the BAW delay become impractical.

As pointed out in this paper, BAW delays are a new, less expensive way to implement recursive notch filters. They can be considered for delays on the order of 1-5 microseconds and provide filter performance comparable to the superconducting notch filters.

ACKNOWLEDGMENTS

I would like to acknowledge the contributions of Dave Peterson, Ernie Buchanan, Wes Mueller, Don Poll and Pete Seifrid for their valuable contributions in realizing and commissioning the hardware.

REFERENCES

[1] R. J. Pasquinelli, "Superconducting Notch Filters for the Fermilab Antiproton Source", Proceedings of the 12th International Conference on High Energy Accelerators, 1983, pp. 584-586.

[2] R. J. Pasquinelli, Kells, Peterson, Marriner, "Optical Correlator Notch Filters for Fermilab Debuncher Betatron Stochastic Cooling", IEEE Proceedings of the 1989 Particle Accelerator Conference, pp. 694-696.

[3] Kuchnir, McCarthy, Pasquinelli, "Superconducting Delay Line for Stochastic Cooling Filters", IEEE Transactions on Nuclear Science, Vol. NS-30, No. 4, August 1983, pp 3360-3362.

[4] Teledyne Microwave Inc., BAW delay short form catalog.

[5] G. Jackson et al., "A Test of Bunched Beam Stochastic Cooling in the Fermilab Tevatron Collider", Proceeding of this conference.

Design and Operational Results of a "One-turn-delay Feedback" for Beam Loading Compensation of the CERN PS Ferrite Cavities

F. Blas, R. Garoby

PS Division, CERN, CH-1211 Geneva 23

Abstract

The periodic transient beam loading in the CERN PS ferrite cavities was diagnosed in 1989 to be the source of performance limitations for the antiproton production beam [1]. A project was then launched to lower the transient beam induced voltage by a factor of four, with a "one-turn-delay feedback" system reducing the equivalent cavity impedance on the first 3 revolution frequency side-bands around the cavity tune frequency. The design is able to cope with a wide frequency range due to particle acceleration (15 % velocity variation) and choice of harmonic number (h=10 to 20). Loop gain is above 0 dB in the vicinity of revolution frequency harmonics over an instantaneous bandwidth of 3 to 4 times the 3 dB bandwidth of the original RF system. Fast digital electronics is applied extensively, resulting in a very reliable and compact implementation. The various functions are described, closed loop performance of a cavity is shown and measurements with beam are presented.

I. INTRODUCTION

After the implementation of a fast feedback around each high power amplifier and cavity system [2], the full set of RF gymnastics proposed in 1983 for the antiproton production beam in the PS was put into operation in 1988 [1]. Performance was then limited by the periodic transient beam loading induced in the cavity because of the partial filling of the machine with particles. Troubles were especially obvious at transition and whenever the voltage was supposed to be reduced continuously to zero on a cavity (as in figure 1 with $1.7 \cdot 10^{13}$ protons).

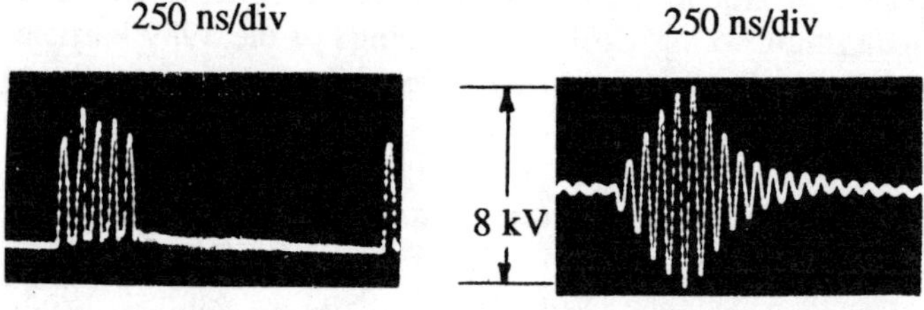

Figure 1 : Beam current and beam induced voltage in a cavity.

The first 3 revolution frequency harmonics of the beam current on each side of the RF are responsible for the cavity voltage shown. The loop gain of the fast feedback having already been pushed to its practical limit, no further increase by a factor of four (12 dB) could easily be expected, so that a complementary system was needed to help reduce the cavity impedance.

II. PRINCIPLE OF ONE-TURN-DELAY FEEDBACK

The beam current spectrum is localized in narrow frequency bands (~10 kHz) centred around revolution frequency harmonics ($f_{RF} \pm 3 f_{REV}$). Moreover, no other feedback loop is active other than at the RF frequency. A total electrical delay of one revolution period is then tolerable in the feedback loop, since high loop gain is only needed over a limited bandwidth and the phase can be correct simultaneously at each revolution harmonic. Such a "one-turn-delay feedback" system has already been designed and applied to wide-band cavities [3]. Figure 2 shows the basic block diagram with the fundamental functions.

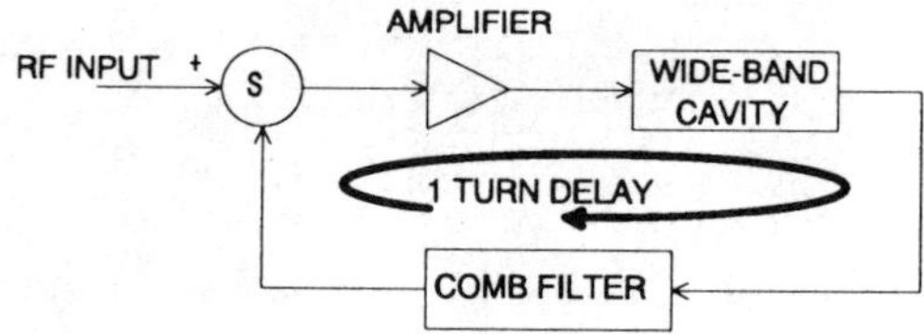

Figure 2 : Block diagram of a one-turn-delay feedback system.

Apart from the requirement for a total electrical delay of one machine turn, a comb filter with gain maxima at harmonics of the revolution frequency is necessary.

III. PRACTICAL REALIZATION

Special features

Table 1 : Feedback specifications.

Loop gain	≥ 12 dB at $f_{RF} \pm f_{REV}$
Revolution frequency	415 kHz to 480 kHz
Harmonic number (h_{RF})	10 to 20 (continuous variation)
RF frequency	4.15 MHz to 9.6 MHz
High power system	~ Cascade of 2 resonators with 1 MHz & 2 MHz 3 dB bandwidth

Table 1 lists the requirements and figure 3 presents the overall system set-up. All the processing is done digitally using ECL circuits after analogue-to-digital conversion (8 bits) at the base-band without heterodyning. The comb filter and the "automatic delay compensation" have a clock frequency of $80 f_{REV}$, so that their useful bandwidth extends comfortably to more (~ h=28) than the highest line to be compensated (h=23). The notch filter, clocked at $4 f_{RF}$, is needed to reduce the loop gain around the RF frequency and to avoid any interference of this new feedback with the many other existing loops (AVC, tuning, beam phase loop, etc.).

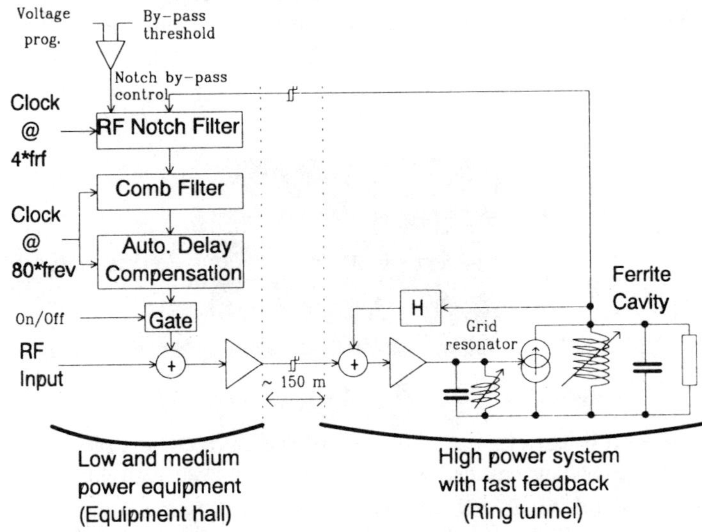

Low and medium power equipment (Equipment hall)

High power system with fast feedback (Ring tunnel)

Figure 3 : Block diagram of the practical realization.

Auxiliary functions are included for on/off control of the loop and to cancel the action of the notch filter when the cavity is left idle with a voltage program at 0 V.

The transfer function of the high-power RF system is shown in figure 4. The 3 dB bandwidth is limited to 1 MHz and the phase shift is 270° over a 3 MHz frequency band, due to the low-Q resonator in the grid of the final tube (figure 3) [2]. Loop stability is preserved by making the overall electrical delay, τ_{total} , smaller than the revolution period, T_{REV}. In fact ,

$$\tau_{total}=T_{REV}(1-1/h_{RF}) \qquad (1)$$

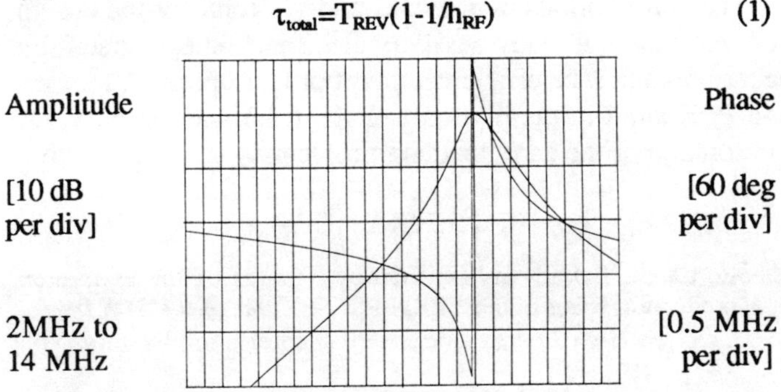

Figure 4 : Transfer function of the high-power RF system.

Comb filter

This is realized as a single-coefficient recursive filter (figure 5) [3]. Its transfer function is given by the z transform

$$C(z) = \frac{a}{\left(1-(1-a)z^{-80}\right)} \qquad (2)$$

with a=2⁻⁴ .

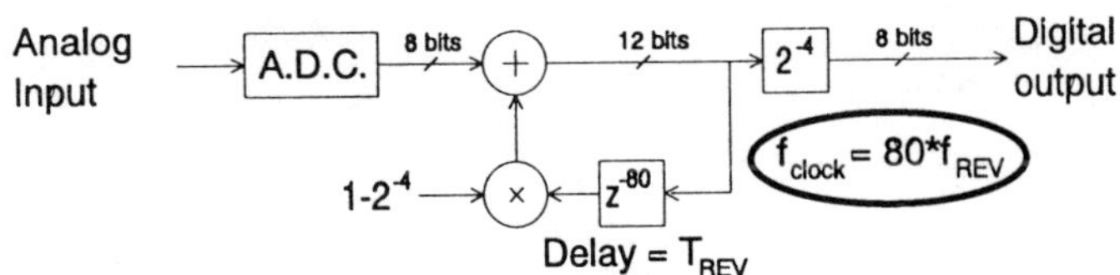

Figure 5 : Comb filter lay-out.

Gain maxima of 0 dB are obtained at all harmonics of the revolution frequency with a 3 dB bandwidth of 15 kHz. Gain minima of -30 dB are located at $(n+1/2) f_{REV}$. The 8-bit output is fed directly in digital form into the "automatic delay compensation unit".

Notch filter

The notch filter is also recursive (figure 6) and its transfer function is given by :

$$N(z) = \frac{\left(1-z^{-4}\right)}{\left(8/7-z^{-4}\right)} \qquad (3)$$

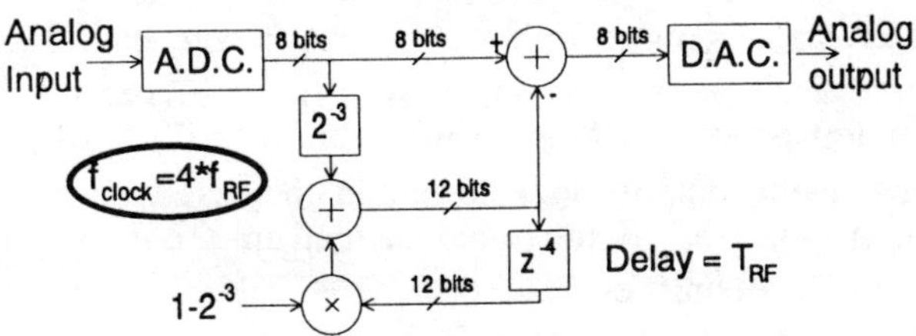

Figure 6 : Notch filter layout.

The output is converted back into analogue form before being sent to the next modules. The open-loop gain resulting from the cascade of the comb and notch filters with the high-power RF system is shown in figure 7.

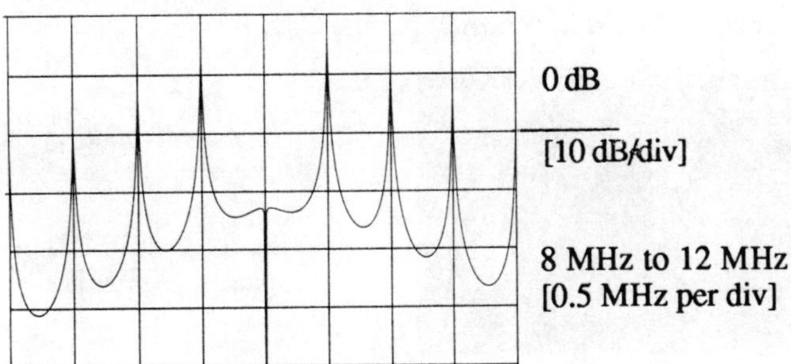

Figure 7 : Open-loop gain

Automatic delay compensation.

The overall electrical delay of the full loop is stabilized at

$$\tau_{total}=T_{REV}(1-1/h_{RF}) \qquad (4)$$

by the action of the automatic delay compensation. The system is based on a "first-in-first-out" register (FIFO), whose clocks are connected as described in figure 8.

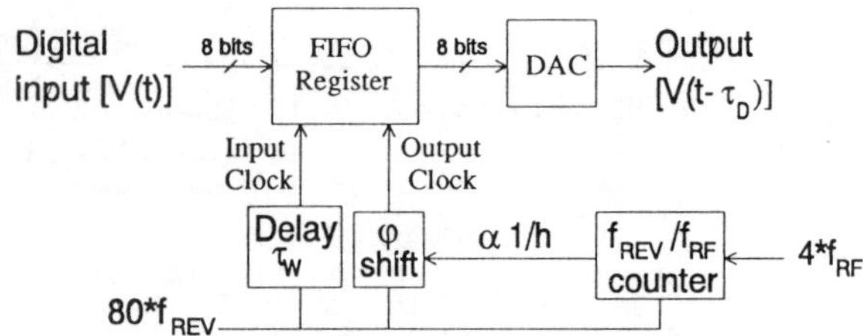

Figure 8 : Block diagram for delay compensation.

The FIFO is preset with n_0 cells at a clock frequency f_{C0} (period T_{C0}) and the delay in the unit is

$$\tau_{D0}=\tau+n_0 \cdot T_{C0} \qquad (5)$$

where τ is a constant group delay. At constant RF harmonic number h_{RF} when the clock is at f_C, the number of cells in the FIFO has changed by

$$\Delta n=(f_{C0}-f_C)\tau_W \qquad (6)$$

and τ_D has become

$$\tau_D=\tau+(n_0+f_{C0}\cdot\tau_W)T_C-\tau_W. \qquad (7)$$

Noting that $(n_0+f_{C0}\cdot\tau_W)T_C$ is a constant phase shift for all harmonics of the revolution frequency ($T_C=T_{REV}/80$), τ_D behaves as a negative delay $\tau-\tau_W$, which can compensate for the large electrical delay in the full RF system (~2 μs).

On top of this, a phase shift proportional to $1/h_{RF}$ is applied to the output clock to obtain the result quoted in equation (4). $1/h_{RF}$ is measured in real time, counting the length of T_{RF} in units of $T_{REV}/80$.

IV. RESULTS

Closed-loop transfer function

A typical closed-loop transfer function measured between the low-level RF input and the cavity gap, with the feedback on or off, is shown in figure 9.

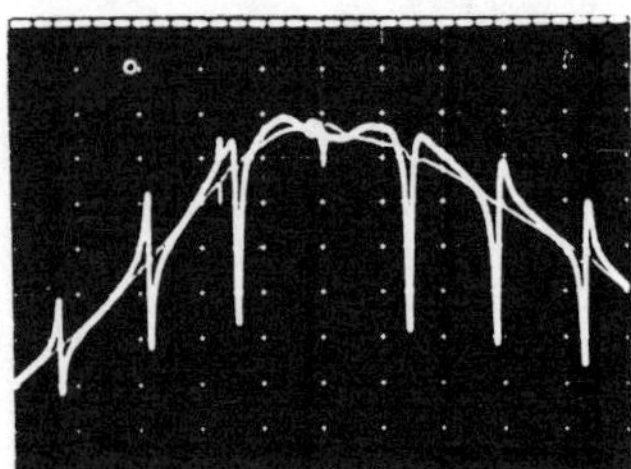

[3 dB/div]

Centre freq.: 8.3 MHz
Span : 2.905 MHz

Figure 9 : Closed-loop transfer function (RF notch active).

The gain decreases as required by 12 dB on the first 2 revolution side-bands, the gain at the following ones being approximately at the same absolute level, so that the beam sees an equivalent resonator with a Q lowered by a factor of four.

Results with beam

A reduction factor of 3.3 is measured, by comparison with the situation without feedback, for the peak voltage induced in an idle cavity (figure 10). The successive bunches now experience the same transient beam loading voltage.

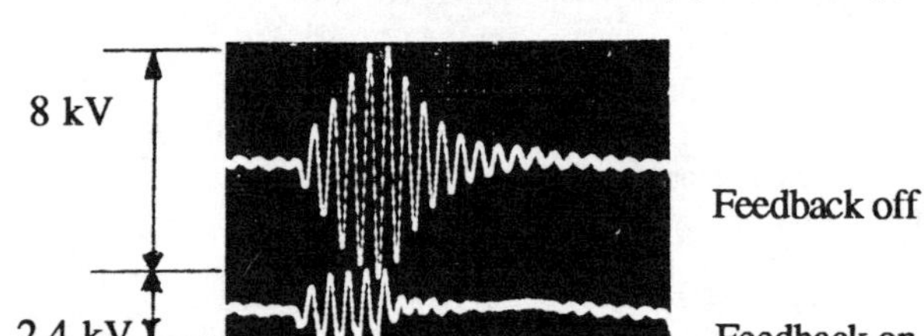

Figure 10 : Beam induced voltage in a cavity.

The full installation on the ten ferrite cavities was put into operation in September 1990. The expected improvements [1] were indeed observed : the 3% loss when crossing transition has almost disappeared as well as most of the bunch shape oscillations triggered by the various gymnastics. However, the coupled bunch instabilities thresholds have not been clearly reduced, which indicates that the disturbing impedance is not due to the cavities. Consequently, the intensity of the antiproton production beam could be raised by 10%. Using the RF dipole in the 1 GeV transfer line between PSB and PS[4], a record intensity of 1.85 10^{13} ppp has been achieved with an acceptable beam quality, while 1.7 10^{13} ppp was routinely obtained in operation.

V. ACKNOWLEDGEMENTS

The contributions of J. Evans and G. Roux for the design and realization of many auxiliary electronic units is gratefully acknowledged. The efficient support of G. Lobeau, P. Maesen and P. Konrad in the high-power RF team was essential for the solution of the early teething problems.

VI. REFERENCES

[1] R. Cappi, B.J. Evans, R. Garoby, "Status of the antiproton production beam in the CERN PS", in *Proc. of the 14th Intern. Conf. on High Energy Acc.*, 1990, Particle Accelerators, Vol. 26, p. 217

2] R. Garoby, J. Jamsek, P. Konrad, G. Lobeau, G. Nassibian, "RF system for high beam intensity acceleration in the CERN PS", in *Proc. of the 1989 IEEE Part. Acc. Conf.*, Chicago, March 20-23, p.135

[3] D. Boussard, G. Lambert, "Reduction of the apparent impedance of wide band accelerating cavities by RF feedback", in *IEEE Trans. Nuc. Sci.*, NS-30, 1983, p. 2239

[4] G. Nassibian, K. Schindl, "RF Beam Recombination ("Funnelling") at the CERN PSB by Means of an 8 MHz Dipole Magnet", in *IEEE Trans. Nuc. Sci.*, NS-32, 1985, p.2760.

Improvement of the Time Structure and Reproducibility of the Bevalac Spill*

C.M. Celata, M.J. Bennett, D.N. Cowles, B. Feinberg,
Robert Frias, M.A. Nyman, G.D. Stover, M.M. Tekawa
Lawrence Berkeley Laboratory
University of California, 64-121
Berkeley, California 94720
and
R. Salomons
RAFAEL, M.O.D. Israel

Abstract

The time structure of the Bevalac beam spill has been measured for spills with and without feedback. A new filter across the magnet has reduced spill ripple near 170 Hz. Low frequency (~a few Hz) spill structure has been improved by removing the scintillator used for feedback from the vacuum chamber, detecting instead radiation generated by collisions of beam particles with a wire chamber. Pulse-to-pulse variation of the circulating beam intensity has been reduced, and continuous tunability of the intensity introduced, by using a new feedback system. This system reduces the rf bucket height to spill beam until the proper intensity is reached.

I. INTRODUCTION

In order to serve the needs of both the nuclear physics and heavy ion medical therapy communities, it has become important to improve the time structure of the beam extracted from the Bevatron. Count-rate-limited nuclear physics detection systems would benefit from a reduction in the ratio of the maximum intensity to the average in a pulse. The UCSF-LBL heavy ion therapy program is beginning 3D conformal treatment, raster scanning tumors with the ion beam. Low frequency (≤ 2 kHz) time structure in the extracted beam would result in spatial inhomogeneities in the dose delivered to the tumor. To investigate these problems, new measurements of spill time structure and its major cause, magnetic field ripple, have been made. These measurements will be presented in Section IIA. In Section IIB, we describe a new feedback detector which has decreased low frequency structure.

The Raster scanner treatment program also requires better control (to within a few percent) of the circulating beam intensity than has been possible, since the dose given is set to be a fixed percentage of the circulating beam intensity. Section III describes a feedback system which has been successful in decreasing the pulse-to-pulse variability of the circulating beam intensity to, in some cases, a few percent.

*This work supported by the U. S. DOE under contract #DEAC03-76SF00098.

II. TIME STRUCTURE OF THE EXTRACTED BEAM

A. Bevatron Spill Time Structure

Beam is extracted from the Bevatron either by using a single sextupole magnet to produce a third order resonance, or by ramping the magnetic field while keeping the particle energy constant. In the case of the resonant spill, the sextupole magnet current may be controlled by a feedback system, which measures the extracted intensity and changes the sextupole current accordingly to keep the intensity constant. Heavy ion therapy time structure requirements mandate the use of the feedback system.

Time structure in the beam spill derives from two sources: (1) ripple in the magnetic field of the synchrotron, and (2) delay in the feedback loop controlling feedback spills. Figure 1 shows the frequency spectrum of the single turn dB/dt pickup loop signal during the time when the main synchrotron field is held "constant" and the beam is extracted. Power to the (weak focusing) magnet is provided by a 12-phase half wave rectified motor-generator (MG). Ripple from this power source is seen at harmonics of the rectifiers' output frequency, which in the case of Fig. 1 is 59 Hz. Also present is low frequency structure at about 2-5 Hz. This has been shown to originate in the regulation loop on the generator. The field ripple shown in Fig. 1 is reduced from the natural ripple provided by the motor generator by two systems. The first is a set of LC filters with central frequencies of 170, 355, 658,

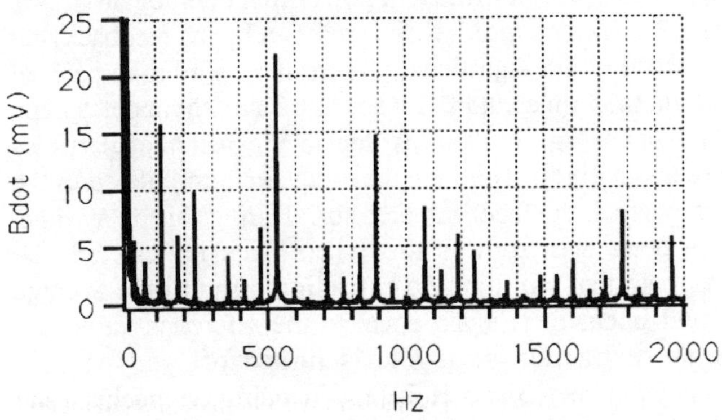

Figure 1.   Frequency spectrum of dB/dt signal. B=2535 G.

1008, and 1345 Hz in parallel with each half of the magnet. The 170 Hz filters were added this year. They have attenuation of 16 and 18 dB at 170 Hz, and full width of 8.2 and 6.7 Hz respectively at the 3 dB points. The second, the "Ripple Reduction" feedback system, reads the dB/dt signal and adjusts pole face windings dynamically to compensate for changing field. This system has unity gain at 10 Hz and 10 kHz.

The frequency structure of the spill is shown in Fig. 2 for the case of a linear ramp for the sextupole current (no feedback). This spectrum is typical of what is seen, but pulse-to-pulse variation in individual spectral lines can be of the order of 50%. The MG harmonics can be seen, as well as line harmonics, but there are also lines at 100, 200, 500, and 1100 Hz. These are due to the fact that the the magnet ramp is digitally constructed using small 10 msec step functions. Thus these components will be relatively easy to eliminate. The largest ripple is in the 2-5 Hz region.

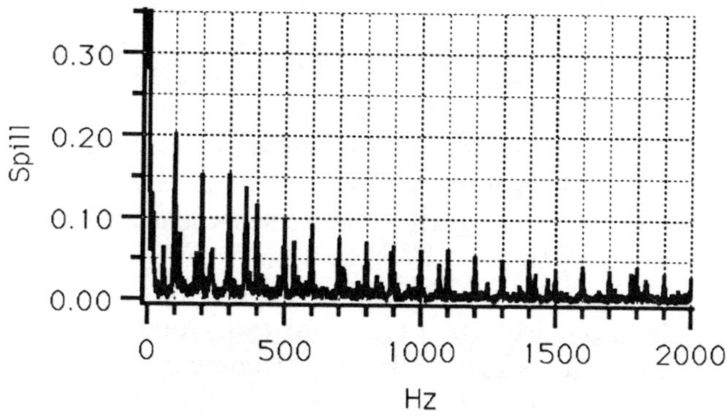

Figure 2. Frequency spectrum of extracted beam, normalized to 0 Hz component. No Feedback. B=2535 G. Ne^{+10} at 4×10^8 particles per 1 second spill.

Figure 3 is a typical spectrum for a feedback spill. Again, pulse-to-pulse variation is significant, but the frequencies of the major spectral components do not change. It is evident from the scale that the feedback system effectively removes most low frequency ripple. What remains are a few MG lines at frequencies without filters, a power line component at 360 Hz, and 2-5 Hz structure, which is reduced by the feedback, but on occasion is still significant. There is also a broad peak of amplitude 0.35 times the DC level of the spill at about 5 kHz, not shown in Fig. 3. This is due to the fact that particles require about 100 μs from the time they are destabilized to the time they reach the feedback detector. During this delay time, enough beam will spill so that when beam is detected, the feedback system will turn the extraction off until the average extracted intensity is again equal to the reference value set. Thus the extraction sextupole is turned off and on at a frequency between 3 and 10 kHz, depending on machine and feedback conditions. If the loop gain of the system is decreased, not as much beam will spill during the delay time. The height of the 5 kHz peaks will decrease, and their

frequency will increase. Thus reducing field ripple, which would permit a decrease in loop gain, is one path to reducing the 5 kHz ripple. The 5 kHz peaks are the largest ripple in the feedback spill, but do not affect raster scanning therapy because of their frequency.

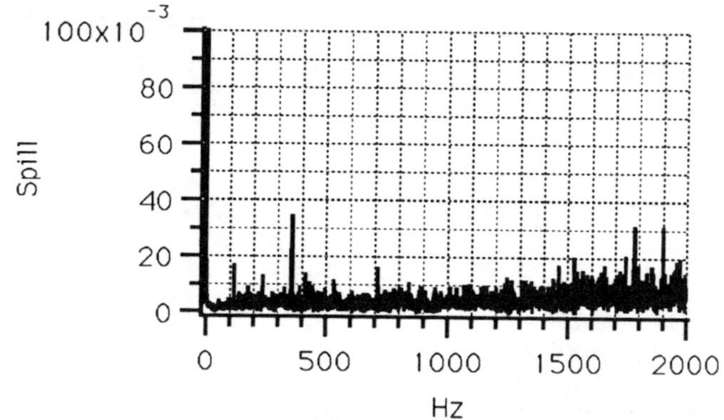

Figure 3. Frequency spectrum of extracted beam with feedback active. B=2535 G. Ne^{+10} at 3×10^8 particles per 1 second spill. Values normalized to 0 Hz component.

B. Improvements in the Extraction Feedback Detector

The extraction feedback system measures the intensity of the extracted beam vs. time, and subtracts from it the requested (reference) level. The extraction sextupole current is then changed, in proportion to this difference, in order to make the extracted intensity equal to the reference. The feedback detector consisted, until recently, of a scintillator and photomultiplier tube. The scintillator was located in the path of the extracted beam, and for high intensity beams (e.g.,≥about 10^9 particles per one second spill for Ne^{+9} at 585 MeV/nucleon). the scintillator browned in a matter of 4-5 hours. The browning was spatially inhomogeneous. Since the beam sweeps across the scintillator during extraction, this browning essentially caused a time-dependent change in the calibration of the feedback detector during the spill. When the beam swept across a brown part of the scintillator, the detector output would decrease, causing the feedback system to increase the sextupole current, and thus the spill rate. This resulted in a variation of order 30-50% in the extracted intensity vs. time.

This situation was corrected by installing, outside the beam pipe, a system consisting of a photomultiplier tube, with a scintillator connected by a short light pipe. The system detects fragments scattered mainly from an upstream wire chamber, as well as gamma rays produced by the fragments. Thus the system is referred to as the Beam Fragment Detector. The light pipe allows the PM tube to reside within a metal pipe, for magnetic shielding, while the scintillator remains outside of the pipe for full exposure to fragments. In this configuration browning is negligible, and the simplicity, high amplification, and fast response of the PM tube are retained. The system response is linear with beam intensity, and is not sensitive to changes in beam position. A further practical advantage is that the whole system is outside of the vacuum, thus making maintenance and changes simple. The system has been used satisfactorily for 1.5 years.

Problems and limits of the system derive from the fact that since the scintillator is not in the beam, it detects 2 to 3 orders of magnitude fewer "particles" than a scintillator in the beam. Thus it is sensitive to "noise" consisting of fragments sprayed from beam scraping in the extraction channel. Normally this scraping does not occur. The low signal amplitude also means that the system cannot be used at low spill rates (e.g., $\leq 1.6 \times 10^8$ particles/s for He^{+2} at 660 MeV). The signal level could be increased by various means, including enlarging the scintillator, but for these low intensity cases the old system with the scintillator in the beam works well. Finally, the system only samples 1 particle for every 10^2 or 10^3 beam particles, so the accuracy of its statistics on the extracted beam is imperfect. Thus we see broadband "shot noise" in the spectrum of the detector signal. For the desired accuracy, bandwidth, and intensities presently desired (1% accuracy below 2 kHz, and intensities $\geq 10^8$ particles per 1 second spill) this is not a problem.

III. FEEDBACK CONTROL OF THE CIRCULATING BEAM INTENSITY

The circulating beam intensity in the Bevatron varies from pulse to pulse by up to 50%, due to variations in the injector output. Moreover, the intensity is controlled by attenuators, and therefore is not easily continuously tunable. A new feedback system has been tested which has been successful in reducing the pulse-to-pulse variation to less than 7%, and which allows continuous tunability. The feedback detector is the "Beam Induction Electrode", or BIE,-- two parallel plates through which the beam passes, which measure the total circulating charge. Because the signal is so small, it is beat against the accelerating rf, which has the same frequency as the beam buckets, in a superheterodyne detector, and the amplitude of the signal at the difference frequency is measured. The feedback system compares this to a requested reference level, and decreases the rf amplitude, dumping beam, if the actual intensity exceeds the reference. A similar system was used at the Bevatron several years ago [1] to spill beam onto an internal target. Results are shown in Fig. 4. Each dot on this graph shows the intensity of one pulse of the synchrotron. Intensities above 1×10^9 are from pulses for which the feedback system was off. As can be seen from the figure, the intensity can be changed continuously over about a factor of 10. Within this range the pulse-to-pulse variation is $\leq 7\%$. The system has been used successfully over periods of hours to hold the intensity constant within this accuracy, and often to within a few percent.

Problems with the feedback system stem from the fact that the output of the superheterodyne detector is sensitive to the input waveform phase and shape, as well as to the amplitude of the signal. As the system dumps particles, the pulse shape

and the phase of the beam with respect to the rf change, giving a false reading of the intensity. The error is often the same, pulse to pulse, over long periods of time, allowing for the success reported above. But it changes with machine and beam parameters. Thus it is impossible at present to predict in advance the exact value of the reference that will produce a given intensity. On occasion we have also seen long term(~60% in 0.5 hour) drift in the intensity. The pulse-to-

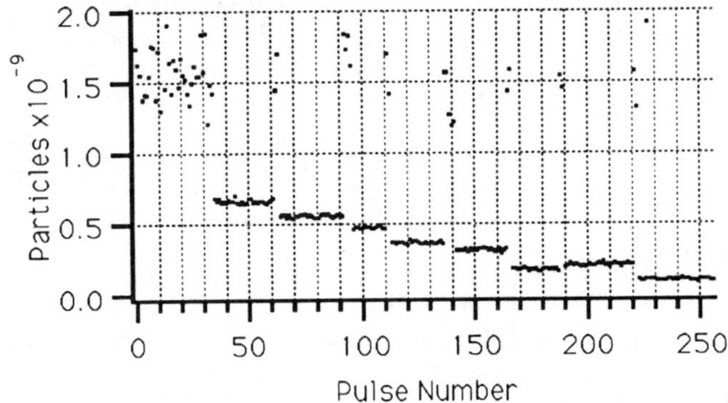

Figure 4. Intensity value from the BIE vs. Pulse Number, for B=2535 G, He^{+2}.

pulse variation in the error accounts for the 7% accuracy limit we have observed. Efforts are underway to design a new detector which would accurately measure circulating charge. We are confident, given the success with the present system, that such a detector would enable accurate intensity control.

IV. REFERENCES

[1] Fred H. G. Lothrop, Rev. Sci. Instr., vol 37, No. 3, pp. 358-361, March 1966.

Prompt Bunch by Bunch Synchrotron Oscillation Detection via a Fast Phase Measurement*

**D. Briggs, P. Corredoura, J. D. Fox, A. Gioumousis
W. Hosseini, L. Klaisner, J.-L. Pellegrin, K. A. Thompson**
Stanford Linear Accelerator Center, Stanford University, Stanford, CA 94309 USA
and
G. Lambertson
Lawrence Berkeley Laboratory, University of California, Berkeley, CA 94720 USA

Abstract: An electronic system is presented which detects synchrotron oscillations of individual bunches with 4 ns separation. The system design and performance are motivated by the requirements of the proposed B Factory facility at SLAC.

Laboratory results are presented which show that the prototype is capable of measuring individual bunch phases with better than 0.5 degree resolution at the 476 MHz RF frequency.

INTRODUCTION

Many accelerator facilities have incorporated feedback systems to suppress the growth of coupled-bunch oscillations [1-5] . The proposed SLAC-LBL-LLNL B Factory design presents several challenges in the design of transverse and longitudinal feedback systems. To achieve a design luminosity of 3*10**33, the B Factory will contain 1658 bunches per ring (total current 1.5 A for the high energy ring, 2.1 A for the low energy ring), spaced every other RF bucket at the 476 MHz RF frequency. This large number of bunches, plus the short (4.2 ns) inter-bunch period, forces difficult constraints on the detection, processing and energy correction stages of the feedback system. Our proposed feedback design uses a bunch by bunch (time domain) strategy that treats each bunch as an independent oscillator . [6-11] This approach has the additional advantage that it damps not only coupled bunch oscillations, but any disturbance, such as injection timing errors, that may induce energy oscillations.

PRINCIPLE OF OPERATION

Figure 1 presents a system constructed to test and evaluate various signal detection schemes for the longitudinal feedback system. In this laboratory prototype we have implemented a two bunch system with a time structure similar to the planned accelerator. The master oscillator of this model uses a 102 MHz oscillator that is harmonically multiplied by a factor of 28 to generate a phase synchronous 2856 MHz reference. The 102 MHz oscillator provides the excitation for two step recovery diodes which generate 60 ps FWHM impulses to simulate beam bunches. Two analog phase shifters are used to adjust the spacing of the bunches relative to the 102 MHz master oscillator, and these phase shifters allow simulated synchrotron oscillations to be impressed on the 60 ps impulses. We space

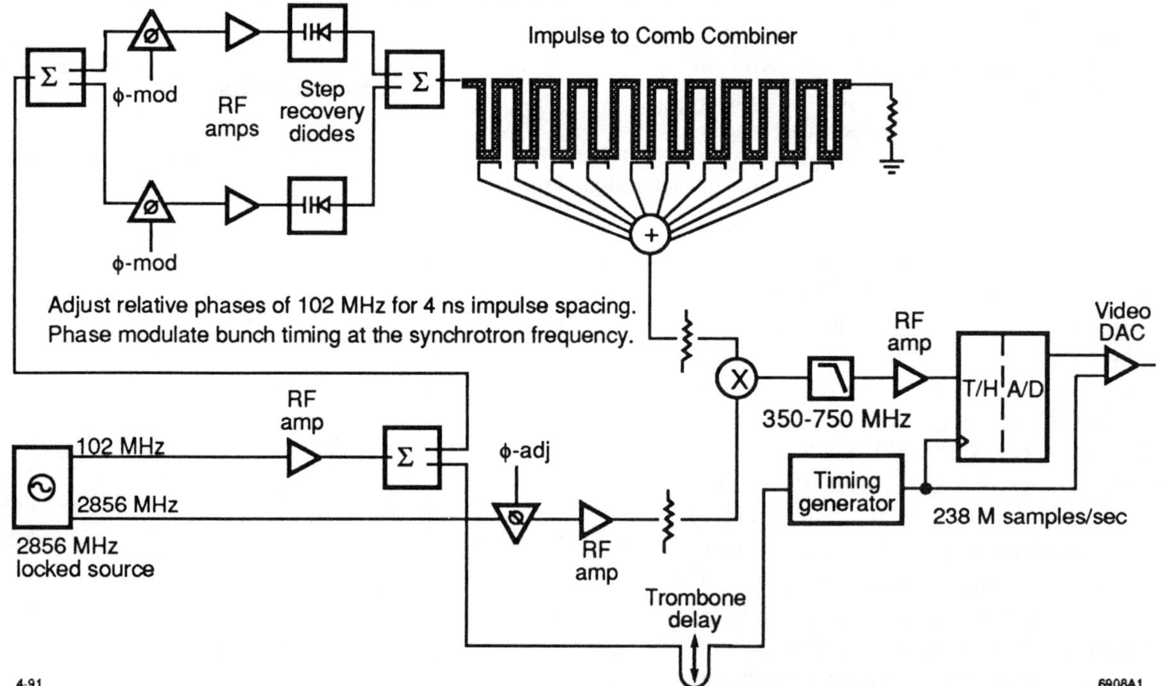

Figure 1. Block diagram of the prototype front end circuitry. This system converts a beam signal impulse into a 2856 MHz 8-cycle tone burst, and compares the phase of the burst against a reference oscillator. A 250 MHz A/D digitizes the phase each bunch crossing. The proposed B Factory design has 1658 bunches with a 4.2 ns bunch interval. The laboratory prototype uses 2 step recovery diodes and phase shifters to simulate bunches with independent synchrotron oscillations.

⋆ Work supported by Department of Energy contract DE–AC03–76SF00515.

the nominal positions of the bunches with a 4.2 ns interval which corresponds to the B Factory ring design, however in our test there is an extra 2 ns gap following the second bunch, after which the sequence repeats. With this system we can simulate independent synchrotron oscillations of the two bunches at frequencies up to 100 kHz, and study the performance of the detection system.

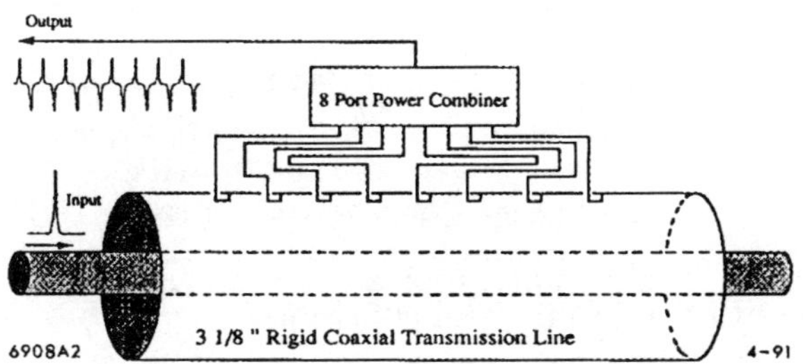

Figure 2. The eight quarter wave couplers are aligned 10 cm apart along the transmission line. Their outputs are connected to the power combiner with semi-rigid cables of equal length.

Figure 1 also shows the detection circuitry. A periodic coupler is used to generate a short eight cycle tone burst from each bunch. The phase of each burst is compared against the 2856 MHz reference oscillator in the double balanced mixer. The output of the phase detector is filtered with a 750 MHz low pass filter to remove the second harmonic and to limit the bandwidth for noise reduction. A fast analog to digital converter (with an internal track and hold) digitizes the phase error signal at a 238 MHz rate [12]. This process provides a unique error signal for each bunch crossing. In a complete feedback system this digital data stream would undergo further digital processing, but in our laboratory model we reconstruct the digitized values with a fast digital to analog converter which allows us to use traditional lab instruments to study the system performance [13]. A digital timing system, constructed with 100 K ECL logic, is clocked by the 102 MHz master oscillator and provides the timing for the A/D and D/A stages. This system allows us to measure the resolution, noise, and inter-bunch isolation of the bunch phase measurements.

RESULTS

The beam pick-up generates a short (less than 4 ns) tone burst at the 2856 MHz frequency. We use a periodic coupler, rather than a tuned resonant structure, to avoid coupling between adjacent bunches. We have studied two possible configurations. We have constructed a comb generator in a coaxial geometry, using an array of quarter wave couplers aligned 10 cm apart along a coaxial transmission line. All the coupler outputs are then combined coherently in a power combiner as shown in Fig. 2. As an alternative, we have also fabricated and measured a structure constructed of periodically coupled stripline circuits. This latter approach minimizes the number of RF connections, and in the case of a beam coupled structure, would also minimize the number of vacuum feedthroughs.

The measured response of the coaxial generator is shown in Fig. 4. We see the signature of the time of arrival of each bunch as an eight cycle tone burst. There is a small amount of residual ringing (approximately 5–10%) evident in the time domain measurements. In comparison, the stripline fabricated generator produced fewer reflections, but due to its larger design coupling value and skin effect losses displayed less amplitude uniformity during the output burst.

LAB MEASUREMENTS OF COMBINED SYSTEM

This periodic coupler output has been used to study the performance of the electronic system. Fig. 4 presents the analog output signal of the mixer for the two bunch system; also shown in the figure is the reconstructed analog output of the D/A converter. We can see that the bunch spacing alternates between 4 and 6 ns, as produced by our beam simulator. The response shows the bandwidth required for independent measurements on bunches with the 4.2 ns spacing.

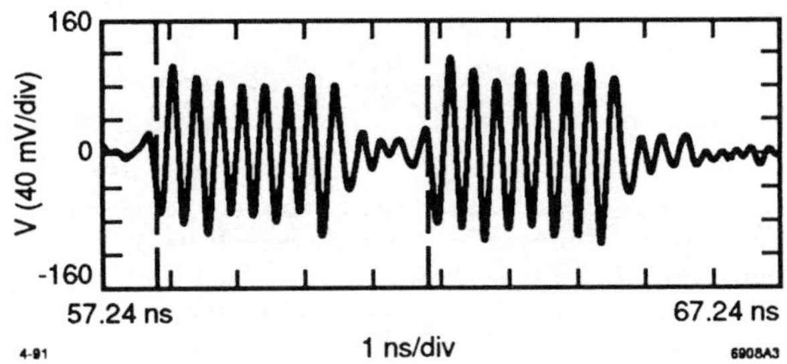

Figure 3. Measured response of the coaxial comb generator for two simulated beam signals with 4 ns bunch spacing.

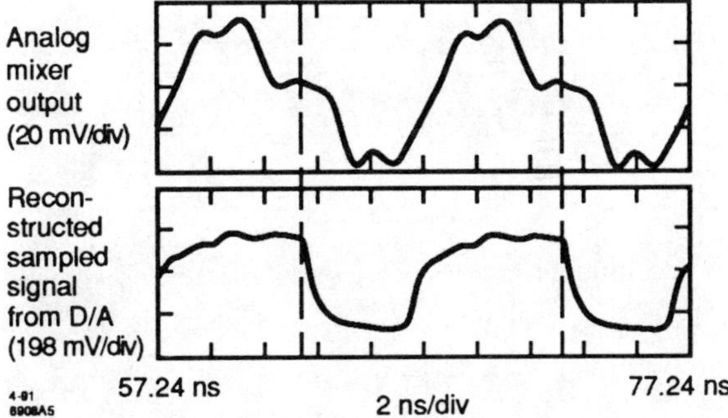

Figure 4. Measured response of the mixer (phase detector) for two bunches nearly 180 degrees apart. Also shown is the digitized phase signal after reconstruction in the D/A.

Figure 5 shows the system response for six independent DC phase offsets of the bunches. The top traces show the analog mixer response for 3 distinct DC phases of bunch A, with bunch B left undisturbed. Similarly, the lower pair of traces show three DC phases of bunch B, with bunch A left undisturbed. Again, the lowest traces are the reconstructed D/A output showing excellent isolation of the bunches.

Figure 6 is an oscilloscope photograph which shows the AC response of the system. In these measurements a 10 KHz AC sine wave is impressed on the control signal of the step recovery diode of a selected bunch. The AC modulation exactly simulates a synchrotron oscillation. In the figure bunch A is modulated with an amplitude corresponding to 0.7 degrees p/p (+/-0.006 radians) at the 476 MHz ring RF frequency. We note the prompt in-phase detection of this oscillation, which corresponds to a +/- 2ps time displacement of the bunch. The rms noise floor in this measurement is found to be better than 0.1 degrees at the 476 MHz RF frequency within the 10 kHz oscillation bandwidth.

The bunch to bunch coupling in our system is measured using a variant of the technique used in Fig. 6. By driving bunch A with a large oscillation, and detecting the bunch B output signal at the modulation frequency we can measure the coupling from A to B.

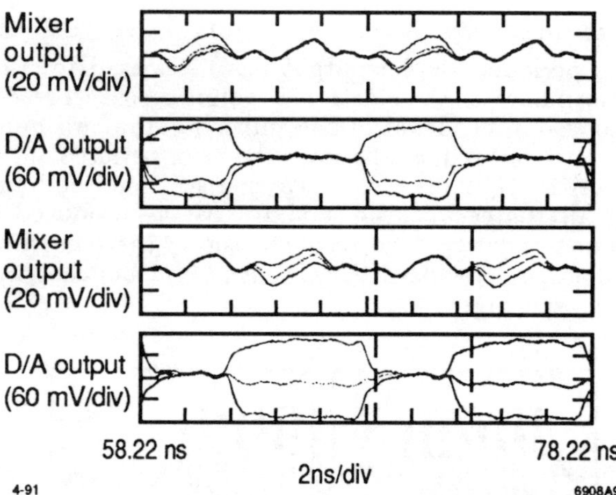

Figure 5. The upper trace shows the analog output signal of the phase detector (mixer) for three phase offsets of Bunch "A", while the next trace shows the digitized and reconstructed waveform of the upper trace. The lower traces are the same measurement of three phase offsets of bunch "B" with bunch "A" undisturbed.

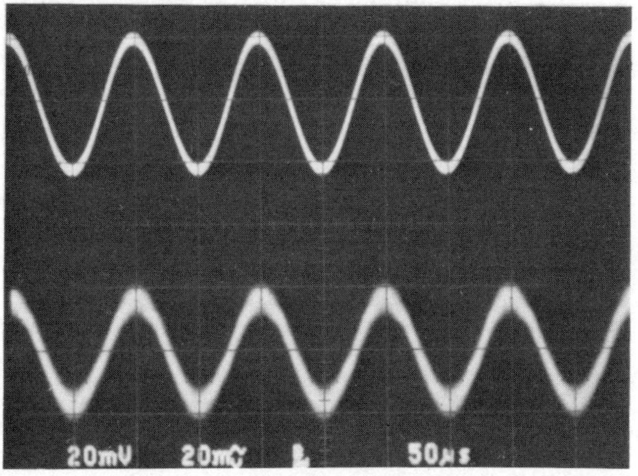

Figure 6. The upper trace shows the signal applied to a step recovery diode phase shifter, and represents a synchrotron oscillation amplitude of 0.7 degrees (+/- 0.006 radians) at the 476 MHz RF frequency. The lower trace shows the reconstructed D/A output.

Table 1 presents a summary of the measured performance of our various system components for the configuration of Fig. 1. We see that the performance of the two comb generators is similar. We also note that the coupling between the bunches for the A to B and B to A cases is very similar, notwithstanding the extra 2 ns present for the B to A case. The noise performance of the system is consistent with the eight bit quantization of the digitizer, and suggest that a system using this technique could easily achieve measurement resolutions of better than 0.5 degrees at the 476 MHz RF frequency.

Summary and Conclusions

We have demonstrated a signal processing system designed to detect longitudinal oscillations of stored bunches in a B Factory like storage ring. Our prototype system has been shown to be capable of measuring the phase of individual bunches separated by 4.2 ns with better than 0.5 degree resolution (at 476 MHZ). We have shown that the periodic coupler is capable of generating isolated tone bursts from the simulated bunches, and that the detection

Table 1
Isolation, Resolution, and Noise Measurements

Comb Generator	Configuration	Isolation
Coaxial	A to B	25.9 dB
Coaxial	B to A	28.5 dB
Microstrip	A to B	26.7 dB
Microstrip	B to A	29.4 dB
Phase Detector Range		±15° at 476 MHz
Phase Detector Resolution		1.3 mRad at 476 MHz
	or	0.08° at 476 MHz
Phase Detector Noise		1.55 mRad rms at 476 MHz
	or	0.09° rms at 476 MHz

of individual bunch phases for a large number of bunches (1658) with a 4.2 ns interval is feasible.

Acknowledgments

The authors would like to thank J. Dorfan, A. Hutton, and M. Zisman for their encouragement and support for this project, and H. Schwarz and J. Judkins for the loan of specialized laboratory equipment.

References

[1] D. Heins, R. D. Kohaupt et al., "Wideband Multibunch Feedback Systems for PETRA," DESY 89–157, 1989.

[2] P.L. Corredoura, J.-L. Pellegrin, H.D. Schwarz and J.C. Sheppard, "An Active Feedback System to Control Synchrotron Oscillations in the SLC Damping Rings," in Proc. Particle Accelerator Conf., 1989, vol. 3, p. 1879.

[3] Toshio Kasuga, Masami Hasumoto, Toshio Kinoshita and Hiroto Yonehara, "Longitudinal Active Damping System for UVSOR Storage Ring," Japanese Jour. Applied Physics, vol. 27, no. 1, 1988, p. 100.

[4] M. A. Allen, M. Cornacchia, and A. Millich, "A Longitudinal Feedback System for PEP," in IEEE Trans. Nucl. Sci. NS–26, 1979, no. 3, p. 3287.

[5] E. Higgins, "Beam Signal Processing for the Fermilab Longitudinal and Transverse Beam Damping System," in IEEE Trans. Nucl. Sci. NS–22, 1975, no. 3., p. 1581.

[6] "An Asymmetric B Factory based on PEP," Conceptual Design Report, SLAC 372, 1991.

[7] K. A. Thompson, "Simulation of Longitudinal Coupled-Bunch Instabilities," B Factory Note ABC–24, 1991, SLAC.

[8] B Factory Accelerator Task Force, S. Kurokawa, K. Satoh and E. Kikutani, Eds., "Accelerator Design of the KEK B Factory," KEK Report 90–24, 1991.

[9] C. Pellegrini and M. Sands, "Coupled Bunch Longitudinal Instabilities," SLAC Technical Note PEP–258, 1977.

[10] R. F. Stiening and J. E. Griffin, "Longitudinal Instabilities in the Fermilab 400 GeV Main Accelerator," IEEE Trans. Nucl. Sci. (1975), NS–22, no. 3, p. 1859.

[11] D. Briggs et al., "Computer Modelling of Bunch by Bunch Feedback for the SLAC B Factory Design," Proc. Particle Accelerator Conference, 1991.

[12] Tektronix Corporation, TKADC–20C 250 MSample/Sec Hybrid Analog to Digital Converter.

[13] Brooktree Corporation, BT108BC Digital to Analog Converter.

Computer modelling of bunch-by-bunch feedback for the SLAC B-factory design*

D. Briggs, J. D. Fox, W. Hosseini, L. Klaisner,
P. Morton, J.-L. Pellegrin, K. A. Thompson
Stanford Linear Accelerator Center, Stanford University, Stanford, CA 94309 USA

and

G. Lambertson
Lawrence Berkeley Laboratory, University of California, Berkeley, CA 94720

Abstract

The SLAC B-factory design, with over 1600 high current bunches circulating in each ring, will require a feedback system to avoid coupled-bunch instabilities. A computer model of the storage ring, including the RF system, wake fields, synchrotron radiation loss, and the bunch-by-bunch feedback system is presented. The feedback system model represents the performance of a fast phase detector front end (including system noise and imperfections), a digital filter used to generate a correction voltage, and a power amplifier and beam kicker system.

The combined ring-feedback system model is used to study the feedback system performance required to suppress instabilities and to quantify the dynamics of the system. Results are presented which show the time development of coupled bunch instabilities and the damping action of the feedback system.

I. INTRODUCTION

The large average current in the SLAC B-factory design is distributed into many bunches of sufficiently small charge to minimize the beam-beam interaction and single-bunch instabilities. Although the cavity higher-order-modes (HOMs) will be strongly damped ($Q < 70$), there will still be significant coupling of the longitudinal and transverse motion of adjacent bunches via wakefields. Furthermore, the high-Q accelerating mode can also strongly couple the bunches longitudinally. The resulting instabilities will be controlled via wideband, bunch-by-bunch feedback. Such a feedback system can handle disturbances to the bunch motion arising from any source, including but not limited to wakefields and injection errors.

The feedback systems to control the longitudinal and transverse coupled-bunch instabilities will be similar in architecture. Since the signal detection and kicker requirements are more stringent for the longitudinal system and for the high energy ring (HER), we shall concentrate our discussion on this case. Basic longitudinal-feedback system specifications are shown in Table 1. The proposed system implementation, its block diagram and description, and hardware tests are discussed elsewhere [1].

Table 1: Basic feedback system specifications

RF freq.	476 MHz
Max. mode amplitude	10 ps = 0.03 rad
Injection scheme	1/5 bunch at 60 pps
$\frac{\delta E}{E}$ injection error	0.002
δt injection error	100 ps

* Work supported by Department of Energy contract DE–AC03–76SF00515.

II. SIMULATION MODEL

The simulation model (see Fig. 1) consists of a model of the feedback system electronics, combined with a model of the dynamics of the bunches in the ring.

The feedback system model simulates the transfer function of the feedback system and includes: (1) the electronic properties of the phase detector, mixer, low-pass filter, and A/D converter, (2) input noise, gain and offset errors, bandwidth limitations, and dynamic range of the analog components, and (3) the algorithm running in the set of digital signal processors (DSP farm), that takes as input the digitized bunch phases and calculates the longitudinal kick.

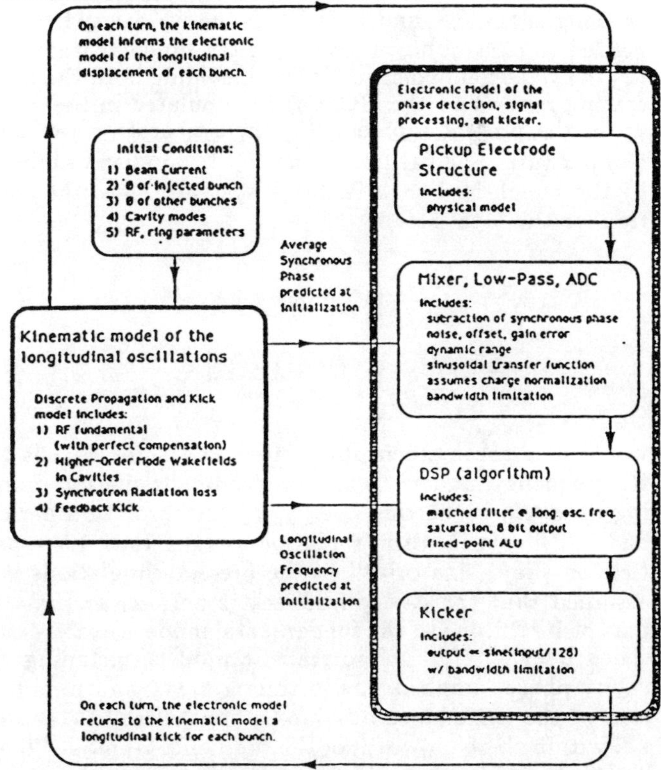

Figure 1. Block diagram of the feedback simulation model.

The model of the DSP farm emulates a 20-tap finite impulse response (FIR) matched filter. The coefficients ξ_j of the taps comprise a sinusoid with the synchrotron period of 19.3 turns, that is,

$$\xi_j = A_{DSP} \sin\left(2\pi\frac{j-1}{19.3}\right) \ , \qquad (1)$$

where $j \in 1, ..., n_{samples}$, and we take $n_{samples} = 20$. Given measures $\tilde{\varphi}_i(k) \equiv \varphi_i(k) - \varphi_s$ (where φ_s is the synchronous phase) of the phase of bunch i on successive turns k as input, the result of the DSP algorithm is the output:

$$y_i(k) = \sum_{j=1}^{n_{samples}} \xi_j \tilde{\varphi}_i(k-j) \quad . \qquad (2)$$

This result is clipped to an 8-bit signed integer and used to set the phase of the kicker oscillator for that bunch on that turn. The kicker model is implemented as a phase-modulated RF kicker with a nominal 4 keV maximum output amplitude.

The measurement of the phase of a bunch is assumed independent of its charge, i.e., it is assumed that a separate measurement of bunch charge is available for normalization. A propagation delay of at least one turn ($7.33\ \mu s$) is enforced in the feedback transfer function.

In the ring simulation model, a discrete kick is given to each bunch at a single point in the ring; that is, the system is modelled as though there were a discrete change in energy at a single point on each turn. This simplification is justified since the synchrotron frequency is small compared to the revolution frequency. The kick given to bunch i is comprised of several components: (1) the RF cavity voltage $\hat{V}_g \sin \varphi_i + V^{cav.fbk.}$, where $\hat{V}_g$ is the peak generator voltage, φ_i is the phase of bunch i with respect to the zero crossing of the RF, and $V^{cav.fbk.}$ is the RF cavity feedback needed to control beam loading in the fundamental mode, (2) the wake field voltage V^{wake} (including both the accelerating mode and the HOM's) accumulated in the cavity up to the present moment, (3) the synchrotron radiation loss per turn, and (4) the voltage V_i^{fdbk} applied to bunch i by the bunch-by-bunch feedback system. Thus, the equation for the total kick is

$$\Delta \dot{\varphi}_i = -\frac{\alpha \omega_{rf}}{E_0/e} \left[\hat{V}_g \sin \varphi_i + V^{cav.fbk.} \right.$$
$$\left. - U_0/e + V^{wake} + V_i^{fdbk} \right] - \frac{2T_0}{\tau_E} \dot{\varphi}_i \quad , \qquad (3)$$

where α is the momentum compaction factor, ω_{rf} is the RF frequency, and τ_E is the longitudinal radiation damping time; E_0 is the ring energy, T_0 the revolution period, and U_0 the synchrotron radiation loss per turn, for a particle on the design orbit. In the present simulations it is assumed that the cavity feedback is perfect, so that the part of V^{wake} due to the fundamental mode is exactly cancelled by $V^{cav.fbk.}$. More realistic models, including cavity phase, amplitude, and tuning loops, and modification of the impedance at coupled-bunch frequencies that fall within the bandwidth of the fundamental mode [2] are under study [3].

III.   SIMULATION RESULTS

Parameters used in the simulations are shown in Table 2. The harmonic number of the HER is 3492, with every other bucket filled except for a 5% gap; the total current is 1.48 A. The initial conditions used were $\varphi_5 = 0.2915$ rad, that is, bunch 5 starts at an 0.1 rad offset from the remaining bunches i, which were started at the synchronous phase $\varphi_i = 0.1915$ rad. The feedback gain in the examples described here was set so that a 5 mrad sinusoid on the input (at the synchrotron oscillation frequency) corresponds to a 4 keV sinusoid on the kicker output.

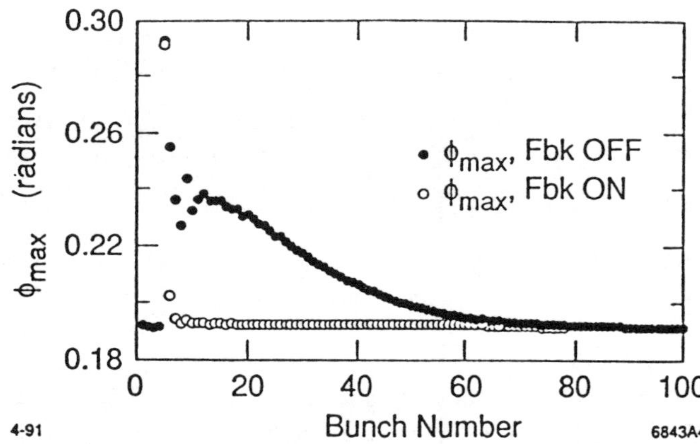

Figure 2. Plot of the maximum bunch offset reached in 3000 turns for the first 100 bunches after the gap, with and without feedback.

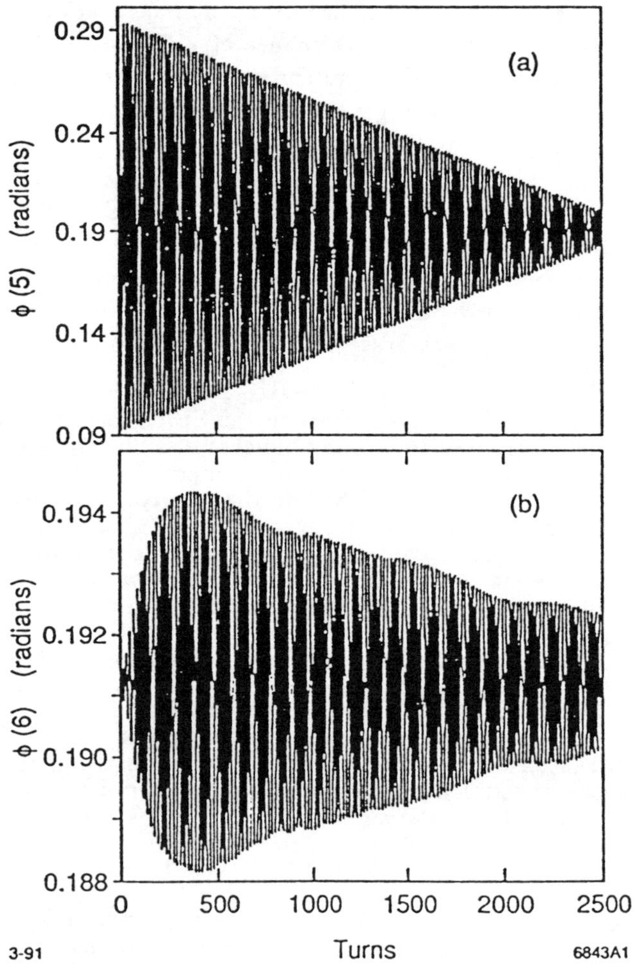

Figure 3. Plots of the longitudinal phases of (a) the injected bunch (#5) and (b) the bunch immediately following (#6), vs turn number, in the presence of feedback. Note the expanded vertical scale in (b).

In Fig. 2, we show a plot of maximum bunch offsets reached after 3000 turns, with and without feedback. Note that the time between injection pulses is 1/60 second, which is about 2300 turns. In the absence of feedback, the disturbance shown would grow even larger and propagate further back in the bunch train. With the feedback system

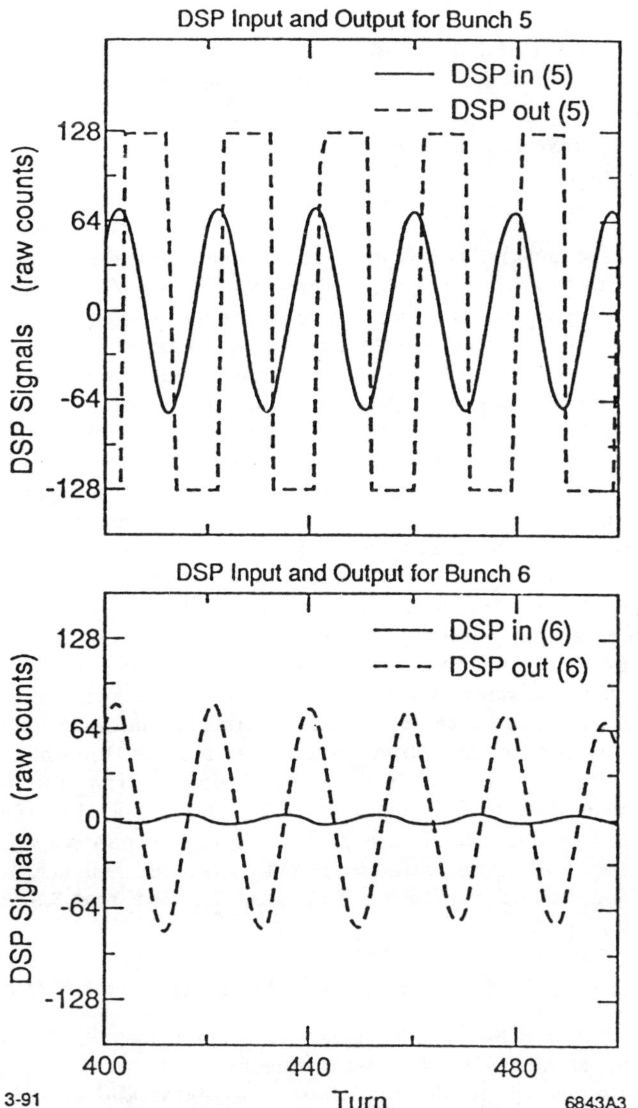

DSP Input and Output for Bunch 5

DSP Input and Output for Bunch 6

Figure 4. Plots of the input and output of the DSP model for (a) bunch #5, and (b) bunch #6, shortly after the injection of bunch #5.

Table 2: Simulation parameters for HER

Bunch charge	4×10^{10}
Number of bunches	1658
Bunch interval	4.2 ns
Number of cavities	20
Freq. of strongest HOM	750 MHz
Q of HOM	70
R/Q of HOM (per cav.)	33 Ω
$\hat{V}_g$	18.5 MV
α	0.00241
U_0	3.52 MeV/turn

turned on, the coupled bunch excitation does not extend beyond a very few bunches.

Fig. 3 shows the phases of the injected bunch (#5) and the immediately following bunch, vs turn number. The envelope of the phase of the injected bunch damps linearly, reflecting the fact that the kicker saturates, and the phase of the following bunch grows quickly to a maximum and then slowly damps. The excitation of subsequent bunches is strongly suppressed.

Fig. 4 shows the input and output of the DSP model for bunches #5 and #6 shortly after injection. The DSP output saturates for bunch #5, but maintains the proper 90° phase lag. Such benign saturation behavior is difficult to realize with conventional analog approaches.

Fig. 5 compares the amplitude of the injected bunch #5 and following bunches, first without (Fig. 5a) and then with (Fig. 5b) a 10% bunch-to-bunch coupling in the front-end electronics and a 3% coupling in the kicker. With coupling, bunch #6 suffers a greater disturbance, but still damps, while subsequent bunches suffer only slightly. Thus the system is tolerant of a reasonable amount of bunch to bunch coupling in the analog components.

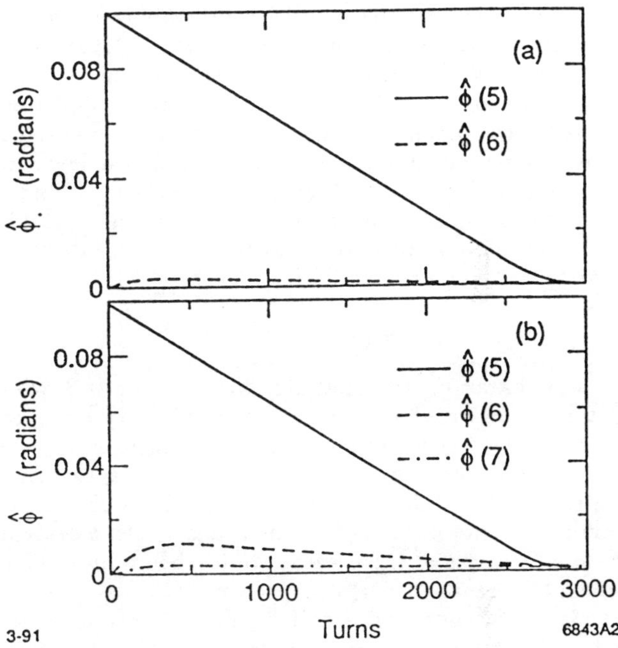

Figure 5. Plots of the phase-space error amplitude for the injected bunch (#5) and following bunch(es), (a) with no coupling, and (b) with 10% bunch-to-bunch coupling in the front-end electronics and 3% coupling in the kicker.

In conclusion, our simulations indicate that the present conceptual approach to bunch-by-bunch feedback is satisfactory. Simulations to support the detailed design effort are in progress.

We thank D. Boussard, J. Galayda, Q. Kerns, F. Pedersen, and P. Wilson for reviewing this work, and J. Dorfan, A. Hutton, and M. Zisman for their interest and encouragement.

REFERENCES

[1]   D. Briggs, et.al., "Prompt Bunch-by-Bunch Synchrotron Oscillation Feedback via Fast Phase Measurement", these proceedings.

[2]   D. Boussard and G. Lambert, IEEE Trans. Nucl. Sci, NS-30, (1983), p. 2239.

[3]   P. Corradoura and F. Volker, private communications.

A Longitudinal Multibunch Feedback System for PEP*

H.-D. Nuhn, Y. Sun, H. Winick, W. Xie† R. Yotam
Stanford Synchrotron Radiation Laboratory, P.O. Box 4349, Bin 69, Stanford, CA 94309-0210
H. Schwarz
Stanford Linear Accelerator Center, Stanford University, Stanford, CA 94309-0210
and P. Friedrichs
Siemens Medical Laboratories, Inc.

Abstract

A bunch by bunch feedback system to suppress longitudinal multibunch instabilities for PEP has been under development by the Stanford Synchrotron Radiation Laboratory (*SSRL*) and *SLAC*. The system is designed to operate on up to 18 equally spaced electron bunches for synchrotron radiation applications or up to 9 electron by 9 positron bunches for colliding beam operation. As an initial step in developing a very capable multi-bunch stabilizer, the system, which is based on a de-Qed 850 MHz 3-cell cavity, should enable stable storage of a total current of about 36 mA in about 18 bunches at 7 GeV in low emittance mode for synchrotron radiation research. This is based on observed single bunch limits of about 2 mA in the low emittance mode at 7 GeV. At higher electron energy, and/or with modifications to the low emittance lattice, it is likely that even higher currents could be achieved, ultimately limited by transverse instabilities.

I. Introduction

PEP has characteristics that are unique in the U.S. which give it extreme capabilities as a high luminosity e^+e^- colliding beam facility as well as a high brightness X-ray synchrotron radiation source [1][2]. Detailed characterization of the ring and undulator beams were carried out during a low emittance run in 1987 [3]. This run gave clear evidence that the PEP ring and the present photon beam lines offer a significant increase in performance over other sources. In particular, emittances of 4 and 6 nm-rad were measured at 7.1 GeV at low current. Stable currents of 10-15 mA in many bunches were achieved during this run. Both single and multi-bunch instability limits were observed. The thresholds for these instabilities must be increased in order for PEP to reach its full performance potential as a synchrotron radiation source.

The consensus of a workshop [4], held to review the results of the low emittance run and to plan future improvements on PEP, was that a wide-band longitudinal feedback system was the most effective approach to raising the levels of stable stored current in PEP in dedicated low emittance operations towards the desired 50-100 mA. It may be necessary to also implement a transverse feedback system to reach the highest levels.

II. Longitudinal Multibunch Instability

Particle bunches generate electromagnetic wake fields when traveling through a vacuum chamber. In cavity-like structures the Fourier components of these fields that correspond to resonant frequencies oscillate for some time depending on the Q of the resonances. Fields generated by a bunch that passes through the cavity with its center of charge not at the synchronous phase, i.e. performing coherent synchrotron oscillations, can excite coherent synchrotron oscillations in trailing bunches, thus leading to instabilities. The growth rate of these instabilities increases with the bunch current. For low currents natural damping mechanisms like synchrotron radiation keep the oscillations from growing. The excitation process leads to an instability above the threshold current for which the growth time of the multibunch instability τ_{inst} is equal to the damping time τ_{rad}. Theoretical calculations for τ_{inst} in PEP using the ZAP code as well as numerical simulations give unsatisfactory low confidence levels since they depend on the data for the resonances in the 120 main PEP cavity cells. Neither the shunt impedances nor the tunes of these resonances are very well know. During the low emittance run a threshold above 5 to 10 mA has been observed at 7.1 GeV ($\tau_{\mathrm{rad}} = 37$ msec). By scaling conservatively from this number we estimate growth times at 100 mA total beam intensity between 2 msec at 7.1 GeV and 3.6 msec at 13 GeV.

III. Detection and Phase Measurement

The bunch-by-bunch feedback system, described in this paper, measures the instantaneous phase of each bunch with respect to the ring master oscillator and provides a correction voltage in each bunch via a cavity. There are two possible approaches to the measurement of longitudinal motion: (1) direct measurement of the phase of each bunch and (2) measurement of the transverse displacement of each bunch in a region of the storage ring with nonzero dispersion. The second approach requires signal processing to separate the betatron motion from the transverse displacement due to longitudinal motion. The direct phase measurement approach has been selected. The system is designed to handle up to 18 equal separated electron bunches or up to 9 electron by 9 positron bunches for a colliding beam mode of operation, proposed by M. Donald and J. Paterson in 1990. This implies a minimum time interval between bunches of 408 nsec. The block diagram of the feedback electronics is given in fig. 1.

A. Pickup Device

The pick up device is a Four Button Beam Position Monitor (BPM), of the type that is used in the PEP ring. The BPM produces a signal every time an electron or positron bunch passes by. The induced signal amplitude is proportional to the distance of the beam from the button and to the beam intensity. If all four buttons are used and the sig-

---

*Work supported by the Department of Energy, Office of Basic Energy Sciences, Division of Material Sciences.

†Now at the SSC Laboratory, 2550 Beckleymeade Avenue, Dallas, TX 75237.

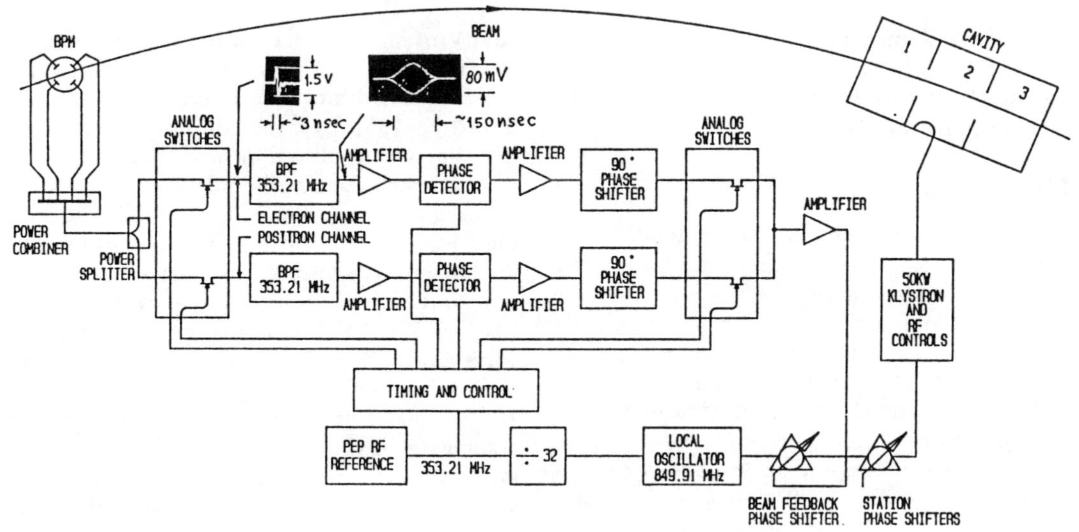

Figure 1

Block Diagram of the Feedback Electronics

nals are transferred through cables and combined into one signal, the signal amplitude will be fairly independent of the beam position. However, each button may contribute a different phase depending on its cable length and phase errors produced by the combiner. It is desired to have a stable signal both in amplitude and phase. The required cable length is 40 m, and the maximum measured phase difference due to the existing variations in the cable length and the combiner phase errors is 5.2°. A special feedback loop is implemented to reduce constant or slowly varying phase offsets.

The peak voltage at the button is given by

$$V_{peak} = \tilde{V}_{peak} \frac{4\alpha}{360°} = \frac{Z_0 \lambda\ q\ e^{-1/2}}{\sqrt{2\pi}\ \sigma^2 v} \frac{4\alpha}{360°} \qquad (1)$$

$Z_0 = 50\ \Omega$	Characteristic Impedance.
$\lambda\ =\ 0.01$ m	Length of the Beam Position Monitor.
$\sigma\ = 40$ psec	Length of the Electron Bunch.
$\alpha\ = 30°$	Height of a BPM Button.
$q\ = 14.7$ nC ≈ 2 mA	Charge per Electron Bunch.
$v\ =\ c$	Speed of the Electron Bunch.

The induced high voltage ($V_{peak} = 1200$ V) at the button is attenuated significantly and the narrow signal is widened at cable end. This is mainly due to the coaxial-cable skin effect.

B. Phase Detection

A direct measurement of the time of arrival is not practical because the required resolution is in the range of picoseconds. Converting from time domain to frequency domain, the BPM signal contains harmonics of the beam revolution frequency of 136 kHz well into 1 GHz. If the 2596 harmonic is chosen, a direct phase comparison with the PEP ring RF of 353.21 MHz is possible and will yield the desired beam phase. Since the feedback system is designed to suppress longitudinal oscillations of both electron and positron bunches, two separate channels which include analog switches, input band pass filters, phase detectors,

and amplifiers had been built. The electron and positron bunches can be singled out by switching the BPM signal with high speed analog switches. The conversion from time domain to frequency domain, after the analog switches, is done by applying the signals to an eight stage Bessel type band pass filter. This filter has been chosen because of its characteristics in the time domain. The band pass filter was designed with a center frequency of 353.21 MHz and a bandwidth of 10 MHz. Using SPICE, the time and frequency responses of the filter to the BPM signal have been simulated. The response to a ≈ 3 nsec BPM input signal is a ringing signal which decays after ≈ 150 nsec and its center frequency is 353.21 MHz. This signal is then used as input to the phase detector. The filters have been tested with inputs from the BPM and performed as expected.

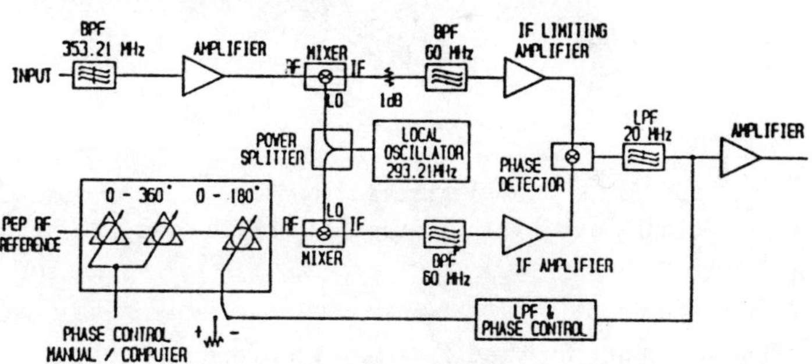

Figure 2

Schematic of the Phase Detector

The requirements for the actual phase detector are: phase resolution of better than 0.5°, and high relative phase stability and accuracy over the range of at least a factor of 10 in beam intensity. The most suitable approach was selected to be a heterodyne system with an IF at 60 MHz, limiting amplifiers at the IF frequency with minimum phase variation over the limiting range, and an analog phase detection scheme at the IF frequency using mixers.

The phase measurement circuit includes a special loop to eliminate constant (or slowly changing) phase offsets at the output of the phase detector. Such offsets are the consequence of the mixers' characteristics, a change in the phase between the PEP master oscillator and the BPM signal, or a change in phase due to the contribution of the cables and other components (fig. 2).

IV. Signal Processing

The damping of the coherent oscillations is done by applying an energy correction to each bunch, i.e. damping the energy oscillations. The energy oscillations are the time derivative of the phase oscillations whose amplitude is measured by the phase detector. In our case the time derivative is equivalent to a 90° phase shift. The signal processing includes the implementation of a 90° phase shifter. The intention is to use a digital finite response filter (FIR), which provides some noise reduction on the raw signal in addition to the 90° phase shift. A 90° phase shift is also required from the system stability point of view.

V. Energy Correction

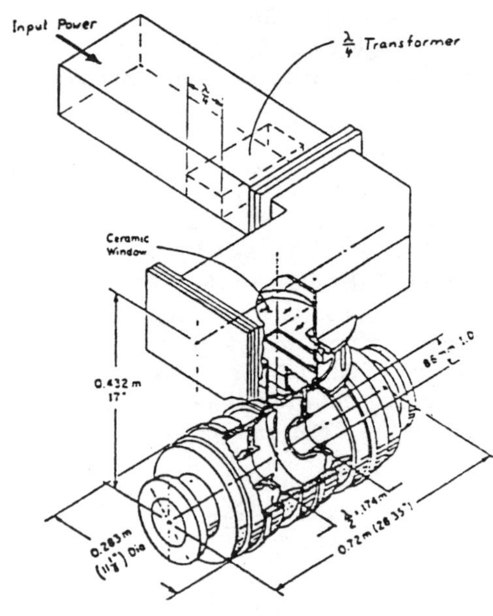

Figure 3
3 Cell Cavity with Waveguide Transformer

A. RF System

The frequency for the RF system was chosen to coincide with the 6237th harmonic of the PEP revolution frequency at an operating frequency of 849.91 MHz. Major components are used from the PEP six bunch longitudinal feedback system[5]. As amplifier a commercial TV klystron is utilized which can provide up to 50 kW cw output power and a 1 dB bandwidth of 7.5 MHz. A circulator directs the considerable reflected power form the over-coupled cavity into a 40 kW load. Typical feedback loops guarantee long term stability of amplitude and phase in the cavity without interfering with the fast feedback signals, for which the system is being designed.

B. Feedback Cavity

To provide the longitudinal kick, the three cell aluminum

cavity of the PEP six bunch longitudinal feedback system is reused. It was originally built for a bunch spacing of 1.2 μsec for which the loaded Q of the assembly was reduced to 1350 by over-coupling through a large coupling iris. For the application now under consideration the minimum bunch spacing is only 408 nsec and the Q had to be lowered further. Since the coupling iris was not accessible because it is blocked by the ceramic window, a quarter wavelength transformer was placed in the feed waveguide outside the window one wavelength away form the iris (see Fig. 3). By choosing the impedance of the quarter wavelength transformer to be half the impedance of the waveguide the loaded Q of the cavity could be lowered to about 330 as expected. Under this loading condition the resonance of the π-mode gets broad enough to start overlapping with the $2\pi/3$-mode and thus puts a limit to the attempt to broaden the bandwidth of the cavity. The inter-

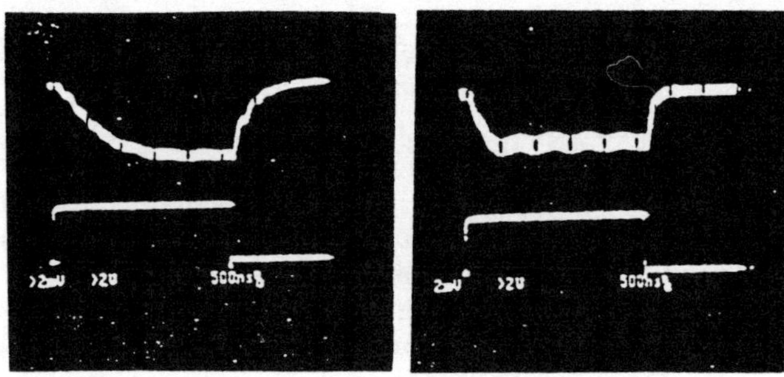

Figure 4
Response of the field in cell 3 for a step function drive into cell 2, .5 μsec/div., a) Q=1350, b) with waveguide transformer, Q=330

cell coupling would have to be increased, if one wanted to reduce the Q further, which is not practical in a completed cavity. A measurement of the response of the cavity field in cell 3 to a step function in the drive to cell 2 indicates a response time (10 – 90 %) of about 300 nsec (Fig. 4) which is sufficient for the feedback to affect individual bunches with a 408 nsec spacing.

VI. Status

The phase detector has been fabricated and tested, partially with beam. The energy correction system is ready to go. Work has been done in developing the signal processing and timing system.

This work has been put on hold with the SLAC decision to not operate PEP during FY 1991 due to a large budged cut. Work will resume when the future operation of PEP is clarified. In the meantime we are shifting our attention to feedback systems on SPEAR, which is now a fully dedicated light source with a dedicated 3 GeV injector synchrotron.

VII. References

[1] A. Bienenstock et al.,. **60**, 1393-1398 (1989).
[2] J. Paterson,. Proc. of the 1989 Part. Acc. Conf.
[3] S. Kramer et al.,. Proc. of the 1989 Part. Acc. Conf.
[4] H.-D. Nuhn, H. Winick edt., 1988. SSRL Rep. 88/06.
[5] M. Allen et al.,. IEEE, NS-28, No. 3, 2317.

Energy Feedback System for the SSRL Injector Linac*

L. Emery[†]

Stanford Synchrotron Radiation Laboratory, P.O. Box 4349, Bin 99, Stanford, CA 94309-0210

Abstract

The energy feedback system for the SSRL Injector linac is presented. The feedback is implemented as a computer program in the SSRL Injector control system, and has been a valuable tool during the commissioning of the injector.

I. Introduction

The SSRL Injector microwave gun, linac, and booster were commissioned last year[1]. For efficient injection into the booster the linac beam energy must be stable to within about 1%. Unfortunately, fluctuations in the line voltage supplying the unregulated klystron modulators cause fluctuations in the linac beam energy, thus making the operation of the booster potentially very difficult. Without any corrections, the linac energy may drift by as much as one percent per minute or it may jump suddenly by one percent. Fortunately, the linac beam energy can be kept close to the desired value by using feedback on the low level controls of the klystrons feeding two of the linac sections. The energy of the linac is sampled at a beam position monitor[2] (BPM) downstream of the first magnet following the linac. The feedback loop is closed using a low-pass filter that noise from the BPM processing electronics. The feedback is implemented as a computer program (energy_feedback) written in C for the SSRL Injector control system[3], and is robust against various fault conditions, as a result of the many changes and adjustments suggested by the commissioning team who are mentioned in the acknowledgements.

II. Linac Beam Energy Fluctuations

The linac is described elsewhere[4]. Briefly, it is composed of three 10-ft 2856 MHz SLAC-type sections, each powered by a $\sim$30 MW klystron pulsed for $2\mu s$ at 10 Hz. The energy gained by a particle travelling on the crest of the RF wave is

$$E_{\text{gain}}[\text{MeV}] = \sum_{i=1}^{3} 10\sqrt{P_i[\text{MW}]}\cos\phi_i \qquad (1)$$

An attenuator and a phase shifter control the low level RF to the klystrons feeding sections 1 and 3. The modulators supplying the voltage to the klystron cathodes are not regulated. The voltage at which the pulse-forming network of the modulators charge varies with the line voltage supplied by the local utility. Typically the charging voltage is set at about 20 kV. A slight variation in klystron cathode voltage can cause a much greater variation in RF amplification since klystrons are very nonlinear devices.

III. Beam Energy Detection

The energy of the particle is measured using the first Linac-to-Booster beamline (LTB) bending magnet as a spectrometer. Since $\gamma = E/mc^2 \approx 250$, the term energy will used interchangably with momentum. Magnet B1 of the LTB line bends the beam of the nominal energy E_0 horizontally by an angle of 41.57 degrees. A beam position monitor (BPM) is located downstream from the bending magnet. If the energy of the beam is increased by ΔE, then the beam position detected by

the BPM will change by

$$\Delta x = \eta\frac{\Delta E}{E_0} \qquad (2)$$

where η is the dispersion at the BPM position. Since there are no quadrupoles in between B1 and the BPM, the dispersion function can be calculated accurately to be 0.75 m.

IV. Closing the Loop

The variable RF attenuators upstream of the klystron RF drive input ports can be used to control the variations of the electron beam energy based on energy measurement of past pulses. For instance, if the high-voltage (HV) from the modulators start to drop a bit, then the attenuators can be adjusted to attenuate less. This prevents the klystron RF output power from dropping when the cathode HV drops. Although this is a nonlinear system, it is treated as linear to keep the feedback algorithm simple.

The amount of the attenuation can be set from the control system. Because of the nature of the attenuator[5], the attenuation is a highly nonlinear function of the programming voltage. (The attenuator operates by a DC current which is provided by a voltage programmable current driver.) The attenuation characteristics for linac section 1 was measured as a function of programming voltage. Using equation 1, one can calibrate the energy gain due to one linac section as a function of the attenuator setting. The voltage dependence on energy gain is plotted in figure 1 for the klystron of section 1. The data was taken

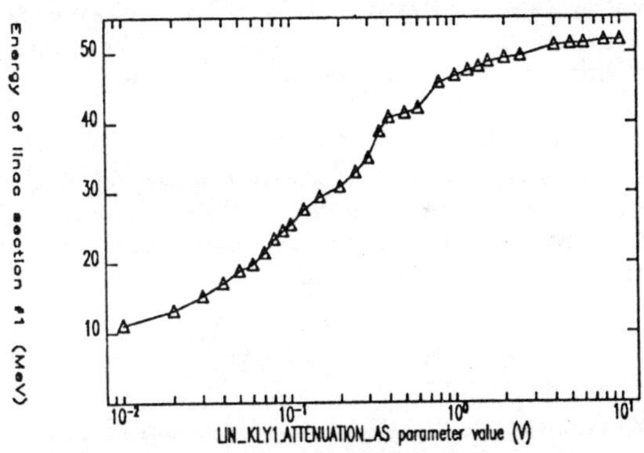

Figure 1

Calibration of attenuator of section 1

with a maximum klystron RF power of 26.7 MW. Variations between klystrons will modify the vertical scale of the plot.

Since the electron beam enters the linac section 1 with an energy of $\sim$2 MeV, there will be a small amount of bunch compression in the first few cells. Varying the RF power delivered to this section may upset the optimal RF phasing between the gun and section 1. Therefore the energy correction is split between section 1 and 3 by a ratio of 1 to 5. The klystron of section 2 is left untouched because it also controls the gun RF power.

The basic feedback algorithm steps are:

1. Measure the position of the beam x at the BPM following B1 relative to a set point x_{ref} selected by the operator.

2. Determine the absolute energy deviation from the set point:

$$\Delta E = E_0(x - x_{\text{ref}})/\eta \qquad (3)$$

*Work supported by the Department of Energy, Office of Basic Energy Sciences, Division of Material Sciences.

†Now at Argonne National Laboratory

">

3. Calculate the necessary attenuator change ΔV_{att} for linac sections 1 and 3 using the calibration data in Figure 1:

$$\Delta V_{\text{att}} = \frac{dV_{\text{att}}}{dE} \Delta E \qquad (4)$$

4. Change the attenuator settings in the database according to the above. The control system then sets the attenuator programming voltage to its new value.

5. Wait for the next beam position measurement and repeat.

For simplicity, the interpolation in step 3 is done by fitting straight line segments to the energy vs log of attenuator voltage curve.

Ideally a correction should be effected for every beam pulse, that is at a 10 Hz rate. However the hardware scans a different BPM along the LTB line at every beam pulse. In addition, the control system database, from which the BPM position data is gleaned, is updated approximately every 0.8 seconds, depending on the system CPU usage. Therefore the energy correction loop can be working at a maximum rate of 1.2 Hz.

V. Stability of loop

The feedback system is susceptible to unstable oscillations if noise is not sufficiently filtered and if the gain (i.e. change in attenuator signal per sampling interval) is not correctly calculated.

The energy deviation signal includes noise from the BPM processing electronics which should be filtered out using a low-pass filter in a computer program. For convenience, the beam position difference

$$(x_{\text{diff}})_k = (x_{\text{ref}})_k - x_k = -\frac{\Delta E}{E_0}\eta \qquad (5)$$

is the quantity that is filtered instead of the energy deviation. Here k means the kth sample of a quantity. The latter value can be set interactively at an operator's console. The filter is of the form

$$(x_{\text{fil}})_{k+1} = A(x_{\text{fil}})_k + B(x_{\text{diff}})_k \qquad (6)$$

where the coefficients A and B are optimized for desired response time and minimum overshoot. The attenuator setting change $(\Delta V_{\text{att}})_{k+1}$ is calculated so that the $(x_{\text{fil}})_{k+1}/\eta$ portion of the energy deviation is corrected.

$$(\Delta V_{\text{att}})_{k+1} = \frac{dV_{\text{att}}}{dE}\frac{(x_{\text{fil}})_{k+1}}{\eta} \qquad (7)$$

Different running conditions such as klystron and modulator settings, and relative phase between linac sections may cause unintended changes in the feedback loop characteristics. The coefficients A and B can be readjusted easily to recover an optimal perfomance.

A low-pass filter allows the feedback system to track the energy well when a slow drift of the charging voltage of the modulators occurs with the high-frequency noise of the BPM electronics.

However, beam energy changes by 2% can suddenly occur because of line voltage changes. Because of the low-pass filter, the feedback loop may take as much as 10 seconds to bring the beam position back to its set point. The response of the feedback can be accelerated by temporarily switching to a faster low-pass filter (higher B value), which can be producing normally unused values in parallel all along. The faster low-pass filter simply assigns a larger weight to the current difference signal. When the beam position is close enough to the set point, the slower low-pass is reinstated.

The program determines automatically the threshold for the above filter-switching. A high-pass filter working in parallel is used to determine the RMS amplitude of the noise which the low-pass filter removes from the signal. The threshold is set to twice the noise level.

VI. Feedback System Performance

Figure 2 shows the drift in beam position at the BPM during a five minute period. In order to test the feedback system a

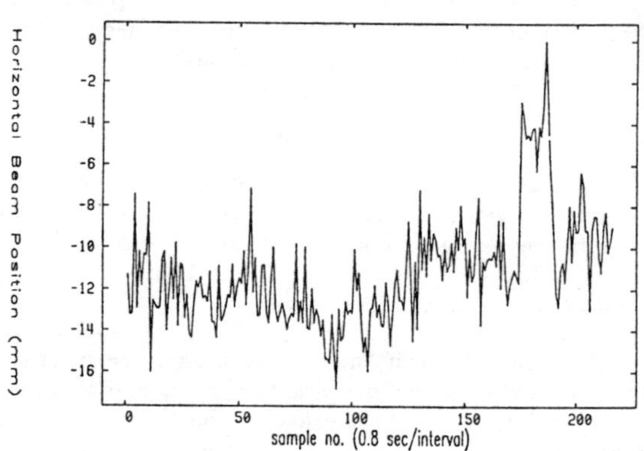

Figure 2

Beam position drift without feedabck

few parameters were recorded simultaneously for a period of 40 minutes. Figure 3 shows a voltage proportional to the charging voltage of the modulator. The sharp downward spikes in the Figure can be ignored; they are due to transients during the

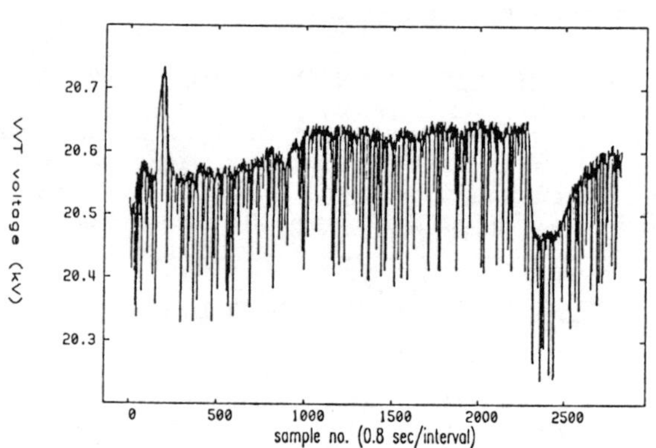

Figure 3

Charging voltage of modulator

recharging of the modulator. That is, the envelope of the curve indicates the variation which should be compensated by the feedback. The beam position during the same period is plotted in figure 4. Note the steady value of the beam position during the high-voltage drift. The spikes on this plot are not spurious. They indicate how suddenly an energy change can occur, and how quickly the energy is corrected. The attenuator signal in Figure 5 basically reflects the error signal. The features of the plot correspond to the same features as the modulator charging voltage, the source of the error.

VII. Fault conditions

The computer program that implements the feedback must be able to distinguish between various error conditions, and hold the loop until some conditions are met. For instance, the sum signal of the BPM strips are continually monitored for possible absence of beam. If the feedback loop was not put on hold, the program would assume that the beam is centered exactly in

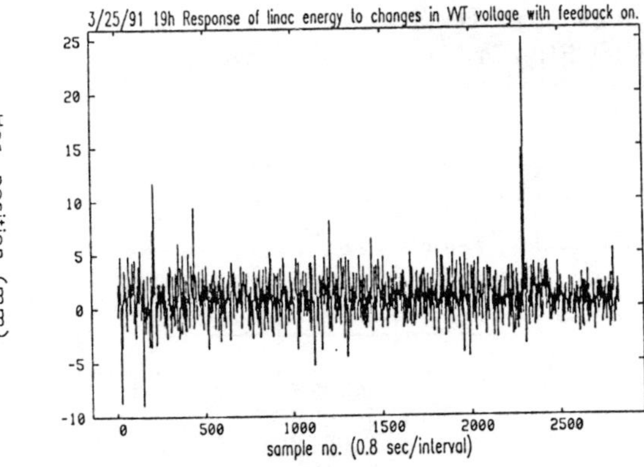

Figure 4
Beam position during feedback

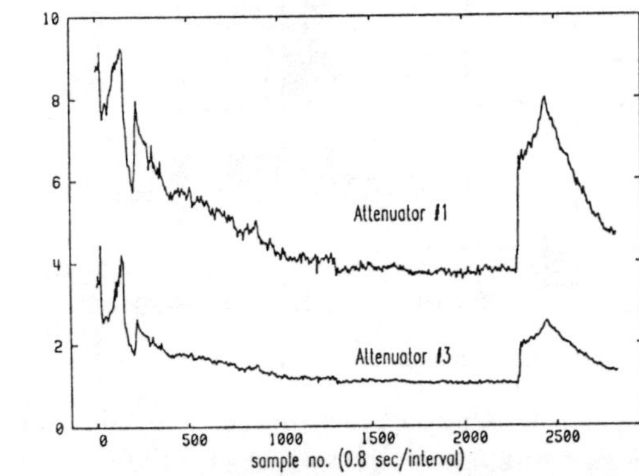

Figure 5
Attenuator signal during feedback

the BPM. If the set point was not 0 mm, then the attenuator settings would become unstable and reach their limits.

An occasional glitch in the BPM processing equipment causes the BPM database value to be exactly 0mm or exactly ± 25 mm (database software limits defined for BPM's) for a single sample. These values are ignored by the feedback, and the program waits a sampling interval before reading the next beam position. When the reading persists for two consecutive samples, then either the beam is lost or the beam energy jumped out of the range of the BPM. The former case was mentioned in the previous paragraph. For the latter case the feedback system considers the beam to be exactly at the limit and switches on the fast feedback.

When the most "active" attenuator (linac section 3) reaches its limits, that is, it cannot provide more power to a drooping klystron, the other attenuator continues working until it too reaches its limit. At this point, the line voltage determines the operating energy of the linac, thus forcing a reduction of the LTB beamline magnet strengths and the booster injection energy.

Occasionally, due to personnel protection system faults, the RF powering the microwave gun and linac sections is turned off. Recovering the previous beam energy and the small energy spread (.1%) takes a few minutes of waiting after the RF is turned back on. The transient effect and the temporarily large energy spread (which can be viewed on a removable phosphorescent screen following the BPM) can be traced to the nature of the microwave gun. The feedback system is capable of turning itself back on as soon as the BPM detects a sufficient amount of beam. It is interesting to watch the beam on the phosphorescent screen as the beam energy is quickly returned to its original value while the energy spread is still stabilizing.

VIII. POSSIBLE IMPROVEMENTS

At large beam offsets the BPMs become nonlinear. The dependence can be derived analytically[6]. If the individual BPM button signal strengths were available to the control system, it would be a simple matter to include the correct beam position in the computer program.

The attenuators of section 1 and 3 could be calibrated automatically as an option when the feedback program is invoked. Assuming that the klystron voltage remains stable during the procedure, the attenuator could be swept through a range of values while the BPM detects the beam position within its limits. The data points could be stored as a fitted function which the feedback algorithm could call. This scheme would have the advantage that it would removes possible daily variations of klystron properties and other imponderables.

IX. CONCLUSION

A linac beam energy feedback system was found to be necessary for injection into the SSRL booster during periods of line voltage fluctuations. The system was implemented by writing a computer program for the control system. The feedback performance for a 40 minute period was described.

X. ACKNOWLEDGEMENTS

I thank James Sebek for my introduction to feedback systems. The graphs were made using software written by Micheal Borland. William Lavender taught me most of what I know of the SSRL Injector control system. Additional thanks go to Simon Baird, Heinz-Dieter Nuhn, James Safranek, Richard Garret, and Bob Genin who all gave me valuable "feedback" on the operation of the system.

XI. REFERENCES

[1] J. Safranek, and S. Baird, "Commissioning the SSRL Injector," in *Proceedings of the 1991 IEEE Particle Accelerator Conference*, 1991.

[2] W. Lavender et al., "The SSRL Injector Beam Position Monitoring Systems," in *Proceedings of the 1991 Particle Accelerator Conference*, 1991.

[3] C. Wermelskirchen et al., "The SSRL Injector Control System," in *Proceedings of the 1991 IEEE Particle Accelerator Conference*, 1991.

[4] J.N. Weaver et al., "The linac and booster rf systems for a dedicated injector for spear," in *Proceedings of the 1991 IEEE Particle Accelerator Conference*, 1991.

[5] H. Schwarz *IEEE Trans.*, vol. NS-32, pp. 1847–51, October 1985.

[6] J. Safranek. Private Communication.

DESIGN OF VAX SOFTWARE FOR A GENERALIZED FEEDBACK SYSTEM*

F. Rouse,[a] S. Castillo,[b] T. Himel, B. Sass, and H. Shoaee

Stanford Linear Accelerator Center, Stanford University, Stanford, CA 94309 USA

Abstract

Fast feedback in the Stanford Linear Collider (SLC) not only works, but is necessary. We have several examples of currently running systems that have greatly improved the performance of the accelerator. In order to increase the number of feedback loops, it has become necessary to redesign the system to allow a database description of any feedback loop. We use digital control theory to formally describe each feedback loop in terms of a matrix equation. Then a new feedback loop requires only an update to the database, and perhaps the installation of a inter-micro communications link. This paper details the design of the VAX software required to implement the new system.

INTRODUCTION

The SLAC Linear Collider (SLC) is a novel accelerator designed to produce e^+e^- collisions at center-of-mass energies up to 100 GeV, i.e., around the mass of the neutral intermediate vector boson Z^0. The collisions occur between electrons and positrons produced on every crossing as opposed to being stored for an extended time, as in electron-positron storage rings. Currently, the SLC has feedback loops that stabilize the energy of the machine, the orbit through a set of collimators near the end of the linear accelerator, and one that maintains the beams in collision. These feedback loops are essential to the operation of the SLC. The software for these feedback loops resides on both a VAX 8800 and a series of INTEL 80386 microprocessors. The 80386 processors actually control the devices that accelerate and control the beam.

We have designed a new system that replaces the current software with generic, database-driven software. We rely on the SLC database to specify each different loop. This is possible because the action of any feedback loop can be cast into a series of matrix equations in the formalism of digital control theory [1].

The problem of closed-loop feedback can be described by a series of matrix equations. We therefore use the term vector to refer to the vectors of measurements, state variables, and control elements used in the feedback loop. We use the term matrix to refer to the matrices that connect the vectors together into an equation [1].

The SLC database specifies the matrices and describes the vectors the matrices act on. The database also contains the complete description of what needs to be measured and how to affect the actuators (usually magnets) to carry out the changes required to stabilize the loop. We design the matrices and specify the loop in the database, add the hardware for the network linking the different micros in the loop, and reboot the micros to start up a new feedback loop in this new system.

---

* Work supported by Department of Energy contract DE–AC03–76SF00515.

(a) Present address: University of California at Davis.

(b) Present address: Apple Computer Corporation.

1416

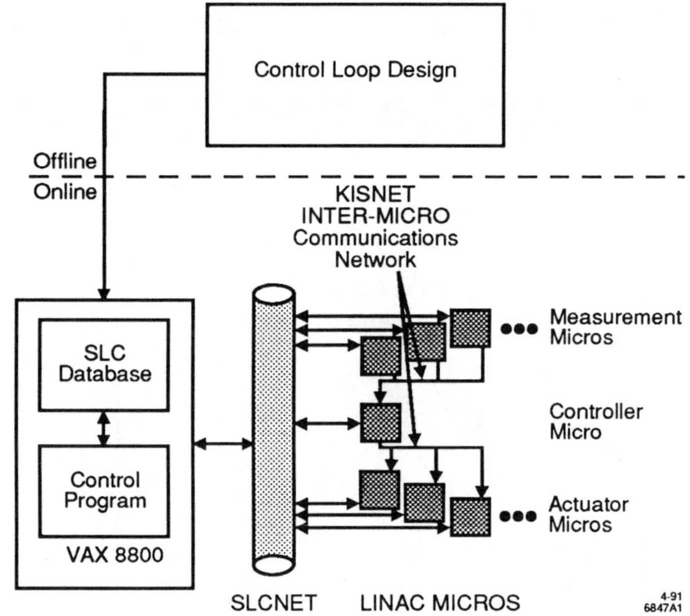

Figure 1. Overview of the components for one feedback loop.

Figure 1 shows the basic components needed for one loop. Matrix design is done offline [1]. The VAX orchestrates how each and every feedback loop works. The INTEL 80386 microprocessors actually do the work of the feedback loop: measure, compute the corrections for and control the hardware devices for the beam. The microprocessors communicate among themselves via a new network called KISNet [2].

Primarily, the VAX orchestrates the new feedback by providing the initialization of all microfeedback jobs, the user interface for analysis and operator diagnostics, and the management of the feedback database.

ELEMENTS OF THE VAX SYSTEM

The new feedback system is first and foremost a database driven system. We built a linked-list system that classifies relevant information together. The objects chosen are shown in Figs. 2 and 3. The feedback structure is the highest ranking linked list. Each linked list consists of pointers to the lower ranking lists, along with the name (used as a key) and other pertinent information. The microstructure mainly points to descriptors associated with each micro attached to the loop. There are also plotting and analysis structures used by the user interface.

The model hierarchy shown in Fig. 3 stores the information associated with the vectors and the matrices. Ring buffers store pulse-by-pulse data from the microprocessors. Data mainly consists of vectors and the status of each vector element. We store information describing each element in the vector and the matrices used by the feedback loop.

We separated the design of the hierarchy of the feedback loop and the interaction with the SLC database from application code. This allows changes in the precise representation of the feedback hierarchy or changes

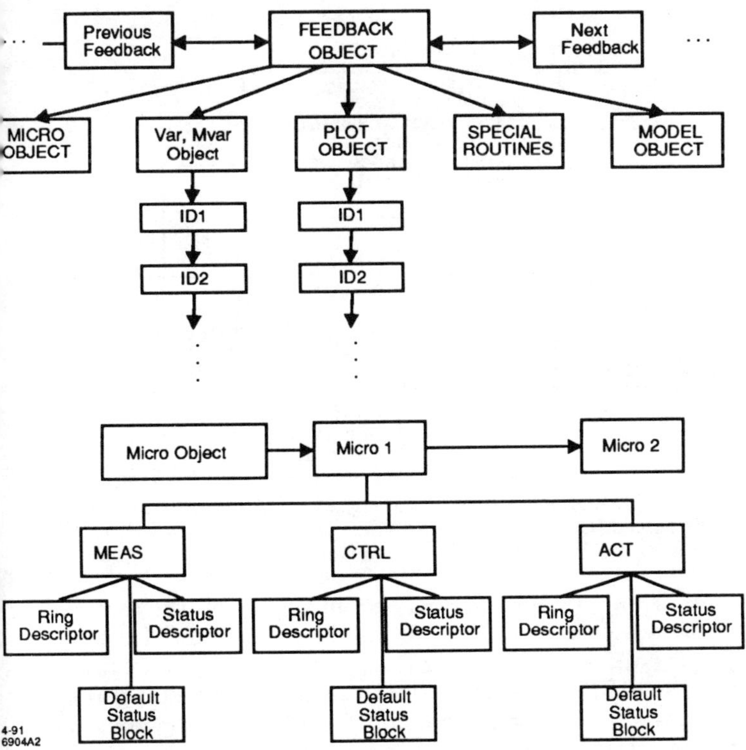

Figure 2. Diagram of feedback hierarchy.

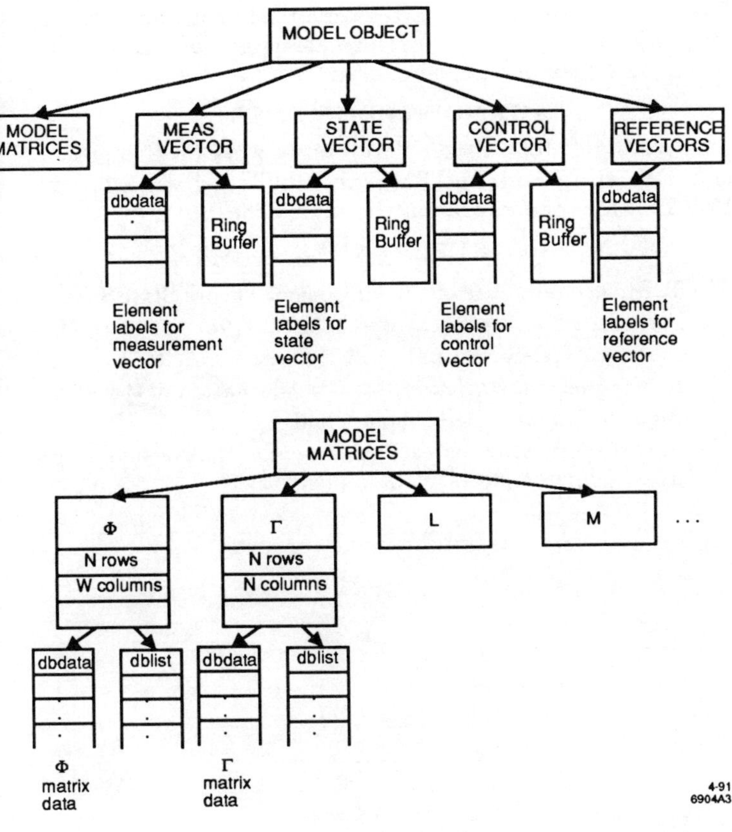

Figure 3. Diagram of model hierarchy.

in the SLC database without having to change the application code. The feedback code is layered as shown in Fig. 4. The outermost layer is the *SCP* layer. This layer manages all user interfaces. We request actions via touch panels or terminals, the SLC's standard interface devices. Actions include loop control and calibration, diagnostic interventions, display of recent feedback data (measurements, states or actuator settings) as a function of time, and listing of pertinent loop information. Displays are shown on a color graphics monitor.

The next layer is called the *application utility* layer. This layer buffers calls from the application code layer to the *kernel* layer. The most important layer is the *kernel*. Here the feedback hierarchy is stored. The particular representation of the hierarchy is a series of linked lists.

The user interface in the new system allows for arbitrary numbers of feedback loops, micros, and plots. Additionally, the user interface handles all common interface functions, such as changing the state of a feedback loop; turning on or off a particular device; listing the status of a feedback loop; analysis and histograms; and listing the feedback loops, micros, plots, or ring buffer elements.

SLC DATABASE

The information in the database for the feedback system consists of two classes: feedback loop information and display information. Feedback loop information includes a loop name, descriptions of the micros carrying out the measurement, controller and actuator tasks of the loop and the communication links between them, the feedback matrices, and the vectors the matrices act upon. We also specify the state vector that the controller uses to compute the actuator settings. The display information consists of the plot names, windowing for specified plots, and variables. We key off of the feedback loop name for all information pertaining to the loop.

The matrices are generated offline by modeling the action of the feedback loop along with the model of the accelerator. The matrices are then loaded into the SLC database by the offline program. They are stored in a sparse format.

The vectors must have specific device information, since the measurement and actuation drivers need CAMAC control words and locations in order to read out or set their respective devices. Typically, feedback routines only need a pointer to specific device information. The device information is already in the database to allow control of the accelerator by preexisting applications. Each vector element has a corresponding label that includes the keywords required for unique database access. Finally, the database also describes physical and display units, tolerances, axis labels, etc., for each vector element

OBJECTS

The basis of the categorization system is the linked list. Each linked list contains different information, but movement from node to node along

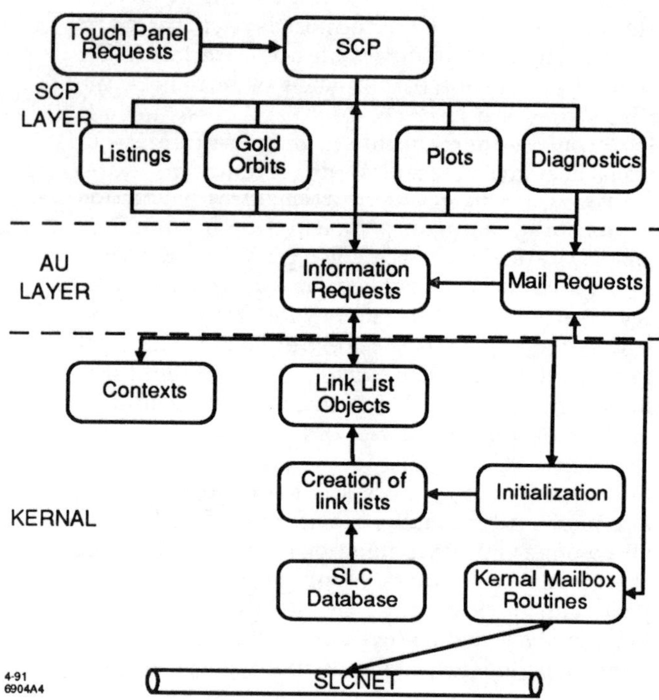

Figure 4. Layering of the VAX feedback code

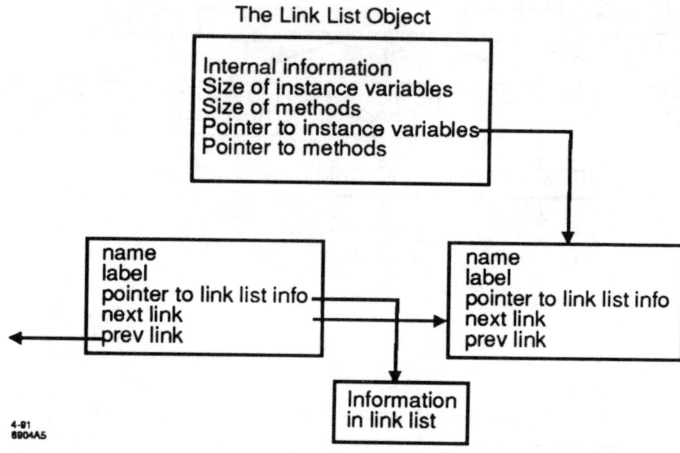

Figure 5. The structure of an object. This example shows the linked list object. The instance variables are also shown. The methods are the common actions for all linked lists.

the link list is a common action. Additionally, the major user interface is through programmable CRT touch panels. Often we do not know how many buttons will be required on the panel. The action initiated by pushing any one button may, of course, be different. The action of pushing a feedback loop button is different than pushing a button that describes the element in a vector. Yet the act of selection or deselection of a button, or the printing of a screen can be described by common code.

The use of *"objects"* [3] can elegantly represent our problem. We allow our objects to inherit instance variables and internal methods to give a slightly different behaviour to our various linked lists and button actions [3].

Let us consider our linked list class in particular. We have linked lists of feedback loops, micros, plots and plotting variables within a feedback loop (see Fig. 2). This linked-list class is an example of a container class [3]. The class handles movement from node to node along the list, listing each node in the class, adding nodes to and removing nodes from the list, and creation and deletion of the entire linked list. Each subclass implements its own particular behavior.

The actual structure of the linked list object that we implemented is shown in Fig. 5. The description of the object is only required in a fully objective system, though we use the SIZEOF fields to create copies (new instances) of objects. The important part of the object are the pointers to the instance variables and to the methods.

CONCLUSIONS

We have described the VAX code used by a generalized feedback system at the Stanford Linear Collider. The system categorizes various classes of information by the use of linked lists. The lists use a primitive form of an objective-C object in which a container class describes all linked lists. This allows us to have an arbitrary number of feedback loops whose behaviour can be globally modified by changing common code.

ACKNOWLEDGMENTS

We thank John Zicker for his early work on this problem. We also thank Lee Patmore, Phyllis Grossberg and Bob Hall for their efforts on the VAX code.

REFERENCES

[1] T. Himel et al., "Use of Digital Control Theory State Space Formalism for Feedback at the SLC," Proc. 1991 IEEE Particle Accelerator Conf., San Francisco, CA, 1991.

[2] K. Krauter and D. Nelson, "SLC's Adaptation of the ALS High Performance Serial Link," ibid.

[3] B. Cox, *"Object Oriented Programming, The Evolutionary Approach,"* (Addison-Wesley, 1986.)

GENERAL, DATABASE-DRIVEN FAST-FEEDBACK SYSTEM FOR THE STANFORD LINEAR COLLIDER*

F. Rouse,[a] S. Allison, S. Castillo,[b] T. Gromme, B. Hall,[c] L. Hendrickson,
T. Himel, K. Krauter, B. Sass, and H. Shoaee

Stanford Linear Accelerator Center, Stanford University, Stanford, CA 94309 USA

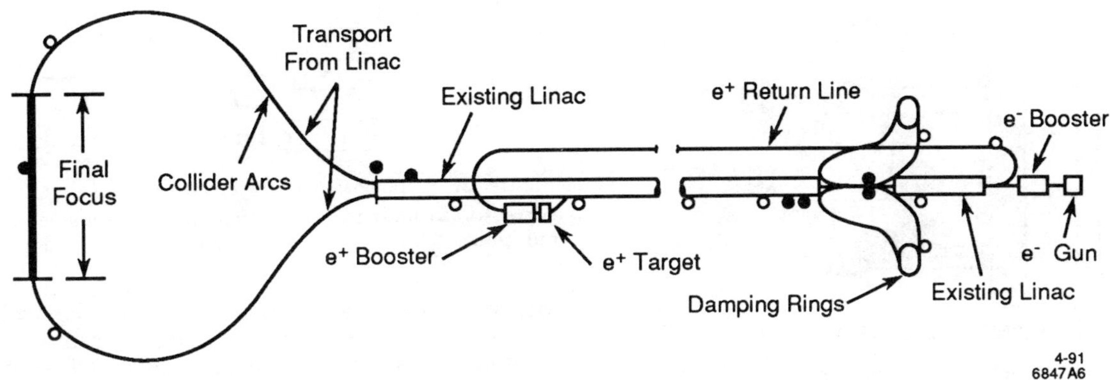

Figure 1. The layout of the SLC. Locations of the presently existing feedback loops are shown with a solid dot. Implementation planned in the next six months are shown with an open dot.

Abstract

A new feedback system has been developed for stabilizing the SLC beams at many locations. The feedback loops are designed to sample and correct at the 60 Hz repetition rate of the accelerator. Each loop can be distributed across several of the standard 80386 microprocessors which control the SLC hardware. A new communications system, KISNet, has been implemented to pass signals between the microprocessors at this rate. The software is written in a general fashion using the state space formalism of digital control theory. This allows a new loop to be implemented by just setting up the online database and perhaps installing a communications link.

INTRODUCTION

The SLAC Linear Collider (SLC) is a novel accelerator designed to produce e^+e^- collisions at center-of-mass energies of up to 100 GeV, i.e., around the mass of the neutral intermediate vector boson Z^0. The collisions occur between electrons and positrons produced on every beam crossing and then thrown away, rather than stored for an extended time as in electron-positron storage rings. Currently, the SLC has feedback loops that stabilize the energy of the machine, stabilize the orbit through a set of collimators near the end of the linear accelerator, and one that maintains the beams in collision. These feedback loops are essential to the operation of the SLC. The software for these feedback loops resides on a VAX 8800 plus a series of INTEL 80386 microprocessors (micros). The micros actually control the devices that accelerate and control the beam. The success of the three feedback loops has led us to redesign the system to allow a more unified and automatic loop specification.

We have replaced the specialized software with generic, database-driven software We rely on the SLC database to specify each loop. This is possible because the action of any feedback loop can be cast into a series of matrix equations in the formalism of digital control theory [1]. The SLC database specifies the matrices and describes the vectors the matrices act upon. The database also contains the complete description of what sensors to use (usually beam position monitors), and how to control the actuators (usually magnets) to carry out the changes required to stabilize the loop. We design the matrices and specify the loop in the database, add the hardware for the network linking the different micros in the loop, and reboot the micros to start up a new feedback loop in this new system.

The biggest constraint on the new feedback system comes from the topology of the SLC. The accelerator consists of several major instruments: an injector, damping rings, positron target, transport lines, arcs, and final focus as shown in Fig. 1. The major accelerating portion of the accelerator is the LINAC itself. It is divided into *30 sectors*.

A single micro controls all devices in one geographical region; for example, a single transport line, a single sector of the LINAC, a damping ring, etc. Correctors and beam position monitors spread out over several micros are required to measure and control the beam position and angle. Additionally, several feedback loops may need to use devices in the same micro. Hence, a feedback system is required to have multiple loops executing multiple tasks in a set of micros.

Figure 2 shows the basic components needed for one loop. Matrix design is done offline [1]. The VAX orchestrates how each feedback loop works and provides users with timely analysis and status information. The INTEL 80386 microprocessors carry out the processing required for feedback: measurement, computation of the corrections needed and control of the appropriate hardware devices. The microprocessors communicate among themselves via a new network called KISNet which is based on the design and hardware of the Advanced Light Source (ALS) [2].

An individual feedback loop may be distributed over several micros. We break the task of feedback into three discrete tasks: measurement, controller, and actuator. The measurement tasks read beam derived information, the controller carries out the matrix arithmetic and determines the next value for the actuators, and the actuator tasks cause the actuators to be set to the designated values.

*Work supported by Department of Energy contract DE–AC03–76SF00515.
[a] Present address: University of California at Davis.
[b] Present address: Apple Computer Corporation.
[c] Present address: General Electric, Consulting.

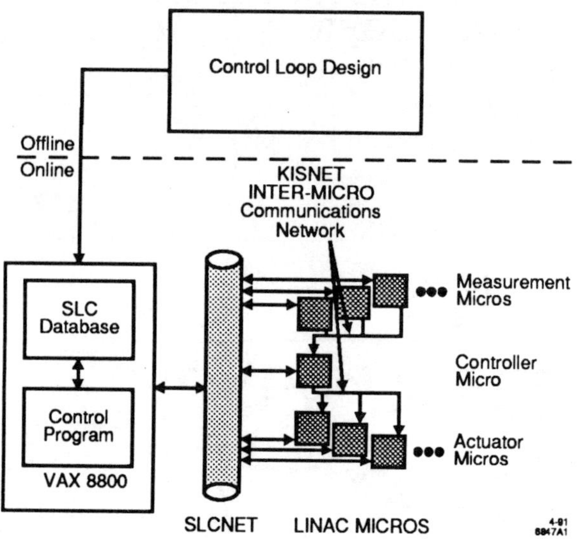

Figure 2. Overview of the components for one feedback loop.

State space formalism used by the controller

Any continuous linear system can be described by a set of first order differential matrix equations [1]. We can change from continuous time to discrete time by solving this equation and integrating over our sampling intervals. If we had perfect knowledge of the accelerator, we could calculate the exact correction to bring the SLC to any desired state. Unfortunately, this is not possible. Instead, we must estimate the state and use the measurements to correct our estimate. The predictor-corrector formalism of state estimation is

$$\hat{\mathbf{x}}(n+1) = \mathbf{\Phi}\hat{\mathbf{x}}(n) + \mathbf{\Gamma}\mathbf{u}(n)$$
$$+ \mathbf{L}(\mathbf{y}(n) - \mathbf{H}\hat{\mathbf{x}}(n)) + \mathbf{M}\mathbf{r} \qquad (1)$$
$$\mathbf{u}(n) = -\mathbf{K}\hat{\mathbf{x}}(n) + \mathbf{N}\mathbf{r} , \qquad (2)$$

where $\hat{\mathbf{x}}$ is the vector of estimated states of the system, $\mathbf{y}$ is the vector of measurements of the system output and $\mathbf{u}$ is a vector of actuation values. The matrices $\mathbf{\Phi}$, $\mathbf{\Gamma}$, and $\mathbf{H}$ represent the system dynamics, account for the state changes caused by the actuators, and connect the current state of the system to the output of the system respectively. The elements of vector $\mathbf{r}$ are the setpoints of the system, and the $\mathbf{M}$ and $\mathbf{N}$ matrices can be chosen by the feedback designer [1]. A pictorial representation of the predictor corrector formalism is shown in Fig. 3.

The $\mathbf{\Phi}$, $\mathbf{\Gamma}$, and $\mathbf{H}$ matrices come from the model of the SLC. Therefore, we need only concern ourselves with the design of the two matrices $\mathbf{K}$ and $\mathbf{L}$. They are chosen to optimize the response of a feedback loop with respect to response time, overshoot, recovery time, etc., of the loop in response to expected disturbances in the accelerator.

COMPONENTS OF THE FEEDBACK SYSTEM

VAX software

A detailed description of the VAX software can be found elsewhere [3]. We only give an overview of the software here.

The VAX is central to the operation of the feedback, since only the VAX has access to the entire SLC database. Each micro has only a copy of the database germane to itself. The VAX must therefore form the signal routing map between micros and download this map, along with other pertinent information, at initialization time to the micro.

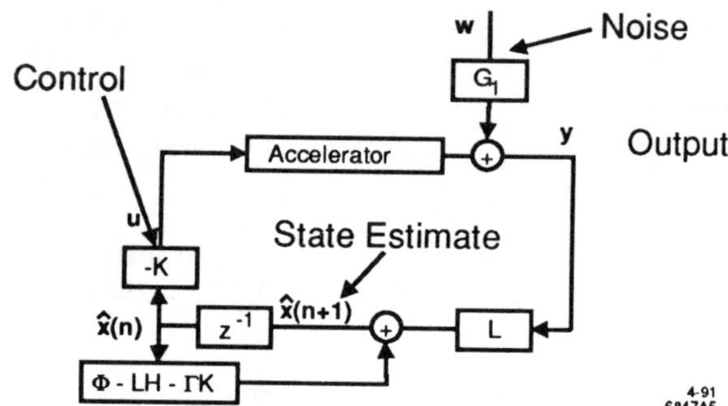

Figure 3. A pictorial representation of the basic predictor-corrector formalism. The operator z^{-1} represents a delay by one pulse. Omitted from the picture are external references.

Additionally, the VAX carries out the functions of information retrieval and display, loop control functions of the system, and user initiated actions. The VAX communicates with all micros involved in the system via the bi-directional communications network, SLCNet. User actions supported by the VAX include loop control and calibration, diagnostic interventions, display of recent feedback data (measurements, states, or actuator settings) by accelerator pulse, and listing of pertinent loop information.

MICRO software

We view the feedback loop as consisting of beam measurements being carried out on a series of micros, with the information being transmitted to the controller micro. The controller micro then uses the state space formalism detailed in the previous section to compute the required actuator settings to restore the beam. Finally, the actuator settings are transmitted to a series of micros that control the actual devices. A status return is routed back to the controller micro. On any one micro, one feedback job called FBCKMAIN is created that oversees all three task types: a measurement, controller, and actuator. Each feedback loop that has a requirement for a particular task type on this micro is treated as a separate task of that particular type (measurement, controller, or actuator).

For example, if one feedback loop needs measurements from sectors 27 and 28, and controls actuators in sectors 26 and 27, and another feedback loop needs measurements from sectors 28 and 29, and controls actuators in sectors 27 and 28, we would need to create two separate measurement tasks in the micro for sector 28, one measurement task each in sectors 27 and 29, two actuator tasks in sector 28 and one actuator task each in sectors 26 and 28. These example feedback loops, along with their KISNet connections, are shown in Fig. 4.

The purpose of the measurement task (FMES) is to assemble measurement information from all input devices and transmit the values to the controller. FMES communicates with the data acquisition drivers [currently, the Beam Position Monitor (BPM) job] for each class of device. Various classes of devices are handled, in addition to beam position monitors. We gather all information from all sources for a feedback loop on one micro before transmitting the entire subvector to the controller.

The controller task (FCTL) waits until all measurement subvectors are assembled before taking action. Once all subvectors from each micro have been received, the controller task implements Eq. (1) to compute the estimated current state of the machine. It then applies Eq. (2) to the estimated state to obtain the next actuator settings required to stabilize the machine. The controller is capable

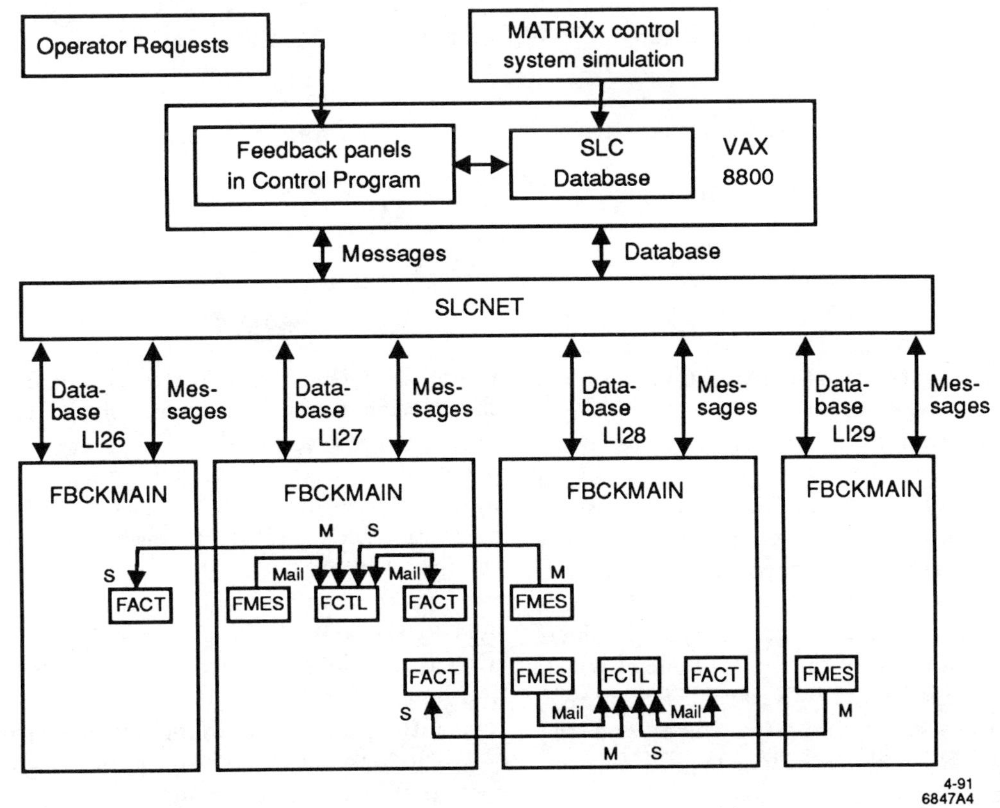

Figure 4. An example of two feedback loops in common micros of the LINAC. Each separate fast-task box corresponds to a separate task under the main task FBCKMAIN. Included in the drawing are the connections both intramicro (via RMX mailboxs and denoted Mail), and intermicro via KISNet. KISNet master and slave ports are denoted M and S respectively in the figure.

of handling nonlinear devices such as phase shifters used to control the beam energies. We expect that in the future this calculation will include state information fed forward from upstream feedback loops.

The actuator task (FACT) receives the new device settings transmitted by the controller. Each destination micro only receives the subvector of information for devices controlled by that micro. The actuator task then sets the device and reports a status code back to the controller.

Communications system

A new intermicro communications network based on the Advanced Light Source(ALS) hardware was built for the feedback system and is described in detail elsewhere [2]. We configure it as a point-to-point network with a *master* port communicating with a *slave* port. Only one master port can be on any one wire.

The time critical communications, namely measurement to controller and controller to actuator, are implemented by having a master port write to a slave port. Each micro involved in a measurement must therefore have a separate master port for each controller to which it must deliver the information. Finally, since only one master can be on a wire, the controller must have one slave port for each measurement micro.

Status information must be returned from the actuators to the controller. This information is not time critical. Instead of running another wire from each actuator to the controller and therefore creating the necessity of adding one port per actuator micro to the controller, we allow the actuator slaves to write the status information back to the

controller master. A master must poll each actuator micro in order to even determine if there is status data.

The software is designed to separate the physical transmission of data from higher level functionality. This allows us to change the physical media of transmission (a follow-on network) from the conceptual task of transmitting a block of data. For example, some information is passed within the same micro. The lowest level routines use mailboxes provided by the operating system instead of communications ports, if the destination is the same micro.

CONCLUSIONS

We have described a general feedback system for the Stanford Linear Collider. This feedback system allows us to control the accelerator beam with standard software. We need only make database entries and connect a limited amount of communications hardware to create a new feedback loop anywhere in the machine.

ACKNOWLEDGMENTS

We thank John Zicker for his early work on this problem. We also thank Lee Patmore and Phyllis Grossberg for their efforts on the VAX code.

REFERENCES

[1] T. Himel et al., "Use of Digital Control Theory State Space Formalism for Feedback at the SLC," Proc. 1991 IEEE Particle Accelerator Conf., San Francisco, CA, 1991.

[2] K. Krauter and D. Nelson, "SLC's Adaptation of the ALS High Performance Serial Link," ibid.

[3] F.R. Rouse et al., "Design of VAX software for a Generalized Feedback System," ibid.

THE TRANSVERSE DAMPER SYSTEM FOR RHIC*

J. Xu, J. Claus, E. Raka, A.G. Ruggiero and T.J. Shea

Brookhaven National Laboratory
Upton, NY 11973, USA

I. INTRODUCTION

If the beam is injected with errors x_c, x'_c (or y_c, y'_c) with respect to the closed orbit or disturbed by transverse instabilities, it will execute coherent oscillations and will be diluted in betatron phase space within a time interval of about $1/\Delta\nu$ turns, even if it is properly matched to the focusing characteristics of the lattice, unless there is an effective damper system to prevent this. Here $\Delta\nu$ is the tune spread in the beam. Such a damper will not prevent dilution due to mismatches. Without such a damper the emittance of the beam will ultimately develop to a properly centered matched ellipse with an area ϵ in phase space that is larger than that of the injected one ϵ_0 which is also matched but off–centered by x_c and x'_c.

Let the equations of the centered ellipse and the injected off–centered but matched ellipse be[1]

$$\epsilon = \gamma x^2 + 2\alpha x x' + \beta x'^2 \tag{1}$$

and

$$\epsilon_0 = \gamma \left(x - x_c\right)^2 + 2\alpha \left(x - x_c\right)\left(x' - x'_c\right) + \beta \left(x' - x'_c\right)^2 \tag{2}$$

respectively, where α, β, γ are the Twiss parameters at the injection point.

The dilution factor ϵ/ϵ_0 can be expressed as follows:

$$\frac{\epsilon}{\epsilon_0} = \left[1 + \sqrt{\frac{\epsilon_c}{\epsilon_0}}\,\right]^2 \tag{3}$$

where $\epsilon_c = \gamma x_c^2 + 2\alpha x_c x'_c + \beta x'^2_c$. We have similar relations for the y–direction.

If we require an emittance growth less than 20% in both directions, the maximum allowable values are x_c = 0.25 mm, x'_c = 0.03 mrad, y_c = 0.6 mm and y'_c = 0.012 mrad. These tolerances may be difficult to meet and a feedback system to damp the coherent oscillations induced by the injection errors should be considered for RHIC.

The damper system consists of a beam position monitor, signal processing electronics, an amplifier station and a kicker device. The beam position signal is amplified and transported to another location where it is applied across the kicker device for the correction. Each kicker is made of a pair of striplines of length ℓ. If $\pm V_k$ is the voltage applied to each plate then the effect of the kicker on the ion motion can be expressed as follows:

$$Am_0\gamma\ddot{x} = \frac{2ZeV_k}{d_k}\,\eta\alpha \tag{4}$$

where Z is the charge state, A the mass number, m_0 the specific mass at the rest of the ion, d_k the plate separation and γ the relativistic energy factor. The factor η describes the interaction between the kicker and the beam. There are contributions from both electric and magnetic fields. If the kicker plates are terminated at the upstream end with the characteristic impedance, the two contributions add and $\eta \simeq 2$. The average electric field in the kicker is approximately $2V_k/d_k$. The factor α measures the enhancement or depression of the electric and magnetic fields at the center of the vacuum chamber due to the geometrical configuration of the kicker design.

The kick angle received by an ion going through the system of kickers is given by

$$\theta = \frac{2Zen_k\ell V_k}{AE_0\gamma\beta^2 d_k}\,\tau\eta\alpha \tag{5}$$

where n_k is number of kickers each of length ℓ, $E_0 = m_0c^2$, β is the relativistic velocity and τ is a transit–time factor which in the frequency domain ω is expressed as

$$\tau = \frac{\sin\frac{2\omega\ell}{c}}{\frac{2\omega\ell}{c}} \tag{6}$$

In the following we have assumed a square voltage pulse applied to the damper kicker, with a duration long enough so that $\tau = 1$.

For a given damper system we will consider two modes of operation: (i) the kicker voltage is proportional to the beam position signal and (ii) the kicker voltage is adjusted to a constant value having a sign which depends on the sign of the signal detected from the beam position.

II. PROPORTIONAL MODE[1]

In the proportional mode the correction effect is proportional to the beam center displacement. As the damping proceeds, the signal and therefore the correction are reduced. The power required at any time is then proportional to the square of the beam displacement, and is the highest at the beginning.

Let $X_{p,n} \equiv \left(x_{p,n}, x'_{p,n}\right)$ be the vector representing the coherent oscillation at the pick–up location during the n–th crossing. The kicker voltage V_{kn} during n–th crossing of the kicker is proportional to x_{pn},

$$V_{kn} = -Gx_{p,n} \tag{7}$$

* Work performed under the auspices of the U.S. Department of Energy.

then the vector during the (n+1)-th crossing is

$$X_{p,n+1} = M_{kp}\, M_k\, M_{pk}\, X_{p,n} \qquad (8)$$

where M_{pk} and M_{kp} are the 2×2 transfer matrices respectively from pick–up to kicker and from kicker to pick–up and M_k is the matrix representing the effect of the kicker. We have

$$M_k = \begin{pmatrix} 1 & 0 \\ 0 & 1 \end{pmatrix} - k_0 \begin{pmatrix} 0 & 0 \\ 1 & 0 \end{pmatrix} M_{pk}^{-1}$$
$$= I - k_0 R M_{pk}^{-1} \qquad (9)$$

where

$$k_0 = \frac{2 Z e n_k \ell G}{A E_0 \gamma \beta^2 d_k} \qquad \eta \alpha = \frac{\theta_n}{x_{p,n}} \qquad (10)$$

is a damper parameter that can be obtained by combining Eqs. (5,7).

We have

$$M = M_{kp} M_k M_{pk}$$
$$= M_0 - k_0 M_{kp} R \qquad (11)$$

where $M_0 = M_{kp} M_{pk}$ is the unperturbed transfer matrix at the pick–up location.

Let

$$M_0 = \begin{pmatrix} m_{11} & m_{12} \\ m_{21} & m_{22} \end{pmatrix} \qquad (12)$$

and

$$M_{kp} R = \begin{pmatrix} a & 0 \\ b & 0 \end{pmatrix} \qquad (13)$$

with

$$a = \sqrt{\beta_k \beta_p}\, \sin \psi_{kp}$$
$$b = \sqrt{\frac{\beta_k}{\beta_p}}\, (\cos \psi_{kp} - \alpha_p \sin \psi_{kp}) \qquad (14)$$

where ψ_{kp} is the phase advance from kicker to beam monitor. The damping rate is given by the eigenvalues of the total transfer matrix M, which are the solutions of the quadratic equation

$$\lambda^2 - \lambda \left(2 \cos \mu_0 - k_0 a\right) + \mathrm{Det}\, M = 0 \qquad (15)$$

where μ_0 is the unperturbed phase advance per revolution. If, finally, we let

$$k = k_0 \sqrt{\beta_p \beta_k} \qquad (16)$$

then

$$\lambda = \cos \mu_0 - \frac{k}{2} \sin \psi_{kp} \pm i \sqrt{\left(\sin \mu_0 + \frac{k}{2} \cos \psi_{kp}\right)^2 - \left(\frac{k}{2}\right)^2}. \qquad (17)$$

The case of practical interest is when the phase advance ψ_{pk} from monitor to the kicker is adjusted close to an odd number of $\pi/2$ radians and the gain k is small. In this case, apart from a phase factor of no consequence, the damping rate of the coherent oscillations is given by

$\frac{1}{2} k \sin \psi_{pk}$ per turn or $\frac{1}{2} \frac{\theta}{x_p} \sqrt{\beta_p \beta_k} \sin \psi_{pk}$ per turn. Since $\sin \psi_{pk} \sim \pm 1$ the damping rate is simply given by

$$D_{pr} = \frac{1}{2} \frac{\theta}{x_p} \sqrt{\beta_p \beta_k}\ \text{per turn} \qquad (18)$$

III. CONSTANT VOLTAGE MODE[2]

In the constant voltage mode the beam position signal is used only to trigger the correction by applying a constant kicker voltage. It also controls the sign of the voltage according to the actual sign measured of the instantaneous beam displacement. This method still provides effective damping but with a considerably reduced peak power requirement.

The Courant–Snyder invariant of the coherent oscillation at the n–th turn, at the kicker location, is given by

$$A_n^2 = \gamma_k x_{k,n}^2 + 2 \alpha_k x_{k,n} x_{k,n}' + \beta_k x_{k,n}'^2 \qquad (19)$$

where $x_{k,n}$ and $x_{k,n}'$ are the x and x' values during n–th crossing of the kicker. Let θ_n be the kick applied, then assuming the position is unchanged, we have a change in A_n given by

$$\Delta A_n^2 = 2 \left(\alpha_k x_{k,n} + \beta_k x_{k,n}'\right) \theta_n + \beta_k \theta_n^2 \qquad (20)$$

If the correction is small we can neglect the quadratic term. Using explicitly equations for $x_{k,n}$ and $x_{k,n}'$, we obtain

$$\Delta A_n = \beta_k^{\frac{1}{2}} \theta_n \cos \psi_{k,n} \qquad (21)$$

where $\psi_{k,n}$ is the phase of the coherent oscillation during n–th crossing of the kicker which varies from turn to turn.

For the correction kicker, we take

$$\theta_n = -\theta_0 \frac{|\cos \psi_{k,n}|}{\cos \psi_{k,n}} \qquad (22)$$

where θ_0 is a constant. We have finally

$$\Delta A_n = -\beta_k^{\frac{1}{2}} \theta_0 |\cos \psi_{k,n}| \qquad (23)$$

which gives the variation of the Courant–Snyder invariant of the coherent oscillation during n–th crossing of the kicker. Then, A_N at the n–th turn can be expressed as

$$A_N = A_0 - \sum_{n=1}^{N} \beta_k^{\frac{1}{2}} \theta_0 |\cos \psi_{k,n}| \qquad (24)$$

We can replace $\sum_{n=1}^{N} |\cos \psi_{k,n}|$ with $N \overline{|\cos \psi|}$ and obtain the following expression:

$$A_N = A_0 - N \frac{2}{\pi} \beta_k^{\frac{1}{2}} \theta_0 . \qquad (25)$$

The average fractional reduction of the coherent oscillation amplitude per turn when it reduces from x_0 to $x_0 e^{-1}$ is

$$D_c = \frac{2}{\pi} \frac{\sqrt{\beta_p \beta_k}\, \theta_0}{x_0 (1 - e^{-1})} \simeq \sqrt{\beta_p \beta_k} \frac{\theta_0}{x_0} \qquad (26)$$

Comparing Eqs. (18) and (26), one finds that the effective damping rate with the constant voltage mode is about twice that obtained with the proportional mode. The constant voltage mode is proposed to allow a reduction by a factor of 4 of the total power requirement. When the residual coherent oscillation amplitude becomes small enough one may switch to the proportional mode.

IV. RHIC PARAMETERS

As determined in ref.[1], the most demanding requirements are set by injection of the beam of gold ion (Z=79, A=197) at 10.5 GeV/u. An initial beam center displacement translated at the location of the pickup, as large as 2 mm can be expected from injection errors. It is required that damping take place in about one hundred turns ($\Delta \nu \simeq 0.01$), which is fast enough to avoid emittance dilution due to the beam tune spread. We take $\sqrt{\beta_p \beta_k} \sin \psi_{pk} = 50$ m, from Eq. (23) one gets $\theta_0 = 0.4$ μrad, which can be maintained with

$$ n_k \ell V_k = 200 \text{ V} \cdot \text{m} \qquad (27) $$

here we take $\eta = 2$, $\alpha = 1$, $\tau = 1$ in Eq. (5).

The total power required is

$$ P_k = 2n_k \frac{V_k^2}{R_k} \qquad (28) $$

where R_k is the termination impedance which we take here to be 50Ω. A reasonable solution is one set of kicker $(n_k = 1)$ $\ell = 2$ meter long. On each plate a voltage 100 volts is applied which requires a total power of 400 watts.

The rise and fall times of the damper system should be short enough to allow individual damping of coherent motion of up to 3 × 57 bunches.

V. REFERENCES

[1] J. Xu, J. Claus and A.G. Ruggiero, BNL AD/RHIC–74, July 1990.
[2] J. Xu and A.G. Ruggiero, BNL AD/RHIC–76, August 1990.

The Stochastic-Cooling System for COSY-Jülich

P. Brittner, R. Danzglock, H.U. Hacker, R. Maier, U. Pfister, D. Prasuhn,
H. Singer, W. Spieß and H. Stockhorst

Forschungszentrum Jülich
Postfach 1913, D-5170 Jülich, FRG

Abstract

The stochastic-cooling system is under development. Cooling characteristics have been calculated. The tanks are similar to those of the CERN-AC. But the COSY parameters have required changes of the tank design. Active RF components have been developed for COSY. Measured results are presented.

I. OVERVIEW OVER THE SYSTEM

The cooling in the Cooler Synchrotron COSY [1] will work in the ranges: Band I: 1 to 1.8 GHz, Band II: 1.8 to 3 GHz. The system allows cooling in the energy range of 0.8 to 2.5 GeV. A parameter list is given in Tab.1.

There will be separate systems for both transverse planes (Fig.1). The longitudinal cooling will be performed using Thorndahl filters in the sum paths added to the transverse RF-signal processing.

Different cooling paths are envisaged for low and high energies during the test operation. A band-I cooling path will be built as a first step. The diagonal ways (Fig.1) will be used in the nominal operation phase.

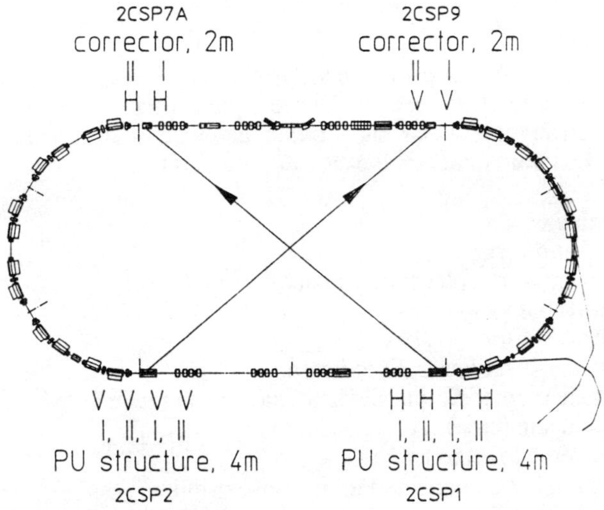

Fig. 1: Stochastic-cooling paths in COSY for nominal operation mode
H (V) : 1-m section of horizontal (vertical) cooling tank;
I,II : Band I: 1 to 1.8 GHz, Band II: 1.8 to 3 GHz.

The time required for transverse cooling from emittances of $5\ \pi$ mm mrad to about $1\ \pi$ mm mrad for 10^{10} protons at energies > 1.5 GeV will be in the order of 30 s. The signal-to-noise ratio and the variable power gain of the system allows cooling times of about 1 s for $1 \cdot 10^8$ stored protons. Similar times will be necessary to reduce the longitudinal phase space by a factor of about 3.

II. THEORETICAL INVESTIGATIONS

The cooling parameters have been calculated using the Fokker-Planck equation [2], including an equilibrium emittance [3], and taking hardware formulae corresponding to [4], [5]. The model has been verified using CERN-AC data.

The diagonal ways (Fig. 1) provide small antimixing in the energy range of 0.8 to 2.5 GeV at the given working point. The resulting phase errors are also tolerable over the whole RF band ($\pm$ 20 - 30 deg.). The transition energy (1.15 GeV) can be shifted if a shorter cooling time is required for experiments around 1 GeV using proton numbers N in COSY of N > 10^9. Further, the beam can be heated longitudinally in order to reduce the cooling time if the experiment tolerates the increased longitudinal phase space. The overall power gain of the cooling chain has been optimized for minimum transverse cooling times for $10^8 \leq N \leq 10^{10}$. The cooling rates will be constant for $N \leq 10^8$, and will be proportional to N for $N \geq 10^{10}$.

The cooling times listed in Table 1 are given in the case of operation of band-I and band-II systems. The expected cooling times are longer by a factor of about 3 during the beginning phase (band I only). The longitudinal cooling times in Table 1 have been calculated under the assumptions that the sum signals of both horizontal and vertical cooling systems are combined and that the accuracy of the delay time is better than 50 ps.

III. MECHANICAL TANK DESIGN

Figure 2 shows a partial view of a stochastic-cooling tank. Many of the tank details are being adapted from the CERN AC tanks [6], [7]. The differences of COSY and AC:

- beam parameters (emittances, energy),
- mechanical and space conditions in the ring,
- cooling-frequency ranges

have caused changes concerning the following items:

Table 1: Stochastic Cooling in COSY Jülich

kinetics				
kinetic energy	T /GeV	0.85	1.50	2.50
momentum	p / GeV/c	1.52	2.25	3.31
rel. mass factor	γ	1.91	2.60	3.66
rel. proton velocity	β	0.85	0.92	0.96
transverse stochastic cooling				
total emittance before cooling	ε_0 / π mm mrad	10	5	5
total emittance after cooling	ε_0 / π mm mrad	1	1	1
cooling time for 10^8 p	$t_{transv.}$ / s	2	1	1
cooling time for 10^{10} p	$t_{transv.}$ / s	30	50	25
longitudinal stochastic cooling				
total mom. spread before cool.	$\Delta p/p$ / 10^{-3}	1.0	1.0	0.5
total mom. spread after cool.	$\Delta p/p$ / 10^{-3}	0.4	0.4	0.2
cooling time for 10^8 p	$t_{long.}$ / s	1	2	2
cooling time for 10^{10} p	$t_{long.}$ / s	10	15	20
ion optics				
orbit length	s_{orbit} /m	183.5		
working point	$Q_{hor.}$ / $Q_{vert.}$	≈ 3.37/3.39		
focussing in bending sections	(horizontal plane)	D-F-F-D		
transition energy	$T_{trans.}$ / GeV	1.15		
transition mass factor	$\gamma_{trans.}$	2.22		
momentum spread at 40 MeV (unbunched)	$\Delta p/p$ /10^{-4}	± 23		
frequency ranges				
frequency range band I	f_I / GHz	1.0 - 1.8		
frequency range band II	f_{II} / GHz	1.8 - 3.0		
tank mechanics				
beam aperture minimum	$w_{min.}$ / mm	20		
beam aperture maximum	$w_{max.}$ / mm	140		
duration of moving cycle	$t_{up/down}$ /s	3		
lifetime of mech. components	n_{cycle}	10^6		
length of tanks	l_{pickup} / l_{kicker} / mm	4355 / 2355		
tank section length	$l_{section}$ /mm	2050		
length of coupler support bar	$l_{support}$ / mm	1950		
inner diameter of tank	d_{tank} /mm	500		
beam pipe diameter	$d_{beam\ pipe}$ /mm	150		
temperature of bars	T_{pickup} / T_{kicker} /K	25 / 300		
water cooling of kicker tanks	P_{water} /kW	1		
vacuum on beam axis	p_{vacuum} / nPa	10		
pickup refrigeration				
temp. of cryo stages	T_1 / T_2 /K	45 / 25		
heat input (cond., radiat.)	P_1/P_2 /W	43 / 6.5		
cool. power (4 RG 580)	P_{c1}/P_{c2} / W	240 / 20		
cooling down time	t_{cool} / h	≈40		
tank RF-components				
beam coupling structure		stripline coupler		
charact. impedance	R_W / Ω	50		
pickup RF length (bands I+II)	l_{pickup} / mm	3840		
kicker RF length (bands I+II)	l_{kicker} / mm	1920		
structure section length	$l_{structrure}$ / mm	960		
RF feed throughs per section	n_{FT}	8		
couplers per feed through	n_{cI} / n_{cII}	6 / 8		
dimensions of couplers	$(b \cdot h)$ I/II	20 · 32 / 20 · 22		
distance of couplers to beam axis	$d_{c \to b}$ /mm	10 → 70		
RF signal processing				
noise temp. of strip terminations	T_N /K	25		
noise temp. of preamplifiers	T_{N_pre} / K	50		
temp. of preamplifiers	T_{pre} / K	300		
location of preamplifiers		outside of tanks		
total amplifier gain	g_{amp} / dB	150		
tolerance of total gain	Δg / dB	2		
signal delay time of amplifiers	t_{amp} / ns	30		
signal delay time of filters	t_{filter} / ns	30		
tolerance of total delay time	Δt / ps	50		
power per kicker feed through	P_{FT} / dBm	48		
total peak power (H+V+L)	p_{tot} / dBm	63		

- single electrodes instead of super-electrodes,
- size and number of electrodes,
- range of distance between beam and electrodes,
- geometry of movable electrode supports,
- drive of movable electrode supports,
- tank lengths,
- division of tank sections in 1-m subsections for the 2 RF bands,
- arrangement in the RF-feed-through domes,
- steps of signal combining within the tanks,
- cryogenic cooling power.

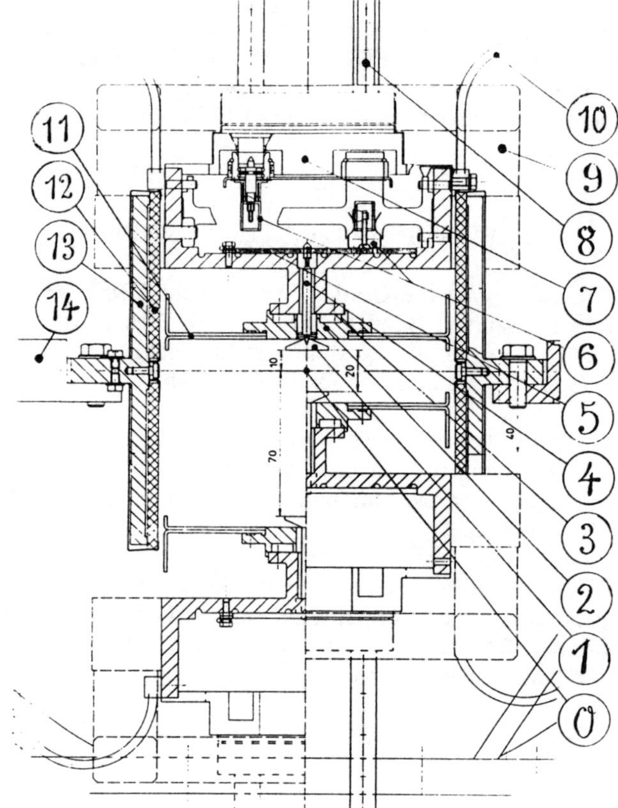

Fig. 2: Stochastic-cooling tank, partial cross-section view around the beam. Three quadrants show the arrangement for the cooled beam; The lower left quadrant indicates the case of uncooled beam.

0: beam center,
1: stripline coupler,
2: stripline-coupler fixing plate,
3: movable support bar,
4: 50-Ohm transmission line,
5: power-combiner/splitter board,
6: connectors to 5, intermediate lines (rectangular coaxial), connectors to 8,
7: traverse for coaxial lines,
8: coaxial line connected to movable stripline,
9: support-bar pivot connected to actuating rams,
10: heat-conducting band (pickup only),
11: RF screening sheets,
12: ferrite tile for attenuation of waveguide modes,
13: ferrite-tile support bar, springs for fixing the tiles,
14: fixed ferrite-bar pillar.

The PU structures will be cooled down to ~ 25 K using the cryopanels of two-stage Gifford-McMahon He cryopumps [8]. The corrector tanks will be similar to the PU tanks. But water cooling will be used instead of He cooling. The modularity of the electrode support allows the construction of 2 m tanks having 1 m of band-I and band-II structure each. The vacuum requirements of COSY (10 nPa) will be fulfilled applying additional cleaning and pumping technics. A large part of the design work of the cooling tanks has been performed up to now. So, the first tanks will be constructed previously during 1992.

IV. RF-SIGNAL PROCESSING

The signals of the single electrodes will be combined for electrode numbers corresponding to 1/4-m PU structure. The further combination steps must be outside the cooling tanks to allow the large energy range of 0.8 to 2.5 GeV. That results in a number of RF vacuum feed-throughs of 8 per 2-m electrode support bar.

2-stage Wilkinson hybrids will be used in the power-combining/dividing networks in order to get the required bandwidth (2-way and 3-way hybrids). The signal of each RF vacuum feed-through will be preamplified in a low-noise HEMT amplifier. Prototypes operating at room temperatur have been developed for both RF bands. The following characteristics of the band-II preamplifier have been measured (the band-I values are slightly more advantageous):

- noise figure 0.7 dB, corresponding to 50 K, ripple ± 1 dB,
- input return loss > 11 dB,
- amplification gain ~ 30 dB, ripple ± 1 dB,
- output return loss > 11 dB,
- group delay 0.9 ns, ripple ± 0.15 ns.

The intermediate-level electronic components are not separated in 2 RF bands.

A total gain of all amplifiers of 150 dB is foreseen. Digitally controled attenuators will enable to minimise the cooling time for the actual number of protons. Filtering, gain and phase equalization will be done at levels around 0 dBm. Ferrite tiles and resistively coated ceramic tubes will avoid wave propagation in the beam pipes.

Prototypes of 1-to-3-GHz components have been fabricated for COSY:

- low-power amplifiers,
- medium-power amplifiers,
- digitally controled attenuators.

The following global characteristics have been measured:

- input return loss > 11 dB,
- amplification gain ~ 30 dB   or
  attenuation 0 ... 33 dB, ripple ± 1 dB,
- output return loss > 11 dB,
- group delay ~ 1 ns, ripple ± 0.15 ns.

The power amplifiers will be built in 2 RF bands.One driver module will provide the power for 2 end-stage modules. Each end-stage module will have a nominal peak power of 35 W. The output of 2 modules each will be combined and fed into one RF feed-through of the kicker tanks. Design work for band-I and band-II end stages has been made (2 FETs NE345L-10B each). Power gains of the end stage of 9 dB and 5 dB have been calculated. The characteristics have been examined for a band-I prototyp.

A complete prototype RF-signal path will be constructed previously during 1991.

V. ACKNOWLEDGEMENTS

We are greatly indebted to thank our colleagues of CERN and GSI, espcially Mr L. Thorndahl, S. Milner, F. Caspers, A. Schwinn and F. Nolden for giving us comprehensive information about the CERN cooling systems and help at designing the COSY cooling system, Mr. D. Möhl and S.v.d.Meer for helpful discussions about cooling and system parameters. We also like to thank the committee members for their advisory work.

VI. REFERENCES

[1] R. Maier, U. Pfister, R. Theenhaus, "The COSY-Jülich project, May 1990 status",Proc. EPAC 90, Nice, June 1990, p. 131.

[2] D. Möhl, G. Petrucci, L. Thorndahl, S. van der Meer, "Physics and technique of stochastic cooling", Phys. Rep. 58,2 [North Holl. Pub. Co.,Amsterd., Feb.80], pp. 73...119 and CERN/PS/AA/79-23 [23.Jul.1979]

[3] K. Bongardt, S. Martin, D. Prasuhn, H. Stockhorst, "Theoretical analysis of transverse stochastic cooling in the Cooler Synchrotron COSY", Proc. EPAC 90, Nice, June 1990, p. 1583.

[4] J. Bisognano, C. Leemann, "Stochastic cooling", Summ. Sch. Hi. En. Part. Acc. 1 [FNAL, Jul.81], AIP Conf. Proc. 87 [1982], pp. 583...655

[5] FNAL, "Design report Tevatron 1 project" [Sept.1984]

[6] B. Autin,Bruno, G. Carron, F. Caspers, L. Thorndahl, "Application of microwaves to antiproton control", Alta Frequenza LVI, N10 [Milano,1987], pp. 381...91

[7] B. Autin, G. Carron, F. Caspers, S. Milner,L. Thorndahl, S. van der Meer, "ACOL stochastic cooling systems", PAC 87, IEEE Catalog No. 87CH2387-9, pp. 1549...51

[8] P. Lebrun, S. Milner, A. Poncet, "Cryogenic design of the stochastic pick-ups for the CERN Antiproton Collector ACOL", Cryog. Eng. Conf. 1985, Adv. Cryog. Eng. 31 [1986], pp. 543....550

Beam Position Monitoring in the AGS Linac to Booster Transfer Line*

T. J. Shea, J. Brodowski, R. Witkover
Brookhaven National Laboratory
Upton NY, 11973

Abstract

A beam position monitor system has been developed[1] and used in the commissioning of Brookhaven's Linac to Booster transfer line. This line transports a chopped, RF modulated H- beam from the 200 MeV Linac to the AGS Booster. Over a 15dB dynamic range in beam current, the position monitor system provides a real-time, normalized position signal with an analog bandwidth of about 20 MHz. Seven directional coupler style pickups are installed in the line with each pickup sensing both horizontal and vertical position. Analog processing electronics are located in the tunnel and incorporate the amplitude modulation to phase modulation normalization technique[2]. To avoid interference from the 200 MHz linac RF system, processing is performed at 400 MHz. This paper will provide a system overview and report results from the commissioning experience.

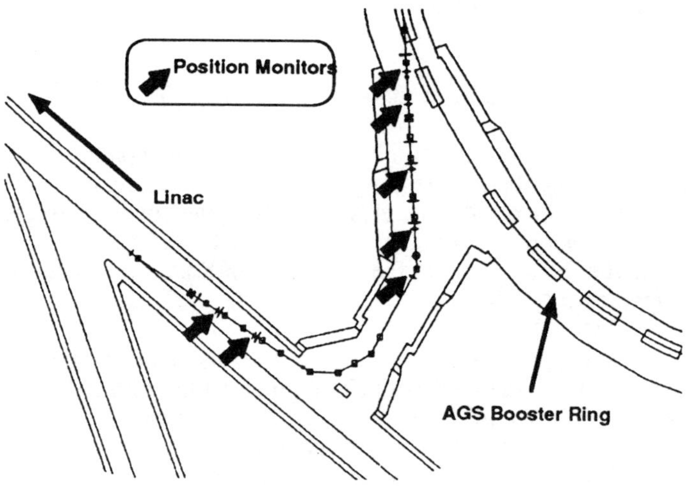

Figure 1. Site Plan.

I. INTRODUCTION

In April of 1991, commissioning of the AGS booster synchrotron commenced with the successful transport of an H-beam from the 200MeV linac to the first sextant of the Booster ring. During this exercise, the beam position monitoring system was used in conjunction with the loss monitor, current transformer, and multi-wire systems. Only the position monitoring system will be described here. This system consists of stripline style pickups with analog processing electronics located in the tunnel. The analog processing modules normalize the position signal by applying the amplitude modulation to phase modulation technique. These signals are then routed to an equipment building for digitization and analog distribution. During commissioning, both digitized and analog signals were available.

II. SYSTEM DESCRIPTION

A. Position monitors

As shown in figure 1, seven position monitors are distributed throughout the transfer line. Each position monitor contains four stripline electrodes to allow position measurement in both the horizontal and vertical planes. To avoid coupling, the horizontal and vertical pairs of electrodes are offset axially from each other. Furthermore, the length of each stripline is chosen such that the first peak in the frequency response occurs at 402.5 MHz, which is a harmonic of the 201.25 MHz bunching frequency. In this way, much of the interference from the linac RF system can be avoided. A cross-section of the monitor is shown in figure 2.

B. Electronics

As illustrated by figure 3, the analog processing modules are located in the tunnel. Each module provides a

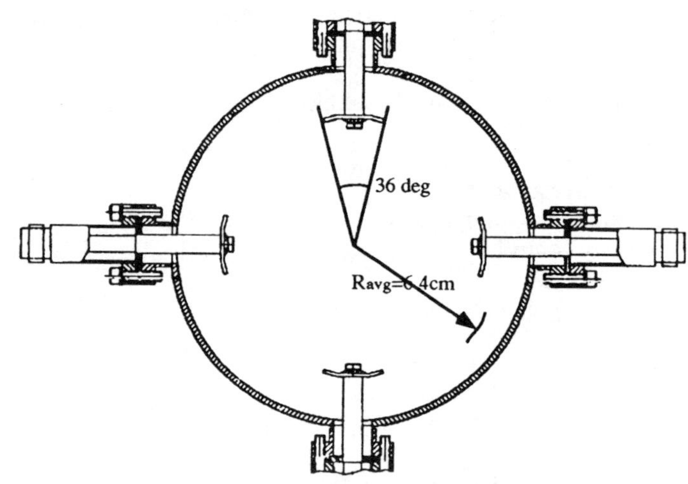

Figure 2. End view of beam position monitor.

*Work supported by the US Department of Energy

"

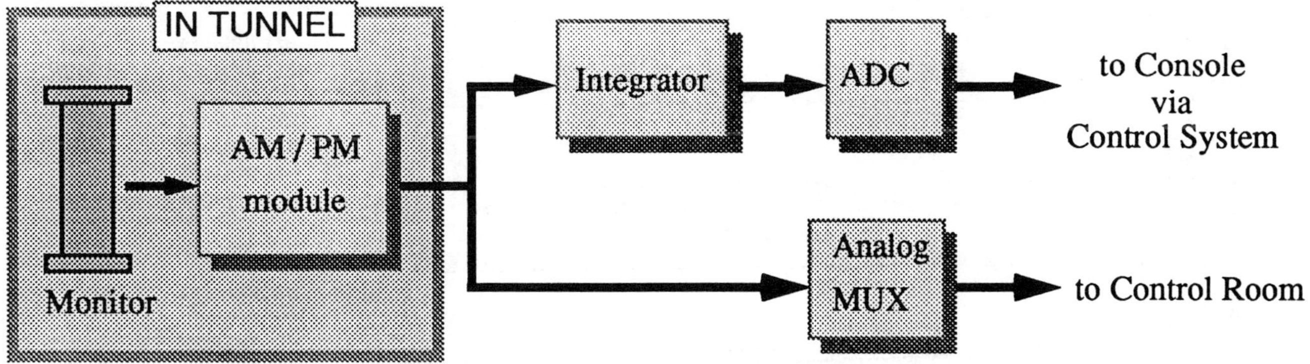

Figure 3. Block diagram of beam position monitoring system.

synchronously detected sum signal and two normalized position signals. For simplicity, all processing is performed without downconversion at the carrier frequency of 402.5 MHz. The normalization is achieved via the amplitude modulation to phase modulation technique.

All signals from the tunnel are transformer isolated and routed on coax to a nearby equipment building. Here the signals are split with one part going to an analog multiplexer and the other to a gated integrator. After integrating over a programmable window time (minimum of about 15µs), the signal is then held until it can be digitized.

III. BENCH TESTS

CW Calibration

Before installation, each module is calibrated on the bench. This calibration is accomplished with an automated testing system that maps the module's transfer function point by point over its full range of beam position and current. The transfer functions for several intensities are then averaged and the

result is fitted in three ranges to transcendental functions. An example transfer function is shown in figure 4 with the three fits overlaid. The control system uses these fits to correct for the nonlinearities that arise when measuring large beam displacements.

Transient Response

During normal operation, the beam in the transfer line will be chopped into pulses to allow efficient capture in Booster RF buckets. The chopper, operating at a frequency of about 2.5 MHz, produces pulses that are a few hundred nanoseconds long. Therefore, the transient response of the AM/PM module is of interest. Also, for diagnostic purposes, a measurement of position variation within a pulse could be worthwhile. Tests were arranged to simulate both a short burst of beam current and a rapid change in beam position. Results from the latter are shown in figure 5. For this test, amplitude unbalanced, phase coherent signals were injected into two input ports on the module. The ratio of the two amplitudes at the input ports was then suddenly changed for a duration of 200ns, thus simulating a beam that suddenly moves transversely by one centimeter and then back again.

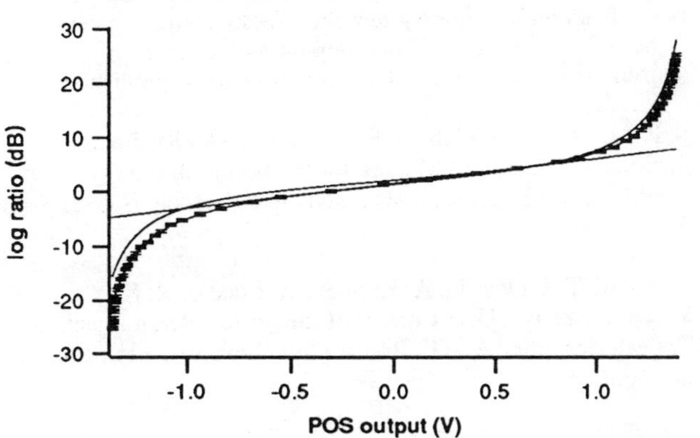

Figure 4. Example of fits to average transfer function.

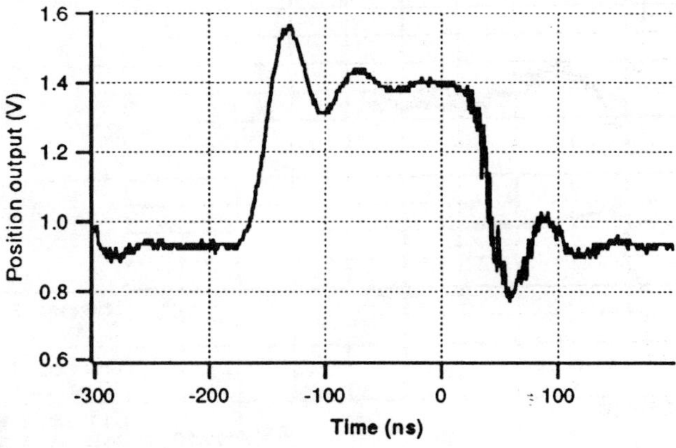

Figure 5. Module's response to simulated beam displacement.

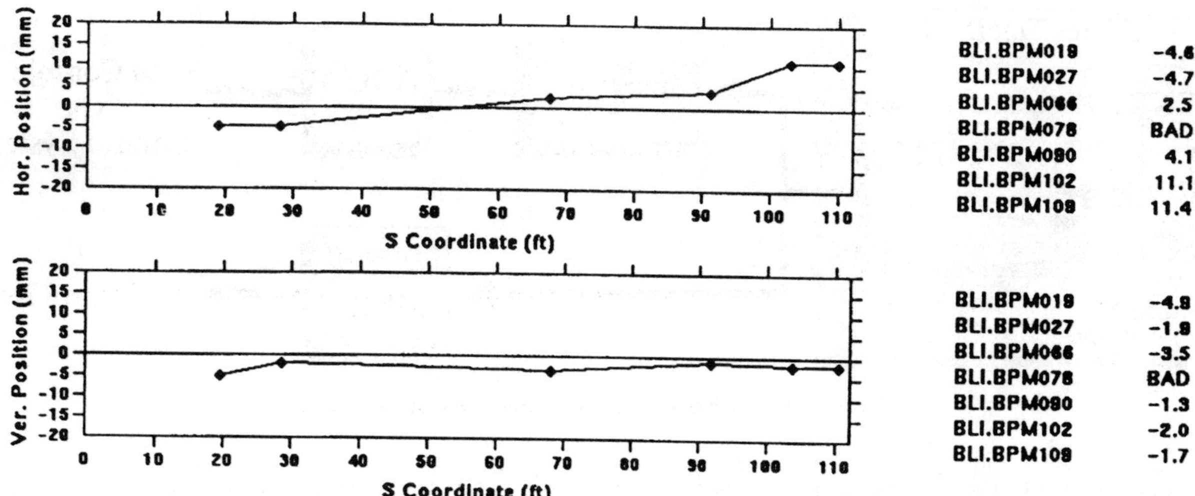

Figure 6. Beam position as displayed on control room console.

IV. COMMISSIONING EXPERIENCE

After the system performed well on the bench, the true test came during use with actual beam. Figure 6 shows a portion of a console display as it appeared during the beamline commissioning in April 1991. Before building the display, the control system utilizes the transfer functions of the monitors, AM/PM modules, and integrators to provide position information as accurately as possible. The information here is also used by an automated steering algorithm that calculates appropriate corrector settings[3]. During the first stage of commissioning, this program corrected the vertical trajectory to ±0.5mm in a single pass. Horizontal correction required two passes, possibly due to an inaccuracy in the last position channel for large displacements. Intensity information was also available and this helped the commissioning team identify valid position measurements.

In addition to the fully calibrated readings displayed on the console, the raw analog signals were also available in the control room via an analog multiplexer. Figure 7 shows a typical digital oscilloscope display of one module's output. The availability of these signals in the control room was crucial at early stages of commissioning because it allowed verification of integrator timing and control system operation.

V. ACKNOWLEDGMENTS

The authors are indebted to the following people who made significant contributions to the position monitor system : R. Bossart, V. Chiampou III, D. Ciardullo, C. Degen, T. D'Ottavio, D. Gassner, V. LoDestro, R. Thomas, and F. Usack.

VI. REFERENCES

[1]   T. J. Shea, C. M. Degen, D. M. Gassner, V. LoDestro, "A Beam Position Monitor System for Brookhaven's Linac to Booster Transfer Line", Workshop on Accelerator Instrumentation, Fermilab, October 1990, to be published.

[2]   S. P. Jachim, R. C. Webber, R. E. Shafer, "An RF Beam Position Measurement Module for the Fermilab Energy Doubler", IEEE Transactions on Nuclear Science, NS-28, No. 3, June 1981.

[3]   J. Alessi, T. D'Ottavio, A. Kponou, A. Luccio, R. K. Reece, J. Skelly, "User Control of the Proton Beam Injection Trajectories into the AGS Booster Synchrotron", see paper in this conference.

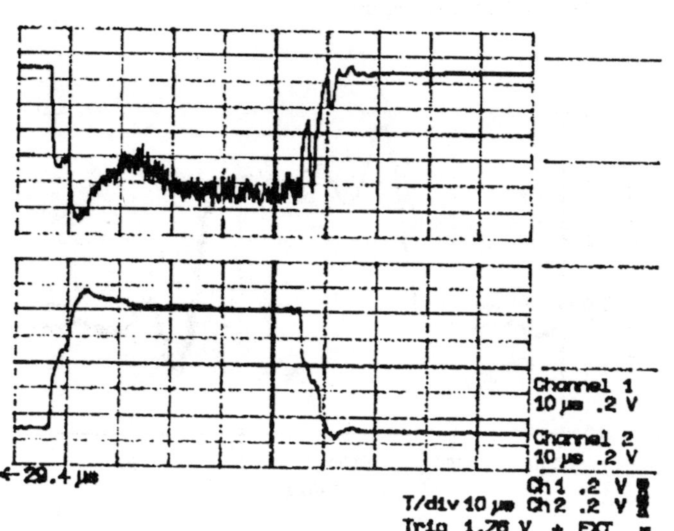

Figure 7. Analog signals: horizontal on top, vertical on bottom. 100mV corresponds to a 1mm displacement.

DESIGN OF THE AGS BOOSTER
BEAM POSITION MONITOR ELECTRONICS[*]

D.J. Ciardullo, G.A. Smith, E.R. Beadle
Brookhaven National Laboratory
Upton, New York 11973

Abstract

The operational requirements of the AGS Booster Beam Position Monitor system necessitate the use of electronics with wide dynamic range and broad instantaneous bandwidth. Bunch synchronization is provided by a remote timing sequencer coupled to the local ring electronics via digital fiber-optic links. The Sequencer and local ring circuitry work together to provide single turn trajectory or average orbit and intensity information, integrated over 1 to 255 bunches. Test capabilities are built in for the purpose of enhancing BPM system accuracy. This paper describes the design of the Booster Beam Position Monitor electronics, and presents performance details of the front end processing, acquisition and timing circuitry.

Front End Processing Module

The interface between the detector and the bunch signal integrators is the Front End Processing (FEP) module. Expected bunch widths (at half max) of from 3750 ns to 50 ns require a processing bandwidth of 40 KHz to 20 MHz. The need for wide dynamic range over this bandwidth [1,2] make the use of gain switching necessary. The module provides the frequency compensation needed to preserve the temporal profile of the bunch signals, and can supply gain or attenuation in three ranges (x 0.1, x 1 or x 10). In addition, the FEP takes the real-time sum and difference of these signals and delivers them to the integrators. Each FEP has a set of auxiliary

outputs which can be configured to provide either buffered bunch signals or their sum and difference. These auxilliary outputs are currently being used as a commissioning aide, and will eventually be utilized by the Tune Meter, RF radial and phase control loops.

Each FEP input channel contains a shunt bleeder resistor whose purpose is to prevent excessive charge accumulation on the electrode and signal cable [3]. This resistor, together with the electrode and cable capacitances defines the low end of the FEP frequency response, as well as the amplitude of the bunch signals. The cables are matched to their characteristic impedance at high frequencies via series RC to prevent VSWR ripple. A series 200 ohm resistor at the detector end of the cable provides rolloff at the high end of the response as well as isolation of the PUE from signal reflections. Peaking at the top end of the response is reduced with a series RLC which shunts the input of the buffers. The center frequency and Q of this notch filter are adjusted during module alignment to maximize common mode rejection at the high end of the band.

A difference signal output which is -50 dB relative to the sum gives an FEP accuracy of $\pm$ 0.3 mm. This requires that both channels be matched to within 0.027 dB over the entire operating bandwidth. AD9610 current feedback amplifiers provide a flat response from DC to 10 MHz, lessening the tuning effort needed to meet this specification. It was found necessary, however, to match pairs of AD9610s due to small response deviations above 10 MHz. These particular wideband

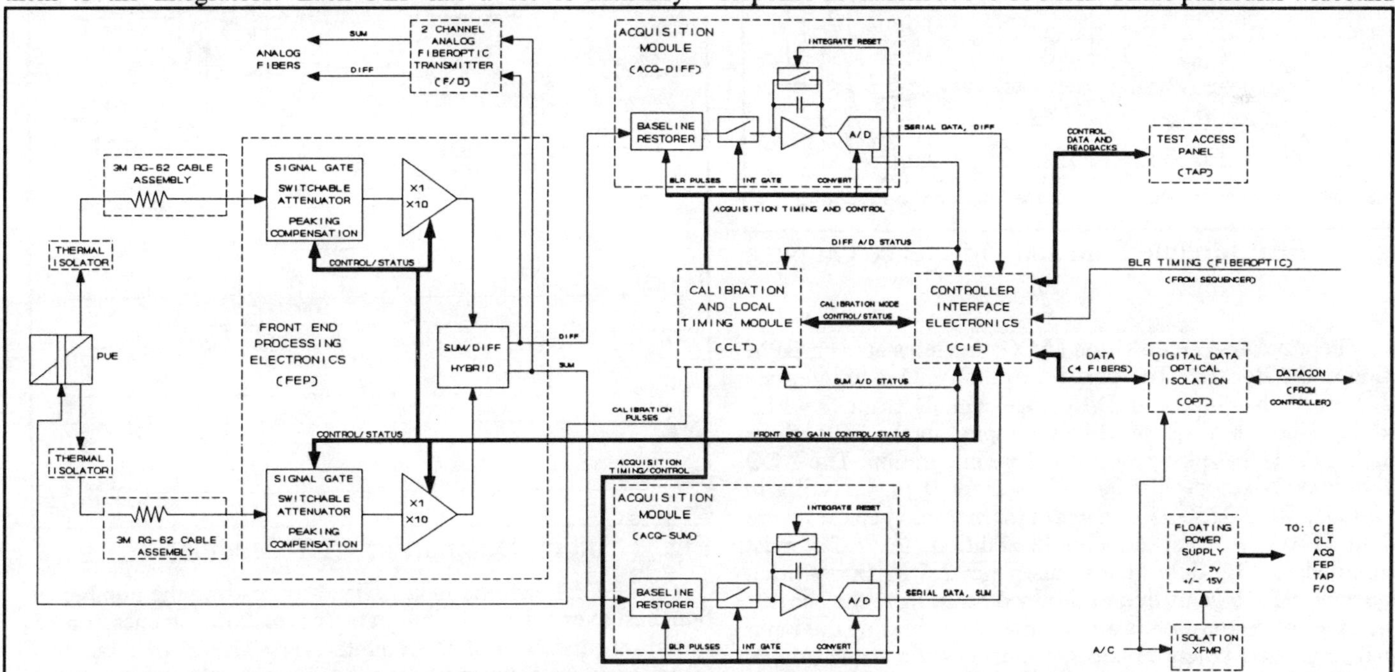

Fig. 1   Block Diagram of Local BPM Electronics.

[*]Work performed under the auspices of the U.S. Department Of Energy.

"

devices were chosen for their low input noise voltage (0.7 nV/√Hz), high non-inverting input impedance (200 KΩ) and low input capacitance (2 pF). Since the AD9610 is a current feedback amplifier, unity non-inverting gain is obtained when its inverting input is left open. Shunting a resistor of appropriate value across this input results in a voltage gain of 20 dB.

Attenuation is accomplished by setting the AD9610 to unity gain and loading the PUE with an appropriate amount of additional capacitance. The additional capacitance also has the effect of widening the bottom end of the bandwidth, resulting in a flat frequency response of 1 KHz to 20 MHz. All of the switching which occurs in the FEP module is done using Clare DSS3105 long life expectancy $(2 \times 10^8$ operations) relays.

A trifilar winding hybrid transformer with a common mode performance of better than -70 dB from 10 KHz to 30 MHz is used to take the sum and difference of the bunch signals. Signals out of the hybrid are buffered by HA5002 high speed video unity gain buffers, then sent to the Acquisition modules and auxilliary FEP outputs.

Figure 2 shows a network analyzer plot from a typical FEP module in the Booster. The calibration ring of a PUE detector is driven by the network analyzer, and the response measured from the AUX SUM and AUX DIFF outputs. The Sum output of all three gains are shown on the same reference line to simplify comparison of common mode performance between ranges.

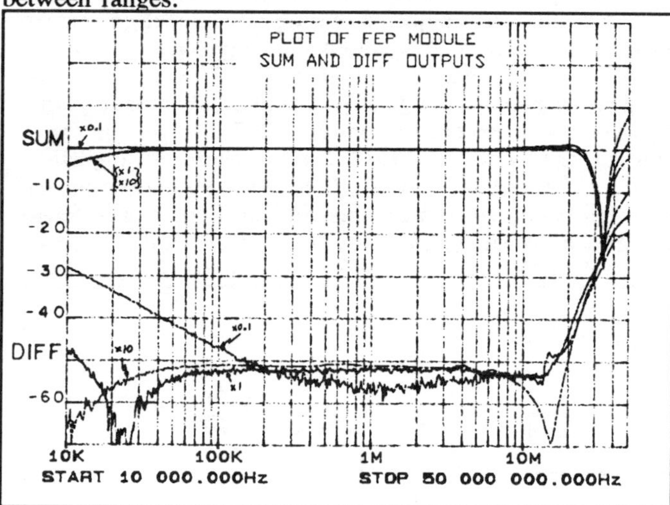

Fig. 2   FEP Module Sum and Difference Outputs.

Acquisition Module

There are two Acquisition (ACQ) modules at each BPM station in the Booster. The purpose of the ACQ is to integrate and digitize the Sum and Difference signals from the FEP module. Since they are reactively coupled in the FEP, these signals must be baseline restored before integration. The ACQ is required to accept bipolar pulses from 0 to 4.5 volts in amplitude, 50 to 3750 nsec in width (at half max) spaced from 240 to 4500 nsec, respectively. In addition, the ACQ must accommodate 75% duty factor pulses at a 2.5 MHz repitition frequency (FEP signals due to chopped beam from the Linac).

Baseline restoration is accomplished by shorting the input signal to ground potential between bunches via a quad diode array arranged as a shunt switch. The diode bridge contains matched components to minimize offsets, and is biased on and off through a balun transformer to prevent unequal + and - bias voltages. The balun also provides a bias voltage gain of x2, since the diode bias must be larger than the amplitude of

the signal whose baseline is being restored. The pulses which drive the baseline restorer (BLR) circuitry are 60 nsec in width to accomodate the minimum spacing between Sum and Difference signals of 100 nsec. Since a 60 nsec pulse width does not allow sufficient time for complete restoration of the baseline, 12 BLR pre-conditioning pulses are received prior to the start of integration.

The baseline restored signal is fed to a high performance gated integrator with two sensitivity ranges (remotely selectable). Following receipt of 12 pre-conditioning BLR pulses, the signal is gated to the integrator for the period of time necessary to allow a pre-set number of bunches to pass through the detector. Once this time has elapsed, the gate is closed and the integrated output is held and digitized. After A/D conversion is complete, the integrator is reset. The gating and reset timing, as well as the A/D conversion start signals are all generated by the Calibration and Local Timing module. Signal gating and reset functions are accomplished with Analog Devices ADG201HS analog switches, chosen for their high switching speed and frequency independent charge injection characteristics.

System specifications require 13 bits of information (10 bits position and 3 bits dynamic range for each gain mode of the FEP module). An ADC71 digitizes the integral into a 16 bit word within 50 usec. The Sum (or Difference) information is then sent serially to the Controller Interface Electronics.

Calibration and Local Timing Module

The Calibration and Local Timing (CLT) module takes external timing bursts from the Sequencer and decodes them into the proper timing and control signals needed by the ACQ modules. A block diagram of the CLT is shown in Figure 3.

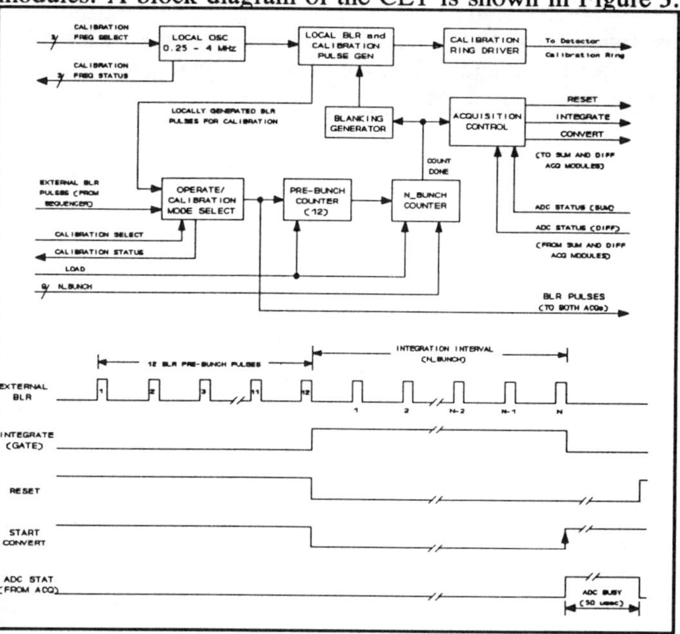

Fig. 3 Block Diagram of CLT Module.

The CLT module is initialized by loading the number of bunches over which to integrate (up to 255 bunches, or 85 turns) into the N_BUNCH counter. The arrival of a burst of 60 nsec pulses from the Sequencer initiates the measurement cycle. At the falling edge of the twelfth BLR pre-conditioning pulse, the CLT opens the integrator gates to both ACQ modules, and removes their reset signals. Analog Sum and Difference signals are integrated by the ACQ modules until N more BLR pulses are counted. The analog signal gates are then closed and a CONVERT command is issued to the A/D converters. Only after the ADCs on both Aquisition modules have completed conversion does the CLT reset the integrators, completing the measurement cycle until the next burst from the Sequencer.

In addition to its normal mode of operation, the CLT has the ability to drive the Calibration ring of the detector with 12 volt pulses, enabling the BPM system to be tested without beam. In this mode, a 16 MHz clock is divided down to generate calibration pulses of programmable frequency. One burst of BLR and calibration pulses is generated each time the host computer sends a CAL_SEL (calibration select) command. Five possible frequencies (0.25, 0.5, 1.0, 2.0 and 4.0 MHz) are available, each with a 0.25 duty factor.

BPM Timing Sequencer

Timing for the BPM system is generated external to the ring by the Sequencer. Digital fiberoptic links enable low skew, precision timing pulses to be transmitted to each BPM station in the ring. The Sequencer divides the Booster ring into 48 equally spaced phases of the revolution frequency (f_{rev}). The

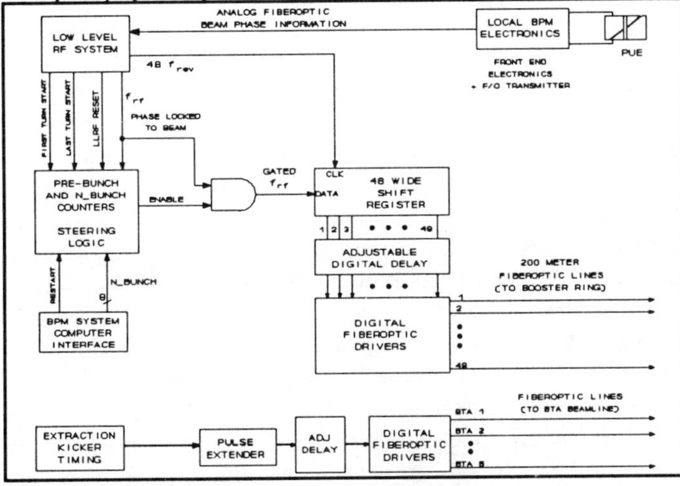

Fig. 4 Block Diagram of Sequencer.

Low Level RF (LLRF) system provides the Sequencer with the RF frequency (f_{rf}) and 48 x f_{rev}, both phase locked to the beam [4] (knowing both of these frequencies simultaneously makes operation of the Sequencer independent of harmonic number). The phase reference is sent to the LLRF system via high speed analog fiberoptic link, which sends a real-time bunch signal from an FEP module in the ring.

RF phase locked to the beam is gated into the data input of a 48 wide shift register, which is clocked by 48 x f_{rev}. Prior to a request for timing from either the BPM system computer interface (RESTART) or the LLRF (First_Turn_Start or Last_Turn_Start), logic 0 is being clocked through the SR. Upon receipt of a timing burst request, the gate opens and allows f_{rf} to be clocked through the SR. The gate is closed after counting 12 + N_BUNCH rf cycles (set by the BPM system computer), at which point the data input to the shift register is set back to a logic 0. The high speed clock to the SR continues, allowing the Sequencer output to "follow the beam" around the Booster ring. The transit time of all fiberoptic lines from the Sequencer to the Booster ring are matched using adjustable delays.

First_Turn_Start (FTS) and Last_Turn_Start (LTS) are high resolution pulses from the LLRF system, and are timed

to occur exactly 12 rf cycles (plus transit delays) before the position measurement is to begin. RESTART is a relatively low resolution pulse originating from the BPM computer interface, and is used to make multiple position measurements within a Booster cycle.

Booster To AGS (BTA) transfer line timing is accomplished by sending an integration gate, rather than BLR pulses, through the fiberoptics. There are always three bunches traveling through the BTA line, eliminating the need for baseline restoration of the sum and difference signals before integration.

Support Electronics

As part of the effort to maximize signal to noise ratio, the local processing electronics are optically isolated from all external devices. In addition, crate power is provided by linear supplies which are isolated from true ground and allowed to float with the beampipe.

The BNL DATACON standard is used for communication between the local processing electronics and the BPM instrument controller. Bipolar DATACON signals are converted to fiberoptics at the Digital Data Optical Isolation unit (DDOI, see Fig. 1). Two transceiver pairs link the DDOI with the Controller Interface Electronics (CIE), which resides in the local processing electronics crate, and is powered by the crate supply. The CIE board fiberoptically receives and decodes commands from the controller, and transmits back status and data readbacks. In addition, the CIE board receives the fiberoptic timing signal from the Sequencer and converts it to a differential TTL signal for use by the CLT module.

The Test Access Panel (TAP) provides access to all of the backplane signals in the crate. A switch and I/O connector on the front panel of the TAP board enable the crate to be put under local control. In this mode, the CIE board is tri-stated and all commands, data and readbacks normally passing through the CIE are routed through bidirectional buffers to the front panel connector of the TAP board.

Analog fiberoptic transmitter modules are resident in several of the local electronics crates within the ring. These modules can be moved from station to station during the commissioning process, allowing real-time FEP signals to be monitored in the Main Control Room. Selectable gains of x1, x0.1, x5 and x50 are available for scaling the input signals to maximize the link SNR. Gain is realized using low noise AD9610 wideband amplifiers and is selected through the BPM system computer, with one bit slaved to the FEP gain. Low distortion transmission [5] is achievable with up to a 1 V_{pp} input. The noise floor of the link is 1 mV_{rms} for over 30 MHz.

Acknowledgements

The advice and help of E. Schulte of CERN in the early stages of this project are gratefully acknowledged, as is the continuing support of J.M. Brennan and R.L. Witkover of BNL. The authors would also like to express their appreciation to J. Cupolo for his high speed analog prototyping talents.

References

[1] E. Beadle, et al, "Design of the AGS Booster Beam Position Monitor System," Proceedings of the 1989 IEEE Particle Accelerator Conf.

[2] D.J. Ciardullo, private communication.

[3] R.E. Shafer, "Beam Position Monitoring," AIP Conf. Proc. 212, E.R. Beadle, V.J. Castillo, Ed., 37 (1989).

[4] T. Hayes, A. Zaltsman, "Indirect Phase Locking of RF Clock to the Beam for the BNL Booster BPM System," this conference.

[5] E.R. Beadle, "Fiber Optics in the BNL Booster Radiation Environment," this conference.

Modeling in Control of the Advanced Light Source *

J. Bengtsson, E. Forest, H. Nishimura, and L. Schachinger
Lawrence Berkeley Laboratory
Berkeley, CA 94720

Abstract

A software system for control of accelerator physics parameters of the Advanced Light Source (ALS) is being designed and implemented at LBL. Some of the parameters we wish to control are tunes, chromaticities, and closed orbit distortions as well as linear lattice distortions and, possibly, amplitude- and momentum-dependent tune shifts. In all our applications, the goal is to allow the user to adjust physics parameters of the machine, instead of turning knobs that control magnets directly. This control will take place via a highly graphical user interface, with both a model appropriate to the application and any correction algorithm running alongside as separate processes. Many of these applications will run on a Unix workstation, separate from the controls system, but communicating with the hardware database via Remote Procedure Calls (RPCs).

Introduction

The ALS is a 1-2 GeV third-generation synchrotron light source under construction at Lawrence Berkeley Lab. The machine consists of an electron gun, followed by a 50 MeV linac, a 1.5 GeV booster ring, and a storage ring with 10 straight sections for insertion devices. The demands placed by the booster on correction algorithms are much less stringent than those of the storage ring — the booster orbit will probably only have to be corrected to 1mm *rms*, and chromaticity correction might not be needed. The storage ring, on the other hand, will require a residual closed orbit distortion of the order of 0.15mm *rms*. This is due to the fact that linear lattice distortions caused by feed-down from orbit offsets in the sextupoles break the symmetry of the lattice, and reduce the dynamic aperture. Also, fast (10-20Hz) correction of the orbit in the insertion devices is needed to maintain beam stability for users. Good chromaticity control is necessary as well. The distortion of the linear lattice by insertion devices also needs correction.

At present we have accelerated beam to 50 MeV in the linac, and are preparing for the imminent injection of the first beam into the booster. Storage ring commissioning is scheduled to begin in 1992.

*Work supported by Director, Office of Energy Research, Office of Basic Energy Sciences, Materials Sciences Division, U.S. Department of Energy under Contract Number DE-AC03-76SF00098.

We have developed tools for linac and transfer line commissioning, two of which are mentioned in the next section. We describe two tools being developed for booster commissioning; one for orbit correction and another for adjusting the tunes. One storage ring application that corrects for linear lattice distortions caused by insertion devices is also included. First we describe our present vision of an ideal system, then report on a sample of applications.

Planned System Structure

Each application will be composed of four parts — the graphical interface, the model of the machine, the correction algorithm, and the hardware database. These four pieces are seen as independent, with well-defined interfaces. In the system as it will eventually be implemented, each will be a separate process. The communication mechanism between them will probably be RPCs, perhaps coordinated by a sequencing language such as Glish [1].

Our current graphical interfaces are written for the X window system in C++ using InterViews[2], a freely available toolkit. In our interfaces, we want to display any relevant physics that will aid in adjusting a parameter; for example, we display resonance lines directly on an application that shows a measurement of the tunes and also allows tune adjustment (see Figure 1).

The model is a description of the accelerator that varies from application to application. In the case of orbit correction, our model is the ideal betas and phases at the beam position monitors (BPM's) and correctors. In setting the tune, the model is a map from tune changes to quadrupole strength changes, or a collection of maps, one for each area of the tune plane. Our models will be simple, though often created by doing studies with the full-blown simulation codes Tracy2[3] or COSY Exterminator[4], both based on the Pascal-S compiler/interpreter system[5] and therefore developed in Pascal. Tracy2 is a tracking code using 4x5 matrices and thin kicks, or alternatively, 2nd or 4th order symplectic integrators. COSY Exterminator generates maps, and is basically a generalization of a matrix code to arbitrary order. It is a cleaner, more powerful implementation of the idea behind COSY-Infinity[6]. Part of its power comes from the ability to write input files in Pascal, rather than COSY-Infinity's special-purpose, but

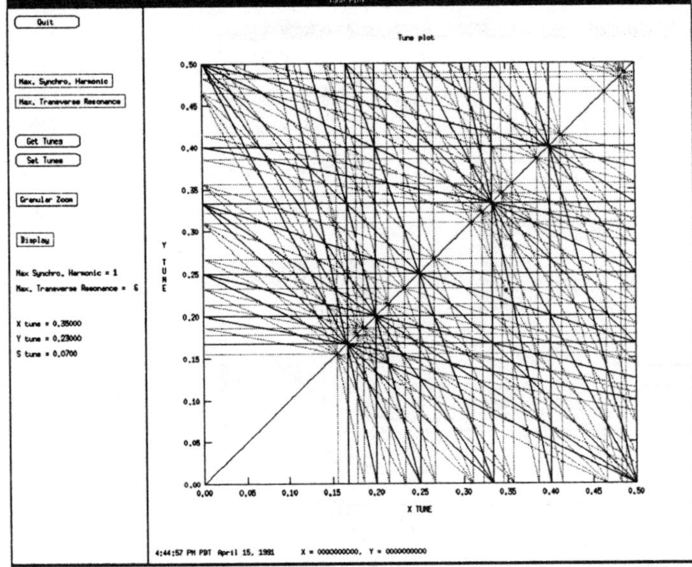

Figure 1: Graphical interface to the tune measurement and correction algorithm.

inappropriate input language. COSY Exterminator has been used to calculate maps to allow independent focusing of the beam in X and Y at waists in a transport line. For another application, MACSYMA[7] was used to simplify the steering in the solenoidal fields downstream of the electron gun by decoupling the effect of dipole correctors. This model also contains the solenoidal field, calculated by POISSON[8] from the solenoid currents.

The correction algorithm uses input from the model and the accelerator to compute new values of accelerator parameters, resulting in a change in a physics parameter of the machine. An example of a correction algorithm is the code which uses ideal betas and phases at the BPM's and correctors from the model and readings from the BPM's to compute new corrector settings, improving the orbit.

The hardware database is described in detail elsewhere [9], but for our purposes it can be viewed as a collection of readable and writable "values" of devices in the machine.

Booster Applications

A simple application is the one that sets the tunes. Our interface is as shown in Figure 1. The model is a map produced by COSY Exterminator, and the correction algorithm simply applies the model to the requested tune changes, producing new quadrupole settings. The correction algorithm iterates, applying the model until the measured tunes equal the requested tunes. The desired tune is selected by pointing and clicking with a mouse, or by typing. The maximum order of resonance lines can be selected using a menu item on the left. Resonance lines identify themselves (e.g., $\nu_x + 2\nu_y = 3$) when pointed at.

The orbit correction application we will use for the booster is based on earlier work done at the SSC Central

Design Group [10]. The correction algorithm uses overlapping local bumps of three correctors, and can correct the orbit both on the first turn and for circulating beam. Minimization is done using the simplex method, so each corrector and monitor may be assigned a weight which need not be a continuous function. This is used to prevent the program from setting the corrector strengths above the maximum value — corrector weights are a step function which becomes very large for strengths above the maximum. The goal orbit can be any arbitrary trajectory.

The interface is highly graphical (see Figure 2). The user can zoom in on either the ring on the right, or the linear display of BPM readings on the left. Zooming in, then hitting the "correct" button causes just the zoomed region to be corrected. The individual monitors and BPM's are displayed under the plot on the left, and pointing at one with a mouse causes its value to be displayed.

Originally this algorithm was implemented in the context of a simulation; we have replaced the modeling program which generated BPM readings with RPC calls to the ALS hardware database. This turned out to be straightforward, due to the modular design of the original program.

Storage Ring Applications

Due to lack of space for dedicated correction quads, local correction of insertion devices is not possible for B^2L greater than approximately $10T^2m$, which is not enough to correct the strongest insertion devices. Therefore, a global correction scheme has been worked out using only the maps in Floquet space between sextupoles[11]. By individually retuning either a subset or all of the existing 48 quadrupoles, it is possible to restore the symmetry of the bare linear lattice. We have used Tracy2 and the simplex method for this 48-dimensional non-linear optimization. This almost restores the dynamical aperture.

A complication in the implementation of this correction scheme is the fact that some of the insertion devices will be scanning. This makes it impossible to do the non-linear optimization in real-time. Therefore, a linear approximation to the relation between the insertion device and quadrupole settings is derived for each type of insertion device. This allows superposition of the contributions from several insertion devices, which is a good approximation for the cases we have studied. Tables of quad strengths as a function of insertion device setting will be used by the correction algorithm, which then only needs to do the superposition. At present, there is no user interface for this algorithm.

To improve the Touschek lifetime, we are working on an algorithm to reduce the amplitude-dependent tune shifts and non-linear chromaticity by retuning all 48 sextupoles in the ring. Analytical expressions for these tune shifts, which depend on properties of the bare lattice that have been calculated using DA techniques[12], have been calculated using MACSYMA[7]. These expressions could be used on line to adjust the sextupole strengths.

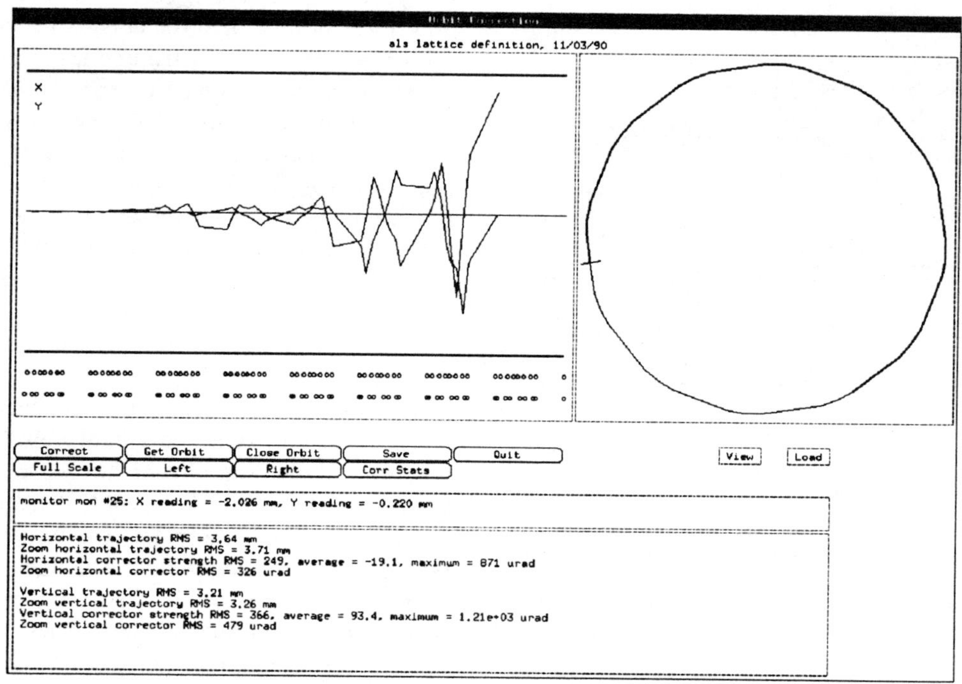

Figure 2: Graphical interface to the orbit correction algorithm. The bird's-eye view showing the entire ring is on the right; the window on the left shows the area currently zoomed-in on.

Future Work

We hope that in the process of commissioning the booster we will learn much that will be of use in developing tools for the storage ring. For example, what sort of interfaces are really convenient to use? How do we want to view and manipulate the machine state? How do the components of the system function together? What other tools or interfaces might be useful?

We are also exploring the potential of other software development tools and languages for possible use in storage ring commissioning applications. For instance, a prototype model is being developed using Eiffel, and with it, a class library for accelerators. We are exploring other graphics toolkits such as *glistk*[13] which is being developed at SSC.

Acknowledgments

Vern Paxson created the interface for orbit correction while at SSC/CDG. The interface for setting the tune is an outgrowth of one he and one of us designed for studying tracking. Many thanks to Chris Timossi for implementing the RPC calls to the hardware database and for supporting our efforts in general.

References

[1] V. Paxson *et al.*, "A Language, Server and C++ Class Library for Event Sequencing," NIM **A**293 (1990), p. 356-362.

[2] M. Linton *et al.*, "Composing User Interfaces with InterViews," IEEE Computer, Feb. 1989, pp. 8-22.

[3] J. Bengtsson and H. Nishimura, unpublished.

[4] J. Bengtsson, "Applications Based on Tracy and COSY Exterminator," LBL LSAP-116, 1991.

[5] N. Wirth, "Pascal-S: A Subset and Its Implementation, Pascal – The Language and Its Implementation," Ed. D. W. Barron, pp. 199-260, Wiley 1981.

[6] Cosy-Infinity Reference Manual, LBL-28881 (1990).

[7] "MACSYMA Reference Manual", Symbolics Inc. (1983).

[8] C. Kim, private communication.

[9] S. Magyary et al., "The Advanced Light Source Control System," NIM **A**293 (1990), p. 36-43.

[10] V. Paxson, *et al.*, "Interactive First Turn and Global Closed Orbit Correction in the SSC," Proceedings of the European Particle Accelerator Conference, Rome, August 1988.

[11] J. Bengtsson and E. Forest, "Global Matching of the Normalized Ring," LBL ESG Tech Note 125.

[12] M. Berz, "Differential Algebraic Description of Beam Dynamics to Very High Orders," Particle Accelerators **24**, 109 (1989).

[13] M. A. Kane *et al.*, The GLISTK Manual, unpublished.

ACCELERATOR AND FEEDBACK CONTROL SIMULATION USING NEURAL NETWORKS*

D. NGUYEN,[†] M. LEE, R. SASS, H. SHOAEE

Stanford Linear Accelerator Center Stanford University, Stanford CA 94305

Abstract

Unlike present constant model feedback systems, neural networks can adapt as the dynamics of the process changes with time.

Using a process model, the "Accelerator" network is first trained to simulate the dynamics of the beam for a given beam line. This "Accelerator" network is then used to train a second "Controller" network which performs the control function. In simulation, the networks are used to adjust corrector magnets to control the launch angle and position of the beam to keep it on the desired trajectory when the incoming beam is perturbed.

Introduction

Both fast and slow feedback control are used in the SLC control system to stabilize the beam and optimize collisions. Fast feedback control executes at or near the beam rate in the 80386 micros, each of which controls all devices in one geographical region[1], while slow feedback control operates in the VAX over a period of many seconds to a few minutes.

All of these systems use a constant model of the accelerator in the region to be controlled. While known noise characteristics can be incorporated into the model, transient conditions or unknown systematic changes over time can reduce the effectiveness of the control.

Adaptive control systems however, can change their behavior in response to changes in the dynamics of the process they control. Like constant model systems, adaptive systems contain an embedded model of the process they control but they also have the ability to modify this embedded model and the control algorithm with time as the process dynamics change. One way to do adaptive control is by the use of multiple neural networks.

Neural Network Architecture

For this simulation, two neural networks were trained with the back propagation algorithm[2]. Even though a network with only a single hidden layer can implement any

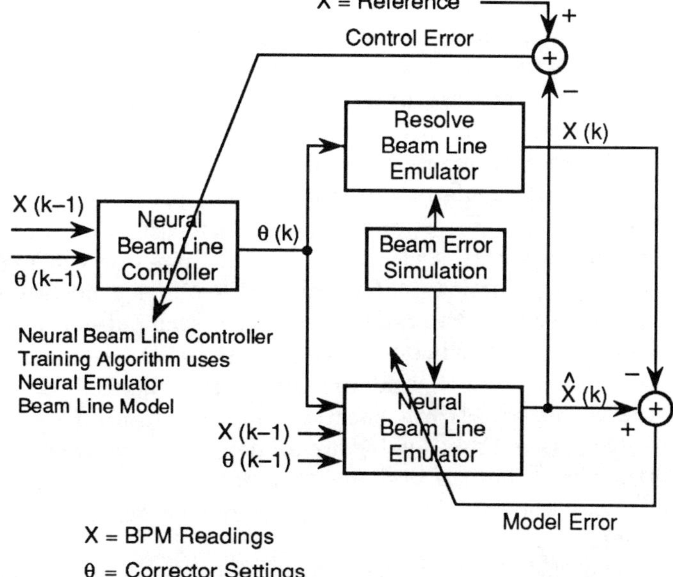

Figure 1. Block Diagram of Neural Network Architecture

nonlinear function[3], multiple networks with more layers often provide a better way to decompose a problem.

Figure 1 shows the neural network architecture used for this simulation which consists of two feed forward networks. The Neural Beam Line Emulator is the machine model, the equivalent of the plant model in control theory, and the Neural Beam Line Controller is the equivalent of the control algorithms used to keep the beam at the desired setpoint.

Because the plant to be controlled is linear, that is, the Beam Position Monitor (BPM) readings for a pulse is a linear function of the corrector settings, both the Neural Emulator and the Controller need only be one layer networks i.e., no hidden layer is required. In a beam line containing 16 BPMs, the Neural Controller is trained to adjust two correctors in each plane near the beginning of the beam line to center the beam positions at two consecutive BPMs near the end of the line.

Training the Networks

The learning process involves two stages. The first stage trains the Neural Emulator to act like the beam line model

*Work supported by the Department of Energy, contract DE-AC03-76SF00515

[†]Present address: ARGO Systems.

"

and "understand" the beam dynamics in this section of the accelerator. It uses the RESOLVE[4] beam line modeling program as its "teacher". The second stage enables the Neural Controller to learn to control the beam by using the Neural Emulator as a model. It is assumed that the factors which affect the trajectories are varying slowly with time and change little from pulse to pulse, enabling the Neural Controller to use BPM and corrector information from the previous pulse to control the next pulse.

Training the Neural Emulator is similar to plant identification (plant modeling) in control theory, except that the plant identification is done automatically by a neural network. The Neural Emulator has 40 inputs and 32 outputs. The inputs are the 32 BPM readings (16 in each plane) and the four corrector settings (two in each plane) of the previous pulse and the four corrector settings of the current pulse. The Neural Emulator then outputs the 32 BPM readings which are a prediction of the trajectory of the next pulse. The current pulse is then launched, and the BPM readings generated by RESOLVE are compared with the Neural Emulator's prediction. The prediction error is used to train the neural network using the back propagation algorithm so that the Neural Emulator will make a better prediction. After training over many pulses with the correctors set to random values for each pulse, it learns to be a good predictor of the next pulse's trajectory. By this process, the Neural Emulator "gets the feel" of how the corrector settings affect the pulse's trajectory. This process is roughly analogous to the steps that would be taken by a human designer to model the beam line. In this case, however, the modelling is done automatically by the network.

The Neural Controller is also a one layer (no hidden units) network with 36 inputs and four outputs. The inputs are the BPM readings and the corrector settings from the previous pulse, and the outputs are the corrector settings to be used in controlling the next pulse. The Neural Controller's outputs are connected to the RESOLVE beam line model and the Neural Emulator.

Before launching each pulse, the BPM readings and the corrector settings from the previous pulse are obtained and fed to the Neural Controller. Because the adaptive weights of the controller are initially at random, it outputs an erroneous set of corrector settings for the next pulse. The pulse is then launched, and the BPM readings at the two BPMs of interest are obtained. The error, which is the difference between the actual beam positions and the desired beam positions at these two BPMs, is computed, and the controller is trained to minimize the magnitude square of this error. However, to train the controller we need to know the error in the controller's outputs, but we have instead the error in the BPM readings shown in Eq. 1.

$$\epsilon = x_{\text{reference}} - x(k) \qquad (1)$$

This is where the Neural Emulator is useful. It essentially acts as an error translator, converting the errors in

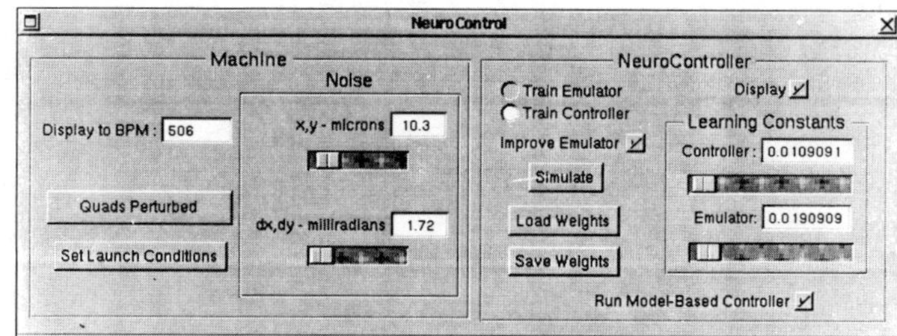

Figure 2. Simulation Control Panel

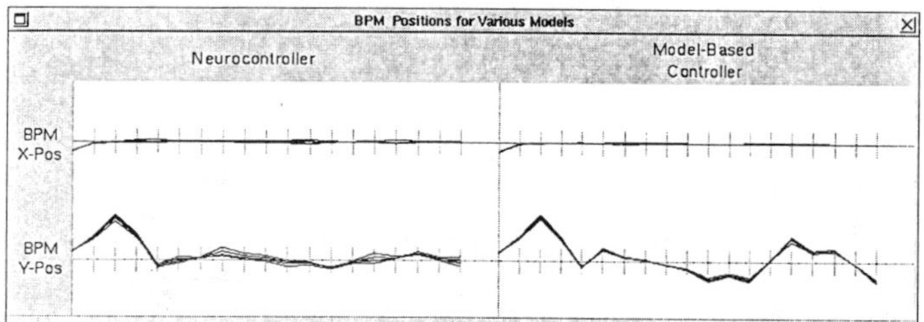

Figure 3. Standard Beamline Environment

BPM readings to errors in corrector settings, θ. The error in the BPM readings is back propagated through the Neural Emulator toward the Neural Controller, and then back propagated through the Neural Controller, updating its weights, W, according to the learning rate μ at the same time as shown in Eq. 2.

$$\Delta W = -\mu \frac{\partial \epsilon}{\partial W}^2 = 2\mu\epsilon \frac{\partial x(k)}{\partial \theta(k)} \frac{\partial \theta(k)}{\partial W} \qquad (2)$$

The Neural Controller training process is repeated for many pulses, until the mean square error decreases to an acceptable value. The training of the Neural Emulator may continue at the same time as the Neural Controller, thus enabling the whole system to continually adapt. If unexpected changes, such as malfunctions or drifts due to temperature fluctuations, occur in the beam line, the Neural Emulator will adapt to the changes, and then the Neural Controller will adapt to the Neural Emulator, enabling the overall system to continue to function.

Simulation Results

Figure 2 shows the control panel for doing the simulation. The learning constants for both of the networks, noise values of beam position and angle, which of the networks are to be trained and comparison with the model based controller can all be selected from this panel. In addition, the "Quads" button allows us to alter predesignated quadrupole strengths and thus invalidate the machine model.

Figure 3 shows a comparison of the trajectories of a series of consecutive pulses controlled by the neural networks and the model based controller. Note how the pulses come into the beam line section off center and at an angle, and are centered by the end of the beam line section. With all devices matching the model, both the model based controller and the neural networks work about equally well.

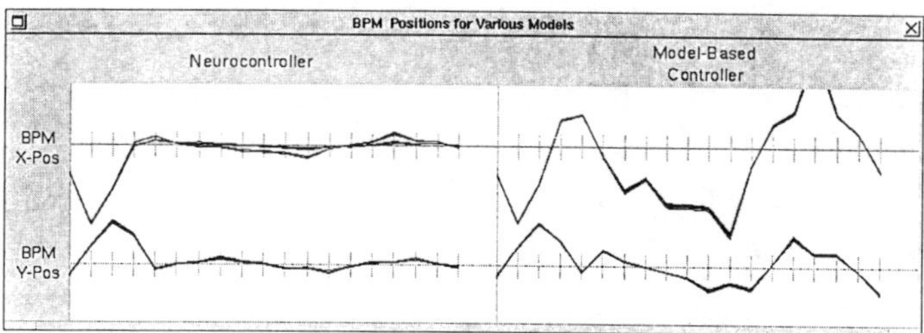

Figure 4. Malfunctioning Beamline Environment

Figure 4 shows the simulation in a beamline environment which contains two quadrupoles whose field strength values have been substantially decreased from their nominal values, simulating a malfunction. Because the model used does not match the actual beam line due to the simulated malfunction, the beam is not well controlled at all by the model based controller. The neural network based controller however has adapted to the change and still controls the beam as well as before, after it has learned how the real machine has changed. When the quadrupoles have been restored to their nominal values, the neural network based controller can again relearn the correct model and control the beam like the model based controller.

Future Plans

Neural network based feedback can be used to compensate for changes in the energy profile caused by Klystron population changes, Klystron phase drift and diurnal effects not compensated for by existing feedback. Less likely though still possible are hardware changes like power supply instabilities and changes in focusing properties of magnets due to hysteresis. All of these dynamic changes to some degree invalidate the static model used by current feedback systems.

To further explore the possibilities for the use of neural networks in the real control environment we anticipate attempting one or more of the following efforts:

1. Train one or more emulators offline using the models generated for the fast feedback system presently running at the SLC[1]. Then run these neural network models in the 80386 micros that presently run the fast feedback system and compare the results with the existing feedback.

2. Use a neural network to control the residual errors of the existing feedback systems. In this mode, the network acts to "touch up" i.e. make small, dynamic corrections to the existing feedback system.

3. In parallel with this work, experiment with a variety of extensions and enhancements to the basic back propagation algorithm to improve performance and study network instability.

Conclusions

We have been able to train multiple neural networks to be both an emulator and controller of a beam line section. This indicates that it would be feasible to use multiple neural networks to compensate for transient or slowly varying beam instabilities or correct for uncertainties in our knowledge of the machine model.

Acknowledgements

Special thanks to Professor Bernard Widrow of Stanford University for his early inspiration on this project.

References

[1] F.R. Rouse et al., "General, Database Driven Fast Feedback System for the Stanford Linear Collider," *Proceedings of the 1991 IEEE Particle Accelerator Conference,* San Francisco, CA (May 1991).

[2] G.E. Hinton, D.E. Rumelhart and R.J. Williams. *Learning internal representations by error propagation,* Vol. **1**, Ch. 8. MIT Press, Cambridge, MA, (1986).

[3] B. Irie and S. Miyake. "Capabilities of three-layered perceptrons," *Proceedings of the IEEE International Conference on Neural Networks,* pp. I–641, (1988).

[4] Cf. M. Lee for information on RESOLVE.

Optimal, Real-Time Control – Colliders*

J.E. Spencer

Stanford Linear Accelerator Center, Stanford University

Stanford, California 94309

Abstract: With reasonable definitions, optimal control is possible for both classical and quantal systems with new approaches called PISC(Parallel) and NISC(Neural) from analogy with RISC(Reduced Instruction Set Computing). If control equals interaction, observation and comparison to some figure of merit with interaction via external fields, then optimization comes from varying these fields to give design or operating goals. Structural stability can then give us tolerance and design constraints. But simulations use simplified models, are not in real-time and assume fixed or stationary conditions, so optimal control goes far beyond convergence rates of algorithms. It is inseparable from design and this has many implications for colliders.

Introduction

Prediction is very difficult, especially about the future
— Niels Bohr

Predicting the future isn't difficult until one demands a high correlation between events and their predicted times of occurrence. For deterministic systems having models with fast predictive cycles compared to their time scales, stable control should be possible. The choice between closed and open loop control then revolves around which is better, the model or the feedback system. Whether either can be done fast enough or accurately enough in real time depends on resources relative to problem demands.

Because there are other stochastic effects beyond the known dynamic nonlinearities, we need both feedback and feedforward for optimal control. We use *causal feedforward* for the special combination that takes optimal advantage of the collapse of the probability distribution. External noise effects are shown in Fig. 1. An example is line noise in magnet supplies that maintain constant current via feedback but leave random field errors from eddy currents and hysteresis. Knowing if tolerances are violated from above or below improves the speed and quality of correction.

New stochastic effects are expected as the energy and luminosity of colliders increase associated with quantum effects and the growth of complexity (degrees of freedom) e.g. it becomes harder to define the system, its variables or their constraints even assuming the system is isolated. Further, even if the dynamics and control model are linear, their physical realization with measurement, roundoff or overflow errors can be nonlinear with chaotic regions. The meaning, possibility and implications of optimal real-time control under such circumstances are discussed.

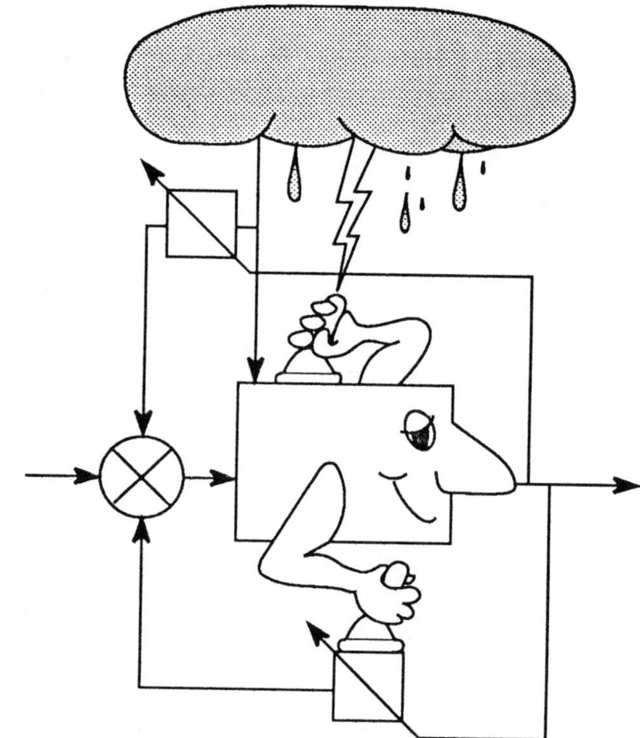

Figure 1: Real-Time Control in the Real World

Problems with Complexity

Despite books and conferences[1] relating it to entropy, order and information, there is still no universal measure of complexity. Real-time control of colliders is a good place to explain why. The largest machines ever are proposed to learn everything about the smallest distances in the least possible time. We consider whether this is consistent and, if so, at what cost and with what techniques?

Figure 2 compares the growing complexities of several systems where the vertical axis can be thought of as the bits of information needed to specify the system state. For ring colliders this is proportional to their radius (or diameter for two rings) which also relates to cost. For single chip DRAM and Intel micros, it is the number of bits or transistors. The time axis gives the first available date of the various products – physics or chips. Thus, the LHC and SSC are located by the lower bounds on their detector dates. The vertical scale is limited above by the human brain with a dozen billion neurons[2] and below by quanta such as photons in pure helicity states with one bit of information classically when we know the wave vector.

---

*Supported by U.S. Dept. of Energy contract DE-AC03-76SF00515.

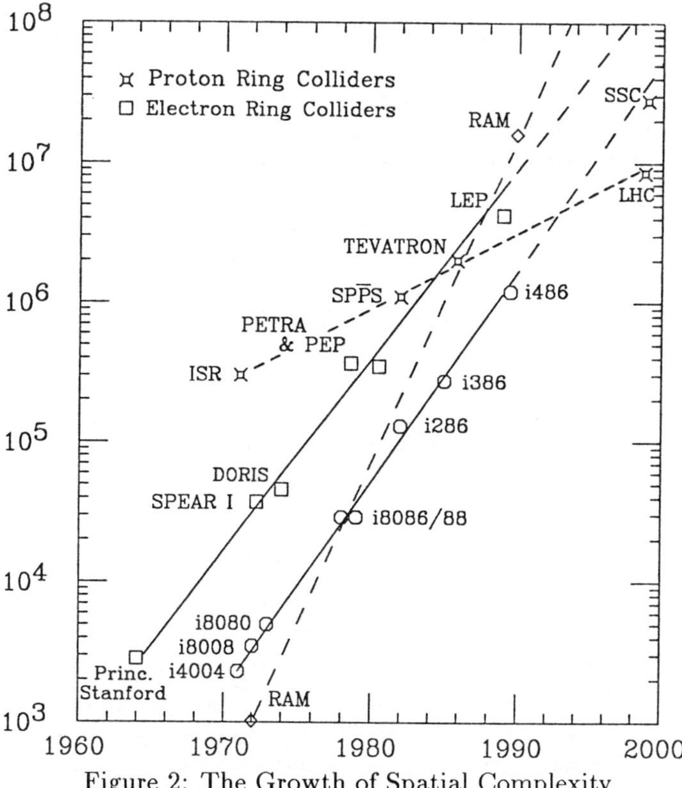

Figure 2: The Growth of Spatial Complexity

Excepting SSC, each class shows an exponential increase in spatial complexity i.e. the number of elements due to increases in size or density. Extrapolating, one can make predictions e.g. it's hard to see how SSC can produce any physics before 2005. Trends within a class help us predict capabilities e.g. what the i586-i786 chips will provide and when[3]. LEP uses some 2000 micros at the i286/386 level. Including memory, LEP's system approaches the human brain in complexity and the SSC may surpass it[2].

While LEP appears inefficient it provides flexibility and redundancy that's applicable for LHC. To compare to SLC, we could, in principle, count their control program statements and calculate execution times to determine the more complex system assuming both were optimal. But storage rings don't need a program most of the time so they must have lower temporal complexity. However, LEP is *much* larger or more spatially complex so the two emphasize complementary resources to solve the same problem. In either case the growth in complexity makes reliability, adaptability and flexibility increasingly important.

Real-time computation, as part of real-time control, faces similar but easier problems. In $n \times n$ bit multipliers, you can interchange space-time complexity going from pure memory methods with as little as one memory cycle but n^2 memory locations to serial multiplication with order n^2 machine cycles but no memory. The neural net is an ideal theoretical tool to study such problems e.g. we can use it as an associative memory or optimal parallel multiplier[3] or to simulate *any* complex process whose K-entropy is large or infinite. Neural nets are also practical and needed for complex systems such as colliders – possibly in chips like the i686 or i786 by the year 2000 for the LHC[3,4].

Comparison of Colliders

Our ability to define our system, identify the variables and conditions they must satisfy in terms of a consistent measure of merit is itself a measure of problem complexity. Current linear collider (LC) designs for the same energy and luminosity differ wildly even before any consideration of how one finds or holds the optimal system state.

Some questions are: 1) What is the optimal collider for physics, 2) What is optimal for electrons, 3) How does one optimize time complexity for LC's and 4) Have we really explored all the possibilities? Some of these were explored by Richter[5], Rubbia[6], Panofsky[7] and by Palmer[8]. Restating 1): Why not build a 2 TeV electron machine rather than a 20 TeV SSC? Also, for 2): Why not build an LC since rings become dominated by radiative losses?

$$Cost = Fixed + Capital + Power$$
$$\approx C_0 + C_1 R + C_2 \frac{\gamma^4}{R} \rightarrow C_0 + 2\sqrt{C_1 C_2}\gamma^2$$

Both cost and radius scale with the square of the energy and the 'optimal' ring approaches an LC asymptotically. SLC and Tevatron scale linearly with length or energy. From Fig. 2 and the fact that the Tevatron has ten times the top energy of LEP[8], it is preferred as long as its cost or complexity per unit length is $C_1^{TeV} < 2C_1^{LEP}$. For SLC, the equivalent path length is less than one-fourth LEP which also roughly measures their relative capital costs.

But no one can really say what C_1 or C_2 are until the physics is done since they depend on integrated luminosity, detector capabilities and how many Z's are needed i.e. the underlying problem complexity. LEP, SLC and the Tevatron with their various detectors are different algorithms for this problem. Each has produced comparable results for different costs in time and other resources. One can estimate the enhanced reliability (and costs) needed for future machines from earlier ones[4,9]. When reliability and temporal complexity of LC's scale linearly, they are preferred and 0.5 TeV is competitive with LHC.

One approach to these problems is via correlations:

$$\langle \cdots f_{i_\pm}(t) f_{j_\pm}(t - \tau) \cdots \rangle \leftrightarrow \langle \cdots f_{i_\mp}(t) f_{j_\mp}(t + \tau) \cdots \rangle$$

with $f_{i_\pm}$ a state variable whose vector is $\mathbf{f}(t) = \{f_i(t) : i = 1 \cdots 2d\}$, with time series $\{\mathbf{f}(t_j) : j = 0 \cdots n\}$ where $t_j = t - j\tau$. We can predict $f_i(t_j > t)$ for chaotic systems when $n \geq 2d$[10] because we can measure and study the dimensionality d_i. We can also control and optimize f_i and thus $\mathbf{f}$. The correlations can be interpreted at fixed time as giving instantaneous envelop equations to any order. Similarly, from the time dependence at fixed location we can monitor the K-entropy which measures the average rate of information loss or phase space deformation i.e. the predictability with time. To keep all correlations bounded we must supply information at least as fast as it is lost through both design and control[11,12]. As K $\rightarrow \infty$ the problem becomes purely statistical but still allows optimal prediction when we know the statistics. However, adaptive prediction is still only worse by log n with n the number of steps in advance time for LMS algorithm.

Model-Driven, Adaptive Control System

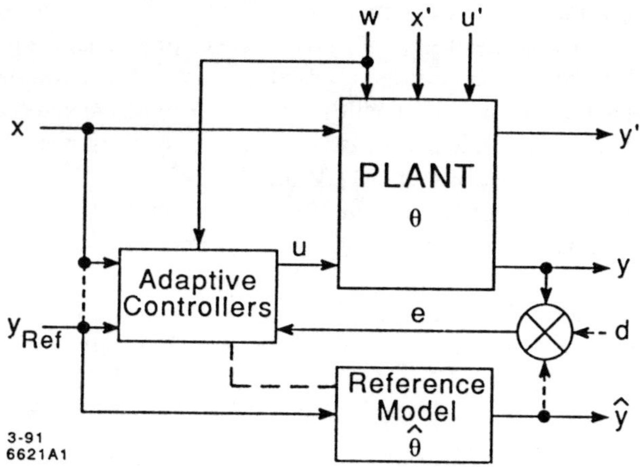

Figure 3: A General-Purpose, Real-Time Control System. It can cancel noise (w), control a plant, given the model, or find a model for a given plant and optimize it i.e. $\hat{\theta} \to \theta$. It is useful for complex systems where u', w, or x' are variables that may influence output but aren't controlled or monitored or when the system is changing and we can't set tolerances or respond fast enough. Ref.[3] gives a case where x and u get contaminated and have to be translated into hardware syntax while also learning.

Optimal, Real-Time Control

Although independent of space-time bounds, our most important postulate here is that of Turing and Church[4]. Also, we assume that complex systems need comparable or greater complexity in their controllers consistent with Shannon's measure of information. We can't drop degrees of freedom or the highest order of a plant or subsystem. Lastly, we can't have too much computing power because it improves the ability to adapt to and control unforeseen, unstable or chaotic conditions e.g. the K-entropy.

We begin by assuming a 'program' can be written that will simulate *any* physical process. It is also assumed that for any specific problem, one can translate this program into 'matched' hardware solutions such as ASIC's. This combination defines an algorithm for the problem. The classic example is the von Neumann computer but it can not be optimal because of its serial hardware and software.

Considering the program as an equivalent description of the problem (or process), information theory tells us the efficiency of any encoding or most efficient algorithm. But real-time control differs from computational algorithms in the importance of external, asynchronous effects that can be more important than the calculation. For optimal solutions, we therefore assume that we must use the process itself even though we may not be able to produce an equivalent software procedure. The hardware, in a very real sense, is the optimal 'program' or problem algorithm that we need to realize in our control system.

We still lack optimal control until we obtain the model and optimize it for the *existing* plant. But we know how to do that[12,3,4], at least in principle. Furthermore, ANN's or artificial neural nets can replace computers in all ways[3] and don't suffer their bottlenecks[4]. They are also the paradigm for the ultimate, parallel, pipelined processor that handles concurrent, parallel data naturally. They apply to computers, colliders and detectors so they provide both the controller and model needed in Fig. 3. Some of these capabilities could be available in the i686 chip[3].

This is in direct contrast to conventional, model-based systems that lack any explicit reference to control. Since they are also used for design, we lack optimal designs as well as the tools to achieve them which can increase control costs C_{ic} drastically. In contrast to ring colliders, the beam characteristics in LC's have tighter tolerances and ultimately depend on source performance. Another paper addresses the source design from this perspective[11].

References

[1] *Measures of Complexity*, L.Peliti & A.Vulpiani (Eds.), Springer-Verlag, Berlin 1988. See also Ref.[10].

[2] *The Brain*, R.S.Thompson, W.H.Freeman and Co., New York 1985. The information to specify a state is different from that required to set it. Number and connectivity between neurons is then more appropriate because the complexity of one neuron can then be $> 10^4$. Likewise, there are large differences between bits and transistors in micros and DRAM[4].

[3] A.Corneliusen et. al., Computation and Control with Neural Nets, NIM A293(1990)507-516.

[4] J.E.Spencer, Real-Time Applications of NN, IEEE Trans. Nuc. Sci., NS36, No. 5(1989)1480. Information capacity of a 'forward' weight can be taken as 1 bit. To specify a simple net of two equal layers with n neurons apiece requires an $n \times n$ connection matrix but a linked list requires only $10^4 n$ while a planar, nonintersecting circuit takes at most $6(n-1)$ or $4(n-1)$ without interconnects within a layer i.e. order n.

[5] Burton Richter, Very High Energy Colliders, 1985 IEEE Part. Accel. Conf., Vol. NS-32, No.5(1985)3828.

[6] Carlo Rubbia, The 'Future' in High Energy Physics, 1988 European Part. Accel. Conf., Rome 1988, 290.

[7] Wolfgang K.H.Panofsky, A Perspective on Lepton-Photon Physics, XIVth Int'l. Symp. on Lepton-Photon Interactions, Stanford, CA(1989).

[8] R.B.Palmer, Prospects for High Energy Colliders, Ann. Rev. Nucl. Part. Sci., Vol. 40(1990)529.

[9] J.E.Spencer, Accel. Diag. and Cont. by NN, 1989 IEEE Part. Accel. Conf., 89CH2669(1989)1642.

[10] *Exploring Complexity An Introduction*, Gregoire Nicolis and Ilya Prigogine, W.H.Freeman and Co., New York 1989. Also: N.H.Packard et al., Phys. Rev. Lett., 45(1980)712 and J.D.Farmer and J.J.Sidorowich, ibid, 59(1987)845. We use ANN's to define models and to optimize. Parts of this paper were given in BAPS, Vol.35(1990)1047.

[11] J.E.Spencer, High Brightness Sources, This Conf.

[12] *Adaptive Signal Processing*, Bernard Widrow and S.D.Stearns, Prentice-Hall, NJ 1985.

Accelerator Simulation and Operation Via. Identical Operational Interfaces*

J. Kewisch, A. Barry, R. Bork, B. Bowling, V. Corker, G. Lahti, K. Nolker, J. Sage, J. Tang
Continuous Electron Beam Accelerator Facility
12000 Jefferson Avenue
Newport News, VA. 23606

ABSTRACT

The CEBAF accelerator contains approximately 2500 power supplies, 340 klystrons, and 800 beam monitors. The operation of such a complex machine requires a control system which can provide a high degree of automation with strong support by simulation and modeling programs.

We present the architecture and first results of a control system which allows one the use of identical operation procedures and interfaces for operation of the real accelerator and high-level accelerator simulation programs. The interfaces were developed using TACL (Thaumaturgic Automated Control Logic) control software, developed at CEBAF for accelerator control. This setup provides the capability to: (1) test and debug the various operation procedures before the completion of the accelerator, (2) execute machine simulations under realistic environmental conditions, and (3) preview and evaluate the effectiveness of operational procedures during run time. The optimized simulation program adds only two seconds to the normal TACL operational cycle.

INTRODUCTION

The CEBAF control system consists of a programming environment referred to as TACL[1] and a large number of user applications. TACL offers the following services to the user:

- Easy access to the hardware. TACL contains a database for the computer and CAMAC configuration. The user only needs to know the signal names for an access.
- Logic transformations of signals. TACL computes signal transformations in the "logic array" which is user generated with a graphic editor. This signal processing is used for example to convert units (*i.e.*, bits into amperes), to combine signals (*i.e.*, magnet current and beam energy into deflection angle) and for feedback loops (cryogenic controls).
- User processes for more complicated operations. TACL provides methods to execute user written programs which may or may not be synchronized to the logic cycle of TACL. These programs are used for the calculation of optical (Twiss) parameters, orbit corrections, etc.
- User interface. TACL handles the operator interaction: input via mouse, knobs or keyboard and output via numbers, bar graphs, dials and icons. User written programs can be executed in a window for non-standard presentations. A graphic editor exists for the layout of screens.
- Networks. TACL provides a two layer network for the transport of signals between computers.

These capabilities make TACL the ideal environment for an accelerator simulation system which runs in parallel to the real accelerator. The concept of this system is described in this paper.

THE TACL SYSTEM

The CEBAF control system uses two levels of computers. The supervisor level employs graphic workstations used for operator interaction and running the high level operation procedures. The local level performs hardware operations and CAMAC access using rather inexpensive computer hardware.

The TACL software is identical for both types of computers. A set of system files determine the configuration of TACL for each computer. Therefore it is also possible to perform all software tasks on a single computer. Figure 1 shows a complete TACL system for one computer.

The heart of TACL is a shared memory segment which is used for all communication between the different processes. The data in this segment are inputs and outputs to and from the operator or operations procedures. They represent not only readings and settings of the accelerator, but also control signals for sequencing of procedures. RESOLVER is the set of function calls used in all TACL and user programs to access this shared memory.

Four TACL processes are active at run time:

- RUN (RUNtime) program is the main operator interface. It has various built-in basic features to display data from the shared memory, and allow for input of data from keyboard, mouse and knobs. For more complicated problems, user-written programs can be executed in a "JWindow" on the RUN screen which allows shared use of the input and output devices. The layout of the operation screens is defined by DISPLAY FILES produced with the DISPLAY EDITOR.
- LOGIC is the main operation program. The actions of LOGIC are controlled by the logic array. Each node of the logic array has a value and a function to generate this value. The large number of built-in functions include reading from and writing to CAMAC and the shared memory, and can range from simple mathematical functions to complex operations like PID-control loops. The "user process" function allows activating user written programs to handle nonstandard devices and complicated operation sequences. The logic array is generated manually with the LOGIC EDITOR, or from INGRES database information, and is stored in the SYSTEM FILES.
- LLAN (Local LAN) and SLAN (Supervisor LAN) handles the communication on the local and supervisor net, respectively.

*Supported by D.O.E. contract #DE-AC05-84ER40150

                    1443

The DISPLAY FILES and SYSTEM FILES can be created not only with the TACL editors, but also by special programs using information from the INGRES database.

TACL AS A ENVIRONMENT FOR CONTROLS AND SIMULATION

The unique capability of the TACL system to perform controls and simulations within one system lies in the architecture of the data flow. The LOGIC program acts as a data switchyard. Data can easily be redirected by changing functions of the logic array. The structure of the logic array remains unchanged for this purpose.

This feature allows an easy substitution of the hardware interface by a simulation program as shown in Figure 2. Figure 2a shows the setup of a supervisor computer in the operation of the real accelerator. The INGRES database contains a detailed description of the accelerator and is used for the assembly of the accelerator and the generation of the logic array. For the operation of the accelerator TACL writes the settings of the hardware to CAMAC and receives the readings on the same path.

Figure 2b shows the system in the simulation mode. The read and write CAMAC functions in the logic array are substituted by operator input and output functions. This automatically generates a new section in the shared memory. The simulation program is implemented as a user process and can access hardware settings via RESOLVER calls. Together, with the description of the machine (from INGRES), the simulation calculates the behavior of the beam and writes the simulated readings of the instrumentation back to the shared memory.

Except for a simple redefinition of the data flow in LOGIC, no change is necessary to convert the control system into a simulation system and vice versa. All operation procedures can be used in the same way. Besides providing theoretical support for the machine physicist, this system offers two major advantages:
- New operation procedures can be tested in a realistic environment.
- New operators can be trained without interrupting the operation of the machine.

SIMULATION OF THE CEBAF MACHINE

A simulation system as described above has been implemented on our simulation computer system. The simulation program is based on the optics code PETROS[2]. PETROS calculates the central orbit of the beam including second order nonlinear terms. A special routine is added to handle the linac cavities and the energy dependance of the beam motion exactly.

The simulated system describes the CEBAF accelerator from the 5 MeV point to the end of the fifth pass and includes 376 main dipoles, 1612 correction dipoles, 806 quadrupoles, 72 sextupoles, 806 beam position monitors and 1690 cavities. Multiple passes through the same element are counted as multiple elements.

The logic array for our system allows the operator to use deflection angles, focal length and energy gains as inputs. Since the beam energy in a linear accelerator depends not only on the settings of the cavities but also on the path length of beam trajectory (which depends on the beam energy), LOGIC has been set up which follows the beam line element by element, calculating beam energy and magnetic fields for each element.

The system is implemented on a single Hewlett-Packard 720 computer. The following computing times were observed:
- Setting all magnets 0.3 sec
- Calculation of new orbit 1.5 sec
- Reading beam position and display 0.1 sec

Using a faster computer like an HP 730 for example would reduce the simulation time to a total of one second which is close to the expected response of the control system with actual machine hardware.

FUTURE EXPANSION

For use in the control room, expansion of the simulation system is planned as shown in Figure 3. Parallel to the real machine control, a simulated system will be available on a separate computer. A modified version of the SLAN program will allow initialization of the settings on the simulation system. The operator can use this system to test changes of the machine settings before they are applied to the real machine.

Although the simulation will predict changes to the machine correct in the first order, the absolute results will be different because of unknown machine errors (tolerances in fabrication and alignment). These errors cannot be measured directly; however, one can develop an approximate description from the effect of the errors on measurable beam parameters as beam position, beam size, etc.

This information will be collected during the commissioning and operation of the machine, and stored in the measurement database. An analyzer program could then be used to extract the information about machine errors from a large number of measurements.

This procedure can again be tested by substituting the real machine with a simulation program: A set of machine errors is generated for the real machine simulator using a random generator. The operation of the simulator produces a measurement database. The analyzer is then used to reconstruct the machine errors, which can be compared with the original random numbers.

CONCLUSION

The data flow concept of TACL makes it for the first time possible to operate a simulation with the same operator interface and operation procedures that are used to control the real accelerator.

REFERENCES

[1] R. Bork: CEBAF's control system, CEBAF-PR89-013.
[2] J. Kewisch: Diplomarbeit Universität Hamburg (1978).

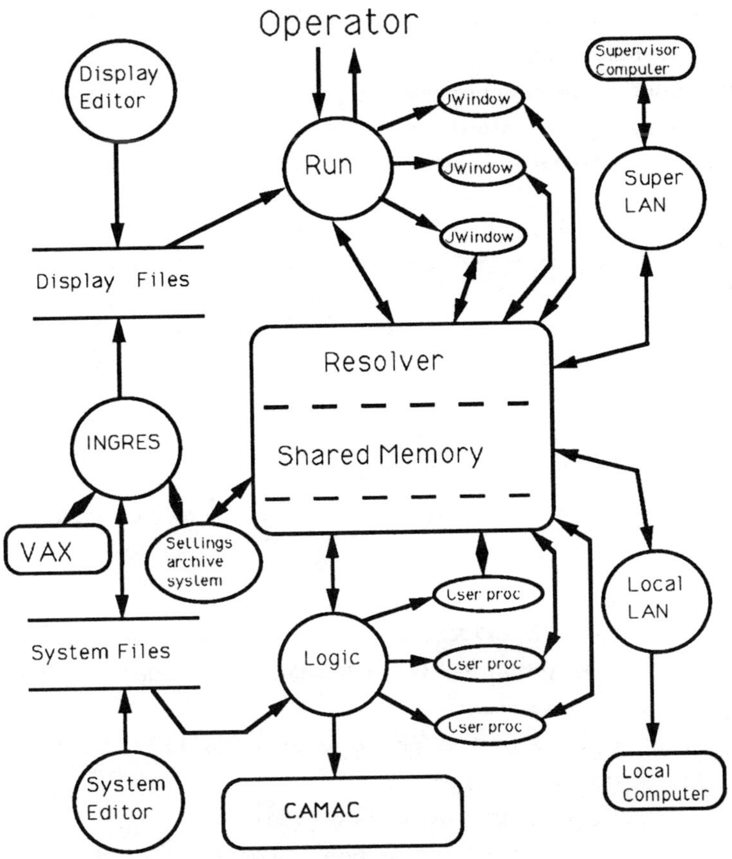

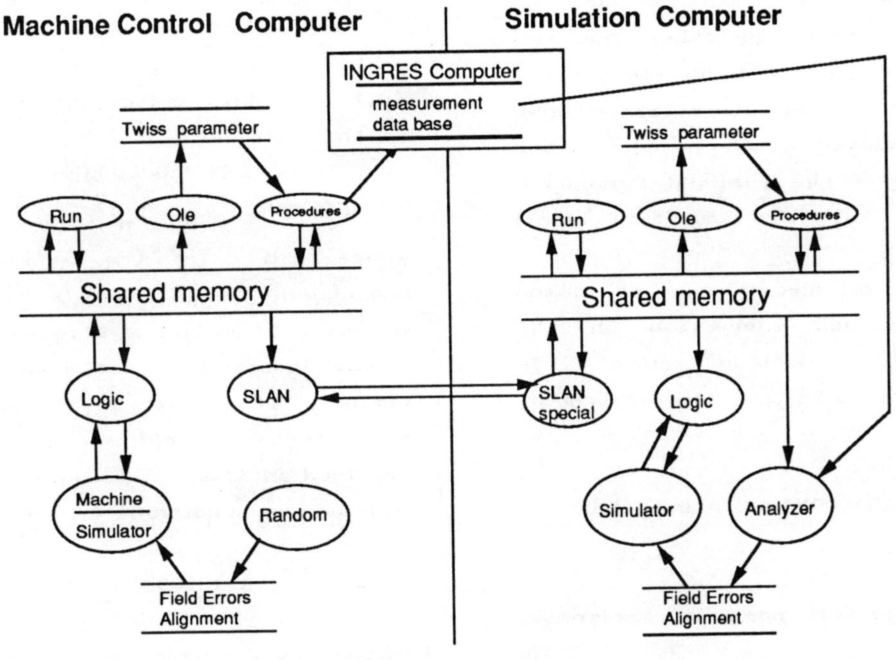

Fig. 1: TACL system overview

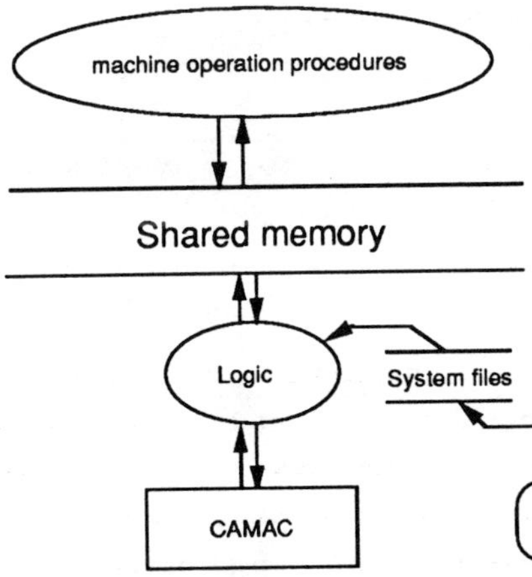

Fig 2a: TACL data flow for machine operation

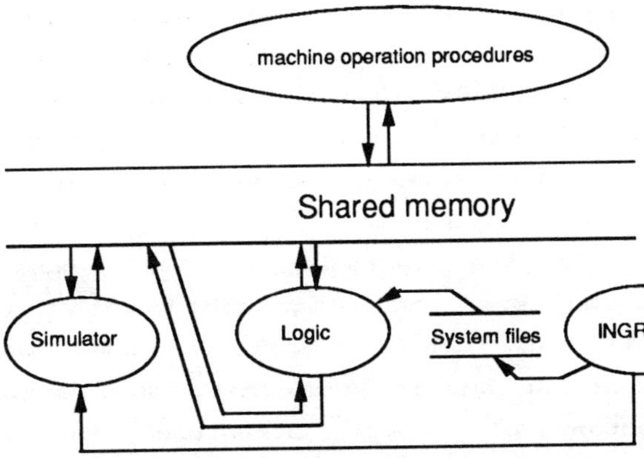

Fig 2b: TACL data flow for simulation

Fig. 3: Parallel operation and simulation with TACL

The Estimation and Control of Closed Orbit in Fast
Cycling Synchrotron

S.H.Ananian,R.H.Manukian,A.R.Matevosian

V.Ts.Nikogossian,A.R.Tumanian

Yerevan Physics Institute,Armenia,U.S.S.R.

The quality of the closed orbit regulation and optimization in synchrotron depends on the composition and the quality of the beam parameters measurements and the methods of the control action development based on this measurements.In the existing closed orbit correction systems the beam position monitors are used as an information source and the dipole corrector magnets are used as the control devices.The algorithms of the control actions are well known in this kind of systems /1/.

In the upgrading of the existing accelerators and equipping with the automatic control systems the most important and in some cases the decisive factor of their structure is the minimization of the extra equipments with maximum usage of existing control system.

In Yerevan synchrotron the beam observation system consists of the beam intensity monitors and the affairs of making precise BPM were failed because of bad noise immunity.From the other side the closed orbit correction 2 bump scheme /2/ doesn't give pure local correction,without distorting the closed orbit in the other part of the ring (the residual distortion is about 25% of the maximum local displacement /3/).

Here a set of algorithms of closed orbit estimation and correction on the magnetic field injection level is presented,which do not need the system of BPM.

1.THE ESTIMATION AND CONTROL ALGORITHMS.

The equations of particle horizontal motion are /3/

$$R_k(s)=R_{co}(s) + \phi(s)C^{k-1}\phi^{-1}(0)(R_o-R_o^{co}), \quad (1)$$

$$R_{co}(s)=\phi(s)[\int_o^s \phi^{-1}(t)BF(t)dt+$$

$$+(C^{-1}-E)^{-1}\int_o^L \bar\phi^1(t)BF(t)dt] \quad (2)$$

$$R_o =R_{co}(0)=\phi(0)(C^{-1}-E)^{-1}\int_o^L \phi^{-1}(t)BF(t)dt,(3)$$

here

$R(s)=[R_1(s),R_2(s)]$ is the phase vector of the particle (coordinate and angle) on the s position;

$R_k(s)$ is the phase vector on the K turn on the s position;

$R_{co}(s)$ is the closed orbit;

R_o is the initial phase vector;

R_o^{co} is the closed orbit on the beam injection position;

$\phi(t)$ is the fundamental matrix of the differential equation of the particle motion;

C is the rotation matrix;

$B=[0,1]^T$;

$F(s)$ is the distorting force,which acts on the particle.

If $F(s)$ is the action of all bumps
$$F(s)=U=[u_1,u_2,...,u_n]^T,$$

where U_i is the action of the i bump,then from (1)-(3) one has the following beam tuning strategy:

- the independence of beam transfer line tuning from linac to the synchrotron and main magnetic field of the synchrotron is provided in the case when the control U fullfils the condition

$$R_o^{co}(U)=0; \quad (4)$$

- the ideal input condition $(R_o=R_o^{co})$ is provided by the beam line tuning.

As it was mentioned any U must fulfil the condition (4).

Here algorithms for the closed orbit estimation and correction are described.

PROBLEM 1 The 'ideal' bump forming .

The algorithms of 'ideal' bump forming is the result of solution of the problem of finding such control vector U which minimizes the functional

$$I=\sum_{l=1}^{n}||R_{co}(l)||^2 \qquad (5)$$

with the constraints

$$R_{co}(j)=R^o, \qquad (6)$$

$$R_o^{co}=0, \qquad (7)$$

$$\sum_{i=1}^{m} U_i=0, \qquad (8)$$

The solution of this problem forms local distortion of the closed orbit R^o in the middle of the foccusing blocks after j period of the magnetic lattice with the closed orbit minimal distortions in the rest part of the ring and without distorting the input condition and with the special conditions on the horizontal bumps power supplies (the constraint (8) /2/).

The beam trajectories are shown on fig.1.Dashed line is the trajectory formed by the 'ideal' bump and the full line is the closed orbit formed by the usual bump (with the condition (8)).The trajectories show that the maximum closed orbit displacement in the rest part of the ring for the ring for the 'ideal' bump is less by the factor of 3.5 than the one for the usual bump.

PROBLEM 2 The algorithm for the beam diagnostic (estimation) of the closed orbit and the input condition on the injection on the magnetic field injection level.

Here the algorithm 1 is used for every j=1,2,...N positions and

$$R_-^{oj} = [R_{1-}^{oj} ,0]^T , \qquad R_+^{oj} = [R_{1+}^{oj} ,0]^T$$

are calcuiated for the cases when the beam is lost on the inner and outer side of the vacuum chamber on the j position

$$R_1^j=(R_{1+}^{oj} - R_{1-}^{oj})/2, \quad R_o-R_o^{co} =R_m-(R_{1+}^{oj} +R_{1-}^{oj})/2$$

give the horizontal closed orbit estimation on the j position and the beam centre oscillation amplitude with respect to the closed orbit.R_m is the horizontal width of the vacuum chamber.It is obvious that $R_o-R_o^{co}$ shows the beam optimal input condition is satisfied.

PROBLEM 3 Closed orbit tuning algorithm.

This algorithm is the solution and realization of control algorithm

$$U_{opt}(s)=\sum_{j=1}^{n} U_{jopt}$$

where U_{jopt} is the solution of PROBLEM 1 for $R_{co}(j) =- (R_1^j,0)^T$,and R_1^1 is the closed orbit displacement on orbit displacement on the N positions.

Algorithms 1-3 are solved with the least square method and on this final state one calculates only matrix multiplications.

REFERENCES

1.J.P.Koutchouk, "Trajectory and closed orbit correction", CERN LEP-TH/89-2.

2.O.A.Gusev and et al."The system of correction of Yerevan synchrotron magnetic field " First All-Union Conference of Particle Accelerator,1970 v.1, p.221.

3.A.G.Agababyan, S.G.Ananian, A.A.Kazarian et al. "The System of Yerevan Synchrotron Closed Orbit Regulation and Estimation",
Preprint YPI 1169(46)-89,Yerevan,1989.

4.Ch.Louson,R.Henson "Least Square Method", Nauka,Moscow,1986.

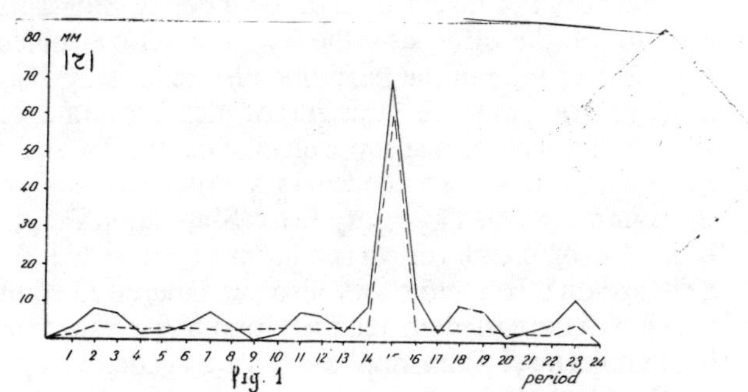

Fig. 1

MARCO - Models of Accelerators and Rings to Commission and Operate

L. Catani, G. Di Pirro, C. Milardi, A. Stecchi, L. Trasatti.
INFN, Laboratori Nazionali di Frascati
P. O. Box 13, 00044 Frascati (Rm) ITALY
M. J. Lee
Stanford Linear Accelerator Center
P.O. Box 4349, Stanford, CA 94309

Abstract

MARCO is a knowledge-based user interface for commissioning and operating modern accelerators and storage rings. Its purpose is to provide a model-referenced graphical interface system between the users and the machine. It allows access to modeling and simulation codes that are used in the design of the machine. It can be used to predict the effects of a change of parameters on the beam, or to compare the predicted with the measured effects. The design and prototype development of MARCO using Hypercard will be described in this paper.

I. DESIGN CODES

Two types of programs are used in machine design:

(1) Lattice Modeling - to define the location and strength of the beamline elements in order to obtain the desired lattice function values (e.g., the transport matrix and the Twiss parameter values). Examples of the Lattice calculation programs are COMFORT[1], LEDA[2], etc.

(2) Error Simulation - to find the location or the strength of the beam monitors and correctors to change the beam errors (e.g., the beam trajectory and shape). Examples of the Correction programs are RESOLVE[3], etc.

MARCO is an attempt to build a common graphical and interactive user interface to these programs, hiding the complexity of the many existing input-output data formats. The integration of such a tool in the control environment of an accelerator will greatly enhance the possibility of understanding and predicting machine behaviour for the operators, during both commissioning and normal operation.

Control Applications

MARCO will be an interface to both types of programs: Lattice Modeling and Error Simulation. Using these programs, it is possible to operate the machine more intelligently. For instance, Lattice Modeling Programs can be used to see the effects on the Twiss function values before making a change in the beamline elements; they also can be used to compute the strength of the beamline elements required for making a specific change on the Twiss functions. Both applications are needed in a Procedure to "Setup the Beamline". Also, Trajectory Error Simulation Programs can be used to predict the effects on beam trajectory before making a change on a corrector; they also can be used to compute the strength of the correctors required for making a specific change in the trajectory. Both applications are needed in a procedure to "Correct the Beam Error".

Commissioning Applications

The goal in commissioning is to find and correct the errors in the beamline. The Error Verification Procedure involves two steps: (1) Identify the good regions by finding the largest regions where the prediction from the Simulation Codes agrees with the measurement; (2) Identify the errors by searching for the most likely candidate in the bad region (a bad region usually lies between two adjacent good regions).

For example, it will be possible to use MARCO to find field errors due to misalignment or miscalibration of the beamline elements by analyzing the measured trajectory of a test beam. The measurement usually involves changing the beam trajectory by kicking the beam with some trajectory correctors and measuring the beam trajectory at the beam position monitors (BPMs). In practice, (1) Focusing errors are found by analyzing a multiple set of beam oscillation data (An oscillation is defined as the difference between two sets of beam trajectory data); (2) Bending errors are found by analyzing a set of trajectory data (not their differences). Both applications are needed in a procedure to "Verify the Beamline." This Procedure has been used successfully to find focusing errors and bending errors in SLC and to find the alignment errors in PEP and SPEAR at SLAC.

II. THE SYSTEM REQUIREMENTS

Recently, a user interface, GENI [4], has been developed for modeling and simulation programs such as COMFORT and RESOLVE. These modeling and simulation programs are normally used to control and commission any machine.

To use these programs, the user has only to read the beamline dataset which contains the Element Definitions and Beamline Definitions. The name, type, strength, etc. of every element are described in the Element Definition data, and the position along the beamline of the elements is contained in the Beamline Definition data. After reading in the beamline dataset, it is possible to use GENI to set up the beamline configuration and to analyze measured beam trajectories.

We would like MARCO to expand the range of GENI. It should offer a standard interface to all of the modeling and simulation programs that are commonly used for lattice design and error study. Since GENI was implemented on a MicroVAX Workstation, its availability will be limited. MARCO is required to operate on a Macintosh II in order to be available to many more users. In particular, the use of HyperCard [5] on a MAC can offer an simple way to maintain and upgrade MARCO.

The HyperCard Solution

We have designed an interface system for MARCO that can meet these requirements. Figure 1 shows a block diagram for the layout of the interface system for MARCO.

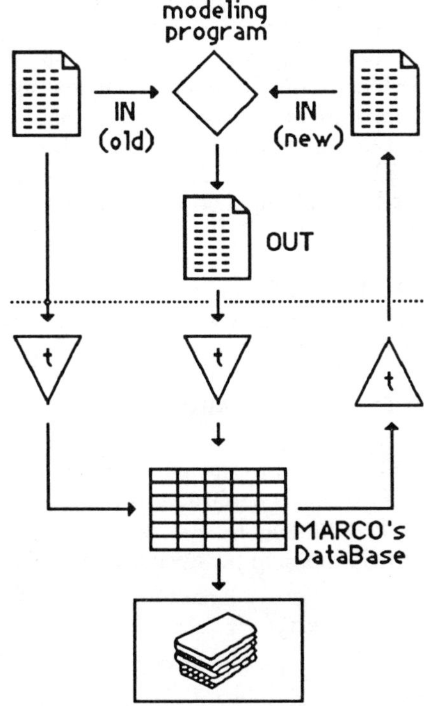

Fig. 1. A Block Diagram showing the layout of an Interfacing System to MARCO for Beamline Set-up. A dotted line separates the modeling program from the translators (t) and MARCO.

For Beamline Set-up, the Lattice Modeling code interfaces to the input beamline description data and to the output data package. Two copies of the input data are provided--an original copy containing the element strength values of the design solution, and a second copy containing the strength values of a new solution. The user can run the modeling code to find the strength of the elements (FITTING) that produce the desired values of the TWISS functions at some specific points (FIT POINT). The output data include a Table of the Twiss Functions at every element and the strength of the elements.

MARCO uses three translator codes: one for translating the Input Data from the data format of the modeling code, the second for translating the MARCO's Output Data to the format of the modeling code, and the last one to translate the Output of the modeling code to the MARCO interface standard. After the input-output data have been translated, the user can make changes or select display interactively using buttons, menus, graphs and windows.

At Frascati, most existing codes are working on a VAX

computer. It is not necessary to implement them to work on the MAC. It is possible using HyperCard to run MARCO on the Mac and to run the modeling code on the VAX simultaneously. Alternatively, for a code that has been implemented to run on the MAC, both MARCO and the modeling code can run on the MAC concurrently. The use of HyperCard offers us a simple way to connect new programs to MARCO with minimum effort.

III. FIRST IMPLEMENTATION

The first two programs to be connected to MARCO are the lattice design codes LEDA and COMFORT. LEDA was originally written to run on a VAX using Top Drawer as a graphical interface. All of the I/O functions provided by Top Drawer are replaced by Hypercard .

Using MARCO it is possible to run Leda in two modes:

- Importing the code to the Macintosh and recompiling it (using MPW 3.1[6] and the Language System FORTRAN[7]).

- Using an RS/232 interface to the VAX (the input and output files are transferred between Macintosh and VAX). In this mode, Leda runs directly on the VAX CPU. In the future the use of the Apple Communication Toolbox will allow faster and more flexible communications with external computers.

These two modes of operation reflect different needs which may arise for different programs: While running a program on the machine for which it was originally intended is usually the easiest way to proceed, it may sometimes be convenient to be able to work in a standalone environment by running it on the Macintosh.

As far as COMFORT is concerned, only the first method has been implemented, due to the high level of complication of the program.

To run LEDA or COMFORT three separate windows have been developed using HyperCard 2.

A) General parameter setting window (see Fig. 2).

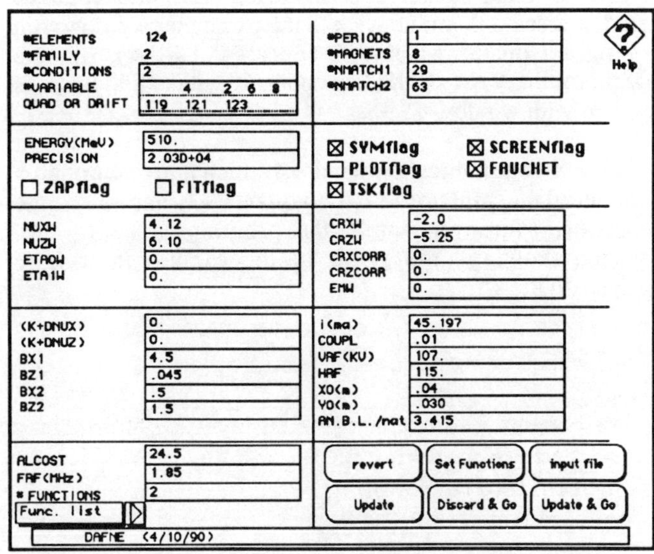

Fig. 2. General parameter setting window for LEDA

When this window is selected, a menu becomes available to load an input data file and to translate it to window B.

B) Element index and histogram window (see Fig. 3). The elements are selected by the upper scroll bar which permits rapid scrolling through the entire machine. Once an element is selected by clicking on the corresponding icon, it is possible to change its parameter values, or to assign/delete a parameter from the list of elements to be used in Fitting (for LEDA, Fitting is the process of calculating the values of the variable element parameters to obtain the desired Twiss Function values at some specific points along the beamline). These points are called the Fit Points. The output is displayed in graphic form. Multiple plots can be shown in the same window, including the output from a previous run of the program.

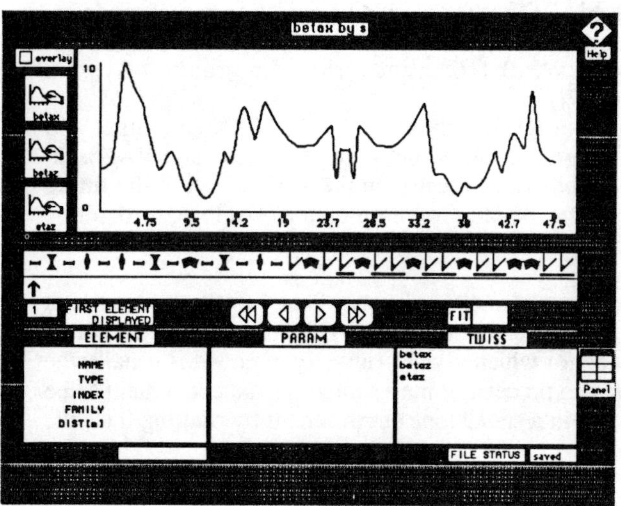

Fig. 3. Element index and histogram window

On window B, the user can click the buttons to: reset the machine to the original configuration; save the changes entered or discard them; go back to the parameter window; run the program on the Macintosh (only for LEDA); run the program on the VAX. If the last option is selected, the user is presented with window C.

C) VAX communication window, which allows through a simple terminal emulator to log unto the VAX and to change directories. Moreover, while the program is being run, messages showing the status of the execution may be displayed on the screen.

IV. SUMMARY

The experience with Hypercard 2 has proven extremely positive. We have demonstrated that a user interface for a model-based control system can be implemented in a highly modular way that can ease debugging and improve maintainability, as can be expected from the use of an object oriented language (HyperTalk).

We have found that development time using HyperCard can be reduced to 1/5 or 1/10 of the development time using UIS graphics (VAXworkstation). In particular, the number of lines of code has been reduced from 5000 (GENI) to less than 2000 (MARCO). We believe that the reduction in the number of lines will also lead to a reduction in the effort to maintain/upgrade MARCO.

The success obtained in building a common interface to programs as different as LEDA and COMFORT encourages us to continue to implement other modeling and simulation programs into MARCO. Eventually, MARCO will be the user interface for the model-based control system for the new PHI-factory DAΦNE under construction at the LNF.

V. ACKNOWLEDGEMENTS

We would like to thank the Accelerator Group of the LNF for continuing discussions and encouragement. We are particularly grateful to T. Vignola, C. Biscari, and Steve Kleban (SLAC) for their help in the implementation of LEDA with MARCO.

VI. REFERENCES

[1] M. D. Woodley, M. J. Lee, J. Jäger, A. S. King, Control of Machine Functions, or Transport Systems, SLAC-PUB-3086, March 1983.

[2] G. Vignola, private communication.

[3] M. Lee, private communication.

[4] S. Kleban, M. Lee and Y. Zambre, "GENI: A graphical Environment for Model-based control," N.I.M. A293 (1990)

[5] HyperCard v2.0v2, Apple Computer Inc, © 1987-90.

[6] Macintosh Programmer Workshop v3.1, Apple Computer Inc, © 1985-89.

[7] Language System FORTRAN, Language System Corporation, © 1988.

USE OF DIGITAL CONTROL THEORY STATE SPACE FORMALISM FOR FEEDBACK AT SLC*

T. Himel, L. Hendrickson, F. Rouse,[a] **and H. Shoaee**
Stanford Linear Accelerator Center, Stanford University, Stanford, CA 94309 USA

Abstract

The algorithms used in the database-driven SLC fast-feedback system are based on the state space formalism of digital control theory. These are implemented as a set of matrix equations which use a Kalman filter to estimate a vector of states from a vector of measurements, and then apply a gain matrix to determine the actuator settings from the state vector. The matrices used in the calculation are derived offline using Linear Quadratic Gaussian minimization. For a given noise spectrum, this procedure minimizes the rms of the states (e.g., the position or energy of the beam). The offline program also allows simulation of the loop's response to arbitrary inputs, and calculates its frequency response.

INTRODUCTION

The SLAC Linear Collider (SLC) is a novel accelerator designed to produce e^+e^- collisions at center-of-mass energies up to 100 GeV, i.e., around the mass of the neutral intermediate vector boson Z^0. The collisions occur between bunches of electrons and positrons, which are produced, accelerated, collided, and dumped at a maximum rate of 120 Hz. When the feedback project described here was started, the SLC had fast-feedback loops that stabilized both the energy of the machine [1] and the orbit, through a set of collimators near the end of the linear accelerator. These feedback loops were essential to the operation of the SLC.

We have designed a new system that replaces the current software with generic, database-driven software [2]. We rely on the SLC database to specify each different loop. This is possible because the action of any feedback loop can be cast into a series of matrix equations in the formalism of digital control theory [3].

This paper briefly introduces the state space formalism of modern control theory, and then describes how it has been applied to the SLC feedback system. Some of the tradeoffs involved in the design of a loop are discussed.

STATE SPACE FORMALISM

The state space formalism provides a simple, elegant method to describe the dynamics of a linear system in a single matrix equation. We now give that equation, and an example of its application to modeling the response of a section of the SLC. Define the following vectors:

- **y** is a vector of outputs. For our example elements of this vector are readings of beam positions from Beam Position Monitors (BPMs).

- **x** is a vector of states. A typical state is the position, the angle, or the energy of the beam at a point. Note that these are not necessarily directly measurable with a sensor. Other states are the angular kicks given to the beam by devices (typically magnets) in the beam. These may be devices directly

---
*Work supported by Department of Energy contract DE–AC03–76SF00515.
[a] Present address: University of California at Davis.

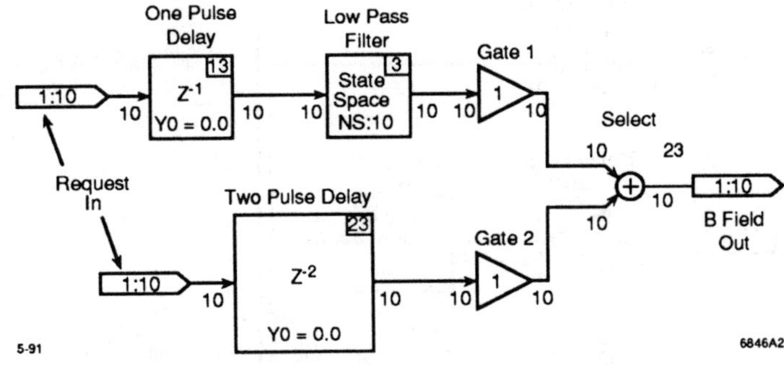

Figure 1. Diagram of actuator dynamics. Normally, either *gate 1* has a gain of 1 and *gate 2* a gain of 0 or vice versa. In the first case, the actuator is assumed to have an RC time constant due to filtering in the power supply. In the second case, the response is modeled as a delay of two beam pulses. This accounts for computation time of the feedback loop and other minor delays.

controlled by the feedback loop, or devices upstream of the loop whose perturbations must be corrected by the loop. The state vector should completely describe the system.

- **u** is a vector of inputs. A typical input is the current requested for a magnet.

These vectors are then related with a general first-order matrix equation using the four matrices **A**, **B**, **C**, and **D**:

$$\dot{\mathbf{x}}(t) = \mathbf{A}\mathbf{x}(t) + \mathbf{B}\mathbf{u}(t)$$

$$\mathbf{y}(t) = \mathbf{C}\mathbf{x}(t) + \mathbf{D}\mathbf{u}(t) .$$

These two equations can be combined into a single one by defining the system matrix, **S**:

$$\mathbf{S} = \begin{pmatrix} \mathbf{A} & \mathbf{B} \\ \mathbf{C} & \mathbf{D} \end{pmatrix} .$$

This gives the continuous state space equation:

$$\begin{pmatrix} \dot{\mathbf{x}}(t) \\ \mathbf{y}(t) \end{pmatrix} = \mathbf{S} \begin{pmatrix} \mathbf{x}(t) \\ \mathbf{u}(t) \end{pmatrix} .$$

In the above equation, **x**, **y**, and **u** are all considered to be continuous functions of time. The equation below is used for the discrete (sampled) case. The subscript k indicates the kth sample:

$$\begin{pmatrix} \dot{\mathbf{x}}_{k+1} \\ \mathbf{y}_k \end{pmatrix} = \mathbf{S} \begin{pmatrix} \mathbf{x}_k \\ \mathbf{u}_k \end{pmatrix} .$$

There are multiple advantages to using this formalism. It is a standard formalism, so there are books and computer tools that can be used to aid in the development. Secondly, the software tends to be general and streamlined.

APPLICATION TO THE SLC

Modern control theory was developed to handle dynamical systems. A satellite attitude control system, for example, involves an object with a finite moment of inertia, and its state (spin velocity) at a certain time is related by a differential equation to its state at a

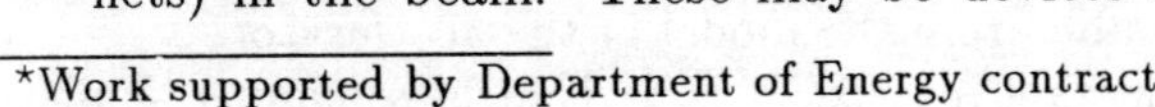

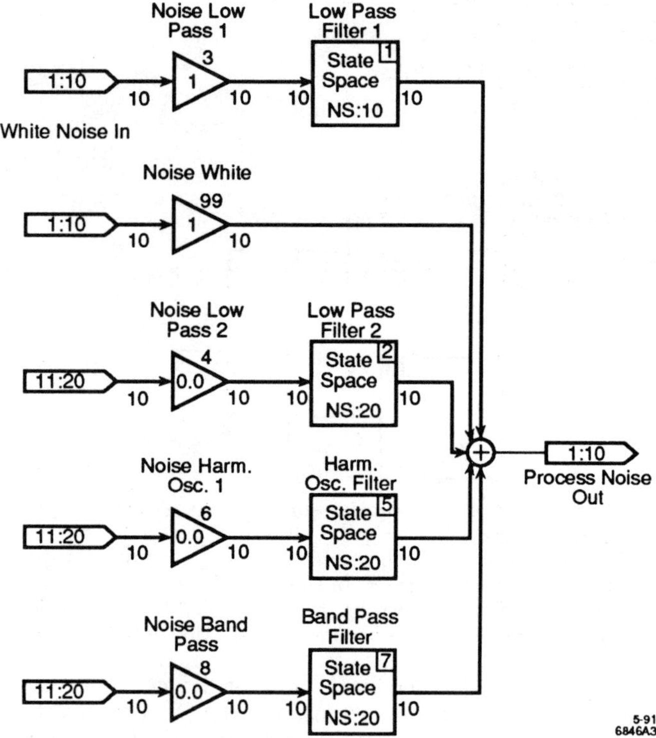

Figure 2. Diagram of beam noise dynamics. Normally only two or three of the gain blocks on the left have nonzero gains. The time constants of the low-pass filters, and the frequency and quality factor of the harmonic oscillator filter, can be adjusted to get the modeled noise spectrum to match the measured one.

previous time. At first glance, the SLC accelerator has no dynamics, no *inertia*. There is a bit of dynamics in the actuators, typically correction magnets, that are used by feedback to steer the beam. The model used for this is shown in Fig. 1. However, the typical magnet is fast on the scale of the 1/120 second period between beam pulses, so this dynamics is not essential to the problem.

The important part of modelling the dynamics of the system is the description of what is causing the beam to move. For example, there may be an oscillating or slowly varying upstream magnet power supply, causing the beam to move. The dynamics of the ensemble of all such power supplies must be modeled. In practice, we measure the spectrum of disturbances to the SLC beam, and then try to match that by adjusting model parameters such as low-pass filter time constants and oscillation frequencies. Figure 2 shows the model used for this dynamics of the noise source.

FEEDBACK ALGORITHM

Having described the model of the accelerator in the state space formalism, we can now go on to design the feedback algorithm. For this, the predictor-corrector formalism of optimal control theory [3] is used. The first controller equation is used to estimate the present value of the state vector:

$$\hat{\mathbf{x}}_{k+1} = \mathbf{\Phi}\hat{\mathbf{x}}_k + \mathbf{\Gamma}\mathbf{u} + \mathbf{L}(\mathbf{y} - \mathbf{H}\hat{\mathbf{x}}_k) \,,$$

where

$\hat{\mathbf{x}}_k$ is the estimate of the state vector on the kth pulse.

$\mathbf{\Phi}$ is the system matrix, and describes the dynamics of the accelerator model.

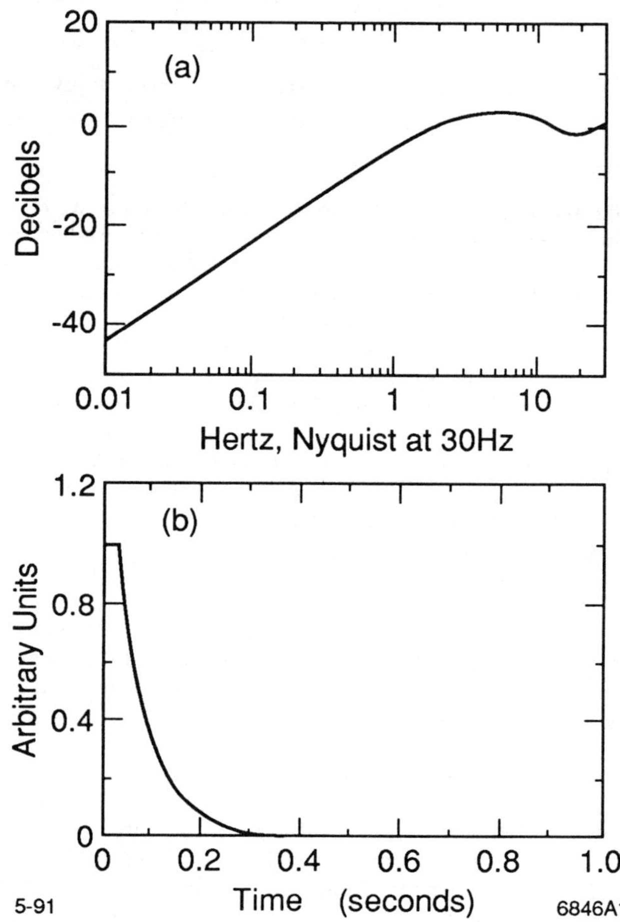

Figure 3. Simulation results: (a) shows frequency response for a typical loop. (b) shows the response to a step change in the incoming states. Typical states recover in 0.1 sec (for a 60 Hz sample rate).

$\mathbf{\Gamma}$ is the control input matrix. It describes how changes in the actuators should affect the state.

$\mathbf{H}$ is the output matrix. It maps the state vector to the output vector. That is, given an estimate of the states, it gives an estimate of what the sensors should read.

$\mathbf{L}$ is the Kalman filter matrix. Given an error on the estimate of the sensor readings, it applies a correction term to the estimate of the state vector.

The matrices $\mathbf{\Phi}$, $\mathbf{\Gamma}$, and $\mathbf{H}$ are obtained from the model of the accelerator. The $\mathbf{L}$ matrix is derived from the other matrices, and is designed (via the Linear Quadratic Gaussian method) to minimize the rms error on the estimate of the state.

The second controller equation calculates the actuator settings from the estimate of the state vector.

$$\mathbf{u}_{k+1} = \mathbf{K}\hat{\mathbf{x}}_{k+1} + \mathbf{Nr}$$

where

$\mathbf{r}$ is the reference vector that contains set points for the loop.

$\mathbf{N}$ is the controller-reference input matrix. It maps the reference vector to actuator settings and is directly derivable from the model of the accelerator.

$\mathbf{K}$ is the gain matrix. It is derived in a manner similar to $\mathbf{L}$. It is designed to minimize the rms of selected state vector elements.

DESIGN TRADEOFFS

There are several design goals for a feedback loop. It should minimize the rms of the states it is trying to control, have a good response to a step change in the incoming beam, have a good frequency response, have a good DC bias rejection, and continue functioning well even if the accelerator is slightly different than the model used by feedback. These design goals are conflicting, so tradeoffs must be made.

The tuning of feedback is done using an offline simulation program. This program is built on top of the Matrix_x control system CAD package written by Integrated Systems Incorporated. All the figures in this paper are outputs from this simulation program. The program takes parameters (such as filter cutoff frequencies) from the on-line database, calculates all the matrices needed by the feedback loop, saves them for use by the loop, and then calculates and plots the frequency response, the response to a step function, and so on. The user can vary the parameters and rerun the simulation until the desired response functions are obtained. Two examples of these plots for a typical loop setup are shown in Fig. 3.

CONCLUSIONS

At present there are seven feedback loops running at the SLC that are designed and implemented using the state space formalism of digital control theory described above. These loops control a total of 52 beam parameters. They work quite well, and there is demand for more. We expect to have about 12 feedback loops operating by the end of the year.

As more loops have been designed and implemented, the power and generality of a database-driven feedback system based on the vectors and matrices of modern control theory have become clear. Implementing new loops requires little or no new on-line software.

ACKNOWLEDGMENTS

We thank John Zicker for originating the concept of applying modern control theory to our feedback problem, and we thank Robert McEwen of Integrated Systems Incorporated for his help in applying that theory to our problem.

REFERENCES

[1] G.S. Abrams et al., "Fast Energy and Energy Spectrum Feedback in the SLC Linac," Proc. 1987 IEEE Particle Accelerator Conf., Washington, D.C, 1987.

[2] F.R. Rouse et al., "General, Database-Driven Fast-Feedback System for the Stanford Linear Collider," this conference.

[3] Gene F. Franklin and J. David Powell, Digital Control of Dynamic Systems, (Addison–Wesley, 1980).

A DATABASE FOR MODELING THE BROOKHAVEN AGS BOOSTER[*]

E. H. Auerbach

AGS Department, Brookhaven National Laboratory
Upton, NY 11973

Abstract

A Database has been developed for Brookhaven's AGS Booster and its ancillary lines which contain both design and measured physical data for the accelerator lattice and its magnets and other components. This database may then be connected to programs for modeling, permitting the programs to use any degree of desired detail from the available data with a view toward increasingly model-based control of the accelerator. These methods are expandable to include the other machines in the AGS complex.

INTRODUCTION

We have used the InterBase (TM) database soft-ware in use on the APOLLO network at the Brookhaven AGS complex to provide a database for computer modeling of the AGS Booster. The software used was that chosen for the Booster control system (for reasons of compatibility) and does not represent an individual evaluation for this purpose.

The AGS Booster project consists of a main ring (one-fourth of the size of the AGS) together with three transfer-lines: Linac-to-Booster (for proton injection), Tandem Transfer Line-to-Booster (for heavy ion injection), and Booster-to-AGS (for injection to the AGS). The needs for modeling and control of these systems provided the primary motivations for the specifications of the database.

COMPONENTS DESCRIBED

The principal components of the accelerator necessary to modeling are its magnetic elements, the rf systems and the instrumentation. The database relations describing the various properties of these components are structured to separate properties according to how generic or specific they are; thus, in the case of a magnet, some properties are maintained under the design-type (generic) while misalignments are treated as a geometric property (location-specific) and measured field errors are treated as specific to individual magnet.

Relations defined are: one database relation defines the geometry of the ring and lines--but only those properties associated with the "lattice locations" are included here; particular components in this relation are referred to by their design-type and serial-number where more specific properties are stored. Another relation contains the properties apposite to the design-type--e.g., a particular type of quad, position-monitor, etc. A set of relations, one for magnets and one for each type of instrument, contain the properties which describe a particular physical device; thus, instrument calibrations, magnetic field errors, etc. are stored here. Magnet strings--magnet-groupings operated under a single power supply and not controllable separately--are described in another relations. Finally, we have added a "model" relation to contain descriptions particular to programs such as MAD or SYNCH use to describe lattice elements.

The advantage of such a structure is that the modeler can use data available to whatever depth of detail deemed appropriate to a particular calculation--from the generic description of the design type, through inclusion of measured misalignments and detailed field measurements where the results of such a program are available. In addition, this has provided a framework for storing the results of these measurement programs in an easily accessible form.

MODELING USE

Routines developed by the Modeling and Algorithms Group utilize the data stored in these relations to produce computer models of the machine and its lines and to provide programs for the interaction between measurement devices and control of correction magnets, etc.; other routines can produce an input set to programs such as MAD or SYNCH from the lattice relations and the relations describing the various elements. These form the beginnings of a system of model-based controls.

An additional database relation containing physics constants and general machine parameters was added to ensure consistent and compatible calculations among the various modeling and control routines.

EXTENSIONS AND FUTURE EXPANSION

As this system has provided for recording and detailed description of the AGS Booster, one can extend this mechanism to the original AGS to the extent that detailed data are available (and where detailed measurements are missing by using the more generic relations for those components). The future extension of the AGS complex toward its use as an injector for the Relativistic Heavy-Ion Collider (RHIC) can easily be included in this framework with the necessary extensions.

[*] Work performed under the auspices of the U.S. Dept. of Energy.

"

A 256 Channel Digital Filter for a Data Acquisition System

W. Roberts and B. Aikens

TRIUMF, 4004 Wesbrook Mall, Vancouver, B.C., Canada, V6T 2A3

Abstract

The TRIUMF Central Control System (CCS) employs several data acquisition systems to monitor its operational parameters. Each system multiplexes 256 analog channels into one analog to digital converter. Space constraints on the multiplexer cards prohibit the installation of adequate anti-alias filters which allows 60 Hz and other noise to corrupt the measurements. The new system overcomes this problem by sampling each channel at a 160 samples/second rate and using a DSP microcomputer to lowpass filter the data. The multiplexer and analog-digital converter operate at 256 times the channel sample rate. The channel filter bandwidth is restricted to approximately 1 Hz due to the rate at which the CCS reads the data. One DSP microcomputer is able to filter the 256 channels in a multiplexed system at a cost less than that of the anti-alias filters which would otherwise have been required.

I. INTRODUCTION

The TRIUMF Cyclotron requires many channels of low speed data acquisition for control and supervisory purposes. The CCS acquires these signals though systems of multiplexed analog-digital converters (ADC). The multiplexer preceding the ADC has 2 stages of 16 to 1 multiplexing for a total of 256 differential channels. In this system the desired signals are essentially dc but are accompanied by noise with a much greater bandwidth. Since this is a sampled data system the possibility of the noise being aliased into the signal bandwidth must be considered.

Sampled data systems acquire data at discrete time intervals of $1/f_s$ where f_s is the sampling frequency. Sampling is equivalent to modulating an impulse function by the input signal. The input signal is replicated around the multiples of the impulse frequency [1]. All signal and noise components at the input to the sampler are translated into the frequency band 0 to $f_s/2$ (the Nyquist frequency) unless filtered out before they are sampled. If the input signal bandwidth is greater than $f_s/2$ the signal components greater than $f_s/2$ are added to those between 0 and $f_s/2$. Figure 1 shows the effect of sampling the signal $1 + 0.1 \sin(2\pi 60 t_s)$ at a sampling frequency of 3 samples per second (SPS) with a uniformly distributed jitter of 1% of the sample frequency. The result is a noisy looking signal with an mean value of 1. The noise which was originally a 60 Hz sinewave is now a random signal which may be difficult to filter out to the

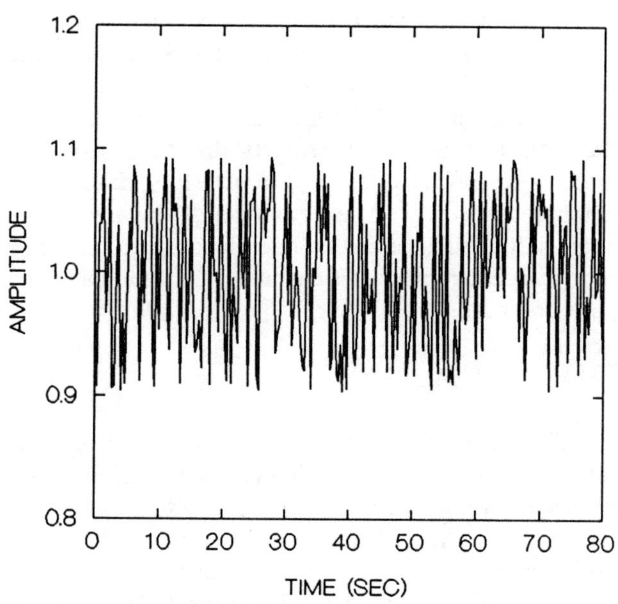

Figure 1. Sampling 60 Hz at 3 +/- .03 SPS

required degree. The worst case occurs when the sampling frequency is an exact submultiple of the signal frequency. In this situation each sample is at the same point on the waveform and thus leads to a dc offset at the sampler or ADC output.

This demonstrates the need to prevent aliasing. Noise which may be easily separated from the signal before sampling may be difficult or impossible to filter out after sampling. In the present CCS multiplexer system the input filtering on each channel consists of a single pole RC lowpass filter with a cutoff frequency of 3.6 Hz. Although the multiplexer system is differential up to the ADC it is inevitable that 60 Hz and other noise will enter the system at different points. The upgrade to the present system replaces the 12 bit ADC with a 14 bit device for increased resolution of low level signals. A 14 bit ADC has a dynamic range of ± 78 dB, therefore, even a small 60 Hz component at the input can mask a large part of the ADC dynamic range. The RC input filters attenuate 60 Hz components by a factor of 0.06 before sampling and digitizing occur but this is not good enough in a high resolution system sampled at 3 SPS.

An anti-aliasing filter is required to prevent this translation of the noise frequencies into the measurement frequency range. It is not possible to construct adequate filters for each channel on the multiplexer cards due to space and cost constraints. The alternative described here implements the anti-aliasing filter after the ADC conversion. In this configuration the ADC is required to sample at a rate much higher than the final sample rate of 3 SPS and then filter the sampled signal to a bandwidth of about 1 Hz. This meets the anti-aliasing requirements for the CCS sampling rate. The front-end RC filter protects the ADC from aliasing since its cutoff frequency is much less than the ADC sampling frequency. The advantage of this method over a purely analog method is that one digital filter can be shared among the 256 multiplexer channels at a significant saving in space and cost.

II. DIGITAL FILTER

The digital filter is implemented in a single Analog Devices ADSP-2105 DSP microcomputer [2]. This device has sufficient internal program memory to contain the filter algorithms. External data memory contains the ADC data and intermediate results. An external dual-port memory is used to transfer the final results to the CCS. The filtering is done in 2 stages in order to reduce data memory requirements and the computational load. The first stage sample frequency is much higher than the final system bandwidth. This is to allow the input RC filter to provide adequate anti-alias filtering for the ADC. This high sample frequency, however, requires many coefficients to reduce the bandwidth to the required value. A low cutoff frequency is more easily attained by sampling at a lower frequency but this cannot be done by simply throwing away unwanted samples since is the same as sampling at a lower rate and aliasing will occur. This 2 stage approach uses finite impulse response (FIR) filters of 22 and 26 coefficients whereas the equivalent single stage FIR filter would use more than 128 coefficients.

The function of the first stage filter is to provide anti-alias filtering for the second stage which can then sample the output of the first stage at a lower sampling rate. This lowpass filtering and resampling at a lower rate is referred to as decimation [3]. A reduction in the decimation stage computational load is attained because the output of the stage only needs to be computed at the second stage sampling rate not at the first stage sampling rate. The function of the second stage is to limit the system bandwidth to a value consistent with the CCS sampling rate.

The first stage filter is equivalent to 3 cascaded n point moving average filters where n = 8. The desirable characteristic of this filter are the notches at intervals of f_s/n in the

frequency domain. In this case the sampling rate is 160 SPS giving notches at multiples of 20 Hz. Figure 2 shows the frequency responses of the 2 filter stages. One function of the notches is to eliminate 60 Hz and its harmonics on the assumption that these are main discrete noise sources. 120 Hz and 180 Hz are greater than the Nyquist frequency but they are aliased to 40 Hz and 20 Hz respectively where notches exist. The other function of the multiple notch response is to provide anti-alias filtering for the second stage. This is shown in Figure 2 where the first stage notches coincide with the second stage passbands and eliminate any noise that would be translated into the second stage passband. The only unprotected regions are around multiples of the first stage sampling frequency which coincide with passbands of the second stage but these are narrow regions and are attenuated at higher frequencies by the input RC filter.

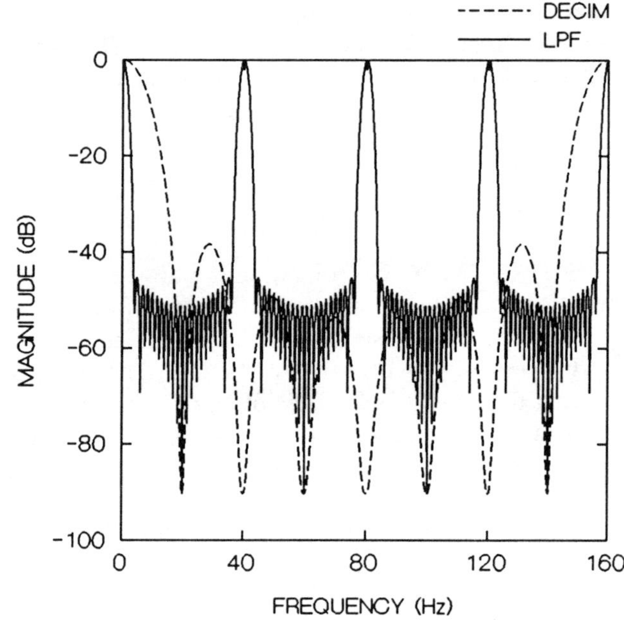

Figure 2. DECIMation, LowPass Filter responses

The second stage is a lowpass filter with a cutoff frequency of 1 Hz. The sampling rate is 40 SPS or 1/4 of the first stage sampling rate. The primary reason for reducing the sample rate is that a 1 Hz cutoff can be achieved more efficiently, in terms of the number of filter coefficients, than would be possible with a 160 SPS sampling rate. A secondary reason is the gain in computing efficiency since only every fourth output from the first stage needs to be calculated. The remaining three output values are not used although all the input data samples are needed to calculate the output values. The lower limit on the sample rate of this stage is the requirement to update the output data as fast as the CCS samples it. A conflicting constraint is to reduce the time delay of the filter, which is product of half the filter length and the sample interval, to a value which will not be noticed by the machine operators. This requires a higher sample rate for the same length filter. For the filter length and sample rate chosen in this case the delay through the filter is equivalent to 2 CCS sample periods.

A characteristic of digital filters is that the frequency response repeats around multiples of the sampling frequency as is seen in Figure 2 where the second stage response is shown out to 4 times the second stage sampling frequency. This shows the need for the preceding anti-aliasing stage.

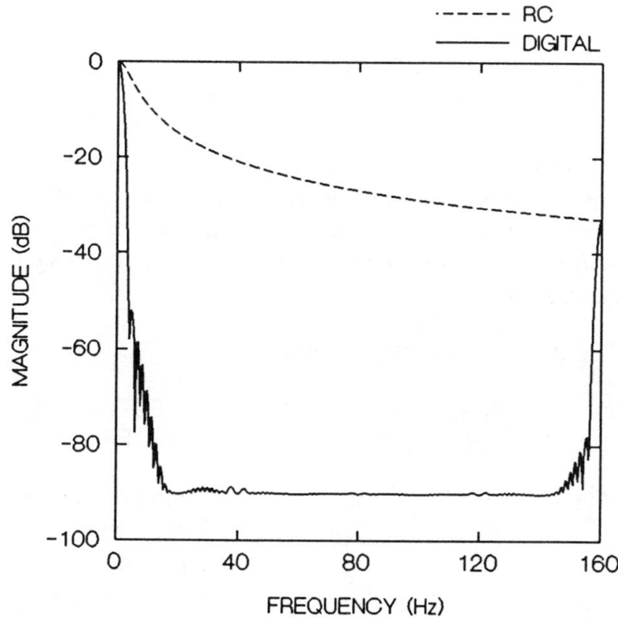

Figure 3. Digital, RC filter responses

The system bandwidth has now been reduced by the second stage to 1 Hz which meets the anti-aliasing requirements for the CCS data acquisition system. Figure 3 shows the combined response for the RC and digital filters compared with the RC filter response alone. There is an improvement of 19.5 dB in noise rejection over the 0 to 160 Hz range. The anti-aliasing effect of the RC filter is seen in the attenuation of the passband at 160 Hz where the envelope of the digital filter response follows the RC filter response. The stopband floor of -90 dB is due to noise added to the plot to simulate the limits of the 16 bit arithmetic results of the DSP processor. Intermediate results in each filter stage are calculated to 32 bits before rounding to the 16 bit result. Higher precision arithmetic would lower the noise floor but to no advantage since the CCS is a 16 bit system.

III. DESIGN AND TEST

The filters were designed and evaluated using the Signal Processing Toolbox for PC-MATLAB [4]. This allowed comparisons of different filter orders and architectures. A prototype system consisting of ADSP-2101 and Crystal Semiconductor CS5016 ADC evaluation boards verified the frequency response and multichannel operation of the filter system at full speed. Less than 25% of the available DSP processor cycles are needed for this system.

IV. SUMMARY

Sampled data systems are subject to a form of signal corruption termed aliasing. The bandwidths of the system and of the anti-alias filter are related to the use of the resulting signal. An operator looking at the data at a rate of once every few minutes is also sampling it. In this situation the bandwidth should theoretically be restricted to a fraction of a Hertz. The same operator watching the data while adjusting a machine parameter wants a wider bandwidth to follow the results of the adjustments more closely. The choice of sampling rate and bandwidth must be based on the bandwidths of the signal and noise, the required time response of the data acquisition system and the uses of the information. In this case the CCS sampling rate and dc nature of the signals determined the system bandwidth. Space constraints on the anti-aliasing filters in the multichannel system led to the design of a digital anti-alias filter based on initial oversampling and subsequent decimation and lowpass filtering.

V. REFERENCES

[1] E. Oran Brigham, *The Fast Fourier Transform*, Englewood Cliffs: Prentice-Hall, 1974.

[2] *Analog Devices, ADSP 2105 DSP Microcomputer Data Sheet*, Norwood: Analog Devices, March 1990.

[3] DSP Applications Staff of Analog Devices, *Digital Signal Processing Applications*, Engelwood Cliffs: Prentice-Hall, 1990.

[4] John Little and Loren Shure, *Signal Processing Toolbox User's Guide*, South Natick: The MathWorks Inc., 1988.

Proposed Data Acquisition System for the Fermilab Booster

V. Bharadwaj, S. Peggs, G. Wu
Fermi National Accelerator Laboratory*
P.O. Box 500 , Batavia, Ill 60510
C. Saltmarsh
Superconducting Supercollider Laboratory*
2250 Beckleymeade Rd., Dallas, Tx 75237

ABSTRACT

At present, studies involving the FNAL Booster (or in fact most accelerators) depend on knowing exactly what detector one has to look at and at what time. Because of this, most studies are done "on-line" and involve looking for repetitive effects using a limited number of detectors. In this paper we propose to design a Booster Data Acquisition System (BDAQ) for the FNAL Booster. In essence this system consists of a large number of digitizers with circular memory buffers. After a machine cycle of interest, these buffers are frozen and then read out into a mass storage device. This paper discusses the hardware and software capabilities needed to make such a data data acquisition system a powerful tool for doing accelerator physics studies and improving machine performance.

I. INTRODUCTION

In many ways the Fermilab Booster is an interesting machine to consider when looking at aspects of accelerator data acquisition systems. Because of its rapid cycling nature, it is very difficult to study using traditional methods. The machine is also relatively small and compact and hence a good candidate for demonstrating the feasibility of accelerator data acquisition systems. Table 1 list some of the relevant parameters of the Fermilab Booster [1].

Table 1. Fermilab Booster Parameters

H^-	multiturn injection (< 10 turns)	
E_{inj}	200	MeV kinetic
E_{ext}	8.0	GeV kinetic
circumference	474.2 meters	
harmonic #	84	
Q_x , Q_y	6.8	
gammat	5.4	
f_{RF}	30.3 - 52.8 MHz	
V_{RF}	950 kV , 17 RF cavities	
transverse aperture	20 pi mm-mrad (normalized)	
longitudinal dp/p	~ +/- 0.6 %	
cycle rate	15 Hz	
max accelerated intensity	$3.0 * 10^{12}$	

Accelerator control and data acquisition are two separate problems. Because they are designed for completely different functions we do not want the design parameters of one system to adversely affect the other. There obviously has to be a data link between the control system and the data acquisition system, but this can be relatively slow and simple.

II. DIGITIZERS AND RATES

The list of digitizers and rates needed for the Booster is shown in Table 2. f_0 is the revolution frequency (which varies from 360 kHz at injection to 628 kHz at extraction). The memory depth per channel that is needed to record one Booster acceleration cycle is also given. In addition to these channels we will also need to integrate test equipment (eg. spectrum analyzers) into the system.

TABLE 2:   Booster Digitizers

RATE	MEMORY DEPTH	#CHANNELS	ACCELERATOR DEVICES
30 kHz	1K	250	magnet currents
f_0	16K	400	BPMs, BLMs RF fanbacks, profile monitors beam current
84*f_0	1400K	10	beam dampers RF parameters
> 4 GHz	120,000K	2	bunch structure

At the higher digitization rates we would not normally store information for the entire acceleration cycle. One would think of only taking partial information at 4 GHz and at the RF frequency. But even so we will easily generate a total of 20 Mbytes of data per machine cycle. Only a few digitizing rates are used in order keep the hardware configuration simple. On an event of interest a trigger pulse will freeze all the circular buffers and the data acquisition will read out all the data. The above table gives the buffer memory needed for one acceleration cycle. Twelve Booster cycles are needed to fill the Main Ring at Fermilab. Although we believe that Booster acceleration cycles behave independently, enough memory depth for 2 Booster acceleration cycles should be considered in order to study cycle-to-cycle correlations.

* operated by the Universities Research Association under contract with the U.S. Department of Energy

Options exist for all the above digitizers. Commercial companies such as Lecroy and Analtek offer digitizing systems with appropriate data rates and memory depths. A design also exists for a VME based "Quickdigitizer" at Fermilab that has 4 channels, 1 MHz per channel digitizing rate and a memory buffer of 64K samples per channel. We propose to use this "Quickdigitizer" for all frequencies up to the revolution frequency, and the Lecroy 6841 for the RF frequency. The >4 GHz digitizers are commercial oscilloscopes and will have allowable memory depth limitations.

The exact hardware used will of course depend on detailed analysis of cost, performance and ease of software support.

III. COMPUTER HARDWARE

The digitizers for the acquisition system will sit at approximately 6 to 8 locations around the Booster. At each location we assume that there will one or more VME crates. The digitizer memories are mapped into VME memory space. Each crate has a controlling microprocessor card (such as a card based SPARC workstation) and all of the crates are connected by a vertical bus to a central processing node. This node can be a real computer or a crate full of VME based processors. How much local (crate-level) processing of information should there be before it is sent to the central node? If the vertical bus has enough bandwidth and the central node has sufficient resources to process incoming data, then the simplicity implicit in the local processor software outweighs any advantages one might reap from having local processing power. The needed bandwidth depends on how fast one need to read all the data for one acceleration cycle. This is obviously a rather flexible number. The fastest desired response is to collect and process the data in 33 milliseconds, ie. one Booster machine cycle. A less capable system that takes a few seconds will still have a "real-time" feel. Given present computer and network hardware trends, the latter is not an unreasonable design goal.

Initially we are considering using VME based SPARC workstations connected by ETHERNET to a central SPARC workstation for a prototype system.

IV. ACCELERATOR PHYSICS NEEDS

The list of foreseeable accelerator physics investigations that would be made possible (or made much more efficient) by a fully fledged BDAQ is long. Although the list is presented here as a piecemeal set of objectives, ordered by increasing data acquisition requirements and complexity, it can be conceptually broken into "phase I" and "phase II" tasks. The former are tasks that are either easy to perform, or are of high enough priority that they deserve concentrated attention as soon as possible. They may be performed with a relatively modest subset of of the final system. "Phase II" tasks are those that, although of less immediacy, are nonethless important if the full performance of the Booster is to be achieved. These aspects require the complete implementation of BDAQ hardware and software.

Although it cannot be listed, a third phase of physics investigations exists. After the system has been commissioned, and initial issues resolved, evolving topics will be recognized as important and interesting. This is in the nature of the physics program that a successful Booster Data Acquisition System will make possible.

Turn-By-Turn data from a few Beam Position Monitors, some status information

A minimal arrangement in which Physics can be performed requires
a) a few channels of Turn-By-Turn (TBT) digitization of Beam Position Monitors (BPM's),
b) a few 30 kHz channels for status logging of various parameters
c) the 25 kHz "pinger" now under construction.

This would enable the measurement of tune and chromaticity though the ramp, in essentially the same way that it is being done using the UDAS [2] setup in the Main Ring. Some investment in additional analysis software is necessary, and the system software is already in place.

The same configuration would allow for the measurement of smear, and the observation of resonances, much as these measurements have been successfully performed in the Tevatron. With the use of a somewhat larger number of TBT signals from BPM's, so that a reasonable average can be calculated, it will also be possible to track the RF offset versus time.

Complete Acquisition of BPM's, Loss Monitors, and Status Information

The next configuration assumes that TBT signals are available from all BPMs, and from all Beam Loss Monitors (BLMs). It also assumes that the full complement of 30 kHz status channels, such as from magnet power supplies, is available.

This enables an initial investigation of beam current losses throughout the Booster cycle. The commonly observed loss of 20% to 30% of intensity at injection into the Booster is a particularly important effect. Full coverage of the BPM's allows an analysis of the evolution of the closed orbit with time. This, correlated with the pattern of losses recorded by the BLMs (in time and space), should give strong indications of the underlying Physics of the losses. However, a complete picture of the role, for example, of space charge interactions, needs more detailed information.

Complete BPM acquisition also enables the measurement of transverse transfer functions. The simplest example of this is the measurement of beta functions and phases, as already attempted in the Tevatron. The only attempt so far failed, apparently because of systematic errors in BPM readbacks. Because of the much larger availability of beam time in the Booster, there is a much greater chance of making this promising scheme work there - with benefits to all Fermilab accelerators. Other examples of transverse transfer functions include the measurement of coupling effects, and nonlinear sextupolar effects, both globally and locally.

Ion Profile Monitors

A complete outfitting would have two horizontal Ion Profile Monitors (IPMs) and one vertical IPM. Each would have 32 channels, giving a measurement of the beam profile once per turn (or possibly faster).

One IPM is already built and is in the process of being commissioned. The detailed information from these devices will enable us to look at many effects such as the evolution of the transverse distribution just after injection into the Booster. This study is important to the study of space charge blowup at injection into low energy proton synchrotrons. It should also be possible to use the two horizontal IPM's to reconstruct the distribution of the beam in horizontal phase space.

A corollary of the depth of potential uses promised by the IPMs is the fact that a relatively large amount of analysis software will have to be written, and a lot of experimentation done, before the IPMs will be fully operational and fully utilized. It is in this kind of situation that the pre-packaged data processing tools, such as MATLAB, show their power and versatility for playing with the data, in order to learn how best to manipulate them.

Inter-bunch monitoring, several channels at RF frequencies

Longitudinal Coupled Bunch Instabilities (LCBI's) at present cause the longitudinal emittance in the Booster to blow up during each cycle. Preliminary experimental data suggests that the main culprit is the large impedance due to RF cavities [3]. A tracking digital longitudinal beam damper [4] is being built and installed in the Booster to counteract the LCBI.

A data acquisition system operating at this level would provide much more accurate information about the longitudinal mode that is driving the emittance growth. It would provide the diagnostics required for monitoring the performance of the damper, and for performing experiments on the behavior of the instability. Here, again, the exploratory nature of the investigation is greatly aided by on-line tools such as MATLAB, at least until a standard operational analysis can be defined and coded.

Transition crossing performance, and transition jump analysis

At least one channel of the 4 GHz digitizer is necessary in order to be able to study the detailed behavior of the Booster as it jumps across transition. A mountain range display of the evolution of a longitudinal bunch profile can be generated from this data and used for direct comparison with results produced by the simulation code ESME [5]. In addition to the important desire to experimentally validate ESME, the study of the efficiency of the Boosters transition jump is useful for other reasons, for example, how the various quantities change with time in the vicinity of transition: horizontal and vertical tunes, the closed orbit, RMS(Dp/p), horizontal beam size (and emittance), dispersion, and beam loss distribution. The Ion Profile Monitors have an essential role to play here, as well. None of this information is available at present.

Booster transition crossing studies are important for designing a transition crossing scheme for the Main Injector [6]. While some transition crossing studies aimed at the Main Injector - such as investigation of higher harmonic cavities - can be done in the Main Ring, the availability of the Booster and its gamma-t jump make transition studies there very desirable.

If it is feasible, it is desirable to not only measure the alpha-1 parameter in the Booster, but also to design a scheme to show that its control is possible and understandable.

Intra-bunch monitoring, a couple of channels at about 5 GHz

Intra-bunch monitoring is extremely valuable for other studies. The efficiency of the process of beam bunching at injection into the Booster is not well diagnosed or understood. This process might well be important, in addition to space charge effects, in causing the large intensity losses at injection. The Linac Upgrade will accentuate the importance of this effect, by making space charge effects less important.

A combination of IPM data and intra-bunch data makes possible a presentation of the evolution of the horizontal, vertical, and longitudinal emittances through the cycle. Although the acquisition system required is quite demanding, the analysis to get this information is not, and the result is a shorthand summary which would present clear signatures for various ailments of the Booster.

CONCLUSION

Eventually, with the whole of the proposed system in place - or a configuration not too far beyond it - it will be possible to study the evolution of the internal distribution of the charge in a bunch, on a Turn-By-Turn basis. In fact, the individual elements of such a scheme of "whole bunch tomography" are fairly straitforward. Two things remain to be done. The first is to integrate all the various pieces required for a Booster Data Acquisition system together. The second is to face up to the learning task that will be necessary to perform the experiments, and understand the results

REFERENCES

[1] E.L. Hubbard et al., "Booster Synchrotron", FNAL, TM-405, Jan 1973.
[2] I. Kourbanis et al., "Unix Data Acquisition System", proceedings this conference
[3] K. Harkay et al., "Studies of Coupled Bunch Modes in The Fermilab Booster", proceedings this conference
[4] I. Haberman, I.Rypshtein, "Longitudinal Damping System for the Fermilab Booster", proceedings this conference
[5] S. Stahl, J. MacLachlan, "User's Guide to ESME v7.1", FNAL, TM-1650, Feb 1990
[6] S.Peggs et al , "Comparison of Transition Jump Schemes in the Main Injector", proceedings this conference

Arbitrary Function Generator
for APS Injector Synchrotron Correction Magnets[*]

Oscar D. Despe
Argonne National Laboratory
9700 South Cass Avenue
Argonne, Illinois 60439

Abstract

The APS injector synchrotron has eighty correction magnets around its circumference to provide the vernier field changes required for beam orbit correction during acceleration. The arbitrary function generator (AFG) design is based on scanning out encoded data from a semi-conductor memory, a first-in-first-out (FIFO) device. The data input consists of a maximum of 20 correction values specified within the acceleration window. Additional points between these values are then linearly interpolated to create a uniformly spaced 1000 data-point function stored in the FIFO. Each point, encoded as a 3-bit value is scanned out in synchronism with the injection pulse and used to clock the up/down counter driving the DAC. The DAC produces the analog reference voltage used to control the magnet current .

I. INTRODUCTION

In order to obtain the precision of the magnetic field required for positron acceleration from 450 Mev to 7.7 Gev with low beam loss, correction magnets are used around the ring. Because of mechanical imperfections in the construction as well as installation of real magnets, the exact field correction required at each magnet location is not known until a beam is actually accelerated. It is therefore necessary to generate a correction field that is a function of the beam energy during acceleration. The fairly large number of correction magnets needed requires that the design of the AFG be as simple as possible yet provide the required performance. An important performance feature is that the function can be changed "on the fly", to provide the operator with a real-time feel during the tune-up process.

II. CONCEPTS USED IN FUNCTION GENERATION

The AFG design is based on scanning out encoded data from a first-in-first-out (FIFO) semi-conductor memory. The data input consists of a maximum of 20 correction values that are specified within the injector synchrotron acceleration window. Additional points between these values are then linearly interpolated to create a uniformly spaced 1000 data-point function stored in memory. Each point is scanned out in synchronism with the acceleration cycle and clocks an up/down counter driving the digital-to-analog converter (DAC). Since the function data is first stored before use, its effect is a feedforward correction rather than a real-time feedback action. The optimum function for each magnet is determined by an iterative process based on the beam quality during machine tune up.

The design of the AFG uses the staircase approximation of line segments to represent the desired arbitrary function correction. In the APS power supply system the DACs that generate the reference voltages to the current regulators are driven from the outputs of up/down counters. (See Fig. 1). The consequence of this design philosophy is that the control of a reference voltage is effected via pulse trains only.

The design presented here differs from other AFG designs [1]. Information derived from the points describing the arbitrary function rather than the points themselves, are stored in a FIFO memory. Each point is encoded in only a three bit value instead of the 11-bit value normally required for a number in the range of ±1000. A readout clock synchronized to the injection pulses of the injector synchrotron scans out data from each FIFO. The data are used to gate a clock to the up/down counter that drives a reference DAC. Circuitry is also provided that permits the "on the fly" update of the input function.

The simplest way to generate an arbitrary waveform for this system is to feed pulses to the counter as in Fig 1.

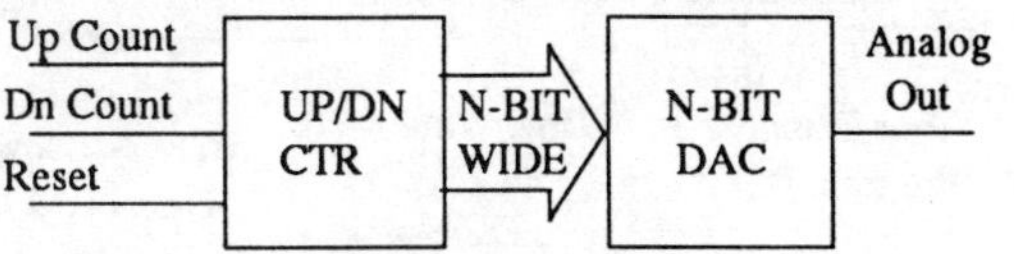

Figure 1. Analog Reference Block

If the pulse train can, in effect, be made to vary its rate, the DAC can produce any arbitrary waveform. For simplicity, the clock is fixed in rate at its maximum value and lower rates are generated by gating off a fraction of the clock pulses. Figure 2 shows an example of how a line segment AB can be approximated by incrementing or not its value at each clock time (using a fixed step size).

* Work supported by US Department of Energy, Office of Basic Energy Sciences under contract No. W-31109-ENG-38

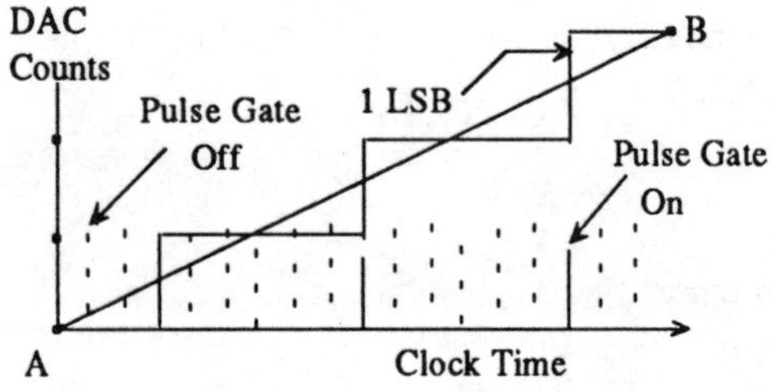

Figure 2 Staircase Line Approximation

II. REQUIREMENTS

The AFG requirements are as follows:

- Polarity   Bipolar
- Resolution (Amplitude & Time) 1×10^{-3}
    10 bits for magnitude
    1 bit for sign
- Repetition Period   0.5 s
- Function Duration (Acceleration)   0.25 s
- Number of points specified $< = 20$
- Number of points supplied   1000
    (Linear Interpolation
    used between specified points)
- Dynamic update of function data

III. THE DESIGN OF THE AFG

Figure 3 shows the correction magnet timing in an injector synchrotron cycle.

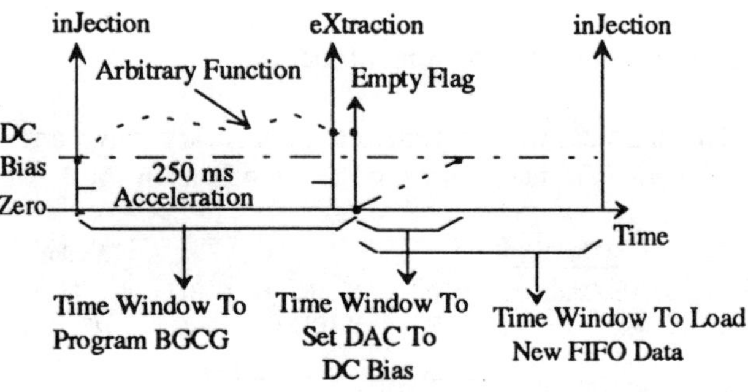

Figure 3. Arbitrary Function Timing in a Synchrotron Cycle

The injection pulse J is used as the reference pulse. The extraction pulse X is at the midpoint of the injector synchrotron cycle. The FIFO empty flag signal is generated when the last data point is output. This occurs after X when the DAC output and the FIFO read pointer are reset to zero.

Since the acceleration interval is divided into 1000 points and the function duration is 250 ms, the required clock period is 250 µs. The points that define the function to be generated are sent by the host computer to the power supply control unit (PSCU) where the pulse enable states, the DIR signal, and other initial parameters are determined prior to loading a 3-bit wide, 1000-bit deep FIFO memory.

3.1 Hardware Blocks

Figure 4 shows the Arbitrary Function Generator.

The Main FIFO block provides temporary data storage for the arbitrary function before transfer to the PS FIFO, so it eases programming and synchronization. A hardware clock is used to read this data and load the addressed PS FIFO.

Each PS FIFO stores arbitrary function data for its power supply. It can receive update data from the main FIFO and provides sync signal for the update process from its empty flag. Since all PS FIFOs are scanned in synchronism with the injection pulse, only one empty flag signal is used.

The Main Gated Count Generator (MGCG) supplies the required number of clock pulses (20kHz) to read the main FIFO and load the addressed PS FIFO. It is gated on by the presence of an update request signal at time the PS FIFO becomes empty.

Each Bias Gated Count Generator (BGCG) unit is programmed to deliver the required number of pulses (100kHz) to set the DAC to its DC bias level after its reset to zero following extraction.

The Input And Output Logic blocks generate other signals needed to synchronize the sequences of events and the selection of clock sources. PLDs (programmable logic devices) are used to implement the logic to reduce the IC count.

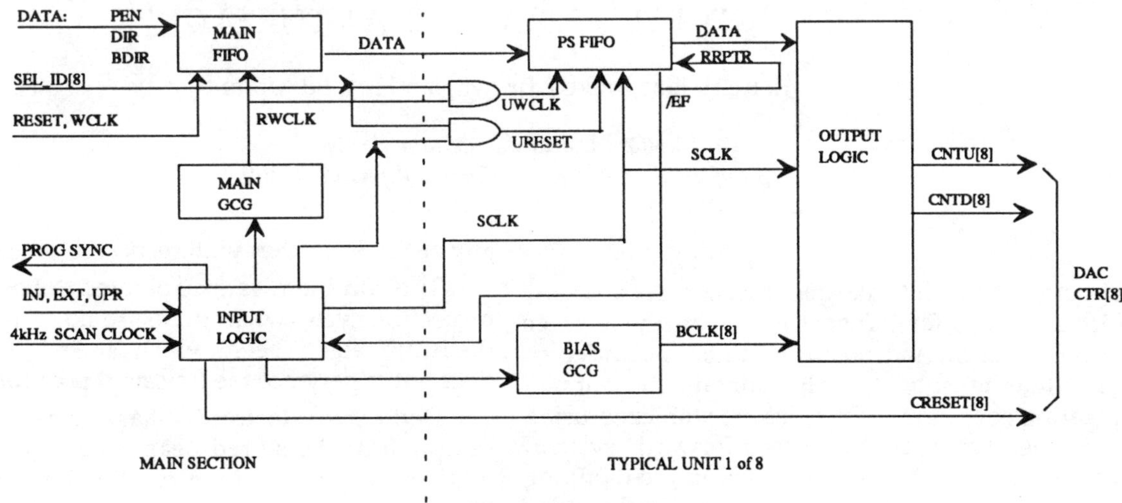

Figure 4. Arbitrary Function Generator

3.2 Operation

The PSCU receives from its host computer the data points describing the arbitrary function. The PSCU maintains the latest function data (20 points) for each power supply it services. From these data it makes calculations of the slope (DIR) and pulse enable signals (PEN) associated with the FIFO readout pulses for each line segment. The software also determines the number of pulses required to set the DAC counter to its DC bias for the new function. This number is programmed into the BGCG following injection if an update request is outstanding. The software generates update request and ID select signals to address the specific FIFO involved.

On the next empty flag signal after the update request, the addressed FIFO and DAC are reset and the addressed BGCG is activated delivering the programmed number of pulses (100kHz) to the selected DAC counter to set it to the new DC level. At the same time, the MGCG is started, reads out the main FIFO and loads the addressed FIFO. The update request latch is then reset. The new function is therefore in place when the next injection pulse occurs.

In its normal mode of operation, each function generator is repetitively scanning its FIFO in synchronism with the acceleration cycle. At every injection pulse, the 4kHz clock is selected to scan the FIFO and drive the DAC counter in accordance with the unit's direction DIR and unit's pulse enable PEN signals. At each empty flag time, the DAC counter and the read pointer of the addressed FIFO are reset to zero. The BGCG is then enabled to set the DAC to its DC bias value in preparation for the next cycle.

If at injection time an update request is present, the following additional actions are performed within the injector synchrotron cycle as indicated in Fig. 3: 1) injection timing signal is sent to the microprocessor for synchronization, 2) the BGCG is programmed with new DC bias data, and 3) addressed FIFO is loaded with new data. These actions are transparent to the normal mode of operation of the function generator as described above.

IV. CONCLUSION

The design of an arbitrary function generator for the APS injector synchrotron correction magnets is based on the staircase approximation of line segments that describe the desired function. The attributes of the segments rather than the point values themselves are stored in first-in-first-out (FIFO) memories This scheme results in a simple hardware implementation that is appropriate for applications requiring a large number of these units.

V. REFERENCES

[1] Kuninori Endo, " Minicomputer Assisted Function Generator for the Dynamic Control of Chromaticity Correction Sextupoles", IEEE Transactions on Nuclear Science, Vol. NS-26 No. 3, June 1979

ENERGY FEED FORWARD AT THE SLC*

R. Keith Jobe, Mike J. Brown, Ian Hsu, Ed Miller

Stanford Linear Accelerator Center
Stanford University, Stanford, California 94309

Abstract

The energy of the SLC scavenger electron beam used to produce positrons for the next cycle and the average energy of the electron and positron bunches collided in the linear collider must be stable to maintain efficient machine operating conditions. Energies are stabilized using a newly installed hardware-based feed forward[1] system. The three bunches are stored and cooled in the damping rings, and co-accelerated in a single machine cycle. Prior to extraction of the stored bunches, the energy gain of the accelerator is set by re-phasing of klystrons to correct for anticipated beam loading effects. We will discuss the hardware associated with determining the intensities and the appropriate energy corrections to compensate for variations in energy due to beam loading effects. We will discuss machine tuning procedures, diagnostics, and operational experience.

I. INTRODUCTION

When operated in its design 3-bunch mode, the positron source for the Stanford Linear Collider[2] (SLC) is sensitive to variations in bunch intensities of not only the scavenger electron bunch (used to produce next cycle's positrons) but of the preceding luminosity positron and electron bunches as well.

There are 3 bunches, each with its own beam loading function, which contribute to the depletion of the available rf fields in the accelerator. The scavenger bunch (bunch number 3) will arrive at the extraction transport line "off energy" if there is any change in the charge of any of the 3 bunches with respect to their nominal values

This problem is exasperated following a machine protection cycle, when there are no stored positrons and the full intensity beam which is used to create positrons does not experience the (missing) positrons' loading.

Dedicated electronics has been built to allow the sampling of the stored beam intensity several milliseconds prior to extraction from the damping rings; the intensities of the three bunches are used to predict the energy error of the next scavenger electron bunch. Two kilometers remote from the damping rings, supporting electronics has been installed which takes the predicted error, and corrects the energy profile of the accelerator just upstream of the extraction point in the linac.

Two sectors (out of 31 total) are used to adjust the energy of the extracted beam. Low power phase shifters are set with coefficients derived from the slower feedback[3,4] system to "kink" the phase of the two sectors of klystrons.

II. ELECTRONICS

There are 4 discrete types of electronics used in the feed forward system: An intensity sampler, a sampling module which predicts the energy error, a computing module which calculates appropriate phase settings, and some rf hardware to affect the correction.

A. Intensity Sampler

In the SLC, the beam in the damping rings is delivered to the linac with very little loss. Sampling of the

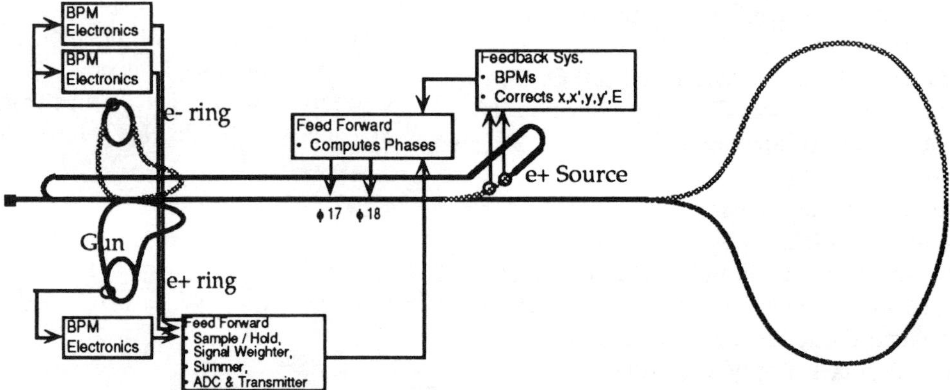

fig 1. The system diagram of the Feed Forward correction system superimposed on the familiar SLC beamline diagram. The extraction line is central and just upstream of the "e+ source". Shown is both the Feed Forward and Feedback components for this system.

* Supported by the Department of Energy, Contract DE-AC03-76SF00515

beam position and intensity is done for steering and diagnostics using a specialized beam position monitor (BPM) electronics package[5]. For this feed forward application, the standard BPM module is used with a stripline position monitor to determine only the intensity of the stored beam.

B. Sampling Module

The beam loading of a nominal 3-5 10^{10} particles in each of the three bunches on the scavenger electrons is different due primarily to the following effects:

• Positrons: The fundamental loading of this bunch is fractionally "washed-out" by the timing of the third bunch due to the the 130 nS bunch timing separation and the 825 nS accelerator fill time .

• Electrons: As for the positrons, there is a wash-out effect with strength of only 60 nS.

• Scavenger electrons: The median energy of this bunch is stabilized, therefore a) only half the charge is considered, and b) since this bunch experiences its own self-loading in the accelerator, the fundamental and all higher modes contribute to self loading.

There is a CAMAC sampling module[6] which interfaces to both the SLC timing system and the BPM electronics. Interfacing to the other modules, a set of gates are provided to the beam position electronics and signals are available for monitoring and diagnostics. Inputs to the module are timing signals from the timing system and the "Sum" signal from the BPM electronics. Internally there are appropriate gain and offset levels. These signals are processed as follows:

1 Offset (or pedestal) value is subtracted independently for each channel,

2 Gain Coefficient is applied to each channel. This data is summed as follows and is sampled with a GADC,

3 Three outputs are summed, with an additional offset term,

4 Result is sampled and digitized. Sampled values are monitored by GADC and a normal ADC, and digital format is sent as differential data to the feed forward computer module. Data transmission is via a 32 kilobaud differential serial link.

Results from this module are sent to the computing module (below) located near the positron extraction line 2 km distant.

C. Computing Module

The resultant action from the predicted energy offset is the re-phasing of two sectors of high power klystrons. Hardware phase shifters allow the control of these rf devices, with the two sectors (acceleration 2 GeV each) to be symmetrically mis-phased, or "kinked". This can allow quick correction for energy control without the introduction of significant changes in energy spread.

Corrections from feed forward are complicated by the simultaneous corrections by slower "fast" energy feedback systems. The feed forward corrects for anticipated errors due to accelerator beam loading effects, while the feedback system repairs errors due to anomalous drifting and changes of accelerator rf, transport components, and any errors introduced by the feed forward system. Communications between the two systems is achieved by locating the phase computer CAMAC module in a feedback controlled crate, and having the feedback process continually adjust the feed forward coefficients to the newest optimum value.

The computing module is a CAMAC module[7] which receives the digital loading correction data from the sampler module, and applies pre-programmed constant, linear and quadratic coefficients to compute new phase shifter settings.

For simplification from the control software standpoint, both this module and the sampler module are designed to identically replicate the SLC standard DAC module used by the SLC, allowing standard software support to be used in the setting and updating of the coefficients. All "DAC" values, whether used internally as digital multipliers or analog offsets, are available from the front panel and read by a multi-channel ADC, again to allow the control system the standard feature of readback of these virtual components.

D. Associated RF Hardware

The phase setting determined from the computing module is sent as an analog voltage to two rf phase shifters[8]. These low power phase shifters consist of a pair of varactor diodes and a 3 dB splitter. Installed just prior to the rf multiplier, they allow a ±180° phase control of a sector of klystrons each. Similar hardware is used by a fast feedback process to stabilize the linac energy (see ref. 3).

III. OPERATIONAL EXPERIENCE

Recent operation of the accelerator has allowed for a brief pre-conference commissioning effort on this new system. Accelerator physics programs have verified that:

• The beam loading intensity dependence is near expected theoretical values.

• The installed hardware has been commissioned, passing expected performance tests.

• The energy deviations of a single bunch with deliberately introduced intensity variations can be reduced.

A. Beam Loading Effects:

The loading of the accelerator was measured using the extraction transport line as a spectrometer while the intensity of the scavenger bunch was statically varied (see fig. 2). The results are in agreement with expectation (*c.f.*: Ref. 1), with resulting energy dependence of -159 MeV / 10**10 particles. Static tests varying the intensity of the

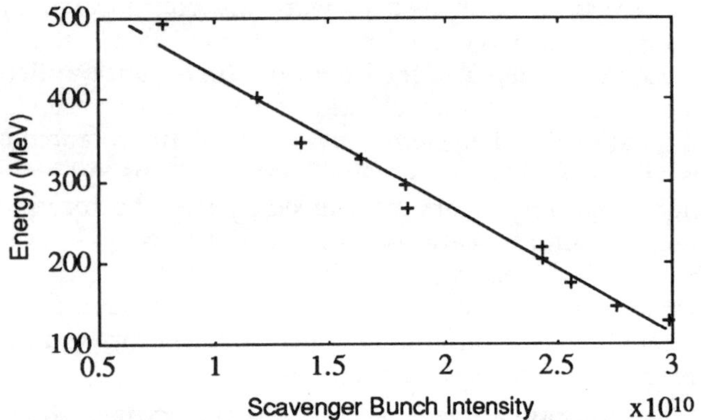

fig. 2. **The energy of the scavenger bunch correlated against intensity measured in the Damping Ring. Slope of −159 MeV/10**10 electrons is in agreement with the expectation value of −172.**

positron current showed similar agreement with expectations. Definitive measurements of the beam energy on a pulse-to-pulse basis is difficult, since the accelerator's extraction line is not equipped with sufficient BPMs that can be read simultaneously (un-multiplexed) to cleanly separate launch errors in x, x' from $\Delta E/E$ errors.

B. Hardware

Initial results suggest that the hardware is operating as intended. While there is insufficient operational experience to report on the machine tuning procedures, it is clear that the calibration of the sampling system (the dependence of energy loading on beam intensities) will require some attention.

Missing from the design is any way to separate the calibration of the detectors from the assignment of beam loading coefficients. The authors intend to rectifying the situation by the addition of a set of three gain stages which will allow the operations support personnel to independently verify that the module reports the same beam current as the machine diagnostics report, and will

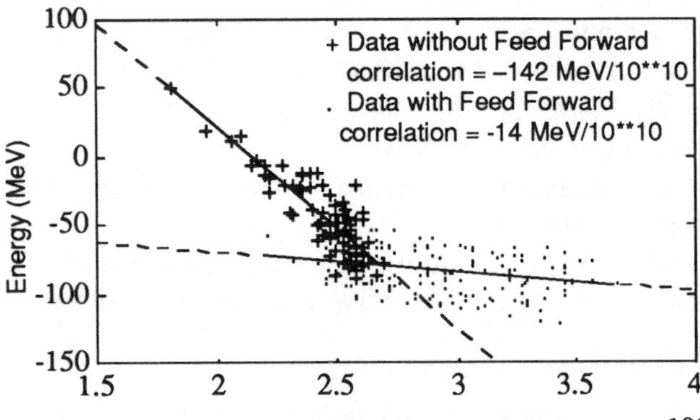

fig. 3. **Shown is the energy jitter of the beam with and without feed forward. The energy of the beam can be estimated on a pulse-to-pulse basis using sampled data and the known dispersion value for the BPM.**

allow the separate assignment of beam loading weights.

C. Commissioning Results:

The Energy Feed Forward system has undergone initial commissioning in early May 1991. For some of these tests, the machine was deliberately mis-tuned to generate larger intensity fluctuations, and therefore larger intensity-dependent energy variations from the scavenger beam's self loading. BPM analysis allows a rough determination of the energy jitter through the knowledge of the system's dispersion of $\eta = -26$ cm and energy of 35 GeV. Shown in fig. 3 is the sampled data from consecutive machine pulses in the SLC. Superimposed on this plot is the energy jitter from the machine both with and without the feed forward system operational.

Initially, the energy intensity dependence as observed was -140 MeV for every 10**10 scavenger electrons; feed forward reduced this by a factor of 10.

Further commissioning will require additional attention paid to the specific intensity dependence the energy has to each of the three bunch currents. Additionally, further verification of the stability and performance of the intensity detectors is required.

IV. ACKNOWLEDGMENTS

The authors would like to thank Artem Kulikov for the suggestion to stabilize scavenger energy by a energy feed forward system based on stored charges.

V. REFERENCES

1 Ian Hsu, *et al.*, "The Energy Stabilization for the SLC Scavenger Beam," 1990 Linear Accelerator Conf., 1990, p. 662. SLAC-PUB-5310.

2 John Seeman, "The Stanford Linear Collider," 1990 Linear Accelerator Conf., 1990, p. 3.

3 R. K. Jobe, *et al.*, "Position, Angle and Energy Stabilization for the SLC Positron Target and ARCs," 1987 IEEE Particle Accelerator Conf., 1987, p. 713. SLAC-PUB-4232.

4 K. A. Thompson, *et al.*, "Feedback Systems in the SLC," 1987 IEEE Particle Accelerator Conf., 1987, p. 748. SLAC-PUB-4217.

5 J.-C. Denard, *et al.*, "Monitoring of the Stanford Linac Microbunches' Position," IEEE Transactions on Nuclear Science (Particle Accelerator Conf., 1983), Vol. NS-3, No. 4, Aug. 1983, p. 2364. SLAC-PUB-3058.

6 M. J. Browne, "Feed Forward Summer", SLAC Drawing Package 125-471.

7 M. J. Browne, "Feed Forward Controller", SLAC Drawing Package 125-472.

8 Heinz D. Schwarz, "Computer Control of RF at SLAC," invited talk, IEEE Transactions on Nuclear Science (Particle Accelerator Conf., 1985), Vol. NS-32, No. 5, Oct. 1985, p. 1847.

BPM DATA ACQUISITION SYSTEM FOR THE BATES PULSE STRETCHER RING

O. Calvo, T. Russ, J. Flanz
Massachusetts Institute of Technology
21 Manning Rd., Middleton, Ma. 01949

Abstract

A beam positioning monitoring system is being developed at Bates. It will include approximately 30 stripline monitors distributed around the South Hall Ring [3], These BPMs will be sampled by flash analog-to-digital converters operating at 40 MSPS. Each ADC is connected to a FIFO that can be read by the CPU (SparcStation in a single VME slot, running SunOS) and transferred via Ethernet to the central control system, after some local processing. Hardware and software operation of the prototype are described

I. INTRODUCTION

MITs Bates Accelerator Center is being upgraded to provide high duty factor CW electron and photon beams of excellent quality, with energies up to 1 GeV, for use in nuclear physics. This pulse strecher ring will extend the beam duty factor to 85 %, with a maximum extracted current of 50 uA. It will also initiate a program of internal target studies when operating in storage mode. Operation is based on repetitive injection (up to 1000 times a second) of a short high peak current pulse (40 mA) during two turns, followed by time-uniform extraction of the stored beam during the interpulse period.

Beam position monitors will be used in different stages of the operation of the machine. In the initial turn-on, they will be used to steer the beam clear of all apertures during injection, and to find the closed orbit, measure the integer and fractional part of the tune, obtain the beta functions and identify instabilities during storage and tune-up. During extraction, they will facilitate the measurements of dynamic aperture, the adjustment of the non-linear elements and the fractional tune, and the optimization of the extracted beam with fast-feedback and feed forward loops. In normal operation, the BPMs are mainly used to determine beam position, phase space and fractional part of the tune.

II. BEAM PICK-UPS

There are thirty one BPMs in the ring (roughly four per betatron wavelength). They have a resolution of +/- 0.1 mm with circulating currents ranging from 1 to 80 mA. We are currently testing a prototype detector made of two pairs of 50 ohms, 3/4 lambda striplines with a position sensitivity of 400 mV/mm. The signals from all detectors are processed by amplitude-to-phase modulation (AM/PM) modules,

downconverting the RF signals to 50 MHz with an analog bandwidth around 20 MHz. These detectors also provide a current output delivering 40 mV per mA of beam current. A button detector is also under development. It has lower output signal levels but is far less disturbing to the beam and could be easier to manufacture[2].

III. ANALOG TO DIGITAL CONVERSION

Each BPM has three channels: two corresponding to vertical and horizontal positions and a third one indicating the beam current. (Fig. 1).

Analog signals coming from the detectors are amplified, offset and sampled by flash analog-to-digital converters, operating at 40 MHz. After testing four different units we settled for the Analog Devices AD9012, which offers a maximum frequency of 75 MHz and is TTL compatible. Each ADC is connected to a high speed FIFO that can be read by the local processor. Information on the FIFOs is read into main memory by a real -time processor residing in the VMEbus. This data is locally processed and the results are sent to the central control system. By using PLCC packages we can integrate nine channels in a 6U, B size VME card. The goal is to put as many as 150 channels in a single VME crate.

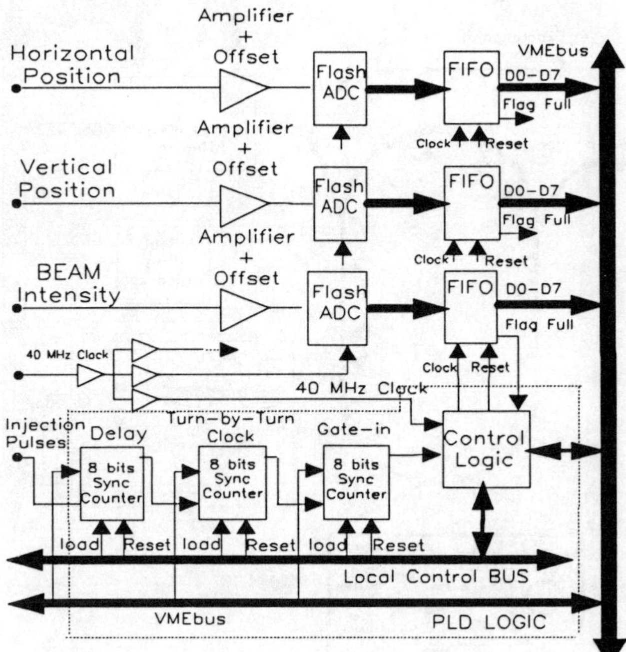

Figure 1. ADC channels and triggering logic

IV. Triggering Logic

The A/D converters are free running at maximum speed, and it is the responsibility of the triggering logic to decide when to latch the data on the FIFOS. Any BPM can be programmed independently of the others to operate in different modes. The main trigger pulse that activates all the channels is also synchronized to and delayed from the injection pulse. A separate timing card is in charge of this function and distributes the timing pulses to all the channels. This board has interrupt generation capabilities, and also provides the timing to the VME processor unit. The main 40 MHz clock is derived from the main RF frequency (2856 MHz) through a PLL divider/multiplier.

Different lengths between cables, and beam delay, can produce time mismatches between channels as large as 800 ns (more than one turn). These delays are compensated in two steps. First, the cables are trimmed so that the delay between channels is always an integer multiple of 25 ns. Then, 6 bit programmable digital delays are set from the control system to any number of 25 ns. intervals between zero and 64 (zero to 1.6 usec) (figure 2). The turn by turn (TBT) clock is obtained by dividing the 40 MHz sample signal by 25. Each bpm has its own set of counters implemented with high speed PLDs (Xilinix 3020). This solution presents more flexibility than using conventional logic: changes in trigger circuitry can be done by merely reprogramming the PLDs.

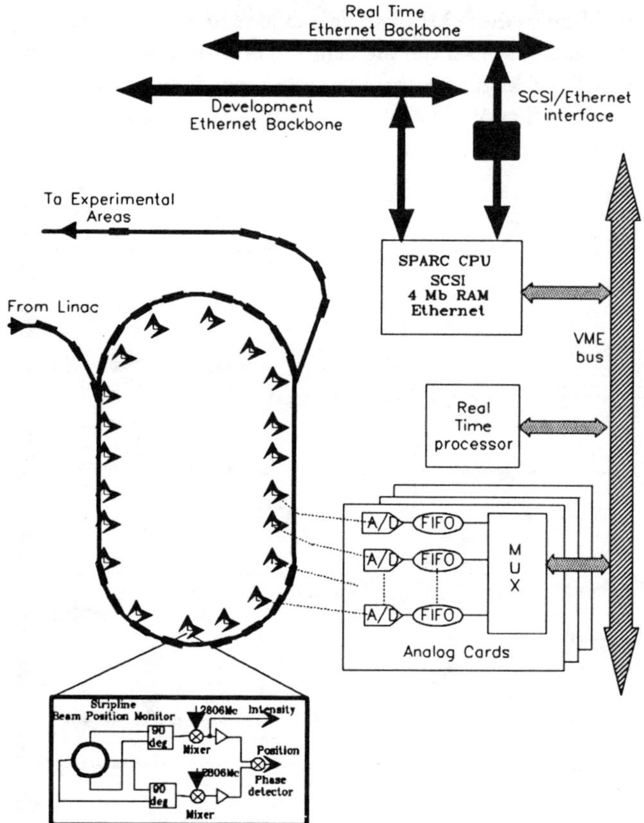

Figure 2. Overall Layout

By avoiding LC delay lines and making all data capturing synchronous, the design of the timing generation functions is greatly simplified.

V. Data Processing

The main processor is based on the SPARC 1E, VME based CPU. This microcomputer engine can run SunOS (Unix). It operates at 20 MHz, delivering 12.5 MIPS with a floating point capacity of 1.33 MFLOPS, and has 4 Mbytes of RAM, a SCSI and an Ethernet port. It can be easily linked to our instrumentation computer, a SUN SparcStation two [3].

Since both computers, target and workstation are basically identical, all the software development can be done either in the workstation and then ported to the target, or in the target itself, across the network. During the software development stage, a private Ethernet connection will be used to link both systems. If future application require it, this "private network" can be used for "heavy" data logging without saturating the main control system Ethernet network.

Having the Workstation "right on the bus", controlling the instruments directly, significantly increases the power of analysis. Otherwise, only relevant data would be available to the instrumentation computer and most of the raw data would be thrown away.

VI. Software

All the software is being written in ``C'' and runs under SunOS. Benchmarks run on the SparcEngine indicate that a hardware interrupt, requested from the VMEbus, can be serviced in 16 usec . A very small routine decodes the command issued in the main console of the control system, encoded in a low level Ethernet packet. A SCSI to Ethernet adapter, developed in house,allows us to transfer data to the network only during a certain time window, minimizing collisions and improving the effective bandwidth.

The software is also responsible for setting the channels to its proper mode of operation., loading all the delays, arming the trigger, transferring the data from the FIFOs to main memory, averaging out the data and performing mathematical computations, such as FFTs.

VII. BPM Measurement Modes

As stated above, the bpms will operate in very different modes depending on the status of the machine:

A. Integer Tune - All BPMs at one time

The purpose of this measurement mode is to obtain a snapshot of the trajectory of the beam during one complete turn. Thirty one BPMs (93 channels), distributed around the ring are sampled at the same time, allowing us to determine

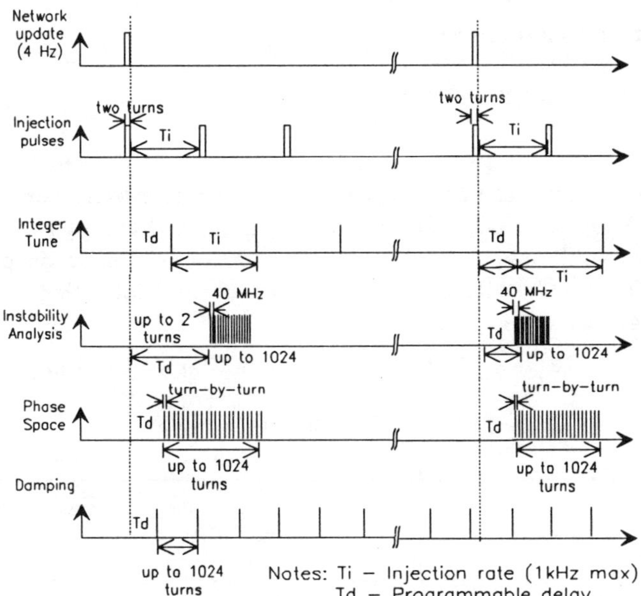

Figure 3. BPM timing diagram

the betatron oscillations. The sampling process is triggered by the injection pulse and delayed a number of turns specified by the operator (up to 1800 turns for 1 KHz injection rate). This is applicable to all the measurement modes. The previous process can be repeated at injection rate for data averaging. Only averaged data will be sent to the network.

B. *Instability analysis and fractional part of the tune.- One BPM vs time*

In this mode of operation, selected BPMs are sampled at 40 MSPS until the FIFOs are full. An FFT will be calculated on these points. The sampling frequency was chosen to measure frequency components as high as seven times the revolution frequency (1.58 MHz). To be conservative, the sampling frequency is 40 MHz, providing an theoretical analog bandwidth of 20 MHz. This means that there will be 25 samples per turn.

C. *Phase Space - Two BPMs, turn by turn.*

In this mode, position and angle are of interest. Two BPMs are monitored, once every turn (turn rate) until the FIFOS are full (1024 turns). The results of this measurement are updated at 4 Hz rate until a new mode is selected.

D. *Damping - One BPM, turn by turn.*

This mode is similar to the phase space measurement, except that only one monitor is necessary and the samples are

stored in main memory every N turns, where N is user selectable. Injection rate is very slow and the number of samples stored is also user selectable.

E. *BPMs in the Injection line*

Five BPMs are located in the injection line. They work with a pulsed beam and operate in two different modes.

1. *Normal operation - Injection rate*

Under normal operations, these BPMs are sampled at a rate of one per injection pulse. The sampling point must fall in the middle of the injection pulse.

2. *Fast Plot mode*

In this mode, all the BPMs are sampled at maximum speed during two turns. This is done only once, and the data is sent to the network for plotting. No averaging is necessary. FIFOs will be refreshed at their maximum allowable speed rate.

VIII. CONCLUSION

The system described is currently under development and will be operational in 1992. A prototype of the flash ADC was built and will be tested soon with the pick-ups located in the Bates Linac. A video matrix switch was also developed to allow the selection of 16 out of 256 signals to be brought to the control room, simplifying the analog monitoring of bpm signals.

The Unix machine on the VME bus reduces enormously the development time and the use of a standard operating system facilitates the communication with the host

The combination of a flash ADC, FIFO and flexible trigger multiplies the capabilities of the acquisition modes. Once the ring enters in operation, new measurement modes could be requested, which can be easily inplemented through reprogramming of the PLD logic.

IX. REFERENCES

[1] R. Shafer et al, "The Tevatron beam position and beam loss monitoring ," Proceedings of the 12th International Conference on High Energy Accelerators, Batavia, Il, August 1983. pp 609-615.

[2] Bates E/E group, "Prototype Stripline BPM - Description and Measurements", Bates internal report, Nov. 1989.

[3] T. Russ, et al "South Hall Ring Control System Design", Bates Internal report, May 1990.

[4] J. Flanz et al "The MIT-Bates South Hall Ring", Proceeding of the 1989 Particle Accelerator Conference, Chicago, Il, March 1989, pp 34-36.

The CEBAF Frequency Distribution System*

A. Krycuk, J. Fugitt, K. Mahoney, S. Simrock
Continuous Electron Beam Accelerator Facility
12000 Jefferson Avenue
Newport News, VA. 23606

INTRODUCTION

The RF system for the CEBAF accelerator requires that two frequencies, 70.000 and 499.000 MHz, be distributed throughout the accelerator site. Proper operation of the multipass beam requires an energy spread of less than 2.5×10^{-5}. This imposes stringent constraints on the allowed slow phase error, in turn implying that there must be active regulation of the phase of the distributed reference frequencies.

There are several methods already in use that regulate the phase of a distributed signal[1,2]. A SLAC type pulsed microwave interferometer was not readily applicable, as this method uses the off time in a pulsed accelerator to perform its phase correction and measurement; CEBAF is a continuous device with every RF bucket filled. Since the CEBAF system requires amplifiers in the distribution line, a diplexed reference signal such as that used at LANL would require a series of discrete control loops, tied together with an overall control loop. It was felt that a less complex system would be desirable.

To accomplish this, the frequency distribution system uses temperature regulated coaxial lines to provide high level signals to the RF control modules in the klystron galleries, combined with a lower power reference signal carried via phase stable fibreoptic cable.

MASTER OSCILLATOR

The master oscillator serves as the RF reference source used throughout the accelerator complex. It is composed of an indirect multiple output frequency synthesizer, a 10 MHz reference source, and power conditioning circuitry. Three output frequencies, 70, 499, and 1497 MHz are derived from the 10 MHz reference. Each frequency is synthesized using a multiple phase locked loop architecture.

Reference phase noise is the largest single contributor to uncorrected errors in the accelerating cavity RF control system. As such, it is imperative that the phase noise of the master oscillator be held to as low a value as possible. The phase noise requirements for the master oscillator are given in Table 1.

A unique requirement for the master oscillator is that all frequencies remain in the same, albeit arbitrary, phase relationship for the life of the oscillator. Normally a frequency synthesizer may lock to a modulo pi phase of its reference. If this were true for the CEBAF MO, then each time the system was brought up it could potentially change

the reference phase of the entire accelerator by 180 degrees. A known phase relationship is maintained through the use of subloops which compare the 499 and 70 MHz outputs. The 1497 MHz output is derived by multiplication of the 499 MHz output.

Table 1. SSB Phase Noise of Master Oscillator

OFFSET FROM CARRIER	SSB PHASE NOISE, dBc/Hz		
	70 MHz	499 MHz	1497 MHz
10 Hz	−90	−85	−75
1 kHz	−130	−135	−120
100 kHz	−150	−155	−140

HIGH LEVEL DISTRIBUTION

This portion of the system distributes the two RF reference frequencies to the RF control system and the klystrons themselves (Figure 1). The oscillator output at both frequencies is amplified to +30 dBm and carried via 7/8" foam filled coax to the accelerator service buildings. In the Injector and RF Separator buildings, the 499.000 MHz signal is fed directly to the RF control modules which regulate those subharmonic cavities.

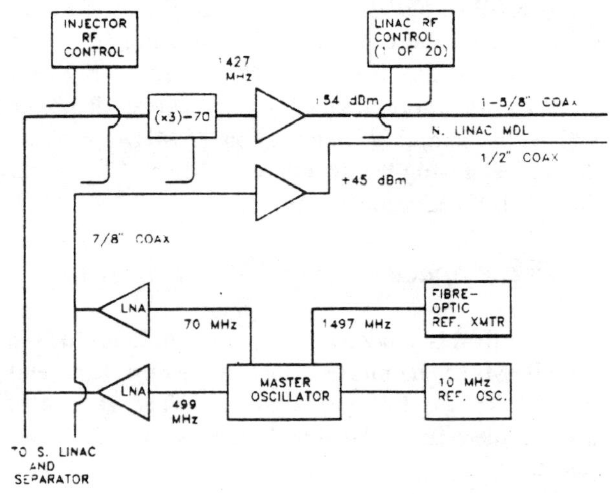

Figure 1. High Level Coaxial Distribution System

The klystrons in the north and south service galleries, however, operate at 1497.000 MHz. The RF controllers for these cavities operate using 70.000 and 1427.000 MHz [3].

*Supported by D.O.E. contract #DE-AC05-84ER40150

Therefore, prior to distribution to the klystron galleries, the 499.000 MHz signal is multiplied by three and mixed with the 70 MHz signal to obtain 1427.000 MHz. It is this frequency which, along with the 70 MHz, is distributed to the control modules.

MAIN DRIVE LINE

The 1427.000 and 70.000 MHz signals are amplified to +54 and +45 dBm respectively, prior to introduction onto the main drive line (MDL). The MDL consists of a pressurized 1-5/8" rigid coax line and a 1/2" foam filled coax line encased in a temperature regulated jacket (Figure 2). The 1-5/8" coax is used as the transmission medium for the 1427 MHz signal; the 1/2" carries the 70 MHz signal. As the 750 ft. long 1-5/8" coax is phase sensitive to temperature changes (6.5° phase/°C @ 1427 MHz), it is imperative to maintain temperature regulation to within ±.1 °C to meet the required phase stability. The 1-5/8" rigid coax is pressurized to 4 psig with dry nitrogen to provide a desiccating atmosphere and control any change in dielectric length due to pressure variations. A similar effect could have been obtained by evacuating the coax, though this approach is not as attractive in view of the large amount of power lost as heat on the inner conductor which must be removed.

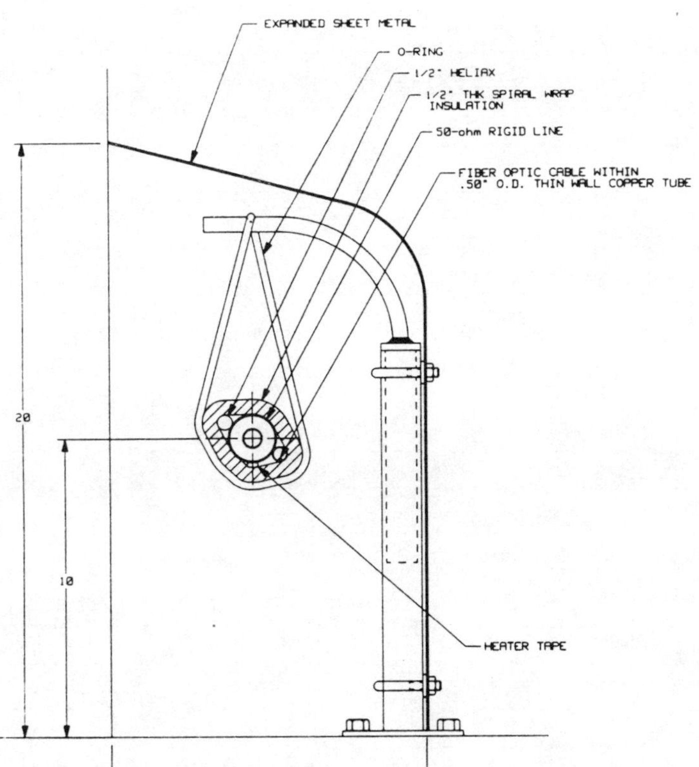

Figure 2. Cross-Section of Main Drive Line

The MDL is fabricated in 10 meter segments, with directional couplers provided for both lines at the end of each section. The 70 MHz (1/2" coax) is fitted with a simple 20 dB coupler, as attenuation does not result in a large change in signal level from one end of the klystron gallery to the other. Line loss in the 1-5/8" coax causes a significant decrease in the 1427 MHz signal, from an input of +53 dBm to +40 dBm at the end. In order to maintain the desired coupled value of +30 dBm at each RF module, it is necessary that the coupling value be adjustable. This could also be accomplished by maintaining a constant coupling value with attenuative pads as required, but this would result in the need for an even larger amplifier, with a concomitant increase in cost. In order to isolate the MDL from any microphonic noise, the entire 750 ft. length is suspended from O-rings spaced every 5 feet.

FIBREOPTIC REFERENCE LINE

Whereas the MDL provides the requisite degree of phase stability in the north and south linac galleries, there is no active temperature regulation of the remainder of the coaxial distribution system. Left uncorrected, this could result in the linacs drifting in phase with respect to each other and the rest of the accelerator. To correct for this, an overall phase reference/correction system was designed.

The method used makes use of the inherent phase stability of fibreoptic cable. Sumitomo Corporation has developed an ultra-phase stable fibre that has been tested with great success at KEK and JPL [4,5]. This fibre, run the entire circuit of the CEBAF accelerator, approximately 1.5 km., would exhibit a thermal shift of .12° phase/°C, as compared to the unregulated coax shift of 5 ° phase/°C. A 2 mW laser diode operating at 1.3 μm was obtained from BT&D. The diode and its associated receivers are capable of being modulated at 6 GHz, far beyond the needs of CEBAF.

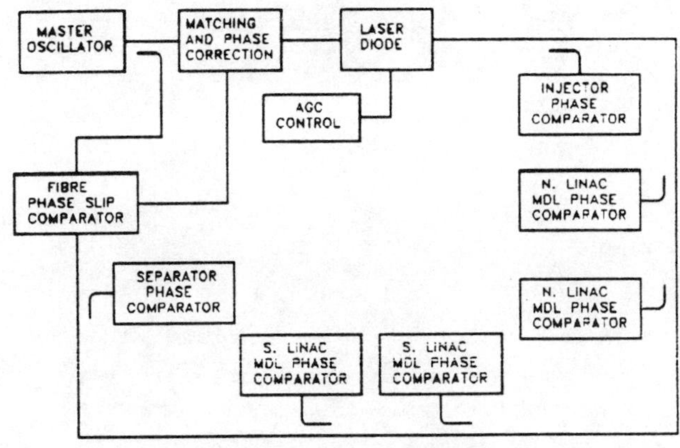

Figure 3. Fibreoptic Reference System

The usual drawback to RF transmission via optical fibre is the high system insertion loss, resulting in a received RF signal which must be amplified 30 or 40 dB to attain its original signal level. Such amplification is not only expensive, but as can be seen in Figure 3, it would be extremely difficult to obtain amplifiers which tracked in phase for each fibreoptic receiver. The source of most of the system insertion loss is the technique used to match the RF source (50 ohm) to the fibre transmitter (7 ohm) and receiver (high impedance). Normally, a 50 ohm load is placed in parallel with the fibreoptic device. This technique is appropriate for communication systems where the devices must be matched over wide bandwidths. The RF reference system, operating at one frequency, has no such bandwidth requirement. Therefore, a reactive matching network[6], which trades wide band response for a reduction in narrow band insertion loss, has been incorporated for both the transmitter and receivers.

The fibreoptic signal is coupled off at selected locations throughout the accelerator (injector, beginning and end of north and south linac MDL's, and separator) and compared with a coupled portion of the distributed high level signal. The phase difference between these two is assumed to originate in the drift of the coax system, and an appropriate correction will be applied. In order to correct for any phase slip or change in the fibre itself, the optical cable is routed back to the control room to be compared in phase with the original master oscillator signal.

CURRENT STATUS

The system design has been completed and components are under test. Two 20 ft. sections of 1-5/8" rigid coax have been encased in a prototype thermal jacket regulated by an in-house constructed temperature controller, with the interior pressure maintained by a commercially-obtained absolute pressure regulator. Preliminary results indicate that temperature regulation to $\pm.1°C$ is achievable. The laser diode source for the fibreoptic reference system has been characterized and a matching circuit is being fabricated.

REFERENCES

[1] Schwarz, H. and Weaver, J., "The RF Reference Line for PEP," IEEE NS-26, No. 3, 3956 (1979).

[2] Jachim, S., Regan, A., Gutscher, D., "A Phase-Stable Transport System," NPB Technical Symposium, May, 1990.

[3] Simrock, S., "RF Control System for CEBAF," presented at this conference.

[4] Tanaka, S., Murakami, Y., *et al.*, "Precise Timing Signal Transmission by a New Optical Fiber Cable", KEK Report 90-5, May 1990.

[5] Primas, L., Lutes, G., Sydnor, R., "Stabilized Fiber-Optic Frequency Distribution System", TDA Progress Report 42-97, vol. January-March 1989, Jet Propulsion Laboratory, Pasadena, California, pp. 88-97.

[6] de La Chapelle, M., "Computer-Aided Analysis and Design of Microwave Fiber-Optic Links", Microwave Journal, pp. 179-186, September 1989.

GENERALIZED EMITTANCE MEASUREMENTS IN A BEAM TRANSPORT LINE[*]

J. Skelly, C. Gardner, A. Luccio, A. Kponou, and K. Reece

AGS Department, Brookhaven National Laboratory
Upton, NY 11973

ABSTRACT

Motivated by the need to commission 3 beam transport lines for the new AGS Booster project, we have developed a generalized emittance-measurement program; beam line specifics are entirely resident in data tables, not in program code. For instrumentation, the program requires one or more multi-wire profile monitors; one or multiple profiles are acquired from each monitor, corresponding to one or multiple tunes of the transport line. Emittances and Twiss parameters are calculated using generalized algorithms. The required matrix descriptions of the beam optics are constructed by an on-line general beam modeling program. Design of the program, its algorithms, and initial experience with it will be described.

INTRODUCTION

Measurement of the transverse emittance and Twiss parameters for a particle beam transport line is commonly accomplished using one or several profile monitors to determine the width of the beam.[1] The subsequent calculation of the desired beam properties is then performed by a computer program that can calculate and apply the beam transport matrices for this problem; conventionally the detailed properties of the transport line are represented in the program code. We report here the development of an application program for the Brookhaven AGS accelerator complex that incorporates several recent computing advances to provide a generic program that can be used with any transport line. Our immediate motivation was the need to commission three new transport lines for the AGS Booster. In addition, the program can be applied to existing beam transport lines.

CONCEPTUAL DESIGN

We formulate the problem in the usual way, as one of determining the properties of a beam at some initial point S_0, by measuring the widths of the beam profiles at one or more profile monitors farther down the line at points S_1, S_2, etc. The Twiss parameters α_i, β_i, γ_i at each point S_i are related to those at S_0 by the beam transport matrices for the line.[1] Then if the beam has an emittance ϵ, the measured widths w_i are given by

$$w_i^2 = \epsilon \beta_i = f_i(\epsilon \alpha_0, \epsilon \beta_0, \epsilon \gamma_0) \qquad (1)$$

for functions f_i determined by the transport matrices (neglecting dispersion). The three unknowns $\epsilon \alpha_0$, $\epsilon \beta_0$, and $\epsilon \gamma_0$ can then be determined if three of more widths are measured. If more than three widths are measured, a least squares approach is used to minimize the sum of the squares of the residuals R_i:

$$R_i = w_i^2 - \epsilon \beta_i \qquad (2)$$

The error matrix for the resulting beam parameters can also be determined from the sum of the squares of the residuals.[1]

If the beam transport line in questions does not have enough profile monitors to satisfy the requirements set forth above, an alternate approach is to vary the tune of the transport line (that is, the focusing strengths of the quadrupoles) and measure the widths of the profiles for each tune used in the beam line. If the transport line has only one profile monitor, this approach is the only one possible; in this case, one attempts to vary the tune of the transport line enough that the β function at the profile monitor exhibits a minimum as a function of the quadrupole strength being varied, and the phase advance at the profile monitor varies through a range comparable to $90°$.

In a transport line instrumented with multiple profile monitors, one can improve the quality of the measurement of the beam parameters by measuring the profile widths as the tune of the transport line is varied.

PROGRAM FEATURES

We have implemented this approach in an emittance program developed at the AGS. We incorporate several features which maintain generality in the program, permitting it to be used with any beam transport line.

The calculation of the beam transport matrices in our approach is carried out by a generalized modeling program for beam transport lines, the program MAD as developed at CERN.[2] We require the user to describe the beam transport line in MAD vocabulary; focusing strength are parameterized in terms of currents supplied to the quadrupole magnets, and these currents are interactively obtained from the control system. For each tune of the transport line, MAD is instructed to perform a TWISS analysis of the line, and the results of the MAD analysis are made available to the emittance program. The emittance program then calculates the beam transport

[*] Work performed under the auspices of the U.S. Dept. of Energy.

matrices from the MAD results. MAD requires initial values for the beam properties at the entrance to the transport line (emittance ϵ, Twiss parameters α and β), but the resulting transport matrices do not depend to first order on the initial values. These calculational techniques were validated in an earlier test on the proton injection line for the AGS.[3,4]

The emittance program is designed and written in a modern object-oriented language, C^{++}. The result of this object-oriented approach is that details of beam-line models, and of beam-line hardware (magnets and instrumentation) are concealed within object-oriented library routines, and are neither visible nor of concern to the emittance program proper; generality of the emittance program is enhanced and much easier to implement.

The selection of options within the emittance program (e.g. choice of beam line, segment within the beam line, measurement strategy) is governed by descriptions of the possibilities which are maintained in a directory structure of data files. Adding new options, e.g. new beam lines or new measurement strategies for existing beam lines, is accomplished by describing these options in the appropriate data files; the emittance program code need not be changed. The emittance program itself offers options for creating new measurement strategies (which quadrupoles to vary, which profile monitors to employ) and tactics (detailed choice of setpoints for each magnet for each profile measurement).

The emittance program offers an extensive graphical interface, which is implemented using an object-oriented graphics package. The object-oriented approach to the graphical interface preserves generality of the program and offers considerable efficiency in program development.

The emittance program offers a convenient graphical interface to the MAD modeling results. A potential emittance measurement strategy can be modeled and the results examined without actually using beam time. The strategy can be revised as appropriate, until the user is satisfied that the strategy will yield a successful emittance measurement; only then is beam time needed to obtain the final measurements.

Subsequent to a successful emittance measurement, the MAD modeling program may again be invoked, this time using the measured beam parameters to describe the initial beam, and using the on-line control values for the currents in the focusing beam elements; this final model of the beam line can be used to predict the beam properties at any location in the line. Locations upstream or downstream of the segment in which the measurement was performed can also be modeled with such a technique.

PROGRAM IMPLEMENTATION

The emittance program is implemented in the AGS Distributed Control System (AGSDCS). The AGSDCS comprises a network of approximately 50 Apollo workstation nodes on a Domain token-ring network which spans the AGS accelerator complex. The AGSDCS is interfaced to some 5800 accelerator devices via some 100 so-called device-controllers. Version 7.2 of the MAD program[2] has been implemented on this computer network. Apollo workstations situated in the AGS Main Control Room Consoles offer a windowing high-resolution (1280x1024) graphical color display, with a multiprogramming environment; each workstation may have several programs concurrently available via separate windows on the display. The emittance program employs a standard AGSDCS user interface, with interactive pull-down menus, a library of pop-up displays, and a standardized graphics package. Access to the control system interface to accelerator devices is accomplished in object-oriented tools which employ standard library routines and a database of available accelerator hardware.

APPLICATION TO BOOSTER PROTON INJECTION LINE

The proton injection line is some 35 meters in length and comprises 13 quadrupoles and 4 bending dipoles producing a net horizontal bend of 126°. The emittance program was used in commissioning this line to measure beam parameters in two straight sections of the line which each have a single profile monitor. Figures 1 through 3 are copies of on-line displays from the program. Figure 1 illustrates graphical displays of these two sections of the beam line. Figure 2 exhibits a typical set of profiles, and Fig. 3 plots the widths of these profiles (widths defined as enclosing 98% of the profile area) against a quadrupole current. The parameters resulting from a least squares fit to these data are shown in Table I.

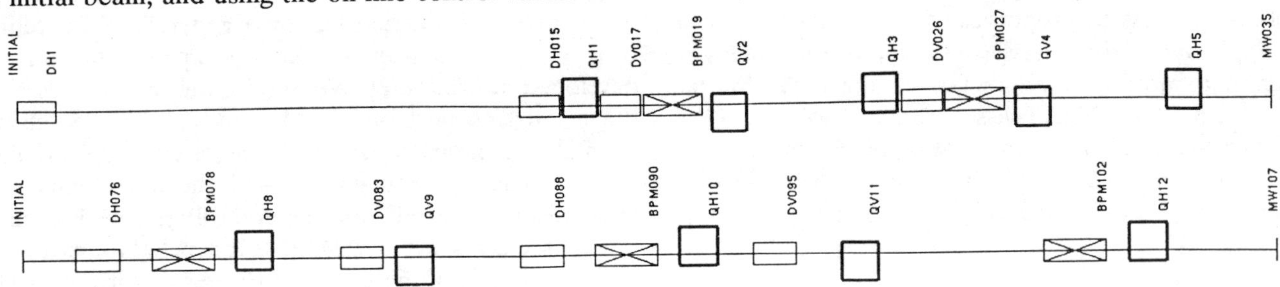

Fig. 1. Graphical layouts of two of the sections of the Linac-to-Booster (LTB) proton injection line; upper layout is LTB1, lower layout is LTB3.

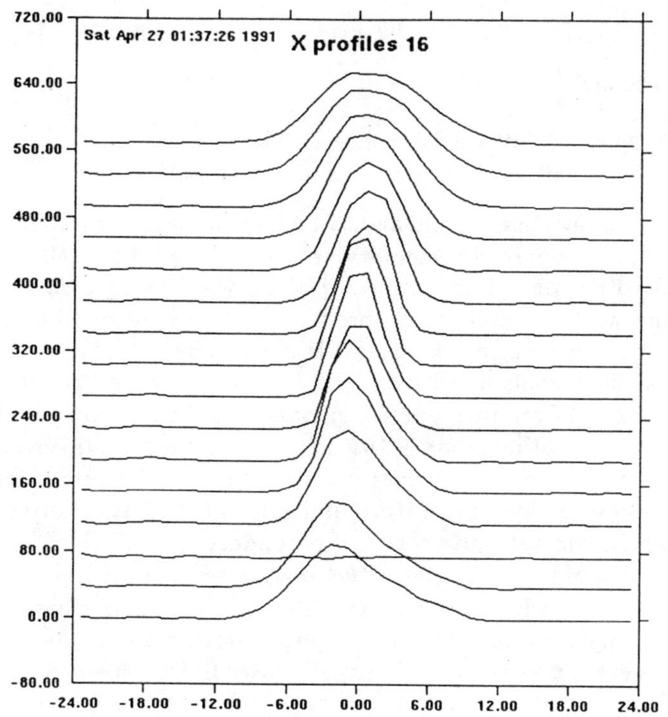

Fig. 2. Horizontal profiles from multiwire MW035 in LTB1, as quadrupoles QV2, QH3, and QV4 are varied together as a triplet.

APPLICATION TO HEAVY-ION TRANSPORT LINE

The basic optical cell of the HITL consists of a drift, followed by a quadrupole doublet, followed by another drift. The doublet is tuned so that the cell images a waist to a waist in both the horizontal and vertical planes. Profile monitors provide beam widths near the waists. Emittance and twiss parameters are determined from profile widths acquired as the current in the doublet is swept through a series of setpoints. A least squares fit to the profile widths yields the desired parameters.

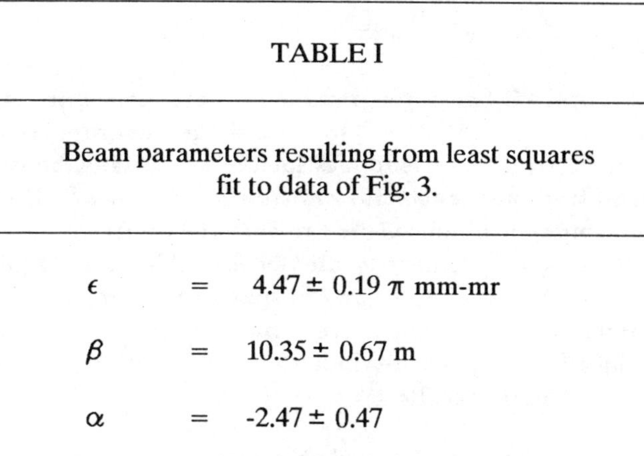

TABLE I

Beam parameters resulting from least squares fit to data of Fig. 3.

$$\epsilon = 4.47 \pm 0.19 \ \pi \ \text{mm-mr}$$

$$\beta = 10.35 \pm 0.67 \ \text{m}$$

$$\alpha = -2.47 \pm 0.47$$

OTHER APPLICATIONS

The emittance program will be used in commissioning the Booster extraction line and Booster Heavy-Ion injection line later this year. There are a number of other beam transport lines at the AGS in which we anticipate useful application of this program as well.

REFERENCES

1. K. Ebihara, M. Tejima, T. Kawakubo, S. Takano, Z. Igarashi, and H. Ishimaru, Nucl. Instr. and Methods 202 (1982) 403-409.
2. F.C. Iselin and J. Niederer, "The MAD Program," CERN/LEP-TH/88-38.
3. A. Luccio, private communication.
4. A. Kponou & A. Luccio, private communication.

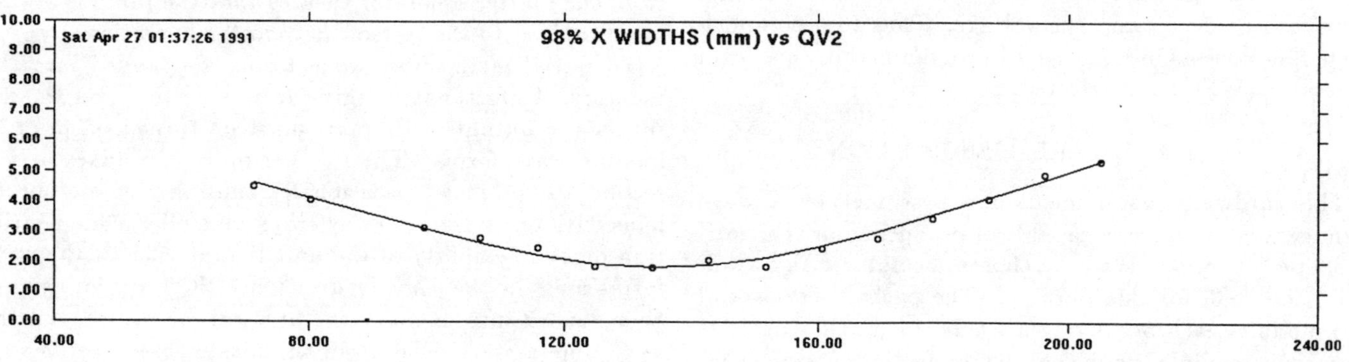

Fig. 3. The width from the profiles in Fig. 2 are plotted (open circles) against the current in QV2. The curve through the points is the result of the least squares fit to these widths.

The Control and Operation of the Programmable Wave Form Generator for the SSRL Injector*

S. Brennan, S. Baird‡ W. Lavender, H.-D. Nuhn, C. Wermelskirchen‡ J. Yang
Stanford Synchrotron Radiation Laboratory, P.O. Box 4349, Bin 69, Stanford, CA 94309-0210

Abstract

A CAMAC-based programmable wave form generator is described which is used to control the quadrupole and sextupole trim power supplies for the new 3 GeV booster synchrotron injector for the electron storage ring SPEAR. This same type of unit is used to vary the cavity voltage of the RF system. Memory located in the CAMAC module is loaded with a wave form and output to the supplies with an internal clock. The wave form can be loaded either graphically using the X-windows-based control system or with an ASCII text file.

I. INTRODUCTION

We have recently completed the commissioning of a new 3 GeV booster synchrotron injector adjacent to the SPEAR storage ring[1]. The injector consists of a 120 MeV linac followed by a booster whose magnets are ramped using a White circuit. [2] The principle of the White circuit is that the dipole and quadrupole magnets in the booster are placed in series with capacitors so that an L-C oscillating circuit is produced. For the case of the SPEAR injector the resonant frequency is 10 Hz. With the dipoles and quadrupole currents being ramped in hardware the only magnet supplies which must be ramped using a programmable power supply are the quadrupole and sextupole trim coils. After 9 months of operation the sextupole magnets have not been utilized, so they will not be discussed further. The purpose of the quadrupole trim coils is to fix minor problems in magnetic field within the quadrupoles during the ramp. The specific shapes of the ramps for the focusing and defocusing quadrupoles are empirically determined. The other component of the injector which needs an external ramp is the radio-frequency (RF) cavity voltage. For this purpose one of several analytic functional forms are used to ramp the voltage. Thus to accomplish these functions a CAMAC-based programmable wave form generator (WFG) was selected.

II. HARDWARE DESCRIPTION

The hardware requirements are best met by a wave form generator which is capable of producing an arbitrarily shaped function. We have chosen a commercially available module[3] for this purpose. The package consists of two modules each occupying a single slot in the CAMAC crate. One module has the output connections and the timing input connections and is addressed by the computer. The second is the memory storage module and talks to the first via a buss connector along the top back edge of the units. The unit can produce 1, 2 or 4 output signals simultaneously using a 12-bit D-A converter. The analogue output signals are driven by two op-amps which are each capable of producing $\pm 5V$ signals with respect to ground, thus in floating mode $\pm 10V$ is possible, but for our system only a single $\pm 5V$ driver is used. Short circuit protection is achieved by temperature limitation of the output driver, which can get quite toasty if grounded.

One of the surprises of the module was a transient voltage spike which occurs each time the voltage is changed during a ramp. Since our application uses the module to generate a wave form that is 100 msec in length a low-pass filter with 1 kHz time constant is sufficient to smooth out the signal and has no effect on operation. It might be a problem for higher frequency operations.

The memory module contains 32K of static memory which can be parsed according to the number of signal channels. For our application a 1000 sample wave form was used, updating the wave form every 0.1 msec.

III. INITIALIZATION

The proper timing of the booster is critical to its functioning. Rather than have the WFG operate in free-running mode a master oscillator is used to synchronize all of the critical events. The WFG can operate either in continuously repeating mode or can output 1-15 repetitions of the form and then stop. As we have implemented the module it is set up in continuous mode with a start pulse sent by the master oscilator. In this mode the unit only completes a full wave form and restarts itself if a start pulse is missed by the master oscillator. The start pulse from the master oscillator ensures that the ramp is always synchronized to the rest of the booster operations.

To initialize the unit wave forms are loaded into the memory. Unfortunately this can not be done on the fly, rather the output of the unit must be turned off prior to loading wave forms. The loading procedure takes over a second to complete, occasionally causing time-out problems with our serial highway crate controller. One restriction on the flexibility of the unit is that each of the wave forms must be the same length. Once the wave forms have been loaded into memory the unit can be armed and the next time a start pulse from the master oscillator is seen the wave form will be unloaded from memory. A software start is also an option after the module has been armed.

*Work supported by the Department of Energy, Office of Basic Energy Sciences, Division of Material Sciences.

‡visiting SSRL from CERN

‡Now at Gesellschaft für Mathematik und Datenverarbeitung (GMD), Bonn, Germany

IV. Wave Form Creation and Loading

The user interface to the WFG is through the X-windows-based graphical user interface of the injector control system.[4] For the focusing and defocusing quadrupoles the inputs to the final wave form are 20 amplitudes adjustable on screen using sliders (a graphical input *widget* provided by X-windows). Each of the 20 sliders represents a different time with respect to the injection of the electron beam into the booster from the linac. There are additional sliders for adjusting the phase of the wave form, a constant offset and an overall amplitude factor. The program divides the 50 msec ramp time into 19 intervals and assigns the 20 slider values to the 20 interval borders. The total number of points in the wave form is 1000, so the first 500 are calculated as a linear interpolation between the slider values. The second 500 values are a smooth interpolation between the last and the first slider value while trying to keep the wave form amplitude low. We plan to replace the linear interpolation by a 3-point spline interpolation in the near future. The phase, offset and overall amplitude are applied to the total wave form before it is written to the data base and loaded into the WFG. A new wave form is calculated and loaded whenever one of the input parameters is changed.

The RF cavity voltage depends on a number of machine parameters such as the AC and DC offset voltages for the White circuit and the linac energy. There are also parameters for the rate of change of the magnetic field in the dipoles ($\dot{B}$) and overall phase, offset and scale parameters. The program converts the White circuit voltages into energy equivalents which are used to estimate the phase angle ϕ of the white circuit at which injection occurs. The 1000 waveform points (to cover the 100 msec period) are calculated by

$$f(t) = \dot{E}\, T_{rev}\, \dot{B} + E(t)^4\, U_0$$

where

$$
\begin{aligned}
E(t) &= ENERGY_{DC} + ENERGY_{AC}\, SIN(\omega t + \phi), \\
\dot{E} &= \omega\, ENERGY_{AC}\, COS(\omega t + \phi), \\
U_0 &= \text{Energy Loss per Turn}, \\
T_{rev} &= \text{Revolution time}.
\end{aligned}
$$

Again, the phase, offset and scale factors are applied to the 1000 waveform points before it is written to the data base and loaded into the WFG. A new waveform is calculated and loaded whenever one of the input parameters is changed. As an alternative, the program can accept as parameters the maximum booster energy and the ratio of the AC to the DC component of the magnetic field which it converts to the AC and DC voltage setpoint by using the calibration factors.

The control system for the injector includes provision for saving to an ASCII text file a particular parameter or set of parameters. These text files can be reloaded into the data base. Thus an arbitrarily shaped wave form can be calculated and written into a text file with the proper format and then loaded into the data base and the WFG module.

V. References

[1] H. Wiedemann, "3 GeV Injector Synchrotron for Spear." At this Conference.

[2] R. Hettel et al., "The 10 Hz Resonant Magnet Power Supply for the SSRL 3 GeV Injector." At this Conference.

[3] Model 5910 Function Generator, Bi Ra Systems, Inc. Albuquerque, N.M. 87107.

[4] C. Wermelskirchen et al., "The SSRL Injector Control System." At this Conference.

Triggers and Timing System for the SSRL 3 GeV Injector[*]

R. Hettel, D. Mostowfi, R. Ortiz, J. Sebek

Stanford Synchrotron Radiation Laboratory, Stanford, CA 94309-0210

Abstract

The electron beam for the SSRL 3 GeV Injector facility is produced in an RF gun, chopped by a stripline deflector to form a 1 nsec long beam bunch, injected into and extracted from a single 358 MHz RF bucket in a 10 Hz booster synchrotron, and injected into a preselected 358 MHz bucket in the SPEAR storage ring. The systems that generate 10 Hz triggers for the linac, beam chopper, and the pulsed and cycling magnets are described.

I. INTRODUCTION

The SSRL Injector facility consists of a 120 MeV S-band linear accelerator and a 10 Hz booster synchrotron designed to provide single bunch filling of 3 GeV electrons for the SPEAR storage ring [1]. The linac beam originates in an RF electron gun [2] driven with power tapped from the waveguide connecting one of the linac sections to its modulated klystron. The gun sources a string of 2856 MHz electron bunches during much of the 1 usec modulator macropulse. A stripline deflector is used to sweep the gun beam past a slit at a rate that only permits 3 of the S-band bunches to enter the linac [3]. These microbunches are accelerated to 120 MeV and are injected into a single 358 MHz bucket in the booster where they are ramped to 3 GeV. The microbunches coalesce into a single bunch before being ejected and transported to a single 358 MHz bucket in SPEAR.

Two classes of trigger timing systems have been implemented to achieve the Injector timing requirements. One system, shown in figure 1, generates 10 Hz triggers for the for the booster resonant magnet power supply system (White Circuit) [4], the focusing and defocusing quadrupole trim tracking supplies, the booster RF gap voltage ramp, and the pulsed ejection septum magnet. This system also provides injection and ejection energy gate signals for the other more precise RF synchronized system, shown in figure 2, used for triggering the linac modulators, beam chopper, booster injection kicker, and booster ejection and SPEAR injection kickers.

II. 10 Hz CLOCK AND ENERGY TIMING

10 Hz Clock and Delayed Triggers

The Injector 10 Hz clock is phase-locked to the 60 Hz line frequency to minimize the impact of power supply ripple on machine stability. The phase-locked loop response is limited to < 1 Hz so that the 10 Hz clock does not track fast line frequency phase jitter. The undelayed 10 Hz clock is used to trigger the White Circuit pulser

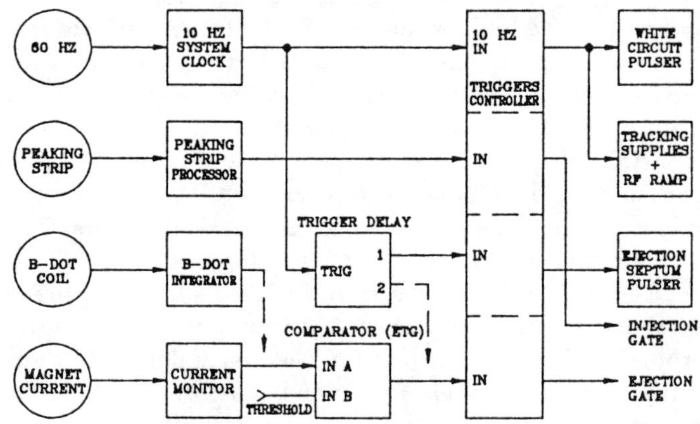

Figure 1. 10 Hz clock and energy gate timing system

network and the waveform generators that drive the tracking amplifiers for the two quadrupole families and the RF cavity gap voltage [5]. The frequency stability and timing precision required for these devices is on the order of 0.1% to achieve a comparable level of machine functional stability. This corresponds to a timing stability requirement of order 100 μsec which is easily met with this system.

The 10 Hz clock signal is delayed by ~80 msec to provide a trigger for the pulsed ejection septum magnet [6]. The magnet pulse width is 17 msec, and the pulse peak timing must be stable to within 100 μsec of the time the booster magnet current reaches the ejection amplitude to maintain 0.1% ejection energy stability. This timing precision over 80 msec is readily achieved with a commercial trigger delay unit; the ejection septum pulser is also stable to this level.

Peaking Strip Injection Energy Timing

Another task for the 10 Hz trigger system is to generate a beam energy-dependent timing gate for the RF synchronized timing system that is used to trigger the linac, beam chopper, and injection kicker magnets. The injection energy acceptance of the booster is ~0.5%; the desired injection energy stability is 0.1% to maximize beam capture and transmission efficiency during ramping. If this stability were to be achieved with a system that generated the injection timing trigger by simply delaying the 10 Hz clock signal, the ring magnet power supply system, which provides a peak current of 630 A at 3 GeV, would have to be accurate to 25 mA at the 25 A injection current.

This stability requirement for the power supplies is relieved by instead deriving the trigger from a permalloy peaking strip located in a solenoid magnet coil situated in the gap of one of the booster dipoles. The peaking strip produces a voltage pulse in the bias coil, which acts also as a sense coil, when the magnetic field in which it is

* Work supported by Department of Energy, Office of Basic Energy Sciences, Division of Material Sciences.

immersed changes sign. The bias field is· adjusted to exactly cancel the dipole field at the proper injection energy. The Peaking Strip Processor develops a trigger signal from the detected pulse and also sources the coil bias current. The pulse width is inversely dependent on the slope of the changing magnetic field and is ~50 μsec at injection when the magnet current is fully biased. The timing stability is ~2 μsec, well within the 5 μsec tolerance needed for 0.1% injection energy stability.

Ejection Energy Timing

The ejection energy timing gate that is used for generating triggers for the booster ejection kicker and the three SPEAR injection kickers is presently produced by delaying the 10 Hz clock by ~85 msec. This technique requires that the magnet power supply system be stable to 0.1% of its full current capability to achieve that level of ejection energy stability. The consequent timing stability is of order 100 μsec. In practice we achieve this degree of supply stability over the short term (1 or 2 hours), but the supplies needed to be adjusted over longer periods.

To improve the ejection energy stability, we are presently implementing an Energy Timing Generator that detects either 1) the magnet current, or 2) the dipole magnetic field and produces trigger signals when preprogrammed levels are reached. The magnet current is sensed by a high accuracy (< .01%) transductor; the magnetic field is calculated by integrating the the signal from a B sense coil situated in the gap of one of the booster dipoles.

Triggers Controller

All of the above trigger signals pass through a Triggers Controller unit that permits computerized signal on/off control. This unit also automatically switches the triggers for the injection and ejection energy timing gates from their magnet field-and current-derived sources when the White Circuit is cycling to the 10 Hz clock when the White Circuit is off. This permits the linac and kickers to be operated without running the booster magnets and ensures that they will not inadvertently lose triggers, which could be disruptive to stable operation, when the White Circuit turns on or off.

III. RF SYNCHRONIZED TIMING SYSTEM

In order to inject beam into a single 358 MHz bucket in SPEAR, several precise timing events must take place: 1) the three linac klystron modulators the pulsed S-band drive amplifier must be triggered so that the 120 MeV beam arrives at the booster injection point within the 0.1% injection energy window; 2) the injection kicker near the injection point must be triggered so that the incoming beam arrives at the kicker flat-top just prior to the falling edge so that the kicker is fully off by the time the beam completes its first revolution in the booster 445 nsec later; 3) the beam chopper must be triggered within the timing aperture created by the linac macropulse, the injection kicker waveform, and the RF bucket acceptance window in the booster; and 4) the booster ejection kicker and SPEAR injection kickers must be triggered within the ejection

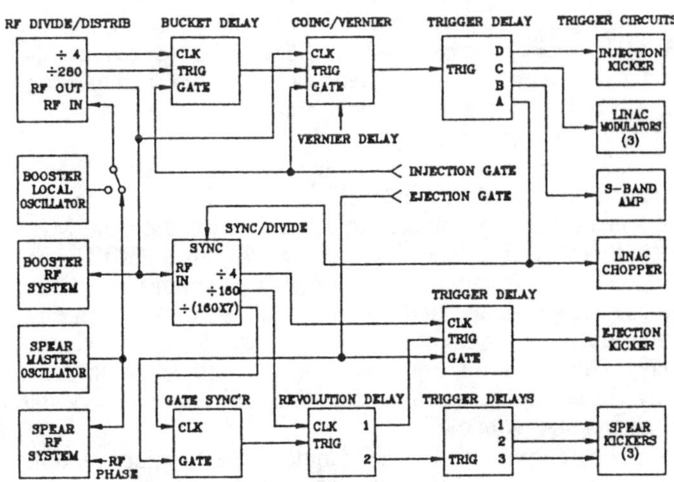

Figure 2. RF synchronized timing system

energy timing gate at a time that will cause the booster bunch to be captured in a preselected SPEAR bucket.

Single Bucket Injection

Two basic system features make single bucket filling feasible: 1) the linac beam chopper creates a short enough (~1 nsec) beam burst with a timing jitter of < 300 psec so that the beam can be reliably captured in single 358 MHz booster bucket; and 2) the booster and SPEAR RF systems are driven by the same 358 MHz master oscillator so that a high level of phase stability between booster and SPEAR buckets is attained.

The second feature guarantees that, because the booster and SPEAR bucket harmonic numbers are 160 and 280 respectively, 7 booster revolution periods will be precisely equal to 4 SPEAR revolution periods. This harmonic relationship ensures that if a booster bucket is extracted at time t_e from the booster, it will hit the same SPEAR bucket as it would if it were extracted at a time $t_e + 7NT_B$, where N is an integer and T_B is the booster revolution period of 445 nsec.

The basic principle of the RF synchronized timing system is thus to 1) create a SPEAR bucket-dependent timing signal that triggers the linac, chopper and injection kicker through appropriate fixed delays; 2) use the chopper trigger signal to reset and synchronize a 358 MHz divider unit that produces timing signals on every 7th booster revolution period; and 3) generate triggers for the booster ejection and SPEAR kickers through appropriate fixed delays from the first such $7T_B$ divider signal occurring within the ejection energy gate.

Bucket Selection Timing

The SPEAR bucket timing signal is obtained by delaying the SPEAR revolution clock signal by integral number of bucket periods. The SPEAR revolution clock is generated by dividing the 358 MHz master oscillator frequency by 280. The computer controlled bucket delay unit counts a programmed number of 89 MHz (358 MHz divided by 4) clock signals; single bucket delay resolution is achieved with a programmable delay line included in the delay unit

[8]. The SLAC designers of this module chose to use the 89 MHz clock, as opposed to 358 MHz, to reduce the need for sensitive high frequency components and packaging methods. The subsequent timing accuracy of this ECL unit is ~500 psec.

The timing accuracy for the beam chopper trigger is increased to the 100 psec level by resynchronizing the output signal from the bucket delay unit with the 358 MHz clock in the Coincidence/Vernier unit [8] using MECL III components. This unit also contains a voltage-controlled varactor diode delay line so that the signal delay can be continuously adjusted over a 4 nsec range. This vernier delay is used to fine-tune the chopper timing so that injected beam arrives within the 300 psec optimal booster RF acceptance window.

The output from the Coincidence/Vernier module triggers a 4-channel programmable delay unit. This unit is used to establish the proper fixed timing relationships between klystron modulator, S-band amplifier, injection kicker and chopper triggers. The relative timing stability requirement between these units is ~10 nsec to achieve 0.1% or better energy spread the linac and to make sure the beam arrives reproducibly at the kicker flat-top [7]; timing jitter on the order of 5 nsec is observed for the modulators and kicker and is caused mainly by the thyratron pulsers.

The critical chopper timing stability of less than 300 psec with respect to the 358 MHz RF drive frequency is maintained by the delay unit and by the MOSFET chopper pulser [3], both which have timing jitters of ~100 psec.

When a different SPEAR bucket is selected, the trigger from the Coincidence/Vernier unit changes by a discreet number of RF bucket periods. The chopped beam timing tracks accordingly with a measured accumulated jitter of ~200 psec (figure 3).

Ejection Timing

If the beam that is injected into the booster were to be extracted the very first time it passed the ejection point, which is a fixed distance from the beam chopper, it would traverse the fixed distance from the ejection point to the SPEAR RF cavity (ignoring the fact that the energy would be wrong!) with the same 300 psec timing precision obtained at the booster cavity. To be captured in the corresponding SPEAR bucket, a vernier delay might be needed so that the beam would arrive within the RF acceptance window. The booster bunch would hit the same SPEAR bucket if it remained in the booster for any integer number of 7 booster revolution periods. Any additional fixed timing delays imposed on the system by component response times or for the purpose of adjusting their relative timing only result in shifting the bucket that will be filled by a fixed number.

The key component in preserving the single bucket timing requirement throughout the ~45 msec between booster injection and ejection is the Sync/Divide unit which performs the synchronized divide-by-seven of the booster revolution period. This unit first divides the 358 MHz clock by 160 using MECL III and ECL components to obtain the booster revolution period. The division is performed by counters that are reset each time the chopper

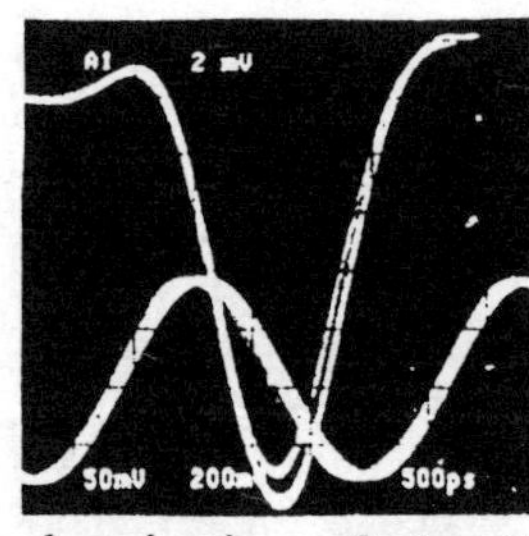

Figure 3. Timing jitter of first turn beam bunch in booster (upper trace) with respect to 358 MHz RF drive (lower trace) is ~200 psec.

is triggered so that the proper synchronization of the divide-by-seven clock and the booster bunch is maintained. When a different SPEAR bucket is selected by shifting the Coincidence/Vernier trigger with respect to the SPEAR revolution clock by a discreet number of buckets, the divide-by-seven clock phase is shifted by precisely the same number of buckets. The timing jitter of the Sync/-Divide output with respect to a synchronizing trigger is measured to be ~200 psec.

The first divide-by-seven clock that falls within the ejection energy timing gate generates triggers for the booster ejection and SPEAR kickers though programmable fixed delay units. Although a 4-channel delay unit identical to the one used for injection timing could have been used here, the booster and SPEAR delay units are located in different buildings for historical reasons. In the case of the booster ejection kicker, this delay is set by cascaded pair of counters: one that counts booster revolution periods for coarse adjustment, followed by a bucket delay unit that provides 2.8 nsec resolution. This timing accuracy is much better than the timing stability of order 10 nsec required for the kickers, also determined by flat-top requirements.

The actual timing precision between the booster and SPEAR buckets is defined by that of the booster beam passing the ejection point, which in turn corresponds to the longitudinal stability of the accelerated booster beam within its own bucket. This precision is better than 100 psec. Vernier timing adjustment is accomplished by adjusting the phase of the 358 MHz drive for the SPEAR RF system using a voltage-controlled 360° phase shifter.

V. References

[1] H. Wiedemann, *et al.*, *3 GeV Injector Synchrotron fo SPEAR*. In these proceedings, 1991.

[2] J.N. Weaver, *et al.*, *The Linac and Booster RF System for a Dedicated Injector for SPEAR*. In these proceedings, 1991.

[3] M. Borland, *et al.*, *Design and Performance of the Traveling-Wave Beam Chopper for the SSRL Injector*. In these proceedings, 1991.

[4] R. Hettel, *et al.*, *The 10 Hz Resonant Magnet Power Supply System for the SSRL 3 GeV Injector*. In these proceedings, 1991.

[5] S. Brennan, *et al.*, *The Control of the Programmable Wave Form Generator for the SSRL Injector*. In these proceedings, 1991.

[6] J. Cerino, *et al.*, *Extraction Septum for the SSRL Injector*. In these proceedings, 1991.

[7] H.-D. Nuhn, *et al.*, *The SSRL Injector Kickers*. In these proceedings, 1991.

[8] K. Slattery, *Hardware Upgrade of the SPEAR Timing System*, SLAC internal technical note, 1985.

MAIN CYCLE CONTROLS FOR THE AGS BOOSTER SYNCHROTRON*

B. B. Culwick, S. Yen

Brookhaven National Laboratory
Upton L.I., N.Y. 11973

ABSTRACT

The AGS Booster is a separated function synchrotron with the main excitation coils of the dipoles and quadrupoles connected electrically in series. This circuit is driven by a complex modular power supply with current and voltage reference functions to obtain the desired magnetic fields as a function of time. The dipole cycle is defined by algebraic functions specifying the desired field profile as a function of time. These functions are processed through successive phases to convert to the signals needed to provide the power supply with one current and six voltage references. The user interface and algorithms to derive the control variables are described.

INTRODUCTION

The main magnet of the AGS Booster synchrotron is designed to operate over a wide range of frequencies from 7.5 Hz for protons to <1 Hz for heavy ions. To accommodate this range while preserving precision of magnetic field value as a function of time and good power factor on the incoming power line requires a complex power supply [1]. The design selected for the power supply makes use of six 24-phase SCR controlled modules one of which is supplied with a current reference and all of which are supplied with voltage references. In addition to these controls, any modules not required at a given time can be by-passed by additional SCRs to eliminate their effect on power factor and ripple voltage.

The current and voltage references to the power supply modules are sequenced by vector function generators which drive digital to analog converters [2]. The function generators are composed of two logical entities referred to as the timing module which provides common timing to all generators, and the amplitude modules, one of which is dedicated to each function and is located at the relevant power supply. Vectors are defined by a starting value, a slope and a time duration. Binary controls also permit multipliers on the slope and enable the by-passes when required.

CONTROL STRATEGY

Considerations of accelerator physics, operational convenience and simplicity led to a decision to define the main magnet cycle in terms of magnetic field versus time. To facilitate suitable parameterization, the user interface allows graphical or numeric definition of field values and derivatives with respect to time. For reasons of tradition the units selected for the operator interface are kilogauss and milliseconds although inter-program communication is in tesla and seconds. The magnetic field versus time function is then defined as a series of zones of particular algebraic form. Zone specifications range from linear through cubic with defined derivatives or derivatives propagated from previous zones. Figure 1 shows a typical magnetic field function with a zone specification menu superimposed. From such a function it is intended that further processing be automatic and manual intervention is confined to verification of program actions and entry of parameters for which defaults are provided. Manual procedures are also invoked if errors or incompatibilities are encountered.

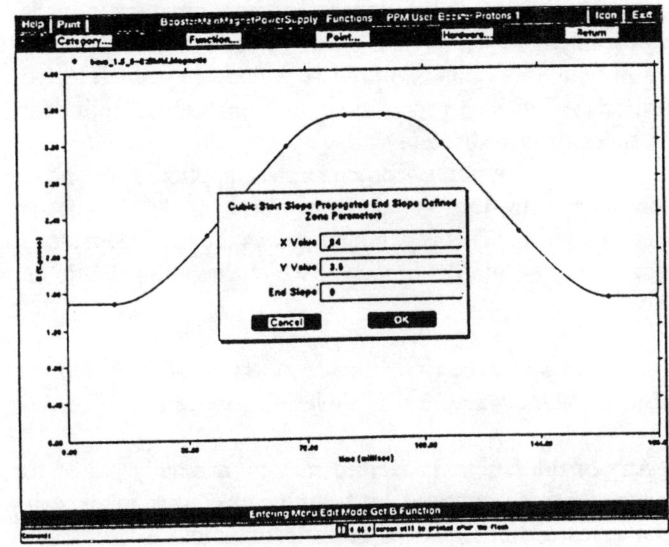

Figure 1 Magnetic Field Plot with Zone Menu

---

*Work performed under the auspices of the U. S. Dept. of Energy

Detailed Processing

The procedures for converting to hardware format pass through the following steps:

1) Conversion of magnetic field values to current. This procedure makes use of a calibration function expressed as a cubic spline fit to the calibration data. This facilitates subsequent differentiation of the function (see 2). No provision for hysteresis is made in the present code.

2) Calculation of the total voltage required to achieve the desired current waveform. This calculation makes use of a simple model in which the magnet is represented by a series combination of resistance and inductance. The voltage is calculated as

$$V = Ri + L \frac{di}{dt} \qquad (1)$$

At present the resistance and inductance are treated as constants but consideration has been given to making them functions of i or di/dt. The time derivative of the current is calculated by the "chain rule"

$$\frac{di}{dt} = \frac{dB}{dt} \times \frac{di}{dB} \qquad (2)$$

to avoid non-analytic procedures which produce discontinuities in the voltage. Finally the voltage waveform is advanced by a predetermined time to compensate for delays in the power supply. This is equivalent to introducing a term in dV/dt in the voltage, a procedure justified in reference [3].

3) Division of the voltage between modules is made using parameters such as maximum voltage per module. The order in which modules are used is defined by user selected priorities. A two step procedure avoids enabling supplies for short times or low absolute voltages.

4) A "rounding" procedure is applied to remove discontinuities in the voltage waveform at the divisions between modules. This is parametrized as the maximum slope change permitted and the time over which rounding should be applied.

This completes the algorithmic processing of the function. A typical voltage waveform is shown in Figure 2.

Any of the functions derived may be manually edited to introduce features required by testing procedures or experimental investigations. These modifications may be for one-time usage or may be preserved for future use. [4] A manually modified function is shown in figure 3.

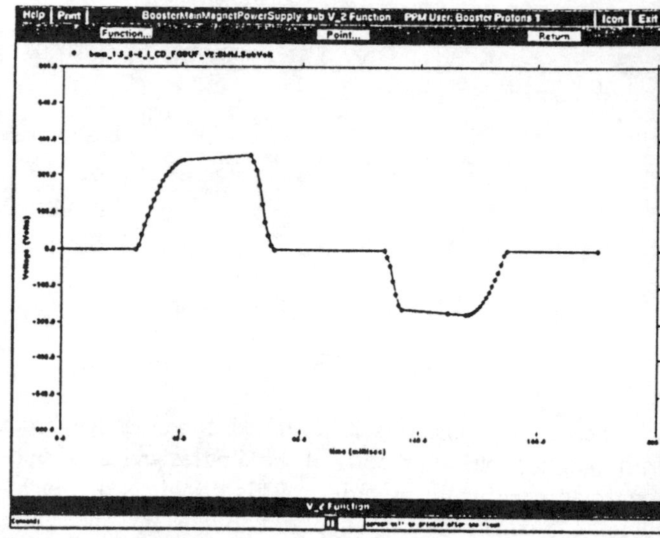

Figure 2 Typical Voltage Plot

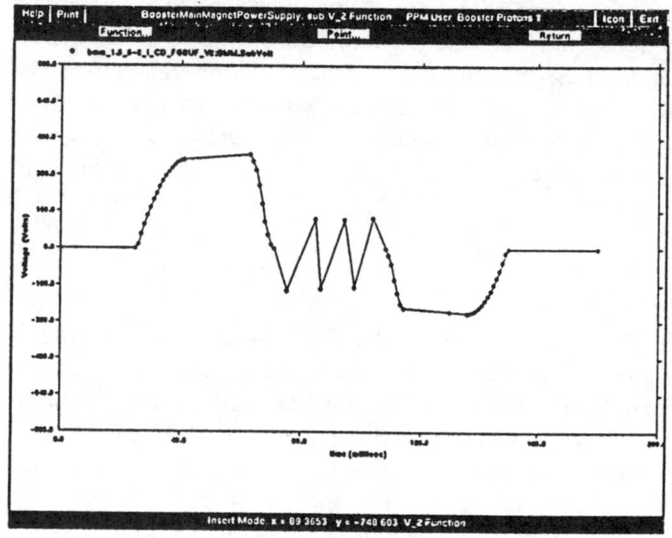

Figure 3   Manually Modified Voltage Plot

Execution

Before loading a function to the Device Controller which operates the power supply in real time, the data must be converted to the vector format. The procedure consists of calculating a series of vectors of increasing length until the average value of the absolute magnitude of the deviations of the vector from the function exceeds a preset value. The previous vector is then accepted and a new vector begun. Any vectors of zero magnitude and slope are flagged to indicate that the by-pass is to be enabled for this vector. This fitting procedure preserves the area under the curve and hence the voltage-time integral. When all functions have been fitted, the common time divisions are derived by or-ing the individual time points. This produces the common timing required by the hardware implementation. The vectors are loaded to buffers in the Device Controller. Interaction between the

Device Controller and the application program then allows execution of the function in a one-shot or repetitive cycle.

POWER SUPPLY MONITORING

In addition to conventional status readbacks, the power supply is equipped with 28 transient digitizers which may be triggered by faults or a timer to preserve data on power supply parameters. The control program provides for retrieval of the data from these recorders as well as from the buffers or function generator tables. These data may be presented numerically or graphically to the operator.

PERFORMANCE

The program has been used quite extensively in tests of the Booster Main Power Supply, the most recent with it connected to about 60% of the ring magnets. Results have been satisfactory The user interface is quite convenient and transient recorder data has been of value in recording power supply parameters. A significant modification to the magnet cycle can be accomplished in about 30 seconds.

When the Booster ring is complete, the program parameters will be optimized to model the full complement of magnets and to prepare for injection, acceleration and extraction studies.

REFERENCES

1. A. V. Soukas et al., "AGS Booster Main Ring Power Supply System," Particle Accelerators, Vol. 29, pp. 121-126, 1990.

2. N. F. Schumberg, "Dipole Power Supply Interface Specification," Private communication.

3. S. Y. Zhang, Private Communication.

4. S. Mandell, Private Communication.

Smart Rack Monitor for the Linac Control System

S. Shtirbu, R. W. Goodwin, E. S. McCrory, M. F. Shea
Fermi National Accelerator Laboratory
P. O. Box 500
Batavia, IL 60510

Abstract

The Smart Rack Monitor (SRM) is a low-cost, one board, data-acquisition module for the upgraded Linac control system at Fermilab. The SRM is based on the Motorola MC68332 microcontroller in the Business Card Computer (BCC) configuration. It is connected to the Linac local control station (to be referred to as Local Station) by an Arcnet LAN, and can be located close to the controlled hardware. Each Local Station is connected to several SRMs. The SRM has 64 A-D channels, sixteen D-A channels, and eight bytes of digital I/O on the mother board. Software Components Group's pSOS is the SRM's kernel. The SRM's software is cross-developed on VAX/VMS in C. The SRM does not have an attached console and is fully controlled by the Local Station. It performs data acquisition and settings, as directed by the Local Station. Its existence is transparent to the rest of the control system. The SRM supports code updates downloading from the VAX, through the Local Station.

I. INTRODUCTION

The 400MeV Linac upgrade project at Fermilab has spawned an upgrade of the Linac local control stations. The major change is the upgrade from a Multibus MC68000 system to a VME MC68020 system. The Local Stations are connected to each other and to the rest of the accelerator control system by a Token-Ring LAN. Since the new Local Stations are more powerful than the previous ones and can control more devices, fewer of them are needed. Reducing the number of Local Stations requires extensive wiring to reach one Local Station from several RF stations.

The Smart Rack Monitor is based on the Rack Monitor designed for DZero [1]. It has its own processor and controls the local devices. It is connected to Local Stations by an Arcnet LAN, which eliminates the need for the extensive wiring. Since such a design also frees the Local Station's CPU from data acquisition tasks that are performed in parallel, more devices can be controlled by one Local Station.

II. GENERAL DESCRIPTION [2]

A. Business Card Computer

The SRM is based on the Motorola MC68332 microcontroller (MCU). This new processor is available in a small subsystem called Business Card Computer (BCC), a 2.25 by 3.5 inch board that has 128 kbytes of PROM, 64 kbytes of RAM, and an RS-232 serial interface. (The SRM is not normally connected to a terminal). The BCC attaches to the motherboard by two 64-socket DIN connectors.

B. Local Area Network Interface

Standard Microsystems Corporation's COM90C65 Arcnet adapter is integrated into the SRM's motherboard.

C. Memory

In addition to the memory on the BCC, the SRM has 64 kbytes of non-volatile RAM. This memory is used for the SRM's code and tables.

D. Packaging

The SRM is a self-powered, 2U (3.5 inch) rack mounted chassis. It has analog and digital I/O connectors on the lower 1U of the rear panel, with additional connector space available on the upper part for connectors used by daughterboards.

III. ON BOARD I/O CAPABILITY [2]

All I/O devices on the SRM board are memory mapped, to allow easy access from the controller.

A. Analog Input

There are 64 channels of differential A-D input. Four connectors on the rear panel of the SRM are used to connect the analog inputs to eight 8-channel differential input multiplexers. Outputs from these multiplexers are combined by a ninth multiplexer that drives an instrumentation amplifier followed by the digitizer. The family of Crystal Semiconductor converters used is pin-compatible with twelve, fourteen or sixteen bit resolutions. The Crystal Semiconductor A-Ds are CMOS devices and include an onboard microcontroller to recalibrate the device at power-on, or by an external command. The converters include an input sample-and-hold function.

B. Analog Output

Sixteen D-A output channels are provided using four Analog Devices' AD664 quad D-A circuits. The D-A registers are accessible for both read and write. Output ranges of these converters are programmable.

Operated by the Universities Research Association, Inc. under contract with the U.S. Department of Energy.

C. Digital I/O

Eight bytes of digital I/O are on the SRM board. Normally, the 74ALS652 chips are used. These chips offer non-inverting bidirectional octal latch-buffer circuit with high active totem pole outputs. All digital I/O lines are pulled to +5V with 1 kohm pullup resistors. The data direction of the digital I/O is determined for each byte by a front panel piano DIP switch.

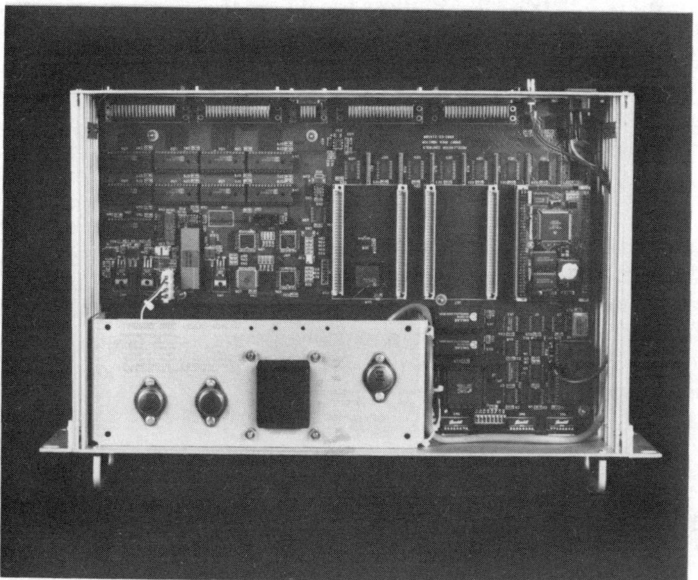

Photograph of a Smart Rack Monitor

IV. DAUGHTERBOARDS [2]

One of the major advantages of the SRM design is the ease of adding custom interfaces where needed. These interfaces are implemented as daughterboards with the same footprint and pinout as the Motorola BCC module, so that all the facilities of the MCU are available. Such boards can be stacked with virtually no limit (there are three sockets for BCC-pinout boards on the motherboard). Several such boards are already implemented.

A. Interface to old Linac hardware

Two daughterboards are needed to interface with the existing Linac hardware. One of them handles nine bytes of digital I/O, and the other interfaces to multiplexed D-A, A-D chassis and three digital I/O bytes. These daughterboards have the same interface as the Multibus boards used in the original Local Stations.

B. Four bytes of digital I/O

This board is an extension to the digital I/O capabilities on the SRM motherboard (eight bytes). These daughterboards can be plugged-in and used for input/output digital I/O when more devices need to be connected.

C. Four Channel Predet Timer

This board uses the Actel-1010A field programmable gate array. It features four run-time programmable 16-bit delay registers (interfaced as D-A channels). It is driven by an external free running clock and a trigger pulse. After a trigger is received, it counts clock pulses. When the count matches the value in one of the delay registers, a one-clock-wide pulse is produced.

V. LINAC LOCAL STATION SUPPORT [4]

SRMs are fully controlled from the Local Station. The Local Station code now supports Arcnet, while using Token-Ring to communicate with the rest of the control system.

The Local Station makes the existence of SRMs transparent to the outside world. When a setting is desired, the Local Station sends the request to the SRM and forwards the reply back to the requestor, as if the devices were connected to the Local Station itself. The Local Station responds to data requests with data from its own data pool.

Every Linac cycle (15 Hz) the Local Station broadcasts a cycle message to all the SRMs in its network. Each SRM acquiresthe current readings and sends a reply message back. The Local Station then analyzes the reply, updates its data pool, and scans for alarms. It takes five to twenty milliseconds for the data-acquisition reply to come back from an SRM, depending on the number of devices connected to it (including devices connected via daughterboards).

New versions of the SRM code are downloaded from a Local Station via Arcnet (instead of downloading the code through the serial port in each SRM, using the M68332BUG interface). The new version is first downloaded once from the host to a Local Station (via a serial port); then it can be sent to all the SRMs that are connected to any Local Station.

Since SRMs do not have a monitor connected to them, the Local Station console can be used instead. A special application program monitors an SRM's memory and enables changes to its contents from a Local Station. This feature is useful for setting up the SRM, telling it how many devices to access, and of what types, and for debugging.

VI. NETWORK INTERFACE [5]

The SRM is connected to the Local Station by an Arcnet local area network. Each Linac Local Station has a separate network with three or more SRMs. The physical media is coaxial cable in most cases. Fiber optics is used to connect to SRMs located in the Cockroft-Walton high voltage domes of the pre-accelerator. Arcnet was chosen because of its simplicity, low cost and the industrial support available. The limited message size (508 bytes) is sufficient in this implementation.

SRMs talk only with their Local Station and no messages are exchanged among them. SRMs do not initiate any message to the Local Station. The Local Station polls its SRMs for data acquisition buffers at the beginning of every 15 Hz Linac cycle. The SRMs poll the hardware and send back the up-to-date values in a single Arcnet frame. SRMs also respond to multiple-reply requests, that require a reply every cycle. One shot requests and settings are also supported. A subset of the Acnet protocol [6] (developed in Fermilab) and a special "number 4"* protocol are used. The Acnet protocol is designed for data acquisition tasks and handles requests, replies and multiple-reply requests. The #4 protocol is an application layer that specifies the action to be taken by the SRM. The general-purpose Acnet protocol fits so well that no #4 protocol header is needed for replies from SRMs to Local Stations.

VII. SRM Software [5]

The software for the SRM is written in C and cross developed on a VAX/VMS environment. The Microtec cross development tools are used.

A. Kernel

The kernel is Software Components Group's pSOS. pSOS is used in the Local Stations and therefore is the natural choice.

B. M68332BUG

The Motorola M68332BUG is an evaluation and debug tool for the MC68332. It comes with the BCC. It uses the BCC's serial interface, and allows the user to download and debug code for the MC68332. The M68332BUG was integrated into the SRM code (it is kept in PROM). The bug program has no knowledge of the kernel and the SRM code. At reset, it tests the MCU and the memory and transfers control to a pSOS initialization routine. One can easily switch to the bug program (when an SRM is running) by toggling a DIP switch which is checked every cycle. When in the bug program, one can download a new version of the SRM code, check and modify tables in the non-volatile memory, and even set chip selects and timers. When done, one can give the bug program a *go* command, and the SRM resumes running its code from the exact point where it was when the bug program took control (unless new code was downloaded). The ability to switch back and forth from the SRM code to the bug program is extremely useful for debug purposes.

The interface to the serial port provided by the bug program is also used to write values of selected channels to the serial port. This is useful in SRMs that are not close to a Local Station. Obviously, one has to connect a terminal or a portable computer to the serial port to see those readings.

C. Local Cycle

Normally the SRM polls the hardware for current values every time it receives a cycle message from the Local Station (at 15 Hz). In order to support data acquisition even when the connection to the Local Station is broken, the SRM has an internal 12.5 Hz cycle that is activated if the external cycle

message fails to arrive. The acquired readings can be seen by connecting a terminal to the serial port (see above).

D. Readings/Settings

The SRM code supports A-D readings, D-A settings, and digital I/O reading, settings, pulses, and motor control of the various A-D, D-A and digital I/O devices on the SRM's motherboard and on the various daughterboards.

VIII. Commissioning

Sixty Smart Rack Monitors are on order. The first group is to be used with the upgraded Linac control system, expected later this year. The rest of the SRM modules are for the 400MeV Linac upgrade (for which the same Local Stations are used), several of which are to be used during testing and commissioning of the new accelerator.

IX. Conclusion

The Smart Rack Monitor designed in Fermilab is an example of a low-cost (both in hardware cost, and in engineering time for design and development) data acquisition module. Nevertheless, it is a very flexible device and can serve as platform for a variety of applications.

X. References

[1] R. Goodwin, et al, "Initial Operation and Current Status of the Fermilab D0 VME-Based Hardware Control and Monitor System" in *Proceedings of the International Conference on Accelerator and Large Experimental Physics Control Systems*, Vancouver, Canada, October 1989, pp. 125-129.
[2] M. Shea, "Smart Rack Monitor", internal note, 1990.
[3] E. McCrory, et al, "Upgrading the Fermilab Linac Control System", in *Proceedings of the 1990 Linear Accelerator Conference,* Albuquerque, New Mexico, September 1990, pp. 474-477.
[4] R. Goodwin "SRM Message Protocols (from the local station)", internal note, 1991.
[5] S. Shtirbu "Smart Rack Monitor Software", internal note, 1991.
[6] C. Briegel, et al, "The Fermilab Acnet Upgrade", in *Proceedings of the International Conference on Accelerator and Large Experimental Physics Control Systems*, Vancouver, Canada, October 1989, pp. 235-238.

---

* This is the fouth data communication protocol currently supported by the Linac local control stations.

SLC'S ADAPTATION OF THE ALS HIGH PERFORMANCE SERIAL LINK*

K. Krauter

Stanford Linear Accelerator Center, Stanford University, Stanford, CA 94309 USA

Abstract

Keep It Simple Network (KISNet) is a very high speed, low overhead, primarily point-to-point network adapted from the LBL Advanced Light Source (ALS) control system. KISNet implements the ALS system in the SLAC Linear Collider (SLC) environment.

Commercially available communication hardware and protocols do not meet the per-small-packet transmission time requirements of the initial application that motivated the development of KISNet. Limited resources restricted what could be developed from scratch.

In the SLC environment, the communications system is used in a slightly different way than at ALS. The software is layered to enable insertion of future hardware as it becomes available. Several enhancements for the future of KISNet are under consideration.

KISNet REQUIREMENTS

The Fast Feedback Project (FFP) [1] for the SLAC Linear Collider (SLC) includes a new high speed micro-to-micro communication path. The microprocessors along the length of the SLC communicate with each other only via a single centralized VAX,® with the exception of a few micros currently in communication with each other over low-performance broadband CAT/CAR links. This intervention of the centralized VAX® during micro-to-micro communications could not be afforded by FFP, so a micro-to-micro communication path concept was born, and the architecture requirements were drafted.

Most important:

- Performance had to be very high for small packet throughput, on the order of a millisecond, in order to operate on the scale at which SLC beam operates (120 Hz).
- The new network had to be operational when FFP was scheduled for use, within about eight months of when the new network decision process was begun.
- The new network had to fit within the modest FFP budget.
- The new network had to be general enough that it would support more clients than just FFP.
- The new network had to be structured enough such that a minimum amount of software and hardware needed to be re-implemented in the future event of replacing the new network with yet a newer one.

Less important but still desirable:

- The new network should be easily extendable as new needs for it are identified in the future; this ties into the nature of FFP which will expand a few micros at a time.
- Budgetary and schedule constraints would be more easily met if the new network consisted of standard commercially available hardware and software, but the performance constraint makes this unlikely.

The single requirement which encompasses most of the above requirements may be stated in the well-known design principle of "Keep It Simple," which led to choosing the name K-I-S-Net for the new network.

DECIDING ON KISNet

General SLC-wide network architectures were almost all eliminated from consideration because of the expense of stringing cable the entire length of SLC. FFP required only a small minority of the micros be connected at first, anyway. Further, most of the general purpose networks have significant overhead processing per data packet, exceeding the FFP performance requirement. In addition to these general drawbacks, each proposed network had its own drawbacks.

An Ethernet connecting all the micros seemed a poor choice due to the nature of the communication expected by FFP: namely in between each beam, all the micros would want to talk to each other simultaneously, resulting in significant packet collision recovery overhead.

A Shared Memory architecture was proposed which involved relocating all the micros in one room where they would talk to each other via common memory, and talk to their respective devices via wire links from the central room to each device's location in SLC. This proposal exceeded the FFP budget and schedule requirements.

The choices were narrowed down to point-to-point communication protocols specifically between those micros that needed to talk to each other. Cable would only be strung between those micros that needed to talk to each other, which only becomes more costly than one SLC-wide general cable as many micros are involved. Only two nodes on each cable need be supported, minimizing collision errors and possibly packet overhead. Extension of the intramicro communication paths requires more cable be strung up, which was a penalty offset by the advantage of being minimal of the initial installation.

Intel's iNA® Ethernet product installed on a point-to-point basis removed the drawbacks mentioned above for an SLC-wide ethernet. However, performance numbers supplied by Intel (on the order of six milliseconds small data packet throughput) exceeded FFP's requirements.

LBL's Advanced Light Source (ALS) [2] project developed a special-purpose network applicable to a point-to-point installation, with many of the same performance and cost constraints as KISNet. Throughput time is on the order of a hundred microseconds per 100 bytes.

An agreement was made to pattern KISNet after the ALS network, buying into their hardware and firmware. Our own driver code would be necessary because of differences between the ALS and SLC environment.

* Work supported by Department of Energy contract DE–AC03–76SF00515.

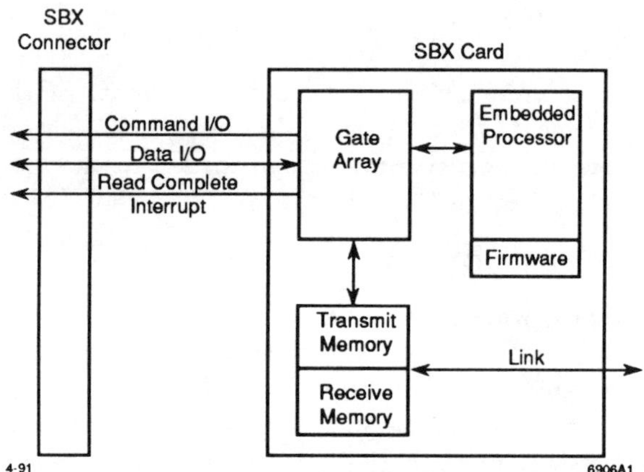

Figure 1. Node architecture.

ARCHITECTURE SUMMARY

Here is a summary of the ALS network architecture after which KISNet is patterned. The hardware at each node consists of an SBX® card with a gate array that handshakes with the microprocessor, some memory that is hooked to the link via DMA, and an embedded processor with firmware that implements RS485 SDLC link protocol. Refer to Fig. 1.

The SBX card's memory is divided into two parts: half for buffer data received, and half for buffer data to be transmitted. The buffer size limits the size of a data packet to 100 bytes. It takes thirty microseconds to download transmit data to the hardware. It takes 100 microseconds to transmit the data (the firmware will timeout a transmission after one millisecond). These times do not include interrupt overhead or other software execution times.

As used by KISNet, this network is one link with a master node on one end and a slave node on the other. The primary difference between these node types is that a master node may transmit data through to the slave node without solicitation. A slave node must hold data until the master sends a request, at which time the slave will transmit the data. In other words a slave must be polled for its data.

DIFFERENCES IN TARGET APPLICATION

ALS's network application has a single master "cluster controller" which talks to many slave device controllers. The master is a relatively slow IBM-PC® running Presentation Manager,® while the slaves are speedy Intel 80386 multibus based microprocessors running RMX.® Because the master is almost guaranteed to run slower than the slaves, and the communication direction is mostly in the polled direction from the slaves to the master, slave data can never overrun the master. The master polls the slaves for the data only as fast as the master can process it. The slaves wait for the poll from the master without losing data.

SLC's network application has a single master which talks to a single slave. Both master and slave are speedy Intel 80386 multibus based microprocessors running RMX.

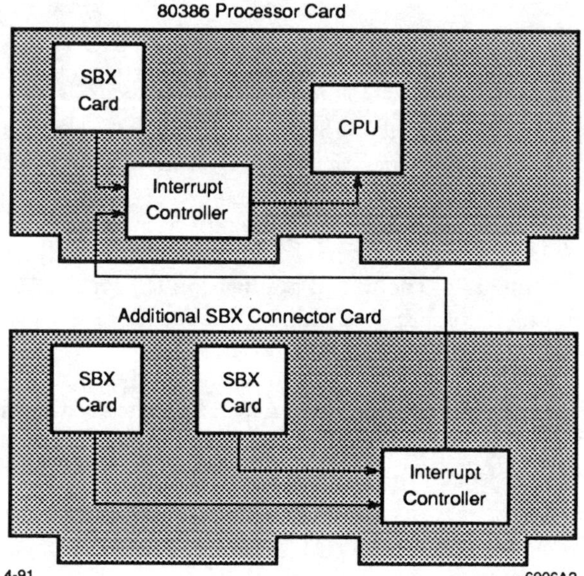

Figure 2. Target architecture.

Data transmitted is very time-critical to the receiving end, and the fastest way to have data appear at the receiver is to send the data in the unpolled master-to-slave direction. This eliminates the overhead of a software polling cycle, although it does introduce the overhead of interrupt handling. The risk of receiver overruns is also introduced since the data is being transmitted asynchronously.

At SLC, the polled slave-to-master direction is also supported, but is only used for non-time-critical transmissions. Multiple slaves per-master is supported, but not used yet.

Because KISNet is used point-to-point only, more than one KISNet network is required if a micro needs to talk to more than one micro. The KISNet hardware connects to one of SLC's 80386 microprocessors through its SBX connector, and so a micro needs as many SBX connectors as it has micros to talk to. SLC has older revisions of Intel's 80386 CPU card which have only one SBX connector on it; newer revisions of this card, such as are used by the ALS project, have more than one SBX connector. SLC uses a separate card with additional SBX connectors to support multiple KISNet ports. There is a slight performance degradation in the non-CPU-card KISNet ports since they must communicate with the CPU over the multibus backplane instead of directly. Refer to Fig. 2.

The FFP application sometimes must transmit data which is larger than the maximum memory size supported by the KISNet hardware. In this case the data is sent in multiple packets, which imposes a performance hit on the order of a millisecond per extra packet.

PROTOTYPE DEVELOPMENT

The ALS project was of a great deal of assistance during KISNet's design phase. Source code for the ALS high-level drivers and firmware documented the order in which I/O must be issued. ALS diagnostic source code demonstrated polled and unpolled communication. Discussions were held about applying the ALS network hardware to the SLC environment, with issues such as interrupt-driven I/O overhead, necessitated by unpolled communication, and firmware error recovery. Advice was given about parking at LBL.

Documentation was generated on the basis of the information received from ALS, including diagrams of the handshaking between the microprocessor and the firmware, flowcharts of the firmware source code, and block schematics of the gate array.

An IBM-PC platform was set up which held two KISNet ports. This platform was invaluable for quick experimentation, and verified some of the differences between KISnet and the original ALS network. Code developed on this platform was not directly transferable to the target systems on SLC; this was acceptable only because the turnaround of the development cycle on the PC platform was so much quicker than on the target system and because the PC platform had user interface support.

Once the KISNet design was moved from the prototyping platform onto the target system, debugging was needed only in those areas where the target system differed. These areas included interrupt handling, receiver overruns resulting from the asynchronous nature of the transmitting and receiving end, and the effect of higher priority interrupts interfering with both the interrupt-driven input and the atomic output of KISNet packets.

DIFFERENCES IN IMPLEMENTATION

The high-level driver code from the ALS system defined the way in which the hardware could be accessed. There are only a couple of areas in which the KISNet high-level driver code varies from the ALS drivers.

The primary difference is in error detection and recovery. This difference results from SLC responding to errors differently from ALS: the physically larger SLC machine makes it impractical to physically visit the hardware if an error occurs, so the hardware is given every possible chance to recover by itself. Potential infinite loops in the ALS drivers were given escape routes, indicating hardware malfunction and need for retry. When an error occurs, an attempt is made to automatically clear the hardware to determine if the error occurs again with the next

operation. An unsolved firmware problem (which generates a firmware error whenever communication switches direction) was worked around in the drivers by anticipating the error, clearing it, and automatically trying again.

Other implementation differences exist in the application interface. Buffers and mailboxes allow the application to use the same interface to send data to a remote destination over KISNet, or to a local destination in another task on the same micro. Support for sequence numbers in the data packets was necessary for multi-packet data.

RESULTS

FFP is now running on the SLC using KISNet. The system works with only minor problems remaining. These are being addressed at the time of this writing.

While working with the firmware developed by ALS, a wish list of desirable additions has been generated. These features would make KISNet a more general-purpose network.

The software which attempts to clear the hardware is inadequate and needs to be replaced by a true hardware reset which may be stimulated by a software command to the firmware.

KISNet needs to query the firmware about the status of the previous write operation submitted to the firmware. Currently the only means by which this status may be obtained is to submit a new operation to the firmware; a workaround may be had by double buffering all write-data so the data from the previous write operation is not lost when the new data is written out and reveals the previous data to have failed.

A firmware retry mechanism is currently available for the write, however none is available for the master's poll operation. Currently, polls are manually retried from the driver.

REFERENCES

[1] S. Magyary et al., "Advanced Light Source Control System," in Proc. 1989 IEEE Particle Accelerator Conf., Floyd Bennett and Joyce Kopta, eds.

[2] T. Himel et al., "General, Database Driven Fast Feedback System for the Stanford Linear Collider," these proceedings.

Use of Ethernet and TCP/IP Socket Communications Library Routines for Data Acquisition and Control in the LEP RF System

E. Ciapala, P. Collier and P. Lienard

CERN, 1211 Geneva 23, Switzerland

Abstract

A general move is being made at CERN towards the direct connection of intelligent equipment and device controllers to the control room consoles by the use of local Ethernet segments bridged to the main Token Ring networks. Communications is based on standard TCP/IP protocols which allows immediate use of standard software packages. The Data Managers which control the LEP RF accelerating units and transverse feedback systems have recently been connected. The implementation of Ethernet and TCP/IP socket communications routines for RF data acquisition and control is described. The adaptation of almost all of the existing software for RF system control, data aquisition and diagnostics to make use of this means of communication has proved straightforward. Furthermore the transparent transfer of data in the form of 'C' structures from the Data Managers to the the control centre workstations and other computers has considerably simplified the software required for remote surveillence and data logging with a corresponding increase in speed and reliability.

I. INTRODUCTION

The RF system of LEP is made up of individual RF units. Each consists of 16 accelerating cavities and their klystron power sources, high voltage power supply, low level electronics and controls. For the LEP phase 1 eight units using copper coupled accelerating/storage cavity assemblies have been installed near interaction points 2 and 6. For the upgrading of LEP to 90 Gev the construction of an additional 12 units using superconducting (SC) cavities is under way, eight of these for installation at points 4 and 8.

II. LEP RF SYSTEM CONTROL

A VME based 'Data Manager' (DM) running the OS9 operating system provides overall control of the RF unit [1]. The access to the equipment making up the unit is over IEEE488 (GPIB) bus to G64 based 'Equipment Controllers' (ECs) which contain the hardware interfaces [2]. There is one EC for each element of the unit i.e. one for each cavity, one for each klystron, one for HV equipment etc. The DM executes global control procedures, carries out surveillance, provides facilities for local control and the connection to the control system for remote operation and data acquisition. The DM is situated in the units control racks which are underground in a klystron gallery running parallel to the LEP machine tunnel. The connection to the general LEP control system originally installed is by a MIL-1553 standard bus which links all the DMs at an interaction point to a Process Control Assembly (PCA) in the surface building. The PCA provides the connection to the LEP token ring networks and to the Apollo workstations used for LEP operation. Another DM in the main Prevessin Control Room (PCR) with a small group of ECs providing control of RF synchronization and timing equipment is connected in a similar manner via a local PCA and MIL-1553 bus.

Communication between DM and the ECs of the RF units or PCR is based on command response transfer of simple ASCII strings. In the interests of simplicity and compatibility the same approach has been implemented for communication between the PCA and the DMs over the MIL-1553 bus. For communication between the control room workstation and the PCA command response string transfer is implemented by a remote procedure call (RPC) client on the workstation with a server running on the PCA.

Applications programs, both for the PCR workstations and for the local DMs, use make use of equipment functions built up as a set of simple macro definitions. These set up the command, call the communication routine and use the appropriate type conversion routine on the reply. The large amount of software required at both levels for LEP operations and RF specialist use has been built up entirely on this basis At the level of the PCR workstations a single communications routine 'lrf_eqacc()' is called whenever the RF system has to be accessed.

III. THE LEP CONTROL SYSTEM

Since the commissioning and startup and of LEP the controls groups of the CERN accelerators have agreed on a common architecture for control systems [3]. This takes advantage both of the increasing computational capacity of microprocessor systems and the availability of off-the-shelf communications packages which can be run on such systems. The PCA will be replaced by a 'Device Stub Controller' (DSC) which is the interface between the control system nework and the accelerator equipment. The general network configuration agreed upon is one which will use TDM based Token Ring networks similar to those of LEP [4] with connection by dedicated bridge units to clusters of DSCs and other intelligent equipment interconnected by short Ethernet segments. TCP/IP communications protocols will be used.

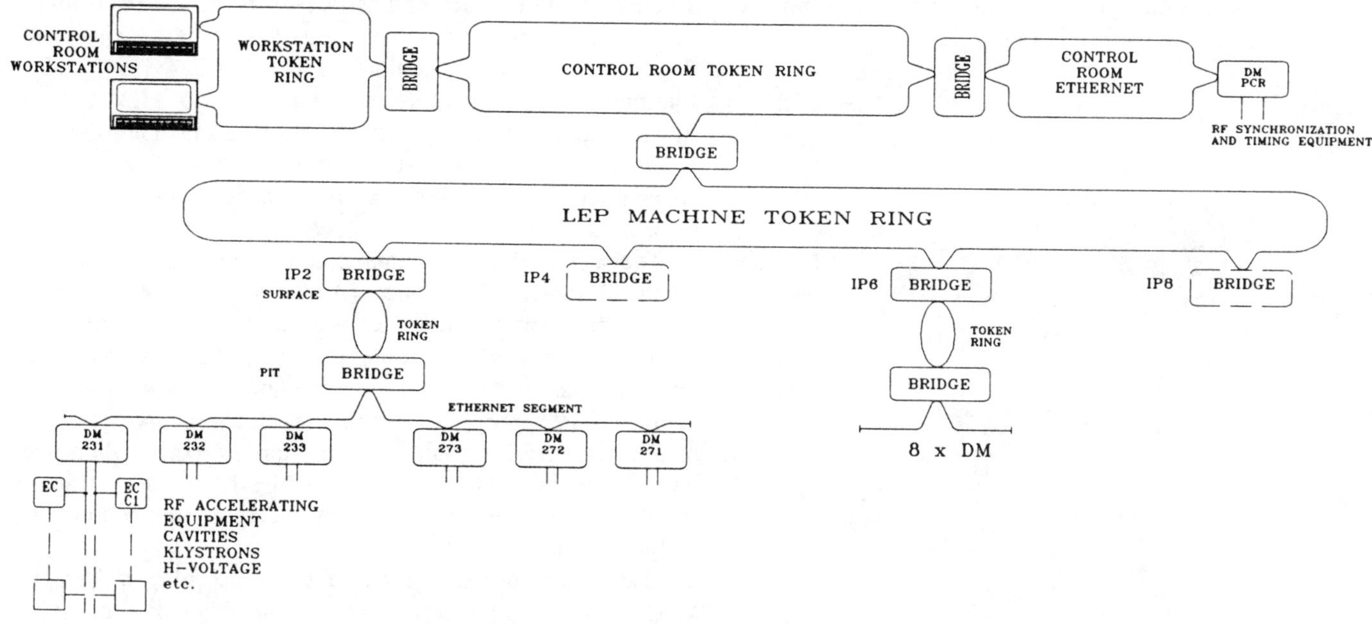

Figure 1. Overall configuration for RF controls with direct Ethernet network connections

IV. ETHERNET CONNECTION OF RF DATA MANAGERS

The DMs of the RF system are functionally equivalent to the DSCs. Ethernet hardware interfaces for the 68020 CPUs used and TCP/IP software have been recently installed. Ethernet segments link the RF units at each of the two points presently installed with RF including the first superconducting unit at point 2. The LEP transverse feedback system near point 2 is also included. A local token ring network, bridged in the SR surface buildings to the main token ring, extends as far as the bottom of the pit. Here an Ethernet to Token Ring bridge provides connection to the RF segment. RF equipment in the PCR is connected via a bridge in a similar manner. Installation of TCP/IP software and setting up of the bridges and routing tables was straightforward, providing communication between all DMs and direct connection to any of the control network computers and workstations. The overall network configuration and the interconnections which concern the RF system are shown in Figure 1.

V. COMMUNICATIONS AND USE OF TCP/IP SOCKET LIBRARY ROUTINES

Command response communication between the PCR workstation and the DM, which is the basis of RF system control, can be implemented directly using the socket library routines which are part of the TCP/IP software package. Hence the process of RPC between the Apollo workstation and the PCA and tranfer via the MIL-1553 can be replaced by direct communication.

Implementation of command response involves use of the socket library routines. Both the Apollo and OS9 Internet system use the concept of sockets as the basis for network communications between tasks. A socket is defined as an end points for communication. A command response server runs permanently on the DM. It initially creates a socket using the socket() library routine, associates a port number with it using the bind() routine and then waits for a connection using the accept() routine. A calling process on the PCR apollo workstation creates a socket to which the name of the DM to be accessed and the port number are bound and opens it as an active socket. It then uses connect() to connect to the listening socket of the DM. When the command response server accepts the opening of the connection the connect routine in the calling process terminates. Since the accept() routine creates a new socket which is used to transfer data the original socket is liberated for a new connection. The calling process sends command data in the form of a 1024 byte single packet using the send_output() routine then immediately calls recv() which awaits return data. The command data is read by the DM server using recv(). After command processing a reply is sent back to the calling process by the send() routine. The return of multiple packets is allowed for. The packet definition contains information on the type of packet and transmission of packets continues till an 'END' type packet is transferred. The calling process terminates and the server loops back to the accept() routine to handle the next connection request.

VI. DATA MANAGER SOFTWARE

At the level of the DM input command data arriving directly over the network is from the calling process is handled in a the same way as are commands from the MIL-1553 multidrop. Use of the same command message format simplifies the changeover of software and keeps compatibility.

The communications, message handling and command interpreter processes are shown in Figure 2.

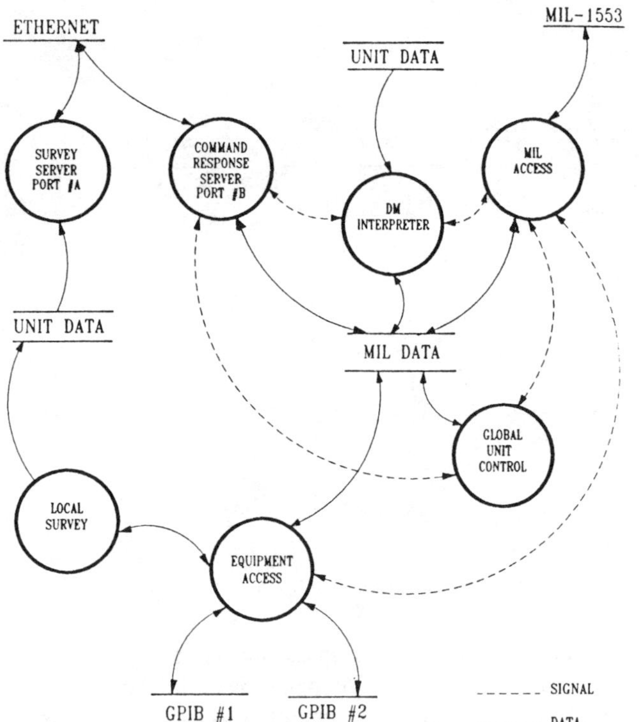

Figure 2. Data Manager Processes.

The system allows access both from the MIL-1553 and from Ethernet. Three processes can be called up : one permits direct equipment access via the GPIB, a second allows the execution of global RF unit control procedures and a third called DM interpreter accepts system calls and provides access to stored RF unit data. These processes are normally asleep and are woken by a signal from either the command response server or the MIL-1553 access process. Command parameters and reply data are passed via the 'MIL DATA' structure. This also stores process identities and handshake flags which are used by the various processes to ensure that signals are not sent to active processes.

VII. RF APPLICATIONS SOFTWARE

At the level of the calling workstation one cycle of command response using the above procedure is invoked by a new lrf_eqacc() function residing in a new communications library. Re-linking applications programs with the new library therefore results in incorpoation of the new communications system. No major modifications were required to convert all control, data acquisition and diagnostics software.

VIII. SURVEILLANCE

Another application of communication based on TCP/IP socket routines is the transfer of data for RF system surveillance. A DM background local surveillance program gathers the most important states settings and readings from the equipment of the RF unit and stores the data in the form of a structure containing approximately 200 integers, strings and floating point values. Remote acquisition based on ASCII command response requires the conversion of all data to a string equivalent and concatenation to form a long reply string. Use of this information remotely for logging or status display requires decomposition of the message and the correct type conversions.

The socket library communication system can be used for the direct transfer of structures. A new 'survey server' process, using a different port number, runs in the DMs to return the survey data. Instead of defining the data to be transfered as an array of characters as is done for command response, it is defined as the appropriate RF unit data structure type. Returned data in memory is treated as a structure of exactly the same type. Since the data format of variables in the OS9 systems and the UNIX operating system of the Apollo are compatible no further conversion is needed.

Use is made of this to provide a global display of the state of all RF units in the control room. The acquisition program runs on the control room DM which displays data on a dedicated monitor. The same data will be used for logging programs, freeing the workstation presently used. The individual RF unit data structures can be formed into an overall RF system data structure. This can be transferred in a similar manner to remote points of the site allowing DM type systems to provide the same overall display facility as in the control room.

IX. PERFORMANCE AND RELIABILITY

Command response communication between the workstations and the DMs has been increased in speed by up to a factor 50. The main speed limitation in accessing equipment is the GPIB access to the ECs which is typically around 20 to 50mS for read operations. Overall a factor two has been gained in speed which produces a noticeable increase in the execution time of PCR application programs.

Reliability of the network, bridges and local connections is good. No problems have been encountered since the new communications system was made operational for the LEP startup in March.

X. REFERENCES

[1] E. Ciapala and P. Collier, "A Dedicated Multi-Process Controller for a LEP RF Unit," in Proceedings of the 1989 Particle Accelerator Conference, Chicago, March, 1989, pp. 1583-1585.

[2] E. Ciapala and M. Disdier, "A Dedicated Multi-Purpose Digital Controller for the LEP RF System," in Proceedings of the 1989 Particle Accelerator Conference, Washington, March, 1987, pp. 520-522.

[3] CERN PS and SL Controls Groups, "PS/SL Controls Consolidation Project," CERN SL 91-12.

[4] P.Lienard, "SPS and LEP Control Network Architecture," in Proceedings of the 1989 Conference on Accelerator and Large Experimental Physics Control Systems, Vancouver, Canada, November 1989, pp. 215-220.

FIBER OPTICS IN THE BNL BOOSTER RADIATION ENVIRONMENT*

E.R. Beadle

AGS Department, Brookhaven National Laboratory

Upton, NY 11973

ABSTRACT

The Booster instrumentation uses analog and digital fiber optic links, designed to withstand at least 50 krads without performance degradation. The links use inexpensive and commercially available components that operate at a center wavelength of 820 nm. The analog link operates to 30 MHz over a 200 m fiber and can provide insertion gain. The digital link provides 60 ns timing pulses without the dispersive effects of coaxial cables. The optical fiber is a step-index hard clad silica type with a 200 micron core. This paper presents the component selection criteria, link design, installation, testing and performance for the optical links in the Booster instrumentation systems.

INTRODUCTION

Fiber optic links used in the Booster Beam Position Monitor (BPM) System satisfy timing and analog signal transmission requirements as well as provide electrical isolation.[1] The signals are sent between a rack in an equipment bay outside the tunnel and electronics subracks located underneath the dipole magnet supports inside the Booster. Each subrack contains a pair of analog transmitters, including switchable gain pre-amps, and digital receiver circuitry for timing. The fibers are in an unshielded cable tray five feet above the beam pipe. Flexible conduit is used to protect the cables where they transition from the overhead tray to the electronics on the floor.

The radiation exposure on the Booster floor has been estimated using CASIM software.[2] The cumulative dose estimate per year for unshielded locations on the floor are 5-27 krad (Si) for ionizing radiation and $3.4\text{-}17 \times 10^{12}$ n/cm^2 for neutrons. The ranges show the difference between typical and worst case points, but exclude the dump and septum locations. The objective was to design a link capable of operating for at least 5 years at a typical point in the Booster. It is expected the magnet supports will provide some shielding for the electronics. During commissioning, radiation detectors will be arrayed on all the devices to monitor absorbed dose.

*Work performed under the auspices of the U.S. Department of Energy.

FIBER SELECTION

The fiber selected for the links is the Ensign-Bickford (Avon, CT) HCR series fiber. The fiber is a 200 micron core step-index hard clad silica (HCS) type and was selected for its low induced loss, large numerical aperture (0.37), and low intrinsic attenuation (5 db/km maximum). The large numerical aperture and core size combine to reduce the mismatch loss at the transmitter. The HCS material is radiation resistant and recovers quickly from high dose rate exposures. The high attenuation, relative to telecom grade fibers, is because the dopants useful for reducing attenuation often cause an intolerable sensitivity to radiation. Typical dopants are germanium, boron, and fluorine.[3] Step-index fibers generally have a lower length bandwidth product than graded-index fibers however, using less than 0.3 Km over 50 MHz of bandwidth is achieved. This exceeds the bandwidth requirements of the transmit/receive pairs. Crimp and cleave style connectors are used because these connectors do not require polishing or epoxying as is typical of many optical connectors. Termination is simple and can be performed easily in the field within 30 minutes. SMA connectors are used because they are inexpensive, repeatable, and provide adequate alignment at this core size. Fiber lengths up to 100 m have been radiation tested in the AGS up to 50 krads without degradation.

ANALOG LINK

Meret, Inc. (Santa Monica, California) MDL288TV components have been selected as the transmit/receive (T/R) pair for the analog fiber optic link. An LED with a center wavelength of 820 nm and dc output of 100 μW is used in the transmitter. A PIN photodiode with a responsivity of 2 mV/μW is used in the receiver. The T/R pair are separately packaged in 10.8 cm x 4.5 cm x 3.3 cm enclosures housing the conversion between electrical and optical signals. Each enclosure has a BNC and SMA 905 connector and dc power terminals. Also, there is a gain adjustment on both components.

The transmitter input is ac coupled with a lower 3 db point of 6 Hz. Up to a 1 V (p-p) signal can be transmitted by the link with better than 2% differential linearity. Larger signals do not damage the transmitter, but significant distortion in the received signal will result. The noise floor of the received output is 1 mV (rms) in a 30 MHz bandwidth. However, as the transmitter input signal level increases, the harmonic levels in the receiver output increase. A two-tone test over a peak envelope range of 200 μV to 1 V showed that the second order intermodulation products are at least -20 dbc. Also, < 4% THD exists for link outputs up to 0.5 V (pk).

The peak value of the transmitter input is expected to vary over a range of tens of millivolts to several volts; therefore, a preamplifier/attenuator is used prior to the optical transmitter. It is used to enhance the received SNR for low level signals or attenuate high level signals to lower the distortion. The gain selections are 1/10, 1, 5, and 50. The preamp uses an Analog Devices 9610 current feedback amplifier as the gain element (Figure 1), providing a selection of 1 or 50. The attenuator is a resistive divider and provides either 1 or 1/10 gain. The gain is selected via the computer controls system. Two preamp/attenuators and transmitters are packaged within one chassis in a 3 U x 14 HP x 220 mm eurocard format.

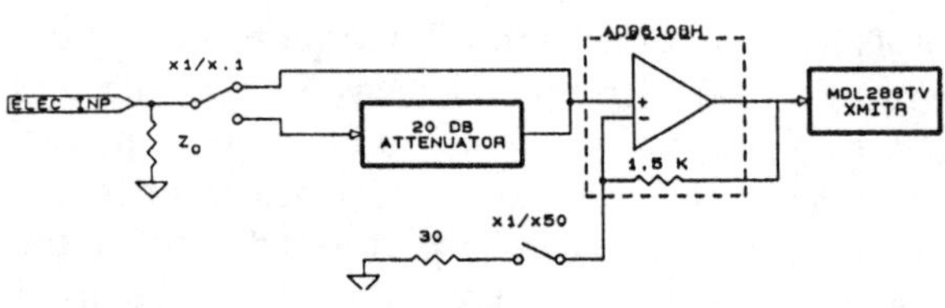

Fig. 1.   Block diagram of pre-amplifier/attenuator.

It is expected that the 9610 will provide the required radiation tolerance because many opamps operate into the 0.1-1 Mrad range. Additional care has been exercised in designing the remainder of the circuit. To improve the circuits radiation resistance, low current relays driven by TTL peripheral drivers have been used to switch the gain. TTL components have demonstrated radiation tolerance into the megarad range.[4] Semiconductor switches are a potential source of analog and digital failures due to radiation. Diodes (1N4446) used to limit the relay coil EMF during switching should pose no threat to radiation induced failure and should provide reliable operation to at least 10^{14} n/cm^2 which is > 6 years of operation at a worst case point on the Booster floor.[5]

Radiation tests on the Meret components were performed in the AGS over a 2 month period. In that time, the link was exposed to 45-50 Krads (Si) as measured by an array of TLDs. This corresponds to a 10 year dose on the Booster floor. No degradation in either power output or bandwidth were noticed in the lot of transmitters tested. The receivers were also tested with no noticeable effects. Considering the dose rate for this test greatly exceeds the dose rate expected in the Booster, we expect to have a link lifetime exceeding 10 years.

The receiver circuitry consists of the Meret receiver and separate buffer/driver electronics. These circuits are located outside the radiation environment. The optical receiver outputs interface to the buffer electronics through a patch panel. The receiver units are packaged in a separate enclosure that allows access to all of the units for gain adjustment. The buffer (Figure 2) circuit provides unity gain and can drive 50 Ohm cables with a 2 V p-p signal at 40 MHz. If required, the circuit can provide either high frequency peaking or increased gain. The buffer circuits are packaged with four channels on a card in a 3 U x 14 HP x 220 mm eurocard package.

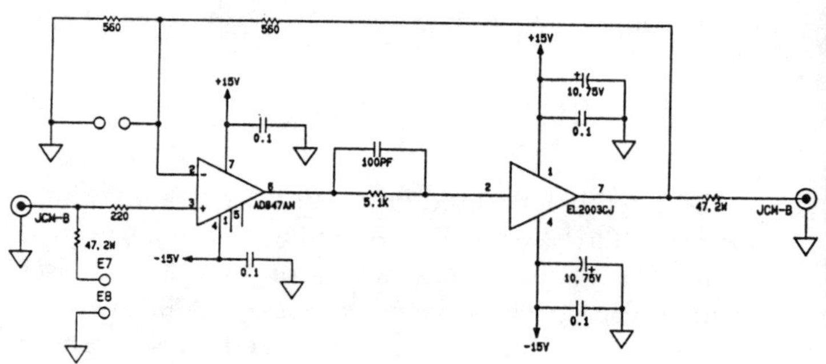

Fig. 2.   Schematic diagram of a buffer/driver channel.

The system was installed in the Booster and used during commissioning. The transmission systems were calibrated for a gain of unity from the BPM front end electronics to the Main Control Room operator stations. To maximize the received SNR, the optical transmitter gain was adjusted to the maximum and the receiver gain set to compensate for link-to-link variations in received signal. The system bandwidth is > 25 MHz at a gain of 50. The primary limitation of the system bandwidth is in the optical T/R pair.

DIGITAL TIMING LINK

The timing link is used to simultaneously place a 60 ns TTL level pulse at multiple locations radially distributed in the Booster. The signal is generated at a central location and transmitted in a star topology. The period of this signal tracks the rf frequency. The differential delay between the pulse edges and width of the pulse must be tightly controlled at each station. Additionally, electrical isolation of the receiving circuitry must be maintained. To solve the problem, a digital fiber optic link was constructed. The link uses Hewlett-Packard photodiodes (HFBR 1402/2404) operating at 820 nm for the transmit/receive pair and the same fiber as the analog link. The link accepts ECL level pulses, translates them to TTL, transmits over 300 m of fiber to the receivers located in the tunnel, and regenerates the pulse with a TTL pulse generator. The generator accepts the TTL pulse output from the receiver as a trigger and produces a 60 ± 1.5 ns pulse. To control the differential delay between stations, the fibers have been cut to a 3 ns tolerance, and in each channel is a digital delay line. The resolution is 2 ns with a ± 2 ns accuracy and compensates for delay variations in each channel. The HP diodes were selected because of the commercial availability, high speed (up to 30 Mbd with fast rise/fall times) and a low cost of $45 per pair. It is expected that because of the AlGaAs material, they would perform well in radiation environments.[6]

The transmitter diode forward current has been set at 20 mA, but can be increased up to 60 mA if necessary. At 20 mA, the link operates at full speed and with sufficient power margin, but some optical power has been reserved to compensate for radiation effects. The transmit and receive circuits were adapted from an HP application note.[7] The transmitter circuit (Figure 3) consists of a ECL/TTL converter, delay line, and a TTL driver to switch the LED on and off. The receiver (Figure 4), consists of HFBR-2404 analog amplifier followed by a differential video amplifier to drive a comparator. The comparator output pulses are regenerated by the pulse generator to precisely 60 ns. Tests on the link have shown it to operate to 20 MHz, although including the pulse generator, the maximum frequency required is below 15 MHz.

The transmit and receive diodes have been radiation tested in the AGS. Over a period of approximately 2 months they were exposed to 28 Krad. During the test, the transmitter and receiver characteristics were monitored and found to be unaffected by radiation at this level. The main thrust of this test was to confirm published data that commercial grade AlGaAs LEDs will perform well in radiation environments. This is significant because the link budget does not require additional margin for receiver/transmitter degradation. In addition, commercial grade components are less expensive than components specifically manufactured as rad-hard. The only concern with the radiation tolerance of the link is the tolerance of the video amplifier and TTL pulse generator. The video buffer amp should not cause a failure for several years because it is used to overdrive the comparator and degradation of the gain will not cause significant problems until it is greatly reduced. The TTL components are not expected to cause radiation induced failures.

REFERENCES

[1] D. Ciardullo, et al., The AGS Booster Beam Position Monitor System, this proceedings.

[2] A. Stevens, private communication.

[3] E.J. Friebele, et al., Overview of Radiation Effects in Fiber Optics, SPIE Vol. 541, p. 70-88, 1985.

[4] Messenger and Ash, The Effects of Radiation on Electronic Systems, Chapt. 10, Van Nostrand Reinhold Co., 1986.

[5] Ricketts, Fundamentals of Nuclear Hardening of Electronic Equipment, p. 231-233, Wiley, 1972.

[6] J. Wiczer and C. Barnes, Optoelectronic Data Link Designed for Application in a Radiation Environment, IEEE Trans. NS-32, 4046-4049 (1985).

[7] Hewlett Packard Application Handbook, Ap Note 76.

Fig. 3. Block diagram of the digital fiber optic transmitter.

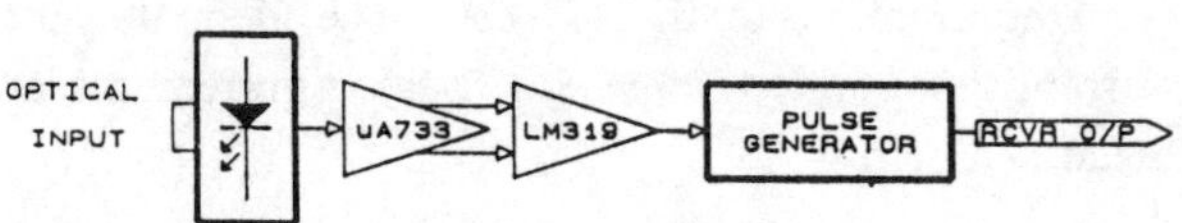

Fig. 4. Block diagram of the digital fiber optic receiver.

I/O Subnets for the APS Control System*

N.D. Arnold, G.J. Nawrocki, R.T. Daly, M.R. Kraimer, W.P. McDowell

Argonne National Laboratory

Advanced Photon Source

9700 South Cass Avenue

Argonne, Illinois 60439

Abstract

Although the Advanced Photon Source Control System allows for microprocessor–based Input/Output Controllers (IOCs) to be distributed throughout the facility, it is not always cost effective to provide such capability at every location where an interface to the Control System is required. I/O subnets implemented via message passing network protocols are used to interface points and/or equipment to a somewhat distant IOC, thereby reducing the number of required IOC's and minimizing the field wiring from the equipment to the Control System. For greatest flexibility, the subnets must support connections to equipment that requires several discrete I/O points, connections to GPIB and RS232 instruments, and a network connection to custom designed intelligent equipment. This paper describes an approach that supports all of these interfaces with one subnet implementation, BITBUS^{TM}. In addition to accommodating several different interfaces on a single subnet, this approach also circumvents several limitations of GPIB and RS232 which would otherwise restrict their use in a harsh, industrial environment.

I. THE APS CONTROL SYSTEM

The APS Control System provides for VME–based Input/Output Controllers to be distributed throught out the facility and interconnected via Ethernet to one another and also to Unix–based Operator Interface consoles.[1] Although this distributed architecture allows for intelligent processors near the major subsystems, I/O subnets are frequently required to interface directly to the equipment and communicate I/O information to the nearest IOC.

II. CURRENTLY SUPPORTED SUBNETS

A. *Allen Bradley Remote I/O*

The Allen Bradley Remote I/O Subnet allows the use of Allen Bradley's 1771 Series I/O modules. These units are inexpensive, rugged, and well proven in harsh industrial environments. This subnet allows field cabling to be kept to a minimum by providing the computer interface close to the equipment. A fiber optic network is available to provide isolation to areas where potentially dangerous voltages and/or EMI/RFI interference may exist. Unfortunately, since this network and the I/O chassis are proprietary designs of Allen Bradley, it cannot be customized to meet some of the unique requirements encountered in accelerator control. Any requirement other than typical process or industrial control operations currently supported by Allen Bradley modules cannot be supported with this implementation.

B. *GPIB*

Sophisticated test and measurement instruments frequently used in laboratories routinely provide GPIB as their interface to external equipment. Utilizing GPIB extensively in an industrial environment presents some serious challenges and potential problems. The distance limitation of 20 meters implies that the IOC must be located relatively close to the instruments. Extenders (both fiber optic and twisted pair) are available, but most efficiently used when a *cluster* of GPIB instruments are far away from the controller. If several GPIB instruments are located far away from each other, the use of multiple extenders is prohibitively expensive. Another concern is the noise susceptibility of the non–balanced signals within the GPIB cable. Industrial environments are potentially very noisy with respect to EMI/RFI, ground spikes, and power line transients. Since GPIB originated as a laboratory instrument interface, it was not designed with such an environment in mind. Differential signal lines, ground isolation between nodes, and error detection capability (such as parity or CRC's) are techniques that are frequently used to improve the reliability of a network in a harsh environment, but have not been included in the GPIB interface standard.

III. REQUIREMENTS FOR ADDITIONAL SUBNET SUPPORT

Several requirements for the APS cannot be adequately met by the currently supported subnets. These requirements are briefly discussed below.

A. *Power Supply Control Units*

The numerous power supplies throughout the APS facility will be controlled by small intelligent Power Supply Control Units (PSCU's). In the Storage Ring alone it is planned to have

---

*Work supported by U.S. Department of Energy, Office of Basic Energy Sciences under Contract No. W–31–109–ENG– 38.

200 PSCU's. Each IOC in the Storage Ring (20 are planned) will need to interface to 10 PSCU's. The basic requirements for this "subnet" are:

– Multi–drop network that supports at least 10 nodes over a distance of 50 m
– Immunity from switching power supply noise
– Ground isolation between nodes
– Low cost per node (due to the large number of nodes)

B. APS Custom Equipment with Embedded Controllers

Several pieces of equipment that are custom built per ANL specifications are sophisticated enough to use embedded controllers for equipment control. Interfacing this equipment to an IOC via a subnet would allow for a clean and inexpensive connection. An example of this is the five Klystron/Modulators located in the Klystron Gallery of the LINAC area. Each has an embedded controller and requires approximately 50 parameters to be passed to/from an IOC. The Modulators are approximately 50 feet apart. The basic requirements for this "subnet" are:

– Multi–drop network that supports at least 5 nodes over a distance of 70 m (200 ft)
– Immunity from EMI/RFI environment
– Ground isolation between nodes (preferably fiber optic)

C. Distributed RS232 Instruments

Several instruments that will be specified for use may only provide an RS232 interface. Residual Gas Analyzers, Stand–alone Single–Loop Controllers (PID control), Ion Gauge Controllers and Digital Display Panels are examples of some things that will require a simple RS232 connection to an IOC. A subnet that would provide multi–drop support over a large distance with appropriate error detection and noise immunity would allow a clean interface to these distributed RS232 instruments.

D. Distributed GPIB Instruments

Advanced Test and Measurement equipment will routinely be used for monitoring and controlling different parameters of APS operation. Examples include Digital Oscilloscopes, Network Analyzers, Precision Timing Generators, RF Power Meters, and Frequency Counters. Ultimately it will be required to interface this equipment to an IOC, even if it is not close to an IOC. A subnet that would accommodate GPIB devices but allow for long, noise immune cabling between instruments would be a significant advantage during the project's lifetime.

E. Single Point I/O

Since the APS facility covers more than 40 acres, it is likely that I/O points to be interfaced will be quite spread out. Although such instances can well be accommodated by field wiring, having the capability to (cost effectively) interface to a handful of I/O points would provide greater flexibility in the design of a subsystem. Attaching an inexpensive module to a subnet that would provide 16 I/O points at a particular location is likely to be a requirement.

IV. NEW SUBNET PROPOSAL

Each of these requirements for additional subnet support has a possible solution. The Power Supply Subnet can be one of many networks, including Ethernet, GESNET, BITBUS, or simple RS232. Distributed RS232 instruments can be supported with Multi–drop RS232 networks available from many vendors. Equipment with embedded controllers can be specified to provide GPIB or RS232 interfaces. Distributed GPIB instruments should be grouped together in racks and an extender used when the distance from the IOC exceeds the recommended distance. Allen Bradley chassis can currently provide small amounts of I/O in random areas, but this approach is not cost effective for just a handful of points.

Providing unique solutions to each of the above requirements dramatically increases the hardware and software that will need to be developed and/or maintained throughout the lifetime of the project. It also increases the unique hardware and software for a given subsystem, making it less likely that anyone other than the engineer who implemented the system could support it.

Ideally, a single subnet could be implemented that would support all of the above listed requirements. With a common solution, cost would probably be less (due to the quantities required) and maintainability would be significantly enhanced.

A. Subnet Architecture

The requirements previously discussed could be implemented with the *single* subnet architecture illustrated in Figure 1. The subnet network is distributed from the IOC throughout the area of interest. Three types of "gateways" can be connected to this subnet: Single Point I/O Gateway; RS232 Gateway; and a GPIB Gateway. In addition, any intelligent node (such as embedded controllers) that implemented this interface could be connected directly to the subnet. This network should have the following characteristics:

– based on a non–proprietary commonly accepted network standard, so components are commercially available
– multi–drop network that supports > 25 nodes
– immune to random electrical noise
– error detection and recovery
– provide ground isolation between nodes
– allow fiber optics for total isolation when required
– fast enough to support a reasonable number of nodes

With this approach, the IOC would communicate to GPIB instruments over the same network it uses to communicate to RS232 devices. The data to be sent to the devices would be encapsulated within the defined subnet protocol. The individual gateways would extract (and possibly buffer) the data and

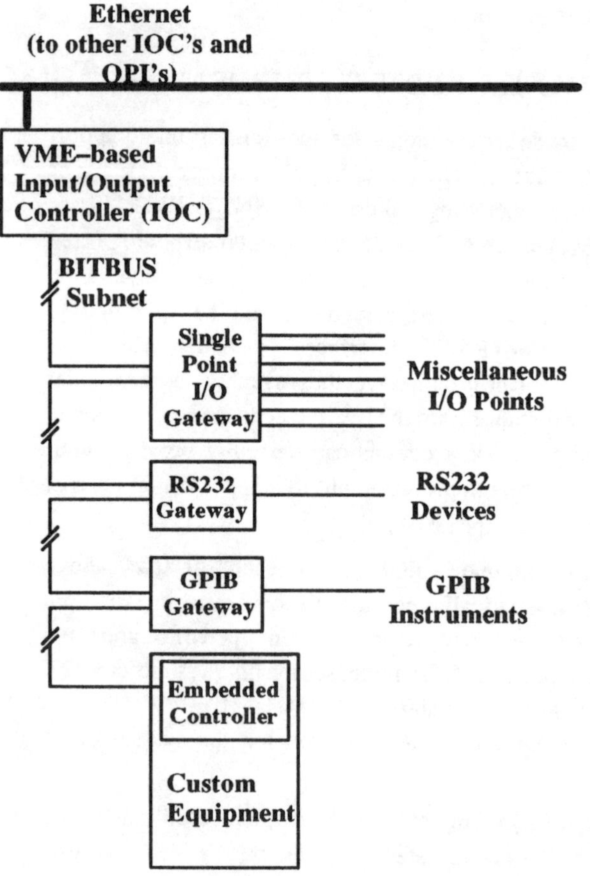

Figure 1. New Subnet Proposal

complete the transmission over the particular interface. This approach also solves a potential problem of slow devices "tying up the network" that may exist in other implementations.

B. BITBUS as the Backbone

Several networks come to mind as possible alternatives for implementing this idea: Ethernet, BitBus, Arcnet, MilStd 1553, or even RS232. When evaluating these networks in the areas of cost, performance, and available products, BitBus seems to emerge as the most promising. In fact, three of the four types of interfaces shown in Figure 1 are currently available from commercial sources for BitBus.

The BitBus Interconnect Serial Control Bus was introduced by Intel to provide high speed transfer of short control messages in hierarchical systems. It quickly became popular in industrial applications as a "field bus" that allowed distributed I/O equipment to be connected to a single host. It is likely to be accepted as an IEEE standard (IEEE P1118), which will further increase its popularity. The goal of the BitBus Interconnect is to provide a message passing interface between tasks at the master node and tasks at multiple slave nodes. The salient features of the specification include:

– Electrical Interface: RS485 (differential twisted pair)
  Self–clocked 62.5Kb/s @ 1200m or 375Kb/s @ 300m
  Fiber Optic tranceivers available

– Data Link Protocol: subset of SDLC
  Single master/multiple slave topology
  Provides CRC error detection and message sequencing
  Provides automatic retry on detection of errors

– Message Protocol
  Destination address includes Node and Task
  Predefined messages for I/O and memory commands
  Data length up to 248 bytes per message

C. Product Availability

Almost all of the components required to implement the capabilities illustrated in Figure 1 are commercially available (discussed below). Although a GPIB Gateway product has not been identified, vendors have expressed an interest in providing such a product. Besides commercially available products, Intel markets a microcontroller (8044) that implements the BitBus protocol on a chip that can be used to provide a Bitbus Interface to custom designed equipment. This, of course, is an advantage of basing the subnet on a commercial standard rather than a proprietary product.

D. 'Proof of Concept' Test Stand

A Test Stand was constructed to demonstrate the feasibility of the concept illustrated in Figure 1. Only commercially available hardware was used (custom software was required). The products used were :

VME Bitbus Card: Xycom XVME–402 Bitbus Controller

Single Point I/O Gateway: Phoenix Contact Interbus–C IBC–EU (Bitbus Interface) and IBC EDIO (Digital I/O)

RS232 Gateway: Micronetics International, Inc; Bitbus/RS232 Gateway

GPIB Gateway: Micronetics International, Inc Analog Data Manager; National Instruments GPIB SBX card; custom software by Micronetics

Embedded Controller: GESPAC G64 chassis with 68000 processor card and FILBUS controller < *this part of the test stand was not fully implemented at the time of this writing* >

V. CONCLUSION

BITBUS is currently being considered as a viable approach to attaching equipment to an IOC in the APS Control System. With the appropriate gateways (either developed or purchased), BITBUS could be used to connect distruted GPIB instruments, RS232 instruments, intelligent controllers, and single point I/O panels via the same data network, thereby greatly enhancing maintainability of the system. Distance limitations and noise susceptability of GPIB are circumvented, as well as the point–to–point nature of RS232.

VI. REFERENCES

[1] M. J. Knott, "The Advanced Photon Source Control System" these proceedings.

The Trajectory Control in the SLC Linac*

I. C. Hsu, C. E. Adolphsen, T. M. Himel, and J. T. Seeman

Stanford Linear accelerator Center
Stanford University, Stanford, CA 94309

Abstract

Due to wake field effects, the trajectories of accelerated beams in the Linac should be well maintained to avoid severe beam break up. In order to maintain a small emittance at the end of the Linac, the tolerance on the trajectory deviations become tighter when the beam intensities increase. The existing two beam trajectory correction method works well when the theoretical model agrees with the real machine lattice. Unknown energy deviations along the linac as well as wake field effects can cause the real lattice to deviate from the model. This makes the trajectory correction difficult. Several automated procedures have been developed to solve these problems. They are : an automated procedure to frequently steer the whole Linac by dividing the Linac into several small regions ; an automated procedure to empirically correct the model to fit the real lattice and eight trajectory correcting feedback loops along the linac and steering through the collimator region with restricted corrector strengths and a restricted number of correctors.

I. INTRODUCTION

Due to the transverse wake field effects in the 3 km linac of the Stanford Linear Collider (SLC), it is necessary to keep the beam trajectory rms deviation less than a few hundred microns in order to avoid large emittance growth. Because the transverse wake kick is linearly proportional to the beam intensity, the tolerance on the trajectory deviations become tighter when the beam intensities increase. The existing two beam trajectory correction method works well when the theoretical model agrees with the real machine lattice. Unknown energy deviations along the linac as well as wake field effects [1] can cause the real lattice to deviate from the model. This makes the trajectory correction difficult. In this paper we present several automated procedures which have been developed to solve these problems. The algorithms of two beam steering of the SLC linac in which e⁻ and e⁺ bunches must be steered simultaneously appear elsewhere.[2]

The SLC linac consists of 30 sectors, *ie*. sector 1 to sector 30. Each 100 m sector contains eight girders ; at the end of each girder of a typical sector is one quadrupole magnet of the FODO lattice. The first sector after the damping rings, sector 2, has four times as many quads, and sectors 3 and 4 have twice as many in order to provide stronger focus which is needed because of lower energy beams are more sensitive to wake fields. The phase advance per cell is 90° in sectors 2 - 4 and 76° in sectors 5 - 14, then tapers to 45° at sector 30 as the quadrupoles saturate.

II. AUTOMATED PROCEDURES

A. Auto-Steer Macro

The disagreement between the real lattice and the model accumulates as the length of the region which we try to steer increases. Therefore, we developed a Button Macro[3] which automatically divides the whole linac into four regions with about equal phase advances. The linac is then steered region by region from upstream to downstream with iterations of steering. All we need to do is just push the button once. This button macro has been successfully tested. The advantage of this button macro is that it frees the operators' attention and steers the linac in pieces precisely the same way every time. It can also be implemented in an automatic procedure which executes every few minutes in a manner similar to slow feedback loops along the linac.

B. Model Updating Macros

When the unknown energy deviations along the linac as well as wake field effects cause the real lattice to deviate from the model too much, the trajectory steering will become very difficult even after dividing the linac into several pieces. A "lattice diagnostic" program[4] has been developed to find the discrepancies. The existing "Linac Energy Management (LEM)" program[5], was then used to implement the results to adjust the real lattice (quadrupole magnets) such that it agrees with the model.

There are two reasons to change the above technique. First, since both e⁻ and e⁺ beams use the same set of quadrupoles, we can only adjust the lattice to fit either the electron model or the positron model. To make it right for both the electron beam and the positron beam, instead of adjusting the lattice, we need to adjust the models of both beams. The second reason is that in the high intensity regime as we operate in SLC now (3 - 4.5 x 10¹⁰ particles per bunch), the discrepancies between the real lattice and the model is usually dominated by wake field effects. Thus, from the optics point of view, we would also like to adjust the models to ease steering while keeping the lattice unchanged so that the beam's

* Work supported by the Department of Energy, contract DE-AC03-76SF00515.

Twiss parameters will be well matched with the lattices' Twiss parameters. This is easily understood by imagining slicing the beam longitudinally, the transverse dipole wake kick will not change the beam's Twiss parameters of each slice and therefore, we like to keep the matched lattice unchanged.

Previously to adjust each model , it took about 50 manual operations and was very difficult to do it correctly each time. That software was not capable of simultaneously adjusting both electron model and positron model correctly. The model updating macros were created to accomplish the model updating by only pushing three buttons and they are capable of adjusting both beam's models simultaneously. They have been successfully tested and are now used routinely.   Each time after we update the models we can steer the whole linac in one piece. The models remain good on the order of one week or when the beam currents are significantly changed.

A measured example of these procedures is included to illustrate their effectiveness.   Fig. 1 shows the result of an incorrect model. There is a big phase discrepancy between the measured oscillation (solid line) and the curve which is predicted by the model (dashed line) by fitting the earlier part of the oscillation.   As we pointed out previously, the bad fitting is due to the discrepancies between the real lattice (measured oscillation) and the model (fitting curve). Fig. 2 shows a good agreement in phase advance between the measured oscillation and the fitting curve after the adjustment of the model. The amplitude discrepancy is due to the decoherence of the beam signal. This data was taken after we used the model updating macros to update the models.

C. Trajectory Correcting Feedback Loop

Once good beam trajectories have been established using the steering techniques described above we want to keep the beams on those trajectories. To accomplish this there are eight trajectory correcting feedback loops [6] spaced along the linac. They are centered at sectors 2, 3, 4, 6, 11, 18, 23, and 27. Each loop reads the positions of the electrons and positrons from 12 to 16 beam position monitors which are spread out over 360 to 720 degrees of betatron phase advance. From these readings a loop calculates the position and angle of the beams and then sets eight correction magnets to keep the beams on the desired orbit. These measurements, calculations and corrections are presently repeated at 20 Hz and we plan to increase that rate to 60 Hz. The loops quickly correct orbit changes caused by klystrons turning on or off or caused by changes of magnet power supplies.

D. The Collimator Steering

A special steering algorithm is required near the end of the linac where the beam passes through two sets of collimators, each of which consists of two pairs of collimators in both planes. Each pair is separated by 90 degrees in phase advance, and electrons and positrons are collimated in alternate pairs. For the collimation to be effective, both beams have to be centered in the collimator jaws to the 100 micron level. For this purpose, eight corrector magnets are used for each set of collimators to steer both beams based on the readings from four BPMs located near the collimators.

In principle, the eight correctors can be used to zero the four BPM values for both beams since there are eight constraint conditions. In practice, the phase advance between the correctors is such that magnet strengths would exceed their maximum limits to correct the orbits given the approximately 100 micron quadrupole misalignments in that region. As an alternative, a steering algorithm was developed which does a least squares minimization of both the orbit deviations and the magnitude of the corrector strengths to determine the corrector settings. The relative weighting of the two constraints was adjusted empirically. The result is that both beams can be steered to within 100 microns of each other and to within 200 microns of zero at all eight BPMs without exceeding the maximum corrector magnet strengths. After steering, the collimator jaws can be centered about the average position of the two beams to better optimize the collimation. In the operation of the linac, the steering is done every few days while the collimator alignment is done a few times a year.

III. Conclusion

Using the above techniques, we were able to control the SLC production electron and positron beam trajectories. However, the scavenger beam trajectory is not so well controlled. A three beam steering method is under developed now.   The trajectory control is vital for the issue of the emittance preservation of the linear collider. The experience, we learned, from controlling the SLC beam trajectories should be helpful for further development of the next generation linear collider.

IV. Acknowledgement

The success of the trajectory control is a result of the efforts of many people. We would like to thank all of them.

V. References

[1] C. Adolphsen *et al.*, "Effect of Wakefields on First Order Transport in the SLC Linac," these proceedings.

[2] K. A. Thompson *et al.*, "Operational Experience with Model-Based Steering in the SLC Linac," *Proceedings of the 1989 IEEE Particle Accelerator Conference*, Chicago, Illinois, March 1989, pp. 1675-1677.

[3] S. Moore, "Button Macro User's Guide, " SLC internal report, July 1989.

[4] T. Himel and K. Thompson, "Energy Measurements from Betatron Oscillations," *Proceedings of the 1989 IEEE Particle Accelerator Conference*, Chicago, Illinois, March 1989, pp. 1529-1530.

[5] M. Woodley, "Design Specification of a Linac Energy Management Facility," SLC internal report, January 1988.

[6] F. Rouse *et.al.* , "A General, Database Driven Fast Feedback System for the Stanford Linear Collider," these proceedings.

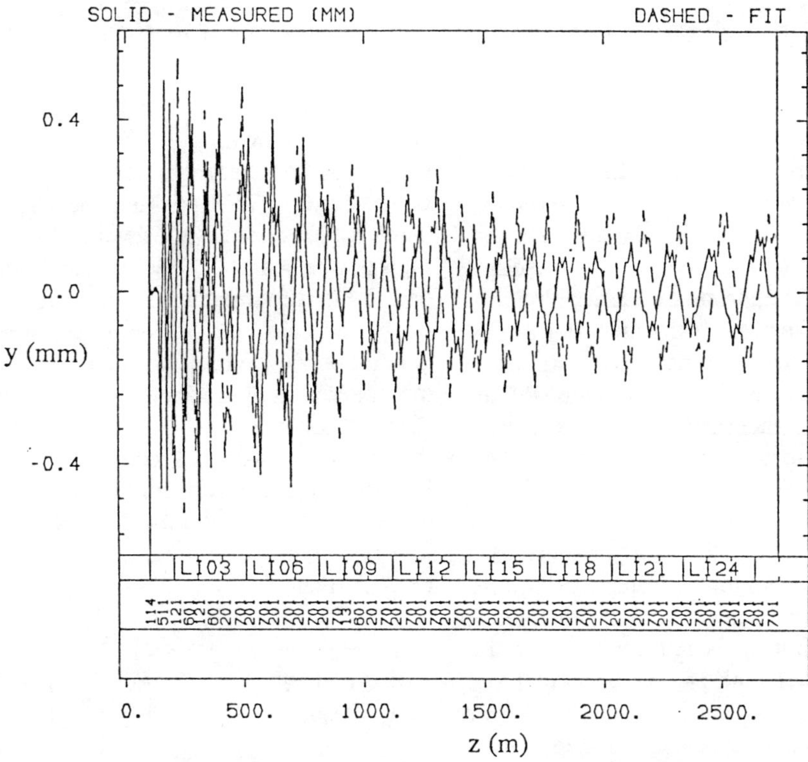

Fig. 1   Trajectory difference from an induced dipole oscillation.   Data are taken before updating the model.   [I = (2.0 ± 0.2) x 10^{10} e+]

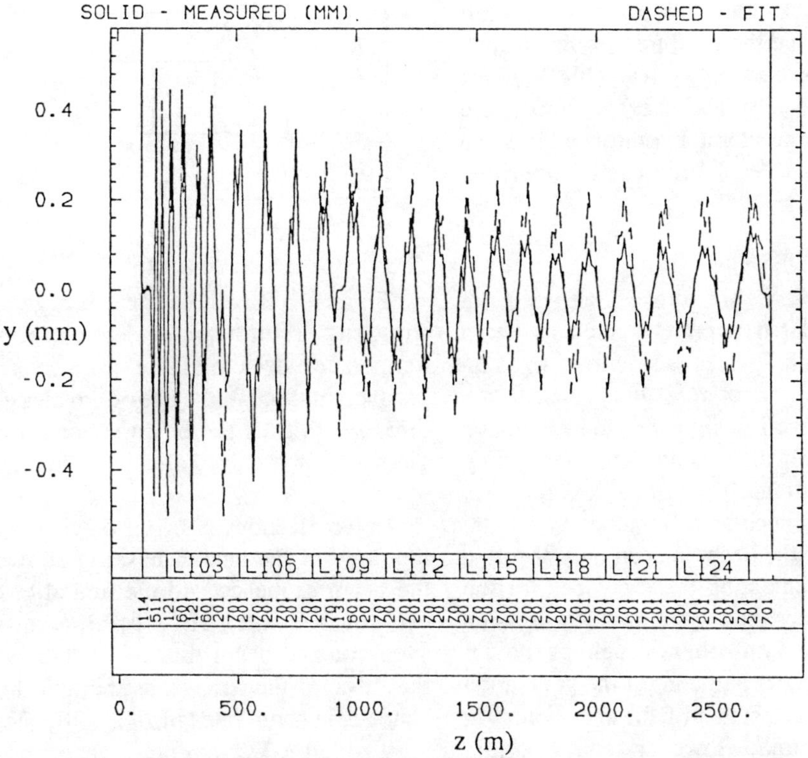

Fig. 2   Trajectory difference from an induced dipole oscillation.   Data are taken after updating the model.   [I = (2.0 ± 0.2) x 10^{10} e+]

Machine Protection Schemes for the SLC

M. C. Ross, Stanford Linear Accelerator Center, Stanford, Ca. 94309

ABSTRACT*

The beamline components of a linear collider must be protected from high power beams in a way that is both reliable and has a minimum impact on integrated luminosity. When an upstream accelerator component fault occurs, the machine protection system suppresses the appropriate beam pulses and restores them when the fault clears or is compensated for. If an unacceptable localized beam loss is detected, without an accompanying component fault that is a likely cause of the loss, the system must provide identical, lower rate (lower average power), beam pulses to be used for diagnosis. This must not be done at the expense of any upstream beam stabilization system since fault diagnosis and recovery may take some time. Since the SLC beam pulse sequence is a regenerative one, i.e. correct function on a given pulse requires that several preceding pulses have been successfully completed, beam pulse repetition rate limiting is not trivial. Smooth, rapid, recovery from this type of fault is very important and can have a significant impact on luminosity. This paper provides an overview of the beam suppression and repetition rate limiting schemes used at the SLC.

INTRODUCTION

Several next generation particle accelerators will control beams of very high effective power which are capable of causing significant damage to beamline components in a short time. While these machines share the need for a sophisticated, highly integrated machine protection system (MPS), their specific requirements differ radically. The extent to which these systems are integrated with the accelerator controls will have a great impact on their operation. Very little has been written about such systems in general.

MPS DESIGN

The MPS we address here are those that rely on measurements of a beam related parameter such as beam intensity, position or loss. Such devices are referred to as an 'errant beam detectors' (EBD).

Generally speaking, in a well designed system, damage from the beam cannot occur if all systems are functioning properly and at their design settings. The primary exception to this are beam defining devices, such as collimators, which are intended to cut the extremes of the beam but can only absorb a fraction of its total power. Neglecting these devices for the sake of argument, one could design a system which drew its input primarily from device controllers, such as power convertor or RF system controllers and control the generation of beam pulses according to the status of these systems. In this case, EBD are not required. However, under certain conditions, loss monitors indicate a problem and all device controllers indicate good status. Under these circumstances beam may be required to assist the diagnosis of the root cause of the failure. This is often the case with collimators. The beam required for diagnosis must be a low power 'copy' of the high power beam. Otherwise it will not be a useful diagnostic and the frustrating situation may result where the low power beam indicates no problem, yet the EBD will not allow generation of the high power beam. Furthermore, these trips may be caused by transient events that escape detection (or are detected too late) by the device controllers so the transition to the low power state and recovery of the high power beam must be done quickly and smoothly and must not involve the generation of any substantially different beam pulses. At SLC, low power diagnostic beams are produced by lowering the repetition rate.

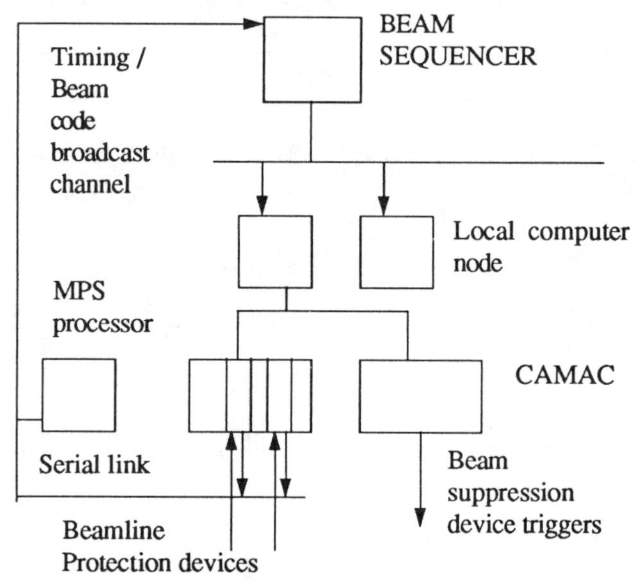

Figure 1: Block diagram of a generic MPS. At SLC, the front end errant beam detectors are usually ion chambers or temperature detectors. The beam sequencer controls the beam production through the use of broadcasted beam codes [1]. It also directly controls the beam suppression devices and controls globally synchronized data acquisition.

Figure 1 shows a block diagram of a generic MPS. The topology of the beam lines, the allowed modes of operation, the beam sequence and the limitations of pulsed devices are critical for the SLC MPS. Additionally, a globally synchronized beam diagnostic data acquisition scheme, like that used at the SLC, must be able to cleanly cope with the changes in beam rates instigated by the MPS.

Typical SLC average beam power is 150KW in a $0.05mm^2$ area. When an electron beam of this power strikes some device the bulk of the energy is dissipated in a relatively

* supported by DOE contract DE-AC03-76SF00515

large volume several radiation lengths behind the point of initial entry. However, if the beam is small enough, significant damage from a single pulse may occur near the point where the beam enters the material. At several places in the SLC, the beams are small enough so that this is a concern. The use of a thin upstream spoiler increases the beam size enough to prevent this type of damage.

To illustrate the problems posed by the above constraints, we will first briefly describe the SLC (figure 2) and some of the relevant subsystems.

SYSTEM CONSTRAINTS

Beam suppression in the upstream part of the SLC is accomplished near the gun. The nominal 120Hz SLC e+/e- bunch production scheme has the following features:

1) Three bunch co-acceleration is required in 2/3 of the main linac (1.1GeV - 30GeV) and the injector region (210 MeV to 1.1GeV). The co-acceleration results in an energy dependence of a trailing bunch on the intensity of the leading bunch(es). This dependence is caused by fundamental beam loading in the disk loaded waveguide structure.

In the SLC this leads to about 1% energy loss per 3 x 10^{10} leading particles. At nominal currents this effect is large compared to the typical beam energy spread and downstream energy acceptances and therefore must be considered. A feedforward system [2] that compensates for this in the main linac is presently being commissioned, but its role in the SLC MPS will not be considered here.

2) In the positron damping ring, the positrons must damp to about 1% of their initial emittance, about 5 damping times. Because the interpulse time is less than 3 damping times at 120Hz, the e+ must be allowed to damp more than one interpulse period. In order to achieve this, the bunches are left in the SDR for two interpulse periods. They are injected and extracted from the ring one at a time and are equally spaced around the ring circumference.

3) The SDR kicker pulses are fast enough to extract a single bunch without disturbing the remaining one. In contrast the NDR kickers use a long flat-topped pulse to inject and extract both bunches at once. Any rate limit scheme must keep the average power dissipated in the kickers and their power supply system constant. This is done using 'standby' pulses.

There is a logical boundary at the beginning of the arcs beyond which the e+ bunch can no longer affect the energy of the e- bunch that follows it. A pair of beam suppression devices, known as 'single beam dumpers', (SBD's) at this location can allow one beam or the other to continue on to the final focus. Any fault detected by EBD downstream of this location can be handled easily. If beam pulses are required to diagnose the problem the single beam dumpers may be programmed to allow a full current, low rate beam through.

RATE LIMITING

In order to provide an equivalent beam of lower power as required to diagnose an errant beam detector fault, the cycle must be broken because each rate limited pulse must produce and deliver positrons in just the same manner as the full rate pulses do. The positron storage time must therefore be made arbitrarily long. Since the kickers continue to fire, only one SDR bunch will survive. This means that the transition from full to low rate must also be a transition from 'n -> n+2' to 'n -> n+1' e+ production (and vice versa). Thus at those transitions a single e+ only (rate drop) or an e- only (rate increase) pulse is produced in the main linac. Because the e-

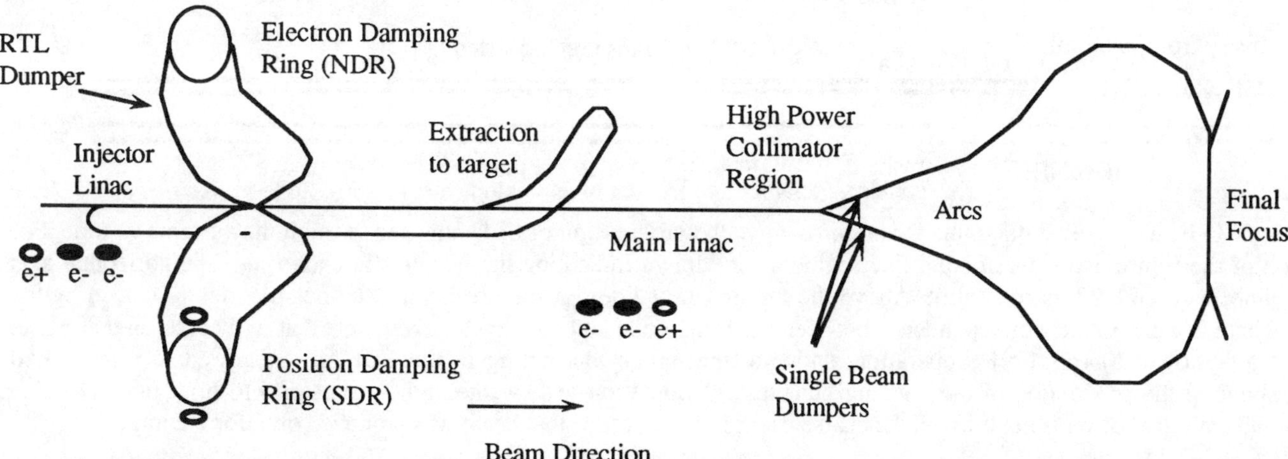

Figure 2: SLC Layout. The sequence begins when an electron bunch pair is extracted after damping 1/120s and follows a leading e+ bunch through the main linac where the trailing bunch is deflected onto the positron target. The newly created positrons are brought back to the injector and inserted behind a new electron bunch pair. The key features of this cycle from an MPS point of view are that the e+ are both delivered and generated on the same pulse and that the e+ generated on pulse *n* are destined for pulse *n+2*. The extraction line feeding the 30GeV beam to the positron target, the region surrounding the high power linac background collimators, the entrance to the arcs and the final focus are critical for MPS and most EBD trips originate in these areas.

only pulse consists of unloaded e-, some nuisance trips will result.

The behavior of the NDR is somewhat different. Since it has a long kicker pulse, no bunches will survive longer than the nominal interpulse time. These considerations result in the scheme shown in figure 3.

Since the e+ trail the e- through the injector, an additional pair of electron pulses is required in rate limited mode to provide the proper beam loading. This beam must not be allowed to continue into the main linac and is disposed of with a small dump in the ring to linac transport line. For extended rate limited periods, such as during machine development, the 'loading bunches' are turned off and an appropriate energy compensation is made.

DATA ACQUISITION

Finally, we must consider the impact of rate changes on the machine data acquisition system. Data is continuously acquired from position monitors for feedback purposes and, less often, for accelerator studies. This is accomplished by dropping the rate of the acquisition codes and assigning high priority first to operator driven acquisition and then to the feedback process. A more complex problem is the control of synchronized data acquisition across region containing the SBD's which may be rate limiting the beam. The beam sequence controls this by broadcasting data acquisition codes to the appropriate data collection processors on a user by user basis. Since it also controls the SBD's, it is able to tag the pulses which will be allowed through.

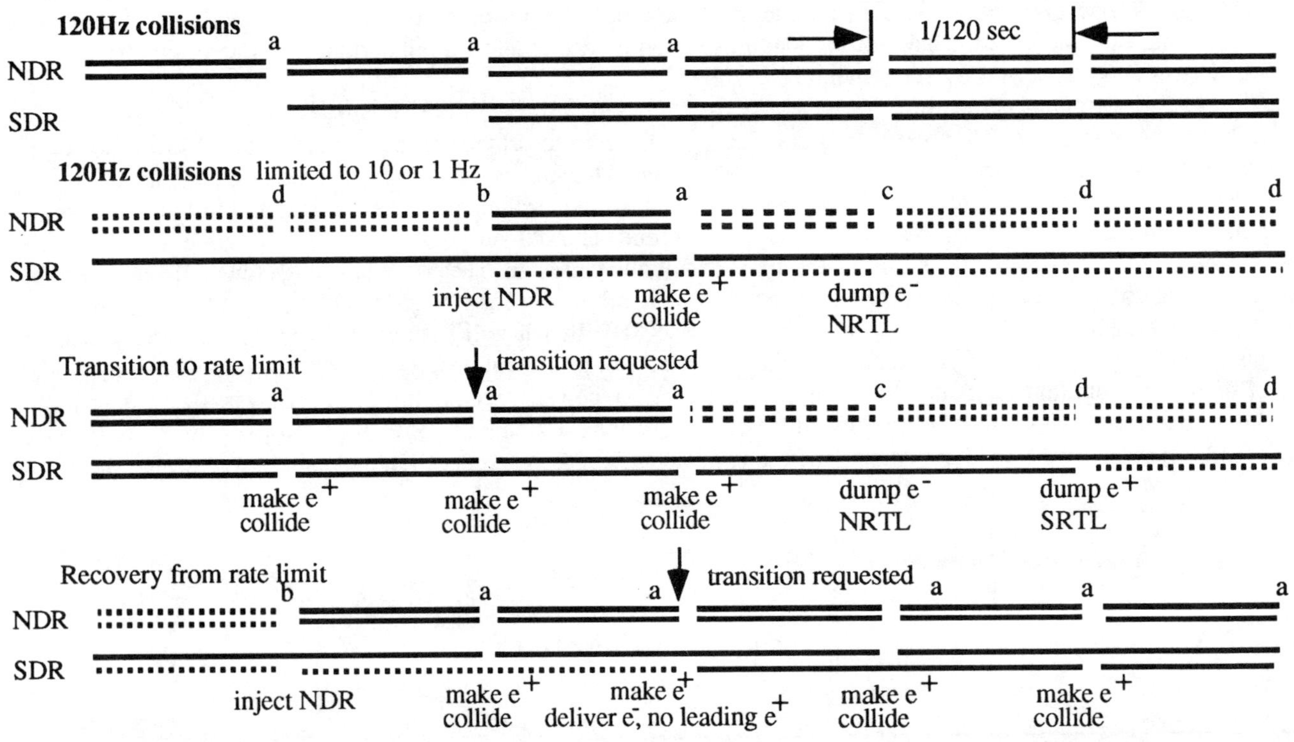

Figure 3: SLC Beam timeline diagram. This figure shows the progression of SLC bunches through the systems vs time. For each section of the figure, two sets of solid line segments are drawn indicating the bunches in each ring. The horizontal axis represents time, with an 8.33ms basic unit. Above the top group of lines, at the breaks, are the beam codes produced by the sequencer. There is a one to one correspondence between the beam code and the bunch movements that will occur on that pulse. Four levels are shown: a) full 120Hz operation, showing the bunch alternating in the SDR, b) Steady state rate limited operation, showing the production of the 'loading bunches', c) transition to low rate and d) transition to high rate. The fine dashed line indicates that beam is not present, the coarse dashed line indicates that beam was produced only for loading.

---

[1] K. Thompson and N. Phinney, 'Timing System Control Software in the SLC', IEEE Trans. Nucl. Sci. NS32-5:2123, 1985.

[2] R. K. Jobe et. al., 'Energy Feed Forward System at the SLC', Proceedings of this conference.

A Programmable Beam Intensity Display System for the Fermilab Accelerators

S. Johnson, D. Capista
Fermi National Accelerator Laboratory *
P.O. Box 500, Batavia, Illinois 60510

Abstract

A programmable system has been constructed to provide an easy means to display beam intensities and efficiencies from the various Fermiblab accelerators. The purpose of these displays is to provide an alternative display method for this information. The system consists of a central microprocessor, a data acquisition subsystem, an interface to the Fermilab control system, and a link driver to drive multiple display boxes. Each display box has eight displays, with a display consisting of the name of the parameter being displayed, the units for the parameter displayed, the time the data was taken, and the actual data. The final system will have a display box at each control console in the Main Control Room. The system is also a backup way to view accelerator performance when the accelerator control system is down for any reason.

Introduction

At Fermilab there are approximately eighty intensity parameters from the accelerators and transport lines. The programmable beam intensity display system gives the user a hardware display to view the intensity parameters of interest, usually to aid in making adjustments to machine parameters. Each display box will be programmable through any control console.

The system consists of a VME crate which contains the following cards: CPU, Token ring interface, MADC, scalar , display link driver, and an accelerator clock monitor. The CPU uses the link driver to communicate with the displays. The token ring interface is for communication with the accelerator control system. The CPU uses the other three cards for data acquisition.

Available Signals

There are several types of signals available which this data acquisition system uses. One type of signal is a voltage which is proportional to the beam intensity. The Linac and circular accelerators at Ferimilb use a voltage type signal to provide real time beam intensity data. Some of the beam transport lines also use a voltage signal to represent the integrated intensity which passes through the line during a machine cycle. Another type of signal is the TTL level pulse

---

*Operated by the University Research Association under contract with the U.S. Department of Energy.

train or scalar. Pulse train signals are found in some of the beam transport lines, particularly in the Switchyard experimental areas. These signals represent the total integrated intensity an area receives by the number of pulses the area returns.

System Hardware

The system hardware consists of a VME crate and several display boxes. The VME crate contains several cards which are commercially available and several cards which have been built in house. The display boxes and the VME crates' scalar, link driver, and accelerator clock monitor cards are in house design and construction. The CPU, MADC, and token ring interface cards in the VME crate are commercial cards.

The Fermilab Accelerator Clock contains information about the status of the various accelerators such as: injection, start of acceleration, end of acceleration, and extraction [1]. The accelerator clock card in this system monitors these eight bit clock events so that the CPU will know when to sample the various signals. The CPU tells the clock card which events to monitor. When one of these events occur, the clock card stores the event data in a register and interrupts the CPU (fig. 1).

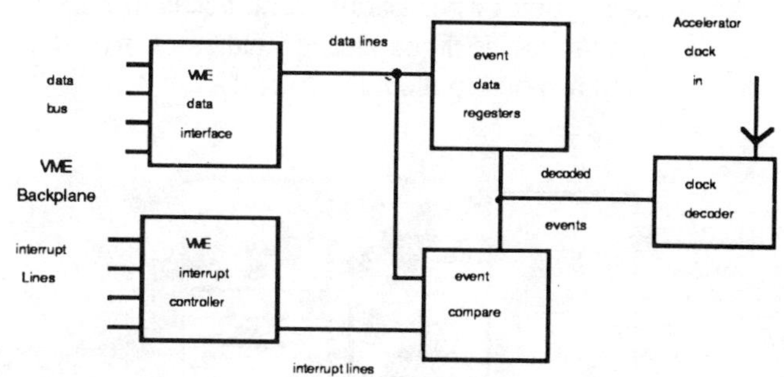

Figure 1. Accelerator Clock Card

The CPU then processes the interrupt and reads the event data. The clock card is capable of four priority levels of interrupts with a storage register for each interrupt. These four levels of interrupts are necessary since several clock events may occur close to each other in time.

The CPU communicates with the display boxes through a parallel link. Each transmission consists of sixteen bits of address and sixteen bits of data. When the CPU communicates with the link driver, the address and data information are moved into latches. The link driver then clears communication with the CPU to allow it to perform other tasks while the link driver transmits the information on the link.

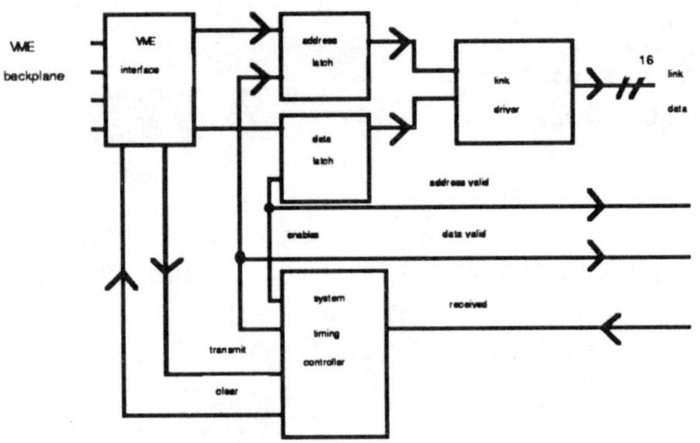

Figure 2. Link Driver Card

After the VME interface moves the address and data information into the latches, it signals the system controller to begin transmission. The system controller then moves the address information onto the sixteen bit link and drives the address valid line true. Next the system controller waits for the display box to respond to the address by driving the received line true. When the system controller views the received line true, it drives the address valid line false and waits for the received line to go false. The system controller then moves the data to the link and drives the data valid line true. The display box which responded to the address processes the data and responds to the link driver by again driving the received line true. To complete the transmission, the system controller drives the data valid line false and issues a clear to the VME interface to allow further transmissions. Each transmission requires about 1us.

Each display box on the parallel link needs to repeat the link information to the next box, decode addresses, pass data to its displays, and respond to the link driver (figure 3).

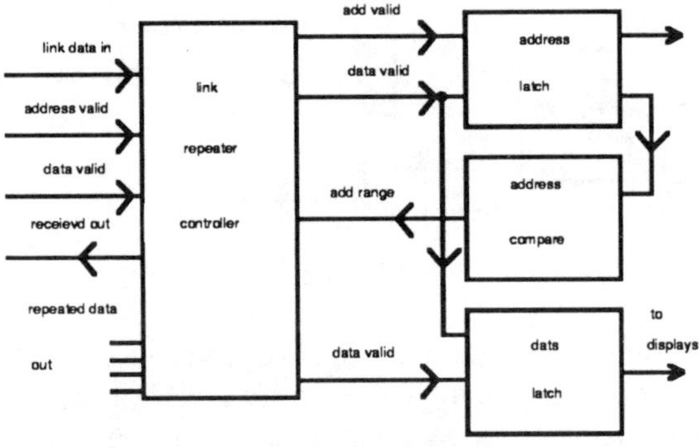

Figure 3. Link Repeater/controller

Once the link driver places the address on the link and drives address valid true, each display box will receive the address through its link repeater/controller and move the information into the address latch. The address compare unit then looks at the most significant byte to see if the address is for this box and if the compare is true, the address range line will be driven

true to notify the link repeater/controller. Upon receiving address range true, the link repeater/controller will drive the received line true to notify the link driver it has the address. The link driver then puts the data on the link and drives the data valid line true. Next the link repeater/controller receives the data and with the address range line still true, transfers the data to the data latch and routes the address and data information to the displays. To complete the link communication, the received line is driven true by the link repeater/controller.

Each of the eight displays in a display box consists of four numeric and sixteen alphanumeric displays (figure 4).

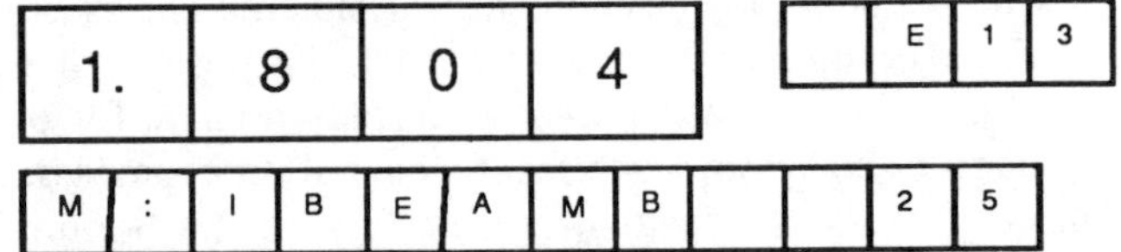

Figure 4. Display layout.

The four alphanumeric displays to the right of the numeric displays represent the units of the data. The twelve alphanumeric displays under the numeric displays represent the the name of the parameter and the accelerator clock event the display updates on.

Software Considerations

The software for the system consists of four major components. These components consist of an operating system, networking software, user interface, and the actual data acquisition and display software. The system software services are provided by the MTOS operating system. This package is a commercial product supplied by Industrial Programming Inc.. The other software components are layered on top of MTOS.

MTOS is designed as a real time operating system for use in embedded applications. MTOS provides multitasking abilities, along with intertask communication facilities. Examples of the services provided by MTOS are semaphores, controlled shared variables, event flags, signals, message buffers, and mailboxes. A prioritized scheduling algorithm is used by MTOS to make sure that critical tasks are run on time. Since the operating system design is for embedded applications, MTOS itself does not support program development. All development work is done on a separate machine, and the finished code is downloaded to the system.

The second part of the software is the network interface. This allows the system to communicate with the Fermilab ACcelerator NETwork or ACNET. The physical connection to the network is through a Token Ring interface. The network interface is provide for convenience in changing the data acquisition setup. This part of the software was written in house at Fermilab.

The user interface task allows the user to change the setup of the system from an attached terminal. This same task is also called by the network software when it changes the data acquisition setup. This task runs at a very low priority and will only make changes when all other tasks are idle. This task also provides debugging routines through the attached terminal for both the hardware and software.

The last part of the software is the actual control and display software that makes up the system. This part of the software can be viewed as four different tasks. Each of these tasks provide some function that can be preformed independently of the other tasks. The data is then passed between the different task via software mailboxes.

The data acquisition is controlled through a central data structure. This structure is an array of linked lists, with each list corresponding to one reset on the Accelerator Clock system. A node on this list will contain the following information: index of the disired parameter, raw data, scaled data, and one entry for each box on the link. The lower eight bits of each box entry correspond to the eight displays on each box. This information is used to determined which displays are updated for this node. When adding or deleting nodes from the linked list, the user interface task will first lock all other tasks from the list to prevent corrupting their pointers to the linked list. Figure 5 shows an example of a node on the linked list.

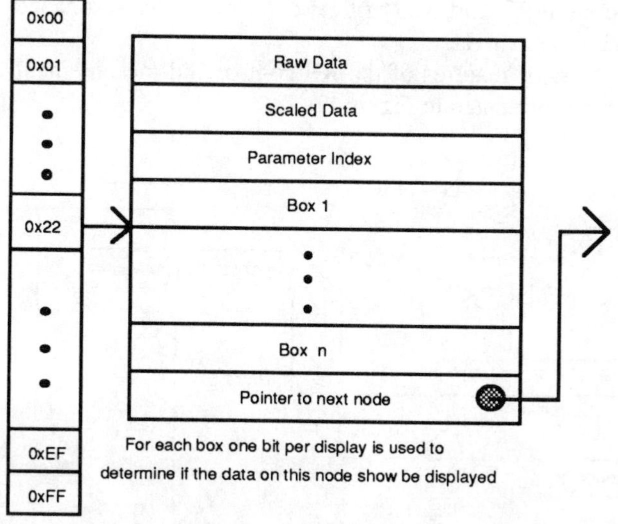

Figure 5. Node on Clock Event Linked List

A very simple database is also maintained in the system. There is one entry for each data parameter, and the database is accessed using the parameter index stored in the node on the linked list. The database provides information on what card and channel to read to get the raw data, how to scale the raw data, and in what units the scaled data is displayed.

The first task reads the clock event to be serviced off the Accelerator Clock card. This task interfaces with the interrupt handler in the system software to provide this information. Once it has read the clock event it passes the information on to the next task.

The second task takes the clock event provided by the first task and traverses a linked list associated with that clock event.

Each node on this linked list is a channel on the scalar or MADC card that should be read. The task reads each channel, stores the raw data in the linked list, and the passes onto the next task the clock event to be processed.

The third task again takes the clock event being passed to it and traverses that link list. This time the task will take the raw data that was collected by the last task, and scale that data and put it into a displayable form.

The last task takes the clock event data and again traverses the linked list. Encoded on each node is the box and display number for where the information should be sent. Figure 6 shows a view of each task's input and output.

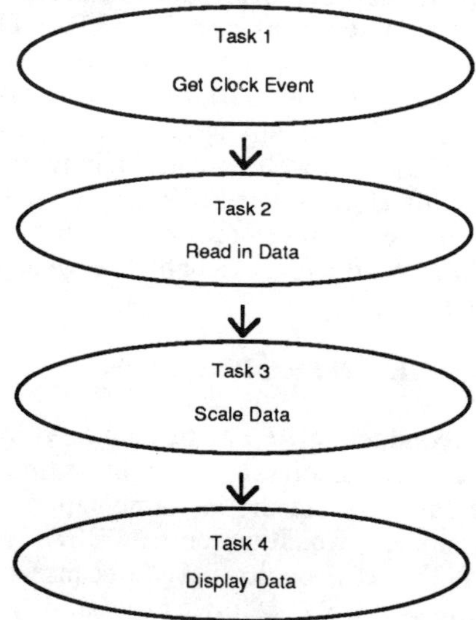

Figure 6 Task Input and Output

Summary

The display system is currently up and running in the Main Control Room at Fermilab. The interface between the network and the data acquisition software still needs to be improved. This should be completed in the next couple of mouths. Additional display boxes are being built so that a box will exist at each console.

Acknowledgements

We would like to thank Linda Klamp who came up with the original ideas for the system, and who was involved with the early programming efforts on the system. We would also like to acknowledge the efforts of James Morgan who is doing the actual display box construction, and Charles Robertson who designed the scalar card.

References

[1]   R. J. Ducar, "Tevatron Clock Event Assignments", Fermilab controls hardware release No. 17.35, February 1991.

An Automated RF Control and Data Acquisition System for Testing Superconducting RF Cavities

C. Reece, T. Powers, and P. Kushnick
Continuous Electron Beam Accelerator Facility
12000 Jefferson Avenue
Newport News, VA 23606

Abstract

An integrated RF control and data acquisition system has been developed to perform acceptance testing of superconducting RF cavities for CEBAF. The system provides options for automated resonance locking, VCO frequency and PLL phase optimizations, CW cavity coupling measurements, pulsed RF coupling and Q measurements, and cavity field regulation as well as data logging and retrieval. The system has dynamic range sufficient to stably operate CEBAF SRF cavities at accelerating gradients in the range 0.05 - 20.0 MV/m and is switchable between six cavity testing positions.

I. INTRODUCTION

As superconducting RF cavities are used in increasing numbers in particle accelerators, an increased degree of automation is called for in the instrumentation used to test cavity performance. Construction of CEBAF requires the installation of 338 SRF cavities.[1] Because of the large investment in assembly effort, provision is made to performance test all cavities prior to insertion into the horizontal cryostats used in the accelerator. A testing system for SRF cavities has been designed and constructed at CEBAF to accommodate the systematic testing of a large number of cavities and to provide a durable facility available indefinitely for possible rework of cavities. By providing automated control of all RF parameters, enhanced testing efficiency is obtained relative to the hardware and instrumentation styles commonly used in R&D applications.

II. RF HARDWARE CONFIGURATION

The SRF cavities to be tested have a fundamental accelerating mode frequency of 1497.0 MHz, are equipped with variable input coupling, and have a fixed transmission port coupling Q_{ext} nominally equal to 2E11. Unloaded Q-factors of the cavities are typically >3E9.[2] The cavities are tested with the input RF port near critical coupling so the resonance bandwidths are < 1 Hz.

The major features of the RF control and measurement system are represented in Figure 1.

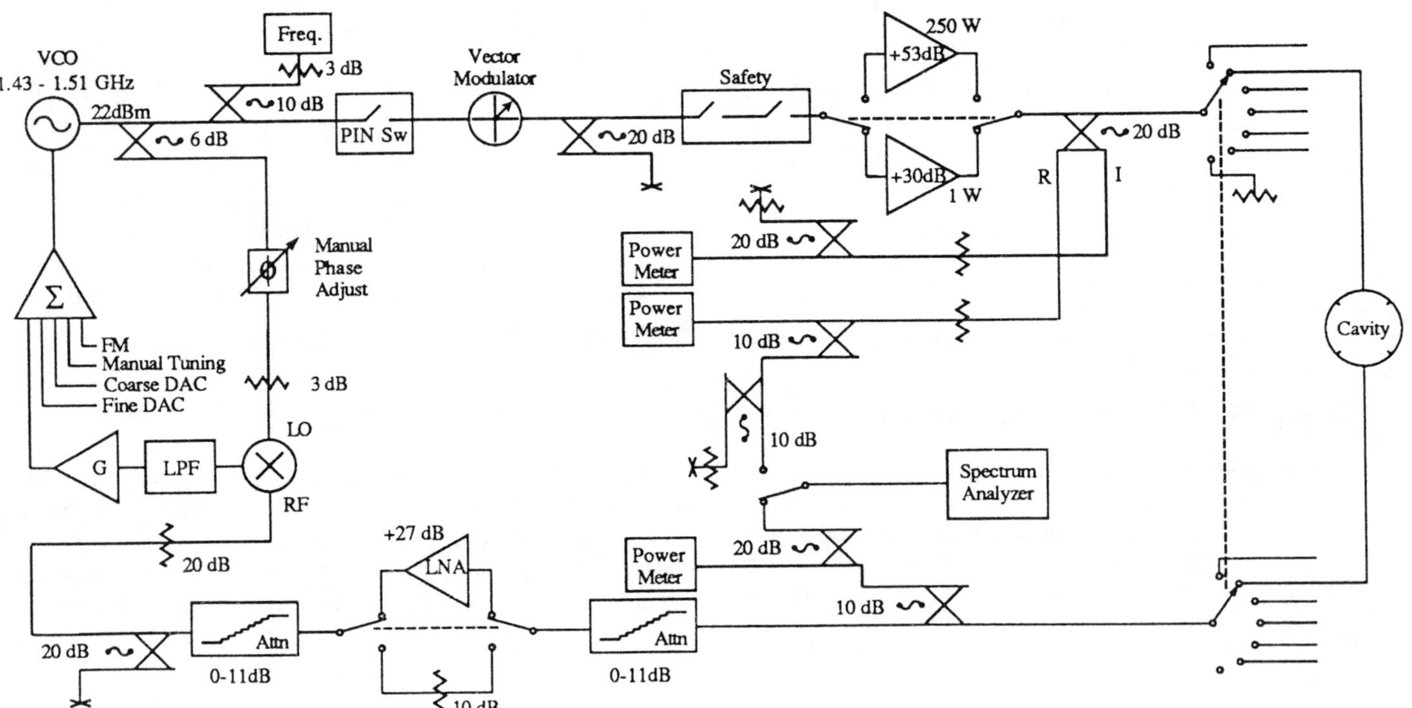

Figure 1.   RF Network for Performance Testing CEBAF Cavities.

This work was supported by the U.S. Department of Energy under contract DE-AC05-84ER40150.

A pair of cavities sharing a common internal vacuum are individually RF tested at 2.0 K in one of two vertical cryostats. In addition, there is the ability to test individual cavities in a third cryostat. Because the cryogenic cycle time for a dewar and cavity pair is long compared with the time required for RF testing, a common RF system is shared between the three cryostats.

A linearized voltage controlled oscillator provides a 22dBm RF signal electrically tunable over the range 1430 to 1510 MHz, which spans the fundamental pass band of CEBAF cavities. A PIN switch with TTL driver and 10 nsec switching time permits pulsed RF testing of the SRF cavities. The output of the PIN switch is conditioned by a vector modulator manufactured by Vectronics Microwave Corp. This vector modulator is implemented with 9-bit TTL selected PROM programmed linearized attenuation having 0.125 dB steps in both the I and Q hybrid legs. With a bi-phase switch in each leg, the 20 TTL lines enable selection of an arbitrary phase shift across the vector modulator with $< 0.5°$ resolution and simultaneous amplitude control over a 40 dB range with 0.125 dB resolution.

Safety interlock RF switches (operation discussed below) allow for interruption of the drive signal prior to routing to one of two solid state amplifiers. A 1 watt amplifier is used for low power testing, and an amplifier, manufactured by Power Systems Technology, Inc. capable of 250 watts output with 53 dB gain, is used for high power operations. The output of the selected amplifier is routed to the cavity under test with a SP7T coaxial switch which is electrically ganged with a similar switch that selects the transmitted signal from the cavity. These switches, manufactured by Loral Microwave-Wavecom, are rated for reliable operation to over one million actuations per position.

The amplitude of the transmitted RF signal is trimmed or boosted with a combination of programmable switch selected step attenuators and a low noise amplifier. This conditioning of the transmitted signal is done to maintain a -20 dBm level at the RF port of the phase detecting mixer while the power transmitted out of the cavity ranges from -25 to +35 dBm. This regulated power level, together with the phase lock loop electronics described below, maintains a consistent locking range of 50 kHz and a capture range of about 5 kHz over a potential testing range of accelerating gradients of CEBAF SRF cavities of approximately 0.03 to 25 MV/m.

For power measurements, portions of the incident, reflected, and transmitted RF signals are coupled into calibrated HP437 power meters and Schottky diode detectors. In addition, samples of the reflected and transmitted signals may be manually switched into a spectrum analyzer.

III. CONTROL ELECTRONICS

a) *RF Controls*

The VCO control electronics consist of a phase lock loop amplifier, a two channel D/A converter module and a function generator module. These modules along with one 4 channel low noise amplifier module and a computer-to-vector modulator interface module were fabricated using a 3U 220 mm eurocard standard and low noise techniques. These modules are interconnected and interfaced with one another, Nubus I/O boards in a Macintosh II computer, and the vector modulator using a small back-plane printed circuit board. Ground-plane printed circuit boards and 0.141 hard line are used throughout this section. All sensitive analog circuitry are housed in shielded enclosures.

The phase error signal is applied to the input of the PLL amplifier where it passes through a 100 kHz, LRC, low pass filter which presents a constant 50 Ω impedance to the mixer to 50 MHz. The phase signal is then amplified using a variable gain amplifier. It is also summed with the output of the function generator (FM drive) and either the manual tuning input or the outputs of the D/A converter module which have been heavily filtered. The composite signal is buffered and then connected to the VCO after passing through a 25 kHz single pole low pass filter. The output noise of the amplifier with the input shorted is 30 μVp-p which equates to 240 Hz of residual FM noise at the VCO output.

Manual tuning of the frequency is accomplished by adjusting two front panel potentiometers, one of which passes through a 150:1 resistive divider network. The function generator provides a triangle waveform of varying amplitude which is used to sweep the frequency.

Computerized tuning is accomplished using a dual D/A converter module. This module consists of a dual 12-bit DAC, associated amplifiers, and drive logic. The two 0-10 V signals are applied to the input of the VCO loop amplifier module where both DAC channels pass through 3 Hz low pass filters, are scaled, and applied to the summing amplifier. The frequency step for the coarse DAC is 16.5 kHz, and for the fine it is 30 Hz. The low frequency filters force the RF frequency slewing to be slow enough to permit PLL acquisition as the incremented frequency steps across the narrow cavity resonance.

Four channel low noise preamplifiers were constructed to provide user and computer interface to various crystal detectors and other signals throughout the system. Each channel consists of a three op-amp instrumentation amplifier circuit using low noise amplifiers, metal film resistors and ground plane circuit techniques. Channels used for crystal detector signals typically operate with 50 dB of gain (Av=316), which scales the top of the detector square-law range to a convenient 8 volts, while miscellaneous signals are buffered by unity gain channels. Signal outputs are buffered and routed to the front panel and to the A/D converter on the computer I/O card.

b) *Safety Systems*

Safety interlock RF switches in the amplifier drive line serve three major functions: they inhibit introduction of high power RF to a cavity unless radiation shielding is in place, they ensure that RF routing switches are not switched live, and they disable all RF whenever significant radiation levels are detected outside shielding. Radiation interlocks were implemented using simple relay logic, while the dewar selection and high power amplifier selection logic was implemented using programmable logic devices.

Movable massive shielding lids surrounding the tops of the recessed cryostats provide attenuation adequate to reduce the possible 2000 R/hr inside to < 0.1 mR/hr outside. Closure of a lid, which is detected using a pair of microswitches, is fed into the radiation interlock system.

In addition to the provision of interlocks on the primary drive circuit, there is logic which drives the switching of the input/output of the power amplifier which inhibits high power operation unless the lid of the selected dewar is closed. Appropriate delays are imbedded into the dewar selection and high power amplifier selection logic to ensure that no RF signal is applied unless the selected RF switches are closed and that no switch changes are made until the RF energy stored in the cavity has had time to decay.

IV. SOFTWARE

a) Programming Environment

All aspects of the software controlled data acquisition, RF control, and analysis are performed on a Macintosh II computer in the environment provided by LabVIEW 2®, published by National Instruments. LabVIEW 2® programming provides a convenient, integrated graphical user interface both for program development and user interaction. The programmer graphically constructs custom virtual instruments which interact with I/O hardware in the desired manner. Since the execution sequence is driven by data flow, these virtual instruments can subsequently be arranged in logically straight forward hierarchical combinations. Modular data acquisition, control, and analysis routines are thus easily integrated for various applications and may be dressed for efficient user interaction.

b) Data Acquisition

The following parameters may be read by the RF system computer via GPIB control of commercial instruments: RF frequency, incident, reflected, and transmission CW power levels, and field emission current. A buffered 12 bit A/D converter channel is provided for each crystal diode detector in the system and also the phase error signal from the mixer. Time dependent RF levels may thus be monitored at sample rates to 100 kHz. The A/D channels may be auto-scaled, making the detector output directly readable over the full square-law range from -53 to -20 dBm. An additional A/D channel is used to log the intensity of any x-radiation that may be induced by field emission in a cavity. Communication over an RS-422 line with the cryogenics control computer permits logging of dewar pressure, He bath temperature, and liquid level.

c) Control

As mentioned above, the VCO frequency may be controlled by two 12-bit DAC's, and RF phase and amplitude are controlled via a 20-bit signal applied to the vector modulator. One program module can set the uncorrected VCO frequency with 1 kHz resolution. Another provides the operator or another module with the ability to independently control phase and attenuation of the vector modulator. These are integrated with monitoring of the transmitted power crystal detector in a resonance search routine which systematically searches the frequency-phase plane near a desired mode, locates the transmitted power maximum, and nulls the phase error signal in 15 - 30 seconds.

Another module trims the mixer RF input to the optimum level and automatically compensates path length differences as attenuators are switched in or out by adjusting the vector modulator phase setting. This maintains a nulled phase error signal in the PLL as the operating power level is changed.

Programmatic control of the PIN switch is provided via GPIB communication with a pulse generator. One shot operation is available as well as user specified repetition rate and duty cycle pulsed mode.

d) Data Analysis

Several software modules have been developed which integrate the control of RF parameters and data acquisition with real time data analysis, logging, and display. Analysis of coupling parameters, calculation of the required cavity performance parameters, (unloaded Q-factor, accelerating gradient, and transmission probe external Q-factor), and error analysis is performed automatically with each measurement. The time dependence of the logarithmic slope of the transmitted decay signal is used to measure the loaded Q-factor directly as a function of stored energy. This feature increases the reliability of high field Q measurements when non-linear loss mechanisms become significant.

Another module enables pulsed RF testing and analysis, employing the computer directly as a four-channel digital storage scope which automatically scales to the selected RF pulse repetition rate and performs coupling and decay analysis.

VI. ACKNOWLEDGMENTS

We are grateful for the very important contributions to low level software drivers which were provided by J. Brown and B. Almeida. I.E. Campisi, P. Kneisel, and J. Mammosser made early contributions to system layout and specifications.

VI. REFERENCES

[1] H.A. Grunder et al., "The Continuous Electron Beam Accelerator Facility," in *Proceedings of the 1987 IEEE Particle Accelerator Conference*, Washington, D.C., March 1987, pp.13-18.

[2] P. Kneisel et al., "Performance of Superconducting Niobium Cavities for CEBAF," contributed to this conference.

A New Data Acquisition and Control System for the Power Amplifier Test Station

Mark S. Champion
Fermi National Accelerator Laboratory*
P. O. Box 500, MS 306
Batavia, Illinois 60510

Abstract

The RF Group at Fermilab maintains approximately 50 power amplifiers that provide RF power for particle acceleration in the Booster and Main Ring accelerators. These amplifiers must be highly reliable since a failure necessitates shutting down the accelerator to allow for technician access to the beam tunnel. To ensure reliability after repairs, each amplifier is tested on an automated test station prior to being placed back in service. By using automated testing we ensure that all amplifiers are tested under the same conditions. Hence comparisons may be made and trends observed. Until recently the test station was controlled and monitored via a Z80 based microcomputer. The system was inflexible, user unfriendly, and increasingly unreliable. An IBM PC/AT with A/D and D/A cards has been installed and programmed using C language. The new system is easily modified and continues to evolve as needed. This paper describes the test station hardware and software.

I. INTRODUCTION

This paper describes the recent upgrade of the data acquisition and control system of the power amplifier (PA) test station. The PA's will be described in section II followed by a description of the test station in section III. The upgrade will be described in section IV, including discussions of the hardware, software, and tests performed. A conclusion is given in section V.

II. POWER AMPLIFIER DESCRIPTION

The RF power for a single accelerating station in the Booster and Main Ring accelerators is provided by a three-stage 100 kW power amplifier [1]. The first stage is the distributed amplifier which makes use of six 4CW800F tetrodes to raise the RF power level from 1 W to 100 W. The second stage is the lower half of a cascode arrangement based on fourteen paralleled 4CW800F tetrodes. This stage raises the RF power level to 2500 W. The final stage and upper half of the cascode depends on a Y567B tetrode to provide 100 kW of RF power to the accelerating cavity. The output power may be varied through programming of the cascode grid bias voltage and of the power tube anode voltage. Maximum output power is approximately 140 kW at an anode voltage of 25 kV. The input level to the distributed amplifier is held constant regardless of the desired output power. The range of frequencies required for the Booster and Main Ring systems is 30-53 and 52.8-53.1 MHz, respectively. Cavity tuning is achieved via horizontal ferrite tuners mounted adjacent to the cavity structure. A wide dynamic power range is required due to varying ferrite losses and gap voltage requirements.

III. TEST STATION DESCRIPTION

The PA test station may be described as consisting of three major subsystems: (1) the physical PA test stand including the PA, the anode resonator, and RF load, (2) the power supplies providing AC and DC voltages and RF drive, and (3) the control and data acquisition system. A functional diagram of the test station is shown in Figure 1.

Instead of mounting the PA directly onto an accelerating cavity, a water-cooled 50 Ω load is used to dissipate the output power. The power amplifier is mounted on a coaxial cavity resonator that is resonant at about 51 MHz with a Q of 60. Power is coupled out of the resonator via a tap on the center conductor and feeds into a 6 inch coaxial line which is terminated with the 50 Ω RF load.

The 30 kV anode supply feeds into the anode modulator which is used for programming the anode voltage on the power tube. The modulator cabinet also houses filament, screen, bias and other supplies necessary for operating the PA. The RF drive is obtained from a programmable frequency synthesizer and 3 W amplifier.

The data acquisition and control is handled by a computer system with digital-to-analog (D/A) and analog-to-digital (A/D) converter cards. The D/A card is used to program the power tube anode voltage, the cascode grid bias, and the RF drive level. (Although the RF drive level is constant in the Booster and Main Ring accelerating stations, the test station RF drive level is adjustable.) The A/D card is used for monitoring the operating parameters of the test station. Most of the inputs to the A/D card come from a monitor box mounted above the PA. This box contains voltage dividers, Hall Effect current sensors, and RF diode detectors. Isolation amplifiers are used to adjust the signal levels to ±10 V referenced to system ground.

Monitoring of the PA output power is accomplished via calorimetry. The water flow through the RF load and the temperature differential between the inlet and outlet are used to calculate the power dissipation. In addition, several RF monitors are built into the PA so that the RF waveforms may be observed.

Until recently the computer system in use was a Z80 based microcomputer equipped with an 8-channel D/A converter and

* Operated by Universities Research Association, Inc., under contract with the United States Department of Energy.

a 32-channel A/D converter. There was no disk storage of any type, and the program resided entirely in read only memory. There were a number of code patches that had to be typed in each time the computer was restarted. The system was completely inflexible and was becoming increasingly unreliable, hence it was decided to install a new computer system, which is described in the next section.

IV. UPGRADE DESCRIPTION

A. Hardware

The main component of the upgrade is an IBM PC/AT personal computer. Accessories include a 20 M-byte hard disk, 3.5-inch and 5.25-inch floppy drives, 640 k-byte base memory and 2944 k-byte extended memory, an enhanced color display, and a dot-matrix printer. The Burr-Brown PCI-20000 instrumentation system was chosen for handling data acquisition and control. This system offers modularity via a carrier card and up to three instrument modules. The carrier card provides interfacing to the PC bus, provides power and mounting for the instrumentation modules, and provides for inter-module communication. We are using the PCI-20041C-3A carrier which supports direct memory access data transfer and supplies 32 bits of digital input/output (I/O). The three instrumentation modules include the PCI-20021M-A 12-bit, 8-channel D/A converter, the PCI-20019M-1A 12-bit, 8-channel A/D converter with sample-and-hold, and the PCI-20031M-1 32-channel analog expander. The expander module works in conjunction with the A/D module to provide 32 channels for monitoring analog signals in the range of ± 10 V. The D/A converter is configured to produce analog signal in the range of ± 10 V.

B. Software

The software is written entirely in C language and is compiled using Microsoft C version 6.0. The test program begins execution by initializing the instrumentation system and setting up the data structures used throughout the program. It also prompts for the name of the operator and the serial numbers of the PA being tested. After initialization, the main control loop is entered. This loop awaits keyboard input and passes control to the appropriate routines when activity is detected. Control is accomplished via the function keys and the numeric keypad. The function keys initiate major tests such as the Q test and the phase test, provide a help screen, allow for system reset, and provide for exiting the program. The numeric keypad provides cursor movement in the control section of the display (lower left corner of Figure 2). The RF drive frequency and level, the power tube anode program, and the cascode grid bias program can all be adjusted in this section.

Data acquisition occurs in one of two ways depending on the RF duty factor. If the duty factor is less than 100%, the duty cycle generator interrupts the computer when the RF turns on to initiate data acquisition. If the duty factor is 100%, no interrupts are generated and the program acquires data periodically as determined by an internal timer.

C. Tests Performed

The first test performed after a PA has been mounted and the filaments have warmed up is to check the system Q. A single keystroke initiates this test, which is performed at an anode voltage of 20 kV, a cascode cathode current of 1.75 A, and a cascode grid RF level of 35 V_{p-p}. The input RF is swept from 49-53 MHz while monitoring the power tube anode RF. A least squares fit is performed on the measured Q curve, and the resulting fit and the measured data are plotted on the computer display. The system Q and resonant frequency are also displayed. Subsequently, the RF input frequency is automatically set to the resonant frequency.

A phase test is performed to check that the "time-of-flight" through the PA stages is correct. Setup of this test requires disabling the DA grid bias so that the RF power levels are kept low. Two test sequences are performed: (1) the phase difference is measured between the RF drive and the cascode grid RF, and (2) the phase difference is measured between the RF drive and the power tube cathode RF. During each sequence the phase difference is measured at 30, 40, and 50 MHz at power tube anode currents of 2, 4, and 6 A. Performing the phase test requires only that the operator swap coaxial cable connections at the PA as prompted by the computer.

Assuming success with the first two tests, the PA is subjected to power testing. It is operated at anode voltages of 20, 22.5, and 25 kV. Output power levels range from roughly 85 to 140 kW. Power testing is performed either with a 50% duty factor with a period of 100 ms, or with a 67% duty factor with a period of 16 seconds. It is desirable to run the power amplifier at power levels of 90-100 kW for several days to ensure reliability. Operating parameters are compared with nominal parameters so that abnormal conditions may be easily noted. High or low readings are displayed in different colors on the computer display. An example of the computer display during a power test is shown in Figure 2.

V. CONCLUSION

As a result of this upgrade, the PA test station now has a flexible data acquisition and control system. The software is slowly evolving as we gain experience with the system. A goal for the future is to develop an on-line database of test results so that it is easier to monitor trends, make comparisons, and generate reports.

VI. ACKNOWLEDGEMENT

Thanks to Jim Crisp for his contributions to this project.

VI. REFERENCES

[1] Q.A. Kerns and H. W. Miller, "100 kW RF POWER AMPLIFIER," IEEE Transactions on Nuclear Science, vol. 18, pp. 246-248, 1971.

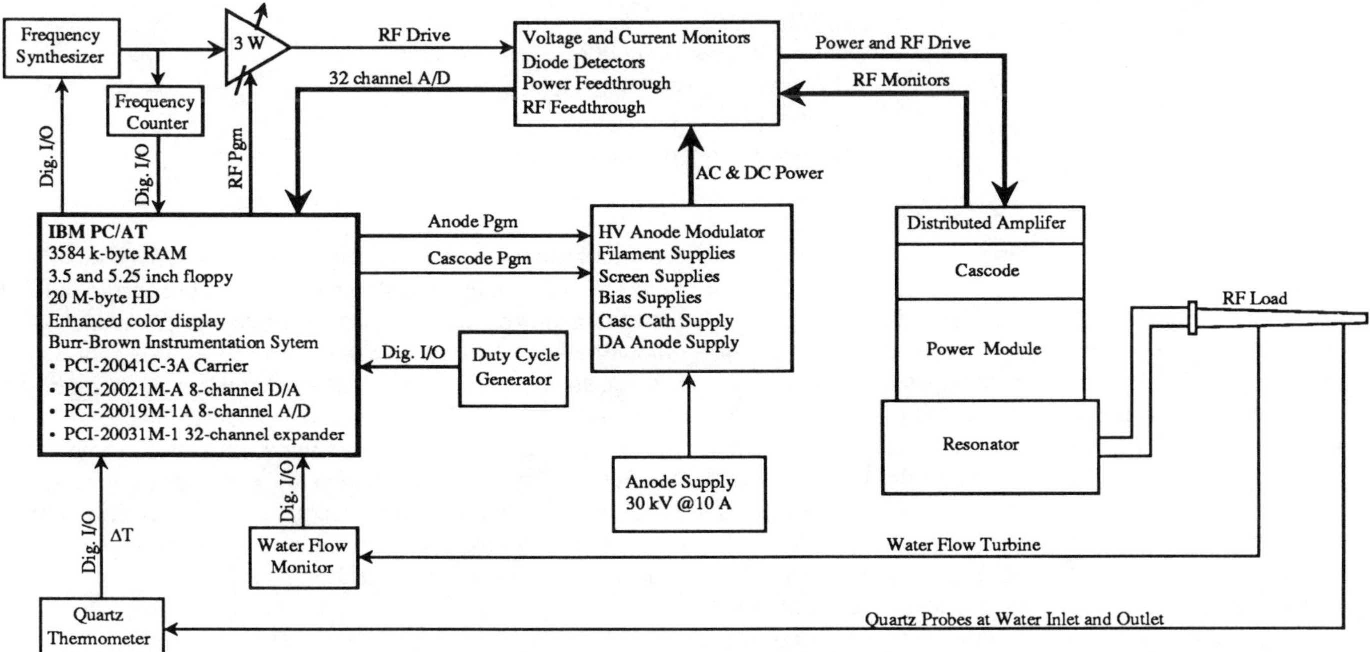

Figure 1. Functional diagram of power amplifier test station.

```
 0  PA anode V              20.581 kV        1  PA Anode I              7.520 Amps
 4  CASC cathode V        -704.102 V         5  CASC cathode I         -9.985 Amps
26  DA anode V             559.082 V        27  DA anode I              2.795 Amps
30  DA bias V               33.740 V         1  DA bias I               0.004 Amps
 6  PA screen V            474.121 V         5  PA screen I             0.088 Amps
 2  CASC screen V          295.898 V        27  CASC screen I           0.428 Amps
28  DA screen V            251.953 V         1  DA screen I             0.081 Amps
22  PA filament V          405.273 V         5  PA filament I           7.920 Amps
24  DA/CASC filament V     466.309 V        27  DA/CASC filament I      1.453 Amps
 8  CASC bias             -11.621 V          1  CASC error              3.125 V

33  Frequency               51.464 MHz      16  Phase Detector          6.504 deg
33  Water flow              17.300 GPM      99  MOD HV on                 YES
33  Delta Temp               9.480 degC     99  MOD ready                 YES
33  Power                   86.591 kW       99  Spark                      NO
33  Efficiency              56.053 %        99  Duty Factor                50 %

13  PA anode RF              4.137 Vpp       10  DA grid RF              1.233 Vpp
12  PA cathode RF           2.328 Vpp       11  RF Drive               16.081 Vpp
 9  CASC grid RF            3.342 Vpp

                                            Program Mode       MANUAL
88  Frequency|        |     51.465 Mhz      DA, Casc, PA #     15,15,15
87  Anod Prog|        |     20.000 kV       Tested by          M. Champion
86  Casc Prog|6.000   |      6.000 V        Date               4/09/91
85  RF Prog  |        |        4.4 V        Time               11:54:18
```

Figure 2. Example of operating parameters display during power testing of PA.

Vacuum Control System for the MLI Model 1.2-400 Synchrotron Light Source

S. Pugh and B. Ng

Maxwell Laboratories, Inc., Brobeck Division
4905 Central Avenue
Richmond, California 94804

Abstract

A new vacuum control system has been developed for the MLI Model 1.2-400 Synchrotron Light Source. The choice of industrial programmable controllers provides expansion capability and the ability to adapt to the profusion of off-the-shelf components. The MLI synchrotron vacuum control system can easily expand to include this diversity, as the ring and user beamline work-stations are expanded. Downtime and repairs are minimized by using readily available modules. The MLI synchrotron vacuum control system can interface to any main accelerator control system, reducing the main computer software loading, and maintaining operator response speed at a much lower cost than a custom system.

I. Requirements

Vacuum control systems in industry and university research facilities have historically used mechanical relay logic, relay patch panels, or home built solid state logic controllers. Linking the signals from all the valve position switches, logic circuits, and pressure gauge controllers to the computer system is simple in concept but difficult in application. Software integration with vacuum equipment from a variety of vendors is a Herculean task.

The Model 1.2-400 Synchrotron Light Source needs a reliable vacuum control system that is inexpensive to build and maintain, while being expandable as beamlines, insertion devices, and experimental sections are added to the ring. For system reliability, each beamline and insertion device must be able to operate rapidly and independently, even if the main computer control system or connected subsystems fail. Because valve and instrumentation signal cables are difficult to shield, the vacuum control system needs some degree of EMI and environmental tolerance.

For flexibility, the vacuum control system must be able to interface with components from a variety of vendors, without major software modifications. The software interface to the main computer control system must not compromise the other critical functions (e.g., magnet power supplies, cooling systems, beam diagnostics). Users must be able to interface their own computers and instrumentation to the vacuum control system and still be able to operate reliably.

II. Design Solution

A relay patch panel was the only option available twenty years ago, and is still in common practice. It is a rugged, easily repaired system immune to electrical noise. However, modifications are difficult, and relays have a limited life before failure. MLI pulse power technology experience and confidence with Allen Bradley programmable logic controllers (PLCs) provided a good solution for magnet interlock and vacuum interlock control decisions. PLC hardware and software experience inhouse allowed configuration and programming decisions to be made without the long lead development times of other options. Allen Bradley PLCs had already been chosen for several magnet subsystems, making it an easy decision to incorporate vacuum control into the existing Allen Bradley network.

III. Application

A. Hardware

The vacuum control subsystem communicates with the main computer database along with all other subsystem I/O, including CAMAC, GPIB, and custom application programs. The PLC network scanner link resides as a card in the MicroVax Qbus; all communication originates from there to each PLC-based subsystem. As each PLC is responsible for its own data acquisition, logic, and I/O, the host computer is ignorant of how the individual controller goes about its business, nor it the controller knowledgeable about other controllers on its network, unless they are involved in a specific data request.

To reduce signal noise and enhance signal reliability, the fiber optic communication is only 110 Kbaud. As the volume of actual data transferred is small, the slow speed is a reasonable exchange for the extra error checking. Data I/O operations inside the PLC are typically less than 10 milliseconds for every 1,000 instructions — much faster than the response through the network and host computer, but slower than through custom hardware.

B. Software

The PLC/VAX software package was supplies by the PLC vendor and runs as a separate process that communicates to the synchrotron control program and database. Although the main control system, VISTA, was designed only for CAMAC I/O, the additional programming to incorporate the PLC data acquisition was minimal.

From the operator's view, the hardware interface is invisible. Each Allen Bradley PLC was programmed by a subsystem engineer. In each case, the subsystem engineer was not familiar with the main computer system control nor with other PLC subsystems. Each PLC was programmed in the vendor's "ladder logic." Ladder logic programming proved to be easy to learn, and finished programs were quickly loaded in the PLC permanent memory. The system programmer provided the link between the PLC scanner table and the main control system database.

IV. Future

During this project, our confidence in PLCs has increased. We would now consider them as first choice for most of our accelerator projects, leaving the particular and exotic control applications in CAMAC, GPIB, and specialized computer interfaces.

The New Vacuum Control System for the SPS

Detlef SWOBODA

AT Division, CERN

1211 Geneva 23, Switzerland

Abstract

The Super Proton Synchrotron (SPS) accelerator at CERN was put into service in 1976. The vacuum control equipment is located mainly in six surface buildings (BA) which are at distances of about 1.5 km from each other. Maintenance and repair of major parts of the vacuum control system have become very expensive if not impossible. The paper describes the hard- and software choices which have been made to replace the existing control system, taking the limitations of the already installed control equipment into account and being based on available and proven technologies.

I. INTRODUCTION

The general data communications system for the control of the SPS accelerator is in operation since the commissioning of the SPS in 1976[1]. A new communication system, relying on the same architecture as in LEP[2] has now been installed in the SPS tunnel and will gradually replace the original system at the rate at which the control equipment is being converted to the new standards. The control equipment for the vacuum system has been developed throughout the life of the SPS under various standards. To obtain the same state of the art level of control for all equipment, it is necessary to upgrade substantially most of the control systems. The replacement of obsolete and onerously maintainable equipment is being undertaken on a building by building basis, system by system, over the years 1990 - 1993.

II. NEW CONTROL SYSTEM ARCHITECTURE

A. Accelerator Network

The system is built from three hierarchical levels which follow, where possible, widely used communications standards. The top layer (*accelerator network*) is used to link the distant buildings. Each building is equiped with a local control computer which is connected to the network and which in turn provides the necessary communications links with the control instrumentation. The totality of the equipment will be interfaced through RS 232 connections in order to obtain a uniform and versatile communications interface at the instrumentation level.

The general transmission system between SPS sites, which is already in operation, complies with the Token-Ring protocol[5]. Inside the BA buildings, this backbone network is isolated from the local communication systems, i.e. Token-Ring and Ethernet (IEEE 802.3), through gateways and bridges but which are transparent to the application software.

B. Equipment Control Computer

The presently installed NORSK frontend minicomputers (GPx) in the BA buildings are being replaced by so-called Device Stub Controllers (DSC)[6].

One *System Equipment Control Computer* (DSC) per building is connected to the general communications network and provides the X.25 links for the control of the vacuum equipment. It is the task of the DSC to supervise the control instrumentation in order to guaranty the integrity of the vacuum control. The DSC are VME based systems, running a real-time operating system (OS9). The communications protocol over the accelerator network conforms to the TCP/IP protocol suite, whereas X.25 [7] is used to exchange information with the vacuum control equipment.

C. Equipment Communications Link

At present, virtually all of the equipment is connected to the MPX system[3]. The vacuum system does not require sophisticated control strategies and algorithms. The required tasks are essentially

- the execution of user driven commands
- the surveillance of status information
- the signalling of exception conditions
- the reading and conversion of analog values
- the inhibition or resetting of commands.

This information can be transported by a "slow" communications link. The MPX system is, therefore, replaced by RS 232 point to point links which connect the equipment to RS 232 concentrator modules with 16 ports each. These are connected to the DSC via X.25 links (Fig. 1)

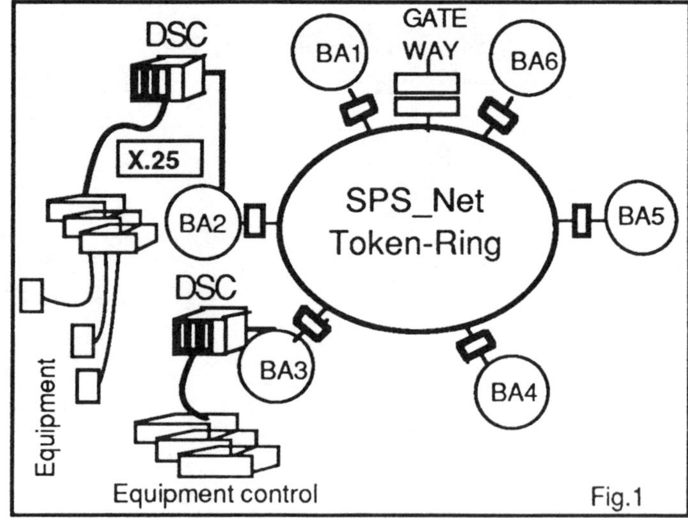

The architecture provides the following advantages

a) Easy conversion of the presently installed instrumentation to the new communications system, e.g. virtually all equipment is provided with a serial interface or can be equiped with such.
b) The unique connection standard largely simplifies modifications or configuration changes of the control system.
c) The required hardware and communications software consists of standard tested industrial equipment and uses internationally accepted communications standards.

III. CONTROLS HARDWARE ADAPTION

A. *Sputter Ion Pump Supplies*

The adaption of the power supply controls to the new communication system is being undertaken in the course of a general overhaul and modification of the presently installed supplies.

The two major improvements are:
- higher precision of the pump current as a measure of the actual pressure in the beampipe.
- simplification of the necessary controls cabling for easier maintenance and reduced costs.

In order to simplify cabling and control of the power supplies, the rearpanels of the units are replaced by a mounting which includes a small printed circuit (addressing unit)[8] and is connected to a flat cable which acts as a parallel bus. Up to 15 supply units can be connected to each of the 8 ports in a 19" G64 controller chassis.

B. *Sublimation Pumps*

Approximately 1300 sublimation pumps are distributed around the SPS accelerator. Groups of up to 10 pumps are chained over a single control cable which is connected to controller crates in the surface buildings.

Newly designed controller crates similar to those for the sputter ion pumps[9] are replacing the presently installed CIM-bus controller crates.

Each G64 crate can handle up to 8 chains of sublimation pumps. The functionality of the new equipment corresponds essentially to that of the CIM bus system. One important improvement is the hardware interlock to switch the 380 Volt mains off in case of malfunction of the system. They are divided into a proper interlock crate which controls the mains over a relay switch and into a watchdog circuit board in each sublimation pump controller crate which monitors the operation of the crate, each of which may be isolated from the interlock chain to allow sublimation despite the failure of this controller crate.

C. *Gauges*

200 new gauge control units have been received to replace the presently installed instruments. The new control units are equiped with serial interface connectors according to RS 232 and are directly connected to the RS 232 concentrators.

D. *MACBUS systems*

The MACBUS is a backplane bus [10], similar to G64 and has been implemented for the control of Roughing Pump stations, Vacuum Valves and the Sector Vaclve Interlock Chain. At present, the MACBUS crates are controlled via the MPX. However, the available CPU cards, with 8 Bit microprocessors are equiped with serial line interfaces, according to RS 422 and may, consequently, be directly connected to RS 232 lines.

Those vacuum valves which are not connected to the MACBUS are readily equiped with a MPX interface connector where all the signals are provided in form of open contacts. The addressing interface which has been developed for the control of the sputter ion pump supplies is being adapted for this purpose. As a consequence, both kinds of equipment could be connected to the same parallel bus.

IV. SOFTWARE

A. *Communication*

Configuration data is transferred to the local control computers using TelNet. The equipment is accessed , conform to the LEP vacuum message format[11]. Requests to a DSC are send in form of *Remote Procedure Calls* (RPC).

The replacement of the MPX by a serial link modifies largely the method of equipment control. More intelligence is close to the instrumentation, thus relieving considerably the load on the control computers. As a consequence, however, the control routines at the instrumentation level have to be modified correspondingly and the information exchange between control computers and remote equipment controllers must be added. In addition, software in the control computers is required to manage the instrumentation busses and the equipment database. The communication process between control computer and equipment is mainly of the command response type. An equipment with an outstanding message must at present wait until being polled by the control computer. File transfer is only required in case of a configuration update or rebooting of the equipment controller.

B. *Database*

A powerful database is one of the key requirements for efficient operation and maintenance of as large a system as the SPS vacuum control. The ORACLE database is already in widespread use at CERN (i.e. LEP vacuum system) and available on the SPS network[12].

The kind of information which is contained in the database is essentially of three types:

- The description of the SPS accelerator defines the physical layout of the machine, such as sector numbers and names, machine positions and such like. This information will in principle only vary in case of major modifications in the SPS tunnel.

- The description of the communications path contains the necessary correspondence tables between equipment names and positions in the accelerator as well as the absolute network addressof the item. These tables are relatively dynamic, since they change as soon as equipment is being moved to other locations. The address of an item is fully described by the "family-member" which identifies equally the RS 232 port to which it is connected.

- The description of the equipment contains the static information about each kind of equipment. A "family" defines the characteristics of an equipment . Each "family" includes an Equipment Data Module (EDM) which consists of the parameters, commands and functions that this kind of equipment may recognize and execute.

The cabling list database provides the correspondence between the equipment, installed in the accelerator tunnel and the related control instrument.

The inventory database of the vacuum equipment of the SPS is linked to this list. The database is used to keep track of the movement and state of the equipment and to establish the "Family-Member" relationship.

C. Application programs

The control computer (DSC) in each building (BA) is responsible for the information routing and checking between equipment controllers and application program, e.g. interactive or batch program. The dynamic pipe-allocation for each transfer session allows multiple equipment access simultaneously (Fig. 2).

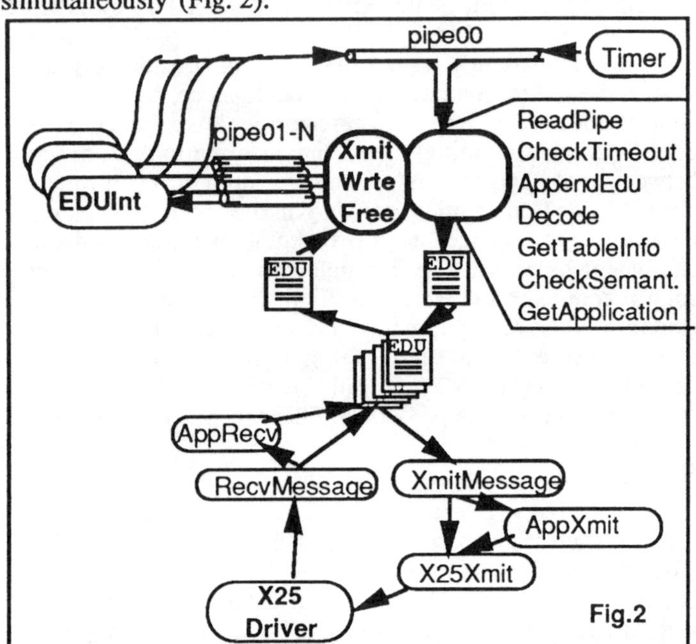

Each DSC runs a Data-Server process which is the default message handler. The Data-Server accesses the local database for address verification and checking of command and equipment type before forwarding the request to the Equipment Directory Unit (EDU). The EDU is responsible for the transmission over the X.25 Interface. Depending on the type of access, different application programs may be called before the transmission, for example to read the pressure of an entire accelerator sextant. The EDU can be directly accessed via the communications interface. In this case a pipe will be opened prior to the connection.

The Equipment Controllers interact with the equipment, such as pumps, valves, etc.. Most devices can accept messages in the standard LEP message format and execute the requested action. The CPU has full control over the equipment and exception conditions can be signalled to the DSC at any time.

Commercial instruments, e.g. vacuum gauge control units, terminals, etc., however, require a corresponding control program in the DSC.

D. User Interface

The general introduction of workstations and PC based consoles provide the tools for graphical user interfaces. Highlevel languages, like C, and multitasking operating systems are replacing the present environment, i.e. Nodal and Sintron[4]. The user access to the vacuum control system will follow the presentation which was chosen for LEP. Since the message format and the addressing scheme are similar, many of the user interface programs may be easily ported to the SPS environment.

In the first stage the user interface which has been developed on the Apollo Domain network is being used while the screen representations, written under the X11 standard, are still under development.

V. REFERENCES

[1] M.C.Crowley-Milling,The Design of the Control System for the SPS,CERN 75-20

[2] D.Swoboda, The LEP Token Ring Network Architecture, SPS/ACC/86-16

[3] C.Michaud,R.Rausch,J.M.Sainson,R.Wilhelm, Système de Multiplexage de Données du SPS, SPS/ACC/77-15

[4] M.C.Crowley-Milling, The Nodal System for the SPS,CERN 78-07

[5] Token Ting Access Method Ansi/IEEE 802.5 - 1985

[6] P.G.Innocenti,The LEP Control System,Proc. Europhysics Conf. on Control Systems for Experimental Physics,Villars 87

[7] CCITT, Data Communication Networks, Vol. VIII, Recommendation X.25, 1980

[8] G.Prat, D.Swoboda, Technical Note : Instrumentation bus for the Sputter Ion Pump Supplies of the SPS, unpublished

[9] G.Prat, D.Swoboda, The new Sublimation control equipment for the SPS, unpublished

[10] F. Ragris, Documentation MACBUS, Déscription du système de communication en serie - MACBUS

[11] The SPS/LEP controls group, LEP controls note 54 (internal publication)

[12] M.Vanden Eynden, Distributed Software for the LEP vacuum control, CERN SPS/89-19

VACUUM AND BEAM DIAGNOSTIC CONTROLS FOR ORIC BEAM LINES

B. A. TATUM

Oak Ridge National Laboratory*
Oak Ridge, TN 37831-6368

ABSTRACT

Vacuum and beam diagnostic equipment on beam lines from the Oak Ridge Isochronous Cyclotron is now controlled by a new dedicated system. The new system is based on an industrial programmable logic controller with an IBM AT personal computer providing control room operator interface. Expansion of this system requires minimal reconfiguration and programming, thus facilitating the construction of additional beam lines. Details of the implementation, operation, and performance of the system are discussed.

INTRODUCTION

Beam lines from the Oak Ridge Isochronous Cyclotron, ORIC,[1,2] contain numerous components related to vacuum and beam diagnostics. A new system for controlling these components and monitoring beam line vacuum has been implemented. It provides capabilities which did not previously exist, standardizes those which did exist, and facilitates expansion associated with new beam lines. An industrial programmable logic controller is the basis of the system, which is a deviation from the more traditional standards used in accelerator facilities, such as CAMAC. The system is dedicated in that it is not interfaced to a host computer and is used solely for the purpose of vacuum and beam diagnostic control.

SYSTEM HARDWARE

From a hardware standpoint, the system consists of an industrial programmable logic controller (PLC), a personal computer (PC), digital vacuum gauge controllers (DGCs), vacuum gauges, valves, and beam viewers. Figure 1 shows a block diagram of the system.

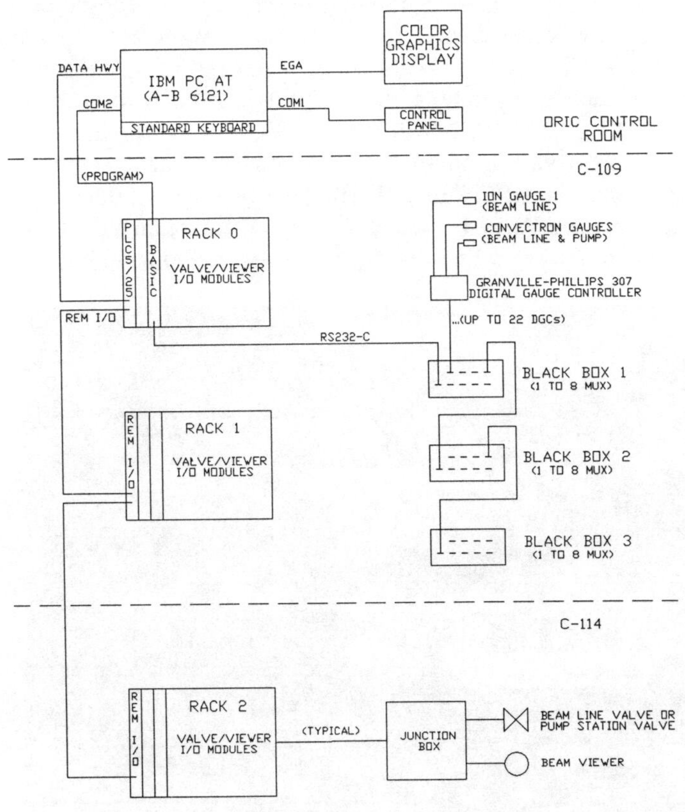

Fig. 1. Vacuum and beam diagnostic controls block diagram.

An Allen-Bradley (A-B) PLC 5/25 processor provides the logic handling capability of the system. The processor resides in a chassis containing I/O modules through which the beam line valves, pump station valves, and beam viewers are interfaced. Granville-Phillips (G-P) 307 DGCs are interfaced through a BASIC module which resides in the A-B chassis and transfers information to the PLC. Each DGC operates one Bayard-Alpert ionization gauge and two G-P Convectron gauges. Additional I/O chassis are networked to the processor chassis via the A-B Remote I/O Link operating at a 57-kbaud transmission rate.

* Research sponsored by the U.S. Department of Energy under contract DE-AC05-84OR21400 with Martin Marietta Energy Systems, Inc.

A distinct advantage of the PLC is the ease in which the valves and viewers are interfaced to the modules. Wiring is connected to terminal strips which swing out of the way for easy module removal. Thus complicated interface boxes and connectors are not required.

Operator interface to the PLC network is provided through a dedicated IBM AT Personal Computer located in the ORIC control room. The A-B Advisor PC Color Graphic System is utilized to provide color graphics displays of the beam lines. Pages may be selected to show overviews of the beam lines with beam line valve or beam viewer status shown by the color of the respective symbols; this is OPEN/CLOSED for valves and WITHDRAWN/INSERTED for viewers. A red symbol refers to a device positioned to block beam passage, and green represents an OPEN or WITHDRAWN state. Additional vacuum pages may be selected to show elevation views of sectors of the beam lines. A typical sector, as shown in Fig. 2,

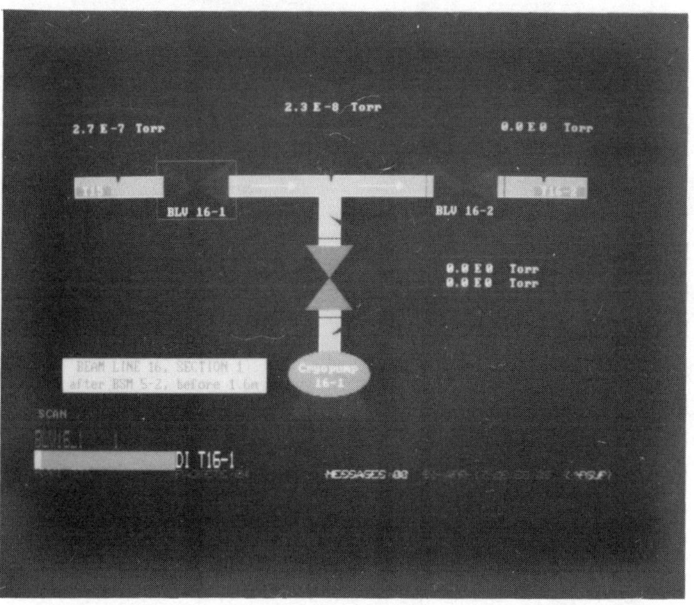

Fig. 2. Typical beam line sector.

consists of a pump station, a pump valve, and two beam line valves to isolate the pump station. These views provide beam line and pump station valve status, pressure readout, and the capability to control valves and ionization gauges in the displayed sector using a programmable A-B keyboard. Available ionization gauge controls are ON/OFF and electron bombardment DEGAS. The keyboard sector select keys are arranged so that an operator can sequentially step down any given beam line to a target station. Valve, viewer, and DGC control keys are also conveniently grouped. Individual components do not have separate keys. Instead, an operator selects the device he wishes to operate and presses the appropriate operation key.

SOFTWARE

Software for the system is divided into three major segments: PLC ladder logic for valve/viewer handling, a BASIC program for DGC handling, and operator interface configuration.

Communication to the PLC is accomplished via A-B PLC-5 programming software running on the control room PC. All of the PLC code is written in relay, or ladder logic. Different subroutines were written for each available operator interface display. The subroutines differ primarily in the block transfers to the BASIC module. To improve the update time of pressure readings, only the DGCs associated with a given sector display are scanned, thus block lengths vary. The block transfers are synchronized in read-write pairs. For simplicity, all valves are scanned within the main program. Viewers are only scanned when the single viewer display is selected.

The purpose of the BASIC module is to provide ASCII communication to the DGCs and to convert ASCII pressure readings to a PLC compatible data format. It is possible to communicate directly with the BASIC module from the control room PC via a terminal emulator. This permits easy program modifications, diagnostics, or uploading and downloading of program code. The BASIC program resides in the 16-kB battery backed RAM of the module. If expansion of the program to more than 16 kB becomes necessary, a 32-kB EPROM is also available.

Operator interface configuration takes on several aspects. Graphics displays are created with a mouse graphics package in which tags are applied to different symbols, e.g. valves, viewers, or pressure readout boxes. Information related to these tags is stored in a database. For example, each valve in the system has an associated tag which contains the PLC status and control bits pertaining to that valve. Thus a tag may be referenced by more than one display. Each graphics display has a corresponding background program which informs the PLC and BASIC module of the current display. The viewer page background program also saves the selected viewer such that if an operator leaves and returns to the page he will not have to restep through the control list. Graphics screens are preloaded into RAM when the system is booted to facilitate rapid screen drawing. The programmable membrane keyboard is easily defined in a database and a single communications table is used to define each of the communication links.

The software operates in the following way:

The advisor PC informs the PLC and BASIC module which display page is active and scans the PLC data tables. Additionally, it writes to the data tables if an operator has requested toggling of a valve, viewer, or ionization gauge state. The PLC and BASIC module branch to subroutines relevant to that page, with the BASIC module transferring DGC pressure readings to the PLC data tables and toggling ionization gauge status as necessary. BASIC module communications to the DGCs is via RS-232C. Switch codes are sent to three Black Box 1-to-8 RS-232C multiplexer boxes which are daisy-chained together and to which the DGCs are connected. The Black Box switches are located adjacent to the DGCs.

In the event of a power failure, the entire system will restart itself. The system is designed to operate in the "fail-safe" mode. If a power failure occurs, valves close, viewers withdraw, and they remain de-energized after power is restored until an operator changes their state.

SYSTEM PERFORMANCE

Programmable logic controllers are traditionally not known for high speed data transfers and there were initially a few problems with moderately slow pressure update rates. However, accelerator beam line vacuum monitoring does not require millisecond updates. What an operator needs is to know the pressure scale and to be able to see changes in pressure. Nevertheless, the pressure update rate has been enhanced to about twice per second, which is more than adequate for normal operation. This was accomplished by some programming adjustments and a direct 57-kbaud A-B Data Highway interface between the PC and PLC. Previously an RS-232C 19.2-kbaud link was used. Presently, the slowest aspects of the system are communications between the BASIC module and the G-P DGCs (limited to 9600 baud), and the fact that the BASIC program is interpreted rather than compiled.

An early problem which had to be overcome was the use of a color CRT in the ORIC control room in which there may be up to a 20 gauss magnetic field when ORIC is operating. Several different types and sizes of CRT were tested. It was apparent that Trinitron technology showed the least distortion. However, some shielding was still necessary. A mu-metal shield and a protective aluminum box were fabricated to house a 13" Sony Trinitron monitor. The assembly was mounted on a tiltable, swivel base and bolted to the top of the ORIC console. Operationally, there have been been no complaints except that it slightly obstructs the view of a portion of the ORIC main panel.

Operators are pleased with the improved information that the system provides. They have contributed numerous ideas to make the system as user-friendly as possible. This is the first color graphics based control system for the Holifield accelerators. Thus there were the usual ergonomic discussions on colors, glare, page layout, symbols, and location. At present, the operator's primary enhancement request is for the replacement of the programmable keyboard with a trackball. The Holifield tandem accelerator control system presently utilizes a trackball. This was not an option with earlier versions of the operator interface software but it is now available and the possibility of such a conversion is being studied.

Portions of the system have been in operation for more than a year and a half and the only hardware failure was, oddly enough, with a defective PLC chassis. Otherwise, the only maintenance required has been the routine replacement of burned out ionization gauge filaments.

The system has proven to be easy to expand. Originally, the beam viewers were not to be part of the system. However, they operate exactly like a valve and thus became a logical addition to the system. Their inclusion was basically as simple as creating a graphics screen, adding them to the database, and wiring them to new modules. Other chassis and modules may also be added with ease. This is attractive for further expansion of Holifield beam lines.

SUMMARY

Vacuum and beam diagnostic equipment on beam lines from the ORIC is now controlled via a new system based on an industrial programmable logic controller with a personal computer operator interface. It has proven to be a viable alternative to the more traditional hardware implementations. PLCs offer simple interfacing, reasonable cost, and they fit well into modular control systems. As PLCs continue to improve in the areas of data transfer and number crunching, they will likely be used in ever increasing areas of accelerator control.

REFERENCES

1. R. S. Livingston and F. T. Howard, Nucl. Instr. and Meth. 6, 1 (1960).
2. J. A. Martin et al., "Operational Experience for Coupled Operation of the Holifield Tandem Electrostatic Accelerator and Isochronous Cyclotron," Proc. 11th Int. Conf. on Cyclotrons and their Applications, Ionics, Tokyo, Japan, p. 38 (1987).

An Automated Vacuum System*

W. H. Atkins**, G. D. Vaughn, C. Bridgman***
MS-H820, Los Alamos National Laboratory, Los Alamos, NM 87545

Abstract

Software tools available with the Ground Test Accelerator (GTA) control system provide the capability to express a control problem as a finite state machine. System states and transitions are expressed in terms of accelerator parameters and actions are taken based on state transitions. This is particularly useful for sequencing operations which are modal in nature or are unwieldy when implemented with conventional programming. State diagrams are automatically translated into code which is executed by the control system. These tools have been applied to the vacuum system for the GTA accelerator to implement automatic sequencing of operations. With a single request, the operator may initiate a complete pump-down sequence. He can monitor the progress and is notified if an anomaly occurs requiring intervention. The operator is not required to have detailed knowledge of the vacuum system and is protected from taking inappropriate actions.

I. INTRODUCTION

The approach of the GTA control system (EPICS, for Experimental Physics and Industrial Control System) is to provide a collection of tools which can be used to implement computerized controls. These tools can be utilized without intimate familiarity of the system. Reducing the programming expertise required of the implementer and the amount of code required increases the system's reliability and reduces its long-term cost.

II. THE SEQUENCER

One of the EPICS software tools, the *Sequencer*, allows a controls problem to be expressed in an accelerator-user's terms and nearly automatically produces programming code to implement the algorithm. The Sequencer tool is comprised of a *State Notation Language* (SNL) compiler which uses a Finite State Machine paradigm, and run-time code to assist in implementation of the program.

A state diagram is often more useful for expressing control problems, as the description is in terms of the problem and is not constrained by programming-language conventions. An example state diagram is shown in Figure 1.

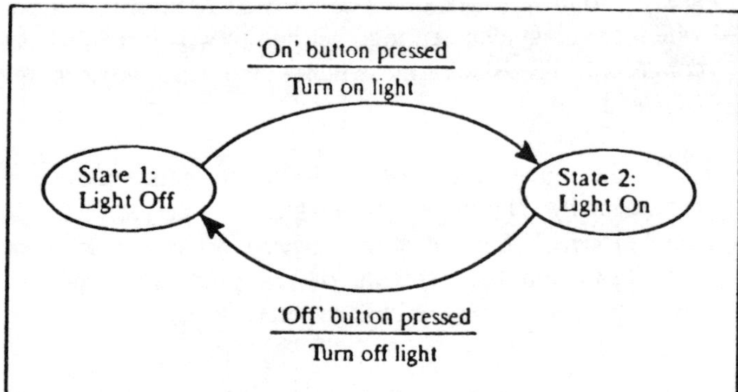

Figure 1. Example State Diagram.

In this example there are two states defined, corresponding to two states of a light. The connecting lines show that the 'system' can transition from one state to the other. The notation next to the state-transition lines describes (above the line) the condition for the transition occurring, and (below the line) actions that occur as a result of the transition. In the example, the condition for transitioning from state 1 to state 2 is that the 'On' button is pressed. The condition for the opposite transition is that the 'Off' button is pressed. No action is taken if the system is in state 2 and the 'On' button is pressed, as there is no transition describing that case.

A state diagram translates directly (not yet automatically) to a State Notation Language program. Figure 2 shows an excerpt of the corresponding SNL program for the example in figure 1.

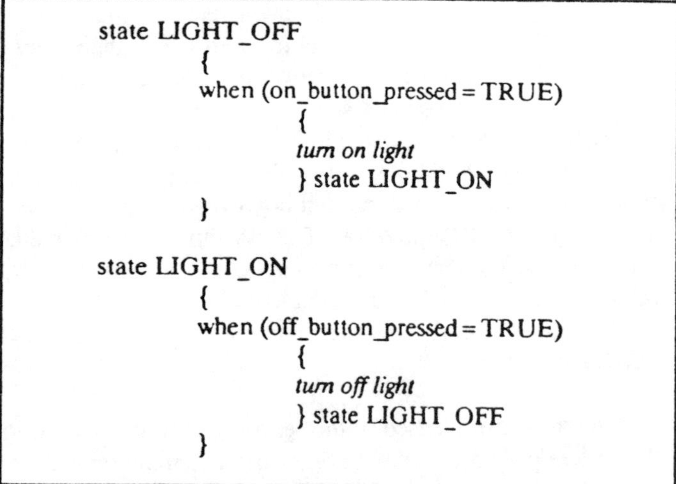

```
state LIGHT_OFF
    {
    when (on_button_pressed = TRUE)
        {
        turn on light
        } state LIGHT_ON
    }

state LIGHT_ON
    {
    when (off_button_pressed = TRUE)
        {
        turn off light
        } state LIGHT_OFF
    }
```

Figure 2. Excerpt from State Notation Language Program.

The statements in italics are not part of the program syntax. The outer-most sets of braces delimit states. Within a state, the transitions are described as *when* clauses, read *"when some condition occurs, take some action(s), and transition to another state."* The condition is specified as the argument to the *when* statement. The actions to take when

---

*Work supported and funded by the United States Department of Defense, Army Strategic Defense Command, under the auspices of the United States Department of Energy.

** Oak Ridge National Laboratory

*** Grumman Corporation, Space Sciences Division.

the condition is true are specified inside the next set of braces. Following these braces the destination state is specified. More than one *when* clause would be present in a state if there were more than one transition out of the state.

The expressions shown in the *when* statement are truly as simple as shown. Accelerator parameters such as temperatures, pressures, etc. are made available to the SNL program as program variables. SNL facilities provide this interface with other parts of the control system.

Without detailing the SNL syntax any further, it is clear that a SNL program can be directly produced from a state diagram with no logic or algorithmic translation. This makes for a quick transition from problem definition to implementation.

III. A VACUUM SYSTEM EXAMPLE

The vacuum system for the GTA Radio–Frequency Quadrupole (RFQ) utilizes two cryogenic pumps and one turbomolecular pump. Figure 3 shows how one cryo system is configured.

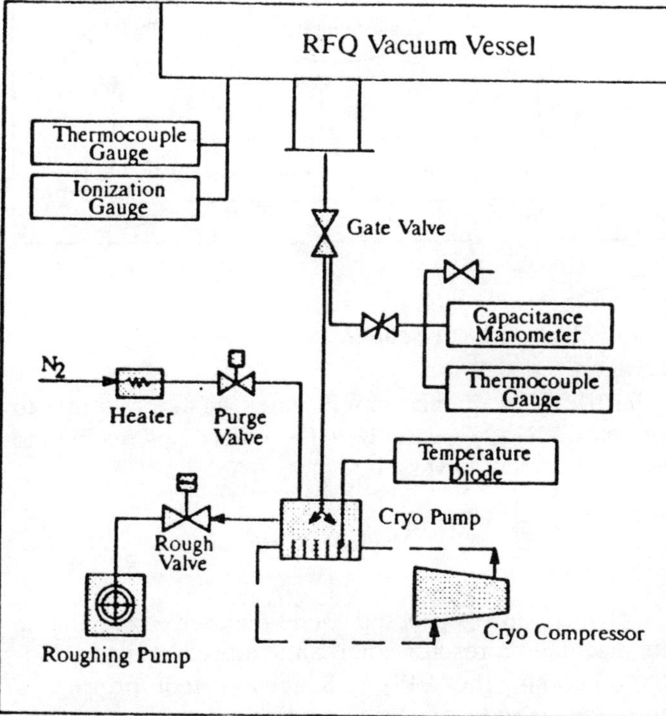

Figure 3. Partial RFQ Vacuum System Schematic.

Each pump and its associated instrumentation can be treated somewhat independently, i.e. a cryo pump can be roughed–out and cooled down independently of what the turbo pump is doing. Coordination of the three pump systems is required when they must cooperate in pumping the vessel itself.

A procedure for preparing the cryo pump for on–line pumping would be as follows:

1. Close gate valve, close purge valve, open rough valve.
2. Start roughing pump.
3. When cryo pump pressure reaches 50 millitorr, close rough valve, start compressor and cold head.
4. If pressure increases to 65 millitorr in less than 1 minute, open rough valve until pressure decreases to less than 50 millitorr, then close rough valve.
5. When cryo pump temperature is below 20°K, the pump is ready to be placed on-line.

Figure 4 shows this procedure expressed as a state diagram.

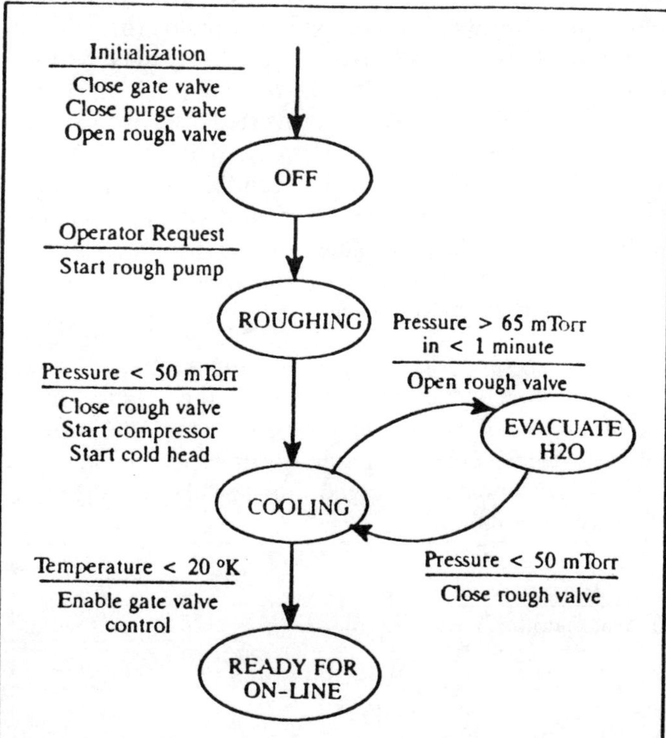

Figure 4. Cryo Procedure as a State Diagram.

The procedure is broken into 5 states. At initialization the system is put into a known state. On an operator request, the system begins evacuating the cryo pump with the roughing pump. This is the ROUGHING state. Note that this means roughing as pertains to the cryo pump, not the RFQ vessel.

Once the pressure in the cryo pump is less than 50 millitorr, the pump is ready to be cooled. The roughing valve is closed, and the cryo compressor and cold head are switched on (state COOLING).

During the first minute of cooling, if water vapor has not been adequately evaporated, it will condense and cause a pressure rise. If this is detected, the roughing valve is opened and the system transitions to state EVACUATE H2O. When the pressure is again low enough, the roughing valve is closed and a transition is made back to state COOLING. Note how transitions in and out of this state implement a hysteresis feature.

Cooling proceeds until the cryo pump temperature is below 20 °K. When this occurs, the state diagram calls for a transition to state READY FOR ON-LINE. The action corresponding to this transition is to enable control of the gate valve. It is not opened automatically because control must be coordinated with the other pumping systems.

This state diagram is not complete. It does not show paths for system shutdown or error handling. Error paths are needed to accommodate problems such as devices not operating as expected, e.g. the roughing valve not opening. Errors that can be anticipated can be handled automatically. Others require operator intervention.

The state diagram is easier to understand because much of the information is visual. It is quick to see what conditions cause state changes without reading through text or program code. A logic error is more likely to be realized through the state diagram than by reading program code.

This state diagram can be converted directly to SNL as described previously. The SNL provides an interface with other EPICS facilities so that conditions and actions specified by the sequence can be monitored and controlled easily. For example, C-language-like subroutine calls are used to test the cryo pump temperature or to operate a valve.

IV. COOPERATIVE SEQUENCES

It was noted that this cryo pump sequence is not completely independent of the rest of the RFQ vacuum system. To treat the RFQ vacuum system as a single entity, a higher-level sequence is appropriate, one specified in terms of the system, not in terms of individual components. Figure 5 shows an example.

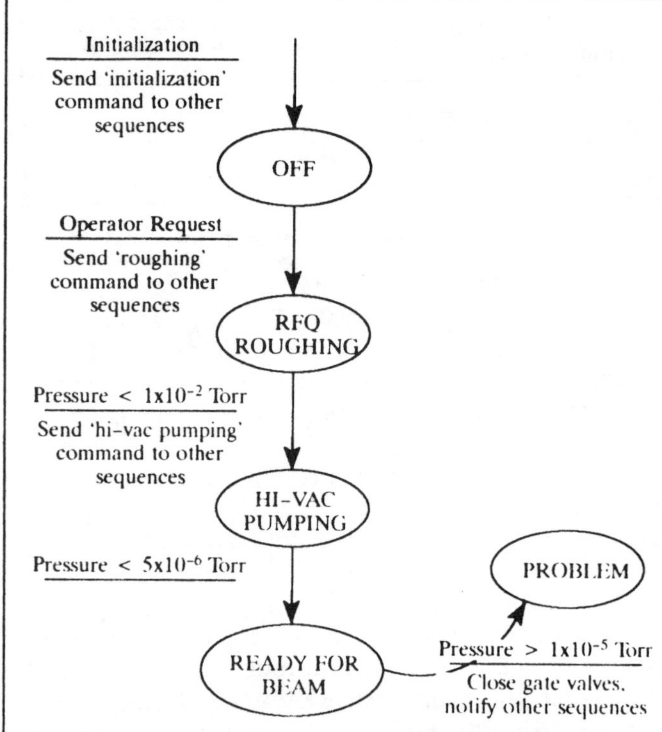

Figure 5. High-level Vacuum System State Diagram.

This diagram is expressed in terms of the RFQ vacuum system, hence 'roughing' means roughing of the RFQ vessel, not one of the pumps. Again, this is not a complete diagram as there is no shutdown path or error paths.

Note the PROBLEM state, which illustrates how an interlock may be implemented as part of the state machine. This interlock is more easily implemented here because the condition of pressure exceeding 1×10^{-5} Torr is only a problem if the system was previously READY FOR BEAM.

Actions specified by this diagram are instructions to lower-level sequences, not operations on particular vacuum components. In turn, the lower-level sequences would accept instructions from this sequence instead of directly from an operator. This hierarchy is illustrated in figure 6.

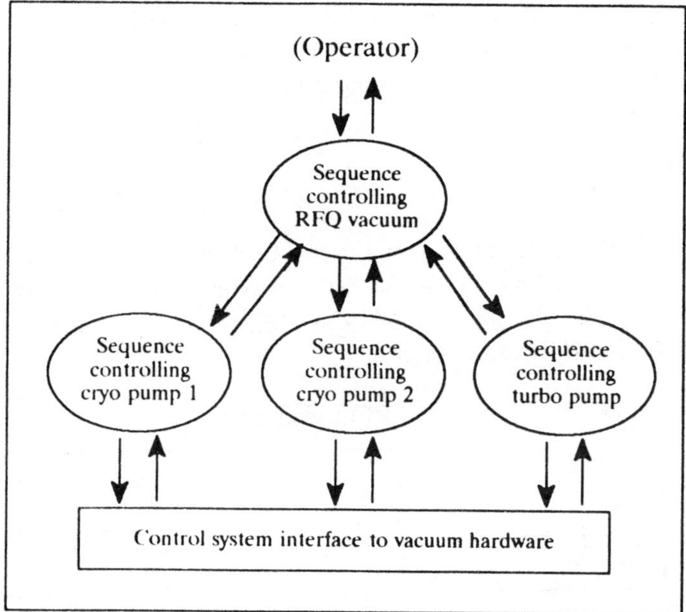

Figure 6. Hierarchical Sequences.

A third-level sequence might later be appropriate to coordinate RFQ vacuum with other upstream and downstream vacuum vessels.

V. SUMMARY

Some controls problems lend themselves better to a state machine representation than to other paradigms. For such problems, the EPICS Sequencer tool provides a convenient and faster implementation method. Hierarchical sequences may be built which provides a basis for higher-level accelerator control.

VI. REFERENCES

[1] A.J. Kozubal, D.M. Kerstiens, J.O. Hill, and L.R. Dalesio, "Run-time Environment and Applications Tools for the Ground Test Accelerator Control System", Proceedings of the International Conference on Accelerator and Large Experimental Physics Control Systems, Vancouver, BC, Canada, Oct 30 – Nov 3, 1989.

Open Loop Compensation for the Eddy Current Effect in the APS Storage Ring Vacuum Chamber*

Y. Chung, J. Bridges, L. Emery, and G. Decker
Argonne National Laboratory
9700 S. Cass Ave., Argonne, IL 60439

Abstract

In the third generation synchrotron light sources, closed orbit stabilization against external vibrations is critical to ensure low emittance and high brightness. The Advanced Photon Source (APS) will use a large number (678) of correction magnets to create local bumps and to achieve global orbit stabilization. In this paper, we will present the result of the effort to counter the effect due to the finite inductance of the magnet and the eddy current in the 1/2"-thick aluminum storage ring vacuum chamber. The amplitude attenuation and the phase shift of the correction magnet field inside the APS storage ring vacuum chamber were measured. A circuit to compensate for this effect was then inserted between the signal source and the magnet power supply. The amplitude was restored with an error of less than 20% of the source signal amplitude and the phase shift was reduced from 80° to 12° at 10 Hz. Incorporation of this circuit in the closed loop feedback scheme and the resulting beneficial effect in the closed orbit stabilization will be discussed.

I. INTRODUCTION

A number of correction dipole magnets will be installed around the APS storage ring to stabilize the beam against the low frequency vibration below 25 Hz from various sources. The correction magnet system consists of 80 main dipole trim coils, 280 sextupole dipole coils, 240 vertical field dipole magnets for horizontal corrections and 78 vertical/horizontal field magnets for horizontal and vertical corrections. A detailed description of the design specification parameters for these magnets can be found in Ref. [1]. The displacement of the quadrupole magnets due to vibration has the most significant effect on the stability of the positron closed orbit in the storage ring. A small displacement of the quadrupole magnet leads to a large distortion of the closed orbit, and hence, the growth of the emittance.

This article describes the measurement of the eddy current effect on the externally applied correction magnet field to counter the low frequency (< 20 Hz) vibration. The correction magnets are controlled individually through the feedback loop comprising the beam position monitoring (BPM) system. We will concentrate on compensation for the eddy current effect using an open loop circuit. Incorporation of this circuit in the closed loop feedback scheme and the beneficial effect in the closed orbit will be also discussed.

II. MEASUREMENT SETUP

For the measurement of the amplitude attenuation and the phase shift, sine waves of various frequencies up to 25 Hz were used. Square waves and triangular waves were also used to measure the rise time of the magnetic field inside the vacuum chamber or to demonstrate the performance of the compen-sation circuit inserted between the function generator and the power supply.

Figure 1 shows the measurement setup, with a section of the APS storage ring vacuum chamber in the magnet bore. Since the bore is too small for the entire vacuum chamber to fit in, the antechamber was cut away, leaving only the positron beam chamber and the photon beam channel. This will have negligible effect on the measurement because of the large aspect ratio of the photon beam channel. The magnet was one of the dipole magnets used in the electron cooling ring experiment at Fermilab[2].

With the inductance L = 10 mH and the coil resistance R_c = 25 mΩ, the time constant of the magnet τ_c is equal to L/R_c = 400 ms. In order to reduce the time constant, a water-cooled stainless steel resistor of R = 0.2 Ω was connected in series to the magnet, as shown in Fig. 2. This reduced the time constant τ to 44 ms.

The unperturbed field is measured with a current transducer (V_c), while the field inside the vacuum chamber is measured simultaneously with a Hall probe (V_g) and a gaussmeter (G).

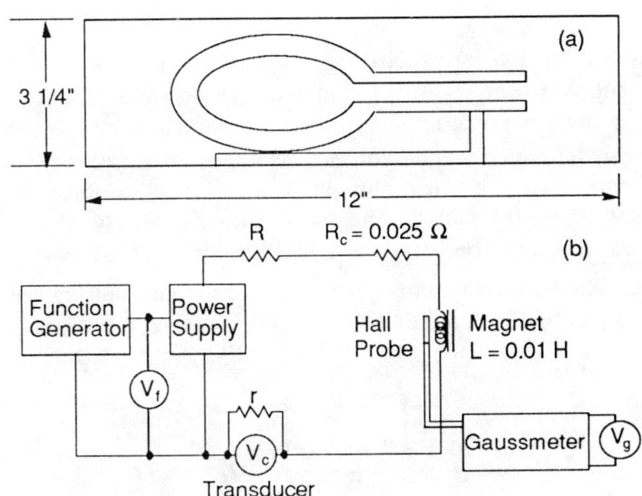

Fig. 1: (a) A section of the vacuum chamber cut short to fit into the magnet bore. (b) Electrical connection.

*Work supported by the U.S. Department of Energy, Office of Basic Energy Sciences, under Contract No. W-31-109-ENG-38.

The field attenuation and the phase shift can then be calculated by comparing V_c and V_g.

III. COMPENSATION CIRCUIT

In order to restore the magnetic field inside the vacuum chamber to the same shape as the driving signal, it is necessary to amplify the attenuated frequency components and to advance their phases accordingly. Consider the following relation.

$$V_c(t) = \int d\omega \, V_c(\omega)e^{-i\omega t},$$

$$V_g(t) = \int d\omega \, a(\omega)e^{i\Delta\phi(\omega)}V_c(\omega)e^{-i\omega t}. \tag{1}$$

Here, $V_c(t)$ is the unperturbed field and $V_g(t)$ is the field inside the vacuum chamber, and $V_c(\omega)$ and $V_g(\omega)$ are the Fourier transforms. Let a_v and a_m be the attenuation factors due to the vacuum chamber and the magnet, respectively, and let $\Delta\phi_v$ and $\Delta\phi_m$ be the phase shifts. Then the overall attenuation factor a and the overall phase shift $\Delta\phi$ will be

$$a = a_v a_m \qquad \text{and} \qquad \Delta\phi = \Delta\phi_v + \Delta\phi_m. \tag{2}$$

If $\Delta\phi(\omega)$ is close to 90° and if $a(\omega)$ behaves like ω^{-1} within the frequency range where $|V_c(\omega)|$ is appreciably large, then $V_g(t)$ is simply the time integration of $V_c(t)$ except for a real multiplication factor. In this case, the compensation can be achieved by feeding the differentiation of the driving signal to the magnet. Let $V_i(t)$ be the source signal and let $V_i(\omega)$ be its Fourier transform. The ideal compensation circuit would then modify the source signal such that

$$V_c(\omega) = \frac{1}{a(\omega)} e^{-i\Delta\phi(\omega)} V_i(\omega). \tag{3}$$

This can be partially achieved by using a simple circuit element shown in Fig. 2. A simple argument shows that this circuit always gives attenuation which decreases monotonically with frequency and frequency dependent phase advance at all frequencies. The maximum phase advance achievable with this circuit is less than 90° and depends on the ratio R_1/R_2. It occurs at some frequency ω satisfying $R_1 < 1/\omega C < R_2$. Assuming harmonic time dependence $e^{-i\omega t}$ and putting $r = R_1/R_2$ and $\tau = R_1 C$, the exact relation between V_i and V_f is

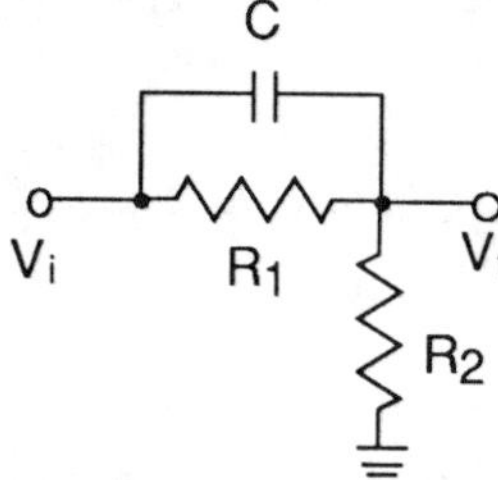

Fig. 2: An element of the open loop compensation circuit.

$$\frac{V_f}{V_i} = a'(\omega) \, e^{i\Delta\phi'(\omega)} \tag{4}$$

where

$$a'(\omega) = \frac{[\{1 + r + (\omega\tau)^2\}^2 + (r\omega\tau)^2]^{1/2}}{(1 + r)^2 + (\omega\tau)^2},$$

$$\Delta\phi'(\omega) = -\tan^{-1}\left(\frac{r\omega\tau}{1 + r + (\omega\tau)^2}\right). \tag{5}$$

This circuit has two disadvantages. One is that the low frequency components are attenuated significantly due to large r. Secondly, it does not have the isolation property needed to shift the phase by more than 90°. These problems can be solved easily by adding an op-amp at the output terminal. Two such circuits were used in combination to compensate for the magnet inductance and the eddy current effect. The parameters r and τ were chosen such that the low frequency behavior of $\Delta\phi'$ follows the measured values as closely as possible up to 10 Hz while maintaining $aa' \approx 1$. If we let a'_v, $\Delta\phi'_v$, a'_m and $\Delta\phi'_m$ be the amplification and the phase correction by the first and the second stages, then the overall compensation will be represented by

$$a' = a'_v a'_m \qquad \text{and} \qquad \Delta\phi' = \Delta\phi'_v + \Delta\phi'_m. \tag{6}$$

From Eqs. (2) and (6), the attenuation and phase shift after compensation will be given by

$$a_c = aa' \qquad \text{and} \qquad \Delta\phi_c = \Delta\phi + \Delta\phi'. \tag{7}$$

IV. MEASUREMENTS AND RESULTS

The current transducer and the Hall probe used for the field measurement were first checked for linearity in their response to DC and AC fields. For DC linearity check, the field was increased up to $G = 1,200$ gauss and V_c/G and V_g/G were measured. The results were:

$$V_c/G = 0.608 \pm 0.010 \text{ (mV/gauss)},$$

$$V_g/G = 0.382 \pm 0.004 \text{ (mV/gauss)}. \tag{8}$$

These numbers are the calibration constants for field measurements. For AC, V_c/V_g was measured for frequencies up to 25 Hz. The DC bias level was set at 1,200 G and the power supply was modulated at various frequencies. The AC field amplitude was set at approximately 25% of the DC bias level, i. e., 300 G. No attenuation or phase shift in the Hall probe response was observed as a function of the driving current in the magnet. After calibration using Eq. (1), we found:

$$V_c/V_g = 0.973 \pm 0.010. \tag{9}$$

After checking the linearity in the Hall probe response, it was placed at the center of the positron beam chamber.

Two separate measurements were made to determine a_v, $\Delta\phi_v$ and a_m, $\Delta\phi_m$. To measure a_v and $\Delta\phi_v$, the current in the magnet, which was converted to voltage V_c by the transducer, was used as reference and the Hall probe signal V_g was compared with V_c. Similarly, to measure a_m and $\Delta\phi_m$, the driving signal V_f from the function generator was used as reference and the signal V_c was compared against V_g. Figure 2 shows the results of these measurements, (a) for the attenuation factors and (b) for the phase shifts.

Figure 3 summarizes the measurement results. It shows that compensation for amplitude attenuation is very good up to 20 Hz. The remaining phase shift is 12° at 10 Hz and 35° at 25 Hz. This deviation comes mostly from the vacuum chamber. This result is obvious since the eddy current effect can shift the phase by more than 90°, while the compensation circuit works only up to 90°. However, unless the source signal, e. g., the vibrational motion that needs to be corrected, has a broad-band spectrum reaching beyond 10 Hz, these deviations will not pose a serious problem.

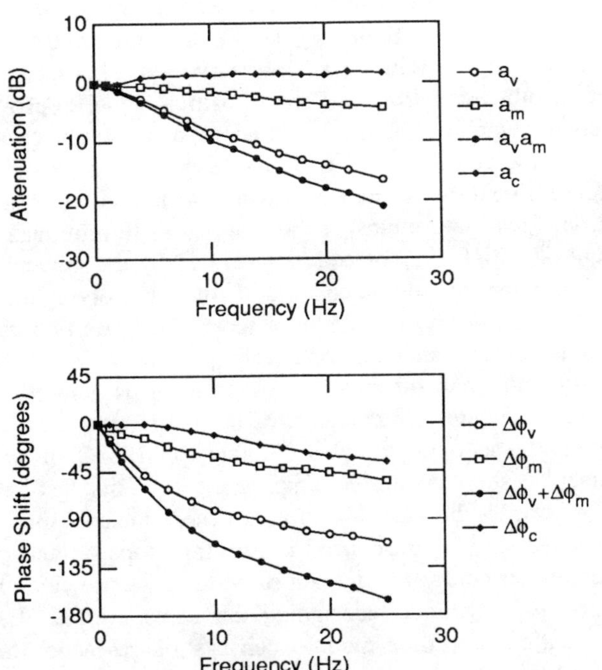

Fig. 3: Amplitude attenuation and phase shift due to the eddy current effect.

V. CLOSED LOOP COMPENSATION

The present work considered only the open-loop compensation without any feedback from the resultant magnetic field inside the vacuum chamber. For better compensation for the eddy current effect in a broader frequency range, a local closed loop feedback scheme[3-4] must be used. In the APS storage ring, the local loop will employ photon beam position monitors to detect the position and the angle of the beam. Let A be the open loop gain. Then, with the loop closed, the overall compensation will be given by

$$\frac{V_f}{V_i} = \frac{aa'A}{1 + aa'A} \, . \tag{10}$$

From Eq. (10), we find that a large gain ($|aa'A| \gg 1$) is needed for the loop to work. However, by Nyquist's theorem, this will cause instability or amplitude growth if the phase delay due to the eddy current effect (a) increases to near 180°, and hence, decrease of the bandwidth of stable operation. This problem can be alleviated by incorporating the open loop circuit (a') in the loop, as shown in Fig. 4. Measurements similar to those described in this paper have been made using this scheme and near perfect compensation for the eddy current effect was achieved ($|a_c| \approx 0.95$ and $\Delta\phi_c \approx -7°$ at 25 Hz). The results will be published separately in the near future.

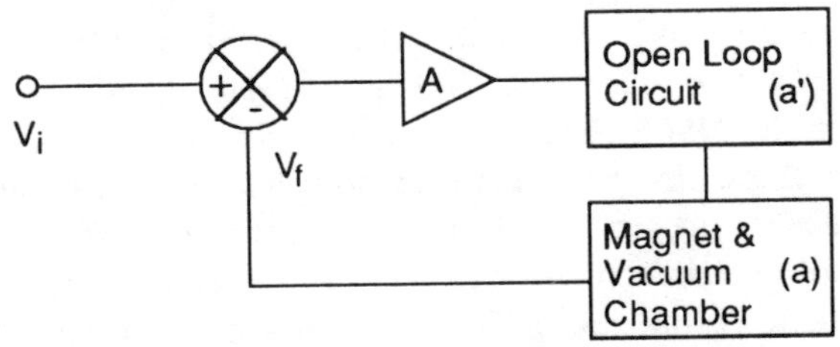

Fig. 4: Incorporation of the open loop circuit in the closed loop feedback.

V. SUMMARY

The measurements performed on the field attenuation and the phase shift due to the eddy current in the APS storage ring vacuum chamber were described. The finite inductance of the magnet adds to this effect. As a solution to correct this problem, a simple circuit was designed and inserted between the signal source and the magnet power supply. The compensation was almost complete below 10 Hz, and between 10 and 20 Hz, more than 85% of the phase shift was corrected. Incorporating the open loop circuit in the closed loop feedback shows much improved compensation.

VI. REFERENCES

[1] APS Design Handbook, Argonne National Laboratory, 1989
[2] Fermilab Electron Cooling Ring Experiment, Design Report, FNAL, 1978
[3] R. O. Hettel, "Beam Steering at the Stanford Synchrotron Radiation Laboratory," IEEE Trans. Nucl. Sci. Vol. NS-30, 2280, 1983
[4] R. J. Nawrocky et al., "An Automatic Beam Steering System for the NSLS X-17T Beam Line Using Closed Orbit Feedback," *Proceedings of 1987 IEEE Particle Accelerator Conference*, 512, 1987

Automatic Local Beam Steering Systems for NSLS X-Ray Storage Ring - Design and Implementation*

O.V. Singh, R. Nawrocky, J. Flannigan
National Synchrotron Light Source
Brookhaven National Laboratory
Upton, New York 11973

Abstract

Recently, two local automatic steering systems, controlled by microprocessors, have been installed and commissioned in the NSLS X-Ray storage ring. In each system, the position of the electron beam is stabilized at two locations by four independent servo systems. This paper describes three aspects of the local feedback program: 1) design, 2) commissioning and 3) limitation. The system design is explained by identifying major elements such as beam position detectors, signal processors, compensation amplifiers, ratio amplifiers, trim equalizers and microprocessor feedback controllers. System commissioning involves steps such as matching trim compensation, determination of local orbit bumps, measurement of open loop responses and design of servo circuits. Several limitations of performance are also discussed.

I. INTRODUCTION

During the year 1990, two local feedback systems were commissioned and are now fully operational: one for hybrid wiggler (HBW) at X25 beamline and the other for LEGS experiment at X5 beamline. These systems employ 4 trim magnet bumps for beam deflection and two beam position monitor detectors, which are all placed in the straight section. Previous work on stabilization of photon beams in a storage accelerator is described in [1] - [3] and a general introduction to all NSLS local feedback systems is presented in [4].

II. DESIGN

The block diagram of NSLS beam steering system is shown in Fig. 1. This beam steering system can be subdivided into three sections: 1. Beam position monitoring (PUE and RF detectors) 2. Beam actuators (Power supplies and Trim magnets) 3. Local feedback electronics. Sections 1 and 2 are discussed in details in [5] and [6] respectively. The last section, local feedback electronics, is discussed in this paper.

The local feedback electronics consist of the following modules: CPU/Communication, signal processors, ramping (on/off), compensation amplifiers, ratio amplifiers, summing and equalization amplifiers. All of these modules are housed in a VME crate which is linked to the host computer using CPU/Communication module. The signal processor module consists of differential receiver circuits to receive signals from the beam position RF detectors (located near the beam pipe) and the reference or target signal which is provided by the computer using a digital to analog converter. This reference signal is subtracted from the beam position signal to provide the error signal which then is used to drive compensation amplifier (servo) loops. It is desirable that the value of this reference signal be as close as possible to the value of beam position signal before closing the loop, so as to minimize the amount of initial trim correction generated by the system.

The ramping module consists of four multiplying digital to analog converters, one for each loop. There are two basic functions for this module; one is to provide a "soft" turn on/off function and other is to provide some dc gain control from the computer, if required. The "soft" turn on/off is achieved by sending digital set points to MDAC in small increments, so it creates a ramping effect. This feature is highly desirable because it minimizes the transients into the beam which can be introduced by turning loops on and off. The computer controlled dc gain provides a way to fine tune the loop gain without redesigning the servo circuits. Each compensation amplifier module consists of cascaded three stage amplifiers with each stage providing circuits to incorporate 2 poles and 2 zeros. This amplifier at X 25 beamline has 5 poles at 1.5 hz, 4 hz, 200 hz, 400 hz (2) and two zeros at 15 hz and 40 hz and a dc gain of 100 (40 db). This combination provides system bandwidth of 50 - 60 hz with a gain and phase margin of about 15 db and 45 degrees respectively.

The ratio amplifier module consists of four multiplying digital to analog converters for each loop. The output values of these converters are controlled by computer set points K_{1j} and K_{2j} for the two loops (j = 1 to 4) which then provides the proportional signals to the corresponding trims. The ratio values are critical to the performance of feedback system and are based on two conditions; the four magnet bump should be local and this bump should provide an angle bump such that while this bump moves the beam at one PUE location, it has

*This work was performed under the auspices of the U.S. Department of Energy.

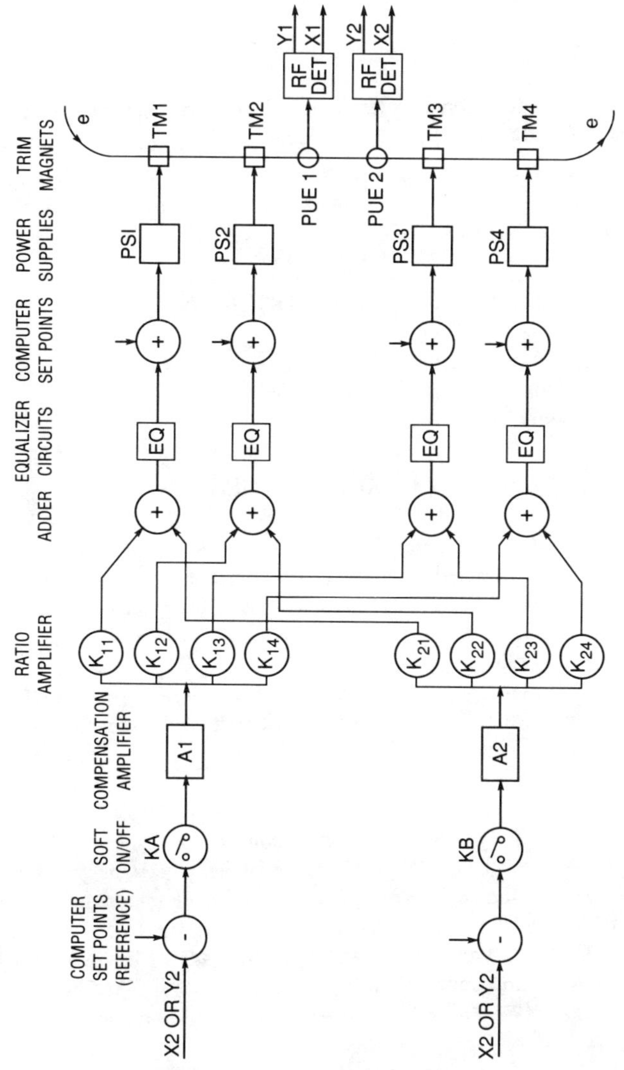

Fig. 1. Block Diagram of NSLS Beam Steering System

no significant effect on the second PUE location. In other words, the angle bump crosses at zero point at the second PUE. The output of four ratio amplifier signals from each detector is added correspondingly into the adder circuits. The equalizer amplifier circuits consist of cascaded two stage amplifiers with provisions of up to four poles and four zeros. The equalizer amplifiers are used to match the beam phase response of four trim magnets. This is necessary so that there is no phase distortion at higher frequencies, which could lead to an unstable feedback system. This module also provides high current drivers which drive these signals to power supply locations.

Since at NSLS storage ring, all the trim power supplies are used to set proper beam orbit by sending set points by the computer (see Fig.1), the feedback trim power supplies have summing junction to accept second input from the feedback electronics. However, computer set points for feedback trims should be kept as small as possible so as to allow large

strengths to be utilized for automatic feedback correction system. This, however, is not always possible.

III. COMMISSIONING

The matching of beam phase responses for all trims can be quite important and sometimes difficult. Once all of the four phase responses are measured, the one which rolls off the fastest is used as a reference, while others are matched to this reference response utilizing the equalizer amplifiers.

As stated earlier, there are two independent loops for each plane, one designated for each detector. Each loop is to provide a stable beam at its detector. For these two loops to be decoupled, it is necessary to develop two sets of four magnet bumps such that while each loop moves the beam at its own detector it causes little or no disturbance at the other one. To achieve this, the first three magnet local bumps are determined by special programs [7]. Using this program, two sets of local bumps are determined, first by using trims 1,2,3 and then by using trims 2,3,4. From these two sets of three magnet bumps, two sets of four magnet bumps are developed as follows:

Say, K_{1j} and K_{2j} represent the bump coefficients for above mentioned two sets of three magnet bumps, where $j = 1$ to 4 (note $K_{14} = K_{21} = 0$). Using these bump coefficients in the ratio amplifier, the beam motion at detectors 1 and 2 are determined when a signal is injected at the input of ratio amplifier (say these are D_{11} and D_{12} due to 1st bump and D_{21} and D_{22} due to second bump). The two decoupled four magnet bumps (K_{1j}' & K_{2j}') can be written as

$$K_{1j}' = K_{1j} - (K_{2j} \cdot D_{12}/D_{22})$$

$$K_{2j}' = K_{2j} - (K_{1j} \cdot D_{21}/D_{11})$$

These calculated bump coefficients are inserted into the ratio amplifiers and the open loop system responses are measured. For X25 beamline, the dc response is about 0 db for vertical plane and about -8.5 db for horizontal plane while the phase response crosses 180° between 400 to 600 hz for horizontal and vertical planes.

IV. RESULTS

Two local feedback systems have been employed successfully and are fully operational on the NSLS X-ray ring beam lines X5 and X25. The results from X25 beam line are shown in Fig.2. Two RF detectors are connected to PUE 30 and PUE 31 and are used for this local feedback. The top trace in Fig 2 shows the chart recorder data of the vertical beam motion in time domain with vertical feedback on and off. From this trace, we see that the beam motion induced by the NSLS Booster (at approximately 0.66 Hz) is clearly reduced, when loops are turned on. The bottom trace is the frequency spectrum of the vertical beam motion at PUE 31 with the loop on and off. The improvement at the second

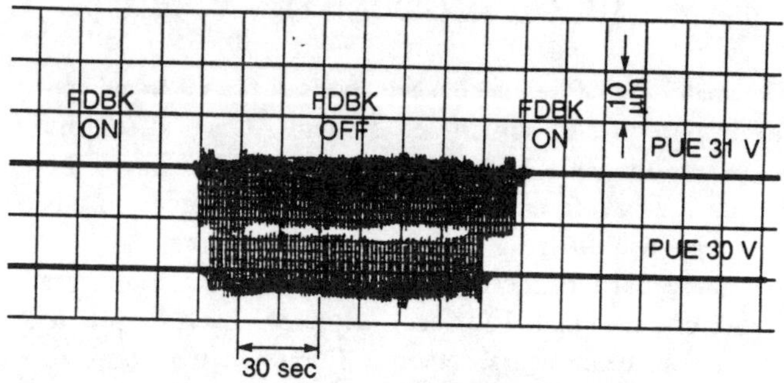

X=1.334 Hz
Y_a = -42.115 dBVrms
Y_b = -76.708 dBVrms
POWER SPEC1

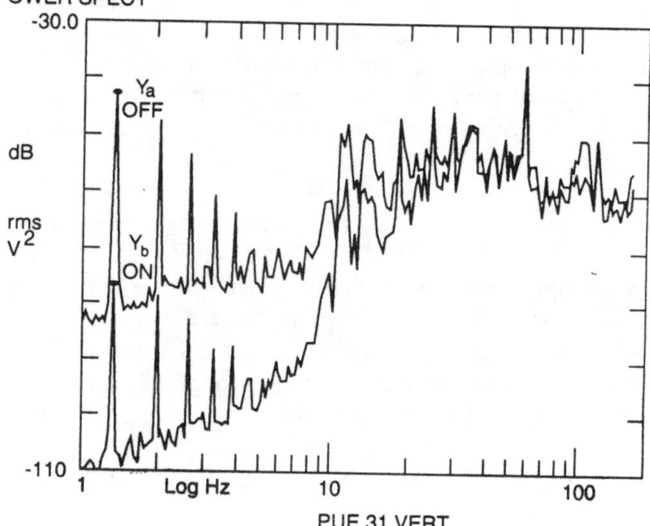

Fig. 2. Beam Motion Reduction with Feedback On.

harmonic of the NSLS Booster is better than 33 db as shown by peaks Ya and Yb which translates to an improvement from 4.4 μm (p-p) to 0.08 μm (p-p). (20 db = 5.6 m p-p).

V. LIMITATIONS

Since the feedback trim power supplies also are used for orbit correction, the full trim strength may not be available for the correction system.   Thus this reduces the feedback correction range. Further, if the initial setting of a trim is too high, and a large amount of correction is required, it is possible that that trim will saturate while other three trims are still in the active range. This can result into a non local bump and thus cause undesirable global beam motion. This condition of high initial setting does occur for one trim in NSLS ring and this problem has been recently solved by providing two

seperate power supplies for this trim, thus increasing its strength.

In the horizontal plane, the beam motion range for a given fill (200-500 μm) is far greater than the range of the feedback system (100 μm). At present time, several global orbit corrections are made during each fill so that the feedback correction demand is reduced.

V. ACKNOWLEDGEMENTS

G. Frisbie was involved in the development and construction of all the feedback circuit boards and in the assembly, testing and commissioning of the overall system.

VI. REFERENCES

[1] R.O. Hettel, "Beam Steering at the Stanford Synchrotron Radiation Laboratory", IEEE Trans. Nucl. Sci., Aug. 1983.

[2] R.J. Nawrocky et al., "An Automatic Beam Steering System for the NSLS X-17 Beam Line Using Closed Orbit Feedback", BNL 39595.

[3] R.J. Nawrocky et al., "Automatic Beam Steering in the NSLS Storage Rings Using Closed Orbit Feedback", Nucl. Instr. and Meth. in Phys. Res., Vol. A266, 1988.

[4] R.J. Nawrocky et al., "Automatic Steering of X-Ray Beams from NSLS Insertion Devices Using Closed Orbit Feedback", 1989 IEEE Particle Accelerator Conference, Chicago, Il.

[5] R. Biscardi and J.W. Bittner, "Switched Detector for Beam Position Monitor", 1989 IEEE Particle Accelerator Conference, Chicago, Il.

[6] O.V. Singh and R.J. Nawrocky, "Upgrade of Beam Steering System Components and Controls for the NSLS Storage Rings", 1989 IEEE Particle Accelerator Conference, Chicago, Il.

[7] L.H. Yu and L. Ma, "Local Bump by Real Time Measurement Iteration Procedure", BNL 38698, 1986.

Controls and Interlocks for a Prototype 1MHz Beam Chopper

G. Waters, D. Bishop, M.J. Barnes, G.D. Wait
TRIUMF, 4004 Wesbrook Mall, Vancouver, B.C., Canada V6T 2A3

Abstract

A prototype 1MHz beam chopper for the proposed KAON Factory at TRIUMF has been constructed. The chopper is an electric field device, driven by a tetrode based pulser, for deflecting a charged particle beam. Associated with the tetrode used in the prototype design are high voltage power supplies for the electrodes. We use an FET based grid pulser and a sequencer capable of accurate digital control of pulse timing to 0.4ns. A safety interlock and control system using a programmable controlller with fibre optic links has been built. This has given us the versatility required in a prototype system.

I. INTRODUCTION

For ring-to-ring transfers in the KAON factory at TRIUMF Kicker magnets will be required. The TRIUMF cyclotron will be used as an injector for the synchrotron. To prevent beam spill at the transfer points, gaps must be introduced into the injected beam of sufficient duration to allow the kicker magnets to switch on and off. These gaps are produced using an electric kicker[1] operating at up to 15Kv. The repetition rate will be 1.022 MHz with alternate pulses of 46ns and 92ns. The field between the deflector plates must rise and fall in less than 39ns.

The electrical pulses that produce this field are stored in a low loss transmission line with a one-way propagation time of 1μs. One end of the transmission line is terminated with the deflector plates while the other is terminated with a short circuit. The line is centre fed by a tetrode whose cathode is at a high negative voltage. The complex reflections from the open and short circuited ends allow a single tetrode to function both as a charger and as a clipper depending upon the polarity of the reflected pulses. This device is described in a separate paper at this conference[2].

In order to operate the equipment safely, proper procedures as well as a safety interlock and control system were required. In anticipation of frequent changes to the prototype we chose to use a programmable controller (PLC) for the interlock system. PLCs are easily programmed using ladder diagrams and feature relay logic, timers and counters. Only a modest number of I/O points are required so the decision was taken to use an Allen Bradley SLC 150[3]. It has the advantage of low cost and small panel size and provides 20 optically isolated inputs and 12 isolated relay outputs.

The pulse sequences required to drive the tetrode were generated using reprogrammable PALs and programmable delays to produce pulses of arbitrary duty cycle and repetition rate (fig 1). The sequencer output was connected to a FET based grid pulser using fibre optics giving the required high voltage isolation and good noise immunity.

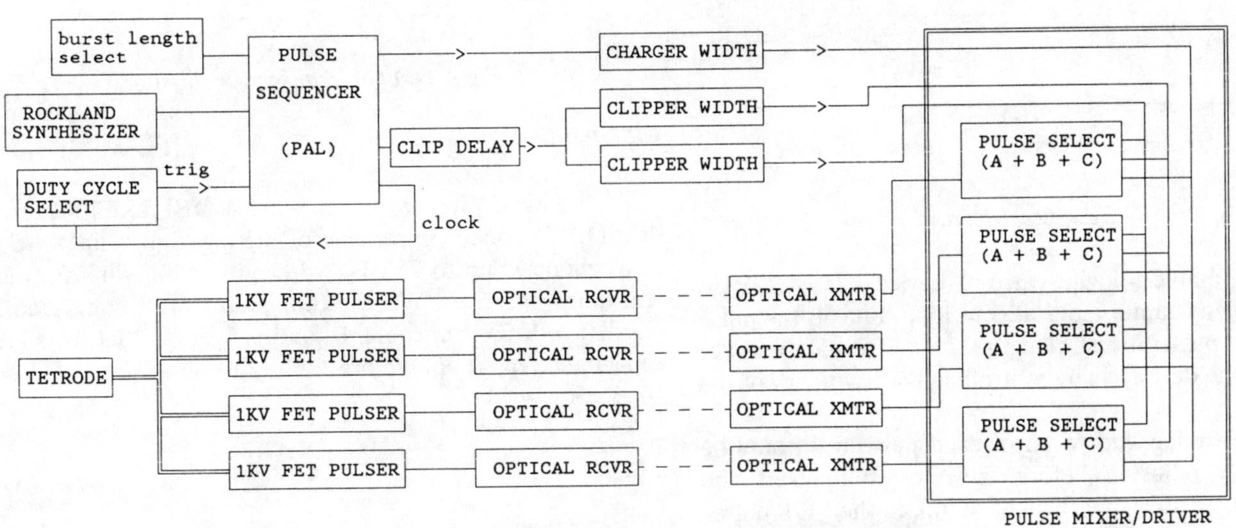

fig 1 Block Diagram of the Grid Pulse Sequencer

II. GRID PULSER

The pulser is required to produce a pulse of magnitude up to 800v at the control grid with a rise time in the 10ns range. The quality of the pulse within the centre fed transmission line is extremely sensitive to the phase and frequency of the grid pulse. A variation of 3ns in timing typically added 10-15ns to the rise time of the stored pulse. It was therefore extremely important to maintain a stable pulse train capable of adjustments in sub-nanosecond increments. To realise this we used a Rockland frequency synthesizer as the rf source and digitally programmable delay lines for pulse timing. The high precision delay lines used (AD9501) are capable of delay increments as small as 10ps.

Sequencer

One operating mode requires a sequence of alternating charger pulses and clipper pulses. This is shown in the lattice diagram fig 2. We can launch a second chopper pulse at the null points when the inverted reflection from the short circuited end of the transmission line cancels the pulse reflected from the open circuit end. This effectively doubles the amplitude of the pulses in the transmission line. In this mode we require a pulse train of two charger pulses followed by two clipper pulses fig 3.

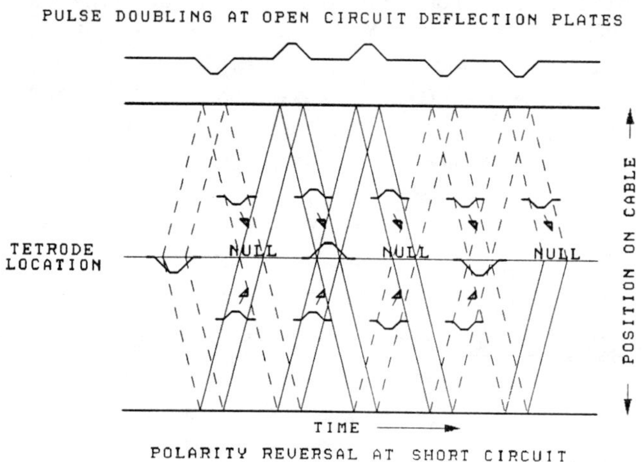

fig 2 Lattice Diagram

A programmable logic array device (PAL22V10) incorporating a counter and state machine controls the pulse mode and pulse burst length. An external counter determines the duty cycle by controlling the frequency of the burst trigger.

The timing and width of each pulse train may be independently adjusted before they are muliplexed and transmitted by a 100Mb optical fibre link to the FET pulser units. Fig 4 shows the transmitter pulse (top trace) and

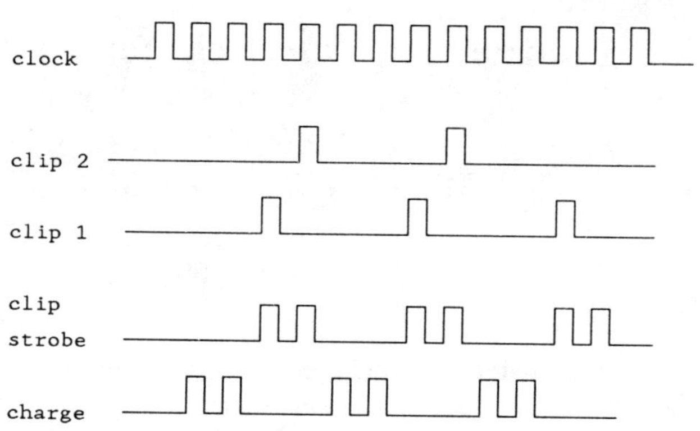

Fig 3 Sequencer output

optical receiver output pulse (bottom trace). We use DIP switches to independently select incremental delays and pulse widths with a precision of 0.4 nanoseconds.

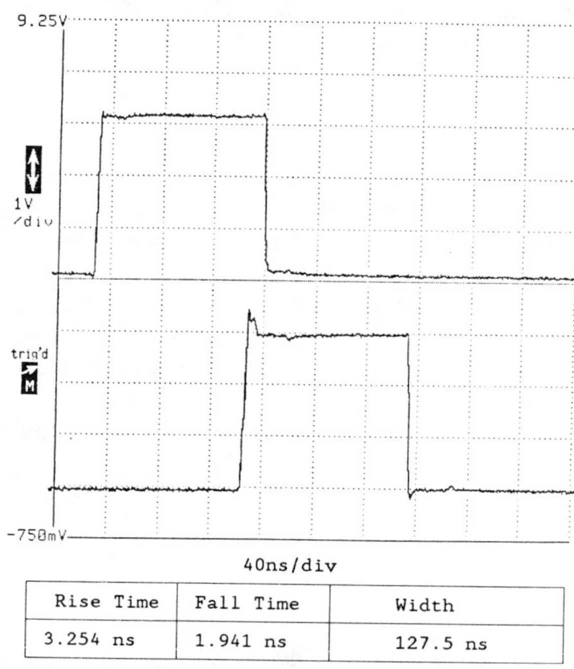

Rise Time	Fall Time	Width
3.254 ns	1.941 ns	127.5 ns

Fig 4 Optical Transceiver Pulses

FET Pulser

This device is built around a DEI 150FPS mosfet driver[4]. This board is capable of driving high voltage FETs at frequencies up to 25MHz with pulse widths of 25ns and rise times of 3-5 ns. We use four pulser units connected in parallel with 50 ohm cable. Equation 1 shows that for a grid capacitance of 700pf we are limited to a rise time of about 10ns at the grid.

$$zc = \frac{50\Omega \times 700pf}{4} \tag{1}$$

The four pulser units operate with a common but variable source voltage to a maximum of -800 volts. In addition the drain voltage of one of the units is variable while the remainder are fixed at ground. The pulser outputs are connected to a summing point close to the grid by coaxial cable. The pulse mixer and the four FET pulser units allows us to produce a composite grid pulse at the control grid of the tetrode.

III. INTERLOCKS

The Programmable Controller.

The traditional way to implement a safety interlock system is with relay logic, a time consuming and very inflexible method. For larger applications we have used a variety of embedded processors and a high level language. This approach while providing sufficient flexibility does not provide for rapid changes. For this reason we chose the SLC 150. As this compact unit already provided sufficient I/O capability for our purpose no further investment in hardware was required.

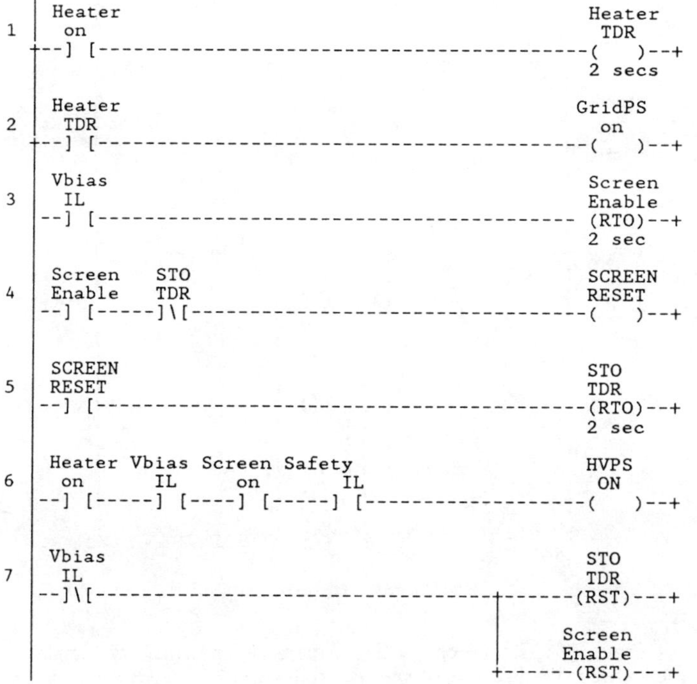

Fig 5 Power on sequence

Programming

The PLC was initially programmed using a PC-AT. The supplied software allowed off-line program development as well as downloading, on-line monitoring and storage. It was not always possible or desirable to permanently connect the PC to the system. For such circumstances a hand held keypad was used. This allowed rapid program changes to be made in the field. The PLC also had an interlock defeat feature and password protection.

The Program

The basic program insures that no high voltage power supplies may be energized until all cabinet doors are closed, the tetrode filament is on, the screen and control grids are biased and the cooling system functioning. Internal counters simulate time delay relays to control the startup and shutdown sequence. Fig 5 illustrates the use of these timers to control switch-on of the screen grid.

A time delay relay (TDR) insures a 10 minute warm up period for the tube heaters before the grid power supplies are energized (rungs 1 and 2). When the control grid bias voltage stabilizes a 2 second delay (rung 3) is initiated before a screen trip reset is generated (rung 4). This in turn starts a second TDR (rung 5) which gates off screen reset after 2 seconds (rung 4). At this stage the main high voltage power supply is enabled (rung 6). Loss of grid bias resets the TDRs ready for a new power up sequence. Should at any time a screen trip occur, the operator is forced to switch off the control grid and to go through the grid supply start up procedure in order to recover from the trip.

IV. CONCLUSIONS

Resources available to us during chopper prototype development were limited so we elected to use commercially available sub-systems wherever feasible. This approach has been vital in commissioning and testing the chopper. It allowed new functionality to be added with minimal disturbance. The components and techniques employed will be used in the design and development of a fully integrated chopper and kicker magnet control system at KAON.

V. REFERENCES

[1] M.J. Barnes, D.C. Fianders, C.B. Figley, V. Rodel, G.D. Wait, G. Waters. "A 1 MHz Beam Chopper for the KAON factory", Proceedings of European Particle Accelerator Conference, 1990.

[2] M.J. Barnes, G.D. Wait, G. Waters, D. Bishop. "Prototype Studies of a 1 MHz Chopper for the Truimf KAON factory", "Proceedings of IEEE Particle Accelerator Conference 1991.

[3] Allen-Bradley, Milwaukee, Wisconsin 53204, "Publication 1745-1.0", January, 1987.

[4] Direct Energy,Inc, 2301 Reseach Blvd.,Ste. 101, Fort Collins, Colorado 80526.

BEAM DIAGNOSTICS USING TRANSITION RADIATION

PRODUCED BY A 100 MEV ELECTRON BEAM

M. Jablonka, J. Leroy, X. Hanus, J.C. Derost
CEN-Saclay, 91191 Gif sur Yvette, FRANCE

L. Wartski
IEF Laboratoire associé au CNRS, 91400 Orsay, FRANCE

Abstract

We report on several experiments using the optical transition radiation (OTR) produced by a 100 MeV electron beam. In using a sensitive video camera coupled with a digital image processing system an accurate and simple beam profile monitor has been devised.

In measuring with a photo-multiplier the radiation emitted in a small solid angle around the direction of the OTR emission, a signal very sensitive to beam energy variations has been obtained.

These experiments have been carried out on the Saclay ALS linac.

Introduction

More than fifteen years ago, one of us (L.W), studying the properties of the optical transition radiation (OTR) produced by 40-70 MeV electrons, demonstrated the interest of this phenomenon for precise and powerful particle beam diagnostics [1,2].

As a matter of fact, OTR, wich is produced whenever a charged particle crosses the boundary separating two media with different dielectric constants (e.g. vacuum and metal) allows, not only precise beam profile imaging, but also measurements of emittance or energy thanks to its high energy-dependant directivity. Moreover, enhanced precision can be obtained by the use of a pair of foils as radiator instead of a single one, leading then to an interference pattern.

Resuming these ideas, recent experiments have been performed on the FEL's electron beams of Los Alamos and Boeing [3].

At the Saclay Linac (ALS) we were primarily interested in devising a simple tool (single foil) for beam profile imaging and small energy variations monitoring.

Theoretical background

In the simplest case, when a relativistic particle of velocity $v = \beta c$ and reduced energy γ travelling in a vacuum impinges on a metallic surface, the radiated energy per unit solid angle $d\Omega$ and unit frequency interval $d\omega$ is given by :

$$\frac{d^2 W}{d\omega \, d\Omega} = R \frac{e^2}{4\pi^2 c} \frac{\sin^2 \theta}{(1 - \beta \cos \theta)^2}$$

where R is the optical reflection coefficient of the material (R = 1 for most of noble metals as Al, Ag, Fe) and θ the angle of emitted rays with respect to the direction of specular reflection of the pseudo-photons (electrons) (Fig 1).

For a highly relativistic particle ($\gamma \gg 1$) θ is very small and the density of radiation becomes :

$$\frac{d^2 W}{d\omega \, d\Omega} = \frac{e^2}{\pi^2 c} \frac{\theta^2}{(\gamma^{-2} + \theta^2)^2}$$

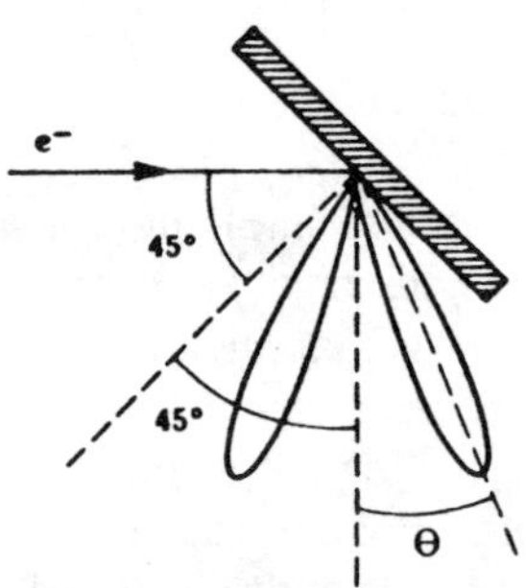

Figure 1 : Schematic illustration of OTR emission.

The angular distribution given on figure 2 represents the signature of OTR. The maximum intensity is peaked in a direction $\theta_M = \gamma^{-1}$.

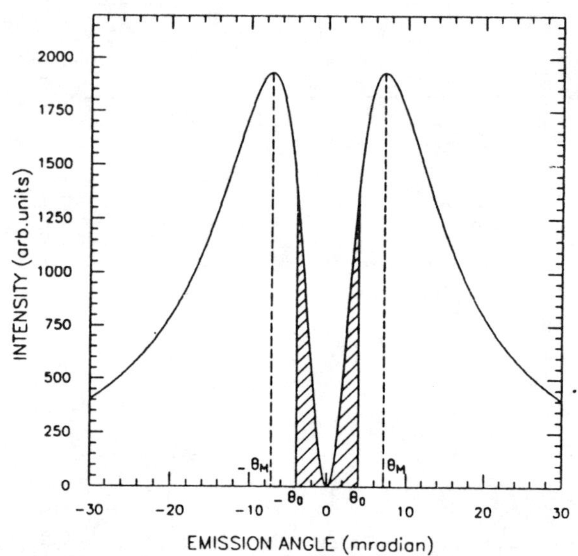

Figure 2 : Theoretical OTR intensity profile vs angles measured from direction of specular reflection.

Integrating the radiation in a cone of apex angle θ_0 one obtains, putting $u = \gamma \theta_0$:

$$u \ll 1 \qquad \frac{dW}{d\omega} = \frac{e^2}{2\pi c} u^4 \tag{1}$$

$$u \gg 1 \qquad \frac{dW}{d\omega} = \frac{e^2}{\pi^2 c} (2 \ln(u) - 1)$$

Of particular interest is the latter case for sensitive energy monitoring of a beam of very small emittance. When the condition $u \ll 1$ is not well satisfied the angularly integrated intensity varies less rapidly than the fourth power of g as indicated above. The power variation u^n where $n < 4$ is listed in table I for different collimating apertures.

u	5.7	2.1	1.3	0.65	0.47	0.28
n	0.8	1.5	2.1	3.6	3.7	4

<u>Table I</u> : Correspondance between a fraction of the maximum emission angle ($u = \gamma\theta_0$) and the exponent n of OTR intensity as a function of the beam energy.

The photon yield in a large frequency interval (i.e. $400\ nm < \lambda < 800\ nm$) is easily shown to be of the order of the fine structure constant :

$$\alpha = \frac{e^2}{\hbar c} = \frac{1}{137}$$

Experimental arrangement

The Saclay ALS (a 700 MeV high duty cycle electron linac) offers after its 6th section a 2 m long drift space in which 1 m was available. At this point, the beam energy is in the 60-120 MeV range, according to the machine regime. We have installed here a retractable radiator consisting of a circular plate of stainless steel 2 mm thick. Stainless was chosen for its ability to be polished at optical standards (2λ) despite its disadvantage of producing much radioactivity. It is placed at a 45° angle with respect to the beam direction.

Because of the high directivity of the RT emission, it was indispensable to carefully align the detection devices. We have used a laser that was itself positionned in reflecting its beam on the center of the rear face of the radiator (also polished), towards the center of an already existing viewscreen at a distance of 2 meters. The camera could then be aligned on the laser beam thanks to a 5 degree of freedom support assembly (Fig. 3).

Camera

First, we have used a low cost camera Thomson TAV 570. It works with a 2/3" ($8,8 * 6,6\ mm^2$) "Newicon" image tube, which differs from a Vidicon only by a more sensitive photoconductive area (0.5 lux). In these conditions, good images were obtained with the following pulsed beam : 5 mA peak current, 20 ms, 6.25 Hz. The image size formed on the sensitive area was approximately 3 mm. When the image was spread over a 10 times larger size we had to use a silicon intensified target camera (S.I.T.) of sensitivity 10^{-4} lux, manufactured by SOFRETEC [4].

Image processing

We have used a VME system equipped with 3 "Imaging Technology" modules [5], namely one analog/digital interface (ADI), one frame buffer (FB), one arithmetic logical unit (ALU) ; a C-library was provided by the manufacturer performing usual elementary functions. It was then possible to develop a Fortran user level.

A VME Fortran software including digital processing and appropriate treatment subroutines has been written in order to study both the beam cross section and the RT angular distribution.

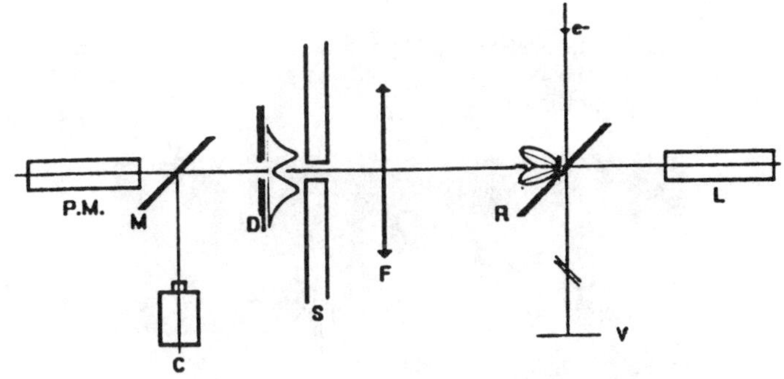

<u>Figure 3</u> : Schematic of the experimental set-up.
R- OTR radiator
L- alignment laser. F- focusing lens f=0.9 m
S- linac shielding   D- diaphragm
C- S.I.T. camera   P.M- photomultiplier
V- viewscreen   M- splitter

Beam cross section imaging

In imaging the radiator itself with the camera, we observe the image of the electron beam cross section via the OTR photons. The advantages of such a viewscreen, consisting in a simple metallic foil,over a classical one (phosphor deposit or chromium doped alumina) are multiple :
- no vacuum degradation under beam irradiation ;
- no discrepancy between the image size and the actual beam size, as is often observed ;
- good linearity, i.e. the number of emitted photons is proportionnal to the number of the incident electrons ;
- no remanence. Moreover, it should be kept in mind that OTR being a very rapid phenomenon, in the picosecond scale, time resolved measurements can be performed ;

From the digitized beam image, the beam size can be extracted by the use of an isointensity curve (Fig. 4). Three-dimension profiles and pseudocolors representations are also possible and very convenient.

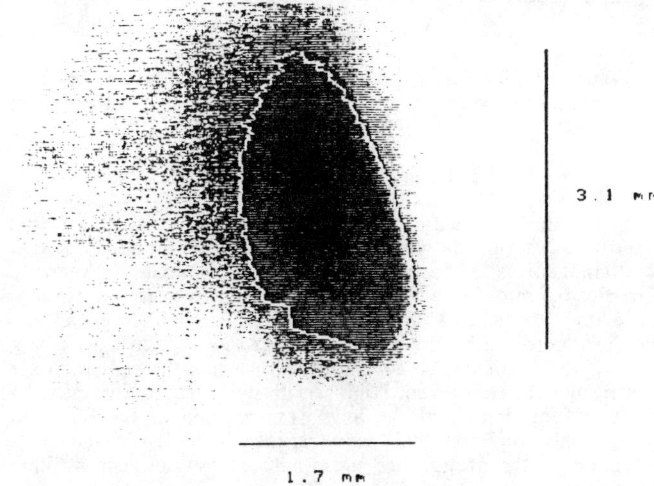

<u>Figure 4</u> : Beam cross-section image with contour and dimensions determined by the computer.

RT angular distribution imaging

In adjusting the camera on infinity, we observe an image of the RT angular distribution (Fig. 5). It can be scanned along a diameter and thus provide an experimental plot of the radiation intensity vs angles of emission. The useful part of this plot is then fitted to the theoretical expression of the OTR intensity convoluated with a gaussian beam divergence (Fig. 6).

As described in reference [1] and demonstrated in [3] one can thus determine the beam energy and the gaussian divergence s of the beam angular distribution.

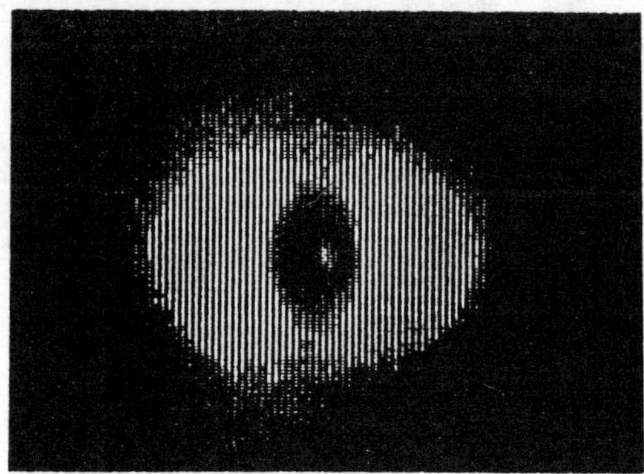

Figure 5 : Image of the RT angular distribution. The small luminous spot inside the dark central region is the image of the linac electron gun cathode, 40 m away.

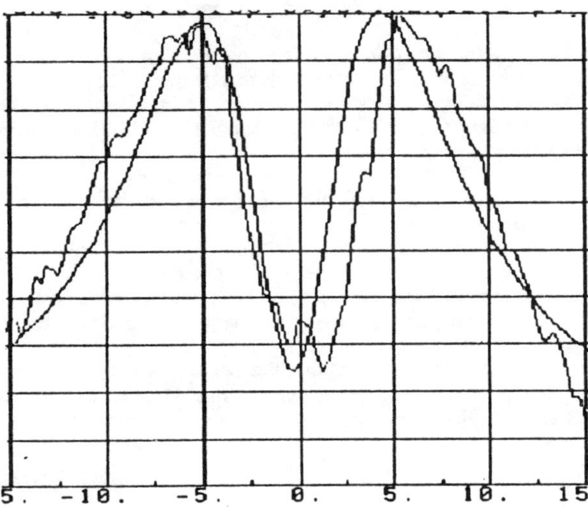

Figure 6 : Fit of an experimental angular distribution by Rule formula [3].The beam energy is found equal to 78 MeV and $\sigma = 1.5$ mrad.

R.T. intensity in a small angle

In order to verify formula (1) we have installed the experimental set-up shown on figure 3. The lens with a focal length of 900 mm forms in the focal plane a circle of maximum intensity of radius 6.5 mm for a 70 MeV beam. At this place, a circular collimator of radius 1.85 mm selects a small fraction of the emission, detected by the photomultiplier ($\theta_0 \gamma = 0.28$). Centering of the distribution can be accomplished in using the camera while steering the electron beam. High sensitivity is required because of the long focal length of the lens. Transmission of the RT light through the shielding wall is indispensable to avoid the P.M. saturation by the background noise induced by the beam striking the radiator.

The verification consists in plotting variations of the P.M. signal amplitude vs variations of the beam energy. Variations of the beam energy are made in varying the input power of the last accelerator klystron. Knowing the relative variation of its RF power, one can calculate (as we had no spectrometer available) the relative variation of its contribution to beam energy :

$$\Delta E/E = 1/2 \, (\Delta P/P)$$

Hence, the initial total energy being known with enough accuracy, one can calculate its relative variation. Plot on figure 7 shows that the P.M. signal varies 4 times faster than the beam energy.

Such a signal could be used for beam energy stability monitoring or as an error signal for a feedback loop.

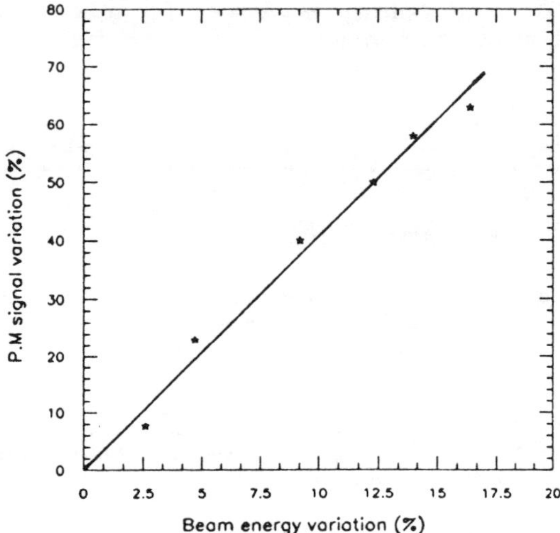

Figure 7 : Experimental plot of the OTR intensity variation in a small angle vs beam energy relative variation.

Aknowledgements

The authors are greatly indebted to J.F. Gournay and F. Gougnaud for their contribution to the image processing development.
The members of the FEL group of Bruyères-le-Chatel are also aknowledged for their interest and support to our experiments.

References

[1] L. Wartski, J. Marcou, S. Roland, "Detection of optical transition radiation and its application to beam diagnostics", IEEE Transactions on Nuclear Science, NS-20, No. 3, June1973.
See also L. Wartski, "Etude du rayonnement de transition produit par des électrons de 30 à 70 MeV", Thèse de doctorat, Orsay, Avril 1976.

[2] L. Wartski, S. Roland, P. Brunet, "Transition radiation monitors in accelerators equipments", 1st International Symposium on transition radiation, Erevan (URSS), Mai 1977.

[3] D.W.Rule, "Transition radiation diagnostics for intense charged particle beams",N.I.M., B24/25 (1987) 901-904.
See also A.H. Lumpkin et.al., "Optical transition radiation measurements for the Los Alamos and Boeing F.E.L experiment", N.I.M. A285 (1989), pp. 343-348.

[4] Sofretec. 95870 Bezons France.

[5] Imaging Technologies Inc., Woburn, Massachussets 01801.

Intelligent Power Supply Controller

R.S. Rumrill, Alpha Scientific Electronics
1868 National Ave. Hayward, California 94545

D.J. Reinagel, JCIL Engineering
293 MacArthur Ave. San Leandro, California 94577

ABSTRACT

We have developed a new power supply controller which would combine 20-bit precision, simple interfacing, and versatile software control. It performs many tasks internal to the power supply and also communicates with an external host computer. Parameters can be entered and/or read over a serial link using one of the 82 command words [1]. In addition, an optional remote control panel can be located up to thousands of feet away. This new controller will reduce the software development time normally spent by the user, while increasing the reliability of the system. The cost is less than buying the equivalent separate CAMAC system. Nonvolatile memory remembers all configuration data; one generic controller can thus be programmed to use anywhere from the smallest power supply to the largest.

HARDWARE

A. General

The controller generates the necessary drive signals for both the transistor regulator and the preregulator. It has inputs for up to 24 interlocks. It controls the main contactor as well as an optional load reverse switch.

The controller is built into a standard 5.25 inch high chassis. An aluminum divider runs vertically down the center axis. The resulting two halves form separate analog and digital sections. Each section has its own motherboard. The motherboards are passive, containing only the 150 pin connectors for the individual boards and several other smaller connectors.

For each major function a new board was developed. These include boards for:

1. CPU Circuits
2. Interlocks Circuits
3. RS-422 Serial Ports
4. DAC's and ADC's
5. Amplifiers
6. Configuration PROM
7. RS-422 Control Panel

Each board is designed to be either a digital or an analog board and to mount into its half of the chassis. The DAC/ADC board is able to span the two halves, picking up the digital half, opto-isolating it, converting it to analog signals, then dropping it down to the analog half of the chassis. CMOS circuits were used wherever possible to reduce power consumption, and therefore, the dissipated heat.

We are presently building 22 power supplies using these controllers. They will be used at the Clinton P. Anderson Meson Facility at Los Alamos (LAMPF) [2].

B. CPU

An NEC V30 microprocessor is used. This is similar to an Intel 8086, however the V30 has a better arithmetic logic unit. 128k bytes of EPROM memory is used, with expansion capability to 512k bytes possible. 64k of battery backed RAM is used, expandable to 256k. A Maxim MAX695 watchdog circuit is used in the reset circuit. Three PAL's are used to provide the various logic decoding, etc.

C. Interlock Circuits

The interlock circuits work on 24 volts DC. Each circuit connects to the LED half of an optical isolator. The output half then connects into the microprocessor circuits. Each interlock one can be individually configured for the following parameters:

1. Interlock name
2. Normal state (open or closed)
3. Fast off or ramp down to off
4. Monitor or ignore

The interlock name is displayed on the front panel during a fault condition and is available to read over the serial link.

The choice was given to allow certain interlocks to ramp down before turning off, thereby, reducing stresses in both the magnets and the input power system. Events such as a transformer overheating may have required many minutes to occur, so a slow ramp down to off is appropriate.

D. RS-422 Serial Circuits

Four serial channels we desired for the following:

1. Host Computer Link
2. Local Control Panel
3. Optional Remote Control Panel
4. Private Diagnostic port

The control panels communicate at 57.6 kbaud. National NS16550 UART's are used, these contain internal FIFO's, and are able to minimize the CPU overhead required for communication. The host computer link can operate at baud rates between 300 and 38.4k. Via this port, one can control virtually any internal function using simple 3 letter commands. The diagnostic port was added for programing and debugging and is not for public use. The two ports which connect to outside equipment are opto-isolated from the microprocessor circuits. A 5 volt DC-DC converter is used to generate to required voltages on the other side of the isolation.

E. DAC/ADC Circuits

The output from the current transductor is measured by an Analog Devices AD624 instrumentation amplifier. The resulting signal applied to a Crystal Semiconductor CS5503 Analog-to Digital Converter. This ADC is a 20 bit delta-sigma device with several features which make it ideally suited for our application. Likewise, the level of the output voltage, transistor voltage, and ground fault current are each digitized by a 16 bit CS5501. The input stage of these ADC's have a 10 Hz lowpass filter with about 50dB of rejection at 60 Hz. The serial output stage is perfect for optical isolating and multiplexing. Because the input range of the device is ±2.5 volts, the 10 volt signals used in the controller are divided 4:1 right at the ADC using precision Vishay resistor networks.

The Current Reference voltage is generated by an 18 bit Sipex SP9380 Digital-to-Analog Converter. The resulting ±10 volt signal is used by the Linear Technology LT1007 error amplifier. Several Burr-Brown DAC's are used to produce the voltage reference and the preregulator reference. A 12 bit DAC is used to provide a fine, or vernier adjustment into the Sipex's summing junction. This accomplishes two things. It produces apparent resolution of greater than 20 bits and allows for a fine correction feedback loop.

This "drift correction loop" was implemented because the Crystal ADC actually has better specs than the best DAC we could find. Because almost all DAC's use a R-2R ladder structure, the ultimate drift is always limited the specs of the resistors used inside the DAC. The ADC, on the other hand, uses no resistors to digitize the signal. We simply try to keep the current measuring ADC reading constant by adjusting the fine vernier DAC as needed. Only the fine DAC is controlled by this loop, the 18 bit Sipex DAC is used "open loop". The current control DAC's are therefore "inside the loop" and their errors reduced to virtually nil. The algorithm used was selected to be dead slow, the desire being to take out drift without causing any loop instabilities, even on the most inductive magnets imaginable.

SOFTWARE

In the design of this controller, it was desirable to use a high-level communications protocol for host communication as well as inter-task communication [3]. This allows for supporting multiple command sources and differing physical communication medium. It was also desirable to have each sub-system in the power supply controller, respond to commands in exactly the same way, independent of the source of the command. To achieve these objectives, the software program was written in an object oriented programming language, C++ [4].

Objects are software entities within the processor. For example, the set of all interlocks form a class; the set of all analog-to-digital converters form another class. However, the most significant class in this controller is not one that corresponds to a physical entity, but one which is used for communication between the objects and application programs. This class, named cmdFrame, forms a generalized data packet which was sufficient for all inter-process communication. This class will be detailed below. Since a number of these objects co-exist and are independent of each other, the system behaves as though several programs were running concurrently under the control of a multi-tasking operating system. No such operating system was used.

The communication class, cmdFrame, has a group of public parameters, listed below with their data type:

int	command_number;
int	done;
int	application_scratch[SIZE_APPLICATION_SCRATCH];
unsigned	next_crc_should_be;
int	error_code;
int	wait_flags;
int	need_current_value;
int	argcount;
float	float_value;
float	pot_scale_factor;
char*	response_units;
int	response_flags;
char*	strings_out[SIZE_STRINGS_OUT];

The command_number indicates which process, or application, is to receive and act on this object. Other entries in this object will give specific information as to what is being requested. The Boolean, done, is set by the receiving process when it has completed the requested task. For a number of commands, as with the reading of the interlock conditions, communication between the command source and the application must repeat until all requested information has been sent. To allow for parameter passing between applications and for intra-application data storage, a small array, called application_scratch, is provided; its usage is very application dependent. The next_crc_should_be is used for error checking when a command source is sending a block of data to the controller. When an application detects an error condition, whether due to the passed parameters or due to the state of the power supply, it signals the error through the error_code parameter. When the cmdFrame object needs to wait for some event before being passed along, the wait_flags parameter is set appropriately. This saves valuable processor resources by eliminating the need for busy-waiting. When one application needs to read another application's current setting, then the Boolean need_current_value is set. If the application is sending a floating point response, then the argcount parameter is set and the value is place in float_value. If there is a unit name associated with this floating point response, e.g., Amps/sec, then the response_units is set to point to the appropriate name. For those applications which have their values adjusted at a control panel, through the use of a digital pot, the pot_scale_factor parameter is set to indicate the magnitude of the effect of the digital pot. When the application is returning a response, the parameter response_flags is used to indicate which parts of the data in the cmdFrame is to be sent back to the control source and in which order. If the application is returning a text message, then the array, strings_out, is set to point to entries in the word dictionary which form the message.

To manage the cmdFrame class, there are a number of subroutines, known as member functions, which will coordinate the loading and parsing of the object. These member functions with their return types are:

```
            cmdframe(const int);
int         get_source_ID();
int         active_controller();
int         set_active_controller(int value);
char*       getbuf_text_out();
char*       gets_text_in();
char        putc_text_out(char);
char*       puts_text_out(char*);
char        getc_text_in();
void        ungetc_text_in(char);
int         check_for_no_args();
void        clear();
void        init();
```

When a cmdFrame object is created, then its constructor, cmdframe(), is invoked. This initializes key private parameters in the object, as well as clearing most of the other parameters by calling another member function, init(). Init() is invoked each time a communication transaction has completed as indicated by the setting of the Boolean done. Another member function, clear() is called by any application when a command is not yet done but all the data in the object has been operated upon and the application wants a clear slate to work on. Although it is ideal for all applications to respond in exactly the same way to all command sources, there are some idiosyncrasies associated with the different command sources. For example, to enter the super-user mode, a password is needed; from the serial communication link, an ASCII string is entered, but from a control panel, a combination must be entered using the digital pot. In order for the SUN application to know which parameter to examine, it must call the member function get_source_ID(). When adjusting power supply parameters, it is necessary to have exclusive access to these parameters for a period of time. To determine if the command requester has been granted this exclusive access, the member function active_controller() is called and compared to the source_ID. If they match, then the application makes the requested change; otherwise, an error_code is returned. The member function set_active_controller() is called by the REQ application to inform all the users of the class cmdFrame who is the active controller at that instant.

Often times in object communication, there is need to pass ASCII characters and strings using the cmdFrame as the medium of transfer. To coordinate this, a set of member functions are provided; for reading characters out of a cmdFrame, the functions getc_text_in() and ungetc_text_in() are provided. For fetching the entire remaining stream of characters, the function gets_text_in() is provided. In some cases, the application expects there to be no input at all, and to verify this condition, the member function check_for_no_args() is provided. For writing characters into the cmdFrame, the function putc_text_out() is provided; this function ensures that character array lengths are not violated. When it is necessary to write characters directly into this array, the function getbuf_text_out() is provided to pass a pointer to the location where the next character should be placed. However, care must then be taken by the application that it does not exceed the character array length. When writing a character string into the array, strings_out, the function puts_text_out() places that string in the next available location. There are a number of private parameters within the cmdFrame class to affect these member functions, but these will not be detailed.

The use of this communication object has proven to provide a clean interface for the entire controller program and makes it simple to locate and solve programming problems. The modularity of this approach also makes it possible to add other communication mediums without requiring significant changes to the other parts of the program, as well as adding other application commands.

REFERENCES

[1] For a detailed listing of the command words contact the authors.
[2] S. Cohen and R. Stuewe, "Magnet Power Supply as a Network Object," in this conference proceedings.
[3] L. J. Chapman, "Object-Oriented Communications," in Proceedings of the 1989 IEEE Particle Accelerator Conference, pp. 1631-2.
[4] B. Stroustrup, The C++ Programming Language, Addison Wesley, 1986.

History Data Facility in the SLC Control System

Ralph G. Johnson   Gregory R. White

Stanford Linear Accelerator Center, Stanford University, Stanford CA 94305 [*]

Abstract

Two major enhancements to the SLC History Data Facility [1] are described separately. First the internal design and procedures used for saving and using long term history data. Second the user interface, facilities and application of the History Data Comparisons sub-system, which is used for analyzing and correlating two or more accelerator device histories.

Overview and Rationale

The history data facility of the Stanford Linear Collider (SLC) records device and feedback loop parameter values. We use the term 'device' to mean any machine unit, such as a monitoring instrument, magnet, power supply, etc. A 'parameter' of a device then, is the current value of one of its properties. Most devices report more than one parameter. A typical example would be a Toroid device for which an important parameter is its detected beam current.

Extensive software has been developed for displaying recorded history data. Much can be learned from looking at the locus of any one machine parameter over time, such as effects predicted by diurnal variation, or the correspondence of a device with a particular accelerator configuration. The comparison of two or more parameters can be used for failure diagnosis, machine tuning, experimentation and the like.

The parameters of any device which is defined in the control system database, may be saved. Currently we are recording over 23,000 different device parameters at a typical interval of 6 minutes. New developments to support historical correlations and long term analyses is presented. Our objective is to promote the use of historical data in analyzing machine characteristics for increasing beam luminosity.

Long Term Histories

As outlined in a preceeding paper [1], a data collection process saves data in a ring buffers - the 'Daily' data files. Each time that process is executed it reads and saves the values for a set of device parameters. File activity is minimized by sequentially saving data in a single buffer. Data in these files covers a period of one day.

We now also have history files which contain data for one week. These 'Weekly' files are organized by device. That is, there is a separate ring buffer for each device parameter. All ring buffers are of the same length. There is also a single ring buffer containing the time-stamps of all the recorded data. For any given time-stamp, the value of each parameter at that time is located at the same offset within each data ring buffer. This format permits easy data management and efficient data retrieval.

Shortly after midnight each day, the data in each Daily file is translated to the Weekly format. Thus we have one process saving data every 2 to 6 minutes and another moving that data to week long ring buffers once each day.

Archiving

Since the Weekly data files are ring buffered, data older than a week is lost each midnight. To provide long term data saving there is another set of files named 'Archive' (also commonly called 'Yearly' files). These are of the same internal format as the Weekly files. Once each week, or whenever needed, a process makes a snapshot copy of each of the current Weekly files. Each of these is given a name which includes the date. To conserve disk space this version is not a complete copy; rather we preserve only every nth data point, in order to achieve a desired file size. Although each history file could be compressed by a different ratio, all are currently compressed by 10:1. The primary consideration has been disk space.

As a consequence of our normal disk backup to tape, these 'Archive' files are preserved in secure storage. At some point we could delete older files from the disk as we would still have that data on tape. We have not yet implemented a procedure or software for autonomously retrieving any data from the backup. However, one could use

[*]Work supported by the Department of Energy, contract DE-AC03-76SF00515

the standard system file restore procedures for any file, whereupon it would then be available automatically.

Consequently, we always have the past week's history in "fine structure" form (at most every 6 minutes) - used typically for diagnosing current problems; and older data in "coarse structure" (each hour) for looking at trends and past setpoint values.

Data Retrieval

We have attempted to minimize the total disk accesses required to retrieve all the parameter data for a given time span. First, any data available in the Weekly file is gathered. Then, if more recent data is also needed, it is obtained from the Daily file by a request to the history process. If older data is needed, the list of available Archive files is obtained. Since the names of these files includes the date of the last data point in each we can then sequentially retrieve data only from those files pertinent to the requested time span. Also, by first using all available Weekly data we insure the presentation of data with the greatest number of values within the requested time period. As each request for new data is processed, any data already available in memory from prior requests is utilized, again minimizing file accesses.

Correlating Parameters over Time

This section describes the History Data Comparisons subsystem of the History Data facility, used to plot and analyze the recorded parameter values of two or more devices. The analyses fall roughly into two classes; *qualitative* - by-eye examinations of the loci of some devices, and *quantitative* - correlation charts, statistics and expressions.

This distinction is interesting when the user community of each class and their respective applications is considered. First we describe the interface and displays available to the users, and then consider the applications to which various users put them.

User Input Interface

The interface to the History Data facility is via the SLC Control Program (SCP) - a large multi-image software suite used to control most of the system wide functions necessary to run the SLC accelerator. Users navigate through the sub-systems of this program via a hierarchical arrangement of push-button 'panels' [5][these proceedings]. There are thirty or so panels in the SCP dealing with History Plots, each oriented toward one facility in the accelerator. From all of these, one can enter the History Buffer Comparisons panel.

With 23,000 or so devices available for inspection, some help is given to the user in selecting the device and parameter they are interested in. This help makes use of the fact that all devices and their parameters are identified by their database entry name, which is moderately formal. A device and each parameter under it, is uniquely identified by a tuple composed of four, or sometimes five, domains. Reading the tuple from left-to-right, each successive domain identifies the device more closely. The help system attempts to satisfy a partially instantiated device name and displays all the possibilities for the most significant missing domain. The user may assign up to 3 devices to *variable labels* A, B and C. All plots are referred to as plots of these variables, either against each other, or vs. time.

Default values for each domain are also maintained, chosen from consideration of the previous assignment of the variable, the assignment of the other variables, and any partial tuple already entered.

Users may also elect to assign to a variable label, a device they had examined when using one of the other History Data facilities.

Device names, however assigned, are reflected on the panel screen. The time range for which History Data Comparisons is to base its plots, is input from any of the panels dealing with the History Data facility as a whole. Once set, all plots for all devices are generated for that range, though the range can be changed at any time and the plot will be re-drawn.

All plots can be scaled manually or automatically. In general, when auto-scaling is selected, the entire range of data values gathered for the device will be charted and used in computing statistics. The function of manual-scaling is slightly different depending on whether a qualitative or quantitative plot is requested; see below.

Qualitative Plots

To compare devices over a common time range, their loci can be plotted either one-above-another (a 'Stripchart') or superimposed (an 'Overlay'). Manual-scaling for these plots refers simply to the upper and lower bounds of their ordinate axes. Major applications of qualitative plots have been in diagnosing errors and investigating unexpected machine behaviour, very much as an *ad hoc* problem solving tool.

Quantitative Plots

Quantitative charts explicitly treat the parameter values of the devices over time as a data vector. Before the vectors are used in expressions and, more significantly, before statistics are calculated on them, we make them congruent. That is, we make sure they all have the same number of data points, that they range over exactly the same time and that all corresponding data points refer to the same delta time from the stamp of the first point. This is achieved by linear interpolation, which is always performed assuming device variable A is the *independent*, and therefore B is "aligned" to it.

There are two groups of quantitative plots. Those plotting the locus of an arithmetic expression in the data vectors, vs time, named 'expression' plots; and one displaying the scatter-plot of the data vectors, named 'A vs B'.

At the time of writing only two expressions are available, being those requested by users to fulfil specific applications, discussed below. They are:

$$L[i] = A[i] + (r * B'[i]) \qquad (1)$$

$$L[i] = A[i]/(r + B'[i]) \qquad (2)$$

where L is the resultant vector, r is a real value entered from one of the buttons, and i ranges from 1 to the number of original data points taken for A (before, if applicable, A was 'bound'). For both expression plots and scatter plots, the user may elect to bind, or 'cut' the data value vectors before they're used. This they do by by manually scaling the device they wish to limit. In this case, original data vector elements outside the range will be replaced by linear interpolation between the previous and next elements whose values are inside the manual-scaling range. This is slightly different to the manual-scaling function when used for the Stripchart and Overlay plots.

In addition, the plot of the resultant locus of an expression plot can be manually scaled.

A significant portion of the systems analysis time was spent looking at anomalies arising from interpolations and their effect on statistics in the data. For instance, there may be periods when little or no data is recorded for one of the devices while the other is recorded normally. The exact effect of this on the plot and on the correlation statistics is different depending on whether it was the independent, or dependent variable for which data was 'missing '.We concluded that it was sufficient to draw attention to these effects in the help and documentation rather than employ more sophisticated data smoothing methods.

Correlation Statistics

The linear correlation coefficient (by Pearson) between A and B is displayed on the A vs B plot.

Pearson's r measures only the extent of correspondence between variables known to have a significant association. That is, it should not be used to decide whether two variables have an association in the first place. It is not yet clear whether users of History Data Comparisons are using the facility informally, to 'find' correlations.

Given some assumptions regarding r, for instance that many calculations of r, over a very long time frame, for the same variables, would yield a symmetric distribution curve, then we can for any one evaluation of r decide roughly how 'reliable' it is. This is quantified by a corrected Complimentary Error Function. That figure is also given with every chart.

Linear interpolation is used to bound the input data vectors on manual-scaling and "align" the vectors in preparation for a plot (this changes the data values most when significantly more data points are taken for one device than the other). All these will make the correlation coefficient figure artificially high.

Quantitative plots are used in machine tuning and accelerator physics experiments. For instance, the positron yield is a simple plot of the ratio of electrons onto the target and resultant positrons, and can be plotted for any one of the current monitoring devices down beam of the target.

Current Development

The simple expressions currently available were those requested by control room operators with specific applications in mind. We are now extending this function to plot 'arbitrarily' complex expressions on two or more device vectors. To do this we are using the Free Software Foundation's (GNU) implementations of two popular Unix[1] program development tools, Lex and Yacc. The GNU versions are named 'Flex' and 'Bison' [3] respectively. Lex is a lexical analyzer, which we use to pass the tokenized expression to a Yacc generated function which parses and evaluates the data. This is a recognized, virtually textbook application [2].

We are also studying the utility of non-linear methods in the normalization of data for the quantitative plots and giving more sophisticated statistics. Operators wishing to analyze the data in depth have the option to down-load the history data and look at it with their own tools. There has therefore been the argument that the on-line control system should provide only a minimum, intuitive, package. We are collecting comments from users before implementing, for instance, polynomial interpolation.

The History Data Facility in general is a major software component of the Control System for the SLC. It is continually upgraded and extended to accommodate tracking and analysis for new sub-systems in the SLC.

References

[1] Johnston, R. Stanford Linear Collider History Data Facility. *Proceedings of the 1989 IEEE Particle Accelerator Conference* 1989. 1716-1718

[2] Kernighan B. W. and Pike P. The Unix Programming Environment. Prentice-Hall. 1984. Chapter 8.

[3] Corbett R. and Stallman R. BISON The YACC-compatible Parser Generator. Free Software Foundation. 1988,1989,1990.

[4] Press W.H., Flannery B.P, Teukolsky S.A., Vetterling W.T. Numerical Recipes, The Art of Scientific Computing. Cambridge University Press. 1989.

[5] Flores M., Hendrickson L., Phinney N., Sanchez-Chopitea L. Correlation Plot Facility in the SLC Control System *Proceedings of the 1991 IEEE Particle Accelerator Conference* 1991.

---

[1] Unix is a trade mark of Bell Laboratories, ok

A Beam Diagnostic System for ELSA

M. SCHILLO, K.H. ALTHOFF, W.v. DRACHENFELS, T. GOETZ, D. HUSMANN,
M. NECKENIG, M. PICARD, F.J. SCHITTKO, W. SCHAUERTE, J. WENZEL

Physikalisches Institut der Universität Bonn

Nußallee 12

D-5300 Bonn 1

Germany

Abstract

A beam diagnostic system, which is based on capacitive beam-position monitors combined with fast electronics, has been developed for the Bonn **EL**ectron **S**tretcher **A**ccelerator ELSA. The position signal of each monitor is digitized at an adjustable sampling rate (max.: 10 MHz) and the most recent 8192 position and intensity values are buffered. This allows a wide range of different beam diagnostic measurements. The main purpose is the closed-orbit correction, which can be carried out on various time scales. To optimize the duty factor of the extracted beam, the system can also be used as a fast relative intensity monitor resolving the intensity distribution of the bunches or of the injected beam. It is designed to support betatron tune and phase measurements with very high accuracy, offering the choice to select any of the beam position monitors. This enables the measuring of many optical parameters. Furthermore any pair of suitable monitors can be used for experimental particle tracking or phase space measurements.

I. INTRODUCTION

The Electron Stretcher Accelerator ELSA, commissioned in 1988, is designed to produce electron beams with a high duty cycle at energies up to 3.5 GeV [1]. The 2.5 GeV synchrotron, in operation since 1967, is now used as an injection-booster. There are three different operation modes for ELSA. In the stretcher mode electrons are injected every 20 msec from the booster. Then the extraction of the electrons is started using a third integer resonance. Above 2 GeV ELSA has to be used in the post accelerator mode. After the injection of several pulses from the booster-synchrotron the electrons are ramped to the required energy (max.: 3.5 GeV). To obtain a satisfactory duty cycle in this mode as well, the extraction time has to be extended. In a third mode ELSA is used as a storage ring for synchrotron-radiation experiments. Due to this multi-purpose usage of ELSA arose the need for a flexible and powerful beam-diagnostic system.

II. THE MONITOR SYSTEM

The ELSA beam position monitor system is based on capacitive pickups[1] with a button-type electrode. A single BPM consists of four electrodes positioned at a 45° angle in a vacuum chamber with a nearly elliptical profile.

---

[1] Same type as used at DESY II

A. RF - Electronics

For the signal processing a narrow bandwidth approach was chosen working at a center frequency of 500 MHz with a bandwidth of ± 10 MHz. This is the main frequency component of the beam current in ELSA, since each bucket is filled.

The sensitivity of the pickups was estimated by measuring the amplitude of this frequency component at one pickup button with a spectrum analyser. With 50Ω cable termination this yields an amplitude of 1.1 mV per mA beam current at each button.

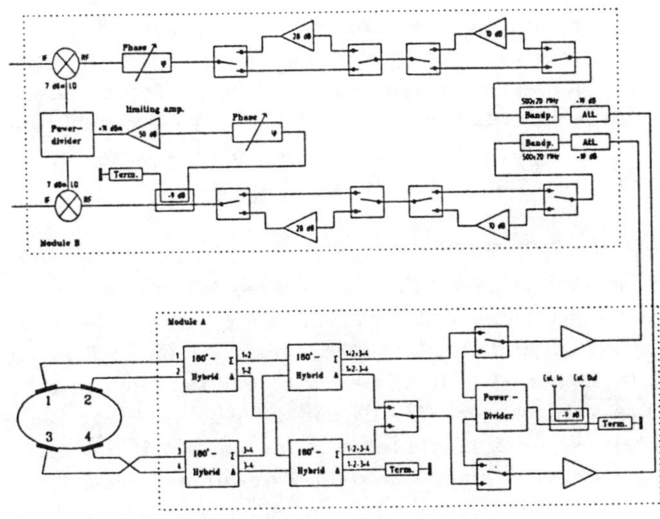

Figure 1

The monitor rf electronics

The signal-processing electronics consists of three different modules. Module A generates 500 MHz horizontal or vertical position signals and an intensity signal. Module B contains amplifiers and the detector circuit for the two 500 MHz signals; the third module carries out the digitization of sum and difference signals. An overview of the whole RF circuit is shown in figure 1. The bottom half of the figure corresponds to the the first rf module. To keep phase errors small the connections between the electrodes and the module have to be short, thus it has to be positioned in the vicinity of the BPMs. The four buttons of the monitor are directly connected to the 180°-hybrids. At the output of the four hybrids we obtain the intensity signal $V_s = V_1 + V_2 + V_3 + V_4$ and the position signals $V_{hor} = V_1 - V_2 + V_3 - V_4$ and $V_{ver} = V_1 + V_2 - V_3 - V_4$.

To reduce the number of channels for the further signal-processing, the horizontal or the vertical signal has to be selected with a GaAs RF-switch, leaving one difference and one sum channel. At this point a calibration signal can be switched to both channels with equal phase and amplitude to enable the correction of phase differences or unequal attenuation in the following components. The signal level at the output of module A is increased by an amplifier stage ($+13$ dB) in each channel.

The two output signals then are carried by two double-shielded coaxial cables to a central station where the other two modules for each monitor are placed. Due to different cable length a special attenuator has to be added for each monitor to obtain an equal overall attenuation for all monitors. At the input of the rectifier-module a bandpass filter with a center frequency of 500 ± 10 MHz is used to remove noise and distortions. A variable amplifier stage, which is realized with two switchable MMIC amplifiers, follows the filter. The detection of the 500 MHz signals is carried out by a homodyne circuit [2] with a dynamic range of 30 dB, which is mainly determined by the limiting amplifier. The two voltage-controlled phase shifters will be used in addition to the calibration signal to correct phase difference in the two channels.

At the output of the mixers we obtain the rectified signal with a bandwidth of 10 MHz and a maximum amplitude of 100 mV. This amplitude range now has to be matched with the input range of the ADC of 0-2 V. A variable gain amplifier, which allows the addition of an voltage offset to the input signal, is used for this purpose.

B. Data Acquisition

The digitization of the signals is carried out by a special sampling module with two 8-bit flash ADCs. The values of the sum and the difference signals are buffered in two static memories of 8 KByte each. The sampling rate can be selected between 78 kHz and 10 MHz. A trigger pulse generated by a microprocessor module (MACS^2) of the control system starts the data acquisition process for all monitors. It is stopped when the buffer is completely filled.

Control signals for the RF-switches, control voltages for the phase-shifters, for offset and for gain inputs are also generated by this module. Five of this sampling modules can be handled by one module of the control-system.

C. Calibration

The position dependency of the difference signals was measured for each monitor with a movable antenna. A central range of -25..25 mm horizontal and -12..12 mm vertical was covered by the calibration with a step size of 1 mm. For each point a set $(\frac{\Delta_{hor}}{\Sigma}, \frac{\Delta_{ver}}{\Sigma}, x, y)$ was recorded. This data was fitted by a polynomial of 8th degree: $x = P_x(\frac{\Delta_{hor}}{\Sigma}, y)$ and $y = P_y(\frac{\Delta_{ver}}{\Sigma}, x)$. To obtain (x, y) from the position data $(\Delta_{hor}, \Delta_{ver})$ an iteration algorithm is used, usually with less then 10 iterations.

Online calibration of the phase shifters and the data acquisition can be performed by the MACS-modules on request, thus cancelling the effects of temperature or long

^{2}Microprocessor Aided Control System, developed at Bonn.

term drift.

III. Software for Data Analysis

A. The ELSA Physics Operating System

For ELSA beam diagnostics and orbit correction in combination with accelerator control, a program named "E-POS" (**ELSA P**hysics **O**perating **S**ystem) was implemented based on the existing control system. EPOS combines BPM data aquisition, analysis and automation of measurement and steering tasks in an interactive, workstation-based environment.

The system provides a simple programming language for ELSA control and data analysis. This language gives access to all accelerator and BPM parameters and can manipulate data objects by a variety of different tools, for instance algorithms dealing with closed orbit correction.

EPOS allows the use of expressions, which may include transcendental or special functions, operations on arrays, matrices or optics of the machine, like measured orbits or computed β–values. The digital signal processing module contains different kinds of timedomain–filters, fourier analysis, power spectrum estimation methods and techniques for peak detection, trend removal and data smoothing. Additionally instruments for curve fitting, interpolation and histogramming are available. The graphical user interface of EPOS is based on `X-WINDOWS` and `GKS` and can generate plots and printouts for all supported data types.

B. Closed Orbit Correction

The closed orbit correction is performed via steering dipoles. Their settings are calculated by the program COCPIT (**C**losed **O**rbit **C**orrection **P**rogram for **I**nteractive **T**asks), which is integrated into EPOS.

First the quadrupole field gradients are evaluated from the machine tune, since the tune can be measured with high accuracy. Then the optical functions are determined. A matrix $\mathcal{M}$ is evaluated which gives the expected beam positions at the BPMs for a given set of steerer stettings. With this a least-square correction can be performed by inversion of $\mathcal{M}$, but harmonic correction is implemented as well. Due to the resonant structure of the betatron motion, high frequency components (with respect to the tune) of the closed orbit are expected to be strongly suppressed for a random distribution of field errors. On the other hand when correcting with LSQ method, high frequency components of the closed orbit produce extreme steerer settings. COCPIT allows to select a frequency range and perform correction only for frequency components inside this range (harmonic correction). The Fourier transformation and necessary filtering can be put together to one matrix $\mathcal{M}'$, given by

$$m'_{ij} = \frac{\sqrt{2}}{\sqrt{\beta_i \beta_j}} \sum_{k \in range} \frac{Q^2}{Q^2 - k^2} \sin(\frac{\pi}{4} + k\Delta\phi_{ij}),$$

with i and j denoting the positions of BPM and steerer respectively. Thus the calculations can be performed in complete analogy to the LSQ method and in the same time. Single monitors can be weighted with errors or completely left out of the calculations, when they are supposed to be erroneous. Also single steerers can be switched off or set to a current limit for the calculations to avoid overloading.

IV. Measurements

A. Closed Orbit

To test the accuracy of the monitors the dispersion versus the detuning of the accelerating rf was measured. For different frequencies the beam's position was taken at all monitors. A straight line was fitted to each monitor's data and this line's maximum deviation from the data points was taken as an upper limit of the monitor errors. This yields an accuracy better than 0.2 mm.

Closed-orbit corrections were carried out with both correction schemes. The LSQ correction seems better to cope with the huge deviations of the uncorrected orbit while the harmonic correction produces the best results when used as the second correction. With both algorithms a final rms-value of 0.15 mm and maximal values of ±0.25 mm can be obtained.

B. Betatron Tune

The transverse tune of the machine is measured by analyzing the beam position data taken from a coherent betatron oscillation. The oscillation can be observed after the injection or can be induced by a kicker magnet. A Fourier transformation is applied on the position data, the fractional tune value is derived from the measured betatron sideband frequency, related to the ELSA revolution frequency of $\approx$1.8 MHz. The loss of coherence limits the useful beam observation time of the BPM system to $\approx 200 \mu sec$, which yields a tune resolution in the range of $\frac{\Delta Q}{Q} \approx 10^{-3}$. For slow resonant extraction (as required by the stretcher principle) near a third integer resonance, however, this accuracy of $\frac{\Delta Q}{Q}$ is not sufficient. To increase accuracy, a special correction method was implemented in the EPOS spectral analysis software. By using an interpolation recipe for the main bins of a detected spectral peak, it is possible to get a $\frac{\Delta Q}{Q} \approx 10^{-4} - 10^{-5}$. By combining this method with the computation of the weighted mean for selected peak bins, the tune measurement has been made possible with a reliable resolution of $\frac{\Delta Q}{Q} \leq 10^{-4}$. Thus, a fine-adjustment of the controlled tune-shift (by fast quadrupoles) during the extraction period can be performed very easily.

All tune measurements are automatically performed by the EPOS system, which is also used to apply more diagnostic methods to the obtained position spectra.

C. Phase Space

First measurements of the phase space for different tunes were carried out by recording the coherent betatron oscillation in the horizontal plane at two neighbouring BPM stations. To keep calculations simple two BPMs were chosen where $\beta_i = \beta_j$ and $\phi_i - \phi_j \approx 90°$. For these measurements the four signals of the two monitors were recorded by digital oscilloscopes with a sampling rate of up to $100\ MHz$. The position value of one bunch ensemble is extracted with a software filter for each revolution out of the recorded signals. Plotting the position values at the second BPM versus the position values at the first BPM shows the phase space in relative coordinates at the position of the first BPM.

Figure 2 shows the evolution of the phase space plot for a horizontal betatron tune close to $4\frac{2}{3}$ and extraction sextupoles in ELSA driving a third-integer resonance. After

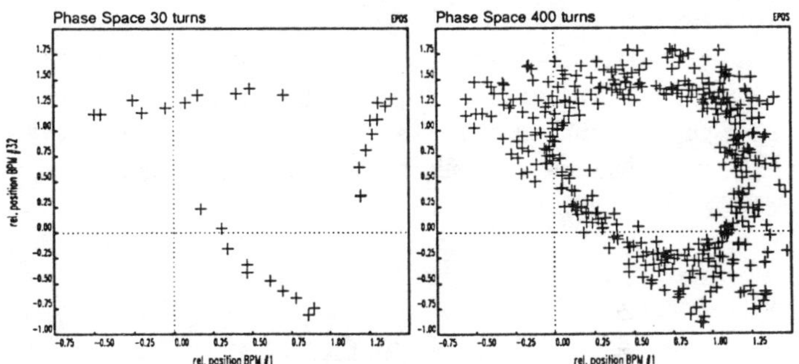

Figure 2
Phase space plot

30 turns (left figure) the shape of the characteristic phase space triangle appears. After 400 turns the loss of coherence pretends a damping of oscillations.

The formula:

$$\cos 2\pi Q = \frac{1}{2} \frac{x_n - x_{n+1} + x_{n+2} - x_{n+3}}{x_{n+1} - x_{n+2}}$$

can be used to measure the fractional part of the horizontal betatron tune Q on a short time scale of four turns ($2.192\ \mu s$). This method was used to analyse the dynamical evolution of the horizontal tune for some hundred turns after the injection. Due to energy oscillations a modulation in the kHz-range on the fractional part of the betatron tune can be observed. When the RF in ELSA is switched off a constant rise in the horizontal tune can be seen, that is caused by the energy loss according to synchrotron light emission.

V. References

[1] K.H. Althoff et al., "ELSA - One Year of Experience with the Bonn Electron Stretcher Accelerator," in *Particle Accelerators, Vol. 27*, pp. 101–106, 1990.

[2] R. Bossart, J. Papis, and V. Rossi, "Synchronous RF-Receivers for Beam Position and Intensity Measurement at the CERN SPS Collider," in *IEEE Trans. Nucl. Sci. NS-32*, p. 1899.

A Table Driven Database and Applications Generator for Accelerator Control Systems

Anthony Carter , Coles Sibley and Tomás Russ
MIT - Bates Accelerator Center
21 Manning Rd., Middleton, MA. 01949

Abstract

Accelerators and related hardware are dynamic entities. Similarly, the control systems that are designed for them must be dynamic. This paper describes our attempt to define a fundamental part of the control system for a pulse stretcher ring that adapts easily to the changes inherent in a system of this complexity.

I. INTRODUCTION

The Bates Pulse Stretcher Ring [1] control system architecture is a distributed system of computers distinguished as either data processing or data acquis-ition/control [2]. The data processing computers serve as either display stations or control stations. The display stations are low-cost workstations or personal computers that provide continuous update of the status of pertinent Ring data. The control stations are high performance workstations that give the operator real time control of Ring elements. Also, there are various other computers that serve specific functions such as Ring modeling and orbit correction. The data acquisition/control computers acquire raw data from the Ring hardware at periodic intervals (roughly 4 Hz) and multicast the data on the Ring Ethernet. They also listen for requests from the data processing computer and perform the necessary control functions.

II. GOAL

The common aspect of all these applications is that they all need certain system-wide information in order to function properly. This information could be constants (which are not necessarily very constant), hardware address information, scaling information, or any information that is not application specific. Typically, this information is hardcoded into an application making it very efficient but very hard to maintain. Our goal was to develop a system whereby an application could be written to query a central database for this class of information (see figure 1).

Writing an application to incorporate system-wide constants is a relatively simple operation. Once the information has been retrieved, the application should proceed with no significant overhead. However, designing an application where scaling information is incorporated at run-time proved to be a much more serious task.

III. DESIGN

Our first attempt at designing a system where the scaling of raw data could be defined in a database was to have a third-order (or higher) polynomial describe the transformation. The database would contain the coefficients of the polynomial and the raw data would be the independent of the polynomial. The computing of the polynomial could be hard coded into an application preserving much of the efficiency. Unfortunately, this proved to be not flexible enough to describe the data in our system.

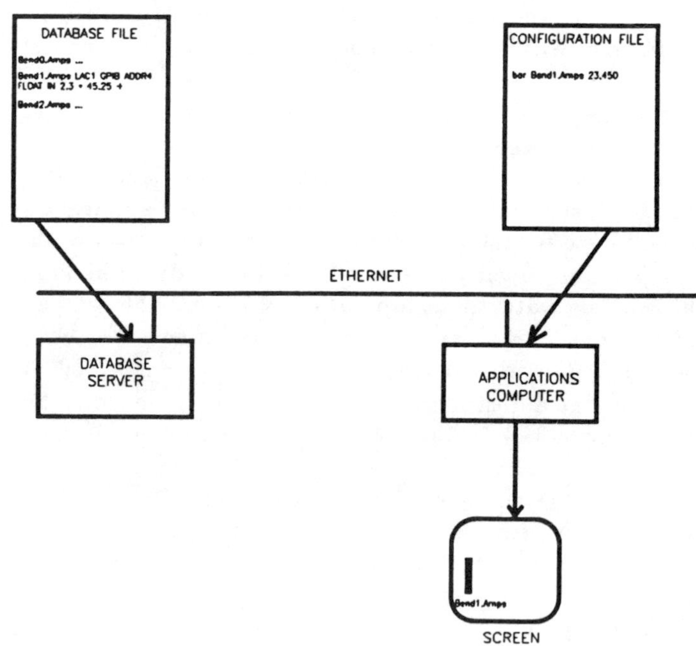

Fig. 1 - Server / application architecture.

Table 1.

Operators	Description
+ - * / %	simple math
! && || > >= <= == !=	boolean
sin, cos, pow, log, ln, sqrt	complex math
sto	memory store
rcl	memory recall
in	input
out	output

Our next step was to throw simplicity out the window and look for a system that could describe virtually any transformation. We settled on a system where the transformation was described by a postfix equation (also known as Reverse Polish Notation or RPN). The supported operators and functions are taken from the C programming language where possible (see table I). A few functions were added to provide for special operations.

The IN function was added to describe how to extract raw device data from the multicast Ethernet packets. It takes four arguments: source computer, source port, offset into the packet and the data type of the raw value. The data type can be any of the following: char, unsigned char, short, unsigned short, long, unsigned long, float or double.

The OUT function defines how to send a request to a data acquisition/control computer. It takes the same four arguments as the IN function, adding a fifth for a command. Each data acquisition/control computer supports a set of commands, such as READ, WRITE, INCREMENT, RAMP, CYCLE, and so on.

In addition the STO and RCL functions provide the memory store and recall capabilities of a calculator. These are used in a special way. When you convert a raw signal to some engineering unit, the IN function effectively holds the place of that raw data in the equation. However, when you want to reverse the process, so that you can SET something for instance, the data you want to convert comes from within the application program. The way we handle this is as follows: the equations for setting things are written with RCL's holding the place of the data. The application programs are written to STO to memory the data to be output before computing the equation. As the equation is computed, the actual data will be inserted anywhere a RCL is found.

IV. IMPLEMENTATION

In our implementation of this system, we have one machine that acts as a central database server. Configuration information is broken into two distinct groups, application dependent and application independent. The former would be things like where on a particular screen a bar graph is placed, and the latter would be things like limits on a particular power supply. Only application independent information is kept in the central database. All programs are written to query the server for this data at initialization time.

For example, one display computer program reads a configuration file with a format like:

BAR CB1.ADC.GAUSS x1 y1 x2 y2

This tells it to put a bar graph from coordinates x1 y1 to x2 y2, computing the data with the algorithm known as CB1.ADC.GAUSS. The equation for converting amps to gauss for that power supply might be

$$B_0 + B_1X + B_3X^3 + B_7X^7 + B_{11}X^{11} \quad (1)$$

where B0, B1, B3, B7 and B11 are constants derived from magnet mapping and X is the signal in normalized amps. This equation would be entered into the central database. Figure 2 shows what would be a typical excerpt from the database, with some names altered for clarity.

At initialization time, the applications read their own configuration files, which specify which signals are needed. The server is queried for the definition of each signal and the equations are parsed and stored in an internal form for processing. How the equations are parsed and stored is application dependent, but a standard parse/compute engine has been written and should work well for most applications.

Since configuration information is only read at initialization time, the overhead of querying the server is insignificant. The overhead of repeatedly computing each of the equations is not. However, given the ever increasing computational power of today's workstations, and some intelligent software design, this should not be a problem.

V. CONCLUSIONS

This system of describing the Ring data should make it easy for the user to do high level application development by hiding details behind logical naming schemes. Also, it will guarantee system-wide configuration data integrity by keeping it defined in one place, in the central database. The decrease in dependence on skilled computer personnel to maintain this information is well worth the added cost to the computer hardware required to support it.

Signal Name	Definition
LAC8	8
IOPORT	1
USHORT	2
B0	2.83236E-2
B1	7.498663
B3	-2.3996E-2
B7	1.00004E-2
B11	9.9411E-3
CQ1.AMPS.MAX	600.00
CQ1.ADC.RAW	LAC8 IOPORT 0x10 USHORT IN
CQ1.ADC.AMPS	CQ1.ADC.RAW CQ1.AMPS.MAX * 65535 /
X	CQ1.ADC.AMPS 150.0 /
CQ1.ADC.GAUSS	X 11 ^ B11 * X 7 ^ B7 * X 3 ^ B3 * X B1 * B0 + + + +

Fig. 2 - Sample database excerpt

VI. REFERENCES

[1] J. Flanz et al. "The MIT- Bates South Hall Ring", Proc. of the
1989 Particle Accelerator Conf., Chicago, Mar. 1989, pp.34

[2] T. Russ et al. " The Bates Pulse Stretcher Ring Control System
Design", Proc. of the 1989 Particle Accelerator Conf.,
Chicago, Mar. 1989, pp.85.

Grid Scans: A Transfer Map Diagnostic*

P. Emma
W. Spence
Stanford Linear Accelerator Center
Stanford, California 94305

ABSTRACT

A beam line transfer map diagnostic is described which uses induced betatron oscillations to search for focusing errors and geometric aberrations. A grid is produced graphically in *normalized phase space* coordinates with the beta match quantified from this grid. Application to the SLC electron damping Ring-To-Linac (RTL) transport line is presented.

I. DATA ACQUISITION

Two horizontal (or vertical) dipole corrector magnets are scanned in *nested loop* fashion to induce cosine-like and sine-like betatron oscillations in the RTL. Beam Position Monitors (BPM) of the RTL are sampled for each corrector setting and this data is saved to disk. BPM readings are averaged over 5 beam pulses to reduce noise levels. The corrector scan range is selected to produce oscillations which explore ~8 times the monochromatic RMS beam size.

II. ANALYSIS

The linear transformation used to obtain normalized phase space coordinates from a pair of BPMs is

$$\vec{u}_1 \equiv \begin{bmatrix} u_1 \\ v_1 \end{bmatrix} = \begin{bmatrix} 1/\sqrt{\varepsilon\beta_1} & 0 \\ \alpha_1/\sqrt{\varepsilon\beta_1} & \sqrt{\beta_1/\varepsilon} \end{bmatrix} \begin{bmatrix} 1 & 0 \\ R_{11}^{(1:2)} & R_{12}^{(1:2)} \end{bmatrix}^{-1} \begin{bmatrix} x_1 \\ x_2 \end{bmatrix}, \quad (1)$$

where $\mathbf{R}^{(1:2)}$ is the 2×2 transfer matrix from BPM-1 to BPM-2, α_1 and β_1 are design Twiss parameters at BPM-1, ε is the nominal RTL beam emittance (6800 μm-μrad), u_1, v_1 are the desired normalized phase space coordinates at BPM-1, and x_1, x_2 are the position readings from a BPM pair. These BPMs are separated by one to optimize phase advance (Fig 1). The mean of all trajectories is subtracted to produce difference trajectories.

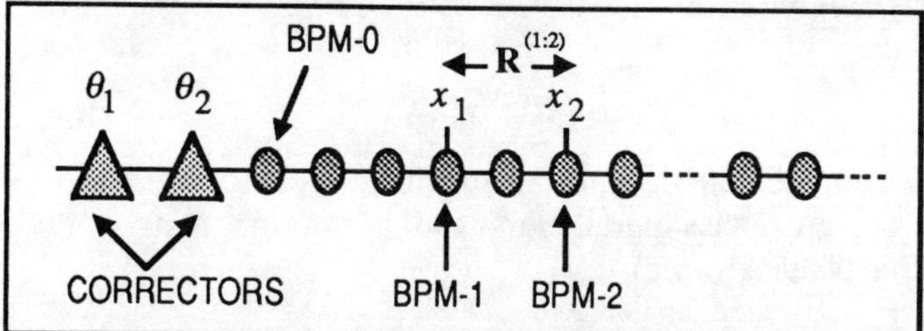

Fig 1. Corrector/BPM relationship (the first BPM in the system, BPM-0, and an arbitrary BPM pair is shown as an example - x_1, x_2)

Including the emittance in the normalization conveniently scales the u,v coordinates to units of nominal RMS beam size. This transformation is applied to each point per BPM pair and

v_1 is plotted versus u_1 to form a grid of points. The grid construction is repeated for each BPM pair in the beam line.

The transfer matrices are taken from the SLC Data Base after an on-line COMFORT [1] run. The Twiss parameters used are the propagation of some chosen initial Twiss parameters, α_0 and β_0 through the modelled transfer matrix elements.

$$\begin{bmatrix} \beta_1 \\ \alpha_1 \\ \gamma_1 \end{bmatrix} = \begin{bmatrix} R_{11}^2 & -2R_{11}R_{12} & R_{12}^2 \\ -R_{11}R_{21} & 1+2R_{12}R_{21} & -R_{12}R_{22} \\ R_{21}^2 & -2R_{21}R_{22} & R_{22}^2 \end{bmatrix} \begin{bmatrix} \beta_0 \\ \alpha_0 \\ \gamma_0 \end{bmatrix} \quad (2)$$

Here the $\mathbf{R}$ matrix transports from some chosen initial fixed point, BPM-0 in Fig 1, to the first BPM of the current pair of interest (BPM-1). The initial Twiss parameters are chosen so that the first BPM pair produces a square grid (see section III). The choice of correctors is then less restricted. It is only necessary that they sufficiently span the space ($\Delta\psi \neq n\pi$). A linear grid is fitted to the transformed data points using the known corrector kick angles θ_1 and θ_2 as a parametrization.

$$u_1(t) = a\theta_1(t) + b\theta_2(t)$$
$$v_1(t) = c\theta_1(t) + d\theta_2(t) \quad (3)$$

The fit coefficients (a, b, c, and d) are used only to evaluate a fitted u_1 and v_1 at each of the grid points, therefore the absolute scale of the kick angles is irrelevant. The transformed data points and the fitted points are then superimposed on the same plot. The fitted points are connected by lines to form the grid, and the transformed raw BPM data points are represented as dots (Fig 2).

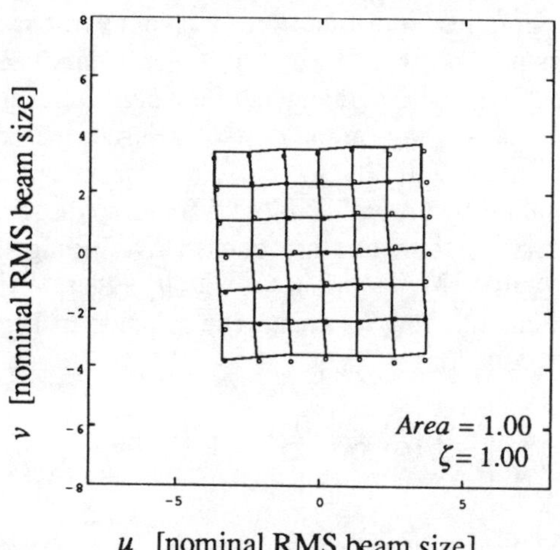

Fig 2. Normalized phase space grid for an RTL BPM pair. Points are BPM data to which the grid is fitted, and linearity is clear.

* Work supported by Department of Energy contract DE-AC03-76SF00515

The area of the grid is calculated (section III) and normalized to the area at BPM-0. With no acceleration or X-Y coupling, the case for this data, the area should remain constant and the grid will rotate but remain square as different BPM pairs are selected down the beam line. A focusing error in the line will linearly distort the grid from its initially square orientation with no area change (quantified in section III).

An example of a geometric aberration which is clearly distinguished is shown in Fig 3. The data is from a horizontal grid scan across the damping ring extraction septum.

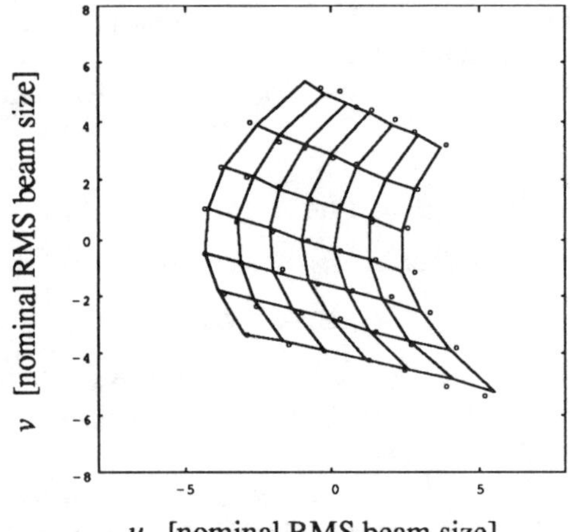

Fig 3. Horizontal grid scan across the damping ring extraction septum showing geometric aberration (fitted to second order).

III. CONCEPT

If we start with the initial transformation from x_1, x_2 to x_1, x_1' $(=dx/ds)$ already applied and begin in these phase space coordinates, then the normalized phase space transformation is straight forward, with the exception of the initial grid 'squaring' at BPM-0. Normalized phase space at BPM-0 is obtained as in (1) by the transformation

$$\vec{u}_0 = \left(\sqrt{\varepsilon} \mathbf{A}_0\right)^{-1} \vec{x}_0 \; ; \quad \mathbf{A}_0 \equiv \frac{1}{\sqrt{\beta_0}} \begin{bmatrix} \beta_0 & 0 \\ -\alpha_0 & 1 \end{bmatrix} \; ; \quad \vec{x}_0 \equiv \begin{bmatrix} x_0 \\ x_0' \end{bmatrix} . \quad (4)$$

For ease of graphic interpretation it is convenient to **choose** the Twiss parameters at BPM-0 in order to produce a square grid there. The Twiss parameters of the *beam* may also be used, however, depending on the correctors used, the grid will then not be square initially so that downstream optical distortions, when they occur, will be more difficult to discern.

The initial Twiss parameters are chosen by finding the transformation matrix $\mathbf{M}$, from x_0, x_0' which 'squares' the initial grid while maintaining its area. The left side of Fig 4 suggests the three equations

$$\mathbf{M}\vec{p}_0 = a \begin{bmatrix} 1 \\ 0 \end{bmatrix} \; ; \quad \mathbf{M}\vec{q}_0 = a \begin{bmatrix} 0 \\ 1 \end{bmatrix} \; ; \quad \det\!\left(\mathbf{M}\right) = 1 . \quad (5)$$

Here a is a scale factor adjusted so that the area is unchanged, and the vectors p_0 and q_0 (Fig 4) are extracted from the data at BPM-0. The transformation $\mathbf{M}$ is calculated as

$$\mathbf{M} = \sqrt{\left|\det\!\left(\mathbf{G}_0\right)\right|}\, \mathbf{G}_0^{-1} \; ; \quad \mathbf{G}_0 \equiv \begin{bmatrix} \vec{p}_0 & \vec{q}_0 \end{bmatrix} = \begin{bmatrix} p_1 & q_1 \\ p_2 & q_2 \end{bmatrix} . \quad (6)$$

The grid area S_0 is calculated as the cross product magnitude

$$S_0 = \left|\vec{p}_0 \times \vec{q}_0\right| = \left|\det\!\left(\mathbf{G}_0\right)\right| . \quad (7)$$

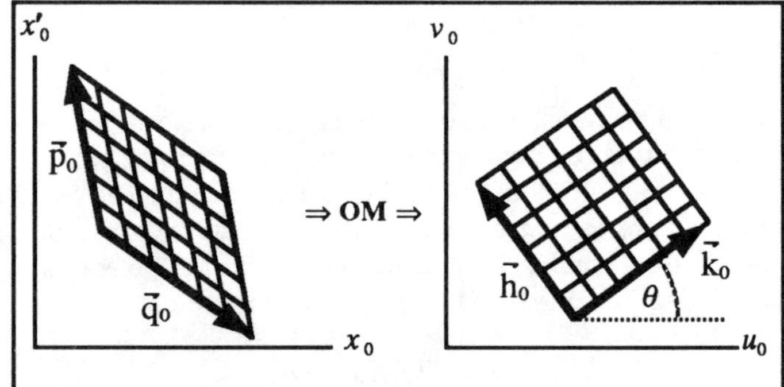

Fig 4. Grid *squaring* at the initial BPM (BPM-0)

Substituting $\mathbf{M}$ for $\mathbf{A}_0$ into (4) provides a square grid with unchanged area (except for the factor of ε) at BPM-0. However, $\mathbf{M}$ is not necessarily lower triangular and therefore identification of the elements of $\mathbf{M}$ as the chosen Twiss parameters is difficult. To convert $\mathbf{M}$ into a lower triangular matrix, we simply rotate it through the angle θ (Fig 4) with the orthogonal matrix $\mathbf{O}$ (equivalent to defining the initial betatron phase advance).

$$\mathbf{OM} = \begin{bmatrix} \cos\theta & \sin\theta \\ -\sin\theta & \cos\theta \end{bmatrix} \mathbf{M} = \begin{bmatrix} \beta_0^{-\frac{1}{2}} & 0 \\ \alpha_0 \beta_0^{-\frac{1}{2}} & \beta_0^{\frac{1}{2}} \end{bmatrix} \quad (8)$$

The angle θ is solved using the lower triangular condition $(\mathbf{OM})_{12} = 0$, and the initial Twiss parameters follow.

$$\tan\theta = -\frac{\mathbf{M}_{12}}{\mathbf{M}_{22}} \; ; \quad \beta_0 = \left(\mathbf{OM}\right)_{11}^{-2} \; ; \quad \alpha_0 = \frac{\left(\mathbf{OM}\right)_{21}}{\left(\mathbf{OM}\right)_{11}} \quad (9)$$

Propagating these Twiss parameters to the other BPMs then reduces to application of (2). The transformation to normalized phase space at a downstream BPM pair follows as in (4) with subscript 0 replaced by 1. Now express phase space at BPM-1 in terms of that at BPM-0 as

$$\vec{x}_1 = \mathbf{R}^{(0:1)} \vec{x}_0 \; , \quad (10)$$

and decompose the design transfer matrix $\mathbf{R}$ into initial and final design Twiss parameters and a rotation matrix Ψ (betatron phase advance).

$$\mathbf{R}^{(0:1)} = \mathbf{A}_1 \Psi_{01} \mathbf{A}_0^{-1} \quad (11)$$

The design Twiss matrices $\mathbf{A}_1$ and $\mathbf{A}_0$ are as defined in (4), and $\Delta\psi_{01}$, implicit in Ψ_{01}, is the design betatron phase advance from BPM-0 to BPM-1.

A focusing error between BPM-0 and BPM-1, will change the final Twiss parameters and the phase advance between these points. The effect of a focusing error on the transfer matrix is written as

$$\widehat{\mathbf{R}}^{(0:1)} = \widehat{\mathbf{A}}_1 \widehat{\Psi}_{01} \mathbf{A}_0^{-1} , \tag{12}$$

where the 'hatted' (perturbed) matrices are introduced to distinguish them from the design. Combining (4), (10), and (12)

$$\vec{u}_1 = \left(\sqrt{\varepsilon}\mathbf{A}_1\right)^{-1}\widehat{\mathbf{A}}_1 \widehat{\Psi}_{01}\mathbf{A}_0^{-1}\vec{x}_0 = \left(\mathbf{A}_1^{-1}\widehat{\mathbf{A}}_1\right)\widehat{\Psi}_{01}\vec{u}_0 . \tag{13}$$

From (13) it is clear that with no focusing errors ($\widehat{\mathbf{A}}_1 = \mathbf{A}_1$) the grid at BPM-1 is simply a clockwise rotation of the grid at BPM-0 through $\Delta\psi_{01} > 0$. Furthermore, since the determinants of all matrices in (13) are unity, the grid area is invariant and independent of the choice of initial Twiss parameters.

The quality of the beta match is indicated in the 'squareness' of the grid. Much like a matched beam remaining circular in normalized phase space, the grid should remain square for a perfect lattice. The 'squareness' of each grid is defined as the sum of the squares of the lengths of the grid spanning vectors h_1 and k_1 at BPM-1 (as is shown for BPM-0 in Fig 4) which are taken from the grid fit to the data. Writing the grid as a 2×2 matrix of these two vectors, the 'squareness' can then be conveniently written as the trace of $\mathbf{H}_1$ times itself transposed.

$$\mathbf{H}_1 \equiv \begin{bmatrix} \vec{h}_1 & \vec{k}_1 \end{bmatrix} , \tag{14}$$

$$\mathrm{tr}(\mathbf{H}_1\mathbf{H}_1^{\mathrm{T}}) = \left|\vec{h}_1\right|^2 + \left|\vec{k}_1\right|^2 \tag{15}$$

Since h_0, k_0 are simply two identically scaled orthonormal vectors rotated through θ, $\mathbf{H}_0$ can be defined as

$$\mathbf{H}_0 \equiv \begin{bmatrix} \vec{h}_0 & \vec{k}_0 \end{bmatrix} = \sqrt{\left|\det\!\left(\mathbf{G}_0\right)\right|} \; \mathbf{O} . \tag{16}$$

Transformation from $\mathbf{H}_0$ to $\mathbf{H}_1$ is the same as transformation from u_0 to u_1 in (13)

$$\mathbf{H}_1 = \left(\mathbf{A}_1^{-1}\widehat{\mathbf{A}}_1\right)\widehat{\Psi}_{01}\mathbf{H}_0 \equiv \mathbf{F}_1\mathbf{H}_0 , \tag{17}$$

where the definition of $\mathbf{F}_1$ is here implicit. Rewriting (15) using (16) and (17) gives the 'squareness' in terms of $\mathbf{F}_1$.

$$\mathrm{tr}\!\left(\mathbf{H}_1\mathbf{H}_1^{\mathrm{T}}\right) = \mathrm{tr}\!\left(\mathbf{F}_1\mathbf{H}_0\mathbf{H}_0^{\mathrm{T}}\mathbf{F}_1^{\mathrm{T}}\right) = \left|\det\!\left(\mathbf{G}_0\right)\right|\mathrm{tr}\!\left(\mathbf{F}_1\mathbf{F}_1^{\mathrm{T}}\right) \tag{18}$$

Substituting $\mathbf{A}_1$ and its perturbed counterpart into the definition of $\mathbf{F}_1$ produces a beta match parameter, ζ, in terms of the Twiss parameters at BPM-1 of the current pair, which is invariant until a second focusing error is encountered.

$$\zeta \equiv \frac{1}{2}\mathrm{tr}\!\left(\mathbf{F}_1\mathbf{F}_1^{\mathrm{T}}\right) = \frac{1}{2}\mathrm{tr}\!\left\{\left(\mathbf{A}_1^{-1}\widehat{\mathbf{A}}_1\right)\widehat{\Psi}_{01}\widehat{\Psi}_{01}^{\mathrm{T}}\left(\mathbf{A}_1^{-1}\widehat{\mathbf{A}}_1\right)^{\mathrm{T}}\right\} =$$

$$\frac{1}{2}\mathrm{tr}\!\left\{\left(\mathbf{A}_1\mathbf{A}_1^{\mathrm{T}}\right)^{-1}\left(\widehat{\mathbf{A}}_1\widehat{\mathbf{A}}_1^{\mathrm{T}}\right)\right\} = \frac{1}{2}\mathrm{tr}\!\left(\sigma_1^{-1}\widehat{\sigma}_1\right) =$$

$$\frac{1}{2}\left\{\frac{\beta_1}{\widehat{\beta}_1} + \frac{\widehat{\beta}_1}{\beta_1} + \beta_1\widehat{\beta}_1\left(\frac{\alpha_1}{\beta_1} - \frac{\widehat{\alpha}_1}{\widehat{\beta}_1}\right)^2\right\} \tag{19}$$

The $1/2$ is introduced so the matched case will result in $\zeta = 1$, and the beam matrix, σ, is introduced, which follows from

$$\mathbf{A}\mathbf{A}^{\mathrm{T}} = \begin{bmatrix} \beta & -\alpha \\ -\alpha & \gamma \end{bmatrix} \equiv \frac{\sigma}{\varepsilon} . \tag{20}$$

The 'hatted' quantities in (19) represent the perturbed beam parameters, while those without the 'hats' represent the design. From (15), (18), and (19), the beta beat amplitude, ζ, at BPM-1 of each BPM pair (indicated at the bottom of the grid in Fig 2), is a direct result of the fitted grid vectors h_1 and k_1.

$$\zeta = \frac{1}{2}\frac{\left|\vec{h}_1\right|^2 + \left|\vec{k}_1\right|^2}{\left|\det\!\left(\mathbf{G}_0\right)\right|} \tag{21}$$

A short proof of the invariance of ζ is shown by calculating ζ_i at a point farther downstream, if we transport both the design beam, σ, and the perturbed beam, $\widehat{\sigma}$, to point-i through the same design $\mathbf{R}$ and use common matrix properties.

$$\zeta_i = \frac{1}{2}\mathrm{tr}\!\left\{\left(\mathbf{R}\sigma\mathbf{R}^{\mathrm{T}}\right)^{-1}\left(\mathbf{R}\widehat{\sigma}\mathbf{R}^{\mathrm{T}}\right)\right\} =$$

$$\frac{1}{2}\mathrm{tr}\!\left\{\left(\mathbf{R}^{-1}\mathbf{R}\right)^{\mathrm{T}}\sigma^{-1}\left(\mathbf{R}^{-1}\mathbf{R}\right)\widehat{\sigma}\right\} = \frac{1}{2}\mathrm{tr}\!\left\{\sigma^{-1}\widehat{\sigma}\right\} = \zeta \tag{22}$$

As a final point, note that although the choice of initial Twiss parameters at BPM-0 does not affect the invariance of ζ or of grid area, this choice does however affect the step size of ζ when a focusing error is encountered. Therefore, the Twiss parameters of the beam itself should be used at BPM-0 (rather than the grid squaring parameters used in Fig 2) if a directly qualitative beta matching study is of prime interest.

IV. CONCLUSIONS

The grid technique has advantages over simply comparing a measured oscillation with a model. The oscillation comparison does not assume an area preserving map and, for example, may confuse a BPM calibration error with a focusing error. Furthermore, since the grid scan does not force linearity, the technique will reveal nonlinearities over the scan range and therefore is useful for qualitatively evaluating geometric effects.

Application to the SLC RTL beam line shows only some small focusing errors along the RTL with no measurable geometric aberrations downstream of the extraction septum. This scheme helped eliminate one potential source for a suspected RMS emittance growth within the RTL itself.

V. ACKNOWLEDGEMENTS

We thank Max Cornacchia for suggesting grid scanning as a diagnostic tool, and Thomas Lohse for some initial analysis.

VI. REFERENCES

[1] M.D. Woodley, et.al., "Control of Machine Functions or Transport Systems", SLAC-PUB-3086, March 1983

Fiber Optic Communications Links for the Main Ring Control System Upgrade

Robert J. Ducar
Fermi National Accelerator Laboratory *
P.O. Box 500
Batavia, IL 60510

Abstract

Fiber optic communication links have been implemented throughout the 6.5 km circumference of the Fermilab accelerator facility in support of the upgrade of the Main Ring Control System. The choice of optical fiber as a communication medium was based on cost and limited plant capacity for additional conventional copper based medium, rather than by the inherent advantages of high bandwidth and low loss. Details of component selection and performance of the custom designed repeater facilities will be discussed.

I. RATIONALE FOR FIBER OPTIC MEDIUM

As plans for replacement of the the Fermilab Main Accelerator control system began to materialize in early 1988, it was clear that additional links would be necessary to support the planned upgrade. The original coaxial cabling supporting Main Accelerator controls was routed through the accelerator enclosure and surfaced at each of the thirty service buildings of the 6.5 km ring. Periodic measurements had revealed a gradual degradation of the cable's characteristic impedance to the point where it was not reasonable to expect support of the new links which were to be an order of magnitude higher in bit rate. Five ring-wide cable paths were deemed necessary for the new control system. Some excess capacity remained on the installed nineteen conductor Heliax cable already in use for the Tevatron control system. However, these five circuits were available only in two-thirds of the ring circumference. Recapture of needed capacity from the nineteen conductor could have been made possible by installation of additional coaxial cable and by consolidation of several remote consoles.

A detailed analysis was performed to compare costs associated with implementation of conventional copper/coaxial links versus fiber optic links. The cost of a fiber optic installation was projected to be $168K, which compared quite favorably to the $187K cost of conventional links. A decision was made to install a twenty-four fiber trunk cable and to proceed with necessary engineering work to implement the desired 10 Mbit/sec links.

II. FIBER OPTIC CABLE INSTALLATION

A multimode graded index fiber (62.5/125 micron) was selected for the installation. Attenuation properties were not a significant issue since most runs were calculated to be approximately 300 meters in length. Bandwidth was also judged sufficient for anticipated applications. Multimode fiber was also considered to be easier to splice and better suited to the economical optical emitters and receivers available from component manufacturers.

Nearly nine kilometers of trunk cable was necessary for the installation. The installation involved thirty-one distinct pulls, ranging in length from 150 to almost 600 meters. Most of the pulls were routed in recently installed communication ducts buried below the Main Accelerator access road. This routing handily avoided exposure of the fiber optic cable to any radiation from the operation of the accelerator. Innerduct was utilized in these communication ducts. Innerduct was used sparingly once the cable was inside the accelerator service building, typically at only those points of high congestion or multiple bends. Cable ends were always routed directly to a rack mounted splice enclosure.

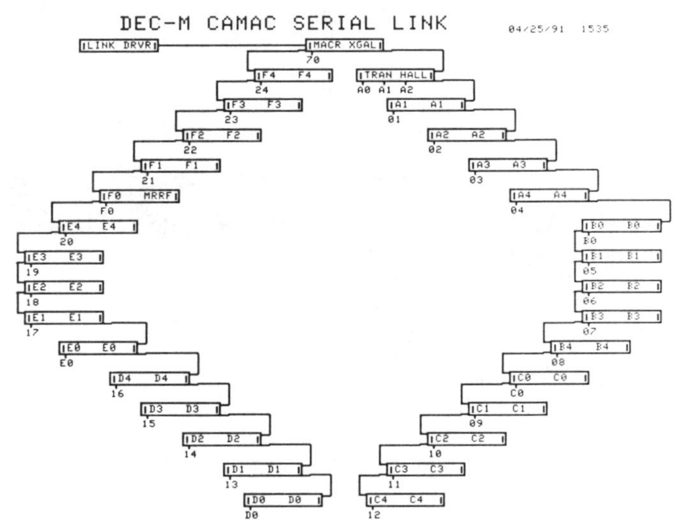

Figure 1. Diagram of Main Ring communications links for the installed CAMAC system. Boxes represent repeaters at service buildings. Numbers outside of the repeater boxes represent the addresses of connected CAMAC crates.

The selected cable contains four loose tube buffers, each holding six optical fibers. Connections to each of the five new links were required at each cable terminus. At the time of installation, no plans existed for use of the remaining nineteen fibers. Seven of these fibers were classed as building to building links and were fusion spliced to ST-connectorized pigtails at each terminus, as were the five necessary link fibers. Six of the remaining fibers were classed as longer sector to sector circuits. These were fusion spliced to pigtails at the sector zero buildings and series fusion spliced at numbered service buildings. The last six remaining fibers were left unspliced at each terminus. The splice enclosures housed six or seven splice trays, each tray accommodating six pigtail

---

*Operated by the Universities Research Association, Inc. under contract with the U.S. Department of Energy.

splices, or six series splices. Two splice trays were installed for the buffer tubes containing unterminated fibers. Pre-terminated and tested ten meter ST jumper cables were purchased and subsequently were cut and prepared as two separate pigtails.

Trunk cable installation and splicing was performed under contract with non-Fermilab personnel. Almost one thousand fusion splices were necessary to complete the installation. Nearly four hundred terminated fiber paths were certified with OPM and OTDR equipment at 850 and 1300 nm wavelengths.

III. FIBER OPTIC REPEATER DEVELOPMENT

The Main Ring Control System Upgrade [1] called for replacement of original (circa 1972) interface hardware with CAMAC crates and modules. Though some new designs were developed and implemented, much of the necessary hardware was already available. The Tevatron Serial Crate Controller [2] was re-engineered to a two-wide module. The Main Ring Power Supply Link was completely redesigned with new transmitter and receiver hardware. All necessary links for the Upgrade were based on the 10 MBit/sec serial protocol already in wide use in the Fermilab accelerator control system. Extant links are based on frequency detection of a 50 MHz carrier over coaxial cables, with the repeaters for this system being housed in 5 1/4" NIM bins [3].

Given the significant time of flight delays in the link cabling system, either coaxial or fiber, there was no significant motivation to develop faster links. There was, however, incentive to rework the nature of the repeater unit itself. The older "rf" repeaters were expensive and time consuming to build as well as difficult to adjust and troubleshoot in place. Local inputs or outputs of individual "rf" repeaters were limited to four. Branch points in implemented links required two slots of NIM space. These disadvantages were overcome with the new repeater design.

Figure 2. Fiber optic repeater chassis.

A repeater system was developed as a single chassis with individual printed circuit boards, rather than modules, serving as link repeaters. Fiber optic connections are made at the front of individual repeaters and are adequately protected behind a swing-down front panel. Local coaxial connections to the repeaters are accommodated at the back of the repeater chassis. Lemo style connectors are mounted directly at the rear of the repeater board and protrude through the repeater chassis back panel when the repeater is inserted. When inserted, the repeater board also connects to a small mother board that serves to

distribute five volt dc power sourced from a power supply board assembly at the right end of the repeater chassis. Each repeater chassis can support up to sixteen individual repeater boards. Fiber optic pigtail cables from the adjacent splice center enclosure enter the chassis from the rear and are channeled at the left end of the repeater chassis to individual repeaters.

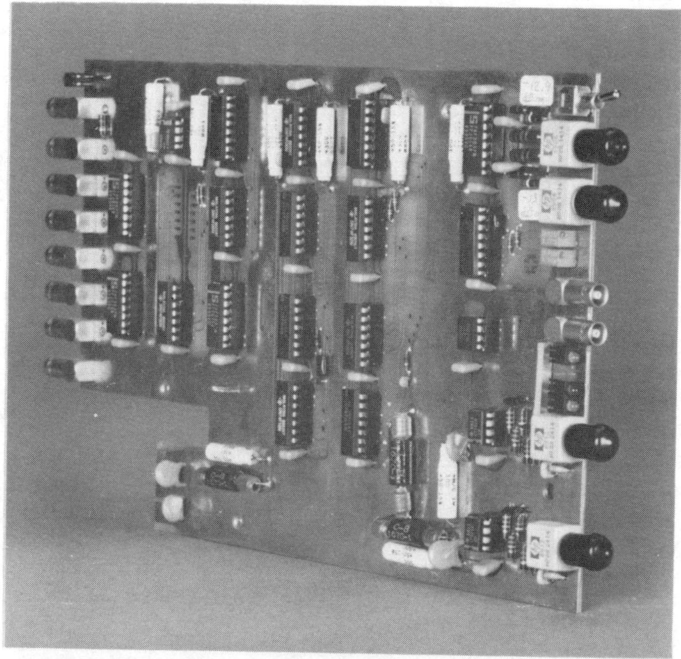

Figure 3. Typical fiber optic repeater. Pairs of optic transmitters and receivers are mounted at the front top and bottom of the repeater board. Local TTL 50 ohm connections are made at the rear of the repeater by means of Lemo style connectors.

The repeater board design is common to fan-in and fan-out configurations, determination made by selective component loading. Provision is made for two optic transmitters and for two optic receivers to facilitate link branch requirements at a single repeater station. Local coaxial connections at the rear of the module provide for six fan-in or fan-out connections as well as separate direct in and direct out connections. Separate LEDs at the back panel indicate the type of repeater as being fan-in or fan-out. Timing adjustments and diagnostic LEDs and signals are facilitated at the front of the repeater. A separate repeater board was designed to support the Main Ring abort link.

The design of the repeater circuit utilized low cost HP optic components (HFBR-1414/2414) connectored for ST. The LED transmitter was more than adequately driven by a single 74F3037 gate. The receiver circuit was developed from HP application notes [4] and employed the LT1016 comparator. The repeater processes 20 MBd data as 50 and 100 nanosecond light pulses that arrive in data frames of two to four microseconds in duration. The repeater's function is to pass leading edges of incoming signals with as little induced time skew as possible. Repeated pulse widths are adjusted locally at each repeater. This approach, also employed by the "rf" repeaters, avoids the necessity of reclocking data, but can present difficulties as minor time skews at an individual repeater accumulate after the signal is processed by multiple repeaters. Given the relatively short distances between service buildings, receiver sensitivity was really not an issue in the design.

Commissioning of the repeater system surfaced several problems with the repeater board design. Leading edges at certain positions in the data stream were being skewed by more than three nanoseconds per repeater. After four or five repeats, the transmissions were beyond the range of acceptance for local hardware. This problem was traced to a RC discharge problem at the input of the LT1016 comparator and was largely solved by reducing the coupling capacitor value from 39 pf to 12 pf. That action reduced sensitivity at the receiver. Because the skewing problem was aggravated at lower levels of input optical power, compensation was made at the transmitter by increasing driver current to near maximum. This lead to a second problem of overdriving the receiver. In analyzing this difficulty, measurements were taken of the output power of the transmitter at maximum drive. Output power of numerous individual transmitters was measured at -10 dBm, the maximum stated on the data sheet, rather than the expected -12 dBm typical specification. Reduction of transmitter drive current from 60 ma to 40 ma solved the overdrive problem. Subsequent measurements of the installed repeater system showed leading edge skew to be less than one nanosecond per repeater. Link signals repeated approximately fifteen times arrived at the far end of the repeater system with sufficient quality to operate error free.

IV. CONCLUSIONS

With more than eighteen months of operational experience, the new fiber optic link system has performed remarkably well. After the gross problems of leading edge skewing were solved, transmission errors are at such a low level that the links are considered error free. There have been no failures or problems with the installed plant cable, nor with any components of the installed repeater system.

Expenditures for the repeater system have been very reasonable. The cost of a powered repeater chassis is less than $500. Individual repeater modules cost approximately $170. Total repeater parts costs for 31 stations and some 150 individual repeater cards totalled less than $40k.

Predictably, spare fibers have been used for other purposes. Two fiber circuits are currently supporting 802.5 Token Ring around the accelerator, using commercially available optic hardware. Other fibers support video and Ethernet facilities. Several fiber circuits were used to bypass coaxial "rf" repeater circuits that were temporarily disconnected during a construction period. The common transmission protocols made the bypass very easy to implement.

V. ACKNOWLEDGEMENTS

The author wishes to thank Mr. Terry Hendricks for his valuable assistance and untiring efforts in the development and installation of the fiber optic communication links. The author also wishes to thank Mssrs. Rupert Crouch, David Jarvis, and Nicholas Visser for their valuable skills, persistence, and patience in bringing the new links to reality.

VI. REFERENCES

[1] K. Seino et. al., "New Main Ring Control System," *Proceedings of the International Conference on Accelerator and Large Experimental Physics Control Systems*, Vancouver, BC, Canada, October 1989, pp. 267-270.

[2] R. Ducar, "A CAMAC Serial Crate Controller for the Tevatron Accelerator," *IEEE Transactions on Nuclear Science*, vol. NS-28, No. 3, June 1981, pp. 2303-2304.

[3] R. Ducar, "Tevatron Serial Data Repeater System," *IEEE Transactions on Nuclear Science*, vol. NS-28, No. 3, June 1981, pp. 2301-2302.

[4] Hewlett-Packard Application Bulletin 73, "Low-Cost Fiber Optic Transmitter and Receiver Interface Circuits," 5954-8415(6/87), Figure 12.

The Improvement of TRISTAN Timing System

J.Urakawa and T.Kawamoto
KEK, National Laboratory for High Energy Physics
Oho 1-1, Tsukuba-shi, Ibaraki-ken 305, Japan

Abstract

The TRISTAN timing system is divided into fast and slow timing systems. In the fast timing system, the 50Ω coaxial cables for the timing signal transmission were replaced with special single-mode optical fiber cables. New fast timing signal transmission system achieved the timing accuracy within $11psec$ over the temperature range from 10 to $35°C$. In the slow timing system, a dedicated microprocessor(LSI-11) for the generation of slow timing signals was ejected and the software was rewritten and installed into one minicomputer in the computer network for the TRISTAN accelerator control. This modification gives more reliability and more flexibility. In this paper we describe the improvement of the TRISTAN timing system.

I. INTRODUCTION

The TRISTAN accelerator complex comprises four accelerators: a 200MeV electron linac for positron production, a 2.5GeV linac, an 8GeV accumulation ring(AR) and a 30GeV main ring(MR). The AR and MR are controlled by a highly computerized control system[1]. Twenty-five minicomputers are linked by optical-fiber cables to form an N-to-N token-ring network. The transmission speed through the cables is 10Mbits/sec. From each minicomputer, a CAMAC serial highway extends to the controlled equipment. The TRISTAN timing system is divided into fast and slow timing systems[2]. The fast timing system supplies timing signals(fast timing) for devices whose operation is synchronized with bunched beams from either the linac or the AR. These signals are also used in various beam monitors and beam feedback systems. The slow timing system generates trigger signals(slow timing) in order to achieve synchronization between the magnetic field and the rf accelerating voltage of the AR or MR. These triggers are also used for the automatic operation of machines. The slow timing system manages the operation mode of the AR and MR with both flexibility and extensibility.

For the precise timing signal transmission, a new optical fiber cable system was developed and installed between the 2.5GeV linac gun room and the TRISTAN control room, and their performance in transmitting the AR RF signal(508MHz) has been evaluated[3].

Triggers for changing the magnetic field are sent out by the slow timing system, which receives pulse signals which are transmitted at the end of the magnetic field change through a twisted-pair cable. The slow timing system distributes the trigger signals from the main control room to all subcontrol stations using MIL/STD-1553B with an 8-bit event code.

The slow timing signals are required for managing synchronization among the magnet system and the rf system as well as for automatic operation of the machines.

Each timing system was controlled through the CAMAC by minicomputers (HIDIC 80E or 80M) and a dedicated microprocessor (LSI-11). However, in order to simplify the software management and to increase the reliability and the flexibility, the LSI-11 was deserted.

In next section we describe the improvement of the fast timing transmission system with minimum transmission delay drift, and E/O, O/E converters. The improvement of the slow timing system is explained in section 3.

II. NEW OPTICAL FIBER TRANSMISSION SYSTEM

A. New Optical Fiber Cable

Optical fiber transmission system has advantages of low attenuation (0.35dB/km), wide bandwidth(more than 1GHz over dozens of kilometer) and immunity to electromagnetic interference. This fiber cable showed the reduced thermal transmission delay change less than 10psec/km in the temperature range from -20 to 30 $°C$, which is 100 times smaller than that of any other existing coaxial cables and conventional optical fiber cables.

B. E/O,O/E Converters

The E/O converter consists of a wide bandwidth laser diode module, a driver circuit, and an auto-power control circuit in order to stabilize the emitted power. The timing signal is fed to the driver circuit with 50Ω impedance and equalized to achieve the flat frequency response. The O/E converter consists of a PIN-AMP module and a commercially available low noise amplifier. The O/E and E/O converters achieved the timing accuracy within 11psec over the temperature range from 10 to 35 $°C$[4]. Another investigation about the timing jitter width clarified that the origin of the phase jitter is derived from the thermal noise in the optical receiver circuit. Hence, an appropriate bandpass filter should be used to improve the signal to noise ratio.

C. Improvement of Fast Timing Transmission System

The new optical fiber cable was installed from the linac gun room to the main control room of the TRISTAN AR in April 1989. The cable has 800m length and contains 6 fibers in it. 300 meter of the total cable was layed even in the underground from beam switching yard to the AR where the cable was subjected to the irradiation. Since the core material of this fiber is pure silica, this fiber cable is much resistant to irradiation than conventional fiber cables.

One of the six fibers in the cable was connected to the E/O and O/E converters at both ends to replace the existing coaxial system of transmitting 508MHz RF timing signal. This coaxial cable, although it had been designed to have a thermally stabilized electrical length, required phase adjusting several times a year. On the contrary, the new optical fiber cable system has been operating without any phase delay adjusting to April 1991. The characteristics of this system has not been degraded during the last two years. The superiority of the new system is demonstrated.

The confirmed high stability suggests that the use of this cable system will effectively simplify and improve the "main drive lines" in the acceleration systems, where large diameter coaxial cables are used in the special conduit with sophisticated temperature control.

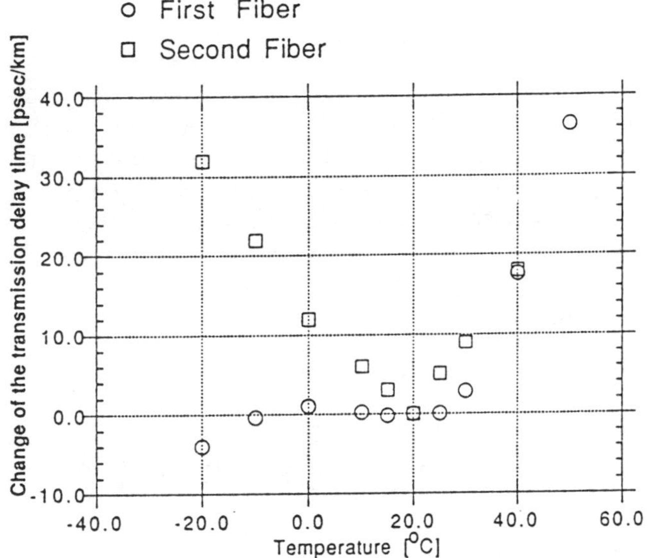

Fig.1 Measured transmission delay time change against temperature.

Since precise bunch oscillation and beam profile monitors require the precise synchronization signals for bunched beam, the synchronization signals will be transmitted to 4 site buildings through the optical fiber cables from the main control room in this summer. Figure 1 shows the temperature dependence of the transmission delay time of the fibers which was used before and will be used. The difference is due to the variations in the process of manufacture. The thermal coefficient for delay of less than $0.5\text{ppm}/^{\circ}C$ is guaranteed.

III. IMPROVEMENT OF THE SLOW TIMING SYSTEM

This system supplies the timing for devices which are operated in synchronization with the magnetic field of the AR or MR. Since there are many kinds of operation modes for TRISTAN, it must be easy to change the timing for the operation mode and to extend the system. It is also required that the timing signals be transmitted to many points distributed over the large area of the accelerator sites. The generation, transmission and receiving of timing signals are carried out through the CAMAC system by means of a serial bus using Manchester II code, based on the standard MIL/STD-1553B.

Since the Manchester II code (biphase level code) is transmitted as a bipolar signal, low noise susceptibility can be realized by using transformer-coupled transmission lines, which allow half-duplex transmission and transmission of information from numerous sources through one cable. The non-word format of the MIL/STD-1553B consists of a twenty-bit field for data or command and one bit for parity. The discrimination of the synchronization signal polarity distinguishes two kinds of word formats: command words and data words. These words are utilized to transmit both clock and event signals, respectively.

There are three kinds of slow timing signals for the TRISTAN control:

(1) a 100Hz clock signal;

(2) 255 event signals coded by eight bits with a dummy identification bit, meaning that the event is not in operation;

(3) time code signals of year, month, day, hour, minute and second.

These signals are converted to the Manchester II code and transmitted to CAMAC timing modules on serial multidrop buses through coaxial cables and repeaters. The command synchronization of the 1553B bus code is used as the clock signal, and the time-code data are divided into three command words. The event signal is attached to the data word. The next clock following an event code gives its exact time.

As old slow timing system which was controlled by the dedicated microprocessor (LSI-11) was described in ref.[2], we explain the reason which the LSI-11 was deserted and new slow timing system. In the old timing system, we must manage two software system and maintain the communication system between the HIDIC 80E and the LSI-11. We think that the unification of the software system and the simplification of the hardware system are important. Though the response speed for the event request is degraded, one of the new timing system is enough for the accelerator operation. We decided to eject the LSI-11 for reliability, flexibility and maintainability. Figure 2 shows the relationships between the software modules and the hardware modules which they control in new slow timing system. In the main control room there is a CAMAC crate with the timing master generator module for generating and transmitting the slow timing signal.

The following programs have been developed.

(1). A Management program to produce slow timing signals runs on the HIDIC 80E (GP0). This program (TRLSI) has the data-table for timing event codes.

(2). A program which is activated by a LAM directly produced from several hardwares requests slow timing signals to TRLSI. This program (EVSIL) runs on the GP0.

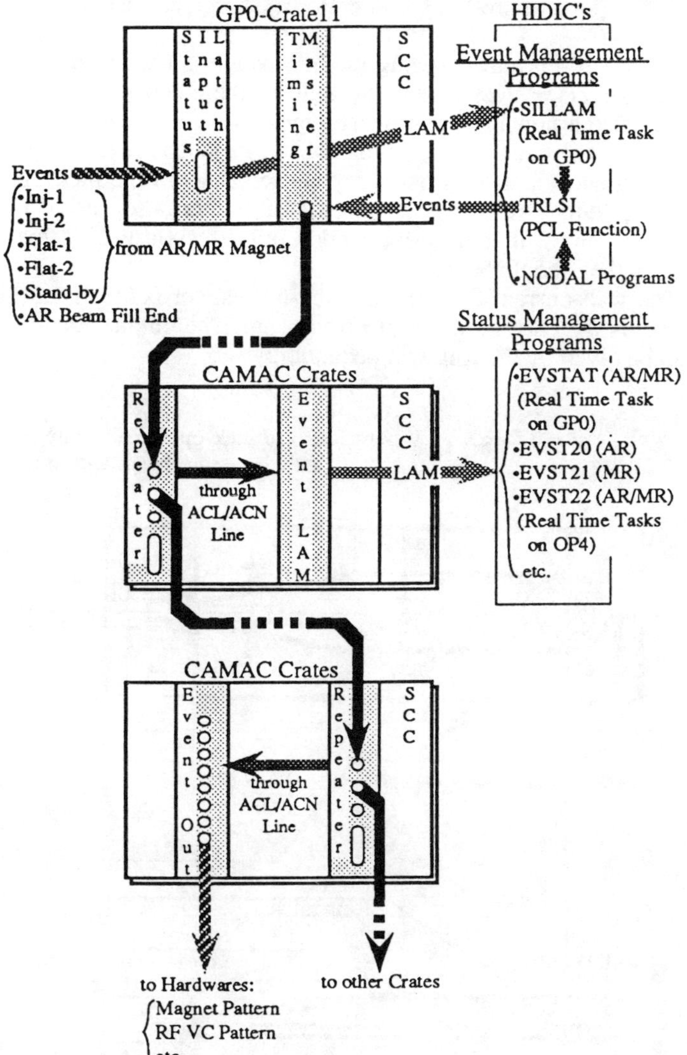

Fig.2 The relationships between the software modules and the hardware modules which they control in new slow timing system.

(3). Programs which are activated by LAM produced by timing modules using a code of slow timing signals run on several HIDIC minicomputers to display the timing status and to activate other programs.(EVSTAT, EVST20, etc.)

(4). A program which manages the bunched beam trans-

fer runs on the GP0. (EVBMC)

(5). Program for setting data into the timing modules run on HIDIC 80's.

(6). Many function programs for the timing system.

(7). A cheking program for the slow timing system.

Program(1) attaches the event signal to the slow timing signal through the timing master module by a CAMAC command.

The slow timing system manages the operation mode of the accelerators. The HIDIC-80E (GP0, OP4) can monitor the operation mode through the operation-mode events of the slow timing signal, using the event LAM module. If one wants to activate a program following the operation mode, the program is queued by a timing event registered in the GP0. This program is either initiated after the operation mode, or activated by the event management program on the HIDIC-80.

IV. SUMMARY

A new optical fiber cable with highly stabilized transmission delay time and improved resistivity against irradiation was installed for the transmission of 508MHz timing signal between the linac gun room and the TRISTAN AR. The new optical system eliminated the frequent phase adjusting in the former coaxial transmission system required from the temperature change in a year. The use of this phase stabilized optical fiber system will be effective in the precise transmission of timing signals in the acceleration systems.

The new slow timing system has managed the operation since January 1990. It gives more reliability and more flexibility. In order to increase the response speed for the event request, we are preparing for the replace of the HIDIC-80E to a 32-bit computer.

V. ACKNOWLEDGEMENTS

The authors would like to thank Professors Y.Kimura, K.Takata, Shigeru Takeda and Seishi Takeda for their support during the work. They also wish to thank Mr. Y.Sato of KEK-PF, Dr. S.Tanaka and Mr. Y.Murakami of SUMITOMO Electric Industries for the collaboration in the construction and the test of the optical fiber transmission system.

VI. REFERENCES

[1] S.Kurokawa et al., Nucl. Instr. and Meth. A247 (1986) 29.

[2] J.Urakawa et al., Nucl. Instr. and Meth. A293 (1990) 198.

[3] J.Urakawa, Particle Accelerators, Vol.29 (1990) 251.

[4] S.Tanaka et al., KEK Report 90-5 (1990)

Bunch Monitor for an S-Band Electron Linear Accelerator

Yuji Otake and Kazuo Nakahara

KEK, National Laboratory for High Energy Physics
1-1, Oho, Tukuba-shi, Ibaraki-ken, 305, Japan

Abstract

The measurement of bunch characteristics in an S-band electron linear accelerator is required in order to evaluate the quality of accelerated electron beams. A new-type bunch monitor has been developed which combines micro-stripline technology with an air insulator and wall-current monitoring technology. The obtained time resolution of the monitor was more than 150 ps. This result shows that the monitor can handle the bunch number of an S-band linac. The structure of the monitor is suitable for being installed in the vacuum area, since it is constructed of only metal and ceramic parts. I can therefore easily be employed in an actual machine.

1. Introduction

An advanced electron linear accelerator, such as a linear collider or a linac for a free-electron laser, is usually required to produce good quality electron beams, thus requiring both a low-energy spread and a low-emittance. For this reason, the measurement quality of each electron bunch is very important. The energy spread of the accelerated electron beam by a linac depends directly on its own bunch length. Bunch length measurements are necessary for the above-mentioned reasons.

A part of KEK, National Laboratory for High Energy Physics, includes the Photon Factory 2.5-GeV linac (PF Linac) which uses the S-band RF; it also has future plans, such as the JLC (Japan Linear Collider) which will use the S-band linac for both pre-acceleration and positron generation.

The bunch length of the S-band linac at the Photon Factory is less than 10ps. If we want to evaluate the electron bunch quality of the above-mentioned accelerators, the time resolution of the bunch monitor must be more than 10ps; the time resolutions of the beam monitors which already exist, however, are not sufficient for bunch length measurements. From above-mentioned reasons, a bunch monitor must be developed.

The development[1] of a bunch monitor was started using an epoxy-glass material (usually called G-ten) for the monitor's base material; a pre-experiment was carried out in the atmosphere by extracting electron beams from the positron generator of the PF Linac. The bunch monitor using G-ten achieved more than a 300ps pulse response. After obtaining reasonable results in the pre-experiment, Designing a new bunch monitor to be installed in vacuum was started.

2. Coupling Mechanism of Bunch monitor[2]

The electromagnetic field of a relativistic electron beam reduces the longitudinal electromagnetic field in proportion to $1/\gamma$. The transverse electromagnetic field of an electron beam projects its own bunch length in the extremely relativistic case, where the transverse electromagnetic field at a radial position r from the beam center can be expressed as

$$E_T = \frac{1}{2\pi\varepsilon_0} \frac{\lambda}{r} ,$$

where λ represents Beam Charge/unit length.

If a wall-current coupling method which is used to detect the transverse electromagnetic field of the electron beam is applied to a bunch-length measurement, we can theoretically measure the bunch length of an S-band linac. However, extra distributed inductances and stray capacitances in the pick-up resistor and transmission circuit of the monitor usually prevent circuit matching for a good pulse response. For this reason they should be reduced.

The wall current through the pick-up resistor is ideally the same as the current of the electron beam. The output voltage can be given by the following equation:

$$V = I * R, \qquad \text{where R is the pick-up resistor value,}$$
$$\text{and I is the beam Current.}$$

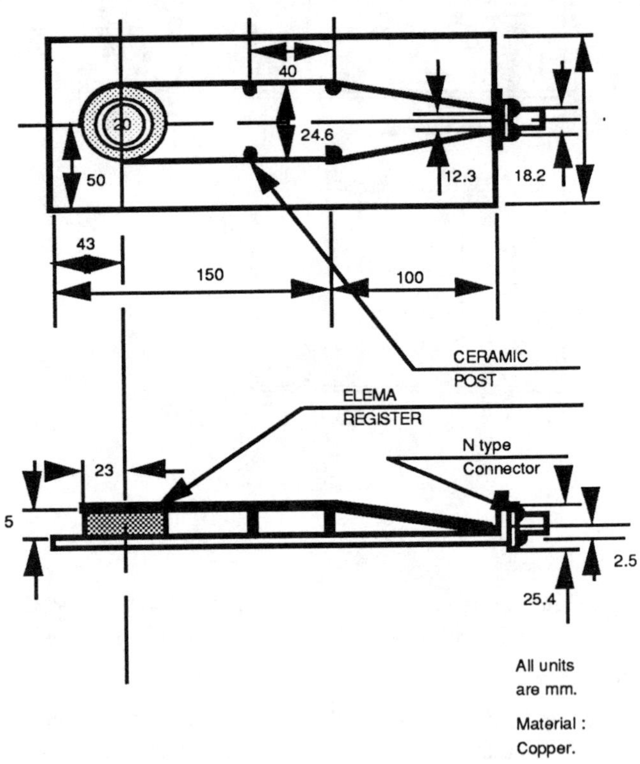

Fig. 1 Stripline-Type Bunch Monitor
with an Air Insulator

3. Design and Evaluation of a Bunch monitor

A bunch monitor using stripline technology was designed, as shown in Fig.1; the matched transmission line into a characteristic impedance of 50 ohm is necessary in order to smoothly transmit the beam signal. We previously made a bunch monitor without any taper transformer in order to omit the complicated part; then, the frequency response of the monitor (looking at the output connector) was good and the main reflection component was from the resistor.

The characteristic impedance of the stripline can be calculated by [3]

$$Z_0 = \frac{120\pi}{\sqrt{\varepsilon_{\text{eff}}}}\left[\frac{w}{h} + 1.393 + 0.667 \ln\left(\frac{w}{h} + 1.444\right)\right]^{-1}$$

and
$$\varepsilon_{\text{eff}} = \frac{\varepsilon_r + 1}{2} + \frac{\varepsilon_r - 1}{2}\left(1 + 12\frac{w}{h}\right)^{\frac{1}{2}} \quad \text{for } w/h > 1,$$

where w is the width of the stripline, and h is the height of the stripline from the ground plane. This equation shows that the w/h ratio is important in determining the characteristic impedance.

The monitor requires a taper transformer which gradually decreases in height while maintaining the characteristic impedance for obtaining a good matching between the stripline and the output connector. The taper electrically rejects the big dimensional difference between the stripline and the output connector.

The value of the pick-up resistor is determined according to the following procedure. The evaluation circuit[2] of the pick-up part is shown in Fig. 2, and the time constant of the bunch monitor can be expressed by

$$\tau = C * R, \qquad$$
where R is the pick-up resistor,
C is the stray capacitance ,
and L is the distributed inductances.

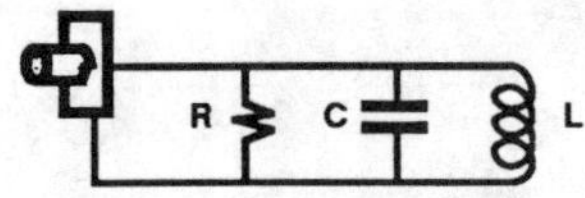

Fig. 2 Evaluation Circuit for Bunch Monitor

At first, a 50-ohm resistor was chosen, since the matching of the stripline can be easily evaluated in the opposite direction from the output connector. Furthermore if the transmission circuit of the bunch monitor has reflection at any point, the reflection signal is absorbed by the resistor. On the other hand, if a better pulse response is required and the transmission line has a good frequency response, the value of the resistor can be reduced for decreasing the time constant of the pick-up circuit.

Frequency domain measurements (scattering parameters) using a microwave network analyzer and time domain measurements using TDR (Time Domain Reflectometer) were carried out in order to evaluate the stripline circuit of the bunch monitor. Fig. 3 shows the frequency response of the monitor; time domain reflection data are shown in Fig, 4. Fig. 4 shows that the 50-ohm pick-up resistor (elema resistor) of the monitor is not very good. The frequency response shown

in Fig.3 is also not very good, since the stripline width of the monitor is not optimized, and the resistor does not have a good frequency response. The main effect of the reflection loss might be attributed to the resistor.

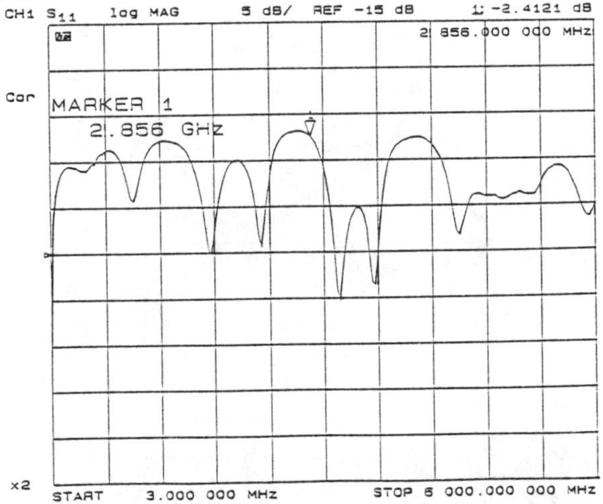

Fig. 3 Frequency Response of Bunch Monitor from the Output Connector

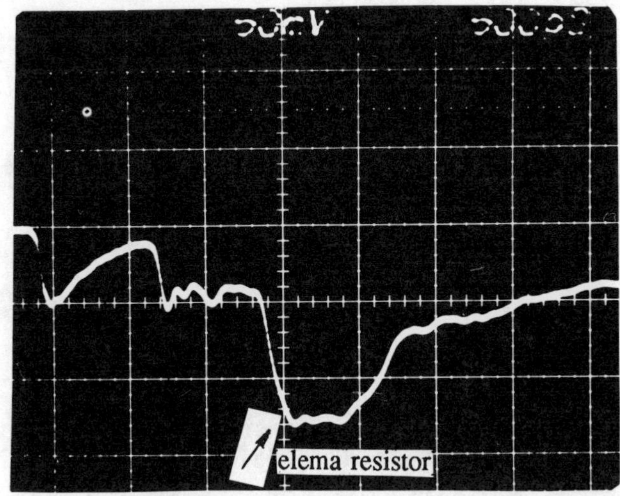

Fig. 4 Data of Time Domain Measurements from the Output Connector

4. Experiment and Result

The beam signal from the bunch monitor with a 50-ohm pick-up resistor has been observed at the end part of the KEK positron generator. The experimental setup is shown in Fig.5.[4] The primary high-intensity electron beam for generating positrons was guided to the monitor. For measurements of the fast signal (more than 100 ps), a 6-GHz transient digitizer (Tektronix, 7250) was used. The RF cable corresponding to 20D was provided between the monitor and the digitizer in order to prevent any distortion of the signal waveform.

Fig. 6 shows the signal from the bunch monitor; the bunch number of the electron beam can be counted by a streak camera, as shown in Fig. 7. The signal of the streak camera was saturated in order to obtain a small bunch signal;

therefore, the signal amplitude is not proportional to each bunch intensity. The rise time of the monitor's signal is almost 100 ps; we can count 7 bunches from Fig. 6. This bunch number is the same as data obtained from the streak camera.

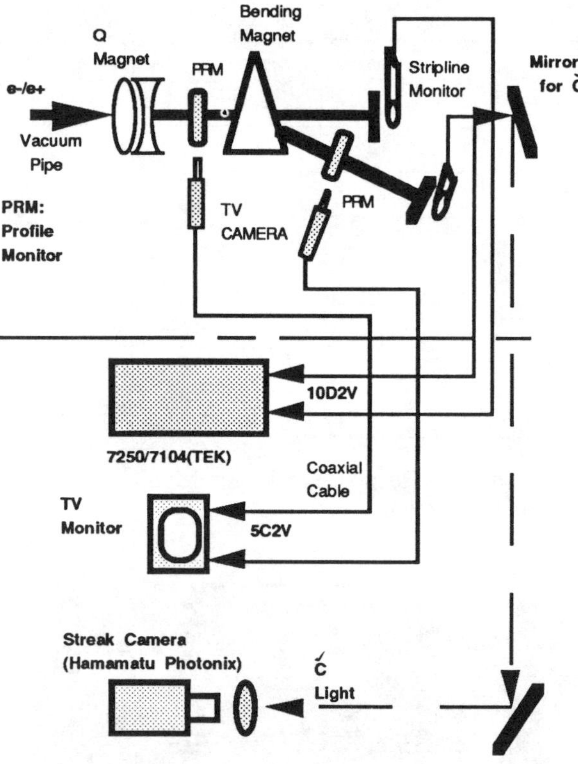

Fig. 5 Experimental Setup for Stripline Typre Bunch Monitor

5. Conclusion

The bunch monitor almost achieved a rise time of 100ps. This value is not sufficient for bunch-length measurements in an S-band linac, although the bunch monitor can count the bunch number of the electron beams. Data from the bunch monitor are still useful for evaluating the characteristics of the accelerated electron beams in S-band linacs.

We think that the characteristics of the bunch monitor can be even more improved by optimizing the stripline width, decreasing the resistor's value and using a good pick-up resistor which has a wide-band frequency response.

6, Acknowledgments

We appreciate Dr. Atsushi Enomoto's help in the experiment of the bunch monitor, and Dr. Takao Urano's help in setting up the trigger system.

Reference

1, M. Washio et.al., "Bunch Monitors for S-Band Electron Linac (in Japanese)", KEK-Report- 79-24, October 1979.

2, "OHO '86", Text of Accelerator Summer School in Japan.

3, T.C. Edwards, "Foundations for Microstrip Circuit Design", John Wiley & Sons Ltd, New York.

4 , T.Shidara et.al., "Improvements on Monitor System in the KEK 2.5-Gev Linac", KEK-Report-88-11, January 1989.

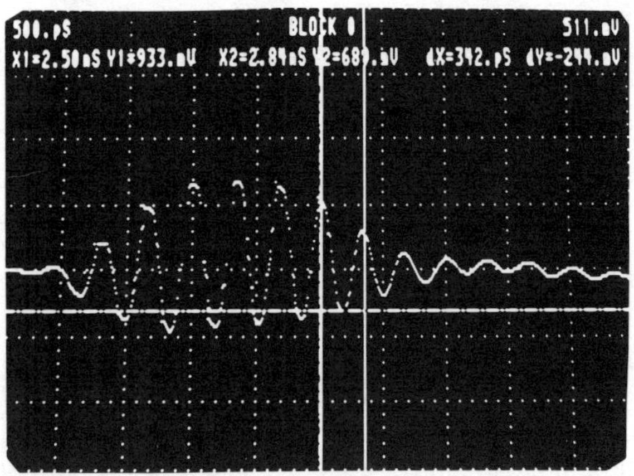

Fig. 6 Signal of Bunch Monitor

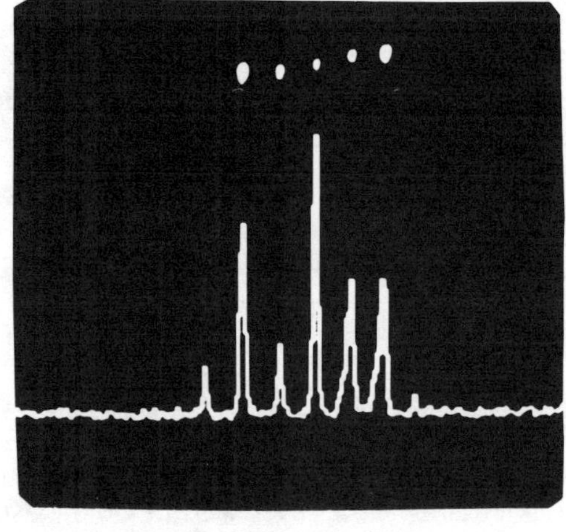

Fig. 7 Signal of Streak Camera

C++ Objects for Beam Physics

LEO MICHELOTTI

Fermilab*, P.O.Box 500, Batavia, IL 60510

1   Introduction.

Let us begin by admitting that the movement of physicists from FORTRAN to C++ has been less than a groundswell. At the same time, let me restate my opinion that scientific programming can be facilitated by an extensible language – such as C++ , Objective-C, or ADA – one in which we can define, easily and naturally, new variable types that behave, in all respects, like *fully functional variables of the language.* C++ is tailored to suit programmers' needs by creating "classes," which specify (a) structures of data, (b) the functions and operators which act upon them, and (c) rules[1] for creating and annihilating them. Among the advantages of working within such a language are:

• **user-friendliness**   Type checking, operator and function overloading, and data hiding make it easy to build user-friendly C++ classes that behave as close to "expected" as possible. For example, operator overloading means that arithmetic on algebraic classes can be implemented with the usual tokens: `+`, `-`, `*`, and `/`. Function overloading means a statement like "`y = cos(x)`" will work regardless of whether `x` and `y` are of type `double`, `complex`, `matrix`, `quaternion`, or another type for which the statement makes sense. In addition, if data conversions are appropriate, such as might arise in mixed mode arithmetic, the class constructors can handle this detail themselves, and other routine bookkeeping tasks can be imbedded within class implementations, freeing the application program(mer) from them.

• **inheritance and extensibility**   The class `DA` , which is discussed below, is an implementation of differential algebra in C++ . Having built it, one can easily go on to define other classes, such as `DAmatrix` – matrices whose elements are `DA` variables – with arithmetic operators overloaded in the obvious way. Should it be useful to do so, we could also have `DAquaternion` or `DAcomplex` variables, and one could go on to develop toolkits for `rational`, `group`, or any mathematical object that might be useful.

• **language support**   The advantages of working within a supported language should not be undervalued. Once defined, classes have *the full functionality of any other variable type within the language.* Suppose that a class "`zlorfik`" has been defined. This functionality means that an applications programmer can: (a) declare `zlorfik` variables just as easily as `double`, `int`, `char`, or any other type of variable, (b) write functions which return a value of type `zlorfik`, (c) declare `zlorfik` aggregrates – multidimensional arrays, lists, trees, or whatever – and (d) apply class operators not only on explicitly declared `zlorfik` variables but also on expressions which evaluate to type `zlorfik`. One gets all this for free – not even the class designer has to sweat the details – by working within a language *designed to be extensible.*

But enough missionary work. We shall describe below a few C++ classes and applications that have been written recently at Fermilab for problems in accelerator physics.[2]

2   MXYZPTLK

In a paper for the previous IEEE PAC I described a C++ class, `nstd`, which directly and very straightforwardly implemented automatic differentiation. Although it could (and did) do simple calculations, it was designed more for pedagogy than practical work. MXYZPTLK was designed to correct serious inadequacies of `nstd`. In it are defined two classes, `DA` and `DAVector` , which implement the "prolonged numbers" of `nstd` as dynamically allocated (and deallocated) doubly linked lists — that is, they are derived from the container class `dlist` — with attributes defined at runtime.

The `DA` and `DAVector` classes possess a unary operator, `.D` and a function `.derivative`, which correspond to performing and evaluating a derivative, as shown below.

---

*Operated by the Universities Research Association, Inc. under contract with the U.S. Department of Energy.

[1] The constructor and destructor functions which bring variables into and out of scope.

[2] The new limit of three 8.5 × 11 pages has made it impossible to provide a bibliography, so there will be no citations of connected works. However, the two or three of you who are actually reading this already know who the major players are in the games these tools are meant for: Alex Dragt, Etienne Forest, Martin Berz, Filippo Neri, Johannes van Zeijts, Ron Ruth, Robert Warnock, Yiton Yan and so forth.

```
DA     u, v;
double x;
int    m[] = { 2, 1 };
   ...
v = u.D( m );           // Line A
x = u.derivative( m );  // Line B
```

Line A corresponds to a functional equation,
$v = \partial^3 u / \partial x_0^2 \partial x_1$, while Line B merely loads the value of
this derivative into the variable x. It is the operator .D
that makes DA objects into a "differential algebra," in the
sense of Berz.

In addition to the unary operator $.D : DA \to DA$ there is a
binary operator $\hat{} : DA \times DA \to DA$ which implements Poisson brackets. For example, the program fragment below is
used to evaluate the Poisson bracket of the two expressions,

$$a = x_1 x_2^2 p_1 p_2^3, \quad b = \sin(x_1 p_2^2 x_2^3) .$$

```
DA x1, x2, p1, p2, a, b, pb;
   ...
a   = x1 * (x2*x2) * p1 * (p2*p2*p2);
b   = sin( x1 * (p2*p2) * (x2*x2*x2) );
pb  = a^b;
cout << pb.standardPart();
```

The last line prints the value of the bracket to the screen.
Because instances of C++ classes are fully functional variables, this could have been done in one step:

```
cout << ( ( x1 * (x2*x2) * p1 * (p2*p2*p2) )
              ^
          ( sin( x1 * (p2*p2) * (x2*x2*x2) ) )
        ).standardPart();
```

which applies .standardPart to the *expression* obtained
by taking the bracket of the two *expressions* formerly
loaded into a and b. In this same vein, the Jacobi identity
can be tested among three expressions, a, b, and c with
the line,

```
( a^(b^c)) + (b^(c^a)) + (c^(a^b)  ) .peekAt();
```

which prints all non-zero members of the expression to the
screen. Indeed, the arguments to the bracket operator (or
any DA function) may also include *functions which return
a value of type* DA .

Each DA variable keeps track of its own attributes, such
as accuracy and reference point. Because the order of
derivatives kept in the list is necessarily truncated, the
DA value resulting from an invocation of .D or the Poisson
bracket operation is not as accurate as the arguments that
went into it (its highest derivatives are missing). If this
variable is used later in calculations, the results are also less
accurate. Such information is carried along and upgraded
automatically across computations. If the application program tries to differentiate a variable too many times, to
access derivatives which are not accurate, or to multiply
two DA variables that have different reference points, an
error message will be printed.

Concatenation is implemented in MXYZPTLK via a binary operator, *, applied to variables of type DAVector .
Although it superficially acts like an array of DA variables,
in fact a DAVector has extra structure.

The first version of MXYZPTLK was released in January, 1990. Source code can be obtained upon request,
and a User's Guide has been written. It is worth mentioning that MXYZPTLK implements automatic differentiation by simple forward-mode algorithms and would
probably not be suitable for large-scale problems requiring hundreds of coordinates. These may be handled better
by reverse-mode methods incorporated into ADOL-C, for
example.

3   beamline

The very name of the class beamline suggests what it is.
beamline is derived from two parents: dlist , a container
class which implements a beamline variable as a doubly
linked list, and bmlnElmnt, a base class that contains information common to all beamline elements, such as geometry, pointers to circuits, or conversion factors (e.g., amperes to Tesla). That beamline is derived from bmlnElmnt
makes it easy to insert one beamline variable into another.

A C++ program fragment that builds a lattice consisting of five identical FODO cells may look as follows.

```
double   length, focalLength;
            ...
drift   O (   length );
thinQuad F (   focalLength );
thinQuad D ( - focalLength );   // Line 5

beamline A ( &F );
A.append    ( &O );
A.append    ( &D );
A.append    ( &O );             // Line 10

beamline B;
for( int i = 0; i < 5; i++ ) B.append( &A );
```

Done in this way, any subsequent adjustment of the focal
length of the F quad will take place simultaneously in all
"five" cells of the lattice. An alternative to building the
beamline variable element by element is to declare it with
a string argument, which is interpreted as a lattice file.
The statements,

```
beamline Tevatron      ( ''lowBeta.synch'' );
beamline mainInjector ( ''mi_17.flat'' );
```

declare two beamline variables, the first defined in a
SYNCH file, the second in a FLAT format file.

The beamline class interface contains the critical lines,

```
virtual void propagate( double* );
virtual void propagate( DAVector& );
```

which establish that every specific beamline element must contain two **.propagate** member functions. The first, which accepts an array of (six) real variables as its argument, does straightforward, element-by-element tracking through the beamline; the second, which accepts a **DAVector** as its argument, constructs the polynomial map corresponding to concatenating the maps of its elements.

The **beamline** class itself has **.propagate** member functions that do the same. One of the extraordinarily useful features of the **virtual** statement is that we can do this sort of thing so easily. The entire source code for the **beamline::propagate** function consists of two declarations and one executable line.

```
void beamline::propagate( DAVector& x ) {
dlist_iterator getNext ( *(dlist*) this );
bmlnElmnt* p;
while ( p = (bmlnElmnt*) getNext() )
        p -> propagate( x );
}
```

The beamline is being told to go element by element and propagate **x** through each one. There is no sequence of decisions to determine what to do based on type. Each type of variable knows itself what it is supposed to do.

And how does it know? One feature of the **bmlnElmnt** class is that *all the physics is isolated and contained in a collection of .physics files.* For example, the **.propagate** implementation for a thin quadrupole element contains the line,

```
{ #include ''thinQuad.physics'' }
```

which file is to contain all the physics associated with passage through a thin quadrupole, the rest being boilerplate and logistics. This file may contain only the lines

```
upr = upr - ( u / f );
vpr = vpr + ( v / f );
```

where **f** is a part of the private data of a **thinQuad** beamline element, representing the focal length of the quadrupole. Now, someone may very well object to using this, as it makes no mention of longitudinal momentum. He has the option of using an alternative file from the **thinQuad.physics** library, say,

```
upr = upr - ( u / ( f*( 1.0 + wpr ) ));
vpr = vpr + ( v / ( f*( 1.0 + wpr ) ));
```

where **wpr** is to be interpreted as $\delta p/p$. In addition, he can tinker with the files on his own. Suppose a user wants to do something unforeseen, such as call on a new symplectic numerical integrator to go very carefully through a thick element. He has the freedom to create his own **.physics** file, and the changes would be completely transparent to any application software using **beamline** . The brackets surrounding the **#include** statement assure that *variables declared or dynamically created by such tinkering can never interfere with variables of the same name in other parts of the class source code.* The C++ scoping rules will prevent it – another feature one has for free when working within the language. Our hypothetical tinkerer need never look at boilerplate and logistics.

A generic type of **bmlnElmnt** can exist in different flavors. For example **quad** is a derived **bmlnElmnt** , as are **thinQuad** , **DAQuad**, and **DAthinQuad**. The implication of **thinQuad** is obvious; the other two refer to a quadrupole whose properties – length and strength – are themselves **DA** variables, enabling them to be used as control coordinates in optimization calculations or to appear in polynomial expansions.

The above was written as though the design were already implemented. Some is; more is not. The first working version of **beamline** should be ready near Fall, 1991, with documentation by All Saints Day.

4   AESOP and canvas

AESOP[3], was introduced and demonstrated at the last IEEE PAC, so we shall not dwell on it here, as space is getting short. Suffice it to say that AESOP is a Phigs-based, prototype graphics shell, programmed in C++ , for implementing "exploratory orbit analysis." Its objective is easy, interactive exploration of four-dimensional phase space maps. AESOP's "four-dimensional cursor," implemented late last year, greatly facilitates the finding and tracking of four-dimensional separatrices through bifurcations. (See *Resonance Seeding of Stability Boundaries in Two and Four Dimensions*, this Proceedings.) A script (VAX/VMS DCL command file) is provided for linking AESOP with any four-dimensional mapping routine, which may be written in C++ , C, or FORTRAN. It has been used at Fermilab to explore the offset beam-beam interaction in the Tevatron, space charge in the booster, and in the Main Ring.

AESOP was written originally for the Evans and Sutherland PS390, but I plan to port it to the Sun environment "soon." The PS390 is a powerful, sophisticated graphics engine which is largely underutilized. The reason for this lack of enthusiasm is that no scientist of sound mind would read, much less assimilate, its seven-volume set of manuals. To respond to this problem of making it easier to view two- three-, and four-dimensional data on the PS390, a C++ class, **canvas**, was built. By simply declaring **canvas** variables application programs are provided with objects that accept (scatterplot, wireframe, or vector-field) data and display them automatically. The "real-time" transformation capabilities of the PS390 are activated in one step by ".connecting" its external devices, the dials and the puck, to the desired canvas. A **rasterCanvas** class is also available for scanning two-dimensional regions (ala Mandelbrot, for example).

---

[3] Analysis and Exploration of Simulated Orbits in Phasespace, or some such thing.

Beam Steering Using Quadrupoles As Position Monitors

Jean-Yves Hémery and Thomas Pettersson
CERN
CH 1211 - Geneva 23

Abstract

An algorithm is proposed to center the beam in the quadrupoles of a transfer line, by changing the focussing strength of a quadrupole and observing the resulting position shift on a monitor downstream. The observed position shift depends linearly on the beam position offset in the quadrupole, provided the phase advance between the quadrupole and the monitor is not a multiple of π. The same monitor may thus be used to center the beam in several places of the beam line.

The centering accuracy depends on the resolution of the monitor and on the quadrupole current increment which can be set in such a way that the beam profile remains sufficiently peaked to determine its mean position.

I. INTRODUCTION

The tight lay-out of the CERN LEAR experimental area (ref. [1]) is such that an extra means of beam position monitoring was necessary. Quadrupoles are required to transport the beam and can be used as diagnostics when they are associated to one real position monitor. Thus, we inherited a number of extra monitors, which are to be added to the 24 real ones.

A simple software program based on an iterative process is used to center the beam. It implies very simple arithmetic and decision logic criteria to minimize the dipolar effect of a quadrupole on a non-centered beam.

A. Why not another method?

Solving the problem analytically would have required a larger amount of work. Namely:

- matrix manipulations to solve the system of linear equations.
- data base initialization, 68 magnets + 24 monitors.
- data base maintenance since the transfer lines lay-out are regularly modified to suit the requests of the experiments.

The authors would like to thank Thys Risselada for helpful discussions and valuable suggestions, and Daniel Jean Simon for his interest and permanent support.

With the method being used, the influence of the following parameters is easy to see:

drift space, magnetic length, focussing power, fringe field, alignment error and parasitic field.

The latter is not negligible on a 20 MeV/c momentum beam (1 G over 1 m gives 1.5 mrad deflection).

The number of measured points required to solve the system of equations would be larger than the necessary number of steps to converge with the present method used.

II. DESCRIPTION OF THE ALGORITHM

Let us take (going from upstream) a pair of correctors, a quadrupole and a classical beam position monitor. Only one plane is treated here, but the same logic remains valid for the two of them.

Focussing strength scan sequence:

1: observe beam position on monitor
2: change quadrupole setting
3: observe beam position shift relative to '1:'
4: restore quadrupole nominal setting

How many pieces of information are available from a non-centered beam in a quadrupole?

When scanning, a beam position shift is observed at the monitor located further downstream. It is up to the operator to choose whether a focussing or defocussing is more adequate, depending on the monitor acceptance and resolution, relative to the beam size.

The signs of the (x, y) position shifts tell in which quadrant the beam is passing in the quadrupole.

From this, one can apply a correction using the 2 upstream correctors in the considered plane and re-iterate the loop until the observed shift is zero and the beam centered at the monitor.

The sign and amplitude of the beam shift compared to the ones observed at the previous iteration tell whether the process converges.

Figure 1 gives a steering in 3 steps. Correctors $_{a,b}$ are tuned according to the beam position on M1 monitor and the shift observed while scanning Q1.

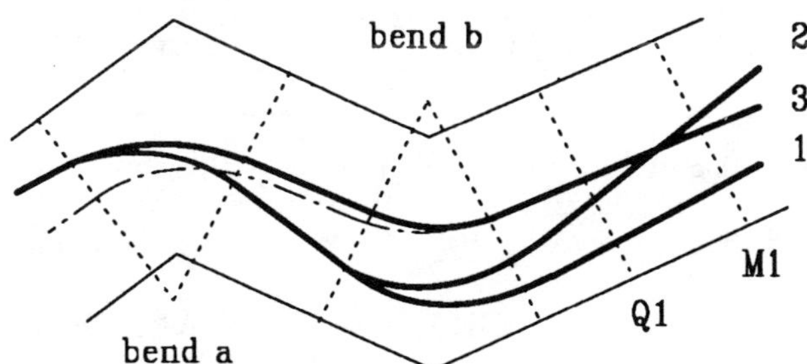

Figure 1: steering in 3 steps

step 1   : beam position > 0 at monitor, and
< 0 shift when scanning
$\Rightarrow$ increase bending angle b

step 2   : beam position < 0 at monitor, and
> 0 shift when scanning
$\Rightarrow$ decrease bending angle a and b

step 3   : beam position $= 0$ at monitor, and
no shift when scanning
$\Rightarrow$ the steering has finished

III.   DECISION TABLES

A.   Example with 1 quadrupole

Table 1 supplies the set of decisions to be taken. Each decision is quoted as a numerical value and corresponds to a change listed below.

The beam position measured at the monitor (while facing downstream the line) before scanning gives one P1 row selected.

Conventions for P1 rows:

$> 0 \Rightarrow$ beam in the right hand or upper side
$= 0 \Rightarrow$ beam in the center
$< 0 \Rightarrow$ beam in the left hand or lower side

The sign of the beam position shift when scanning gives one S1 column selected.

Conventions for S1 columns:

$> 0 \Rightarrow$ shift to the right or up
$= 0 \Rightarrow$ no shift observed
$< 0 \Rightarrow$ shift to the left   or down

warning : the sign of the beam shift has to be inverted in case the quadrupole scan consists in decreasing the focussing or increasing the defocussing strength.

This pair of values supplies the row and column coordinates, and leads to a unique decision.

$P1 \setminus S1$	< 0	$= 0$	> 0
< 0	1	4	4
$= 0$	3	0	4
> 0	3	3	2

Table 1: 1 quadrupole + 1 monitor

Meaning of the decision numerical values:

0: *last iteration, the beam is centered in the quadrupole and at the monitor*

1: $v_{b_{j+1}} = v_{b_j} + i_b \times d_b \times s_b$

2: $v_{b_{j+1}} = v_{b_j} - i_b \times d_b \times s_b$

3: $v_{b_{j+1}} = v_{b_j} - i_b \times d_b \times s_b$
$v_{a_{j+1}} = v_{a_j} + i_a \times d_a \times s_a$

4: $v_{b_{j+1}} = v_{b_j} + i_b \times d_b \times s_b$
$v_{a_{j+1}} = v_{a_j} - i_a \times d_a \times s_a$

With

v: *bending magnet setting value*
v_a and v_b are respectively for the upstream and downstream bending magnets
j is the iteration index number

i: *bending magnet increment value*

d: *+1 for deflection to the left or down,*
-1 for deflection to the right or up

s: *respectively the sign of v_{a_j} or v_{b_j}*

B.   Example with 2 quadrupoles

This case is very similar to the previous one. The meaning of the decision numerical values as well as the conventions remain identical. Only the signs are changed for the S2 rows of table 2 compared to the P1 rows in table 1.

$S2 \setminus S1$	< 0	$= 0$	> 0
> 0	1	4	4
$= 0$	3	0	4
< 0	3	3	2

Table 2: 2 quadrupoles + 1 monitor

IV. COARSE AND FINE STEPS

A. Justification and criteria

⊙ The process does not converge and the results are oscillating (fig.2) with poor damping around the solution and justifies to diminish the step size increments to speed up the convergence.

criterion 1: the two entries in table 1 corresponding to the states 1 and 2 are not neighbours.

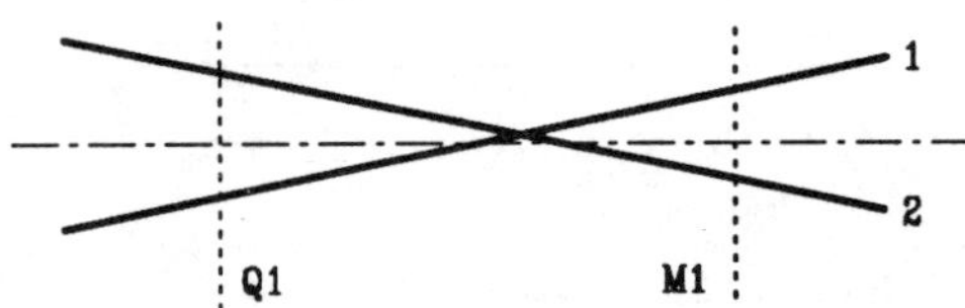

Figure 2: see-saw convergence

⊙ The program may converge so slowly (fig.3) that it is worth expanding the step size increment on the correctors to speed up the convergence.

criterion 2: the two entries in table 1 corresponding to the states 1 and 2 are the same, and converging.

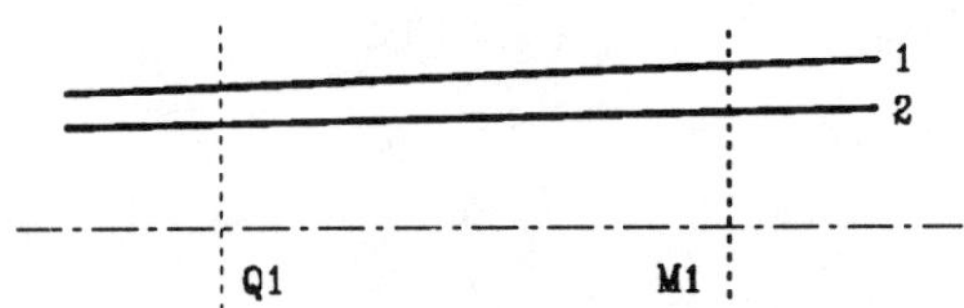

Figure 3: turtle convergence

⊙ The program may diverge if the transition is from state 2 to 1 (fig.3). The increment $i_{a,b}$ are to be negated.

criterion 3: the two entries in table 1 corresponding to the states 2 and 1 are the same, but diverging.

B. Neighbouring check Implementation

Each entry of table 3 contains its (row,column) coordinates:

1,1	1,2	1,3
2,1	...	2,3
3,1	3,2	3,3

Table 3: (m, n) coordinates used for 'neighbouring check'

By evaluating:

$$N = \mid m_j - m_{j-1} \mid + \mid n_j - n_{j-1} \mid \qquad (1)$$

in which j is the iteration index number.

The step size increments are altered, depending on the value of N:

N	action	comment
$= 0$	$i_{a,b} = i_{a,b} \times 1.1$	speed up
$= 1$	$i_{a,b} = i_{a,b}$	keep going
> 1	$i_{a,b} = i_{a,b} \times 0.9$	slow down

Table 4: criteria for step size increments alteration

V. NO POSITION SHIFT WHILE SCANNING

The reasons of an undetectable position shift on the monitor while scanning may come from the resolution of the monitor, the weak scan amplitude or limited phase advance close to a multiple of π between the quadrupole and the monitor.

The above statement is simply deduced from the well-known expression:

$$\Delta x_M = x_Q \, \Delta K \, l \, \sqrt{\beta_Q \, \beta_M} \, sin(\mu_Q - \mu_M) \qquad (2)$$

Where

x_Q : *beam position at the quadrupole*

ΔK : *quadrupole strength increment*

l : *quadrupole length*

β, μ : *lattice function values at Q and M positions*

About 60% of the 28 quadrupoles in the LEAR experimental area have betatron phases which allow them to produce a measurable displacement on the associated monitor without provoking beam loss on the vacuum chamber.

VI. RESULTS

The procedure is successfully applied to center the beam position along the LEAR transfer lines, and on the target of the experiments with zero angle. An accuracy of 0.1mm and 20μradian on a target is obtained in 6 iterations per plane.

VII. REFERENCES

[1] D. Dumollard, J-Y. Hémery and D.J. Simon,
    The LEAR Experimental Areas, Present Lay-out.
    CERN/PS/91-13(PA).

OPTICAL FIBER CHERENKOV DETECTOR FOR BEAM CURRENT MONITORING

I.V.Pishchulin, N.G.Solov'ev, O.B.Romashkin

Moscow Engineering Physics Institute, Moscow, USSR

ABSTRACT

The results obtained in calculation of an optical fiber Cherenkov detector for accelerated beam current monitoring are presented. The technique of beam parameters monitoring is based on Cherenkov radiation exitation by accelerated electrons in the optical fiber. The formulaes for calculations of optical power and time dependence of Cherenkov radiation pulse are given. The detector sensitivity and time resolution dependence on the fiber material characteristics are investigated. Parameters of a $10\,\mu m$ one-mode quartz optical fiber detector for the free electron laser photoinjector are calculated. The structure of a monitoring system with the optical fiber Cherenkov detector is considered. Possible applications of this tecnique are discussed and some recomendations are given.

INTRODUCTION

Cherenkov detectors are widely used for energy measurements of high energy beams [1]. There are many types of Cherenkov radiators which were tested in particle energy registration. It's known that Cherenkov radiation is emitted in a cone with the Cherenkov angle

$$\theta = cos^{-1} \frac{1}{n\beta}$$

where n is Cherenkov detector medium index of reflaction, β is the beam velocity normalized to the vacuum speed of light c.

In this paper a version of Cherenkov detector for the low energy beam current monitoring is described. The main component of the detector is a quartz optical fiber. It is inserted into the beamline. The angle between the beam line and the fiber is determined by the fiber core index of refraction. Cherenkov radiation is generated in the fiber core and transmitted along the optical fiber. In the present work the goal of the authors was to calculate the sensitivity and the time parameters of the optical fiber detector which would be used for current measurements of the single pulse bunched beam generated by photoinjector [2]. Such measurements require the detector time resolution less then 25 ps. It was proved that the radiation output power is sufficient to be registered by a high speed photochronograph.

OUTPUT POWER OF AN OPTICAL FIBER CHERENKOV DETECTOR

Let's assume that the beam crossing the optical fiber has the charge density ρ_s, which is uniform over the beam cross-section. The beam diameter is If a fiber with square crosssection is placed at the angle θ to the beam line one can be find out the fraction of the beam charge q_f that crosses the optical fiber with the diameter d_f

$$q_f = \frac{4 q_B d_f}{\pi d_B sin\,\theta}$$

where $sin\theta = \sqrt{1 - 1/(n\beta)^2}$

The Cherenkov radiation power is given by [1]

$$P = P_0 f(n)$$

$$P = 32 \frac{d_f^2}{d_B^2 \sqrt{1 - \frac{1}{n^2 \beta^2}}} \frac{\lambda_2^2 - \lambda_1^2}{\lambda_1^2 \lambda_2^2} I_B^2 t_B \;,$$

where I_B is the beam current amplitude, t_B is the beam pulse duration, $\lambda_1 = \bar{\lambda} - \Delta\lambda$, $\lambda_2 = \bar{\lambda}_1 + \Delta\lambda$ and $\bar{\lambda}$ is the wavelength in the centre of the band, $f(n) = 1/\sqrt{1 - 1/(n^2 - \beta^2)}$.
For the beam energy 200 keV, the current amplitude 500A, the pulse duration 250ps, the beam diameter 0,01m and for the index of refraction value of 1,5 in the wavelength band of 0,3-0,7μm the maximum power value is equal 162W. In practice the registration in such wide wavelength band is very difficalt. Taking account of the dependence $P(\lambda)$, the dependence the optical power losses in fiber on wavelength λ and the selectivity of photoreceiver the real value of optical power will be much less. In any case the output power depends on the photoreceiver wave length band, on the optical power losses and on the wave length. The middle $\bar{\lambda}$ is chosen on the basis of requirement of needed bandwidth and sufficient detector output power.

Fig.1 shows the one mode optical fiber output power P as a function of the wavelength λ with $\Delta\lambda = 0,1\mu m$ at various beam normalized velocities β .

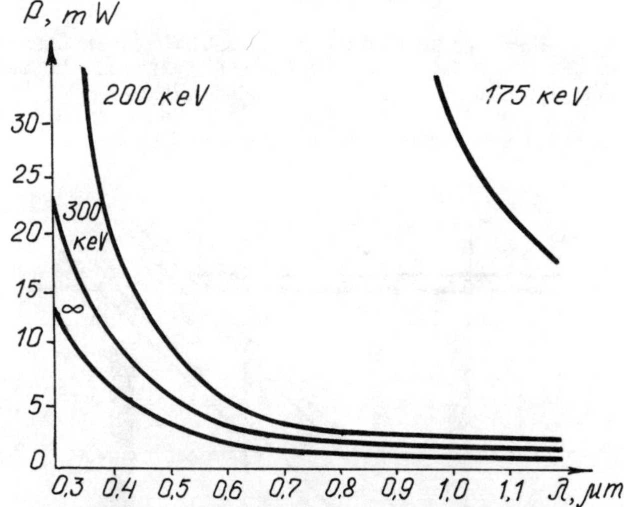

Fig.1. Cherenkov radiation power dependence on wave length.

$d_B = 1sm$, $d_f = 10\mu m$, $I_B = 100A$, $t_B = 250ps$
$n = 1,5$

Fig.2 shows the detector output power as a function of the index of refraction n .

It was expected , that the output power would be decreased with increasing n . Minimum n for given β is determined from the Cherenkov radiation generation threshold. For Cherenkov radiators with high value of n the energy range is widened at the low energy end.

For example it is possible to carry out current measurement of the beam with energy less than $50 keV$ by using GaAs fiber ($n = 3,34$).

The output optical radiation is registed by a high speed photochronograph, which requires very low

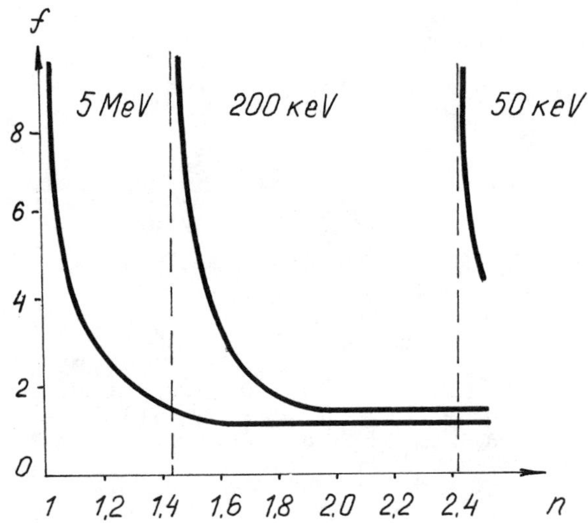

Fig.2. Cherenkov radiation power dependence on the index of the refrection.

power level for its operation. For the chosen photo-chronograph "Agat" the required incident power density is about 100 W/cm^2 . So if the fiber diameter d_f is equal 10 μm than Pmin=78,5·10^{-6}W. In practice detector output power levels are higher than Pmin (see Fig.1,2).

TIME RESOLUTION OF OPTICAL FIBER CHERENKOV DETECTOR.

The time resolution of an optical fiber Cherenkov detector is determined by the nonsynchronizm between the optical radiation and electron beam and by the fiber time distortion. Fig.3 illustrates a beam and a fiber sizes and typical time moments.

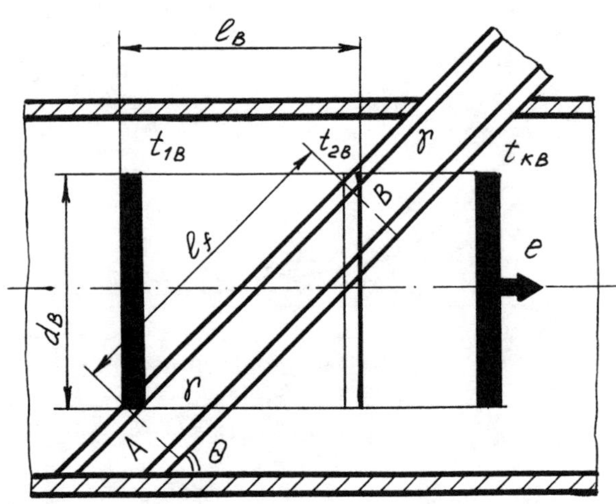

Fig.3. A beam and a fiber sizes and typical time moments.

For calculations the following terms are used:
Δt_1 is the delay time, which depends on the fiber material dispersion. It is essentially the difference between the time required to travel from the surface "A" to the surface "B" to the first photon emitted on the surface "A" at the moment t_{1B} and that for the last photon;
t_1 is the fly time of the fastest photon to the surface "B";
t_2 is the fly time of the slowest photon to the surface "B";
Δt_B is the fly time to the beam front from the surface "A" to the surface "B".
The calculations were carried out on the basis of the following formulas

$$\ell = \frac{d_B}{\sin\theta} \ ; \ \ell_B = \frac{d_B}{n_\beta \sin\theta} \ ; \ U_B = \beta c \ ;$$

$$\Delta t_1 = \frac{d\tau}{d\lambda}\Delta\lambda \ ; \ t_1 = \frac{d_B n}{c \sin[\cos^{-1}\frac{1}{n\beta}]} \ ;$$

$$t_2 = t_1 + \Delta t_1 \ ; \ \Delta t_B = \frac{\ell_B}{U_B} = \frac{d_B}{c \sin[\cos^{-1}\frac{1}{n\beta}]} \ ;$$

where ℓ_f is the fiber length, ℓ_B is the beam length, $\frac{d\tau}{d\lambda}$ is the fiber   material dispersion and $\Delta\lambda$ is the spectral line width.
We considered three possible cases:
a) $\Delta t_B > t_2 > t_1$, in this case the optical pulse distortion is $\Delta = \Delta t_B - \Delta t_1$;
b) $\Delta t_B = t_1 < t_2$, $\Delta = t_2 - \Delta t_B$;
c) $\Delta t_1 = \Delta t_B$, $\Delta = \Delta t_1$.
The results of Δt_1, t_1 and Δt_B calculations show that for any wave length
$\Delta t_B < t_1$. So that the optical distortion being expressed as the difference between the optical pulse duration and the beam pulse duration can be written

$$\Delta = t_1 + \Delta t_1 - \Delta t_B$$

This distortion depends on the angle θ to the beam line and the fiber material dispersion $d\tau/d\lambda$. The calculations results show that Δ is less then 35ps and it can be seen from Table 1. Table 2 illustrates the optical pulse time distortion in the one mode optical fiber which has the length 10m.
Results of Table 1,2 show that total pulse time distortions are less than 30ps by wave length band 0,6-0,95 μm, $\Delta\lambda$ is less than 10$^{-2}\mu m$ and β is less than 0,8.

Table 1. The calculations results of Δ , ps.

β \ $\lambda,\mu m$ / $\Delta\lambda$	0,4 0,2	0,6 0,2	0,8 0,2	0,95 0,35	1,3 0,2
0,7	19,6	17,6	16,2	16,8	15,0
0,8	30,2	29,1	28,35	28,7	27,7
0,9	35,7	34,8	34,1	34,4	33,6

Table 2. The calculation results of Δt, in one mode optical fiber, ps.

$\Delta\lambda$, μm \ λ , μm	0,4	0,6	0,8	0,95
$5 \cdot 10^{-3}$	35	20	10	8
10^{-2}	70	40	18	16
$3 \cdot 10^{-2}$	210	120	54	48

CONCLUSION

An optical fiber Cherenkov detector was investigated. It has the advantage of combining the functions of a detector and transmission line. The results of this investigation lead to the following conclusions.

1. Quartz optical fibers can be used as Cherenkov radiators for low energy beam current monitor.
2. The optical fiber detector are able to measure the beam pulse duration with the time resolution of 30ps.
3. If the beam energy is less then 175 keV it is necessary to choose the fiber material with high index of refraction values. With GaAs fibers it is possible to measure currents of beams with energies less than 50 keV.

REFERENCES

[1] Zrelov V.P. Izluchenie Vavilova-Cherenkova i ego primenenie v phizike vysokikh energiy.- M., Nauka, 1968.
[2] Airapetov A.Sh. et.al. Photoelectronnaya subnanosekundnaya pushka. - Tezisy dokladov XII Vsesoujznogo soveshchaniya po uskoritelyam zaryazhennykh chastits. M., 1990. - p.179.

ELECTRON BEAM PUMPED SEMICONDUCTOR LASER FOR PARTICLE BEAM DIAGNOSTICS

O. V. GARKUSHA, I. V. PISHCHULIN, N. G. SOLOV'EV, O. V. ROMASHKIN

Moscow Engineering Physical Institute, Moscow, USSR

ABSTRACT

Some characteristics of a new technique for accelerated particle beam monitoring are presented. This technique for beam current and pulse duration measurement is based on the electron beam pumped semiconductor laser (EBP-laser) principle of operation. First experimental investigations of $300\,\mu$m GaAs primary probe were carried out. Accelerated electron beam pulses with current range 50-500 A, pulse duration $5\,\mu$s and energy 200 keV were registered. The system sensitivity in this current range was $5\star10^{-3}$ V/A. It was shown that it's possible to apply the thechnique to monitoring of picosecond electron bunches in free electron laser photoinjectors.

INTRODUCTION

High power generation of the ultra-short optical pulse radiation is based on electron beam pumping of a semiconductor laser [1]. Such lasers generate the optical pulses with power 10-50 W, duration 5-10 ps by pumping current density above 10^5 A/m² and electron energy 50-200 keV. The free electron laser photoinjector is worked and developed [2]. The photoinjector electron bunches with current range 50-500 A, bunch duration range 40-200 ps and electron energy range 100-400 keV is achieved. The bunch-to-bunch interval is 10-20 ns with beam pulse duration 200 ns. In present work, the goal of the authors was to research the use of EBP-laser for photoinjector bunches diagnostics. This paper reports results of the first experiments on diagnostic equipment with EBP-laser monitor on the high current accelerator in MEPhI.

SUMMARY OF THE THEORY

In our calculations of radiation power, threshold current and resolution time of EBP-laser we used the monitor structure which is shown in fig.1.

In this structure electron beam is normal to surface of the GaAs crystal, which has thickness H. The optical radiation is normal to side surface. The distance between them is L. For transmission optical radiator the optical fiber with diameter d_f is used. The longitudinal dimention of radiation surface is l_s. In our case it's determined by opening in the collimator. D and d are thickness of active and radiation layer accodingly.

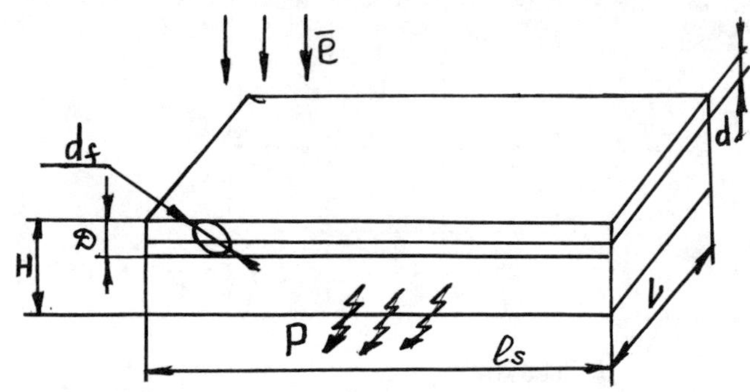

Fig. 1. Monitor structure

The optical power of the EBP-laser is defined as

$$P_{opt} = P_{el} \cdot l \qquad , (1)$$

where

$$P_{el} = \frac{S_c}{S_B} \cdot \frac{E I}{t_i} \qquad , (2)$$

S_B is the area of beam cross section and S_c is the area of opening in the collimator.

The laser efficiency can be expessed as [3]

$$l = l_{in} \frac{1 - exp(-\alpha L)}{\alpha L} \cdot \frac{1 - R}{1 - R\,exp(-\alpha L)} \qquad , (3)$$

where l_{in} is the internal efficiency (l_{in}=0.33) α is the dispersive losses in the active layer (α =30 m⁻¹), R is the resonator mirror reflection index (R=0.3). For E=50 keV, I=200 A, t_i=100 ps, S_B=3$\star$10⁻⁴m², S_c=7.5$\star$10⁻⁷m² it's shown, that l =0.05, P_{el} =2.5$\star$10⁴ W and P_{opt} =1.25$\star$10³ W. The real value P_{opt}^R is less than P_{opt} by three orders because the electron energy distribution is ingomogenious in the active layer. Therefore the layer efficiency must be decreased to 0.01. Moreover P_{opt} is decreased 2 times because the optical radiation is generated from 2 resonator edges. At last, there are the fiber edge deflection losses and the value l_s is much greater then fiber diameter d_f (l_s=0.5 mm, d_f=10 m). So P_{opt}^R isn't more then 1.25 W.

Using the condition of a laser generation the expression for the threshold current density J_t may be written as [4]

$$J_t = \frac{8\pi n^2 \Delta \nu e E D}{\lambda_o^2 E S\, l_q} \left[\frac{d}{D}\alpha_a + (1 - \frac{d}{D})\alpha_p + \frac{1}{L} ln\left(\frac{1}{R}\right) \right] , (4)$$

where λ_o is the optical radiation wavelength (λ_o=9.0$\star$10⁻⁷ m), l_q is the internal quant efficiency (l_q= 0.7), n is GaAs refraction index (n=3.34), $\Delta\nu$ is the spontaneous amplification bandwidth ($\Delta\nu$=1.5$\star$10¹³s⁻¹), e is the electron charge, S is the reverse dissipation factor (S=0.68), $\mathcal{E}$ is the average energy for one electron - hole pair creation ($\mathcal{E}$=4.5 eV), d_a=12 sm^{-1} and d_p=210 sm^{-1} are dispersive losses in the active and radiation layers, accordingly. From (4) for E=50 keV, L=1.5 mm, the threshold current density is equal $1.2*10^4$ A/m^2. For measuring the electron beam current it is necessary to determine the laser time resolution. The laser generation occurs if the pumping current pulse duration is more then the generation delay time t_d. The delay time t_d can be estimated as

$$t_d = \tau_e \, \ell n \left[\frac{\mathcal{J}}{\mathcal{J} - \mathcal{J}_{t'}} \right] \qquad , (5)$$

where $\tau_e = 10^{-9}$ s for GaAs, $\mathcal{J}_{t'}$ slightly exeeds $\mathcal{J}_t$. The generation delay time t dependence on the electron current density is shown in Fig.2. where $\mathcal{J}_{t'}=1.5*10^4 A/m^2$ and $\mathcal{J}_t=1.2*10^4$ A/m^2.

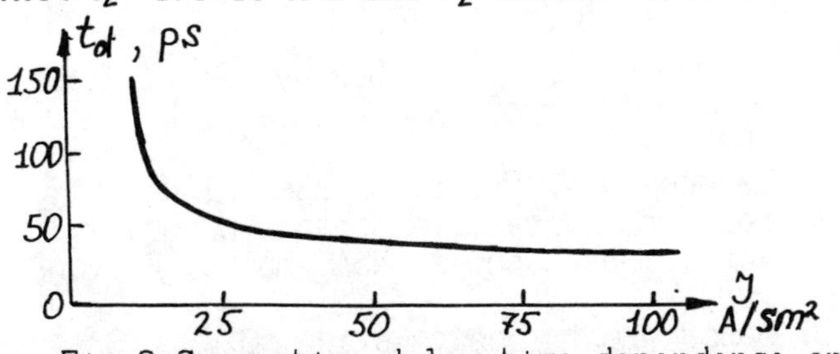

Fig.2. Generation delay time dependence on electron current density.

The dependence in Fig.2. indicates that t_d isn't more then 20 ps at high $\mathcal{J}$. It's known that delay time t_d may be decreased by additional constant pumped current $\mathcal{J}_c$. $\mathcal{J}_c$ must be not more then $\mathcal{J}_t$.

To minimize the difference between the optical radiation pulse form and the beam current pulse form, it's necessary, that $t_8 < T$. The time T characterises the electron-photon relaxation period and can be determined as

$$T = \frac{2\tau_e}{Y+1} \; , \quad Y = \frac{\mathcal{J}}{\mathcal{J}_t} \qquad , (6)$$

The results of calculation are shown in Fig.3. It is seen that for high Y , the value of T is small (T=20-25 ps).

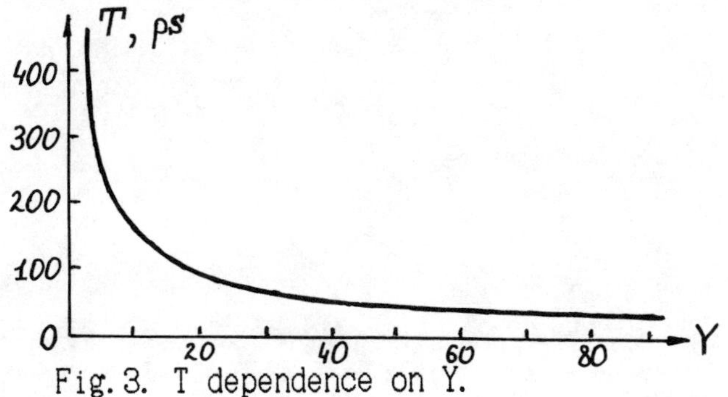

Fig.3. T dependence on Y.

So, if $\mathcal{J}$ exceeds $\mathcal{J}_{t'}$ significantly and the electron beam pulse duration is enough for laser generation, the pulse form distortion isn't creat. The maximum of pulse distortion should not exceed 20 ps.

MONITOR CONSTRUCTION

The main component of the monitor is GaAs semiconductor crystal. The crystal substrate is placed in a special hollow of a massive metallic substrate (Fig.4). The optical radiation

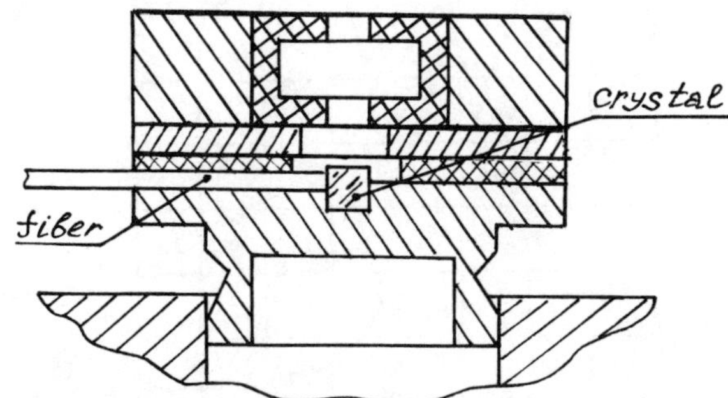

Fig.4. Monitor construction

output is carried into the optical fiber end. The fiber is placed in the V-type channel in a metallic substrate. The fiber is fixed in the channel by a dielectric elastic laying, a metallic plate and a supplementary clamp. The electron beam is transported to the surface of the semiconductor substrate through a collimator. The collimator consists of an aluminium block and a graphite insert with a through opening. To change the electron pumping area on the substrate surface a few graphite inserts with different opening sizes are used. The minimal size of the pumping area is limited by 0.5*1.5 mm^2. To decrease the influence of the secondary electrons a special cavity is inserted into the graphite. The metallic substrate, the metallic plate and the collimator are united in an indivisible construction with 2 screws. The monitor is placed on Faraday cup opening in the vacuum chamber of the accelerator. The monitor construction is provids the optical radiation transmission from a semiconductor laser to the optical fiber. It forms the pumping area on the semiconductor substrate surface and enables to insert different crystal substrates, fibers and graphites.

EXPERIMENT

The experimental work was performed with the high current electron accelerator in MEPhI laboratory of high current beam physics. The samples of the semiconductor laser were prepared on GaAs substrate. The substrate thickness was 300 μm. The vacuum output of the optical fiber is realised on a flat rubber seal. Experiments under vacuum conditions proved that for achiving the limit pressure (10^{-6} mm Hg) such a seal was sufficient. The monitor samples were studied with the high current electron accelerator with electron energy 200 keV, current density range (15-150)$\star 10^{4}$ A/m^{2}, pulse duration 5 μs. First experiments results are presented in Fig. 5.

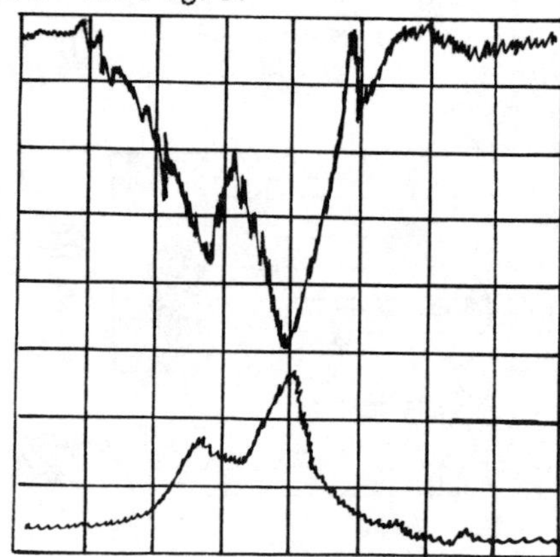

Fig. 5. Experimental results: the upper trace is the beam current pulse (80 A/div, 1 μs/div) ; the lower trace is the EBP-laser radiation (50 mV/div, 1 μs/div).

The optical radiation was transmitted from the laser to a receiver by the 30 m optical fiber. This optical radiation was measured with the avalanche photodiode with sensitivity 27 A/W at wavelength 0.85 μm and was registered by memory - type oscilloscope. The photodiode load resistance is equal 510 Ohm. From Fig. 5. it is seen that the system sensitivity with the electron beam pumped semiconductor laser monitor is $5\star 10^{3}$V/A.

CONCLUSION

According to the obtained results the electron beam pumped laser is a useful diagnostic tool for a picosecond electron bunches in free electron laser photoinjector. The pulse duration of photoinjector current can be measured on the basic of this technique. Due to the high time resolution and sufficient sensitivity this detector can be used for particle beam diagnostics.

REFERENCES

[1] Bogdankevich O.V., Darznek S.A., Eliseev P.G. Poluprovodnikovye lazery, M., Nauka, 1976.
[2] Lebedev A.N. Raboty po LSE PhIAN. - II Mezhdunarodnoe soveshchanie po novym metodam uskorenija. Non-Amberd, Erevan, 1989.
[3] Basov N.G., Bogdankevich O.V., Grasynk A.Z., Semiconductor Laser with Radiating Mirrors, IEEE J.Quantum Electron., 1966, QE-2, N 9, p.594.
[4] Hunsperger R.G. Solid State Electr., 12,215 (1968).

METHOD AND APPARATUS FOR MULTIFUNCTIONAL
NONPERTURBING DIAGNOSTICS OF H$^-$ BEAMS

A.S.Artiomov, N.G.Vaganov, A.K.Gevorkov,
V.V.Limar, V.P.Sidorov
I.N.Vekua Institute of Physics and Technology
Sukhumi 384914, USSR

Abstract

A method of nonperturbing diagnostics for negative ion beams based on electrons resulting from their near-threshold photodetachment is considered. A compact apparatus for realizing this method in the Moskow Meson Factory Linac (MMFL) and measuring various parameters of H$^-$ ions beam is proposed.

Method

For high current ion accelerators, it is important to obtain information on various parameters of a beam not affecting whem appreciably during measurements (nonperturbing diagnostics). In the linear areas of the transport line for negative ions beam such information may be obtained via the method based on the electrons resulting from the near-threshold ($\varepsilon_\text{п}$) photodetachment of some negative ions. A kinematic analysis of a elementary single-photon photodetachment acts shows that accuracies of determination of the energy and momentum direction of the ion by means of electron parameters are found

$$\frac{\Delta E_e}{\overline{E_e}} = \frac{2 \cdot \beta \cdot \gamma}{(\gamma - I) \cdot M_e \cdot C} \sqrt{2 \cdot \mu_{eo} \cdot (\hbar\omega - \varepsilon_\text{п})} ;$$

$$\Delta\theta_e \ [\text{рад}] = 2 \cdot \frac{\sqrt{2 \cdot \mu_{eo} \cdot (\hbar\omega - \varepsilon_\text{п})}}{\gamma \cdot \beta \cdot M_e \cdot C} ; \qquad (1)$$

where $\mu_{eo} = M_e \cdot M_o / (M_e + M_o) \approx M_e$, $\gamma = (1-\beta^2)^{-0.5}$, $\beta = V_i / C$, V_i is the velocity of an ion with the energy E_i and the mass M_i, $\omega = \omega_o \cdot \gamma \cdot (1 - \beta \cdot \cos\eta)$, ω_o is the photon frequency in a laboratory frame, η is the angle between ion and photon momenta, $\overline{E}_e = E_i \cdot M_e / M_i$. For fixed ω_o and η values, the $\Delta E_e / \overline{E}_e$ and $\Delta\theta_e$ accuracies may be improved nearly five and nine times, respectively, with the photon polarization suitably chosen [1]. When forming a stripping target using laser radiation with highly monochromatic and directed photons the nonperturbing diagnostic accuracy for high energy ion beams is mainly limited by energy ($\Delta\beta$) and angular ($\Delta\theta_i$) spreads of ions in a beam due to a strong dependence of near-threshold photodetachment cross section $\sigma(\omega) = 2 \cdot I0^{-16} \cdot (\hbar\omega - \varepsilon_\text{п})^{1.5} \cdot (\hbar\omega)^{-3}$ [2] on photon energy in the ion reference system. To obtain the best accuracies for $\Delta\beta$ and $\Delta\theta_i$ given, the minimum value of $\hbar\omega - \varepsilon_\text{п}$ in (1) is chosen so that photodetachment cross sections $\sigma(\omega)$ are the same for all H$^-$ ions. This condition take place when

$$\hbar\omega - \varepsilon_\text{п} \gg \gamma \cdot \hbar\omega_o \cdot [\Delta\beta \cdot [\beta \cdot (I - \beta \cdot \cos\eta) \cdot \gamma^2 - \cos\eta] +$$

$$\beta \cdot \sin\eta \cdot \Delta\theta_i]. \qquad (2)$$

From this e.g., for the MMFL with the energy of H$^-$ ions 600 Mev, $\Delta\beta/\beta \approx \pm I$ mrad and suitably polarized photons with $\hbar\omega_o = I.I7$ ev, one may obtain accuracies of correspondence between electron and ion distributions in a beam $\approx 3 \cdot I0^{-2}\%$ in energy and $\approx 2 \cdot I0^{-4}$ rad in angle. The photon target power necessary for measurements is defined by the conditions of information extraction from the total flux of electrons from the target and residual gas.

Apparatus

In this paper a compact multifunctional apparatus for realization of this nonperturbing diagnostical method is proposed. A schematic layout is shown in figure. A dipole magnet with a homogeneous field (MA) is used to extract electrons from the ion beam and to analyze the information carried by them. The interpolar distance Π_m

is chosen to be sufficient to let ions pass through the analyzer unhindered.

Ion energy spectrum and longitudinal emittance measurements are performed according to a scheme (a) well known for magnetic analyzers where, using the laser radiation, instead of a diaphragming slit of the analyzer, a band type target (0) is formed with the required spatial localization ΔX_o along the X_o axis. The energy spectrum of ions is reproduced according to the spatial distribution of the electron flux density along the X axis measured by the detector $Д_{е1}$ with the its spatial resolution d taken into account. In order to measure an electron momentum with the accuracy of $\delta P/P \approx 2\cdot 10^{-4}$ for $\Delta X \approx (5\div 10)\cdot 10^{-2}$ mm, $d\approx (4\div 5)\cdot 10^{-2}$ mm, $\Pi_m = 40$ mm and for the expected angular spread of ions $\Delta\theta_i \approx \pm 1$ mrad, one may choose, e.g., magnetic analyzers with $\phi = \pi/2$, R= 200 mm, $L_1 = 240$ mm and $L_2 = 160$ mm, or with $\phi = \pi$, R= 230 mm and $L_1 = L_2 = 0$. The latter choice ensures the minimum effect of the ion beam space charge on electrons. Electrons with energies needed for phase analysis are separated via diaphragming when the analyzer magnetic field sign and value are changed. The diaphragmed electron beam may be dispersed in phase at the detector $Д_{е2}$ in a cavity with a circularly polarized rf-field [3] or with rf-transverse deflector [4]. The accuracy of transformation of the ion beam phase structure into an electron flux at the entrance of the cavity dispersed in phase (CDF) is mainly defined by the projection ΔZ_m of the target area (where the electrons are collected from) onto the Z-axis and by the difference of electron trajectory lengths in the magnetic analyzer due to the ion angular spread in the beam and its space charge. E.g. to ensure the phase resolution of $\approx 1^{\circ}$ for a beam of H$^-$ ions with the energy of 160 Mev (f$\approx$ 200 MHz), in the second type of analyzer, one must provide $\Delta Z_m \approx 1$ mm. When using a photon target inclined towards the ion beam, with a sufficiently small transverse dimension along the Z axis, the required operating region of the target may be localized by choosing the diaphragming slit (S) correspondingly limited along the Y axis.

While measuring the ion transverse emittance and profile, the band-type photon target is localized within the $X_o Z$ plane, moves in parallel along the Y_o axis and has a spatial localization $\Delta Y_o \approx (5\div 10)\cdot 10^{-2}$ mm required for measurements (b). The analysis of electron distributions at the detector $Д_{е1}$ within the plane of (Y'Y) obtained as a result of computer experiments have shown that for ion energies $E_{H^-} = 600$ Mev, $\Delta\beta/\beta \leqslant 10^{-2}$, $|X_o, Y_o| \leqslant 0.5$ sm, $|X'_o, Y'_o| \leqslant 1$ mrad effects of boundary fields and the analyzer geometry may be taken into account via expression

$$Y = A\cdot Y_o + B\cdot Y'_o \qquad (3)$$

where A and B are defined according to the type of the analyzer chosen, and equal to A= 1.47 and B= 0.086 sm/mrad for the first choice, and A= 1 and B= 0.072 sm/mrad for the second one, respectively. The Y'_o distribution is reproduced with the accuracy of $\delta Y'_o \approx 2\cdot 10^{-4}$ rad from the measured spatial distributions of electron fluxes at the detector along the Y axis for controllable characteristics of the target (defining the probability of an electron generating) and its position in the space (Y_o). At the same time, a functional dependence of the integral electron flux at the detector upon the target position defines the ion beam profile along the Y_o axis.

For a short time interval (e.g. during a pulse of the ion beam) a certain information on the ion distribution over the (Y'_o, Y_o) plane may be obtained by means of several band-type targets fixed in space, created and separated from each other along the Y_o axis by diaphragming the laser radiation. The distance between them is defined by the condition of the electron distributions overlapping at the detector along the Y axis and can be estimated as $\delta Y_o \approx 1.5$ mm in the both types of analyzers for the ion beam of 600 Mev energy.

For measuring energy spectra and transverse emittance of the H$^-$ ions beam, with accuracies sufficient for the MMFL $\Delta Z_m \approx 2$ sm may be use. For the first, simpler in realization, type of the analyzer the ion beam space charge does not practically reveal itself within the accelerator area with ion energies 600 Mev and it exert an essential influence on the accuracy of beam parameter measurements in the area with $E_{H-}=$ 160 Mev. Inaccuracies in adjusting and manufacturing of the magnetic dipole result in systematic errors of measurement may be taken into account by preliminary calibration of the apparatus by means of a testing electron beam and experimental definitions of A and B values. For the precision operation of the apparatus, spatial positions of the band-type photon target should be controlled with accuracies of $\delta(X_o) \approx \delta(Y_o) \approx (5 \div 10) \cdot 10^{-2}$ mm, $\alpha(X_o) \approx \alpha(Y_o) \approx (2 \div 5)$ mrad. Moreover, background magnetic fields H_b should be well shielded off as well as high accuracies of the required magnetic field magnitude H in the analyzer should be achieved ($\delta H \approx H_b \approx 3 \cdot 10^{-4} \cdot H$). Expected relative losses of the H^- ions beam during the process of measurement are $\approx 10^{-7}$.

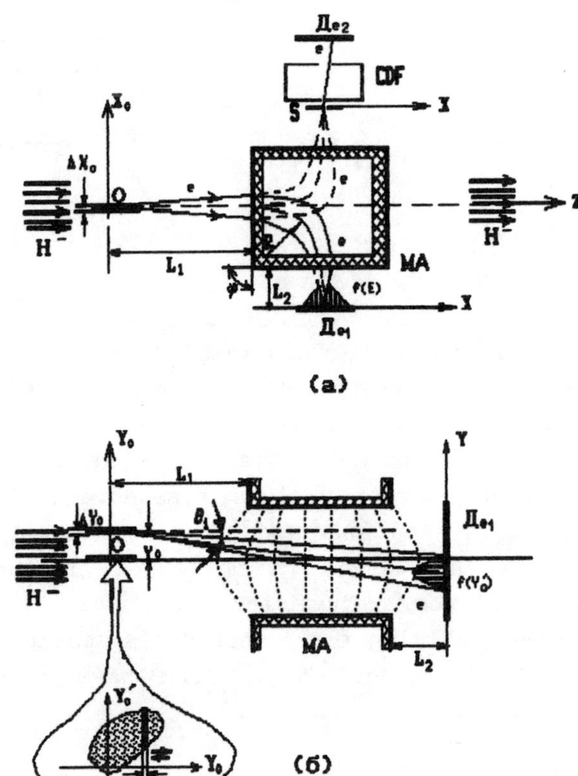

References

1. A.S.Artiomov, A.A.Avidzba. Sukhumi Inst. Phys. and Technol. Preprint № 17, 1990.

2. B.H.Armstrong "Phys.Rev.",131, 1132, 1963.

3. E.S.Zlunitsyn, A.I.Zykov, G.D.Kramskoy, V.A.Kushnir. "VANT. Ser. Tekn. Phys. Eksper.", № 2(28), 37, 1986.

4. A.A.Stepanov, A.V.Feshchenko, A.G.Chursin. "VANT. Ser. Tekn. Phys.Eksper.", № 3(34), 43, 1987.

CORRELATION METHOD OF NONPERTURBING
MEASUREMENTS OF ION BEAM ENERGY SPECTRA

A.S.Artiomov

I.N.Vekya Institute of Physics and Technology,
Sukhumi, 384914, USSR

Abstract

A correlation method of nonpertubing
control of ions energy spectra is suggested.
The method is based on measurements of mutual
correlation function between a fluxes of
particles or target photons, pseudorandomly
modulated in time and that of fast neutrals
formed at the target and detected on the
drift distance. Characteristics of the
apparatus realizing the proposed diagnostical
method in a source of H⁻ ions have been
evaluated.

Introduction

In order to control acceleration processes
in modern accelerator complexes it is
necessary to carry out nonperturbative
measurements of ion beam parameters in the
tpansport line areas with external
electromagnetic fields, e.g., with a bending
magnet. For these purpose, fast neutrals may
be used which are generated as a result of
ion destruction or charge excharge process at
a specially shaped and practically
transparent for a beam target (see,
e.g.,[1,2]). The target is formed so that
information carrying neutral particles
(IN-particles) should follow the ion velocity
in magnitude and direction with accuracies
required for measurements. At the transport
line bending area, they leave the ions beam
without perturbing the information carried
and they may be analyzed in special devices
not affecting the ions. E.g., in analyzing
the energy spectra of IN-particles, a time of
flight method is usually used, or their
ionization and consequent magnetic analysis
is performed.

For ion sources, the probability of
IN-particles generation on the residual gas

(η_r) may be considerable. For this case, it
is impossible to extract the information
directly from a flux of background
IN-particles at the detector at any density
of the target used. In particular, it takes
place when the nonperturbing diagnostical
method proposed in [2] is applied to sources
of H⁻ ions ($\eta_r \approx 0.3$).

Method

In this paper the correlation method for
measuring energy spectra of ion beam
using test IN-particles detected on the
drift distance L is proposed. For an ideal
case, under the test particles, we assume
IN-particles having their autocorrelation
flux function $\psi_T(t)$ equal to a periodical
δ-function (a pseudorandom flux):

$$R_{TT}(\tau) = \int_{-\infty}^{\infty} \psi_T(t) \cdot \psi_T(t-\tau) \cdot dt =$$

$$= \sum_{K=-\infty}^{\infty} \delta(\tau - K \cdot T) . \qquad (1)$$

For generating such particles, at the initial
area of the transport line bend a target is
formed with the density $\psi_t(t)$ pseudorandomly
modulated in time. For the ion beam current
being unchanged during measurement, the
target is spatially localized so that the
flux of the particles generated at the target
should adequately reproduce the target time
modulation, $\psi_T(t) = const \cdot \psi_t(t)$. Pulse
characteristic h(t) of the drift distance
from the target to the detector is related to
the velocity V_o distribution of IN-particles
($t = L/V_o$) and, hence, with the energy spectra
of ions. The flux of IN-particles ψ_o at the
entrance of the drift distance consists of
fluxes of background (ψ_r) and test (ψ_T)
IN-particles generated in the course of
interaction of ions with residual gas

components and the target, respectively. Since the drift distance is a linear system the IN-particle fluxes at the detector $f_o(t)$ and at the entrance of the path length $\psi_o(t)$ are interrelated by the convolution

$$f_o(t) = \int_0^\infty h(\tau) \cdot \psi_o(t-\tau) \cdot d\tau \ . \qquad (2)$$

We consider that the contribution of background IN-particles generated at the bending area after the target may be neglected. Taking the independence of ψ_r and ψ_T into consideration, we find that measurements of a mutual correlation function between the fluxes of target particles or photons and IN- particles at the detector allows us to obtain periodical reproductions of pulse characteristics of the drift distance

$$R_{io}(\tau) = \int_{-\infty}^{\infty} \psi_t(t) \cdot f_o(t-\tau) \cdot dt =$$

$$= const \cdot \int_0^\infty h(t) \cdot R_{TT}(\tau+t) \cdot dt = \qquad (3)$$

$$= const \cdot \sum_{K=0}^{\infty} h(\tau - K \cdot T) \ .$$

In reality, the flux of test IN-particles must be such that the convolution of $h(t)$ and R_{TT} does not change the $h(t)$ function. In accordance with [3] this condition means that a periodically replicating element of the autocorrelation function of the $\psi_t(t)$ signal must have a sufficiently narrow shape in time, with the width of $\Delta \ll \tau_{max}$, where $h(\tau) = 0$ for $|\tau| \geqslant \tau_{max}$, and its period T must satisfy the condition $T > 2 \cdot \tau_{max}$.

Correlation methods allow to perform measurements in condition when background signal exceeds by several orders of magnitude the required one. Hence, in measuring $R_{ro}(\tau)$ using correlators the energy spectra of the ion beam may be controlled nonperturbatively.

Apparatus

The considered method of energy spectra measurements may be used, e.g., in surface plasma sources of H^- ions [4] according to the scheme shown in figure. H^o atoms, generated from the stripping of H^- ions on the residual gas or the photon target, reproduce the ion energes with high accuracy ($\approx 2 \cdot 10^{-4}$) and can be used as IN-particles. The stripping target 1 is formed at the beginning of the bend area by diaphragming the radiation with the wavelength of $\lambda = 10600$ Å ($\hbar\omega = 1.17$ ev) from the Nd^{+3}:YAG laser with synchronized modes. Duration of series of pseudorandom picosecond radiation pulses is $T_s \approx 100$ ns and their autocorrelation function has a shape close to a triangular one with the width $\Delta \approx 50$ ps [5]. Such a photon target with H^- ions stripped at it may serve as an efficient generator of H^o test-atoms allowing to measure pulse characteristics of the drift distance $h(\tau)$ which are fairly short in time. To generate them in accordance with the target autocorrelation function the interaction area length should not exceed $\Delta \cdot V_{H-} \approx 10^{-2}$ sm. At present, potentialities of the above measurement method are limited mainly by the fast acting capability of correlators.

The mutual correlation function $R_{\gamma o}(\tau)$ between the target photon flux and that of H^o-atoms at the detector 2 may be measured by means of time-integrating correlators in charge-coupled (CC) devices similar to that suggested in [6]. Some laser radiation after the interaction area is spatially modulated by means of a GaAS CC-linear structure 3, in which an spatial distribution of pixel charges corresponds to a discrete-in-time representation of the shape of a current signal from the H^o-atom detector. The radiation 4 modulated by the photoelectric absorption is detected by the silicon CC-linear structure 5. Spatial distribution of the charge accumulated there during the measurement time corresponds to the correlation function $R_{\gamma o}(\tau)$. The function may be read out, e.g., during intervals between the target switchings or the ion beam pulses and then it may be transformed into the time scale, the charge transfer velocity along the modulating CC-linear structure being known. Assuming that for a source of a band-type ($\approx 0.1 \times 1$ sm^2) beam of H^- ions with energy $E_{H-} \approx 20$ keV and $\eta_r \approx 0.3$ measurements are possible for $\psi_r/\psi_T \approx 10^2$ we obtain the needed power of the photon target $P_\gamma \approx 350$ W. The target required, having the transverse dimensions $\approx$ $10^{-2} \times 0.1$ sm^2 within the interaction region, may be formed by diaphragming the compact laser radiation with an average power within duration of series pulses being $P_l \approx$ 200 kW. Fairly large target photon fluxes make it possible to use a waveguide propagatin of the radiation in the CC-linear structure 3 and to reach the maximum dynamical range of modulation ($\approx$ 100%). Moreover, this fact allows to provide the detector pixels with the sufficient charge and to carry out energy spectra measurements during the time $T_s \approx$ 100 ns when needed in the course of the pulse of ion beam. Fast acting capability of the corellator and the apparatus as a whole is determined by the guiding frequency of the modulator CC-linear structure which may amount to $\approx$ GHz for GaAs CC-devices [7]. Thus operating one may measure h(τ) in detail with the minimum value of $\tau_{max} \approx$ 10 ns. For the source considered, with the path length for H$^\circ$ atoms L$\approx$ 100 sm, the apparatus suggested permits to carry out nonperturbative measurements of energy spectra of H$^-$ ions with accuracies $\approx$ 0.4 %.

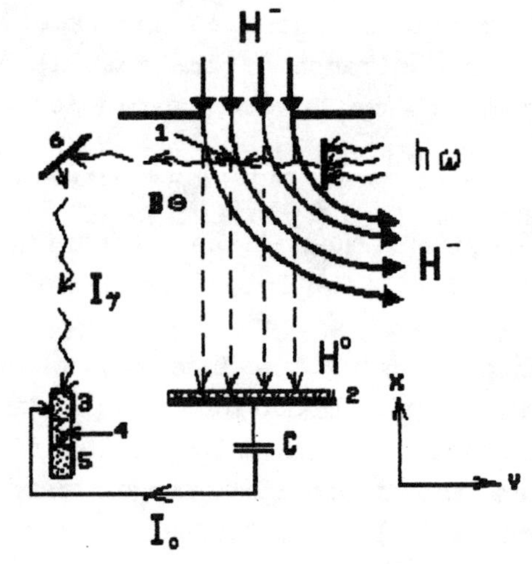

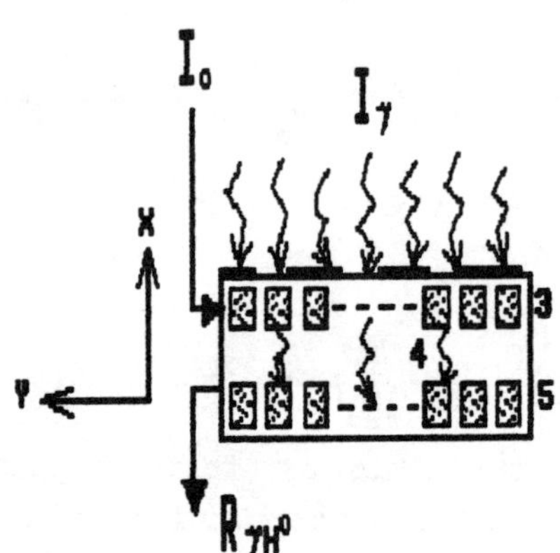

References

1. Stephen L.Kramer, D.Read Moffet, IEEE Trans.on Nucl.Science, NS-28, 2174 (1981).

2. Cottingame W.B. et al, IEEE Trans.on Nucl. Science, NS-32, 1871 (1985).

3. Methodes et techniques de traitement du signal et applications aux mesures physiques. Tome 1, par J.Max, Paris, 1981.

4. Belchenko Y.I. et al, Proc. XIII Intern. Conf.High Energy Accel., V.2, Novosibirsk, 276 (1987).

5. Ultrashort Light Pulses (picosecond techniques and applications), Edited by S.L.Shapiro, New York, 1977.

6. Kingston R.H. Proc.IEEE, 72, 954 (1984).

7. Deyhimy I.et al,IEEE Electr. Device Lett., EDL-2, 70 (1981).

Specialized Microprocessor Modules for
the Synchrotron Automatic Control System

A.G.Agababian,   S.H.Ananian,   A.A.Kazarian,

M.Yu.Khoetsian,A.R.Matevosian

Yerevan Physics Institute,Armenia,U.S.S.R.

At present a modernization of the Yerevan synchrotron control system is being performed with the aim to create multilevel architecture of computing means with the use of micro and personal computers.

To operate at the lower level, there is elaborated a set / 1-5 / of specialized modules based upon INTEL 8080 microprocessors that permit to solve the following problems :

- continuous measurement, tolerable check of parametrs and formation of actions that control the accelerator subsystems;
- buffering, preliminary processing, information conversion and its transfer to a higher-level computer;
- feedback local control;

The connection of the modules with the computer is realized via the RS 232 standard interface of a "Vector" design.

The main cause for elaboration of specialized modules and rejectoin of using standard nuclear electronic equipment is the intention of apparatus minimization with suffisient versatility of its functional possibilities.

In the present report we consider peculiarities of the construction of the microprocessor modules elaborated and also give specifications and functional diagrams.

1.   Module of Measurement and Processing of the Time Intervals /1/

The module performs measurement and preliminary processing of 50 time intervals in a fixed measurement cycle under two possible modes :

- measurement of 50 time intervals with begining at the same moment and termination at the different time moments in one cycle of measurements;
- measurement of 16 time intervals with begining and termination at arbitrary time moments in one cycle of measurements.

The module diagram is given in Fig.1 where the following main units are shown:

- central processor (CPU) based upon a INTEL 8080 microprocessor;
- ROM and RAM with 2 Kb each;
- RS 232 communication line adapter based upon INTEL 8251A chip which performes the connection of the module with the central computer in asynchronous mode of information exchange;
- control and synchronization circuit (CSC);
- array of counters (AC) based upon INTEL 8253 timer chips.

The accuracy of measurement of time intervals is 0.5 μs, the maximum duration of the measurement cycle interval is 32 ms.

The given module is used for measuring, check and diagnosis of the driving magnetic field and basic synchronization pulses of the Yerevan synchrotron.

2.   Timer Module /2/

The Timer Module has the following functional possibilities :

- programmed distribution of the basic synchronization pulse in different channels depending on the operator-preset succession;
- formation of controllable delay in fixed channels;

time selection of the basic synchronization pulse, check of its timely arrival and travelling.

A structural diagram of the module is shown in Fig. 2.

The tolerable check unit (TCU) passes the basic synchronization pulse only in the case when it falls to the tolerable "gates" whose parameters are formed by user's program.

In the commutation channel code memory (CCCM) the processor records the succession of channels' numbers according to which the comparision circuit (CC) commutes the channels with each arrival of the basic synchronization pulse. The time delay former (TDF) delays pulses in two channels by values determined by user's program. The hardware of TCU and CC as well as the introduced additional memory CCCM enabled to completely release CPU from direct control over pulse distribution process and tolerable check and to use it for solving other problems if necessary.

The main specifications of the timer module are as follows :
- the number of pulse distribution channels - 8;
- the number of time delay distribution channels - 2;
- the range of programmed time delay of pulses - from 1 μs to 32 ms;
- accuracy of delay formation - 0.5 μs;
- the length of the pulse distribution periodicity over channels is arbitrary, up to 256.

The timer module is used for synchronization of all output devices and physical experimental equipment with the operation cycles of the synchrotron in all modes of exploitation.

3. Universal Measurement-Control Module /3/

The Universal Module (UM) has the following functional possibilities :
simultaneous measurement of the shape of two analog signals and realization of tolerable check;
- formation of 6 pulses with controllable time delays;
- measurement and tolerable check of duration of 16 time intervals.

A structural diagram of the UM is given in Fig. 3. Here CSC is a control-and-synchronization circuit for the array of counters AC; ADC1 and ADC2 are 10-bit analog-digital converters for the measurement of values of analog signals with repeated triggering; LVM1 and LVM2 are memories for ADC1 and ADC2, respectively to store tolerable values; CC are circuits of comparison of measured and tolerable signals.The TDF timers work on arrivals of the external triggering pulse irrespective of the CPU, which permits the UM to operate simultaneously both in the tolerable check or measurement modes and in the mode of information exchange with the central computer.The number of discretization points in the measurement of analog signals is up to 256, the pitch of discretization and measurement - 55 μs.

The UM is the basic module for all control subsystems of slow beam ejection from the Yerevan synchrotron /4/.

4. Controlable Function Generator

The Controlable Function Generator (CFG) represents a program multichannel function generator intended for formation of controlling analog signals of arbitrary shape for 6 channels. The minimum value of signal time discretization is 1 μs, the number of programmed values of one signal is not more than 1024. The structural diagram of the CFG is presented in Fig. 4.

12-bit digital-analog converters (DAC) form program values of functions loaded by the CFG processor to the memory of values of generated curves (DTM). The control circuit (CC) provides time fixation of the shaped curves to the process, AC are address counters. The CFG is intended for formation of the modulation curve of the RF-system and power supply control of magnetic elements of the Yerevan synchrotron output systems.

5. Check and Diagnosis Module of State Signals /5/

The Check and Diagnosis Module (CDM) is intended for cyclic detection of state signals of "on-off" type, for the establishment of the fact of absence (presence) of changes, and for execution of diagnosis by realization of logic algorithms for situation recognition according to the state signals changes.

The CDM structural diagram is given in Fig. 5. Here M is an 8-bit multiplexer, CVR is a 16-channel 8-bit multiplexer and currentvalue register of state signals, CC are comparision circuits. The current values of state signals from the CVR outputs and the corresponding basic values from the RAM outputs arrive and are compared in the CC. In case of their anticoincidence the situation is analyzed by the processor. Further on the process is repeated.

The CDM is realized on 128 input channels.

REFERENCES

1. A.G.Agababian,S.H.Ananian,A.A.Kazarian et al."Microprocessor device for the magnetic field check system of the Yerevan synchrotron".
Preprint YerPI-1112(75)-88,Yerevan,1988.

2. A.G.Agababian,S.H.Ananian,N.A.Zapolsky et al."Timer synchronization unit for output subsystems of the Yerevan synchrotron".Preprint YerPI-1252(38)-90, Yerevan,1990.

3. A.G.Agababian,S.H.Ananian,V.G.Grigorian et al."Control unit of γ-beam extraction channel of the Yerevan synchrotron". Preprint YerPI-1113(76)-88,Yerevan,1988.

4. A.G.Agababian,S.H.Ananian,V.G.Grigorian et al."Structure and Organization of automatic control subsystem of beam extraction from the rapid-cycling synchrotron".Proceedings of the 11-th All-Union Conference on Charged Particle Accelerators. Dubna, 1989.

5. A.G.Agababian,S.H.Ananian,A.A.Kazarian et al."Unit for Check and Diagnosis of state signals for basic subsystems of the Yerevan synchrotron". Preprint YerPI-1093(56)-88, Yerevan, 1988.

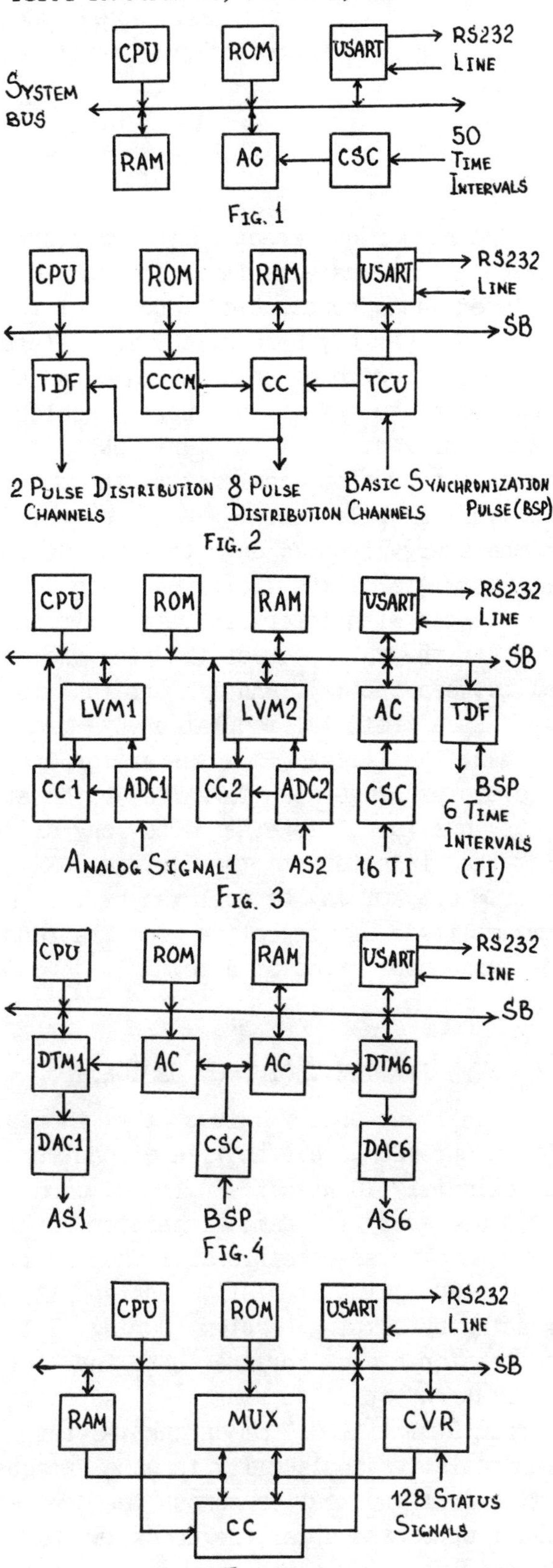

THE PULSERS OF THE DAMPING INJECTION OSCILLATION SYSTEMS FOR THE UNK I-STAGE

I.N.Ivanov, N.A.Malachov, V.A.Mel'nikov, N.V.Pilyar, A.S.Shcheulin
Joint Institute for Nuclear Research. Dubna.
V.V.Akimov, V.B.Ermakov, T.A.Latypov, G.L.Mamaev.
Moscow Radiotechical Institute, USSR Academy of Sciences, Moscow.

Summary

The hard requirement to the emittance of the beam is a characteristic feature of the designed and constructed accelerators for high and super-high energy. It is determined by working regime of an accelerator as a collider or buster for the next acceleration stage. The I-stage of the IHEP UNK[1] is a proton synchrotron accelerating the beam with a common particle number $5 \cdot 10^{14}$ in the beam at the energy from 70 GeV to 400 GeV. The beam consists of the 12 trains each of 4.6 μsec length with intervals between trains 0.6 μsec. At the full pack of the ring the interval between the 12th and 1st trains is 5.2 μsec. Each train is modulated by frequency 200 MHz. The I-stage is a buster for the 2nd accelerator stage at 3 TeV. The both stages occupy one tunnel together. The long distance from the injector to the I stage UNK (the order $\approx$ 6.5 km) may be followed by the maximum amplitude of injection oscillations to the consequent error of 5 mm.

I. THE CHOICE OF DESIGN.

The linear theory gives the increasing time about 120 μsec (the time of one turn is approximately 70 μsec). In these conditions it is necessary to damp injection oscillations as soon as possible till the amplitude could keep against resistive instability in creasing by damping system during all the accelerator cycle. For the UNK the initial oscillation amplitude reduce is supposed to be from 5 mm till 0.5 mm during 1-3 turns and damp resistive instability in the frequency region from the most dangerous low one (about 4 kHz) up to the upper frequency, where Landau damping is expected for the given energy spread. In our case this frequency is 500 kHz.

Due to this the solution was taken to perform separate systems for injection oscillation damping and limitation of resistive instability increasing. Every of these systems uses 2 kickers in X- and Z- directions. Every pair of kickers is placed on the accelerator ring in such a way, to make the phase advance of betatron oscillations between them close to $\pi/2$. The supply system for injection oscillation damping includes the pulsers, which must provide necessary time characteristics and accuracy requirements to the current pulses in the kickers. In our case the maximum oscillation of the generator pulse top must be less than $\pm 3\%$. At such a pulser design, in principle, two approches are possible:

1) to design a pulser, which provides to damp the oscillation during one turn only;

11) to design a pulser which permits to influence the beam during turns.

II. KICKER

We agree that the kicker constructed in IHEP for the 1 UNK[2] stage may satisfy us. The theorethical data have shown that the intergral magnetic field at the level of 0.014 Tm is required. The kicker is a vacuum box with current lines surrounded by ferrit yokes inside. The aperture of the kicker is 60x60 mm. The current lines are divided into two parts, each of them has got an independent high voltage exit. It gives a possibility to provide the magnetic field pulse of the different sign within each kicker. The inductivity of the every half of kicker current lines is 2-3 μH. It means that the maximum current of 500 A corresponds to the necessary magnetic field.

The kickers are connected by cable lines of the supply systems placed in a special building on the ground (the total connection lines length $\geqslant$ 300 m).

III. ONE TURN DAMPING PULSER[3]

The general idea of this generator is as follows: the U_0 summary capacity form line charged preliminary till the constant voltage is discharged during the time $\leqslant$ 50 μsec (that time is determined by the time one turn minus the time for control system work) to the necessary level corresponding to the pick-up signal. The scheme of the generator is represented in Fig.1.

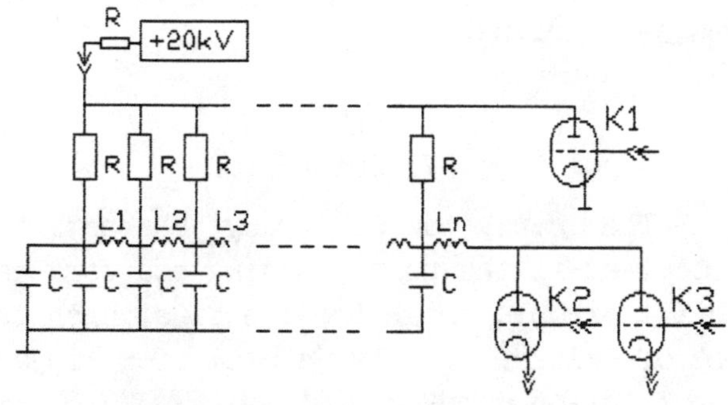

Fig.1 One turn pulser scheme.

The effective meaning of the resistances R1-RN is chosen to reduce the voltage up to the level of 0.03 Uo during 50 μsec while the commutator K1 is switched on (the line is discharged like a concentrated capacity) The commutator K2 starts functioning in time To after the measuring moment. While switching on the K1 the law of the transformation is chosen to have the voltage by the moment of K1 working at the line corresponding to the current necessary for A_{meas} damping. The higher A_{meas} - the later K1 is switched on.

The curves illustrating the principle of the scheme work are shown in Fig.2, and the current on the loading has satisfietied the beam time characteristics in Fig.3.

The usage of the third commutator K3 switched on at the kicker with different current line allows one to get magnetic field pulses of different polars from one generator with one polar charged voltage.

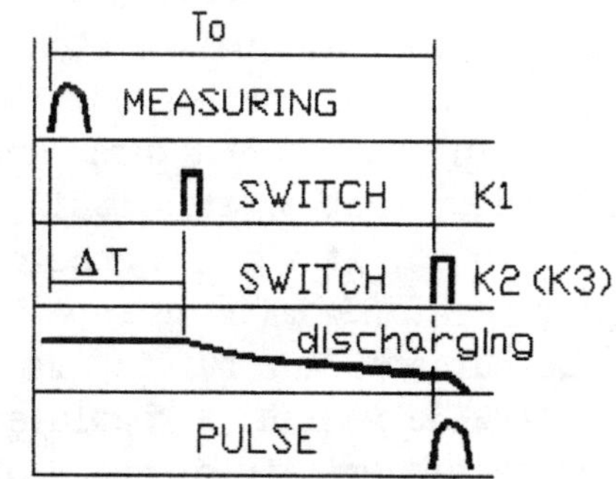

Fig.2   Time diagram of one turn pulser work

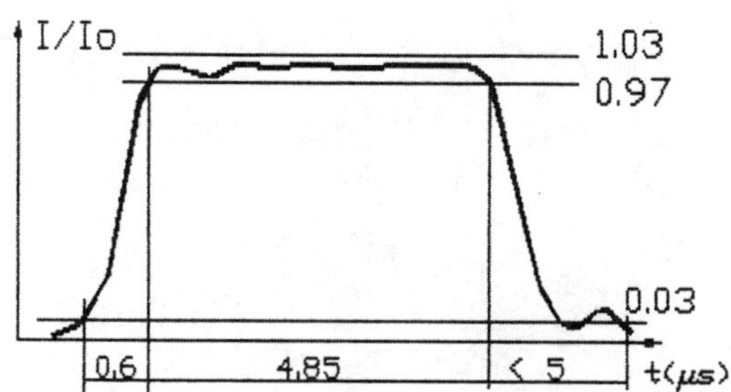

Fig.3   Current curve on the loading

The constructed line possesses the effective wave resistance 17 Om and provides its sequence with a feeder. The sequence of the feeder with the kicker was done by including it into LC-chain with the same wave resistance. Model experiments have shown the correspondence between the real time, amplitude, accuracy parameters of the output pulse and the calculations in all the range of changes (current pulse is changed in the loading for 32 times due to the change of switching delay for 50 μsec).

IV. THE PULSER FOR MULTI TURN ACTION AT THE BEAM.

Developing the one turn damping system for the UNK initial betatron oscillations we cannot but take in account the reasons due to which it could turn out not effective enough. The main reasons are error amplitude oscillation measuring and partial instability rising during one turn. So, we must provide the possibility of multi turn action at the beam.

On this purpose the scheme of fast form line generator with a regulated amplitude of charging is being designed. The repetition frequency of the same pulses should not be less than 14 kHz, the rest parametrs of the pulse are similar to the one turn damping generator. The current on the loading at the level of 500-100 A. At the ETA accelerator[4] a MAG tipe generetor work with repetition frequency of the pulses 5 kHz but it is not supposed for the wide region of regulated voltage. The investigated principle scheme of the generator is represented in Fig.4.

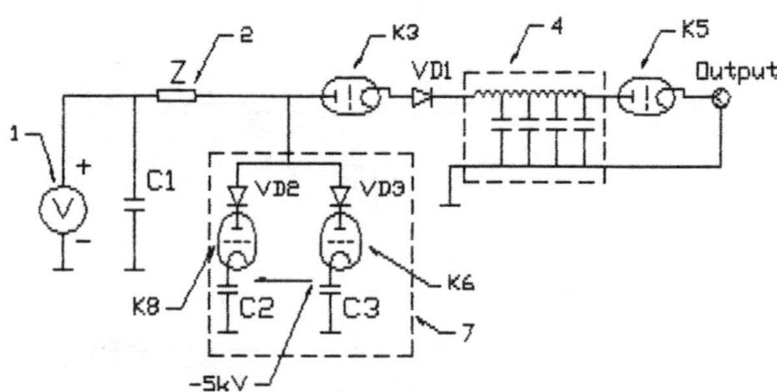

Fig.4 The principle scheme of a multi turn pulser.

The artificial form line 4 is charged by commutator K3 and one-port 2 from the constant voltage source 1. The commutators K6 and K7 are included into a symmetrically double-pole switch 7. The length of charging and, therefore, the magnitude of the voltage on the line 4 is determined by the time interval between commutator K3 and K6 (or K8) switching. A double-pole switching scheme was taken to decrease reguirements to the commutator K6 and K8 recovery time. The availability time for work of each arm switch 7 must include the recharging time the storage capacities C_2, C_3 in thyratron catode chain. One-port 2 accelerates the current break in commutators K3, K6, K8. It provides the principle posibility of the scheme work with necessary frequency repetition pulses 14 kHz. The voltage on the storage capacity line curve is shown in Fig.5.

The total time of line charging up to 15 kV is not more than 30 μsec. On the back front of the voltage curve one can see the leap, connected with the discharging line for the sequence loading. The investigations on the generator have shown the principle possibility of work with high frequency repetition

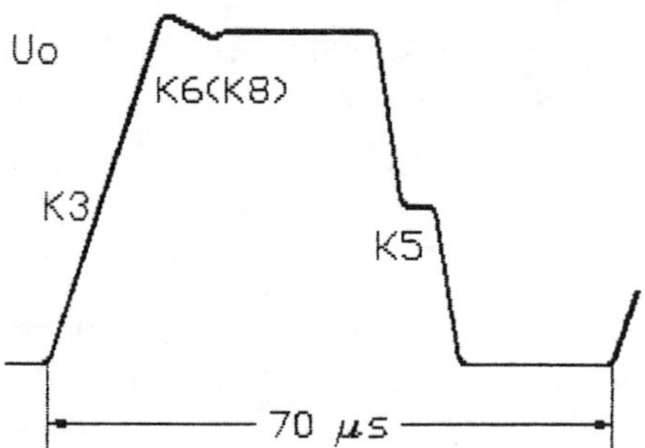

Fig.5 The line voltage

of bursts containing several tens pulses. The limitation for a number of pulses is the middle current of a thyratron. Every generated pulse may be regulated inside the 0-15 kV amplitude region. The amplitude setting must be done not later than 30-40 μsec before the pulse starting.

V. CONCLUSION

The study on accuracy characteristics received at the pulse loading and work reliability should determine the final architecture of kicker supply system for injection oscillation damping.

REFERENCES:

[1] A.I.Ageev et al. Proceeding of the X All-Union conferens on particles accelerators, JINR, D9-87-105, Dubna,v.2,p.430.
[2] V.N.Andreev et al. Proceeding of the IX All-Union conferens on particles accelerators, JINR, Dubna, v.2, p.395.
[3] Zhabitskij et al. 'Simulation of one-turn damping system of initial betatron beam oscillation for UNK I stage'. P9-90-548, JINR, 1990.
[4] M.Newton et al. 'Energy flow and high repetition rate issue for ETAII magnetic modulator system'. Lawrence Livermore National Laboratory, Livermore, California, 9450.

Particle Amplitude Growth Due to Single
or Repetitive Resonance Crossing

S. R. Mane and W.T. Weng

Brookhaven National Laboratory, Upton, N.Y. 11973

We study the problem of the growth of the amplitude of a particle executing betatron oscillations in a synchrotron or storage ring, due to the crossing of a resonance line. The resonance can be crossed either only once, or repeatedly. We treat only one degree of freedom in this paper, and assume the particles in the beam are independent. We consider some simple models of amplitude-dependent tunes, to discern the effects of detuning away from the resonance as the particle amplitude grows. The basic equation of motion we shall treat is the harmonic oscillator equation

$$\ddot{x} + \omega_x^2 x \;=\; f(t) \tag{1}$$

where the frequency $\omega_x = \omega_x(x,t)$ may depend on x and t, and also other parameters, and f is a driving term, which may also depend on t and other quantities. A more complete description of our analysis and results, including numerical results, is given in Ref. [1]. We only present the main points here. We shall assume that the driving term f has a sinusoidal time dependence, i.e. $f \propto F\cos(\omega_f t)$, or sometimes $f \propto Fe^{i\omega_f t}$. First, let us suppose that $\omega_x = \omega_{x0}$, a constant, and

$$\ddot{x} + \omega_{x0}^2 x \;=\; \frac{F}{2\tau_{rev}} e^{i\omega_f t}, \tag{2}$$

where τ_{rev} is the revolution time around the ring. The solution for the driven part of the motion is easily found to be

$$x \;=\; -i\,\frac{F}{2\tau_{rev}}\,\frac{e^{i\omega_f t}}{\omega_{x0}^2 - \omega_f^2} \tag{3}$$

and, on resonance, i.e. $\omega_{x0} = \omega_f$, the solution is

$$x \;=\; -i\,\frac{F}{2\tau_{rev}}\,\frac{t\,e^{i\omega_f t}}{2\omega_f}. \tag{4}$$

This grows without bound as $t \to \infty$. Suppose, however, there is an amplitude dependent tuneshift, given by

$$\omega_x \;=\; \omega_{x0} + \mu I_x, \tag{5}$$

* Work performed under the auspices of the U.S. Department of Energy.

where μ is a constant, and $I_x = [(\omega_{x0}x)^2 + \dot{x}^2]/(2\omega_{x0})$ is the action. The above model is an octupole tuneshift, since it is quadratic in the particle amplitude, i.e. linear in I_x. Suppose also that the motion is on resonance in the absence of the tuneshift, i.e. $\omega_f = \omega_{x0}$. Then the motion is resonant for small values of I_x, but as I_x increases, the motion is detuned, i.e. it goes off resonance. Putting $x = u\,e^{i\omega_f t}$, we find that, from the above definition, $I_x = \omega_{x0}u^2/2$. We also find that the motion does not grow to infinity, but reaches a peak value, given by [1]

$$u_{max} \;=\; \left(\frac{F}{2\tau_{rev}\mu\omega_{x0}^2}\right)^{1/3}, \tag{6}$$

$$I_{max} \;\propto\; u_{max}^2 \;\propto\; \left(\frac{F}{\mu}\right)^{2/3}. \tag{7}$$

We have verified the above result by tracking [1].

We now treat the next case in our investigation. We now suppose that ω_x is a constant, without any amplitude dependence. Suppose also that the frequency of the driving term increases linearly with time, i.e. $\omega_f(t) = \omega_1 + \dot{\omega}t$, where ω_1 and $\dot{\omega}$ are constants. Without loss of generality, we may suppose that $\omega_1 = \omega_x$ and that $\dot{\omega}$ is positive. In that case, the frequency of the driving term starts off below ω_x, then becomes resonant with ω_x, and then again becomes nonresonant. We again expect that the particle amplitude will grow, to a large but finite asymptotic value, but the expression for the peak amplitude will be different from the previous case. The problem can in fact be solved analytically. We write

$$\phi(t) \;=\; \int_0^t \omega_f(t')\,dt' \;=\; \omega_1 t + \frac{\dot{\omega}t^2}{2} \tag{8}$$

and, using a Green function approach, we can show that

$$\begin{aligned}
x \;&=\; \frac{1}{\omega_x}\int_{-\infty}^t \sin[\omega_x(t-t')]F\sin[\phi(t')]\,dt'\\[4pt]
&=\; -\frac{F}{2\tau_{rev}\omega_x}\left(\frac{2\pi}{\dot{\omega}}\right)^{1/2}\cos(\omega_x t + \frac{\pi}{4}),
\end{aligned} \tag{9}$$

after some algebra. The peak amplitude is therefore

$$|\omega_x^2 x^2 + \dot{x}^2|^{1/2} \;=\; \frac{F}{2\tau_{rev}}\left(\frac{2\pi}{\dot{\omega}}\right)^{1/2} \;\propto\; \frac{F}{\dot{\omega}^{1/2}}. \quad (10)$$

Lysenko [2] also found, independently, that the parameter which characterizes the amplitude growth is $F/\dot{\omega}^{1/2}$. We can offer a heuristic derivation of the above result. In the vicinity of the resonance, the amplitude of x grows linearly with time, $x \propto t$, so the amplitude growth is approximately proportional to the time spent near the resonance, say T. Now the particle will be close to the resonance when $\omega_f \simeq \omega_x$, so that the phase of the particle oscillation, $\omega_x t$, and of the driving term, $\int \omega_f \, dt'$, are in phase. However, ω_f increases with time, so after a while $\omega_x t$ and $\int \omega_f \, dt'$ will get out of phase. This will happen, roughly speaking, when the phase gap between them reaches 2π. Hence the time T is approximately given by

$$\int_0^T \omega_f \, dt' \;\simeq\; \omega_x T + 2\pi. \quad (11)$$

Putting $\omega_f = \omega_x + \dot{\omega} t'$, we obtain

$$\frac{\dot{\omega} T^2}{2} \;\simeq\; 2\pi \quad (12)$$

or $T \propto \dot{\omega}^{-1/2}$. As argued above, the particle amplitude growth is proportional to T, hence $\dot{\omega}^{-1/2}$. The factor of F is obvious because the equation of motion is linear, hence $x \propto F/\dot{\omega}^{1/2}$.

We now turn to the third case to be studied in this paper, viz. multiple, or repetitive, crossing of a resonance. This type of situation is more pertinent to the "storage mode," where the revolution frequency, betatron frequency, and driving frequency are fixed for long periods of time. Because of synchrotron oscillations and chromaticity, the betatron frequency of a particle may oscillate sinusoidally and cross a resonance line repeatedly. Let us therefore suppose that

$$\omega_x \;=\; \omega_{x0} + \lambda \cos(\omega_m t) \quad (13)$$

where λ is the amplitude of the tune modulation and ω_m is the modulation frequency. We refrain from writing ω_s instead of ω_m because the above modulation may be caused by other sources, e.g. power supply ripple, not only synchrotron oscillations. We do, however, assume that $\omega_m \ll \omega_{x0}$ and $\omega_m \ll \omega_{rev}$, the (angular) revolution frqeuency. The solution for x, using a Green function approach, can be written as

$$x \;=\; \frac{F}{\omega_{x0}} \int_{-\infty}^t \sin\left(\int_{t'}^t \omega_x \, dt''\right) \sin(\omega_f t') \, dt'. \quad (14)$$

We use the Bessel function identity

$$e^{ir\sin\psi} \;=\; \sum_{n=-\infty}^{\infty} e^{in\psi} J_n(r). \quad (15)$$

We also average over the betatron oscillations, and treat only the centroid $\bar{x}$. The solution can then be written as [1]

$$\bar{x} \;=\; \int_{-\infty}^t \frac{F(t')}{\omega_{r0}} e^{-\bar{\mu}(t-t')} \sin\Big[\omega_{x0}(t - t') \\ + \frac{\lambda}{\omega_m}[\sin(\omega_m t) - \sin(\omega_m t')]\Big] \frac{dt'}{\tau_{rev}} \quad (16)$$

A convergence factor $e^{-\bar{\mu}(t-t')}$ has been introduced into the above expression, to model the decoherence caused by an amplitude dependent tune. After some algebra, the solution is

$$\bar{x} \;=\; \frac{F}{2\omega_{x0}\tau_{rev}} \mathrm{Re}\Big\{ e^{i(\lambda/\omega_m)\sin(\omega_m t)} \times \\ \sum_n e^{-in\omega_m t} J_n\left(\frac{\lambda}{\omega_m}\right) \times \\ \Big[\frac{e^{i\omega_f t}}{\bar{\mu} - i(\omega_{x0} + n\omega_m - \omega_f)} \\ - \frac{e^{-i\omega_f t}}{\bar{\mu} - i(\omega_{x0} + n\omega_m + \omega_f)} \Big]\Big\}. \quad (17)$$

In other words, there are resonances at the betatron frequency ω_{r0} plus "sidebands" $\omega_{x0} + n\omega_m$, $n = 0, \pm 1, \pm 2, \ldots$

There are now two cases to consider. If the sidebands are very close, i.e. $\omega_m \ll |\omega_f - \omega_{x0}|$, then we find [1]

$$\bar{x} \;=\; \frac{F}{2\omega_{r0}\tau_{rev}} \mathrm{Re}\Big\{ \frac{e^{i\omega_f t}}{\bar{\mu} - i(\omega_{x0} - \omega_f)} \\ - \frac{e^{-i\omega_f t}}{\bar{\mu} - i(\omega_{x0} + \omega_f)} \Big\}. \quad (18)$$

This is exactly as if there had been no tune modulation at all, as can be verified by putting $\lambda = 0$ in the integral for $\bar{x}$. Hence if the driving frequency lies outside the "cluster" of the betatron frequency and its sidebands, then the cluster can be treated as one single frequency.

If the sidebands are widely separated, so that ω_m is comparable to $|\omega_{r0} - \omega_f|$, then we approximate by assuming that only one sideband (only one value of n) is important, in which case

$$\bar{x} \;\propto\; \frac{F}{2\omega_{r0}\tau_{rev}} \frac{J_n(\lambda/\omega_m)}{\bar{\mu} - i(\omega_{x0} + n\omega_m - \omega_f)}. \quad (19)$$

If the tune modulation is due to synchrotron oscillations, then we must average over the amplitude and phase of the synchrotron oscillations. The details are given in Ref. [1] and are not relevant here.

In conclusion, we have presented three case studies of situations where the amplitude of a betatron oscillation can grow, due to crossing a resonance. We have presented expressions for the amplitude growth caused by the resonance, in the various cases. Further investigations, e.g. involving two degrees of freedom, or other types of driving terms and other types of resonances, will be presented elsewhere.

References

[1] S.R. Mane and W.T. Weng, Nucl. Instr. and Meth., in press.

[2] W.P. Lysenko, Part. Accel. **5**, 1 (1973).

CONSTRUCTION OF HIGH ORDER MAPS FOR LARGE PROTON ACCELERATORS[*]

JOHN IRWIN

Stanford Linear Accelerator Center Stanford University, Stanford, California 94309

Abstract

We propose a simple perturbation method using a truncated CBH Theorem to calculate the nonlinear generator for one turn maps, and argue that such maps should give accurate long turn dynamic apertures for proton colliders such as the SSC and LHC.

I. INTRODUCTION

A promising application for maps is the determination of long term dynamic apertures. Since the exact machine is not known, a statistical ensemble must be studied to establish confidence in design parameters and specifications. This requires studying a large set of statistically generated machines. To use maps to study dynamic apertures one must be able to rapidly track them and construct them. The rapid tracking of maps may be done through a kick factorization [1] or a Fourier developed generating function. [2] Here we address calculation of maps.

The present method of calculating high order maps requires tracking a Taylor series element-by-element around the ring to determine the one turn Taylor series map. [3] At interesting orders the CPU time to construct the map in this way becomes prohibitive for a large machine. For example a ninth order SSC map requires more than two hours of Cray CPU time.

We propose here a method by which these maps can be calculated rapidly. The method also provides insight into the physical sources of terms in the map.

II. DESCRIPTION OF THE METHOD

A. The Lattice Representation

Most element-by-element tracking programs begin by approximating the lattice by a series of linear transformation and nonlinear kicks. Canonical integrators prescribe the magnitude and sequence of kicks for thick elements. [4] The linear transformations may be analytically represented by matrices and the kicks may be represented by exponential Lie operators. [5] There are also ways to represent a thick element by two linear transformations with an exponential Lie operator acting at the center. [6] Errors in placement and strength as well as fringe fields can be represented this way. Consequently, a product of linear transformations and nonlinear exponential Lie operators is a very general starting point for a lattice representation.

Using a similarity transformation,

$$Re^{:f:}R^{-1} = e^{:Rf:},$$

where R is a linear operator, one can rearrange operations to obtain a product of linear transformations times a product of Lie operators.

$$\prod_{1 \leq n \leq N} e^{:f_n:} R_n = \prod_{1 \leq n \leq N} R_n \prod_{1 \leq n \leq N} e^{:\bar{R}_n f_n:}$$

where $\bar{R}_n = \prod_{1 \leq m \leq n} R_m$.

[*] Work supported by Department of Energy contract DE–AC03–76SF00515.

B. CBH Theorem

The nonlinear problem becomes a problem in concatenating the Lie operators. Two exponential Lie operators may be combined into one by using the Cambell-Baker-Hausdorf (CBH) Theorem.

$$e^{:a:}e^{:b:} = e^{:c:}$$

with

$$c = a + b + \frac{1}{2}[a, b] + \frac{1}{12}[a - b, [a, b]] + \frac{1}{24}[b, [a, [a, b]]] + \dots$$

To combine several exponential Lie operators one can start at one end and proceed stepwise. At the nth step,

$$e^{:f_{n+1}:}e^{:F_n:} = e^{:F_{n+1}:}$$

where F_n is the generator corresponding to the product of the first n Lie operators.

The CBH theorem can be understood as a perturbation series. The initial sum of generators, $c \approx a + b$, gives the correct result when there is no interaction between the nonlinearities. The first Poisson bracket (PB), $\frac{1}{2}[a, b]$, gives the first order (in generator strength) modification as the action of one operator is altered by the presence of the other, corresponding to the modification of one kick that results from a previous kick in a ring. The double PB gives second order effects and so on. Thus one may view the CBH theorem for the product generator as a perturbation series in the nonlinearity strength. If each nonlinearity is small, which is indeed the case with proton colliders, the higher order terms in the perturbation series are quite small.

It is important to note that this perturbation series is not the same as truncating at some polynomial order. The individual nonlinearities can contain high order monomials.

C. Separation of Orders

There may be occasion or interest in keeping track of contributions that come from different orders. In the method described above terms of different order become mixed. This is not necessary, and with a bit more work, terms of different order can be kept separate. For this purpose we will replace f_n by εf_n and expand F_n.

$$F_n = \varepsilon F_n^1 + \varepsilon^2 F_n^2 + \varepsilon^3 F_n^3 + \dots$$

Expanding the CBH theorem in powers of ε and identifying terms of equal power one finds:

$$F_{n+1}^1 = f_{n+1} + F_n^1$$

$$F_{n+1}^2 = F_n^2 + \frac{1}{2}[f_{n+1}, F_n^1]$$

$$F_{n+1}^3 = F_n^3 + \frac{1}{2}[f_{n+1}, F_n^2] + \frac{1}{12}[f_{n+1} - F_n^1, [f_{n+1}, F_n^1]]$$

$$F_{n+1}^4 = F_n^4 + \frac{1}{2}[f_{n+1}, F_n^3] + \frac{1}{12}[f_{n+1}, [f_{n+1} - F_n^1, F_n^2]]$$
$$+ \frac{1}{24}[F_n^1, [f_{n+1}, [f_{n+1}, F_n^1]]].$$

Using these relationships it is possible to write the F_n's directly in terms of the f_n's. For example,

$$F_m^1 = \sum_{n \leq m} f_n$$

$$F_m^2 = \sum_{p \leq n \leq m} \frac{1}{2} [f_n, f_p].$$

III. Estimating and Calculating Poisson Brackets

A. Estimating Poisson Brackets (PBs)

From the basic definition of the exponential operator,

$$z' = e^{:a:} z = z + [a, z] + \frac{1}{2} [a, [a, z]] + \cdots$$

it follows that for a weak nonlinearity $\delta z \approx [a, z]$. For an action variable $J \equiv 0.5(x^2 + p_x^2)$, $\delta J \approx [a, J]$.

For a a polynomial of order p, which we denote by a_p, let us take as an estimate of the strength of a_p the expression

$$a_p^\# \equiv \max_{0 \leq \theta \leq 2\pi} \frac{|\delta J|}{2J}$$

We have included the factor 2 because J is quadratic in phase space variables x and p_x, and defined this way the estimate is related to the definition of smear [7]. Writing x and p_x in terms of J and θ we find

$$a_p = \sum_k \alpha_{pk} J^{\frac{p}{2}} e^{ik\theta}$$

where α_{pk} is complex, $\alpha_{p,-k}$ is the complex conjugate of α_{pk}, and only even or odd values of k occur according to whether p is even or odd. Thus

$$\frac{\delta J_a}{J_a} \approx \sum_k ik\alpha_{pk} J^{\frac{p}{2}-1} e^{ik\theta}.$$

As a rough estimate of the sum, we replace each quantity by its average value and multiply by the number of turns. Then

$$\frac{|\delta J_a|}{J} \approx \frac{p^2}{2} \langle |\alpha_{pk}| \rangle J^{\frac{p}{2}-1}$$

Likewise for b_q

$$b_q = \sum_j \beta_{qj} J^{\frac{q}{2}} e^{ij\theta}$$

and

$$\frac{|\delta J_b|}{J} \approx \frac{q^2}{2} \langle |\beta_{qj}| \rangle J^{\frac{q}{2}-1}$$

To estimate $\delta J_{[a_p, b_q]}$ we first calculate $[a_p, b_q]$.

$$[a_p, b_q] = J^{\frac{p+q}{2}-1} \sum_{j,k} \alpha_{p,k} \beta_{q,j} i \frac{pj - qk}{2} e^{i(j+k)\theta}$$

From $\langle |(pj - qk)(k + j)| \rangle \approx \max(p,q)pq/3$ and the same estimation procedure as above, we get

$$\frac{|\delta J_{[a,b]}|}{J} \approx J^{\frac{p+q}{2}-2} \frac{\max(p,q)(pq)^2}{6} \langle |\alpha_{p,k}| \rangle \langle |\beta_{q,j}| \rangle$$

from which we deduce

$$\frac{|\delta J_{[a,b]}|}{J} \approx \frac{2}{3} \max(p,q) \frac{|\delta J_a|}{J} \frac{|\delta J_b|}{J}.$$

or

$$\left(\frac{1}{2} [a_p, b_q] \right)^\# \approx \frac{2}{3} \max(p,q) a_p^\# b_q^\#$$

Another way to look at this is to make use of the Jacobi identity to write

$$\delta J_{[a,b]} \approx [[a, b], J] = [a, [b, J]] - [b, [a, J]] = [a, \delta J_a] - [b, \delta J_b]$$

where it is now clear that the change of J due to $[a, b]$ is the change of a change, and hence of higher order. The factor $2 \max(p, q)/3$ comes from the fact that the derivative of the change comes in, not just its magnitude. We have used the fact that for polynomials of known order we can estimate the derivative in terms of the function.

For the SSC, at values of J near the long term dynamic aperture $\frac{|\delta J_f|}{2J} \approx .001$ for an f generating the nonlinearities of a single dipole. As one proceeds around the ring the cumulative effects of these nonlinearities grows as $\sqrt{N}$, where N is the number of dipoles. For the generator representing the nonlinearities of the full ring $\frac{|\delta J|}{2J} \approx .06$ at J near the long term dynamic aperture. We note that the long term dynamic aperture is very near what is referred to as the linear aperture, the latter being defined as that aperture with smear equal 10%.

The connection between the nonlinear generator we are calculating and the generator giving the smear is to be found in Normal Form theory. [8] Normal Form theory attempts to combine the linear transformation and the nonlinear factors into one grand generator which would be the pseudo-Hamiltonian of the ring. This is a risky procedure which works in some situations and not in others. Resonance denominators are involved. Such problems do not arise in the formation of the nonlinear generator we are constructing.

B. Calculation of Poisson Brackets

A PB in two degrees of freedom can be calculated by taking four derivatives of each function and performing four multiplications. In the general situation it is preferrable to use the PB for the basic monomials:

$$[\alpha_{abcde} x^a p_x^b y^c p_y^d \delta^e, \beta_{pqrst} x^p p_x^q y^r p_y^s \delta^t] = \alpha_{abcde} \beta_{pqrst} \delta^{e+t}$$

$$\{ (aq - bp) x^{a+p-1} p_x^{b+q-1} y^{c+r} p_y^{d+s}$$
$$+ (cs - dr) x^{a+p} p_x^{b+q} y^{c+r-1} p_y^{d+s-1} \}$$

The number of operations involved is somewhat greater than two multiplications if the monomials are scanned to optimize computation time. When one of the functions is a kick the calculation can be reduced to two multiplications. This follows from the relationship

$$\left[\sum \alpha_{nm} x^n y^m, F \right] = \sum n\alpha_{nm} x^{n-1} y^m [x, F]$$
$$+ \sum m\alpha_{nm} x^n y^{m-1} [y, F].$$

C. Comparison with Power Series Tracking

When carefully optimized the number of polynomial multiplications per dipole in power series tracking is about twice the order retained in the error multipoles of the dipole. For 5th order multipoles (through dodecapole) this is 12 multiplications per dipole. For the procedure we are suggesting the number of PBs will be one or two per dipole corresponding to approximately two or five multiplications, independent of maximum multipole order. For ninth order multipoles the improvement in computational time could be almost a factor of ten. Some computations will be needed to establish the exact improvement because there are other effects: the order of the generators is larger by one than the order of the map, though they begin at third order for the first PB and at fourth order for the second PB and many pairs are absent.

IV. Choice of Truncation of the CBH Theorem

A. Truncation Order for Ensembles of Maps

In concatenating the many weak nonlinear generators into one generator it is crucial to maintain at least one of the PBs in the CBH theorem. This conclusion rests on the fact that one PB is needed to include the effects of sextupoles acting on sextupoles. Such terms produce an octupole like term that gives an important contribution to the phase shift with amplitude.

However there is an indication that statistically this may be sufficient. Statistical analytic calculations of tune shift and smear, including one PB for the tune shift and none for the smear, have been compared with tracking over 400 seeds. The tracking results were Fourier analyzed so that one could compare smear details. The results agreed [9] to within the accuracy of the tracking analysis.

One might further argue that if you have the correct tune shift with amplitude, which determines the position of resonance lines in the particle phase space, and the correct smear amplitudes, which determine the driving strengths of resonances, the long term dynamic apertures would also agree. We believe this will be especially true when one compares ensembles of machines because the coefficient of any particular monomial in the map will vary more widely with the change in the first order terms than with modification by higher order effects. If this is indeed the case, the limits on our ability to know and measure the high order multipoles in a single magnet will limit our ability to predict the long term dynamic aperture.

B. Statistical Generation of Maps

If it can be established that the maps constructed with one PB produce the same long term dynamic aperture, or even that statistically they give the same long term dynamic aperture, then it is possible to generate these maps analytically. [9] This can be done extremely fast. In this case it should be possible to identify the important coefficients in the map and their source. This would provide important insight into the determination of the long term dynamic aperture.

C. Truncation Order for Individual Maps

Though we believe it is enough to include one PB of the CBH theorem when studying the properties of an ensemble of machines, we recognize it would build confidence in the method to show that the map for a single particular machine generated in the way we propose gives exactly the same dynamic aperture as the element-by-element tracking of that machine. For this purpose we conjecture that one more PB may be enough. This belief rests on our estimation of the accuracy of the approximation at this level, that the tune shift with amplitude and the resonant

strengths are determined accurately, and also on the fact that the next order PB, if it were to be included, is even an order smaller than would be expected at first sight. In other words including one more PB includes all effects that occur in the next two orders of approximation. We now present the argument for this result.

Denote by f_n the nth nonlinear generator, and by F_n the generator that combines the first n generators. For the SSC the result of summing up terms $[f_{n+1}, F_n]$ is about $[F_N, F_N]$, and so is approximately of magnitude $(.06)^2$ at the long term dynamic aperture. The result of summing a term like $[F_n, [f_{n+1}, F_n]]$ will be down by another factor of .06 because it involves an additional PB. However the term $[f_n, [f_n, F_n]]$ will be smaller yet by a factor of 1/50 because F_n grows like $\sqrt{n}$ compared to f_n, and half way around the ring $\sqrt{n} \approx 50$. Thus at any given order of PB, terms which have two f's will be smaller than terms with one f by more than one order.

Since the only third order PB occurring in the CBH theorem is

$$[f_{n+1}, [F_n, [f_{n+1}, F_n]]]$$

and since it has two f's, effects from this order will be more than two orders smaller than the second order PB terms.

V. Conclusion

We have described a method to calculate one turn maps that offers insight into the structure of the maps and is substantially faster than standard power series methods. If the program suggested here is carried out and the conjectures proven correct, the nonlinear generators for one turn map maps could be calculated directly using analytic expressions containing random numbers. Substantial insight would be gained into the physics behind the determination of long term dynamic apertures.

References

[1] J. Irwin, A Multi-kick Factorization Algorithm for Nonlinear Maps, SSC-228 (1989)

[2] J. Berg and R. Warnock, Symplectic Full-turn Maps in a Fourier Representation, SLAC-PUB-5458 (1991)

[3] M. Berz, "Differential Algebraic Description of Beam Dynamics to Very High Orders", Part. Accel. **24**, 109 (1989)

[4] R. Ruth, "A A Canonical Integration Technique", IEEE Trans. Nuc. Sci, NS-30, 2669 (1983); H. Yoshida, Phys. Lett. A, **150**, 262 (1990)

[5] A. Dragt, F. Neri, G. Rangarajan, D. Douglas, L. Healy, and R. Ryne, "Lie Algebraic Treatment of Linear and Nonlinear Beam Dynamics", Ann. Rev. Nucl. Part. Sci. **38** 455 (1988)

[6] J. Irwin, "The Application of Lie Algebra Techniques to Beam Transport Design", Nuc. Inst. and Methods, **A298**, 460 (1990)

[7] M. Furman and S. Peggs, "A Standard for Smear", SSC-N-634 (1989)

[8] E. Forest, M. Berz, and J. Irwin, "Normal Form Methods for Complicated Periodic Systems", Part. Accel. **24**, 91 (1989)

[9] J. Bengtsson and J. Irwin, "Analytical Calculations of Smear and Tune Shift", SSC-232 (1990)

Refinement of the Hamilton-Jacobi Solution using a Second Canonical Transformation*

W. E. Gabella†
Department of Physics,
University of Colorado,
Boulder, CO 84309

R. D. Ruth and R. L. Warnock
Stanford Linear Accelerator Center,
Stanford University,
Stanford, CA 94309

Abstract

Two canonical transformations are implemented to find approximate invariant surfaces for a nonlinear, time-periodic Hamiltonian. The first transformation is found from the non-perturbative, iterative solution of the Hamilton-Jacobi equation. The residual angle dependence remaining after performing the transformation is mostly eliminated by a second, perturbative transformation. This refinement can improve the accuracy, or the speed, of the invariant surface calculation. The motion of a single particle in one transverse dimension is studied in a storage ring example where strong sextupole magnets are the source of the nonlinearity. The refined transformation to action-angle variables, and the corresponding invariant surface, can attain accuracy similar to that of a good non-perturbative transformation in half the computation time.

I. The First, Non-Perturbative Canonical Transformation

We assume that we have solved the Hamilton-Jacobi equation iteratively to find an approximate invariant torus and the canonical transformation to an approximate set of action-angle variables for the full, nonlinear Hamiltonian[1]. In this paper, we discuss a refinement of this solution.

We start with the time-periodic Hamiltonian that describes the transverse motion of a single charged particle in a storage ring with sextupoles,

$$H(\Phi_0, I_0, s) = \frac{I_0}{\beta(s)} + \frac{1}{3!} S(s)(2\beta I_0)^{3/2} \cos^3 \Phi_0 \quad , \quad (1)$$

where $S(s)$ gives the strength in m^{-3} and distribution of the sextupoles around the ring. We have used the action-angle variables of the linear part of the Hamiltonian. The arclength around the storage ring $s \in [0, C]$ serves as the 'time' variable in the problem. In principle, the method for finding the non-perturbative solution is applicable to an arbitrary nonlinearity, as long as it is time-periodic. For single particle motion in storage rings, the dominant nonlinearity often comes from sextupole magnets.

The canonical transformation from the variables (Φ_0, I_0) to (Φ_1, I_1) is generated by $F_2(\Phi_0, I_1, s) = \Phi_0 I_1 + G^1(\Phi_0, I_1, s)$. The three different sets of action-angle variables will be discriminated by subscripts. For the two different generators G we use superscripts. The non-perturbative, numerical solution of the Hamilton-Jacobi equation described in Reference [1] yields the Fourier coefficients of G^1 on a finite mode set. The Fourier series is

$$G^1(\Phi_0, I_1, s) = \sum_{m \in M_0} g^1(m, I_1, s)\, e^{im\Phi_0} \quad . \quad (2)$$

The summation is over a finite set of modes indicated by M_0.

The numerical solution for G^1 serves as a perfectly good canonical transformation even if it does not yield the action-angle variables for the problem. It is a well-defined function of the variables (Φ_0, I_1, s), and it is very accurately periodic in s. We can implement the canonical transformation, allowing for a possible residual angle dependence, finding the Hamiltonian in the new variables:

$$H_1(\Phi_1, I_1, s) = \frac{I_1}{\beta(s)} + A(I_1, s) + V_1(\Phi_1, I_1, s) \quad . \quad (3)$$

The V_1 is defined such that its average over the new angle is zero. If G^1 were a perfect solution of the Hamilton-Jacobi equation then the residual angle dependence, V_1, would be zero.

The two terms are most easily calculated as integrals over the original angle variable Φ_0 using the transformation $\Phi_1 = \Phi_0 + G^1_I(\Phi_0, I_1, s)$ and its Jacobian. We group all terms in the Hamiltonian transformation equation that might have angular dependence into

$$F(\Phi_0, I_1, s) = G^1_s(\Phi_0, I_1, s) + \frac{G^1_\Phi}{\beta(s)} + V(\Phi_0, I_1, s) \quad . \quad (4)$$

Then A is the average of F over the new angles; this can be written in the original angles, by using the Jacobian, as

$$A(I_1, s) = \int_0^{2\pi} \frac{d\Phi_0}{2\pi} \, \mathcal{J}(\Phi_0, I_1, s)\, F(\Phi_0, I_1, s) \quad . \quad (5)$$

where $\mathcal{J}(\Phi_0, I_1, s) = 1 + G^1_{I\Phi}(\Phi_0, I_1, s)$. In a similar fashion, the Fourier coefficients of V_1 with respect to the *new* angles Φ_1 are calculated as

$$V_1(m, I_1, s) = \int_0^{2\pi} \frac{d\Phi_0}{2\pi} \, \mathcal{J}(\Phi_0, I_1, s) \times$$
$$e^{-im(\Phi_0 + G^1_I)}\, F(\Phi_0, I_1, s) \quad . \quad (6)$$

The Φ_1 dependence can be found using the Fourier series $V_1(\Phi_1, I_1, s) = \sum_{m \in M_1} V_1(m, I_1, s)\, e^{im\Phi_1}$, where M_1 is a mode set bigger than the original M_0. For the second perturbative transformation it is the Fourier coefficients themselves that are interesting.

* Work supported by the Department of Energy, contracts DE-AC03-76SF00515 and DE-FG02-86ER40302.

† Current mailing address is SLAC, Bin 26, P. O. Box 4349, Stanford, CA 94309.

TABLE 1. The results for refined solutions and their comparison with non-perturbative solutions.

action I_2	with refinement				cpu	non-perturbative only			cpu
$(10^{-5}$ m)	M_0	N_{RK}	M_1	δ	(secs)	M_0	N_{RK}	δ	(secs)
2.2	31	$6\rightarrow28$	127	$1.79\cdot10^{-2}$	244	63	$14\rightarrow28$	$4.10\cdot10^{-2}$	468
2.0	31	$6\rightarrow20$	127	$2.09\cdot10^{-3}$	177	63	$10\rightarrow20$	$3.67\cdot10^{-3}$	314
1.0	31	$6\rightarrow20$	127	$5.52\cdot10^{-4}$	156	63	$10\rightarrow20$	$5.41\cdot10^{-4}$	436
0.1	7	$2\rightarrow12$	31	$4.98\cdot10^{-5}$	5.3	31	$6\rightarrow12$	$5.05\cdot10^{-5}$	40.5
0.01	3	$1\rightarrow4$	31	$9.45\cdot10^{-6}$	2.0	7	2	$1.78\cdot10^{-4}$	0.4

II. The Second, Perturbative Canonical Transformation

Using the new Hamiltonian (3) with the residual angle dependence, we can write the Hamilton-Jacobi equation that gives the transformation to a new set of variables (Φ_2, I_2). If the G^1 canonical transformation was an approximate solution to the original Hamilton-Jacobi equation then the residual perturbation will be much smaller than the original perturbation. It then makes sense to solve the new Hamilton-Jacobi equation with perturbative techniques.

The Hamilton-Jacobi equation for the generator $G^2(\Phi_1, I_2, s)$ is

$$H_2(I_2, s) \simeq G_s^2 + \frac{I_2 + G_\Phi^2(\Phi_1, I_2, s)}{\beta(s)} +$$
$$A(I_2, s) + \partial_I A(I_2, s) G_\Phi^2 + V_1(\Phi_1, I_2, s) \quad . \quad (7)$$

In Equation (7) we have expanded to first order in the presumed small parameters V_1 and G^2. The above equation can be solved by using a Fourier series for G^2 of the same form as Equation (2) with Fourier coefficients $g^2(m, I_2, s)$ and summing over $m \in M_1$.

The Hamiltonian (3) defines a betatron phase advance that is slightly different from the linear phase advance. We define the new phase as

$$P(s) = \int_0^s \frac{d\sigma}{\beta(\sigma)} + \int_0^s \partial_I A(I_2, \sigma)\, d\sigma \quad . \quad (8)$$

Then the Fourier coefficients of $G^2(\Phi_1, I_2, s)$ can be written as

$$g^2(m, I_2, s) = -e^{-i\,m\,P(s)} \{ \frac{1}{e^{i\,m\,P(C)} - 1} \times$$
$$\int_0^C e^{i\,m\,P(\sigma)} V_1(m, I_2, \sigma) d\sigma +$$
$$\int_0^s e^{i\,m\,P(\sigma)} V_1(m, I_2, \sigma) d\sigma \} \quad , \quad (9)$$

recalling that $m \in M_1$ and that I_2 is a constant parameter.

The original action I_0 can be specified as a function of the original angle Φ_0 by chaining through the two canonical transformations. After expanding all functions of $I_1 = I_2 + G_\Phi^2(\Phi_1, I_2, s)$ around the 'best' constant action I_2, we find the compact form

$$I_0(\Phi_0, I_2, s) = I_2 + G_\Phi^1(\Phi_0, I_2, s) +$$
$$\det\left(\frac{\partial I_0}{\partial I_1}\right) G_\Phi^2(\Phi_0 + G_I^1, I_2, s) \quad . \quad (10)$$

This is the refined invariant torus. It gives the distortions to the linear Courant-Snyder actions under the effects of the nonlinearity.

III. Numerical Results

The accuracy of the invariant torus can be estimated by finding its deviation from numerically computed trajectories. The trajectories are computed by a fourth order symplectic integrator[2] . The deviations of the actions as each trajectory crosses the $s = 0$ point (the point at which we are studying the torus) from the corresponding actions of the torus are found. The normalized deviation is calculated as

$$\delta = \max \frac{\sum_{k=1}^{1000} |I_0(\Phi_0^T(kC)) - I_0^T(kC)|}{\sum_{k=1}^{1000} |I_0(\Phi_0^T(kC)) - I_2|} \quad , \quad (11)$$

where superscript T indicates points on the trajectory, and where the summation is over the turn number. The maximum is over 16 trajectories with different initial conditions starting on the torus. The δ parameter measures the worst agreement between a trajectory and the torus. Notice it is normalized by the distortion of the torus and not by the action value itself.

In Table 1 we compare the refined solution to a good non-perturbative solution. We give the mode sets and the number of integration steps, N_{RK}, for the fourth order Runge-Kutta integration used. The final number of integration steps plus one is the number of knots used for the cubic spline interpolation of the s dependence of the Fourier coefficients g^1 and g^2 in a sextupole. The s dependence outside the nonlinear elements is trivial. The cubic spline is found using the 'not-a-knot' condition, and it is used to evaluate the integrals in Equation (9). The 'CPU' is the computation time to calculate the solution on the SLAC IBM 3090. The time under 'with refinement' gives the time for finding the poor non-perturbative canonical transformation, for estimating $\partial_I g^1$, and for completing the perturbative calculation. The time under 'non-perturbative only' is that for calculating just the good non-perturbative solution.

The numerical results are for the ideal single cell of the Berkeley Advanced Light Source (ALS)[3] . The largest amplitude in Table 1 corresponds to an approximate x offset of 22 mm, and the smallest to 1.5 mm. The x tune is $\nu = 1.18973$. There are four 0.20 m long sextupoles symmetrically placed in the cell with strengths of -88.09, 115.615, 115.615, and -88.09 m^{-3}.

In Figure 1, we display three representative invariant curves. The top curve corresponds to $I_2 = 2 \cdot 10^{-5}$ m and the bottom to $I_2 = 10^{-6}$ m. On this plot, the difference between the original and the refinement to an invariant torus is not discernible. If there were no nonlinearity, the curves would be flat lines at their respective values of I_2.

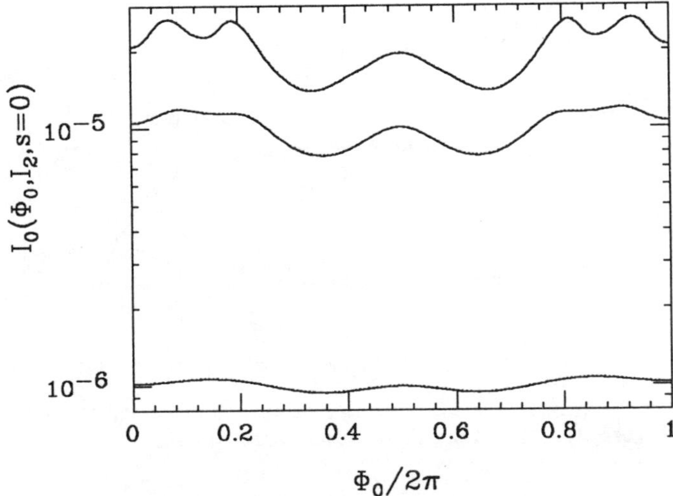

Figure 1. Invariant curves shown at $s = 0$ for the constant actions $I_2 = 2 \cdot 10^{-5}$ m, 10^{-5} m, and 10^{-6} m. The refinement is not noticeable on this plot.

From the first non-perturbative transformation, we calculate the function A which defines the nonlinear phase advance, $P(s) - \Psi(s)$, where $\Psi(s) = \int_0^s d\sigma/\beta(\sigma)$ is the linear phase advance. Notice that it varies only in the nonlinear elements. In Figure 2, we show the nonlinear part of the phase advance in each of the four sextupoles. On the horizontal axis in the plot, 1 to 2 corresponds to the first sextupole, 2 to 3 the second, and so on. The offset from zero at the end of the last sextupole is 2π times the nonlinear tune shift.

In Figure 2, notice that the shape of the function $P(s) - \Psi(s)$, is very similar for the three cases shown even though the magnitudes are different. This is typical for this accelerator example. The three cases have constant actions $I_2 = 10^{-6}$ m to $2 \cdot 10^{-5}$ m and correspond to the invariant curves shown in Figure 1.

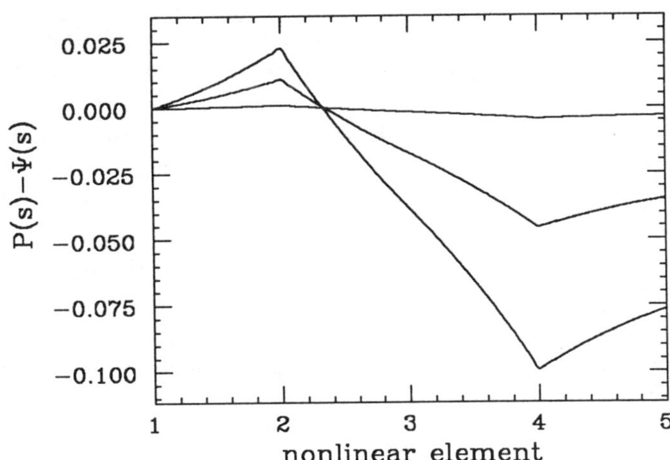

Figure 2. Nonlinear phase advance in the sextupoles for $I_2 = 10^{-6}$ m, 10^{-5} m, and $2 \cdot 10^{-5}$ m. The horizontal axis is 1 to 2 for the first sextupole, 2 to 3 for the second and so on. The small offset at the end of the last sextupole is 2π times the nonlinear tune shift. The linear phase advance is $\Psi(s) = \int_0^s d\sigma/\beta(s)$.

IV. Conclusions

We found that the second canonical transformation used to refine a poor non-perturbative solution gave more accurate tori than the poor solution. However the refinement of very good non-perturbative solutions did not increase the accuracy of the solutions. From Table 1 we see that in most cases the refinement gives tori with similar accuracy in about half the computation time. From the implementation of the non-perturbative transformation, we were able to calculate the nonlinear phase advance in the sextupole magnets. The form that these functions take seems to depend on only a relatively small number of parameters as the constant action I_2 is changed.

V. References

[1] W. E. Gabella, R. D. Ruth, and R. L. Warnock, in the Proceedings of the US-Japan Workshop on Nonlinear Dynamics and Particle Acceleration, Tsukuba, Japan, October 22-25, 1990, edited by T. Tajima and Y. Ichikawa, also SLAC-PUB-5414; W. E. Gabella, R. D. Ruth, and R. L. Warnock, presented at the IEEE Particle Accelerator Conference, Chicago, Illinois, March 20-23, 1989, edited by F. Bennett and J. Kopta.

[2] R. Ruth and E. Forest, Physica **43 D**, 105 (1990).

[3] *1-2 GeV Synchrotron Light Source, Conceptual Design Report—July, 1986*, Lawrence Berkeley Laboratory, PUB-5172 rev., Berkeley, California.

Tracking Studies for the Oxford Instruments Compact Electron Synchrotron

Chas. N. Archie
IBM Research, Thomas J. Watson Research Center, Yorktown Heights, N.Y. 10598
Jan Uythoven
Oxford Instruments Ltd., Synchrotron Division, Osney Mead, Oxford, England OX2 0DX

Abstract

Simulating electron trajectories (tracking) in a realistic compact ring introduces special problems because the magnetic bending radius can be as little as twenty times the horizontal displacement. Common simplifying approximations in tracking codes may not be appropriate. Results are presented from a thin lens tracking code which correctly handles the small bending radius virtual forces and real magnetic forces to octupole level. Nonsymplectic behavior is controlled by highly segmenting the Oxford superconducting dipoles. A qualitative comparison with experimental results is made.

Introduction

A superconducting electron storage ring intended primarily for X-ray lithography has recently been built by Oxford Instruments for IBM. This machine is currently being commissioned at the East Fishkill, New York IBM facility. The machine, called HELIOS, includes two 180° superconducting dipole magnets with a typical magnetic field of 4.5 T at an electron energy of 700 MeV. This gives a synchrotron light critical wavelength of 8.4 Å. The bending radius is 0.52 m and the total circumference is 9.6 m with about one third of this length within the dipole magnets.

Tracking an electron in this machine requires dealing with special computational problems not encountered in larger rings with conventional magnets. In particular, the usual procedure of switching to a coordinate system tied to the design energy closed orbit (a non-inertial reference frame) introduces *virtual* forces which are comparable to the higher order magnetic forces coming from a Taylor series expansion of the local magnetic field. Several of these virtual forces are examined in the case of HELIOS in the next section after which some tracking results are presented.

The Equations of Motion

In this paper there is room enough only for a brief outline of the derivation of the equations of motion to third order in the transverse coordinates and velocities.[1] Figure 1 shows the curvilinear coordinate system tied to the closed orbit of a design energy electron. Assuming that the design orbit is totally in the median plane, time derivatives of the unit vectors are

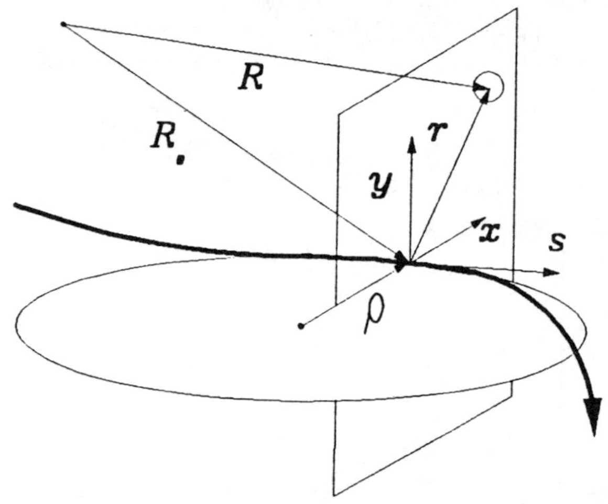

Figure 1. The Curvilinear Coordinate System

$$\dot{\vec{x}} = h\dot{s}\vec{s} \qquad \dot{\vec{y}} = 0 \qquad \dot{\vec{s}} = -h\dot{s}\vec{x} \tag{1}$$

where $h(s) = 1/\rho$. The position relative to a fixed arbitrary point, velocity, and acceleration are given by

$$\vec{R} = x\vec{x} + y\vec{y} + \vec{R}_0 \tag{2}$$

$$\vec{v} = \dot{x}\vec{x} + \dot{y}\vec{y} + \dot{s}(1 + hx)\vec{s} \tag{3}$$

$$\dot{\vec{v}} = \left(\ddot{x} - h\dot{s}^2(1 + hx)\right)\vec{x} + \ddot{y}\vec{y} \\ + \left(\ddot{s}(1 + hx) + 2h\dot{x}\dot{s} + h\dot{x}\dot{s}\right)\vec{s} \ . \tag{4}$$

These relations should be substituted into the Lorentz equations for the particle motion

$$\dot{\vec{v}} = \frac{e}{m}\left(\vec{v} \times \vec{B}\right) \ . \tag{5}$$

Time derivatives in the equations for the x and y directions can be rewritten in terms of s derivatives, denoted by ′, with the help of the equation of motion for the s direction:

---

[1] C. N. Archie, D. J. Orrell and J. A. Uythoven, "Particle Tracking in Small Radius Electron Storage Rings," to be published.

$$x'' - \frac{h'x + 2hx'}{1 + hx} x' - h(1 + hx) =$$
$$\frac{e}{p} \sqrt{x'^2 + y'^2 + (1 + hx)^2} \left[\frac{x'y'}{1 + hx} B_x - (1 + hx)\left(1 + \frac{x'^2}{(1 + hx)^2}\right) B_y + y' B_s \right] \tag{6}$$

and

$$y'' - \frac{h'x + 2hx'}{1 + hx} y' =$$
$$\frac{e}{p} \sqrt{x'^2 + y'^2 + (1 + hx)^2} \left[(1 + hx)\left(1 + \frac{y'^2}{(1 + hx)^2}\right) B_x - \frac{x'y'}{1 + hx} B_y - x' B_s \right] . \tag{7}$$

The magnetic fields can be written down in more familiar accelerator physics terms. The expansions are limited to third order for B_x and B_y but only second order for B_s since B_s always enters in the equations of motion multiplied by a transverse velocity component.

$$\frac{e}{p_0} B_y = h + kx + \frac{1}{2} m_1 x^2 - \frac{1}{2} m_2 y^2 + \frac{1}{3} n_1 x^3 - n_2 xy^2 \tag{8}$$

$$\frac{e}{p_0} B_x = ky + m_1 xy + n_1 x^2 y - \frac{1}{3} n_2 y^3 \tag{9}$$

$$\frac{e}{p_0} B_s = h'y + (k' - hh')xy \tag{10}$$

Upon introducing these expressions and expanding the square roots we obtain

$$x'' = -(k + h^2)x - (\frac{1}{2} m_1 + 2hk + h^3)x^2 + \frac{1}{2} m_2 y^2 + h'(xx' + yy') + \frac{1}{2} h(x'^2 - y'^2)$$
$$+ (\frac{1}{3} n_1 + m_1 h + h^2 k)x^3 + (n_2 - hm_2)xy^2 - hh'x^2x' - (\frac{3}{2} k + 2h^2)xx'^2 + kx'yy' - \frac{1}{2} kxy'^2 + k'xyy' \tag{11}$$

and

$$y'' = ky + (m_1 + 2hk)xy + h'xy' - h'x'y + hx'y'$$
$$+ (n_1 + 2hm_1 + h^2 k)x^2 y - \frac{1}{3} n_2 y^3 - k'xx'y - hh'xx'y - (k + 2h^2)xx'y' + \frac{1}{2} kx'^2 y + \frac{3}{2} kyy'^2 . \tag{12}$$

The higher order terms on the right hand sides of Equations (11) and (12) partition into those which stem solely from the higher order magnetic fields present and *virtual* forces which result from the transformation to a noninertial reference frame, the curvilinear coordinate system. Clearly all forces on the real particles are magnetic forces but this artificial grouping into *virtual* forces and magnetic forces is instructive for showing the importance of keeping certain terms in the truncations of the equations of motion in the curvilinear coordinate system. We focus on the relative strengths of the sextupole order forces as these largely determine the dynamic aperture.

With these terms dependent on various powers of x, x', y and y', it is not obvious how they compare. One way to do this relies on the higher order force terms being generally small compared to the dipole and quadrupole field terms as is certainly true for small excursions about the design orbit. Hence the pseudo-harmonic solution to the strong focussing terms alone describes the motion fairly well over a few revolutions in the lattice. It is therefore meaningful to use time averages of the pseudo-harmonic solution to make relative comparisons of the higher order terms. This assumption is only used to gain an appreciation of the relative magnitudes of the forces in the problem.

For simplicity the discussion is restricted to comparing force terms for a purely horizontal deviation of the particle from the closed orbit. The solution to the linear part of the equation of motion is given by

$$x = \sqrt{\varepsilon_x \beta_x} \cos \mu_x \tag{13}$$

where β_x is the usual accelerator physics horizontal betatron function, ε_x is the conserved horizontal emittance and μ_x is the betatron phase. Let the brackets $< >$ denote time averaging at a given lattice position. Then it can be shown

$$<x'^2> = \tfrac{1}{4}(4 + \beta_x'^2)/\beta_x^2 <x^2> \tag{14}$$

and

$$<xx'> = \tfrac{1}{2}\beta_x'/\beta_x <x^2> . \tag{15}$$

Returning now to Equation (11), the purely horizontal sextupole order force terms on the right hand side can now be compared. Figure 2 shows the lattice parameters β_x, β_y and the dispersion η for a possible operating tune point of the HELIOS lattice. Using the above time averages and the calculated dipole fields,

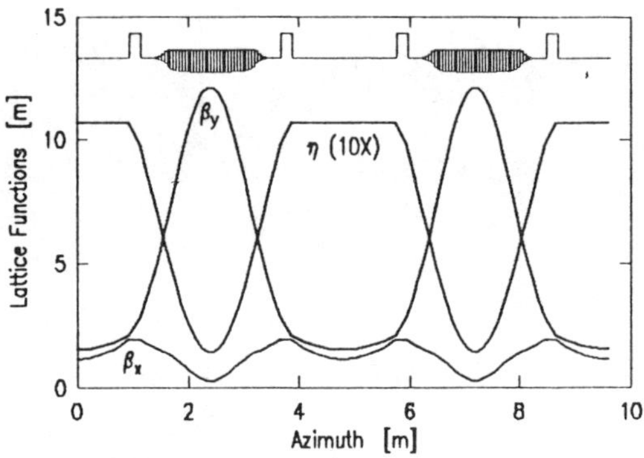

Figure 2.   HELIOS Lattice Parameters:

the various coefficients of second order force terms are shown in Figure 3. Several features are worth noting. The magnetic field contributes kick-counterkick pairs at each end of the dipole magnet and hence there is partial cancellation. By contrast, the three *virtual* force terms, some of which extend throughout the body of the magnet, do not similarly pair up.

Tracking Application to HELIOS

The previous section demonstrates that significant higher order forces extend throughout the HELIOS dipole. This, together with the relatively large length of the dipoles, implies that a thin lens kick code can only be used provided the dipoles are highly segmented. The results reported here come from a thin lens kick code in which gradient dipoles are treated exactly by a matrix formalism and all higher order force are represented by kicks between the magnet segments. Because HELIOS is a compact ring with few actual components in the ring, the dipoles can be

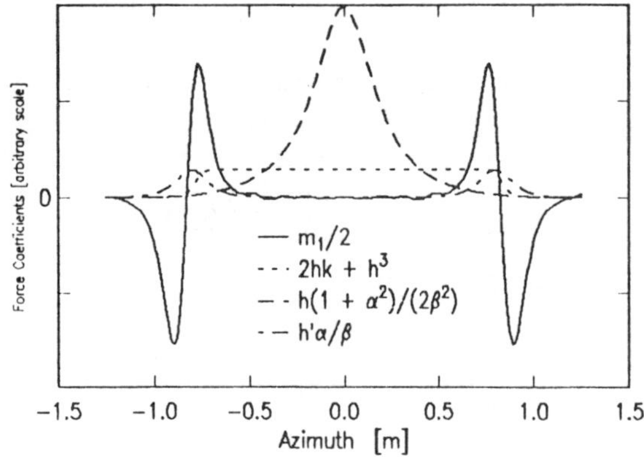

Figure 3.   Time Averaged Sextupole Order Force Comparison: Note $s = 0$ at the dipole center and $\alpha = -\frac{1}{2}\beta'$.

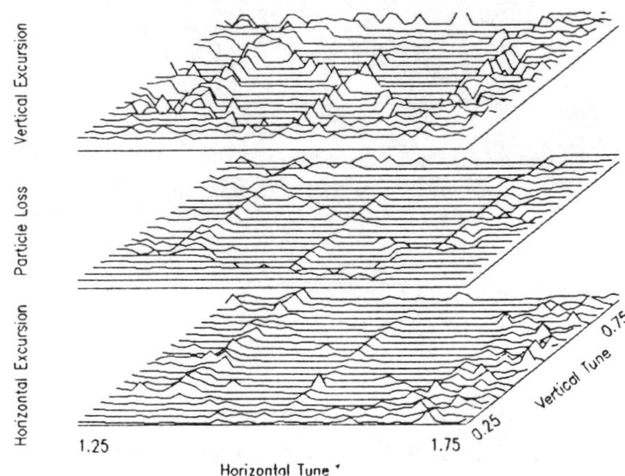

Figure 4.   Stored Beam Tune Scan

highly segmented without serious computational difficulty.

In Figure 4 a scan over tune space for HELIOS is presented. Adjustable sextupoles in the dipoles and one straight are set to zero the chromaticities at each tune point. The initial excursion of the particle at the middle of the ring straight section is calculated at each tune point to be ten times the natural beam size ($10\sigma_x$) with 10% coupling between horizontal and vertical motion. The particle is tracked for a maximum of 200 revolutions and the maximum vertical and horizontal excursions are noted. Doubling the number of maximum revolutions has little effect on the topology of the figure. If the excursion exceeds 50 mm in either of the two transverse directions, the particle is considered lost. The middle graph is a quantity which is proportional to 200 minus the number of successful revolutions of the particle. In some sense this graph is a composite of the other two graphs. If the particle is lost then this quantity is zero.

Of course, these studies do not take into account collective effects which may have significant effects on beam stability at any tune point. Nevertheless, preliminary commissioning experience on HELIOS provides some qualitative confirmation of these tracking results in that sizable currents have been injected and stored in many of the quiet regions of Figure 4. It should be emphasized that the comparison of tracking results and commissioning experience reported is preliminary and that considerably more experience is necessary in order to go beyond qualitative remarks.

The authors wish to thank particularly Mark Barton for his suggestions and constructive criticism and the people at Daresbury Laboratory and Oxford Instruments who designed HELIOS.

Suppression of Single Bunch Beam Breakup by Autophasing[*]

R.L. Gluckstern and J.B.J. van Zeijts

Department of Physics, University of Maryland, College Park, MD 20742

F. Neri

AT Division, Los Alamos National Laboratory, Los Alamos, NM 87545

I. INTRODUCTION

The increased interest in using intense, narrow, short bunches in accelerators has triggered an interest in the phenomenon of single bunch beam breakup[1]. Several analyses of this phenomenon have been made[2-4], including the effects of energy spread in the bunch. In a recent paper[4] we developed an analysis of single bunch beam breakup, including the effect of a linear variation of the transverse focussing force with longitudinal position within the bunch (leading to what is known as BNS damping[5]). A subsequent suggestion by Balakin[6] to shape the variation of transverse force with position within the bunching, called autophasing, makes it possible, in principle, to eliminate growth due to beam breakup.

The physical picture of autophasing is easy to understand and leads to a rough approximation for the necessary force gradient. But the required shaping is technically very difficult and one will probably have to be satisfied with approximate suppression.

In Section II we review the theory of single bunch beam breakup with BNS damping for a coasting beam and in Section III show the implications of autophasing. In Section IV we present a reliable way to estimate the growth due to beam breakup with linear variation of the transverse force in terms of two universal parameters, and include comparison with simulations.

II. SINGLE BUNCH BEAM BREAKUP

The equations which govern the displacement ($\xi(N,M)$) and excitation of the deflecting mode (proportional to $z(N,M)$) for the Mth "macroparticle" of the bunch as it enters the Nth cavity are

$$\frac{\partial^2 \xi(N,M)}{\partial N^2} + \mu^2(M)\xi(N,M) = z(M,N) \qquad (1)$$

and

$$z(N,M) = \omega\tau \int_o^M d\ell(M-\ell)\, r(\ell)\, \xi(N,\ell), \qquad (2)$$

where $\mu(M)$ is the phase advance of the transverse oscillation per cavity. We have approximated the cavity wakefield, which is proportional to
$r(\ell)\, sin(M-\ell)\, \omega\tau$, by its linear approximation for a bunch whose length is short compared to an r.f. wave length. Here

$$r(\ell) = \frac{e\sigma(\ell)}{2W}\frac{Z_\perp T^2}{Q}\, L, \qquad (3)$$

where $r(\ell)$ is a measure of the charge in the ℓth macroparticle and its influence on the transverse motion. The bunch has an energy W and a total charge $N_p e = \int d\ell\sigma(\ell)$. The cavities, separated from each other by a distance L, have a transverse mode frequency $\omega/2\pi$, a quality factor Q, and a shunt impedance parameter $Z_\perp T^2/Q$, where T is transit time factor. For uniform longitudinal bunch density we have $\sigma(\ell) = N_p e/M_o$, where the bunch is divided into M_o macroparticles. One readily includes several modes by the replacement

$$\omega(Z_\perp T^2/Q) \;\rightarrow\; \sum_j \omega_j(Z_\perp T^2/Q)_j. \qquad (4)$$

Two derivatives of Eq. (2) with respect to M lead to

$$\frac{\partial^2 z(N,M)}{\partial M^2} = \omega\tau\, r(M)\,\xi(N,\ell). \qquad (5)$$

Equations (1) and (5) are the starting point for our analysis. Our notation is very similar to that used in an earlier formulation of cumulative beam breakup.[7]

III. AUTOPHASING

In autophasing[6], one looks for a solution in which $\xi(N,M) = \xi(N)$ is a function only of N, that is, the bunch moves as a rigid body. This is accomplished by writing

$$\mu^2(M) = \mu_o^2 + \delta\mu^2(M), \qquad (6)$$

where

$$\frac{\partial^2 \xi}{\partial N^2} + \mu_o^2\xi = 0, \qquad (7)$$

leading to

$$\delta\mu^2(M)\xi(N) = z(M,N). \qquad (8)$$

[*]Work supported by the Department of Energy.

From Eq. (5) we obtain the autophasing condition

$$\frac{d^2}{dM^2}[\delta\mu^2(M)] = \omega\tau r(M), \qquad (9)$$

that is, the second derivative of the focussing force with respect to position within the bunch must match the transverse wakefield.

If we assume a symmetric Gaussian-like bunch with 90% of the bunch between $M = M_1$ and $M = M_2$, Eq. (9) can be integrated to obtain

$$\delta\mu^2(M_2) - \delta\mu^2(M_1) \cong \frac{\omega\tau(M_2 - M_1)}{2}\int_{-\infty}^{\infty} d\ell\; r(\ell). \qquad (10)$$

This is then the change of the focussing force from one "end" of the bunch to the other required to suppress beam breakup.

IV. UNIVERSAL PARAMETERS

In view of the difficulty in shaping the variation of the transverse force with position within the bunch, we shall treat the case of a linear variation for a coasting beam bunch of uniform density. Equations (1) and (5), for constant $\mu(M) \equiv \mu$ and $r(M) \equiv r$, have the approximate solution[4]

$$\frac{\xi(N, M)}{\xi_o} \cong \frac{1}{\sqrt{4\pi}} Re\left(\frac{e^{i\mu N + ue^{-i\pi/6}}}{\sqrt{ue^{-i\pi/6}}}\right), \qquad (11)$$

where ξ_o is the initial bunch offset, and where

$$u = \frac{3}{2}\left(\frac{\omega\tau}{\mu}\right)^{1/3} M^{2/3}\, N^{1/3} \qquad (12)$$

is a universal parameter assumed to be large compared to 1 in Eq. (11). Beam breakup is associated primarily with the term in u in the exponential.

In Fig. 1 we plot $\ln(\eta^s) \equiv \ln(\xi_{max}/\xi_o)$, obtained from numerical simulations of the original difference equations[7], as a function of u, and compare it with the analytic value predicted by Eq. (11):

$$\ln(\eta^a) \equiv \ln(\xi_{max}/\xi_o) = \frac{u\sqrt{3}}{2} - \frac{\ln\, 4\pi u}{2}. \qquad (13)$$

Here ξ_{max} is the amplitude of the oscillation with respect to N. Specifically, we show $\ln(\eta^s)$ and $\ln(\eta^s) - \ln(\eta^a)$ in Fig. 1. The agreement between η^s and η^a is excellent (within $\pm4\%$) for values of $u > 3$.

If we now allow $\mu(M)$ to have a linear dependence on M:

$$\mu(M) = \mu + \alpha M, \qquad (14)$$

such that Eq. (1) can be approximated by

$$\frac{\partial^2\xi(N, M)}{\partial N^2} + \mu^2 + 2\alpha M\mu \cong z(N, M), \qquad (15)$$

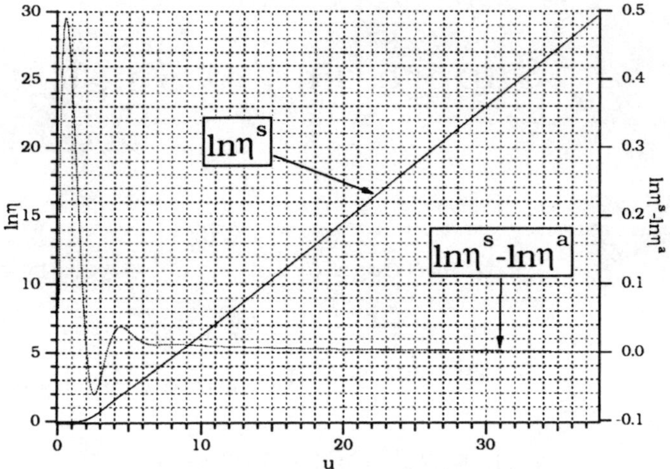

Figure 1: $\ln(\eta^s)$ and $\ln(\eta^s) - \ln(\eta^a)$.

we can obtain the solution for ξ/ξ_o in the limit $u \gg 1$:

$$\frac{\xi(N, M)}{\xi_o} \cong \frac{1}{\sqrt{4\pi}} Re\left(\frac{e^{i\mu N + uf(v) + g(v)}}{\sqrt{ue^{-i\pi/6}}}\right). \qquad (16)$$

The second universal parameter v is defined as

$$v = \alpha\left(\frac{\mu}{r\omega\tau}\right)^{1/3} M^{1/3}\, N^{2/3} \qquad (17)$$

and the functions $f(v)$ and $g(v)$, whose defining equations are given in Reference 5, are shown in Figs. 2 and 3. In

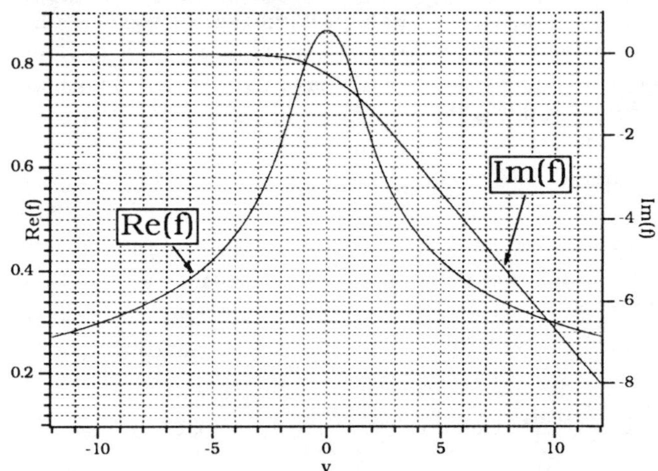

Figure 2: $Re\; f(v)$ and $Im\; f(v)$.

Fig. 4 we plot $\ln(\eta^s)$, obtained from numerical simulations, as a function of v and compare it with the analytic value predicted by Eq. (16):

$$\ln(\eta^a) = u\, Re\; f(v) + Re\; g(v) - \frac{1}{2}\ln\, 4\pi u + \Delta(u), \qquad (18)$$

where

$$\Delta(u) = \ln\,(\eta^s)\,|_{v=0} - \left(\frac{u\sqrt{3}}{2} - \frac{\ln\, 4\pi u}{2}\right) \qquad (19)$$

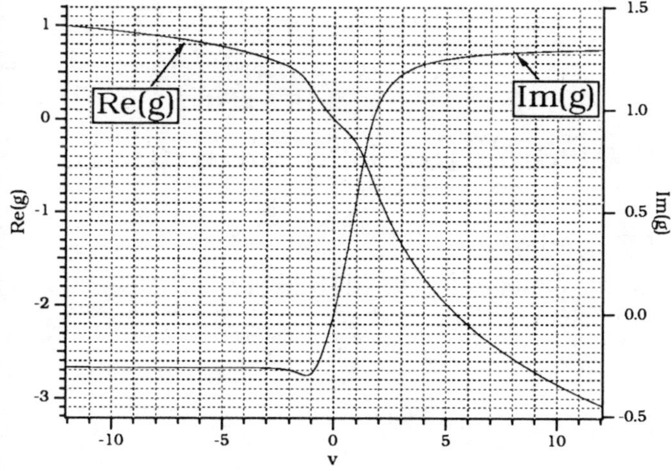

Figure 3: $Reg(v)$ and $Img(v)$.

corrects for the small inaccuracy of Eq. (11) for $v = 0$ shown in Fig. 1. Once again, the agreement between the simulations and the predictions is excellent, particularly over the range $-2 < v < 10$.

Figure 4 contains the necessary information to determine the gradient of u necessary to suppress beam breakup. Specifically, one calculates u from Eq. (12), and uses the appropriate curve to determine the value of v for which $\ln(\eta) \cong 1$. The desired α is then obtained from Eq. (17).

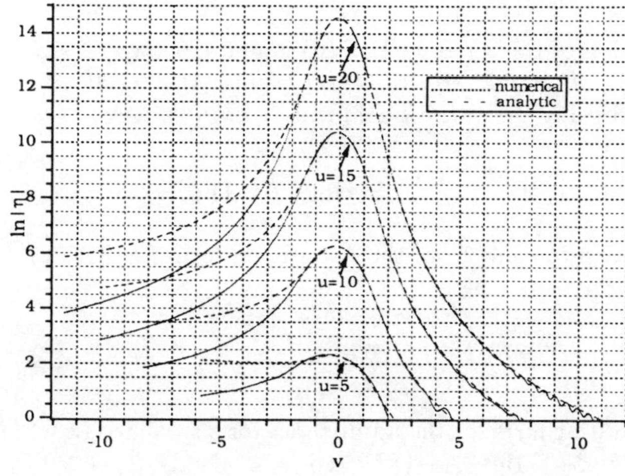

Figure 4: $\ln(\eta^s)$ and $\ln(\eta^a)$.

The relation between the accurate results in Fig. 4 and the approximation in Eq. (10) can be obtained by observing that suppression occurs at relatively large values of v. Considering only the term $u\,Re\,f(v)$ in the exponential in Eq. (16) and using the large v limit, $f(v) \cong (8/9v)^{1/2}$, we find

$$u\,Re\,f(v) \cong (8u^2/9v)^{1/2} = (2M\,r\,\omega\tau/\mu\alpha)^{1/2}. \quad (20)$$

If we assume that suppression takes place when $u\,Re\,f(v)$

is of order 1, we predict

$$\delta\mu^2(M) = (\mu + \alpha M)^2 - \mu^2 \cong 2\alpha\,M\,\mu \sim M^2 r\,\omega\tau, \quad (21)$$

in close agreement with the simplified autophasing version in Eq. (10).

Finally, we reiterate that our analysis is for a coasting beam bunch of uniform density, with a linear variation of focussing force with position within the bunch. These restrictions can be removed by using numerical simulations. It is also necessary to include the longitudinal wakefield in determining the focussing force within the bunch. Another concern is that our formulation assumes that each macroparticle transverses the cavity before the next one enters. This can be justified by averaging the displacement of the macroparticles over each cavity.[8]

V.   REFERENCES

[1] A.W. Chao, B. Richter, and C-Y Yao, Nucl. Instr. and Methods 178, 1 (1980).

[2] K.L.F. Bane, A.I.P. Conference Proceedings of the Accelerator Summer School, FNAL, pp. 971 ff (1987).

[3] D. Chernin and M. Mondelli, Particle Accelerators 24, 177 (1989).

[4] R. Gluckstern, F. Neri, and J.B.J. van Zeijts, "Suppression of Single Bunch Beam Breakup by BNS Damping", Proceedings of the Linac Conference, Albuquerque, NM, Sept. 1990.

[5] V.E. Balakin, A.V. Novokhatsky and V.P. Smirnov, Proceedings of the 12th International Conference on High Energy Accelerators, FNAL, p. 119 (1983).

[6] V.E. Balakin, Proceedings of the 1988 Workshop on Linear Colliders, SLAC, p. 55.

[7] R.L. Gluckstern, R.K. Cooper and P.J. Channell, Particle Accelerators 16, 125 (1985).

[8] R.L. Gluckstern, "Compensation of Single Bunch Transverse Beam Breakup in a Chain of Long Periodic Cavities", CERN Report: CLIC Note 128, October 17, 1990.

Longitudinal Coupling Impedance of a Thin Iris Collimator*

R.L. Gluckstern and W.F. Detlefs

Department of Physics, University of Maryland, College Park, MD 20742

I.   INTRODUCTION

We present here the results of numerical calculations of the longitudinal coupling impedance of a thin iris collimator in a beam pipe. The calculations are performed using an analytic result derived in a previous note[1]. The impedance is calculated as a function of frequency $(kc/2\pi)$ and inner and outer radius $(r = b$ and $r = a)$. The results are presented by plotting the admittance as a function of $k(a - b) = kax$ for different $x \equiv (a - b)/a$. Comparison is made with the expected behavior of the admittance in the small iris limit $(x \ll 1)$.

II.   ANALYTIC RESULT

We now summarize the derivation of the analytic result given in the previous research note[1]. We began by writing the electromagnetic fields as the sum of a source term plus transmitted and reflected modes in the beam pipe. The fields were then matched at $z = 0$ (the axial location of the iris) to obtain an integral equation for the discontinuity in the axial electric field from one side of the iris to the other. Finally the following variational form was constructed for the admittance

$$Z_o Y(k) = \frac{\int_b^a rdr \int_b^a r'dr'\, g(r)\, g(r')\, K(r,r)}{[\int_b^a rdr\, g(r)\, \ell n(a/r)]^2}, \qquad (1)$$

where $g(r)$ is a trial function proportional to the discontinuity in the axial electric field, and $K(r, r')$ is the kernel of the integral equation, given by

$$K(r, r') = \frac{2\pi}{k} \sum_{i=1}^{\infty} \frac{\beta_i J_o(p_i r/a)\, J_o(p_i r'/a)}{J_1^2(p_i)}. \qquad (2)$$

Here p_i are the zeros of $J_o(u)$ and

$$\beta_i a = (k^2 a^2 - p_i^2)^{1/2} = -j(p_i^2 - k^2 a^2)^{1/2}. \qquad (3)$$

The solution to the integral equation was obtained by expanding the discontinuity in magnetic field in terms of the complete set $F_1(\sigma_m r)$ in the interval $b \leq r \leq a$,

where

$$F_1(\sigma_m r) = Y_1(\sigma_m r) - \frac{Y_o(\sigma_m a)}{J_o(\sigma_m a)} J_1(\sigma_m r) \qquad (4)$$

and where the eigenvalues σ_m satisfy $F_1(\sigma_m b) = 0$. The final result for the impedance was

$$\frac{Z(k)}{Z_o} = \sum_m \sum_m (P^{-1})_{mn}, \qquad (5)$$

where $Z_o = 120\pi$ ohms is the impedance of free space and the m, n element of the symmetric matrix P is given by

$$P_{mn} = \frac{2\pi b^2}{ka^2} \sum_i \frac{J_1^2(p_i\frac{b}{a})}{J_1^2(p_i)} \frac{\sigma_m^2 a^2}{\sigma_m^2 a^2 - p_i^2} \cdot \frac{\sigma_n^2 a^2}{\sigma_n^2 a^2 - p_i^2}. \qquad (6)$$

Truncation of the series in Eq. (6) is not expected to significantly affect numerical accuracy, since the result was obtained using the variational form in Eq. (1).

III.   SMALL IRIS LIMIT

It is easy to show from the variational form for $Z_o Y(k)$ that[1]

$$Z_o Y(k) = Z_o G(k) + j Z_o B(k) \cong \frac{\pi}{2}ka - \frac{j\, Aa}{k(a - b)^2} \qquad (7)$$

is the limiting form for the admittance for a small iris $(x \ll 1$, or $a-b \ll a)$. Here A is expected to be weakly dependent on ka and of order 1.

The small x approximation in Eq. (7) has two interesting features. The first is that $G(k)$, the real part of the admittance, is independent of the geometry of the obstacle. The second is that $B(k)$, the imaginary part of the admittance, is inversely proportional to the cross-sectional area in the $r - z$ plane where the field is significantly disturbed by the iris (approximate dimensions $a-b$ by $a-b$). These features are the same as those encountered in the case of a small convex obstacle in a beam pipe[2], where the corresponding result was

$$Z_o \tilde{Y}(k) \cong \pi\, ka - \frac{2\pi\, ja}{k\Delta}, \qquad (8)$$

---

*Work supported by the Department of Energy.

where Δ is the cross-sectional area of the obstacle. We speculate that these features are part of a more general result which will also apply for a small iris of small but finite thickness and possibly even general shape.

Finally we can also examine the high frequency limit for a small iris where kax may be of order 1 or greater. The form of the result can most easily be seen by taking the corresponding limit in Eq. (6) where $J_1(p_i b/a)/J_1(p_i)$ goes over to $\cos p_i x$, the equivalent 2-D result. The prediction is then that $xZ_oY(k)$ will be a function of the universal variable kax. The small kax limit is that given in Eq. (7), namely

$$xZ_oY(k) \cong \frac{\pi}{2} \, kax - \frac{j\,A}{kax} \, , \ kax \ll 1 \qquad (9)$$

and the large kax limit can be obtained via Eq. (6), and is

$$xZ_oY(k) \simeq \pi - \frac{2j}{kax} \, , \ kax \gg 1. \qquad (10)$$

IV. CONVERGENCE BEHAVIOR

The convergence of the analytic method was examined with respect to truncation of the series in Eq. (6) and with respect to the size of the matrix P. To check the convergence of the series, we set $b = 0.9, a = 1.0$ and chose values for ka and n, where $n \times n$ is the matrix size. (Note: throughout the rest of this paper the length scale is chosen so that $a = 1.0$). The values of ka used here were between 10 and 100, and $n = 20, 25$, or 30. We then calculated the admittance for different values of i_{max}, the maximum value of the index in Eq. (6). The real and imaginary parts of the admittance were then plotted as a function of $1/i_{max}$. This procedure was repeated for different values of ka and n. The resulting plots indicated that the convergence was linear in $1/i_{max}$, which was the expected behavior. This allowed for a simple linear extrapolation to estimate the error in the admittance due to the truncation of the series. We determined that a value of $i_{max} = 5000$ would give an error due to truncation of approximately one part in 10^5 , which is negligible in comparison with the error due to finite matrix size (discussed below). All calculations referred to throughout the rest of this paper were performed using $i_{max} = 5000$.

To determine the dependence of the calculated admittance on matrix size, we let $b = 0.9$ and chose a value for ka. We then calculated the admittance for $n = 10, 15, 20, 25$, and 30. The real and imaginary parts of the admittance were then plotted as a function of $1/n$. This procedure was repeated for different values of ka. The resulting plots revealed a strong linear dependence on $1/n$ as well as a dependence on higher-order terms in $1/n$. We determined that a quadratic interpolation would give the best results for the admittance, resulting in an error of no more than one part in 10^2.

V. NUMERICAL RESULTS

The numerical results are presented in Figures 1-4. Equation (9) suggests that $xZ_oY(k)$ is primarily a function of kax for $kax \ll 1$. We therefore plot $xZ_oG(k)$ (Figure 1) and $kax^2Z_oB(k)$ (Figure 2) versus kax for $x = .05, .1, .3$. The plots do show the behavior expected from Eqs. (9) and (10). In Fig. 1, we plot the line with slope $\pi/2$ and the horizontal line with ordinate π; it is then easily seen that the calculated values for $G(k)$ are consistent with Eqs. (9) and (10).   The fact that the curves in Figs. 1 and 2

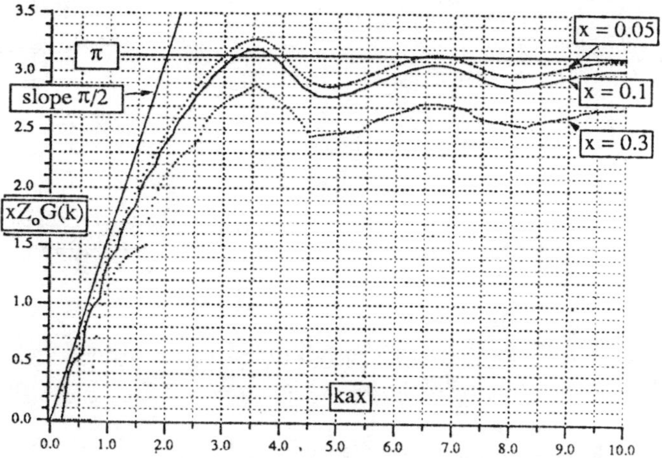

Figure 1: Real part of the admittance (multiplied by x) vs. kax for small x.

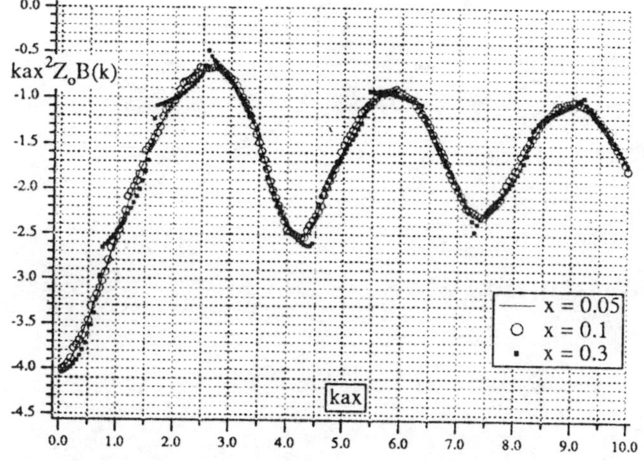

Figure 2: Imaginary part of the admittance (multiplied by kax^2) vs. kax for small x.

lie very close to each other suggests that for small x, the calculated values for the impedance are universal, that is, a function only of kax (except for an overall factor x). It should be noted that Fig. 2 suggests the value of A in Eq. (9) is close to 4.

Figs. 3 and 4 explore the departure from Eq. (9) as x becomes of order 1. In Fig. 3 we plot $xZ_oG(k)$ versus

kax for $x = .05, .5, .7,$ and $.9$. The curves show roughly the same oscillatory behavior as in Fig. 1, in the sense that the minima and maxima occur at the same locations. As expected, as x increases the curves depart significantly from the result for small x.

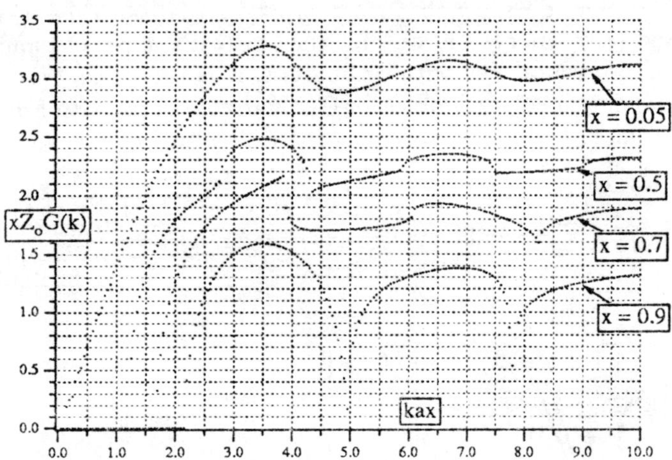

Figure 3: Real part of the admittance (multiplied by x) vs. kax for large x.

In Fig. 4 we plot $xZ_oB(k)$ versus kax for $x = .05, .5, .7,$ and $.9$. We see roughly the same oscillatory behavior as in Fig. 2. It should be noted from Eq. (10) that for x close to 1 and large kax, $B(k)$ will be small compared to $G(k)$ and the impedance will be approximately real. The calculated values are consistent with this conclusion.

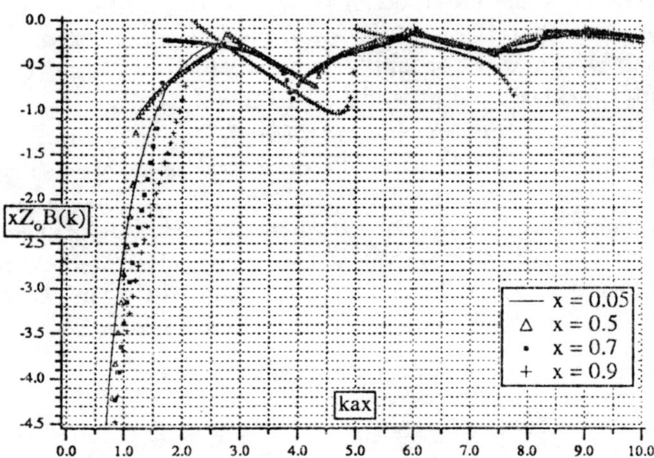

Figure 4: Imaginary part of the admittance (multiplied by x) vs. kax for large x.

VI.   ACKNOWLEDGMENT

One of the authors (WFD) is grateful to Johannes van Zeijts for helpful discussions.

VII.   REFERENCES

[1] R.L. Gluckstern, Longitudinal Coupling Impedance of an Iris in a Beam Pipe, CERN research note CERN SL/90-113 (AP).

[2] R.L. Gluckstern and F. Neri, Longitudinal Coupling Impedance of a Small Obstacle, Proceedings of the Particle Accelerator Conference, Chicago, IL, p. 1271 (1989).

Study of Loss Factor for Slots in the Vacuum Chamber*

Yong-chul Chae** and Lee C. Teng
Advanced Photon Source
Argonne National Laboratory
Argonne, IL 60437

Abstract

Using 3-D wakefield calculation program MAFIA [1], the longitudinal loss factor (energy lost by the beam) and the transverse loss factor (transverse kick experienced by the beam) are systematically studied with the depth and the length of a slot as parameter values. When the bunch is longer than the slot, we find that simulation results agree with the early theoretical prediction by M. Sands [2], the loss factor decreases exponentially as the slot gets deeper. However, when the bunch is shorter than the slot, we find that the longitudinal loss factor decreases but the transverse loss factor does not change very much as slot gets longer. Finally, we compare the loss from a long slot with the loss from a number of holes whose diameter is equal to the slot width and covering roughly the same area as the slot.

I. INTRODUCTION

Energy loss by the beam from small holes and slots in the vacuum chamber was investigated by M. Sands [2] using H.A. Bethe's small hole theory [3] where a hole was replaced by the electric and magnetic dipole to calculate the field diffracted by the hole. In Sands' paper, the energy lost by the beam, i.e. the longitudinal loss factor (K_z) is equal to the total radiated energy half of which is diffracted through the depth of the hole to the outside where the vacuum pumps are located while the other half is scattered back into the beam chamber. Sands' results may be summarized as follows:

1) If the bunch length (σ_z, rms bunch length of Gaussian distribution) is larger than the slot length (slot is parallel to the beam direction), as the depth of the slot, namely the thickness of the vacuum chamber wall, gets larger, the evanescent mode in the slot is damped and consequently the power diffracted through the slot will decrease exponentially resulting in an exponential decrease in K_z.
2) If a slot is longer than the bunch length, K_z increases slowly and becomes independent of slot length.
3) For multiple holes whose radius is smaller than σ_z, the total K_z is given by gN times the K_z for a single hole where N is the number of holes and g is the coherence factor, approximately equal to the number of holes within the bunch length σ_z.

II. GEOMETRY USED IN MAFIA SIMULATION

The beam chamber and vacuum pump chamber configuration considered in this study is the rectangular waveguide coupled by a slot on the common wall (Fig. 1). The size of waveguide used is 8cm by 4cm and the slot length (l), width (w) and depth or thickness (t) are varied as parameter values. Since the vacuum pump chamber has a lot of complicated pump elements, our coupled waveguide model is only an approximation to the real situation. In order to reduce the number of mesh point used in MAFIA, we used the symmetry condition whenever applicable. For example, to simulate the configuration such as Fig. 1-a, instead of using two full waveguides, we used the symmetry condition at the central plane of the slot neck. By specifying the condition that the tangential electric field $E_{tan}=0$ at the central plane, we are simulating the situation where there is an electron beam in the beam chamber and a positron beam in the vacuum pump chamber, and we denote the longitudinal loss facor as $K_{z,E}$. Similarly we can obtain $K_{z,H}$ for Fig. 1-c. Finally, K_z for the configuration of Fig. 1-a can be obtained from $K_z=(K_{z,E}+K_{z,H})/2$.

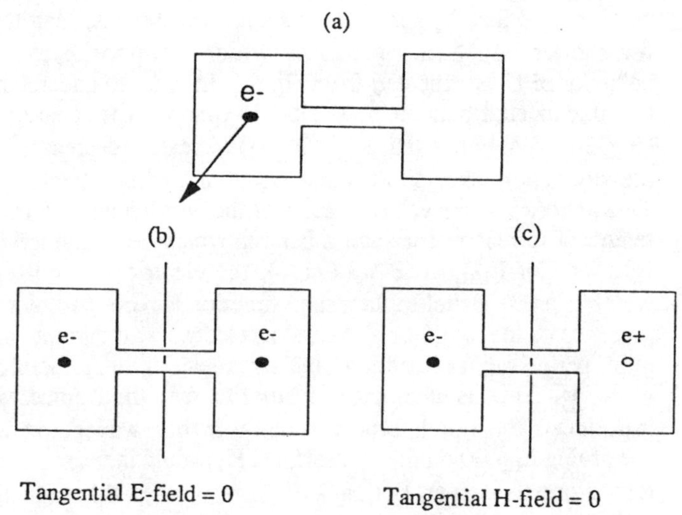

Fig. 1 Coupled waveguide configuration used in MAFIA simulation. (a) Geometry to be simulated (b) $E_{tan}=0$ at the central plane (c) $H_{tan}=0$ at the central plane

---

*Work supported by the U.S. Department of Energy, Office of Basic Energy Sciences, under Contract No. W-31-109-ENG-38.
** Graduate student from Univ. of Houston, Physics Dept.

III. SLOT THICKNESS EFFECT

When the slot is infinitely thin, the energy left in the field diffracted through the slot to the outside or the vacuum pump chamber, U_{out}, and the energy scattered back into the beam chamber, U_{in}, will each be equal to $U_0/2$ where U_0 is the total radiated energy. However, for the slot whose thickness is finite, the evanescent field diffracted through the slot, assuming that the dominant frequency in the bunch spectrum is well below the cut-off frequecy of the slot, will decrease as $\exp(-2\pi t/\lambda_c)$ where t is the slot thickness and λ_c is the lowest cutoff wavelength. Then total diffracted energy loss may be expressed as

$$U=U_{out}+U_{in}=U_0[(1+a)+(1-a)\exp(-2\pi t/\lambda_c)]/2 \quad (1)$$

where a is the portion of bunch spectrum above the cutoff frequency of the slot, and is determined by the bunch length and the slot length ($l>w$ in this study). Note that when $t=0$, U is equal to U_0. In M. Sands' paper, U is identified as K_z. This is a good approximation for short bunches. Contrary to this, Eq. (1) is valid for a bunch longer than the slot. Therefore, it may not be a good approximation to identify Kz with the total radiated energy U given by Eq. (1). Nevertheless it is interesting to compare U with the loss factor Kz as given by MAFIA. One may expect that the qualitative dependence on parameters are the same for these two quantities.

We computed K_z as a function of the thickness (depth) of the slot. The slot dimensions are taken to be $l=1$cm and $w=0.4$cm, and values of K_z normalized to unity at a slot thickness of 0.5mm, are plotted in Fig. 2-a. The simulation result shows that K_z for longer bunch decreases less than that for shorter bunch as t increases which is opposite to the behavior of U as expected from Eq. (1). In order to understand this unexpected behavior, we plot $K_{z,E}$ and $K_{z,H}$ separately (Fig. 2-b). We notice that $K_{z,E}$ ($K_{z,H}$) increases (decreases) as the slot gets thicker and become equal at a certain thickness. This trend was observed regardless of the bunch length and slot length. The reason for such a behavior may be explained as follows. For $E_{tan}=0$ (E-wall case), the electric wall current which is anti-parallel to the beam direction is perturbed by the presence of the slot. For a thick slot, electric wall currents are more perturbed at the edge of the slot resulting in an increase in $K_{z,E}$. This is also clear from the fact that, since we introduced the notch type resonator in our waveguide by specifying $E_{tan}=0$ on the slot surface, K_z would increase as the resonator gets deeper. For $H_{tan}=0$ (H-wall case), the magnetic wall current exists only on the surface of the slot and current intensity decreases as the thickness increases. Thus magnetic wall current is more perturbed for a thin slot resulting in the decrease in $K_{z,H}$ as the slot gets thicker. However, the behaviour in Fig. 2-a still remains to be explained.

Finally, transverse loss factor K_y was obtained by offsetting the beam in the direction perpendicular to the slot surface (the y-direction) and we found that K_y decreases as the slot gets thicker but more slowly than K_z.

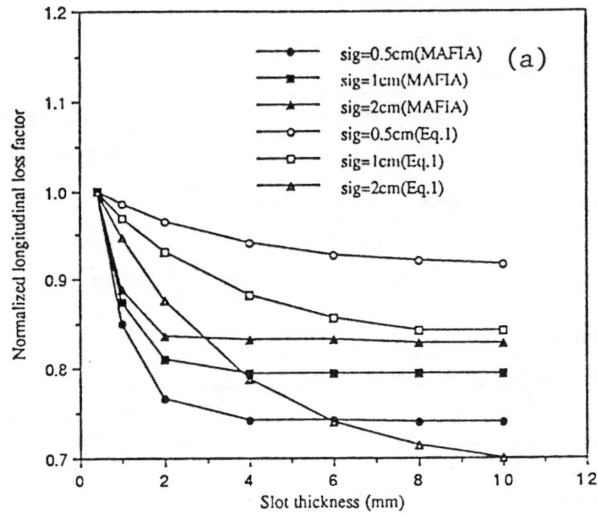

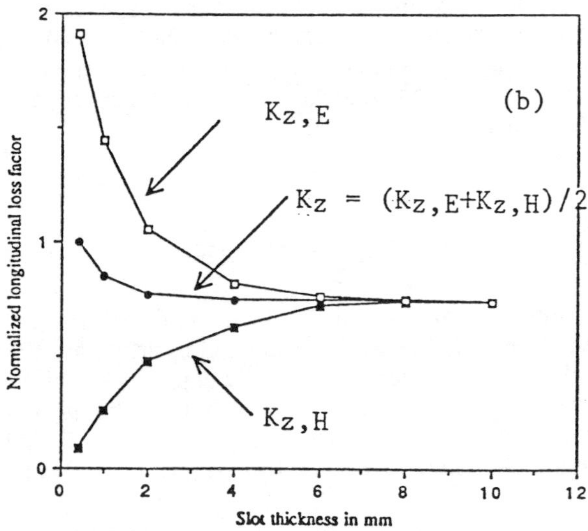

Fig. 2 Normalized loss factors for various bunch lengths as functions of slot thickness.

IV. SLOT LENGTH EFFECT

We studied the slot length effect in cases where the slot is longer than the beam bunch. In this study, the slot has a width $w=0.4$cm, a thickness $t=0.4$cm and a length which is varied as a parameter. According to M. Sands, K_z would increase slowly with the slot length regardless of the bunch length. However, Fig. 3-a shows three different behaviors depending on the bunch lengths $\sigma_z=0.5$cm, 1cm or 2cm. We also found that when we reduced the beam chamber size by a half, i.e. to a 4cm by 2cm waveguide, a similar trend was observed for $\sigma_z=0.25$cm, 0.5cm and 1cm. However, in the case of K_y, it does increase slowly as shown in Fig. 3-b.

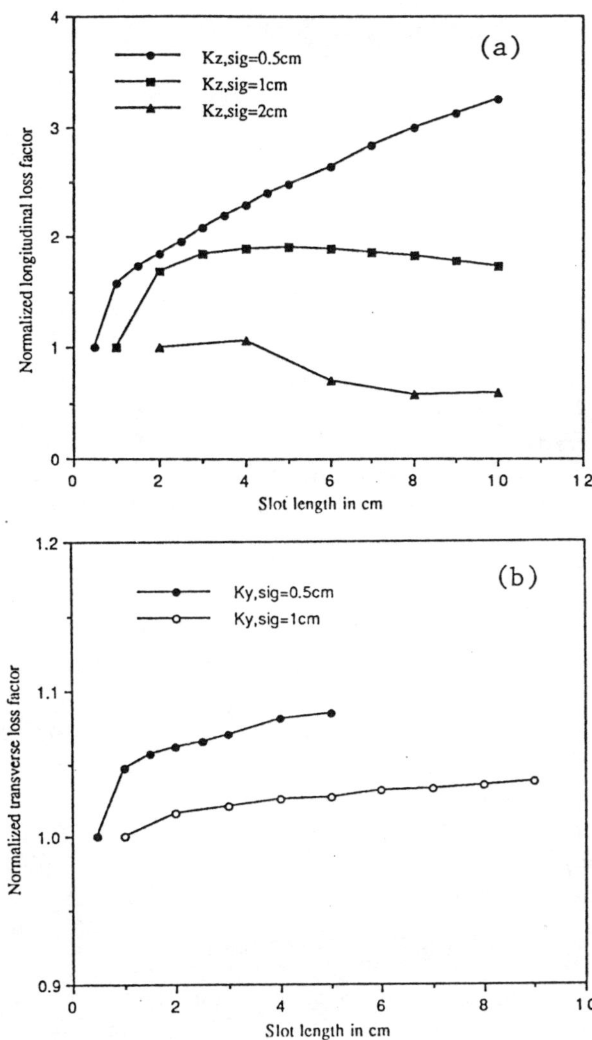

Fig. 3 Normalized loss factors as functions of slot length. The values for $\sigma_z = l$ are used as reference values.

V. SINGLE SLOT VERSUS MULTIPLE HOLES

Slots in the vacuum chamber wall are provided for vacuum conductance. For this purpose the slots may equally well be replaced by holes as long as they give the same opening area. It may in fact be easier to drill holes than to fabricate slots. Therefore, we compare the loss factors for two configurations. One is a long slot with a width of $w=0.4$cm and various lengths. For given slot length l, we compare with the second configuration of a number $N=l/w$ square holes with side equal to w. However, when $l=2$cm, we took $N=4$ instead of $N=5$ for neighboring holes are separated by 1mm. Thus, the opening areas for the two configurations, although not exactly the same, are quite similar. Fig. 4 shows K_z for these two configurations where $\sigma_z=0.5$cm. We can easily see that K_z for the multiple holes is proportional to the number of holes, whereas K_z for a single slot increases only slowly as the slot length increases. In addition, we also investigated how effectively one could reduce the loss factor by reducing the slot width. We replaced a single slot with $w=0.4$cm by two slots with $w=0.2$cm each. Suprisingly, K_z is reduced almost by a

factor of ten. This is also true when we replaced a hole of $r=0.2$cm by four holes of $r=0.1$cm. All we discussed for K_z applies equally to K_y.

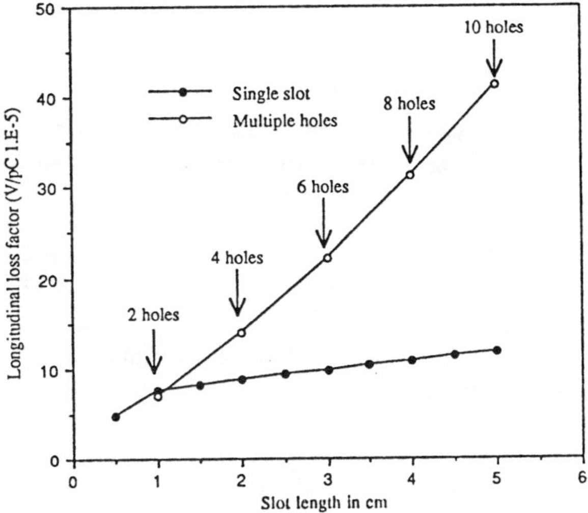

Fig. 4 Longitudinal loss factors for a single slot and for multiple holes as functions of slot length.

VI. CONCLUSION

In this study, we investigated the effect of slots in the vacuum chamber wall on the loss factor of the beam. We found that the contribution to the longitudinal loss factor is quite small, typically of the order of $10^{-6} \sim 10^{-4}$ V/pC/slot. But its behavior is sensitive to the variation of beam bunch length, slot size and vacuum chamber size. The contribution to the transverse loss factor for the slot is also marginal, typically $0.1 \sim 1$ V/pC/m/slot and is rather insensitive to the variation of parameter values. We believe that, in order to understand the effects correctly, we need to calculate the impedance semi-analytically in a manner similar to the semi-analytical study of the pill-box cavity. Finally, for either the longitudinal or transverse effect, the proper geometry of the opening which yields the smallest loss factor is that which causes the least interruption in the wall current, namely long slender slots instead of multiple holes.

VII. REFERENCES

[1] R. Klatt et al., "MAFIA -A Three Dimensional Electromagnetic CAD System for Magnets, RF Structures, and Transient Wake-field Calculations," *Proceedings of the 1986 Linear Acclerator Conferences,* June 1986, pp. 276-278.

[2] M. Sands, "Energy Loss from Small Holes in the Vacuum Chamber," PEP-253, September 1977.

[3] H.A. Bethe, "Theory of Diffraction by Small Holes," Physical Review, vol. 66, pp. 163-182, October 1944.

Bunched Beam Longitudinal Stability

R. Baartman

TRIUMF, 4004 Wesbrook Mall, Vancouver B.C., Canada V6T 2A3

Abstract

Instabilities driven by narrow-band impedances can be stabilized by Landau damping arising from the synchrotron frequency spread due to the nonlinearity of the rf waveform. We calculate stability diagrams for various phase space distributions. We find that distributions without tails are unstable in the 'negative mass' regime (inductive impedance below transition or capacitive impedance above transition). We also find that longitudinal instability thresholds of the (usually neglected) higher order radial modes are lower than expected. For example, the next to lowest dipole mode has a lower threshold than the lowest sextupole mode even though the latter has the larger growth rate in the absence of Landau damping.

I. INTRODUCTION

It is difficult to calculate coupled-bunch instabilities for the case of arbitrary impedance functions. Also, it is usually not very illuminating because little insight is gained and strategies for stabilization are not easily deduced. In this paper, we find thresholds and growth rates arising from parasitic narrow-band resonances. We closely follow the formalism developed by Balbekov [1], but use a different longitudinal phase space distribution function. In particular, we use a distribution which, as a function of synchrotron amplitude, is parabolic near the centre. Such a shape agrees with measured bunch profiles and is moreover expected from thermodynamic considerations.

II. THRESHOLD

Lebedev [2] showed that the Vlasov equation for longitudinal phase space can be cast into the form of an eigenvalue problem for the beam current perturbation harmonics. Assume there is a parasitic resonator which is narrow-band in the sense that the quality factor is large compared with the ratio of resonator to beam frequency ($=n/h$ where h is the harmonic number). In that case, the impedance can couple to a coupled-bunch mode at only one frequency so the eigenvalue matrix reduces to 1×1. The dispersion equation for the harmonic n ($\omega = n\omega_0 + \Omega$, where ω_0 is the revolution frequency) is given by

$$\frac{Z_0}{Z_n(\Omega)} = Y_n(\Omega) \equiv -2i\omega_{s0} \sum_{m=-\infty}^{\infty} m \int_0^{\mathcal{E}_0} \frac{F'(\mathcal{E})|I_{mn}(\mathcal{E})|^2}{\Omega - m\omega_s(\mathcal{E})} d\mathcal{E} \tag{1}$$

where

$$I_{mn}(\mathcal{E}) = \frac{1}{2\pi} \int_{-\pi}^{\pi} \exp\left[im\psi + i\frac{n}{h}\phi(\mathcal{E}, \psi)\right] d\psi \tag{2}$$

and

$$Z_0 = n\frac{\gamma\beta^2\eta E_0}{eI}\left(\frac{\Delta p}{p}\right)^2 = \frac{n}{h}\mathcal{E}_0\frac{V\cos\phi_s}{I}. \tag{3}$$

($\Delta p/p$ is the half-width of the momentum spread at base, I is the average current, V is the rf voltage.) $\omega_s(\mathcal{E})$, ψ and $\mathcal{E}$ are the frequency, angle and action of synchrotron oscillations, $\mathcal{E}_0$ is the maximum action in the bunch: for linear oscillations, the rf phase of a particle w.r.t. the synchronous phase is $\phi = \sqrt{\mathcal{E}}\cos\psi$. $F(\mathcal{E})$ is the distribution function of a bunch; it is normalized according to $\int(F(\mathcal{E})/\omega_s(\mathcal{E}))d\mathcal{E} = \mathcal{E}_0/\omega_{s0}$.

The threshold curve $Y_n^{\text{th}}(\Omega)$ can be found from (1) by letting the imaginary part of Ω approach zero. We consider only cases of weak nonlinearity, where the function $I_{mn}(\mathcal{E})$ (2) can be replaced by the Bessel function $J_m(\frac{n}{h}\sqrt{\mathcal{E}})$ because $\phi \approx \sqrt{\mathcal{E}}\cos\psi$. This restricts us to cases where the tune spread is less than around 20%.

It is clear from (1) that distribution functions F with large slopes are less stable. We consider bunches populated according to the density function

$$F(\mathcal{E}) = K(1 - \mathcal{E}/\mathcal{E}_0)^\mu \tag{4}$$

($K \approx \mu + 1$ for small synchrotron frequency spread). For $\mu < 1$, F' is infinite at the beam edge so the threshold impedance is zero. This is illustrated in Fig. 2 where we have plotted stability diagrams in the impedance plane for the elliptic distribution (parabolic line density) $\mu = 0.5$, the parabolic distribution (parabolic in ϕ, linear in $\mathcal{E}$) $\mu = 1$, and $\mu = 1.5$. These are for $\sqrt{\mathcal{E}_0} = 95°$ in the $\mu = 0.5$ case, and the other cases had $\mathcal{E}_0$ scaled to maintain the same peak line density (see Fig. 1). The two diagrams are for resonators at $n/h = 2$ and $n/h = 0.5$ [1]. It can be seen that indeed for $\mu = 0.5$, the stability boundary passes through the origin. It also does so for $\mu = 1$ and the reason is more subtle: since F' remains finite at the beam edge, the integral in (1) diverges when Ω is exactly $m\omega_s(\mathcal{E}_0)$ because in that case only one side of the singularity is integrated over.

The stability boundaries in Fig. 2 are for the dipole mode ($m = 1$). The other modes lead to additional loops in the impedance plane. An example is shown in Fig. 3 where up to $m = 4$ is shown for the case of $\sqrt{\mathcal{E}_0} = 55°$ with distribution $\mu = 1.5$ for a parasitic resonance at $n/h = 5$. Other such curves are summarized in Fig. 4 where reciprocal threshold shunt impedances are plotted as a function of frequency.

---

[1] No significance should be attached to the fact that in general we use round numbers for n/h. In fact, if n/h is exactly an integer or half integer (always the case if h is 1 or 2), the present stability analysis is not correct because frequencies (of either sign) separated by the rf frequency contribute to the same coupled-bunch mode. Effectively, this means that the present analysis ignores Robinson stability.

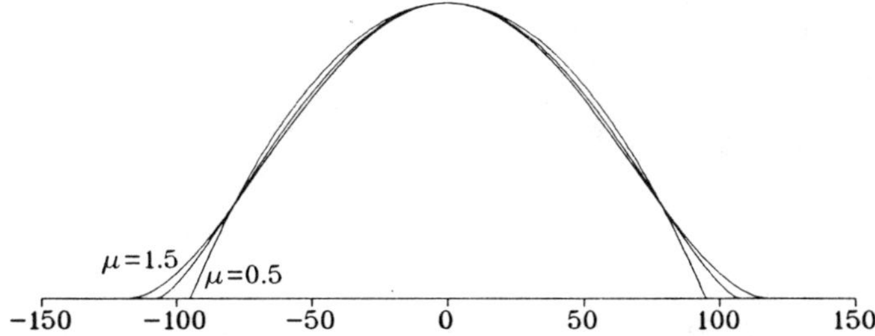

Figure 1: Line density profiles for the three distributions $\mu = 0.5$, 1.0 and 1.5 (see Eqn.4). With the x-axis in degrees, we get the stability diagrams below.

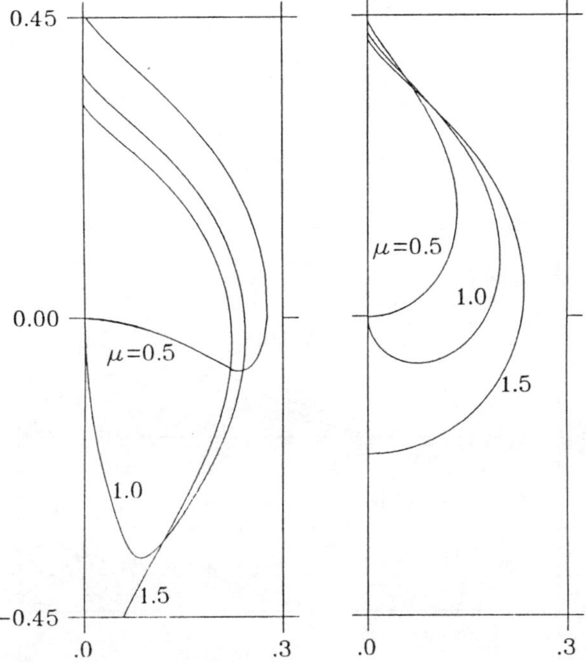

Figure 2: Stability boundaries in the impedance plane for the dipole mode for the three line densities of Fig. 1. The left plot is for $n/h=2$ and the right plot is for $n/h=0.5$. The impedance is in units of $V\cos\phi_s/I$. Scaled in this way, the diagram depends only upon the frequency (n/h) and the bunch length.

III. GROWTH RATE

An upper limit on growth rate ($1/\tau_m$) of azimuthal mode number m is found from the mth term in (1) by ignoring the synchrotron frequency spread and replacing Ω by $m\omega_s + i/\tau$. We get

$$\frac{1}{\tau_m} = -\frac{2m\omega_s}{n/h}\frac{R_{\text{sh}}I}{\mathcal{E}_0 V\cos\phi_s}\int_0^{\mathcal{E}_0} F'(\mathcal{E})J_m^2(n/h\sqrt{\mathcal{E}})d\mathcal{E}. \quad (5)$$

For the distribution family (4), Satoh [3] has solved this integral as an infinite sum and he has shown moreover that the individual terms in the infinite sum correspond to the different radial modes belonging to the azimuthal mode m:

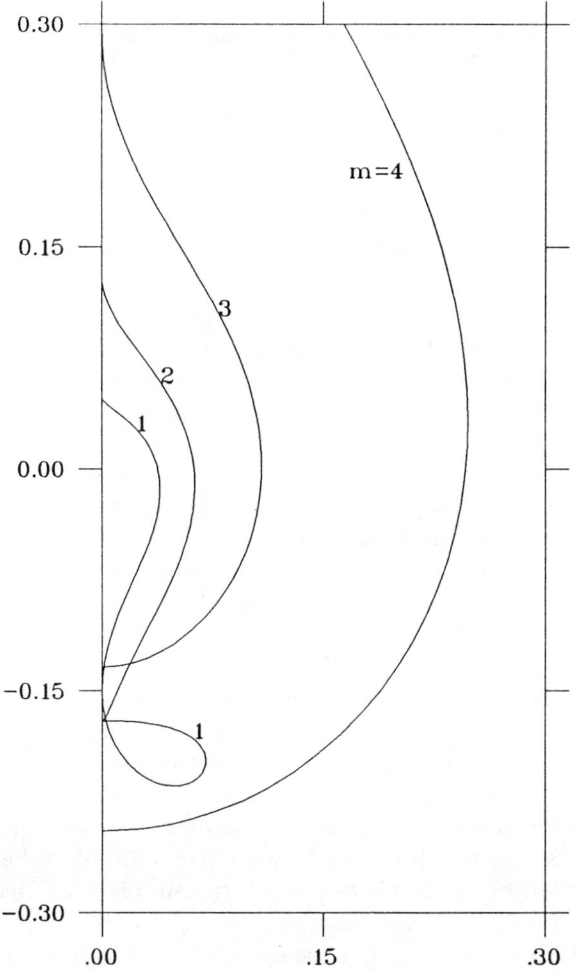

Figure 3: Stability diagram in the impedance plane for a parasitic near 5 times the rf frequency, $\sqrt{\mathcal{E}_0}=55°$. Notice that the dipole mode is peculiar in that it has an extra loop. This arises because there is more than one oscillation of the square of $J_1(5\sqrt{\mathcal{E}})$ for $\mathcal{E} < \mathcal{E}_0$, so there are two dipole modes; ordinary and extraordinary. Inside the extra loop, the extraordinary mode is stable and the ordinary dipole mode is unstable. Both of these dipole modes are linear combinations of the radial modes (k) discussed below.

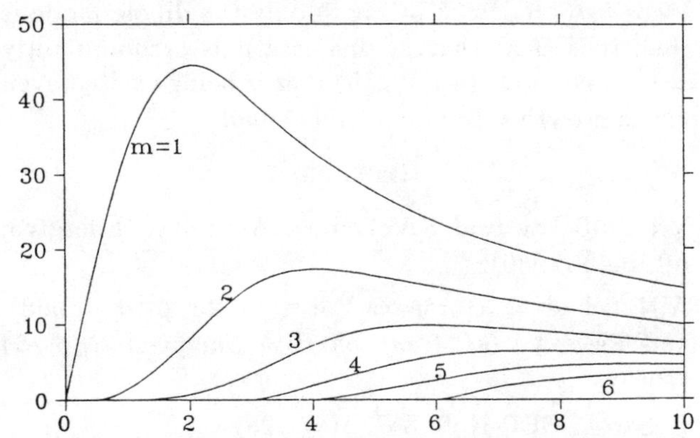

Figure 4: Reciprocal of the threshold R_{sh} (in units of $I/(V\cos\phi_s)$) vs. n/h for $\sqrt{\mathcal{E}_0}=55°$ with distribution $\mu=1.5$. Thresholds up to the dodecapole are shown.

$$\frac{1}{\tau_m} = \sum_{k=0}^{\infty} \frac{1}{\tau_{mk}} = m\omega_s \frac{n}{2h} \frac{R_{\mathrm{sh}}I}{V\cos\phi_s} \mu(\mu+1) \left(\frac{2}{X}\right)^{2\mu+2} \times$$

$$\times \sum_{k=0}^{\infty} \frac{(m+2k+\mu)\Gamma(k+\mu)\Gamma(m+k+\mu)}{k!(m+k)!} J_{m+2k+\mu}^2(X) \tag{6}$$

(X is an abbreviation for $\frac{n}{h}\sqrt{\mathcal{E}_0}$).

Both the individual terms of (6) and their sums have been plotted in Fig. 5. Take particular note of the rigid ($k=0$) mode for each m. These correspond to the form factors given by Sacherer [4] (Sacherer picture). The bottom plot of Fig. 5 is a summary plot of the sums $1/\tau_m$. This can be thought of as the Laclare [5] picture. The Sacherer picture is relevant for calculating growth rates from resistive impedances when a short-range wake like space charge lifts the degeneracy of radial modes and dominates in determining the thresholds. The Laclare picture is relevant when the short-range wake effects are small compared with those due to the narrow-band resonator.

IV. Discussion

A resonator can be represented by a circle tangent to the imaginary axis in the impedance plane. If no other impedance is present, the circle is centred on the real axis and it is clear (Fig. 2) that distributions with $\mu<1$ are unstable for any finite R_{sh}. If there is also a negative imaginary impedance (eg. due to inductive wall effect below transition or space charge above transition) then distributions with $\mu\leq1$ are unstable. In other words, tails are essential in this case. On the other hand, with capacitive impedance below transition, bunches can have line densities which are close to parabolic. This is consistent with the measured bunch shapes of the CERN PS Booster [4].

It is interesting to compare Figs. 4 and 5. Counter to intuition, it is not the mode with the fastest growth rate (in the absence of Landau damping) which has the lowest threshold. In fact, thresholds at any frequency go monotonically with azimuthal mode number and the dipole mode always has the lowest threshold. For example, at $n/h=5$ and $V\cos\phi_s/(IR_{\mathrm{sh}})=23$ in Fig. 4 only the dipole mode is unstable. It is clear that in this case it is predominantly the $k=1$ radial mode (see Fig. 5) that is being excited even though its growth rate is relatively small.

V. References

[1] V.I. Balbekov and S.V. Ivanov, Atomnaya Energiya, **60**, p.45 (1986).

[2] A.N. Lebedev, Atomnaya Energiya, **25**, p.100 (1968), and *Proc. VI Int. Conf. on High Energy Acc.*, p.284 (1967).

[3] K. Satoh, PEP-Note 357, May 1981.

[4] F.J. Sacherer, IEEE Trans. Nucl. Sci., **NS-20**, p.825 (1973), and IEEE Trans. Nucl. Sci., **NS-24**, p.1393 and p.1396 (1977).

[5] J.L. Laclare, CAS Proc., CERN 87-03, p.264.

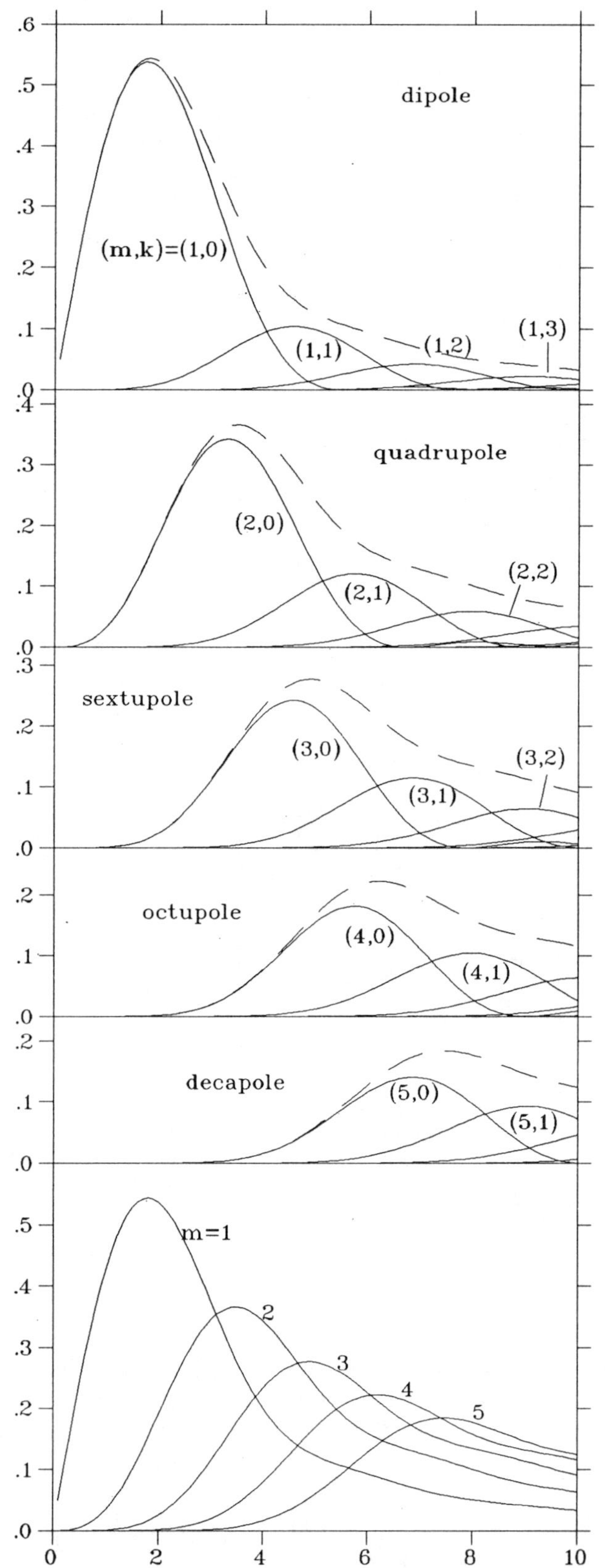

Figure 5: Growth rates (in units of $\omega_s R_{\mathrm{sh}}I/(V\cos\phi_s)$) vs. n/h for the modes indicated. The dashed curves and the curves in the summary plot at the bottom are the sums over k for any given m: they apply to the case of narrow-band impedance only, i.e. no splitting due to space charge.

REDUCTION OF BEAM BREAKUP GROWTH BY
BLEEDING CAVITIES IN LINEAR ACCELERATORS

D. G. Colombant and Y. Y. Lau
Beam Physics Branch, Plasma Physics Division
Naval Research Laboratory, Washington, DC 20375-5000

ABSTRACT

We show that, by coupling several dummy (bleeding) cavities to the accelerating cavities in a linear accelerator, beam breakup growth may be reduced significantly, by factors of hundred in some examples. This growth reduction results from the sharing of the deflecting mode energy in the accelerating cavities with the bleeding cavities. Some issues on the viability of this novel method of BBU control are discussed.

I. INTRODUCTION

There are three known types of rf cure to control beam breakup (BBU) growth: (1) Lowering of the quality factor Q of the deflecting modes, such as by adding transverse slots to the cavities [1] or by loading the cavities with ferrite material [2,3], (2) Reducing the transverse shunt impedance that the beam experiences, e.g., by the use of a shielded gap [4,5], and (3) Stagger tuning the cavities, so that the breakup mode frequencies are off-tuned from one to the next [6-9]. Here, we consider a fourth possibility: (4) By simply attaching to the main accelerating cavities several identical dummy (bleeding) cavities so that the main cavities' deflecting mode (that causes the beam to deflect sideways), has its mode energy shared by the dummy cavities to which it is coupled. This method of BBU control, in some sense, is similar to mechanism (2), but it has the advantage that all deflecting modes can be controlled, as identical cavities are employed. It is different from mechanism (1), as it does not involve any dissipative process.

We stumbled on the above mentioned mechanism while we assessed the effects of cross-coupling among cavities [10] in a recirculating accelerator known as the Spiral Line Induction Accelerator (SLIA) [11]. By unwrapping the SLIA into a linac, we show here that the addition of bleeding cavities would reduce BBU growth in a linac geometry.

II. MODEL AND RESULTS

For simplicity, consider a continuous coasting beam of current I, relativistic mass factor γ, pulse length τ, in a linac which consists of N identical accelerating cavities. Each accelerating cavity supports a deflecting mode of frequency ω_o, quality factor Q, and transverse shunt impedance $Z_\perp$. At the ith accelerating cavity, the deflecting mode amplitude is $f^{(i)}(t)$, and the beam's transverse displacement there is $x^{(i)}(t)$. Each accelerating cavity is coupled to several dummy cavities [Fig. 1]. The deflecting modes in these dummy cavities are denoted by $g_1^{(i)}(t)$, $g_2^{(i)}(t)$, etc. For simplicity of notation, we shall drop the superscript (i) unless specified otherwise.

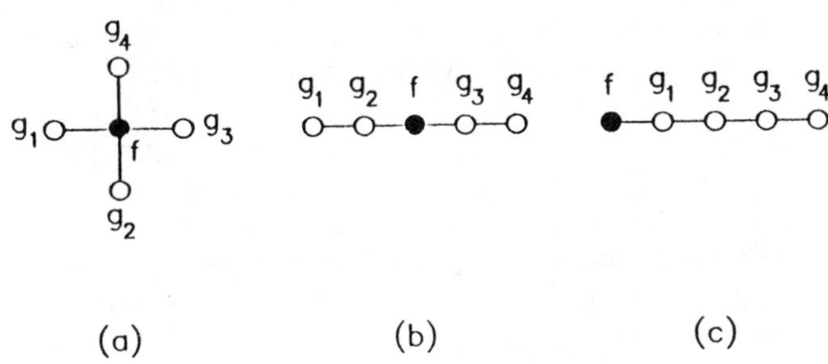

Fig. 1. Various coupling configurations between a typical accelerating cavity (solid circle) and the bleeding cavities (open circle).

The accelerating cavity can be coupled to the dummy cavities in many different ways. Figure 1 shows three configurations where an accelerating cavity (solid circle), is coupled to four dummy cavities (open circles). We assume that the coupling constant between each pair of cavities is κ [10]. The governing equations for configuration b, for instance then read

$$Lf(t) = \varepsilon_r h(t)x(t) + \kappa g_2(t) + \kappa g_3(t)$$

$$Lg_1(t) = \kappa g_2(t)$$

$$Lg_2(t) = \kappa f(t) + \kappa g_1(t) \qquad\qquad (1)$$

$$Lg_3(t) = \kappa f(t) + \kappa g_4(t)$$

$$Lg_4(t) = \kappa g_3(t)$$

where L denotes the operator

$$L \equiv \frac{d^2}{dt^2} + \frac{\omega_o}{Q}\frac{d}{dt} + \omega_o^2,$$

$\varepsilon_r = (I/17\gamma kA)Z_1(\Omega)/30Q$ is the dimensionless coupling constant, and $h(t)=1$ when the beam is present within the accelerating cavity and $h(t)=0$ otherwise. The term $\varepsilon_r hx$ in Eq. (1) describes the excitation of the breakup mode in the accelerating cavity by the beam's transverse displacement. The remaining terms in the RHS of Eq. (1) describe the cavity excitation as a result of the coupling (κ). Similar models have been used to describe the mode coupling in two identical cavities, e.g., in a conventional two-cavity klystron oscillator, where κ equal to a few per cent has been routinely used [12].

Upon crossing the ith accelerating cavity at time t, a beam slice would pick up an incremental transverse momentum $\Delta p^{(i)}(t) = f^{(i)}(t)$; its transverse displacement remains unchanged for a narrow gap. We assume that the motion of this beam slice, upon exiting the ith cavity, will be advanced to the (i+1)th cavity via a 2×2 transport matrix.

As an example, we unwrap a 7-turn, 70 cavity recirculating accelerator, that models a SLIA Upgrade, into a linac with 70 accelerating cavities, each of which is connected to four dummy cavities as shown in Fig. 1. The parameters for the SLIA Upgrade were given in Ref. 13; briefly, 10 kA beam current, 35 ns pulse length, 25 MeV energy, $\gamma\varepsilon_r = 0.245$.

In Fig. 2 we show the BBU amplitude gain, $x(70)/x(1)$, at the 70th cavity as a function of κ for configurations (a), (b), (c). Note from this figure that reduction of BBU growth up to a factor of 300 is possible if $\kappa \to 0.1$.

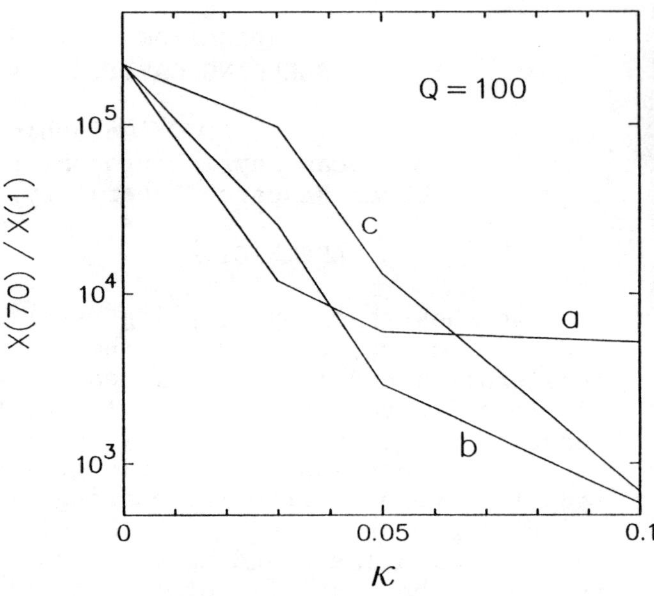

Fig. 2. BBU amplitude gain as a function of κ for the three configurations shown in Fig. 1.

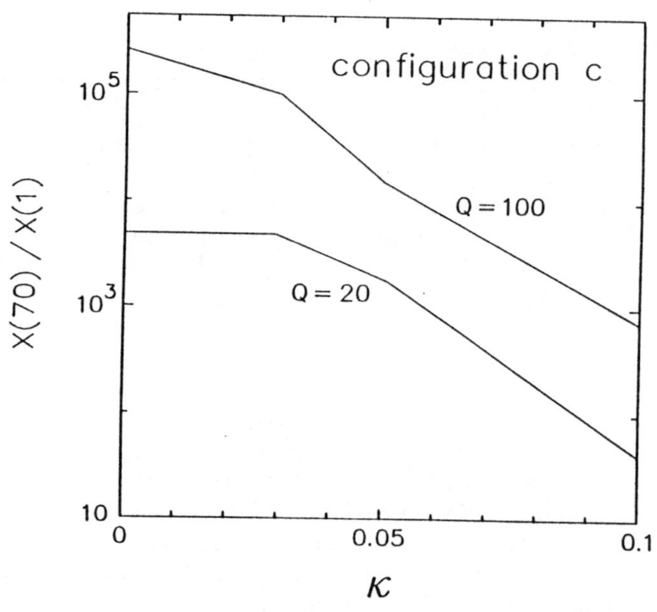

Fig. 3. Comparison between the cases of Q=100 and Q=20 for configuration c.

We have set Q = 100 in Fig. 2 and we include the data for Q = 20 in Fig. 3, for configuration C. In Fig. 4, we eliminate the "bends" in the SLIA Upgrade so that, when it is unwrapped into a linac, all seventy cavities have equal separation. Here, we see that the reduction of BBU growth is only moderate.

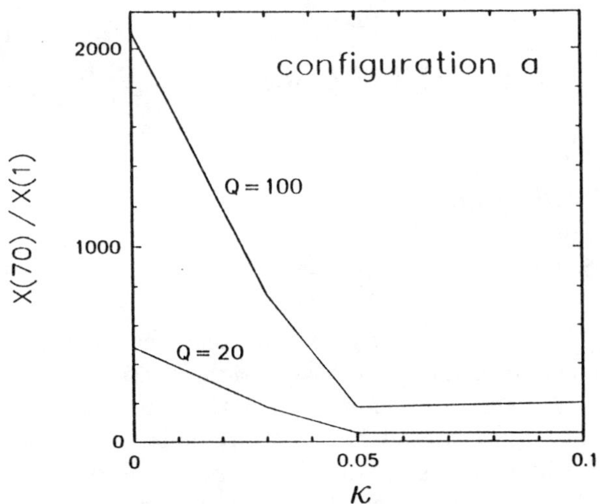

Fig. 4. Comparison between the cases of Q=100 and Q=20 for configuration a. Here, the accelerating cavities have equal separation.

The above data supported the general conclusion that bleeding cavities can reduce BBU growth. The amount of reduction depends on the geometrical arrangements of the accelerating cavities and the bleeding cavities. In general, the amount of BBU reduction increases with Q, as in our previous studies [10].

III. SOME ISSUES

Many questions may undoubtedly be raised regarding this method of BBU control. (1) Would the dummy cavities actually worsen BBU growth, as their presence introduces asymmetries to the accelerating cavity, where the beam passes through? (2) Would the dummy cavities lower the acceleration gradient, as they may also provide leakage of the power intended for particle acceleration? (3) What is the proper way to control the degree of cavity couplings? Would coupling loops suffice? (4) The method of couplings may vary. For example, should mode g_1 be coupled to mode g_4 in Fig. 1, and what determines the optimum configuration? (5) In the above formulation we have assumed that all κ's are the same. How would different phasings in the cavity coupling influence BBU growth? (6) Should this method of BBU control be applied to linear colliders or more suitably to high current induction accelerators [14]? (7) How about the cost of this method of BBU control in comparison with other rf cures?

Although these questions cannot be answered by the present cursory study, some of them can be addressed by small scale experiments, especially those concerning the methods of cavity couplings and their purported tendency of reducing BBU growth. The usefulness of this seemingly novel method of BBU control may then be further assessed, once such small scale experiments are performed [15].

ACKNOWLEDGMENT

This work was supported by the Strategic Defense Initiative Organization, managed by the Office of Naval Research. It was also supported by the Defense Advanced Research Project Agency under ARPA Order 7781 and monitored by the Naval Surface Warfare Center.

REFERENCES

[1] R. B. Palmer, SLAC Publication 4542 (1988); H. Deruyter et al., Proc 1989 IEEE Part. Accel. Conf. p. 156.

[2] D. Birx, Lawrence Livermore Natl. Lab. Report UCID 18630 (1980).

[3] R. J. Briggs et al., Part. Accel. 18, 41 (1985).

[4] M. Wilson, IEEE Trans. NS-28, 3375 (1981).

[5] R. B. Miller et al., Sandia Report # SAND87-1389J (1987).

[6] R. Helm. and G. Loew, in Linear Accelerators, Eds. P. M. Laposelle and A. L. Septier, p. 173 (1970).

[7] R. L. Gluckstern et al., Part. Accel. 23, 53 (1988).

[8] D. G. Colombant and Y. Y. Lau, Appl. Phys. Lett. 55, 27 (1989).

[9] K. A. Thompson and J. W. Wang, in these proceedings (1991).

[10] D. G. Colombant, Y. Y. Lau and D. Chernin, in Part. Accel. 35, No. 4 (1991).

[11] V. Bailey et al., 1987 IEEE Part. Accel. Conf., p. 920; also in Proc. SPIE 1407, Ed., H. E. Brandt (1991).

[12] See, e.g., H. J. Reich et al., Microwave Theory and Technique, (Van Nostrand), p. 539 (1953).

[13] D. G. Colombant, Y. Y. Lau and D. Chernin, in Proc. SPIE No. 1407, Ed., H. E. Brandt (1991).

[14] Y. Y. Lau, Phys. Rev. Lett. 63, 1141 (1989) [Errata: 63, 2433 (1989).]

[15] R. Gilgenbach, private communication (1991); also P. Menge et al., in Proc. SPIE 1407 (1991).

Observation of Magnetized Cooling in the IUCF Cooler

Timothy JP Ellison

The Indiana University Cyclotron Facility
2401 Milo Sampson Lane
Bloomington, IN 47405 USA

Abstract

Measurements of the longitudinal drag rate (R_D, the rate at which the electron beam can change the energy of the proton beam) for 45 MeV protons having longitudinal velocities relative to the electron beam, $v^*_{p\parallel}$, in the range from $1 \times 10^{-7} < v^*_{p\parallel}/c < 1 \times 10^{-2}$, are summarized along with measurements of the *effective longitudinal electron beam energy spread*, $\delta E_{e,spread}$, as function of electron beam current, I, and angular misalignment between the electron and proton beams, $\theta_{misalignment}$. It was found that $\delta E_{e,spread}$ increases approximately linearly with $\theta_{misalignment}$; this appears to be consequence of the combination of the electron beam space charge depression and the fact that the equilibrium proton beam size increases approximately linearly with $\theta_{misalignment}$. For well aligned beams, $\delta E_{e,spread}$ is consistent with electron beam intrabeam scattering, and in this regime, the normalized drag rate is enhanced considerably over the predictions of the nonmagnetized theory. These results are compared with measurements made at other laboratories.

I. INTRODUCTION

The initial measurements of the longitudinal drag rate [1] for the IUCF 270 keV electron cooling system were in close agreement with the nonmagnetized theory, although there was a larger than expected effective longitudinal electron beam energy spread, $\delta E_{e,spread}$. The nonmagnetized longitudinal drag rate as a function of the longitudinal proton velocity relative to the electron beam for a transversely cooled beam, $R_D (v^*_{p\parallel})$, can be simply expressed as:

$$R_D\left(v^*_{p\parallel}\right) = R_{D,\mathrm{max}} \left(\frac{\Delta_{e\perp}}{v^*_{p\parallel} + \Delta_{e\perp}} \right)^2 \qquad \left(v^*_{p\parallel} > \Delta_{e\parallel}\right) \quad (a)$$

$$= R_{D,\mathrm{max}} \frac{\delta E_{e,error}}{\delta E_{e,spread}} \qquad \left(v^*_{p\parallel} < \Delta_{e\parallel}\right) \quad (b) \tag{1}$$

where

$$R_{D,max} = \frac{4I\eta\left(r_e mc^2\right)^2}{e\gamma r_b^2 kT_c} \Lambda_c \tag{2}$$

where $\pm \Delta_{e\parallel} = \pm \beta c \left(\delta E_{e,spread}/2W\right)$ is the width of the electron beam longitudinal velocity spread corresponding to

This work is supported by The National Science Foundation under Grant NSF PHY 90 15957.

$\delta E_{e,spread}$; $\Delta_{e\perp}$ is $(kT_c/m)^{\frac{1}{2}}$ where k is the Boltzmann constant and m the electron mass; r_e is the classical electron radius; e the electron charge and γ the usual relativistic parameter. T_c, η, r_b, and Λ_c and W are defined in Table I. Since we will be discussing the *electron beam effective energy spread*, $\delta E_{e,spread}$, it is more illuminating to express the drag rate as a function of the *equivalent electron energy error* $(\delta E_{e,error} = 2W(v^*_{p\parallel}/\beta c))$, i.e. the drag rate experienced by a cooled proton if the electron beam energy were to be instantaneously stepped by $\delta E_{e,error}$.

The nonmagnetized model [Eqs. (2) and (1) using the parameters for the IUCF electron cooling system tabulated in Table I and $I = 1$ A] is shown along with early

Electron Energy Deviation for 24.4 keV Electrons (eV)

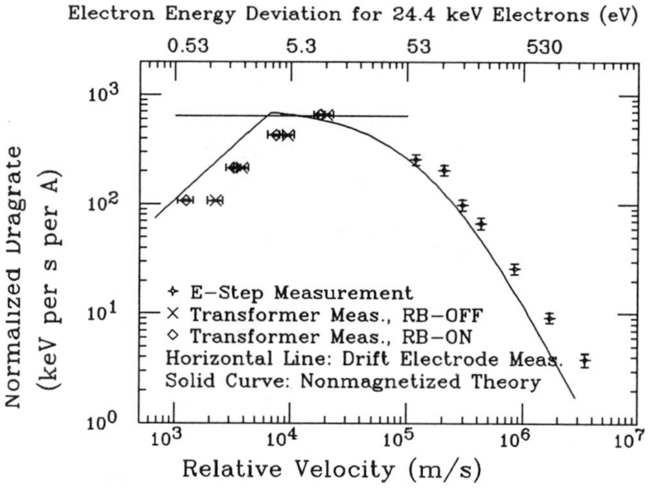

Figure 1. Measured normalized drag rate (R_D/I) in the IUCF Cooler.

measurements [1] in Figure 1. The upper scale shows the drag rate as a function of $\delta E_{e,error}$.

As can be seen in Fig. 1, the model assumes an $\delta E_{e,spread}$ of ± 3.6 eV, corresponding to the region where the drag rate increases linearly with $\delta E_{e,error}$ (i.e. $\delta E_{e,error} < \delta E_{e,spread}$). The width of the linear region for the *measured* drag rate, however, is consistent with a $\delta E_{e,spread}$ of about 10 eV; a value much larger than expected from the cathode potential regulation (± 1 V). There are a number of other mechanisms which can also cause a real or effective electron beam energy spread: longitudinal-longitudinal ($\parallel$-$\parallel$ IBS) and transverse-longitudinal ($\perp$-$\parallel$ IBS) intrabeam scattering [2], and effects arising from the electron beam space charge depression [3].

Table I. Parameters of the IUCF 270 keV cooling system during measurements.

Parameter	Symbol	Value	Units
Cooling Region Length	L_c	2.2	m
Ring Circumference	C	87.8	m
L_c/C	η	0.025	
Coulomb Logarithm	Λ_c	10.7	
Electron Beam Radius	r_b	1.27	cm
Cathode Temperature	T_c	1300	K
E-Beam Kinetic Energy	W	24.3	keV

II. MEASUREMENTS OF THE EFFECTIVE ELECTRON BEAM ENERGY SPREAD

One can attempt to determine the predominant cause of $\delta E_{e,spread}$ by determining how it varies with I and $\theta_{misalignment}$: $\delta E_{e,spread}$ should vary as $I^{1/6}$ if primarily due to ‖-‖ IBS, $I^{1/2}$ if primarily due to ⊥-‖ IBS [2], and as I if primarily due to the electron beam space charge [3].

A. Effect of Electron Beam Current

The measured *normalized drag rate* (R_D/I) is shown in Figure 2 as a function of $\delta E_{e,error}$, for constant ring and electron cooling system settings and I = 0.1, 0.2, 0.4, and 0.8 A. Two features are readily apparent from these data:

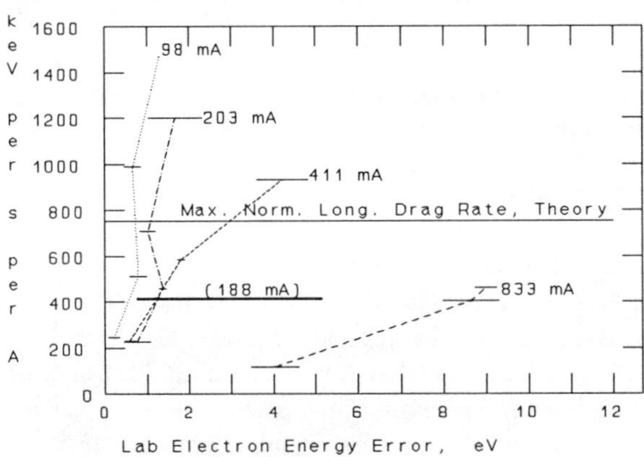

Figure 2. Normalized drag rate (R_D/I) vs. lab frame electron energy error for varying electron beam currents.

firstly, $\delta E_{e,spread}$ increases $\approx$ linearly with electron current pointing to the electron beam space charge as a predominant source. This can be seen from the width of the region over which the drag rate increases to its maximum value. Secondly, the $(R_D/I)_{max}$ decreases with I.

B. Effect of Electron/Proton Beam Alignment

An angular misalignment, $\theta_{misalignment}$, between the proton and electron beams in the cooling section causes the equilibrium proton beam emittance to increase substantially; for example, we have seen that a 0.37 mrad misalignment

between the beams increases the equilibrium proton beam emittance from about 0.05π to 1π μm, such that the rms divergence approximately equals $\theta_{misalignment}$ [3,4]. Consequently one expects to observe $\delta E_{e,spread}$ to increase linearly with $\theta_{misalignment}$ due to the combination of the electron beam space charge and both the increase in proton beam size and transverse movement with respect to the equipotential electron beam axis:

$$\delta E_{e,spread} = \pm \frac{eIr\Delta r}{2\pi\epsilon_o \beta c r_b^2} \qquad (3)$$

where

$$\Delta r = \theta_{misalignment} \cdot \left[\frac{L_c}{2}, \beta_{x,cooler} \right]_{max}$$

where r is the average offset between the electron and proton beams, $\beta_{x,cooler}$ the beta function in the cooling region, βc the proton beam velocity, L_c the length of the cooling region, and ϵ_o the permittivity of free space. Assuming (Eq. 1b) that the drag rate increases linearly in the region $\delta E_{e,error} < \delta E_{spread}$, one would expect to observe $\delta E_{e,error}$ to increase linearly with the product of I and $\theta_{misalignment}$ for a fixed value of R_D/I. The apparent linear increase in $\delta E_{e,error}$ with I is observed in Figure 2. In Figure 3 one observes an approximately linear increase in $\delta E_{e,error}$ with $\theta_{misalignment}$ for fixed I (0.188 A) and R_D (78 keV/s). The model (solid line) assumes r = 3.3 mm. These tests should be repeated with known values of r.

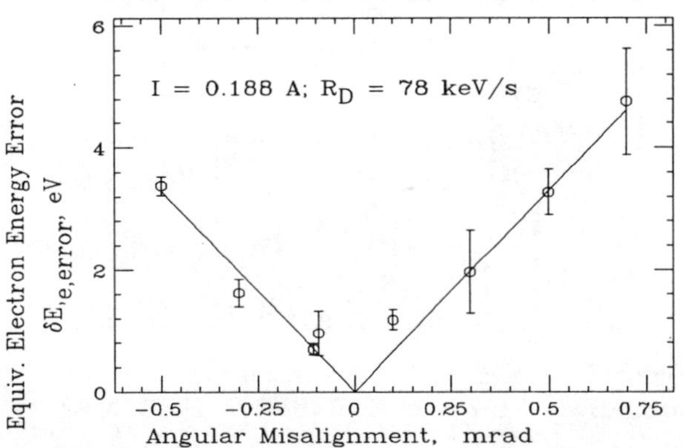

Figure 3. Measured $\delta E_{e,error}$ as a function a angular alignment between the electron and proton beam with a fixed normalize drag rate.

The author believes that the sensitivity of the equilibrium emittance and δE_{spread} to small changes in the $\theta_{misalignment}$ provides a measurement of the straightness of the field lines in the main solenoid.

III. OPTIMIZATION OF COOLING AND OBSERVATION OF MAGNETIZED EFFECTS

The cooling was optimized by using the measurement of $\delta E_{e,spread}$ as a function of $\theta_{misalignment}$ as a diagnostic to minimize $\theta_{misalignment}$. Using a 0.208 A electron beam, the

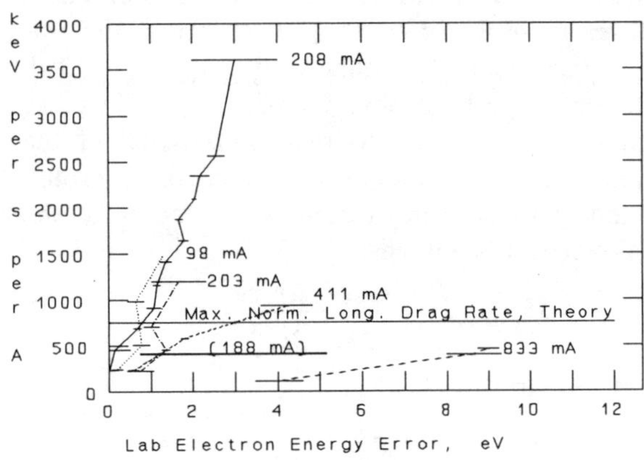

Figure 4. (R_D/I) vs. $\delta E_{e,error}$. The solid curve (0.208 A) is data after minimizing $\delta E_{e,spread}$.

maximum measured longitudinal drag rate increased to a value about 5 times higher than the predictions of the nonmagnetized theory (Figure 4). The effective $\delta E_{e,spread}$, however, was not reduced below about 2 - 3 eV, a value which appears to be consistent with $\|$-$\|$ and $\perp$-$\|$ IBS.

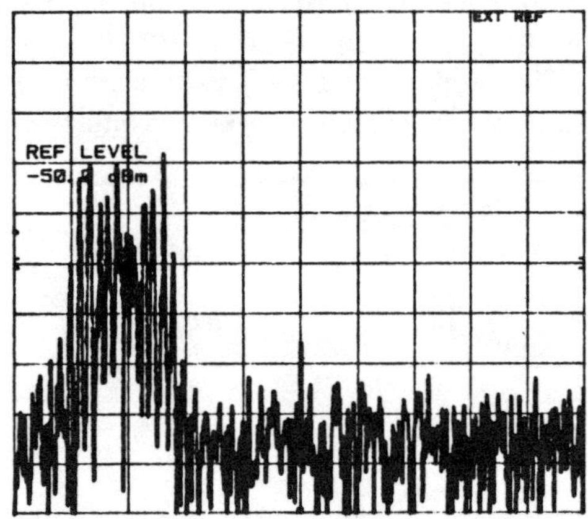

Figure 5. 45 MeV proton beam $\|$ Schottky signal. (CF 54.7 MHz; RBW = VBW = 300 Hz; 5 dB/div; 2 kHz (0.1 s)/div.

Figure 5 is a Schottky signal spectrum obtained while operating with this high drag rate. The sharp spikes occurring each 1/60th s in sweep time are characteristic of a pure tone being frequency-modulated by a 60 Hz triangle waveform with a deviation of $\pm$ 3.1 kHz. My interpretation of this signal spectrum is that the proton beam energy is being coherently modulated by:

$$\pm\ 2.9\ \text{keV}\ [\ =\ \pm\ \tfrac{1}{2}\beta^2\gamma M_p c^2 \eta^{-1}(3.1\ \text{kHz}/54.6\ \text{MHz})]$$

The normalized drag rate necessary for this is about 3500 keV/s-A, consistent with measurements shown in Figure 4.

IV. MAGNETIZED COOLING

These high drag rates are consistent with the theories of cooling with a magnetic field[5]. In this regime, cooled protons experience adiabatic collisions (taking place over a time period long compared to the electron gyroperiod) with the electron Larmor circles. Since the cooling solenoid magnetic field straightness in on the order of 0.3 mrad (rms) (about 7 times smaller than the intrinsic angular divergence of the electron beam due to the 0.12 eV cathode temperature) the electrons appear about 50 times "colder"; however, the coulomb logarithm for such adiabatic collisions is about 10 times smaller, yielding an $\approx 500\%$ increase in the drag rate (see Eq. 2). In Figure 6 the measured *scaled* drag rate ($\gamma R_D/\eta j_e$, where $\eta = L_c/C$ and j_e is the electron current density) is shown along with data from other laboratories as well models showing the contribution to the drag rate from the "fast" (nonmagnetized model) and "adiabatic" (magnetized model) collisions.

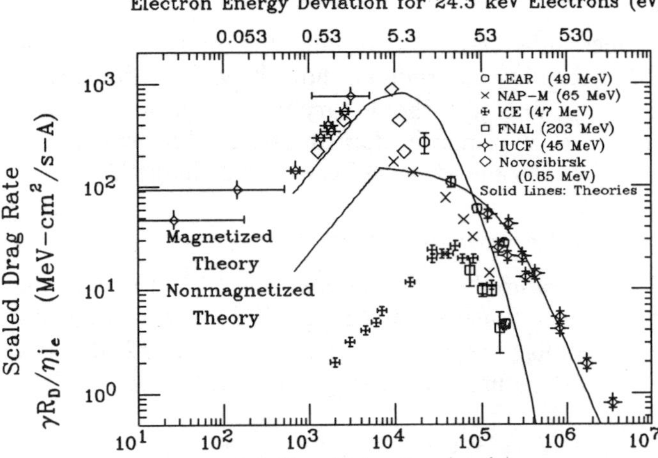

Figure 6. Scaled drag rates from IUCF and other laboratories along with magnetized and nonmagnetized model.

Acknowledgements

These measurements would not have been possible without the help of a myriad of IUCF personnel, especially M. Ball, B. Brown, J. Collins, V. Derenchuk, D. Friesel, B. Pollock, and T. Sloan. I also owe thanks to L. Lebedev (I.N.P., Novosibirsk) for explaining to me, in an intuitive manner, the basics of electron beam intrabeam scattering.

References

[1] Timothy JP Ellison, Dennis L. Friesel, and Robert J. Brown, "Status and performance of the IUCF 270 keV electron cooling system", *Proc. 1989 IEEE P. A. C., Acc. Eng. and Tech.*, (Chicago, IL, 20-23 March 1989).

[2] N.S. Dikansky, V.I. Kudelainen, V.A. lebedev, I.N. Meshkov, V.V.Parkhomchuk, A.A. Sery, A.N. Skrinsky, B.N. Sukhina, "Ultimate possibilities of electron cooling", preprint 88-61 (I.N.P., Novosibirsk, 1988).

[3] T. Ellison, "Electron Cooling" (Ph.D. Thesis, Indiana University, 1990).

[4] M. Ball, D. Caussyn, J. Collins, D. Duplantis, V. Derenchuk, G. East, T. Ellison, D. Friesel, B. Hamilton, B. Jones, M.G. Minty, S.Y. Lee, and T. Sloan, "Beam property measurements in the IUCF Cooler", these proceedings

[5] Ya. S. Derbenev and A.N. Skrinsky, "The effect of an accompanying magnetic field on electron cooling, *Part. Accelerators*, **8**, 235-243 (1978).

TUNE SPLITTING IN THE PRESENCE OF LINEAR COUPLING*

G. Parzen

Brookhaven National Laboratory

Upton, NY 11973, USA

I. INTRODUCTION

The presence of random skew quadrupole field errors will couple the x and y motions. The x and y motions are then each given by the sum of 2 normal modes with the tunes ν_1 and ν_2, which may differ appreciably from ν_x and ν_y, the unperturbed tunes. This is often called tune splitting since $|\nu_1 - \nu_2|$ is usually larger than $|\nu_x - \nu_y|$. This tune splitting may be large in proton accelerators using superconducting magnets, because of the relatively large random skew quadrupole field errors that are expected in these magnets. This effect is also increased by the required insertions in proton colliders which generate large β–functions in the insertion region.

This tune splitting has been studied in the RHIC accelerator (the Relativistic Heavy Ion Collider proposed at Brookhaven National Laboratory). For RHIC, a tune splitting as large as 0.2 was found in one worse case. A correction system has been developed[1] for correcting this large tune splitting which uses two families of skew quadrupole correctors. It has been found[1] that this correction system corrects most of the large tune splitting, but a residual tune splitting remains that is still appreciable.

RHIC has to operate within a box in tune space whose width is $\Delta \nu = 33 \times 10^{-3}$ in order to avoid all resonances which are tenth order or less. It appears desirable to correct the tune splitting to a level which is much smaller than 33×10^{-3}. The residual tune splitting that remained after correction with the 2 family tune splitting correction system was found to be about 18×10^{-3}.

The residual tune splitting appears to be due to higher order effects of the random a_1 multipole. Loosely speaking, one may say that the residual tune splitting is associated with the nearby linear sum resonance, $\nu_x + \nu_y =$ integer. A skew quadrupole correction system has been developed that appears able to correct a large part of the residual tune splitting. This consists of skew quadrupole correctors near the high–β quadrupoles in the insertions, which are excited so as to generate the harmonics that are near $\nu_x + \nu_y$. An analytic result for the ν shift, that includes higher order terms in a_1, indicates that the important harmonics are the harmonics near $\nu_x + \nu_y$.

II. CALCULATION AND CORRECTION OF THE NORMAL MODE TUNE

The normal mode tunes ν_1 and ν_2 can be computed in the following way. The effect of the a_1 multipole in each magnet is represented by putting point a_1 multipoles at the entrance and exit of each magnet, with half

the strength of the a_1 multipole. The one turn transfer matrix around the ring can be computed by multiplying the transfer matrices of all the magnets and all the point a_1 multipoles in the ring. The 4×4 one turn transfer matrix can be written as

$$T = \begin{bmatrix} M & n \\ m & N \end{bmatrix} \qquad (2.1)$$

where M, N, m, n are each 2×2 matrices. The tunes ν_1 and ν_2 can then be computed from the following equations[2]

$$\cos \mu_1 + \cos \mu_2 = \frac{1}{2} \, Tr \, (M + N)$$

$$\cos \mu_1 - \cos \mu_2 = \left\{ \left[\frac{1}{2} Tr \, (M - N) \right]^2 + |m + \overline{n}| \right\}^{\frac{1}{2}}$$

$$\overline{n} = \begin{bmatrix} n_{22} & -n_{12} \\ -n_{21} & n_{11} \end{bmatrix}, \mu_1 = 2\pi\nu_1, \mu_2 = 2\pi\nu_2 \, . \qquad (2.2)$$

Treating the random a_1 as a perturbation one can find an approximate result which is valid close to the resonance line $\nu_x - \nu_y = p$, p being some integer.

$$\nu_1 = \overline{\nu}_x \pm \left[(\nu_x - \nu_y - p)^2 / 4 + |\Delta \nu \, (\overline{\nu}_x, \overline{\nu}_y)|^2 \right]^{\frac{1}{2}}$$

$$\nu_2 = \overline{\nu}_y \pm \left[(\nu_x - \nu_y - p)^2 / 4 + |\Delta \nu \, (\overline{\nu}_x, \overline{\nu}_y)|^2 \right]^{\frac{1}{2}}$$

$$\Delta \nu \, (\nu_x, \nu_y) = (1/4\pi\rho) \int ds \, (\beta_x \beta_y)^{\frac{1}{2}} \, a_1 \exp \left[i \, (-\nu_x \theta_x + \nu_y \theta_y) \right]$$

$$\overline{\nu}_x = (\nu_x + \nu_y + p) / 2, \overline{\nu}_y = (\nu_x + \nu_y - p) / 2,$$

$$\theta_x = \psi_x / \nu_x, \theta_y = \psi_y / \nu_y \qquad (2.3)$$

$\Delta \nu$ may be regarded as the width of nearby coupling resonance. ν_1 is the ν value that goes to ν_x when $a_1 \rightarrow 0$, and similarly for ν_2 and ν_y. For the $\pm$ sign, the top sign is used when $\nu_x > \nu_y + p$ and the bottom sign when $\nu_x < \nu_y + p$. ρ is the radius of curvature in the main dipoles and a_1 is defined by $B_x = B_0 a_1 x$ where B_0 is the main dipole field.

Equations (2.3) suggest that if the a_1 correctors are set so as to make $\Delta \nu = 0$, then the tune splitting is minimized. This is not a practical way to set the a_1 correctors as the $\Delta \nu$ and the a_1 errors are not known for a real accelerator. Usually the a_1 correctors are set so as to minimize the tune splitting which can be measured. The results from this way of setting the a_1 correctors is discussed in Section III. It is instructive to see what happens when the a_1 correctors are set to make $\Delta \nu = 0$. These results are shown in Table 1 for 10 distributions of

*Work performed under the auspices of the U.S. Department of Energy.

a_1 errors and for a RHIC lattice having $\beta^* = 2$ at all 6 crossing points.

The results in Table 1 were found using the RHIC lattice with the expected[3] random quadrupole errors, a_1 and b_1. The tune of the normal modes, ν_1 and ν_2, are computed using the exact result, Eq. (2.2). The tune shift due to the random b_1 has been corrected using the QF and QD quadrupoles in the regular cells. A correction system has been developed[1] for correcting the tune splitting which uses two families of skew quadrupoles as correctors. Results are given for ν_1, ν_2 before and after the application of the tune splitting correction, and are listed as the uncorrected and corrected cases. The tune splitting correction was set so as to make $\Delta\nu = 0$, where $\Delta\nu$ is the $\nu_x = \nu_y$ coupling resonance width given in Eqs. (2.3).

The unperturbed ν_x, ν_y for RHIC for Table 1 is $\nu_x = 28.826$, $\nu_y = 28.821$. Only the fractional part of the ν–values are listed in Table 1. The largest tune splittings found in Table 1 are $|\nu_1 - \nu_2|/10^{-3}$, $= 228$ uncorrected and 23 corrected. One sees that the tune splitting remaining after correction is considerably larger than the 5×10^{-3} predicted by Eq. (2.3). The remaining tune splitting after correction is called the residual tune splitting in this paper. It appears to be due to higher order terms in a_1 which are omitted in Eq. (2.3). An extension of Eq. (2.3) to include higher order terms in a_1 has been found, and is given by the two solutions for $\nu_{x,s}$ and $\nu_{y,s}$, $\nu_{x,s} = \nu_{y,s} + p$, of the equations

$$\left(\nu_{x,s}^2 - \tilde{\nu}_x^2\right)\left(\nu_{y,s}^2 - \tilde{\nu}_y^2\right) = 4\nu_x\nu_y |\Delta\nu\left(\nu_{x,s}, \nu_{y,s}\right)|^2$$

$$\tilde{\nu}_x^2 = \nu_x^2 + \Delta_x, \tilde{\nu}_y^2 = \nu_y^2 + \Delta_y$$

$$\Delta_x = 4\nu_x\nu_y \sum_{n \neq -p} |c_n|^2 / [(n+p)(n+\nu_x+\nu_y)]$$

$$\Delta_y = 4\nu_x\nu_y \sum_{n \neq p} |b_n|^2 / [(n-p)(n+\nu_x+\nu_y)]$$

$$b_n = (1/4\pi\rho) \int ds\, a_1 (\beta_x\beta_y)^{\frac{1}{2}} \exp\left[i\left(-(n+\nu_y)\theta_x + \nu_y\theta_y\right)\right]$$

$$c_n = (1/4\pi\rho) \int ds\, a_1 (\beta_x\beta_y)^{\frac{1}{2}} \exp\left[i\left(\nu_x\theta_x - (n+\nu_x)\theta_y\right)\right]$$

$$\tag{2.4}$$

For $p = 0$, the $\nu_x - \nu_y = 0$ resonance, Eqs. (2.4) can be solved to give

$$\nu_1^2 = \left(\tilde{\nu}_x^2 + \tilde{\nu}_y^2\right)/2 \pm \left[\left(\tilde{\nu}_x^2 - \tilde{\nu}_y^2\right)^2/4 + 4\nu_x\nu_y|\Delta\nu\left(\nu_1, \nu_1\right)|^2\right]^{\frac{1}{2}}$$

$$\nu_2^2 = \left(\tilde{\nu}_x^2 + \tilde{\nu}_y^2\right)/2 \mp \left[\left(\tilde{\nu}_x^2 - \tilde{\nu}_y^2\right)^2/4 + 4\nu_x\nu_y|\Delta\nu\left(\nu_2, \nu_2\right)|^2\right]^{\frac{1}{2}}$$

$$\tag{2.5}$$

For $\Delta\nu\left(\nu_1, \nu_2\right)$, $\Delta\nu\left(\nu_x, \nu_y\right)$ is evaluated for $\nu_x = \nu_y = \nu_1$, where ν_1 is given by Eq. (2.3), and similarly for $\Delta\nu\left(\nu_2, \nu_2\right)$. Eq. (2.5) reduces to Eq. (2.3) when higher order terms in a_1 are dropped. The above equations show that when $\Delta\nu\left(\bar{\nu}, \bar{\nu}\right) = 0$ for $\bar{\nu} = \left(\nu_x + \nu_y\right)/2$ and $\nu_x = \nu_y$, there is a residual tune splitting given by $\nu_1 - \nu_2 =$

Table 1: Results for the correction of the tune splitting for a $\beta^* = 2$ RHIC lattice using a 2 family tune splitting correction system set to make $\Delta\nu = 0$.

Error Field Dist.	—Uncorrected—			—Corrected—		
	ν_1	ν_2	$\begin{array}{c}\|\nu_1 - \nu_2\| \\ /10^{-3}\end{array}$	ν_1	ν_2	$\begin{array}{c}\|\nu_1 - \nu_2\| \\ /10^{-3}\end{array}$
1	0.796	0.854	59	0.828	0.823	6
2	0.707	0.935	228	0.838	0.819	19
3	0.869	0.783	86	0.825	0.829	4
4	0.772	0.883	111	0.831	0.823	7
5	0.779	0.872	93	0.836	0.820	16
6	0.848	0.805	43	0.832	0.821	11
7	0.840	0.847	7	0.852	0.834	18
8	0.742	0.895	153	0.838	0.818	20
9	0.785	0.866	81	0.828	0.823	6
10	0.749	0.891	142	0.823	0.827	5

$\left(\Delta_x - \Delta_y\right)/2\nu_x$, and $\left(\nu_1 + \nu_2\right)/2 = \bar{\nu} + \Delta_x/4\nu_x + \Delta_y/4\nu_y$. Thus after a tune splitting correction that makes $\Delta\nu \simeq 0$, the harmonics of a_1 near $\nu_x + \nu_y$ contribute most to the residual $|\nu_1 - \nu_2|$. This result is useful in designing a correction system for the residual tune splitting (see Section III). The derivation of the above result will be given in a future paper.

III. CORRECTION OF THE RESIDUAL $|\nu_1 - \nu_2|$

A first step in reducing the residual $|\nu_1 - \nu_2|$, is to set the a_1 correctors of the two family tune splitting correction system to minimize the residual $|\nu_1 - \nu_2|$, instead of setting them to make $\Delta\nu = 0$ as indicated by Eqs. (2.3). The results of doing this are given in Table 2. In Table 2, the results for the residual $|\nu_1 - \nu_2|$ listed in Table 1 using the $\Delta\nu = 0$ correction procedure are compared with the results found using the minimizing $|\nu_1 - \nu_2|$ procedure for the same 10 distributions of random a_1 and b_1 errors and for the $\beta^* = 2$ lattice for RHIC. Table 2 shows that the minimizing $|\nu_1 - \nu_2|$ procedure considerably reduces the residual $|\nu_1 - \nu_2|$ for some field error distributions. A possible explanation is that for some error distributions, the setting of the correctors to make $\Delta\nu = 0$ excites the harmonics in a_1 that are near $\nu_x + \nu_y =$ integer, and it is these harmonics that are believed to be responsible for the residual $|\nu_1 - \nu_2|$.

One may note that after correction, not only is there a considerable residual $|\nu_1 - \nu_2|$ for some field error distributions, but also the average tune $\nu_{av} = \left(\nu_1 + \nu_2\right)/2$ may be considerably off from $\left(\nu_x + \nu_y\right)/2$. It was found that the location of ν_{av} may be corrected using the normal tune correction system which in RHIC consists of the QF and QD quadrupoles in the regular cells of the arcs which are corrected in two families. This is indicated by Eqs. (2.3). However the normal tune correction system will not change the residual $|\nu_1 - \nu_2|$ by much.

Table 2: Residual tune splitting resulting from two different methods of tune splitting correction using the 2–family tune splitting correction system.

Error Field Distribution	$\|\nu_1 - \nu_2\|/10^{-3}$ $\Delta\nu = 0$ Correction	$\|\nu_1 - \nu_2\|/10^{-3}$ minimize $\|\nu_1 - \nu_2\|$ Correction
	$\beta^* = 2$	
1	6	5
2	19	18
3	4	3
4	7	7
5	16	8
6	11	11
7	18	7
8	20	16
9	6	6
10	5	4

To further reduce the residual tune splitting requires additional a_1 correctors so that one can control not only the harmonics in a_1 that drive the difference resonance, $\nu_x - \nu_y$, but also the harmonics in a_1 that are near $\nu_x + \nu_y$, which for RHIC are principally the 58 and 57 harmonics in a_1. The most effective place to put these correctors is near the quadrupoles in the insertions that have high β–functions. In RHIC which has 6 beam crossing points, there are 12 such high β locations. The two family tune splitting correction system[1] has a_1 correctors at 6 of the high β locations. The proposed enlargement of the a_1 correctors is to have a_1 correctors at all 12 of the high β locations, and all 12 a_1 correctors are separately powered.

A computer study was done to evaluate the performance of the enlarged a_1 correction system in reducing the residual $\|\nu_1 - \nu_2\|$. The results are given in Table 3 for one particular method of using the a_1 correctors which is described below. One may note that the unperturbed tune of the accelerator in this study has a tune splitting of $\|\nu_x - \nu_y\| = 5 \times 10^{-3}$, and the results listed in Table 3 for the residual $\|\nu_1 - \nu_2\|$ using the enlarged a_1 correction are good.

The method used to reduce the residual $\|\nu_1 - \nu_2\|$ is the following: the 6 a_1 correctors that make up the 2–family tune splitting correction system are corrected in the usual way as described in Ref. 1. This provides two variables which are primarily used to correct the effects of the difference resonance $\nu_x = \nu_y$. In addition the 12 a_1 correctors near the high β quadrupoles are excited according to the equation

$$a_1 = A \cos(\psi_x + \psi_y) + B \sin(\psi_x + \psi_y),$$

where ψ_x, ψ_y are the betatron phases at the correctors and A and B are two so far unknown variables. This provides 2 more variables used primarily to correct the $\nu_x + \nu_y$. Using the above hook up of the a_1 correctors, the 4 available variables are adjusted to minimize $\|\nu_1 - \nu_2\|$. This gives the results in Table 3.

The hook up of the a_1 correctors described above does not use the full flexibility of the enlarged a_1 correction system, but it appears to do a fairly good job of reducing the residual $\|\nu_1 - \nu_2\|$.

There appears to be some advantages to having some a_1 correctors in the arcs of the accelerator. The a_1 correctors in the arcs cannot usually entirely correct the difference resonance as they tend to have about the same value of $\psi_x - \psi_y$ if they are put near the quadrupoles in the arcs. However if the phase is defined so that $\psi_x - \psi_y \simeq 0$ at the quadrupoles, then the arc a_1 correctors can be used to correct the real part of $\Delta\nu$, the width of the difference resonance $\nu_x = \nu_y$, which is often a large part of $\Delta\nu$. The strength required of the a_1 correctors in the insertions is then reduced, and there is some improvement in the performance of the correction system. In RHIC, it is proposed to have a_1 correctors near most of the defocusing quadrupoles in the arcs which are connected in series in one family, which would be used to correct the real part of $\Delta\nu$, the width of the difference resonance. For the results given in Table 3, the arc a_1 correctors were also used.

Table 3: The reduction in the residual tune splitting obtained using the enlarged a_1 correction system, compared with the results obtained with the 2–family correction system, minimizing $\|\nu_1 - \nu_2\|$.

Error Field Distribution	$\|\nu_1 - \nu_2\|/10^{-3}$ 2 Family Correction System	$\|\nu_1 - \nu_2\|/10^{-3}$ enlarged a_1 Correction System
	$\beta^* = 2$	
1	5	5
2	16	6
3	3	3
4	5	5
5	5	5
6	9	3
7	14	4
8	10	2
9	6	3
10	1	3

The random skew quadrupole errors also produce an appreciable change[4] in the $\beta-$ functions of the normal modes, β_1 β_2. The change in β_1, β_2 can be quite large for some distributions of random a_1 errors, as much as 100% has been found for RHIC. The above described enlarged a_1 corrector system has been found to be able to also correct most of the change in β_1, β_2. It was also found[5] that correction of the tune splitting also corrected the change in the β–functions to a considerable extent.

IV. REFERENCES

[1] G. Parzen, BNL Report AD/RHIC–AP–72 (1988).
[2] E.D. Courant and H.S. Snyder, Ann. Phys. 3, 1 (1958).
[3] G. Parzen, BNL Report AD/RHIC–82 (1990).
[4] G. Parzen, BNL Report AD/RHIC–73 (1990).
[5] G. Parzen, Beta functions in the presence of linear coupling, these proceedings.

The WKB Approximation and the Travelling-Wave Acceleration Cavity

David C. Carey

Fermi National Accelerator Laboratory *

Batavia, Illinois 60510

Abstract

The equations of motion of a charged particle in a travelling-wave accelerating cavity are those of a harmonic oscillator with a varying restoring constant. This restoring constant is negative for transverse motion, resulting in diverging trajectories. The restoring constant is positive for longitudinal motion. For many years the only solution to the equation in the program TRANSPORT [1] was for massless particles. The WKB approximation of quantum mechanics yields a very accurate solution for massive particles. It allows both the amplitude and the wave number of the trajectory to vary as the cavity is traversed. Solutions are approximately 30 times more accurate than previously published. Comparisons are made with numerical integration.

1 Introduction

The original derivation of the transfer matrix for the travelling-wave accelerator cavity was for electrons. It assumed an ultra-relativistic beam in which the mass of any particle was negligible compared to its kinetic energy. For many years this ultra-relativistic approximation was the only option in the computer program TRANSPORT. [1] People who wanted to design proton or ion beams containing accelerating cavities had to work with a very bad approximation.

In the 1987 edition of this conference, Hurd and McGill [2] (henceforth referred to as simply "Hurd") pointed out the need for a representation of an accelerating cavity for massive particles. They also provided a representation which was a substantial improvement over what was previously available.

In the travelling-wave accelerator cavity, both the amplitude and the wavelength of the transverse motion changes as the particle is accelerated. The wavelength of the transverse motion is not the same as the wavelength of the accelerating field. Hurd approximated the solution by holding the wavelength constant and letting the amplitude vary. About the same time the present author produced a solution where the wavelength varied and the amplitude was held constant. The two solutions were of comparable accuracy. An exact solution was also derived by this author using Runge-Kutta integration. The two solutions differed from the exact solution by about one percent.

It is desirable to have a solution which satisfied two criteria: (1) It is accurate to the number of decimal places which are printed out in the transfer matrix by TRANSPORT. (2) It longitudinally segments to the accuracy that the transfer matrix is printed by TRANSPORT. Previously, the accelerating cavity for massive particles was the only element in TRANSPORT for which longitudinal segmentation was not exact.

The WKB approximation of quantum mechanics [3] allows both the amplitude and wavelength of the motion to vary as the particle is accelerated. It is an approximation which may be iterated to any level of accuracy. We found that a single iteration was sufficient to produce the desired accuracy. The discrepancy from the exact solution was reduced from the two previous approximations by a factor which ranged from 20 to 200, depending on the numbers involved and the specific matrix element. Below we give the analytic derivation of the expressions and make comparisons with numerical solutions produced by Runge-Kutta integration.

2 Transverse Motion

The transfer matrix elements are referred to the standard six coordinates as used in TRANSPORT. They are x, x', y, and y' in the transverse plane, and ℓ and δ in the longitudinal direction. The quantity ℓ describes the longitudinal separation between two particles as a function of distance along the reference trajectory. The quantity δ is the fractional momentum deviation from the reference particle.

In the description of the equations of motion we shall follows Hurd's paper. [2] [4] [5] We are unable to improve on his presentation. We shall also employ his notation.

*Operated by the Universities Research Association, Inc. under contract with the U.S. Department of Energy

The electric and magnetic fields are assumed to be constant over the length of the accelerating cavity. They are approximated then by:

$$E_z = \tfrac{1}{q}\Delta E \cos\phi, \qquad E_r = \tfrac{\gamma_s}{q}\left(\tfrac{\pi r}{\gamma_s \beta_s \lambda}\right)\Delta E \sin\phi \qquad (1)$$

$$B_\theta = \tfrac{\gamma_s \beta_s}{qc}\left(\tfrac{\pi r}{\gamma_s \beta_s \lambda}\right)\Delta E \sin\phi$$

Here ΔE refers to the maximum possible energy gain of the cavity, and ϕ to the synchronous phase of the particle. The energy gain of the synchronous particle is then $\Delta E \cos\phi$. The wavelength of the rf field is λ. The quantities γ and β are relativistic factors and c is the speed of light. The subscript "s" refers to the synchronous particle. The charge of the particle being accelerated is q.

In the derivation we shall also use a quantity η which is the product of γ and β. The additional subscripts "o" and "f" refer respectively to the initial and final values of the quantities to which they are attached. The length of the accelerating cavity will be L. The rest energy of the particle is mc^2.

The expressions for the fields lead to the first-order equation for the transverse motion of the particle:

$$\frac{d}{dz}\gamma_s \beta_s \frac{d}{dz}x = \left(\frac{\pi \Delta E \sin\phi_s}{\lambda L mc^2}\right)\frac{1}{\gamma_s^2 \beta_s^2}x \qquad (2)$$

The WKB approximation yields solutions for the transfer matrix elements as follows:

$$R_{11} = R_{33} = \left(\frac{\eta_{sf}}{\eta_{so}}\right)^{\frac{1}{4}}[\cosh(I_t) \qquad (3)$$

$$-\tfrac{1}{4}\frac{\eta_{so}'\eta_{so}^{\frac{1}{2}}}{\sqrt{Q}}\sinh(I_t)]$$

$$R_{12} = R_{34} = \frac{\eta_{so}^{\frac{3}{2}}}{\sqrt{Q}}\left(\frac{\eta_{sf}}{\eta_{so}}\right)^{\frac{1}{4}}\sinh(I_t)$$

$$R_{21} = R_{43} = \tfrac{1}{4}\left[\frac{\eta_{sf}'}{\eta_{sf}^{\frac{3}{4}}\eta_{so}^{\frac{1}{4}}} - \frac{\eta_{so}'\eta_{so}^{\frac{1}{4}}}{\eta_{sf}^{\frac{1}{4}}}\right]\cosh(I_t)$$

$$+[(\tfrac{\eta_{sf}}{\eta_{so}})^{\frac{1}{4}}\frac{\sqrt{Q}}{\eta_{sf}^{\frac{3}{2}}} - \tfrac{1}{16}\frac{\eta_{sf}'\eta_{so}'\eta_{so}^{\frac{1}{2}}}{\eta_{sf}^{\frac{3}{4}}\sqrt{Q}}]\sinh(I_t)$$

$$R_{22} = R_{44} = \tfrac{1}{4}\frac{\eta_{so}^{\frac{1}{2}}}{\sqrt{Q}}\frac{\eta_{sf}'}{\eta_{sf}^{\frac{3}{4}}\eta_{so}^{\frac{1}{4}}}\sinh(I_t) + (\tfrac{\eta_{so}}{\eta_{sf}})^{\frac{1}{4}}\cosh(I_t)$$

$$\text{where} \qquad I_t = \int_0^z \frac{\sqrt{Q}}{\eta^{\frac{3}{2}}}d\zeta \qquad (4)$$

and

$$Q = \frac{\pi \Delta E \sin\phi_s}{\lambda L mc^2} \qquad (5)$$

We are assuming that the rate of energy gain with distance is uniform over the accelerating cavity. This means that γ changes uniformly with distance and its derivative for the synchronous particle is given by:

$$\gamma' = \frac{\Delta E \sin\phi_s}{L mc^2} \qquad (6)$$

The integral I_t may then be evaluated by a binomial expansion. Taking out the constant factor $\sqrt{Q}$, we have:

$$\int_0^z \frac{d\zeta}{\eta^{\frac{3}{2}}} = \frac{1}{\eta_{so}^{\frac{3}{2}}\gamma'}[\Delta\gamma - \tfrac{3}{4}\frac{\gamma_{so}}{\eta_{so}^2}(\Delta\gamma)^2 \qquad (7)$$

$$-\tfrac{1}{4\eta_{so}^2}(\Delta\gamma)^3 + \tfrac{7}{8}\frac{\gamma_{so}^2}{\eta_{so}^4}(\Delta\gamma)^3]$$

$$\text{where} \qquad \Delta\gamma = \gamma_{sf} - \gamma_{so} \qquad (8)$$

3 Longitudinal Motion

We first formulate the equations of motion in the longitudinal direction in terms of the particle energy W, and its phase ϕ. The synchronous energy is W_s and the synchronous phase is ϕ_s. The deviations of a given particle from these quantities are given by ΔW and $\Delta\phi$. The first-order equations then become:

$$\frac{d\Delta W}{dz} = -\frac{\Delta E \sin\phi_s}{L}\Delta\phi \qquad (9)$$

$$\frac{d\Delta\phi}{dz} = \frac{2\pi}{\lambda}\frac{\Delta W}{\gamma_s^2 \beta_s^3 mc^2}$$

The quantities used are related, in first order, to the standard TRANSPORT variables by:

$$\ell = \frac{\beta_s \lambda}{2\pi}\Delta\phi \qquad \text{and} \qquad \frac{\Delta p}{p_s} = \frac{1}{\beta_s^2}\frac{\Delta W}{W_s} \qquad (10)$$

The WKB approximation yields solutions for the longitudinal transfer matrix elements as follows:

$$R_{55} = \frac{\beta_{sf}}{\beta_{so}}[(\frac{\eta_{so}}{\eta_{sf}})^{\frac{3}{4}} cos(I_l)$$

$$+\frac{3}{4}\eta'_{sf}\sqrt{\eta_{so}}(\frac{\eta_{so}}{\eta_{sf}})^{\frac{1}{4}}\frac{1}{\sqrt{Q_l}}sin(I_l)]$$

$$R_{56} = -\frac{\beta_{sf}\beta_{so}\lambda}{2\pi}\frac{Lmc^2}{\Delta E sin\phi_s}\{\frac{3}{4}[(\frac{\eta_{so}}{\eta_{sf}})^{\frac{1}{4}}\eta'_{sf}$$

$$-(\frac{\eta_{so}}{\eta_{sf}})^{\frac{3}{4}}\eta'_{so}]cos(I_l)$$

$$-[\frac{9}{16}\frac{\eta'_{so}\eta'_{sf}}{\sqrt{Q_l}}\frac{\eta_{so}^{\frac{3}{4}}}{\eta_{sf}^{\frac{1}{4}}} + \sqrt{Q_l}\frac{\eta_{so}^{\frac{1}{4}}}{\eta_{sf}^{\frac{3}{4}}}]sin(I_l)\}$$

$$R_{65} = -\frac{\eta_{so}^{\frac{3}{4}}}{\eta_{sf}^{\frac{1}{4}}}\frac{1}{\beta_{so}\beta_{sf}}\sqrt{Q_l}\,sin(I_l)$$

$$R_{66} = (\frac{\eta_{so}}{\eta_{sf}})^{\frac{1}{4}}\frac{\beta_{so}}{\beta_{sf}}[cos(I_l) - \frac{3}{4}\frac{\eta_{so}^{\frac{1}{2}}\eta'_{so}}{\sqrt{Q_l}}sin(I_l)]$$

Here $\qquad I_l = \int_o^z \frac{\sqrt{Q_l}}{\eta^{\frac{3}{2}}}d\zeta \qquad (12)$

and $\qquad Q_L = 2Q \qquad (13)$

References

[1] K.L. Brown, F. Rothacker, D.C. Carey, and Ch. Iselin, "TRANSPORT, A Computer Program for Designing Charged Particle Beam Transport Systems", SLAC Report SLAC91.

[2] J.W. Hurd and J. McGill, "Modification of Acceleration Element in TRANSPORT, IEEE Particle Accelerator Conference, p. 1198, 1987.

[3] J. Mathews and R.L. Walker, <u>Mathematical</u> <u>Methods</u> <u>of</u> <u>Physics</u>, Benjamin/Cummins Publishing Company, 1970.

[4] K. Mittag, "On Parameter Optimization for a Linear Accelerator," Kernforschungszentrum Karlsruhe GmbH, Report KfK 2555, Jan. 1978.

[5] M. Stanley Livingston and John P. Blewett, <u>Particle</u> <u>Accelerators</u>, McGraw-Hill, New York, 1962.

4 Comparison of Results

Numerical comparison of the values of the transfer matrix elements derived from Runge-Kutta integration and from the old and new versions of TRANSPORT. The example used is from Hurd's article. The initial energy is 100 MeV, the energy gain is 3.19 MeV over a length of 290 cm, and the synchronous phase is 30 degrees. There is some slight discrepancy with the exact values as printed in Hurd's article, which is probably attributable to rounding error in taking the parameters of the accelerating cavity directly from his article.

Element	Numerical Integration	Transport (Old)	Transport (New)
$R_{11} = R_{33}$	1.197	1.0	1.197
$R_{12} = R_{34}$	0.307	0.286	0.307
$R_{21} = R_{43}$	1.386	0.0	1.386
$R_{22} = R_{44}$	1.179	0.973	1.179
R_{55}	0.653	1.000	0.653
R_{56}	2.059	0.0	2.060
R_{65}	-0.281	-0.026	-0.281
R_{66}	0.624	0.997	0.624

Kick Factorization of Symplectic Maps*

A.J. Dragt and I.M. Gjaja
Department of Physics, University of Maryland, College Park, MD 20742

G. Rangarajan
Lawrence Berkeley Laboratory, Berkeley, CA 94720

I. INTRODUCTION

There are two sanguine (perhaps almost to the point of naivete) and yet remarkably tacit assumptions made in the design and construction of accelerators and storage rings. The first, made by machine builders, is that if two machines are nearly the same, their performance (including long-term orbit stability) should be nearly the same. Without this assumption, it would be impossible to proceed since construction errors are unavoidable at some level. The second, made by accelerator theorists, is that analytical/numerical models of machine behavior, despite their approximate nature, still have relevance for real machines. Without this assumption, it would be impossible to design machines. From a mathematical perspective, what is being assumed in either case is that if two symplectic maps are *close* (in some not yet precisely defined sense), then the behavior (including long-term behavior) of systems described by these maps should be nearly the same. This note explores in a mathematical context, and for a simple example, some aspects of this assumption.

Consider a nonlinear symplectic map characterizing a nonlinear Hamiltonian system. In a Taylor series approximation, one truncates this map at a given order in phase-space variables. Such a truncated map typically produces spurious damping or growth when used to analyze the long-term behavior of trajectories. This undesirable behavior arises from the fact that the truncated map violates the symplectic condition. There are at least two ways of coping with this problem. The first consists of replacing the Taylor series with a generating function whose effect is identical to that of the Taylor series through some order (perhaps the order to which the Taylor expansion is known). The second is to replace the Taylor series by a sequence of polynomial symplectic maps that can be evaluated exactly and whose net effect is again identical to that of the Taylor series through some order. In either case, a given symplectic map is being replaced by some nearby symplectic map, and one hopes this replacement still leads to valid conclusions.

II. A SIMPLE SYMPLECTIC MAP

Let $\mathcal{M}$ be a map acting on two dimensional phase space q, p and suppose $\mathcal{M}$ is written in the product form

$$\mathcal{M} = \mathcal{R}(\phi)\mathcal{N}. \tag{1}$$

Here $\mathcal{R}(\phi)$ denotes a linear map correspondng to rotation by angle ϕ (a simple phase advance) in the q, p plane; and $\mathcal{N}$ is a *nonlinear* map defined by the equations

$$q_f = \mathcal{N}q_i = q_i(1 - p_i)^2, \tag{2}$$

$$p_f = \mathcal{N}p_i = p_i \sum_0^\infty (p_i)^n = p_i/(1 - p_i). \tag{3}$$

Evidently $\mathcal{N}$ has a nonterminating Taylor expansion. It is also symplectic. Indeed, it has been selected because it has the simple Lie representation

$$\mathcal{N} = \exp : qp^2 : . \tag{4}$$

Figure 1 shows the dynamic aperture (tracking data) for $\mathcal{M}$ [1]. This figure was obtained by viewing $\mathcal{M}$ as a one-turn map, and tracking for 1000 turns the initial conditions $p_i = 0, q_i = .1, .2, \cdots .9$. Evidently, trajectories are stable if $q_i \leq .6$, and are unbounded if $q_i \geq .7$. This is a remarkably large dynamic aperture considering that, according to (3), $\mathcal{N}$ is singular at $p = 1$.

Suppose the Taylor series (3) is truncated beyond terms of degree 8. Figure 2 shows the result of tracking for the truncated map. The truncated map is not symplectic. In the figure this violation is first evident for the fourth "ring" out from the center where one sees spurious damping and on which $p \simeq .4$. The first neglected term in the Taylor series is p^9, and $(.4)^9 = 2.6 \times 10^{-4}$. In the course of 1000 turns this error could in principle accumulate to $\sim 10^3 \times 2.6 \times 10^{-4} = .26$. From the figure one sees that the accumulated error is somewhat less, but of this general order of magnitude. Consider the third ring out from the center. On this ring $p \simeq .3$ and $(.3)^9 = 2 \times 10^{-5}$. One would expect that this ring would show error effects after 10^4 turns since $10^4 \times 2 \times 10^{-5} = .2$. This has been found to be the case. We conclude that the Taylor approximation to $\mathcal{M}$ is not satisfactory for long-term tracking studies unless terms well beyond 8th order are retained.

---

*Work supported in part by the U.S. Department of Energy.

III. GENERATING FUNCTION APPROXIMATION

Suppose the map $\mathcal{N}$ is approximated by using a polynomial generating function $F(q_i, p_f)$, and the results of this generating function are required to agree with the Taylor series through 3rd order [2,3]. This procedure approximates $\mathcal{N}$ by a map $\mathcal{N}_{gen}$ that is exactly symplectic. Indeed, by construction $\mathcal{N}_{gen}$ has a factored Lie product expansion of the form

$$\mathcal{N}_{gen} = \exp : qp^2 : \ \exp : f_5 : \ \exp : f_6 : \cdots, \qquad (5)$$

where the homogeneous polynomials $f_5, f_6, \cdots$ are in general not zero, but are hoped to have negligible effect for trajectories of interest.

Figure 3 shows the dynamic aperture for the map $\mathcal{R}\mathcal{N}_{gen}$. Evidently the topological features of figures 1 and 3 are similar. However, the 4th ring is slightly deformed, and the rings successively farther out are successively more deformed. Indeed, there is even a somewhat ragged 7th ring. These deformations are presumably the effect of the terms $f_5, f_6 \cdots$.

To study the effect of these terms, consider the map $\mathcal{N}^{1/4} = \exp : qp^2/4 :$. Approximate this map by a generating function map $\mathcal{N}'_{gen}$. For this map we have the relation

$$\mathcal{N}'_{gen} = \exp : qp^2/4 : \ \exp : f'_5 : \ \exp : f'_6 : \cdots, \qquad (6)$$

where $f'_5 \cdots$ are now at least a factor of $(1/4)^3$ smaller. This procedure allows us to construct the improved approximation

$$\mathcal{M} \simeq \mathcal{R}(\mathcal{N}'_{gen})^4. \qquad (7)$$

Figure 4 shows the dynamic aperture for the improved approximation (7). Evidently figures 1 and 4 are remarkably similar. The only noticeable difference is a slightly different nonlinear phase advance as is evident for the 5th ring. This agreement is all the more remarkable when one considers that the nonlinear term $1/(1-p)$ in (3) is very significant for the outer rings.

IV. KICK APPROXIMATION

Although, as illustrated, the generating function approximation works well, it is somewhat awkward computationally since it involves the use of a Newton's method procedure, and this procedure may sometimes not converge. An alternate approach is to factor the nonlinear part of a map $\mathcal{N}$ into a product of special Lie maps called *kick* maps. Here a kick map is defined to be one for which the exponential series $\exp :*:$ terminates [3,4,5]. For the problem at hand we have constructed a kick factorized map $\mathcal{N}_{kf}$ of the form

$$\mathcal{N}_{kf} = \prod_{i=1}^{5}(\mathcal{R}_i e^{:\beta_i q^3 + \gamma_i q^4:} \mathcal{R}_i^{-1}) \qquad (8)$$

in such a way that one has the approximation

$$\mathcal{N}_{kf} = \exp : qp^2 : \exp : g_5 : \exp : g_6 : \cdots, \qquad (9)$$

where the polynomials $g_6 \cdots$ are again hoped to have negligible effect. Here the maps $\mathcal{R}_i$ are phase-space rotations whose angles, along with the coefficients β_i, γ_i, are chosen in such a way that (9) is satisfied. Kick maps have at least two advantages: They can be evaluated directly (no Newton procedure) and rapidly to machine precision; they can be inverted exactly since their inverses are also kick maps.

Figure 5 shows the dynamic aperture for the map $\mathcal{R}\mathcal{N}_{kf}$. Evidently, the first 3 rings agree well with those in figure 1. For the 4th ring and beyond the agreement is not as good as the generating function approximation (figure 3). What is notable about this result is not that the kick factorization does not work well [indeed the nonlinearities associated with $1/(1-p)$ are already large on the 3rd ring where the kick factorized approximation still works], but that the generating function works so remarkably well.

As in the generating function example, the effect of the terms $g_5 \cdots$ can be studied by kick factorizing $\mathcal{N}^{1/4}$ instead of $\mathcal{N}$ and then making the approximation $\mathcal{M} \simeq \mathcal{R}(\mathcal{N}'_{kf})^4$ where $\mathcal{N}'_{kf}$ is the kick factorized approximation to $\mathcal{N}^{1/4}$. Figure 6 shows the dynamic aperture for this map. Evidently, figure 6 is nearly as close to figure 1 as figure 4 is. Thus, the kick factorization approximation also works well.

V. CONCLUSIONS

Based on this brief exploration we draw the following tentative conclusions: First, map approximations that violate, even in small amount, the symplectic condition, such as truncated Taylor expansions, are not well suited to long-term tracking studies. Second, if two symplectic maps are sufficiently close, then at least gross features of their long-term behavior are in fact similar. Indeed, the symplectic approximations $(\mathcal{N}'_{gen})^4$ and $(\mathcal{N}'_{kf})^4$ probably differ more from $\mathcal{N}$ for a single turn than does the truncated Taylor expansion. Yet they give far better predictions of long-term behavior. Finally, the use of kick factorization deserves further study. Indeed, it can be shown that for the full 6-dimensional phase space symplectic maps can be approximated through 3rd order using only 9 kicks, and through 11th order using only 68 kicks [5]. Thus, kick factorized approximations should be ideal for realistic long-term tracking studies.

VI. FOOTNOTES AND REFERENCES

[1] For all the tracking studies of this note we have used the nonresonant phase advance $\phi = 66$ degrees.

[2] This is the method currently used in MARYLIE 3.0. For MARYLIE 5.0 f_5 and f_6 are required to vanish as well. See the MARYLIE manual and reference 3.

[3] A.J. Dragt et al., Ann. Rev. Nucl. Part. Sci. $\underline{38}$ (1988).

[4] J. Irwin, SSC Note 228 (1989).

[5] G. Rangarajan, Ph.D. Thesis (1990).

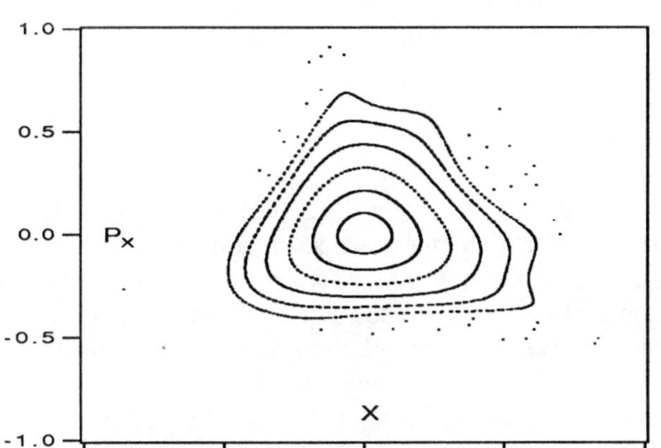

Figure 1: Tracking Results for Exact Map.

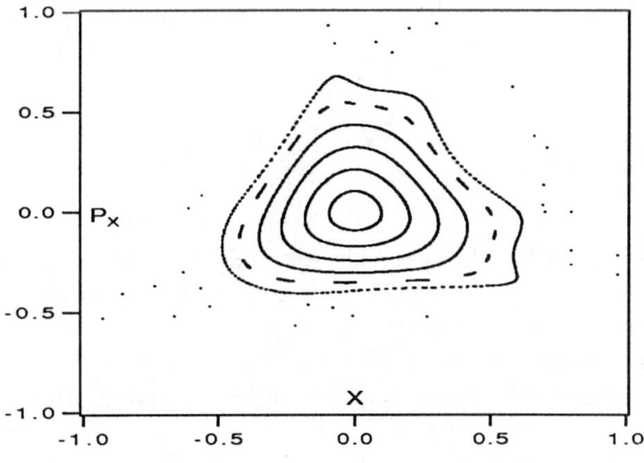

Figure 4: Improved Generating Function Approximation.

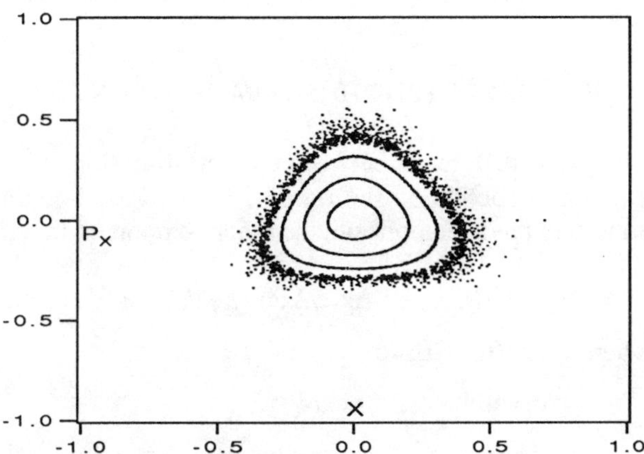

Figure 2: Truncated Taylor Approximation Results.

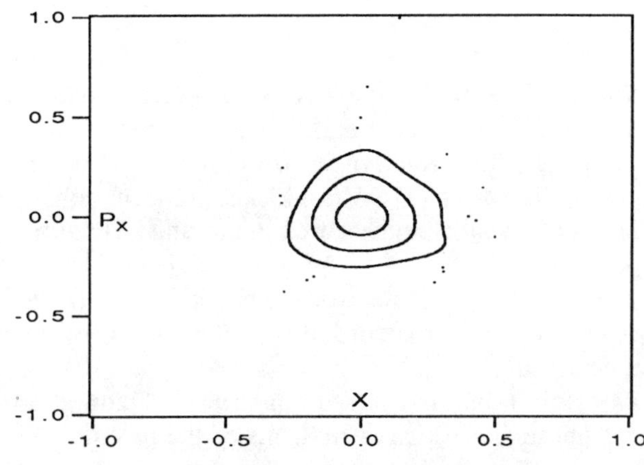

Figure 5: Kick Approximation Results.

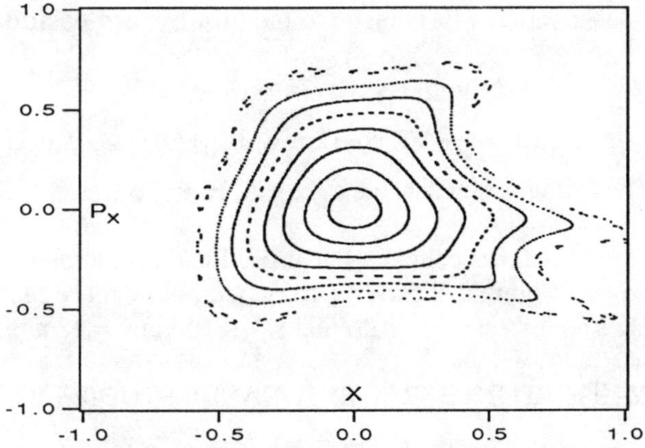

Figure 3: Generating Function Approximation Results.

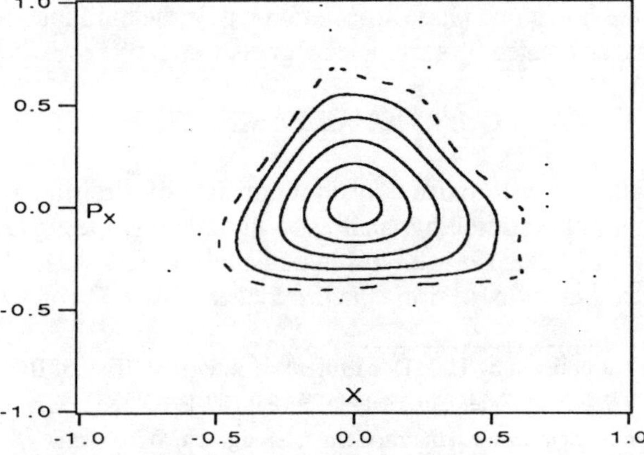

Figure 6: Improved Kick Approximation.

Non-Linear Resonance Studies at the Synchrotron Radiation Center, Stoughton, Wisconsin*

E. Crosbie, J. Bridges, Y. Cho, D. Ciarlette,
R. Kustom, Y. Liu, K. Symon,[1] L. Teng, W. Trzeciak[1]
Argonne National Laboratory
9700 South Cass Avenue
Argonne, IL 60439

Abstract

A single bunch is stored in ALADDIN at 800 MeV. The coherent oscillations produced by a fast kicker are recorded for each turn at two horizontal position detectors separated by $\pi/2$ in phase. Displaying the recorded positions on horizontal and vertical axis produces a phase plot of the coherent motion. Resonances are produced by adding a controlled amount of sextupole field. The results are compared with computer simulations. To adequately describe the behavior near unstable fixed points the finite size of the beam must be taken into account.

I. INTRODUCTION

Non-linear resonance studies are being conducted at the ALADDIN 800-MeV Electron Storage Ring located at the University of Wisconsin Synchrotron Radiation Center in Stoughton, Wisconsin [1]. A fast kicker produces coherent oscillation of a single bunch stored in the ring. The turn to turn positions are recorded at two sets of horizontal and two sets of vertical pickup electrodes spaced at 90° in each plane. The motion is recorded for 4096 turns for each experimental run.

The stored data is used to construct horizontal and vertical phase space representations of the motion. The experimental results are compared with results from the computer tracking program PACMAN. Since the bunch in the storage ring has a finite size ($\sigma_x = 0.8$ mm), the tracking program has been modified to follow the average phase space motion of several particles. In this paper we describe only the horizontal phase space motion near the third integral resonance created by sextupole magnets.

II. DATA ACQUISITION CAPABILITIES FOR SRC

The data acquisition system consists of circuitry for timing, pulse stretching, analog-to-digital converters, data display, and data storage. It is initiated by a trigger pulse at a selected time prior to firing the fast kicker. The A/O conver-ters acquire and store the data at the beam revolution frequency. Eight channels are used to monitor the signals from 2 sets of horizontal and 2 sets of vertical stripline position pickups. The 1 nsec signals from the stripline are stretched to 100 nsec to match the A/O acquisition times.

The data is then down-loaded into an IBM PC compatible computer and the difference divided by the sum of each pair of signals is calculated. A look-up table can be used to cancel the nonlinear transfer function of the voltage versus position signal. Position as a function of time (turn) can be displayed along with phase-space diagrams of both horizontal and vertical data. Once the experiment is finished the data is transferred to a VAX computer for further analysis.

III. CORRECTED PHASE SPACE PLOTS

The phase difference between the beta-functions at the two pickups depends on the tune of the ring. Choosing $\Delta x_1(n)$ as one phase component, the other component is

$$Px(n) = (R\Delta x_2(n) - \Delta x_1(n) \cos \Delta\phi)/\sin\Delta\phi \quad (1)$$

where $\quad \Delta x_i(n) = x_i(n) - c_i,$

$\qquad c_i = $ orbit positions at i,

$\qquad \Delta\phi = $ phase separations between 1 and 2,

and $\quad R^2 = \beta x_1/\beta x_2.$

The necessary corrections are obtained from linear oscillation data. A small amplitude oscillation is produced, and a least squares fit is made to the turn by turn position data.

$$x_1(n) = c_1 + A \cos (2\pi n\nu_x + \phi_0) \quad (2)$$

$$x_2(n) = c_2 + A/R \cos (2\pi n\nu_x + \phi_0 + \Delta\phi) \quad (3)$$

The fitting parameters are $c_1, c_2, A, R, \nu_x, \phi_0, \Delta\phi.$

Figure 1 shows coherent linear oscillation monitored at position 1. Figure 2 shows the corrected phase space plot. The plots are presented in BPM units. 1 BPM unit ≈ 18 mm.

IV. THIRD INTEGRAL RESONANCE STUDIES

For resonance studies, the ALADDIN ring is tuned to near $\nu_x = 7.33$. A single sextupole is powered to produce the 22nd harmonic. Computer simulations predict the separatrices shown in Figures 4,6, and 8[2].

*Work supported by U.S. Department of Energy, Office of Basic Energy Sciences under Contract No. W-31-109-ENG-38.

[1] At the Synchrotron Radiation Center, Stoughton, Wisconsin.

[2] In order to get agreement with the experimental results it was necessary to add an average octupole tune shifting component. Such a component can come from quadrupole edge fields which are not included in our tracking program.

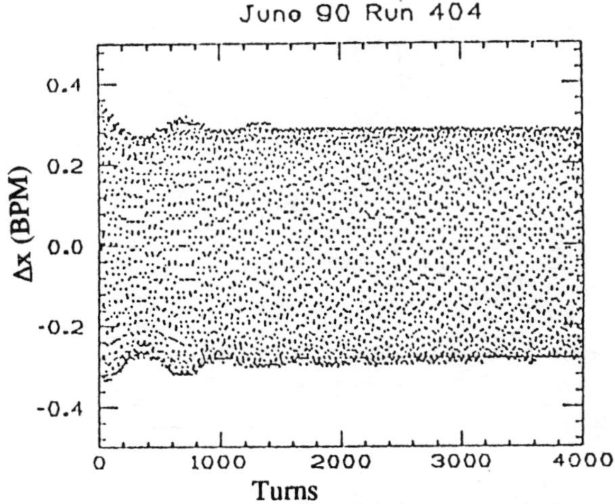

Figure 1. Coherent oscillation at pick up 1.

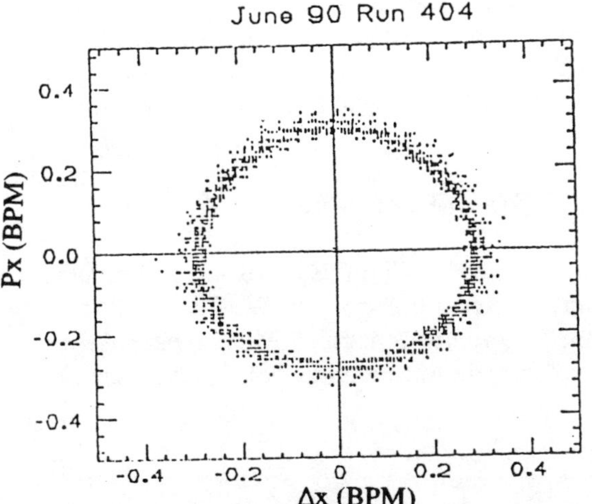

Figure 2. Phase plot derived from pickups 1 and 2.

Figure 3 shows a typical experimental phase plot for a small amplitude oscillation. Because of the finite beam size and because of the proximity of portions of the bunch to the separatrix, there is an apparent damping of the oscillations due to loss of coherence. Figure 4 is a computer simulation of the same behavior. The figure shows the center of gravity motion of 5 particles starting on the vertical spin with initial amplitudes spread over 2 mm starting at about 1 mm from the separatrix.

Figure 5 illustrates the behavior for a larger initial kick. The kick put most, but not all, the beam into the "lobe". In the computer simulation, Figure 6, 4 of the 5 initial amplitudes are beyond the separatrix. The total spread is 2 mm.

When the kick is large enough the whole team oscillates about the stable fixed points. The behavior is illustrated in Figure 7. Again because of the finite size of the beam, the individual electrons get out of phase and produce an apparent damping of the beam to the three fixed points. (The real damping time constant is about 27 ms or 100,000 turns.) The corresponding computer simulation shown in Figure 8 shows the average of 5 particles spread initially over 2 mm.

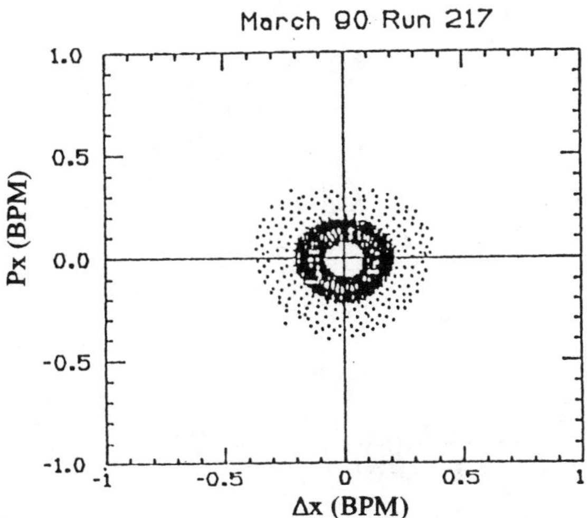

Figure 3. Experimental phase plot for small amplitude oscillation near the resonance $3\nu_x=22$.

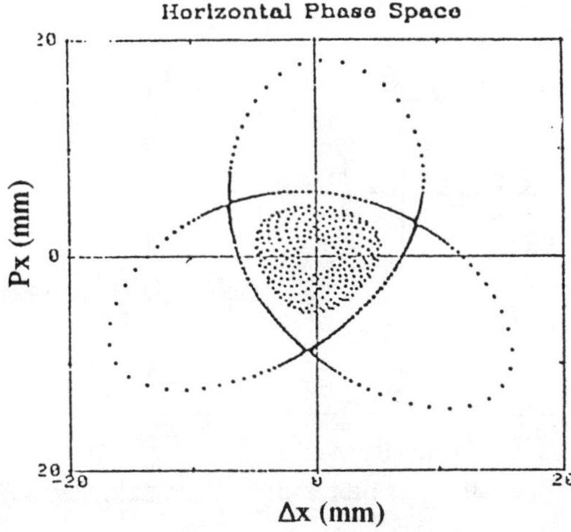

Figure 4. Computer simulation corresponding to Figure 3.

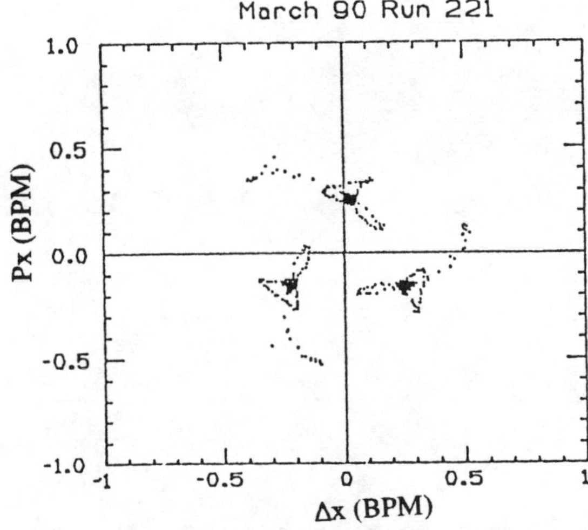

Figure 5. Experimental phase plot with large amplitude oscillations for the same conditions as Figure 3.

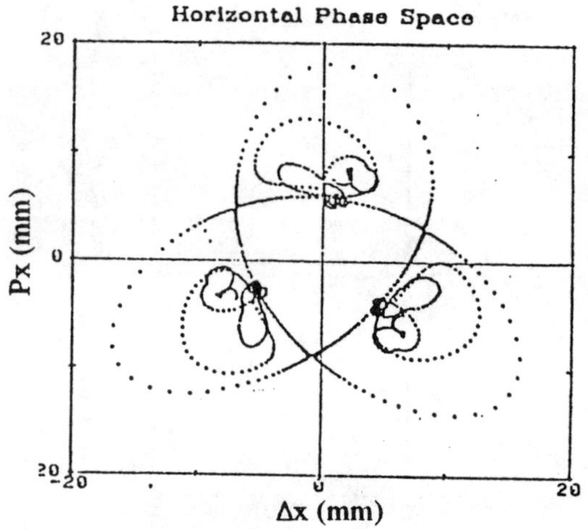

Figure 6. Computer simulation corresponding to Figure 5.

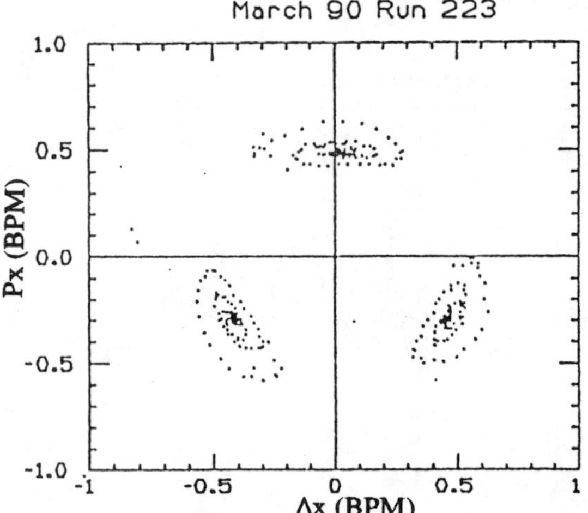

Figure 7. Experimental phase plot with total beam kicked to the stable fixed points.

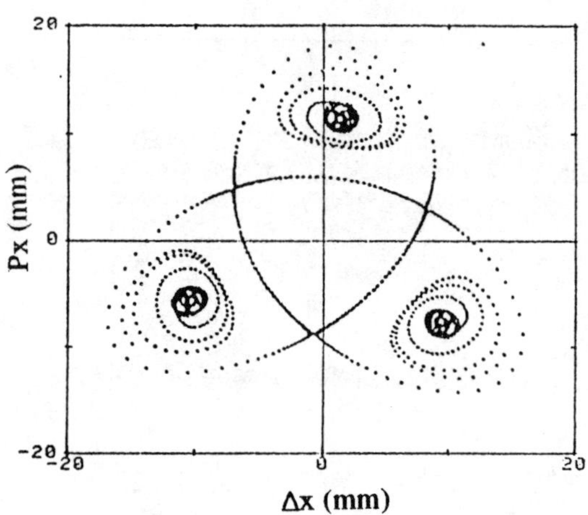

Figure 8. Computer simulations corresponding to Figure 7.

Once the beam is in the "lobes" it stays there with no apparent loss in lifetime. The position monitors show the bunch moving from fixed point to fixed point. To return the beam to the center it is necessary to switch off the sextupole magnet.

V. REFERENCES

[1] J. Bridges, Y. Cho, W. Chou, E. Crosbie, S. Kramer, R. Kustom, K. Kleman, R. Otte, W. Trzeciak, and K. Symon, "Dynamic Aperture Measurements on ALADDIN," Particle Accelerators, vol. 28-29, issues 1-4, pp. 479-487.

CORRECTION OF SKEW–QUADRUPOLE ERRORS IN RHIC*

G.F. Dell, H. Hahn, S.Y. Lee[†], G. Parzen and S. Tepikian

Brookhaven National Laboratory
Upton, NY 11973, USA

Abstract

The correction scheme for skew–quadrupole (a_1) errors in the Relativistic Heavy Ion Collider is presented. The a_1 errors result from fabrication and from installation errors. At the selected betatron tune of $\nu_x = 28.826$ and $\nu_y = 28.820$, the principal resonances driven by a_1 errors are $\nu_x - \nu_y = 0$, the sum resonances near $\nu_x + \nu_y$ which cause residual tune splitting and distortion of beta function, and $\nu_y = 28\&29$ which cause vertical dispersion. Skew quadrupole correctors located in the insertions and arcs will be used to correct the coupling and sum resonances and, if required, separate correctors in the arcs could be used for vertical dispersion correction. The study of the a_1 correction system confirmed that correction is possible at $\beta^* = 2$ m and $\beta^* = 6$ m and that the required corrector strengths are consistent with those specified for RHIC.

INTRODUCTION

The correction system for the skew–quadrupole errors in RHIC, the Relativistic Heavy Ion Collider, is discussed in this paper. The most important effect of these errors is the tune splitting associated with the $\nu_x - \nu_y = 0$ coupling resonance. This effect can be corrected to a large degree with a two–family corrector system that corrects the driving term of the resonance.[1] Two other effects induced by skew–quadrupoles are residual tune splitting that remains after the driving term of the $\nu_x - \nu_y = 0$ resonance has been corrected and distortion of the beta functions.[2–3] These two effects have been found to be associated with the sum resonances near $\nu_x + \nu_y$.[2-4] Skew–quadrupoles also produce vertical dispersion that is associated with the integer resonances at $\nu_y = 28\&29$.

A system of skew correctors will be provided in RHIC to counteract the beam dynamic effects mentioned. The correctors are located in the insertions as well as arcs. Several configurations were investigated and served as a guide in selecting the present version in which the number of cold–to–warm feedthroughs and the number of power supplies required at start–up was minimized.[5,6] In fact, only correction of the coupling resonance will initially be available.

The performance of the skew–quadrupole correction system is evaluated by determining its ability to correct

---

*Work performed under the auspices of the U.S. Department of Energy.

†Present address: Dept. of Physics, Indiana University, Bloomington, IN 47405.

the important stopbands of the resonances associated with the four effects mentioned above. The capacity and flexibility of the system will be demonstrated by correcting the $\nu_x - \nu_y = 0$ and the $\nu_x + \nu_y = 58\&59$ stopbands. However, the system is sufficiently flexible to be operated in ways which correct the beam dynamic effects directly without reference to stopbands.

SKEW–QUADRUPOLE ERRORS

The a_1 errors consist of construction errors from coil tolerances in the dipoles resulting in $a_1 \approx 1.7 \times 10^{-4}$ cm^{-1} rms as well as errors from an anticipated 0.5 mrad rms uncertainty in the orientation of the planes of the quadrupoles that is a combination of field measurement errors and installation errors. Furthermore, solenoids which are part of experimental detectors at the crossing points will be a skew–quadrupole source.

The complex strength of the stopband integrals and driving terms can be approximated by the sums listed in Table I. Estimates of their strength were made with random a_1 errors generated according to a Gaussian distribution that is truncated at $\pm 3\sigma$. A systematic error was added to the arc dipole errors to produce a 0.3σ average a_1. The maximum expected stopband/driving terms were estimated by sampling 10 machines with the results given in Table II.

Sorting of arc dipoles can be used to reduce the driving terms. It is planned to sort in a way which reduces vertical dispersion resulting in an error reduction by a factor of 2-3. It is expected that sorting eliminates the need for vertical dispersion correction, nevertheless the capability will be provided.

Table I: Stopband Integrals and Driving Terms.

$$\Delta_0 = \frac{1}{4\pi} \sum \frac{a_1\ell}{\rho} \sqrt{\beta_x \beta_y} \exp i \left(\varphi_x - \varphi_y \right)$$

$$\Delta_{57} = \frac{1}{4\pi} \sum \frac{a_1\ell}{\rho} \sqrt{\beta_x \beta_y} \exp i \left(\frac{28.5}{\nu_x} \varphi_x + \frac{28.5}{\nu_y} \varphi_y \right)$$

$$\Delta_{58} = \frac{1}{4\pi} \sum \frac{a_1\ell}{\rho} \sqrt{\beta_x \beta_y} \exp i \left(\frac{29}{\nu_x} \varphi_x + \frac{29}{\nu_y} \varphi_y \right)$$

$$\Delta_{28} = \frac{\nu_y}{4\pi} \sum \frac{a_1\ell}{\rho} X_p \sqrt{\beta_y} \exp i \left(\frac{28}{\nu_y} \varphi_y \right)$$

$$\Delta_{29} = \frac{\nu_y}{4\pi} \sum \frac{a_1\ell}{\rho} X_p \sqrt{\beta_y} \exp i \left(\frac{29}{\nu_y} \varphi_y \right)$$

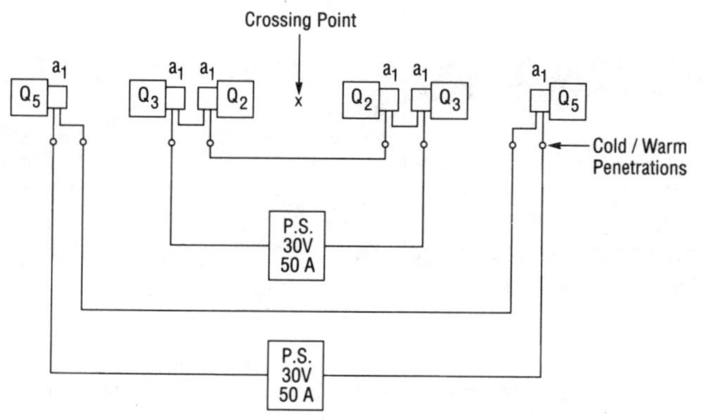

Fig. 1: Insertion corrector configuration.

Table II: Estimates of Correction Requirements: Stopband/Driving Term Strengths.

| | $\beta^* = 2$ m | | $\beta^* = 6$ m | | |
	Re	Im	Re	Im	Units
"0"	7.6	4.4	7.2	1.0	$\times 10^{-2}$
"57"	3.1	4.8	2.2	2.6	$\times 10^{-2}$
"58"	3.6	2.0	2.3	1.7	$\times 10^{-2}$
"28"	17.8	8.5	17.8	9.4	$\times 10^{-2}$ m$^{1/2}$
"29"	9.6	13.6	8.9	13.6	$\times 10^{-2}$ m$^{1/2}$

CORRECTOR SYSTEM

The RHIC lattice consists of three superperiods each having an inner arc of 12 cells, an In–Out insertion, an outer arc of 12 cells, and an Out–In insertion. The insertions are identified by their location on the face of the clock as 2:, 6:, 10: and 4:, 8:, 12:, respectively. The phase advance per cell is approximately 90°. The location of the a_1 correctors in the insertions is shown in Fig. 1 and of the arc correctors in Fig. 2. The standard–aperture correctors, i.e. in the arcs and in the insertion at Q5, have a strength of 1.5 T. The large–aperture correctors are weaker, but the combination of one corrector each at Q2&Q3 also provide 1.5 T.

The correction scheme adopted assumes families of correctors, representing a group of correctors with separate power supplies, which are software coupled. Considerable effort went into finding families which are orthogonal due to their hardware configuration.

The two families required for correction of linear coupling are obtained in the case of:

$\beta^* = 2$ m with (Q3 @ 2: 6: 10: + Q3 @ 4: 8: 12:) and (Q3 @ 2: 6: 10: − Q3 @ 4: 8: 12:) producing real and imaginary terms respectively;

$\beta^* = 6$ m with (Q3 @ 2: 6: 10: + Q3 @ 4: 8: 12:) and (Q5 @ 2: 6: 10: − Q5 @ 4: 8: 12:). On day–one only the linear–coupling correction system will be powered.

The two families for correction of the "57" stopband involve the 3×8 correctors each in the outer arcs (QO) and in the inner arcs (QI). Within one arc there are 4 positive and 4 negative correctors cancelling a contribution to "0". (However, the wiring maintains the option of locally correcting an average a_1 in the arc.)

The correction scheme for "58" takes advantage of the 120° phase shift between superperiods which leads to $1 - \frac{1}{2}e^{i120} - \frac{1}{2}e^{-i120} = 0$ assuring orthogonality with "0" and "57". The two families are then (Q3 @ 2: $-\frac{1}{2}$6: $-\frac{1}{2}$10:) and (Q3 @ 4: $-\frac{1}{2}$8: $-\frac{1}{2}$12:).

The remaining 12 independent corrector pairs in the arcs will be available for vertical dispersion correction. Each pair is counterphased and thus orthogonal to "0". These correctors could be combined into four families to correct "28" and "29" or alternatively to correct directly the vertical dispersion at the crossing points.

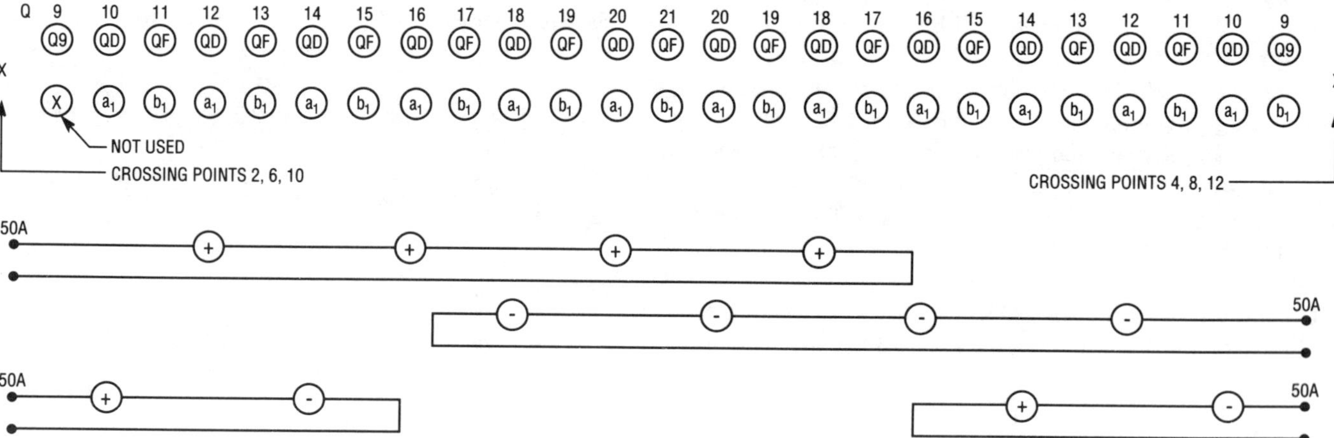

Fig. 2: Arc corrector configuration.

Table III: Stopband Matrix (Δ in units of 10^{-2}).

$$
\begin{pmatrix} \mathrm{Re0} \\ \mathrm{Im0} \\ \mathrm{Re57} \\ \mathrm{Im57} \\ \mathrm{Re58} \\ \mathrm{Im58} \end{pmatrix}
=
\begin{pmatrix}
33.1 & 1.1 & 0 & 0 & 0 & 0 \\
0.7 & -54.9 & 0 & 0 & 0 & 0 \\
1.8 & 32.6 & -5.9 & -3.6 & 0 & 0 \\
-1.1 & 54.7 & 3.9 & -6.0 & 0 & 0 \\
0 & 0 & 0 & 0 & -4.3 & 15.7 \\
0 & 0 & 0 & 0 & 15.4 & -3.0
\end{pmatrix}
\begin{pmatrix}
\mathrm{Q3}\,(2:6:10:+4:8:12:) \\
\mathrm{Q3}\,(2:6:10:-4:8:12:) \\
\mathrm{QO} \\
\mathrm{QI} \\
\mathrm{Q3}\,(2:-\tfrac{1}{2}6:-\tfrac{1}{2}10:) \\
\mathrm{Q3}\,(4:-\tfrac{1}{2}8:-\tfrac{1}{2}12:)
\end{pmatrix}
$$

STOPBAND MATRIX AND CAPABILITIES

The impact of each corrector family on all stopbands can be represented in matrix form which indicates the real and imaginary stopband produced by full–strength excitation of the correctors in a family. For example, the matrix for $\beta^* = 2$ m in all insertions is shown in Table III.

The correction capability for each stopband was obtained by inverting a related matrix which connects the individual corrector to the stopbands, and by determining the maximum modulus $|\Delta|$ that can be achieved with 1.5 T corrector strength. In cases where one corrector is required for two stopbands, a strength of 1.5 T/$\sqrt{2}$ was allotted to each stopband. The resulting correction capabilities for $\beta^* = 2$ m are shown in Fig. 3 and are compared to the error estimates. The capability of the system would seem sufficient to correct the skew–quadrupole errors with a 95% confidence level. The correction requirements for $\beta^* = 6$ m are less stringent.

SIMULATION RESULTS

The operation of the a_1 correction system was simulated using the *PATRICIA* tracking program for ten sets of random errors. The "0", "57" and "58" stopband integrals were evaluated and used to obtain the required corrector strength. With the correctors powered, the stopbands were reduced by a factor of 10^3.

In order to study tune splitting, the tune was set to $\nu_x = 28.826$ and $\nu_y = 28.820$ when skew–quadrupoles are absent. Introduction of skew–quadrupole errors produces tune splitting, $\nu_1 - \nu_2$. The tunes, before and after stopband correction, were obtained by Fourier analysis of 2000 turn tracking runs. The rms tune splitting was determined for the 10 sets of random errors to be $\Delta\nu \approx 124 \times 10^{-3}$ rms for $\beta^* = 2$ m and 110×10^{-3} rms for $\beta^* = 6$ m. Correction of the "0" stopband reduced the tune splitting by more than one order of magnitude.

In conclusion, the simulation studies confirmed that the a_1 correction system has the strength and flexibility to cope with the expected skew–quadrupole errors.

REFERENCES

[1] G. Parzen, BNL Report AD/RHIC–AP–72, (1988).

[2] G. Parzen, BNL Report AD/RHIC–82 (1990).

[3] G. Parzen, BNL Report AD/RHIC–73 (1990).

[4] G. Parzen, *Tune splitting in the presence of linear coupling*, (these proceedings).

G. Parzen, *Beta functions in the presence of linear coupling*, (these proceedings).

[5] H. Hahn and G. Parzen, BNL Report AD/RHIC–RD–20 (1990).

[6] S.Y. Lee and S. Tepikian, BNL Report AD/RHIC–75 (1990).

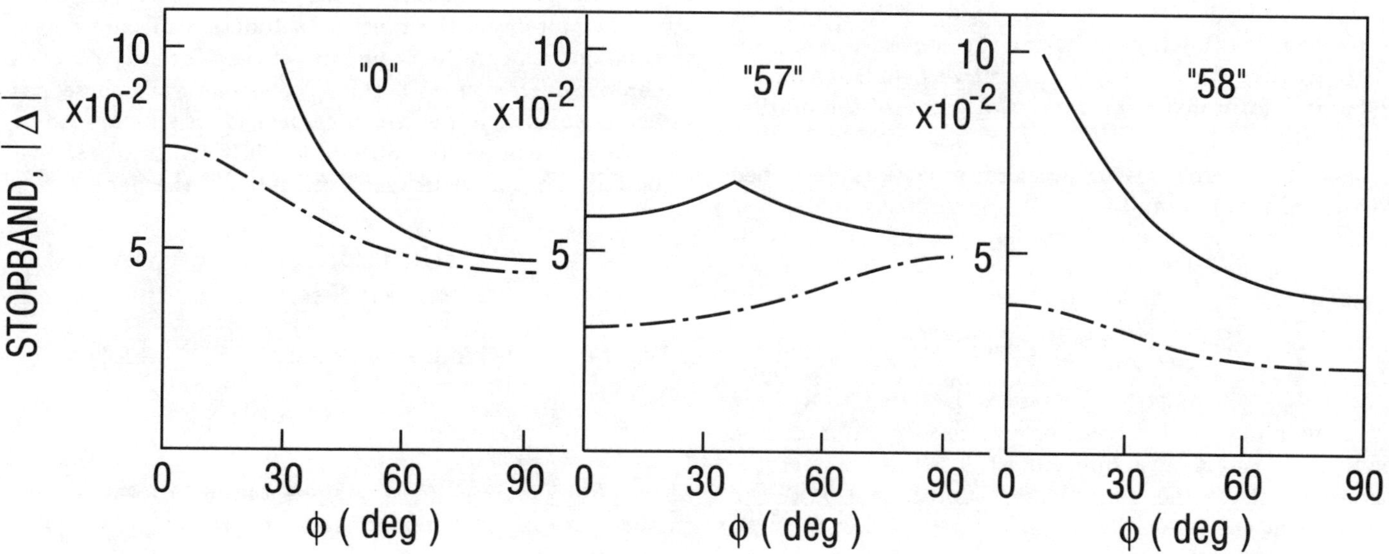

Fig. 3: Stopband corrector capability (solid line) and requirements (dashed lines) for $\beta^* = 2$ m.

Invariant Metrics for Hamiltonian Systems*

Govindan Rangarajan
Lawrence Berkeley Laboratory, University of California, Berkeley, CA 94720
Alex J. Dragt
Center for Theoretical Physics, University of Maryland, College Park, MD 20742
Filippo Neri
AT Division, Los Alamos National Laboratory, Los Alamos, NM 87545

Abstract

In this paper, invariant metrics are constructed for Hamiltonian systems. These metrics give rise to norms on the space of homeogeneous polynomials of phase-space variables. For an accelerator lattice described by a Hamiltonian, these norms characterize the nonlinear content of the lattice. Therefore, the performance of the lattice can be improved by minimizing the norm as a function of parameters describing the beam-line elements in the lattice. A four-fold increase in the dynamic aperture of a model FODO cell is obtained using this procedure.

I. INTRODUCTION

Given an accelerator lattice, various correction schemes (lumped correctors, shuffling of magnets etc.) can be used to improve its performance. However, to be able to implement these schemes, it is essential to have a "merit function" (depending on the parameters describing the beam-line elements) that can be minimized to produce the optimal lattice. This merit function should be a reliable measure of the nonlinearilty of the lattice since it is nonlinear effects that degrade the performance of the lattice and it is these effects that have to be minimized. In this paper, we propose a merit function satisfying the above criteria. This function will turn out to be a positive definite symmetric bilinear form invariant under the action of the unitary group U(3).

We restrict ourselves to accelerator lattices described by a (nonlinear) Hamiltonian in a six-dimensional phase space. Given such a system, an equivalent description is provided by the following one-pass or one-period symplectic map[1]

$$\mathcal{M} = M e^{:f_3:} e^{:f_4:} \ldots e^{:f_m:} \ldots \tag{1}$$

Here, the 6×6 matrix M characterizes the linear part of the map and the Lie transformations $e^{:f_m:}$ characterize the nonlinear part. The operator $:f_m:$ is the Lie operator corresponding to the homogeneous polynomial $f_m(z)$ of degree m in the phase-space variables z_i ($i = 1, 2, \ldots 6$).

The polynomial $f_m(z)$ can be expanded as follows:

$$f_m(z) = a_\alpha^{(m)} P_\alpha^{(m)}(z) , \tag{2}$$

where we have used Einstein's summation convention. Here $P_\alpha^{(m)}(z)$ denotes the following mth degree basis monomial

$$P_\alpha^{(m)}(z) = q_1^{r_1} p_1^{r_2} \ldots q_n^{r_5} p_n^{r_6}, \quad r_1 + \cdots + r_6 = m. \tag{3}$$

The monomials are ordered using the index α[2]. The summation over α in Eq. (2) extends from 1 to $N(m)$ where $N(m)$ is given by the relation[3]

$$N(m) = \binom{m+5}{m}. \tag{4}$$

In this paper, we will construct a symmetric positive definite bilinear form on the space spanned by homogeneous polynomials of degree m in phase-space variables. This will enable us to define a norm on this space. This norm will also be invariant under the action of the unitary group U(3). Since the nonlinear part of a symplectic map is specified by homogeneous polynomials, a norm defined on the space of homogeneous polynomials can be used to quantify the nonlinear content of the map (or equivalently the lattice). Moreover, the norm is a function of parameters specifying the beam-line elements of the accelerator lattice under consideration. Therefore, one can vary these parameters so as to minimize this norm. This should lead to improvements in performance of the lattice. This is shown to be true for a model FODO cell later in the paper.

II. CONSTRUCTION OF INVARIANT METRICS

We start by defining a bilinear form $g_{\alpha\beta}^{(m)}$ as follows

$$g_{\alpha\beta}^{(m)} \equiv (P_\alpha^{(m)}(z) , P_\beta^{(m)}(z)) \tag{5}$$

where $(P_\alpha^{(m)}(z) , P_\beta^{(m)}(z))$ denotes a bilinear form defined on the space of basis monomials of degree m. We require the bilinear form to be symmetric and positive definite so that it can be used to define a norm on this space.

*Work supported in part by the U.S. Department of Energy

Ideally, we would also require $g^{(m)}_{\alpha\beta}$ to be invariant under the action of all symplectic maps. Then, we would obtain a metric as unique as possible. However, this turns out to be impossible. The set of all symplectic maps forms a non-compact Lie group[1]. It can be shown that such groups can not have invariant metrics[4]. Only compact groups can have such metrics. Therefore, we are forced to impose a more modest requirement that the metric be invariant under the action of the largest compact subgroup of the symplectic group. We will require that $g^{(m)}_{\alpha\beta}$ be invariant under the action of the compact unitary group U(3).

We define the bilinear form $(P^{(m)}_\alpha(z)\,,P^{(m)}_\beta(z)\,)$ as follows[3]:

$$(P^{(m)}_\alpha(z)\,,P^{(m)}_\beta(z)\,) \equiv \frac{1}{r^{2m}} \int d\Omega_5\, P^{(m)}_\alpha(z)\, P^{(m)}_\beta(z)\,,\tag{6}$$

where $d\Omega_5$ is the solid angle for the 5-sphere and $r^2 = q_1^2 + p_1^2 + \cdots + p_3^2$. We show that it is invariant under the action of U(3). Consider the following expression

$$(\hat{U}P^{(m)}_\alpha(z),\hat{U}P^{(m)}_\beta(z)) = \frac{1}{r^{2m}} \int d\Omega_5\, P^{(m)}_\alpha(Uz)P^{(m)}_\beta(Uz).\tag{7}$$

Here, $\hat{U}$ is the Lie transformation corresponding to the element U belonging to U(3)[3]. We change to a new variable z' defined to be equal to Uz. Since the solid angle $d\Omega_5$ is invariant under the action of U(3), we obtain the relation

$$(\hat{U}P^{(m)}_\alpha(z)\,,\hat{U}P^{(m)}_\beta(z)\,) = \frac{1}{r^{2m}} \int d\Omega_5\, P^{(m)}_\alpha(z')P^{(m)}_\beta(z')\tag{8}$$

Since U was an arbitrary element of U(3), we get the desired result

$$(\hat{U}P^{(m)}_\alpha(z)\,,\hat{U}P^{(m)}_\beta(z)\,) = (P^{(m)}_\alpha(z)\,,P^{(m)}_\beta(z)\,)\tag{9}$$

$$\forall\, U \in \text{U(3)}.\tag{10}$$

It is easily seen from the definition that this bilinear form is symmetric. Obviously, it is also positive definite. Hence, Eq. (6) gives a valid invariant metric.

This invariant metric can be evaluated as follows. Consider the following equation

$$\int d^6z\, e^{-r^2} P^{(m)}_\alpha(z)\, P^{(m)}_\beta(z) = \int dr\, r^{2m+5} e^{-r^2}$$
$$\times \frac{1}{r^{2m}} \int d\Omega_5\, P^{(m)}_\alpha(z)\, P^{(m)}_\beta(z)\,.\tag{11}$$

This is seen to be correct since we have merely reexpressed the infinitesimal volume element d^6z in terms of the radius vector r and solid angle $d\Omega_5$. Inserting Eq. (6) in Eq. (11), we obtain the relation

$$(P^{(m)}_\alpha(z)\,,P^{(m)}_\beta(z)\,) = \frac{\int d^6z\, e^{-r^2} P^{(m)}_\alpha(z)\, P^{(m)}_\beta(z)}{\int dr\, e^{-r^2} r^{2m+5}}\,.\tag{12}$$

Both the numerator and the denominator can now evaluated easily[3].

Using the above construction, we obtain the following expression for $g^{(3)}_{\alpha\beta}$ (we do not list entries below the diagonal; we also restrict ourselves to the four-dimensional case due to lack of space):

$$
\begin{aligned}
g^{(3)}_{i\,i} &= c_1 \quad i = 1, 11, 17, 20,\\
g^{(3)}_{i\,i} &= c_2 \quad i = 2,3,4,5,8,10,12,13,14,16,18,19,\\
g^{(3)}_{i\,i} &= c_3 \quad i = 6,7,9,15,\\
g^{(3)}_{1\,5} &= g^{(3)}_{1\,8} = g^{(3)}_{1\,10} = g^{(3)}_{2\,11} = c_2,\\
g^{(3)}_{3\,17} &= g^{(3)}_{4\,20} = g^{(3)}_{11\,14} = g^{(3)}_{11\,16} = c_2,\\
g^{(3)}_{12\,17} &= g^{(3)}_{13\,20} = g^{(3)}_{17\,19} = g^{(3)}_{18\,20} = c_2,\\
g^{(3)}_{2\,14} &= g^{(3)}_{2\,16} = g^{(3)}_{3\,12} = g^{(3)}_{3\,19} = c_3,\\
g^{(3)}_{4\,13} &= g^{(3)}_{4\,18} = g^{(3)}_{5\,8} = g^{(3)}_{5\,10} = c_3,\\
g^{(3)}_{8\,10} &= g^{(3)}_{12\,19} = g^{(3)}_{13\,18} = g^{(3)}_{14\,16} = c_3.
\end{aligned}
$$

Here the indices $1, 2, \ldots 20$ represent monomials q_1^3, $q_1^2 p_1$, $q_1^2 q_2$, $q_1^2 p_2$, $q_1 p_1^2$, $q_1 p_1 q_2$, $q_1 p_1 p_2$, $q_1 q_2^2$, $q_1 q_2 p_2$, $q_1 p_2^2$, p_1^3, $p_1^2 q_2$, $p_1^2 p_2$, $p_1 q_2^2$, $p_1 q_2 p_2$, $p_1 p_2^2$, q_2^3, $q_2^2 p_2$, $q_2 p_2^2$, and p_2^3 respectively. And the constants c_1, c_2, and c_3 have the following values

$$c_1 = 5/64, \quad c_2 = c_1/5, \quad c_3 = c_1/15.\tag{13}$$

III.   CONSTRUCTION OF NORMS

Using the metric defined above, we now define a norm on the space of homogeneous polynomials of degree m. This norm can then serve as a merit function that can be used to minimize nonlinearities of degree m.

Each metric $g^{(m)}_{\alpha\beta}$ gives rise to a norm on the space of homogeneous polynomials of degree m. Consider a general homogeneous polynomial of degree m denoted by f_m. We are interested in obtaining a norm for storage ring lattices. Since the emittances in the three degrees of freedom can be quite different, we normalize them by factoring out the betatron functions. This is achieved by going to the so-called normal form[1] of the linear part M of the map $\mathcal{M}$. Let A be the symplectic transformation that takes M into its normal form N i.e.

$$N = AMA^{-1}\tag{14}$$

where N is a block-diagonal matrix with 2×2 blocks on the diagonal[5]. Applying the transformation A to the map $\mathcal{M}$, we obtain the result

$$\mathcal{N}_2 = A\mathcal{M}A^{-1} = AMA^{-1}Ae^{:f_3:}e^{:f_4:}\ldots e^{:f_m:}\ldots A^{-1}.\tag{15}$$

Using Eq. (14), we get the relation

$$\mathcal{N}_2 = Ne^{:f_3^{\text{tr}}:}e^{:f_4^{\text{tr}}:}\ldots e^{:f_m^{\text{tr}}:}\ldots\tag{16}$$

where[1]

$$f_m^{\text{tr}} = Af_m(z) = f_m(Az).\tag{17}$$

Since A depends on M, the f_m^{tr}'s also now depend on the linear part of the map. These transformed f_m's can be reexpressed in the original basis as follows

$$f_m^{\text{tr}}(z) = b^{(m)}_\alpha P^{(m)}_\alpha(z)\,.\tag{18}$$

We are now in a position to define a norm on the space of homogeneous polynomials

$$\|f_m\| \equiv (f_m^{\text{tr}}, f_m^{\text{tr}})^{\frac{1}{2}} = (b_\alpha^{(m)} P_\alpha^{(m)}(z), b_\beta^{(m)} P_\beta^{(m)}(z))^{\frac{1}{2}}. \tag{19}$$

Using Eq. (5), we get the following result

$$\|f_m\| = (g_{\alpha\beta}^{(m)} b_\alpha^{(m)} b_\beta^{(m)})^{\frac{1}{2}}. \tag{20}$$

From the above equation, we see that the norm $\|f_m\|$ is a function of parameters characterizing the beam-line elements in the accelerator lattice that we started out with (since the coefficients $b_\alpha^{(m)}$ are determined by these parameters). Therefore, one can think of varying these parameters so as to minimize this norm. Since the norm quantifies the nonlinear content of the lattice, this may lead to improvements in the performance of the system. We also note that $\|f_m\|^2$ is a positive definite quadratic function of the strengths of the m-th order multipoles (e.g. $\|f_3\|^2$ is such a function of the sextupole strengths). Hence $\|f_m\|^2$ is guaranteed to have an unique global minimum as a function of these multipole strengths.

IV. EXAMPLE

In this section, we study a model FODO cell with systematic sextupole errors to illustrate the utility of the invariant metric. The FODO cell consists of the following elements: a thin-sextupole corrector, a drift, a focusing-quadrupole with fringe fields and sextupole error, a drift, a thin-sextupole corrector, a drift, a defocusing-quadrupole with fringe fields and sextupole error, a drift, and finally, another thin-sextupole corrector.

First, we turn off the correctors and compute the norm $\|f_3\|^2$ in a four-dimensional phase space. It has a certain value (≈ 400 in our case). Next, we set the corrector strengths by minimizing the norm. The minimum is found to correspond roughly[6] to setting the corrector strengths according to Simpson's rule (i.e. the three strenghts are in the ratio 1:4:1)[7]. For this setting, the value of $\|f_3\|^2$ is reduced (from the uncorrected case) by almost two orders of magnitude. To verify that third order nonlinearities have actually been reduced in magnitude, the dynamic aperture of the FODO cell was computed for these two cases. The dynamic aperture for the corrected case was found to be larger by a factor of four.

V. SUMMARY

In this paper, we constructed invariant merit functions for accelerator lattices described by Hamiltonians. These metrics were used to define norms on the space of homogeneous polynomials of phase-space variables. These norms quantify the nonlinear content of the accelerator lattice. They can be minimized as a function of parameters describing the beam-line elements to improve the performance of the lattice. Finally, we considered a model FODO

cell with sextupole errors. By minimizing the third degree norm using correctors, we obtained a four-fold increase in the dynamic aperture.

VI. REFERENCES

[1] A. J. Dragt, in *Physics of High Energy Particle Accelerators*, edited by R. A. Carrigan, F. R. Huson, and M. Month, AIP Conference Proceedings No. 87 (American Institute of Physics, New York, 1982), p. 147; see also A. J. Dragt *et. al.*, Annu. Rev. Nucl. Part. Sci. **38**, 455 (1988).

[2] One possible indexing scheme can be found in A. Giorgilli, Comp. Phys. Comm. **16**, 331 (1979).

[3] G. Rangarajan, Ph. D. thesis, University of Maryland, 1990.

[4] J. F. Cornwell, *Group Theory in Physics*, volumes 1 and 2 (Academic Press, London, 1984).

[5] We note that in going to the normal form, we have "used up" the non-compact part of the symplectic group Sp(6,R) and we are left only with the compact subgroup.

[6] Due to lack of a proper optimizer routine, we did not make a global search for the minimum. In this case, we already knew what the approximate minimum was supposed to be from theoretical considerations. See Ref. [7] below.

[7] D. Neuffer and E. Forest, Phys. Lett. A **135** 197 (1989).

Further Dynamic Aperture Studies on a Wiggler-Based Ultra-Low-Emittance Damping Ring Lattice *

L. Emery[†]

Stanford Synchrotron Radiation Laboratory P.O. Box 4349, Bin 99 Stanford, CA 94309-0210

Abstract

Further dynamic aperture studies on an ultra-low emittance damping ring lattice are presented. A past conference paper[1] explained how the fast damping rate, the low emittance and the large dynamic aperture are acheived for this lattice. Dynamic aperture improvement with octupole correction was also reported. In this paper the dynamic aperture improvement is emphasized with a more systematic derivation and study of the octupole correction. Also, the modified sextupole proposal of Cornacchia and Halbach[2] is applied to the damping ring lattice.

I. Introduction

The ultra-low emittance damping ring lattice reported in a past conference paper[1] makes use of long dispersion-free straight sections filled with strong wigglers to produce fast synchrotron radiation damping. The lattice also has large radius arcs with strongly-focusing FODO cells to produce low quantum excitation. As Wiedemann points out in a proposal to lower the emittance of PEP[3], both features yield a very low equilibrium emittance. In the damping ring lattice proposed in [1], an emittance of 2.5×10^{-11} m-rad for a beam energy of 4 GeV is acheived ($\epsilon_n = \epsilon\gamma = 2 \times 10^{-7}$m-rad).

To maximize the dynamic aperture (the range of stability for transverse oscillations), FODO cell achromats as defined by Brown and Servranckx[4] with *non-interleaved* sextupoles are adopted. Although an interleaved achromat arrangement can accomodate a larger number of sextupoles, thus reducing the individual strengths, the dynamic aperture suffers greatly. In both interleaved and non-interleaved cases the second-order geometric aberration sextupole terms are made to vanish. Thus, vanishing second-order aberration terms does not guarantee the maximum possible dynamic aperture[5]. In general, sextupoles interact with each other to produce higher-order aberration terms which become important for low emittance lattices. Non-interleaved achromats, as are implemented in the damping ring proposed above, are simply a way to prevent sextupoles from interacting in this way. Further examination shows that the main geometric aberrations produced in these optics modules are due to the lengths of the sextupoles.

In this paper, new analytical formulae for the sextupole length aberrations are derived. Octupoles can be inserted into the lattice to selectively cancel some of these aberration terms, thus enlarging different parts of the dynamic aperture. Numerical tracking of particle trajectories confirms this.

II. Sextupole Length Aberration

The equations of transverse motion inside a sextupole are

$$x'' = -\frac{1}{2}m(x^2 - y^2), \qquad (1)$$

$$y'' = mxy. \qquad (2)$$

where m is the normalized sextupole strength, $(e/cp)\partial^2 B_y/\partial x^2$. These equations will be integrated along the longitudinal coordinate s using an iterative method. One starts with constant initial solutions $x(s) = x_0$, $x'(s) = x_0'$, $y(s) = y_0$, $y'(s) = y_0'$,

*Work supported by the Department of Energy, Office of Basic Energy Sciences, Division of Material Sciences.

[†]Now at Argonne National Laboratory

and iterate the following 4 steps,

$$x'(s) \leftarrow x_0' + \int_0^s \left(-\frac{1}{2}m(x(s)^2 - y(s)^2) \right) ds', \qquad (3)$$

$$x(s) \leftarrow x_0 + \int_0^s x'(s') \, ds', \qquad (4)$$

$$y'(s) \leftarrow y_0' + \int_0^s (mx(s)y(s)) \, ds', \qquad (5)$$

$$y(s) \leftarrow y_0 + \int_0^s y'(s') \, ds'. \qquad (6)$$

until the resulting functions (polynomials in s) are of sufficient accuracy. Keeping only terms cubic in coordinate variables, the sextupole exit coordinates take the form

$$x_i(l) = \sum_{j=1}^{4} R_{ij}(l)x_{j0} + \sum_{j=1}^{4}\sum_{k=1}^{j} T_{ijk}(l)x_{j0}x_{k0}$$

$$+ \sum_{j=1}^{4}\sum_{k=1}^{j}\sum_{l=1}^{k} U_{ijkl}(l)x_{j0}x_{k0}x_{l0} \qquad (7)$$

for $1 < i < 4$ where $x_1 \equiv x$, $x_2 \equiv x'$, $x_3 \equiv y$, and $x_4 \equiv y'$, and l is the sextupole length. The notation and formalism of nonlinear matrix elements is that of K.Brown[4]. The R_{ij} matrix elements represent a drift space of length equal to the sextupole length. The T_{ijk} matrix elements are proportional to the integrated sextupole strength. An achromat is designed so that the T_{ijk} terms for $1 < i, j, k < 4$ contributed by the sextupole pair within the achromat cancel exactly. The U terms are called third-order matrix elements, and are the most important terms for an achromat with long sextupoles, since the contributions of each sextupole of the pair add together. It is therefore sufficient to examine the U terms of one sextupole for local octupole compensation.

A. Largest Third-Order Matrix Elements

Out of 80 possible U_{ijkl} terms (for $1 < i, j, k < 4$), 40 are non-zero. Most of these are small. The strongest terms are found by converting the $x_i x_j x_k$ factors into normalized coordinates. The normalized coordinates are $u = x/\sqrt{\beta_x}$ and $\dot{u} = \sqrt{\beta_x}x' + \alpha_x x/\sqrt{\beta_x}$ for the horizontal plane, and $v = y/\sqrt{\beta_y}$ and $\dot{v} = \sqrt{\beta_y}y' + \alpha_y y/\sqrt{\beta_y}$ for the vertical plane. In a linear lattice, the particle trajectories in normalized coordinate phase space are circles (i.e., $u^2 + \dot{u}^2 = 2J$ is an invariant). Inserting the normalized coordinate definitions into equation (7) gives cubic terms in u, $\dot{u}$, v, and $\dot{v}$, which perturb the value of the linear invariant at every sextupole location. The coefficients of the cubic terms are the U_{ijkl} times some power of $l/\beta_{x,y}$ where l is the sextupole length. Normally in a storage ring, $l \ll \beta_{x,y}$. Therefore the most important nonlinear matrix elements have the lowest power of $l/\beta_{x,y}$.

One can repeat the same integration procedure above for particle motion in an octupole field:

$$x'' = -\frac{1}{6}O(x^3 - 3xy^2), \qquad (8)$$

$$y'' = -\frac{1}{6}O(y^3 - 3x^2y), \qquad (9)$$

where $O = (e/cp)\partial^3 B_y/\partial x^3$. The largest third-order matrix elements for octupoles are the same type as those of sextupoles.

If both sextupole and octupole fields are combined in a magnet, the largest third-order aberration terms are

$$\Delta\dot{u} = \left(-\frac{1}{6}Ol + \frac{1}{12}m^2 l^3\right)\beta_x^2 u^3 + \left(\frac{1}{2}Ol + \frac{1}{12}m^2 l^3\right)\beta_x\beta_y uv^2, \tag{10}$$

$$\Delta\dot{v} = \left(-\frac{1}{6}Ol + \frac{1}{12}m^2 l^3\right)\beta_y^2 v^3 + \left(\frac{1}{2}Ol + \frac{1}{12}m^2 l^3\right)\beta_x\beta_y u^2 v. \tag{11}$$

These two important equations summarize the largest aberration terms for a long sextupole and connect the aberrations from sextupoles with those of octupoles. They are the basis of octupole correction.

III. LOCAL OCTUPOLE CORRECTION

An octupole field with strength O can be superimposed on the sextupole field to cancel one or the other higher-order aberration terms, but not both, unfortunately. However, one of the two terms in each of these equations is much greater than the other because the sextupoles are placed where $\beta_x > \beta_y$ or $\beta_y > \beta_x$ for effective chromaticity correction. The larger of the two terms must obviously be targeted for cancellation using the octupole field strength as an adjustable parameter. The appropriate octupole integrated strength is

$$Ol = \frac{1}{2}m^2 l^3. \tag{12}$$

For instance, if $\beta_x > \beta_y$, this choice of octupole strength cancels the u^3 term in equation (10) and also the small v^3 term in equation (11), but increases the value of the coefficient of the coupling terms uv^2 and u^2v by a factor of four. Equations (10) and (11) become

$$\Delta\dot{u} = \frac{1}{3}m^2 l^3 \beta_x\beta_y uv^2 \tag{13}$$

$$\Delta\dot{v} = \frac{1}{3}m^2 l^3 \beta_x\beta_y u^2 v \tag{14}$$

For oscillations of equal invariant value in both planes (i.e. $v \approx u$) the ratio of the coupling term to the u^3 term is $4\beta_y/\beta_x$, which is of order one for FODO lattices. The coupling effect on the dynamic aperture from tracking doesn't seem to increase that much as a result.

For the other case, $\beta_y > \beta_x$, the largest term is the v^3 term, which vanishes when the same octupole strength $O = -\frac{1}{2}m^2 l^2$ is used. Note that the sign of the octupoles are the same in both cases. The reason is that the main nonlinear effect produced by the interaction of two sextupoles must always scale with the product of the sextupole strengths. In the case of sextupole self-interaction, the strength is squared, and the sign disappears.

A. Dynamic Aperture with Octupolar Correction

The octupole correction described above will be tested by tracking the damping ring lattice with an octupole family correcting each sextupole family. The damping ring uses two sextupole families. To simplify the discussion the sextupoles that correct the horizontal chromaticity are called SF sextupoles and the SF-correcting octupoles, OF octupoles. Similarly, the sextupoles that correct the vertical chromaticity are called SD and the correcting octupoles, OD octupoles. Figure 1 shows schematically where the SF sextupoles are placed relative to the focusing quadrupoles within the 90° FODO cell achromat of the damping ring lattice. The required integrated strengths of the octupoles are listed in Table 1. Ideally the octupolar field should be superimposed on the sextupole field for proper cancellation of aberration terms. In the tracking the octupoles were modeled as a thin-lens inserted in the center of each sextupole. Also, the sextupoles in the tracking were each modeled as two kicks spaced $l/2$ apart. In this approximation the third-order aberration term of the sextupole is reduced by a factor of 3/4. The correcting octupole strengths are adjusted to cancel the third-order aberration terms of the two-kick sextupole.

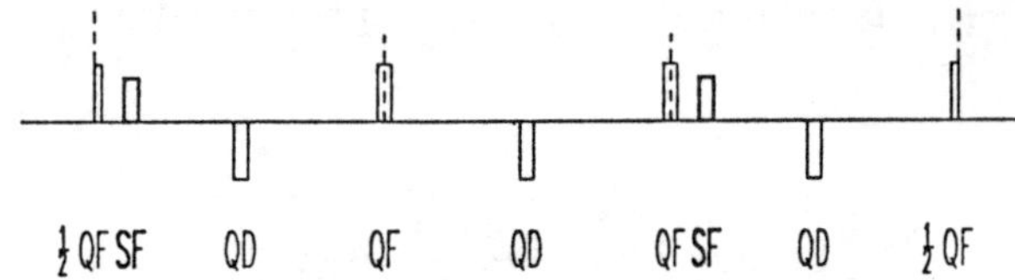

Figure 1

Achromat module for SF sextupoles

Table 1

Required octupole strength for correction

Octupole	Ol (m^{-3})	Associated sextupole	
		ml (m^{-2})	l (m)
OF	41.3	-16.6	0.3
OD	43.8	17.1	0.3

The difference in strength between the two-kick model and the uniform model should not change the conclusions.

One can look at the octupole correction in stages. First the octupoles (OF) that correct the SF's where $\beta_x \gg \beta_y$ are inserted. The calculated dynamic aperture for β_x=12.2 m and β_y=2.2 m is shown in Figure 2. A large improvement is observed in the stability in the horizontal plane, as one would expect from equation (10), and none in the vertical plane since the cancelled v^3 term in equation (11) is very small. The uncorrected SD's are responsible for the vertical aperture limit. The OF octupole also has the effect of correcting the horizontal tune shift with amplitude. When the octupoles (OD) that correct the SD's are inserted where $\beta_y \gg \beta_x$ (see Figure 3) the dynamic aperture in the vertical direction is improved.

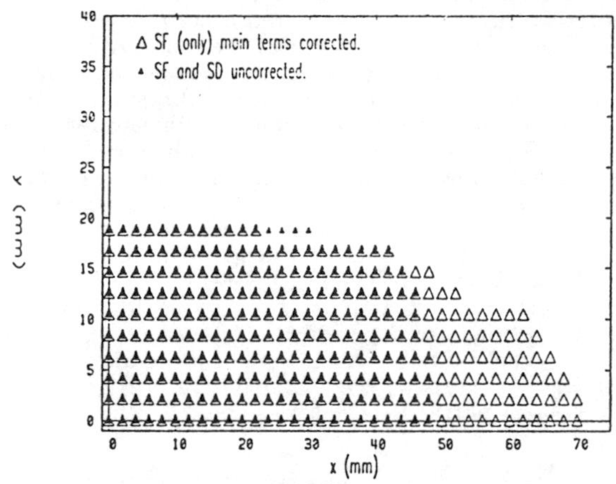

Figure 2

Dynamic aperture with SF-correcting octupole.

When both octupole families are turned on (Fig. 4), the stability along both axes is improved compared to the case with no octupoles. However, the dynamic aperture along both axes is smaller than the best possible values achieved when only one octupole family is used. There is no improvement in the area of the x-y plane where both x and y are large. This is understandable because the octupoles do not remove nonlinear coupling, but in fact increase it.

The dynamic aperture can be improved (up to 40%) and shaped with octupoles according to one's need. For instance if only horizontal aperture is required, then only the octupoles associated with the SF sextupoles should be turned on.

As an aside, a straightforward way of reshaping the dynamic aperture would be to redistribute the number of SF modules

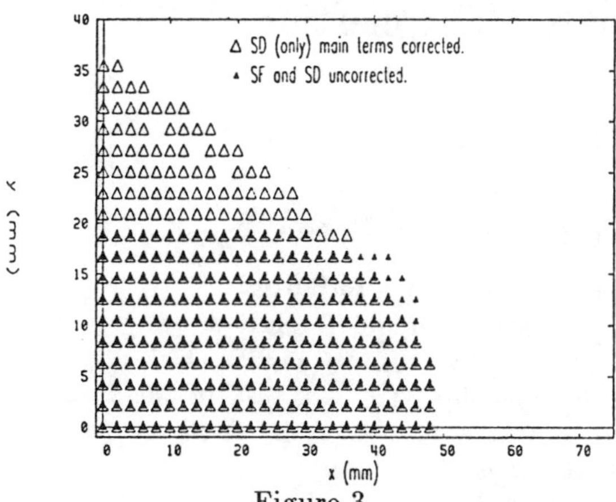

Figure 3

Dynamic aperture with SD-correcting octupole.

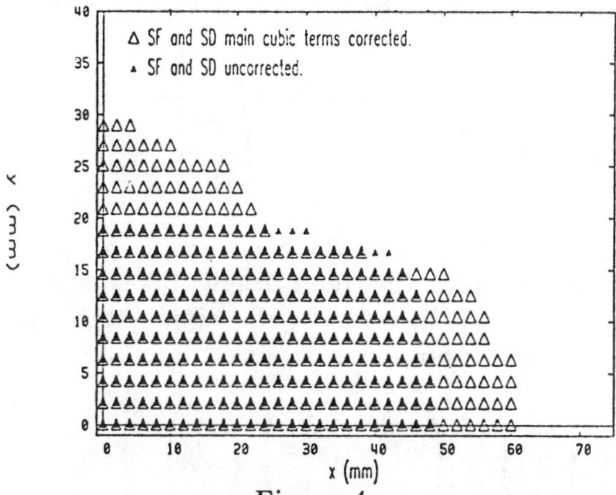

Figure 4

Dynamic aperture with SF- and SD-correcting octupole.

and SD modules. If for some reason the horizontal dynamic acceptance needs to be much greater than the vertical one, then the achromats used for SD sextupoles can be converted over to SF sextupoles. Repartitioning the modules in this way would reduce the individual SF sextupole strength and increase the individual SD sextupole strength. The horizontal dynamic aperture will then increase at the expense of the vertical dynamic aperture.

IV. APPLICATION OF MODIFIED SEXTUPOLES

After the above work with octupole correction was completed, a report on the design of modified sextupole magnets for dynamic aperture improvement in storage rings was published by Cornacchia and Halbach[2]. They propose reshaping the poles of sextupoles in order to reduce the magnetic field along a chosen transverse direction, while preserving the sextupole field symmetry near the axis. Outlying particles of a stored gaussian beam would feel a weaker nonlinear field and follow a stabler trajectory, while the core of the beam feels the regular sextupole field. This has the effect of increasing the dynamic aperture. While the chromaticity of the core of the beam is corrected, the chromaticity of large-amplitude orbits is not. Fortunately, the head-tail instability, for which the chromaticity correction is needed, is a collective effect, and it's not necessary for all of the beam to have zero or positive chromaticity.

Cornacchia and Halbach mention many possible field distributions for modified sextupoles. A modified sextupole field design that is probably the easiest to implement is the following[2]:

$$B^* = B_x - iB_y = -iAz^2 \exp(\kappa z^2) \tag{15}$$

where $z = x + iy$, A is a constant, and κ is a decay parameter

to be adjusted. Because of the $\exp(\kappa z^2)$ factor this sextupole is called a gaussian sextupole.

If $\kappa = 0$ we have an ordinary sextupole field. For $\kappa < 0$, the sextupole field eventually decays along the horizontal axis, but grows indefinitely along the the vertical axis, while the reverse is true for $\kappa > 0$. It would seem that the dynamic aperture would increase in one plane at the expense of the other. However, β_x and β_y are normally very different at sextupoles. One finds that the SF family of sextupoles (where $\beta_x > \beta_y$) should have $\kappa < 0$, and that the SD family of sextupoles (where $\beta_y > \beta_x$) should have $\kappa > 0$. Since the gaussian sextupole is very non-linear and not a simple multipole, its study using matrix elements is difficult. It is therefore difficult to predict which absolute value of κ will improve the dynamic aperture the most for a given lattice. A rough estimate would be to set κ such that $\kappa z^2 \approx$ -1 at the dynamic aperture limits measured at the sextupole position.

I optimized by trial and error the dynamic aperture for the damping ring, treating κ for each of the sextupole families as independent adjustable parameters. The dynamic aperture for the optimal gaussian sextupoles with $\kappa_{SF} =$ -75 m^{-2} and κ_{SD} = 250 m^{-2} is shown in Figure 5. Again, the dynamic aperture

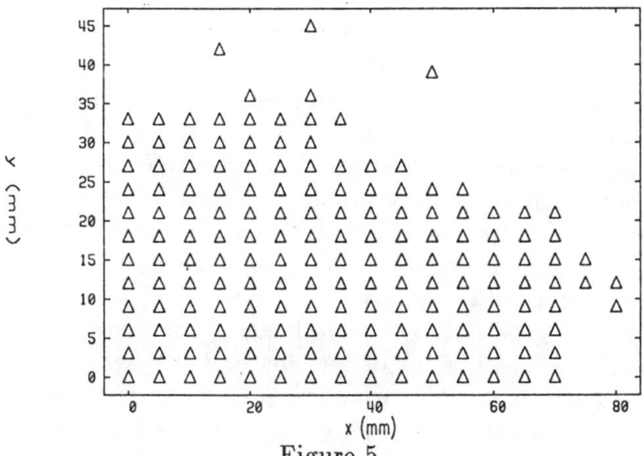

Figure 5

Dynamic aperture with gaussian sextupole.

is limited by sextupole length effects. In this case however, the dynamic aperture improvement along both axes (about 40%) is better than that of octupole correction.

V. CONCLUSION

The large dynamic aperture of non-interleaved achromat-based lattices can be improved further with correcting octupoles. The modified gaussian sextupole was also successfully applied to the low-emittance damping ring, and may be worth studying further.

VI. REFERENCES

[1] L. Emery, "An Ultra-Low Emittance Damping Ring Lattice and Its Dynamic Aperture," in *Proceedings of the 1989 IEEE Particle Accelerator Conference*, p. 1225, March 1989.

[2] M. Cornacchia, and K. Halbach, "Study of Modified Sextupoles for Dynamic Aperture Improvement in Synchrotron Radiation Sources," *Nuclear Instruments and Methods in Physics Research*, vol. A, no. 290, pp. 19–33, 1990.

[3] H. Wiedemann, "An Ultra Low Emittance Mode for PEP Using Dampling Wigglers," in *Nuclear Instruments and Methods in Physics Research*, (Amsterdam), p. 24, North-Holland, 1988.

[4] K. L. Brown and R. V. Servranckx, "First- and Second-Order Charged Particle Optics," SLAC-PUB 3381, SLAC, July 1984.

[5] L. Emery, *A Wiggler-Based Ultra-Low-Emittance Lattice and Its Chromatic Correction*. PhD thesis, Stanford University, August 1990.

Analytic Closed Orbit Analysis for RHIC Insertion*

S.Y. Lee[†]

Department of Physics, Indiana University, Bloomington, IN 47405

S.Tepikian

Accel. Development Department, Brookhaven National Laboratory, Upton, NY 11973

Abstract

Analytic closed orbit analysis is performed to evaluate the tolerance of quadrupole misalignment and dipole errors (b_0, a_0) in the RHIC insertion. Sensitivity coefficients of these errors are tabulated for different β^* values. Using these sensitivity tables, we found that the power supplies ripple of 10^{-4} can cause closed orbit motion of 0.05 mm at the IP in comparison with the rms beam size of 0.3 mm. It is desirable to have the power supply ripple less than 10^{-5}.

1. Introduction

The closed orbit error in the insertion region is of fundamental important to the collider physics. For RHIC, the closed orbit at the high-β triplets may also affect the dynamical aperture. Evaluation of the sensitivity on the quadrupole alignment errors and dipole excitation or rotation angle error gives us a feeling of the alignment tolerance. Some correction schemes may also applied to obtain proper orbit control in the insertion region.

Fig. 1 shows the RHIC inserion layout. There are nine quadrupoles on both sides of the interaction point (IP). The closed orbit can result from (1) quadrupole misalignment, (2) dipole error. etc. This paper discusses an analytic method in the closed orbit analysis.

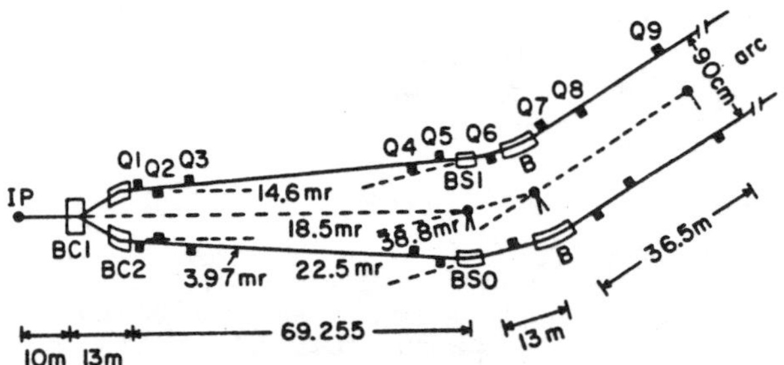

Fig. 1: Schematic Layout of a RHIC insertion.

2. Method of Orbit Error Analysis

For a particle in the accelerator, the equation of motion is given by[1]

$$\frac{d^2 y}{ds^2} + K(s)y = \frac{\Delta B(s)}{B\rho} \tag{1}$$

where y represents either the radial or vertical coordinates, $K(s)$ is the focusing function. For the horizontal closed orbit error, $\Delta B(s)$ arises from the vertical dipole field error due to quadrupole horizontal misalignment and dipole field errors. For the vertical closed orbit error, $\Delta B(s)$ arises from the horizontal dipole field error due to quadrupole vertical misalignments and dipole rotations.

2.1 Orbit kick due to a quadrupole

When a single quadrupole is shifted away from the central closed orbit by Δ_q, the angular kick due to the quadrupole is given by $-(y - \Delta_q)/f_q$ in the thin lense approximation, where f_q is the the focal length of the quadrupole and y is the actual closed orbit. For a focusing quadrupole, $f_q > 0$ and similarly $f_q < 0$ for a defocusing quadrupole. Thus the particle closed orbit after the quadrupole is related to the closed orbit before the quadrupole by,

$$\begin{pmatrix} y_a \\ y_a' \\ 1 \end{pmatrix} = \begin{pmatrix} 1 & 0 & 0 \\ -\frac{1}{f_q} & 1 & \frac{\Delta_q}{f_q} \\ 0 & 0 & 1 \end{pmatrix} \begin{pmatrix} y_b \\ y_b' \\ 1 \end{pmatrix}. \tag{2}$$

2.2 Closed Orbit Kick due to a Dipole

When a particle passes through a dipole, the closed orbit is modified by the dipole field error. Using a thin dipole approximation, we obtain then

$$\begin{pmatrix} y_a \\ y_a' \\ 1 \end{pmatrix} = \begin{pmatrix} 1 & 0 & 0 \\ 0 & 1 & \Delta\theta \\ 0 & 0 & 1 \end{pmatrix} \begin{pmatrix} y_b \\ y_b' \\ 1 \end{pmatrix}. \tag{3}$$

where $\Delta\theta$ is the dipole field error.

2.3 The Closed Orbit Error Propagation

The error propagation of the closed orbit is then obtained from multiplying matrices of orbit kicks discussed in previous sections and matrices of appropriate drift spaces. The procedure is equivalent to integrating Eq.(1) along the beam line. This procedure remains valid in the thick lense calculation. Thin lense approximation however simplify the calculation greatly. The final closed orbit distortion can then be expressed in terms of the angular errors of dipoles and quadrupole misalignments of quadrupoles by substituting the quadrupole strengths correspondingly. The actual result of these calculations[2] for RHIC will be discussed in the next section.

3. Closed Orbit Kicks in a RHIC Insertion

Table 1 lists the sensitivity coefficients for the horizontal closed orbit error at Q5, Q3, Q2, Q1 and IP as a function of the error fields discussed in section 2. As an example, table 1 gives the closed orbit deviations at Q3 and IP (for $\beta^*=2$m) as,

$$x_3 = -7\,x_9 - 123\,x_9' + 2.72\,\Delta_8 - 12.5\,\Delta_7 + 3.5\,\frac{\Delta\theta}{\theta} +$$

$$+4.26\,\Delta_6 + 1.1\,\frac{\Delta\theta_s}{\theta_s} - 8.17\,\Delta_5 + 5.76\,\Delta_4$$

"

and

$$x_{IP} = 1.3\,x_9 + 18.1\,x_9' - 0.97\,\Delta_8 + 3.32\,\Delta_7 - 0.85\,\frac{\Delta\theta}{\theta}$$

$$-0.85\,\Delta_6 - 0.18\,\frac{\Delta\theta_s}{\theta_s} + 1.14\,\Delta_5 - 0.53\,\Delta_4 - 2.62\,\Delta_3$$

$$+3.18\,\Delta_2 - 2.37\,\Delta_1 + 0.30\,\frac{\Delta\theta_{c2}}{\theta_{c2}} + 0.21\,\frac{\Delta\theta_{c1}}{\theta_{c1}}$$

We observe clearly that the closed orbit, x_3, at Q3 (similarly at Q2 and Q1) is very sensitive to x_9'. A 0.1 mrad error in x_9' can give rise to 12 mm error at Q3 location by assuming a perfect machine elsewhere. This sensitivity is due to betatron phase advance between Q3 and Q9 and a large betatron function at Q3 position. At the same time, x_3 is also sensitive to quadrupole misalignment at Q7. A 0.25 mm alignment in Q7, Q6, Q5 and Q4 can cause 3 mm rms closed orbit error at Q3. Similarly, the closed orbit at IP is sensitive to Q1–Q3 quadrupole alignment.

Table 1: Sensitivity Coefficients of the horizontal Closed Orbit Displacement. Here $\delta\theta = \Delta\theta/\theta$ represents percentage dipole field error.

$\beta^* = .5$m	x_5	x_3	x_2	x_1	x_{IP}	x'_{IP}
x_9	-1	-15.5	-8.6	-10.3	3.1	0.5
x_9'	-11.2	-246	-137	-164.	46.4	8.6
Δ_8	1.06	8.57	4.71	5.62	-1.89	-0.3
Δ_7	-3.37	-34.2	-18.8	-22.5	7.19	1.2
$\delta\theta$	.79	8.68	4.8	5.74	-1.8	-0.31
Δ_6	.27	3.85	2.14	2.56	-0.77	-0.13
$\delta\theta_s$	.07	1.63	0.91	1.1	-0.31	-0.06
Δ_5		0	0	0	0	0
Δ_4		5.3	3.03	3.72	-0.65	-0.18
Δ_3			-0.59	-1.23	-2.57	-0.05
Δ_2				0.6	3.16	0.1
Δ_1					-2.42	-0.1
$\delta\theta_{c2}$					0.30	0.02
$\delta\theta_{c1}$					0.21	0.01
$\beta^* = 2$m	x_5	x_3	x_2	x_1	x_{IP}	x'_{IP}
x_9	-0.8	-7	-3.8	-4.6	1.3	0.2
x_9'	-2.5	-123.	-69.1	-83.5	18.1	4.1
Δ_8	1.46	2.72	1.4	1.58	-0.97	-0.1
Δ_7	-3.91	-12.5	-6.7	-7.8	3.32	0.45
$\delta\theta$	0.9	3.5	1.88	2.21	-0.85	-0.12
Δ_6	0.64	4.26	2.34	2.78	-0.85	-0.15
$\delta\theta_s$	0.07	1.1	0.61	0.73	-0.18	-0.04
Δ_5		-8.17	-4.59	-5.56	1.14	0.27
Δ_4		5.76	3.3	4.05	-0.53	-0.19
Δ_3			-0.59	-1.23	-2.62	-0.06
Δ_2				0.6	3.18	0.1
Δ_1					-2.37	-0.1
$\delta\theta_{c2}$					0.30	0.02
$\Delta\theta_{c1}$					0.21	0.01

The table also gives us a guide line for the stability requirement in the power supply. As an example, we find

$$x_{IP} = -0.85\,\frac{\Delta\theta}{\theta} - 0.18\,\frac{\Delta\theta_s}{\theta_s} + 0.30\,\frac{\Delta\theta_{c2}}{\theta_{c2}} + 0.21\,\frac{\Delta\theta_{c1}}{\theta_{c1}}$$

from table 1 at $\beta^*=2$m. A power supply ripple will affect $\Delta\theta/\theta$ for all dipoles coherently. At $\Delta\theta/\theta \simeq 10^{-4}$, we expect the orbit will be shifted by 0.05 mm in comparison with the $\sigma_{\text{beam size}} \simeq 0.3$ mm for heavy ion beam at 100 GeV/u. The effect will be very harmful to the beam life time due to the presence of beam–beam interaction. It is therefore important to achieve the power supply ripple less than $\frac{\Delta\theta}{\theta} \leq 10^{-5}$. At $\beta^* = 0.5$ m, table 1 gives

$$x_{IP} = -1.8\,\frac{\Delta\theta}{\theta} - 0.31\,\frac{\Delta\theta_x}{\theta_s} + 0.30\,\frac{\Delta\theta_{c2}}{\theta_{c2}} + 0.21\,\frac{\Delta\theta_{c1}}{\theta_{c1}}$$

At $\Delta\theta/\theta \simeq 10^{-5}$ ripple will give $\Delta x_{IP} \simeq 0.02$ mm in comparison with the proton beam size $\sigma = 0.08$ mm at 250 GeV/c. Thus the power supply stability is especially important to the proton collision mode.

Table 2: Sensitivity Coefficients of the Vertical Closed Orbit. Here ϕ is the dipole rotation angle.

$\beta^* = .5$m	z_5	z_3	z_2	z_1	z_{IP}	z'_{IP}
z_9	-1.4	3.6	6.7	4.2	-2.5	-0.3
z_9'	18	-79.3	-141.	-87.9	60.5	6
Δ_8	-4.5	17.08	30.71	19.11	-12.5	-1.28
Δ_7	2.44	-8.78	-15.8	-9.86	6.36	0.66
φ	.62	-1.94	-3.53	-2.2	1.35	0.14
Δ_6	-.27	0.55	1.03	0.65	-0.31	-0.04
φ_s	.07	0.02	0.01	0	-0.08	0
Δ_5		0	0	0	0	0
Δ_4		-5.3	-9.27	-5.7	4.49	0.41
Δ_3			0.59	0.53	1.4	0.04
Δ_2				-0.6	-6.05	-0.22
Δ_1					2.42	0.1
φ_{c2}					0.30	0.02
φ_{c1}					0.21	0.01
$\beta^* = 2$m	z_5	z_3	z_2	z_1	z_{IP}	z'_{IP}
z_9	-1.1	2.5	4.7	3	-1.5	-0.2
z_9'	9.4	-40.8	-73.1	-45.5	31.8	3.1
Δ_8	-3.21	9.31	17.05	10.72	-6.34	-0.69
Δ_7	1.77	-4.52	-8.36	-5.27	2.91	0.33
φ	0.51	-0.95	-1.8	-1.15	0.5	0.07
Δ_6	-0.64	0.42	0.94	0.64	0.14	-0.02
φ_s	0.07	0.11	0.15	0.09	-0.17	-0.01
Δ_5		2.07	3.54	2.16	-2.02	-0.17
Δ_4		-5.76	-10.0	-6.2	5.04	0.45
Δ_3			0.59	0.53	1.39	0.03
Δ_2				-0.6	-6.01	-0.22
Δ_1					2.37	0.1
φ_{c2}					0.30	0.02
φ_{c1}					0.21	0.01

Table 2 lists the sensitivity table for the vertical closed orbit by integrating Eq. (1) along the insertion, where the quadrupole vertical misalignment is given by $\Delta_1, \ldots, \Delta_8$ and the dipole rotation is given by φ, φ_s, φ_{c2} and φ_{c1} for the corresponding dipoles B, BSI, BC2 and BC1 respectively.

Similar to that of the horizontal motion, the quadrupole alignment of Q1–Q3 is important to the proper collision at

IP. Power supply ripple is also critical for the operation at $\beta^* \leq 2$ m.

4. Closed Orbit of the Accelerator

We have calculated the closed orbit error propagation from the begining of the insertion to the interaction point, i.e.

$$\begin{pmatrix} y_{IP} \\ y'_{IP} \\ 1 \end{pmatrix} = \begin{pmatrix} a_{11} & a_{12} & a_{13} \\ a_{21} & a_{22} & a_{23} \\ 0 & 0 & 1 \end{pmatrix} \begin{pmatrix} y_9 \\ y'_9 \\ 1 \end{pmatrix}, \qquad (4)$$

where the matrix elements a_{ij} may be obtained directly form Tables 1 and 2. For example, $a_{11} = 3.1, a_{12} = 46.4$ and $a_{13} = -1.89\Delta_8 + 7.19\Delta_7 - 1.8\delta\theta + \cdots$ etc. for the horizontal closed orbit propagation at $\beta^* = 0.5$ m. To obtain a proper closed orbit of the entire circular accelerator, we shall assume that the orbit is propagated through the rest of the machine, i.e.

$$\begin{pmatrix} y_9 \\ y'_9 \\ 1 \end{pmatrix} = \begin{pmatrix} M_{11} & M_{12} & M_{13} \\ M_{21} & M_{22} & M_{23} \\ 0 & 0 & 1 \end{pmatrix} \begin{pmatrix} \tilde{y}_{IP} \\ \tilde{y}'_{IP} \\ 1 \end{pmatrix}. \qquad (5)$$

where 2×2 matrix corresponds to the propagation of betatron motion from the IP through the rest of the accelerator to the end of Q9. The matrix elements M_{13}, M_{23} are related the orbit kicks associated with the rest of the accelerator. The total closed orbit is then obtained by multiplying matrices of Eqs. (4) and (5) with the following constraints,

$$\tilde{y}_{IP} = y_{IP}; \quad \tilde{y}'_{IP} = y'_{IP}. \qquad (6)$$

The 2×2 matrix of the final product corresponds to the one turn betatron transfer map. They are given by

$$a_{11}M_{11} + a_{12}M_{21} = \cos 2\pi\nu + \alpha^* \sin 2\pi\nu,$$

$$a_{11}M_{12} + a_{12}M_{22} = \beta^* \sin 2\pi\nu,$$

$$a_{21}M_{11} + a_{22}M_{21} = -\gamma^* \sin 2\pi\nu,$$

$$a_{21}M_{12} + a_{22}M_{22} = \cos 2\pi\nu - \alpha^* \sin 2\pi\nu,$$

where ν is the betatron tune for either horizontal or vertical betatron motion, α^*, β^* and γ^* are the Courant-Snyder parametrization of the betatron functions at IP. Normally $\alpha^* = 0$ and $\gamma^* = 1/\beta^*$. Solving the closed orbit condition of Eq.(6), we obtain then

$$y_{IP} = \frac{1}{2\sin \pi\nu}\{(a_{11}M_{13} + a_{12}M_{23} + a_{13}) \sin \pi\nu$$

$$+\beta^*(a_{21}M_{13} + a_{22}M_{23} + a_{23}) \cos \pi\nu\}. \qquad (7)$$

$$y'_{IP} = \frac{1}{2\sin \pi\nu}\{-\frac{1}{\beta^*}(a_{11}M_{13} + a_{12}M_{23} + a_{13}) \cos \pi\nu$$

$$+(a_{21}M_{13} + a_{22}M_{23} + a_{23}) \sin \pi\nu\}. \qquad (8)$$

Note here that the orbit error is enhanced by the nearness of the betatron tune to an integer. The sensitivity factor is however still proportional to the tables 1 and 2 through a_{13} and a_{23} coefficients in Eqs. (7) and (8).

5. Conclusion and Discussion

The closed orbit analysis for the RHIC insertion is analyzed in an analytic model in terms of the quadrupole alignment errors and the dipole errors. We calculate the sensitivity coefficients for various β^* values. The closed orbit error becomes large at the high$-\beta$ quadrupoles, Q1–Q3. To minimize these errors, one should properly align the quadrupoles in the insertion. One should also measure x_9, x'_9, z_9 and z'_9 so that a proper orbit correction scheme for the insertion can be established due to the fact that the closed orbit is sensitively dependent on the parameters x'_9 and z'_9 (see Tables 1 and 2).

Using the table, we can also set the tolerance on the power supply ripple. At low β^* value, the power supply ripple of 10^{-5} is critical to obtain a proper beam–beam collision.

Finally the ground motion due to the high tide, local traffic, etc. will shift quadrupole alignment. A shift of 10μ in Q1–Q3 can also cause beam movement at the IP by 0.01 mm, which will affect the performance. Fortunately, the ground motions which does not change the relative motion of the accelerator component do not affect the luminosity. The random noise is however normally small, $\simeq 1 \mu$. Thus the effect should not be important. Realistic measurement of the effect in RHIC tunnel should be confirmed.

The method described in this paper is useful in evaluating the effect of few important magnetic elements on the closed orbit distortion. Another possible application is analyzing the orbit distortion of a beam tranfer line.

---

* Work performed under the auspices of the U.S. Department of Energy.

† On leave of absence from Brookhaven National Laboratory

Reference

1. E.D. Courant and H.S. Snyder, Ann. of Phys. <u>3</u>, 1 (1958)

2. S.Y. Lee and S. Tepikian, AD/RHIC-64, Feb. 1990.

BEAM–BEAM INTERACTION AND HIGH ORDER RESONANCES*

S. Tepikian and S.Y. Lee[†]

Brookhaven National Laboratory
Upton, NY 11973, USA

Abstract

There is experimental evidence from SPS[1], that very high order resonances (an order > 10) can cause particle loss during the beam–beam interaction. These results can be simulated by tracking a single particle against a round beam. If the round beam has an rms size of σ and the particle's initial amplitude is 2σ then no effects are observed due to high order resonances. However, if the particle's initial amplitude is 5σ the beam–beam 16th order resonances are strong. Furthermore, if we add tune modulation (i.e. power supply ripple) then the 5σ particle's motion becomes chaotic (when it is near a 16th order resonance) and can result in particle loss. This is in agreement with the beam–beam experiments performed at the SPS. Implications on RHIC and SSC will be discussed.

I. INTRODUCTION

Beam–beam experiments on the SPS have shown that when colliding a proton beam ($\approx 26\pi$mm·mrad) against an anti–proton beam ($\approx 13\pi$mm·mrad), there is an increase in background proton radiation when the tunes cross through the 16th order resonance. One surprise is that the loss occurred to the larger beam. The reason for this will be made clear as we continue.

Beam–beam effects have been studied both experimentally and theoretically.[2-8] However, most of these studies dealt with the low order resonance effects (an order < 10). In order to study the high order resonance effects we tracked a single particle, with a given initial amplitude, through a collision point against a round beam of rms radius σ. The low order resonances were avoided by choosing the operating tunes of RHIC.[9]

II. THE SIMULATION

The main accelerator is modeled as a linear machine with the following transfer matrix

$$\begin{pmatrix} \cos 2\pi\nu_x & \beta_x^* \sin 2\pi\nu_x & 0 & 0 \\ -\frac{1}{\beta_x^*} \sin 2\pi\nu_x & \cos 2\pi\nu_x & 0 & 0 \\ 0 & 0 & \cos 2\pi\nu_y & \beta_y^* \sin 2\pi\nu_y \\ 0 & 0 & -\frac{1}{\beta_y^*} \sin 2\pi\nu_y & \cos 2\pi\nu_y \end{pmatrix}$$

where β_x^* and β_y^* are beta values at the crossing point and ν_x, ν_y are horizontal and vertical betatron tunes respectively.

After the particle circles the accelerator, it will encounter the round beam at the crossing point. This beam–beam interaction is modeled with the following potential

$$U(r) = \frac{Nr_o}{2\gamma\sigma} V(r) = \frac{Nr_o}{2\gamma\sigma} e^{-r^2/2\sigma}$$

where N is the number of particles, $r_o = q^2 e^2/Am_o c^2$ is the classical radius of the particle with charge qe, mass number A, m_o is the atomic mass unit, γ is the Lorentz factor and σ is the rms beam size. From the above potential, the kicks to the slope become:

$$\Delta x' = 4\pi\, \xi f(r)\, \frac{x}{\beta^*}$$

$$\Delta y' = 4\pi\, \xi f(r)\, \frac{y}{\beta^*}$$

where $\xi = Nr_o\beta^*/4\pi\gamma\sigma^2$ is the linear beam–beam parameter, $\beta^* = \beta_x^* = \beta_y^*$ and

$$f(r) = \frac{2\sigma^2}{r^2} \left[1 - e^{-r^2/2\sigma^2} \right] \underset{r\to 0}{\sim} 1.$$

A particle with an initial maximum amplitude of $x_o = y_o = 2\sigma$ will have an rms amplitude over betatron motion of $<x>_{rms} = <y>_{rms} = \sqrt{2}\sigma$. Hence, the particle is most likely to be found in a box of area $8\sigma^2$. A particle with the same probabilities in a circle has a radius, $r = \sqrt{\frac{8}{\pi}}\sigma$, which occupies the same area. Expanding the potential term into a multipole expansion leaves

$$V(r) = \sum_{k=0}^{\infty} \frac{1}{k!} \left(-\frac{r^2}{2\sigma^2} \right)^k$$

The multipoles in this case are very small. However, when the initial maximum amplitude is $x_o = y_o = 5\sigma$, the corresponding radius is $r = 5\sqrt{\frac{2}{\pi}}\sigma$. This leads to large multipole terms, in particular the 8th term (driving the 16th order resonance) grows to ~ 400 and the 11th term (driving the 22nd order resonance) then drops to ~ 200. This is confirmed with tracking discussed in section III.

When two beams are colliding with different beam sizes, particles of the larger beam are more likely to be at the 5σ position than those of the smaller beam, then the losses will be seen in the larger beam. Additionally,

*Work performed under the auspices of the U.S. Department of Energy.

[†]Present address: Dept. of Physics, Indiana University, Bloomington, IN 47405.

only even order resonances are excited by this multipole expansion. However, when two beams are colliding off center, the odd order resonances will become important.

The final ingredient in the simulation is to include the linear effects of the beam–beam potential. This affects both the betatron tunes and the β^* at the crossing point. When these effects are taken into account, we observe the smear goes to zero as the initial amplitude goes to zero as it should. The simulation code was checked by reverse tracking to reproduce the initial conditions. In 10^4 revolutions, the error in reproducing the initial conditions is $\sim 10^{-13}$, increasing to $\sim 10^{-7}$ in 10^7 revolutions (in double precision on an IBM 3090).

Finally, we also study the effect of tune modulation on the particle motion. The betatron tunes are assumed to be $\nu_x = \nu_x^o + \Delta\nu_x \sin\left[n\overline{f}\right]$ and $\nu_y = \nu_y^o + \Delta\nu_y \sin\left[n\overline{f} + \varphi\right]$ for the n'th revolution and a tune modulation of $\Delta\nu_x$ and $\Delta\nu_y$. Furthermore, $\overline{f} = 2\pi Cf/v$ where f is the frequency of the tune modulation and φ sets up a phase difference between the modulations.

III. THE RESULTS

The simulation, described above, was used to study the beam–beam interaction. In particular the tunes were varied from $\nu_x^o = 28.809$ to $\nu_x^o = 28.830$ in steps of 0.0001 and $\nu_y^o = \nu_x^o - 0.004$. Furthermore, $\xi = 0.02$, $\beta^* = 2$ m, $\sigma = 5.5 \times 10^{-4}$ m, $\gamma = 100$, $C = 3833.845$ m, $x_o = 5\sigma$, $x_o' = 0$, $y_o = 5\sigma$ and $y_o' = 0$. Figure 1 shows the maximum $E_x + E_y$ and minimum $E_x + E_y$ versus the resulting ν_x, (where E_x and E_y are the emittances in πmm·mrad). In this figure we observe 3 major regions: Region (I) shows larger emittance to the 16th order resonances at harmonic 461. Additionally, there are 8 peaks in $(E_x + E_y)_{max}$ and the 8 troughs in $(E_x + E_y)_{min}$ corresponding to the 9 resonances $16\nu_x = 461$, $14\nu_x + 2\nu_y = 461$, $12\nu_x + 4\nu_y = 461$, ..., $16\nu_y = 461$. Also the parametric $16\nu_x$ and $16\nu_y$ resonances are harder to be recognized. Region II shows no resonances; Region III shows some peaks due to the 22nd order resonances with harmonic at 634.

Although we see resonance effects in Fig. 1, long term tracking will not reveal any instabilities or chaotic behavior. When tune modulation effects are included, new effects appear. Consider a tune modulation of $\Delta\nu_x = \Delta\nu_y = 0.001$, $\varphi = 0$ (i.e. tune modulation parallel to the coupling line) and at a frequency of 60 Hz. Note, it takes about 1,300 revolutions for the particle to go through one full cycle of the tune modulation, thus, long term tracking is necessary.

Figure 2 plots $(E_x + E_y)_{max}$ and $(E_x + E_y)_{min}$ versus ν_x with tune modulation on (each point tracks through 10^5 revolutions). In region I, the emittances have greatly increased due to the 16th order resonances. Region III shows no significant effect due to the 22nd order resonance.

The points A and B shown on Fig. 2 are the tunes where we performed long term tracking. A particle at point A can cross a 16th order with results shown in Fig.

3. In 10^6 revolutions the resonance can be crossed 1500 times. The motion seems chaotic with large emittance growth. The total emittance is outside the available aperture. Detailed phase space indicates that the particle is locked on a sum resonance line and then is transported through different resonance to another sum resonance. Through this mechanism the total emittance is increased irreversibly. For comparison, Fig. 4 shows the tracking in Region II at point B (i.e. point B on Fig. 2). Here, the motion appears to be quite regular and bounded.

IV. CONCLUSION

In all large colliders such as RHIC, SSC, SPS, Tevatron, etc. it is very likely that these machines will collide beams of unequal size. In this case, the larger beam can have a significant number of particles at the position of 5σ or larger relative to the smaller beam. Since, these particles may also have tunes near high order resonances, they can get lost leading to increased background radiation which the detectors must be able to handle. However, this will not lead to total beam loss, only losses to particles at the fringes of the beam.

This is especially important to machines such as RHIC which will collide beams of different species. When colliding a proton beam against a gold beam, the gold beam will be larger. As the tune of the machine crosses a high order resonance, there will be an increase in the background of gold ions that the detectors must contend with.

V. REFERENCES

[1] L. Evans, "The Beam–Beam Interaction", in Workshop on the RHIC Performance, BNL Report 41604, pp, 245–251 (1988).

[2] A.W. Chao, "Beam–Beam Instability", AIP Conference Proc. 127, pp. 202– 242 (1983).

[3] A. Gerasimov, F. Izrailev, J. Tennyson, A. Tyomnikh, "The Dynamics of the Beam–Beam Interaction", In Lecture Note in Physics 247, pp. 154–175 (1985).

[4] S. Myers, "Review of Beam–Beam Simulation", ibid, pp. 176–237.

[5] A.W. Chao, P. Bambade, W.T. Weng, "Nonlinear Beam–Beam Resonances", ibid, pp. 77–103.

[6] M. Month, J. Herrera, editors, "Nonlinear Dynamics and the Beam Beam Interaction", AIP Conference Proc. 57, 1979.

[7] S.Y. Lee, S. Tepikian, "Tangent Map Analysis of the Beam–Beam Interaction", Third Advanced ICFA Beam Dynamics Workshop, Novosibirsk, USSR, pp. 62–65 (1989).

[8] M. Cornacchia, L. Evans, "The Effects of Magnetic Nonlinearities on a Stored Proton Beam and there Implications for Superconducting Storage Rings". Particle Accelerator 19, pp. 125–144 (1986).

[9] Conceptual Design Report for RHIC, BNL Report 52195 (1989).

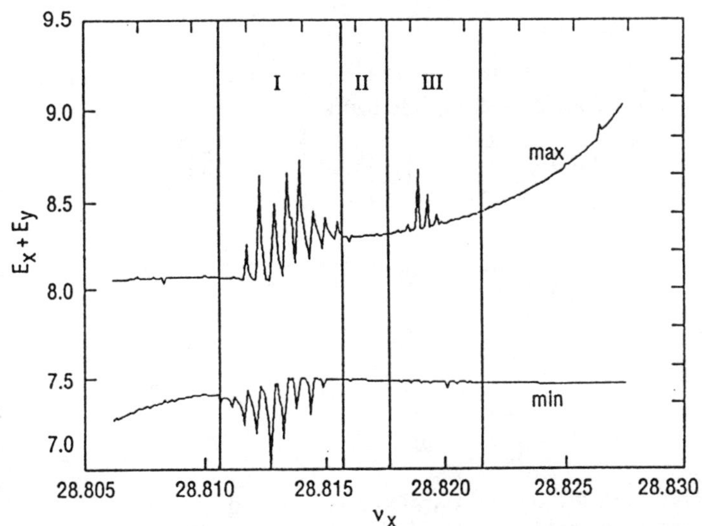

Fig. 1: Plot of $(E_x + E_y)_{min}$ and $(E_x + E_y)_{max}$ versus the horizontal tune ν_x with no tune modulation. Regions I, II, and III refer to the particle crossing different resonances, see text.

Fig. 2: Plotting $(E_x + E_y)_{min}$ and $(E_x + E_y)_{max}$ versus ν_x with tune modulation. Point A and B refer to Figs. 3 and 4 respectively. Note the expanded scale of the total emittance.

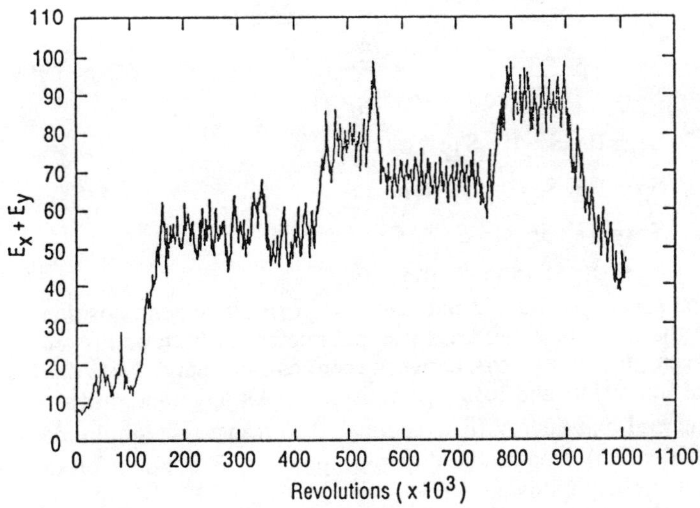

Fig. 3: Long term tracking of particle on a 16th order resonances. The tunes are shown at point A on Fig. 2.

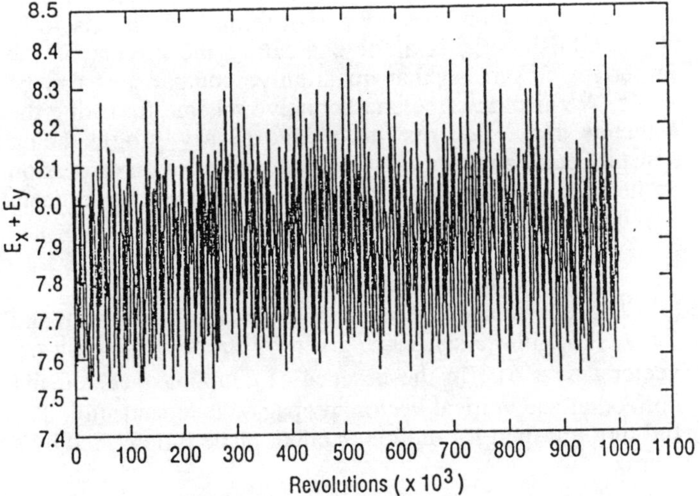

Fig. 4: Long term tracking of a particle away from resonances. The tunes are shown at point B on Fig. 2.

Single Beam Crab Dynamics

T. Chen, D. Rubin
† Newman Laboratory, Cornell University
Ithaca, NY 14853.

Abstract

In an electron-positron machine in which the beams cross at an angle, the characteristics of head on collisions can be preserved by tilting the bunches[1,2,3]. The crab angle is created by driving head and tail in opposite directions at a location of zero dispersion or by introducing a head tail energy difference at a location of finite dispersion[4]. We observe that bunches are tilted at any point in the machine where the dispersion is non zero and the longitudinal phase space ellipse is not erect. A crab angle is therefore generated at the interaction point as long as it is not a symmetry point of the longitudinal focusing and the dispersion is finite. The matrix formalism developed for the analysis of transverse coupling is generalized to the horizontal- longitudinal case. A measure of residual coupling (as distinguished from the unavoidable coupling associated with dispersion) is defined. General expressions for the crabbing angle and residual coupling are given in terms of transfer matrix elements.

I. INTRODUCTION

It has been proposed that synchrobetatron resonances excited by beams crossing at an angle be compensated by introducing a comparable crab angle to the bunches. The bunch is tilted in the x-s (horizontal-longitudinal) plane so that colliding beams overlap exactly. Two schemes have been suggested for rotating the bunches. Transverse crabbing requires introduction of a time dependent transverse force that gives head and tail equal but opposite kicks. Alternatively an equal but opposite longitudinal kick to head and tail in a region of horizontal dispersion can yield the necessary coupling to tilt the beams at the IP. Both schemes employ a closed (or very nearly closed) x-s coupling bump about the interaction region so that the synchrobetatron coupling within the crabbing region does not persist throughout the machine. We establish a criteria for evaluating the effects of the imperfect closure so that quantitative comparisons can be made. We also describe an alternative scheme for tilting the bunches that requires no extraordinary single beam synchrobetatron coupling. The scheme relies on dispersion at the collision point. In such a scheme however, beam beam synchrobetatron coupling is not precluded.

II. MATRIX FORMALISM

The coordinates of a particle in the horizontal and longitudinal phase space are given by the four dimensional vector (x, x', s, δ). In the absence of coupling mechanisms horizontal and vertical vectors propagate independently. The full turn transport for motion in either plane is described by a matrix

$$\begin{bmatrix} \cos\mu + \alpha\sin\mu & \beta\sin\mu \\ -\gamma\sin\mu & \cos\mu - \alpha\sin\mu \end{bmatrix}.$$

The longitudinal mapping is generated from accelerating cavity and arc matrices of the form[5]

RF Cavity:
$$\begin{bmatrix} s \\ \delta \end{bmatrix} = \begin{bmatrix} 1 & 0 \\ \frac{1}{F} & 1 \end{bmatrix} \begin{bmatrix} s_o \\ \delta_o \end{bmatrix}$$

where $\quad \dfrac{1}{F} = \dfrac{2\pi V_{rf}\cos\phi_S}{\lambda E}.$

Arc:
$$\begin{bmatrix} s \\ \delta \end{bmatrix} = \begin{bmatrix} 1 & -L \\ 0 & 1 \end{bmatrix} \begin{bmatrix} s_o \\ \delta_o \end{bmatrix}$$

where $L = \displaystyle\int_{s_o}^{s} \dfrac{\eta(s)}{\rho} ds.$

The longitudinal and transverse motion are coupled by bending magnets, accelerating cavities where dispersion is non zero and transverse deflecting cavities. The propagation of the particle vector is defined by 4X4 matrices. The matrix describing transport through the machine arcs, including bends can be written as:

$$\begin{bmatrix} S_{11} & S_{12} & 0 & S_{14} \\ S_{21} & S_{22} & 0 & S_{24} \\ S_{31} & S_{32} & 1 & -L \\ 0 & 0 & 0 & 1 \end{bmatrix}$$

where $S_{11} = \sqrt{\beta_2/\beta_1}(\cos\mu_{21} + \alpha_1\sin\mu_{21})$

$S_{12} = \sqrt{\beta_1\beta_2}\sin\mu_{21}$

$S_{21} = - \dfrac{(1 + \alpha_1\alpha_2)\sin\mu_{21} + (\alpha_2 - \alpha_1)\cos\mu_{21}}{\sqrt{\beta_1\beta_2}}$

$S_{22} = \sqrt{\beta_1/\beta_2}(\cos\mu_{21} - \alpha_1\sin\mu_{21})$

$S_{14} = \eta_2 - S_{11}\eta_1 - S_{12}\eta_1'$

$S_{24} = \eta_2' - S_{21}\eta_1 - S_{22}\eta_1'$

$S_{31} = \eta_1' + S_{21}\eta_2 - S_{11}\eta_2'$

$S_{32} = -\eta_2 + S_{22}\eta_2 - S_{12}\eta_2'$

The subscripts 1 and 2 indicate start position and end position of the arc. So, when the twiss parameters at both points are given, the 4x4 matrix between them can be found. Note that the upper left and lower right 2X2 blocks have determinant one and that energy (δ) is coupled to transverse amplitude (dispersion). Longitudinal displacement (s) is not coupled to transverse amplitude.

The transverse deflection crabbing cavity matrix is given in terms of crab angle ϕ and the horizontal beta-function at the cavity β_c and at the IP β^*:

---

† Supported by the National Science Foundation.

$$\begin{bmatrix} 1 & 0 & 0 & 0 \\ 0 & 1 & t & 0 \\ 0 & 0 & 1 & 0 \\ t & 0 & 0 & 1 \end{bmatrix}$$

with $t = \dfrac{\tan\phi}{\sqrt{\beta_c\beta^*}}$

In general then the mapping of the particle motion is described by 4X4 symplectic matrices. If the full turn matrix is block diagonal then there is no local coupling of horizontal and longitudinal phase space. Crabbing requires a correlation between longitudinal and transverse displacement at the interaction point and therefore the full turn matrix is not block diagonal. Indeed, the crab angle can be extracted from the elements of the matrix by performing a normal mode decomposition. For any symplectic matrix T there exists a unique block diagonal matrix U related to T by similarity transformation[9].

$$T=VUV^{-1}$$

where $T= \begin{bmatrix} M & n \\ m & N \end{bmatrix}$; $U= \begin{bmatrix} u_1 & 0 \\ 0 & u_2 \end{bmatrix}$, $V= \begin{bmatrix} \gamma I & C \\ -C^+ & \gamma I \end{bmatrix}$, and

M,m,N,n,u_1,u_2 are 2x2 matrices. The normal mode unit determinant matrices u_1 and u_2 can be written in terms of normal mode Twiss parameters. The matrix V relates normal mode and real coordinates. It is convenient to normalize the coordinates such that they are independent of Twiss parameters. The normalized V matrix is:

$$\bar{V}= GVG^{-1} = \begin{bmatrix} \gamma I & \bar{C} \\ -\bar{C}^+ & \gamma I \end{bmatrix},$$

where $G= \begin{bmatrix} G_1 & 0 \\ 0 & G_2 \end{bmatrix}$, with $G_i= \begin{bmatrix} 1/\sqrt{\beta_i} & 0 \\ \alpha_i/\sqrt{\beta_i} & \sqrt{\beta_i} \end{bmatrix}$ The

orientation of the beam ellipse is completely described by the matrix C in the limit of weak coupling and where the longitudinal emittance is much greater than the transverse emittance. The angle of the ellipse (the crab angle) is:[6]

$$\phi = \sqrt{\beta_1/\beta_2}\,\bar{c}_{11}=c_{11}+(\alpha_2/\beta_2)c_{12}.$$

The maximum horizontal excursion of the beam ellipse is given by:

$$x_{max}= \sqrt{\varepsilon_2(c_{11}{}^2\beta_2+c_{11}{}^2\gamma_2-2c_{11}c_{12})(1+\alpha_2{}^2)}\cos(\tan^{-1}(\tfrac{1}{\alpha_2}))$$

The matrix V is equivalent to transport from a point of no coupling in the ring to a point at which T is evaluated. If T corresponds to a point of dispersion but no additional synchrobetatron coupling then c_{12} and c_{22} can be identified with the dispersion η and η'.

When the normal mode emittances are more nearly equal the orientation of the beam ellipse is determined by the C matrix and the emittances[7].

III. CRABBING WITH DISPERSION AT IP

Consider a ring with no cavity induced synchrobetatron coupling. As shown above the beam ellipse in the x-s plane

is tilted by an angle $\phi=c_{11}+\dfrac{\alpha_2}{\beta_2}c_{12}$ where α_2 and β_2 are the

Twiss parameters corresponding to the matrix u_2. If the coupling is weak then $u_2=N$ where the matrix N describes the synchrotron motion. Then $\alpha_2 \sim \alpha_s$ and $\beta_2 \sim \beta_s$. It was also noted that $c_{12}=\eta$ and if there is no additional coupling then $c_{11}=c_{21}=0$ and we find that there is a crab angle at a point of non zero dispersion $\phi=\dfrac{\alpha_s}{\beta_s}\eta$. α_s as usual is related to the orientation of the longitudinal phase space ellipse and is in general non zero except mid way between the accelerating cavities. If the total accelerating voltage V~35MV, f_{RF} =500MHz, the synchrotron tune Qs~0.09, and η^*~.5m and the beam energy is 8GeV, then ϕ=10mr, appropriate to a crossing angle of 10mr[9]. All of the accelerating cavities are necessarily bunched together and near to the IP(see Figure 1) so that β_s is changing most rapidly.

There are many advantages to this crabbing scheme. No deflecting cavities are required. No extraordinary synchrobetatron coupling is introduced to generate the crab angle. In the transverse and dispersive crabbing schemes imperfect compensation due to phase or voltage errors leads to residual synchrobetatron coupling. In a scheme that relies strictly on dispersion at the interaction point there is by definition no coupling to compensate and no residual errors can arise. The effects of such residual coupling in deflection and dispersive schemes remains to be investigated.

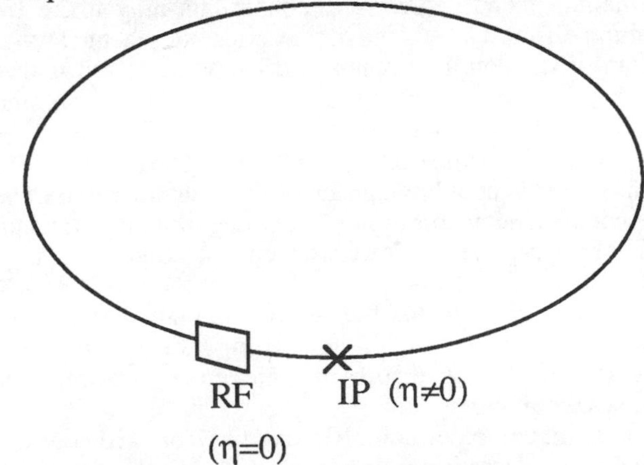

Figure 1 η^* crabbing scheme.

Another advantage of this scheme is the absence of constraints on transverse phase advance. The horizontal phase advance from cavity to interaction point is not relevant to the crabbing angle. The only parameters related to the angle are η^* and RF voltage.

An undesirable characteristic of this scheme is that beam beam interaction induces synchro-betatron coupling due to the finite dispersion at the IP. This is a nonlinear coupling, that may lead to a blow up of the transverse emittances. As a result, the saturated beam-beam tune shift may be reduced. CESR operated successfully with dispersion comparable to that required for crabbing. Consequences to the beam beam tune shift limit of finite dispersion at the IP are not clear but certainly the effects are not intolerable. More study is needed in both theoretical analysis and experimental investigation.

IV. RESIDUAL COUPLING

In both the transverse and dispersive crabbing schemes, an RF cavity initiates the requisite coupling of longitudinal and transverse motion in the incoming beam and a second cavity eliminates that coupling as the beam exits the interaction region. In the case of the transverse crabbing, deflecting cavities are located $\pm 1/4$ horizontal betatron wavelength from the IP. The transverse kick to the beam by the first cavity is compensated by the second a half wavelength away. The relative displacement of head and tail is a maximum at the IP. The dispersive crabbing results from applying equal but opposite energy kicks to head and tail (typical for accelerating cavities) of the bunch in a dispersive region. Because the cavity effects a displacement of the betatron trajectory the peak relative displacement of head and tail appears a full wavelength from the cavity. The betatron oscillation is again eliminated by applying a similar displacement $(n+1/2)$ wavelengths away. The coupling is initiated by a cavity an integer number of wavelengths from the IP and eliminated by a cavity a half integer number of wavelengths from the IP.

In both cases exact compensation requires that there be no phase advance in the longitudinal phase space through the crabbing region and therefore no bending magnets. This condition is quite inconsistent with the dispersive crabbing scheme since it is not possible to go from zero dispersion at the IP to nonzero dispersion in the cavities without intermediate bends. Practical constraints may make that condition difficult to achieve in the deflecting scheme as well. Closure of the coupling bump also is compromised if there are errors in the horizontal phase advance, or in the amplitude and phase of the cavity voltages.

If the compensation is imperfect there is residual synchrobetatron coupling outside of the crabbing region. We note that dispersion constitutes a coupling of longitudinal and transverse phase space that we are prepared to tolerate. Even if the crab coupling compensation is perfect the full turn 4X4 transfer matrix evaluated outside of the crab region is not block diagonal. We need to distinguish the coupling associated with the errors in the compensation from dispersive coupling.

In a machine without RF cavities, or with cavities located where there is no dispersion, the transverse amplitude of the beam depends on the energy spread but is independent of the bunch length and therefore of the accelerating voltage. Furthermore, the full turn matrix evaluated at a point of finite dispersion can be block diagonalized by a mapping based on static elements (bends, drifts and quads). Equivalently the dispersion can be eliminated by transport through such elements and at such a point the motion is decoupled. We can distinguish mechanisms that couple energy to horizontal amplitude from those that couple bunch length to horizontal amplitude. If bunch length is coupled to the horizontal amplitude, then the horizontal motion depends on the cavity voltage.

Again we refer to the matrix V that transforms from normal mode to horizontal and longitudinal coordinates. Matrix elements c_{11} and c_{21} couple longitudinal displacement into the horizontal plane. The amplitude of the longitudinal displacement is determined by the accelerating cavity parameters. Therefore c_{11} and c_{21} are a measure of the residual , nondispersive coupling and can simply be evaluated throughout the machine. We anticipate that such an analysis will permit quantitative study of the single beam effects of the residual coupling in any crabbing scenario.

V. CONCLUSIONS

The η^* crabbing scheme is a relatively simple alternative to the deflection and dispersive cavity schemes. The η^* scheme requries no special RF cavities and introduces no single beam synchrobetatron coupling mechanism so the possibility of residual coupling is precluded. Indeed such crabbing is characteristic of all machines although typically not at the IP. We have applied the transverse coupling formalism to the longitudinal and horizontal planes and have identified elements of the transformation matrix with the angle and height of the beam ellipse. Finally, a correspondence between specific full turn matrix elements and residual synchrobetatron coupling is demonstrated. We hope to further develop these techniques in the analysis of the residual coupling associated with crabbing.

VI. ACKNOWLEDGEMENT:

Authors would like to sincerely thank Prof. M. Tigner for his very helpful advice, and Y. Orlov and D. Sagan for their valuable discussion.

VII. REFERENCES

[1] A. Piwinski, IEEE Trans. Nucl. Sci. NS-24, p1408 (1977).

[2] R. B. Palmer, SLAC Report No. SLAC-PUB-4707 (1988).

[3] K. Oide & K. Yokoya, Physical Review A, Vol. 40, no.1, 315 (1989).

[4] G. Jackson, AIP Conf. Proc. 214, 327 (1990).

[5] R. Littauer, CON-90-9 (1990).

[6] P. Bagley & D. Rubin, "Correction of transverse coupling in a storage ring", in *Proc. of the 1989 IEEE Particle Accelerator Conf.*, p875.

[7] F. Willeke & G. Ripken, "Methods of Beam Optics", *AIP Conf. Proc.184*, p758(1989).

[8] CESR B, CLNS 91-1050, Cornell, (1991).

[9] D. Edwards, L. Teng, IEEE trans. on NS, vol. NS-20, No. 3, (1973).

Emittance Growth due to Beam Motion

King-Yuen Ng
Fermi National Accelerator Laboratory*
P.O. Box 500
Batavia, Illinois 60510

Jack M. Peterson
Superconducting Super Collider Laboratory[†]
2550 Beckleymeade Avenue
Dallas, Texas 75237

Abstract

When a beam undergoes a collective oscillation, its emittance tends to grow because of variations in betatron tune which are due to energy spread and chromaticity and to non-linear lattice characteristics. In this paper we estimate the emittance growth in the SSC caused by the measured ground motions at the SSC site and by beam-beam effects in the Jostlein beam-centering scheme [1].

I. Emittance Growth from Chromaticity

The vector trajectory of a pencil beam kicked through an angle α at the point s_o, where the betatron function has the value β_o, is (using normalized coordinates $\eta \equiv x/\sqrt{\beta}$. and $\eta' \equiv d\eta/d\phi$, ϕ being the betatron phase beginning at s_o):

$$\vec{\eta} \equiv \eta + i\eta' = i\alpha\sqrt{\beta_o}\,e^{-i\phi}. \qquad (1)$$

However, after N turns, the pencil beam has spread into an arc whose rms width [2] in the phase coordinate is $\sigma_\phi = (2\sigma_E \xi/v_s)\sin \pi v_s N$, where σ_E is the rms energy spread, ξ the chromaticity, and $v_s = f_s/f_o$ is the relative synchrotron frequency.

If there is only a single kick, this spreading in phase space is reversed after one half of a synchrotron period, and, in the linear approximation, at integral numbers of turns the pencil beam again collapses to a point, with no net smearing having occurred. However, if the beam is repeatedly kicked at random times, the elementary spreading effects occur in many directions in phase space and will tend to add as in a two-dimensional random walk. The characteristic time for this process is one half of a synchrotron period, which for the SSC at 20 TeV is 0.125 sec, or 430 turns.

A. Non-Linear Tune Variation

If the betatron tune varies with amplitude as

$$v = v_o + \mu x^2, \qquad (2)$$

then any radial locus of points in phase space is continually transforming in a spiral, adding one turn to the spiral every $(\mu x^2)^{-1}$ turns. Thus, a circular beam of phase-space radius A that is kicked into a collective oscillation of amplitude a will eventually smear out into a circle of radius $A + a$. For the SSC μ is expected [3] to be about $-4.8 \times 10^{-5}\,\mathrm{mm}^{-2}$ and the rms beam width 0.12 mm, so that the characteristic time for non-linear smearing will be about 1.4×10^6 turns, or 7 minutes.

B. Amount of Emittance Growth

If the amplitude of collective oscillation is a, Syphers [4] has shown that the emittance dilution factor F at equilibrium is

$$F \equiv \epsilon/\epsilon_o = 1 + (a/\sigma)^2/2, \qquad (3)$$

where σ is the rms beam width. With random excitation, the collective amplitude after N turns is expected to be $a_1\sqrt{N}$, where a_1 is the rms increase in amplitude per turn. (The case of sinusoidal excitation will be treated in a later section.)

II. Amplitude of Collective Oscillations Produced by Ground Motion

If a single quadrupole (of focal length f_Q) at the point s_o in an otherwise perfect machine is suddenly moved by the amount δx, the resulting trajectory of a bunch that had been traveling on the ideal central orbit is given by equation (1) with the angle α given by $\alpha = \delta x/f_Q$. However, the bunch will no longer be on the closed orbit, which now becomes

$$\vec{\eta}_{co} = (\delta x/f_Q)(\sqrt{\beta_o}/2\sin \pi v)e^{-i(\phi - \pi v)}. \qquad (4)$$

The collective oscillation is the difference between equations (1) and (4):

$$\vec{\eta}_{coll} = -(\delta x/f_Q)(\sqrt{\beta_o}/2\sin \pi v)e^{-i(\phi+\pi v)}. \qquad (5)$$

Thus, it can be seen that the amplitude of the collective oscillation induced by the *elementary* motion of a quadrupole is equal in magnitude to the corresponding contribution to the induced closed orbit.

If now the quadrupole is moved again during the revolution time of the beam, the consequent change in the *closed orbit* will add *linearly* to that produced by the first motion. However, the two contributions to the *collective oscillation* must be added *vectorially*, since they will differ in phase by $2\pi v$.

*This work supported by Fermi National Accelerator Laboratory, for the U.S. Department of Energy, under Contract No. DE-AC02-76CH03000.

[†]Operated by the Universities Research Association, Inc., for the U.S. Department of Energy under Contract No. DE-AC02-89ER40486.

Because of the fixed relationships in amplitude and in phase between the contributions to the collective oscillation and to the closed orbit produced by the elementary motion of a quadrupole, we can conclude that the collective amplitude produced by the elementary motion of the 1,000 quadrupoles in each SSC ring is equal to the corresponding increment in the overall closed orbit. We have already computed the closed orbits for each ring and combined them to find the beam separations at the crossing points for several types of ground waves at many frequencies [5]. For each case we can conclude that the rms incremental amplitude of collective oscillation is equal to $(2\sqrt{2})^{-1}$ of the maximum incremental beam separation. Thus, at a typical ground-wave frequency $(\omega_g/2\pi)$ we obtain

$$a_1 = \omega_g\, x_g/(2\sqrt{2}f_o), \qquad (6)$$

where x_g is the beam separation due to ground motion at that frequency at a reference crossing point, and f_o is the revolution frequency in the SSC (3.44 kHz). The resultant emittance dilution factor then is

$$F = 1 + \omega_g{}^2\, x_g{}^2 T/(16 f_o \sigma^2), \qquad (7)$$

where T is the time period of interest.

III. Growth from Measured Ground Motions

Measurements [6] were made at tunnel depth at the SSC site of ground motions due to trains crossing overhead, blasts from nearby quarries, and ambient background. All showed very ragged, irregular wave forms so that we feel that equation (7) should apply.

A. Growth Due to Train Crossings

The maximum beam separation due to trains [5] was $0.10\,\mu\mathrm{m}$ caused by ground waves in the 3-Hz band plus $0.066\,\mu\mathrm{m}$ from a band near 7 Hz. The rms width of the beams at this interaction point is $4.8\,\mu\mathrm{m}$. For a time T of 120 seconds, the dilution factor (adding the two contributions in quadrature) is $1 + 1.1 \times 10^{-3}$, which would not be observable. If the quadrupole displacements should be amplified by a factor of, say, 5 due to the magnet support structures, the emittance growth would be $1 + 2.8 \times 10^{-2}$—still not an important effect.

B. Growth Due to Quarry Blasts

The maximum beam separation due to quarry blasts was found to be $0.93\,\mu\mathrm{m}$ due to ground waves in the 1-Hz band plus $1.12\,\mu\mathrm{m}$ from the 3-Hz region. The geophone detectors showed relatively large ground motion for about 30 seconds in a typical section of the ring. The resulting dilution factor is $1 + 1.14 \times 10^{-2}$, which is not a worrisome level. However, an amplification factor of 5 due to the quadrupole supports would raise this dilution factor to $1 + 0.28$, which would be significant, since the luminosity would drop by the same factor permanently if corrections could not be made promptly.

C. Growth Due to Ambient Ground Motion

Measurements of the ambient ground motion at several tunnel positions around the ring at different times varied over two orders of magnitude. The average amplitude was $0.015\,\mu\mathrm{m}$, the typical frequencies were in the region of 3 Hz, and the wave-forms were ragged and irregular. Since we know little else about these waves, we have tried to make a conservative estimate of their effects by assuming that they are due to random plane waves having an unfavorable direction and negligible attenuation. For such waves our program [5] predicts a maximum separation of about $0.03\,\mu\mathrm{m}$ at a low-beta crossing. For a 12-hour period the resulting emittance dilution factor is $1 + 1.1 \times 10^{-2}$, which is about the same as for the worst-case quarry blast. Again an amplification factor of 5 would raise this dilution factor to a significant level. It would be harder to correct in this case because the growth of the collective oscillation per characteristic smearing period is so much smaller that it would be difficult to measure. More complete data, including wave direction and velocity, are needed.

IV. Growth due to a Beam-Centering Scheme and the Beam-Beam Effect

In the Jostlein [1] beam-centering scheme, one beam is rotated about the other at an interaction point, and the resulting variation in luminosity serves to measure the amount and direction by which the two beam centers miss each other. In this situation each beam acts on the other to first order like a moving quadrupole of focal length $f_Q = \beta_o/(4\pi\Delta\nu_{bb})$, where $\Delta\nu_{bb}$ is the head-on beam-beam tune shift. On the nth turn each beam receives a kick:

$$d\vec{\eta} = ia_n \equiv i\left(b_m\sqrt{\beta_o}/f_Q\right)\sin(2\pi\nu_m n), \qquad (8)$$

where b_m is the amplitude of the sinusoidal beam modulation at relative frequency $\nu_m \equiv f_m/f_o$ cycles per revolution.

Consider a pencil of beam whose vector amplitude at turn zero is $\vec{A}$. At turn N its amplitude will be:

$$
\begin{aligned}
\vec{\eta}_N &= \left(\dots\left(\left((\vec{A}+ia_o)e^{-2\pi\nu_o}+ia_1\right)e^{-i2\pi\nu_1}\right.\right.\\
&\qquad \left.\left. +ia_2\right)e^{-i2\pi\nu_2}+\dots+ia_N\right)e^{-i2\pi\nu_N}\\
&= \vec{A}e^{-i2\pi\sum_0^N \nu k}+i\sum_0^N a_n e^{-i2\pi\sum_n^N \nu k}, \qquad (9)
\end{aligned}
$$

where ν_k is the net betatron tune on the kth turn.

A. Chromaticity Effects

Consider first the variation of betatron tune with energy:

$$\nu_k = \nu_o + \delta E\,\xi\,\sin(2\pi\nu_s k + \theta), \qquad (10)$$

where θ is the phase in synchrotron phase space at turn zero. In this case equation (9) can be summed approximately to give the spreading in phase space, relative to an on-momentum particle, as:

$$\vec{\eta}_N = \left[2\pi b_m \sqrt{\beta_o}\,\delta_E\,\xi/(f_Q \sin \pi v_s)\right] e^{-i2\pi v_o N}$$
$$\left\{\left[\cos(2\pi v_s N + \theta)\sin \pi v_+ N\, e^{i\pi v_+ N}/(2\sin \pi v_+)\right.\right.$$
$$- \sin \pi(v_+ + v_s)N\, e^{i\pi(v_+ + v_s)N + i\theta}/(4\sin \pi[v_+ + v_s])$$
$$\left.- \sin \pi(v_+ - v_s)N\, e^{i\pi(v_+ - v_s)N - i\theta}/(4\sin \pi[v_+ - v_s])\right]$$
$$- [\text{same with } v_+ \longrightarrow v_-]\},\tag{11}$$

where $v_\pm \equiv v_o \pm v_m$. This expression describes a complicated motion that repeatedly goes through zero with a period of $1/v_s$ turns. It represents a breathing motion whose maximum amplitude, evaluated at the collision point, is approximately

$$|x_{\max}| = 2\pi v_m b_m \beta_o \delta E \xi/(f_Q \sin \pi v_s \sin^2 \pi v_o).\tag{12}$$

Thus the motion is bounded. For the case $f_m = 20\,\mathrm{Hz}$, $\beta_o = 0.5\,\mathrm{m}$, $\delta E = \sigma_E = 6 \times 10^{-5}$, $\xi = 5$, $\Delta v_{bb} = 0.004$ ($\longrightarrow f_Q = 10\,\mathrm{m}$), $v_s = 1.16 \times 10^{-3}$, and $v_o = 123.4$, $|x_{\max}| = 0.57 b_m$. Thus, if the amplitude of modulation is a small fraction of the beam size, the increased emittance due to the chromaticity tune spread is negligible, provided that no resonances are involved.

Simulations in which particles are followed according to equations (9) and (10) are in excellent agreement with these conclusions.

B. Non-Linear Tune-Shift Effects

Using the non-linear tune dependence of equation (2) in equation (9) and assuming relatively small quadrupole motion, we obtain for the vector amplitude at the Nth turn the approximate expression:

$$\vec{\eta}_N = e^{-i2\pi N v_A}\left\{\vec{A} + \left[b_m \sqrt{\beta_o}/(2f_Q)\right]\right.$$
$$\left[e^{i\pi v_+ N}\sin \pi v_+ N/\sin \pi v_+\right.$$
$$\left.\left.- e^{i\pi v_- N}\sin \pi v_- N/\sin \pi v_-\right]\right\},\tag{13}$$

where $v_A = v_o + \mu A^2$, and here $v_\pm \equiv v_A \pm v_m$. Thus, all particles of the same absolute initial amplitude $|A|$ have

the same relative motion, so that no smearing occurs, and the beam motion induced by the sinusoidal motion of the "beam-beam quadrupole" averages to zero, again provided that no resonances are involved. Again, simulations using equations (2) and (9) are in excellent agreement with these conclusions.

C. Possible Disagreement with Observations

These results on the effect of sinusoidal modulation of one beam relative to the other at a crossing point may be in disagreement with the experimental observations [7] at CERN, where beam effects were observed at modulation frequencies in the range 2 to $10\,\mathrm{kHz}$, well below the betatron frequency band (13.5–$14.5\,\mathrm{kHz}$) in the SPS collider. However, they also observed in the beams higher harmonics of the modulation frequencies, and they found that the modulation equipment produced higher harmonics of the excitation frequencies. Since the deleterious beam effects could have been due to harmonics that fell in the betatron frequency band, it is still not clear experimentally whether sinusoidal modulation in the presence of the beam-beam effect causes appreciable emittance degradation. No such effects were seen at modulation frequencies below $2\,\mathrm{kHz}$.

V. References

[1] H. Jostlein, "Automatic Beam Centering at the SSC Interaction Regions," Fermilab TM-1253, March 1984.

[2] R. Meller et al., "Decoherence of Kicked Beams," SSC-N-360, May 1987.

[3] Y. Yan, private communication, June 1990.

[4] M. Syphers, "Injection Mismatch and Phase Space Dilution," Fermilab FN-458, June 1987.

[5] K.-Y. Ng and J. Peterson, "Effects from Measured Ground Motions at the SSC," SSCL-277, Dec. 1990.

[6] K. Hennon and D. Hennon, "Field Measurements and Analyses of Underground Vibrations at the SSC Site," SSC-SR-1043, Dec. 1989.

[7] K. Cornelis et al., "Experimental Study of a Beam Excitation in the Presence of the Beam-Beam Interaction," SPS/AMS 89-04, Jan. 1989.

Adiabatic Invariance for Spatially Dependent Accelerating Structures

John R. Cary and David L. Bruhwiler
Astrophysical, Planetary and Atmospheric Sciences Department
University of Colorado
Boulder, Colorado 80309-0391

Abstract

The adiabatic dynamics of charged particles in accelerating structures is significantly altered if these structures vary in space rather than in time. Spatial variation occurs in, for example, free-electron lasers and radio-frequency quadrupoles. The adiabatic invariants for slow temporal and slow spatial variations differ. This causes the longitudinal emittance of adiabatically trapped beams to differ in the two cases. The number of particles trapped in each accelerating bucket also differs. We present analytic and numerical results to clarify these ideas.

I. INTRODUCTION

For a general Hamiltonian $H(q,p,\varepsilon t)$, which depends slowly on the time variable, the adiabatic invariant is the action:

$$I(E,\varepsilon t) \equiv \oint p|_{H=E} \, dq. \tag{1}$$

The action is the phase-space area enclosed by a contour of H at fixed time t. *Slowly* means that the particle executes many oscillations before the parameters of the Hamiltonian change significantly (indicated formally by taking $\varepsilon \ll 1$). Invariance of the action allows one to solve for the energy at one time in terms of the energy at another time. Adiabatic theory has also been used to predict the phase-space area occupied by a beam adiabatically trapped in an accelerating potential.

We analyze the case of adiabatic, spatially varying, accelerating structures. We show that the usual adiabatic invariants for the temporally varying case are not invariant in the spatially varying case. This affects, in particular, the arguments that give the area occupied by a beam that has been adiabatically trapped in a *spatially* varying structure. Our results apply to problems such as beam trapping in the accelerating potential of an RFQ[1] or in the ponderomotive (decelerating) potential of free-electron lasers[2] and to the rf heating of a tokamak plasma.[3]

II. SLOW TEMPORAL VARIATION

The *wave Hamiltonian,*

$$H(q,p,\varepsilon t) = \frac{p^2}{2m} + e\Phi(\varepsilon t) \cos(k[q - u(\varepsilon t)t]). \tag{2}$$

describes the longitudinal motion in an accelerating potential. Here, e and m are the mass and charge of the particle. The amplitude of the potential is Φ. The particles oscillate in a trough of phase velocity u and length $2\pi/k$.

This work is supported by the U.S. Department of Energy and by a grant from the San Diego Supercomputer Center.

For trapped particles the adiabatic invariant is the phase-space area given by the usual closed-loop integral $\oint p \, dq$. However, for passing (or untrapped) particles, the loop integral is over one period (from kq=0 to kq=2π). We are, thus, able to define three different areas, or action functions, $I_{\pm}(E,\varepsilon t)$ and $I_T(E,\varepsilon t)$, which give the loop integrals along the contour of energy E at time t for particles passing above (+) or below (-) or trapped (T) inside the stable region.

This information determines the longitudinal emittance (phase-space area) of a beam that becomes trapped. We suppose for example in Fig. 1 that initially the amplitude of the wave vanishes, and the beam is at momentum $p_i = m(u+\Delta)$. At this time the value of the adiabatic invariant is the area under the curve $p = p_i$: $2\pi m(u+\Delta)/k$. As the amplitude grows (with u remaining constant), the contour of H for these particles distorts, but has below it always the same area. However, eventually there comes a time, called the crossing time t_x, when the area under the more positive part of the separatrix equals the initial value of the adiabatic invariant. Beyond this time there is no passing-particle curve having the same area. Thus, the particles must become trapped. Furthermore, these particles are on and remain on a trapped contour of H containing area $(16m/k)A_x^{1/2}$, where $A_x \equiv A(\varepsilon t_x) = e\Phi(\varepsilon t_x)/m$, since that is the area enclosed by the separatrix at the time of crossing.

These results determine the phase-space area occupied by an adiabatically trapped beam of particles. A beam extending from p=mu to p=m(u+\Delta) occupies an area per wavelength of $2\pi m\Delta/k$. As the wave amplitude increases from zero, first the particles with p=u become trapped. As the amplitude grows, the larger momentum particles become trapped. The largest momentum particles become trapped at an amplitude A_x satisfying $2\pi\Delta = 8A_x^{1/2}$. This result is obtained by equating the initial action with the value of the action on the separatrix.

III. ADIABATIC THEORY FOR SPATIAL VARIATION

For linear structures, such as RFQ's or free-electron lasers, the Hamiltonian (2) is not appropriate. The potential instead has an amplitude that varies spatially:

$$H(q,p,t) = \frac{p^2}{2m} + e\Phi(\varepsilon q) \cos(k[q - u(\varepsilon q)t]). \tag{3}$$

Now the particles see an amplitude growing as they enter the accelerating structure. After being trapped in a stable bucket they may be accelerated or decelerated depending on how the phase velocity u changes with position.

To determine the new adiabatic invariant we introduce the phase-space variational principle, which states that the correct dynamics makes stationary the integral,

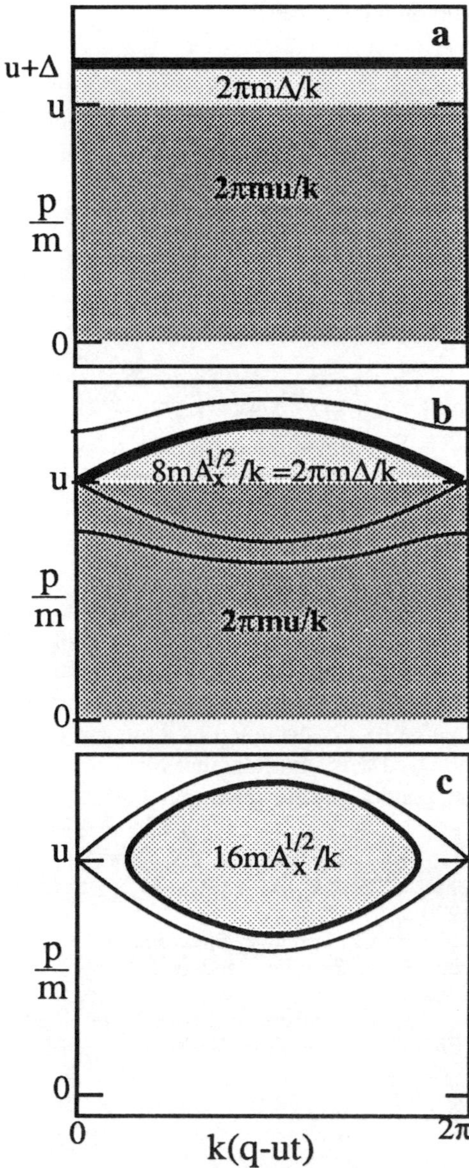

Fig. 1. Adiabatic trapping of a beam in time.

$$\mathcal{C} \equiv \int [p\, dq - H(q,p,\varepsilon t)\, dt], \qquad (4)$$

of the phase-space differential action, $d\mathcal{C}\equiv pdq-Hdt$. For historical reasons the quantity $\mathcal{C}$, which is a line integral along the trajectory, is also known as the action. The other action (1) is related to $\mathcal{C}$ by being the integral around a closed loop of constant H at constant t. The Euler-Lagrange equations for the functional (4) are exactly Hamilton's equations. Adiabatic theory follows when the Hamiltonian H is a slow function of the independent variable (t). In this case, the loop integral of the other conjugate pair (q,p) at constant slow pair (t,H) is the adiabatic invariant.

For the spatially varying case, $\mathcal{C}$ has the following form:

$$\mathcal{C}_q \equiv \int [p\, dq - H(q-ut,p,\varepsilon q)\, dt], \qquad (5)$$

which is not the form required by adiabatic theory. However, we can put $\mathcal{C}_q$ in the appropriate form by first subtracting the total differential $mu^2 dt/2$, then adding and subtracting the term $(pu\,dt)$ to obtain

$$\mathcal{C}_q \equiv \int [(p+K/u)\, d(q-ut) - (K/u)\, dq] , \qquad (6)$$

where

$$K(q-ut,p,\varepsilon q) \equiv \frac{(p-mu)^2}{2m} + e\Phi(\varepsilon q)\cos(k(q-ut)) \qquad (7)$$

is equal in value to the Hamiltonian obtained by transforming to the frame moving with the phase velocity.

The action integrand is now in the form needed to determine the spatial adiabatic invariant. The new momentum $z\equiv p+K/u$ is conjugate to the fast variable $q-ut$. The new Hamiltonian K depends slowly on the independent variable q. Thus, the new adiabatic invariant is given by the integral,

$$J = \int (p+K/u)\, d(q-ut) = \int p\, d(q-ut) + (K/u)\int d(q-ut) , \qquad (8)$$

along a contour of constant K. The last equality in Eq. (8) follows from the fact that K is held constant in the integration.

For trapped particles, the phase (q-ut) begins and ends at the same point. Hence, the second term vanishes and we have,

$$J_T(I) = \oint p\, d(q-ut) \equiv 2I. \qquad (9)$$

The last equality defines the variable I, which for trapped particles is simply half the enclosed area. This factor of 1/2 makes I a continuous phase-space variable.

For passing particles, the phase q-ut does not begin and end at the same value, but changes by $2\pi/k$. Thus, the adiabatic invariant for $\pm$ passing particles is given by

$$J_\pm(I,\varepsilon q) = I + 2\pi K(I,\varepsilon q)/ku , \qquad (10a)$$

where

$$I(\varepsilon q,K) \equiv \int_0^{2\pi} (p-u)\, d(q-ut) \qquad (10b)$$

is the integral along a contour of constant K. The convention chosen here of integrating in the direction of increasing phase q-ut makes I a signed quantity; it is positive (negative) for positive (negative) passing particles.

IV. Trapping in the Spatially Varying Case

Invariants of the motion permit understanding of the dynamics without solving the differential equations. Because the adiabatic invariant $J(I,\varepsilon q)$ (referring collectively to the three functions $J_\pm$ and J_T) is conserved, the trajectory stays on contours of this function. This allows us to determine the value of I as the particle moves through the accelerating structure.

We show in Fig. 2 a contour plot of $J(I,\varepsilon q)$, with contours divided into four regions by the I axis and the heavy curves outlining what appear to be lips. This outline indicates the location of the separatrix, which is given by $I=\pm 8(m/k)A^{1/2}$. The lips vanish at large distances, where the accelerating potential vanishes. Above the lips are the positive passing particles, and below the lips are the negative passing particles.

The trapped particles are in the upper lip, because for trapped particles I is positive according to Eq. (9). The lower lip is an unphysical region.

We first discuss the trajectories in Fig. 2 which do not intersect the lips (i.e. do not become trapped). The top orbit, labeled (a), is an orbit that passes right over the structure. Similarly, the orbit labeled (h) is a left moving orbit that passes over the structure in the opposite direction. The orbit (g) and its mirror reflection on the right begin moving into the potential, but then are reflected. The corresponding particles are reflected by the ponderomotive potential[4] associated with the oscillating electric field. The orbit (f) moves into the accelerating structure, but its velocity is sufficiently negative of the phase velocity that it does not become trapped by the potential.

Now we turn to the trajectories that do become trapped. The trajectory (b) collides with the separatrix, at which point the corresponding particle becomes trapped. For trapped particles, I is the adiabatic invariant, and so I remains constant. Eventually this orbit exits the structure at the other side, becoming untrapped. As indicated by the arrows, this orbit may exit onto a positive passing trajectory (c) or a negative passing trajectory (e) connected by the downward arrow. That is, the trajectory can exit out the top half of the phase-space separatrix or the bottom half. This phenomenon, *beam splitting*, implies that a single beam becomes two beams after trapping and detrapping take place. The time reversed phenomenon is that there are two trajectories, (b) and (d), which trap at the same action.

The trapping areas follow from application of this adiabatic theory. We consider a beam of particles with momentum $p_i = m(u+\Delta)$ far from the accelerating structure. Such particles are in the positive passing state and have an initial value of the adiabatic invariant given by,

$$J_{+i} = (2\pi m/k)\left(\Delta + \tfrac{1}{2}\Delta^2/u\right).\qquad(11)$$

These particles trap when the value of the positive passing adiabatic invariant on the separatrix,

$$J_{+sx} = 8mA^{1/2}/k + 2\pi mA/ku,\qquad(12)$$

equals the initial value of the adiabatic invariant. Equating (11) and (12) shows that these particles become trapped at an amplitude A_x satisfying

$$\Delta + \tfrac{1}{2}\Delta^2/u = 4A_x^{1/2}/\pi + A_x/u,\qquad(13)$$

which implies that these particles are spread throughout an area

$$16mA_x^{1/2}/k =$$

$$(32mu/k\pi)\left(\sqrt{1+\pi^2\Delta/4u+(\pi\Delta/2u)^2/2} - 1\right).\qquad(14)$$

For small values of the momentum difference Δ (and, therefore, trapping amplitude), (14) reduces to the result obtained in the time varying case.

V. CONCLUSIONS

The case of adiabatic, spatially varying accelerating structures differs significantly from that of temporally varying adiabatic structures. The adiabatic invariant for passing particles is not the action (10b), but rather the "new" adiabatic invariant (10a). The phase space area finally occupied by a trapped beam in the temporally varying case is invalid; that result is replaced by (14). In addition, ponderomotive reflection appears.

VI. REFERENCES

1   I. M. Kapchinskii and V. A. Teplyakov, Prib. Tek. Eksp. **119**, 19 (1970).
2   C. W. Roberson and P. Sprangle, Phys. Fluids **1**, 3 (1989).
3   W. M. Nevins, T. D. Rognlien, and B. I. Cohen, Phys. Rev. Lett. **59**, 60 (1980).
4   H. Motz and C. J. H. Watson, in *Advances in Electron and Electronic Physics* (Academic, New York, 1967), Vol. 23, p. 168.

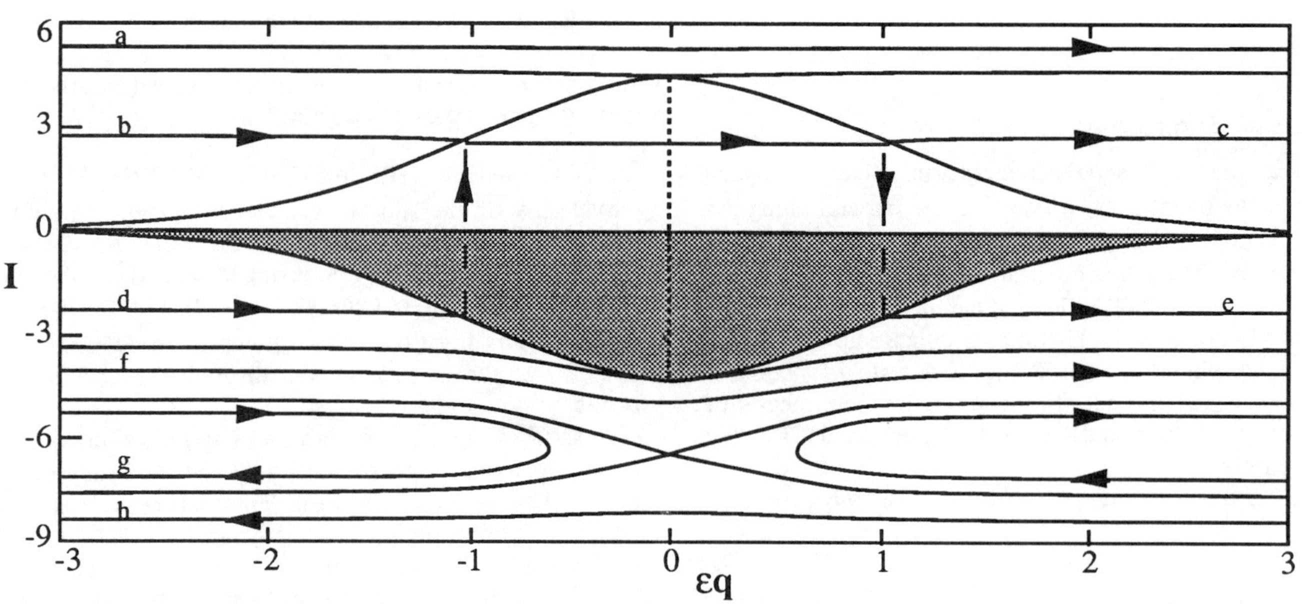

Fig. 2. Contours of the spatial adiabatic invariant for Φ gaussian in shape and $ke\Phi_{max}=0.4mu^2$.

Particle Tracking and Map Analysis for Compact Storage Rings

Michael F. Reusch,

Grumman Aerospace Corporation, 4 Independence Way, Princeton, NJ, 08540,

Etienne Forest,

Exploratory Studies Group, Lawrence Berkeley Laboratory, Berkeley, CA 94720

and James B. Murphy,

National Synchrotron Light Source, Brookhaven National Laboratory, Upton, NY, 11973

Abstract

Particle tracking codes for large storage rings approximate the Hamiltonian by neglecting terms of order x/ρ and higher. For storage rings with small bending radius magnets like the Super Conducting X-ray Lithography Source (SXLS) [1], which make use of combined function bending magnets, these approximations cannot be made. We use an explicit symplectic integrator [2] to construct a tracking code which uses the exact Hamiltonian for drifts and isomagnetic combined function bending magnets, in a manner similar to the Teapot code [3]. Hard edge fringe fields are included in a symplectic manner for dipoles and quadrupoles. The integrator is coupled to the DA package of Berz [4] to provide an arbitrary order map which can be analyzed using the tools of Forest and Irwin [5]. A discussion of the techniques and an application to the SXLS ring at Brookhaven is presented.

I. Introduction

The present note describes our modeling of a compact synchrotron xray lithography source, the SXLS device [1]. The first phase of this device, which incorporates conventional electromagnets, is in operation at Brookhaven National Laboratory and is described in a separate presentation at this meeting. SXLS, which is being developed at BNL in cooperation with Grumman Aerospace Corporation under the auspices of the Defense Advanced Research Projects Agency, [1] is intended to eventually serve as a commercial x-ray source for the lithographic production of computer chips.

The second phase of this device will incorporate 3.87 Tesla superconducting combined function bending magnets which are being constructed by General Dynamics Corporation. The design of these superconducting magnets and their effect on the optics of the phase II device is our principal concern.

SXLS has an 8.5 meter circumference a 60 centimeter bend radius and an unique gradient FODO-like lattice structure. The small size of the device implies that some approximations made in large-machine codes are invalid, in particular, the dropping of terms in the Hamiltonian of order x/ρ and higher, where x is horizontal de-

viation from the closed orbit and ρ is the bending radius of the design orbit. Analysis is further complicated by the strong effect of high-order magnetic field multipoles, the combined function nature of the dipole magnets and the relatively large fraction of the device circumference occupied by fringe fields.

We have not yet addressed all of these problems in our approach to modeling SXLS. For example, our analysis of non-isomagnetic effects is currently done through Marylie 3.0 and a Genmap-like code [6] tailored specifically for SXLS. However, we have begun to develop a computational tool, the Krakpot code, which, we believe, will eventually let us answer many of our unresolved questions.

The principal result of our work is that in devices of this small size it is important to make the high order field multipoles as small as possible. Failure to do so reduces the dynamic aperture. In order to determine this result, we have had to push the state-of-the-art in both magnetic field calculations and tracking code design.

II. Overview of Krakpot

Krakpot, so named because of both because of its origins in Teapot[3], is an analysis code, not a design code. At present, it incorporates • a drift, • a combined function straight element, • a combined function bend element, • a simple RF cavity kick, and • hard-edge fringe elements.

The combined function elements are all of the isomagnetic type. In each of the elements the Hamiltonian is split into one or two exactly soluable pieces as in Teapot [7], and the resulting equations of motion used to implement an explicit canonical integrator.

Quadratic, fourth order [2], and sixth order integrators are all implemented within the code, and higher order integrators are possible. Also present in the code are a routines for the tracking of orbits, determination of the dynamic aperture, and a six dimensional fixed point finder.

The code is intimately linked to the numerical differential-algebra (DA) package [4] of Berz. [2] A "DAified" version of each element and each canonical integrator exists. An arbitrary order Taylor map through a succession of elements can be generated and saved for later analysis.

[1] This work was performed under the auspices of the U.S. Department of Energy and funded by the U.S. Department of Defense

[2] This DA package was developed by Dr. Martin Berz at Lawrence Berkeley Laboratory in 1987 and 1988. It has been considerably modified by the LBL Exploratory Studies Group. The original author cannot be held responsible for its contents.

The map analysis is carried out through post-processor codes based on the LIELIB package of Forest [5]. These extract the normal form of the map, yielding the tunes, chromaticities, geometric tune shift with amplitude and other dynamical lattice parameters.

III. DODECAPOLE ORDER COMBINED FUNCTION DIPOLE

The Hamiltonian for an isomagnetic combined function bending magnet can be written in cylindrical coordinates, with longitudinal distance as the independent variable of integration, as a sum of two explicitly soluable pieces, a drift-like term

$$H_d = -(1 + x/\rho)\sqrt{(1+\delta)^2 - P_x^2 - P_y^2} - \frac{P_t}{\beta_0}, \quad (1)$$

and a kick

$$H_k = -(1 + x/\rho)\frac{A_s(x,y)}{B_0\rho}. \quad (2)$$

In Krakpot, the midplane field expansion for A_s has been carried out to dodecapole order (x^5) with MACSYMA using a stream function method [8] developed for the treatment of non-isomagnetic bends.

An exact solution of the equations of motion of each of these Hamiltonians is straightforward and given by Forest [7]. Incorporation of the solutions into an explicit canonical integrator is straightforward. A quadratic integrator results by applying the drift transformation for half a time step, the kick transformation for a full time step, and then the drift transformation again for half a time step.

One minor difficulty of this method in the combined function bend is that the closed orbit, for which initially x, y, t, P_x, P_y, P_t are all zero and $P_s = 1$, is not sent into the closed orbit. The size of the closed orbit error is given by the order of the integration method. A quadratic integrator applied to the isomagnetic combined function bend yields, to lowest order, closed orbit errors of

$$P_x(s) \approx \frac{s^3}{6\rho^3}, \quad x(s) \approx \frac{s^4}{12\rho^3}. \quad (3)$$

A simple subtractive procedure is adopted in Krakpot to avoid buildup of the closed orbit error over many timesteps.

For a given element, the closed orbit errors from a single time step are calculated and saved. Thereafter, these saved errors are subtracted from the result of every time step for a general orbit. This subtraction procedure is symplectic, since it is only a translation of the origin of phase space, retains the order of accuracy of the integration method, and preserves the closed orbit to machine precision.

In the DAified versions of these transformations, the six canonical dependent variables (x, y, t, P_x, P_y, P_t) are declared as independent DA variables. Arbitrary order derivatives of the canonical transforms with respect to these variables can then be manufactured for the whole lattice by concatenation of transforms from single time steps. An Nth order Taylor map generated in this fashion is automatically canonical up to terms of order $2N + 1$, which is the order of the error term in the Poisson brackets of the Taylor map.

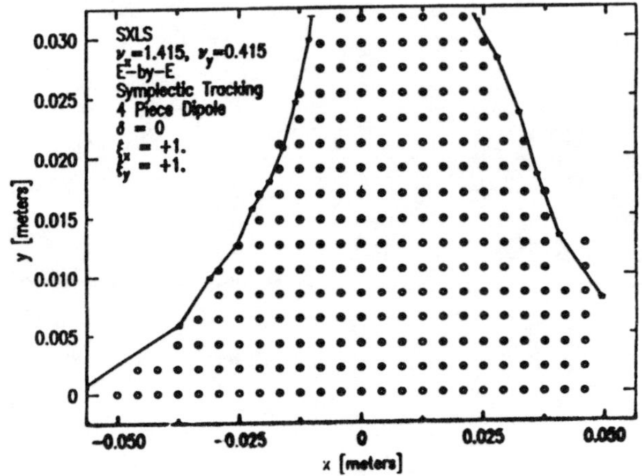

Figure 1: Comparison of Krakpot and Marylie Isomagnetic Apertures

IV. COMPARISON TO MARYLIE

Several codes have been applied to SXLS and disagree in their predictions of even the first-order chromaticity. The Marylie 3.0 code makes minimal approximations and is well suited for comparison with Krakpot.

Figure 1 overlays dynamic apertures found using isomagnetic elements with Marylie and Krakpot. Horizontal and vertical tunes were parked at the design values of 1.415 and 0.415 respectively. Sextupole elements in the lattice were adjusted to yield a chromaticity of 1.0 in both planes. On energy orbits were tracked symplectically for 1000 turns through the lattice. Tracking was done on an element by element basis in Marylie using the circulate command. The dipole was split into four pieces to improve accuracy. Each element in the Krakpot code was divided into as many as one hundred explicit canonical transformations. Even though the tracking methods used differ considerably, the apertures agree. This agreement extends to the values of nonlinear lattice parameters found by a normal form transformation of the one turn maps, which can be carried out to only third order with Marylie.

V. RESULTS FOR SXLS

The chief impetus for the development of Krakpot was to obtain a reliable tool for the modeling of high order effects not treatable with the Marylie 3.0 code. Based on previous code results, we expected that bending magnet multipoles, of octupole order and above, would have a significant effect on the SXLS dynamic aperture. This has turned out to be the case, and has a considerable impact on the design of the bending magnet.

The results presented below were all generated with Krakpot using isomagnetic elements and hard edge fringes. Since tracking was carried out using explicit canonical integration rather than a taylor map, the only approximations in order are made in the expressions for the vector potential.

Figure 2 gives the variation of on energy dynamic aperture as a function of the magnitude of octupole (B_3) in the 180 degree bending magnet.

The improvement of aperture with the addition of a small amount of octupole shown in Fig. 2 may explained

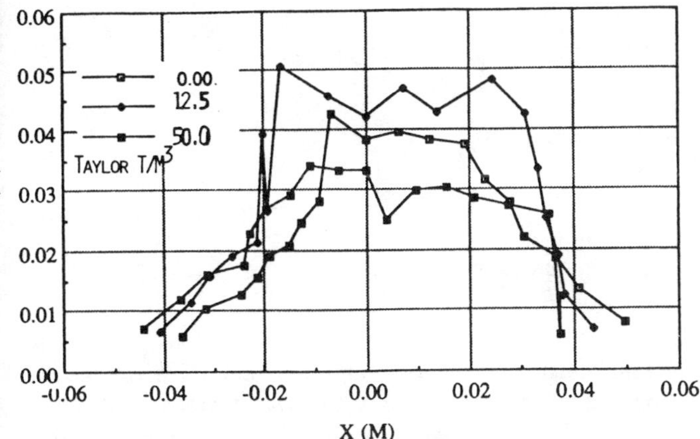

Figure 2: Variation of SXLS Aperture with Octupole

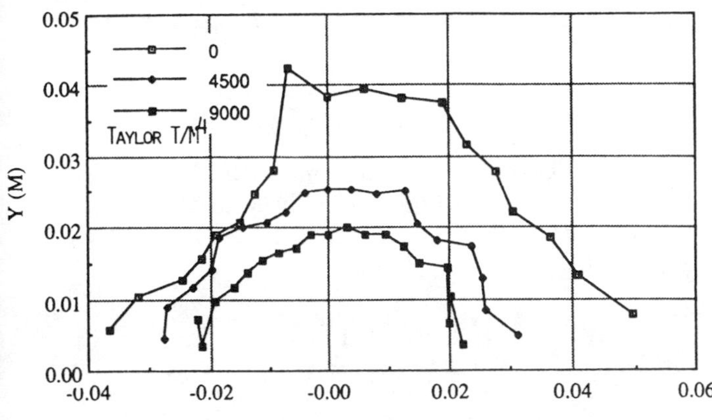

Figure 3: Variation of SXLS Aperture with Decapole

by the form of the vertical tune shifts with amplitude as a function of the octupole strength, $\Pi_{yy} = -28.16 + 7.08\,B_3/B_0$. Vertical aperture can be increased by reducing Π_{yy} to zero with an appropriate choice of $B_3/B_0 \approx 4$ with relatively little effect on the horizontal aperture.

Aperture variation with decapole is given in figure 3. Decapole components on the order of $10^4\,T/M^4$ are quite likely in small radius, combined function, bending magnets designed without regard for higher order multipoles.

Trim windings, mainly for controlling the quadrupole and sextupole moments, have been included in the design of the SXLS phase II magnet because of our concern about some of these effects.

VI. DA Applied to Magnetic Field Calculations

We have written a new code which applies the DA package to the Biot Savart law, in essence, differentiating under the integral sign. The output of the code is not only the three components of the magnetic field at a point but also the arbitrary order spatial derivatives of the field components at the point, that is, the field moments. This procedure gives a more accurate local estimate of the moments than may be obtained from the alternative procedure of fitting a polynomial to the field at several points. We plan to apply this code to fringe regions in the Krakpot code.

A bonus of this approach is that the conductor positions, or combinations of them, may be declared as DA variables, yielding the derivatives of the various moments with respect to conductor position. Mechanical tolerances are immediately established based on the maximum tolerable amplitude of moments. Precise combinations of moments may be sought by automated iterative adjusment of the conductor positions.

VII. Conclusions

Compact electron storage rings, like the BNL Superconducting X-ray Lithography Source, present a significant challenge to the accelerator designer. A novel "Krakpot" code has been described, which has been developed to overcome some of the difficulties presented by these rings. Some of the techniques employed in this code may find an application in other accelerators.

References

[1] J.B. Murphy et alii, The Brookhaven Superconducting X-Ray Lithography Source (SXLS), Proc. 2nd EPAC, p1828 (1990).

[2] E. Forest and R. D. Ruth, Fourth-Order Symplectic Integration, Physica D **43** 105 (1990)

[3] L. Schachinger and R. Talman, TEAPOT - A Thin Element Accelerator Program for Optics and Tracking, SSC Central Design Group SSC-52 (1986)

[4] Martin Berz, The Description of Particle Acclelerators Using High-Order Perturbation Theory on Maps, AIP Conference Proceedings 184, Physics of Particle Accelerators Volume I, American Institute of Physics, New York 1989.

[5] E. Forest et alii, Normal Form Methods for Complicated Periodic Systems, Particle Accelerators, 1989, Vol 24, pp 91-107.

[6] L. N. Blumberg et al., Noisomagnetic Third Order Beam Dynamics in the Compact Storage Ring SXLS, Proc. 2nd EPAC p1825 (1990)

[7] E. Forest, Canonical Integrators as Tracking Codes, AIP Conference Proceedings 184, Physics of Particle Accelerators Volume I, American Institute of Physics, New York 1989.

[8] M. F. Reusch, H. Moser, and J.B. Murphy, Non Isomagnetic Bending Magnet Hamiltonian Expansions for Marylie and their Application to the SXLS X-ray Lithography Source, AIP Conference, Washington, D.C., (1990).

Symplectic Full-turn Maps in a Fourier Representation*

J. S. Berg
Department of Physics, Stanford University
Stanford, CA 94305-4060

R. L. Warnock
Stanford Linear Accelerator Center
Stanford University, Stanford, CA 94309

Abstract

We have developed a method that uses an arbitrary symplectic tracking code to generate an exactly symplectic full-turn or multi-turn map. The map is obtained from a generating function, which is a finite Fourier series in the final angle coordinates, the Fourier coefficients being represented as a B-spline series in the initial action coordinates. We achieve fast iteration of this implicitly defined map, and good accuracy. As a first application, we treat a simplified model of arcs of the SSC.

I. Introduction

This paper presents a method for finding a full-turn or multi-turn symplectic map for an arbitrary accelerator lattice. The map is derived in a direct manner from data supplied by an arbitrary symplectic tracking code, unmodified. In contrast to Irwin's construction of symplectic maps [1], our method does not rely on approximation of the map by a succession of kicks and rotations, and avoids convergence questions associated with the use of Taylor series.

The basic idea is to work in action-angle coordinates, and use tracking data to construct a truncated Fourier series in the angle coordinates, with action-dependent coefficients [2]. If the map is represented explicitly by such a series, it is not exactly symplectic. We give a nonperturbative method to find an associated generating function, also represented as a finite Fourier series, that defines implicitly an exactly symplectic map. The symplectic map is iterated numerically by an efficient application of Newton's method, so that the time for iteration is only a little larger than that for the corresponding explicit map.

On a typical orbit of interest, the action variable has relatively little variation, so that the Fourier coefficients can be described locally (in a neighborhood of a given orbit) as rather simple functions of action. We choose to represent the coefficients as quadratic B-spline series, in the interest of fast evaluation. To evaluate the basis functions at a given point, only three quadratic functions per degree of freedom need be computed. Furthermore, one finds that many Fourier coefficients in the truncated series are negligible. Thus, we have a description of the map that is very simple locally. In general, iteration of the map will involve far fewer function evaluations than are required for a power series map of comparable quality.

* Work supported by the Department of Energy, contract DE-AC03-76SF00515

The method works in any number of dimensions. In this paper, a test is made for two-dimensional betatron motion. An extension to include synchrotron oscillations is possible at moderate cost; the Fourier coefficients would have momentum dependence, also represented by B-splines.

II. Explicit Map

We first present an explicit map that is not exactly symplectic. The tracking code gives us the functions $\Theta(\boldsymbol{\Phi}, \mathbf{I})$ and $\mathbf{R}(\boldsymbol{\Phi}, \mathbf{I})$ such that if $\boldsymbol{\Phi} \mapsto \boldsymbol{\Phi}'$ and $\mathbf{I} \mapsto \mathbf{I}'$ over n turns,

$$
\begin{aligned}
\boldsymbol{\Phi}' &= \boldsymbol{\Phi} + \Theta(\boldsymbol{\Phi}, \mathbf{I}) \quad, \\
\mathbf{I}' &= \mathbf{I} + \mathbf{R}(\boldsymbol{\Phi}, \mathbf{I}) \quad,
\end{aligned}
\tag{1}
$$

where $\boldsymbol{\Phi}$ and $\mathbf{I}$ are action-angle coordinates of some underlying "unperturbed" Hamiltonian system (not necessarily the underlying linear system). There may be anomalous regions of phase space in which action-angle coordinates of the linear system are not suitable for our purposes; for instance, a region where y is much larger than x if the Hamiltonian contains sextupole terms, $x^3 - 3x^2 y$ [3]. In such regions, one can apply a simple, preliminary canonical transformation to find suitable variables.

Our algorithm creates a map that takes the form:

$$
\begin{bmatrix} \Theta(\boldsymbol{\Phi}, \mathbf{I}) \\ \mathbf{R}(\boldsymbol{\Phi}, \mathbf{I}) \end{bmatrix} = \sum_{\mathbf{m}} \begin{bmatrix} \mathbf{t_m}(\mathbf{I}) \\ \mathbf{r_m}(\mathbf{I}) \end{bmatrix} e^{i\mathbf{m} \cdot \boldsymbol{\Phi}} \quad,
\tag{2}
$$

$$
\begin{bmatrix} \mathbf{t_m}(\mathbf{I}) \\ \mathbf{r_m}(\mathbf{I}) \end{bmatrix} = \sum_{ij} \begin{bmatrix} \tau_{\mathbf{m};ij} \\ \rho_{\mathbf{m};ij} \end{bmatrix} B_i^{(1)}\left(\sqrt{I_1}\right) B_j^{(2)}\left(\sqrt{I_2}\right) \quad, \tag{3}
$$

where the $B_i^{(d)}$ are quadratic B-splines [4].

The map is constructed by taking a fixed $\mathbf{I}$ and evaluating $\Theta(\boldsymbol{\Phi}, \mathbf{I})$ and $\mathbf{R}(\boldsymbol{\Phi}, \mathbf{I})$ on a uniform $J_1 \times J_2$ mesh in $\boldsymbol{\Phi}$ using the tracking code. A fast Fourier transform gives the coefficients $\mathbf{t_m}$ and $\mathbf{r_m}$ evaluated at this $\mathbf{I}$. Since the tracking code defines $\boldsymbol{\Phi}'$ only modulo 2π, one must first remove the jumps of Θ that occur just after $\boldsymbol{\Phi}'$ reaches 2π. By adding increments of 2π, we make Θ a continuous function, suitable for Fourier analysis. The number J_i of mesh points is taken to be about four times the largest mode number $\max|m_\alpha|$; i.e., about twice as big as the minimum value required by the Nyquist criterion.

Repeating this procedure, we find $\mathbf{t_m}(\mathbf{I})$ and $\mathbf{r_m}(\mathbf{I})$ on a uniform $n_1 \times n_2$ mesh in $(\sqrt{I_1}, \sqrt{I_2})$.

The B-spline coefficients $\tau_{\mathbf{m};ij}$ and $\rho_{\mathbf{m};ij}$ are then found quickly by solving the systems

$$\begin{bmatrix} \mathbf{t_m}(I_{1;\alpha}, I_{2;\beta}) \\ \mathbf{r_m}(I_{1;\alpha}, I_{2;\beta}) \end{bmatrix} = \sum_{i=1}^{n_1} \begin{bmatrix} \tau'_{\mathbf{m};i}(I_{2;\beta}) \\ \rho'_{\mathbf{m};i}(I_{2;\beta}) \end{bmatrix} B_i^{(1)}\left(\sqrt{I_{1;\alpha}}\right) \; , (4)$$

$$\begin{bmatrix} \tau'_{\mathbf{m};i}(I_{2;\beta}) \\ \rho'_{\mathbf{m};i}(I_{2;\beta}) \end{bmatrix} = \sum_{j=1}^{n_2} \begin{bmatrix} \tau_{\mathbf{m};ij} \\ \rho_{\mathbf{m};ij} \end{bmatrix} B_j^{(2)}\left(\sqrt{I_{2;\beta}}\right) \; , \qquad (5)$$

where α and β index the mesh points. The first system (4) can be solved simply by inverting the $n_1 \times n_1$ matrix $A_{i\alpha} = B_i^{(1)}\left(\sqrt{I_{1;\alpha}}\right)$. A similar inversion is required for system (5). Note that since n_1 and n_2 are typically less than 40 or so, these matrix inversions are easy to perform.

III. Generating Function Map

A. General

Now we shall find a map $(\mathbf{\Phi}, \mathbf{I}) \mapsto (\mathbf{\Phi}', \mathbf{I}')$ induced by a generating function, which is exactly symplectic. Recall that the generating function $G(\mathbf{\Phi}', \mathbf{I})$ is defined so that

$$\mathbf{\Phi} = \mathbf{\Phi}' + G_{\mathbf{I}}(\mathbf{\Phi}', \mathbf{I}) \; , \quad \mathbf{I}' = \mathbf{I} + G_{\mathbf{\Phi}}(\mathbf{\Phi}', \mathbf{I}) \; , \quad (6)$$

where subscripts indicate partial differentiation. This implies the identification

$$\Theta(\mathbf{\Phi}, \mathbf{I}) = -G_{\mathbf{I}}(\mathbf{\Phi}', \mathbf{I}) \; , \quad \mathbf{R}(\mathbf{\Phi}, \mathbf{I}) = G_{\mathbf{\Phi}}(\mathbf{\Phi}', \mathbf{I}) \; , \quad (7)$$

where

$$G(\mathbf{\Phi}', \mathbf{I}) = \sum_{\mathbf{m}} g_{\mathbf{m}}(\mathbf{I}) \, e^{i\mathbf{m}\cdot\mathbf{\Phi}'}$$

$$g_{\mathbf{m}}(\mathbf{I}) = \sum_{ij} g_{\mathbf{m};ij} B_i^{(1)}\left(\sqrt{I_1}\right) B_j^{(2)}\left(\sqrt{I_2}\right) \; . \qquad (8)$$

B. Finding Fourier Coefficients

For this subsection, $\mathbf{I}$ will be fixed. Once $g_{\mathbf{m}}(\mathbf{I})$ has been found at several $\mathbf{I}$ mesh points, a B-spline interpolation defines it at all $\mathbf{I}$. Our mathematical problem is this: we want the Fourier series in $\mathbf{\Phi}'$ of a function $F(\mathbf{\Phi})$, where $\mathbf{\Phi}' = \mathbf{\Phi} + \Theta(\mathbf{\Phi})$. The solution is to to make a change of integration variable in the integral defining the Fourier coefficient:

$$f_{\mathbf{m}} = \frac{1}{(2\pi)^2} \int_0^{2\pi} \int_0^{2\pi} F(\mathbf{\Phi}) \, e^{-i\mathbf{m}\cdot\mathbf{\Phi}'} d^2\mathbf{\Phi}'$$

$$= \frac{1}{(2\pi)^2} \int_0^{2\pi} \int_0^{2\pi} F(\mathbf{\Phi}) \, e^{-i\mathbf{m}\cdot\mathbf{\Phi}}$$

$$\cdot e^{-i\mathbf{m}\cdot\Theta(\mathbf{\Phi})} \det\left(1 + \Theta_{\mathbf{\Phi}}(\mathbf{\Phi})\right) d^2\mathbf{\Phi} \; , \qquad (9)$$

where 1 is the 2×2 identity matrix. Now, make a discrete sum approximation to this integral. The result is

$$f_{\mathbf{m}} = \frac{1}{J_1 J_2} \sum_{\mathbf{j}} F(\mathbf{\Phi_j}) \, e^{-i\mathbf{m}\cdot\mathbf{\Phi_j}}$$

$$\cdot e^{-i\mathbf{m}\cdot\Theta(\mathbf{\Phi_j})} \det\left(1 + \Theta_{\mathbf{\Phi}}(\mathbf{\Phi_j})\right) \; , \qquad (10)$$

where $\mathbf{\Phi_j} = 2\pi\,(j_1/J_1, j_2/J_2)$.

We can evaluate $\Theta_{\mathbf{\Phi}}$ in terms of values of Θ on mesh points, as follows:

$$\Theta_{\Phi_\alpha}\left(\frac{2\pi k}{J_\alpha}, \Phi_\beta\right) =$$

$$\sum_{j=0}^{J_\alpha - 1} \Theta\left(\frac{2\pi j}{J_\alpha}, \Phi_\beta\right)(1 - \delta_{jk})\frac{(-1)^{j-k}}{\sin\left(\pi(j-k)/J_\alpha\right)} \; . \quad (11)$$

This formula would be exact if all Fourier modes of Θ beyond $|m_\alpha| = (J_\alpha - 1)/2$ were absent.

C. Computing $g_{\mathbf{m}}$ from Θ and $\mathbf{R}$

From equations (7) and (8),

$$\Theta_\alpha(\mathbf{\Phi}, \mathbf{I}) = -\sum_{\mathbf{m}} \frac{\partial g_{\mathbf{m}}(\mathbf{I})}{\partial I_\alpha} e^{i\mathbf{m}\cdot\mathbf{\Phi}'} \; , \qquad (12)$$

$$R_\alpha(\mathbf{\Phi}, \mathbf{I}) = \sum_{\mathbf{m}} i m_\alpha g_{\mathbf{m}}(\mathbf{I}) \, e^{i\mathbf{m}\cdot\mathbf{\Phi}'} \; . \qquad (13)$$

We can apply (10) with $F(\mathbf{\Phi}) = R_\alpha(\mathbf{\Phi}, \mathbf{I})$ to get $i m_\alpha g_{\mathbf{m}}(\mathbf{I})$. To get $g_{\mathbf{m}}$ itself, we choose α so that $m_\alpha \neq 0$, and divide by $i m_\alpha$. This is possible unless $\mathbf{m} = (0,0)$, for which case we must employ information from Θ.

The function $\partial g_{00}/\partial \mathbf{I}$ is obtained on the $\mathbf{I}$ mesh from (10) with $\mathbf{m} = (0,0)$ and $F(\mathbf{\Phi}) = \Theta(\mathbf{\Phi}, \mathbf{I})$. To determine the spline representation of this function, we have to account for the identity $\sum_{i=1}^{n} B_i(x) = 1$, which implies that the derivatives $B_i'(x)$ are linearly dependent. Applying the identity, we cast the equations to be solved in the form

$$\frac{\partial g_{00}}{\partial \sqrt{I_1}} = \sum_{i=2}^{n_1} \sum_{j=1}^{n_2} (\Gamma_{ij} - \Gamma_{1j}) B_i^{(1)\prime}(\sqrt{I_1}) B_j^{(2)}(\sqrt{I_2}) \; , (14)$$

where $\Gamma_{ij} = g_{00;ij}$. Evaluating (14) at the points $(I_{1;\alpha}, I_{2;\beta})$, $\alpha = 2, \cdots, n_1$, $\beta = 1, \cdots, n_2$, we get $(n_1 - 1)n_2$ independent equations in a similar number of unknowns $\gamma_{ij} = \Gamma_{ij} - \Gamma_{1j}$, $i \neq 1$. Solving the equations, we obtain the desired coefficients $\Gamma_{ij} = \gamma_{ij}(1 - \delta_{i1}) + \Gamma_{1j}$ expressed in terms of n_2 constants of integration Γ_{1j}, one of which, say Γ_{11}, may be chosen arbitrarily. The remaining constants are determined from equations involving data on $\partial g_{00}/\partial \sqrt{I_2}$ evaluated at the action points not yet used, $(I_{1;1}, I_{2;\beta}), \beta = 2, \cdots, n_2$.

D. Iteration of the map

To compute values of the map defined implicitly by the generating function, we must solve the nonlinear equation $\mathbf{\Phi} = \mathbf{\Phi}' + G_{\mathbf{I}}(\mathbf{\Phi}', \mathbf{I})$ for $\mathbf{\Phi}'$, then substitute the result in $\mathbf{I}' = \mathbf{I} + G_{\mathbf{\Phi}}(\mathbf{\Phi}', \mathbf{I})$. The equation is solved by Newton's method, with a first guess for $\mathbf{\Phi}'$ from the explicit map. The cost of generating an adequate guess is low, since it is sufficient to include only a few Fourier modes in the explicit map. The Newton iteration converges quickly to a solution with machine precision, usually in 4 or fewer steps.

IV. Example: a Model of SSC Arcs

We illustrate by applying a tracking code JJIP by D. Ritson, which provides a highly simplified model of the SSC arcs. It give results similar to Ritson's more elaborate SSC model embodied in SSCTRK [5] , and presumably entails a complexity of transverse nonlinear effects similar to that of a full SSC model (not including beam-beam interaction and interaction point optics). The model consists of 8 FODO cells, with phase advance slightly greater than 90° per cell. For each F-D pair there are three thin-lens multipole kicks, embodying typical systematic and random errors of dipole magnets, up to 14-poles. The kicks are lumped at the D and F quadrupoles, and at the midpoint between F and D. Bends are omitted, as are chromaticity sextupoles, which almost cancel owing to the near 90° phase advance per cell. The tunes are chosen, rather arbitrarily, to be $\nu_x = 2.0314$, $\nu_y = 2.0500$.

We give results for maximum amplitudes $\max(x, y) \approx 2.2$mm (runs A) and 3.5mm (runs B). Figure 1 shows a typical phase-plane plot at 3.5mm. The 10000-turn dynamic aperture along a line $I_x = I_y$ is at about $\max(x, y) = 7.7$mm. Table 1 gives times in milliseconds for iteration of the symplectic map on the IBM 3090 (without vectorization), and the parameter ϵ which measures deviation of the map from the tracking code that produced it. The latter is defined by

$$\epsilon = \frac{1}{4} \left[\left| \frac{\Delta I_1}{I_1} \right| + \left| \frac{\Delta I_2}{I_2} \right| + \left| \frac{\Delta \Phi_1}{\pi/2} \right| + \left| \frac{\Delta \Phi_2}{\pi/2} \right| \right] \quad . \quad (15)$$

We give ϵ_N, the maximum of ϵ over N turns, for $N = 1$ and $N = 1000$. The number of B-splines is the same in each degree of freedom, $n = n_1 = n_2$, and so is the maximum mode number, $M = \max|m_1| = \max|m_2|$.

We explore the variation of results with n and M. After choosing n we increase M until there is no appreciable improvement in the accuracy as judged by ϵ_{1000}. The final M is listed in the table; it is gratifying that it has a rather small value. As was mentioned above, many modes with $|m_\alpha| \leq M$ are negligible, and are discarded so as to optimize the speed of iteration.

The time t to compute one iterate is substantially independent of the number of B-splines; it varies at all in the table only because we have discarded more or fewer small Fourier modes in the various trials. This illustrates the motivation for using B-splines; one can increase the density of interpolation points without increasing the time of evaluation of the interpolated function, since the number of nonzero splines at a point depends only on the order of the splines.

The accuracy over 1000 turns improves rather slowly with the number n of B-splines. One does not expect rapid improvement, since the functions G_Φ and G_I are represented by piecewise quadratic and piecewise linear functions of $\mathbf{I}$, respectively; (our quadratic B-splines have continuous first derivatives). Increasing the number n expands memory requirements during iteration, but does not necessarily increase the cost of constructing the map.

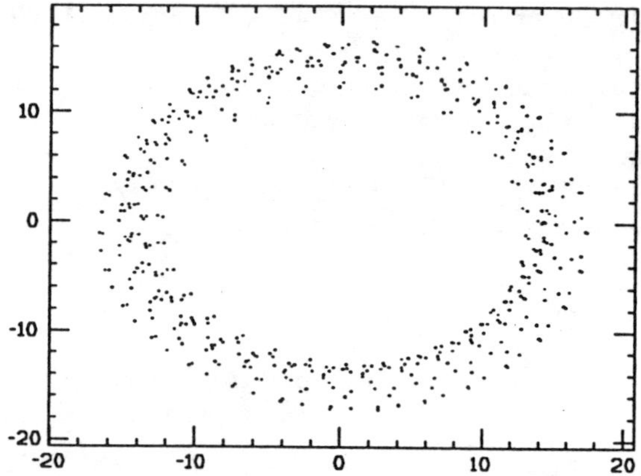

Figure 1: Horizontal phase space plot for Run B, p_x vs. x.

We plan to set up the map construction based on higher-order functions and few interpolation points, then use quadratic B-splines on a finer mesh to calculate the high-order functions during iteration. In a region close to the dynamic aperture, a somewhat higher order representation may be desirable even in iteration.

The speed of iteration that we have achieved is quite encouraging. We estimate that the inclusion of synchrotron motion, with one r.f. kick per turn, would increase the time for iteration by only a factor of two. Accounting for the fact that the iteration lends itself to vectorization, we can then expect at least 10^5 turns per minute in six degrees of freedom on a Cray computer.

We are grateful to David Ritson for offering us his tracking code, and helping us to use it.

References

[1] J. Irwin, SSC-228

[2] R. L. Warnock, Proceedings of the 1989 IEEE Particle Accelerator Conference, p. 1322.

[3] É. Forest, private communication.

[4] C. de Boor, *A Practical Guide to Splines*, New York: Springer-Verlag, 1978, pp.96–164.

[5] T. Garavaglia, S. K. Kaufmann, R. Stiening, and D. M. Ritson, SSCL-268.

Run	n	M	t(ms)	ϵ_1	ϵ_{1000}
A	10	6	1.9	6×10^{-6}	1×10^{-4}
A	20	7	2.1	1×10^{-6}	2×10^{-5}
A	32	7	2.7	8×10^{-7}	1×10^{-5}
B	10	4	1.3	1×10^{-4}	5×10^{-3}
B	20	5	1.6	4×10^{-5}	2×10^{-3}
B	32	7	2.7	4×10^{-6}	1×10^{-5}

Table 1: Iteration time and relative errors on comparing map to tracking code. Run A/B is a map made to track a particle at about 2.2/3.5 mm, n is the number of B-splines used, and $M = \max|m_\alpha|$ is the maximum Fourier mode number. t is the time to iterate the map once on the IBM 3090. ϵ_n is the maximum error over n turns, as described in the text.

Comments on the Behavior of α_1 in Main Injector γ_t Jump Schemes

S.A. Bogacz and S. Peggs
Accelerator Physics Department,
Fermi National Accelerator Laboratory*
P.O. Box 500, Batavia, Illinois 60510

Abstract

Tracking studies of transition crossing in the Main Injector have shown that the Johnsen effect is the dominant cause of beam loss and emittance blow up. To suppress this effect one has to have control over α_1 (dispersion of the momentum compaction factor α). Various γ_t jump configurations are examined and the resulting changes in α_1 are assessed. These results are further validated by comparison between the simulation and simple analytic α_1–formulas derived for a model FODO lattice with full chromaticity compensation in the presence of an eddy current sextupole component. A scheme involving the introduction of a dispersion wave in the arcs of the Main Injector, around transition time, seems to be promising if one regards the strength of the eddy current sextupole family as an external "knob" to control values of α_1.

I. INTRODUCTION

Tracking studies of transition crossing in the Main Injector and other Fermilab accelerators, using the code ESME, have shown that the Johnsen effect is the dominant cause of beam loss and emittance blow up [1, 2]. This effect is rooted in the variation of γ_t the transition gamma, with the momentum offset. A useful parameter characterizing the strength of this effect [3] is the Johnsen time, T_J,

$$T_J = \frac{\gamma_t{}^o}{\dot{\gamma}} \left[\frac{3}{2} + \frac{\alpha_1}{\alpha_o} - \frac{\alpha_o}{2} \right] \frac{\sigma_p}{p} , \qquad (1)$$

where σ_p denotes the rms momentum spread. The Johnsen time, T_J, is directly related to the lattice parameter α_1, which is defined by the following equation

$$\frac{\Delta C}{C_o} = \alpha_o \, \delta + \alpha_1 \, \delta^2 + ... , \qquad \delta = \frac{\Delta p}{p_o} , \qquad (2)$$

where C_o is the nominal closed orbit path length, and ΔC is the increase in path length for an off momentum particle. The coefficients α_o and α_1 are geometrical properties of the lattice, given by

$$\alpha_o = \frac{2\pi}{C_o} \langle \, \eta_o \, \rangle , \qquad \alpha_1 = \frac{2\pi}{C_o} \langle \, \eta_1 \, \rangle , \qquad (3)$$

where angle brackets $\langle \, ... \, \rangle$ denote averaging weighted by bend angle. The quantities being averaged are component dispersions in a momentum expansion of the total dispersion. That is,

$$\eta(s) = \eta_o(s) + \eta_1(s) \, \delta + ... \qquad (4)$$

explicitly showing the dependence of the dispersion on s, the accelerator azimuth.

Clearly, if it is possible to measure and control α_1, then it should be possible to make $T_J = 0$, and ameliorate the damage done by the Johnsen effect [4], by setting

$$\alpha_1 = -\frac{3}{2} \alpha_o + \frac{1}{2} \alpha_o{}^2 . \qquad (5)$$

This may be true especially if α_1 control is combined with RF gymnastic tricks, such as the use of a synchronous phase of 90^0 and a second harmonic cavity, as now being discussed elsewhere [5].

II. DISPERSION, η_1 – DIFFERENTIAL EQUATION

The horizontal closed orbit h(s) is found by solving the differential equation

$$h'' + \frac{K(s)}{1 + \delta} h + \frac{S(s)}{1 + \delta} h^2 = G(1 - \frac{1}{1 + \delta}) \qquad (6)$$

with periodic boundary conditions. A prime indicates differentiation with respect to s, K is the quadrupole strength, S is the sextupole strength, and G is the dipole bending strength. Here, h is expanded in a dispersion function series

$$h = x_{co} + \eta_o + \eta_1 \, \delta + ... \qquad (7)$$

The solution of the lowest order equation is trivial when there are no closed orbit perturbations, $x_{co} = 0$, so that the remaining two equations become

$$\eta_o'' + K \eta_o = G , \qquad (8a)$$

$$\eta_1'' + K \eta_1 = - G + K \eta_o - S \eta_o{}^2 . \qquad (8b)$$

According to Eq.(3), solution of the above equations yields a direct knowledge of α_1, and quantitatively identifies the major factors which affect α_1.

*Operated by the Universities Research Association, Inc., under a contract with the U.S. Department of Energy

III. FODO LATTICE WITH EDDY CURRENT SEXTUPOLES

Suppose that an accelerator like the Main Injector is represented as made up purely of FODO cells. The quadrupoles are thin, and there are no drift spaces. All of the half cell length L is filled with two identical dipoles of bend radius R, which are separated by a thin sextupole representing the field due to vacuum chamber eddy currents, induced during the ramp. There are also two thin chromatic correction sextupoles per half cell, immediately adjacent to the focusing and defocusing quadrupoles. The strength of the (half) quadrupoles is $\pm q$, and of the sextupoles is g_F, g_D, and g_E, given by

$$q = \frac{s}{L} = \frac{\sin(\phi_{1/2})}{L} , \qquad (9a)$$

$$g_I = \frac{(S_I \Delta l)\, \eta_o^I}{q} , \quad I = F, D, E . \qquad (9b)$$

Here, $\phi_{1/2}$ is the half cell phase advance, while S_F and η_o^F (for example) are the sextupole gradient and the lowest order dispersion, at the F chromatic sextupole of thin length Δl. Using these convenient definitions, it can be shown by solving Eqs.(8) that the matched values of the dispersion at F, E, and D are

$$\eta_o^F = \frac{L^2}{R}\frac{2 + s}{2\, s^2} , \qquad (10a)$$

$$\eta_o^E = \frac{L^2}{R}\frac{8 - s^2}{8\, s^2} , \qquad (10b)$$

$$\eta_o^D = \frac{L^2}{R}\frac{2 - s}{2\, s^2} . \qquad (10c)$$

When only the F and D sextupoles are turned on, at a strength to correct for f times the natural chromaticity, it can easily be shown that

$$\alpha_o = \frac{L^2}{R^2}\frac{1}{s^2}\left[\, 1 - \frac{s^2}{12}\, \right] , \qquad (11a)$$

$$\alpha_1 = \frac{L^2}{R^2}\frac{1}{s^2}\left[\, 1 - f + \frac{s^2}{12}\, \right] . \qquad (11b)$$

IV. CONTROLLING α_1 WITH SEXTUPOLES

The middle sextupole, of strength g_E, can be thought of in (at least) two ways. In the first point of view, it represents the sextupole field caused by eddy currents induced in the vacuum chamber of the dipoles. In the second point of view, g_E represents a free knob with which α_1 can be controlled. For the sake of a semi-quantitative interpretation, suppose that the F and D sextupoles have their strengths set to compensate for the sum of the chromaticity induced by an eddy current sextupole of strength g_E, plus f times the natural chromaticity. In this

case it is readily shown that the F and D sextupole strengths are given by

$$g_F = \frac{f}{2} - \frac{2 - s^2}{4}\, g_E , \qquad (12a)$$

$$g_D = -\frac{f}{2} - \frac{2 - s^2}{4}\, g_E . \qquad (12b)$$

Carrying out similar analysis as in the previous section one can generalize Eq.(11b) as follows

$$\alpha_1 = \frac{L^2}{R^2}\frac{1}{s^2}\left[\, (1 - f + \frac{s^2}{12}) - \left(\frac{3}{8}\right)^2 s^3\, g_E\, \right] . \qquad (13)$$

To test the results of Eq.(13) and to gain some insight into the prospect of controlling α_1 in the Main Injector, consider a lattice made up of 80 simple FODO cells. The lattice design code MAD was used to study the variation in closed orbit path-length as a function of $\Delta p/p$, over a range from -0.003 to $+0.003$. Table 1 shows good qualitative agreement between the analytic predictions and simulated values of α_o and α_1.

(f, g_E)	$\alpha_o\ (\times 10^{-3})$		$\alpha_1\ (\times 10^{-3})$	
predicted	predicted	simulated	predicted	simulated
(0,0)	2.956	2.956	3.213	3.332
(1,0)	2.956	2.956	0.129	0.0244
(1,5)	2.956	2.956	−0.638	−0.512

Table 1. Comparison of predicted and simulated α_o and α_1.

If the sextupole family strength parameter g_E is employed to control values of α_1, the inevitable conclusion is that the sensitivity to the family is too weak to reduce the Johnsen time to zero. Investigations are currently under way to find an optical configuration that will significantly increase the orthogonality of the three families beyond the unfortunate results of the FODO lattice. An apparently promising candidate involves the introduction of a dispersion wave in the arcs of the Main Injector, around transition time. This begins to resemble an unmatched γ_t jump scheme [6] − except that the lattice perturbation can be introduced slowly, and that the needed size of the dispersion wave is expected to be relatively modest.

V. BEHAVIOR OF γ_t JUMP SCHEMES

The satisfactory agreement between MAD simulations and analytic predictions reported in the previous section encourages the use of the program to study the behavior of momentum compaction factors for more realistic Main Injector lattices, where analytic results are no longer tractable. Here we consider two families of Main Injector lattices representing matched and unmatched γ_t jump schemes. These schemes are described in detail elsewhere [6]. It is important to check that the resulting change of α_1 does not greatly affect the Johnsen time T_J, extending the variation of transition crossing time for different parts of a bunch.

The simulation places one thin eddy current sextupoles of strength g_E at the middle of each dipole, with a multipole strength [7] of $b_2 = 0.561$ m^{-2}. Two families of chromatic sextupoles are used to compensate for both natural and eddy current chromaticities. It is assumed that the F and D sextupole strengths are not changed while jumping through transition.

Figure 1 summarize the behavior of the *matched* scheme with bipolar ($\Delta\gamma_t = \pm\,0.65$) and unipolar ($\Delta\gamma_t = -1.3$) excitations as well as an *unmatched* unipolar excitation - a bipolar jump is not possible in this scheme. The linear character of $\alpha_p(\delta)$ in the realistic range $\delta = -0.01$ to $+0.01$ is apparent in all cases.

One can conclude that eddy current sextupoles are not expected to significantly affect transition crossing performance. Continuing investigations suggest that a modest dispersion wave significantly improves the orthogonality of three families of sextupoles.

VI. CONCLUSIONS

General analytical expressions make it possible to evaluate how the critical Johnsen time depends on effects like eddy current sextupoles in the Main Injector dipoles, or on transition jump configurations. In the simple case of a FODO lattice representation of the Main Injector, analytic results are in good quantitative agreement with a lattice design code.

Examination of nominal Main Injector transition jump schemes reveals that a *matched* scheme produces little change in the Johnsen time, but that T_J is more than doubled in the *unmatched* scheme.

A third family of sextupoles might be used to deliberately and practically control the Johnsen time, without modifying the nominal transition momentum. Two other sextupole families are used to achieve the desired net chromaticities. Such control, especially when used in conjunction with RF gymnastics, may make transition crossing so innocuous that it becomes unnecessary to include a transition jump in Main Injector designs.

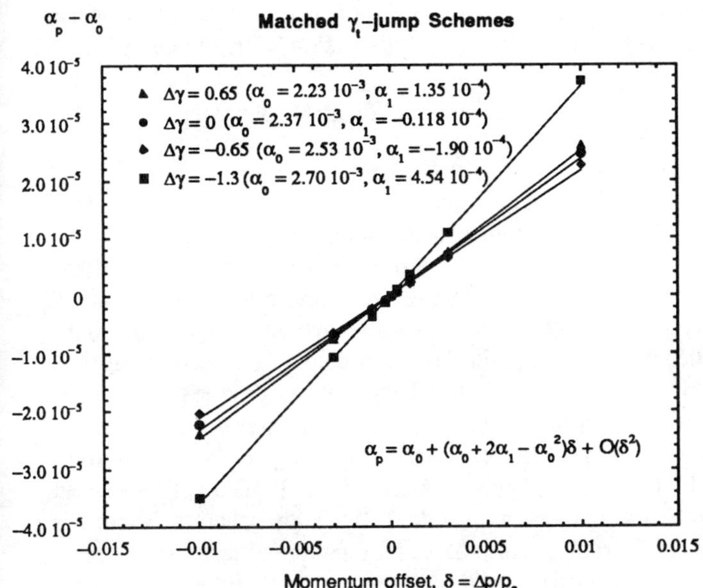

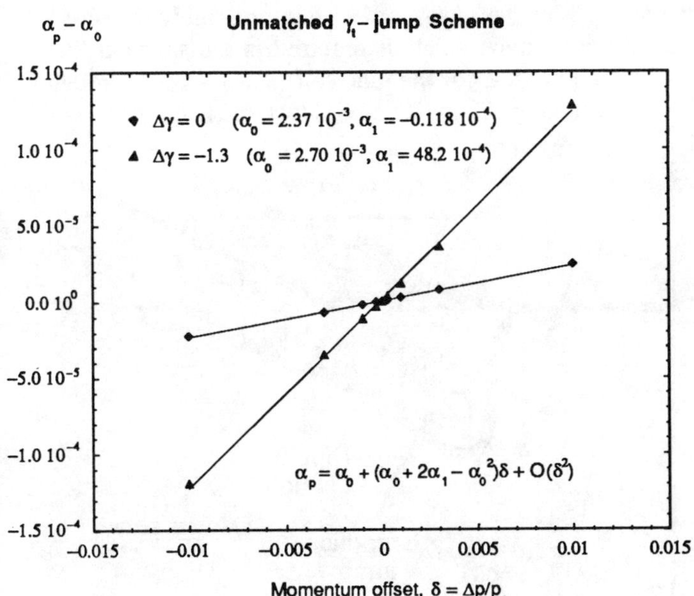

Figure 1. Numerical simulation of α_p versus δ carried out for various γ_t jump configurations.

VII. REFERENCES

[1] S.A. Bogacz, "Transition crossing in the Main Injector - ESME simulation", Proceedings of the F III Instability Workshop, Fermilab, July 1990.

[2] I. Kourbanis and K.Y. Ng, "Main Ring transition crossing simulations", Proceedings of the F III Instability Workshop, Fermilab, July 1990.

[3] K. Johnsen, "Effects of non-linearities on the phase transition", CERN symposium on high energy accelerators and pion physics, Vol. 1, p. 106, 1956.

[4] A. Sorenssen, "Crossing the phase transition in strong focusing proton synchrotrons", Particle Accelerators Vol. 6, pp. 141-165, 1975.

[5] J. Griffin and J. MacLachlan, "Synchrotron phase transition crossing using an RF harmonics", Fermilab, TM-1734, March 1991.

[6] S.A. Bogacz, F. Harfoush, and S. Peggs, A comparison of matched and unmatched γ_t jump schemes in the Main Injector, These Proceedings.

[7] J-F Ostiguy, Eddy current induced multipoles in the Main Injector, Fermilab, MI-0037, October 1990.

A Comparison of Transition Jump Schemes for the Main Injector

S. Peggs, A. Bogacz, F. Harfoush

Fermi National Accelerator Laboratory*, PO Box 500, Batavia, Illinois 60510

I. INTRODUCTION

Transition comes when the differential of circulation time with respect to $\delta = \Delta p/p_0$, the off momentum parameter, is zero. This happens when the relative rate of change of speed β equals the relative increase in path length - when

$$\frac{1}{\beta}\frac{d\beta}{d\delta} = \frac{1}{\gamma^2} \equiv \frac{1}{\gamma_T^2} \tag{1}$$

$$= \frac{1}{C_0}\frac{dC}{d\delta} = \frac{1}{C_0}\int_0^{2\pi} \eta \, d\theta = \frac{2\pi}{C_0}\langle\eta\rangle$$

Here C is the circumference of the closed orbit, η is the dispersion function, and θ is the accumulated bend angle. Angle brackets $\langle\;\rangle$ represent an average over bending dipoles. Significant beam loss and emittance growth are expected in the Main Injector when $\gamma \approx \gamma_T$, if nothing is done to ameliorate transition crossing[1]. Here we consider ways in which γ_T can be modified for a short period of time, so that transition crossing is very rapid. Figure 1 illustrates unipolar and bipolar "transition jump" schemes. In order that the passage is graceful, it is necessary that

$$| \gamma - \gamma_T | > 0.65 \tag{2}$$

for as long as possible[2]. Hence the unipolar scheme shown needs a lattice with $\Delta\gamma_T = -1.3$, and the bipolar scheme needs two lattices, with $\Delta\gamma_T = \pm 0.65$.

A major design issue is, how "gross" are the changes of other single particle properties of the lattice? Another issue is simplicity of design. How strong are the perturbation quadrupoles, how many are there, in how many families? What is the tracking performance? Two prototypical schemes are compared, "matched" and "unmatched", referring to whether or not a large dispersion wave circulates when the perturbation quads are turned on. Results presented from an MI_15 lattice analysis are valid for the contemporary MI_17 lattice, with small changes (except where noted).

II. MATCHED AND UNMATCHED CONFIGURATIONS

A small quadrupole perturbation q shifts the tune by

$$\Delta Q_H = \frac{q\,\beta_{quad}}{4\pi} \tag{3}$$

where β is now a beta function. Downstream from the perturbation there is a free horizontal beta wave

$$\frac{\Delta\beta}{\beta} = -q\,\beta_{quad}\,\sin[2(\phi - \phi_{quad})] \tag{4}$$

with a phase advancing twice as fast as the betatron phase. There is also a free horizontal dispersion wave, given by

$$\frac{\Delta\eta}{\sqrt{\beta}} = -q\,\eta_{quad}\,\sqrt{\beta_{quad}}\,\sin(\phi - \phi_{quad}) \tag{5}$$

advancing in step with the betatron phase.

Most of the Main Injector circumference consists of four FODO cell arcs, with a phase advance per cell close to 90 degrees. No betatron wave escapes one of these arcs if the perturbing quads are arranged in pairs, either with identical strengths 90 degrees apart, or with equal and opposite strengths 180 degrees apart. Similarly, no dispersion wave escapes if identical strength pair members are 180 degrees apart, or if equal and opposite members are 360 degrees apart. Neither wave escapes if identical quads are arranged in groups of four, with 90 degrees of phase advance - one FODO cell - between neighbors. This is the matched γ_T jump scheme. There is a total of 48 such quadrupoles at focussing locations in arc cells, where their effect on γ_T is the greatest. Essentially, these quadrupoles retune the standard arc FODO cell, changing the matched dispersion function. The change in γ_T is first order in the strength of the quadrupoles. Unfortunately, a second family of perturbation quads is required in a dispersion free region, to compensate for the tune shift accumulated through the arc cell retuning. There is no need to avoid dispersion

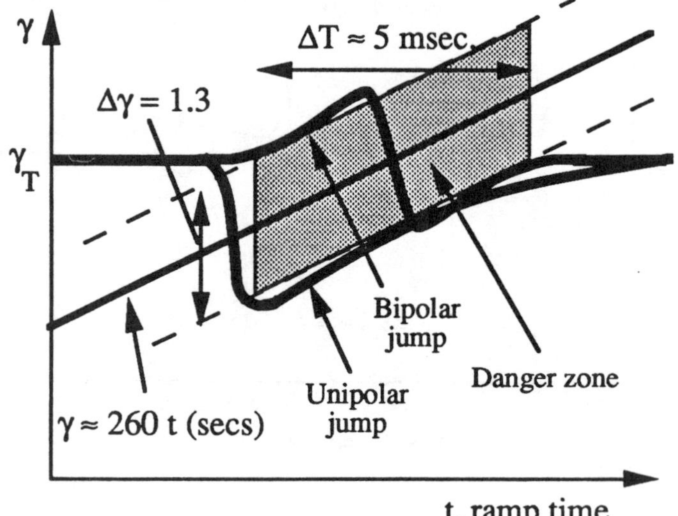

Figure 1   Main Injector passage through transition with nominal parameters, including unipolar or bioplar jumps.

*Operated by the Universities Research Association, Inc., under contract with the U.S Department of Energy.

waves in the second family, so the quadrupoles are arranged in pairs, 90 degrees apart. In the MI_15 lattice there is only space for an effective total of 8 quadrupoles. (In MI_17 there is space for 16 - a significant improvement, as will be seen.) The strength of the second family is about 6 times the strength of the first, and with opposite polarity.

There is only one family of 24 quadrupoles in the unmatched scheme, with opposite polarity members at every other horizontally focussing arc quadrupole[3]. This is similar to the scheme in daily use in the Fermilab Booster [4-6]. There is no global beta wave and no net tune shift, to first order in the perturbation strength. However, there is a large first order global dispersion wave, while the average change in γ_T is only of second order. (In a pure FODO lattice with no straights, flipping all perturbation quad polarities results in exactly the same lattice. The "polarity symmetry" is somewhat broken in MI_15, but the first order term in the variation of $\Delta\gamma_T$ is still negligible.)

Case number	1	2	3	4	5
Matched?		Yes	Yes	Yes	No
$\Delta\gamma_T$	0.0	−1.3	−0.65	0.65	−1.3
Population 1		48	48	48	24
Population 2		8	8	8	—
Strength 1		−.055	−.029	.031	.074
Strength 2		.334	.159	−.157	—
$\beta_{H\,min}$(m)	10.9	1.3	4.68	8.86	8.86
$\beta_{H\,max}$(m)	56.7	432.7	81.0	97.5	65.4
$\beta_{V\,min}$(m)	10.9	10.3	10.9	10.9	10.7
$\beta_{V\,max}$(m)	78.9	83.5	80.1	80.3	82.0
η_{min}(m)	−0.12	−1.75	−0.47	−0.31	−7.71
η_{max}(m)	2.07	4.61	2.64	2.29	9.59

Table 1 Configuration and performance in various schemes. Quad strengths are relative to the regular FODO quad strength.

III. LINEAR OPTICAL PERFORMANCE

Table 1 compares the 5 cases of interest. It is implicit that the net horizontal and vertical tune shifts are negligibly small, of order 10^{-3}, although they are not exactly zero in any of the cases. The table shows that the assumption that first order perturbation theory is adequate is not completely true for either scheme. Case 2 (matched, unipolar, $\Delta\gamma_T = -1.3$) shows the largest strength, with a second family strength of one third of the nominal strength of an arc quadrupole. The optical solution is not acceptable, because the maximum horizontal beta function rises to 433 meters. In all cases the vertical beta functions are negligibly disturbed, since the perturbing quads are at horizontally focussing locations. The matched bipolar scheme (cases 3 and 4) causes much more modest optical perturbations - although they are still not negligible - with a maximum horizontal beta of 98 meters. It is expected that this situation would improve dramatically in the MI_17 lattice, with twice as many quads in the second family, and with none of them placed back to back, as in MI_15.

In the unmatched scheme (case 5), the minimum and maximum horizontal beta functions are disturbed by less than 20%. However, the dispersion reaches extremes of −7.7 and 9.6 meters, leading to a reduced dynamic aperture for off-momentum particles. This is especially painful at transition, when the momentum width of the beam is at its largest. Large dispersion swings also cause a large variation of γ_T with momentum, raising the Johnsen time significantly[7].

IV. TRACKING RESULTS

The dynamic aperture about a displaced closed orbit is found as a function of $\delta = \Delta p/p$ using the code TEAPOT. The δ range used, 0.0 to 0.01, is about twice the large total momentum spread expected for about 10 milliseconds, or 1000 turns, around transition. Successive particles are launched with decreasing amplitude, in steps of 0.5 centimeters, until a particle survives for 1000 turns at the "stable" amplitude. (Amplitudes are initially identical horizontally and vertically, and are scaled to β_{max} in a FODO cell). While it is clearly incorrect to include synchrotron oscillations, such a static model is flawed by not including ramping and transition in all three spatial dimensions. No appropriate code exist at present.

Only those systematic dipole bend magnetic errors listed in Table 2 are included, according to the expansion

$$B_{vert} = B_0 \left[1 + \sum_n b_n \left(\frac{x}{r_0}\right)^n \right] \qquad (6)$$

The reference radius r_0 in Table 2 is 1 inch - 2.54 centimeters. Multipole values come from the Main Injector dipole design calculations[8-10] at a transition momentum of 19.10 Gev/c, and a dipole field of $B_0 = 0.237$ Tesla. They are consistent with prototype measurements[10]. The sextupole field is dominant, due almost completely to eddy currents. Higher order eddy current effects and remanent fields are negligible.

Results are shown in Figures 2 and 3 for two configurations: a matched γ_T jump with $\Delta\gamma_T = -0.65$ (case 3 in Table 1), and an unmatched γ_T jump with $\Delta\gamma_T = -1.3$ (case 5). The net chromaticities are set to zero in both. A straight line is fit to the data, with a slope $dx/d\delta$ analogous to a dispersion if the aperture stop is a "brick wall". The slopes (−2.7 meters in Figure 2 and −7.1 meters in Figure 3)

Source	b_2	b_4	b_6
		$(10^{-4}$ @ 1 inch$)$	
Eddy current	3.405	-.087	-.028
Saturation	.215	.184	.046
TOTAL	3.620	.097	.018

Table 2 Systematic dipole errors used for tracking.

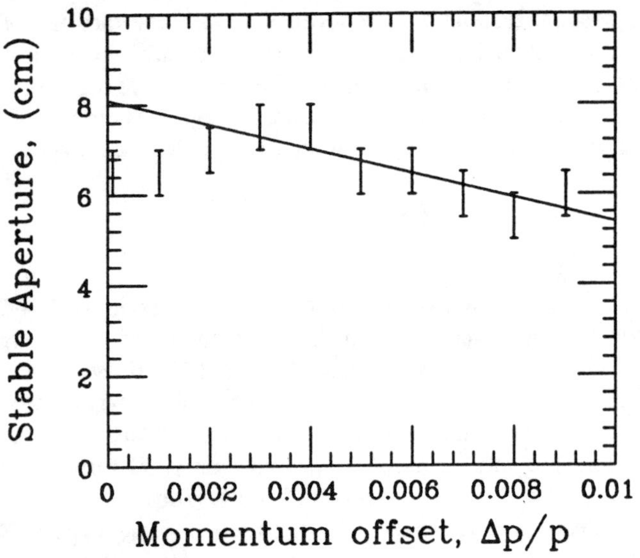

Figure 2 Dynamic aperture versus momentum for a matched bipolar jump, Case 3 in Table 1.

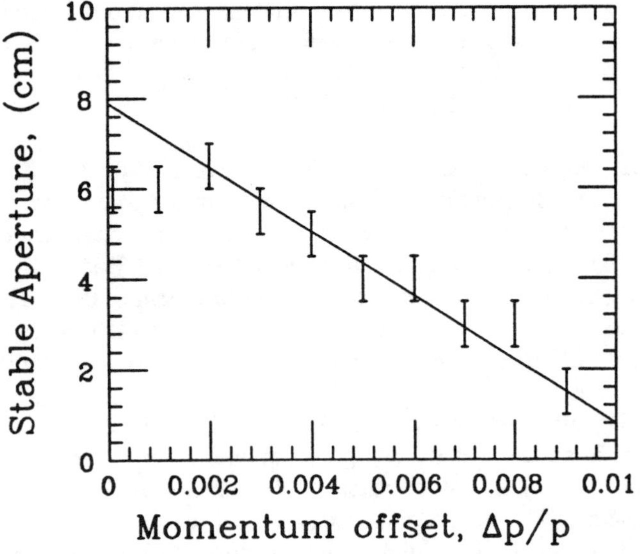

Figure 3 Dynamic aperture versus momentum for an unmatched unipolar jump, Case 5 in Table 1.

are in fair agreement with the corresponding maximum dispersions quoted in Table 1 (2.6 and 9.6 meters). The matched scheme has a definite advantage in improved off-momentum dynamic aperture performance.

V. CONCLUSIONS

While neither the matched nor the unmatched transition jump scheme is entirely satisfactory as presented, either one could be improved and made to work well in the Main Injector.

The unmatched scheme has simpler hardware requirements - only one family of perturbation quadrupoles is required. Betatron functions and tunes are negligibly affected (in an ideal lattice without errors). Its disadvantages stem from the large induced dispersion wave, and from the second order dependence of $\Delta\gamma_T$ on perturbation strength. The 9.6 meter maximum dispersion decreases the dynamic aperture significantly (although not severely) at large momentum offsets.

The matched scheme requires two quadrupole families to keep the betatron tunes unchanged. This leads to a large perturbation strength that, in the worst case of a unipolar $\Delta\gamma_T = -1.3$ jump, distorts the linear lattice almost to the point of instability. This problem is expected to be greatly ameliorated in the MI_17 lattice, with a factor of 4 decrease in maximum quadrupole strength. The behavior of the linear lattice in a bipolar jump of strength $\Delta\gamma_T = \pm 0.65$ is reasonable, even in the MI_15 lattice.

VI. REFERENCES

[1] For a broad discussion of Main Injector transition crossing issues, see S. Peggs, M. Harvey (editors), "Proceedings of the Fermilab III Instabilities Workshop", Fermilab, June 1990.
[2] S.Y. Lee, "Summary Report on Transition Crossing in the Main Injector", Proc. Fermilab III Instabilities Workshop, 1990.
[3] S. Holmes, "Main Injector Transition Jump", Collected reports of the accelerator and beam line options working group, Breckenridge workshop, August 1989.
[4] P. Lucas, J. Maclachlan, "Simulation of Space Charge Effects and Transition Crossing in the Fermilab Booster", p. 1114, IEEE PAC, Washington, 1987.
[5] K.Y. Ng, Space-charge Effects of Transition Crossing in the Fermilab Booster", p. 1129, IEEE PAC, Washington, 1987.
[6] W. Merz, C. Ankenbrandt, K. Koepke, "Transition Jump System for the Fermilab Booster", p. 1343, IEEE PAC, Washington, 1987.
[7] A. Bogacz, S. Peggs, "Comments on the Behavior of α_1 in Main Injector γ_T Jump Schemes", these proceedings.
[8] J-F. Ostiguy, "Main Injector Dipole Magnet: 2 D Field Computations", Fermilab note MI-0036, October 1990.
[9] J-F. Ostiguy, "Eddy Current Induced Multipoles in the Main Injector", Fermilab note MI-0037, October 1990.
[10] D. Harding et al, "Design Considerations and Prototype Performance of the Fermilab Main Injector Dipole", these proceedings.

NORMALIZATION OF THE PARAMETERIZED COURANT-SNYDER MATRIX FOR SYMPLECTIC FACTORIZATION OF A PARAMETERIZED TAYLOR MAP

Yiton T. Yan

Superconducting Super Collider Laboratory*,
2550 Beckleymeade Avenue, Dallas, TX 75237

Abstract

The transverse motion of charged particles in a circular accelerator can be well represented by a one-turn high-order Taylor map [1]. For particles without energy deviation, the one-turn Taylor map is a 4-dimensional polynomials of four variables. The four variables are the transverse canonical coordinates and their conjugate momenta. To include the energy deviation (off-momentum) effects, the map has to be parameterized with a smallness factor representing the off-momentum and so the Taylor map becomes a 4-dimensional polynomials of five variables. It is for this type of parameterized Taylor map that a mehtod is presented for converting it into a parameterized Dragt-Finn facctorization map. Parameterized nonlinear normal form and parameterized kick factorization can thus be obtained with suitable modification of the existing technique.

I. INTRODUCTION

A one-turn Taylor map parameterized for off-momentum transverse propagation relative to the dispersed closed orbit in a circular alternating-gradient synchrotron [2] can be given in the form

$$\vec{x}' = m(\vec{x}, \delta) : \vec{x}$$
$$= M(\delta)\vec{x} + \sum_{i=0}^{n_2} \vec{U}_{2,i}(\vec{x})\delta^i + \sum_{i=0}^{n_3} \vec{U}_{3,i}(\vec{x})\delta^i + \cdots, \quad (1)$$

where the transverse canonical phase space coordinates, $\vec{x} = [x, p_x, y, p_y]$ are deviations relative to the dispersed closed orbit; $\delta = \Delta P/P_0$ is a parameter for the off-momentum, which is usually a smallness factor in the domain of physical interest. Terms in each of the polynomials $\vec{U}_{k,i}(\vec{x})$ for $k = 2, 3, \cdots$ and $i = 0, 1, \cdots, n_k$ are all of the order k of $\vec{x}$. The parameterized Courant-Snyder matrix, $M(\delta)$, is truncated at an order of δ^n and given by

$$M(\delta) = M_0 + \delta M_1 + \delta^2 M_2 + \cdots + \delta^n M_n, \quad (2)$$

where $M_0, M_1, \cdots, M_n$ are all 4×4 and usually coupled matrices. Note that $n \geq n_2 \geq n_3 \geq \cdots \geq n_k \geq \cdots$ and that the dispersed closed orbit can be expressed as

$$\vec{x}_c = \vec{x}_{c,0} + \delta\vec{x}_{c,1} + \delta^2\vec{x}_{c,2} + \cdots, \quad (3)$$

** Operated by Universities Research Association, Inc., for the U.S. Department of Energy under the Contract No. DE-AC02-89ER40486.

which is a polynomial of δ with an order of n or higher.

If the Taylor map is extracted from a symplectic tracking program, then it is symplectic when it is expanded to the infinite order of $\vec{x}$ and δ. We assume the Taylor map given by Eq. 1 is such a type of Taylor map truncated at a finite order of δ and $\vec{x}$. Such a truncated Taylor map, although not symplectic because of the missing of the high-order effects, can be order-by-order converted into a Dragt-Finn factorization map [3]. However, the addition of the parameter δ makes such a factorization process trickier. In this paper, the author presents a method of factorizing such a perameterized Taylor map given by Eq. 1. Only a single parameter, δ, is considered though extension to a multiparameter case is analogous.

II. NORMALIZATION OF $M(\delta)$

The key to the Dragt-Finn factorization of a parameterized Taylor map given by Eq. 1 is the normalization of its associated parameterized Courant-Snyder matrix $M(\delta)$ given by Eq. 2. A method [4] has been presented recently to normalize $M(\delta)$ via order-by-order symplectic factorization such that

$$R(\delta) = A^{-1}(\delta)M(\delta)A(\delta) + \sigma(\delta^{n+1}),$$

where the normalized parameterized rotation $R(\delta)$ and the canonical generation matrix $A(\delta)$ are given as follows.

$$R(\delta) = \begin{pmatrix} \cos\mu_1(\delta) & \sin\mu_1(\delta) & 0 & 0 \\ -\sin\mu_1(\delta) & \cos\mu_1(\delta) & 0 & 0 \\ 0 & 0 & \cos\mu_2(\delta)^{\cdot} & \sin\mu_2(\delta) \\ 0 & 0 & -\sin\mu_2(\delta) & \cos\mu_2(\delta) \end{pmatrix},$$

where

$$\mu_\alpha(\delta) = \mu_{\alpha,0} + \delta\mu_{\alpha,1} + \delta^2\mu_{\alpha,2} + \cdots + \delta^n\mu_{\alpha,n},$$

for $\alpha = 1, 2$. The inverse of $R(\delta)$ is also a decoupled rotation given by

$$R^{-1}(\delta) = \begin{pmatrix} \cos\mu_1(\delta) & -\sin\mu_1(\delta) & 0 & 0 \\ \sin\mu_1(\delta) & \cos\mu_1(\delta) & 0 & 0 \\ 0 & 0 & \cos\mu_2(\delta) & -\sin\mu_2(\delta) \\ 0 & 0 & \sin\mu_2(\delta) & \cos\mu_2(\delta) \end{pmatrix}.$$

The symplectic matrix $A(\delta)$ is of order $\delta^{n(n+1)/2}$ which is a concatenation of a series order-by-order canonical generation matrices given by

$$A(\delta) = A_0(I + \delta^1 A_1)(I + \delta^2 A_2)\cdots(I + \delta^n A_n).$$

Its inverse is given by

$$A^{-1}(\delta) = (I - \delta^n A_n) \cdots (I - \delta^2 A_2)(I - \delta A_1)A_0^{-1}.$$

Note that A_0 and its inverse A_0^{-1} are the familiar generation matrices for normalization of M_0 such that [5]

$$
\begin{aligned}
R_0 &= A_0^{-1} M_0 A_0 \\
&= \begin{pmatrix}
\cos\mu_{1,0} & \sin\mu_{1,0} & 0 & 0 \\
-\sin\mu_{1,0} & \cos\mu_{1,0} & 0 & 0 \\
0 & 0 & \cos\mu_{2,0} & \sin\mu_{2,0} \\
0 & 0 & -\sin\mu_{2,0} & \cos\mu_{2,0}
\end{pmatrix}.
\end{aligned}
$$

For the detailed solution of $\mu_{\alpha,i}$ and A_i for $i = 1, 2, \cdots, n$ and $\alpha = 1, 2$, please refer to Reference [4].

III. DRAGT-FINN FACTORIZATION

For Dragt-Finn factorization of the parameterized Taylor map given by Eq. 1, one first makes a similarity transformation on the map such that

$$
\begin{aligned}
_1m(\vec{x},\delta):\vec{x} &= \mathcal{A}(\vec{x},\delta)m(\vec{x},\delta)\mathcal{A}^{-1}(\vec{x},\delta):\vec{x} \\
&= R(\delta)\vec{x} + \sum_{i=0}^{n_2} {}_1\vec{U}_{2,i}(\vec{x})\delta^i + \sum_{i=0}^{n_3} {}_1\vec{U}_{3,i}(\vec{x})\delta^i + \cdots,
\end{aligned}
$$

where $\mathcal{A}(\vec{x},\delta)$ is the global form of $A(\delta)$. One then make a concatenation of the transformed map $_1m(\vec{x},\delta)$ with $\mathcal{R}^{-1}(\vec{x},\delta)$ such that

$$
\begin{aligned}
_2m(\vec{x},\delta):\vec{x} &= \mathcal{R}^{-1}(\vec{x},\delta)_1m(\vec{x},\delta):\vec{x} \\
&= \vec{x} + \sum_{i=0}^{n_2} {}_2\vec{U}_{2,i}(\vec{x})\delta^i + \sum_{i=0}^{n_3} {}_2\vec{U}_{3,i}(\vec{x})\delta^i + \cdots, \quad (4)
\end{aligned}
$$

where $\mathcal{R}^{-1}(\vec{x},\delta)$ is the global form of the inverse of the rotation matrix $R(\delta)$. Due to symplecticity of the Taylor map (when it is expanded to the infite order), for each of the $_2\vec{U}_{2,i}(\vec{x})$ for $i = 0, 1, \cdots, n_2$ there exists a third-order (every term is third order) polynomial of $\vec{x}$, $f_{3,i}(\vec{x})$ such that

$$[f_{3,i}(\vec{x}), \vec{x}] = {}_2\vec{U}_{2,i}(\vec{x}). \quad (5)$$

Therefore, from Eq. 4, one obtains

$$_3m(\vec{x},\delta):\vec{x} = exp(: \sum_{i=0}^{n_2} f_{3,i}(\vec{x})\delta^i :)\vec{x} + \sum_{i=0}^{n_3} {}_2\vec{U}'_{3,i}(\vec{x})\delta^i + \cdots,$$

where

$$_2\vec{U}'_{3,i}(\vec{x}) = {}_2\vec{U}_{3,i}(\vec{x}) - \frac{1}{2}[f_{3,i}(\vec{x}), [f_{3,i}(\vec{x}), \vec{x}]]$$

Concatenating $_2m(\vec{x},\delta)$ with

$$exp(-: \sum_{i=0}^{n_2} f_{3,i}(\vec{x})\delta^i :),$$

one obtains

$$
\begin{aligned}
_3m(\vec{x},\delta):\vec{x} &= exp(-: \sum_{i=0}^{n_2} f_{3,i}(\vec{x})\delta^i :)_2m(\vec{x},\delta):\vec{x} \\
&= \vec{x} + \sum_{i=0}^{n_3} {}_3\vec{U}_{3,i}(\vec{x})\delta^i + \sum_{i=0}^{n_4} {}_3\vec{U}_{4,i}(\vec{x})\delta^i + \cdots.
\end{aligned}
$$

Again, due to symplecticity, for each of the $_3\vec{U}_{3,i}(\vec{x})$ for $i = 0, 1, \cdots, n_3$, there exists a fourth-order (every term is fourth order) polynomial of $\vec{x}$, $f_{4,i}(\vec{x})$ such that

$$[f_{4,i}(\vec{x}), \vec{x}] = {}_3\vec{U}_{3,i}(\vec{x}). \quad (6)$$

One thus obtains the Dragt-Finn factorization of the fourth order given by

$$exp(: \sum_{i=0}^{n_3} f_{4,i}(\vec{x})\delta^i :).$$

Following the similar steps given above, one can proceed further to obtain the Dragt-Finn factorization of the higher orders given by

$$exp(: \sum_{i=0}^{n_{k-1}} f_{k,i}(\vec{x})\delta^i :),$$

for $k = 5, 6, \cdots$. The parameterized Taylor map given by Eq. 1 can then be represented by a symplectic map given in global form by

$$
\begin{aligned}
\vec{x'} &= m(\vec{x},\delta):\vec{x} \\
&= \mathcal{A}^{-1}(\vec{x},\delta)m_f(\vec{x},\delta)\mathcal{A}(\vec{x},\delta):\vec{x}, \quad (7)
\end{aligned}
$$

where

$$m_f(\vec{x},\delta) = \mathcal{R}(\vec{x},\delta)exp(: \sum_{i=0}^{n_2} f_{3,i}(\vec{x})\delta^i :)exp(: \sum_{i=0}^{n_3} f_{4,i}(\vec{x})\delta^i :)\cdots.$$

IV. DISCUSSIONS

At first glance, one may wonder why Eq. 5 (or Eq. 6), which is one of the important steps for leading to the parameterized Dragt-Finn factorization of a dispersed-closed-orbit Taylor map, is true. Let us assume that the Taylor expansions are all infinite orders of δ, that is, $n = n_2 = n_3 = \cdots = \infty$. If a suitable small factor is substituted for δ into Eq. 1, then one obtains the associated regular closed-orbit non-parametrized Taylor map which is a familiar type of map for Dragt-Finn factorization. Now if we first obtain the linear parameterized normalized rotation $R(\delta)$ and its associated parametrized canonical generation matrix $A(\delta)$ up to the infinite order of δ, and then substitute the same small factor for δ, we would have obtained the same linear non-parameterized normalized rotation R and its associated non-parametrized canonical generation matrix A as those obtained for the associated non-parameterized Dragt-Finn factorization. Therefore, it would make no differnece whether we first obtain Eq. 4 to infinite order of δ and then substitute the small factor for δ or we first substitute the small factor for δ in Eq. 1 to obtain a non-parameterized closed-orbit Taylor map and then obtain the associated non-parameterized equivalence of Eq. 4. Thus, due to symplecticity, there must exist a third-order (every term is third order) polynomial of $\vec{x}$, $f(\vec{x}, \delta)$, such that

$$[f(\vec{x}, \delta), \vec{x}] = \sum_{i=0}^{\infty} {}_2\vec{U}_{2,i}(\vec{x})\delta^i. \tag{8}$$

Since $f(\vec{x}, \delta)$ can always be Taylor expanded as

$$f(\vec{x}, \delta) = \sum_{i=0}^{\infty} f_i(\vec{x})\delta^i, \tag{9}$$

by comparing Eq. 8 and Eq. 9, one can conclude that

$$[f_i(\vec{x}), \vec{x}] = {}_2\vec{U}_{2,i}(\vec{x})$$

for each $i = 0, 1, 2, \cdots, \infty$. Furthermore, since terms with a certain order of δ can only be contributed through concatenation from terms with lower or equal orders of δ, truncation of higher-order terms will not change the outcome for the lower-order Lie operators. Therefore, $f_{3,i}(\vec{x}) = f_i(\vec{x})$ for $i = 0, 1, \cdots, n_2$.

In practice, differential algebra [6] can be used to obtain the Taylor map. Indeed, a truncated Taylor map of the type given by Eq. 1 and its associated dispersed closed orbit given by Eq. 3 can be obtained for the SSC lattice (and others) with the use of a post-Teapot [7] map extraction program Zmap [8]. Implementation of such a parameterized Taylor-map Dragt-Finn factorization can be achieved with some modification of the available non-parameterized Dragt-Finn factorization routine in Zlib [9] and others [10] [11]. Once the parameterized dragt-Finn factorization of the type given by Eq. 7 is obtained, paramterized normal form can be obtained with suitable modification of the existing technique [10] and parameterized kick factorization [12] [13] can be obtained for faster long-term symplectic kick map tracking.

V. Acknowledgments

The author thanks A. Chao for his support and encouragement and J. Ellison, E. Forest, and J. Irwin for many discussions.

VI. References

[1] Y. Yan, T. Sen, A. Chao, G. Bourianoff, A. Dragt, and E. Forest, SSC Laboratory Report SSCL-301 (1990); also to appear in the proceedings of the workshop on nonlinear problems in future particle accelerators, Capri, Italy (1990).

[2] E. Courant and H. Snyder, Ann. Phys., **3**, 1 (1958)

[3] A. J. Dragt and J. M. Finn, J. Math. Phys. 20, 2649 (1979)

[4] Y. Yan, SSC Laboratory Report SSCL-302 (1990).

[5] D. Edwards and L. Teng, in Proceedings of the 1973 IEEE Particle Accelerator Conference, p. 885 (1973); L.C. Teng, Fermi National Laboratory Report FN-229, 1971.

[6] M.Berz, in Floyd Bennett and Joyce Kopta, editors, Proceedings of the 1989 IEEE Particle Accelerator Conference, March 20-23, 1989, IEEE Catalog Number 89CH2669-0.

[7] L. Schachinger and R. Talman, Particle Accelerators, Vol. 22, 1987.

[8] Y. Yan, SSC Laboratory Report SSCL-299, 1990.

[9] Y. Yan and C. Yan, SSC Laboratory Report SSCL-300 (1990).

[10] E. Forest, M. Berz, and J. Irwin, Particle Accel. 24, 91 (1989), and M. Berz, *ibid.* **24**, 109 (1989)

[11] L. Michelotti, in these Proceedings, and Fermi National Accelerator Laboratory Report FN-535, January 31 (1990)

[12] J. Irwin, SSC Laboratory Report SSCL-228 (1990).

[13] A. J. Dragt, to appear in the proceedings of the workshop on nonlinear problems in future particle accelerators, Capri, Italy (1990).

LEP dynamic aperture with asymmetrical RF distribution

Francesco Ruggiero
CERN
SL Division, AP Group
CH-1211 Geneva
Switzerland

Abstract The reduction of dynamic aperture for LEP 200 caused by an asymmetrical RF distribution is investigated by particle tracking. In particular, the dependence of this effect on synchrotron tune is studied under different conditions in case of break down of an RF unit. The localization of the accelerating cavities and the consequent discontinuous replacement of radiated energy are taken into account.

1 INTRODUCTION

Owing to synchrotron radiation, electrons and positrons lose energy almost uniformly along the arcs of the machine, while this radiated energy is restored by a few localized RF-stations. Therefore, even neglecting magnetic imperfections, the orbits of electrons and positrons are different: the maximum orbit separation scales with the cube of the particle energy and is inversely proportional to the number of RF-stations. The consequent gradient distortions induced in quadrupoles and sextupoles lead to optics perturbations and thus to a potential deterioration of the LEP performance.

In Section 2, we discuss the chromaticity correction of the 90° lattice and, in Section 3, the corresponding stability limits for different beam energies, synchrotron tunes and symmetries of the RF-system. For asymmetric RF-configurations, we find an unexpected dependence of the dynamic aperture on the choice of synchrotron tune. In particular, the reduction of dynamic aperture is clearly related to the appearance of instability stop-bands in betatron amplitude, which seem to exist even in the absence of radiation effects.

2 CHROMATICITY CORRECTION

The chromaticity correction of the 90° lattice with 4 sextupole families [1] has been improved by an optimisation of the weights appearing in the program HARMON [2], used to compute the sextupole strengths. In particular, the chromatic variation of the tunes has been globally kept below 0.03 for Q_x and 0.06 for Q_y over the relevant momentum range $\Delta p/p = \pm 1.5\%$ (see Fig. 1), corresponding to 10 times the natural energy spread at 100 GeV. These curves should be compared to those of the 60° lattice (Fig. 3.1 in [3]), where the chromatic variation of Q_y over the same momentum range is almost double.

The stability limits discussed in the next section refer to the standard lattice LEP290H (with $Q_x = 91.385$, $Q_y = 97.285$ and $\beta_y^* = 7$ cm, corresponding to n07h46 with 4 sextupole families) and are much higher then those presented in [4], where the chromaticity correction of the 90° lattice had not yet been optimized. A good correction has also been found for a vertical beta-function at interaction $\beta_y^* = 4.3$ cm and for the modified insertions allowing the installation of super-conducting cavities. It should be stressed that the chromaticity correction is computed for an ideal machine without radiation effects, since the latter would be different for electrons and positrons.

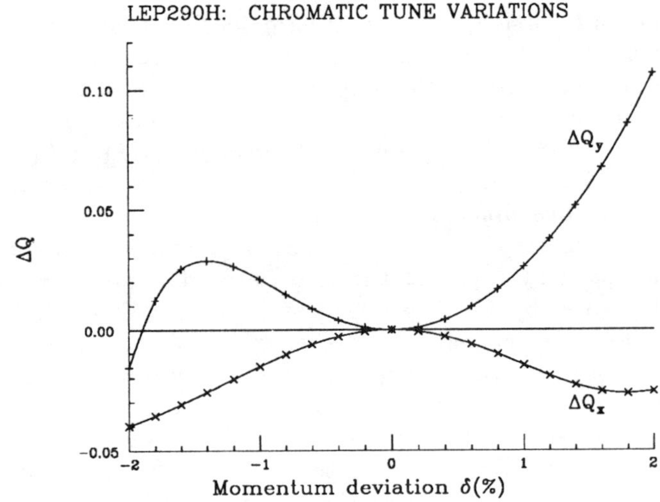

Figure 1: Tune variations with momentum.

3 DYNAMIC APERTURE

In Figs. 2–5, we show different stability limits obtained by particle tracking (with the program MAD [5]) over 400 turns in presence of aperture limiting collimators. In all cases, the effect of radiation on the closed orbit is included (whereas radiation damping and quantum excitation are neglected). Tracking is performed in presence of synchrotron oscillations and we make the pessimistic assumption of full coupling.

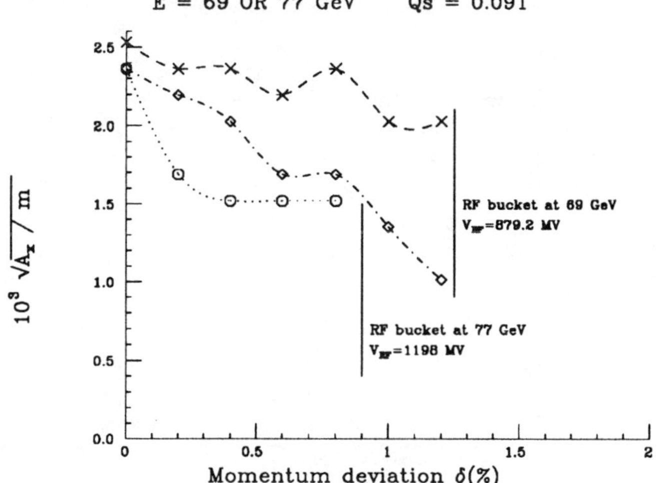

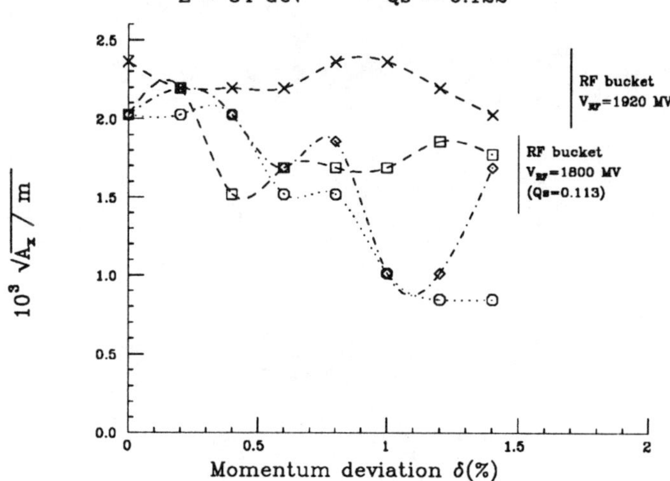

Figure 2: Dynamic aperture versus momentum deviation at 69 GeV with 4 RF-stations (crosses), with 2 RF-stations (diamonds) and at 77 GeV with 2 RF-stations (circles).

Figure 3: Dynamic aperture versus momentum deviation at 84 GeV: with 4 symmetric RF-stations (crosses), with an asymmetry in the voltage at points IP4-IP8 with respect to IP2-IP6 (ratio 2 to 1: diamonds, ratio 3 to 1: circles), with a tripped RF-unit (squares).

The three curves in Fig. 2 show the difference between the case of 2 and 4 RF-stations at an energy of 69 GeV and give the dynamic aperture for the maximum energy (77 GeV) which should be reached with only 2 RF-stations around the interaction points 2 and 6; this aperture turns out to be still sufficient to operate LEP with a good beam life-time. In all cases, the synchrotron tune is assumed to be $Q_s = 0.091$.

In Fig. 3 we consider an energy of 84 GeV, showing the dynamic aperture in the case of 4 RF-stations symmetrically distributed around the interaction points (upper curve) or for less symmetric situations: two of the lower curves refer to the case where the RF-voltage around points 4 and 8 is twice or three times the voltage around points 2 and 6 (always keeping the left-right symmetry around each IP). The synchrotron tune is $Q_s = 0.122$, corresponding to a total accelerating voltage of 1920 MV and to an energy acceptance of the RF-bucket around 1.7%. The remaining lower curve refers to the (originally) symmetric case with a tripped RF-unit: this corresponds to a voltage drop of 120 MV on the left or on the right of one of the interaction points and to a synchrotron tune $Q_s = 0.113$. The energy acceptance of the RF-bucket is reduced to 1.5%.

In Fig. 4 we show similar curves for an energy of 100 GeV and a comparable value of the synchrotron tune $Q_s = 0.119$, corresponding to a total accelerating voltage of 3200 MV and to an energy acceptance of the RF-bucket around 1.2%. It can be seen that even a deviation by a factor two from the 4-fold symmetry leads to a severe reduction of dynamic aperture. In the case of a tripped RF-unit, corresponding to a voltage drop of 200 MV on the left or on the right of one of the interaction points and to a synchrotron tune $Q_s = 0.098$, the energy acceptance of the RF-bucket is reduced to 0.7%, i.e. to less than five times the natural energy spread.

Finally, in Fig. 5, we consider again the case of 100 GeV but for a larger value of the synchrotron tune $Q_s = 0.141$, corresponding to a total accelerating voltage of 3520 MV and to an energy acceptance of the RF-bucket larger than 1.5%. In this case, the reduction of dynamic aperture is tolerable even for a deviation by a factor three from the 4-fold symmetry. The lower curve refers to the (originally) symmetric case with a tripped RF-unit: this corresponds to a voltage drop of 220 MV and to a synchrotron tune $Q_s = 0.127$. The energy acceptance of the RF-bucket is not much reduced, but there is a significant loss of dynamic aperture.

The unexpected dependence of the dynamic aperture on the choice of synchrotron tune (for asymmetric RF-configurations) is currently under investigation, as well as the dependence of these results on the choice of working point. There is a clear relation between the reduction of dynamic aperture and the appearance of instability stop-bands in betatron amplitude, which seem to exist even in the absence of radiation effects (see also [6, 7]). For example, Fig. 6 shows the stability limit obtained by tracking with 2 RF-stations and $Q_s = 0.091$ (corresponding to the lowest two curves in Fig. 2) *in the absence of synchrotron radiation*. The instability stop-band, whose shape and position depends on the choice of 400 turns for the particle survival time, limits the useful dynamic aperture at large momentum deviations down to values comparable to those obtained including radiation effects at 69 or at 77 GeV.

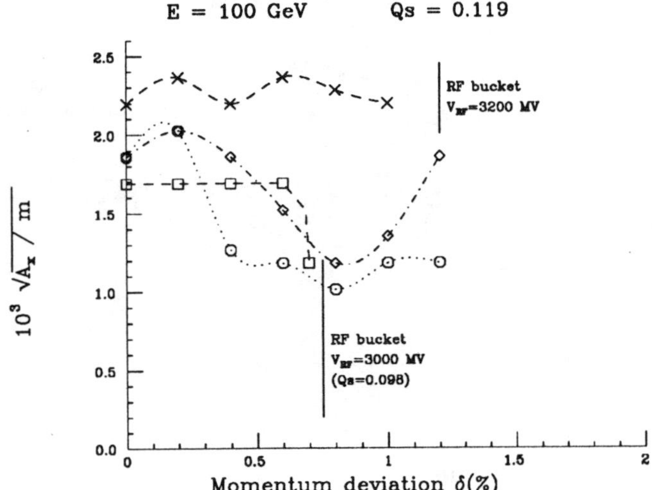

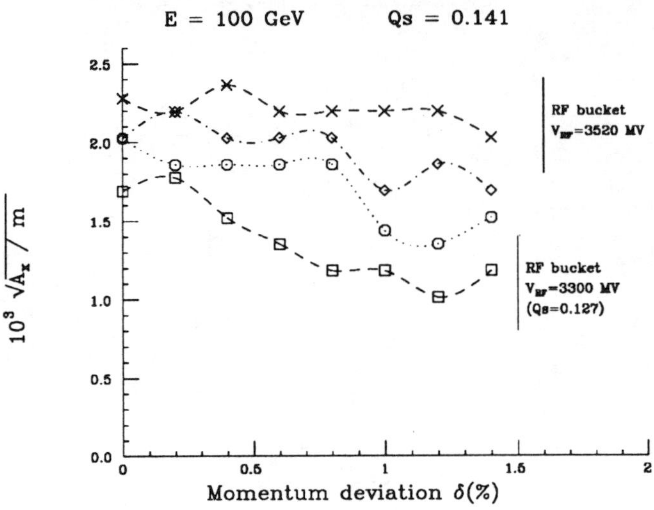

Figure 4: Dynamic aperture versus momentum deviation at $100\,\text{GeV}$ and $Q_s = 0.119$: with 4 symmetric RF-stations (crosses), with an asymmetry in the voltage at points IP4-IP8 with respect to IP2-IP6 (ratio 2 to 1: diamonds, ratio 3 to 1: circles), with a tripped RF-unit (squares).

Figure 5: Dynamic aperture versus momentum deviation at $100\,\text{GeV}$ and $Q_s = 0.141$: with 4 symmetric RF-stations (crosses), with an asymmetry in the voltage at points IP4-IP8 with respect to IP2-IP6 (ratio 2 to 1: diamonds, ratio 3 to 1: circles), with a tripped RF-unit (squares).

By a further optimization of the sextupole strengths, at constant chromaticity, it is possible to substantially increase the survival time in the stop-band. In the case of Fig. 6, for example, the survival time at $\delta = 1\%$ can be increased from 130 to 260 turns by a relative change of $\pm 4\%$ in the strengths of the two focusing sextupole families. Since the betatron damping time at $77\,\text{GeV}$ is about 150 turns, a survival time of 260 turns is sufficient for particle stability.

References

[1] A. Verdier, "Optimisation of the sextupole scheme for the chromaticity correction of LEP13 with 90 degrees per cell", CERN LEP Note 503 (1984).

[2] M.H.R. Donald and D. Schofield, "A user's guide to the HARMON program", CERN LEP Note 420 (1982).

[3] "LEP design report", CERN/LEP/84-01 (1984) p. 22.

[4] A. Fauchet, "Dynamic aperture calculation for LEP 90 deg", CERN/SL-AP Note 90-18 (1990).

[5] H. Grote and F.C. Iselin, "The MAD program (version 8.1): user's reference manual", CERN/SL/90-13 (AP) (1990).

[6] F. Pilat, "Optics for LEP 200 insertions", CERN LEP Note 588 (1987) and "Optics and performance of LEP 200", CERN/LEP-TH/88-29 (1988).

[7] F. Ruggiero, "Consequences of the discontinuous replacement of radiated energy on the performance of LEP 200", Proc. IEEE Particle Accelerator Conference, Chicago, Illinois, Marh 1989, pp. 1298–1300.

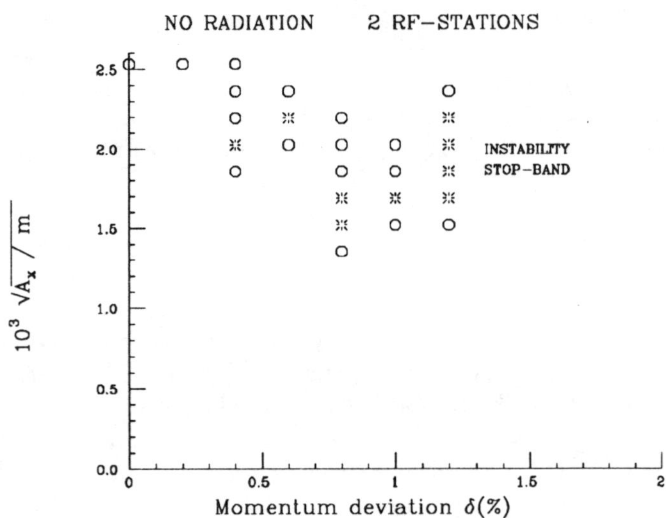

Figure 6: Stability limit with no synchrotron radiation and 2 RF-stations at $Q_s = 0.091$: circles denote particles surviving at least 400 turns, whereas stars in the instability stop-band correspond to particles typically lost after 100–200 turns.

Long-Term Stability Studies for the Large Hadron Collider

F. Galluccio, Z. Guo,* W. Scandale, F. Schmidt and A. Verdier
CERN
CH-1211 Geneva 23

Abstract The dynamic aperture of the next generation of hadron colliders will be strongly limited by the non-linear components of the bending and focusing fields, which are unavoidable in superconducting magnets. In the Large Hadron Collider (LHC) particle stability over about 10 min is required at injection: this corresponds to 7×10^6 turns in the machine. Tracking simulations up to such a high number of turns for a significant number of different cases is impractical, if not impossible, even with the most powerful computer; therefore a different approach must be used to evaluate the LHC long-term dynamic aperture. The most realistic machine has been modelled, including RF cavities, residual closed-orbit distortion after compensation, systematic and random multipolar errors in dipoles and quadrupoles, skew quadrupoles, and lumped multipolar correctors. Long-term computer tracking (4×10^5 turns/$\sim$ 40 s) has been performed for a large number of machine configurations, for particles with different starting coordinates and momenta. The border between stable and chaotic motion has been estimated from the analysis of the medium term tracking results using the Lyapunov exponent method. Only for a small number of selected cases has tracking been resumed and carried on up to 10^6 turns. The results of this study and the estimated value of the long-term dynamic aperture of the LHC at injection are given here.

1   Introduction

No analytical tool is yet available to allow a fast and reliable prediction of the long-term dynamic aperture for conservative systems like the Large Hadron Collider. The reason for this is twofold: first, it is difficult to detect an instability that is caused by the interaction of many weak resonances, and secondly, finding this border of stability, for instance via the Lyapunov method, does not tell us when the actual particle loss will take place. It therefore remains mandatory to do tracking over many turns to estimate the range of amplitudes for which the storage and the acceleration of particles can be ensured. It goes without saying that analytical tools are of paramount importance to understand the results.

In order to make use of the power of today's computers such as the Cray-XMP at CERN, the tracking codes MAD [1] and SIXTRACK [2] have been vectorized. The use of two independent programs, having about the same

computational speed (2×10^5 s of CPU for 20 particles followed over 10^6 turns), allowed these costly investigations to be double-checked. Most of the tracking data have been recorded on tape, to allow further analysis when new aspects arise.

The effect of rounding errors is always a matter of concern when up to 10^{12} operations are performed for 10^6 turns of a full LHC structure. It has been checked that with single precision on the Cray these errors do not invalidate our results up to 10^6 turns.

We therefore feel confident that our tracking studies allow, in the limit of the physical model, a safe estimate of the LHC dynamic aperture.

2   The Machine Model

The layout and the optics of the LHC as used for these studies are described in Ref. [3]. A thin-lens lattice model is adopted for tracking, which has been shown to give results nearly identical to those of the thick-lens model. The machine consists of eight arcs and eight straight sections, with general-purpose insertions in straight sections 4, 6, and 8, experimental insertions in straight sections 1, 2, and 5, a dump insertion in straight section 3 and a halo-cleaning insertion in straight section 7. At injection all general-purpose and experimental insertions are detuned with $\beta_x^* = \beta_y^* = 8$ m; in the dump insertion $\beta_x^* = \beta_y^* = 220$ m, and in the cleaning insertion $\beta_x^* = \beta_y^* = 15$ m.

The selected tunes are $Q_x = 70.28$ and $Q_y = 70.31$; this working point lies inside one of the three best regions of tunes (Fig. 1) and at a distance to the diagonal that minimizes the effects of coupling [4].

The performances of a superconducting machine like the LHC are determined by the multipolar imperfections of the magnetic fields, and by the correction elements introduced to reduce their influence. In order to simulate the machine in the most realistic way, all linear and non-linear perturbations and their respective correction elements are implemented in the model.

The systematic and random multipolar errors of all dipoles and quadrupoles [5], with both normal and skew components, and up to the 20-pole, can be found in Table 1. Moreover, a closed-orbit distortion is simulated by misaligning all quadrupoles.

We have included the following correction elements: four families of skew quadrupoles are located in the insertions to compensate linear coupling, orbit correction dipoles are

---

*On leave of absence from IHEP, Beijing, China.

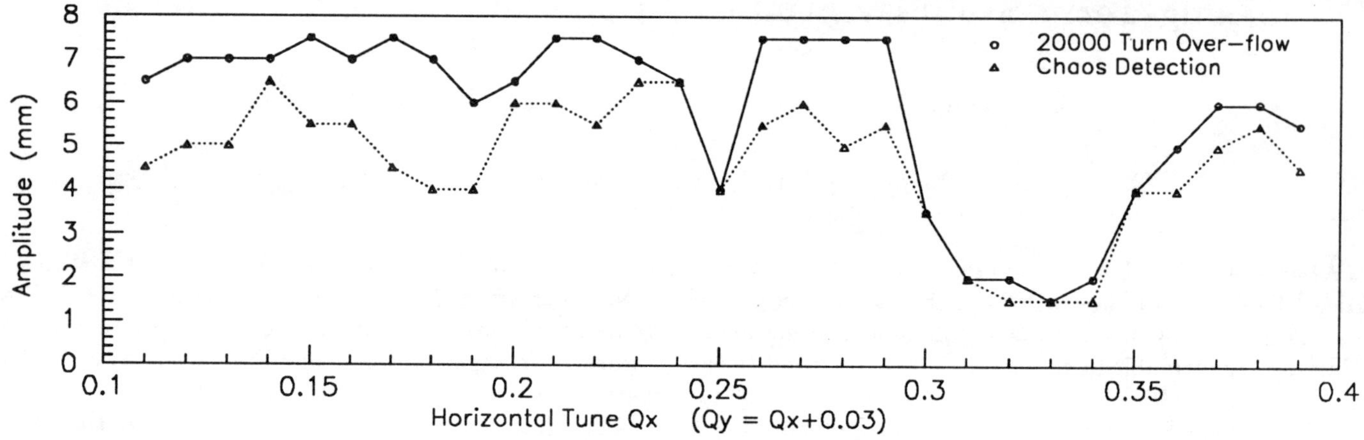

Figure 1: Dynamic Aperture as a Function of the Working Point

Table 1: Coefficients of Systematic and Random (r.m.s.) Multipolar Errors (in units of 10^{-4} at $R_r = 1$ cm) at Injection

| Order | Dipole Errors | | | | Quadrupole Errors | | | |
| | Systematic | | Random | | Systematic | | Random | |
n	b_n	a_n	b_n	a_n	b_n	a_n	b_n	a_n
1	0.0	0.0	0.0	0.0	0.0	0.0	0.0	0.0
2	3.00	0.80	1.2	1.7	0.0	0.0	0.0	0.0
3	-3.35	0.10	1.7	0.5	0.02	0.28	1.06	1.04
4	0.05	0.03	0.15	0.2	0.19	0.01	0.24	0.39
5	0.45	0.03	0.22	0.07	0.05	0.01	0.10	0.09
6	0.0	0.0	0.0	0.0	-0.43	0.0	0.19	0.07
7	0.14	0.02	0.02	0.04	0.0	0.0	0.02	0.03
8	0.0	0.0	0.0	0.0	0.0	0.0	0.01	0.01
9	0.034	0.001	0.005	0.002	0.0	0.0	0.01	0.01
10	0.0	0.0	0.0	0.0	0.01	0.0	0.01	0.0

available next to each quadrupole, and lumped multipolar correctors, placed 'à la Neuffer', are included in every cell in the arcs.

The linear coupling compensation and the closed orbit correction are extensively reviewed in these proceedings [4]. The compensation of the amplitude- and momentum-dependent detuning due to the systematic multipolar errors of the dipoles is accomplished by means of multipole correctors (sextupoles, octupoles, and decapoles). Their strengths are set by minimizing the chromaticity up to third order [6], which also leads to a strong reduction of the amplitude-dependent tune-shift.

Extensive studies have shown that the multipolar corrector in the center of the half-cell is mandatory, except, maybe, for the octupole.

3   Long-Term Tracking

The many parameters that may limit the stability of the LHC make it difficult to provide a reliable estimate of the dynamic aperture. Our first task was therefore to find a small but sufficient set of those parameters to reduce the enormous need of computing time, thereby allowing a repetition of the whole study in case of changes during the design period.

The synchrotron motion is routinely included since tracking with fixed momentum deviation has shown to be too optimistic. We consider transversely round beams and we fix the relative momentum deviation to 10^{-3} ($2\sigma_E$ at injection). Moreover, we can post-process those data to obtain detuning, smear and other quantities of interest. An early detection of chaos via the Lyapunov exponent method provides us with an estimate of the border of infinite stability and the actual particle loss is recorded in survival plots for up to 4×10^5 or even 10^6 turns. A critical issue is how to assign the random errors to each individual element. From 10 sets of random distributions the worst, best, and most realistic ones were selected on the basis of short-term tracking (1000 turns). In order not to mix different effects, this choice of random distributions was done separately for dipoles, quadrupoles and quadrupole alignment errors.

4   Results

Figure 2 shows a typical example of a survival plot, in this case with multipolar errors in the dipoles only. The tracking has been carried on up to 10^6 turns. For the three selected random error distributions we find a steep increase of the loss turn number (above 10^5) when the Lyapunov limit is approached. The unexpectedly large dispersion of the dynamic aperture obtained for different random distributions remains to be understood. It is hoped, however, that it can be reduced by sorting the magnets.

Table 2 gives a summary of the results of our study for the most realistic error distribution: in all cases the border between regular and chaotic particle amplitudes is given, together with the largest stable amplitude for the indicated number of turns.

The rather large b_7 and b_9 field components lead to a reduction of the dynamic aperture by 1.5 mm (cases Nos. 1 and 2) in the presence of dipole errors only. Adding

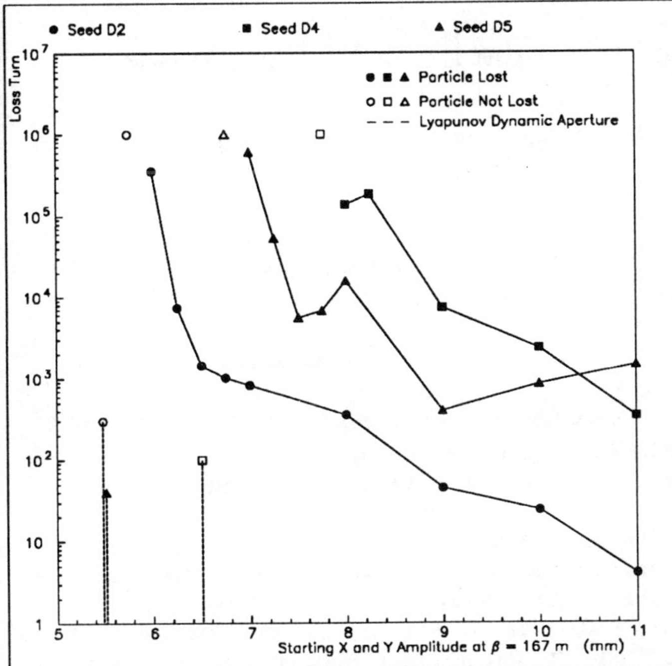

Figure 2: Survival Plot for the LHC with Errors in the Dipoles Only - 3 Random Error Distributions

Table 2: Dynamic Aperture in mm at a Focusing Quadrupole ($\beta = 167$ m)

Machine Configuration	Dynamic Aperture based on:		
	Chaos Detection	10^5 turn overflow	4×10^5 turn overflow
1) Dipole ($\phi = 50$ mm) Err. No a_2 - No Syst. b_7, b_9	7.	9.	
2) Dipole ($\phi = 50$ mm) Err. No a_2 - With Syst. b_7, b_9	5.50	7.00	7.00
3) Dipole ($\phi = 50$ mm) and Quad. ($\phi = 56$ mm) Err. No a_2 - With Syst. b_7, b_9	5.75	7.25	6.75
4) Dipole ($\phi = 50$ mm) and Quad. ($\phi = 56$ mm) Err. All Multipole Errors	5.0	6.5	6.0
5) All Multipole Errors in Dipoles and Quads. + Orbit ($\sigma = 0.6$ mm)	4.75	6.75	6.50
6) All Multipole Errors in Dipoles and Quads. + Orbit ($\sigma = 1.7$ mm)	3.75	5.75	5.25
7) Dipole ($\phi = 56$ mm) and Quad. ($\phi = 56$ mm) Err. No a_2 - With Syst. b_7, b_9	6.25	8.25	8.25

the quadrupole errors does not change the dynamic aperture in the limit of our resolution of 0.25 mm. The skew quadrupole components (a_2) in the dipoles, notably the large random part, also lead to a significant decrease of the dynamic aperture of 0.75 mm (case No. 4). A small closed orbit (case No. 5) does not affect the stability behaviour very much, whilst a sizeable and probably realistic orbit of $\sigma = 1.7$ mm causes an additional drop of more than 1 mm (case No. 6). For a safe operation of the LHC it is believed that an aperture of 4.5σ of the beam is needed. In our case this corresponds to 5.2 mm, which is larger than our overall dynamic aperture of 3.75 mm.

Increasing the dipole bore diameter from 50 to 56 mm, a large fraction of the loss in the dynamic aperture has been recovered (cases Nos. 2 and 7), so that the required stable region is achievable. A similar improvement of the dynamic aperture has been obtained by reducing the b_7 and b_9 components by a factor of 2 and 3, respectively. Another amelioration can be expected from the reduction of the random a_2 component. It remains to be seen which strategy for increasing the dynamic aperture is followed.

Although the performance of the machine at injection is believed to be more critical, we still have to investigate the behaviour of the machine in collision mode.

Acknowledgements

We are very grateful to E. McIntosh from the CN Division of CERN who assisted us in all questions concerning computing, and allowed us to make the best use of our facilities, including data transfer between computers and data convertibility between programs, graphics and the like. F. Zimmermann from DESY helped us modify SIXTRACK to allow the analysis of the correction scheme. We finally would like to thank J. Gareyte for many stimulating discussions and continuous support.

References

[1] H. Grote and F. C. Iselin, "The MAD Program (Methodical Accelerator Design) Version 8.1, User's Reference Manual", CERN–SL 90–13 (AP).

F. C. Iselin, "Improvements in MAD in View of LHC Design", These Proceedings.

[2] F. Schmidt, "SIXTRACK Version 1, Single Particle Tracking Code Treating Transverse Motion with Synchrotron Oscillations in a Symplectic Manner, User's Reference Manual", CERN–SL 90–11 (AP).

F. Schmidt and M. Vänttinen, "Vectorization of the Single Particle Tracking Program SIXTRACK", CERN SL/AP/Note/90–20.

[3] W. Scandale, "The Lattice of the LHC: Version 1", CERN–SL 91–03 (AP), LHC Note 139.

[4] F. Galluccio, Z. Guo, T. Risselada, W. Scandale and F. Schmidt, "Compensation of Linear Lattice Imperfections in the Large Hadron Collider", These Proceedings.

[5] The LHC Study Group, "Design Study of the Large Hadron Collider (LHC)", CERN 91–03, May 1991.

[6] F. Galluccio, "Compensation of Systematic Multipolar Errors in the LHC from the Operational Point of View", CERN SL/AP/Note/90–12, LHC Note 133.

Compensation of Linear Lattice Imperfections in the Large Hadron Collider

F. Galluccio, Z. Guo[*], T. Risselada, W. Scandale and F. Schmidt
CERN, CH-1211 Geneva 23, Switzerland

Abstract Closed-orbit and linear coupling corrections have been studied for the LHC lattice, using different samples of the random error imperfections. Distortions in the closed-orbit are mainly due to alignment and field errors of the main magnets. The main source of the linear coupling is found to be the skew quadrupole component of the dipoles, with its random and systematic part. The linear imperfections expected in the LHC are large enough to prevent the confinement of the beam in the vacuum chamber. Beam position monitors and correction dipoles near each lattice quadrupole are able to reduce the r.m.s. closed-orbit distortion to below 1 mm. Skew quadrupoles located in the insertions and powered in four families are sufficient to handle the linear coupling. The dynamic aperture, with the injection optics and realistic corrections of the linear lattice imperfections, has been evaluated by computer tracking simulations.

I. LATTICE MODEL

The layout adopted is that of Ref. [1]. It consists of eight arcs interleaved with eight insertions. Each arc contains 25 regular FODO cells 99 m long, with 90° phase advance. The insertions have a general purpose (GP) design, except those in point Nos. 3 and 7, dedicated to the beam dump and the halo cleaning respectively. The values of the ß-functions at the interaction points (IP) are shown in Table 1. With collision optics, only the insertions Nos. 1, 2, and 5 are tuned to low-ß.

Table 1
Values of ß* [m] at the interaction points

	GP No. 1, 2, 5	GP No. 4, 6, 8	Cleaning No. 7	Dump No. 3
Injection	8.0	8.0	15	220
Collision	0.5	8.0	15	220

A thin-lens version of the lattice has been used for computer tracking simulations. An aperture limitation of 25 mm, as large as the dipole inner coil radius, has been assumed. To introduce synchrotron oscillations an RF cavity has been added. The beams are separated vertically by $\pm3\sigma$ in all the IPs, and cross horizontally at an angle of ±100 μrad in all the GP insertions.

The choice of the nominal working point $Q_H = 70.28$, $Q_V = 70.31$, has been based on the optimization of the medium-term ($\sim 10^5$ turns) dynamic aperture, using the computational tools of Ref. [2]. An increase by one unit of Q_H has been considered in order to reduce the linear coupling resonances. Larger horizontal-vertical tune separations have not been retained since they imply either an unacceptable mismatch of the optical functions in the insertions, or a heavy modification of the lattice layout.

II. IMPERFECTIONS

Positioning errors with random Gaussian distributions, truncated at 3σ, and with an r.m.s. value of 0.14 mm for the misalignment and of 0.24 mrad for the tilt are envisaged in the LHC, but in order to take into account alignment difficulties related to the twin aperture design of the main magnets, up to four times larger r.m.s. values have been considered in our simulations. Beam position monitor reading errors of 0.6 mm and the field-shape imperfections in Table 2 are also relevant to the linear lattice behavior of the LHC, and have been included in our simulations together with the higher-order multipole components of the field-shape imperfections quoted in Ref. [2].

Table 2
Field errors in units of 10^{-4} at $R_r = 1$ cm

Errors	Relative field error	a_2 systematic	a_2 random r.m.s.
Dipole	5	0.8	1.7
Quadrupole	5	-	-

III. CORRECTORS

The two LHC rings are treated individually as far as corrections are concerned. Orbit correctors and monitors are located next to each twin lattice quadrupole. Directional monitors are installed near the inner triplets of the GP insertions, where the counter-rotating beams are almost collinear. Correctors are excluded there to avoid coupled orbit distortions in the two LHC rings. Additional correctors and monitors are added in the insertions to take care of the vertical separation and of the crossing angle at the IPs, as well as of the possible orbit mismatch at the outer ends of the dispersion suppressors. The integrated strength of the correcting dipoles is 1.5 T·m in the arcs, and 2.2 T·m in the insertions, where the orbit functions are more irregular and a better correction is required. To decouple the horizontal and vertical motions a global correction is implemented: 92 skew quadrupoles, powered in four families, are installed in the insertions, twelve in each GP insertion, and ten in each machine insertion. Two of the families are almost orthogonal in phase for the main diagonal Q_H-Q_V. The correcting magnets are 0.72 m long, with a gradient of 120 T/m, and are of the same design as the arc tuning quadrupoles.

Sextupole, octupole and decapole correctors à la Neuffer are used in the regular cells [2].

---

[*]On leave of absence from IHEP, Beijing, China

IV. ORBIT CORRECTION

The correction of the orbit is very efficiently performed with the MICADO algorithm [3], using the full set of correctors. Satisfactory results have also been obtained with a reduced number of correctors in the arcs. Sophisticated powering schemes have thus been considered to reduce in costs. The most promising of them is the one in which the correctors of two consecutive regular cells are connected in series with those of the next two consecutive cells. In a smoothly aligned machine the results of Fig. 1 can be easily reached, and local large misalignments can still be handled.

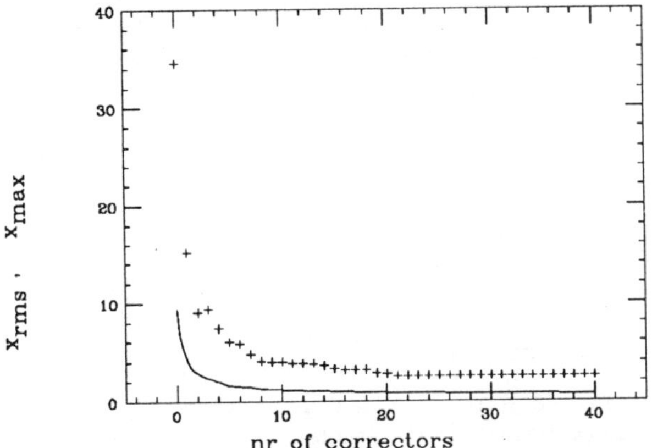

Fig. 1. R.m.s. (continuous line) and maxima (dotted line) of the closed-orbit deviations, as functions of the number of correctors used around the ring

V. LINEAR COUPLING CORRECTION

At injection, the linear coupling of the LHC is mainly due to the systematic and random skew quadrupole component a_2 in the field of the main dipoles. The other sources of coupling, i.e. the tilt of the main quadrupoles, and the vertical orbit in the chromaticity sextupoles, are more than an order of magnitude weaker. This is also true for the collision optics, although in this case the tilt of the inner triplet of the experimental insertions is no longer negligible, and possible axial fields in the experimental devices will need to be locally corrected.

With realistic errors, and the nominal setting of the main quadrupoles gradients, the strength of the main diagonal resonance is as large as almost half a unit in tune, which will make it impossible to operate an uncorrected machine. With one unit separation of the horizontal and vertical tunes, this strength is reduced to 0.02; it can be further diminished with two families of skew quadrupoles orthogonal in phase, by minimizing the value of the closest tune approach with numerical methods. In the example of Fig. 2, where the case of a realistic distribution of a_2 is considered before and after the coupling compensation, tune separations of 0.001 units are reached after a few iterations.

There are two other approaches to globally decouple the machine with the four families of skew quadrupoles, assuming a perfect knowledge of the magnetic imperfections a_2 in all the dipoles. In the first approach, the strength of the correctors is adjusted to cancel the off-diagonal coupling terms of the one-turn transfer matrix computed from IP1. Of course, the results will depend on the choice of the starting point of the computation, but we found this dependence to be almost negligible in practice. In the second approach, the first-order skew resonances $Q_H \pm Q_V = p$ are compensated. At the nominal working points, 70.28/70.31 and 71.28/70.31, with both systematic and random a_2, the two methods lead to a decoupled machine with almost equal compensating strengths.

The proposed decoupling strategies are inadequate to cancel the vertical dispersion around the accelerator: they leave residues of about 0.6 m in the arcs and less than 4 cm at the IPs for the two considered optics. Other correction schemes which could compensate this effect are under investigation.

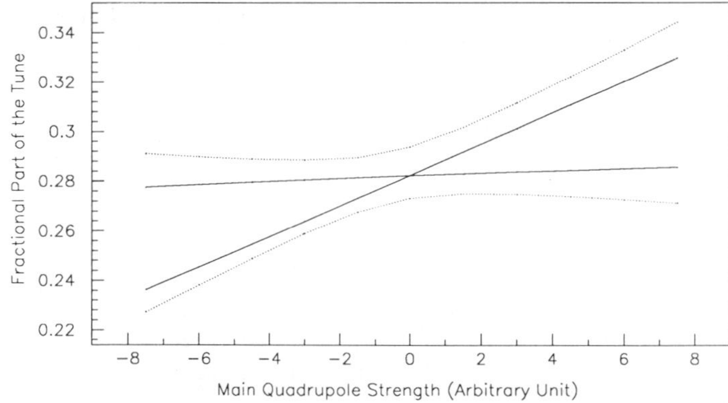

Fig. 2. Fractional part of the tunes as a function of the strength of a tuning quadrupole, before (dotted line) and after (continuous line) the linear decoupling

VI. OPTIMIZATION OF THE WORKING POINT

The choice of the nominal working point of the LHC has been discussed in Ref. [2]. Here we would like to motivate the difference of 0.03 units between the horizontal and the vertical tunes, which has been found to be the best compromise between the necessity for the working point to be far enough from the main diagonal to avoid linear and non-linear coupling resonances, and close enough to keep away from higher-order resonances. Figures 3 and 4 show the medium-term dynamic aperture, i.e. over $2 \cdot 10^4$ turns, as a function of the working point, in the injection configuration, without and with the random and systematic a_2, respectively. Different distances from the diagonal are considered and, in both cases, we see that the maximum beam stability is achieved for $Q_H\text{-}Q_V = 0.03$. Furthermore we observe that the results with and without a_2 are practically the same, from which we conclude that the global decoupling, made in this case with the matrix formalism, is very effective, and that the beam stability is essentially determined by the higher order resonances. Nevertheless it has to be pointed out that in a realistic machine the third order skew resonance $3Q_V = 211$ is larger than when a_2 is neglected. This is likely to be due to the local residual coupling in the chromaticity sextupoles, and in the locations of the sextupolar field imperfections. Results in agreement with the previous ones have also been obtained from the detection of chaos by means of the Lyapunov exponent method.

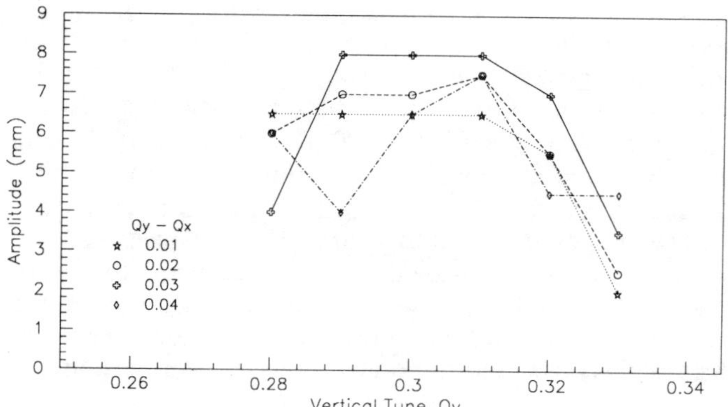

Fig. 3 Medium-term dynamic aperture as a function of the working point, without skew quadrupolar terms

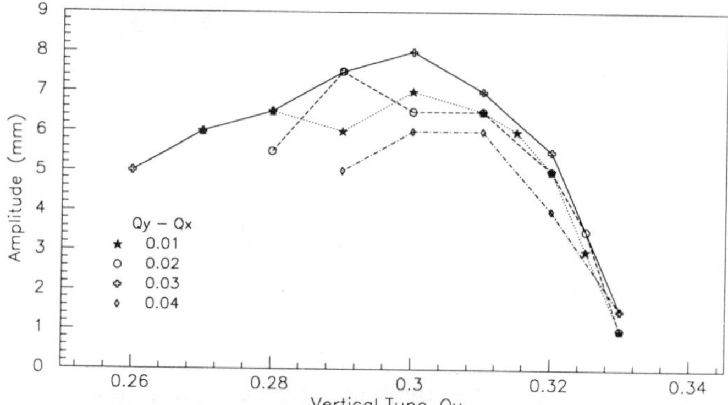

Fig. 4 Medium-term dynamic aperture as a function of the working point, with skew quadrupolar terms

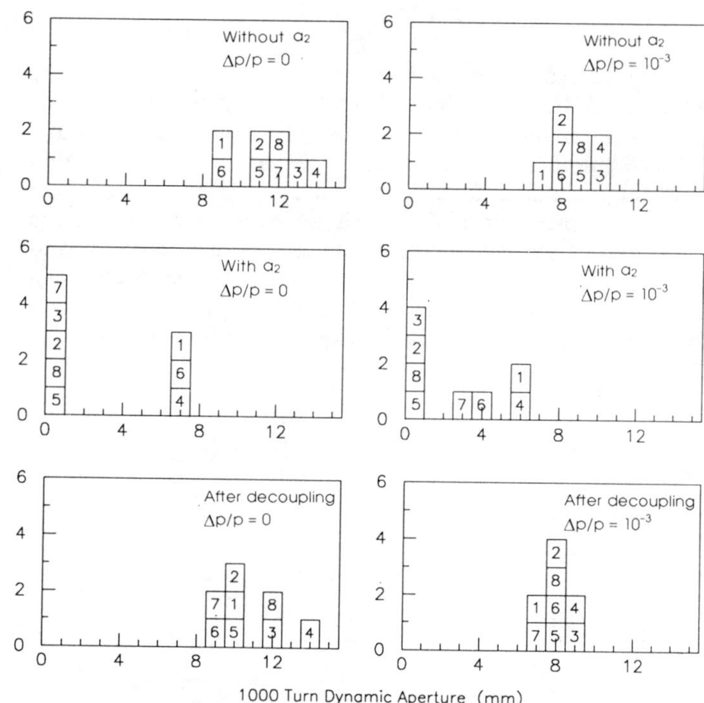

Fig. 5 Short-term dynamic aperture with field-shape imperfections in the dipoles

VII. EFFECTS ON THE DYNAMIC APERTURE

The effect on the short-term dynamic aperture of the a_2 components in the dipoles has been studied with injection optics, and eight samples of the random imperfections in the dipoles. The results are summarized in Fig. 5. The stability region is drastically reduced when uncompensated skew quadrupole errors are added. Most of the reduction is recovered when the global decoupling is applied. For one case, in which the quadrupole field-shape imperfections have also been included, the comparison has been carried out for about 10^6 turns, as shown in Fig. 6. The reduction of the dynamic aperture, with globally compensated a_2 imperfections is of about 1 mm, a result also confirmed by the analysis of the Lyapunov exponent.

The effect of a finite closed-orbit on the dynamic aperture is smaller but still non-negligible. With r.m.s. orbit deviations of 1 mm, a reduction of 0.3 to 0.5 mm has been computed.

VIII. ACKNOWLEDGEMENTS

We would like to thank J. Gareyte, E. Keil, J.P. Koutchouk, and E. Moshammer for helpful discussions.

IX. REFERENCES

[1] W. Scandale, "The Lattice of the LHC: Version 1", CERN SL 91-03 (AP), LHC Note 139, (1991).

[2] F. Galluccio, Z. Guo, W. Scandale, F. Schmidt, and A. Verdier, "Long-Term Stability Studies for the Large Hadron Collider", These Proceedings.

[3] B. Autin and Y. Marti, "Closed-Orbit Correction of Alternating Gradient Machines using a Small Number of Magnets", CERN/ISR-MA/73-17, (1973).

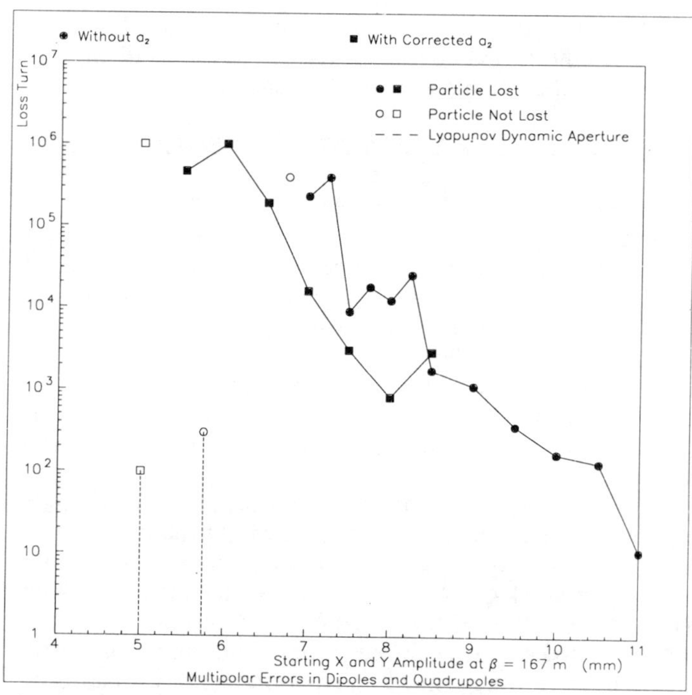

Fig. 6 Survival plot for the LHC with multipole errors in dipoles and quadrupoles, without and with a_2

High-Luminosity Insertion for a B-meson Factory

Bruno Autin

PS Division, CERN, CH-1211 Geneva 23

Abstract

The major characteristics of a very high-luminosity B-meson factory are presented. The machine would consist of two rings lying in the same horizontal plane and using a *crab crossing* scheme to restore the conditions of head-on collisions. Emphasis is put on very low-β values (≤ 1 cm) to relax the constraints imposed by the intensity. This results in a very strongly focusing lattice.

INTRODUCTION

The B-factories [1,4] are characterized by very high luminosities combined with a low level of background. Two orders of magnitude in luminosity have to be gained over existing lepton rings. The solution consists of increasing the number of bunches and avoiding unwanted collisions by making electrons and positrons circulate in two different rings. The separation of the bunches remains a difficulty in the interaction region. If the collisions are head-on, one takes advantage of the different energies of the electron and positron beams to separate them magnetically but the bunch spacing can reach its minimum value only with a geometric separation which has the extra properties of reducing the synchrotron background by eliminating any dipole field and being applicable to machines with symmetric energies. A design using a crossing angle suffers from several handicaps: it reduces the overlap of the bunches and therefore the luminosity, and it is a dangerous source of synchro-betatron oscillations which were a serious limitation in the first DORIS approach [5]. These problems can actually be solved with the *crab crossing* scheme [6,7] which restores the conditions of head-on collisions by using two rf deflectors located $\pi/2$ apart from the interaction point. It is this approach which will be discussed in the context of a study of a B-factory in the CERN ISR tunnel where the two rings, like the old proton rings, would lie in the same horizontal plane.

DESIGN STRATEGY OF A VERY HIGH LUMINOSITY B-FACTORY

Central in the parameterization of a machine is the choice of the beam intensity and of the vertical β^* value at the crossing point. Here, we emphasize the arguments in favor of the smallest β^* value. A high beam intensity has adverse effects in at least two major areas: synchrotron power and higher order modes. The synchrotron power must be maintained within reasonable limits; it is an obvious waste of energy; for the energies and intensities of a B-factory, the synchrotron light beam impinging on the vacuum chamber imposes technical solutions which are costly; furthermore, the background in the experimental areas is directly related to the beam intensity. The higher order modes power per unit length generated by the interaction of a bunched beam with a cavity is given by

$$P_{\text{hom}} = k\,\tau_b\,I^2$$

where k is of the order of 0.15 V/pC/m at 350 MHz and increases with the square of the frequency, τ_b is the time separation between consecutive bunches and I the beam current. This power exceeds 600 kW in the CERN-PSI design [8] and is clearly a major concern both for the development of couplers which have to evacuate this power and for the wake field which drives coupled bunches instabilities.

According to the expression of the luminosity, the intensity can only be reduced by diminishing β^* and the bunch length in the same proportion. A first limit is imposed by the rf voltage required to focus short bunches. Another difficulty lies in the correction of the high chromaticity associated with low-β^* values yet maintaining a sufficient dynamic aperture; this is the realm of non-linear optics where the experience accumulated with recent storage rings gives confidence in the finding of efficient correction schemes. Last, the bandwidth of the feedback systems which have to damp the coupled bunch instabilities has to be increased, but the developments, made especially in the domain of stochastic cooling, let hope that technical solutions exist [9].

Our discussion is limited to flat beams in the interaction area. In spite of the gain of a factor 2 in luminosity which could be theoretically accomplished with round beams, it turns out from more detailed studies on the beam-beam interaction [10] and optics schemes [11] that the gain increase with respect to flat beams may not exceed 40% and that the complications bound to such a scheme prevent the use of much more powerful methods of luminosity upgrading. In these conditions, the beam-beam tune shift, which is the measure of the beam-beam limit, is proportional to the beam density in the normalized horizontal phase-space N/ε_x where N is the number of particles per bunch and ε_x the horizontal emittance. For a given luminosity, the beam-beam tune shift decreases with the number of bunches but one may like nevertheless to work near the beam-beam limit by reducing the emittance. A low emittance has the substantial advantage of limiting the synchrotron radiation in the quadrupoles and is consistent with the short bunches necessary in very low-β scheme.

SCALING LAWS FOR SHORT BUNCHES AND LOW-β VALUES

A collider can work near the beam-beam limit (≈ 0.05 for leptons) if the bunch length is roughly half the β^* value [12,13]. We shall discuss the implication of short bunches on rf voltage and single bunch longitudinal instability then how to achieve low-β^* values. We assume that the regular part of the lattice is composed of FODO cells for which scaling rules are well known. In particular, the horizontal emittance is given by

$$\varepsilon_x = \frac{2R}{Q_x^3}\,\sigma_\varepsilon^2$$

where R is the arc radius, Q_x the horizontal betatron tune and σ_ε the energy spread. Another quantity which plays a central role is the momentum compaction which measures the rate of change of the orbit circumference with the momentum:

$$\alpha_p = \frac{2\pi R}{C}\,\frac{1}{Q_x^2}$$

where C is the machine circumference. Under these conditions, the voltage required to get a bunch of length $2\sigma_z$ is

$$V = \frac{c\,R\,E\,\sigma_\varepsilon^2}{e\,f_{rf}}\,\frac{1}{\left(Q_x\sigma_z\right)^2}$$

where E is the bunch central energy and f_{rf} the rf frequency. In the same way, the maximum impedance permitted for the longitudinal stability of the bunch is proportional to the product $\alpha_p\sigma_z/N$; replacing N by its expression as a function of the beam-beam tune shift and the emittance, it turns out that again this impedance depends on the product $Q_x\sigma_z$. In brief, as far as the lattice structure is concerned, any reduction of the bunch length must be accompanied by a corresponding increase of the betatron tune.

For the low-β insertion, it is proposed that the two beams have independent focusing. Diminishing the constraints improves the simplicity and the performance of the structure. The insertion shown in Fig.1 is made of a single lens to achieve the shape of the beam at the crossing point followed by a quarter wavelength transformer to match the final focus to the regular FODO cell [14].

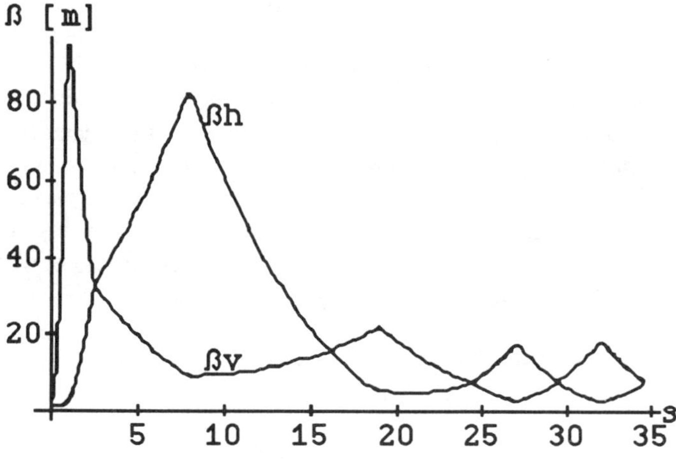

Fig. 1. Model of low-β insertion
($\beta_v^* = 1$ cm, $\beta_h^* = 1$ m, $d^* = 1$ m).

The betatron matching is performed between planes where the β functions are equal and the α functions opposite so that the whole insertion is fully parameterized analytically within the thin lens model. The normalized focal length of the final lens is

$$q = \frac{\beta_h + \beta_v - \sqrt{\beta_h\beta_v\left[4 + (\beta_h - \beta_v)^2\right]}}{\beta_h - \beta_v}$$

All the quantities in the above expression are normalized to the distance from the crossing point to the final lens. As expected, the method can only be applied if β_h and β_v are different. For flat beams, q is practically independent of β_v and roughly equal to 0.8. The measure of the flatness of the beams is given by the coupling coefficient c, ratio of the vertical to horizontal β functions also equal to the ratio of the beam emittances. The scaling of the parameters is conveniently performed at constant c so that β_h is automatically related to β_v. The disdance d which separates the final lens from the transformer is then

$$d = \frac{\sqrt{4 + (1-c)^2\beta_h^2}\left(1 + c - \sqrt{c\left[4 + (1-c)^2\beta_h^2\right]}\right)^2}{2\sqrt{c}\left[4\left(1 + c - \sqrt{c\left[4 + (1-c)^2\beta_h^2\right]}\right) + (1-c)^2(1+c)\beta_h^2\right]}$$

The variations of d and q versus c are very similar except at the limit of very flat beams ($c \approx 0$) where d tends towards infinity whereas q cannot exceed 1.

The length of the insertion is not only determined by d but also by the length of the transformer which varies like $\sqrt{(\beta_0/\beta_v)}$; the insertion can be maintained compact if the β function of the lattice β_0 is small enough. This is one more aspect of the general rule: a small β_v at the crossing point implies a strong focusing lattice.

CROSSING ANGLE

The crossing angle is determined by the room available for the final quadrupole and the crab crossing cavity.

The final quadrupole in the type of insertion which has just been presented has a focal length of the order of 0.8 m and thus an integrated gradient at of 33 T at 8 GeV/c. It must have a width smaller than $l\alpha$ where α is the total crossing angle and l the distance from the crossing point to the quadrupole. As a tentative set of parameters, we assume a coupling coefficient β_v/β_h or $\varepsilon_v/\varepsilon_h$ equal to 0.01, a horizontal emittance of 3×10^{-8} m, a value of β_v^* of 1 cm and a distance l of 1 m so that the beam is round with a standard deviation of 0.25 mm. A permanent quadrupole with an internal radius of 1 cm should therefore be well adapted to accommodate the vacuum chamber, a beam envelope of 10σ, a closed orbit error of a few millimeters and to keep some flexibility for a further reduction of β_v^*. With rare earth permanent magnets, a magnetization of 1 T and therefore a gradient of 100 T/m are typical characteristics; the magnetic length of the high-energy ring quadrupole would therefore be 33 cm.

In order to have a magnet width compatible with a small horizontal crossing angle, slim geometries [15] which reproduce for permanent magnets the properties of figure of eight quadrupoles were investigated. The aperture is circular but the outer contour has an egg shape described by the polar equation

$$r = r_0 \exp\left[k(1 - \cos 2\phi)\right]$$

When k is zero, the contour is simply a circle of radius r_0. When k increases, the contour is elongated in the vertical direction and, for small values of k, takes a quasi-elliptical shape; a transition occurs at k = 0.25 value beyond which the contour has a waste in the horizontal medium plane of symmetry. The shape associated with k = 0.25 is called super-elliptical. The characteristics of two types of quadrupoles are summarized in Table 1.

Table 1 Characteristics of permanent quadrupoles

Outer shape	Gradient [T/m]	Field region with gradient error $<10^{-3}$
Circle	1	63 mm
Super ellipse	1.18	52 mm

The higher gradient of the super ellipse type is at the cost of a reduced good field region. The circular shape is thus chosen as the reference design and a super elliptical shape may become an interesting alternative if one has to increase the inner radius without changing the overall width. In both cases, the residual field in the adjacent low energy ring quadrupole is negligible.

The rf deflector can be located near the second quadrupole of the transformer where β_h is maximum and the horizontal betatron phase close to $\pi/2$. At 500 MHz, the radius of the cavity is 24 cm and is compatible with a position 10 m far from the interaction point should the cryostat be the same for the cavities of the two rings.

A more detailed analysis is needed to determine the complete layout of the interaction region, it should address questions such as the choice of the exact frequency of the rf deflectors, the interest of a third cavity to compensate any phase imperfection between the deflectors, the exact quadrupole aperture for sub-centimeter β^* values and the vertical orbit distortion created by the detector. It seems, however, that 25 mrad would be the right choice for the half-crossing angle.

CONCLUSION

Scaling laws have been presented for the design of the interaction areas of B-meson factories. Luminosities as high as 10^{34} cm^{-2}s^{-1} may be achieved with currents limited to 1.5 A for a 5 mm β value at the crossing point and a betatron tune of the order of 30. This seems to be possible if the beams cross horizontally with an angle such that the focusing is independent for the two beams. Technical solutions for the permanent quadrupoles and for the rf deflectors exist but require developments.

ACKNOWLEDGMENTS

V. Mokhov has studied the permanent quadrupoles. His calculations have been checked with the finite element program Flux2D by F. Rohner.

REFERENCES

[1] R. Siemann, "The accelerator physics challenges of B-factories", CLNS 89/938 (August 1989, in *Proc. of the Int. Conf.e on High Energy Accelerators*, Tsukuba, Japan (August 1989).
[2] T. Nakada (Editor), "Feasibility study for a B-meson factory in the ISR tunnel", CERN 90-02, PSI PR 90-08, 1990.
[3] "Investigation of an asymmetric B-factory in the PEP tunnel", LBL PUB-5263, SLAC-359, CALT-68-1622.
[4] Y. Funakoshi et al., "Asymmetric B-factory project at KEK", KEK Preprint 90-30, 1990.
[5] A. Piwinski, IEEE Trans. Nucl. Sci., NS-24, No. 3 (1977).
[6] R. Palmer, SLAC-PUB 4707 (1988).
[7] K. Oide and K. Yokoya, Phys. Rev. D40, 315 (1989).
[8] Y. Baconnier and D. Möhl, "On the choice of parameters for a B-factory in the ISR tunnel", CERN PS/90-74 (AR), 1990.
[9] F. Pedersen, "Feedback systems", in: *Beam Dynamic Issues of High-Luminosity Asymmetric Collider Rings, AIP Conf. Proc. 214*, AIP Conf. Proc. 214, pp. 246-269 (1990).
[10] K. Hirata, "Round and flat beams in the Phi-factory", INFN-FRASCATI report, ARES-15 (1989).
[11] D. Möhl, "On the low-beta optics for round and flat beams in colliders", *This conference*.
[12] S. Myers, "Review of beam-beam simulations" in: *Non-Linear Dynamics Aspects of Particle Accelerators*, Sardinia 1985, Lecture Notes in Physics, Vol. 247, pp. 176-237, Springer- Verlag.
[13] J. Tennyson, "The beam-beam limit in asymmetric colliders. Optimization of the B-factory parameter base", in: *Beam Dynamic Issues of High-Luminosity Asymmetric Collider Rings, AIP Conf. Proc. 214*, pp. 130-174 (1990).
[14] B. Autin, "Analytical design of low-β insertions for flat beams", Particle World, 2/3(1991).
[15] B. Autin and V. Mokhov, to be published.

Diffusive Transport Enhancement by Isolated Resonances and Distribution Tails Growth in Hadronic Beams.

A.Gerasimov

Fermilab*, P.O.Box 500, Batavia, IL 60505, USA

Abstract

The escape rates and evolution of a distribution of particles are considered for a 2-D model of transverse motion of particles in hadronic storage rings, when nonlinear resonances and external diffusion are present. Dynamic enhancement of diffusion inside separatrices can develop under a certain geometry of resonance oscillations and relatively wide resonances, leading to the fast growth of distribution tails and escape rates . The phenomenon is absent in 1-D.

1 General description

In hadronic colliders, the escape of particles to large betatron amplitudes and associated growth of distribution tails due to the small random modulations of the lattice parameters (predominantly the RF power) is an important practical issue. Experimental evidence indicates that the escape rate has an appreciable magnitude only in the presence of the beam-beam interaction. However, to the present knowledge we cannot expect a fast escape of particles to originate from the beam-beam interaction alone. Therefore, it seems apparent that the external noise and the beam-beam nonlinear dynamics "interfere" somehow to efficiently magnify their respective effects. The present paper (a concize version of /1/) is devoted to the description of one particular mechanism of amplification.

We will discuss the effect of noise on Hamiltonian dynamics, more particularly the following system:

$$\dot{\vec{x}} = \vec{p}$$
$$\dot{\vec{p}} = -\frac{\partial U(\vec{x}, t)}{\partial \vec{x}} + \sqrt{2\eta}\,\vec{\xi}(t) \qquad (1)$$

where η is the diffusion intensity. $\xi_i(t)$ here is the white-noise vector process $\langle \xi_i(t)\xi_k(t+\tau)\rangle = \delta_{ik}\delta(\tau)$. The potential U is supposed to consist of an unperturbed time-dependent part $U_0(\vec{x})$ corresponding to exactly integrable motion and a small perturbation $U = U_0 + \epsilon\delta U(\vec{x}, t)$, time-periodic with frequency Ω. Since in realistic situations, the beam is small relative to the aperture during the entire storage time, we will be concerned with the distribution tails only.

For this model, the situation will be quite different depending on what type of Hamiltonian dynamics one is considering. The simplest case is when the dynamics is exactly integrable. Then, one can envision it as (topologically equivalent to) the trajectotries spiralling around the constant-action tori with a simultaneous diffusion in both phases (position on the torus surface) and actions (tori radii). In many cases however, the diffusion intensity is small, so that the trajectory densly covers the tori surface before moving appreciably in tori radii. One can then perform a suitable averaging on before-said surface, as it is common in the conventional Fokker-Planck problems /2/, and reduce the evolution to a certain diffusion process in the space of actions only. In most cases the dependence of diffusion intensities on actions will be smooth and monotonous, so that the evolution of any initial δ-functional distribution will be at least qualitatively similar to a usual gaussian spreading for coordinate-independent diffusion, without any "structure" to be observed. The tails of distribution, which are responsible for the particle escape to the distant boundaries, can be described in the same way as in the weak-noise asymptotics (WNA) for conventional nonequilibrium systems (with damping) /3/ as $\rho = Z\exp(-\phi/\eta)$ where Z and ϕ are both functions of phase space variables and time while η is the general factor of diffusion intensity (small parameter). The function ϕ, defining the exponential smallness of the transition probability from given initial condition, can be easily shown to depend on time as $\phi = \varphi/t$. The (time-dependent) escape rate r from any given initial condi- tions can be found in weak-noise approximation as $r = R\exp(-G/\eta)$, where G is the minimum of ϕ on the boundary $G = \varphi_{min}/t$. Thus, time-dependent escape in purely integrable Hamiltonian systems under the influence of weak noise can be analytically described through the combined implementation of averaging along Hamiltonian trajectories and WNA, and does not show any peculiarities.

More difficult case is when the Hamiltonian is not exactly integrable, but only nearly so, i.e. consists of an exactly integrable time-independent part and a small perturbatron (with periodic, if any, dependence on time). Such perturbations are known /4/ to drive nonlinear resonances, which constitute in their turn an everywhere dense net of progressively (higher orders)/(narrower widths). The question then is, first, how can each undividual resonance affect the evolution of distribution tails and, second, what is the combined effect of many resonances. From a qualitative considerations one can infer that the answer will be quite different in 1-D (one spatial coordinate) and higher dimensionality. Indeed, in 1-D the only spatial scale, associated with each resonance i is its width, which is proportional to the square root of perturbation strength ϵ. Choosing, to be more particular, the unperturbed action space I, one can also say that the sum of all widths of res-

*Operated by the Universities Research Association, Inc. under contract with the U.S.Dept. of Energy.

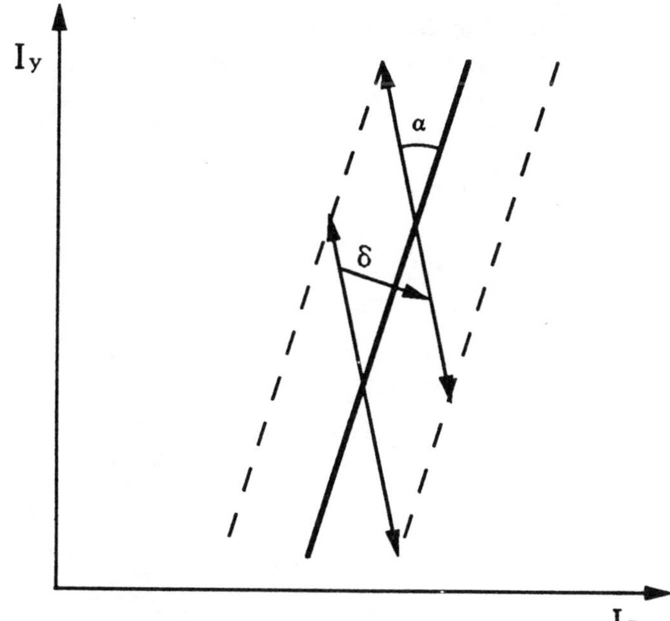

Figure 1: Displacement of resonance oscillation center by transverse kick. Thick solid line is the resonance line. Dashed lines are the separatrix.

onances $\Delta_{int} = \sum_{i=1}^{\infty} \Delta I_i$ is finite and proportional to $\sqrt{\epsilon}$. The smallness of Δ_{int} for small ϵ means, loosely speaking, that the resonances cover only a small submanifold of any given region of I axis. It is quite clear that the perturbation of ϕ by resonances $\Delta\phi = \phi - \phi_0$ (the difference of ϕ's with and without resonances, which we expect to be always negative) in 1-D should be of the order of Δ_{int}, and thus small for small ϵ. Otherwize put, in 1-D the effect of (Hamiltonian) perturbation on escape rate is small as long as its "mechanical" effect is small. In accelerator problems, in most cases the resonance widths are much smaller than the characteristic apertures where particle loss occurs; then in 1-D the transport ability of these resonances is minimal, since their influence is confined to a small region near their separatrices.

In 2 and higher dimensions, resonances are surfaces (or lines) in the action space, and the possibility of the particles to diffuse along them staying inside the separatrices changes the situation completely. The major reason for this is a certain "renormalization" of diffusion inside separatrices, leading to an increase of diffusion intensity along the surface. This can be explained, in 2-D for simplicity, as follows. In the plane of actions I_x, I_y, where resonances are lines, one can draw the arrow of separatrix oscillations, which shows the direction of trapped particle oscillations about the resonance line. Its length Δ is simply the width of the separatrix (or twice the maximum oscillation amplitude) and its center is the resonance line (see Fig. 1). Now consider a small kick δ applied to a trapped particle in the direction orthogonal to the resonance line; it is clear

that the center of oscillations will be displaced a distance $\delta \cot(\alpha)$ along the resonance line. Similarly, if we introduce noise of intensity η in this direction, then the diffusion of the oscillation center along the resonance will have the diffusion coefficient $\eta \cot(\alpha)$. Thus, for small angles α between the resonance oscillations and resonance line, diffusion is enhanced inside the separatrix. This enhancement has been termed diffusive "resonance streaming" and is well known /5/. For escape rate and distribution tail problems, it leads to a very strong effect, since the particles can travel long (as compared to small resonance width) distances along the resonance lines while staying within the regions of enhanced diffusion intensity. The density in the distribution tails and associated escape rates increase exponentially strongly. More particularly, the decrease of the function ϕ by the effect of the resonances is of the order of unity $\phi - \phi_0 \sim \phi_0$ even for small perturbation strength ϵ (as long as $\epsilon \gg \eta$), which is a drastic exponentially strong amplification of the effect as compared to 1-D. The power of the exponential in the escape rate can increase thus several times, making the phenomenon spectacularly strong and supposedly important for applications. One can say that unlike in 1-D, the effect of (Hamiltonian) perturbation on escape rate in 2-D can be large even when its "mechanical" effect is small. The situation is somewhat similar to the escape rate and distribution function behaviour in oscillator with nonlinear resonances, damping and noise /6/, where both damping and diffusion are "renormalized" within the separatrices.

It should be stressed that the WNA description of the system, with the "resonance streaming" emerging as it's ingredient is essentially relying on the condition $\epsilon \gg \eta$ of the resonance being wide enough relative to the noise intensity. When this condition is violated, the situation is getting more involved The basic dynamic process to take into account to evaluate the "macroscopic"(i.e. involving distances much larger then the resonance width) transport rate along the resonance line is the diffusion of particles in transverse direction, so that because of the different longitudinal diffusion intensities inside and outside of the separatrix, the transverse diffusion modulates the longitudinal one. This makes the effective one-dimensional random walk along the resonance line a more complex stochastic process, in fact not even describable by any sort of diffusion process.

When trying to apply the WNA to a generic noisy nearly-integrable oscillator, which posesses an infinite hierarchy of arbitrarily narrow resonances, one immedeately runs into a major difficulty. Indeed, the condition of the resonance being wide enough in respect to noise $\epsilon V_m \gg \eta$ (V_m here is the resonance harmonic amplitude), will break down for all resonances of high enough order. Therefore, the "resonance streaming" diffusion enhancement scenario /5/, implicitly relying on this condition, is essentially incomplete. The fuller description, carried out in /1/ in a certain phenomenological approach gives the overall effect of the resonance as dependent on the width $\Delta \sim \sqrt{\epsilon V_m}$ of the resonance, going to zero as Δ tends to zero. This introduces a certain cutoff of narrow enough resonances and "regularizes" the problem.

2 Concluding remarks

Few observations are in order about the more technical (though very important) question under what conditions can it manifest itself in the real hadronic colliders. The necessary condition of resonance-induced enhancement is the condition (54) of Ref./1/, requiring smallness of angle between the resonance line and resonant oscillations direction. The question then is when this small angle can appear. The important point is that this angle is determined only by the nonlinear tune shifts $\delta\nu_x$, $\delta\nu_y$ dependencies on betatron amplitudes A_x, A_y, and not by the harmonic amplitudes (defining the resonance width). The tune shifts are created by both the multipole components of magnetic fields and the nonlinear beam-beam interaction field. Since the hadronic beams are usually round , the beam-beam interaction is symmetric and preliminary numerical evidence is that the resonant oscillations are always nearly orthogonal to the resonance line. Thus it does not look likely that the phenomenon can manifest itself in the absence of multipole components. Superimposing the latter on the top of beam-beam interactions can however change the situation.

Consider now the effect of the multipoles in the absence of beam-beam force. The lowest order multipole tuneshifts come either from the first-order perturbation term of the octupole component or the second-order one of the sextupole component and have the same functional forms:

$$\delta\nu_x = C_1 A_x^2 + C_2 A_y^2$$
$$\delta\nu_y = C_2 A_x^2 + C_3 A_y^2 \qquad (2)$$

where A_x, A_y are the betatron amplitudes and the coefficients C_1, C_2, C_3 are the integrals of the multipole amplitudes along the ring and can vary in respect to each other in the wide range. It is easy to see that the resonance line is straight in the action variables $J_x = A_x^2$, $J_y = A_y^2$ and that the angle between this line and the resonant oscillations can be varied arbitrarily by varying the constants C_1, C_2, C_3. Thus the phenomenon of the resonance-enhanced diffusion can be more easily observed in the absence of beam-beam interaction and is conceivable when both beam-beam interaction and lattice nonlinearities affect the tune shifts.

References

1. A.Gerasimov "Diffusive transport enhancement by isolated resonances and distribution tails growth in hadronic beams", Fermilab -Conf-90/250, to be submitted to Phys.Rev.A.

2. K.Gardiner, Handbook of stochastic methods, Springer, 1985.

3. R.Graham, in: Fluctuations, Instabilities and Phase transitions, T.Riste Ed., Plenum, 1975.

4. B.Chirikov, Phys. Rep. 52 (1979) 263

5. J.Tennyson, Physica 5D (1982) 123

6. A.Gerasimov, Physica 41D (1990) 89

Evaluation of the Synchrotron Close Orbit.

Yu.A.Bashmakov, V.A.Karpov.

P.N.Lebedev Physical Institute of USSR Academy of Sciences
Leninsky prospect 53, 117924 Moscow,USSR

Abstract.

The computation of the close orbit is reduced to a determination at an arbitrarily chosen azimuth of the eigenvector of the total transfer matrix of the synchrotron ring and to tracing with this vector desired orbit. The eigenvector is found as a result of an iteration.

Introduction.

The knowledge of the closed orbit position is an essential condition for the effective work of any accelerator. Therefore questions of calculations, measurements and controls have great importance [1]. For example, during injection of particles into a synchrotron, the amplitudes of their betatron oscillations may become commensurable with the working region of the synchrotron. This makes us pay attention at the problem of formation of the optimum orbit with use of correcting optical elements. In addition, it is often necessary to calculate such an orbit at the end of the acceleration cycle when particles are deposited at internal targets or removed from the synchrotron. In dedicated synchrotron radiation sources, large e^+e^- colliders or sperconducting accelerators non-linear forces act on the particle and may lead to unstable motion [2]. In these maschines the dynamic aperture is small and the closed orbit distortion may excite losses of particles. It may lead to the transition of superconducting elements in the heating condition [3]. The correction of the closed orbit and of trajectories of particles is require for decreasing of blackgrounds in the experimental detectors.

Calculations of the closed orbit.

In matrix calculations, the result of passing through a periodic element is that initial vector $\vec{Y}_o$ is linearly transformed into vector $\vec{Y}$. This transformation is described by the corresponding transfer matrix M [4]. For the matrix M can be written:

$$M\vec{Y}_\lambda = \lambda\vec{Y}_\lambda \qquad (1)$$

where λ - the eigenvalue of the matrix M and $\vec{Y}_\lambda$ - eigenvector of the matrix M corresponding to λ. To particles moving on the closed orbit meet following conditions:

$$\vec{Y}_{co}(\vartheta+2\pi)=\vec{Y}_{co}(\vartheta), \qquad (2)$$

or:

$$M_t\vec{Y}_{co}(\vartheta)=\vec{Y}_{co}(\vartheta), \qquad (3)$$

where Y_{co} - the vector describing the particle position and M_t - the transfer matrix of one turn. If compare (1) and (3) one can see that the vector $\vec{Y}_{co}$ is eigenvector of matrix M_t for eigenvalue $\lambda=1$.

In principle, calculations of a transient equilibrium orbit reduce to determination of the eigenvector of the complete transition matrix that corresponds to a given azimuth and to construction of the sought orbit [5].

The analyses was done in the Poincare map corresponding to the given form of the motion (radial or vertical). The eigenvector was determined in the following iteration procedure. First, the initial position of a particle is arbitrarily specified, and its motion in the given azimuth is followed during a certain number of revolutions. If the motion of the particle is determinated by linear forces, the phase space vectors are found on the boundary off an ellipse for the motion in one degree of freedom. If a small nonlinearity is turned on, the ellipses will be distorted but will still enclose the same area. The number of revolutions depends on the frequency of betatron oscillations, it is chosen so that the phase-plane points, which image the particle's motion in the given azimuth, fulfill the phase ellipse almost completely.

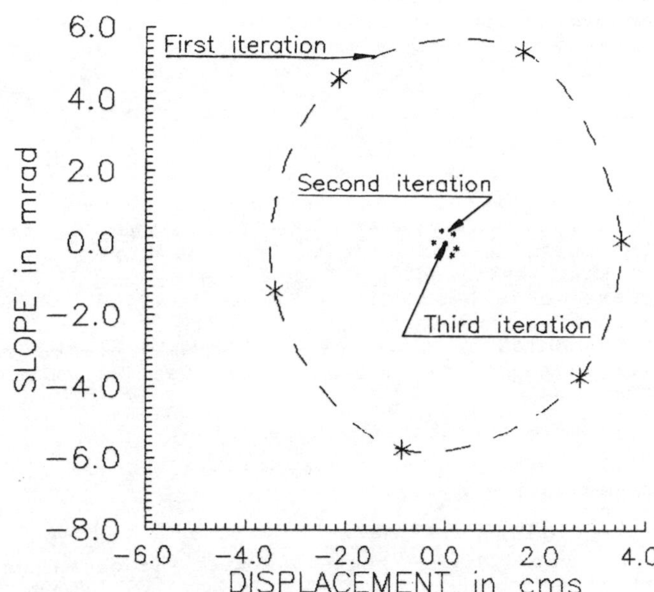

Fig.1. Graphic representation of the iteration process of searching for the closed orbit.

The number of revolutions is given by the following expression:

$$N=E\left(\frac{1}{1-(\nu_{x,z}-E(\nu_{x,z}))}\right) \qquad (4)$$

where $\nu_{x,z}$ is the radial betatron tune or the vertical betatron tune. The function E calculate integer part of the argument.After calculations we obtain a set of N vectors corresponding to passing through the given azimuth in each revolution. Then we find the components of vector which are the arithmetic averaged of the corresponding components

$$X_{km}=\frac{1}{N}\sum_{k=1}^{N}X_k \qquad X_{km}=\frac{1}{N}\sum_{k=1}^{N}X_k'$$

$$Z_{km}=\frac{1}{N}\sum_{k=1}^{N}Z_k \qquad Z_{km}'=\frac{1}{N}\sum_{k=1}^{N}Z_k' \qquad (5)$$

Components of the obtained vector $\vec{Y}_{km}$ are consecutively compared with those of $\vec{Y}_k$ If at least one difference between components of and exceeds the tolerance specified for the given calculation, then the calculation is run again but the initial vector is taken to be equal to $\vec{Y}_{km}$. Calculations are repeated until the eigenvector is determined with necessary accuracy. On fig.1 is shown the Poincoare map for three iteration procedure. So we determine the eigen vector of the complete transition matrix at the given azimuth. Finding of the transient equilibrium orbit reduces to determination of the trajectory of a particle whose initial state is described by the eigen vector of the complete transition matrix.

<u>The code structure.</u>

In order to make calculations we have developed a computer code to calculate both trajectories of individual particles and closed transient orbits. The method in question is based on the fact that the entire magnetic structure of a synchrotron may be represented as a series of discrete elements. All such elements may be classified into three groups:

a) free rectilinear segments,

b) rectilinear segments with strayed magnetic field,

c) magnetic sectors,

d) rectilinear segments with accelerating electric field.

Even element has its own index depending onthe element's position, this allowing information related to an element to be easily filed or retrieved. Thus the parameters of the elements needed for calculations were filed in computer with use of the said notation. This splitting of magnetic structure into homogeneous elements gives us opportunity to make calculation in the matrix form [4], hence

compilation of the computer code and its usage become more convenient. Also, it is easy to modify the code if necessary.

In our calculations particle conditions are specified by vectors in following forms:

$$\vec{Y}=(x,x',z,z',L,E) \qquad (6)$$

where x,x',z,z' - parameters of transversal motion, L - the lengh of the particle path and E - its energy.

The six-dimensional transition matrix may be represented as:

$$M=\begin{bmatrix} M(4,4) & \vdots & M(4,2) \\ \cdots\cdots\cdots\cdots\cdots \\ M(2,4) & \vdots & M(2,2) \end{bmatrix} \qquad (7)$$

where matrix M(4,4) is well-known matrix of transition in transversal motion. Matrix M(2,4) and M(4,2) are zero matrices. Matrix M(2,2) describes the linear transformation of the trajectory length of a particle and its energy when passing through an element. This matrix has the following form:

$$M(2,2)=\begin{bmatrix} 1+\Delta L/L & 0 \\ 0 & 1+\Delta E/E \end{bmatrix} \qquad (8)$$

where ΔL and ΔE are changes in the trajectory length of the particle and its energy in the lattice element. L and E are the path length and the energy of the particle before entering this element. ΔL and ΔE are calculated by corresponding analitical expressions truncated at the second order in small quantities.

Matrix M(2,2) is introduced because it seems to be of interest to study the effect of betatron oscillations on the process of phase oscillations [6,7] which is due to the difference in lengths of the trajectories of particles having different amplitudes of betatron oscillations. It proves to be necessary to introduce the element of matrix M(2,2) describing transformation of energy because the changes in energy due to, say, induction acceleration or synchrotron radiation, must be taken into account when passing through an element.

In our calculations we use a series of subroutines which evaluate the components of the transition matrix for a certain group of elements. Calculations of matrix components is done with due regard for the code of the parameters.

Thus, the code for numeric calculations allows us to determine the coordinates of a particle , the length of its path, and its energy before and after passing through each element. The results of calculations may be presented in graphics, this being very convenient for analyzing the motion of individual particles in an accelerator.

<u>Rezults of calculations.</u>

The method and the code described above have been used for calculations of the closed orbit in the "Pakchra" synchrotron. It was found out that for determining the closed orbit to the accuracy of 10^{-2} cm only two or three iteration steps are needed. The code used includes several orbit correction algorithms. It allowes to make calculations for real magnet lattice including various corrections. The resulting closed orbit is presented in Fig.2.

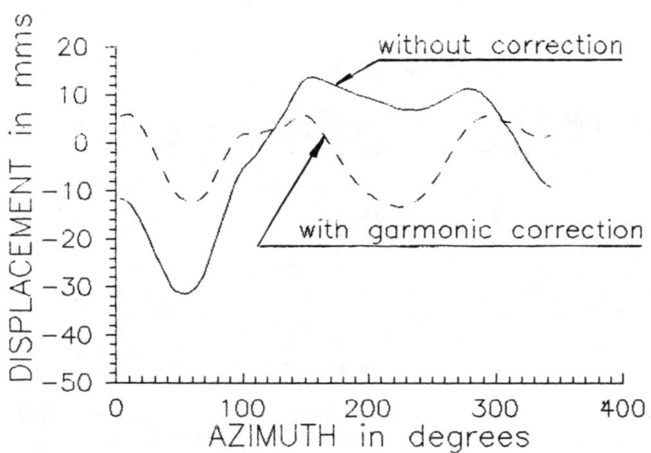

Fig.2. Azimuthal dependence of the radial deviation of the closed orbit.

Also, the orbit obtained after introducing corrections is shown. The data obtained are in reasonable agreement with experiments.

This method may be realized with any program used for particle tracking [7]. One can note that the method may be used for solutions of others problems concerned with finding eigenvectors of a matrix.

References

[1] J.-P. Koutchouk, "Trajectory and closed orbit correction", Frontiers of particle beams; Observation, diagnosis and Correction, Lectures Notes in Phisics 343, Springer-Verlag, 1989,p. 46.

[2] A. Ropert, "Dynamic aperture", CERN Accelerator School,Uppsala University. Sweden, 1989, cern 90-04, p. 26.

[3] M. Harisson, "Resonant extraction at the Tevatron", Fermilab summer Sch. 1987, Cornell summer Sch. 1988, vol.2, New York 1989, p 2009.

[4] H. Bruck, "Acceleraturs circullaires de particules", Paris, 1966.

[5] Yu. A. Bashmakov, V. A. Karpov, "On formation of the instantaneous orbit in the synchrotron", Proc. IX-th All-Union conf. on Charged Particle Acceleratos, Dubna, 198 , p. 190.

[6] D. C. Carey, R. V. Servranckh. K. L. Brown, "Second-order path length terms for a general bending magnet", Particle Accelerators, 1983, v.13 No.3-4, p.199-207.

[7] F. Willeke, "Analysis of particle tracking data", CERN Accelerator School,Uppsala University. Sweden, 1989, cern 90-04, p. 26.

PHASE TRAJECTORY ANALISIS
AT THE NONLINEAR RESONANCES

Yu. A. Bashmakov

P. N. Lebedev Physical Institute USSR Academy of Sciences
Leninsky Prospect 53 USSR 117924 Moscow

Abstract: The phase trajectories close
to the parametric resonance and to the
nonlinear resonances of third and fourth
order, which play an important role in the
cyclic accelerators particle dynamic, are
investigated. The consideration is carried out
in the canonical variables X,Y related by a
simple way to the angle and the displacement
of the circulating particle with respect to
the reference orbit at a given accelerator
azimuth. The problem is reduce to the
construction of the phase trajectories
H(X,Y)=const, that are third or fourth order
curves and determine the mode of motion in the
vicinity of the resonance. The phase
trajectories build-up is performed in the
Klein's perturbation method.

Introduction

The problems of nonlinear dynamics
play an important role in different fields of
the modern physics such as particle physics,
nuclear physics, plasma physics, quantum
electronics and, of course, in accelerator
physics. An excitation of nonlinear resonance
is widely used for particle extraction from
cyclic accelerators [1] , action of nonlinear
resonances determines the dynamical aperture
of large accelerators and storage rings [2,3].
One of the most important problems in study of
the nonlinear oscillations is construction of
the trajectory of motion of representative
points on phase plane and determination of
stable motion regions. In present paper a
geometric method of investigation of particle
motion at nonlinear resonances is developed.

General consideration

The equation of one-dimensional
particle motion in an accelerator in the
presence of a disturbance is the following

$$\frac{d^2 x}{d\varphi^2} + \nu^2 x = \varepsilon\, f(\varphi, x, dx/d\varphi) \quad , \quad (1)$$

where x the transverse displacement of the
circulating particle with respect to the
equilibrium orbit, ν is the betatron
oscillation frequency, $\varphi = \int (ds/\nu\beta)$ is the
generalized azimuthal angle, $f(\varphi, x, dx/d\varphi)$ is
the periodic function in φ with period equal
to 2π, β is the betatron function. Taking the
smallness of the perturbation into account the
solution of the equation (1) can be
represented in the form

$$x = a(\varphi)\, \cos[\nu\varphi + \psi(\varphi)] \quad (2)$$

with the amplitude a and the phase ψ depending
on φ. For further analysis of the motion it is
convenient to make the transformation to new
variables [4,5]

$$X = a\, \cos\psi, \quad Y = -a\, \sin\psi \quad (3)$$

In this variables the equations of particle
motion, take the canonical form

$$\frac{dX}{d\varphi} = \frac{\partial H}{\partial Y} , \quad \frac{dY}{d\varphi} = - \frac{\partial H}{\partial X} , \quad (4)$$

where H is the Hamiltonian. The particle
motion is mapped on the phase plane by the
H(X,Y)=const trajectories.

Third order resonance

The third order resonance $\nu=q/3$ is
excited by the suitable q harmonic of the
quadratic magnetic field $f(\varphi, x, dx/d\varphi) = -A_2 x^2 \cos q\varphi - Bx^3$, B is the constant component
of qubic magnetic field. The Hamiltonian in
this case has the following form [5]

$$H = \frac{A_2}{16}(Y^3 - 3X^2 Y) + \frac{1}{2}(\nu - \frac{q}{3})(X^2 + Y^2) + \frac{B}{2}(X^2 + Y^2)^2 \quad . \quad (5)$$

Let us introduce the new designations

$$s = Y - \sqrt{3}X + \frac{2\sqrt{3}}{3}X_0,$$

$$t = Y + \sqrt{3}X + \frac{2\sqrt{3}}{3}X_0, \quad (6)$$

$$u = Y - \frac{\sqrt{3}}{3}X_0,$$

where $X_0 = \frac{8\sqrt{3}}{3}(\nu - \frac{q}{3})/A_2$, and in accordance
with the sign of the tune shift $\delta = (\nu - \frac{q}{3})$ it
takes positive or negative value. Taking (6)
into account one can cast (5) in the form

$$stu + \frac{8B}{A_2}(X^2 + Y^2)^2 = \frac{16}{A_2}H - \frac{4}{9}X_0^3 \quad (7)$$

In general that is the equation of a fourth
order curve. In absence of the constant
component of qubic magnetic field (B=0) it
defines a third order curve. The phase
trajectories are specified by the equation

$$stu = \eta \quad , \quad (8)$$

where $\eta = 16\, H/A_2 - 4\, X_0^3/9$ is a constant. By
analogy with electrostatic one can call the η
quantity as charge [6]. Then the complete

phase plane splits into two domains: one carrying positive charge $\eta>0$, and other carrying negative charge $\eta<0$. The boundary between these domains is the separatrix

$$stu=0 \quad , \qquad (9)$$

on separatrix $H=\frac{\sqrt{3}}{36}A_2 X_0^3$. This equation is satisfied by the set of three straight lines: $s=0$, $t=o$, $u=o$ forming the equilateral triangle at its intersection. The vertexes of this triangle determine the position of three unstable fixed points

$$1)\ X = 0, \quad Y = -\frac{2\sqrt{3}}{3}X_0;$$

$$2)\ X = -X_0, \quad Y = \frac{\sqrt{3}}{3}X_0;$$

$$3)\ X = X_0, \quad Y = \frac{\sqrt{3}}{3}X_0.$$

The separatrix splits the phase plane into regions of stable and unstable motion. Let us investigate the nature of the phase curves. It is previously worth to note that each of the straight lines $s=0$, $t=0$, $u=0$ divides the plane into two half-plane having positive and negative values of corresponding variables s, t, u. In accord with this fact the sign of the stu production is determined. For a small η value such that $|\eta| \ll |X_0^3|$ the corresponding phase trajectories place close to the separatrix in the domain defined by the η sign. It gives the qualitative technique for graphic build-up of the phase trajectories [7] (Fig.1, where $X_0>0$). There is the largest

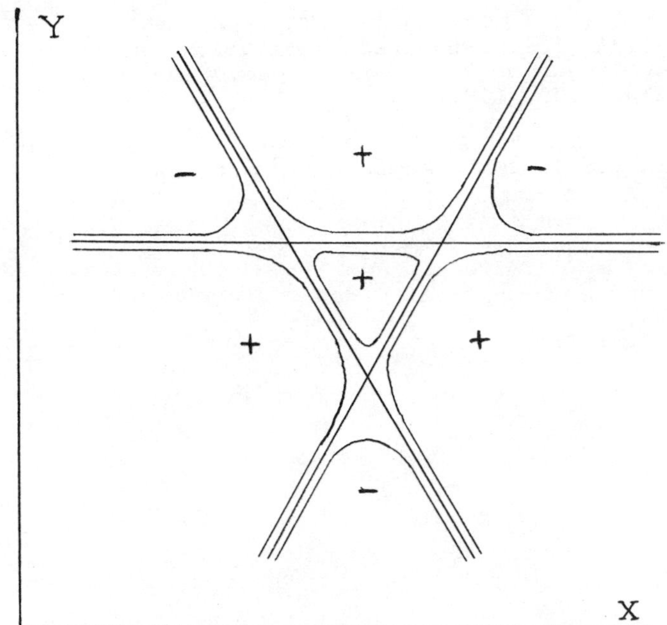

Fig. 1. Phase trajectories at third order resonance, B=0.

detachment of the curves from separatrix near

to the fixed points. In the central region there are the close curves contained the stable fixed points. At $X^2+Y^2=R^2 \gg X_0^2$ the phase curve distance from the separatrix decreases as η/R^2. At B not equal to zero at the large distance from the center of the coordinate system $R^2 \gg X_0^2$ term in (7) proportional to $(X^2=Y^2)^2$ dominates and determines the sign of the left-hand side of the equation keeping it constant everywhere outside of some central circumference of a large enough radius. This gives rise to the coalescence of the separatrix branches. Moreover there is combining of the branches bounded the sectors with η opposite in sign to B.

<u>Fourth order resonance</u>

The fourth order resonance $\nu=q/4$ is excited by the suitable q harmonic of the qubic magnetic field $f(\varphi,x,dx/d\varphi)=-A_3 x^3 \cos q\varphi - Bx^3$. The Hamiltonian in this case has the following form [4,5]

$$H = \frac{A_3}{48}(X^4-6X^2Y^2+Y^4)+ \frac{1}{2}(\nu-\frac{q}{4})(X^2+Y^2)+$$
$$\frac{B}{2}(X^2+Y^2)^2 \quad . \qquad (10)$$

It defines the phase trajectoiries to be the fourth order curves. These curves, as follows from (10), are symmetric about both axes. $H(X,Y)$ in (10) may be represented in the following form

$$f_1(X,Y) \quad f_2(X,Y)=\eta, \qquad (11)$$

$$f_1 = g(X^2 + \frac{1}{k_i} Y^2 + \frac{sgn\ \delta}{(1+k_i)(1+D)} X_0^2) \ ,$$

$$f_2 = g^{-1}((1+D)X^2+(1+D)k_i Y^2+\frac{sgn\ \delta\ k_i}{(1+k_i)} X_0^2) \ ,$$

$$\eta = \frac{k_i}{(1+k_i)(1+D)} X_0^4 + \frac{48}{A_3} H$$

where g is scale factor, $\delta =\nu-q/4$, $D= 24B/A_3$, $X_0=2(6|\delta|/A_3)^{1/2}$, $k_{1,2}=(-3+D\pm\sqrt{8(1-D)})/(1+D)$, $k_1 k_2=1$. The equations

$$f_1(X,Y)=0, \quad f_2(X,Y)=0 \qquad (12)$$

are the second order curve equations. At $D > 1$ there are the unintersecting closed phase trajectories. The intersecting points of the curves are real for $D \leqslant 1$. Hence for such D one has a couple intersecting phase trajectories. The equation

$$f_1(X,Y) \quad f_2(X,Y)=0 \qquad (13)$$

determines the separatrix. And mentioned points define position of the fixed points. At $-1 \leqslant D \leqslant 1$ there are two hyperbolae [4,5,8]. Let us consider in more detail the case $D < -1$, when our curves are two ellipses (Fig.2).

The phase trajectory bild-up may be performed in "the small variation method" explained by F.Klein []. The product $f_1 f_2$ is positive in the piece of the plane lying within or outside both ellipses; in four crescent pieces belonging only to one ellips it is negative. For small η the phase trajectories are arranged close to the separatrix in domains defined by the η sign. This treatment, as one can see, offers to seek the fixed points positions and to analyze its nature.

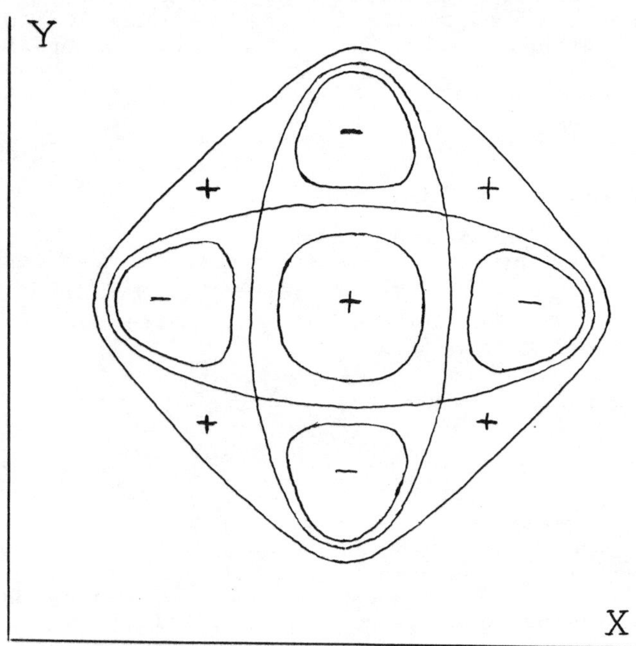

Fig. 2. Phase trajectories at fourth order resonance, D < −1.

References

[1] Yu. A. Bashmakov, K. A. Belovintsev, V. A. Karpov, "An extraction of electrons from a synchrotron on fourth order resonance," Epac'90, Nice June 12-16, 1990, Proc. 2nd European. Particle Accelerator Conf.,Editions Frontieres, V.2, p.1601.

[2] A. Ropert, "Dynamic aperture", CERN Accelerator School,Uppsala University. Sweden, 1989, cern 90-04, p. 26.

[3] F. Willeke, "Analysis of particle tracking data", CERN Accelerator School,Uppsala University. Sweden, 1989, cern 90-04, p. 156.

[4] Yu. A. Bashmakov, K. A. Belovintsev, "Possibility of particle slow extraction from a synchrotron on fourth order resonance, " Kratk. soobshn. Fiz. FIAN SSSR,N1, pp.18-22, 1972 [Sov Phys.Lebedev Inst.Rep., 1972].

[5] Yu. A. Bashmakov, K. A. Belovintsev, Particle dynamic at slow extraction from the "Pakhra" Synchrotron at 3/4 and 2/3 resonances, Preprint FIAN SSSR, N105, M.,1972.

[6] A. N. Kolmogorov, S. V. Fomin, "Elements of the function theory and functional analisis,"[in Russian], Izd. "Nauka", M., 1968, p.496.

[7] Yu. A. Bashmakov, "Search for phase tajectorias near to the nonlinear resonances," Proc. All-Union Conf. on the Development and Practical Application of the Electron Accelelators (September 3-5, 1975, Tomsk) [in Russian], Izd. Tomsk. Gos. Univ. 1975, p.7.

[8] M. Conte, G. Dellavalle, G. Rizzitelli, "Resonant oscillation driven by a nonlinear forcing term," Nuovo Cimento, vol. 57B, p.59, 1980.

[9] F. Klein, "Higher geometry," [in Russian], GNTI, M.-L. 1939, p.400.

Emittance Calculation Using Liouvilles Theorem for a Diagonalized Hamiltonian

H. Heydari

Technical University, EN-2

D-1000 Berlin 10

Abstract

Application of Liouville theorem for the calculation of emittance of particle beams through a special electrostatic arrangement using the separation of phase space by means of a diagonalizable structure of the underlying Hamiltonian.

I. Introduction

It is demonstrated, that in a special electrostatic arrangement the z-dependence (direction of beam) of the Hamiltonian can be neglected. The Hamiltonian separates into two parts describing the degrees of freedom of the perpendicular motion of the particles. This allows using Liouville's theorem to calculate emittance. A formula is easily derived by taking into acount the additive structure of the Hamiltonian (constant partial phase space). A simple method is presented for measuring the emittance.

II. Liouville theorem and the constancy of the phase space

Ions represent a system consisting of n particles corresponding to f degrees of freedom in the phase space. Hence their total mechanical state is described by $F = nf$ generalized configuration coordinates and F generalized conjugate impulse coordinates, therefore by a point in a $2F$-dimensional space in which the fundamental laws of mechanics read:

$$\vec{R} = \omega \, \frac{\partial}{\partial \vec{R}} H \;, \quad \omega = \begin{bmatrix} 0 & +1_{f \times f} \\ -1_{f \times f} & 0 \end{bmatrix} \tag{1}$$

$H = H(q^i, p_k)$ is the Hamiltonian,

$$\vec{R} = [q^1 \ldots q^f \; p^1 \ldots p^f]^T \quad \in R^{2q} \;, \tag{2}$$

$\dot{q}_j = -\partial H/\partial q_j$ and $\dot{p}_j = \partial H/\partial p_j$ are canonical conjugated variables. Given $H = H(\vec{R}, t)$ from (1) follows

$$\omega \, \frac{\partial}{\partial \vec{R}} H(\vec{R}, t) = \vec{V}^H(\vec{R}, t) \;, \tag{3}$$

where $\vec{V}^H$ is the Hamiltonian vector field representing the time derivative along the trajectories in the phase space. Due to the fact that the Hamiltonian vector field $\vec{V}^H$ has always vanishing divergence in phase space R^{2f} for arbitrary phase space function $F = F(\vec{R}, t)$, it follows

$$\frac{\partial}{\partial \vec{R}} \, \vec{V}^H(\vec{R}, t) = 0 \quad . \tag{4}$$

By using Reynold's transport theorem and Gauss theorem results Liouville equation [1]:

$$\frac{\partial \varrho}{\partial t} = \vec{V}^H(\vec{R}, t) \frac{\partial}{\partial \vec{R}} \, \varrho(\vec{R}, t) = -\frac{\partial \varrho}{\partial \vec{R}} \, \omega \, \frac{\partial H}{\partial \vec{R}} = \{\varrho, H\} \;, \tag{5}$$

$$\frac{d\varrho}{dt} = 0$$

$\varrho(\vec{R}, t) \, d^N \vec{R}$ permits the probability of finding N particles in the volume of the phase space $d^N \vec{R}$ at time t. Now eq.(5) for 3 degrees of freedom is represented in the following form

$$\iiint\!\!\iiint \varrho(q_1, q_2, q_3, p_1, p_2, p_3) \, dq_1 \, dq_2 \, dq_3 \, dp_1 \, dp_2 \, dp_3 =$$

$$= \text{const.} \tag{6}$$

The local distribution, i.e. angular and energy distribution, immediately results from the angular distribution of the degrees of freedom, corresponding to each of the 3 space axis, the coordinates x, y and z, and the classical impulses $m\dot{x}$, $m\dot{y}$ and $m\dot{z}$.

Assuming the following that x, p_x and y, p_y are the conjugate canonical variables, secondly that the total energy and the impulse remains constant in z-direction and thirdly $\varrho(x, p_x, y, p_y)$ factorizes as follows $\varrho(x, p_x) \cdot \varrho(y, p_y)$ in the (x, y)-plane, the invariants

$$\iint \varrho(x, p_x) \, dx \, dp_x = const., \iint \varrho(y, p_y) \, dy \, dp_y = const. \tag{7}$$

can be derived by means of (5). But the classical impulses $m\dot{x}$ and $m\dot{y}$ are not always canonical impulses conjugate to the classical coordinates x and y.

Moreover, eq.(2) shows that the conjugate canonical variables enter into the general expression of the Hamilton function

$$H = \sum_i p_i \dot{q}_i - L \qquad (8)$$

related to the corresponding Lagrange function

$$L = T(q_i, \dot{q}_i) - \Pi(q_i)$$

where e.g. L could have the form [2]

$$L = -mc^2 \sqrt{1 - \beta^2} + e(\vec{v} \cdot \vec{A}) - e\Phi \qquad (9)$$

describing the motion of a charged particle in electromagnetic fields with $\Pi =$ potential energy, $T =$ kinetic energy, $\Phi =$ scalar potential, $\vec{A} =$ vector potential. Example (9) shows that the presence of the term $(\vec{v} \cdot \vec{A})$ leads to a more complicated relation between the canonical impulses p_x, p_y (conjugate to x, y) and the 3 classical impulses $m\dot{x}, m\dot{y}, m\dot{z}$ thus preventing us from specifying invariants similar to (7).

Supposing that the total Hamiltonian splits additively [1]

$$H(\vec{R}_{(1)}, \vec{R}_{(2)}) = H_1(\vec{R}_{(1)}) + H_2(\vec{R}_{(2)}) \qquad (10)$$

where $\vec{R}_{(1)}$ and $\vec{R}_{(2)}$ are collections of conjugate canonical variables. The time derivative of the projected volume of the phase space is given as follows:

$$\frac{d}{dt} \left[\int_{p_r(g(t))_{N_1}} d^{N_1} \vec{R}_{(1)} \right] = 0 \quad . \qquad (11)$$

The integration in eq.(11) applies to the projection of the total phase space domain $g(t)$ to the corresponding plane. In addition, the following conditions must be fulfilled

$$g(t) \subset R^N \ , \ \vec{R}_{(1)} \in R^{N_1} \ , \ \vec{R}_{(2)} \in R^{N_2} \ , \ N_{(1)} + N_{(2)} = N \quad .$$

The validity of (11) can be proven in analogy to (4), for example for the projection in the y, p_y-plane, as follows

$$\frac{d}{dt} \left[\int_{p_r(g(t))_{(y, p_y) - plane}} dy \, dp_y \right] = 0$$

To achieve (7) appropriate coordinates with the property

$$\iint \varrho(z, p_z) \, dz \, dp_z = const. \qquad (12)$$

are introduced, so that from equ.(12)

$$\iiiint \varrho(x, p_x, y, p_y) \, dx \, dp_x \, dy \, dp_y = const. \qquad (13)$$

is derived, where the z-component of the impulse of the particle is assumed to be constant.

III. The emittance space and the emittance area

In the previous chapter it is demonstrated that the equation of motion of the trajectories (with time independent fields), relating six independent variables is overdetermined, one of them beeing taken as a parameter.

Now starting from the general form of the canonical conjugated coordinates $p_i = dq_i / dt$ the corresponding impulse in the x-coordinate

$$p_x = \frac{dx}{dt} = \frac{dz}{dt} \frac{dx}{dz} = p_z \, x' \approx p_0 \, x' \qquad (14)$$

is derived in a field free space. When the requirement, which has already been substantiated, $p_z = p_0 = const. = 1$ is inserted, then x' is the canonical coordinate conjugated to x and has the meaning of an angle projection onto the (x, z)-plane. This angle measures then deviation of the path of the particle from the z-axis. Therefore, the (x, p_x)-subspace of the phase space can be replaced by the (x, x')-space and respectively also (y, p_y)-subspace can be replaced by (y, y')-space thus leading to the emittance spaces E_x, E_y.

Fronteau called such spaces "double degenerated phase space" because of $p_z = p_0, p_0 = 1$. Explicitly stated, E_x coordinized by x and x' the emittance space for the x-direction and E_y (coordinized by y and y') for the $y-$direction respectively. The area that arises from both emittance spaces E is named emittance area $F_{xx'}$ and $F_{yy'}$, see also Fig.1d.

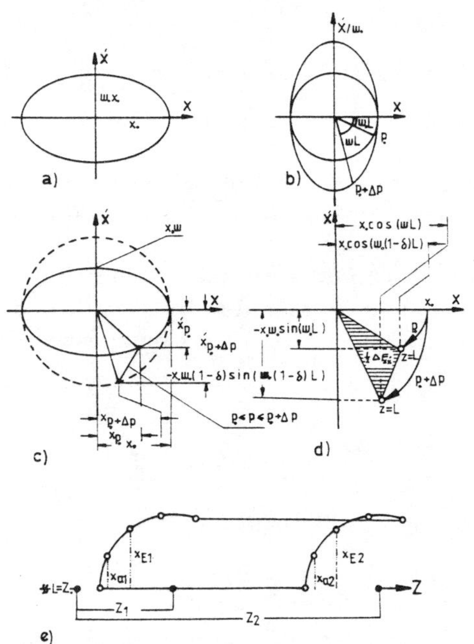

Figure 1: Schematic representation of the emittance area (a-d) and for measuring the emittance (e)

IV. Electrostatic arrangement and emittance area

For the potential of an electrostatic arrangement consisting of three lenses one can use the approach [1]

$$V = \frac{3}{8} U A \frac{\varrho^2}{R^3} \quad , \quad A \ll R \quad . \tag{15}$$

There are U - voltage between two lenses, A - distance between them, R - radius of iris and ϱ - component of cylindrical coordinates. Now the potential is z-independent.

The corresponding Hamilton function of a particle in this system reads

$$H = \frac{p_x^2}{2m} + \frac{p_y^2}{2m} + q U A \frac{1}{R^3}(x^2 + y^2)\frac{3}{8} \quad . \tag{16}$$

Eq.(16) shows that the canonical impulse conjugated to x is independent of y. Therefore the equation of motion in any of those directions is dependent only on its respective direction coordinates and not on the other coordinates. Eq.(16) permits to derive the subsequent equations of motion

$$\frac{\mathrm{d}^2 x}{\mathrm{d}z^2} + \omega_0^2 x = 0 \, , \, \frac{\mathrm{d}^2 y}{\mathrm{d}z^2} + \omega_0^2 y = 0$$

$$\omega_0^2 = \frac{3}{4}\frac{qU A m}{R^3 p_z^2} \, , \, p_z = p_0 \quad . \tag{17}$$

In order to examine the emittance area the initial conditions

$$z = z_0 = 0 \Rightarrow y = y_0, x = x_0, \frac{\mathrm{d}y}{\mathrm{d}z} = 0, \frac{\mathrm{d}x}{\mathrm{d}z} = 0 \tag{18}$$

lead to unique solutions of the subsequent special forms

$$x_{p_0}(z) = x_0 \cos(\omega_0 z) \, , \, y_{p_0}(z) = y_0 \cos(\omega_0 z) \tag{19}$$

$$x'_{p_0}(z) = -x_0\omega_0 \sin(\omega_0 z) \, , \, y'_{p_0}(z) = -y_0\omega_0 \sin(\omega_0 z)$$

and represent ellipses parametrized by z (Fig.1a) characterized by the semi-axis x_0, ωx_0 and y_0, ωy_0 respectively. For further investigations, we present a detailed treatment of the emittance space $E_{xx'}$. It is proved below, if this calculation of emittance area within the beam guiding system is adequate yielding a statement about the constancy of emittance area and the goodness of the used approaches with reference to the measuring apparatus developed for this project (Fig 1e). This device for measuring the emittance consist mainly of a Faraday cup. The cup is movable in φ−direction and picks up the current in z-direction.

In case of varying the impulse p from p_0 to $p' = p_0 + \Delta p$ the new orbital equation using the same initial conditions (18)

$$x_{p_0+\Delta p}(z) = x_0 \cos(\omega_0[1 - \delta]z) \tag{20}$$

$$x'_{p_0+\Delta p}(z) = -x_0\omega_0(1 - \delta)\sin(\omega_0[1 - \delta]z)$$

also describes an orbit of an ellipse with a changed angular velocity $\omega' = \omega_0(1 - \delta)$ (Fig.1).

Defining $\delta = \Delta p/p_0$ as the ratio of the variation Δp and the impulse p_0 at the beginning, the subsequent emittance area (Fig.1d)

$$\Delta F_{xx'} = \begin{vmatrix} \vec{i} & \vec{j} & \vec{k} \\ x_0 & -x_0\omega_0^2 L & 0 \\ x_0 & -x_0\omega_0^2 L(1 - \delta) & 0 \end{vmatrix} = 2|\delta| \, L \, x_0^2\omega_0^2 \tag{21}$$

is evident and the above formula is valid for all p, $p_0 \leq p \leq (p_0 + \Delta p)$ and where $L = 2A$ has been introduced. The term $\omega^2 L$ equals f^{-1}, f beeing the focal length of the electrostatic lense. The increase of emittance area therefore reads

$$\Delta F_{xx'} = \frac{1}{f}2|\delta|x_0^2 = |x_{0(max)}| \, 2 \, |\delta| \tag{22}$$

where $x_{0(max)} = x_0$ and $x'_{(max)} = x_0\omega_0^2 L$ have been used. The difference of the angles

$$\Delta x' = x'_{p+\Delta p}(L) - x'_{p_0}(L) \tag{23}$$

can be expressed as

$$\Delta x' = x_0\omega_0^2 L \, 2\delta \tag{24}$$

so that

$$\Delta x' = \frac{x_{E_1} - x_{E_2} - x_{a_1} + x_{a_2}}{a_2 - a_1} \tag{25}$$

$$x_E = r \cos\varphi_E \quad , \quad x_a = r \cos\varphi_a$$

follows considering the geometry of the measuring apparatus (Fig. 1e).

V. Conclusion

For definite assumptions of technical relevance, the Hamiltonian of electrostatic devices is separable. In that case a good approximation can be achieved with the assumption $z \gg R$ regarding.

The assumption $z \gg R$ is evident, for instance in the accelerator-technique of nuclear physics the component of focussing possesses a greater distance from the target chamber. This statement is demonstrated in eq.(16).

Eq.(25) represents a simple solution for construction of an apertural measurement of the emittance.

References

[1] H. Heydari, "Untersuchungen zum Einfluß der Zustandsfunktionen auf Wirkungsquerschnitte direkter p,γ-Reaktionen bei Energien im 100 keV-Bereich und Projektierung eines Kaskadenbeschleunigers zur Messung von p,γ-Einfangquerschnitten", Doctoral Thesis, TU Berlin, 1988

[2] D. Lennartz, "Phasenraummessungen und Untersuchung der Strahlparameter bei externer Energievariation an einem 20 MeV Protonenzyklotron" Doctoral Thesis, TU Berlin, 1977

Method for Calculating Strong Synchrotron Tune Modulation of Depolarizing Resonances in Storage Rings

S. R. Mane

Brookhaven National Laboratory, Upton, N.Y. 11973

It is well-known that the spins of electrons and positrons circulating in storage rings become spontaneously polarized by the emission of synchrotron radiation; this is now known as the Sokolov-Ternov effect [1]. They predicted that the polarization would build up exponentially to an asymptotic value of $8/(5\sqrt{3}) \simeq 92.4\%$. In practice, the actual value of the polarization can be strongly reduced by so-called "depolarizing spin resonances." The spin resonances occur whenever the condition

$$\nu = m_0 + m_1\nu_x + m_2\nu_y + m_3\nu_s \qquad (1)$$

is satisfied. Here ν is the spin tune, the ν_i are the orbital tunes, and the m_i are integers, including zero. A formalism for calculating the strengths of these resonances was derived by Mane [2]. It expands a suitably defined perturbation series in powers of the orbital actions, followed by an average over the orbital action-angle variables. Effectively, this yields a series in powers of the orbital emittances.

However, the perturbation series can be summed in other ways, in particular for the so-called "synchrotron sideband resonances" [3] – [7], and all the results have been shown to be mathematically equivalent [7]. The expression for the spin resonance strengths is written, not as a power series, but as a series of the form

$$\sum_{m=-\infty}^{\infty} C_m e^{-\alpha} I_m(\alpha) \qquad (2)$$

where the I_m are modified Bessel functions, and the C_m are coefficients whose values are not important here. The perturbation expansion parameter α is not merely $\sigma_\epsilon^2 = (\sigma_E/E_0)^2$, but rather $\alpha = (\gamma a)^2 \sigma_\epsilon^2/\nu_s^2$, where ν_s is the synchrotron tune, γ is the particle energy in units of rest energy, and $a = (g-2)/2 \simeq 0.00116$ for an electron. In a ring like LEP, $\gamma a \simeq 100$ and $\nu_s \simeq 0.1$, so $\alpha \simeq 10^6 \sigma_\epsilon^2 \gg \sigma_\epsilon^2$. Numerical calculations, given in Ref. [7], indicate that $\alpha \simeq 1.4$ in HERA at 30 GeV and $\alpha \simeq 0.8$ in LEP at 50 GeV,

so a power series expansion in α might not converge suitably rapidly for these rings. In this paper, we shall be concerned with the problem of treating cases where α is large, i.e. $O(1)$ or more. We shall call this "strong synchrotron tune modulation."

Let us define the problem more precisely. There would be no problem if synchrotron spin resonances were the *only* type present in storage rings, because formulas, treating only synchrotron spin resonances, have been given in Refs. [3] – [7], for all values of α. There would also be no problem if there were *no* synchrotron sideband spin resonances, because it is known empirically (there is no rigorous proof) that a power series expansion for the other resonances, in powers of the orbital emittances, converges rapidly. There is also no problem if there are both "parent" spin resonances and synchrotron sidebands, for weak synchrotron tune modulation, i.e. $\alpha \ll 1$, because a power series expansion, for all the resonances, will converge rapidly. The problem is: there is no known formalism which deals with both parent spin resonances plus *strong* synchrotron tune modulation, $\alpha \geq O(1)$. This paper proposes to offer an *approximate* method for dealing with the problem.

The polarization is given by the Derbenev–Kondratenko formula [8]

$$P = \frac{8}{5\sqrt{3}} \frac{\left\langle |\rho|^{-3}\left[\hat{b}.\hat{n} - \hat{b}.\vec{d}\right]\right\rangle}{\left\langle |\rho|^{-3}\left[1 - \frac{2}{9}(\hat{n}.\hat{v})^2 + \frac{11}{18}|\vec{d}|^2\right]\right\rangle} \qquad (3)$$

where $\hat{n}$ is the quantization axis of the spin eigenstates and $\vec{d} = \gamma(\partial\hat{n}/\partial\gamma)$ is the spin orbit coupling vector. The definitions of $\hat{b}$, $\hat{v}$, ρ and the angular brackets do not matter here. The term in $|\vec{d}|^2$ has the greatest relevance when calculating the strengths of spin resonances. Suppose the parent resonance is a first order betatron spin resonance, $\nu_p = m_0 \pm \nu_{x,y}$, and $|\vec{d}|^2 = |A_p|^2/(\nu - \nu_p)^2$, i.e. $|\vec{d}|^2 \to \infty$ as $\nu - \nu_p \to 0$. Here A_p is a complex number, which determines the strength of the resonance. Then in Ref. [7] it was shown that the synchrotron tune modulation yields

*Work performed under the auspices of the U.S. Department of Energy.

sidebands given by

$$|\vec{d}|^2 \quad \rightarrow \quad |A_p|^2 e^{-\alpha} \left\{ \sum_{m=-\infty}^{\infty} \frac{1}{(\nu - \nu_p + m\nu_s)^2} \times \right.$$

$$\times \left[I_m + \frac{J_\epsilon}{J_{x,y}} \frac{\alpha}{2} (I_{m-1} + I_{m+1}) \right]$$

$$\left. + \sum_{m=-\infty}^{\infty} \frac{1}{\nu - \nu_p + m\nu_s} \frac{mJ_\epsilon}{\nu_s J_{x,y}} I_m \right\} \quad (4)$$

where I_m means $I_m(\alpha)$ and the J's are damping partition numbers.

Now, in general, $\vec{d}$ contains several resonances, and is of the form

$$|\vec{d}|^2 \;=\; \left| \sum_p \frac{A_p}{\nu - \nu_p} \right|^2$$

$$=\; \sum_p \frac{|A_p|^2}{(\nu - \nu_p)^2} + \text{cross-terms.} \quad (5)$$

We neglect the cross-terms, because there is no formula to obtain synchrotron sidebands for the cross-terms. We introduce synchrotron sideband spin resonances for the diagonal terms via

$$|\vec{d}|^2 \;\rightarrow\; \sum_p \frac{|A_p|^2}{(\nu - \nu_p)^2} \quad (6)$$

followed by

$$|\vec{d}|^2 \quad \rightarrow \quad \sum_{p,m} \frac{|A_p|^2 e^{-\alpha}}{(\nu - \nu_p + m\nu_s)^2} \times$$

$$\times \left[I_m + \frac{J_\epsilon}{J_{x,y}} \frac{\alpha}{2} (I_{m-1} + I_{m+1}) \right]$$

$$+ \sum_{p,m} \frac{|A_p|^2 e^{-\alpha}}{\nu - \nu_p + m\nu_s} \frac{mJ_\epsilon}{\nu_s J_{x,y}} I_m . \quad (7)$$

If the parent resonance is not a first-order betatron spin resonance, e.g. if it is a higher order betatron spin resonance, then the factors of $J_\epsilon / J_{x,y}$ become more complicated, but the basic idea of starting from $|A_p|^2/(\nu - \nu_p)^2$ and generalizing it remains the same. Thus the basic procedure is: (1) use a power series expansion to obtain an expression for $\vec{d}$, hence $|\vec{d}|^2$, (2) neglect the cross-terms, (3) generalize the diagonal terms to include synchrotron sidebands, using modified Bessel functions. There are several obvious approximations in such a procedure. It is not necessarily true that the cross-terms are negligible. Furthermore, a resonance of the form $\nu = \nu_x + \nu_s$ can be visualized as (i) a synchrotron sideband of the parent resonance $\nu = \nu_x$, or (ii) as a betatron sideband of the parent resonance $\nu = \nu_s$. In the above scheme, we neglect case (ii) and do not include the contribution

to the resonance strength coming from the betatron sideband of the parent resonance $\nu = \nu_s$. This is not a priori justified. In addition, the formalisms of Refs. [3] – [7] assume the ring is flat, with all non-flatness due to imperfections. This is not true in a ring with spin rotators, or Siberian Snakes, etc., so the above Bessel function formula is not a priori valid for a ring such as HERA.

Further work is clearly required to see if the above approximate scheme works satisfactorily in practice. Note that, since the power series expansion of a Bessel function converges absolutely for *all* values of the argument, a power series expansion for the synchrotron sidebands should always converge. The problem is that the convergence may be slow. The above idea may at least provide a starting point for the eventual development of a suitably fast algorithm to deal with the problem of strong synchrotron tune modulation.

References

[1] A.A. Sokolov and I.M. Ternov, Dokl. Akad. Akad. Nauk SSSR **153** (1963) 1052 [Sov. Phys. Doklady 8 (1964) 1203].

[2] S.R. Mane, Phys. Rev. A **36** (1987) 120.

[3] Ya.S. Derbenev, A.M. Kondratenko and A.N. Skrinsky, Part. Accel. **9** (1979) 247.

[4] K. Yokoya, Part. Accel. **13** (1983) 85.

[5] C. Biscari, J. Buon and B.W. Montague, Il Nuovo Cimento 81 (1984) 128.

[6] J. Buon, Proc. 8th Int. Symp. on High Energy Spin Physics, Minneapolis (1988) (American Institute of Physics Conf. Proc. No. 187, vol. 2 (1989) p. 963).

[7] S.R. Mane, Nucl. Instr. and Meth. A292, 52, (1990).

[8] Ya.S. Derbenev and A.M. Kondratenko, Zh. Eksp. Teor. Fiz. **64** (1973) 1918 [Sov. Phys. JETP 37 (1973) 968].

First Turn Beam Correction for the Advanced Photon Source Storage Ring*

Y. Qian, E. Crosbie, L. Teng
Argonne National Laboratory
Advanced Photon Source
9700 South Cass Avenue
Argonne, IL 60439

Abstract

A procedure was developed for precise realignment of the quadrupoles in a synchrotron radiation storage ring which can substantially ease the required precision of the initial survey. The procedure consists of first using the injected beam to obtain a closed orbit which is centered on the beam position monitors by the correction dipoles. The strengths of the correction dipoles then give the required fine-adjustment of the quadrupole positions. In this paper we discuss only the algorithm for obtaining the closed orbit.

I. INTRODUCTION

In the Advanced Photon Source (APS) the closed orbit distortion is mainly caused by misalignments of the quadrupoles. Because of the very strong focusing in the storage ring the amplification factor from quadrupole misalignment to orbit distortion is very large. The orbit distortion caused by a random transverse displacement δ of the quadrupoles is about 55 δ. The sextupole magnets in the ring are also very strong. Therefore, orbit distortions lead to large detunings, hence large reductions of the dynamic aperture [1]. An rms orbit distortion of about 5 mm causes a 50% reduction in dynamic aperture. If no orbit correction is applied this implies a tolerance in quadrupole misalignment of about 0.1 mm (rms). This tolerance is rather difficult to achieve with only survey techniques. The correction dipoles (correctors) are, therefore, added to the lattice to ease the survey tolerance.

The ring quadrupole alignment procedure is conceived to be of two stages. In stage one the ring is aligned by survey techniques to a high precision but not necessarily better than 100 μm. The correctors are then employed together with the beam position monitors (BPM) to steer an injected beam in a sequential procedure during the first turn to obtain a closed and corrected orbit. The distortion of the corrected orbit is limited by the resolution of the BPM readings for a single pass beam. Nevertheless, the distortion should be much better than the 5 mm tolerance required by the dynamic aperture consideration. One should be able to store a good beam on this orbit.

In stage two one uses the corrector settings to realign the quadrupoles. The strengths of the correction dipoles needed to produce the corrected orbit give direct information on the misalignment of individual quadrupoles. One can then move each individual quadrupole to correct its misalignment. The precision of this correction is limited by the accuracy with which the supporting mechanisms of quadrupoles can be adjusted. The accuracy of this adjustment is expected to be better than 20 μm. With the ring quadrupoles realigned to 20 μm one should attain an orbit distortion of some 1 mm with all correction dipoles now turned off. If desired the correctors can still be used to further correct the orbit or to stabilize the beam against vibrations through a feedback system to a precision of about 20 μm, the resolution of the BPM readings for a coasting beam.

II. ORBIT CORRECTION PROCEDURE

The APS storage ring consists of 40 Chasman-Green cells joined by long straight sections. In one cell there are 8 horizontal correctors, 9 vertical correctors and 9 BPMs each giving both horizontal and vertical readings. Their locations are shown in Fig. 1. Only seven pairs of correctors and BPMs are used in the simulation. All BPMs except the first and the last ones are located close to sextupoles. When the beam is centered in the BPM, it is then also centered in the sextupole. Hence, the BPM-centered orbit will suffer the least amount of detuning. The positions of the first and the last monitors in the cell have direct bearing on the position of the beam in the insertion device placed in the straight section.

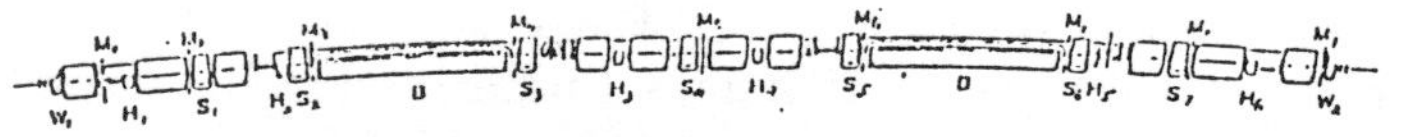

H = Horizontal Correction Dipole
W = Combined Horizontal and Vertical Correction Dipole
S = Vertical Correction Coil in Sextupole
B = Dipole (horizontal) Trim Winding
M = Beam Position Monitor

Figure 1. One cell of the APS ring showing the locations of the correctors and BPMs

In this paper we explore a rather straightforward algorithm for setting the correctors in stage one. The procedure is the following:

---

*Work supported by U.S. Department of Energy, Office of Basic Energy Sciences under Contract No. W-31-109-ENG-38.

1) The quadrupoles are given a set of random Gaussian misalignments Δx_i and Δy_i with a prespecified variance.

2) The beam is injected at angle ϕ_1 through the center of BPM one (M1 of the first cell), tracked through a corrector (horizontal H1 in this case) to the following BPM, M2. In what follows we will only consider the horizontal plane as an example.

3) Corrector H1 is then adjusted to deflect the beam by an angle θ_1 to center it at M2. Even with sextupoles turned on, the dependence of the displacement at M2 on the deflection at H1 is linear enough to make linear interpolation quite good.

4) The corrected beam is tracked through the next corrector H2 to the following monitor M3. The deflection θ_2 at H_2 is then adjusted to center the beam at M3. This process continues all the way around until the last corrector is adjusted to center the beam back at M1.

5) The angle ϕ_2 of the beam at the start of the second turn at M1 will generally not be the same as ϕ_1. The injection angle ϕ_1 is then adjusted and the one-turn monitor-zeroing procedure repeated until $\phi_2 = \phi_1$. We now have a closed orbit which is centered at all monitors.

With a set of random misalignments of standard deviation 0.5 mm in quadrupoles, the uncorrected trajectory is shown in Fig. 2. The corrected closed orbit is shown in Fig. 3 and the required correction angles θ_i are plotted in Fig. 4. Figure 3 shows that the rms distortion of the corrected closed orbit is no more than 1 mm, well within the tolerance of 5 mm dictated by the dynamic aperture criterion.

III. REFINEMENTS, DISCUSSIONS AND CONCLUSIONS

Two refinements were also implemented:

1) The errors in the BPM reading and in the corrector setting can be simulated by adding a random displacement of specified variance to the zero value at the monitor after each step of correction. For the button type BPM and a single pass beam the appropriate variance is estimated to be about 200 μm. With these errors included the results are shown in Fig. 5 and 6.

2) When the required correction angle is larger than the maximum that can be provided by the corrector, the correction is set to the maximum value and proceeds to the next corrector. In this manner, initial quadrupole misalignments as large as 2 mm rms were shown to be correctable. The corrected orbit distortion has a variance of about 4 mm when the BPM resolution is again taken to be 0.2 mm.

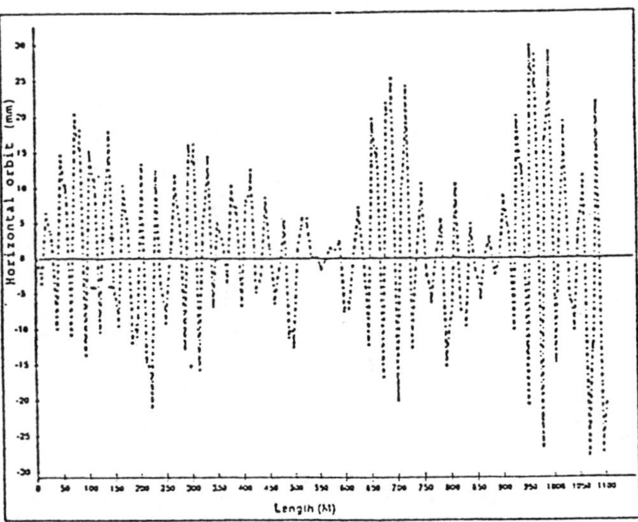

Figure 2. The uncorrected first turn beam with quadrupole misalignments $\Delta X_{qrms} = 0.5$ mm in horizontal plane.

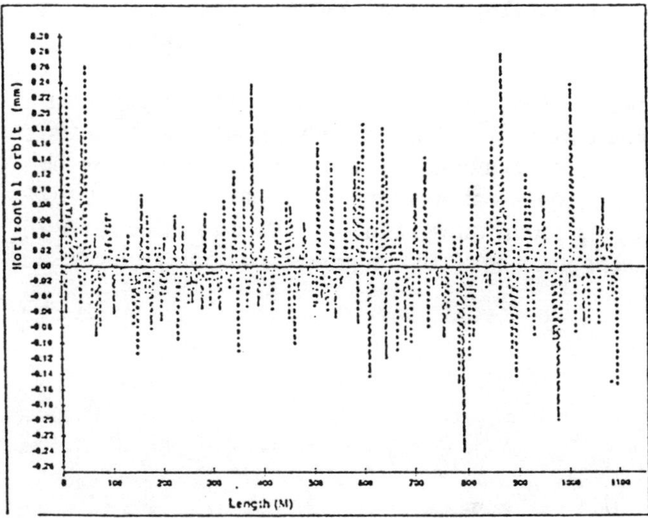

Figure 3. The horizontal closed orbit which is centered at all BPMs with quadrupole misalignments $\Delta X_{qrms} = 0.5$ mm

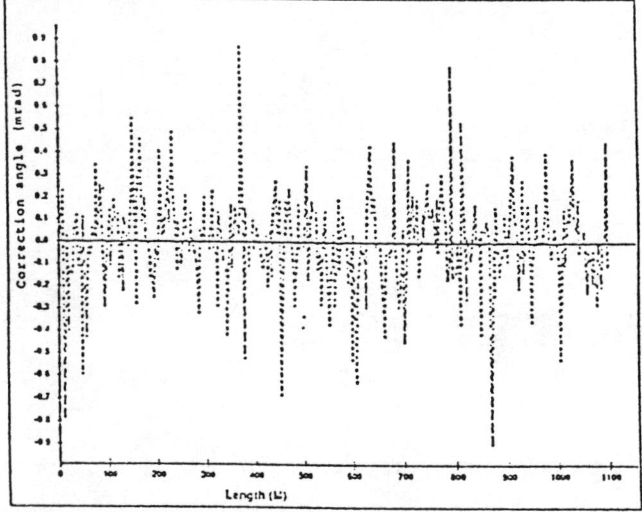

Figure 4. The required correction angles for quadrupole misalignments $\Delta X_{qrms} = 0.5$ mm

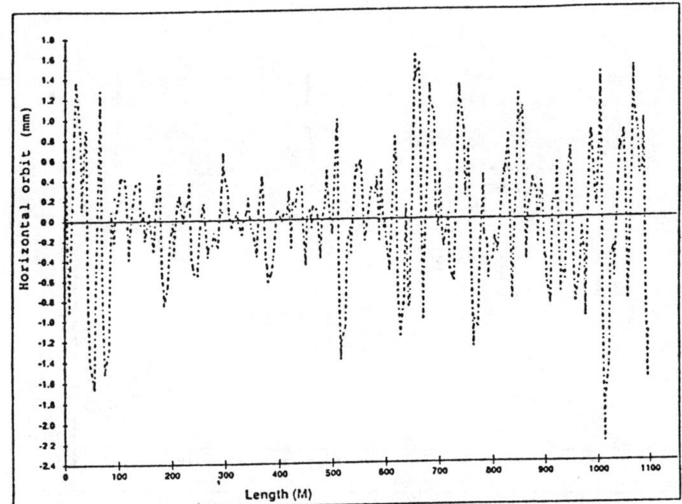

Figure 5. The horizontal closed orbit with quadrupole misalignments ΔX_{qrms} = 0.5 mm and errors in BPM reading ΔX_{mrms} = 0.2 mm

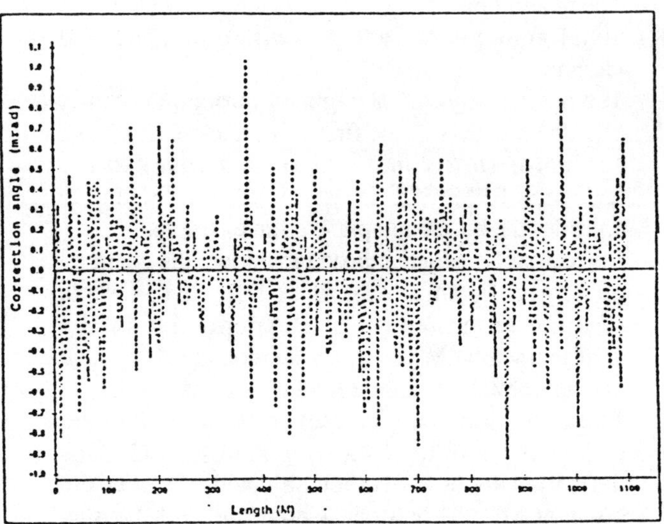

Figure 6. The required correction angles for the same case as in Figure 5

IV. REFERENCES

We conclude from these studies that a fine-realignment procedure as described, using the beam, the correctors and the beam position monitors as aids, can substantially ease the precision required of the initial alignment obtained only by surveying techniques. For the APS an initial survey alignment precision of 0.5 mm proves to be quite adequate.

[1] E. Lessner et al., "Effects of Errors on the Dynamic Aperture of the APS Storage Ring," Proceedings of this Conference.

UNIFORM BEAM DISTRIBUTIONS USING OCTUPOLES*

N. Tsoupas, R. Lankshear, C.L. Snead, Jr.,
T.E. Ward, and M. Zucker
Brookhaven National Laboratory
Department Of Nuclear Energy
Upton, New York 11973

and

H.A. Enge
Deuteron Inc.
Lincoln, Massachusetts 01773

Abstract

The Gaussian beam profile of the BNL 200 MeV
H$^-$ Linac beam at the Radiation Effects Facility target location was transformed into a rectangular profile
with almost uniform distribution by placing two octupole
magnetic elements at particular locations along the beam
line. Experimental results of the beam profile projection
in the horizontal and vertical planes, with and without
octupoles, are presented and compared with third order
calculations.

I. INTRODUCTION

The purpose of the present study was to demonstrate
experimentally that a particle beam which is transported
to a given target location using first order quadrupole
focusing, can have its Gaussian profile distribution transformed into a rectangular profile with almost uniform
distribution. This can be accomplished by using third
order focusing which is provided by octupole magnetic elements placed at particular locations along the beam line.
The concept of introducing third order focusing using octupoles to modify the distribution of the beam profile,
was first proposed by P.F. Meades.[1] Other authors[2,3,4,5,6]
have also discussed the concept and published theoretical results on the subject. The octupole focusing method
to achieving uniform beam profiles was employed in the
present study in which the first experimentally demonstrated uniform beam distributions were obtained at the
Radiation Effects Facility (REF) using the 200 MeV H$^-$
beam delivered by the BNL LINAC. Beam profiles with
uniform distributions are very useful in applications such
as isotope production, radiation effects studies of materials and Nuclear Medicine where it is desirable that
the area under irradiation receive a uniform dose. The
experimental results obtained were compared to the

theoretical third order optics calculations applied to the
REF beam transport line.

II. EXPERIMENTAL RESULTS OF UNIFORM BEAM PROFILING

The TRANSPORT computer code[7] was used for the
first order calculations in order to position the octupoles.
Figure 1 shows the horizontal and vertical beam envelopes
of the first order focussing and the preferred location
of the octupoles for the REF beam line. The REF
beam line is part of the REF/NBTF (Neutral Beam Test
Facility) facility of BNL. The beam emittance used in the
calculations was experimentally measured.

The horizontal beam profile, at the location of the
octupole that affects the distribution of the vertical profile
(Figure 1), has the minimum possible profile that the first
order focussing allows. Such a small profile keeps the
horizontal beam size close to the octupole axis, therefore
limiting the contribution from the aberration coefficients
which mix the horizontal and vertical coordinates through
the octupole field. Similar beam conditions hold at the
location of the octupole that affects the horizontal beam
profile, but with the horizontal and vertical focussing
beam conditions reversed.

When both octupoles are off, first order focussing
dominates and the beam distribution is Gaussian at any
point along the beam line. When the octupoles are turned
on, third order aberrations are included in the beam focussing which partly cancel out the effect of the first order
terms resulting in the uniform beam profile distribution at
the target location. The theoretical calculations were performed using the TRANSPORT code which calculates the
third order aberration coefficients of the beam line. Subsequently, a Monte Carlo program was used which randomly
selects rays from the experimentally measured beam ellipse and then calculates the horizontal and vertical beam
profiles at the target location. The profiles were correct
to third order in aberration coefficients. Figure 2 shows
the theoretical calculations (top graph) of the horizontal

---

*Work performed under the auspices of the Defense Nuclear
Agency through the Weapons Lab., Kirkland AFB, NM.

beam profiles at the REF target location, with octupoles on and off. The bottom set of points shows the corresponding experimental horizontal profiles as taken by the harps. The harps are tungsten multiwire (1 mm spacing) monitors in both vertical and horizontal planes. The wire current is directly proportional to the H$^-$ beam intensity. Similarly, Figure 3 shows the theoretical calculations and the experimental results of the vertical beam profiles for both cases, octupoles on and off.

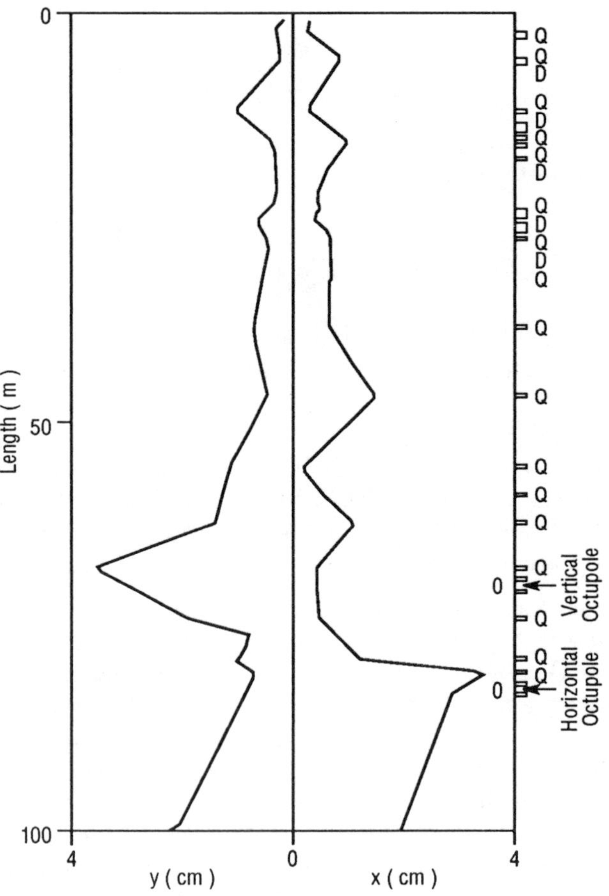

Figure 1. Horizontal and vertical beam envelopes (1σ) for the REF beam line as calculated by the first order optics TRANSPORT code. Definite first order focussing conditions are imposed at the location of the octupoles which introduce the third order focussing.

The theoretical and experimental data shown in Figures 2 and 3 were analyzed with regard to flatness and the "folding" efficiency of the beam in the wings of the Gaussian. The beam tune was not optimized to give the most uniform beam. It was simply our first trial. The octupoles were loaned to us by the acclerator group from the BNL National Synchrotron Light Source, and, although adequate for our purposes to demonstrate the principle, they were not a high enough field to produce optimum beams

for both horizontal and vertical beams. The vertical profile (theory and experiment in Figure 3) has "ears" caused by overcorrecting with the octupole field; yet the theory and experiment are in good agreement on this account. Analysis of the horizontal profile in Figure 2 shows good

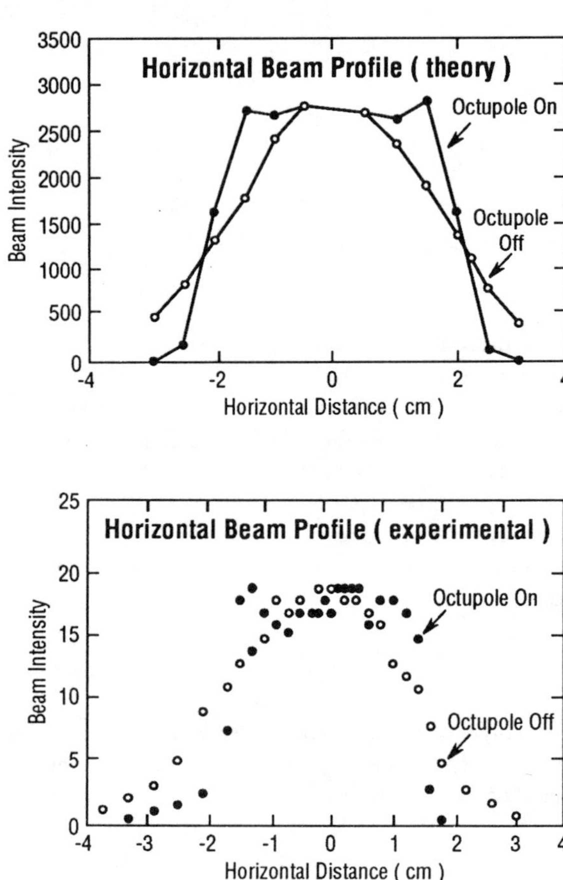

Figure 2. Theoretical calculations (top graph) and the experimental results (bottom graph) of the horizontal beam profiles at the target location in the REF facility, with octupoles on and off.

agreement between theory and experiment with an experimental uniformity of $\pm 7.5\%$ and more than 97% of the beam folded into the central distribution. Theoretically, more than 99.7% of the distribution was predicted to fall within the FWHM of the initial Gaussian profile. This discrepancy is not too surprising as the tune is not our best and the borrowed octupoles were really designed for correcting small third order aberrations, not to purposely introduce them.

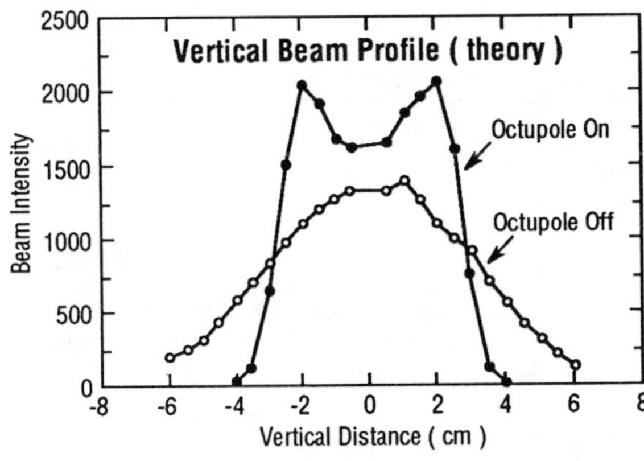

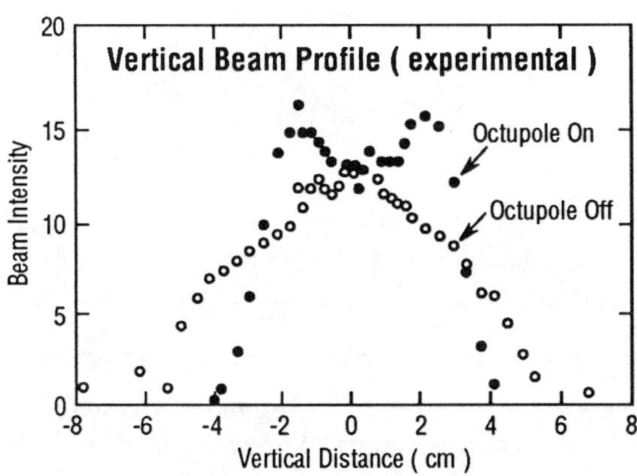

Figure 3. Theoretical calculations (top graph) and the experimental results (bottom graph) of the vertical beam profiles at the target location of the REF facility, with octupoles on and off.

III. REFERENCES

[1] P.F. Meads, IEEE Transactions on Nuclear Science, Vol. NS-30, No. 4, p. 2838, 1983.

[2] H. Wollnick, Optics of Charged Particles, Academic Press, 1987, p. 218.

[3] E. Kashy B. Sherrill, NIM $\underline{B}$26 p. 610, 1987.

[4] E. Kashy and B. Sherril, U.S. Patent No. 4736106.

[5] A.J. Jason, B. Blind, and E.M. Svaton, Linear Accelerator Conference Proceedings, CEBAF Report-89-001, October 1988, p. 192.

[6] B. Sherrill, J. Bailey, E. Kashy, and C. Leakeas, NIM $\underline{B}$40/41, p. 1004, 1989.

[7] K.L. Brown, D.C. Carey, Ch. Iselin and F. Rothacker, CERN Report No. 80-4 (1980).

Sensitivity Reduction against Misalignment of Quadrupole and Sextupole Magnets

K. Tsumaki

Power Electric Machine Design Dept., Hitachi Works, Hitachi, Ltd.
1-1 Saiwaicho 3 chome, Hitachi-shi, Ibaraki-ken, 317, Japan

H. Tanaka and N. Kumagai

RIKEN-JAERI SPring-8 Project Team
2-28-8 Honkomagome, Bunkyo-ku, Tokyo, 113, Japan

Abstract

We studied a magnet alignment method and proposed a new method which treats the quadrupole and sextupole magnets between bending magnets as a unit and aligns their magnetic center precisely before setting it on a design orbit. Alignment between each unit is done by the usual way. The equation of closed orbit distortion was derived and the expected closed orbit distortion was calculated. The results showed that the closed orbit distortion is less than 50 % that of the usual alignment method for the case of SPring-8 storage ring.

I. INTRODUCTION

The emittance goal of third-generation synchrotron radiation sources are $\sim 10^{-9}$ m·rad [1][2]. Electron beams in such a low emittance storage ring are strongly focussed by quadrupole magnets. These strong field quadrupole magnets generate a large chromaticity. To correct this large chromaticity, strong field sextupole magnets are needed, which makes the dynamic aperture a small one. If the quadrupole and sextupole magnets are misaligned, the closed orbit is significantly distorted and the small dynamic aperture is further reduced. The large closed orbit distortion (c.o.d.) and small dynamic aperture makes it difficult to commission a ring and the ring final performance is degraded substantially. To avoid this, precise alignment of quadrupole and sextupole magnets are needed.

Generally magnets are aligned one-by-one on a design orbit using a transit or the like. In this case, alignment errors of magnets are randomly distributed around the design orbit and the expected c.o.d. is the sum of the randomly generated c.o.d.'s which originate from each quadrupole magnet misalignment. Then, we propose a new alignment method in which the expected c.o.d. is not the simple sum of the randomly generated c.o.d.'s but the cancellation works between the c.o.d.'s generated from operation of each quadrupole magnet. The results of the c.o.d. calculation and other effects to the electron beams are described.

II. ALIGNMENT METHOD

The magnet lattice of third-generation synchrotron radiation sources is the Chasman-Green type [1] or the Triple Bend Achromat type [2]. In these magnet lattices, there are several quadrupole and sextupole magnets between the bending magnets. Our alignment method treats these magnets which are placed on a straight line as a unit and aligns their magnetic center precisely before setting it on a design orbit. Alignment between each unit is done by the usual way which results in the error of 0.1-0.3 mm. Then the kicks due to misalignment of the quadrupole magnets within a unit cancel each other and

the expected c.o.d. is reduced substantially. Figure 1 shows a generation pattern of misalignment of a unit and magnets for our alignment method.

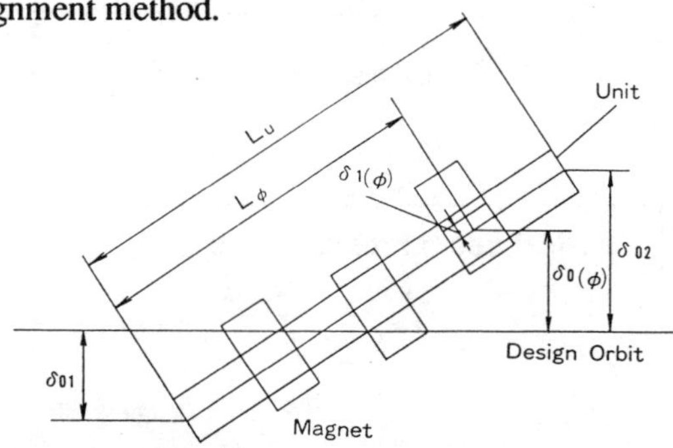

Fig. 1 A generation pattern of misalignment of a unit and magnets.

III. CLOSED ORBIT DISTORTION

A. Equation for C.o.d. in the New Alignment Method

The c.o.d. due to misalignment of quadrupole magnets is[3]

$$\eta(\phi) = \frac{\nu}{2\sin\pi\nu}\int_{\phi}^{\phi+2\pi} f(\xi)\,\cos\nu(\pi+\phi-\xi)d\xi,$$

$$\eta=\beta^{-1/2}y, \quad \phi=\int\frac{ds}{\nu\beta}, \quad f(\xi)=\beta^{3/2}F(\xi), \quad F(\xi)=\frac{\Delta B_0+\Delta B_1}{B\rho}, \quad (1)$$

where ΔB_0 : error field due to misalignment of a unit; ΔB_1 : error field due to misalignment from the unit center. The quantity V (which corresponds to the Courant Snyder invariant for c.o.d.) is

$$V(\phi) = \frac{\nu^2}{4\sin^2\pi\nu}\int_{\phi}^{\phi+2\pi}\int_{\phi}^{\phi+2\pi} f(\xi)f(\chi)\cos\nu(\chi-\xi)d\xi d\chi$$

$$= \frac{\nu^2}{4\sin^2\pi\nu}\int_{\phi}^{\phi+2\pi}\int_{\phi}^{\phi+2\pi}\beta^{3/2}(\xi)\beta^{3/2}(\chi)\cos\nu(\chi-\xi)\cdot$$

$$[\frac{\Delta B_0(\xi)\Delta B_0(\chi)+\Delta B_1(\xi)\Delta B_0(\chi)+\Delta B_0(\xi)\Delta B_1(\chi)+\Delta B_1(\xi)\Delta B_1(\chi)}{(B\rho)^2}]d\xi d\chi$$

$$(2)$$

We separate the contribution of errors into two cases.

(1) Contribution from the magnets in different units.

Since errors are uncorrelated, the expectation value $\langle f(\xi)f(\chi)\rangle=0$.

(2) Contribution from the magnets placed in the same unit.

Since ΔB_1 is randomly distributed, the expectation value of the second and third terms of eq. (2) is zero. Thus, the expectation value of the quantity V is

$$\langle V(\phi)\rangle = \frac{\nu^2}{4\sin^2\pi\nu}\int_\phi^{\phi+2\pi}\int_\phi^{\phi+2\pi}\beta^{3/2}(\xi)\beta^{3/2}(\chi)\cos\nu(\chi-\xi)\cdot$$

$$[\frac{\langle\Delta B_0(\xi)\Delta B_0(\chi)\rangle+\langle\Delta B_1(\xi)\Delta B_1(\chi)\rangle}{(B\rho)^2}]d\xi d\chi$$

$$= \frac{\nu^2}{4\sin^2\pi\nu}\{\sum_{i=1}^m\sum_{j=1}^n\beta^{3/2}(\xi_i)\beta^{3/2}(\chi_{ij})\cos\nu(\chi_{ij}-\xi_i)\cdot$$

$$\langle\frac{\Delta B_0(\xi_i)\Delta\xi\Delta B_0(\chi_{ij})\Delta\chi}{(B\rho)^2}\rangle+\sum_{i=1}^m\beta^3(\xi_i)\langle(\frac{\Delta B_1(\xi_i)\Delta\xi}{B\rho})^2\rangle\}. \quad (3)$$

where i is summed for all quadrupole magnets and j for the magnets in the same unit as i. Using the relation

$$K\delta = \frac{B'L_q\delta}{B\rho} = \frac{\Delta B L_q}{B\rho} = \frac{\Delta B\nu\beta\Delta\phi}{B\rho}, \quad (4)$$

$$\langle V(\phi)\rangle = \frac{1}{4\sin^2\pi\nu}\{\sum_{i=1}^m\sum_{j=1}^n\beta^{1/2}(\xi_i)\beta^{1/2}(\chi_{ij})\cos\nu(\chi_{ij}-\xi_i)\cdot$$

$$\langle K(\xi_i)\delta_0(\xi_i)K(\chi_{ij})\delta_0(\chi_{ij})\rangle+\sum_{i=1}^m\beta(\xi_i)\langle(K(\xi_i)\delta_1)^2\rangle\}, \quad (5)$$

where K: magnet strength; L_q: magnet length; $\delta_0(\phi)$: displacement error of a unit; $\delta_1(\phi)$: displacement error of a magnet measured from the center of a unit. If we set the length of a unit as L_u, displacement errors at the entrance and exit of a unit δ_{01} and δ_{02} and the length from the entrance of a unit to a quadrupole magnet L_ϕ, $\delta_0(\xi_i)\delta_0(\chi_{ij})$ is

$$\delta_0(\xi_i)\delta_0(\chi_{ij}) = \delta_0(L_{\xi_i})\delta_0(L_{\chi_j})$$
$$= [\frac{L_{\xi_i}}{L_u}\delta_{02} + (1-\frac{L_{\xi_i}}{L_u})\delta_{01}][\frac{L_{\chi_j}}{L_u}\delta_{02} + (1-\frac{L_{\chi_j}}{L_u})\delta_{01}]. \quad (6)$$

The expectation value $\langle\delta_{01}\delta_{02}\rangle=0$ and $\langle\delta_{01}^2\rangle=\langle\delta_{02}^2\rangle$. Putting $\langle\delta_{01}^2\rangle=\langle\delta_c^2\rangle$ and using these relationships, we have

$$\langle V(\phi)\rangle = \frac{1}{4\sin^2\pi\nu}\{\sum_{i=1}^m\sum_{j=1}^n\beta^{1/2}(\xi_i)\beta^{1/2}(\chi_{ij})\cos[\nu(\chi_{ij}-\xi_i)]K(\xi_i)K(\chi_{ij})\cdot$$

$$(1 + \frac{2L_{\xi_i}L_{\chi_j}}{L_u^2} - \frac{L_{\xi_i}+L_{\chi_j}}{L_u})\langle\delta_c^2\rangle+\sum_{i=1}^m\beta(\xi_i)\langle(K(\xi_i)\delta_1)^2\rangle\}. \quad (7)$$

From eq. (7), we can obtain the expectation value of c.o.d. y_c as follows.

$$y_c=\frac{\sqrt{\beta}\sqrt{\langle V\rangle}}{\sqrt{2}}=\frac{\sqrt{\beta}}{2\sqrt{2}\sin\pi\nu}\{\sum_{i=1}^m\sum_{j=1}^n\beta^{1/2}(\xi_i)\beta^{1/2}(\chi_{ij})\cos[\nu(\chi_{ij}-\xi_i)]\cdot$$

$$K(\xi_i)K(\chi_{ij})(1+\frac{2L_{\xi_i}L_{\chi_j}}{L_u^2}-\frac{L_{\xi_i}+L_{\chi_j}}{L_u})\langle\delta_c^2\rangle+\sum_{i=1}^m\beta(\xi_i)\langle(K(\xi_i)\delta_1)^2\rangle\}^{1/2}. \quad (8)$$

In a special case where a unit is displaced parallel to the design orbit, c.o.d. becomes

$$y_c=\frac{\sqrt{\beta}}{2\sqrt{2}\sin\pi\nu}\{\sum_{i=1}^m\sum_{j=1}^n\beta^{1/2}(\xi_i)\beta^{1/2}(\chi_{ij})\cos\nu[(\chi_{ij}-\xi_i)]K(\xi_i)K(\chi_{ij})\langle\delta_c^2\rangle$$

$$+\sum_{i=1}^m\beta(\xi_i)\langle(K(\xi_i)\delta_1)^2\rangle\}^{1/2}. \quad (9)$$

B. Numerical Examples

As an example, we applied our method to the SPring-8 storage ring [4]. Assuming $\delta_1=0.025$ mm, the numerical calculation was done using eq. (8). Simulation was also done for 20 machines using the computer code RACETRACK [5]. Evaluation points of c.o.d. are the monitor positions. Results are shown in Fig. 2. For comparison c.o.d. for a case where the quadrupole magnets are aligned one-by-one by the usual method are shown. The expected c.o.d. by our method is reduced to about 40 % that of the usual method if $\delta_c\gg\delta_1$. Numerical calculation and simulation for a special case where a unit is displaced parallel to the design orbit is also shown in Fig. 3. Vertical c.o.d. is reduced drastically than that of the usual method.

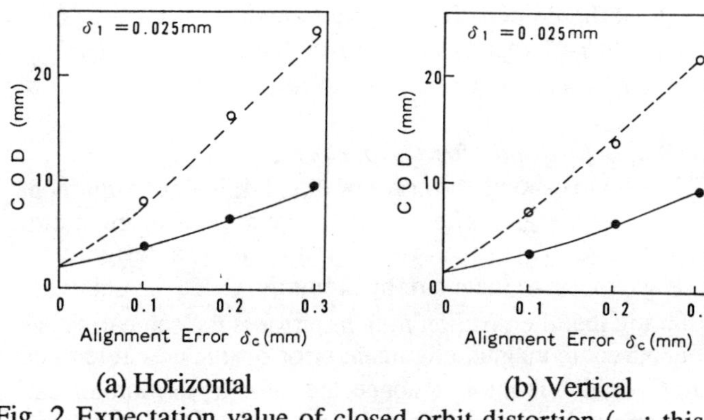

Fig. 2 Expectation value of closed orbit distortion (—: this method (theory), • : this method (simulation),– – –: usual method (theory), ○ : usual method (simulation)).

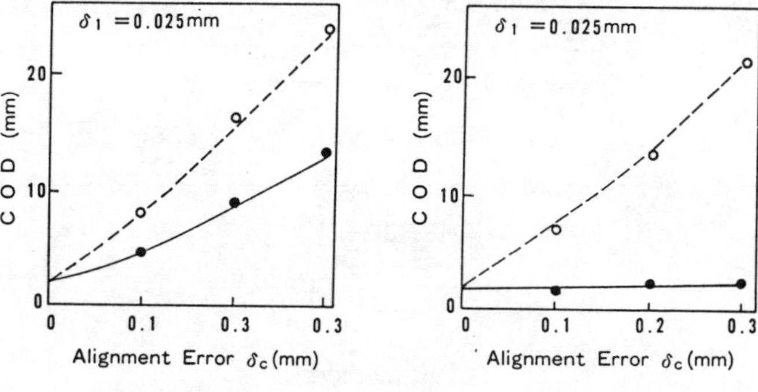

Fig. 3 Expectation value of closed orbit distortion for a special case where a unit is displaced parallel to the design orbit.

The reason for the c.o.d. reduction is as follows. For simplicity we assume that $\delta_c\gg\delta_1$ and the phase difference in a unit is small. Thus c.o.d. by our method is

$$y_c = \frac{\sqrt{\beta}}{2\sqrt{2}\sin\pi\nu}[\sum(\sum_{j=1}^n\beta^{1/2}(\xi_j)K(\xi_j)\delta(\xi_j))^2]^{1/2}. \quad (10)$$

The summation for j is done for a magnet in the same unit. On the other hand, c.o.d. by the usual method is

$$y_c = \frac{\sqrt{\beta}}{2\sqrt{2}\sin\pi\nu}[\sum (\beta^{1/2}(\xi_j)K(\xi_j)\delta(\xi_j))^2]^{1/2}. \tag{11}$$

We can find from eqs. (10) and (11) that the c.o.d. by our alignment method is determined by $(\Sigma\beta^{1/2}K\delta)^2$ and that of the usual method is determined by $\Sigma(\beta^{1/2}K\delta)^2$. Accordingly c.o.d. in our case is less than that of the usual method. The reason that the vertical c.o.d. for the special case is reduced drastically is as follows. If we set K_{Fi} and K_{Di} as the strengths of focussing and defocussing magnets, respectively, the horizontal and vertical c.o.d's are

$$x_c \propto \Sigma(\sqrt{\beta_x}K_{Fi}-\sqrt{\beta_x}K_{Di}) , \quad z_c \propto \Sigma(\sqrt{\beta_z}K_{Fi}-\sqrt{\beta_z}K_{Di}) . \tag{12}$$

In the low emittance ring, K_F is larger than K_D. The betatron function β_x at the focussing magnet is larger than β_x at the defocussing magnet and β_z at the focussing magnet is smaller than β_z at the defocussing magnet, which means $\Sigma(\sqrt{\beta_x}K_{Fi}-\sqrt{\beta_x}K_{Di})$ has a larger value than zero but $\Sigma(\sqrt{\beta_z}K_{Fi}-\sqrt{\beta_z}K_{Di})$ has a value which is almost equal to zero.

C. C.o.d. Before and After C.o.d. Correction

Figure 4(a) shows the concept of c.o.d. for our alignment method before and after c.o.d. correction. Before c.o.d. correction, COD_0 is the orbit distortion from a design orbit C_0. If we measure the c.o.d. by monitors, COD_0' is obtained. Assuming that the position monitor error is the same order as the quadrupole magnet alignment error δ_1, the new reference orbit C_1 which is the line connected monitor-to-monitor can be obtained. After correction of c.o.d., residual COD_1 which is the orbit distortion from the reference orbit C_1 is obtained. Thus the alignment error of a unit after c.o.d. correction has no influence on a beam.

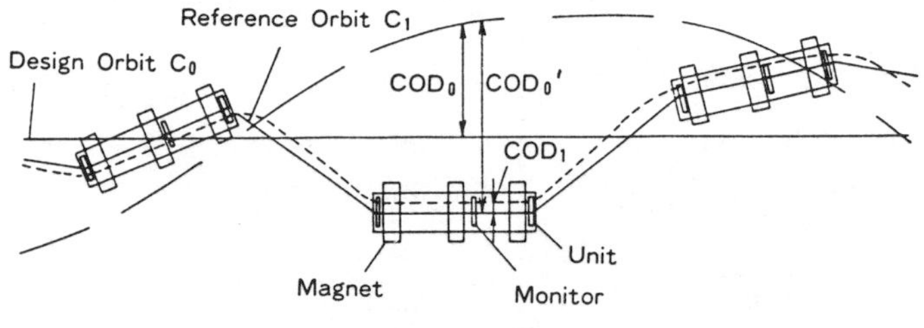

(a) This method

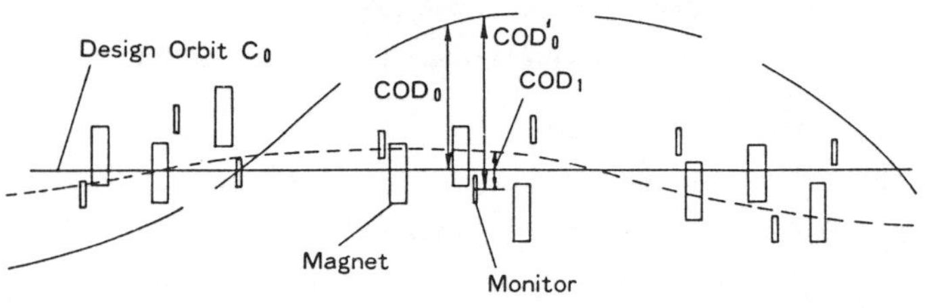

(b) Usual method

Fig. 4 The concept of c.o.d. before and after c.o.d.correction.

IV. OTHER EFFECTS

A. Commissioning

Closed orbit distortion of a low emittance storage ring before c.o.d. correction is so large that the commissioning of such a ring is expected to be difficult. However, since the c.o.d. generated by our alignment method is less than 50 % that of the usual one, application of our alignment method makes the commissioning of a low emittance ring easier.

B. Strength of Correction Magnets

Correction magnet strength is expected to be reduced according to the reduction of the natural c.o.d..

C. Dynamic Aperture

In a low emittance storage ring a dynamic aperture of the ring with errors is much smaller than that of the ideal ring. After c.o.d. correction the dynamic aperture recovers to some extent, but does not recover 100 %. The reason of this is that the small amount of the residual c.o.d. still remains and field error of the quadrupole magnet and misalignment of the sextupole magnet exist. If the error of the quadrupole magnet strength $\Delta K/K$ is about 5×10^{-4} and the misalignment of the sextupole is about 0.1mm, the quadrupole component due to misalignment of the sextupole magnet is three to four times larger than the error of the quadrupole magnet strength in the SPring-8 storage ring. Thus if the misalignment of the sextupole magnet is reduced to 0.025 mm, the dynamic aperture is expected to recover to that of the ideal ring.

D. Spurious Dispersion and Residual C.o.d.

Spurious dispersion and the residual c.o.d. is also expected to be reduced to the ratio of misalignment of the new and usual methods.

V. CONCLUSION

A new alignment method of quadrupole and sextupole magnets was proposed. The method treats the quadrupole and sextupole magnets between the bending magnets as a unit and aligns their magnetic center precisely before setting it on a design orbit. Alignment between each unit is done by the usual way which results in the error of 0.1 to 0.3 mm. We derived the equation for closed orbit distortion in the proposed alignment method and applied it to the SPring-8 storage ring. We found that the closed orbit distortion was reduced to less than 50 %.

Detailed discussion will be published in reference [6].

VI. REFERENCES

[1] H. Kamitsubo et al., in European Particle Accelerator Conference Proceedings, Rome, June 1988, pp.374-376.
[2] M. Cornacchia, in IEEE Particle Accelerator Conference Proceedings, Washington, March 1987, pp.409-413.
[3] E. D. Courant and H. S. Snyder, Annals of Physics, 3, pp.1-48, 1958.
[4] M. Hara et al., in IEEE Particle Accelerator Conference Proceedings, Chicago, March, 1989, pp.476-478.
[5] A. Wrulich, DESY Report-84-026.
[6] H. Tanaka, N. Kumagai and K. Tsumaki, To be published in Nucl. Instrum. & Methods.

Dynamic Aperture of Low Beta Lattices at Tevatron Collider.

Vladimir Višnjić

*Fermi National Accelerator Laboratory**
Batavia, IL 60510, USA

Abstract

A comparative study of the dynamic aperture for several Tevatron lattices is presented with the special emphasis on the new low beta insertions. The results of tracking calculations indicate that the phase I ($\beta_* \approx 50$cm at both B0 and D0) lattice will have about twice the acceptance of the present machine, however the predicted acceptance for phase II ($\beta_* \approx 25$cm) is 40% below that of the present machine.

1 Computer Simulation

In order to increase the luminosity and at the same time provide another interaction region in Tevatron Collider, new low beta insertions are going to be installed in both D0 and B0 straight sections. This is expected to push the Tevatron performance to a new limit ($\beta_* \approx 25$cm, $\beta_{max} \approx 1500$m) and the question is what is the dynamic aperture of the machine going to be. This question is studied here by means of tracking and analytically.

The tracking is done using the code TEVLAT. The Tevatron is modelled as follows. First there are the linear elements of the lattice–the dipoles and the quadrupoles. Their strength is determined from the energy and the tune at which we want to do the tracking. Then the chromaticity sextupoles are turned on in order to compensate for the chromaticity of the linear lattice. Finally, the magnet errors in the form of higher multipole coefficient are read from the actual measurement data for each element.

In order to determine the dynamic aperture I scan a particular subspace of the four dimensional phase space (the xy plane) by tracking for 256 turns. The results of the numerical simulation given in Figs. 1 through 4. What is plotted here is the largest initial amplitude for which the particle survives 256 turns. The amplitude is obtained as $x_0\sqrt{1 + \alpha^2(s_0)}$ and normalized as usual by $\sqrt{\beta(s_0)}$, s_0 the launching point. In other words, we consider the quantities

$$\rho_x = x(s)\sqrt{\gamma_x(s)} \quad \text{and} \quad \rho_y = y(s)\sqrt{\gamma_y(s)} \qquad (1)$$

*Operated by the Universities Research Association, Inc. under contract with the U.S.Dept. of Energy.

which are independent of s and thus invariant. In this formula $\gamma_x(s) = \frac{1+\alpha_x(s)^2}{\beta_x(s)}$ and analog for γ_y.

I define dynamic apertures as maximal values of ρ_x and ρ_y. The dynamic aperture thus defined is the "radius" of the acceptance and is linear in the maximal amplitude the machine can take. In order to obtain the maximal amplitude in mm at a given point of the accelerator from ρ, Figs. 1 through 4, one has to multiply ρ expressed in (microns)$^{1/2}$ with $\sqrt{\frac{\beta[\text{meters}]}{(1+\alpha^2)}}$ at that point. The physical aperture (the beam pipe) is taken to be square with side 70 mm and is represented on the plots by a dotted line rectangle with dimensions $70\text{mm}/\sqrt{\beta_{xmax}} \times 70\text{mm}/\sqrt{\beta_{ymax}}$.

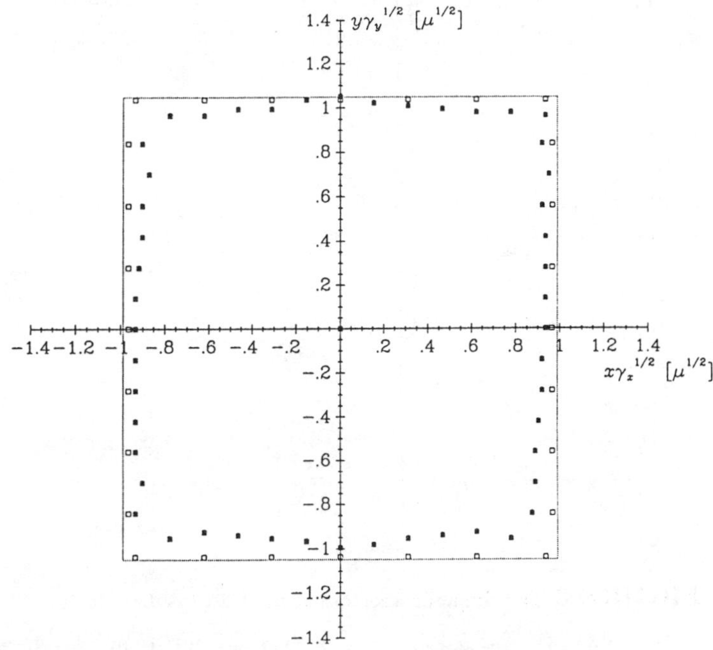

FIGURE 1 The dynamic aperture of the $\beta_* = 70$cm lattice. The open and solid squares denote the linear and the full lattice, respectively. The dotted rectangle is the physical aperture with the dimensions $70\text{mm}/\sqrt{\beta_{xmax}} \times 70\text{mm}/\sqrt{\beta_{ymax}}$.

For each lattice, the tracking is done with and without the nonlinear elements, in particular the higher multipoles in the dipoles and quadrupoles (magnet errors). The linear lattice contains only the ideal linear elements, *i. e.* dipoles

"

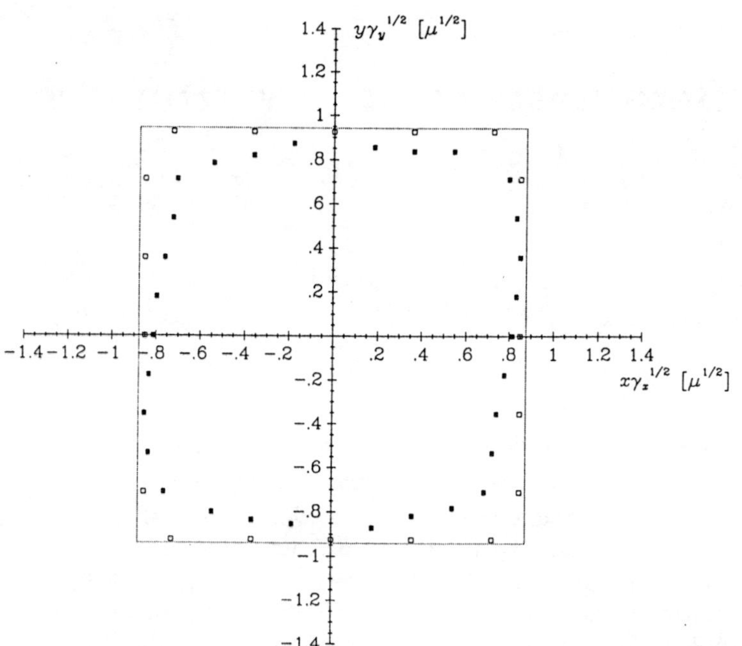

FIGURE 2 The dynamic aperture of the 1988/89 Collider run lattice.

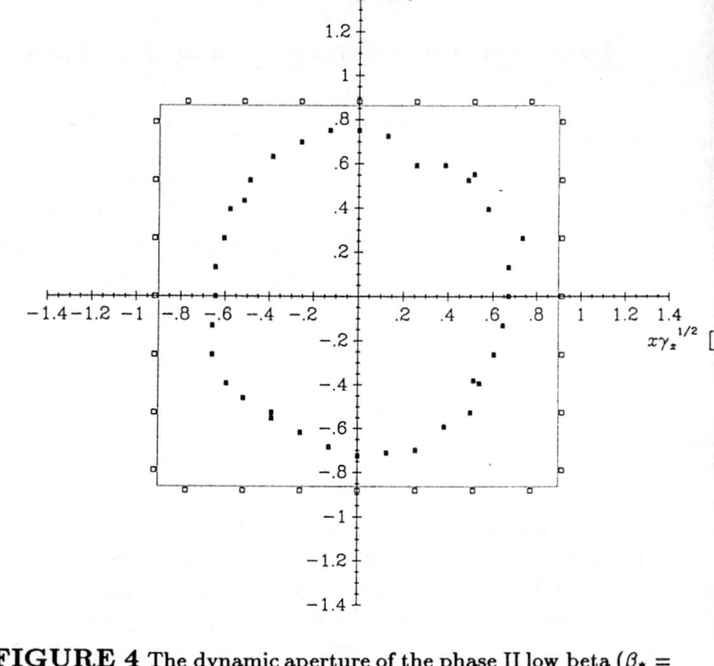

FIGURE 4 The dynamic aperture of the phase II low beta ($\beta_* = 25$cm) lattice.

FIGURE 3 The dynamic aperture of the 1991 low beta lattice.

FIGURE 5 Vertical *vs.* horizontal acceptance for the four Tevatron lattices studied here.

and quadrupoles and its dynamic aperture is identical with the physical one. Note that of the four lattices studied, only the phase II one has the dynamic aperture inside the physical one. Comparing the *horizontal* dynamic apertures of the four lattices normalized to the one used in the 1988/89 run we obtain the ratio

$$1.15 : 1 : 1.46 : 0.82. \qquad (3)$$

The acceptance is obtained by taking the square of the dynamic aperture multiplied by π:

$$A_x = \pi\rho_x^2 \quad \text{and} \quad A_y = \pi\rho_y^2. \qquad (4)$$

In terms of the acceptance, the ratios of Eq. (3) read

$$1.32 : 1 : 2.13 : 0.67. \qquad (5)$$

The graphs A_x *vs.* A_y are shown in Fig. 5.

2 Tune *vs.* Amplitude Calculation

I will use the Hamiltonian approach. The Hamiltonian due to higher multipoles is

$$H = -\frac{RA_3}{B\rho},$$

where

$$A_3 = -B_0 \sum_{m,n}^{\infty} c_{mn} x^m y^n,$$

and the coefficients c_{mn} are given by

$$c_{mn} = \frac{1}{m+n}\binom{m+n}{n}\begin{cases}(-1)^{n/2}\; b_{m+n-1}, & \text{even } n \\ (-1)^{(n+1)/2}\; a_{m+n-1}, & \text{odd } n\end{cases}$$

Consider now the *normal dodecapole, i.e.* b_5 which gives the dominant contribution to $\Delta\nu$. The only nonvanishing c_{mn} are

$$c_{06} = -c_{60} = -\frac{1}{6}b_5 \quad \text{and} \quad c_{24} = -c_{42} = -\frac{5}{2}b_5$$

which gives

$$A_3 = \frac{1}{6}B_1 b_5 (x^6 - 15x^4 y^2 + 15x^2 y^4 - y^6).$$

The amplitudes

$$x = \sqrt{2\beta_x I_x}\sin\phi_x \quad \text{and} \quad y = \sqrt{2\beta_y I_y}\sin\phi_y,$$

where $I = \frac{\epsilon}{2\pi} = \frac{\epsilon_N}{2\pi\gamma}$. Averaging over ϕ gives for $< A_3 >$

$$\frac{5}{12}B_1 b_5 [(\beta_x I_x)^3 - 9(\beta_x I_x)^2 \beta_y I_y + 9\beta_x I_x (\beta_y I_y)^2 - (\beta_y I_y)^3]$$

and for the averaged Hamiltonian

$$< H >= \frac{5}{12}\frac{1}{2\pi}\times$$

$$\sum_{i=1}^{12}\frac{B_1 \tilde{b}_5}{B\rho}[(\beta_x I_x)^3 - 9(\beta_x I_x)^2 \beta_y I_y + 9\beta_x I_x (\beta_y I_y)^2 - (\beta_y I_y)^3].$$

Here the integration over the arc length s is implicit–this is the origin of the summation sign and $\tilde{b}_5^i$ here is $\int b_5 dl$ over the i-th magnet, the actual number given by the measurement. The β functions here are replaced by their average values in given magnets.

The tune shifts are

$$\Delta\nu_x = \frac{\partial < H >}{\partial I_x} \quad \text{and} \quad \Delta\nu_y = \frac{\partial < H >}{\partial I_y},$$

i.e.

$$\Delta\nu_x = A_1 \left(\frac{\epsilon_x}{\pi}\right)^2 + 2A_2 \frac{\epsilon_x}{\pi}\frac{\epsilon_y}{\pi} + A_3 \left(\frac{\epsilon_y}{\pi}\right)^2$$

and

$$\Delta\nu_y = A_2 \left(\frac{\epsilon_x}{\pi}\right)^2 + 2A_3 \frac{\epsilon_x}{\pi}\frac{\epsilon_y}{\pi} + A_4 \left(\frac{\epsilon_y}{\pi}\right)^2,$$

where

$$A_1 = \frac{5}{8\pi}\frac{1}{(2\gamma)^2}\frac{1}{B\rho}\sum_{i=1}^{12} B_1^{(i)} \tilde{b}_5^{(i)} \beta_x^{(i)3}$$

$$A_2 = -\frac{15}{8\pi}\frac{1}{(2\gamma)^2}\frac{1}{B\rho}\sum_{i=1}^{12} B_1^{(i)} \tilde{b}_5^{(i)} \beta_x^{(i)2} \beta_y^{(i)}$$

$$A_3 = \frac{15}{8\pi}\frac{1}{(2\gamma)^2}\frac{1}{B\rho}\sum_{i=1}^{12} B_1^{(i)} \tilde{b}_5^{(i)} \beta_x^{(i)} \beta_y^{(i)2}$$

$$A_4 = -\frac{5}{8\pi}\frac{1}{(2\gamma)^2}\frac{1}{B\rho}\sum_{i=1}^{12} B_1^{(i)} \tilde{b}_5^{(i)} \beta_y^{(i)3}.$$

At the maximal current $\tilde{b}_5$ is -1.74 and -2.9 in^{-3} for the 230 and 130 inch quadrupoles, respectively. The values of γ and $B\rho$ at $E = 900\,$GeV are 959 and 3×10^4 kGm, respectively. The gradients and the beta functions are obtained from SYNCH and we obtain

$$A_1 = 9.31 \times 10^{-7}\text{mm}^{-2} \quad A_2 = -3.79 \times 10^{-7}\text{mm}^{-2}$$

$$A_3 = -5.21 \times 10^{-7}\text{mm}^{-2} \quad A_4 = 1.16 \times 10^{-6}\text{mm}^{-2},$$

and, finally

$$\Delta\nu_x = 9.31\times 10^{-7}\left(\frac{\epsilon_x}{\pi}\right)^2 - 7.59\times 10^{-7}\frac{\epsilon_x}{\pi}\frac{\epsilon_y}{\pi} - 5.21\times 10^{-7}\left(\frac{\epsilon_y}{\pi}\right)^2$$

and

$$\Delta\nu_y = -3.79\times 10^{-7}\left(\frac{\epsilon_x}{\pi}\right)^2 - 1.04\times 10^{-6}\frac{\epsilon_x}{\pi}\frac{\epsilon_y}{\pi} + 1.16\times 10^{-6}\left(\frac{\epsilon_y}{\pi}\right)^2.$$

The values of $\Delta\nu_x$ and $\Delta\nu_y$ for ϵ_x and ϵ_y in the range between 10 and 50 π mm mrad are less than 10^{-6}.

3 Concluding remarks

Empirically, the tracking results tend to overestimate the acceptance of the machine by approximately a factor of two as compared with the measurement. If we revise the predictions of Fig. 7 in view of this uncertainty, we conclude that the normalized acceptance at 900 GeV of the phase I lattice is expected to be 700 π mm mrad and that of the phase II lattice 200 π mm mrad. The tracking results assert that the most important single factor limiting the dynamic aperture is the physical aperture at the point where $\beta = \beta_{max}$ (*i. e.* in low beta quadrupoles), rather than the field quality of those quadrupoles.

The analytical calculation of tune *vs.* amplitude confirms the conclusion from the computer simulation that the new low beta lattice is expected to perform very well.

Adaptive Method of Closed Orbit Correction

Yao Cheng[†] and Chen-Shiung Hsue[†⋆]
†Synchrotron Radiation Research Center
No. 1, R & D Road VI, Science-Based Industrial Park
Hsinchu 300, Taiwan, R.O.C.
⋆ Department of Physics, National Tsing-Hua University
Hsinchu 300, Taiwan, R.O.C.

Abstract

A new adaptive method of closed orbit correction with self-modelling feature is developed for third generation synchrotron light source with many insertion devices. The quasi-Newton method is applied iteratively to minimize the closed orbit distortions and to approach the optimal control. The information from iterative position measurements is used to update the response matrix adaptively. A roughly estimated matrix from optic functions can initialize the correction scheme. The convergence rate is at least superlinear. Hence, the adaptive method may increase the efficiency of fast closed orbit correction. The noise immunity of this method is also considered. Particularly in the case of changing optic functions, this approach provides the searching ability of optimal model.

1   Introduction

The closed orbit correction is one of the basic daily operations in the control routines of every accelerator. The performace of the closed orbit control is important for the beam stability and the flexibility of machine studies. Since the recent progress in the control system supports the on-line modelling control[1][2], an optimal self-modelling control routine free from the working condition will be the interested subject in the future.

There are many different approachs of the closed orbit correction[3]. No matter which method of correction is used, the basic algorithm is to calculate the new corrector settings from the linear response of the machine and the measured orbit distortions. The orbit measurement starts this closed loop iteratively, untill the orbit distortion converges to a stable minimum. If the machine is linear and the model response is exact, the procedure shall terminate at the first iteration. Usually, we need more than one iteration for the correction. It means that the model response is not exact, since the nonlinear effect near the center orbit is negligible. If a improper model response is used, the iterations could even drive the orbit into the nonlinear region and cause the run-away of machine working point. Since the model response estimated from the theoretical optic functions may be different from the actual values in real operation, many laboratories use the measured response in the model function.

If we find a proper model reponse which leads to the convergence of iterations, the convergence rate and the residual decide whether the model is optimal. Particularly in the fast feedback system, faster convergence rate and smaller residual will improve the efficiency of the system. In section 3 of this paper, we will analyze these criteria for a general approach.

In real machine, the response function is a function of working point and working condition. It may take a long time to find the true response for a certain working point. During the commissioning periode or machine studies, frequent change of working point inhibits the search of the true response. An adaptive self-modelling procedure provides the searching ability of optimal model. It should be helpful for the commissioning and machine studies.

2   Model Function

A general model function to be minimized for the closed orbit distortions by iteration in the least square sense may be written as

$$F = (\widetilde{Y + AX})P(Y + AX) + \gamma \tilde{X} X, \qquad (1)$$

where $\tilde{A}$ denotes the transpose of the matrix A and:

- Y is n-dimensional vector and its element Y_i is the measured closed orbit distortion at station i.

- X is m-dimensional vector and its element X_j is the current setting of corrector j.

- A is the model response matrix $(n \times m)$ between current and distortion.

- P is diagonal matrix of the positive weight factor.

- γ is the positive weight connected to the limit of total corrector current.

The $n \times m$ matrix elements of the response matrix A can be calculated from the optic functions or measured from real machine. After the measurements of the closed orbit distortions $Y^{(k)}$ at kth iteration, the current settings $X^{(k+1)} = X^{(k)} + \sigma \delta X^{(k)}$ are determined uniquely by the Newton step,

$$\delta X^{(k)} = -(\tilde{A}PA + \gamma I)^{-1}\tilde{A}PY^{(k)}. \qquad (2)$$

I indicates the identity matrix. $\delta X^{(k)}$ is multiplied by a selected factor σ to increase the convergence rate and reduce the oscillation or to increase the convergence radius. The positive constant γ is used to avoid the limiting cut of high current settings. It also lets the matrix $\tilde{A}PA + \gamma I$ to be positive definite. Therefore, the inverse of matrix involved always exists. The diagonal positive weight factor P provides the freedom to select suitable stations for extra minimization.

This Newton step will be repeated iteratively with a fixed model response A, until a stable minimum of the orbit distortions is reached. In the next section, we formulate the criterion of convergence and the criterion of optimal orbit control.

3 Convergence Rate

Let us assume that the true linear coefficients between stations and corrector currents are T_{ij}. Here, we neglect the nonlinear part. The residual of initial $\tilde{A}PY^{(0)}$ after kth iteration of correction may be written as

$$Y^{(k)} = Y^{(k-1)} + \sigma T \delta X^{(k-1)}, \tag{3}$$

$$\tilde{A}PY^{(k)} = \left[I - \sigma \tilde{A}PT(\tilde{A}PA + \gamma I)^{-1}\right]^k \tilde{A}PY^{(0)}. \tag{4}$$

In the case of $A = T$, a similar transformation can always change the symmetric matrix $\tilde{A}PA$ to the diagonal form. The equation (4) becomes

$$\tilde{A}PY^{(k)} = S_a \left[diag\,(1 - \sigma\lambda_i^a)^k\right] S_a^{-1} \tilde{A}PY^{(0)}, \tag{5}$$

where $\lambda_i^a = \lambda_i/(\lambda_i + \gamma)$ and λ_i are the semipositive eigenvalues of the matrix $\tilde{A}PA$. The norm of the residuals converges linearly to zero for $0 < \sigma < 2/max(\lambda_i^a)$. If there are some $\lambda_i^a = 0$, we can lower the size of the matrix, or select the most efficient correctors.

In the case of $A \neq T$, we transform the matrix $\tilde{A}PT(\tilde{A}PA + \gamma I)^{-1}$ to Jordan form $J(\lambda_i^t)$ with eigenvalues λ_i^t,

$$\tilde{A}PY^{(k)} = S_t \left[I - \sigma J(\lambda_i^t)\right]^k S_t^{-1} \tilde{A}PY^{(0)}. \tag{6}$$

Note that, only the condition $0 < \sigma\lambda_i^t < 2$ will guarantee $X^{(k)}$ to converge linearly to the minimizer X^*. Since λ_i^t are close to unity, σ has to be positive. If there is any $\lambda_i^t < 0$, no σ exists for the convergence. Such a model response matrix A is certainly improper.

The iterations can always find the minimizer succesfully with a proper model response. Then, the question arises here: "Is this result the optimal correction in the $A \neq T$ case?" The answer is "No". We find only the minimun of the model fuction F. For the optimal correction, we should replace the matrix A by the true linear response T in the model function F and then search for the minimun.

4 Quasi-Newton Method

The quasi-Newton method is a modification of the Newton method[4]. It is almost the only method in a modern approach of unconstrained nonlinear optimization problem. We apply this approach on this adaptive method not because of the nonlinear feature of the orbit correction but for the searching of true response matrix T. Futhermore, if we let $\lim_{k\to\infty} A^{(k)} = T$, the equation (6) transfers to the equation (5) and the eigenvalues λ_i^t approach unity, in the case of $\gamma = 0$, after iterations. The convergence rate is increased to be superlinear for $\sigma = 1$. That is to say

$$\lim_{k\to\infty} \frac{\|\tilde{A}^{(k+1)}PY^{(k+1)}\|}{\|\tilde{A}^{(k)}PY^{(k)}\|} = 0. \tag{7}$$

Hence, we propose $\lim_{k\to\infty} \gamma \to 0$ and $\sigma = 1$ in the quasi-Newton approach. The role of σ will be replaced by another constant τ in the later development of this adaptive method.

There is an easy way to understand such an application on the closed orbit correction. We adjust only one corrector current with δX_j. After correction, the vector of position change δY_i divided by δX_j becomes one row vector of the new response matrix. As a matter of fact, we have measured one row vector of matrix. Let the matrix M to be the change,

$$\delta Y_i = (A_{ij} + M_{ij})\delta X_j. \tag{8}$$

This equation (8) is called quasi-Newton condition (the spirit of quasi-Newton method). We say that the feedback information from the measurements is used to correct the response matrix. If several correctors change current simultaneously, then the problem becomes complicated. There are more than one method to update the matrix. However, the quasi-Newton condition must be always hold.

Following, we consider a rank-one matrix $M^{(k)} = u^{(k)}\tilde{v}^{(k)}$ at kth iteration. It is the direct product of a n-dimensional column vector $u^{(k)}$ and a m-dimensional row vector $\tilde{v}^{(k)}$. From the quasi-Newton condition, we have

$$\begin{aligned}
u^{(k)} &= \frac{\delta Y^{(k)} - A^{(k)}\delta X^{(k)}}{\tilde{v}^{(k)}\delta X^{(k)}} \\
&= \frac{Y^{(k+1)} - Y^{(k)} - A^{(k)}(X^{k+1} - X^{(k)})}{\tilde{v}^{(k)}(X^{k+1} - X^{(k)})}, \quad (9)
\end{aligned}$$

where $A^{(k)} = A^{(k-1)} + M^{(k-1)}$ is the updated matrix of last iteration. This form takes any vector v of nonzero denominator. To avoid $\tilde{v}^{(k)}\delta X^{(k)} = 0$, the vector $v^{(k)} = \delta X^{(k)}$ is chosen. This update is indeed the Broyden's formula[5]

$$M^{(k)} = \frac{(\delta Y^{(k)} - A^{(k)}\delta X^{(k)})\widetilde{\delta X^{(k)}}}{\widetilde{\delta X^{(k)}}\delta X^{(k)}}. \tag{10}$$

This is the most successful update when there are no special feature of matrix A to be reflected[6]. The norm of the matrix $(T - A^{(k)})$ converges to zero with a rate as fast as the rate of convergence of $X^{(k)}$ to the minimizer X^*[4]. It is known that $X^{(k)}$ converge superlinearly with Broyden's update. The convergence radius with the initial matrix $A^{(0)}$ is, in principle, unlimited. If the T matrix is exactly linear, there exists only one minimizer.

Considering the onset of noise in the measurements, the update $M^{(k)}$ may be incorrect. Therefore, we insert a positive constant $\tau \in (0,1)$ to multiply the update and let

$$A^{(k+1)} = A^{(k)} + \tau M^{(k)}. \tag{11}$$

This factor plays a role like a digital filter that damps the noise.

In the next quasi-Newton step,

$$\begin{aligned}
\delta X^{(k+1)} &= -(\tilde{A}^{(k+1)} P A^{(k+1)})^{-1} \tilde{A}^{(k+1)} P Y^{(k+1)} \\
&= -H^{(k+1)} \tilde{A}^{(k+1)} P Y^{(k+1)},
\end{aligned} \tag{12}$$

it is not necessary to calculate the time-consuming inversion except for the initial inversion $H^{(0)} = \left(\tilde{A}^{(0)} P A^{(0)} \right)^{-1}$. We change the quasi-Newton condition of equation (8) for the symmetrical inverse matrix $H^{(k+1)}$

$$\delta X^{(k)} = H^{(k+1)} \delta Z^{(k)} \tag{13}$$

with

$$\delta Z^{(k)} = (\tilde{A}^{(k)} + \tilde{M}^{(k)}) P \delta Y^{(k)}. \tag{14}$$

The BFGS formula updates the inverse matrix directly[4],

$$\begin{aligned}
H^{(k+1)} &= H^{(k)} + \tau \frac{\delta Z^{(k)} \widetilde{\delta Z^{(k)}}}{\widetilde{\delta Z^{(k)}} \delta X^{(k)}} \\
&\quad - \tau \frac{H^{(k)} \delta X^{(k)} \widetilde{\delta X^{(k)}} H^{(k)}}{\widetilde{\delta X^{(k)}} H^{(k)} \delta X^{(k)}}.
\end{aligned} \tag{15}$$

The BFGS update preserves the symmetric form of matrix. The updated matrix remains positive definite, if and only if $\widetilde{\delta Z^{(k)}} \delta X^{(k)} > 0$. It means that the positive sequence

$$\widetilde{\delta Z^{(k)}} \delta X^{(k)} = \widetilde{\delta Y^{(k)}} P \delta Y^{(k)} \tag{16}$$

lead to the positive definite sequence of $H^{(k)}$. The inversion of a updated matrix needs $O(m^3)$ floating point computing operation. The BFGS update consumes only $O(m^2)$ floating point computing time. It is the most successful way to update a symmetric inverse matrix[6].

We don't use the same formula uniformly to update $A^{(k)}$ and $H^{(k)}$, because they have the different features. Dennis *et al* showed that both formula are the best update with associated weighting norm in the sense of least change[6]. Since we have the freedom to choose the weight P, there is no contradiction of unequal treatments.

We summarize the whole procedure from the begining as follows

1. Assume that $A^{(0)}, H^{(0)}, Y^{(0)}, X^{(0)}$ are the initial values.

2. From the equation (12), find the $\delta X^{(k)}$ for new current setting $X^{(k+1)}$.

3. After correction, measure the new orbit distortions $Y^{(k+1)}$ and check its convergence for the termination of iterations.

4. use $\delta Y^{(k)} = Y^{(k+1)} - Y^{(k)}$ to calculate the update $A^{(k+1)}$ with formulae (10) and (11).

5. use equation (14) to calculate the update $H^{(k+1)}$ with formula (15) and close the loop here to the step 2.

5 Concluding Remark

The Broyden and BFGS updates have been studied very intensively during the seventies and become slowly the standard algorithms in the nonlinear programming and optimal control. As far as we know, this paper is the first attempt of using this kind of combined algorithms on the closed orbit correction towards an intelligent accelerator control system.

Applying the quasi-Newton method, we can find the true response matrix. The solution of optimizing the model function should be the best correction. Another advantage of this method is to avoid the measurement of A matrix, which is time-consuming and unacceptable for an electron storage ring such as SRRC, if this task has to be executed frequently. We start the correction with the response matrix $A^{(0)}$ estimated from the optic functions. This adaptive method can correct the matrix by itself in the iterations. Even if the optic functions are changed, the method adjusts the response matrix automatically.

The BFGS update keeps the matrix in equation (15) to be positive definite. It is not necessary to pay extra care on the possible sigularity. The BFGS update reduces the time to inverse matrix. The feature of $O(m^2)$ operations instead of $O(m^3)$ operations is benificial and could be essential for the use of the adaptive method in the fast feedback correction.

References

[1] Jan, G.J., "Beam Position Measurement System for SRRC", This conference.

[2] Juan, P., Darpentigny, J., Martin, P.C. and Nghiem, P., "Closed Orbit Measurement and Correction at Super-ACO", Super-ACO/90-27, Orsay, June 1990.

[3] Koutchouk, J.-P. "Trajectory and Closed Orbit Correction" Frontiers of Particle Beams; Observation, Diagnosis and Correction, M. Month, S. Turner(Eds), Springer-Verlag Berlin Heidelberg 1989.

[4] Dennis, J.E. JR, and More, J.J., "Quasi-Newton Methods, Motivation and Theory", SIAM Review, vol 19, No. 1, January 1977.

[5] Broyden, C.G., "A class of methods for solving nonlinear simutaneous equations' Math. Comp., 19, pp. 577-593, 1965.

[6] Dennis, J.E. JR, and Schnabel, R. B., "Least Change Secant Updates for Quasi-Newton Method", SIAM Review, vol 21, No. 4, October 1979.

PERTURBATION TREATMENT OF THE LONGITUDINAL COUPLING IMPEDANCE OF A TOROIDAL BEAM TUBE*

H. Hahn and S. Tepikian

Brookhaven National Laboratory
Upton, NY 11973, USA

Abstract

A simple analytical expression for the longitudinal coupling impedance of a toroidal beam tube below the resonance region has been derived by expanding the electromagnetic fields of the toroidal beam tube in a power series in curvature and substituting directly into Maxwell's equations. The resulting expression consists of the impedance of the straight beam pipe plus a correction terms due to the curvature. It has been verified that this result gives excellent agreement to the exact solution below the first resonance.

INTRODUCTION

The longitudinal coupling impedance of a toroidal beam tube with rectangular cross section was addressed by a number of recent publications.[1] An exact treatment of this problem with many references to earlier work can be found in the paper by Warnock and Morton.[2]

An approximate expression for the coupling impedance below the resonance region, but valid beyond cut off, was derived by Ng and Warnock using the Debye's asymptotic expansions of relevant Bessel functions.[3] In the present paper, a simple analytical expression for the curvature term of the sub-resonant coupling impedance is derived by expanding the electromagnetic fields in a power series in curvature of the toroidal beam tube and substituting directly into Maxwell's equations.

The perturbation treatment of electromagnetic problems is well–known due to the work by Jouguet[4] and Lewin[5] and its application to the present problem conveys considerable physical insight without loss of accuracy. In fact, it has been computationally verified that the results presented here are in better agreement with the exact solution than the Ng–Warnock approximation, although the differences are inconsequential.

FORMAL SOLUTION

A convenient method of obtaining an expression for the longitudinal coupling impedance presented by a smooth beam tube with rectangular cross section (Fig. 1) to a filamentary current involves field matching along a vertical plane common to the inner and outer subregion. In the case of a curved tube, the fields must be expanded in terms of H and E modes. In a straight tube, a pure TM mode would be adequate. A formal solution, valid for both cases, is obtained by always using H/LSE and E/LSH modes.

The centered filamentary current $Ie^{-jn\theta}e^{j\omega t}$ (mode number n, frequency $\omega = vn/R$) is represented by a current sheet in the vertical plane,

$$i = \frac{2I}{h}e^{-jn\theta}e^{j\omega t}\sum_m \cos\xi_m z$$

with $\xi_m = m\pi/h$; $m =$ odd. To achieve notational simplicity, the index m and the common exponential factor will be suppressed in the sequel.

The field components in each subregion have the general form

$$\mathbf{E} = \begin{pmatrix} \mathcal{E}_r\cos\xi z \\ j\mathcal{E}_\theta\cos\xi z \\ \mathcal{E}_z\sin\xi z \end{pmatrix} \text{ and } \mathbf{H} = \begin{pmatrix} \mathcal{H}_r\sin\xi z \\ j\mathcal{H}_\theta\sin\xi z \\ \mathcal{H}_z\cos\xi z \end{pmatrix}$$

The field in each subregion is given by the sum of E and H modes with 2 expansion coefficients (per index m). The four coefficients are determined by requiring continuity of E_θ, E_z, H_θ and application of Ampère's law to H_z at the vertical current sheet. The longitudinal coupling impedance is defined by

$$\frac{Z}{n} = -\frac{R}{nI}\int_0^{2\pi} E_\theta\left(z=0, x=0\right)e^{jn\theta}d\theta$$

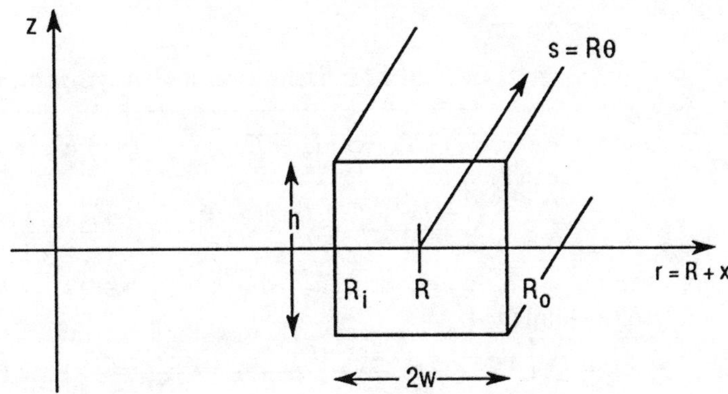

Fig. 1: Toroidal Beam Tube Geometry.

* Work performed under the auspices of the U.S. Department of Energy.

Table I: Field Components in Beam Tube with Rectangular Cross Section.

| H(LSE) mode | | E(LSH) mode | |
toroidal	straight	toroidal	straight
$\mathcal{E}_r = v\Omega^2\frac{R}{r}C(\kappa r)$	$v\Omega^2\cosh\chi(x\pm w)$	$-\frac{\kappa}{\xi}S'(\kappa r)$	$-\frac{\chi}{\xi}\cosh\chi(x\pm w)$
$\mathcal{E}_\theta = -v\Omega\frac{\kappa}{\xi}C'(\kappa r)$	$-v\Omega\frac{\chi}{\xi}\sinh\chi(x\pm w)$	$\Omega\frac{R}{r}S(\kappa r)$	$\Omega\sinh\chi(x\pm w)$
$\mathcal{E}_z = 0$	0	$(1-v^2\Omega^2)S(\kappa r)$	$(1-v^2\Omega^2)\sinh\chi(x\pm w)$
$\mathcal{H}_r = \frac{\kappa}{\xi}C'(\kappa r)$	$\frac{\chi}{\xi}\sinh\chi(x\pm w)$	$-v\Omega^2\frac{R}{r}S(\kappa r)$	$-v\Omega^2\sinh\chi(x\pm w)$
$\mathcal{H}_\theta = -\Omega\frac{R}{r}C(\kappa r)$	$-\Omega\cosh\chi(x\pm w)$	$v\Omega\frac{\kappa}{\xi}S'(\kappa r)$	$v\Omega\frac{\chi}{\xi}\cosh\chi(x\pm w)$
$\mathcal{H}_z = (1-v^2\Omega^2)C(\kappa r)$	$(1-v^2\Omega^2)\cosh\chi(x\pm w)$	0	0

with the formal solution in natural units ($c = \mu_o = 1$)

$$\frac{Z}{n} = j\frac{2}{v}\sum\frac{1}{m\Omega(1-v^2\Omega^2)}\times$$

$$\times\left\{\frac{2\mathcal{E}^E_{\theta i}\mathcal{E}^E_{\theta o}}{\mathcal{E}^E_{ri}\mathcal{E}^E_{\theta o}-\mathcal{E}^E_{\theta i}\mathcal{E}^E_{ro}}-\frac{2v^2\Omega^2\mathcal{H}^H_{ri}\mathcal{H}^H_{ro}}{\mathcal{H}^H_{\theta i}\mathcal{H}^H_{ro}-\mathcal{H}^H_{ri}\mathcal{H}^H_{\theta o}}\right\}$$

where

$$\Omega = \frac{n}{\xi R} = \frac{n}{m\pi}\frac{h}{R}$$

and the field components $\mathcal{E}$ and $\mathcal{H}$ for inner (index i) and outer (index o) subregion to be evaluated at $x = 0$. The present solution equals in substance the Warnock–Morton result, however the different formulation is essential to the subsequent treatment of the problem. Note also, that the distinction between $Z(n)$ and $Z(n,\omega)$ can be ignored in the subresonant region, which is addressed in this paper.

The expressions for the field components of E and H modes are given in Table I. In the toroidal case the functions $C(\kappa r)$ and $S(\kappa r)$ are linear combinations of modified Bessel functions with their definition given in Table II and

$$\kappa = \xi\sqrt{1-v^2\Omega^2}.$$

In the straight case

$$\chi = \xi\sqrt{1+\Omega^2/\gamma^2}$$

with the relativistic $\gamma^{-2} = 1-v^2$. Note that the boundary at conditions R_i and R_o, respectively $x = \pm w$, are fully satisfied.

The coupling impedance of the straight beam tube, the space charge term, follows as ($Z_0 = c\mu_0 = 1$)

$$\frac{Z}{n} = -jZ_0\frac{2}{v\gamma^2}\sum\frac{\text{tgh}\,\xi w\sqrt{1+\Omega^2/\gamma^2}}{m\sqrt{1+\Omega^2/\gamma^2}}$$

which vanishes for $v\sim 1$.

PERTURBATION TREATMENT

The residual curvature term for $v\sim 1$ could be obtained from the above formal solution by using asymptotic expansions for the Bessel functions similar to the Ng–Warnock treatment.[3] An asymptotic expansion is here obtained directly from Maxwell's equations by expanding the field components according to

$$\mathcal{E} = E(x) + e(x)/R$$
$$\mathcal{H} = H(x) + h(x)/R$$

with the zero–order solution given by the straight tube. The resulting set of differential equations is given in Table III. Higher–order solutions are obtained by iteration.

Using the symbolic manipulation program MACSYMA, asymptotic expressions for the field components were determined to second order in w/R for the case of $v\sim 1$. The results for $C_i(\kappa R)$ and $S_i(\kappa R)$ are given as example in Table II.

Table II: Definition and Asymptotic Expansion of Toroidal Functions.

$$C_{i,o}(\kappa r) = \kappa\sqrt{RR_{i,o}}\left\{K_n(\kappa r)I'_n(\kappa R_{i,o}) - I_n(\kappa r)K'_n(\kappa R_{i,o})\right\}$$

$$C_i(\kappa R) \sim \cosh\xi w+$$

$$+\frac{w}{R}\left\{\frac{1-(1-\xi^2 w^2)\Omega^2}{2\xi w}+\frac{w}{R}\frac{\xi w(3-\Omega^2)\Omega^2}{24}+\frac{3-10\Omega^2+7\Omega^4}{32\xi R}\right\}\sinh\xi w+\frac{w^2}{R^2}\left\{\frac{(1-\Omega^2)\Omega^2}{32}+\frac{3-10\Omega^2+7\Omega^4}{64\xi^2 w^2}\right\}\cosh\xi w$$

$$S_{i,o}(\kappa r) = \kappa\sqrt{RR_{i,o}}\left\{I_n(\kappa r)K_n(\kappa R_{i,o}) - K_n(\kappa r)I_n(\kappa R_{i,o})\right\}$$

$$S_i(\kappa R) \sim \sinh\xi w+$$

$$+\frac{w}{R}\left\{\frac{\xi w\Omega^2}{2}+\frac{w}{R}\frac{\xi w(3-\Omega^2)\Omega^2}{24}-\frac{1-6\Omega^2+5\Omega^4}{32\xi R}\right\}\cosh\xi w+\frac{w^2}{R^2}\left\{\frac{(1-\Omega^2)\Omega^2}{32}+\frac{1-6\Omega^2+5\Omega^4}{64\xi^2 w^2}\right\}\sinh\xi w$$

Table III: Perturbation Formulation of Maxwell's Equations.

$$\frac{\partial^2}{\partial x^2} h_z - \left(\xi^2 - \omega^2 + \frac{n^2}{R^2}\right) h_z = +\frac{\xi^2 - \omega^2}{\omega}\left(x\frac{\partial}{\partial x}E_\theta + \omega x H_z + E_\theta\right) - \frac{\xi}{\omega}\frac{\partial}{\partial x}\left(\xi x E_\theta + \omega x H_r\right) + \frac{n}{R}\left(\xi x H_\theta + \omega x E_r\right)$$

$$\frac{\partial^2}{\partial x^2} e_z - \left(\xi^2 - \omega^2 + \frac{n^2}{R^2}\right) e_z = -\frac{\xi^2 - \omega^2}{\omega}\left(x\frac{\partial}{\partial x}H_\theta - \omega x E_z + H_\theta\right) + \frac{\xi}{\omega}\frac{\partial}{\partial x}\left(\xi x H_\theta + \omega x E_r\right) - \frac{n}{R}\left(\xi x E_\theta + \omega x H_r\right)$$

$$h_r = \frac{\omega}{\xi^2 - \omega^2}\left(-\frac{n}{R}e_z + \frac{\xi}{\omega}\frac{\partial}{\partial x}h_z + \xi x E_\theta + \omega x H_r\right)$$

$$e_r = \frac{\omega}{\xi^2 - \omega^2}\left(\frac{n}{R}h_z - \frac{\xi}{\omega}\frac{\partial}{\partial x}e_z + \xi x H_\theta + \omega x E_r\right)$$

$$h_\theta = -\frac{1}{\omega}\left(\xi e_r + \frac{\partial}{\partial x}e_z\right)$$

$$e_\theta = -\frac{1}{\omega}\left(\xi h_r - \frac{\partial}{\partial x}h_z\right)$$

The perturbation expression for the curvature term of the longitudinal coupling impedance of a toroidal beam tube with rectangular cross section was found to be

$$\frac{Z}{n} = -jZ_0\left(\frac{h}{\pi R}\right)^2 \sum (1 - 3\Omega^2)\frac{\frac{1}{2}\sinh 2\xi w - \xi w}{m^3 \cosh^2 \xi w}$$

This perturbation expression has been numerically compared with the exact results obtained by using the SLATEC subroutines for the Bessel functions. It was confirmed that the total coupling impedance in the relativistic case of $v \sim 1$ is given by the sum of the straight beam tube space charge term plus the γ–independent curvature term.[3]

In Fig. 2, the total Z/n from above approximations is compared with the exact results as function of γ for a geometry with $2w = h$ and $\pi R/h = 10^4$. It is noted that the modified Bessel functions revert to ordinary Bessel functions at $\Omega = v^{-1} \sim 1$, the cut–off frequency corresponds to $\Omega \approx \sqrt{2}$ and the first resonance occurs at $\Omega_s = 85.8825$. Complete agreement at frequencies up to and above cut–off, but below the resonance region, was found.

ACKNOWLEDGMENTS

It is a pleasure to acknowledge the stimulating discussions and repeated correspondance with Dr. R. Warnock which contributed to our understanding of this problem.

REFERENCES

[1] H. Hahn and S. Tepikian, Proc. EPAC 90, Nice (1990), p. 1043.

[2] R.L. Warnock and P. Morton, Particle Accelerators, 25, 113 (1990).

[3] King–Yuen Ng and Robert Warnock, Proc. IEEE Particle Accelerator Conf., Chicago, IL, p. 798 (1989); Phys. Rev. D 40, 231 (1989).

[4] M. Jouguet, C.R. Acad. Sci. Paris, 222, 36 (1946); 223, 380 (1946).

[5] L. Lewin, *Theory of waveguides*, (Wiley, New York, 1975), Chap. 4.

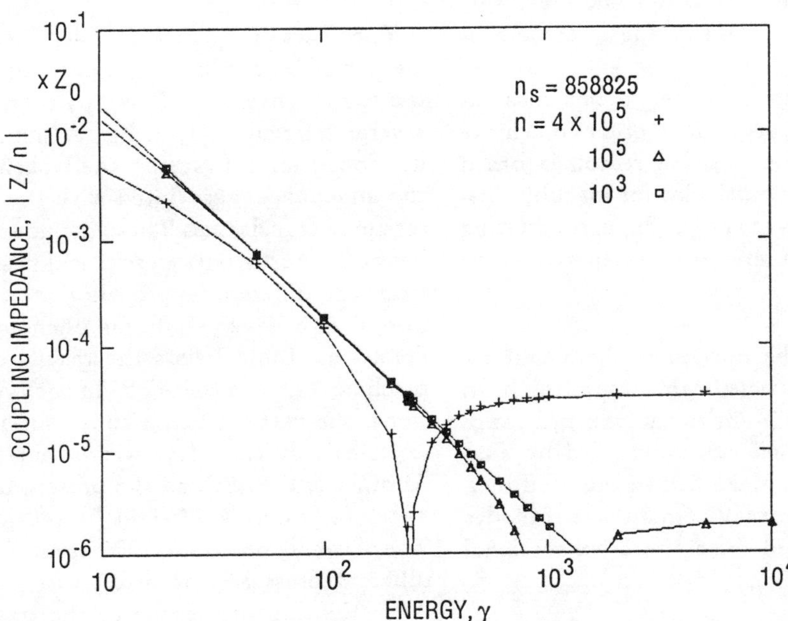

Fig. 2: Coupling impedance as function of γ. The solid curves are obtained from the approximations in this paper and the points from the exact expressions.

Impact of Cross-Sectional Changes in a Beam Tube on Beam Dynamics

Weiren Chou

Accelerator Design and Operations Division

Superconducting Super Collider Laboratory *

2550 Beckleymeade Avenue

Dallas, TX 75237

Abstract

Cross-sectional transitions are seen in accelerators. The beam tube dimensions may vary from one section (of dipole, quadrupole, or undulator, etc.) to another. This paper studies the impact of these discontinuities on single bunch as well as on multiple bunches. The transitions contribute coupling impedances that may significantly affect the single bunch behavior when the beam tube dimensions are small (as in the case of an undulator). They also serve as parasitic rf cavities that have a set of rf modes with high quality factors. When a bunch of particles is passing by, these modes may get excited and lead to instability and/or emittance dilution in the bunches that follow. Several possible cures are discussed.

I. Introduction

In the design of storage rings, a beam tube of uniform size is usually prefered. But sometimes, cross-sectional transitions in a beam tube are inevitable. There are various reasons for this to happen. One is due to some specific applications of a storage ring, such as a synchrotron light source, in which the beam tube height becomes extremely small in the undulator sections in order to achieve a high magnetic field. Another possible reason is based on economic considerations, in particular for machines using superconducting magnets, such as the Superconducting Super Collider (SSC), of which large magnet apertures are expensive.

This non-uniform beam tube may have significant effects on the behavior of the particle beam circulating in it. For instance, the collective instabilities and the emittance dilution. In this paper, these effects are studied by analyzing, in the time domain, the wake fields generated by a bunched beam, and in the frequency domain, the rf modes of the parasitic cavities formed from the cross-sectional variations.

There are two distinctive cases. Firstly, for the synchrotron light sources, the transverse coupling impedance, $Z_\perp$, can be dominated by these cross-sectional transitions because of the fact that $Z_\perp$ is roughly inversely proportional to the third power of the beam tube size. This may become the bottleneck of single bunch current intensity in such machines (e.g., the Advanced Photon Source (APS) at the Argonne National Laboratory.). Secondly, for a machine equipped with superconducting magnets, such as the SSC, the concern is different type. In order to reduce the wall impedance, the beam tube is coated with a thin copper layer. The resistivity of the layer is very low at liquid helium temperature. This means that the quality factor, Q, of the rf modes of the parasitic cavities formed from the cross-sectional transitions will be very high ($\sim 10^5$). Therefore, once these modes get excited by a traversing bunch, they can stay there for a long while before being damped and will have impact on other bunches. The result may be a substantial emittance dilution.

II. Impact on a Single Bunch

The effect of a bunch on itself mediated by the environment can be studied by analyzing the wakefields within the bunch envelope. This has been discussed in detail in several references [1, 2, 3]. A number of tools has been developed for computing the wakefields (or, equivalently, the impedance) associated with the structure of the cross-sectional transitions. These include numerical simulations, boundary perturbation, and scalings. The results obtained from the different approaches are in agreement. Therefore, it is believed that this phenomenon is relatively understood. Table 1 lists the results calculated for the two machines. One is the APS. In each of its 34 undulator sections, the vertical beam tube half-height is 0.4 cm. This explains why their transverse impedance is so dominant. Another is the SSC. In the present design, the quadrupole aperture (4 cm) is different from that of the dipole (5 cm). There would be about 1000 transitions if a beam tube of different cross section sizes were used. Contrary to the APS, the impedance due to the transitions in the SSC is insignificant because of the relatively large beam tube radius in the quads (1.65 cm).

*Operated by the Universities Research Association, Inc., for the U.S. Department of Energy under Contract No. DE-AC02-89ER40486.

Table 1. Impedance of Transitions

Machine	$Z_\parallel/n$		$Z_\perp$	
	Ω	% †	$M\Omega/m$	% †
APS	0.03	3	0.06	40
SSC	0.008	2	0.2	1

† The percentage of the total machine impedance.

III. Impact on Multiple Bunches

The effects of multiple bunches versus a single bunch are different. The beam tube consists of a sequence of large and small pipes, connected by tapered or untapered transitions. The large pipes can be regarded as parasitic cavities with certain eigenfrequencies and eigenmodes. These modes may get excited by a traversing bunch. When the mode frequencies are below the cutoff frequency of the small pipe, they will be trapped. Otherwise, they will propagate along the beam tube. If the quality factor of the trapped modes are high enough, they will stay there for a long time and will affect the behavior of the beam.

The SSC serves as a good example to demonstrate these multiple bunch effects. Assume the radius of the large pipe is 2 cm, the small one 1.6 cm. The conductivity, σ, of copper at liquid helium temperature ($\sim$ 4 K) in a high magnetic field (6.6 T) is taken to be $1.8 \times 10^9\ \Omega^{-1}m^{-1}$ (corresponding to RRR = 30). The first cutoff of the TM wave in the large pipe is 5.7 GHz, and in the small pipe 7.2 GHz. The skin depth and quality factor at these frequencies are, respectively,

$$\delta = \sqrt{\frac{2}{\mu\sigma\omega}} \sim 0.15\ \mu m,$$

$$Q = \frac{V}{S\delta} \cdot G \sim 10^5,$$

in which V is the volume of the cavity, S the total surface area, and G the geometrical factor of the order of unity. The bunch separation in the SSC is 60 MHz. Therefore, an rf mode of a frequency of 6 GHz and a quality factor of 10^5 would be seen by 1000 bunches.

The modes are computed by URMEL [4]. Some of them are shown in Figure 1 and listed in Tables 2 and 3. A question exists whether or not these modes would be excited when a bunch of particles pass by. The answer is not straightforward and depends on whether or not the bunch contains high frequency components. For electron machines, the bunch distribution is Gaussian. One may readily compute the high frequency part from its spectrum. But, unfortunately, for proton machines, the bunches are not really rigid Gaussian. It has been observed in the Tevatron at the Fermilab that 8 GHz lines are seen in the spectrum of a proton bunch which has an rms length of about 60 cm. This may come from the fine structures within the bunch [5]. If, on the other side, one assumes a parabolic distribution for a proton bunch, then the sharp discontinuities at both ends of the bunch will certainly produce high frequency components. In a real machine, the discontinuities in a bunch distribution may well be generated by the filamentation. No matter what the explanation is, that there are high frequency components in a proton bunch is an acknowledged fact.

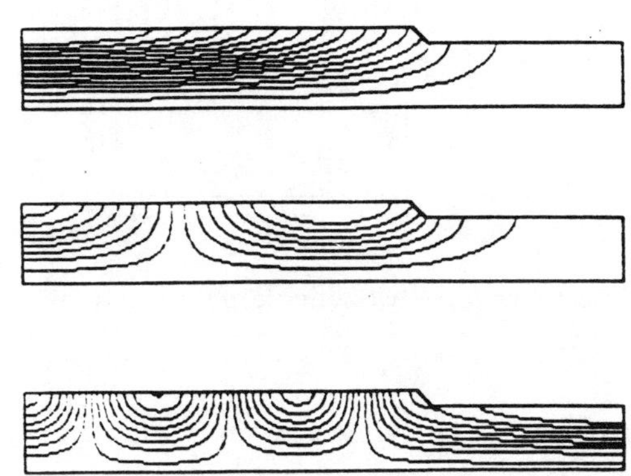

Figure 1: The $\vec{E}$ lines of several TM_{01} modes.

Table 2. Sample Monopole Modes

Mode Type TM_{mnp}	f MHz	R/Q Ω
TM_{011}	5761	0.002
TM_{012}	5852	0.008
TM_{013}	5999	0.037
TM_{014}	6197	0.006
TM_{015}	6436	0.016
TM_{016}	6698	0.045
TM_{017}	7155	8.213
TM_{018}	7230	4.938
TM_{019}	7319	10.17

Table 3. Sample Dipole Modes

Mode Type TE/TM_{mnp}	f MHz	R/Q * Ω
TE_{111}	4551	0.000
TE_{113}	5457	0.001
TE_{115}	6258	0.003
TM_{111}	9156	0.000
TM_{113}	9661	0.001
TM_{115}	10629	0.000
TM_{117}	11437	2.815
TM_{119}	11977	0.859

*Integrated at 3 mm displacement from the axis.

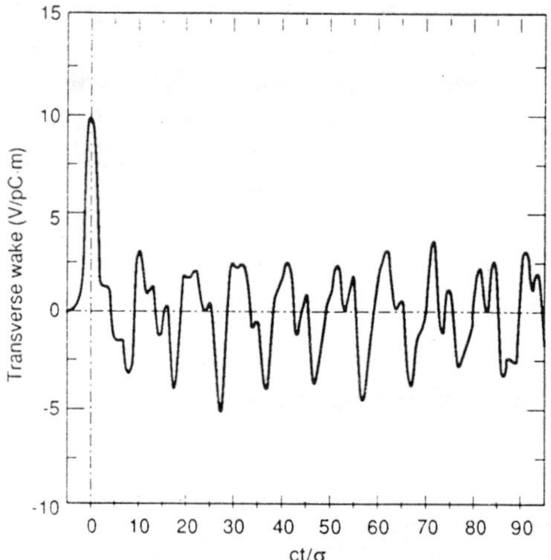

Figure 2: The transverse wakefields generated by a rigid Gaussian bunch.

For the purpose of carrying out the numerical estimation for the SSC, let us assume the bunch distribution is parabolic (approximated by half-cosine) with a full bunch length of a (in seconds).

$$\lambda(t) = \begin{cases} \frac{\pi}{2a}\cos\frac{\pi t}{a} & |t|\leq \frac{a}{2} \ , \\ \\ 0 & |t|> \frac{a}{2} \ . \end{cases}$$

Its spectrum is

$$\tilde{\lambda}(f) = f_0^2 \cdot \frac{\cos\frac{\pi f}{2f_0}}{f_0^2 - f^2} \ ,$$

in which

$$f_0 = \frac{1}{2a} \ .$$

For $a = 0.5$ ns, we have $f_0 = 1$ GHz. Thus, those components in the spectrum which are above $5f_0$ may be called the *high frequency components* and will be able to excite the rf modes in the cavities. These components account for about 10% out of a total of 100%.

Figure 2 shows the transverse wakefields generated by a rigid Gaussian bunch traversing a cavity. This bunch is short enough so that it contains abundant high frequency components (r.m.s. width 8 GHz). The peak of the wakes behind the bunch is about 5 V/pC·m. A very rough estimation is as follows.

For the SSC, each bunch has 0.75×10^{10} protons. This gives 1 nC. The average displacement from the axis is about 1 mm (closed orbit, survey error, etc.). Assume 10% high frequency component in a bunch. Then one bunch will generate $5 \times 10^3 \times 10^{-3} \times 0.1 = 0.5$ V in one cavity. There are 1000 bunches that will see each other. There are 1000 cavities. Therefore, the transverse kick per turn experienced by the protons will be

$$\Delta x' = \frac{0.5 \cdot \sqrt{1000} \cdot 1000}{E} \sim 10^{-9} \quad \text{rad},$$

in which $E = 20$ TeV is the proton energy. These continuous random kicks will result in emittance growth. The growth rate is

$$\begin{aligned} \dot{\epsilon} &= \bar{\beta}\cdot\overline{\Delta x'^2}\cdot f \\ &= 200 \times 10^{-18} \times 3 \times 10^4 \\ &= 6 \times 10^{-12} \quad \text{m}-\text{rad/s} \end{aligned}$$

in which $\bar{\beta}$ is the average betatron function (in meters), $\overline{\Delta x'^2}$ the average square of the kick, and f the revolution frequency (in Hz). When compared to the nominal value of the emittance, 5×10^{-11} m-rad, we see that the emittance would be doubled in approximately about 10 seconds.

IV. CONCLUSIONS

1. There are two possible consequences when adopting a non-uniform beam tube. The first is the single bunch instability due to large transverse impedances of these transitions. This is the case of the APS. The second is the probable transverse emittance dilution due to the multibunch effects, which is demonstrated using the SSC parameters.

2. The best cure is to use a uniform beam tube. When this is impossible, one has to adjust the beam parameters carefully to reach the desired value of the bunch intensity or to apply a feedback system to stabalize the beam.

3. If a concern exists for the trapped rf modes, dampers may be installed. Another conceivable solution is to randomely vary the beam tube lengths by a few centimeters in each section in order for the statistically averaged kicks to become smaller.

References

[1] W. Chou and H. Bizek, Proc. of Impedance and Bunch Instabilities Workshop, ANL, 1989, p. 7, ANL/APS/TM-5 (April, 1990).

[2] W. Chou, Argonne National Laboratory, Light Source Note LS-149, 1990.

[3] H. Bizek and W. Chou, Argonne National Laboratory, Light Source Note LS-150, 1990.

[4] URMEL is a frequency domain simulation code written by T. Weiland.

[5] G. Jackson, private communication.

Coupled-Bunch Instabilities in the APS Ring*

L. Emery

Argonne National Laboratory 9700 So. Cass Ave., Argonne, 60439

Abstract

A study of coupled bunch instabilities for the APS storage ring is presented. The instabilities are driven by the higher-order modes of the fifteen 352-MHz single-cell RF cavities. These modes are modeled using the 2-D cavity program URMEL[1]. The program ZAP[2] is then used to estimate the growth time of the instabilities for an equally-spaced bunch pattern. The cavity modes most responsible for the instabilities will be singled out for damping.

I. INTRODUCTION

One of the eventual goals of the Advanced Photon Source (APS) main ring is the capability to store 300 mA in many bunches[3]. It is therefore valuable to investigate the current limits imposed by coupled-bunch motion instabilities which may jeopardize this goal. Instabilities of bunch motion in the longitudinal or transverse plane may occur when slowly decaying wakefields generated by a bunch passing through vacuum chamber discontinuities are felt by following bunches. Normally, the higher-order modes (HOM's) of RF cavities are the largest contributors of slowly decaying wakefields and, therefore, of coupled-bunch instabilities. The wakefields produced in cavities are usually described as high-Q resonator impedances for each HOM. The design of the APS cavities can be found in reference [3].

Computer calculations of the HOM impedances of the main ring single-cell RF cavity are presented. Then the current limit due to the impedances of the fifteen RF cavities is estimated. The results give a clear indication that cures against the instabilities are required for the operation of the ring. One cure is to attach tuned dampers to the cavity in order to reduce the shunt impedance of the strongest HOM's. To assist the dampers, one may implement systematic geometry differences (slight elongation of the cavities) in all or groups of cavities to spread the HOM resonant frequencies across a number of revolution harmonics. This reduces the chance that many cavity HOM's contribute to the same coupled-bunch mode instability. The required deQing effect of the tuned dampers is estimated.

II. CAVITY IMPEDANCES

In order to predict the longitudinal and transverse coupled-bunch instability limits, one would ideally make impedance measurements on all cavities to be installed in the ring. Since the cavities are not built yet, one can model the monopole and dipole cavity modes using URMEL[1], which uses a rectangular grid on the radial-longitudinal plane. An output file provides the resonant frequency f_r, the shunt impedance R_s, and the quality factor Q for each HOM. Tables 1 and 2 list the strongest modes below the cut-off frequency of the cavity beampipe.

Since the rectangular grid can't fit the spherical contour of the cavity exactly and since the real cavity is not cylindrically symmetric (due to the RF power coupling loop and various

Table 1

Longitudinal coupled bunch motion with one cavity.

f_{HOM} (MHz)	R_s (MΩ)	Q	$1/\tau_{300\text{mA}}$ (sec^{-1})	$I_{\text{thresh.}}$ (mA)
536.7	1.67	41000	790.	80.
922.5	0.62	107000	500.	130.
939.0	0.23	42000	190.	340.
1173.2	0.18	44000	190.	340.
1210.8	0.49	94000	510.	130.
1509.1	0.40	88000	510.	130.

Table 2

Transverse coupled bunch motion with one cavity.

f_{HOM} (MHz)	R_t (MΩ/m)	Q	$1/\tau_{300\text{mA}}$ (sec^{-1})	$I_{\text{thresh.}}$ (mA)
588.7	13.6	68000	390.	80.
761.1	25.6	53000	730.	43.
962.0	6.1	54000	170.	190.
1017.4	2.6	41000	73.	435.
1145.1	2.7	92000	78.	410.
1219.2	3.6	41000	101.	315.

ports), one would expect the measured frequencies of an assembled cavity to differ from the URMEL calculation. The measured frequencies of a prototype cavity are reported in [4]. The URMEL frequencies should therefore be taken as uncertain to some extent. For conservative estimates of the coupled-bunch instability growth rates, the frequencies will be shifted to the closest positive revolution frequency harmonic synchrotron sideband.

In addition, the Q of each HOM may be lower in the real cavity due to the influence of the RF power coupling loop. It is expected however that the measured value of R_s/Q should not change appreciably from the calculated one.

This paper will include only HOM's with frequency lower than the cut-off frequency for the 7-cm radius beam pipe connecting the RF cavities. However, standing wave modes can be trapped in these sections. The TM mode cutoff frequency for 7-cm radius is 1.6 GHz while the cutoff frequency for the regular elliptical vacuum chamber is 4.5 GHz. All modes of intermediate frequencies can be trapped. URMEL can evaluate the impedance contributed by these, and they are by no means negligible.

Absorbing materials placed in these regions will damp these modes. I will assume that strong damping is much more easily acheived for these modes than for the lower HOM's, and no further study is required for the moment.

III. RING PARAMETERS

Table 3 lists the relevant ring parameters for the stability

*Work supported by U.S. Department of Energy, Office of Basic Energy Sciences under Contract No. W-31-109-ENG-38.

calculations. The bunch length used in the calculations takes

Table 3
Parameters used in Stability Calculations

Energy E	7 GeV
Revolution frequency f_0	271.55 kHz
RF frequency f_{RF}	351.931 MHz
RF Harmonic number h	1296
RF voltage V_{RF}	9.5 MV
Natural Bunch length σ_{l0}	5.33 mm
Bunch length σ_l	7.0 mm
Energy Spread σ_ϵ	9.6×10^{-4}
Momentum compaction η	2.28×10^{-4}
Synchrotron frequency f_s	1.49 kHz
Synchrotron tune ν_s	0.00549
Horizontal tune ν_x	35.2
Vertical tune ν_y	14.3
Longitudinal damping rate $1/\tau_\epsilon$	213 $\sec^{-1}$
Longitudinal damping time τ_ϵ	4.7 msec
Transverse damping rate $1/\tau_\epsilon$	106 $\sec^{-1}$
Transverse damping time τ_ϵ	9.4 msec
Total current I	300 mA
Number of bunches n_b	54
Bunch current I_b	5.6 mA

into account single bunch intensity effects for a bunch current of 5 mA according to a pseudo-Green function method[5]. The energy spread is not affected however. Since 5 mA is close to the transverse mode coupling limit[5], the number of bunches for a total current of 300 mA is set at 54, which must be a factor of h for the application of most calculational methods of growth rates of coupled-bunch instabilities.

IV. COUPLED-BUNCH MODES GROWTH RATE

The threshold current for instabilities in electron storage rings is determined by the partially cancelling action of instability growth and the synchrotron radiation damping rate. The instability growth is proportional to the stored current, while the radiation damping rate is constant. The current threshold is then defined as the current at which the instability growth is equal to the damping rate. All growth rates will be calculated for I=300 mA. The current threshold will be extrapolated from

$$I_{\mathrm{thresh}} = 300\mathrm{mA}\left(\frac{\tau_\epsilon}{\tau_{300\mathrm{mA}}}\right). \tag{1}$$

The coupled-bunch modes treated in this section are those produced by a symmetric bunch pattern, i.e. bunches that are equally spaced and equally charged so that the frame of reference of one bunch is equivalent to that of another. The coupled-bunch modes are then rather simple. The amplitude of all bunches is the same, and the phase separation of the motion of each succesive bunch is $\Delta\phi = 2\pi m/n_b$ where n_b is the number of bunches (n_b must be a factor of the RF harmonic number) and m is the coupled-bunch mode number where where $0 < m < n_b - 1$.

For the longitudinal plane the growth rate of each longitudinal mode m is determined by the values of all longitudinal impedances sampled at revolution harmonic synchrotron sidebands

$$f_p = (pn_b + m)f_0 + af_s \tag{2}$$

where p is an integer, f_s is the synchrotron frequency, a is the synchrotron mode number (for dipole mode, a=1), f_0 is the revolution frequency. For a conservative estimate of the current threshold, we consider the dipole synchrotron mode ($a \doteq 1$), since higher order synchrotron modes grow more slowly.

The positive revolution harmonic sidebands ($p > 0$) contribute to growth, and the negative ones ($p < 0$) contribute to damping. It follows from this that if a HOM causes growth for mode m, then it will cause damping for the $n_b - m$ mode.

Similarly for the transverse motion, the relevant quantities are the transverse impedances from dipole HOM's and revolution harmonic betatron sideband frequencies $f_p = (pn_b + m + \nu_\perp)f_0 + af_s$. The horizontal motion is considered here, therefore we set $\nu_\perp = \nu_x$. For zero chromaticity, the growth rate is greatest for the rigid bunch mode, a=0. Here positive p causes damping and negative p causes growth — the opposite of longitudinal motion.

A. Conservative estimates

If the resonant frequency of a monopole HOM happens to equal one of the positive f_p's, then the growth of a particular longitudinal coupled-bunch mode is maximal for that HOM. The growth rate formula in the Wang formalism[6] option of ZAP[2] reduces to approximately

$$\frac{1}{\tau} = \frac{I f_{\mathrm{HOM}} \eta F R_s}{2(E/e)\nu_s} \tag{3}$$

with $F \approx 1$, and similarly for the transverse motion and dipole HOM's:

$$\frac{1}{\tau} = \frac{I c F R_t}{4\pi(E/e)\nu_x} \tag{4}$$

Since one cannot easily control the HOM frequencies, one must assume that some of the cavity HOM's will fall on such a frequency. The widths of the resonances ($\Delta f = f_r/Q$) are much narrower (of order 10 kHz) than the spacing between the harmful revolution harmonics (271kHz). Therefore it would seem that there is a only a small chance that a HOM from no more than one cavity will accidentally fall on or close enough to a harmonic sideband. Tables 1 and 2 show the worst-case growth rate for I=300 mA due to the strongest individual HOM's from a *single* cavity, as calculated by ZAP or equations 3 and 4. Unfortunately the desired 300 mA current is so large that the strongest mode (536 MHz) of a single cavity need only be 36 kHz away on either side of a revolution harmonic to cause instablity. Thus there is a 25% chance that the 536 MHz mode of any one cavity will cause instability. With 15 cavities, the growth rate for 300 mA will very likely exceed the radiation damping rate. Reducing the shunt impedance is necessary.

B. Systematic cavities modifications combined with deQing

If all or some of the resonant frequencies differ among cavities one can reduce the chance of two HOM's adding to the growth of the same coupled-bunch mode[7]. The HOM frequencies need to be at least f_0 apart. The cavities can be modified by inserting shims between the center equatorial ring of the cavity and the two nose cone sections. The proposed cavity elongation is along the beam axis. URMEL was used to estimate the rate at which resonant frequencies decrease as a function of shim thickness. Table 4 summarizes the results. The quantity $1/\tau$ does not include radiation damping, so the values shown should be compared to $1/\tau_\epsilon = -215$ sec^{-1}. Conveniently, the fundamen-

Table 4

Shimming the RF cavity and deQing monopole HOM's

f_{HOM} (MHz)	$d(f_{\mathrm{HOM}})/d(\Delta z)$ (MHz/mm)	$1/\tau$ for $Q = Q_0/10$ (sec^{-1})	Required Q/Q_0
351.93	-0.017	—	
536.7	-0.7	80	4
922.5	-1.1	50	2.3
939.0	-1.1	19	1
1173.2	-1.3	19	1
1210.8	-2.0	45	2.1
1509.1	-2.7	45	2.1

tal mode is hardly influenced. To spread the frequencies of the strongest HOM by a sufficient amount, I assume all cavities are shimmed in increasing thickness by steps of 1mm. This separates the 536 MHz modes by about $2.5 f_0$ from one cavity to the next. All the other mode frequencies are separated by about twice as much.

It would seem that the conservative estimate of growth rate in the previous subsection (shown in Tables 1 and 2) applies exactly to this situation since no two cavities have the same HOM frequencies. However, with 54 possible longitudinal modes for 54 bunches, and $15 \times 6 = 90$ individual HOM's it is likely that many coupled-bunch modes will be influenced by more than one HOM through aliasing, i.e., $f_{r1} = (p_1 n_b + m + \nu_s)f_0$ and $f_{r2} = (p_1 n_b + m + \nu_s)f_0$ with $p_1 \neq p_2$. But it is also likely that some HOM's will cancel each other in their damping and anti-damping effect on many modes. This statistical problem should be further studied. For this report I will assume that in this fifteen cavity system, there is only one HOM fully exciting some coupled-bunch mode.

Installing dampers to lower the shunt impedance and Q of individual modes can greatly reduce the growth rates. However, a lower Q means a wider impedance function, and a larger probability that a HOM will contribute significantly to some coupled-bunch mode growth even if the HOM is not centered on a revolution harmonic sideband. Table 4 lists the maximum growth rates for the cavities with a deQing factor of 10. The HOM's are considered individually. The basic result is that the growth rates are scaled by exactly the deQing factor. Coupled bunch modes associated with adjacent revolution harmonics (not shown) were calculated to have a growth rate of 10% or less than those corresponding to the main excited coupled bunch mode. One can specify the minimum deQing by comparing the growth rate with the radiation damping rate, and one finds that the largest deQing factor for longitudinal HOM's is 4. These low deQing factors may well occur naturally due to the input RF coupling loop without the aid of dampers.

The transverse HOM treatment gives similar results except that the 588 MHz mode does not shift appreciably with shims. This is because the electric fields of this HOM have no longitudinal variation. The spread of the 588 MHz HOM for fifteen cavities with 1mm incremental shim thickness is about 1.2

MHz, enough to cover 5 revolution harmonics. When deQing this mode one should consider each coupled-bunch mode to be fully excited by three cavities. Table 5 shows that the HOM

Table 5

Shimming the RF cavity and deQing dipole HOM's

f_{HOM} (MHz)	$d(f_{\mathrm{HOM}})/d(\Delta z)$ (MHz/mm)	$1/\tau$ for $Q = Q_0/10$ (sec^{-1})	Required Q/Q_0
588.7	-0.08	117.0	11
761.1	-0.7	73.0	4
962.0	-1.2	17.0	2.3
1017.4	-1.7	7.3	1
1145.1	-1.5	7.8	1
1219.2	-1.9	10.	2.1

needing the most deQing is the 588 MHz mode.

Instead of making all cavities different one can make groups of cavities with the same shimming. However this will increase the required deQing of the modes by a factor between 1 and the number of cavities in the group. The exact deQing factor depends on the statistical distribution of frequency shift due to construction tolerances and other imponderables.

V. Conclusion

The growth rate for instabilities were evaluated for reasonably pessimistic cases. The deQing requirements for most of the HOM's are modest if all cavities are shimmed differently.

VI. Acknowledgement

I've greatly benefited from conversations with Bruno Zotter and Steve Kramer on impedances and coupled bunch instabilities in general. I thank Lee Teng for his guidance.

VII. References

[1] T. Weiland *NIM*, vol. 216, pp. 329–348, 1983.
[2] M.S. Zisman, S. Chattopadhyay, and J.J. Bisognano, "ZAP User's Guide," LBL 21270, Lawrence Berkeley Laboratory, December 1986.
[3] APS, "7 Gev Synchrotron Radiation Source Conceptual Design Report," ANL 87-15, APS, 1987.
[4] J. Cook, J. Bridges, and R.L. Kustom, "Measurement on Prototype Cavities (352 MHz) for the Advanced Photon Source (APS)," in *Proceedings of the 1991 IEEE Particle Accelerator Conference*, 1991.
[5] W.Chou and H. Bizek, "Impedance and Bunch Lengthening in the APS," in *Proceedings of the Impedance and Bunch Instability Workshop*, 1990.
[6] J.M. Wang, "Longitudinal symmetric Coupled Bunch Modes," BNL 51302, BNL, December 1980.
[7] R. Kustom. Private communication.

DAMPING OF HIGHER-ORDER MODES IN A THREEFOLD SYMMETRY ACCELERATING STRUCTURE

D. Yu[*]

DULY Consultants, Rancho Palos Verdes, CA 90732

N. Kroll[†]

University of California, San Diego and SLAC

Abstract

We investigate a waveguide-coupled damping structure with threefold symmetry using the 2-D and 3-D MAFIA codes. Within the frequency range considered, all higher order modes except the TM011 and TE111 modes are heavily damped. Possible ways to detrap these by using asymmetric waveguides offset with respect to the accelerating cavity in the direction of the beam are studied. External Qs and resonant frequencies are calculated using recently developed computer methods.

INTRODUCTION

Much work has been done since the first workshop on Future Linear Collider held in SLAC in 1988 on the design of damped accelerating structures using various waveguide loaded cavities. Accelerating structures in which higher order modes (HOM) are damped by coupling to waveguides via radial and circumferential slots have been proposed by R. Palmer[1]. We have studied several of these damped cavities with radial or circumferential couplers and have calculated, in each instance, the external Qs and resonant frequencies of the damped structure using a theory developed by Kroll and Yu[2], based on frequencies and phase shifts for shorted waveguides. A useful extension and variation of this theory has been recently developed[3] and is also used in our evaluation.

We begin by recapitulating some results for cavities with twofold-symmetry radial or circumferential slots which have been reported previously[4]. The calculated Q for the principal transverse mode of the radially slotted cavities is between 8.7 to 13.2, depending on the waveguide cutoff frequency. Results for the Q values using the Kroll-Yu method are highly accurate and convincing, and compare favorably with those obtained with other methods[5]. In addition, the Kroll-Yu method determines the resonance frequencies accurately. A problem with the radial slot cavity is the existence of a slot mode which persistently has a lower frequency than the accelerating mode. Because it has a different symmetry from the accelerating mode, however, it can be strongly coupled out by use of a double ridge waveguide[2] having a lower cutoff frequency than the fundamental mode, without significantly damping the latter.

An advantage of circumferentially coupled cavities is the absence of any slot modes. Not restricted by clearance considerations, use of smaller irises in these cavities also helps preserve accelerating field structure and depress HOM longitudinal coupling. A potential problem of these cavities is that the lowest transverse frequency is very near the waveguide cutoff. It is important to design the structure with extreme care in order to avoid trapped transverse modes. The possibility exists that a transverse mode is trapped even when its frequency, in the absence of damping waveguides, is above the waveguide cutoff. Ideally, the waveguide cutoff frequency should be far enough below the HOM to avoid trapping, and yet be sufficiently high above the fundamental, so that the accelerating mode is well preserved[5].

Recently, Arcioni and Conciauro[6] (AC) proposed a threefold symmetric, circumferentially coupled, waveguide loaded cavity. They showed that the shunt impedance and the Q of the accelerating mode are degraded by 20% and 12%, respectively, from those of a closed cavity, while most of the HOMs were effectively damped. In this research we applied our computer method to further analyze this S-band cavity. The resonance frequencies and external Qs of the HOM for this structure were calculated. Then we studied a similar X-band waveguide loaded accelerating structure with a threefold symmetry, and showed that in the frequency range considered, two HOMs (TM011 and TE111) remain trapped. Finally, we studied methods to detrap these modes.

3-FOLD SYMMETRY CAVITY

The AC cavity shown in Figure 1 has a radius (r) of 3.0 cm and a height of 3.5 cm. The cylindrical cavity is coupled to three square waveguides, 120° apart, each having a width of 3.5 cm. We calculated the frequencies of

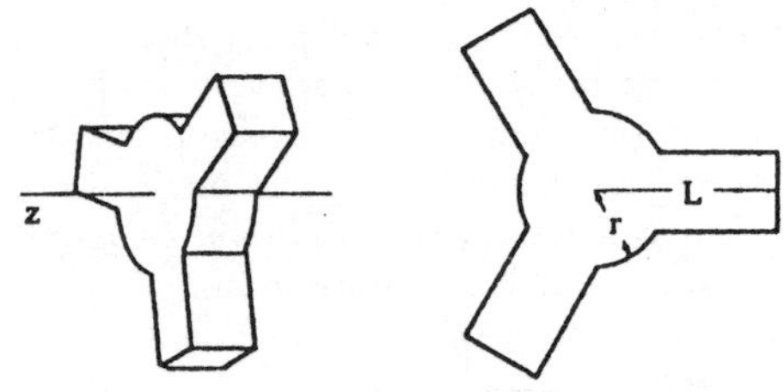

Fig. 1 A threefold symmetry cavity (ref. 6)

the TM and TE modes of this structure with URMEL-T for waveguide lengths (L) of 7.3, 7.9, 8.5, 9.1 and 18.0 cm. For ease of mode identification, only half of the structure was modeled with reflection symmetry imposed on the half plane. A 32000-point mesh is used for the half model with the longest guide length. The number of mesh points for shorter guide lengths was reduced to maintain compatible mesh fineness. The beam pipe was not included in the 2-D model as its effect on the external Q was judged to be negligible. Table 1 lists the frequencies of the TM modes, with principal electric fields along the axis of the resonator, shorted at various guide lengths:

Table 1 Modal frequencies vs guide lengths for AC cavity

Mode	7.3 cm	7.9 cm	8.5 cm	9.1 cm	18.0 cm
TM010	3354.30	3360.94	3352.05	3356.40	3341.75
TM110	4637.98	4603.07	4556.38	4526.22	4335.70
TM020	5104.74	4989.96	4865.69	4785.28	4376.92
TM120	5699.26	5529.12	5374.89	5252.20	4553.26
TM030	6682.62	6367.92	6101.79	5883.56	4664.38
TM130	6955.29	6730.32	6498.38	6292.62	4893.36
TM210	7956.85	7683.32	7448.07	7255.65	
TM040	8371.52	7927.96	7545.96	7203.79	5101.62
TM140				8161.31	5340.50
TM050				8542.02	5640.47
TM150					5865.92

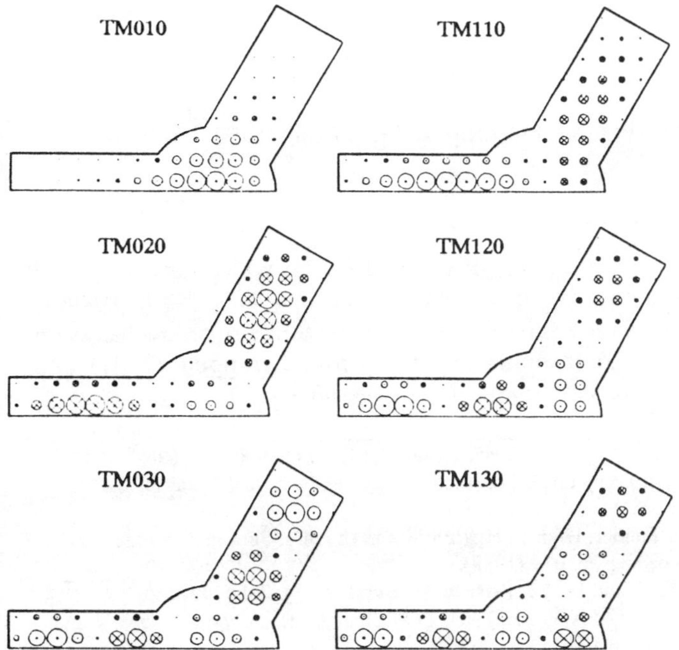

Fig. 2 Electric field plots from URMELT-T

A typical set of axial electric field plots is shown in Figure 2 for the L=9.1 cm case. The phase shifts, defined as:

$$\varphi = \frac{L_g}{c}\sqrt{\omega^2 - \omega_c^2} + m\pi$$

are plotted in Figure 3 for both the monopole (TM0n) and dipole (TM1n) modes as a function of frequency. In the above expression, ω is the angular frequency for a given mode, ω_c is the waveguide cutoff, L_g is equal to L-r, and c is the speed of light. The integer m depends upon the branch on which the mode appears. When properly chosen all the points for a specific symmetry fall on a single curve. This is illustrated in Figure 3, which includes all data from four guide lengths (excluding the longest length). The phase shifts for the monopole modes (represented by data points on the upper curve) undergo a change less than π in the frequency range considered. The data were fitted to a single-resonance 4-point formula[2] with a frequency of 9309 MHz and a Q of 2.3. The data for the dipole modes (lower curve) clearly show that there are two resonances in the same frequency range. The phase shifts in this frequency range undergo a change of 2π. The lower curve on Figure 3 is a theoretical fit obtained with a two-resonance form-ula[2] for the phase shifts:

$$\varphi(\omega) = \tan^{-1}\left(\frac{v_1}{\omega - u_1}\right) + \tan^{-1}\left(\frac{v_2}{\omega - u_2}\right) - \chi(\omega) + m\pi$$

$$\chi(\omega) \approx \chi_0 + \chi_1\omega$$

The resonance frequencies of 4459 MHz and 7350 Mhz, and their corresponding Q of 2.7 and 6.3 were obtained from the six parameters u_1, u_2, v_1, v_2, χ_0 and χ_1 fitted to the data. A high degree of consistency for these results is evident by selecting data points from different crossed branches. We also used the data points from a *single* run with the longest guide length to calculate the dipole resonance frequency and Q. We obtained, using the four-point formula of reference 2, a resonance at 4311 MHz with Q=2.6, or 4385 MHz with Q=2.9, depending on the selection of the four points. These were consistent with the results for the lower resonance obtained from Figure 3 using data from four different runs. Data for the monopole modes using only the longest guide length were insufficient to yield a four-point solution. The extremely low Q values for the single monopole and two dipole resonances lead us to conclude that principal TM mode resonances in the AC cavity are indeed highly damped.

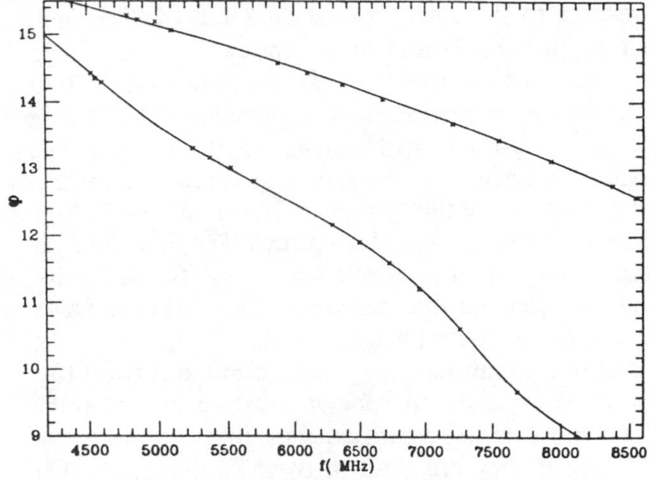

Fig. 3 Phase shift vs frequency Plot

TRAPPED MODES AND DETRAPPING

A review of the AC cavity URMEL-T plots of the electric fields in the plane perpendicular to the axis of the resonator shows that there are two possible trapped modes. These are shown in Figure 4, identified as (a) a TE111 mode, and (b) a TM011 mode in the cavity. To assess possible impact of these trapped modes on the accelerating cavity, we considered a π-mode, X-band accelerating structure with a threefold waveguide symmetry, similar to the AC cavity, using 3-D MAFIA models with up to 460000 mesh points. The 3-D models included long waveguide lengths (up to five times the cavity radius) and a finite cavity and waveguide height ($\sim\lambda$, where λ is the fundamental wavelength). The waveguide width was chosen so that the cutoff frequency is above the fundamental and below the HOM. The MAFIA models confirm that the TE111 and TM011 modes are in fact trapped in the cavity. The TE111 mode may not require damping because its impedance arises only from the probably very small TM contamination. On the other hand, damping is required for the TM011 mode. In order to minimize emittance growth of electron bunches over a very long distance, as in a linear collider, it was desirable to eliminate as much strayed

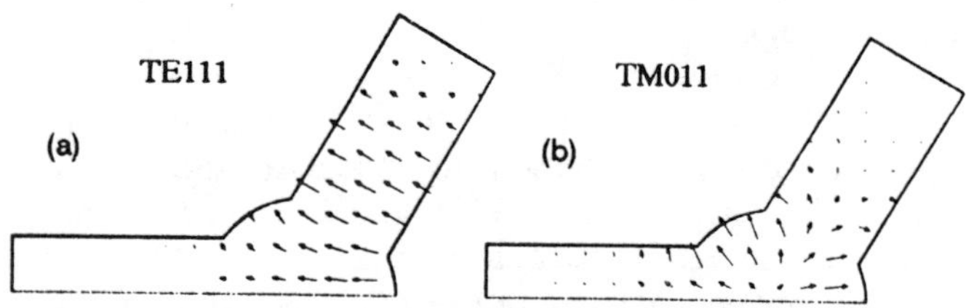

Fig. 4 Trapped modes in the AC cavity

fields in the structure as possible. We therefore considered methods to detrap the TE111 and TM011 modes. In a first unsuccessful attempt we tried to shift each of the three waveguides by a small offset (about 1/10 of the cavity height) with respect to the cavity along the beam direction. This proved insufficient to detrap the modes. In a second attempt, we tried to break the symmetry of the waveguide with respect to the cavity by reducing half of its height. This detrapped the TM011 mode (though not the TE111 mode), but it also distorted the accelerating mode[7]. In a third attempt we increased the waveguide height by 50% in an unsymmetrical way with respect to the cavity. The results of this calculation are shown in Figures 5a and 5b, showing two views of the detrapped TE111 and the TM011 mode, respectively. By making two MAFIA runs at different guide lengths, we were able to use the derivative method[3] to calculate the loaded Q. The TE111 mode is effectively detrapped with a Q of 10.9. The Q of the TM011 mode is substantially reduced, but still over 100. Oversized waveguides preclude installation on adjoining cavities. But damping waveguides may not be needed in every cavity if they are used in combination with other methods[8] of wakefield suppression.

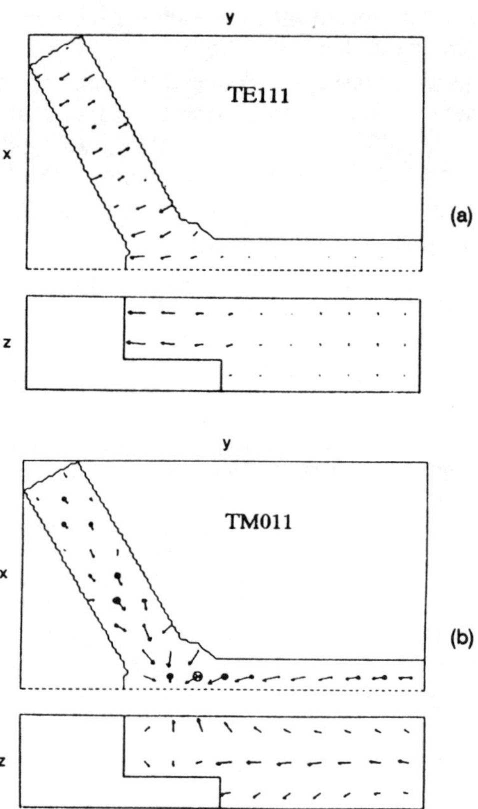

Fig. 5 Detrapped TE111 and TM011 modes

CONCLUSIONS

The problems of trapped modes which have persisted in previous damped structure designs have been partially solved with a threefold symmetry. Some progress has been made in detrapping the newly found trapped TM011 and TE111 modes unique to this geometry.

* Supported by DOE SBIR Grant No. DE-FG03-90ER81080.

† Supported by DOE Contracts DE-AC03-89ER40527 and DE-AC03-76SF00815.

[1] R.B. Palmer, High Energy Physics in the 90's, Snowmass Conference Proceedings, p. 638 (1989).

[2] N. Kroll and D. Yu, Particle Accelerators, 34, 231 (1990).

[3] N. Kroll and X.T. Lin, Proc. 1990 Linac Conf., Albuquerque, NM, June 10-14 (1990) p. 238.

[4] D. Yu and Kroll, poster paper in Linear Accelerator Codes Conf., Los Alamos, NM, January 22-25 (1990); H. Deruyter, et al., Proc. 1990 Linac Conf., Albuquerque, NM, June 10-14 (1990) p. 132.

[5] Y. Goren and D. Yu, Proc. 1989 IEEE Part. Acce. Conf., Chicago, IL, March 20-23 (1989) p. 189.

[6] G. Conciauro and P. Arcioni, European Particle Accelerator Conference, Nice, France (1990).

[7] K. Ko has provided us with ARGUS computations for a similar model considered for this case, with an asymmetric step which reduces the waveguide height by 38%. Using the derivative method of reference 3, we calculate a Q of 74 for the TM011 mode. The TM110 continues to exhibit extremely low Q's.

[8] D. Yu and J.S. Kim, "Wakefield Suppression Using Beatwave Cavities", this conference.

WAKEFIELD SUPPRESSION USING BEATWAVE STRUCTURES[*]

D. Yu and J.S. Kim

DULY Consultants, Rancho Palos Verdes, CA 90732

Abstract

A proposed method of suppressing transverse wake-fields in an accelerating structure makes use of the fact that superposition of long-range wakes excited by an electron bunch traversing a series of accelerating cells with different transverse frequencies can produce interference cancellation, thereby significantly reducing the magnitudes of the harmful wake potentials. Analytic calculations as well as time-domain and modal sum simulations are performed to study the beatwave effects produced by detuned, disk-loaded cavities as a function of their transverse frequency spread and the population density.

INTRODUCTION

A major concern in designing the high gradient accelerating structure for future high-luminosity linear colliders is how to minimize beam instability and energy spread of electron bunches caused by long-range wakefields generated by preceding bunches traversing the accelerating cavities. Several ideas have been proposed in the last few years to suppress the transverse wakefields. While in principle low-Q cavities coupled to external waveguides[1] can be designed to take out higher order modes (HOMs), these designs are quite complicated and can pose many manufacturing problems. Furthermore the problem of discarding the HOMs, once taken out of the accelerating structure, remains challenging. In parallel with this approach, one of us (DY) has proposed a different method of suppressing the long-range HOM wake potentials. In this method, accelerating cells with the same fundamental frequency are detuned for all HOMs. Interference of wakefields at different frequencies for each HOM, made possible by cell-to-cell structural variations, would considerably diminish the amplitudes of all transverse and higher order longitudinal wake potentials.

A simple way to implement this concept is to mix cavities with different iris diameters in a disk-loaded accelerating structure. By properly adjusting other cavity parameters (e.g. the cavity radius), the fundamental frequency of the accelerating mode is identical for all cavities, but the frequency of HOM varies from cavity to cavity. In an early illustration of this idea, the transverse wakes of a three-cell "beatwave" structure were calculated. Figure 1 shows the dipole wakes (from a 2/27/89 TBCI run) as a function of the distance behind the bunch, for a 3-cell, 2.856-GHz, $2\pi/3$-mode structure with an aperture radius to wavelength ratio (a/λ_0) of 0.20, 0.15 and 0.10. The corresponding spread in the lowest transverse frequency is from

3.70 to 4.35 GHz. Two effects were apparent from the results of this simulation. First the interference effects indeed produced a relatively broad null in the combined wake potential. Secondly there was a gradual damping of subsequent peaks of the wake potential.

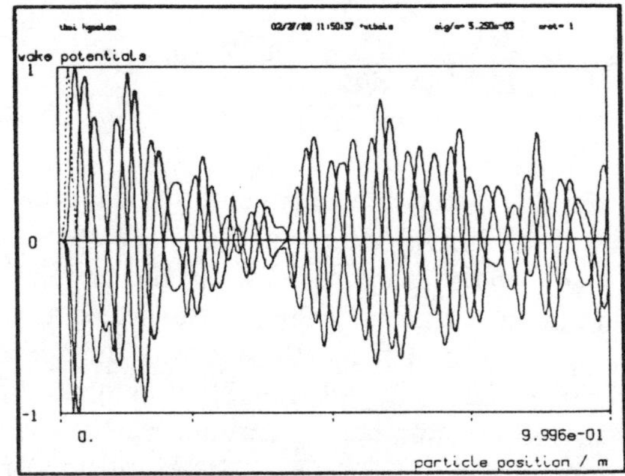

Fig. 1 TBCI dipole wakes for a 3-cell structure

The interference nulls in the wake potentials were further studied in detail by Yu and Wilson[2]. They showed by time and frequency domain calculations that using only two different sets of disk-loaded accelerator cells, relatively broad beating nulls in the transverse wakes could be produced. Since the position of the nulls varies as a function of the transverse frequency spread, it is possible to adjust the amount of detuning required so that the following bunch can be placed at the resulting interference nulls of the long-range transverse wake. For a beatwave structure with $a/\lambda_0 = 0.20$ and 0.15, resulting in a frequency spread of $\approx 8\%$ for the dominant transverse mode, the interference effects reduced the amplitude of the transverse wake by an order of magnitude from its maximum at a distance of 48 cm for 2.865 GHz, or 12 cm for 11.424 GHz. For such a beatwave structure, the wakes are suppressed over a distance on the order of a wavelength.

DAMPED BEATWAVE STRUCTURE

The second effect observed in Figure 1, i.e. damping of the long-range wake and broadening of the region of wake suppression, becomes increasingly pronounced as the number of transverse frequencies participating in the interference beating increases. As shown below, the damping rate of the transverse wake and the length of the wake suppression region are proportional to the population density (i.e. number of different frequencies) in a given frequency spread. The damping effect was recognized recently also in accelerator structure work at SLAC[3], where it was shown that substantial damping of the transverse wake could be sustained over a long distance behind the bunch provided a very large number of cavities with a Gaussian distribution of transverse frequencies were used.

We will first demonstrate the damping effect analytically in the continuum limit. Although the principle of wakefield suppression by interference of detuned cavities applies to any three-dimensional accelerating structure, it is straightforward to illustrate the effect using cylindrically symmetric disk-loaded structures. The dipole wake potential for such a structure is given by[4]:

$$W_d(s) = \frac{2r_0}{a} \sum \frac{k_{1n}}{\omega_{1n} a/c} \, e^{-\omega_{1n}^2 \sigma^2/2c^2} \sin(\omega_{1n} s/c)$$

where s is the distance behind the center of the bunch, r_0 is the offset of the driving charge and a is the aperture of the cavity. ω_{1n} and k_{1n} are respectively the angular frequencies and the loss factors of the nth transverse mode of the cavity. σ is the bunch length, and c is the speed of light. Consider now only the lowest, dominant dipole mode for an infinite number of detuned cavities with a frequency spread of $\Delta\omega = \omega_2 - \omega_1$. Superposition of the transverse dipole wakes gives

$$W_d \propto \int_{\omega_1}^{\omega_2} d\omega \left(\frac{k_1 c}{\omega a^2} \right) e^{-\omega^2 \sigma^2/2c^2} \sin(\omega s/c)$$

for a continuous range of equally weighted frequencies between ω_1 and ω_2; and

$$W_d \propto \int_{\omega_1}^{\omega_2} d\omega \, e^{-\left(\frac{\omega-\omega_a}{\Delta\omega}\right)^2} \left(\frac{k_1 c}{\omega a^2} \right) e^{-\omega^2 \sigma^2/2c^2} \sin(\omega s/c)$$

for a Gaussian frequency distribution centered at $\omega_a \approx (\omega_1+\omega_2)/2$ with a half width of $\Delta\omega$. Define the scaleable dimensionless parameters: $x \equiv \omega/\omega_0$ and $\bar{s} \equiv s/\lambda_0$, where the subscript 0 refers to the fundamental mode. We have calculated values of k_1, ω_0 and ω using the TRANSVRS program for selected values of a/λ_0 ranging from 0.10 to 0.20. For this range of parameters, the combination $k_1 c/\omega a^2$ (for the dominant transverse mode) can be expressed as a linear function in x, i.e. $C_1+C_2 x$. Substituting into the above equations, we find, for $\sigma/\lambda_0 \ll 1$,

$$W_d \propto \frac{1}{\pi \bar{s}} \left[(C_1+C_2 x_1)\sin(\pi\bar{s}x_+)\sin(\pi\bar{s}x_-) - \frac{C_2}{2} x_- \cos(2\pi\bar{s}x_2) \right]$$
$$+ \frac{C_2}{2\pi^2\bar{s}^2} \sin(\pi\bar{s}x_-)\cos(\pi\bar{s}x_+)$$

with $x_+ \equiv x_1+x_2$, and $x_- \equiv x_2-x_1$, for the case of flat distribution; and

$$W_d \propto (\Delta\sqrt{\pi}) \, e^{-\Delta^2\pi^2\bar{s}^2}[(C_1+C_2 x_a)\sin(2\pi\bar{s}x_a)$$
$$+2\pi^2\bar{s}\,\Delta^2 C_2 \, \cos(2\pi\bar{s}x_a)]$$

+ terms containing complex Error functions,

with $\Delta \equiv \Delta\omega/\omega_0$ for the case of Gaussian distribution. At large distance, the wake potential falls off as $1/\bar{s}$ for the flat distribution, and as $\bar{s} \cdot \exp(-\Delta^2\pi^2\bar{s}^2)$ for the Gaussian distribution. These falloffs are further modulated by a sinusoidal function with a period inversely proportional to the frequency spread. The location of the first wake null for the flat distribution is also inversely proportional to the frequency spread. The dipole wake potentials for a flat, continuum, transverse frequency distribution is illustrated in Figure 2a for a frequency spread of 9% centered at $x_a=1.4$. The wake potentials for truncated Gaussian and Gaussian distributions are shown, respectively, in Figures 2b and 2c. Contributions to the wake potentials from higher order modes can be similarly included by summing over these modes.

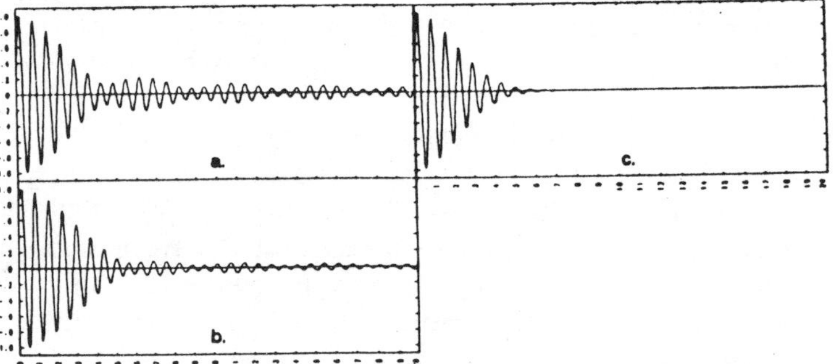

Fig. 2 Transverse wakes from analytic calculations

WAKEFIELD SIMULATIONS

The transverse wake damping rate and the locations of the interference nulls which characterize a beatwave structure with a finite number of discretely detuned cavities are determined with numerical simulations. Using a relatively small number of these cavities, it is possible to design a beatwave structure which provides a suppressed transverse wake profile tailored to match the multibunch requirements of a linear collider.

Extending the work of Yu and Wilson[2], we calculate the wake potential for a train of $2\pi/3$-mode, disk-loaded cells with different irises, using both the time-domain TBCI code, and the modal summation method in the frequency domain. By fixing the fundamental frequency, the cavity radius, b, is calculated for nine values ($a/\lambda_0 = 0.10$ to 0.20, in steps of .0125) of the beam-hole radius, a, using the KN7C program. With no loss of generality, an S-band frequency of 2.856 GHz is used as the basis for scaling. The cavity parameters and the HOM frequencies are completely scaleable with the fundamental frequency of any disk-loaded structure, i.e. S-band, X-band, etc. The corresponding transverse frequencies for the first 25 HOMs are calculated with the TRANSVRS program. Scaling of the wake potentials vs fundamental frequency is given in reference 2.

To study the variation of the wake potential as a function of the population density and the frequency spread, we use TBCI to calculate the combined wakes for several beatwave structures:

case	$(a/\lambda_0)_{min}$	$(a/\lambda_0)_{max}$	no. of cavities
a	0.100	0.200	3
b	0.100	0.200	5
c	0.100	0.200	9
d	0.125	0.175	5
e	0.125	0.175	20

A long beam pipe is appended to the structure in the TBCI models in order to avoid the problem of reflections at the boundaries. The mesh size is chosen so that the bunch length ($\sigma/\lambda_0 = 0.025$) is represented by at least five mesh points. Figures 3a-d show the TBCI transverse wakes ($m=1$) for cases a-d. The wake potential is normalized to its maximum value, and plotted as a function of the dimensionless distance, s/λ_0. Comparison of Figures 3a, b and c shows that for a given frequency spread, the position of the first null remains the same, while the damping improves as the number of cavities increases. The residual wake is damped to 55%, 33% and 25% of the initial peak for 3, 5 and 9 cavities, respectively. Comparison of Figures 3b and 3d shows the relationship between the null position and the frequency spread. For a frequency spread of 16.2%, the first null occurs at $s/\lambda_0 = 3.5$. For a frequency spread of 8.6%, the null occurs at $s/\lambda_0 = 6.5$. The distance of the first null from the center of the leading bunch is thus inversely proportional to the frequency spread. TBCI calculations for the monopole ($m=0$) mode show that the longitudial wakes are well preserved for the accelerating mode. Higher-order longitudinal wakes, as well as quadrupole ($m=2$) transverse wakes are found to be effectively damped.

Longer-range transverse wakes are calculated with the modal summation method. Figures 4a-c show the beatwave effects due to the lowest transverse mode for cases b-d. Figures 4e-g include 25 transverse modes with $\sigma/\lambda_0=0.025$ for the same cases. The effectiveness of long range damping is further enhanced by the inclusion of higher order modes. These figures also show clearly the range broadening of the wakefield suppression as the number of cavity increases. It is possible to construct a region of suppressed wakes broad enough to accommodate the entire train of multibunches (say, about $100\lambda_0$). Figure 4d shows the transverse wake potential from 20 different frequencies with a frequency spread of 9% (case e). For a larger frequency spread of 16% (as in cases a-c), about 30 different frequencies would be needed in order to suppress recurring peaks in the wake potential within a distance of $100\lambda_0$. The amplitudes of the residual wakes decrease exponentially as the number of beating frequencies increases. However unless the number of frequencies is extremely large, a small residual wake persists. The cumulative residual transverse wakes in the beatwave structure can be taken out using waveguide coupled damping structures. Because of the effectiveness of the beatwave method to suppress long range wakes, such damping structure is not required for every cavity. This will simplify the manufacturing process considerably.

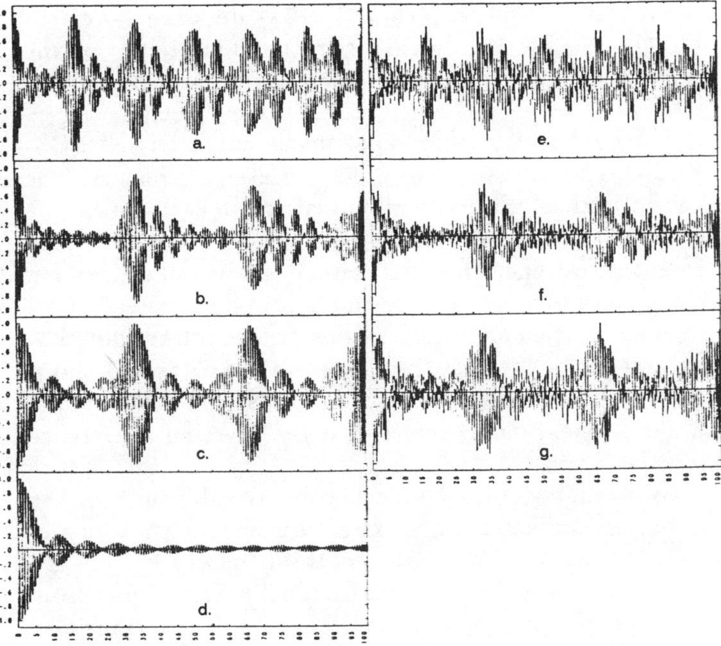

Fig. 4 Transverse wakes from modal sum method

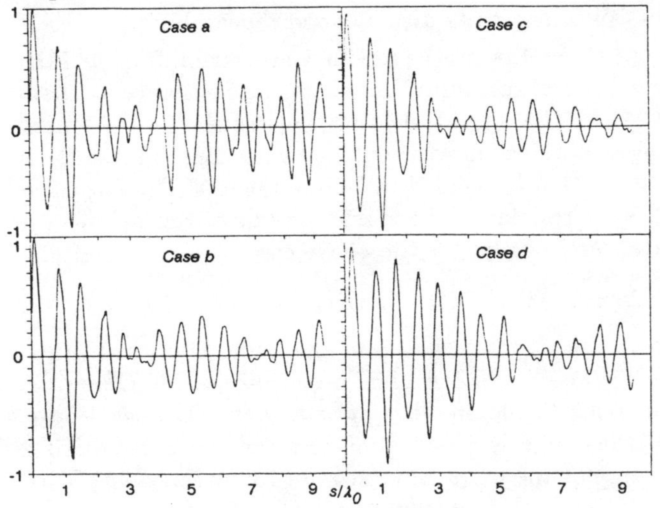

Fig. 3 TBCI transverse wakes for cases a-d

*Work supported by DOE SBIR Grant No. DE-FG03-90ER81080.

[1] R.B. Palmer, H.E. Physics in the 90's, Snowmass Conf. Proc. pp.638-641 (1989); G. Conciauro and P. Arcioni, EPAC, Nice, France (1990).
[2] D. Yu and P. Wilson, SLAC-PUB-5062, September (1989); and Particle Accelerators 30, pp.65-72 (1990).
[3] H. Deruyter, et al., Proceedings of 1990 Linear Accelerator Conference, September 10-14 (1990) p. 132.
[4] See, for example, P. Wilson, SLAC-PUB-4547, January (1989).

Measurement Of Longitudinal Impedance For a KAON Test Pipe Model with TSD-Calibration Method

Y.Yin, C.Oram,TRIUMF, Canada
N.Ilinsky, P.Reinhardt-Nikulin,INR,USSR

Abstract

We report measurements of longitudinal impedances for a KAON factory beam pipe model by means of the TSD-calibration method. The experimental method and the results are discussed. The frequency band is from 48 MHz up to 900 MHz, within which range the method produces measured impedances accurate enough to be useful in indicating whether a test pipe will have a suitably low impedance.

I. Introduction

The coupling impedance of the beam pipe is an important parameter for the accelerator design. Theoretical calculation is not feasible in many cases. So a measurement is necessary. In the KAON Factory Booster and Driver rings the bunch length is about one to two metres, which is much longer than the cut-off wavelength of about 30cm for the KAON Factory pipe. This makes the high-frequency (above the cut-off frequency, 1 GHz) impedance negligible [3]. We use the co-axial wire method [1] [2] for the impedance measurements for a KAON Factory beam pipe model in the low frequency range (below cut-off).

A segment of co-axial pipe can be characterised by S-parameters, the coefficients of the scattering matrix. The error matrices related to all unmatched reflections can be eliminated by mathematically de-embedding the test pipe from the test line. The TSD method is a way of calibrating using Thru pipe, Delay pipe and Short plate, instead of using match, open and short. In this method, the required S-parameters of the beam pipe segment are extracted mathematically, after measuring S_{12}, S_{21}, S_{11} and S_{22} for the T:Thru pipe, S:Short plates, D:delay pipe and a reference pipe [4].

By de-embedding, TSD can solve the difficulty of the measurement which lies in extracting the transmission and phase changes in an experimental set up where there are transition pieces between the co-axial cable coming from the network analyzer and the beam pipe. These transition pieces cannot be perfectly matched, so there are always reflections present in the measurement.

II. Measurements on KAON Test Pipe

Test pipe No.1 is 104 mm wide, 78 mm high and 700 mm long, and made from G-10 board. It is to simulate a section of ceramic chamber in the Booster and Driver rings. The pipe has strips 4mm wide with 1 mm gap between each other pasted on the inner surface of the pipe. The 4 strips in the middle of each plane are cut into 2 sections, forming a 5 mm wide slot with a 5 ohm resistor bridging across the slot externally to make a wall current beam position monitor (BPM) [6]. Test pipe No.2 has the same structure and cross section but is only 25 cm long.

The impedance due to the strips and the slots of the BPM has been measured by the TSD method and by independent measurements using the time domain method [6] [7]. The results from the time domain method are presented for comparison with the TSD method.

A. Measurement with 50 Ohm line impedance

For calibration the TSD method requires, a "through" pipe with length of 370 mm and a "delay" pipe with length of 700 mm, and a "short plate"; all have been used over the frequency range from 48 MHz to 200 MHz [2]. These pipes are made of solid copper with the same cross section as the test pipe, and have a one inch copper pipe passing through them to form a 50 ohm coaxial structure. At both ends there are two cone adapters as transition parts, in order to connect to the 50 ohm coaxial cables. The centre pipe also has two adapters at both ends which are tapered to maintain a 50 ohm impedance. The centre pipe has different lengths, in order to match the length of the"through", "delay" and "short". Every effort has been made to improve the RF contact and the precision of the structure to improve repeatability.

To distinguish the impedance changes caused by the strips, the slots, and the resistors bridging the slot, measurements were done for several different cases. To demonstrate the validity of the calibration, the measurement was also applied to a solid copper reference pipe (with same geometric size as the test pipe) of known impedance. The measurements have been repeated several times for all cases, and the results display good repeatability.

Fig.1 shows that in the resistance spectrum from 48 MHz to 198 MHz of test pipe No.1 there is a resonance at about 170 MHz. The amplitude and the frequency of the peak changed as the slots were patched with copper (case B) or shorted (D) or bridged with 5 ohm resistors (C). The curve "A" is the resistance of the reference pipe. Fig.2 shows the reactance spectrum for the above cases.

B. Measurement with line impedance of 180 Ohm

Both the TSD method and the Time Domain method have been used also for a line impedance of 180 Ohm, as a comparison to the 50 Ohm measurement discussed above. The delay pipe is 10 cm long and the through pipe is 1 cm long, for the frequency range of 180 MHz to 900 MHz.

1722

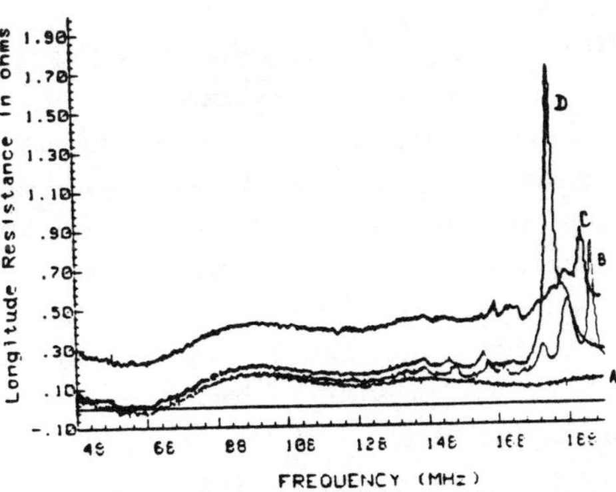

Figure 1: Resistance of Test Pipe No.1 (at 50 ohm)

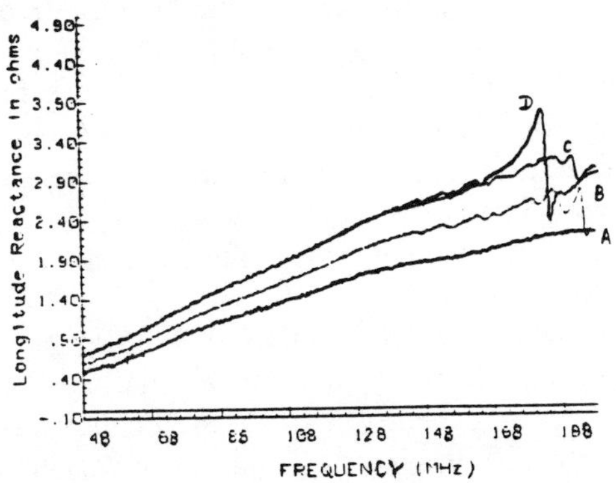

Figure 2: Reactance of Test Pipe No.1 (at 50 ohm)

The centre pipe has diameter of 1/8 inch. The end plates of pipes are 10mm thick and were machined to ensure flatness. The interfacing end of the two adaptors has a 2 mm wide lip to make a good RF contact with the pipes to be attached to it. A special mechanical design was used to keep the centre pipe in tension, to maintain straightness over the 700 mm length of the test pipe. The terminal end of the adaptor has been carefully designed to keep good assembly repeatability. Fig.3 shows results for test pipe No.1: the resistance of the reference pipe(A), the resistance when slots were shorted from externally(B, T1) and when the slots were bridged with 5 ohm resistors(C,T2). T1 and T2 are results from time domain method. Fig. 4 shows the resistance of the pipe with 5 ohm resistors bridging the slots.

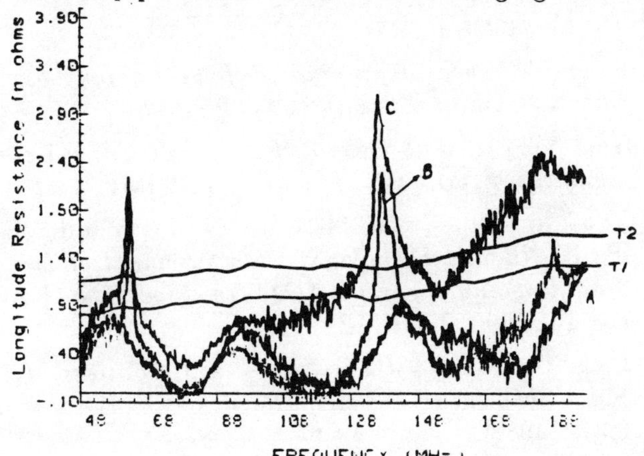

Figure 3: Resistance of test pipe No.1, 48-198 MHz(180 ohm)

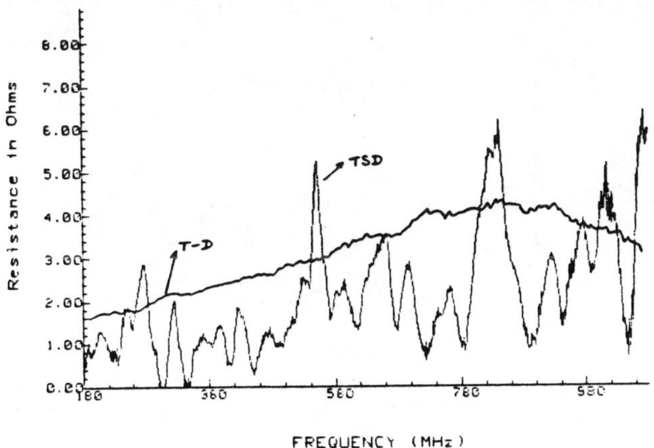

Figure 4: Resistance of pipe No.1, TSD and Time Domain(180 ohm)

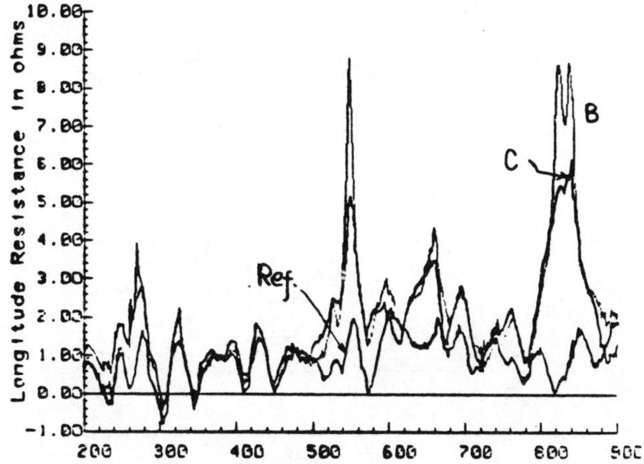

Figure 5: Resistance of test pipe No.1, 200-900 MHz(180 ohm)

The big peaks are due to the slots and the strips. The Time Domain curves T1, T2 are smooth and flat, showing only the trending of the changing impedance. This can also be seen in Fig.4, for the range 180 MHz to 1 GHz. Fig. 5 shows the resistances of test pipe No.1 with 5 ohm resistors(C), and with the slots shorted(B), and the reference pipe.

III. THE ANALYSIS OF THE MEASUREMENT

The longitudinal coupling impedance is a description of the coupling between the beam and vacuum chamber. The theory is that the energy lost by a short current pulse on a central wire is the same as that lost by a bunch of particles having equal time shape. But the presence of the wire obviously modifies the time evolution of the fields after the pulse [8]. So usually a thin wire is preferred for the impedance measurement.

The test pipe we measured has discontinuities: the break in the strips, which will interrupt the wall current. The reflected wall current will superpose with the input signal. When both signals are in phase periodic peaks will appear in the results. The frequencies of the peaks are related to the length from the break to the end of the test line. For the 50 ohm line impedance measurement, the cone adapters are 10 cm long, the total test line is 90 cm, and so the lowest resonant frequency is 170 MHz. The same phenomenon applies to the 180 ohm measurement. The 60 MHz reso-

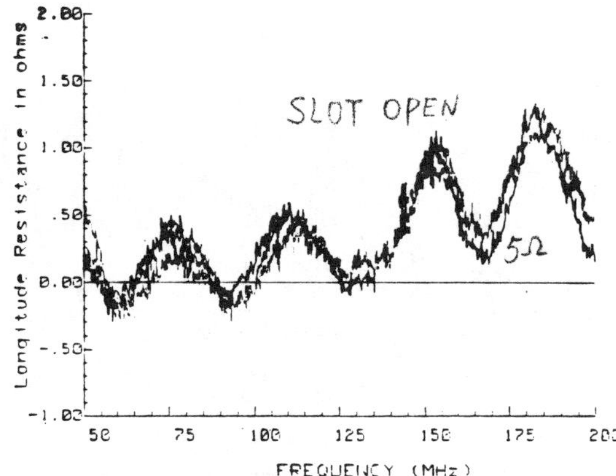

Figure 6: Rsistance, test pipe No.2 (180 ohm)

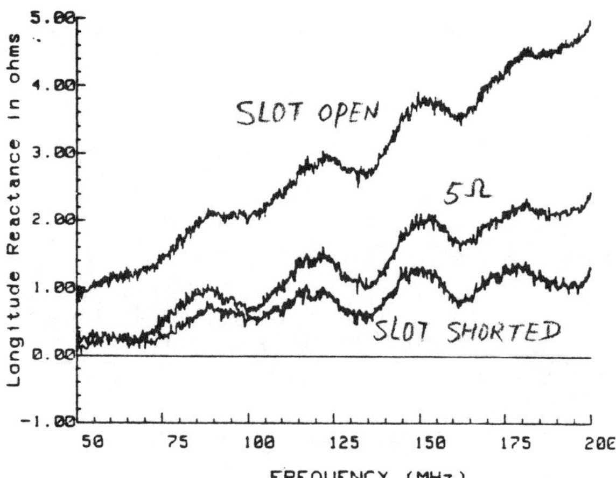

Figure 7: Reactance, test pipe No.2 (180 ohm)

nance is due to the field leakage from the gap between the 2 sides of the G-10 pipe. However, these resonances do not exist in the real machine since the vacuum pipe is a ring. Thus the peaks do not represent the impedance of the pipe in the working condition. But the other parts of the curve do show the real impedance. In order to measure the impedance close to the real situation, the resonances should be eliminated or avoided, because they can not be cancelled by de-embedding. If we ignore the resonances, the impedances of the various cases should have little difference over the frequency range of the measurement. This is more clear in the 180 ohm line impedance measurement, because the big centre pipe of the 50 ohm line impedance changes the field pattern.

Test pipe No.2 is only 25 cm long, this moves the resonances to a frequency of 230 MHz; so in the low frequency range, there are no resonances. The measurement shows in Fig.6,7 that the impedance is as small as from 0.5 to 1 ohm, close to the reference pipe. This is consistent with theory.

IV. CONCLUSION

The stretched wire method is a useful way to measure the longitudinal impedance; but, since a thin wire is needed to improve the accuracy of the measurement, the high reflections will cause the resonances to be related to the length

of the test line set up if there is a discontinuity.

The TSD method requires very careful mechanical design and implementation to ensure good RF signal contact and repeatability of mechanical assembly. Also several measurements with different mechanical setups are necessary, unlike the time domain method. However the TSD method can show narrow resonances, and other detailed structure in the impedance curve (although one has to be careful to analyse it), providing additional information to the time domain method.

Time domain measurements of the same structure do not display the peaks, because the network analyzer simulates the time domain response to an impulse input and mathematically transforms the frequency domain data into the impulse response. [9] The first pulse is kept. The reflection signals which contain the resonance information are therefore gated out. To reduce the sidelobes and improve the dynamic range of a time domain measurement, the windowing filters the frequency domain data prior to converting it to the time domain. This also smeare out the resonance.

The G-10 board model is different from a real ceramic one with no capacitors at the end, a smaller slot will be used to pick up the beam position signal because of the high beam intensity. Hence the impedance will be different in the real situation. However, the measurement shows that Z/n for the whole ring is still small. For instance, at 150 MHz, for the Booster, it is less than 1 ohm.

V. ACKNOWLEDGMENTS

The authors give their thanks to Linda Walling, for use of her TSD measurement program and to Walling and Dr. B. Zotter for the many helpful discussions leading to an understanding of the results and of the method.

VI. REFERENCES

[1] F.Caspers: *Beam Impedance Measurements using the Coaxial Wire Method*, CERN

[2] Linda Walling: *Transmission-Line Impedance Measurements for an Advanced Hadron Facility*, LANL

[3] S.A.Kheifets: *Coupling Impedance in Modern Accelerators* ,SLAC-pub-5297,1990

[4] *HP product Note 8510-8*

[5] Sands: *A Bench Measurement of the Energy Loss of a Stored Beam to a Cavity*,SLAC,PEP-95

[6] Yan Yin: *A wall current BPM built on a Ceramic Chamber*, KAON-DN -144,1990, TRIUMF

[7] Yan Yin, Chris Oram, TRIUMF, N.Ilinsky and P. Reinhardt -Nikulin, INR, USSR: *Measurement of Longitudinal Impedance For a KAON Test Pipe Model*,KAON-DN-142,1990, TRIUMF

[8] Luigi Palumbo, Vittorio G.Vaccaro: *Frontiers of Particle Beams; Observation,diagnosis and correction*, P.312, 1988

[9] *HP 8753 operation manual.*

Calculation of Seed Values for Longitudinal Coupled Bunch Dipole Instability Due To Uneven Bucket Population

S.R. Koscielniak

TRIUMF, 4004 Wesbrook Mall, Vancouver B.C., Canada V6T 2A3

Abstract

Conventionally, the starting amplitude (or 'seed') for bunched beam coherent oscillations is assumed independent of the steady state Fourier components of the beam, and is attributed to imperfect technology or randomness. However, there are occasions when the dominant component of the seeds derive from the transient response of the impedance to the revolution harmonics. Such harmonics are caused by uneven bucket population. In this paper we find the beam response to sweeping the radio frequency through a cavity parasitic resonance in order to calculate the increases in the individual bunch dipole moments and the coupled-bunch oscillation amplitude. In the KAON Booster, an hypothetical first HOM with $Q = 4000$ and $R/Q = 30$ gives a coupled-bunch dipole oscillation of $27°$ of rf phase; which would swamp any injection timing errors.

I. PARASITIC RESONATOR PROBLEM

The design for the KAON [1] Factory Booster synchrotron requires 3 Amps circulating beam current. Because of rf buckets left empty to form a kicker gap, the beam current has components at all multiples of the revolution frequency ω_0. If one falls within the bandwidth of a parasitic resonance of the cavity, significant excitation can occur. This is almost inevitable as there is a large frequency swing during acceleration. Imagine a resonator with bandwidth less than the synchrotron frequency ω_s. Then it is possible for a revolution harmonic $q\omega_0$ to sweep through resonance, and immediately afterward for the lower sideband at $(q\omega_0 - \omega_s)$ to drive a coupled bunch mode. What is worrisome is that the disturbance to the bunches, before the sideband starts to interact with the parasitic, might give rise to large coherent seeds for the coupled bunch modes.

A. Solution Approach

Far from resonance, the steady state electromagnetic fields due to the parasitic impedance are of small amplitude. Hence, at these times the longitudinal stable phase is nominally identical. Thus any observed change in the dipole moment between these times is a real change, and cannot be attributed to different equilibrium conditions. To find the changes, what we do is pretend for a moment that the bunches do not change shape or position while the Fourier component sweeps through the resonance, then we compute the response of the resonator driven by the beam, and then we calculate the effect on the beam of the voltage induced on the resonator. We assume the resonator

bandwidth is small so the resonator interacts with only one revolution harmonic line at a time.

II. BUNCH MOMENTS IN ACTION ANGLE FORM

Let the rf phase be x and the rate of phase-slippage be $y = \dot{x} = (h\omega_0\eta_s/E_s\beta_s^2) \times \Delta E$. E_s is the synchronous energy, ΔE is the energy deviation of a particle, h is harmonic number, and η_s is the usual slip factor. The individual particle Hamiltonian is $H = y^2/2 + \omega_s^2 x^2/2$. We introduce action-angle variables (J, β) defined through:

$$x(J, \beta) = \sqrt{2J/\omega_s}\cos[\omega_s t + \beta] \qquad (1)$$
$$y(J, \beta) = \sqrt{2J\omega_s}\sin[\omega_s t + \beta] . \qquad (2)$$

The equilibrium particle ensemble is $\psi(J)$. Suppose that the system is perturbed. Formally we write

$$J(t) \rightarrow J + \Delta J(t) \quad \text{and} \quad \beta(t) \rightarrow \beta + \Delta\beta(t) \qquad (3)$$

where ΔJ and $\Delta\beta$ are to be determined; and J, β take on the rôle of initial conditions. We now find the increase of the dipole moment. Higher moments may be found similarly, as in Reference [2].

A. Dipole Moment

Two orthogonal measures of the dipole moment are

$$D_x = \langle x(t)\rangle = \sqrt{2/\omega_s}\,\langle\sqrt{J(t)}\cos[\omega_s t + \beta(t)]\rangle \qquad (4)$$
$$D_y = \frac{\langle y(t)\rangle}{\omega_s} = \sqrt{2/\omega_s}\,\langle\sqrt{J(t)}\sin[\omega_s t + \beta(t)]\rangle . \qquad (5)$$

The angle brackets denote an average over the ensemble at time t. Substituting (3) in (4) the result is $D_x \approx$

$$\int\psi(J)\sqrt{\frac{2J}{\omega_s}}\int\left[\frac{\Delta J}{2J}\cos(\omega_s t + \beta) - \Delta\beta\sin(\omega_s t + \beta)\right]d\beta dJ .$$
$$(6)$$

The integral is over initial conditions so ψ is the distribution function for the *unperturbed* ensemble. The orthogonal dipole measure D_y is found similarly. The dipole moment is $M_1 = \sqrt{D_x^2 + D_y^2}$.

III. CHANGES IN ACTION AND ANGLE

We assume the beam-current fundamental and the radio-frequency maintain perfect synchronism. The voltage induced by the beam can be written in a form that tracks with beam frequency. Hence the Hamiltonian is $H =$

$$\frac{y^2}{2} + \omega_s^2\frac{x^2}{2} + \xi\omega_s^2\int^x [\mathcal{A}(t)\cos(ps) + \mathcal{B}(t)\sin(ps)]ds . \qquad (7)$$

The prefactors $\mathcal{A}$ and $\mathcal{B}$ may vary from bunch to bunch. The constant p is the quotient: revolution harmonic q which drives the parasitic divided by h. Let the total rf voltage per turn be $\hat{V}_{\mathrm{rf}}$, the synchronous phase ϕ_s, the circulating beam current I_{dc}, and the shunt resistance of the resonator R; then the scaling factor is

$$\xi = (I_{\mathrm{dc}}R)/(h\pi\hat{V}_{\mathrm{rf}}\cos\phi_s) \ .$$

We assume that the disturbance starts at $t = -T$ and ends at $t = +T$. From canonical perturbation theory, we find the equations of motion for $-\dot{J} = (\partial H/\partial x)(\partial x/\partial\beta)$ and $\dot{\beta} = (\partial H/\partial x)(\partial x/\partial J)$. In a first order approximation we substitute the unperturbed motion into the right hand sides and treat J and β as constants. Then a trivial integration with respect to time gives $\Delta J(T)$ and $\Delta\beta(T)$.

IV. Increase of Single Bunch Dipole Moment

We substitute ΔJ and $\Delta\beta$ into the integral (6) and integrate over β. The result is $D_x(T) =$

$$2\pi\xi\omega_s\int_{\hat{J}}^{\check{J}}\psi(J)\,J_0(pz)dJ\int_{-T}^{+T}\mathcal{A}(t)\sin\omega_s(t-T)dt \ . \quad (8)$$

Here J_0 is the Bessel function of integer order. A similar calculation for $D_y(T)$ shows the two components are in quadrature. We write the sine and cosine transform of $\mathcal{A}$ evaluated at frequency ω_s as $\mathrm{SFT}[\mathcal{A}]_{\omega_s}$ and $\mathrm{CFT}[\mathcal{A}]_{\omega_s}$, respectively. Then the moment M_1 is

$$2\pi\xi\omega_s\int_{\hat{J}}^{\check{J}}\psi(J)\,J_0(pz)dJ\left\{\mathrm{CFT}[\mathcal{A}]_{\omega_s}^2 + \mathrm{SFT}[\mathcal{A}]_{\omega_s}^2\right\}^{1/2}. \quad (9)$$

When there are empty buckets, the components $D_x(n)$, $D_y(n)$ for the nth bunch can be found simply by replacing $\mathcal{A}$ by $\mathcal{A}_n$. The coefficients $\mathcal{A}_n$ and $\mathcal{B}_n$ are defined below.

V. Growth of Coupled Bunch Dipole Modes

Before interaction with the resonator, the amplitude of coupled bunch modes is assumed zero. The collective motion $D_{x,y}(n)$ of each bunch dipole mode can be represented as a phasor $\psi_n = M_1(n)\,e^{j\omega_s t}e^{\phi_1(n)}$ in the Argand diagram. In general the amplitude $M_1(n)$ and phase $\phi_1(n)$ of each phasor will be different. If a pure coupled bunch mode is present there is a fixed phase difference between each of the phasors. For N equally spaced bunches, there are N coupled bunch rigid dipole modes. These are identified by the mode index $m = 0, 1, 2, \ldots N - 1$, and the rotation between consecutive phasors is $\Delta\phi_1(m) = m \times (2\pi/N)$ for a pure mode. Simply adding up the individual phasors by vector addition will *not* give the coupled bunch 'seed'. Instead, to find the amplitude of the mth coupled-bunch mode, we have to rotate each phasor ψ_n backwards by $\theta_1(n,m) = (n-1)\times\Delta\phi_1(m)$ and then add them by vector addition.

VI. Resonator Drive

To find the coefficients $\mathcal{A}$ and $\mathcal{B}$ we have to find the beam current Fourier components which might drive the parasitic, and the resonator response as the revolution frequency sweeps.

A. Beam Current Fourier Spectrum

We assume that bunches are identical in shape. Let us adopt coordinates such that the 1st bunch spans the interval $[-\pi, +\pi]$ and the nth bunch spans the range $[(2n - 3)\pi, (2n - 1)\pi]$. Suppose the beam-shape function $\Lambda(x)$ is normalized so that $\int_{-h\pi}^{+h\pi}\Lambda(x)dx = 1$. Let us represent the function as a Fourier series:

$$\Lambda(x) = \frac{1}{2}\Lambda_0 + \sum_{q>0}\mathcal{R}[\Lambda_q]\cos\left[\frac{qx}{h}\right] - \mathcal{I}[\Lambda_q]\sin\left[\frac{qx}{h}\right] \ .$$

Let the DC component of beam current be I_{dc}. The current components I_q are related to the Λ_q by

$$I_q = (I_{\mathrm{dc}}/\pi h)\Lambda_q \ .$$

VII. LCR Circuit with Swept Drive Frequency

We give *approximate* expressions for the response of a circuit to a swept drive frequency $\omega(t) = \omega[1 + \epsilon_w t]$. We allow for a slowly varying resonance frequency through linear time variations for inductance L and capacitance C: $L(t) = L \times [1 + \epsilon_L t]$ and $C(t) = C \times [1 + \epsilon_C t]$. The circuit shunt resistance R is constant. The fundamental equation is:

$$L \times I_T(t) = \int_T^t V_T(t')dt' + L\frac{d}{dt}[C \times V_T] + V_T\frac{L}{R} \ . \quad (10)$$

A. Frequency Tracking Drive and Response

The current driving the circuit is of constant amplitude with the form:

$$I_T(t) = I_q\exp[j\omega t + \phi(t)] \quad \text{with} \quad \phi(t) = \int_{-T}^t \Delta\omega(t')dt' \ .$$

We assume the frequency variation is linear in time about $t = 0$, that is $\Delta\omega = \omega \times \epsilon_w t$. We will drop the index q for brevity. Now suppose the response can be written in the form

$$V_T(t) = V(t)\exp[j\omega t + \phi(t)]$$

where $V(t)$ is to be determined and may be complex.

B. Relaxation Approximation

Our first approximation is to neglected $\ddot{V}$ compared with $\omega\dot{V}$. Secondly we suppose that $|\epsilon t| < 1$ so that $(1 + \epsilon t)^k$ can be Taylor expanded as $(1 + k\epsilon t)$.

Now introduce $\omega^2 = 1/(LC)$ and $\omega/Q = 1/(RC)$ so that quantities are defined as the values at $t = 0$. After manipulation equation (10) becomes:

$$IR[1 - j\epsilon_L/\omega + (\epsilon_w - \epsilon_C)t] = V[1 + (Q/\omega)(\epsilon_w + \epsilon_l + 2\epsilon_c)$$

$$-j(Q/\omega^2)\epsilon_L(\epsilon_C + \omega/Q)] + (2Q/\omega)\dot{V} + \ldots$$

$+V[jQ(\epsilon_L + \epsilon_C + 2\epsilon_w) + (\epsilon_w - \epsilon_C) + (Q/\omega)\epsilon_w(\epsilon_L + 2\epsilon_C)]t$.

Hence we have an equation of the form:

$$V[A + at] + \dot{V}/\alpha = IR[B + bt] \qquad \text{with} \quad \alpha = \omega/2Q \,.$$

The formal solution is:

$$V(t')e^{[\alpha t'(A+at'/2)]}\Big|_{-T}^{t} = IR\int_{-T}^{t}[B + bt']e^{[\alpha t'(A+at'/2)]}dt' \,.$$

C. Series Expansion in $2Q/\omega$

The right hand side may be repeatedly integrated by parts to give a series expansion in $1/\alpha$.

$$\frac{V(t')}{IR}e^{\alpha t'(A+at'/2)}\Big|_{-T}^{t} = \frac{(B + bt')}{(A + at')}e^{\alpha t'(A+at'/2)}\Big|_{-T}^{t} +$$

$$\left[-\frac{1}{\alpha}\frac{(b-a)}{(A + at')^3} - \frac{3}{\alpha^2}\frac{(b-a)}{(A + at')^5} + \cdots\right]e^{\alpha t'(A+at'/2)}\Big|_{-T}^{t}$$

In fact, $\alpha = \omega/(2Q) >> 1$ and $A \approx 1$ so the series converges rapidly. We now define a time-dependent impedance $Z(\omega, t) = V(t)/I$.

D. Voltage at a Specific Bunch

In the general case, the beam current of the qth revolution harmonic line is:

$$I_q(x)/2\langle I\rangle R = \mathcal{R}[\Lambda_q]\cos(qx/h) - \mathcal{I}[\Lambda_q]\sin(qx/h)\,.$$

The response is:

$$V_q(x)/2\langle I\rangle R = \mathcal{A}_0\cos(qx/h) + \mathcal{B}_0\sin(qx/h)$$

$$+\mathcal{A}_0(t) = \mathcal{R}[Z(q\omega_0)]\mathcal{R}[\Lambda_q] - \mathcal{I}[Z(q\omega_0)]\mathcal{I}[\Lambda_q] \qquad (11)$$
$$-\mathcal{B}_0(t) = \mathcal{R}[Z(q\omega_0)]\mathcal{I}[\Lambda_q] + \mathcal{I}[Z(q\omega_0)]\mathcal{R}[\Lambda_q]\,. \qquad (12)$$

These expressions are fine for the first bunch, in the domain $x = [-\pi, +\pi]$, but not so for other bunches. Our earlier equations assumed the use of local coordinates Δx_n with origin at the centre of the nth bucket. We find a local voltage of the form $V(\Delta x_n)/2\langle I\rangle R =$

$$\mathcal{A}_n\cos[(q/h)\times\Delta x_n] + \mathcal{B}_n\sin[(q/h)\times\Delta x_n] \qquad (13)$$

with the definitions

$$+\mathcal{A}_n = \mathcal{A}_0\cos[q(n-1)2\pi/h] - \mathcal{B}_0\sin[q(n-1)2\pi/h]$$
$$-\mathcal{B}_n = \mathcal{A}_0\sin[q(n-1)2\pi/h] + \mathcal{B}_0\cos[q(n-1)2\pi/h]\,.$$

Comparing (13) with (7) we recognise $p = q/h$.

E. Fourier Transform

The increase in bunch moments is proportional to the moduli of the time-to-frequency Fourier transforms of $\mathcal{A}_0$ and $\mathcal{B}_0$ evaluated at multiples of the synchrotron frequency. All integrals have to be performed numerically.

VIII. Example Cases for Booster Ring

The KAON Booster ring has harmonic number $h = 45$ and bucket occupancy scheme of 40 filled buckets followed by 5 empty. We assume the bunch half-length is $a = 1$ radian, and the shape is $\lambda(x) \propto (a^2 - x^2)$. The rf cavities are of the Los Alamos [3] perpendicular bias type, with a frequency swing 46.1-60.7 MHz. Midway through acceleration $\hat{V}_{\text{rf}} = 550$ kV per turn, the frequency sweep is $\Delta\omega/\omega = 0.01$ and duration $T = 1$ ms. The circulating beam current is $I_{dc} = 3$ A.

A. High Q parasitic

Prior to measurement, it was assumed the parasitics would have quality factors similar to the fundamental. Consider the case of a parasitic at revolution harmonic $q = 85$ with quality $Q = 4000$; the bandwidth is 20 kHz. At the midpoint of the Booster acceleration, the synchrotron frequency is $f_s \approx 20$ kHz, so the cavity bandwidth falls between upper and lower synchrotron sidebands. Below transition, when $N = 45$, the lower sideband at $2N - 5 = 85$ will drive the $m = 5$ coupled-bunch mode. The shunt resistance of the parasitic is $R = 120$ kΩ assuming the HOM has $R/Q \approx 30$ ohm, similar to the fundamental. Since the cavities are identical, the total induced voltage will be 10 times that on a single cavity. Hence we multiply ξ by 10.

The $m = 5$ coupled-bunch dipole mode 'seed' is found to be $M_1^{\text{cb}}(5) = 0.47$ radian or 27°. This is a large oscillation, and could not be tolerated. In fact, the perturbation is sufficiently large that our first-order treatment starts to break down; but still signals a major problem.

B. Medium Q parasitic

Measurement [4] of the cavity parasitics shows that the quality factors of all HOMs are substantially smaller than the fundamental. Likely values are as follows: $R \approx 5.5$ kΩ and $R/Q \approx 32$ ohm, giving a quality factor $Q = 170$. The revolution harmonic which crosses the parasitic might be $q = 75$ at 90 Mhz. The lower sideband of harmonic line $N + 30$ will drive the coupled-bunch mode $m = 15$ when $N = 45$. The coupled-bunch dipole mode 'seed' $M_1^{\text{cb}}(15) = 9.2 \times 10^{-4}$ radian. This seed is insignificant since the value is comparable with the likely statistical 'seed' that would arise if bunches each had random phase-errors of root mean square value 0.4°.

IX. References

[1] *KAON Factory Study Accelerator Design Report*; Triumf, Vancouver B.C., 1990.

[2] S. Koscielniak: "Calculation of Seed Values for Longitudinal Coupled Bunch Instability Due to Uneven Bucket Populations"; TRI-DN-91-K161.

[3] R.L. Poirier: "Perpendicular Biased Ferrite Tuned Cavity for the Triumf Kaon Factory Booster Ring"; IEEE Particle Accelerator Conference, Chicago Illinois, 1989.

[4] T.A. Enegren: "Measurement of RF Cavity Shunt Impedance"; TRI-DN-90-K104.

Simulation of Hollow Beams with Cancellation of
Steady State Non-linear Space-charge

Shane R. Koscielniak

TRIUMF, 4004 Wesbrook Mall, Vancouver, B.C., Canada V6T 2A3

Abstract

This paper describes a technique for making multi-particle computer simulations of coherent instability consistent with conventional bunched-beam longitudinal instability theory. The argument behind the technique exposes an oversight of instability theory: the response of the phase-space distribution to the non-linear steady state wakefields is neglected. As an example, the technique is applied to a beam with the steady-state space-charge fields artificially cancelled. The computer simulations presented are seen to agree with mode-coupling theory. Beam and machine parameters are taken from the proposed TRIUMF-KAON [1] Accumulator, which is a high current proton storage ring.

I. Critique of Stability Analysis

The stability analysis of a particle beam is a two step process: (i) determine the steady state conditions; and (ii) discover the behaviour of small perturbations.

A. Steady State

The single particle equation of motion is:

$$\ddot{x} + \omega_s^2 x = F(x) = \omega_s^2 \xi \sum_p Z_p(0) \Lambda_p e^{jpx} \ . \tag{1}$$

The wakefield F is the product of the complex impedance $Z_p(0) = Z(ph\omega_0)$ evaluated at frequencies $ph\omega_0$ and the steady state beam Fourier components Λ_p. Here ω_0 is the revolution frequency. The wakefield F comprises a linear part $L(x)$ and a non-linear part $N(x)$. The linear term is removed by renormalizing the incoherent frequency. Then (1) becomes:

$$\ddot{x} + \langle \omega_{\text{inc}}^2 \rangle x = N(x) \ . \tag{2}$$

Here $\langle \omega_{\text{inc}} \rangle$ is the ensemble average incoherent frequency.

The behaviour of the steady state ensemble with phase-space distribution function $\Psi(r, \theta)$ is governed by the Vlasov equation. This is written in polar coordinates:

$$\sqrt{\langle \omega_{\text{inc}}^2 \rangle} \frac{\partial}{\partial \theta} \Psi - N(x) \left[\sin \theta \frac{\partial}{\partial r} + \frac{\cos \theta}{r} \frac{\partial}{\partial \theta} \right] \Psi = 0 \ . \tag{3}$$

If $N \equiv 0$ then Ψ has no angular dependence and so $\Psi = \Psi(r)$. We can turn the argument around, and say that if the steady state distribution is a function only of radius, then the non-linear part of the wakefield must have been exactly cancelled by an externally applied counter-field. This is the assumption made in analytical instability theory. Physically, the cancellation is very difficult to achieve; but it is trivial to implement in a computer simulation.

B. Small Perturbations

We write the time dependent part of the phase-space ensemble as $\varepsilon \psi(r, \theta) e^{j\omega t}$; and this generates a wakefield

$$\varepsilon f(t, x) = \varepsilon \, e^{j\omega t} \sum_p Z_p(\omega) \lambda_p \, e^{jpx} \tag{4}$$

where λ_p are the harmonics of the perturbation bunch-shape and $Z_p(\omega) = Z(ph\omega_0 + \omega)$ is the impedance at the perturbation frequency ω. The Vlasov equation becomes:

$$\left[\frac{\partial}{\partial t} - \langle \omega_{\text{inc}} \rangle \frac{\partial}{\partial \theta} \right] \left[\Psi + \varepsilon \psi e^{j\omega t} \right] +$$
$$\frac{\varepsilon f(x, t)}{\langle \omega_{\text{inc}} \rangle} \left[\sin \theta \frac{\partial}{\partial r} + \frac{\cos \theta}{r} \frac{\partial}{\partial \theta} \right] \left[\Psi + \varepsilon \psi e^{j\omega t} \right] = 0 \ . \tag{5}$$

(5) is linearized in ε, and the steady state equation (3) is subtracted giving:

$$\left[j\omega - \langle \omega_{\text{inc}} \rangle \frac{\partial}{\partial \theta} \right] \psi + \frac{f(x)}{\langle \omega_{\text{inc}} \rangle} \sin \theta \frac{\partial}{\partial r} \Psi(r) = 0 \ . \tag{6}$$

C. Consistency Technique

Instability theory seeks solutions of (6). A multi-particle simulation code is easily made to emulate equation (5), and this is a close relative of equation (6). We simply subtract an analytic expression for the steady state wakefield from the wakefield calculated from the Monte Carlo ensemble. When this is done, we should expect agreement on the threshold beam current for coherent instability; because the simulation has been made as unrealistic as the analytic instability theory.

D. Real Beam

Of course, the real beam *will* respond to the steady state wakefield. When the non-linear fields are substantial, then complicated mismatching effects will occur; and though they may mimic an instability these are, in fact, transient response behaviours. In this light, it is no surprise that the calculations and simulation for a space-charge dominated hollow beam reported in reference [2] differ in their conclusions.

II. Mode Coupling Results

An instability analysis for a bunched beam which is hollow in longitudinal phase-space and for which the internal self-forces derive solely from the space-charge force has been carried out by Baartman [2]. Mode-coupling theory shows

"

there to be a dipole instability when the zero amplitude incoherent frequency $\omega(0)$ becomes $\sqrt{2}$ times the zero intensity synchrotron frequency ω_s, as in figure 3.

A further prediction of reference [2] is that there is no quadrupolar instability until the threshold frequency $\omega(0) \approx 1.58\omega_s$; also as in figure 3.

III.   SIMULATION MODEL

The computer experiments were made with the code LONG1D [3]. The rf-cavities produce a linear restoring force. There are 4 cavity crossings per turn, and space-charge is implemented by a second order symplectic mapping with 4 sub-steps between each cavity. The beam bunch is modelled by an ensemble of 6×10^4 macro-particles. A random number generator was used to prepare an approximation to the hollow gaussian distribution:

$$\Psi(r) = (r^2/4\pi\sigma^4)\exp(-r^2/2\sigma^2) . \qquad (7)$$

The parameter σ is measured in radians of rf-phase, and the spacing between bunches is 2π. The finite number of simulation particles implies a statistical jitter in the bunch shape, and this is responsible for seeding any unstable behaviour that may occur.

A.   Cancellation of Steady State Wake

The steady state Fourier components are:

$$2\pi\Lambda_p = \left[1 - p^2\sigma^2/2\right]\exp(-r^2\sigma^2/2) \qquad (8)$$

and these are subtracted from the components found by binning the statistical ensemble and Fourier analysing. The remainder gives the residual λ_p.

The linear part of the wakefield is restored by substituting the ensemble average incoherent tune in place of the zero intensity synchrotron tune.

$$\langle\omega_{\mathrm{inc}}^2\rangle = \omega_s^2\left[1 - C/16\sigma^3\sqrt{\pi}\right] \qquad (9)$$

The bunch length parameter σ has to be adjusted to account for the bunch lengthening consistent with $\langle\omega_{\mathrm{inc}}\rangle$. Let β, γ be relativistic kinematic parameters; Z_0 the impedance of free space; $\hat{V}_{\mathrm{rf}}$ the peak accelerating voltage; $\langle I \rangle$ the dc component of circulating current in Amps; h the harmonic number; and $g_0 = 1 + 2\ln(b/a)$. The constant C is given by

$$C = \left(\frac{g_0 Z_0 h}{2\beta\gamma^2}\right)\left(\frac{2\pi\langle I\rangle}{\hat{V}_{\mathrm{rf}}\cos\phi_s}\right) . \qquad (10)$$

Equations (7–11) are taken from reference [4].

IV.   ENSEMBLE CHARACTERISTICS

The behaviour of the beam coherent motion was monitored by extracting 2 orthogonal measures of the dipole moment, $\langle r\cos\theta\rangle$ and $\langle r\sin\theta\rangle$; and 2 measures of the quadrupole moment, $\langle r^2\cos(2\theta)\rangle$ and $\langle r^2\sin(2\theta)\rangle$. The quadrature sum of 2 measures will give the net amplitude of the dipolar or quadrupolar disturbance. The Fourier transform with respect to time of an individual measure will show the frequencies of the many coherent radial modes which contribute to each polar disturbance.

A.   Ellipse Rotation and Scaling

When longitudinal motion is modelled by difference equations (rather than a differential equation) the matched phase-space circle becomes a tilted ellipse. Whereas the polar moments of a uniformly filled circle are zero, those of an ellipse are finite. Consequently, the ensemble coordinates have to be multiplied by the inverse longitudinal Twiss parameter matrix before the multipole moments are extracted. This procedure is described in reference [5].

V.   TRIAL CASES AND RESULTS

Following Baartman, the trial cases are ordered as a function of the zero amplitude incoherent frequency

$$\omega^2(0) = \omega_s^2\left[1 + C/2\sigma^3\sqrt{2\pi}\right] . \qquad (11)$$

Trial cases for a space-charge compensated beam are summarized in Table 1. The entries in the last column indicate if the simulation behaved as below ($\Downarrow$), or at dipolar threshold ($\rightleftharpoons$), or above ($\Uparrow$). The beam current is given in Amps.

Table 1: Trial cases for beam with hollow phase-space.

Case	Current	σ	$\langle\omega_{\mathrm{inc}}\rangle/\omega_s$	$1 - \omega^2(0)/\omega_s^2$	
A	0.00	0.500	1.00	0.00	n/a
B	1.5867	1.00256	0.99489	-0.05993	$\Downarrow$
C	4.4008	1.00708	0.98599	-0.16399	$\Downarrow$
D	2.2004	0.51459	0.94411	-0.61462	$\Downarrow$
E	3.6673	0.52407	0.91026	-0.96976	$\Downarrow$
F	4.0341	0.52641	0.90218	-1.05257	$\rightleftharpoons$
G	4.4008	0.52874	0.89425	-1.13315	$\Uparrow$
H	5.1343	0.53336	0.87882	-1.28795	$\Uparrow$
I	7.3347	0.54695	0.83569	-1.70617	$\Uparrow$

A.   Dipolar Instability

Figure 1 shows the dipolar amplitude versus time for 3 cases (G, H, I) above threshold and one case (E) below threshold. The threshold frequency shift $\omega(0) \approx 1.4327\omega_s$, inferred from case F, is in excellent agreement with the value predicted by Baartman, $\omega(0) \approx 1.4142\omega_s$.

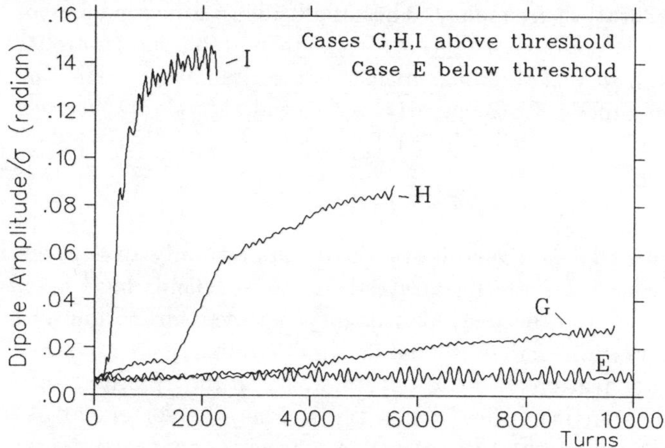

Fig.1 Space−charge compensated hollow beams

Well above threshold, as in cases H and I, the effect of

the instability is to transform the initial two-humped line density into a single-humped gaussian distribution.

B. *Quadrupolar Instability*

There should be no growth in the quadrupolar amplitude for cases A through H. However, for trial case I the zero amplitude frequency is $\omega(0) = 1.645\,\omega_s$; well above threshold and quadrupolar growth is expected. Figure 2 shows the quadrupolar amplitude versus time for cases E, G, H and I; confirming the predictions for the quadrupolar instability threshold.

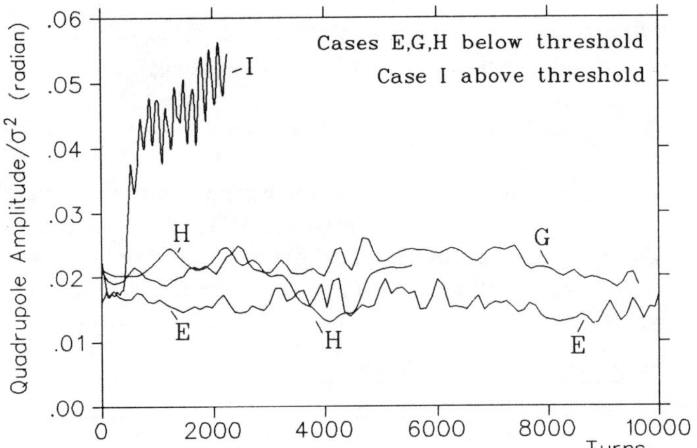

Fig.2 Space—charge compensated hollow beams

C. *Frequency Spectra*

The theoretical frequency spectrum of coherent oscillation modes is given in figure 3, which is adapted from reference [2]. The precise values and number of modes depends on the size of the matrix used to find the eigen-frequencies. Each polar disturbance is split into several radial modes.

It is anticipated that the fast Fourier transform (FFT) of the simulation polar components should show a similar pattern of frequencies. This is verified in the example figures 4a and 4b. The ordinate ω_0/ω is the inverse of the synchrotron tune. The small amplitude zero-intensity tune is $\omega_s/\omega_0 = .04883$. The annotations give the oscillation frequency divided by the ensemble average incoherent synchrotron frequency. The amplitude of a particular frequency component derives from statistics of the ensemble and how well the polar measures discriminate a particular radial mode. Consequently, not all radial mode frequencies need be present in the FFT.

VI. Conclusion

Bunched beam instability theory applies only to a beam in which the non-linear steady state wakefields have been artificially cancelled. Consistency between simulation and calculation demands that a similar, albeit non-physical, compensation of steady state wakes is incorporated into multi-particle codes. When this is done, the agreement between the simulations of a space-charge compensated beam presented here, and the analysis of Baartman leads to a high level of confidence in the computer model.

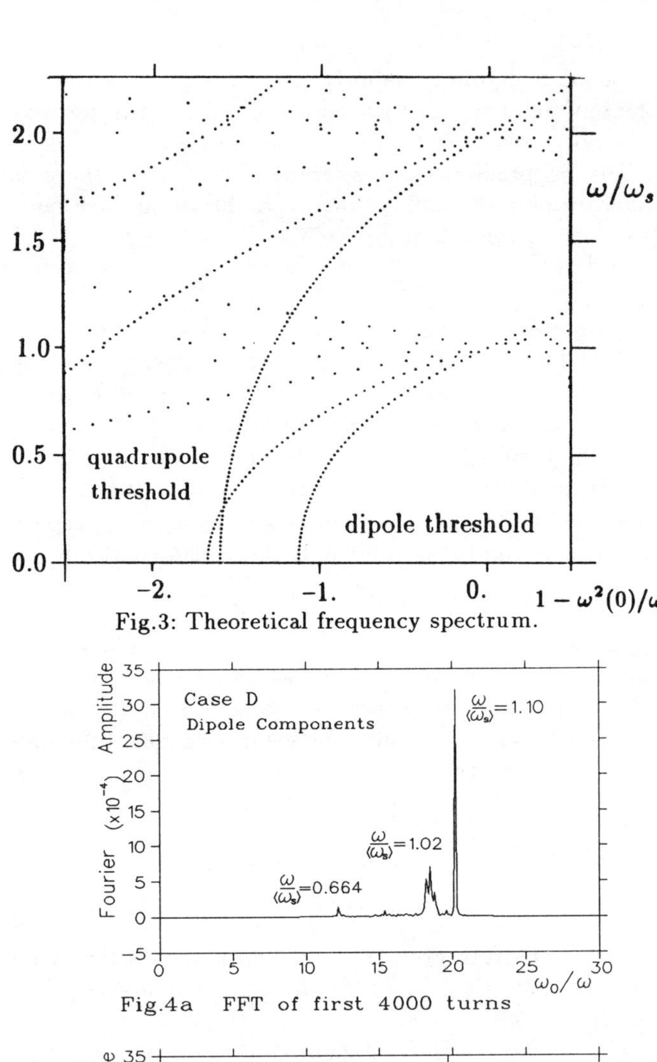

Fig.3: Theoretical frequency spectrum.

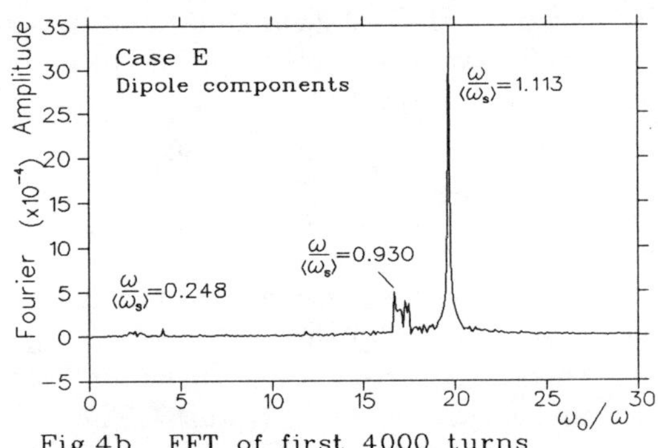

Fig.4a   FFT of first 4000 turns

Fig.4b   FFT of first 4000 turns

VII. References

[1] Kaon Factory Study, Accelerator Design Report; TRI-UMF, Vancouver, Canada, March 1990.

[2] R. Baartman and S. Koscielniak : "Stability of Hollow Beams in Longitudinal Phase Space"; Particle Accelerators, 1990, Vol.28 pp.95–99

[3] S. Koscielniak: "The LONG1D Simulation Code", Proc. *1st European Particle Accelerator Conference*, Rome 1988, pg. 743.

[4] S. Koscielniak: "Phase and incoherent frequency shifts due to steady-state wakefields"; TRI-DN-90-K136.

[5] S. Koscielniak: "Detailed Longitudinal Phase-Space Matching"; TRI-DN-91-K160.

Stationary Longitudinal Phase Space Distributions with Space Charge

R. Baartman

TRIUMF, 4004 Wesbrook Mall, Vancouver, B.C., Canada V6T 2A3

Abstract

An iterative technique has been employed to find longitudinal phase space distributions which are stationary under the influence of a constant, imaginary Z/n. In the positive mass regime (where $\mathrm{Im}(Z/n)$ is positive below transition or negative above transition), the main result is that, irrespective of the starting distribution, as the intensity is raised, the distribution tends towards one which has a parabolic line density. Since the locally elliptic distribution already has a parabolic projection, it is always stationary. In the negative mass regime, where the impedance augments the longitudinal focusing, centrally peaked stationary distributions cease to exist beyond a certain threshold intensity. In this regime, the limiting line density is also a parabola but one which is of opposite sign, i.e. with minimum density at the centre.

I. Introduction

In the study of beam stability, it is important to distinguish between stability and stationarity. Stability should only be defined with respect to a stationary distribution, because if a given distribution is perturbed in some way, it is not sensible to ask whether the perturbation grows unless the unperturbed distribution is stationary. With a short-range wake field such as space charge, the stationary distribution changes with beam intensity. This has recently been emphasized by Oide and Yokoya [1] who showed that mode-coupling calculations which ignore potential well distortion give different instability thresholds and growth rates than those which are self-consistent.

An additional source of confusion between stability and stationarity is the fact that the Keil-Schnell stability criterion, applied locally using the space charge impedance, is formally equivalent to the condition that the bucket area remain finite [2]. Hence, the Keil-Schnell criterion, applied with a predominantly reactive impedance to a bunched beam, pertains more to the stationarity of the distribution than to its stability.

We restrict ourselves to the space charge longitudinal wake given by an imaginary impedance proportional to frequency. It is well-known that then the space charge force is proportional to the derivative of the beam's line density. Hence, for the 'locally elliptic' distribution [2] (whose projection gives a parabolic line density), the space charge force is linear and, with a linear external focusing force, the phase space trajectories lengthen but remain elliptical. Analogously to the Kapchinsky-Vladimirsky distribution of the transverse case, this is the only distribution which does not change shape with intensity. As a result, tune

shifts, bucket areas, and even longitudinal eigenmode frequencies [3] can be calculated by means of simple formulae. However, it is a highly idealized distribution, possessing no tails, and it is not difficult to see that an infinite number of other stationary distributions exist. In the following sections, we investigate three stationary distributions whose individual zero intensity line densities are, respectively, gaussian, hollow, and flat-topped.

II. Theory

This section presents Sacherer's theory [4] for transverse stationary distributions adapted to the longitudinal case. There are two requirements for a stationary distribution. (1) Particle motion must be derivable from a Hamiltonian. (2) There must be one or more integrals of motion with no explicit time dependence; i.e. there must be a relation $J(p,q)=$constant between a particle's momentum p and position q. The stationary distribution $(\rho(p,q))$ is then given by

$$\rho(p,q) = f(J(p,q)) \tag{1}$$

where f is an arbitrary function.

For longitudinal motion in synchrotrons, the effect of the accelerating cavities can be approximated as non-local, since the synchrotron frequency is in general much less than the revolution frequency. In that case, the Hamiltonian itself has no explicit time dependence and we can use H for J. Normalizing in such a way that the unperturbed phase space trajectories are circles, the space charge Hamiltonian is

$$H = \frac{p^2}{2} + \frac{q^2}{2} + I\lambda(q). \tag{2}$$

I is the intensity parameter, included explicitly here because the line density

$$\lambda(q) = \int f(H(p,q))dp \bigg/ \int\int f(H(p,q))dpdq \tag{3}$$

has been normalized to unity. We allow I to have either sign. A negative value corresponds to the 'negative mass' regime where the intensity-dependent part of the potential is attractive and there is potential-well bunch shortening. This can arise for space charge above transition or 'inductive-wall' impedance below transition.

We wish to find how the phase space trajectories ($H=$constant) and the line density deform as the intensity is raised for a given choice of f. The numerical method consists of an iteration where an old λ (as defined on a finite number of points q) is used in H to generate a new

"""

one by (3). In some cases, this procedure diverges and a relaxation technique (where a λ is generated from a weighted mean of old and new) is required.

One undesirable feature of the Hamiltonian (2) is that as I is changed, the average value of H changes as well, and this can change the nature of the distribution as we shall see below. For this reason, a constant is usually added to H and its value is changed along with I to keep some feature of the distribution constant. For example, adding $-I\lambda(0)$ makes invariant the slice $f(p,0)$.

III. GAUSSIAN

The gaussian case, $f(H) = e^{-H}$, is singular in that the momentum dependence can be factored out:

$$\lambda(q) = \lambda(0)\exp(-q^2/2 - I[\lambda(q) - \lambda(0)]). \tag{4}$$

This is the Haissinski equation [5] applied to our particular case. For large positive I, in particular, $I \gg |\ln\lambda|/\lambda$, the line density becomes parabolic:

$$\lambda(q) = \lambda(0) - \frac{q^2}{2I}. \tag{5}$$

This can be seen on the right in Fig. 1[1]. It is analogous to space charge dominated transverse transport [7], where the distribution adjusts to cancel the external focusing force. The concept of Debye-shielding introduced in [7] applies here as well.

In the negative-mass or bunch-shortening regime, no solution is found beyond a certain intensity. This can be seen by Taylor-expanding λ in (4). For $\lambda(q) = \lambda(0)(1 + aq^2 + bq^4 + ...)$, we find

$$a = \frac{-1/2}{1 + I\lambda(0)}, \tag{6}$$

and clearly there is a singularity if $I = -\lambda(0)$. At this intensity, the phase space trajectories at the centre of the distribution are vertical. See the plot on the left in Fig. 1. Increasing the intensity further causes the iteration to quickly diverge with the momentum spread at the beam centre becoming infinite. This case is analogous in some respects to a black hole.

Not surprisingly, the condition $I > -\lambda(0)$, when converted back into unnormalized units, is exactly the same as the condition for avoiding the negative mass instability in a coasting beam with gaussian momentum distribution.

---

[1] The density plots were made with the DENSITY/DIFFUSION command in the graphics package PLOTDATA [6]. This uses a neighborhood halftoning process in which a pixel is turned on, or not, depending both upon the value of the density function at that point and upon a correction factor based on previous data values passed through an error filter. Signal responses are thus diffused over a weighted neighborhood. In this way, statistical fluctuations are avoided, but disadvantages include: correlated artifacts, directional hysteresis due to the raster order of processing, and transient behaviour near edges or boundaries. The last is the reason that the density plots shown here are not symmetric about the q-axis, but have a 'soft' edge in the upper half of the plane and a harder edge in the lower half.

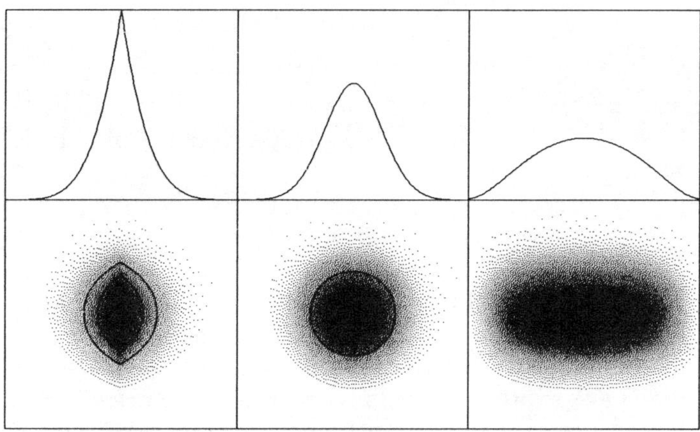

Figure 1: Line densities (top) and density plots (bottom) of stationary distributions for the gaussian case at intensities (left to right) $I = -1.552$, 0, and 20. The left-most plot is at threshold: for $I < -1.552$, no stationary distributions with $f(H) = e^{-H}$ exist. A few iso-density contours have also been plotted. On the contours, H=constant and, therefore, they also represent phase space trajectories. In these plots, as well as those following, the ranges of p and q are from -4 to 4.

IV. HOLLOW

We specify a hollow distribution as

$$f(H) = \exp[-(H - H_b)^2/2], \tag{7}$$

i.e. gaussian in H, *not* in the phase space coordinates. We set H_b=constant$-I\lambda(0)$ as described in section II. to maintain the hollowness as I is raised. The results with a choice of 2 as the constant are shown in Fig. 2 with a positive mass example on the right and a negative mass example on the left. In both cases we see the line density tending toward parabolic, but in the positive mass case an inverted cusp develops in the line density and no stationary distributions can be found for I greater than about 20. This is very similar to the threshold for the gaussian in the negative mass regime. It shows that a hole in the positive mass regime has a negative mass character.

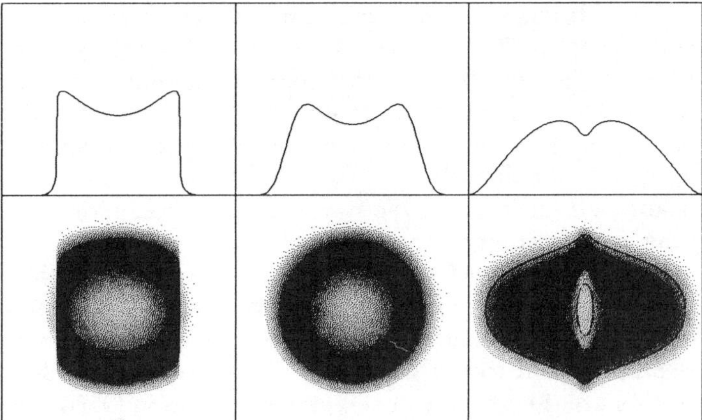

Figure 2: Line densities (top) and density plots (bottom) of stationary distributions for the hollow beam (7). Intensities (left to right) are $I = -8$, 0, and 20.

In the negative mass case, no threshold is found and the line density tends towards a parabola as $|I|$ is raised.

However, the parabola is inverted from the normal case and the slopes at the head and tail become very large. Particles in the bunch travelling towards the head or tail experience a very large space charge force at the edge and are reflected. This is evident from the outer contour (phase space trajectory) in the phase space density plot (Fig. 2, left). The length of the turn-around region is essentially a Debye screening length [7].

V. FLAT-TOPPED

It was pointed out in [2] that a flat-topped line density can be realized by subtracting two locally elliptic distributions. The attraction is that for a given bunch population, flattening the top reduces the peak line density, thereby reducing the *transverse* tune shift.

Specifically, we define the density function as

$$f(H) = \sqrt{H_2 - H}^* - \sqrt{H_1 - H}^* \tag{8}$$

where $\sqrt{x}^*$ is $\sqrt{x}$ for $x > 0$ but 0 for $x < 0$. We set $H_1 = H_0 + I\lambda(0)$. Then with H_2 and H_0 fixed, the line density is independent of I. It is flat to $|q| = q_0 \equiv \sqrt{2H_0}$ and proportional to $H_2 - q^2/2$ from q_0 to $q_2 \equiv \sqrt{2H_2}$. As I is raised, we see in Fig. 3 that 'ears' develop in phase space. Continuing to raise I flattens the 'ears' till they collapse to zero area at

$$I = \frac{2}{3}(q_2^3 - q_0^3). \tag{9}$$

Above this threshold no stationary distributions exist. As a check, we can set $H_0=0$ and we recover the known threshold $I = \frac{2}{3}q_2^3$ above which no bucket exists for the locally elliptic distribution [2].

VI. REFERENCES

[1] K. Oide and K. Yokoya, *Longitudinal Single-Bunch Instability in Electron Storage Rings*, KEK Preprint 90-10 (1990).

[2] A. Hofmann and F. Pedersen, *Bunches with Local Elliptic Energy Distributions*, IEEE Trans. Nucl. Sci., NS-26, p. 3526 (1979).

[3] G. Besnier and B. Zotter, *Oscillations Longitudinales d'une Distribution Elliptique, Couplées par un Résonateur: Application au Calcul de l'Allongement de Faisceaux Intenses*, CERN-ISR-TH/82-17 (1982).

[4] F.J. Sacherer, *Matched Distributions with Non-Uniform Space Charge and No Emittance Growth*, CERN/SI/Int. DL/70-5 (1970).

[5] J. Haissinski, Nuovo Cimento 18B, p. 72 (1973).

[6] J.L. Chuma, **PLOTDATA** *Users' Guide*, TRIUMF Computing Note, TRI-CD-87-03a (1987).

[7] I. Hofmann, *Space Charge Dominated Beam Transport*, CAS Proc., CERN 87-03, p. 327 (1987).

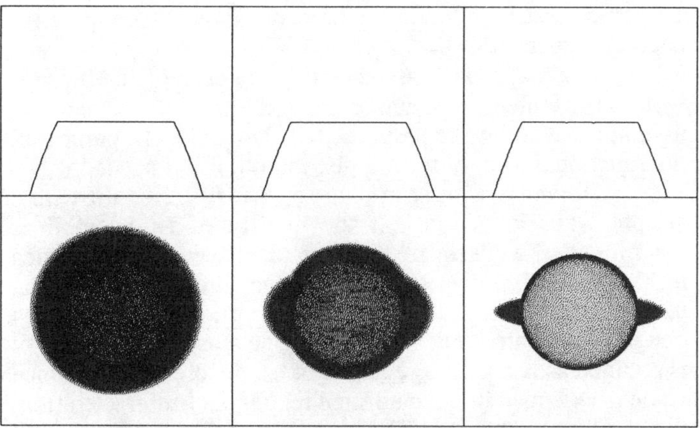

Figure 3: Line densities (top) and density plots (bottom) of stationary distributions for the flat-topped line density case (8) with $H_2=4.5$ ($q_2=3$) and $H_0=2$ ($q_0=2$). Intensities (left to right) are $I = 0$, 8, and 12. For the chosen parameters, we find from (9) that no stationary distributions with the indicated line density exist for $I > 38/3$.

From (9) we see that flattening a line density (raising q_0) reduces the longitudinal threshold. This must be balanced against the gain obtained in the transverse space charge limit.

Characterization and Monitoring of Transverse Beam Tails*

J. T. Seeman, F. J. Decker, I. Hsu, and C. Young
Stanford Linear Accelerator Center, Stanford, California, 94309

Introduction

Low emittance electron beams accelerated to high energy in a linac experience transverse effects (wakefield, filamentation, optics,...) which produce non-Gaussian projected transverse beam distributions. Characterizations of the beam shapes are difficult because the shapes are often asymmetric and change with betatron phase. In this note several methods to describe beam distributions are discussed including an accelerator physics model of these tails. The uses of these characterizations in monitoring the beam emittances in the SLC are described here as well as in Ref. 1.

First, two dimensional distributions from profile monitor screens are reviewed showing correlated tails. Second, a fitting technique for non-Gaussian one dimensional distributions is used to extract the core from the tail areas. Finally, a model for tail propagation in the linac is given.

Beam Profiles from an X-Y Screen

When a beam strikes a fluorescent screen, it produces a two dimensional light distribution which can be observed with a TV camera and monitor or can be digitized and processed [2].

The digitized TV picture can be projected onto the x or y axis and compared to beam shaped measured with other devices such as a wire scanner [3]. Information is lost during the projection process. In addition, size data from fluorescent screens are unique in that an image of a single beam pulse can usually be measured. Whereas, many pulses (~ 20) are needed for a wire scanner measurement.

A two dimensional image is a projection of the x, x', y, y' distributions onto the x,y plane. Information about the other variables must be obtained at locations with different betatron phase advances or by using an adjustable quadrupole upstream. For example, a measured beam projection is shown in Fig. 1 where a transverse beam tail is apparent. The orientation of the tail in phase space is not known unless further measurements are taken. However, because of the longitudinal extent of the bunch and the fact that the back of the bunch tends to be the tail as produced by wakefields, a two dimensional image provides clues to the tail orientation. The x-y view aids in deciphering the head from the tail. The transverse tails observed on this type of screen can be adjusted with upstream variables but solutions which look small usually contain a tail in the angular dimension.

Beam Projections from Wire Scanners

In the SLC, beam sizes (projections) are now routinely measured with wire scanners. When the beams are well behaved, the projected beam size is well represented by a

* Work supported by Department of Energy contract DE-AC03-76SF00515.

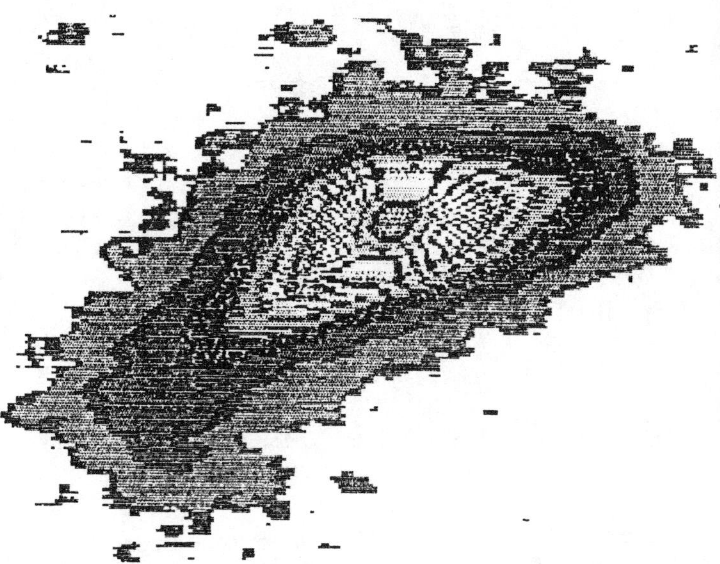

Fig. 1   Beam distribution with transverse tail measured on a screen profile monitor. This normally colored, two dimensional representation of the beam gives more information than a projection alone. The horizontal and vertical Gaussian beam sigmas are about 150 μm.

Gaussian. A straight forward weighted least squares fit yields reliable results. However, a beam with "tails" leads often to a very poor fit. We report here on some techniques for analyzing such cases.

This approach assumes that the core of the beam is well represented by a gaussian, but otherwise makes no assumptions about the properties of the "tail". In particular, the functional form of the tail distribution is not needed.

Given a wire scan data set, we first fit to it with a simple Gaussian form. If the quality of fit is good as determined by a chi square cut, then the distribution is deemed to have no tail. Otherwise, we use the fitted gaussian mean to separate the data into left and right portions, and refit each one independently with Gaussians. The smaller (or better) fit is retained as being representative of beam core. Its functional value is extended into the unfitted region, and subtracted from the scan data. The residual distribution is the "tail". This tail can then be characterized by its moments relative to the fitted Gaussian mean. Two examples of this fitting procedure are shown in Fig. 2.

When the population in the tail is a significant fraction of the core or when it extends very far from the axis, the new Gaussian fit may still be unacceptably poor. This is easily cured by allowing extra iterations during partitioning of the data and refitting.

The benefits of this approach are its simplicity, robustness and independence of tail distribution function. Reliable standard Gaussian fitting routines can be used without custom modification or additional debugging. Its convergence is essentially independent of the properties of the tail distribution function; therefore, the algorithm can be used in many situations. If a specific distribution was assumed, then the algorithm would have to be re-coded for a different functional form for each specific case.

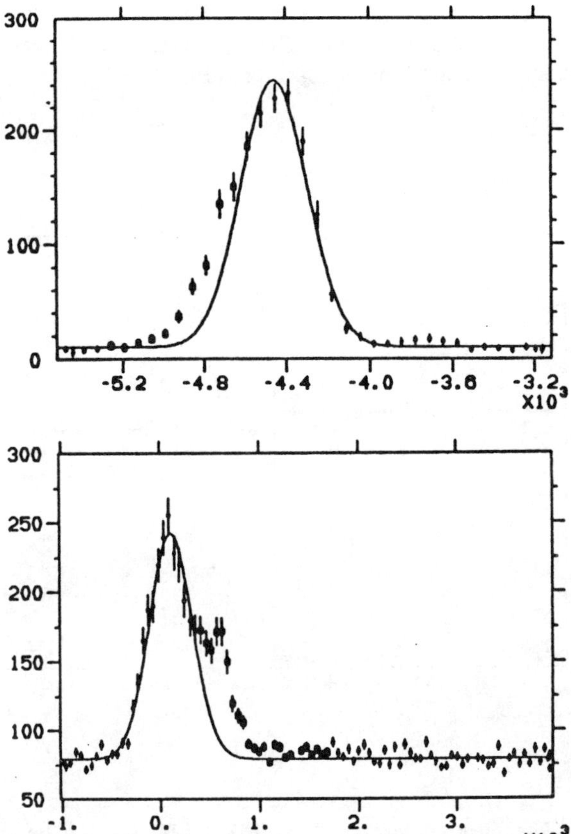

Fig. 2 Profiles fitted with a left or right core gaussian and showing a residual tail.

Accelerator Physics Model for Tail Projection

A bunch executing a betatron oscillation in the quadrupole lattice of the linac experiences transverse wakefields in the accelerating structure. The head of the bunch drives the core and back of the bunch to ever increasing amplitudes producing a non-Gaussian tail. Simulations of this growth have been made where a bunch is divided into longitudinal slices and traced through the linac. In Fig. 3 the centroid positions of these slices are shown for a simulated SLC bunch of 5×10^{10} electrons after oscillating from an initial amplitude of about 100 microns. The (nearly) exponential growth from head to back is apparent.

The transverse particle distribution $\rho(x)$ of each slice is equal to that of the initial phase space distribution. The initial distribution for the SLC is a gaussian with width σ.

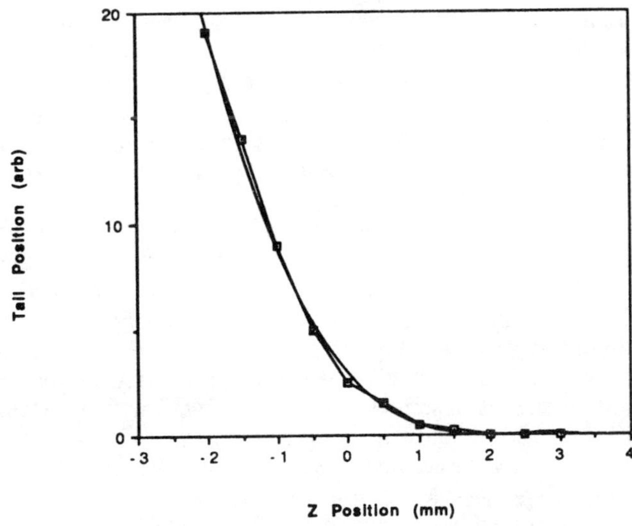

Fig. 3   Transverse slice centroid positions of a simulated SLC beam. The longitudinal head is on the right and the back of the bunch is on the left. Note the (nearly) exponential growth of the transverse position.

$$\rho(x) = \exp(- [x - x_0]^2 / 2\, \sigma^2) / \sigma\, (2\pi)^{1/2} \, , \qquad (1)$$

where x_0 is the mean of the distribution. The overall distribution must be integrated over the slices with different transverse positions x_0.

The position x_0 of each slice is represented by an exponential which initiates at position z_0 along the length of the bunch, as in Fig. 3, and has a growth rate of τ. This tail rotates in phase space with the betatron phase ϕ_i but has an initial phase ϕ_0. The emittance of each slice of the beam is ε and the betatron function at each profile measurement 'i' is β_i. $\sigma_i^2 = \varepsilon\, \beta_i$. The tail extension is scaled locally by σ_i.

$$x_0(\phi_i, z) = \sigma_i\, U(z_0 - z)\, [\, \exp(\, (z_0 - z)\, \tau / \sigma_z) - 1\,]$$

$$X\ \cos(\phi_i + \phi_0) \qquad (2)$$

where U is the unit step function. The bunch length is σ_z. The transverse distribution of each slice is given by :

$$\rho(x, \phi_i, z) = \exp(- (x - x_0(\phi_i, z))^2 / 2\sigma_i^2) / (2\pi)^{1/2}\sigma_i \quad (3)$$

Now, the overall transverse distribution we will call f(x) is given by

$$f(x, \phi_i) = \int_{-\infty}^{+\infty} \rho(x, \phi_i, z)\, h(z)\, dz \qquad (4)$$

where h(z) is the longitudinal profile, usually assumed to be a gaussian as in Eqn. 1 but with length σ_z. By choosing ϕ_0, τ, and z_0, the beam shape can be calculated at any location over a reasonably short (less than a betatron wavelength) region of the linac. In Fig. 4 various calculated beam spots using this formalism are shown. Clearly, many different shapes can be generated.

Conversely, measured beam shapes can be analyzed to determine the tails structure of the beam and measure the effective ϕ_0, z_0, and τ. An oscillation was induced in the SLC electron beam with a dipole magnet and the resulting oscillation is shown in Fig. 5. The associated beam profiles are shown in Fig. 6 which cover a range in betatron phase. The beam shapes observed have a definite tail with a phase. Profiles (a) and (b) show no tails (it is in angles), but profiles (c) and (d) show large tails. Several simulated profiles from Fig. 4 closely resemble the measured shapes.

This analysis breaks down when the tail of the beam becomes quite convoluted after many oscillations or very strong wakefields. Longitudinal shapes go from banana-like into worm-like. Therefore, this analysis handles only moderately enlarged beams which covers the standard running condition of the SLC. If the shapes are much worse than shown, we would stop the program to fix them.

In the near future we plan to implement this algorithm into the online emittance package for eventual use in a feedback system for tails. This technique is properly suited for feedback as the phase of the tail is determined as well as the amplitude.

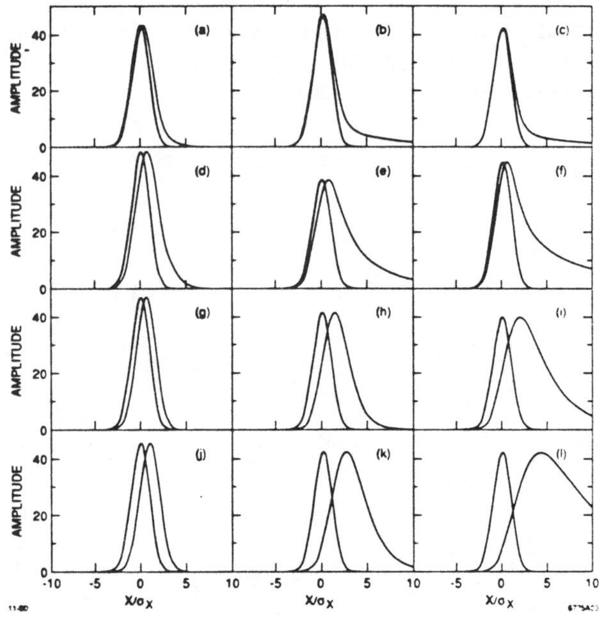

Fig. 4   Calculated transverse bunch projections showing the effects of different input conditions with the rate of tail growth τ and the place z_0 along the bunch where the exponential growth starts. Large τ (horizontal axis) makes long thin tails and a z_0 closer to the front of the bunch (vertical axis) makes a broader shoulder.

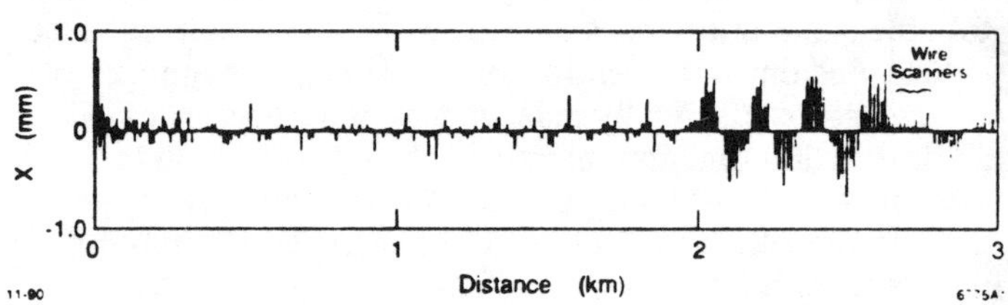

Fig. 5   Induced horizontal oscillation (measured) in the SLC linac which generated the beam tails in Fig. 6

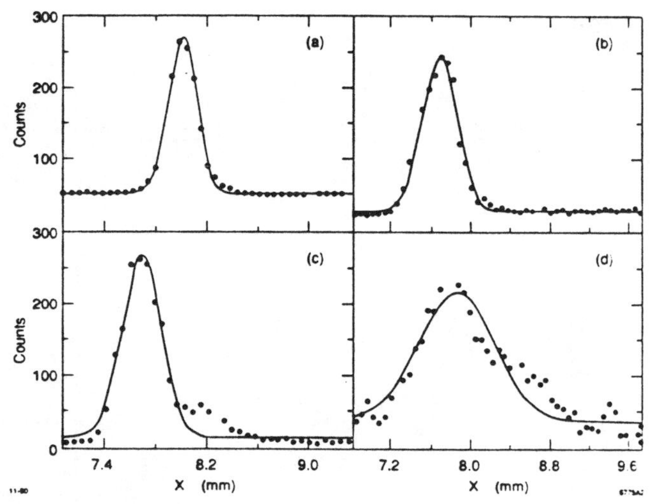

Fig. 6   Measured beam profiles (47 GeV) for the betatron oscillation in Fig. 5. The projections were taken with four wire scanners spaced at 0, 22.5, 90., and 112.5 degrees in betatron phase. Note the similarities of these profiles with those calculated in Fig. 4.

Acknowledgments

Many thanks are extended to the SLC Mechanical and Controls Groups for their help with the beam size measurement systems. L. Brown, D. McCormick, and M. Ross have been very helpful.

References

1) J. Seeman et al., 'Summary of Emittance Control in the SLC Linac', SLAC-PUB-5437 and Proc. of US PAC San Francisco (1991).
2) F. J. Decker, 'Beam Size Measurement at High Radiation Levels', SLAC-PUB-5481 and US PAC San Francisco (1991).
3) M. Ross et al., 'Wire Scanners for Beam Size and Emittance Measurements at the SLC', Proc. of US PAC San Francisco (1991).

Analysis of Resonant Longitudinal Instability in a Heavy Ion Induction Linac*

Edward P. Lee and L. Smith
Lawrence Berkeley Laboratory
University of California, Berkeley, California 94720

SUMMARY

A high current beam of subrelativistic ions accelerated in an induction linac is predicted (in some circumstances) to exhibit unstable growth of current fluctuations at high frequencies ($\nu \sim 100$ MHz). The instability is driven by the interaction between the beam and accelerator modules at frequencies close to a cavity resonance. The extent of unstable growth depends on features of the coupling impedance, beam parameters, and total pulse and accelerator lengths. Transient and asymptotic analysis is presented.

Induction Linac Model

We treat a cluster of beams drifting at velocity v, with line charge density λ and current $I = \lambda v$. It is assumed here that all the beamlets ($N \sim 16$) effectively act in concert so that λ and I are the total values and v is the common velocity. The continuity equation, using laboratory frame variables (z,t) is:

$$\frac{\partial \lambda}{\partial t} + \frac{\partial I}{\partial z} = 0 \ . \tag{1}$$

The beam cluster is treated as a cold, 1-d, non-relativistic fluid. An externally imposed field E^{ex} and a smoothed longitudinal field E, induced by interaction of I with the acceleration modules, acts on v:

$$\frac{\partial v}{\partial t} + v \frac{\partial v}{\partial z} = \frac{q e}{m}\left(E + E^{ex}\right) \ . \tag{2}$$

Neglected in this model are velocity spread and a direct space-charge force proportional to $\partial \lambda / \partial z$. These are significant stabilizing features at the high frequencies treated here, so the calculation may be considered pessimistic. The purpose of the present work is to delineate the phenomena of the longitudinal instability occuring in heavy ion induction linacs, and to guide future study. The analysis is similar to that given by V. K. Neil[1] for relativistic electron beams.

The equilibrium beam drifts at constant velocity v_0, so the total equilibrium field $\left(E_0 + E_0^{ex}\right)$ vanishes. Equilibrium current I_0 and line charge density λ_0 are, in general, functions of the retarded time $\tau = t - z/v_0$. However, they are taken to be constant for the duration of the pulse $(0 < \tau < \tau_p)$.

*Work was supported by the Director, Office of Energy Research, Office of Basic Energy Sciences, Advanced Energy Projects Division, U.S. Dept. of Energy under Contract No. DE-AC03-76SF00098.

Let E^{ex} have a small additional component which acts at z = o and therefore perturbs the beam: $E^{ex} = E_0^{ex} + V(t)\delta(z)$. From eqns. (1) and (2) the resulting perturbed beam variables satisfy

$$\frac{\partial \lambda_1}{\partial t} + \frac{\partial I_1}{\partial z} = 0 \ , \tag{3}$$

$$\frac{\partial v_1}{\partial t} + v_0 \frac{\partial v_1}{\partial z} = \frac{q e}{m}\left(E_1 + V\delta(z)\right) \ , \tag{4}$$

$$I_1 = \lambda_0 v_1 + v_0 \lambda_1 \ .$$

Module Response

The beam-generated field E_1 is induced by by the passage of the return current $(-I_1)$ through the module impedance Z. If we assume the driven form $I_1 \sim \exp(-i\omega t)$, with induced field $E_1(\omega)$, then impedance is

$$Z(\omega) = -E_1(\omega)/I_1 \ . \tag{5}$$

Specifically, we treat an isolated module resonance characterized by a parallel L-R-C equivalent circuit:

$$C \frac{\partial^2 E_1}{\partial t^2} + \frac{1}{R} \frac{\partial E_1}{\partial t} + \frac{E_1}{L} = -\frac{\partial I_1}{\partial t} \ . \tag{6}$$

The circuit parameters are related to measured resonance features; let R be the real impedance peak (units of Ω/m in the smoothed field model), occuring at angular frequency ω_0. and resonance width is $\Delta\omega = \omega_0/Q$ at 71% of maximum $|Z|$. Then we have

$$C = Q/\omega_0 R \ \text{(F-m)}, \quad LC = \omega_0^{-2} \ . \tag{7}$$

The impedance formula for general (complex) ω is

$$Z = \left(\frac{1}{R} - i\omega C - \frac{1}{i\omega L}\right)^{-1} = \frac{R}{1 - iQ\left(\frac{\omega}{\omega_0} - \frac{\omega_0}{\omega}\right)} \ . \tag{8}$$

Typical resonance parameters of interest for the Heavy Ion Fusion application are $\nu_0 = \omega_0/2\pi = 30\text{–}300$ MHz, Q = 10-100, and R = 100-1000 Ω/m.

Beam Frame Equations

It convenient to use the retarded time variable $\tau = t - z/v_0$ and z instead of the laboratory frames variables t and z. Then eqns. (3) and (4) become

$$\frac{\partial I_1}{\partial z} = \frac{\partial}{\partial \tau} \frac{\lambda_o v_1}{v_o} , \qquad (9)$$

$$\frac{\partial v_1}{\partial z} = \frac{q e}{m v_o} \left[E_1 + \delta(z) V(\tau) \right] . \qquad (10)$$

We eliminate v_1 to obtain

$$\frac{\partial^2 I_1}{\partial z^2} = \frac{\partial}{\partial \tau} K^2 C \left[E_1 + \delta(z) V(\tau) \right] , \qquad (11)$$

where
$$K^2 = \frac{q e \lambda_o}{m v_o^2 C} . \qquad (12)$$

Equation (6) becomes

$$\left(\frac{\partial^2}{\partial \tau^2} + \frac{\omega_o}{Q} \frac{\partial}{\partial \tau} + \omega_o^2 \right) E_1 = - \frac{1}{C} \frac{\partial I_1}{\partial \tau} . \qquad (13)$$

The natural scale frequencies for z and τ are K and ω_0. Generally the scale magnitude $\omega_0 \tau \sim 300$ is much larger than $Kz \sim 10$. Eqns.(11) and (13) describe strongly coupled oscillations in both τ and z, which exhibit growth in both variables. We expect large growth in τ will be produced by an external perturbation $V(\tau)$ which contains any appreciable content near the resonant frequency ω_0. In eqns. (11)-(13), K may be a function of τ, however a numerical treatment would then be required. Henceforth K is taken to be constant.

The perturbation $V(\tau)$ is assumed to turn on within the beam pulse at some $\tau = \tau_0 > 0$. The structure of the coupled equations ensures that perturbed quanatities may be consistently assumed to vanish outside the zones $\tau \geq \tau_0 , z \geq 0$.

General Solution by Laplace Transformation

Taking E_1 and I_1 to vanish for $(\tau < \tau_0 , z < 0)$, we find from eqn. (11) the initial conditions at $z = 0+$:

$$I_1 (0+) = 0, \quad \frac{\partial I_1}{\partial z} (0+) = \frac{q e \lambda_o}{m v_o^2} \frac{\partial V}{\partial \tau} . \qquad (14)$$

Performing the Laplace transformation

$$\left(\widehat{I}_1 , \widehat{E}_1 \right) = \int_0^\infty d\tau\, e^{i\omega\tau} \left(I_1 , E_1 \right) , \qquad (15)$$

we get

$$\frac{\partial^2 \widehat{I}_1}{\partial z^2} = - K^2 C i\omega \widehat{E}_1 = \left(\omega \Gamma K \right)^2 \widehat{I}_1 , \qquad (16)$$

with
$$\Gamma = \left(\omega_o^2 - \omega^2 - i\omega\omega_o/Q \right)^{1/2} .$$

The solution $\widehat{I}_1$ satisfying the initial conditions (14) is

$$\widehat{I}_1 = \left(\frac{q e \lambda_o}{m v_o^2} \right) \left(-i\omega \widehat{V} \right) \frac{\sinh(\omega \Gamma K z)}{(\omega \Gamma \kappa)} . \qquad (17)$$

The inverse transformation is $I_1 = \dfrac{1}{2\pi} \displaystyle\int_{-\infty}^{+\infty} d\omega\, e^{-i\omega\tau} \widehat{I}_1 .$ (18)

with all singularities below the real ω axis.

Asymptotic Growth Formulas

The general character of resonant growth can be determined by the saddle point method. The form of I_1, given by eqn (18) is

$$I_1 \sim \int_{-\infty}^{+\infty} d\omega\, f(\omega) \exp(g) , \qquad (19)$$

with
$$g = -i\omega\tau \pm \omega \Gamma K z . \qquad (20)$$

Intrinisic singularities are located at the complex resonance values (poles of Γ^2),

$$\omega = \pm \overline{\omega} - i\omega_o/2Q , \qquad (21)$$

where
$$\overline{\omega} = \omega_o \sqrt{1 - 1/(2Q)^2} .$$

Saddle points are located at the six roots of the equation

$$0 = \frac{\partial g}{\partial \omega} = - i\tau \pm Kz\Gamma \left[1 + \Gamma^2 \omega \left(\omega + i\omega_o/2Q \right) \right] . \qquad (22)$$

The integration contour in eqn (19) can be deformed to pass through the saddle points, and the dominant contributions to I_1 are produced at these locatons for sufficiently large z and τ. To simplify the saddle calculaton let

$$u = \omega / \overline{\omega} + i\varepsilon, \quad \varepsilon = \omega_o/2\overline{\omega}_o Q \qquad (23)$$

assumed small compared with unity. Then eqns. (20) and (21) become (at the saddles)

$$g = -i\overline{\omega}\tau \frac{u(u-i\varepsilon)^2}{(1-i\varepsilon u)} , \qquad (24)$$

$$(u^2-1)^3 = \left(\frac{Kz}{\overline{\omega}\tau} \right)^2 \left(1 - i\varepsilon u \right)^2 . \qquad (25)$$

For mode growth associated with the module resonance we expect $u \approx \pm 1$, which requires $\overline{\omega}\tau \gg Kz$, Eqn. (25) gives

$$u^2 = 1 + r \left(\frac{Kz}{\overline{\omega}\tau} \right)^{2/3} \left(1 - i\varepsilon u \right)^{2/3} , \qquad (26)$$

where r is any cube root of unity. The small quantity $\varepsilon \approx (2Q)^{-1}$ is a fixed constant. However we shall find that at peak growth for fixed z, $(Kz/\overline{\omega}\tau)$ is of order $\varepsilon^{3/2}$. To solve eqn (26) by an expansion in the small parameters we formally define

$$\beta \equiv \frac{(Kz / \overline{\omega}\tau)^{2/3}}{\varepsilon} , \qquad (27)$$

and regard β as of order unity. The resulting expressions for u and g are, assuming $u \approx + 1$,

$$u = 1 + \frac{\varepsilon\beta r}{2} - \varepsilon^2 \left(\frac{\beta^2 r^2}{8} + \frac{i\beta r}{3} \right) + \dots , \qquad (28)$$

$$g = -i\overline{\omega}\tau \left[1 + \varepsilon \left(\frac{3\beta r}{2} - i \right) + \varepsilon^2 \left(\frac{3\beta^2 r^2}{8} - i\beta r \right) + \dots \right] . \qquad (29)$$

For unstable growth the relevant cube root of unity is $r = (\sqrt{3}\, i - 1)/2$. We get, keeping only terms through order ε:

$$g_r = \frac{3\sqrt{3}}{4} \left(\overline{\omega}\tau \right)^{1/3} (Kz)^{2/3} - \frac{\omega_o\tau}{2Q} , \qquad (30)$$

$$g_i = -\overline{\omega}\tau + \frac{3}{4} \left(\overline{\omega}\tau \right)^{1/3} (Kz)^{2/3} . \qquad (31)$$

At specified z the maximum value of g_r is readily found from eqn (30):

$$0 = \frac{\partial g_r}{\partial \tau} = \overline{\omega} \left[\frac{\sqrt{3}}{4} \left(\frac{Kz}{\overline{\omega}\tau} \right)^{2/3} - \varepsilon \right] . \qquad (32)$$

At this point $\beta = 4/\sqrt{3}$, which is of order unity, as assumed. The maximum growth factor is

$$(g_r)_{max} = \left(\frac{\sqrt{3}}{2} \right)^{3/2} \left(\frac{\overline{\omega}Q}{\omega_o} \right)^{1/2} Kz . \qquad (33)$$

Application to Heavy Ion Fusion

The maximum growth is calculated here at a medium energy position in a fusion driver, with ion parameters ($T = 1000$ MeV, $m = 200$ amu, $q = 1$), and the typical pulse parameters ($I_0 = 10^3$ A, $\tau_p = 500$ ns.) For the module response we take $\overline{\omega} \cong \omega_0 = 2\pi \times 10^8\ \text{s}^{-1}$, $Q = 30$, $R = 300\ \Omega/\text{m}$. Then we have

$$v_o = .104c, \quad \lambda_o = 32.2\ \mu C/m ,$$

$$C = 10^{-9}/2\pi\ \text{F-m}, \quad K = .0100\ \text{m}^{-1} .$$

At the pulse end $\overline{\omega}\tau_p = 314$ and the maximum growth point is

$$z = \frac{\overline{\omega}\tau_p}{K} \left(\frac{2}{\sqrt{3}\, Q} \right)^{3/2} = \frac{2.36}{K} = 236\ \text{m} ,$$

$$(g_r)_{max} = \left(\sqrt{3}/2 \right)^{3/2} Q^{1/2} Kz = \omega_o\tau_p/Q = 10.5 .$$

This calculated total growth [$\exp(10.5) = 36300$] is large enough to be of concern, even though the initial disturbance $V(\tau)$ may be very small in the unstable band. A small rms velocity spread $\left(\Delta v/v_o \approx v_o(g_r)_{max}/\omega_o z_{max} \approx .0022 \right)$ would be sufficient to eliminate growth, but would constrain the focal spot radius achievable in a fusion reactor.

Dispersion Relation

A Laplace transformation in both τ and z on eqns (11)-(13) yield

$$I_1 \sim \int d\omega \int d\Omega \frac{F(\omega)}{D} e^{-i(\omega\tau + \Omega z)} , \qquad (34)$$

where the dispersion relation is

$$D(\omega,\Omega) = -K^2\omega^2 + i\omega\Omega^2/CZ(\omega) \qquad (35)$$

$$= \left(\omega^2 - \omega_o^2 + i\omega\omega_o/Q \right) \left(\Omega^2 - K^2 \right) - K^2 \left(\omega_o^2 - i\omega\omega_o/Q \right) . \qquad (36)$$

The latter form of D clearly indicates that a pair of strongly coupled resonances are present, and appear symmetrically when $Q = \infty$. A growth formula, valid for near resonance $\Omega^2 \approx K^2$, may therefore be obtained from eqn. (30) by interchanging $\omega_o\tau$ with Kz. For $Kz \gg \omega_o\tau$ we have

$$g_r = \left(3\sqrt{3}/4 \right) \left(\omega_o\tau \right)^{2/3} (Kz)^{1/3} , \qquad (37)$$

$$g_i = -Kz + (3/4) \left(\omega_o\tau \right)^{2/3} (Kz)^{1/3} . \qquad (38)$$

The roots of the dispersion equation $D(\omega,\Omega) = 0$ can be used to find the growth in z for given real ω; we find for the imaginary part of Ω

$$\Omega_i = \left[\frac{K^2 C\omega}{2} \left(|Z| - Z_i \right) \right]^{1/2} , \qquad (39)$$

with

$$Z = Z_r + iZ_i = \frac{R\left[1 + iQ \left(\frac{\omega}{\omega_o} - \frac{\omega_o}{\omega} \right) \right]}{1 + Q^2 \left(\frac{\omega}{\omega_o} - \frac{\omega_o}{\omega} \right)^2} .$$

The maximum growth formula eqn. (33) may be recovered from eqn. (39) for large Q by maximizing Ω_i with respect to driving frequency ω.

Reference

[1] V. K. Neil, Interaction of the ATA Beam with the TM_{030} Mode of the Accelerating Cells, LLNL report UCID-20456, 1985.

An Investigation of the Source of a Low-Q, Low-Frequency Impedance Disrupting Bunch Coalescing in the Fermilab Main Ring

P. L. Colestock, J. Griffin, X. Lu, G. Jackson,
C. Jensen, and J. Lackey
Fermi National Accelerator Laboratory *
P.O. Box 500   MS 308
Batavia, IL 60510

Abstract

Recent studies of bunch coalescing in the Fermilab Main Ring have indicated the likelihood of a low-frequency, low-Q resonator being responsible for the distortion of the coalesced bunch at high beam intensities. Numerical simulations have shown that a sufficiently high longitudinal impedance is sufficient to account for the observed bunch break-up. A survey of possible offending beam components has been made, and measurements of the longitudinal impedance of these components, including a variety of kicker magnets and coalescing cavities in the Main Ring have been carried out.

I. INTRODUCTION

The coalescing process in the Fermilab Main Ring consists of adiabatically lowering the rf voltage so that a series of bunches effectively debunches, followed by a phase space rotation and recapture in a somewhat larger bucket. In this manner, a high intensity bunch can be constructed out of a series of lower intensity bunches. [1] However, it has been observed that the efficiency of this process degrades as the number of protons in the ensemble increases. In particular, experiments have indicated that charges which arrive at later times in a series of bunches are being pushed out of the buckets, and the overall coalescing efficiency has been observed to decrease nearly linearly with total intensity. [2]

In order to explain these results, the existence of a low-Q, low-frequency resonator has been hypothesized, which can account for several of the details of the observations. [3] In this paper we extend the earlier analyses with a theoretical simulation of the effect, and present the results of bench measurements of various beamline magnets, which were proposed as candidates for the source of the offending longitudinal impedance.

II. RESONATOR MODEL

Since the leading edge of the bunch train is unaffected during coalescing, it is assumed that the resonator Q is such that the resonator fields have decayed away within one revolution period (20.9 μS). Moreover, the resonator period is ostensibly considerably longer than the series of bunches (200 nS), since the deceleration appears to increase roughly linearly

towards the end of the ensemble. A feasible candidate would be a resonance between a few hundred kHz to a few MHz. For such a case, the voltage applied to the beam can be modelled by a simple RLC circuit.

During the brief period of passage of the bunch series, the voltage is primarily due to the buildup of charge on the capacitance C, and analysis of the required voltage puts the estimate of C at about 40 pF, for an average bunch current of 200 mA. Q is expected to be in the range of 1 - 5 and hence R ~ 1 kΩ. In order to make quantitative estimates, we consider the time domain solution of the induced voltage using standard Laplace methods,

$$V(t) = f(t)u(t) - f(t-\Delta)u(t-\Delta)$$

where

$$f(t) = I_o R\left(1 - \cos\left[\omega_r t\right] e^{-\frac{Rt}{2L}}\right) + I_o \omega_r \left(L - \frac{R^2 C}{2}\right) \sin\left[\omega_r t\right] e^{-\frac{Rt}{2L}}$$

u(t) is the Heaviside function, and Δ is the total bunch length. For times short with respect to $1/f_r$, the voltage rises linearly with time.

III. THEORETICAL SIMULATION

Particle tracking simulations have been carried out using the ESME code. [3] In this work, the protons are assumed to be at an intensity of 10^{10} per bunch. In Fig. 1(a) and (b), the results of a coalescing simulation are shown for the case of zero impedance. With increasing time (toward the top of the plot), the bunches first begin to debunch, then rotate and are finally recaptured into a single high intensity peak. In Fig. 1 (c) and (d), a low Q ~ 10 resonator with ω_r ~ 1 MHz, and a shunt impedance of 200 kΩ was assumed. It is clear from this study that such a low Q resonator would have an effect close to that observed, provided its shunt impedance were high enough. As a result of parametric studies, it has been found that total ring impedance levels below about 10 kΩ are insufficient to produce substantial deceleration of the trailing bunches.

Of the various beam components in the Main Ring, the most suspect of containing low Q resonances are the coalescing cavities themselves, or any of the variety of ferrite or ferromagnetic kicker magnets in use. Direct beam loading measurements of the cavities suggest that their combined impedance is less than 100 Ω [4], so attention has been focussed on bench measurements of the kicker magnets' impedance.

* Operated by the Universities Research Association under Contract with the U. S. Department of Energy

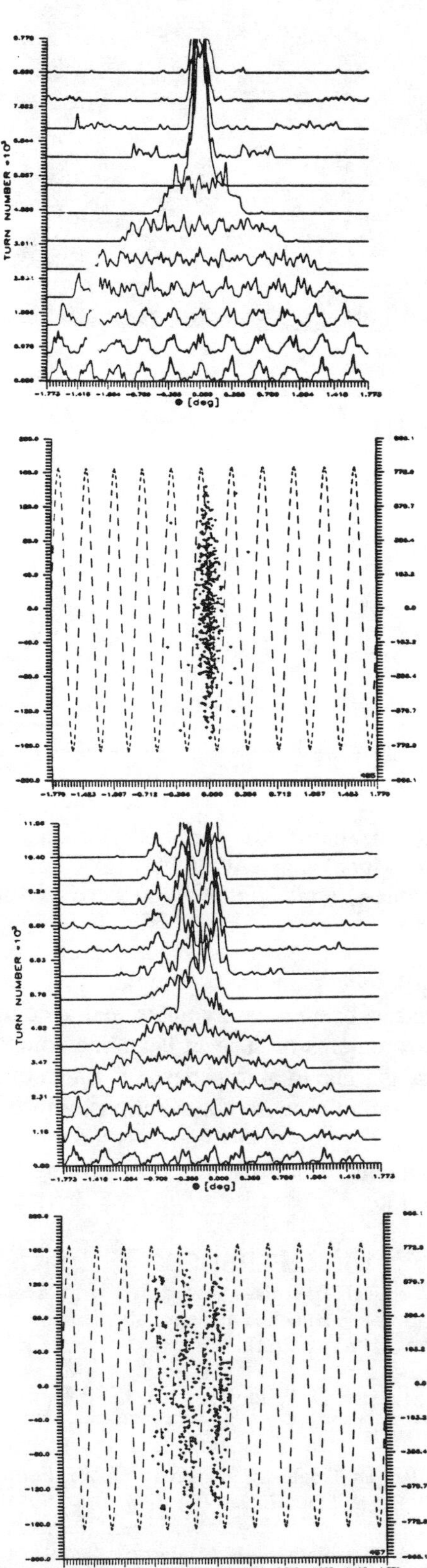

Fig. 1 (a) Plot of bunch current versus arrival time for successive times during the coalescing cycle. Time increases toward the top of the plot. (b) is the corresponding phase space plot. In case (a) and (b), no impedance is assumed. In (c) and (d), a large ~100kΩ impedance is assumed, which causes deceleration of trailing bunches in the ensemble.IV. Kicker Magnet Measurements

IV. KICKER MAGNET MEASUREMENTS

The inventory of kicker magnets in the Main Rng is shown in Table 1. They fall into two classes (a) C-frame magnets and (b) H-frame magnets, with either ferrite or tape-wound-steel as

Table 1. Kicker Magnet Types in the Main Ring

Location	Type	Magnetic Material
B-48	H	Tape-wound-steel
A-11	C	Ferrite
F-14	H	Ferrite
C-48	H	Ferrite
E-17	C	Tape-wound-steel

magnetic materials. The H-frame magnets are particularly suspect because of bus lines that may be closely coupled to ferrite materials. A sketch of a typical H-frame device (the C-48 kicker) with associated busswork is depicted in Fig. 2. The bus bars form a coupled transmission line that is effectively

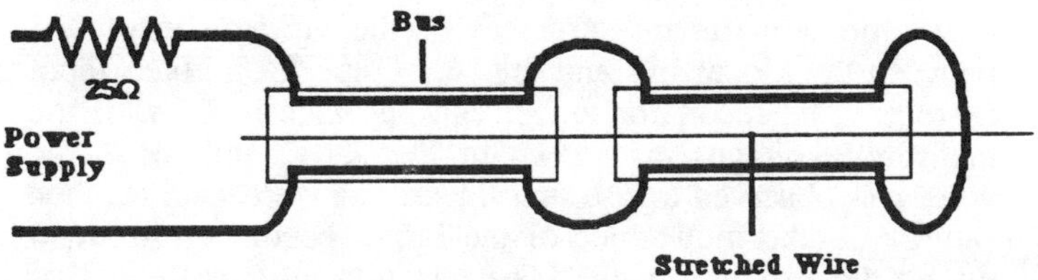

Fig. 2 Sketch of the C-48 kicker magnet. The bus bars form a ferrite loaded transmission line which comprises a low-Q resonator.

terminated in a capacitance at the upstream and downstream ends. In addition, the power feed circuitry applies a short circuit to the input of one bus bar, while 25 Ω occurs across the other. Similar situations occur in the other kicker magnets. The characteristic impedance of the individual lines is measured to be ~ 70 Ω, and due to the presence of the magnetic material, the phase velocity is about 0.7 c. The primary capacitance of interest is at the upstream end, since it can affect trailing bunches. A direct measurement places this value at about 80 pF.

The longitudinal impedance of this kicker was measured with a stretched-wire [5] of characteristic impedance 356 Ω, as shown in Fig. 2. Resistor matching networks at the ends provided a wideband match from DC to about 800 MHz, which permits the extraction of the longitudinal impedance from the transmission loss. The measured frequency response of the kicker, including the power feed circuitry, is shown in Fig. 3.

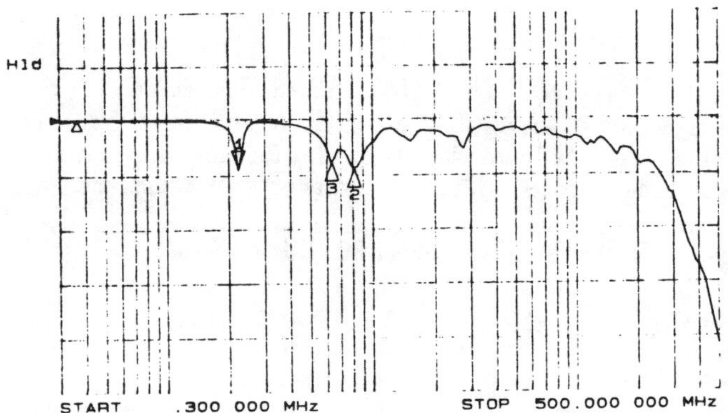

Fig. 3 Transmission response of the kicker from stretched-wire measurements. Absorption peaks indicate impedance maxima. Several low-Q resonances are observed.

The results indicate several low-Q resonances, though due to the requirement that the resonant frequency be less than the inverse bunch length in seconds, it is likely that only the lowest frequency mode can have an appreciable effect. It is noteworthy to observe that the impedance becomes large at high frequencies, likely due to losses in the low conductivity wall coatings and magnetic materials of the kicker, though this impedance is not expected to play a role in the coalescing process.

To determine the quantitative effect such a resonator may have on the beam, we apply the simple resonator model of Sec. II to a time domain measurement of the voltage developed between the kicker bus and the beam pipe for a given input current. The resonator model can be made to fit well the measured voltages, as shown in Fig. 4 (a) and (b). The voltage is observed to rise first linearly on the capacitor, then saturate as the inductance of the ferrite becomes important during the first half cycle. The bunch train exits the kicker before the full voltage of the resonator has built up, giving rise to a maximum voltage of about 15 volts. Other kicker types were found to have a similar order of magnitude response, which leads to the conclusion that about 100 Volts can be accounted for by the kickers.

V. DISCUSSION AND SUMMARY

We have confirmed theoretically that a low-Q resonator of sufficiently high impedance can explain the experimental observation that increasing bunch intensity leads to decreasing coalescing efficiency. In particular, the deceleration of trailing bunches points to a resonant frequency in the MHz range with an approximate shunt impedance of 1 - 5 kΩ. Measurements of the several kicker magnet types employed in the main ring suggest such a resonator may indeed exist, but the magnitude of the measured voltages falls a factor of 10-100 below the expected values based on simulations. While it can be expected that there is some uncertainty in both the measurements and the simulations, it does not appear likely that the kicker magnets are alone responsible for the decrease in coalescing efficiency.

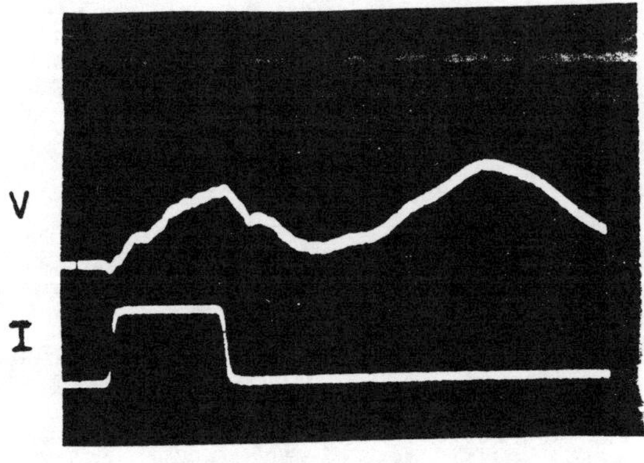

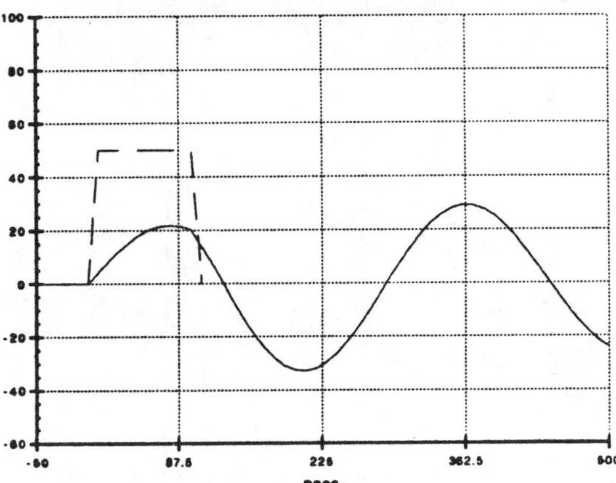

Fig. 4(a) Measurement of induced kicker voltage corresponding to a peak current of 200 mA. and (b) the corresponding resonator model with C=460 pF, L= 5.4 μH and R=7.5 Ω.

In future work, we plan to investigate the source of impedance with direct beam measurements, and to continue to make bench measurements of suspect beamline components. It may be the case that the total impedance is a sum of a large number of small, roughly equal contributions, which would suggest that the most appropriate solution is the development of a feedback system to combat the decelerating voltages around the entire ring.

VI. REFERENCES

[1] J. Griffin, et. al.,"RF Exercises Associated with Acceleration of Intense Antiproton Bunches at Fermilab", IEEE Proc. Nucl. Sci., NS-30, 4, (1983) p.2627

[2] P. Martin,K. Meisner, D. Wildman, Proc. IEEE Particle Accel. Conf., (1987), p. 1527

[3] J. Griffin, "Bunch Coalescing", Proc. of the Fermilab III Workshop on Beam Instabilities (Batavia, Ill., 1990)

[3] J. MacLachlan, "Fundamentals of Particle Tracking for the Longitudinal Projection of Beam Phasespace in Synchrotrons", Fermilab FN-481, 1988

[4] D. Wildman, Private Communication

[5] P. Colestock, et. al., These Proceedings.

[6] X. Lu, G. Jackson, These Proceedings

A Critical Survey of Stretched-Wire Impedance Measurements at Fermilab

P. L. Colestock, P. J. Chou, B. Fellenz, M. Foley, F. Harfoush, K. Harkay,
G. Jackson, Q. Kerns, D. McConnell and K. Y. Ng
Fermi National Accelerator Laboratory
P.O. Box 500   MS 306
Batavia, IL 60510

Abstract

Bench measurements of the impedance of various beam line components, as well as several test structures have been carried out using cavity perturbation and stretched wire methods. Several pitfalls of the wire measurement technique have been uncovered and improvements were developed to resolve ambiguities and improve overall accuracy. In this paper, we summarize the results of our findings concerning measurement techniques, accuracy limitations, and the comparison with numerical simulation of model test structures.

I. INTRODUCTION

A concerted effort is being made at Fermilab to understand and control the contributions of various beamline components to the overall impedance seen by the beam, in order to prevent the onset of various instabilities under normal operating conditions, and at the increased intensity levels expected in the near future. Several methods have been used to determine the effective impedance of a given structure, including (a) cavity perturbation techniques using small, suspended beads [1], (b) beam measurements using a coasting beam in situ [2] and (c) bench measurements employing a thin, coaxial wire stretched along the beam path. [3] Of these methods, the stretched wire techniques are most commonly used because of the ease with which the measurements may be carried out. However, a number of measurement difficulties arise in the use of stretched wires, owing to the fact that reflections may readily occur on a coaxial wire, but not on a high energy beam.

II. STRETCHED WIRE METHODS

The measurement of impedance is divided into two subareas: (a) longitudinal impedance, which can produce a beam energy loss, and (b) transverse impedance, which is responsible for a transverse deflection of the beam. In its most general form, an object in the beam line can be considered to be a 2-port network, involving both shunt and series currents, as shown in Fig. 1, (a) and (b). The two models are equivalent, but energy loss associated with longitudinal fields, i. e. longitudinal impedance, is best determined from the normalized Pi-network, as

$$Z_{||} = \frac{Z_o}{Y_{12}} \qquad (1)$$

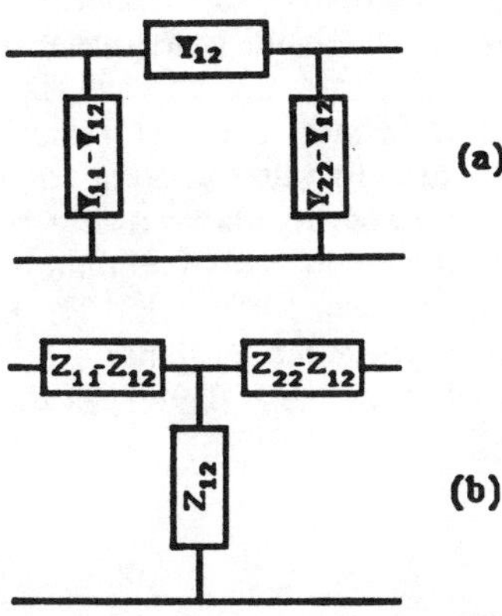

Fig. 1 (a) Pi network model and (b) T network model

while an equivalent shunt impedance is found from the normalized T-network to be

$$Z_S = Z_o Z_{12} \qquad (2)$$

We note the distinction between this perpendicular impedance and the transverse impedance described previously. The above perpendicular impedance is not necessarily associated with a net transverse force on the beam. Using the familiar S-parameters, which are the quantities usually determined from the modern network analyzer, the above impedances are expressed as

$$\frac{1}{Y_{12}} = \frac{1 + S_{11} + S_{22} + \Delta_s}{2 S_{21}}$$

$$Z_{12} = \frac{2 S_{21}}{1 - S_{11} - S_{22} + \Delta_s} \qquad (3)$$

$$\Delta_s = S_{11} S_{22} - S_{21}^2$$

We note that in the limit of negligible shunt losses,

$$Z_{12} \to \infty, \quad S_{11} = S_{22} = 1 - S_{21} \quad \text{and} \qquad (4)$$

$$Z_{||} \to 2 Z_o \left(\frac{1}{S_{21}} - 1 \right)$$

which is the usual result quoted in the literature. [4] However, this condition does not hold in general, and it can be expected that the application of this expression to cases where the shunt

---

* Operated by the Universities Research Association under Contract with the U. S. Department of Energy

"

impedance is non-negligible will be in error. In the case of a high energy beam, no TEM reflected waves can exist, implying $S_{11} = 0$, which is a major distinction from the case of a stretched wire.

As such, the effects of reflections on the stretched wire need to be carefully eliminated from the measured data, which requires a knowledge of the full 2-port scattering matrix for the unknown device. This requirement leads to the necessity of developing an accurate model for the test fixture used to carry out the measurements, which must include any transformers used to make a transition from the 50Ω line to the (typically a few hundred ohm) stretched-wire. Generally, one of two methods for effecting a wideband transformation is used: (a) cone transformers which vary the characteristic impedance gradually along their length and (b) resistive matching networks. The practical length limitation of cone transformers limits their frequency range to above a few hundred MHz, while parasitic reactances of resistive matching networks restrict their usage to below about 1 GHz. The effect of these transformers can be represented in general by linear 2-port networks cascaded with the test device, as shown in Fig. 2.

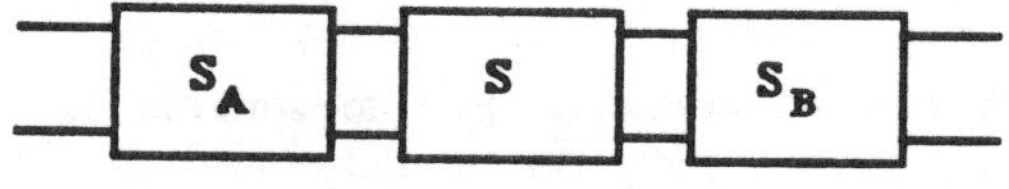

Fig. 2 Cascaded measurement network. S represents the S-parameters of the unknown impedance.

The measurement problem becomes one of de-imbedding the unknown S-parameters from the external measurements of the system.

We have used two techniques to carry out the de-imbedding process. First, the full 2-port S-parameters of the test fixture alone are measured with standard (open-short-load) calibrations to remove the cable contributions. Using a least-squares circuit optimizer [5] with the model circuit shown in Fig. 3,

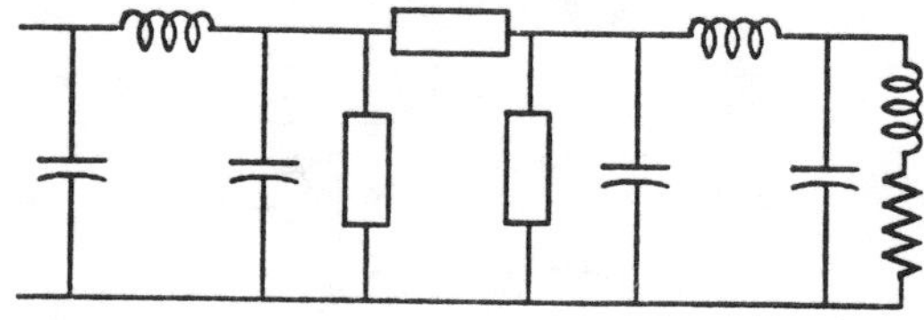

Fig. 3 Lumped element model of the measurement fixture. The unknown forms a 2-port network in the center.

the impedance is deconvolved from the data using standard circuit analysis. Resistor matches have been used at the fixture ends to reduce reflections and facilitate deconvolution. We note that shunt terms have been ignored here. This

approach yields satisfactory results over about a 2-octave bandwidth, until the lumped element end models break down. The second process does not depend on termination standards, but uses two lengths of line of characteristic impedance Z_O, differing by about 1/4 wavelength, as well as two shorts placed at the reference plane. This Through-Short-Delay (TSD) procedure [6] affords the advantage of removing the effects of nonideal transformers and cables in one operation. It also permits a high degree of spatial resolution in the impedance measurement. The effective bandwidth of the method is not limited, but several delay lengths may be required to cover the desired band.

De-imbedding has been carried out using the above procedure for a simple test cavity, as shown in Fig. 4. The

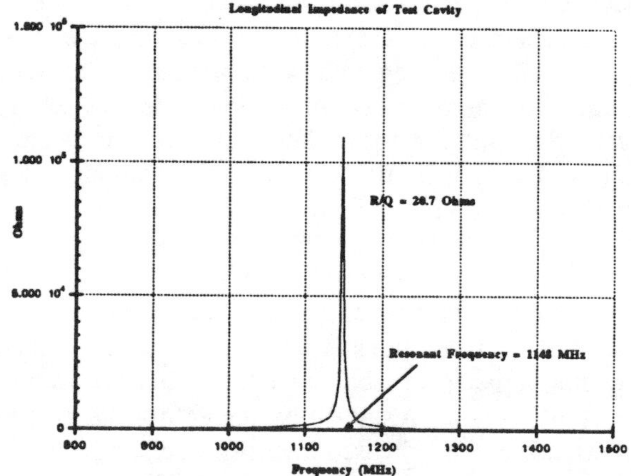

Fig. 4 Real part of $Z_{||}$ for a test cavity, derived from measurements by de-imbedding.

relatively low impedance of the stretched-wire (326 Ω) lowers the Q of the cavity substantially. However, since the R/Q value is preserved, the actual Q can be recovered. The actual Q values compare favorably with weakly coupled probe measurements.

Another problem occurring with wire measurements occurs when more than one source of impedance exists in a given device, such as the dual gap booster cavity of Ref. [7]. Since cavity gap impedances are typically large, significantly large standing waves can occur on the wire between the gaps, introducing spurious resonances into the system. There is no simple means of determining both gap impedances from a single, external S-parameter measurement. One solution of this problem is to only measure one gap at a time, i.e. via S_{11} measurements, using a termination at the unexcited end of the wire. This approach was used effectively in producing the booster data above. An alternative approach consists of inserting a miniature laser/light pipe system into the beam pipe, as shown in Fig. 5. The light pipe does not couple the gaps together, while permitting an intrinsically more accurate S_{21} measurement of the gap impedance. An equivalent, but less perturbing method involves the use of probes instead of a stretched-wire. The probes form a capacitive divider whose shunt loading impedance on the test device can be made exceedingly high, subject to the dynamic range of the system. With the addition of in-situ, low-noise amplification, we have extended the dynamic range of our measurement to 180 dB, facilitating highly non-perturbing measurements.

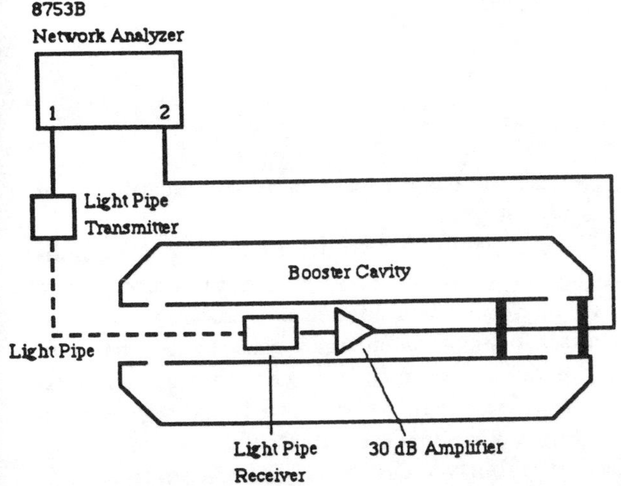

Fig. 5 Single gap S_{21} measurement using a light pipe to avoid inter-gap coupling.

We have compared the results of experimental measurements of the longitudinal impedance with theory for a simple pillbox cavity, where analytic solutions are known. Moreover, the cavity was of a sufficiently small size to force the resonant frequencies to be in a range where careful determination of the transformer impedance was required. We compare the results of the wire method with a lumped circuit model with the theoretical values in Table 1. It is observed that the resonant

Table 1

	Theory	Measurement
f_0 (GHz)	1.134	1.147
	2.621	2.670
R/Q	27	20.7
	21	15.0

frequencies are somewhat shifted and Q values of the unloaded cavity considerably exceed the wire-loaded values, but the R/Q values are in reasonably good agreement with the theory.

III. DISCUSSION AND SUMMARY

We have developed stretched-wire impedance models that include both longitudinal and shunt components, and have studied the effects of non-negligible reflections on measurement interpretation and accuracy. Using the TSD calibration method, we have devised an accurate measurement technique for determining the local impedance of a given

structure, and have used transmission probes to greatly reduce the loading of high-Q devices.

Several problem areas remain, however, which require further study. There is presently no stretched-wire method of determining the individual contribution of physically separated objects to the overall impedance of a distributed structure, owing to the possibility of reflected waves along the wire. A recently proposed approach involving time-resolved wave packets may of use in this case, though the maximum measureable Q must be traded off against spatial resolution. At Fermilab, there is no general approach to impedance measurements above the beam pipe cutoff frequency, where higher modes are observed to occur due to discontinuities in the system. There is a need to develop single-mode wave launchers and receivers to facilitate measurements in this domain. Finally, it is desireable to apply similar methods to the measurement the transverse impedance directly.

IV. REFERENCES

[1] L. C. Maier and J. C. Slater, "Field Strength Measurements in Resonant Cavities", J. Appl. Physics, 23, 1, p. 951

[2] G. P. Jackson,"Review of Impedance Measurements at Fermilab", Proc. Fermilab III Workshop on Instabilities, Batavia, University Research Association, 1990

[3] F. Caspers,"Beam Impedance Measurements using the Coaxial Wire Method", CERN PS/88-59 (AR/OP), Oct. 1988

[4] H. Hahn and F. Pedersen,"On Coaxial Wire Measurements of the Longitudinal Coupling Impedance", BNL 50870, Brookhaven National Lab, Apr. 1978

[5] TOUCHSTONE® Reference Manual, EESOF, (1989), Westlake Village, Ca.

[6] H.-J. Eul and B. Schiek, "A Generalized Theory and New Calibration Procedures for Network Analyzer Self-Calibration", IEEE Trans on Microwave Theory and Techniques, 39, 4 1991, p. 724

[7] K. Harkay, et. al., "Studies of Coupled-Bunch Modes in the Fermilab Booster", These proceedings.

Longitudinal Instability in the Fermilab Accumulator During Slow Transition Crossing

X. Q. Wang and G. Jackson
Fermi National Accelerator Laboratory*
P.O. Box 500
Batavia, IL 60510

Abstract

The Fermilab Accumulator was built primarily as an 8 GeV kinetic energy antiproton storage ring. Recently the Accumulator was modified for use in a medium energy experiment studying charmonium states. The experiment required a circulating antiproton beam with momenta between 6.7 GeV/c and 3.7 GeV/c. In order to decelerate the beam below 5 GeV/c, transition must be crossed. The method adopted for Accumulator transition crossing was to raise the value of γ_t slowly while the antiproton beam circulated in the ring unbunched. During Accumulator transition crossing with this coasting beam, longitudinal instability was observed with a resistive wall current monitor and a dedicated longitudinal Schottky detector. This paper presents the results of measurements in which longitudinal coherent signals and emittance growth were observed. The data is compared to the predictions of a multiparticle computer simulation.

1   Introduction

The Fermilab Accumulator is an 8 GeV antiproton storage ring which collects and stores antiprotons for operation of the Tevatron collider. Since 1987, a medium energy experiment has been conducted in the Accumulator to study the formation of charmonium states ($c\bar{c}$ bound states) [1]. In the experiment, an intense antiproton beam is decelerated to specific energies in the charmonium region between 6.8 GeV and 3.8 GeV, and collided with protons from an internal hydrogen gas jet target.

The design value of the Accumulator transition energy is 5.1 GeV. In order to perform the experiment below that energy, transition must be crossed. Because of the restrictions from the Accumulator RF system and magnet power supplies, the deceleration can only proceed in a rate which is not fast enough for crossing transition without a major disruption of the bunches. Therefor, an alternative method for transition crossing is adopted in the deceleration operation. In this technique, the beam is debunched at an energy slightly above transition, then γ_t is raised to a value such that the beam energy is below transition, and the beam is recaptured.

During the Accumulator transition crossing, longitudinal instability imposes a critical limit on the amount of $\bar{p}$ beam that can be decelerated below transition. The instability causes beam momentum spread to increase and results in the beam intensity loss when beam current is above 10 mA. Measurements in both frequency domain and time domain were made to identify the source of the longitudinal instability. The result from a multiparticle computer simulation is also presented for comparison.

2   Measurements

In transition crossing, the beam was debunched at an energy of 5.286 GeV ($\gamma = 5.634$), and γ_t was raised from either 5.15 or 5.5 to 6.11. The values of γ_t before and after the crossing process were measured by changing the current of the main dipole bus and measuring the corresponding revolution frequency of the beam. The momentum spread of the beam was obtained from the longitudinal Schottky signal, using the relationship

$$\frac{\Delta P}{P} = -\frac{1}{\eta}\frac{\Delta f}{f},$$

where $\eta = 1/\gamma_t^2 - 1/\gamma^2$. The longitudinal Schottky signal was detected by the Accumulator Schottky pickup, a quarter wave length device resonating at 79.3 MHz.

Beam energy loss and momentum blow up was observed during transition crossing. Figure 1 shows the longitudinal frequency distributions of a beam at $h = 127$ before and after a crossing process. The beam was stochastically cooled before the process started. The 10dB width of the momentum spread was 0.25% before the crossing and 0.82% after the process. The energy loss of the beam in the process was determined from the peak shift of the frequency distribution. In figure 1 the frequency shift is 1.36 kHz, from which about 300 Hz results from the change in the orbit length. This gives a result of 14 MeV for the energy loss in the crossing process. Since near transition a small error in the γ_t measurement gives rise a large error in η, these measurement results are subject to about 40% uncertainty.

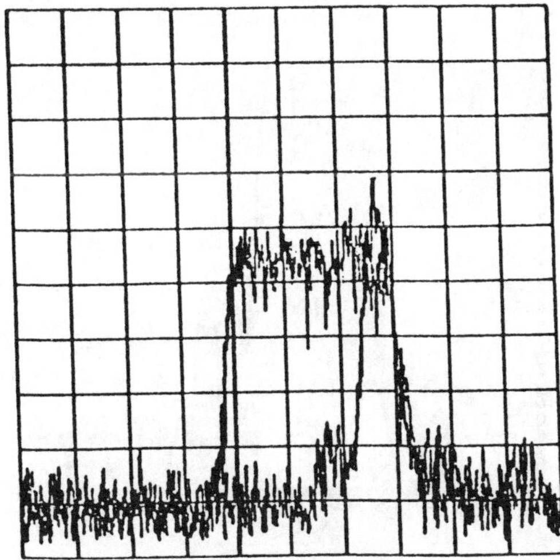

Figure 1: Longitudinal Schottky profiles at $h = 127$ before (narrow trace) and after (wide trace) a transition crossing.

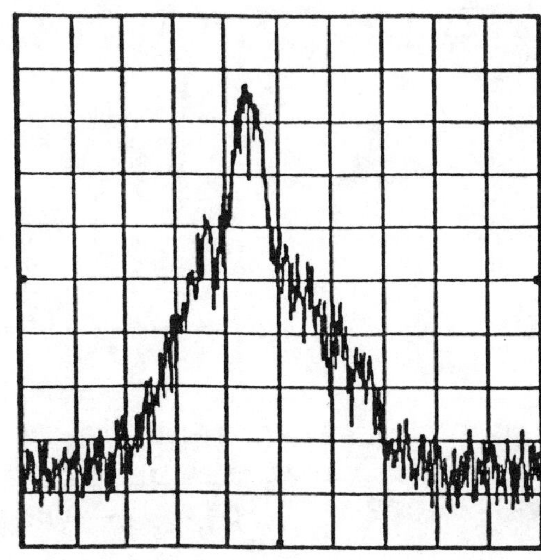

Figure 3: Longitudinal Schottky signal at $h = 127$ in the same transition crossing.

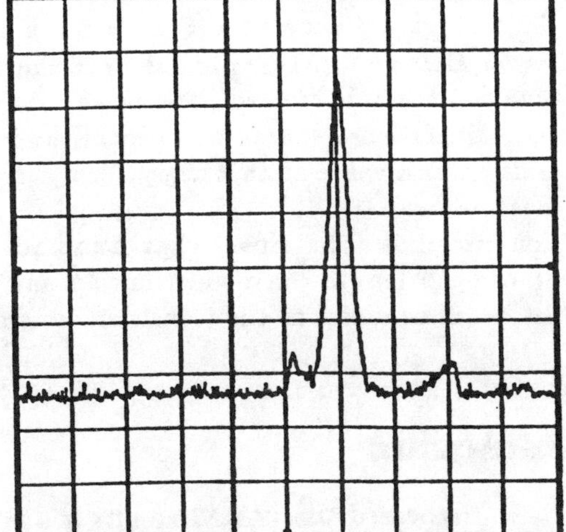

Figure 2: Beam coherent signal at 52.9 MHz in transition crossing, measured by the resistive wall monitor.

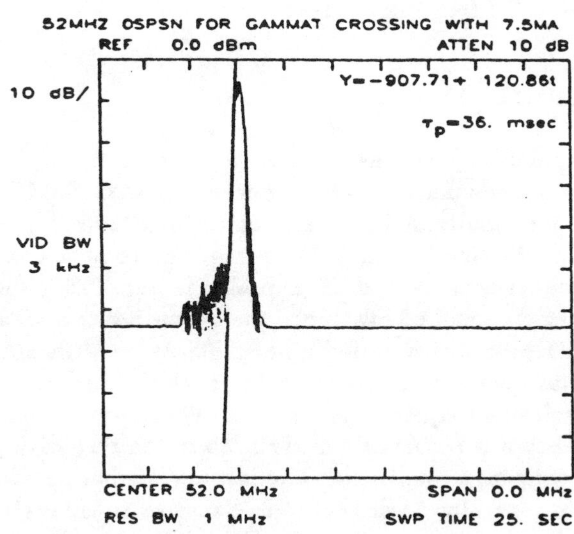

Figure 4: Time evolution of the coherent effect.

In order to discover the sources which caused the momentum blow up and energy loss of a beam in transition crossing, the longitudinal frequency spectrum of the beam was studied as γ_t was being raised. A broadband 3 kHz – 6 GHz resistive wall monitor was used to observe the coherent signal in a wide frequency range. It was found that coherent lines appeared at harmonics of 53 MHz, which was near the resonant frequency of the Accumulator main RF system ARF1. Figure 2 shows the fundamental coherent line at 52.9 MHz. The beam intensity was 5.7 mA. Meanwhile, a sharp peak was observed to rise from the longitudinal Schottky profile, indicating the self-bunching effect of the beam, as shown in figure 3. These observations provided clear evidence that the large shunt impedance of ARF1 was the major source of the beam longitudinal instability in the transition crossing process.

The time evolution of the coherent signal was also measured by using the resistive wall monitor and setting the spectrum analyser to a center frequency near the ARF1 resonant frequency with zero frequency span. The resolution bandwidth of the spectrum analyser was chosen such that at least one revolution harmonic of the beam was covered. Figure 4 shows one measurement result with a beam of 7.5 mA. The sharp rise of the signal occurred when transition was crossed. The rise time of the signal power could be obtained by a linear least-square fit in logarithmic scale. Fitting gave a result of 36 msec.

Transition studies were also done with ARF1 shorted. It was observed that the magnitude of the coherent signal was reduced by at least 50dB, and there was no significant disruption of the beam. Faster transition crossings (6×) also greatly reduced the longitudinal coherent effect, therefore the momentum blow up, of the beam in the crossing process.

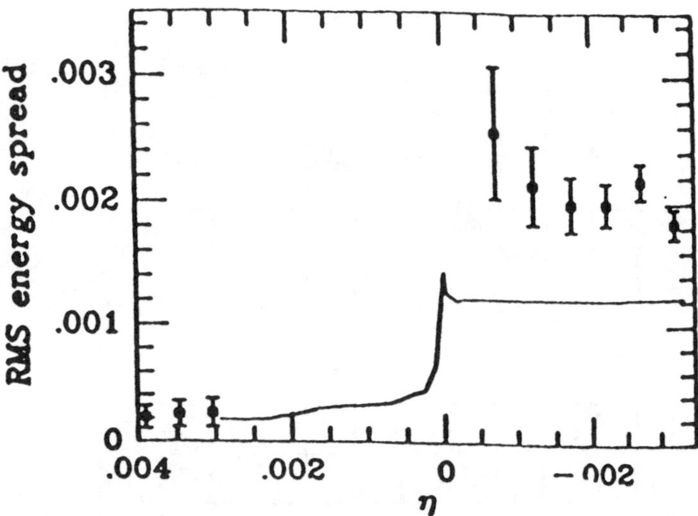

Figure 5: Measured and simulated beam rms energy spread as a function of η. The solid line represents the simulated result. The error bars is determined by the accuracy in measuring the frequency spread.

3   Simulation

Computer simulations of the Accumulator transition crossing process have also been performed with ESME [2] which does multiparticle tracking in longitudinal phase space. In the simulation, ARF1 is modeled to have a resonant frequency at 52.9 MHz, a quality factor of 200, and a shunt resistance of 60 $k\Omega$. Since the beam-induced voltage in ARF1 produces a periodic beam structure with a fundamental frequency of about 53 MHz which is 85 times of the revolution frequency, simulation of the effect of ARF1 on the beam is restricted within 1/85 of the ring with periodic boundary conditions. The initial distribution of the beam is generated such that it is Gaussian in energy and randomly uniform in azimuth. To get the Fourier spectrum of the beam current, the azimuth is divided into 16 bins which will give 8 terms in the current spectrum. The total number of macro-particles used for the simulation is 10000, which results in about 625 particles in each bin and therefore about 4% statistical fluctuation.

Simulations have been done where the effects of space charge and non-linear dependence of path length on momentum are ignored. According to the simulation, in the crossing process there is no real bunching of the beam in the usual sense, i.e., bunching when an external RF voltage is present. When $|\eta| \lesssim 1 \times 10^{-4}$, the periodic coherent structure is continually smeared due to the frequency spread of the beam. Dramatic change in the beam distribution in longitudinal phase space occurs when $|\eta|$ is smaller than 10^{-4}. The energy loss in the crossing process is approximately 9 MeV. In figure 5 the simulated rms energy spread in the crossing process is compared to that from the measurement. The error bars only represents the error in measuring the frequency spread. There is at least additional 40% of uncertainty coming from the η measurement. These results agree with the measurements within

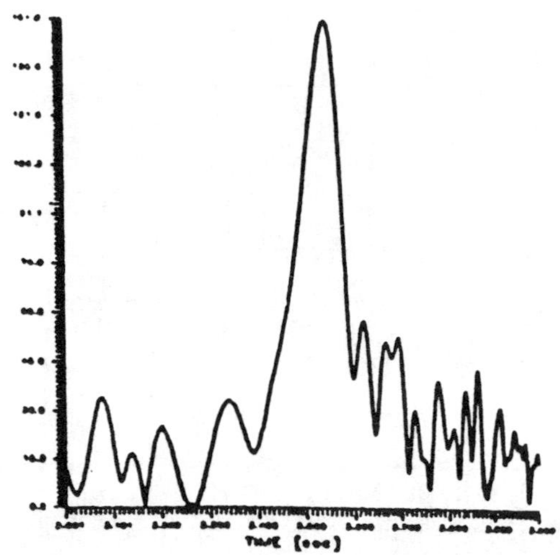

Figure 6: The Fourier component of the beam current at 53 MHz.

the error limit. Figure 6 shows the Fourier component of the beam at 53 MHz as function of time. The sharp peak appears when transition is crossed. The power rise time is obtained by a linear-least square fit in logarithmic scale, and the result gives a value of 36.4 msec. This agrees well with the measurement.

Simulation also shows that space charge and non-linear dependence of path length on momentum do not significantly affect beam behavior in the transition crossing process.

4   Conclusion

It has been discovered that ARF1 results in the major energy loss and energy spread blow up of a beam in the crossing process. The simulation results agree with the measurements. Similar technique will be used in the future to find other impedances that may cause problems at higher intensities.

References

[1] V. Bharadwaj, et al., "Operation of the Fermilab Accumulator for Medium Energy Proton-Antiproton Physics", Proceedings of the 2nd EUROPEAN PARTICLE ACCELERATOR CONFERENCE, June 12-16, 1990, pp.617-619.

[2] S. Stahl and J. M. Maclachlan, "User's Guide to ESME v. 7.1", Fermilab internal note TM-1650, February 26, 1990.

Results from Longitudinal Impedance Measurements in the Fermilab Tevatron

G. Jackson, A. Bogacz, P.J. Chou, P. Colestock, F. Harfoush, X. Lu, K.Y. Ng,
W. Pellico, T. Sullivan, and X.Q. Wang
Fermi National Accelerator Laboratory*
P.O. Box 500   MS 308
Batavia, IL 60510

Abstract

In order to answer questions concerning performance limitations in future accelerator upgrade scenarios caused by coherent instabilities, a considerable amount of effort has been devoted to impedance measurements. This work presents the results from beam-based longitudinal impedance measurements in the Tevatron. Longitudinal transfer function scans with high current unbunched beam at injection is the method used for determining the impedance at every revolution harmonic. Self-bunching instabilities and wave-mixing phenomena occurring during these transfer function measurements are also described. While presenting results, methods to minimize noise and systematic errors are reviewed.

I. EXPERIMENTAL METHOD

By sinusoidally modulating the current of a coasting beam the longitudinal impedance may be measured at any frequency corresponding to a harmonic of the revolution frequency [1,2]. This technique, in the form of a longitudinal closed loop transfer function measurement, has been applied to the Tevatron in an attempt to accurately map the longitudinal impedance as a function of frequency. A sketch of the experimental equipment is displayed in figure 1. The spectrum analyzers are used to measure the longitudinal Schottky distribution and to monitor "ghost" lines (wave-mixing).

After injecting 972 bunches into the 1113 53 MHz RF buckets at 150 GeV the RF voltage was adiabatically reduced by paraphasing two sets of four cavities. When the minimum voltage was attained, the RF power amplifiers were turned off. This beam could then be studied for up to a day.

Because of the problem of time dependent chromaticity in the Tevatron [3], early in these studies the chromaticities were rather large, causing a betatron tune spread which encompassed a number of betatron resonances. With practice it was possible to adjust the tunes and chromaticities during the reduction of the RF to avoid resonance overlap. This is an important consideration since particles whose tune overlapped a resonance were quickly ejected from the accelerator. Since chromaticity generated a correlation between betatron tune and momentum, these losses would generate gaps in the momentum distribution of the beam.

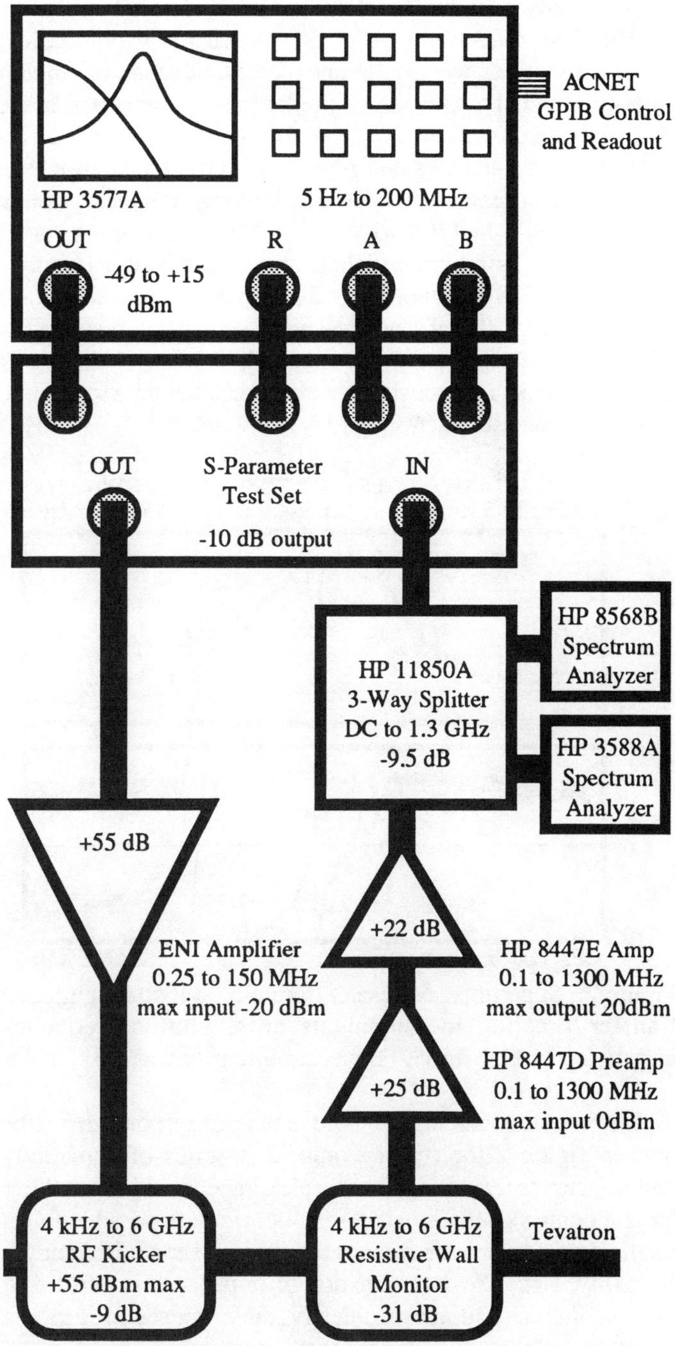

Figure 1: Schematic drawing of the experimental apparatus used to perform the longitudinal closed loop transfer function measurements.

*Operated by the Universities Research Association under contract with the U.S. Department of Energy.

II. IMPEDANCE RESULTS

The first frequency band to be measured in the Tevatron was near the resonant frequency of the RF cavities at 53 MHz. There are 8 RF cavities, each having an unloaded shunt impedance of 1.2 MΩ and an unloaded Q of approximately 7100 [4]. With the cavities loaded to a Q of approximately 4000 [5], the total shunt impedance at harmonic h=1113 should be 5.4 MΩ. This is expected to be by far the largest single impedance in the Tevatron, and easily detectable.

The first measurements of the closed loop longitudinal transfer functions were quite unexpected. Instead of smooth amplitude and phase responses similar to those measured in the Fermilab Accumulator [6], the Tevatron beam responses contained deep notches and regions of remarkable structure. Figure 2 is an example of 6 superimposed response results from harmonic h=1100 to 1105. Using the resistive wall monitor as a longitudinal Schottky detector, and a Hewlett-Packard HP3588A dynamic signal analyzer centered at h=10, it was determined that the notches were actually in response to holes in the momentum distribution of the beam, as discussed earlier. This is a reasonable conclusion, since the structure is common to measurements at every harmonic.

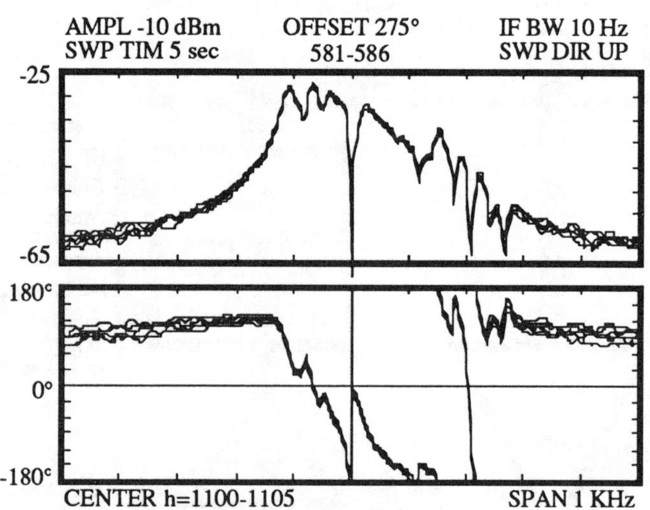

Figure 2: Superimposed results from closed loop longitudinal transfer function measurements at revolution frequency harmonics 1100 to 1105. The beam intensity was 4.3x10^{12}.

If one were to attempt to plot the inverse response from the data in figure 2 the result would be a series of apparently random curves traversing the complex impedance plane. After careful control of tunes and chromaticities, it is possible to get relatively clean inverse response curves, as shown in figure 3. In reality, since the determination of impedance is performed via calculations and not graphically, the existence of response structure is not an impediment to this method.

Analyzing the data from a set of inverse response measurements spanning the region around the RF frequency, from h=1100 to 1125, the real and imaginary parts of the longitudinal impedance can be calculated. The determination

of the absolute shift in the inverse response requires knowledge of the momentum distribution, which was not measured. As a substitute for the preliminary analysis in this paper, revolution harmonic h=1100 was used as a reference (assumed to have zero longitudinal impedance). The data in figure 4 is the result of this analysis, where the estimated error is approximately 20%. Since the water temperature feedback system for the RF cavities was off, the cavity resonant frequencies were shifted away from h=1113. Therefore the nonzero imaginary impedance at h=1113 is reasonable, as is the smaller than expected real part of the impedance (1.4 MΩ).

In the future as time becomes available a systematic program of longitudinal impedance measurements will be pursued. The DC-200 MHz region will be studied first.

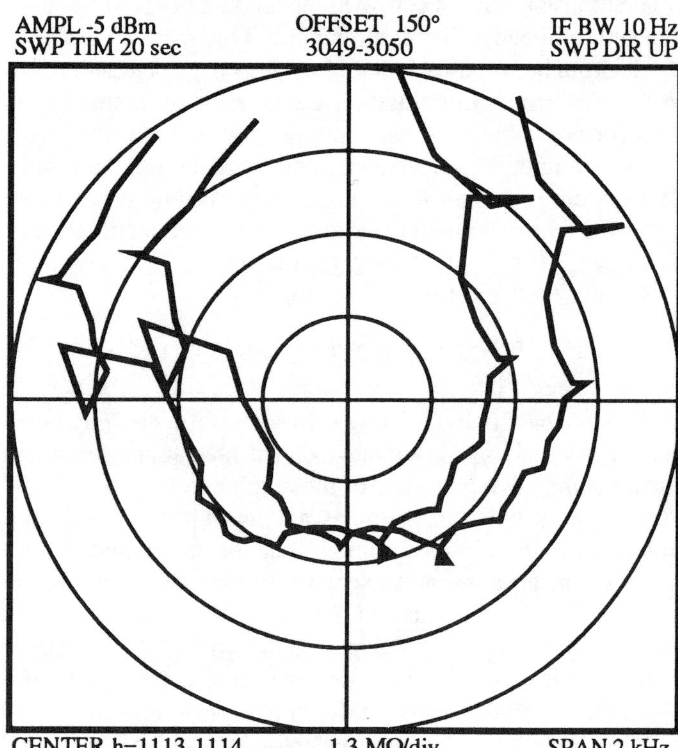

Figure 3: Inverse responses of the beam in the complex longitudinal impedance plane. The harmonics h=1113 and 1114 are plotted. The beam intensity was 5.0x10^{12}.

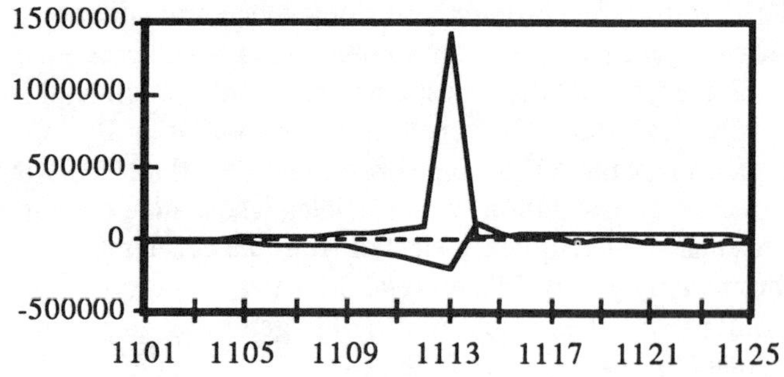

Figure 4: Measured real and imaginary part of the longitudinal impedance (Ω) as a function of revolution harmonic number in the region around the resonant frequency of the RF cavities.

III. OBSERVATION OF NONLINEAR PHENOMENA

Though the measurements of the impedance near the RF frequency show initial success, there are inconvenient and enigmatic phenomenon which have plagued these studies. In the inconvenient category is the impedance of the RF cavities. It is so high that often the coasting beam will spontaneously become self-bunching unstable. This has the effect of mixing the external modulation and the 53 MHz self-bunching modulation, therefore also sampling the impedance at their sum and difference frequencies. In addition, the momentum spread of the beam is increased until the instability disappears.

below the harmonic undergoing study were also excited. Figure 5 contains a typical example of this effect, which was captured by putting the spectrum analyzer in max hold mode. The time development of this phenomenon as a function of harmonic number is presented in figure 6. While studying this phenomenon these lines were dubbed "ghost" lines. When the speed of the network analyzer frequency sweep was drastically reduced, the effect disappeared. Figure 7 shows the resultant spectrum, which shows the residual AM modulation of the current due to mixing of the network analyzer drive and external electrical noise signals from a number of systems. In addition, below a threshold DC current (which depends on the momentum spread) the ghost lines no longer appear.

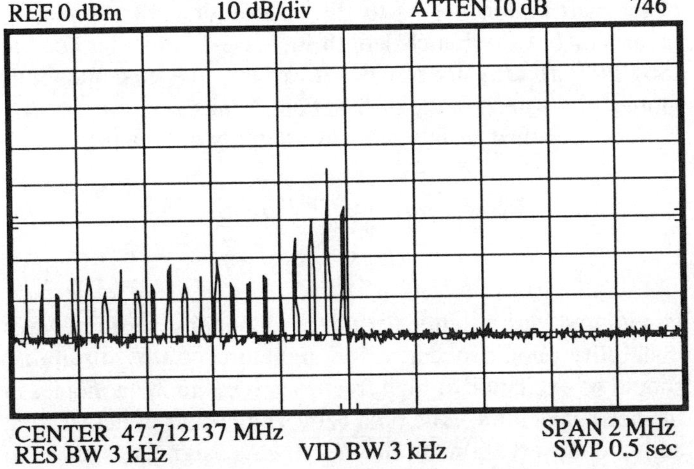

Figure 5: Observation of "ghost" lines during a transfer function measurement at h=1000 (rightmost peak). The spectrum analyzer was in max hold mode, and the beam intensity was 4.3×10^{12}.

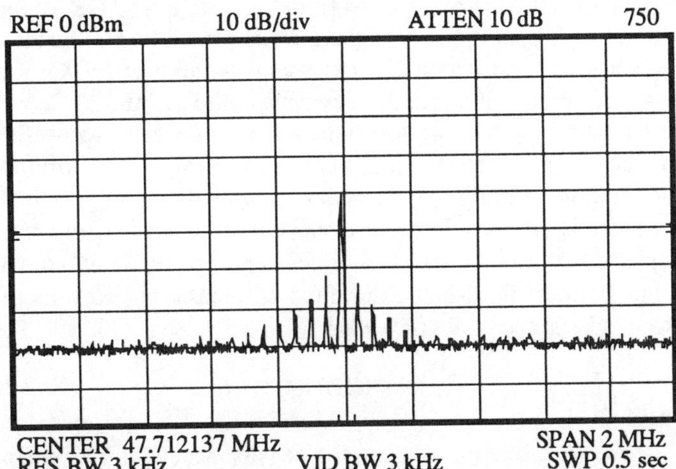

Figure 7: The exact conditions as presented in figure 5, except the transfer function sweep speed was reduced from 40 Hz/sec to 0.25 Hz/sec.

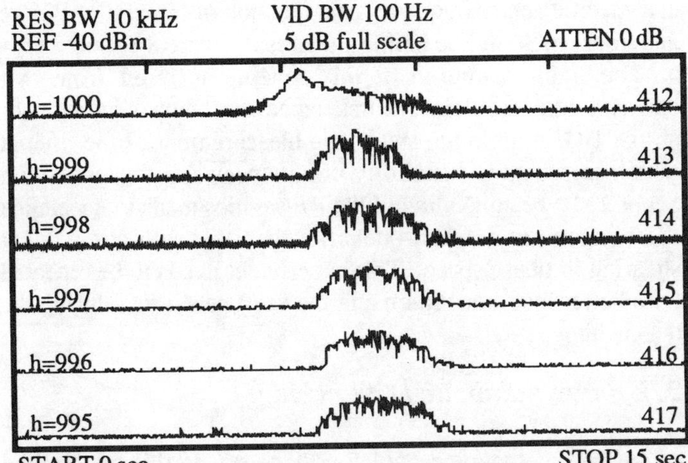

Figure 6: Power vs. time at each harmonic during a coincident longitudinal transfer function measurement at h=1000. The beam current was 10.8×10^{12}. The data was recorded measured by the HP8568B spectrum analyzer in zero span mode.

When the beam current was monitored over a broad frequency range while a transfer function measurement was in progress, it was noted that the revolution harmonic frequencies

IV. REFERENCES

1. A. Faltens, E.C. Hartig, D. Möhl, and A.M. Sessler, "An Analog Method for Measuring the Longitudinal Coupling Impedance of a Relativistic Particle Beam with its Environment", Proc. Int. Conf. on High Energy Acc., CERN (1971), p. 338.
2. A. Hoffman, "Single Beam Collective Phenomena - Longitudinal", Proc. 1st Course Int. Acc. School, Erice (1976): *Theoretical Aspects of the Behaviour of Beams in Accelerators and Storage Rings*, CERN 77-13, Geneva (1977), p. 175.
3. D.E. Johnson and D.A. Herrup, "Compensation of Time Varying Fields in the Tevatron Superconducting Magnets", Proc. IEEE Part. Acc. Conf., Chicago (1989), p. 521.
4. Q. Kerns, et al., "Energy Saver Prototype Accelerating Resonator", IEEE Trans. Nucl. Sci., Vol. NS-28, No. 3 (1981), p. 2782.
5. Q. Kerns, et al., "Fermilab Tevatron High Level RF Accelerating Systems", IEEE Trans. Nucl. Sci., Vol. NS-32, No. 5 (1985), p. 2809.
6. G. Jackson, "Review of Impedance Measurements at Fermilab", Proc. Fermilab III Instability Workshop, Fermilab (1990), p. 245.

Measurement of the Fermilab Main Ring Longitudinal Impedance

X. Lu and G. Jackson
Fermi National Accelerator Laboratory*
P.O. Box 500   MS 308
Batavia, IL 60510

Abstract

The Fermilab Main Ring provides a variety of services to the accelerator complex. High current pulses containing 1000 proton bunches are accelerated from 9 GeV/c to 150 GeV/c for fixed target operations. Single, intense proton and antiproton bunches are produced in the coalescing operation at 150 GeV/c just before injection into the Tevatron collider. At 120 GeV/c the bunch length of proton bunches are shortened using the bunch rotation process just before targeting for antiproton production. In all of these beam manipulations longitudinal stability is crucial. In order to improve present performance and predict beam behavior in future upgrades, the longitudinal impedance of the Main Ring must be measured. Results of such measurements are presented.

I. INTRODUCTION

The performance of a high intensity accelerator is often affected by impedances. In the Main Ring, the offending impedances seem to be both transverse [1] and longitudinal [2]. For example, during Tevatron Collider operations bunch coalescing [3] inefficiencies are responsible for excessively long bunches, lowering the luminosity. It has been hypothesized that the source of this inefficiency is longitudinal impedance generated by a low-Q, low frequency resonator [4].

In the past a number of methods have been applied to the Main Ring to measure either Z_L/n or R/Q. Their results are reviewed, as well as their strengths and weaknesses. The method of closed loop transfer function measurements on a coasting beam [5,6] has recently been attempted on the Main Ring. Preliminary results are presented.

II. TRADITIONAL METHODS

A. Z_L/n by Sudden Debunching

A beam with a few intense bunches is accelerated to a momentum of 150 GeV/c. The RF voltage, which is running at a high enough magnitude to create short bunches, is then turned off abruptly. Particles with different momenta have revolution periods different from the design value by the amount [7]

$$\frac{\Delta T}{T_0} = \eta \, \frac{\Delta p}{p} \qquad , \quad (1)$$

where η is the slip factor (equal to 0.00281 at 150 GeV/c). Therefore, the longitudinal phase space ellipse enclosing the beam distribution begins to tilt. The momentum spread is invariant, but the bunch length increases. In order for the longitudinal area to remain invariant, the instantaneous momentum spread of the distribution shrinks.

If the bunched beam microwave stability criteria [8]

$$Z_L/n \le (2\pi)^{3/2} \frac{|\eta| \, \sigma_p^2 \, \sigma_\tau}{N \, e \, E \, \beta^2} \qquad , \quad (2)$$

is violated during this period of debunching, microwave instability should occur. The signature of this instability should be excitation of high frequency revolution harmonics in the beam spectrum. The time between the start of debunching and the observation of the microwave signals is used to calculate the value of Z_L/n [9].

This experiment was performed in the Main Ring before and after the insertion of bellow shields. A typical measurement used a HP8568B spectrum analyzer in zero span mode to record the power in the specified resolution bandwidth at a specific center frequency as a function of time [10]. Before and after shielding the bellows the results were 8 Ω and 2 Ω.

The implementation of this scheme suffered from two experimental problems. First, since the beam signals at or above 1 GHz are quite small, the measurement of the time at which microwave instability commences is very uncertain. Second, the beam loading of the RF cavities tends to decelerate the bunches in a quasi-bucket, rather than allowing linear shearing of phase space. This experiment needs to be repeated with better instrumentation and the cavities shorted during the debunching period.

2. Z_L/n by Adiabatic Debunching

Instead of snapping the RF voltage off, in this method the RF voltage is adiabatically reduced to different minimum levels, and then adiabatically increased again [11]. In this Main Ring experiment the RF voltage started at and returned to 150 kV while coasting at 150 GeV/c. Photographic double exposures of a bunch profile as detected by a resistive wall monitor before and after the adiabatic debunching were made. If no instability took place during this operation the bunch length would remain unchanged. Figures 1 and 2 show such

---

*Operated by the Universities Research Association under contract with the U.S. Department of Energy.

double exposure photographs for two different minimum voltages. In the case of V_{min}=72 kV (figure 1), no instability was observed. On the other hand, the beam went unstable during the adiabatic debunching process in figure 2.

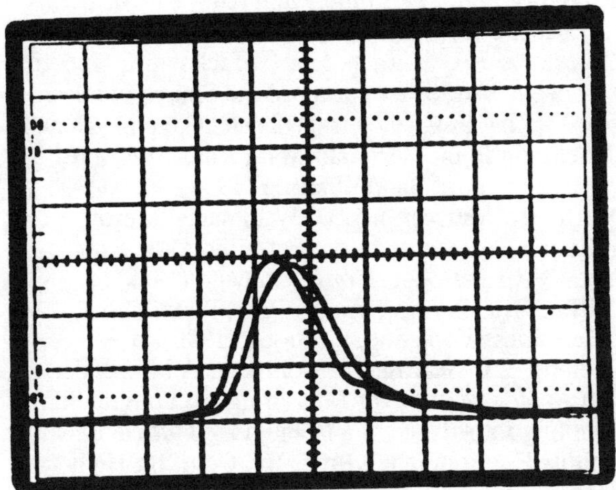

Figure 1: Bunch profile before and after reducing the RF voltage to a minimum value of 72 kV/turn. The scales are 100 mV/div and 1 nsec/div.

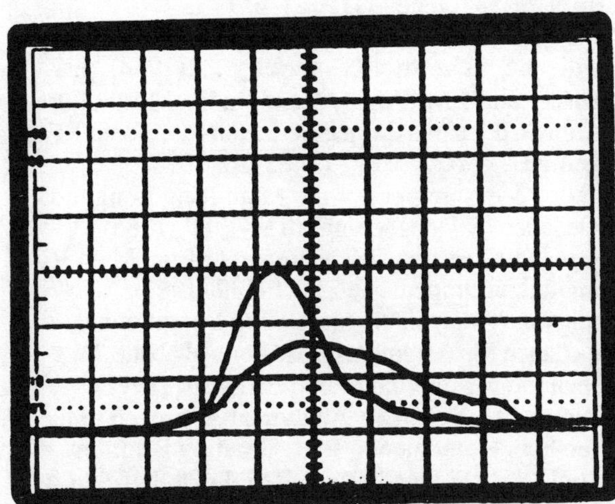

Figure 2: Bunch profile before and after dipping to a minimum RF voltage of 6 kV. Note that after the RF voltage program the bunch length has increased.

To analyze this data, one uses the bunch length at 150 kV to calculate the longitudinal emittance. So now the bunch length and momentum spread are known at all voltages (especially at V_{min}) in the absence of longitudinal emittance dilution. Applying equation (2), limits can be placed on Z_L/n depending on whether or not instability occurred. Since an unstable beam will stabilize after its momentum spread has overshot [12] the criterion in equation (2), by calculating the final longitudinal emittance and extrapolating it back to V_{min} it is possible to place another upper limit on the impedance.

The minimum RF voltages that yielded useful information were 6, 12, and 72 kV. Table 1 contains the above calculations of the longitudinal impedance criterion based on before and after bunch lengths. The conclusion is that the effective Z_L/n for long bunches in the Main Ring falls between 6 and 16 Ω.

RF Voltage (MV)	Z_L/n Before (Ω)	Z_L/n After (Ω)
6	5	36
12	6	17
72	16	16

Table 1: Longitudinal impedance criterion before and after reducing the RF voltage to the specified minimum value.

C. Observations of Resonators

In general the parameterization of the longitudinal impedance in terms of Z_L/n is inadequate due to the existence of prominent high-Q resonators scattered in frequency. These resonators tend to be single objects, like RF cavities and speciallized beam signal pickups. In order to cure an instability or calculate the behavior of the accelerator after future intensity upgrades, it is important to identify and characterize resonators.

A bunched beam method of searching for and identifying resonators was applied in the Main Ring with good success. It involved the observation of the longitudinal coupled bunch oscillation spectrum [13]. After filling the Main Ring with 1×10^{13} protons distributed into 12 equally spaced batches of 84 bunches, the beam was accelerated. A longitudinal coupled bunch oscillation will appear in the longitudinal beam spectrum as pairs of sidebands on either side of harmonics of the RF frequency. The RF frequency in the Main Ring ramps from 52.8 to 53.1 MHz as the beam is accelerated. If the resonator has a fixed frequency, the coupled bunch spectrum will shift during acceleration. By looking for the RF harmonic interval in which the coupled bunch mode frequency is constant, the frequency of the resonator is found. In the case of the Main Ring, two prominent coupled bunch modes generated by resonators at 119 and 130 MHz were observed. Detailed observation of the width of these modes yielded the resonator Q. It turned out that the resonators were higher order modes in the RF cavities [14]. In fact, the cavities were modified to damp the 128 MHz mode because of its adverse effects on the beam during resonant extraction [15].

Because the RF cavities were fitted with gap monitors, it was possible to measure the voltage of these modes. The current at these frequencies was extracted from the spectrum analyzer data. Therefore, the shunt impedance could be calculated. Since the Q was determined from the width of the coupled bunch mode, R/Q was found for each mode:

$$R/Q(119) = 0.5 \text{ k}\Omega \qquad R/Q(128) = 5.0 \text{ k}\Omega$$

In general a gap monitor is not available, so the shunt impedance may not be measurable with this method. On the other hand, it is good for identification of offending resonators.

III. Systematic Impedance Measurement

A more general method of measuring the longitudinal impedance as a function of frequency is based on the external excitation of a coasting beam. Unlike the case in the Tevatron [16] where the beam can be stored for many hours, the maximum storage time in the Main Ring is about 60 seconds. Therefore, techniques which do not require a fixed momentum distribution or long measurement times are required.

Figure 3 is an example of a closed loop longitudinal transfer function measurement in the Main Ring. The apparatus is identical to that used in the Tevatron. After injection and acceleration to the desired energy, the RF must be adiabatically reduced to zero before the voltage is turned off and the cavities shorted. If the beam is not debunched when the RF voltage is turned off, the beam loading voltage decelerates the beam into the aperture, where it is lost. Once the RF is off, the beam must be allowed to distribute itself uniformly around the aperture. This time depends on the momentum spread and the value of η at that momentum.

In the future the longitudinal impedance of the Main Ring will be systematically measured. To prepare, the debunching technique will be modified in order to minimized the extensive momentum distribution structure apparent in figure 3. Alternatively, a new algorithm for determining the impedance from transfer function measurements will be tested [17].

Finally, interest has been expressed in trying to repeat the measurements performed in the Tevatron diagnosing the origins of the ghost lines [16]. Besides having much more available time for studies, the Main Ring has the advantage of being able to store beam both above and below transition. For example, it has been hypothesized that below transition the ghost lines should propagate upward in harmonic number, instead of downward as is the case in the Tevatron.

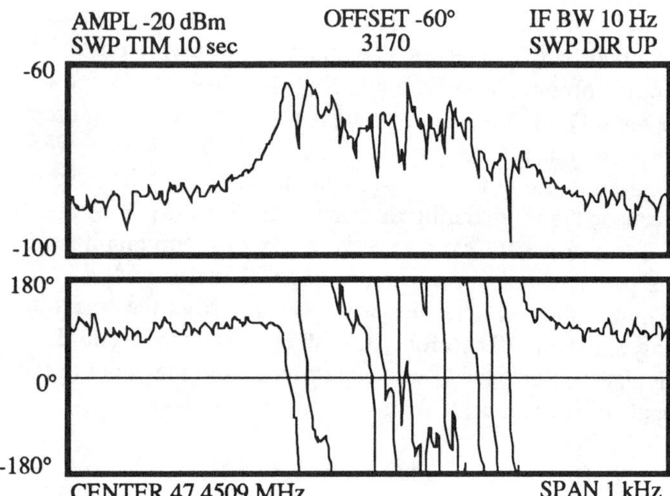

Figure 3: Example of a preliminary closed loop longitudinal transfer function in the Main Ring at 9 GeV/c. The beam intensity was 0.15×10^{12}.

IV. References

1. G. Jackson, "Measurement of the Resistive Wall Instability in the Fermilab Main Ring", Proc. IEEE Part. Acc. Conf., San Francisco (1991).
2. P. Colestock, J. Griffin, X. Lu, G. Jackson, C. Jensen, J. Lackey, "An Investigation of the Source of a Low-Q, Low Frequency Impedance Disrupting Bunch Coalescing in the Fermilab Main Ring", Proc. IEEE Part. Acc. Conf., San Francisco (1991).
3. P. Martin, K. Meisner, and D. Wildman, "Improvements in Bunch Coalescing in the Fermilab Main Ring", Proc. IEEE Part.Acc. Conf., Chicago (1989), p. 1827.
4. J. Griffin, "Bunch Coalescing", Proc. Fermilab III Instabilities Workshop, Fermilab (1990), p. 88.
5. A. Faltens, E.C. Hartig, D. Möhl, and A.M. Sessler, "An Analog Method for Measuring the Longitudinal Coupling Impedance of a Relativistic Particle Beam with its Environment", Proc. Int. Conf. on High Energy Acc., CERN (1971), p. 338.
6. A. Hoffman, "Single Beam Collective Phenomena - Longitudinal", Proc. 1st Course Int. Acc. School, Erice (1976): *Theoretical Aspects of the Behaviour of Beams in Accelerators and Storage Rings*, CERN 77-13, Geneva (1977), p. 175.
7. E. Courant and H. Snyder, "Theory of the Alternating-Gradient Synchrotron", Ann. Phys. 3 (1958), p. 1.
8. S. Krinsky and J.M. Wang, "Longitudinal Instabilities of Bunched Beams Subject to a Non-Harmonic RF Potential", Part. Acc. 17 (1985), p. 109.
9. K.Y. Ng, "Measurement of the Main Ring Longitudinal Impedance by Debunching", TM-1389 (1986).
10. J. Crisp and L. Evans, "Measurement of the Main Ring Longitudinal Impedance", EXP-139 (1987).
11. G. Jackson, "Analysis of Main Ring Longitudinal Impedance Measurements Based on Adiabatic Debunching", Internal memo EXP-174 (1991).
12. Y. Chin and K. Yokoya, "Analytical Approach to the Overshoot Phenomenon for a Coasting Beam in Particle Accelerators" Phys. Rev. D 28, No. 9 (Nov.1,1983), p. 2141.
13. G. Jackson, "Calculation of R/Q for Observed Main Ring RF Cavity Modes at 119 and 130 MHz", Internal Fermilab memo EXP-149 (1987).
14. R. Dehn, Q. Kerns, J. Griffin, "Mode Damping in NAL Main Ring Accelerating Cavities", IEEE Trans. Nucl. Sci. NS-28, No. 3 (1971).
15. Q. Kerns and H. Miller, "Fermilab 500 GeV Main Accelerator RF Cavity 128 MHz Mode Damper", IEEE Trans. Nucl. Sci. NS-24, No. 3 (1977).
16. G. Jackson, et al., "Results from Longitudinal Impedance Measurements in the Fermilab Tevatron", Proc. IEEE Part. Acc. Conf., San Francisco (1991).
17. G. Jackson, "New Methods for Extracting Longitudinal Impedance Information from Coasting Beam Transfer Function Measurements", FN-564 (1991).

Measurement of the Resistive Wall Instability in the Fermilab Main Ring

G. Jackson
Fermi National Accelerator Laboratory*
P.O. Box 500   MS 308
Batavia, IL 60510

Abstract

An instability disruptive to high current Main Ring operations is the vertical resistive wall instability. Tunes, chromaticities, and feedback loop gains must be continuously adjusted to preserve the beam intensity at injection. Results of frequency and time domain measurements aimed at understanding the sensitivity of the beam to this instability are presented. The influence of selected accelerator parameters are studied in detail. In addition, the effectiveness of various cures are reviewed.

I. INTRODUCTION

The existence of the vertical resistive wall instability in the Main Ring has been known for some time now [1]. It is an operational problem during fixed target operations where 12 batches of 81 bunches are injected from the Booster and accelerated each Main Ring acceleration cycle. The intensity per bunch is operationally around 1×10^{10}. For reasons not understood, a severe version of this instability occurs predominantly during the warm summer months.

Two feedback loops presently work to stabilize the vertical motion of the beam. The slow damper loop has a megahertz scale bandwidth, and is the most effective of the dampers. The bunch-by-bunch vertical superdamper is a much weaker loop, but has good diagnostics which are utilized in this work. The slow damper is always off for the work presented in this paper.

II. EXPERIMENTAL METHOD

The measurement of vertical betatron oscillations is accomplished by monitoring the output of the Main Ring bunch-by-bunch vertical superdamper reciever. A block diagram of the system is in figure 1. The spectrum analyzer monitors the transverse position signal, which because of the 53 MHz heterodyning superdamper reciever is only sensitive to dipole beam oscillations. Figure 2 contains a plot of the transverse betatron spectrum in the form of an AM modulation pattern around the 53 MHz RF frequency. Taken while the beam was suffering from this instability, note that the n-Q lines (Q=19.42) have 10x more oscillation power than the n+Q lines, as expected from a resistive wall instability [2].

To perform betatron oscillation damping rate or growth measurements, the spectrum analyzer was placed in zero span mode, centered on the lower n-Q line visible in figure 2. The resolution bandwidth was 3 kHz, which is narrow enough to isolate the line without restricting damping rate measurements.

Injection requires the transfer of 12 batches from the Booster, where the beginning of successive batches are separated by either 95 or 599 RF buckets (h=1113). For these studies the Main Ring remained at the injection energy.

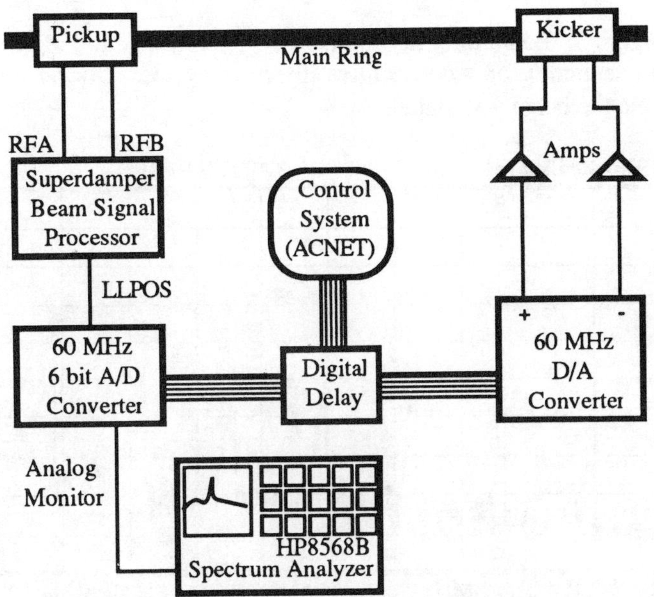

Figure 1: Schematic diagram of the Main Ring vertical bunch-by-bunch superdamper system.

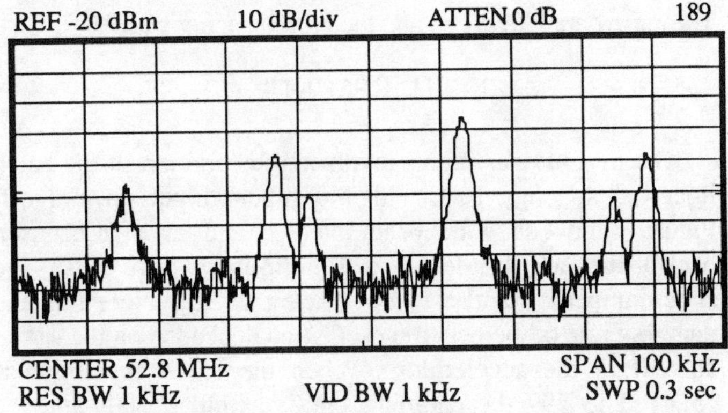

Figure 2: Measured betatron spectrum. The tallest peak corresponds to the RF frequency. The betatron sidebands to either side are visible due to the instability.

*Operated by the Universities Research Association under contract with the U.S. Department of Energy.

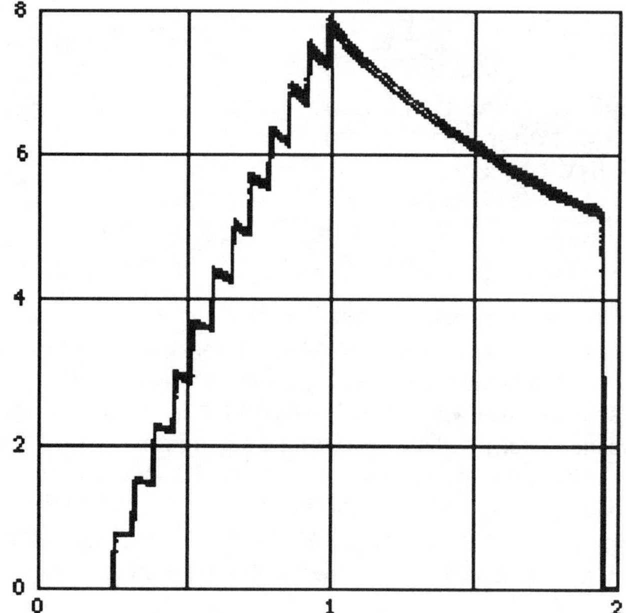

Figure 3: Beam intensity (units of 10^{12}) as a function of time (in seconds) for 4 consecutive injection cycles. During this time the beam was stable.

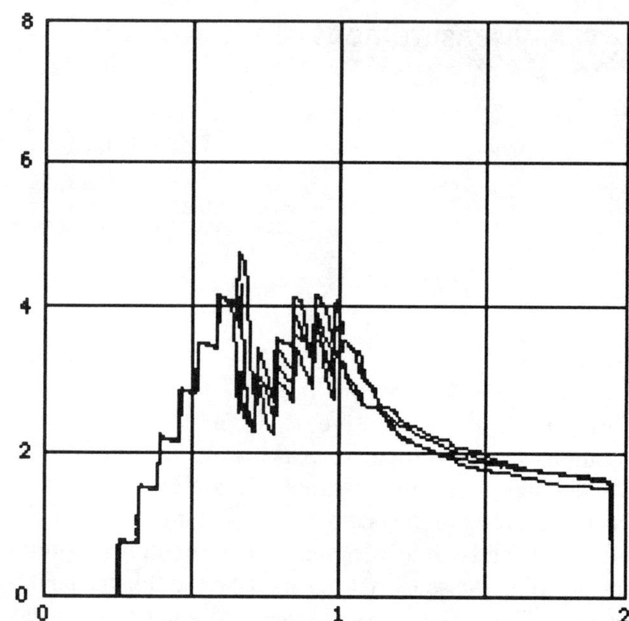

Figure 5: Beam intensity vs. time with the superdamper turned off and the batch spacing set at 599 RF buckets. Under these circumstances the beam clearly goes unstable.

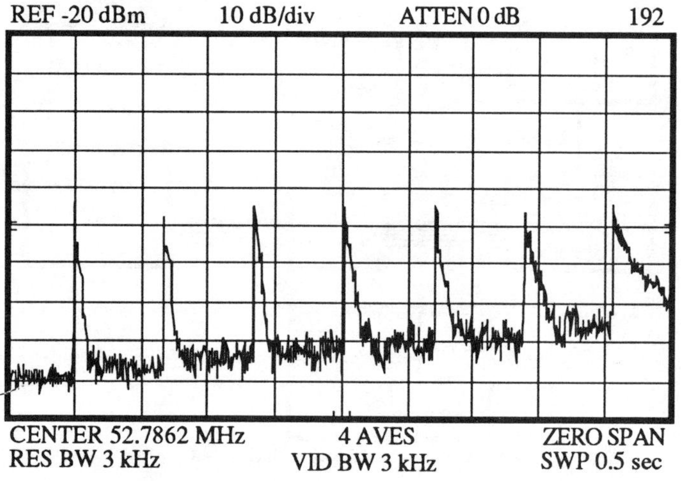

Figure 4: Betatron oscillation amplitude vs. time during the injection cycles displayed in figure 3. Batches 1-7 are shown. The superdamper was on and the batch spacing was 95.

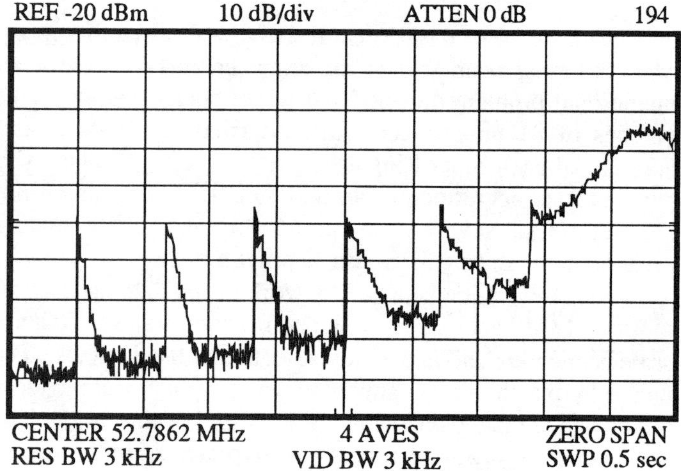

Figure 6: Injection of batches 1 through 7 during the injection cycles displayed in figure 5. Note that batch 6 was marginally unstable, and batch 7 is no longer visible.

III. RESULTS

The first measurements involved the variation of the batch injection spacing, both with the superdamper on and off. Figure 3 and 4 show the beam intensity and average betatron oscillation amplitude as a function of time with the superdamper on and a batch spacing of 95. Note that the damping rate of the betatron oscillation depended on the stored current in the accelerator. When the batch spacing was changed to 599 the damping time was cut in half and the intensity lifetime improved. With the superdamper off injections with either value of the batch spacing were unstable. Figures 5 and 6 show the case of a batch spacing of 599. For a spacing of 95 even batch 5 was unstable.

By studying a number of accelerator parameters it was found that the instability was most sensitive to vertical chromaticity $\xi=\Delta Q/(\Delta p/P)$. With the superdamper on the chromaticity was varied from -40 (stable below transition) to +20. Larger absolute values of chromaticity yielded shorter betatron damping times and intensity lifetimes. Near zero chromaticity the beam was maximally unstable. Table 1 contains a summary of the chromaticity scans.

The RF voltage also played a big role in the damping rate of the injection betatron oscillations. See Table 2 for a summary of the data. The general trend indicates that higher RF voltages cause the beam to be more betatron stable. The superdamper was off during these measurements on batch 6. The batch spacing was 599 RF buckets and $\xi=-20$.

Chromaticity ξ	Growth Rate (sec^{-1})
-40	-83
-30	-37
-20	-23
-10	-3
0	+11
10	0
20	-20

Table 1: Oscillation growth rate as a function of chromaticity at injection.

RF Voltage (MV/turn)	Growth Rate (sec^{-1})
0.5	-6.8
1.0	-10.4
1.5	-16.4

Table 2: Oscillation growth rate as a function of the RF voltage at injection.

In order to further explore the dependence of betatron stability on the condition of longitudinal phase space, the above studies were repeated with the RF voltage off. After the bunches are injected they immediately begin to shear in phase space. For an injection repetition rate of 15 Hz and a 95 bucket batch delay, only about 10% of the previous batch intensity shears into the aperture of the following injection kick. The resultant injection profile is shown in figure 7.

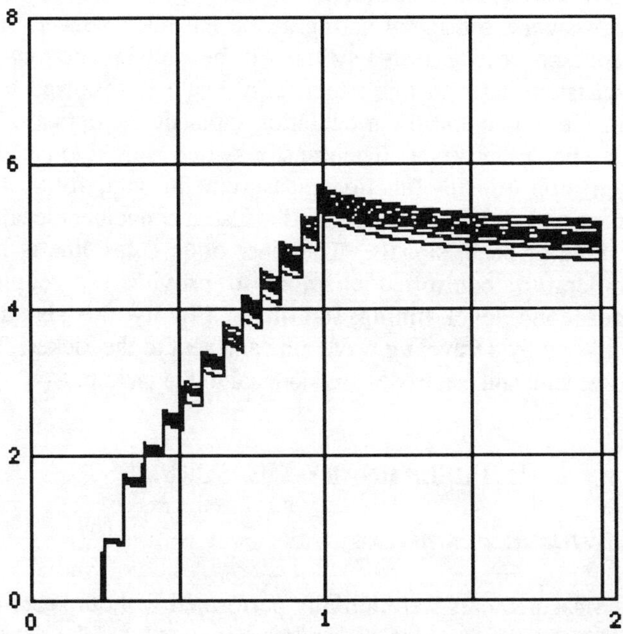

Figure 7: Beam intensity vs. time under the same conditions as the data in figure 5, except the RF is turned off. Because of phase space shearing in the time between transfers, some stored beam is kicked out each batch injection. Therefore, the peak intensity is lower than in the RF-on case.

Even though this data was taken under the same conditions as those in figure 5, note that now the beam is completely stable with a very long lifetime. Unfortunately, since the superdamper reciever depends on the existence of 53 MHz beam structure, no oscillation information was recorded.

IV. PROPOSED CURES

Under normal circumstances the vertical resisitive wall instability can be suppressed by running with large, intensity dependent chromaticities and with the dampers turned on. Unfortunately, the beam intensity lifetime is sacrificed when the chromaticity is run so large. In addition, the dynamic range of the damper recievers are quite small in power and aperture, requiring constant tuning of attenuators and the closed orbit through the beam position striplines.

It has been calculated that the wall resistance in the Main Ring may be the dominant source of transverse impedance at frequencies below 1 GHz [3]. Recently however it has been discovered that the ferrite loaded window-frame type kicker magnets may actually supply the dominant impedance at low frequencies [4,5]. If true, the best cure for this instability is to fix its source and modify these kicker magnets.

In the mean time a set of studies have been initiated to understand the open loop transfer funtion characteristics of the damper systems. For example, at the location labelled LLPOS in figure 1 the signal path was interupted and a network analyzer installed. Preliminary measurements have been successful. For instance, a peak gain of -35 dB was observed.

Finally, a narrow band n-Q feedback loop is presently under construction. In parallel, better injection oscillation correction controls are being designed. The goal is provide a high gain and sensitivity feedback loop on the dominant resisitive wall betatron mode which can operate throughout the cycle. Since it will be a narrow band analog circuit it will not suffer from the same dynamic range deficiencies of the present dampers. Testing will occur during fixed target operations.

V. REFERENCES

1. K.Y. Ng, "Transverse Coupled-Bunch Instability in the Fermilab Main Ring", Internal Fermilab memo FN-482 (1988).
2. F. Sacherer, "Transverse Bunched Beam Instabilities-Theory", Proc. 9th Int. Conf. on High Energy Acc., Stanford (1974), p.347.
3. K.Y. Ng, "An Estimate of the Longitudinal and Transverse Impedances of the Main Ring in the TeV I Project", Internal Fermilab memo TM-1388 (1986).
4. G. Jackson, "Review of Impedance Measurements at Fermilab", Proc. Fermilab III Instability Workshop, Fermilab (1990), p. 245.
5. P. Colestock, J. Griffin, X. Lu, G. Jackson, C. Jensen, J. Lackey, "An Investigation of the Source of a Low-Q, Low Frequency Impedance Disrupting Bunch Coalescing in the Fermilab Main Ring", Proc. IEEE Part. Acc. Conf., San Francisco (1991).

A Test of Bunched Beam Stochastic Cooling in the Fermilab Tevatron Collider

G. Jackson, E. Buchanan, J. Budlong, E. Harms, G. Lee, J. Marriner, D. McGinnis,
R. Pasquinelli, D. Peterson, D. Poll, D. Rohde, P. Seifrid, D. Voy
Fermi National Accelerator Laboratory*
P.O. Box 500   MS 308
Batavia, IL 60510

Abstract

In order to double the integrated luminosity of the Tevatron collider in the next running period, a 4-8 GHz bunched beam betatron stochastic cooling system has been designed. The horizontal and vertical emittances of the protons and antiprotons will be cooled to counteract the effects of power supply noise, beam-beam interaction, and intrabeam scattering. A vertical proton test system has been installed in the Tevatron and tested. In addition to measurement results, details of the hardware and cooling calculations are reviewed. Improvements coming from experience with this test configuration are being incorporated into the construction of the full system.

I. INTRODUCTION

In order to specify the requirements for the stochastic cooling system, the effect of emittance damping on luminosity lifetime [1] was explored [2]. If proton and antiproton horizontal and vertical emittance cooling was implemented with an average time constant of approximately 20 hours, the integrated luminosity of the next collider run should double.

The basic equation describing the optimum emittance cooling time is

$$\tau_\varepsilon = T_0 N_s (M + U) \quad , \qquad (1)$$

where T_0 is the revolution period, M is the mixing factor, U is the noise power to signal power ratio, and N_s is the number of protons per bandwidth sample. Calculations of the bunched and unbunched betatron spectra suggest that the mixing factor should be the same under both conditions. This hypothesis was confirmed by experiments in the Accumulator ring [3].

At an RF voltage of 800 kV/turn the rms bunch lengths and fractional momentum spreads are 2 nsec and 1.5×10^{-4} at 900 GeV. The horizontal and vertical invariant 95% emittances are approximately 20π mm-mrad. Assuming a cooling bandwidth of 4-8 GHz and the expected beam intensity and longitudinal emittance values of 7×10^{10} and 5 eV-sec, the expected average mixing factor and sample intensity are 4.5 and 1×10^9. Since the noise to signal power ratio is negligible and the revolution frequency period is 20.94 μsec, the calculated emittance cooling time is 26 hours.

*Operated by the Universities Research Association under contract with the U.S. Department of Energy.

II. PROTOTYPE SYSTEM DESIGN

A prototype vertical proton stochastic cooling system was designed and installed in the Tevatron near the F0 warm straight section. Due to technical, fiscal, and ecological concerns, it was not feasible to build a classical stochastic cooling feedback loop where the signals from the pickup to the kicker cut a chord across the ring. Instead, this system uses an optical delay line between the adjacent pickup and kicker. The physical separation of the pickup and kicker is approximately 60 m to provide a $\lambda_\beta/4$ fractional betatron tune advance for optimal cooling rate.

The pickup and kicker are composed of two moveable plates, each supporting an array of 16 coplanar loops [4]. The loops are actually etched images on a teflon circuit board. The signals from the 16 loops are combined on the reverse side of the board. The combined signals from each plate are brought outside the tank and are immediately subtracted by a 180° hybrid. After 35dB of gain from a low noise preamplifier (which is the only active circuit element in the tunnel), the difference signal is sent to the F0 RF building.

After more amplification the signal goes through a fast pin diode switch for signal gating at the RF bucket level. The absolute necessity of this switch will become apparent in the discussions later in this paper. This gate is followed by a transfer switch and the modulation input of the optical fiber delay line laser driver. The transfer switch is used to perform open loop transfer function measurements vital for system timing and gain adjustments, and is also a convenient location to monitor beam signals. The fiber optic delay line is in a temperature controlled chamber to provide the required picosecond level timing stability. Finally the signal is amplified by a traveling wave tube and sent to the kicker. The kicker tank and electrodes are identical to the pickup.

III. PRELIMINARY MEASUREMENTS

A. Without Beam

Measurements were initially performed without beam [5]. Spectral harmonics of the spikes generated by the class C power amplifiers of the Tevatron RF cavities [6] were observed to be transmitted down the beam pipe and received by the pickup, which was approximately 10 m away from the nearest cavity. By changing the grid bias to widen the current pulses, it will be possible to eliminate these unwanted signals.

Since the beam pipe acts as a microwave waveguide in the 4-8 GHz frequency band the pickup and kicker are able to close the feedback loop without beam. This caused an instability since the loop gain was greater than unity. The solution to this problem, whose severity depended on the aperture between the opposing arrays, was to install the bucket gating switch. Allowing only the signals from the 6 out of 1113 RF buckets to be amplified, the delay between the kicker and the pickup made positive feedback amplification impossible.

The bucket gating switch also had a beneficial effect on passive spectra and transfer function measurements. By gating the signal after the initial stages of amplification, only a fraction of the thermal noise power makes it to the input receivers of the spectrum and network analyzers. Therefore the signal to noise of those measurements improved by the ratio of the revolution period to six times the gate width, approximately a factor of 100.

B. Passive Spectrum

An example of a passive spectrum measurement is shown in figure 1. There are four distinct features to the spectrum. First and foremost, the vertical Schottky betatron lines are clearly visible as broad distributions rising 7dB above the noise floor. Since the vertical betatron tune was approximately 0.41, the upper and lower sidebands are close together. The observation of these Schottky signals is a very encouraging sign for the success of this project.

At the frequencies corresponding to three revolution harmonics, the very strong coherent beam signals almost wash out the longitudinal Schottky bands, which rise above the noise floor by about 10dB. These coherent signals can be 50dB higher than the betatron Schottky signals, causing a serious dynamic range problem in the amplifier chain.

C. Transfer Functions

A number of transfer function measurements have been performed to time the system in preparation for cooling studies. Though the cooling studies did not occur, much was learned about the response of the system and the beam. Figure 2 contains an example of such a transfer function measurement, where four full revolution harmonic bands are visible. The betatron lines stand out quite clearly, and the phase responds as expected.

The feature which distinguishes a bunched beam response function from a coasting beam transfer function measurement is synchrotron sideband structure. It usually appears as a region of noise in the amplitude and phase near the peak of a betatron resonance. Figure 3 contains a transfer function measurement of a single pair of betatron lines which exhibits this structure. The synchrotron sidebands of the betatron tune appear in this region since the lower order sidebands are relatively narrow and can be resolved independently from their neighbors. Note that these betatron peaks are readily visible at 4 GHz, whereas at 8 GHz they form an almost uniform signal floor with the help of the broader longitudinal Schottky bands.

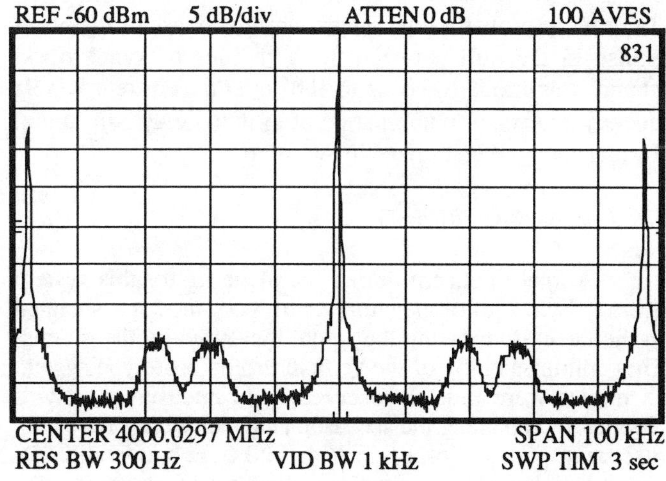

Figure 1: Typical Tevatron vertical bunched beam spectrum as measured by the vertical proton pickup. The intensity was 3.3×10^{10} per bunch and the beam energy was 900 GeV.

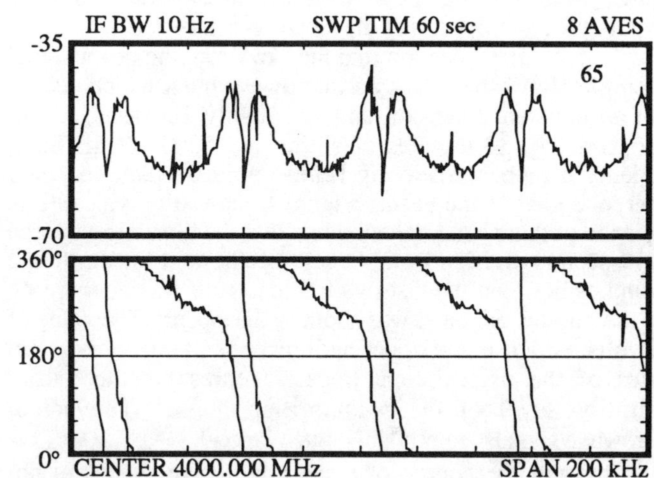

Figure 2: Multiband transfer function measurement of the cooling loop with beam. The bunch intensity was 3.1×10^{10}.

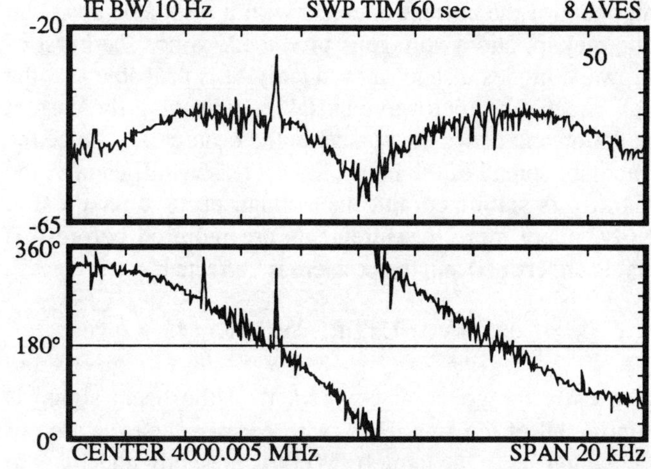

Figure 3: Closeup of a pair of betatron lines. The sharp spike in the left betatron line was caused by a head tail instability when the chromaticity of the Tevatron drifted negative.

In order to properly time the system, the phase of the beam response in the null between these regions of synchrotron sideband structure must be set to 180° over the entire 4-8 GHz frequency octave. A combination of system delay adjustment and equalizers will be utilized in the future.

D. Microwave Instability

A major concern during the planning for this system was the existence of propagating microwave modes associated with the beam dominating the signal measured by the pickup. During commissioning of the vertical proton prototype system such modes were indeed observed. The measurement was performed by replacing the spectrum analyzer with a Tektronix 11802 sampling oscilloscope. With a 20 GHz bandwidth SD-24 sampling head, this scope was capable of observing the beam signal from the pickup in the time domain. One of the more convenient data acquisition modes for monitoring the beam signal was the envelope function, which stored the minimum and maximum voltage in each sampling bin across the screen. Figure 4 contains 3 such traces. The bucket gating system was disabled during these measurements, and the travelling wave tube driving the kicker was turned off.

Just after injection into the Tevatron, the beam signal is completely washed out by a microwave burst which decays with an apparent time constant of roughly 100 nsec. After approximately 30 minutes of sitting at 150 GeV this burst suddenly disappears, leaving behind only the Schottky and coherent signal of the beam. Figure 4 shows the typical time sequence of this phenomenon. Note that the bunch is centered on the second division from the left (see bottom trace). In the 4 minutes between the last two frames, there was no warning that the microwave burst was about to disappear. The value of 0.3 volts, which is the upper and lower limits of the vertical scales of the oscilloscope traces, represents the output saturation level of the preamplifier chain. The actual microwave signals are probably much larger.

At present the source of these microwave modes is not understood. The hypothesis at the moment is that the coalescing process generates bunches with a large amount of high frequency phase space structure. This high frequency component of the current interacts with a resonator upstream of the pickup, and a voltage is produced. Since the burst of microwave modes disappears suddenly, it is probable that the bunch is actually microwave unstable, enhancing the current modulation and hence the voltage in the beam pipe. Once the momentum spread of the bunch has increased sufficiently, the instability is stabilized and the voltage drops dramatically. Clearly, many more measurements are required before this signal is understood and the problem is corrected.

IV. FUTURE WORK

Presently the coherent component of the beam signal is saturating all of the amplifiers (with perhaps the exception of the preamplifier in the tunnel). Work is presently underway to replace the 180° hybrid to improve common mode rejection, and studies are planned to understand the cause of these anomalously large coherent signals associated with such a long bunch. A comb notch filter is being designed to suppress the coherent revolution harmonic power, but it may have serious dynamic range problems itself.

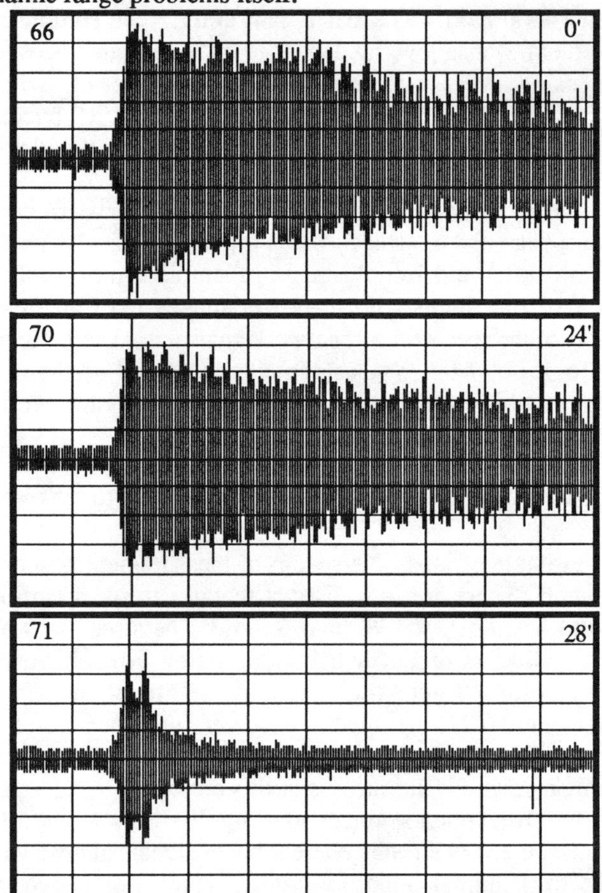

Figure 4: Sampling oscilloscope traces of the pickup signal voltage as a function of time. The scales are ±0.3 volts full scale vertically and 20 nsec/div horizontally. The numbers in the upper right corners is the time since injection.

V. REFERENCES

1. G. Jackson, "Winter 1988 Luminosity Lifetime Studies Analysis Results", internal Fermilab memo EXP-154 (1988).
2. G. Jackson, "Bunched Beam Stochastic Cooling", Proc. IEEE Part. Acc. Conf., San Francisco (1991).
3. J. Marriner, G. Jackson, D. McGinnis, R. Pasquinelli, D. Peterson, J. Petter, "Bunched Beam Cooling in the FNAL Antiproton Accumulator", Proc. Eur. Part. Acc. Conf., Nice (1990), p. 1577.
4. D. McGinnis, J. Budlong, G. Jackson, J. Marriner, and D. Poll, "Design of 4-8 GHz Bunched Beam Stochastic Cooling Arrays for the Fermilab Tevatron" Proc. IEEE Part. Acc. Conf., San Francisco (1991).
5. G. Jackson, J. Marriner, D. McGinnis, R. Pasquinelli, and D. Peterson, "Bunched Beam Schottky Signal Measurements for the Tevatron Stochastic Cooling System", Proc. Adv. Instr. Workshop, KEK (1991).
6. Q. Kerns, et al., "Energy Saver Prototype Accelerating Resonator", IEEE Trans. Nucl. Sci., Vol. NS-28, No. 3 (1981), p. 2782.

The Self-Cooling of Charged Particle Beams in a Straight Line

Ya. S. Derbenev

Randall Laboratory of Physics, University of Michigan

Ann Arbor, Michigan 48109-1120 USA

ABSTRACT

A method of decreasing the beam emittance is considered using intrabeam Coulomb scattering of particles during the processes of beam formation and acceleration. The necessary conditions for efficient heat transfer from transverse degrees of freedom into longitudinal ones are formulated, and the principal limitations on the rate of cooling and achievable temperature are established. An example of the self-cooling is given for the case of a beam formed and accelerated in the space charge regime and in the presence of external accompanying magnetic field.

INTRODUCTION

Cooling of charged particle beams is one of the important topics of modern accelerator physics. Four methods are known and are either used today or have good prospects to be applied for the cooling of circulating beams:

1. radiation damping of ultrarelativistic particles in synchrotrons;

2. electron cooling which is based on the use of a co-moving electron beam, straight or circulating, as in a thermostat;

3. stochastic cooling based on the use of RF feedback systems;

4. laser cooling of ion beams.

Among these methods, only the electron cooling is capable of cooling a straight low-energetic proton or ion beam within an acceptable length of cooling section; it is effective due to a very low value of the longitudinal electron temperature ($T_{\parallel}^e \sim 10^{-4}$ eV) and the transverse motion of electrons being bounded because of the magnetic field. However, in view of the longitudinal heating of electrons by the ions of the beam under cooling, the ratio between the electron's and the ion's current densities must be not less than $T_{\perp}^i/T_{\parallel}^e$. Therefore, the use of electron cooling for the ion beams with current density above ~ 1 mA/cm^2 seems to be problematic.

In this report, we focus on the possibility of using intrabeam scattering of particles for the transverse cooling of a beam, with corresponding heating in the longitudinal direction; in this process, there are no external heat energy transfers from the beam. In such a situation, the total beam entropy is not decreased, but there is redistribution between the degrees of freedom of the beam. The decreasing of the transverse entropy then occurs under the condition $T_{\perp} > T_{\parallel}$; this condition can be maintained for a long time by longitudinally stretching the beam during acceleration and by transversely compressing the beam. The increase of the total entropy is small when these processes are performed slow with respect to the process of temperature relaxation. Note that the adiabatic process should start from a state of $T_{\perp} = T_{\parallel}$, in order to avoid a substantial increase of the entropy during the initial stage of the transverse cooling. Also, if we cannot exclude some intermediate states of the beam when $T_{\perp}$ is not close to $T_{\parallel}$, these states should be passed as fast as possible in order to minimize the resulting transverse phase space volume of the beam.

SELF COOLING IN THE ADIABATIC LIMIT

We describe evolution of the beam by the variables

$$\Gamma_{\perp} = T_{\perp} \cdot \pi\,a^2, \quad \Gamma_{\parallel} = \sqrt{T_{\parallel}} \cdot \frac{\gamma v e}{I},$$

where

$$T_{\perp} = \frac{<(\Delta\vec{p}_{\perp})^2>}{2m}; \qquad T_{\parallel} = \frac{<(\Delta p_{\parallel})^2>}{\gamma^2 m},$$

a is the beam radius, v is the average longitudinal velocity, $\gamma = 1/(1 - \frac{v^2}{c^2})^{1/2}$, e is the particle charge, and I is the beam current. The variables $\Gamma_{\perp}$ and $\Gamma_{\parallel}$ are the dynamical invariants of the beam as an ensemble of particles; they can be considered as the transverse and longitudinal phase space volume of an element of beam length, respectively. The product $\Gamma = \Gamma_{\perp} \cdot \Gamma_{\parallel}$ is the corresponding 6-dimensional phase space volume of the element of beam length. Near the equilibrium state $T_{\perp} = T_{\parallel} = T$ we have

$$\Gamma_\perp = \pi a^2 T, \quad \Gamma_\parallel = \frac{\gamma v}{I} e \sqrt{T}, \quad \Gamma = \gamma \frac{T^{3/2}}{n},$$

where n is the particle concentration. During the adiabatic process, the parameters v, a, and I may change, but the volume Γ remains unchanged and we get the relation between values of $\Gamma_\perp$ at the end to that at the beginning of the adiabatic process as:

$$\frac{\Gamma_\perp}{\Gamma_{\perp o}} = \left[\left(\frac{Ia}{\gamma v}\right)/\left(\frac{Ia}{\gamma v}\right)_0\right]^{2/3} \tag{1}$$

The maximum cooling effect occurs when all the process of beam formation and acceleration is performed adiabatically. In this case, the value v_0 is related to the cathode temperature, T_c, ($T_c = mv_0^2$) and a_0 is the beam size at the cathode, a_c. Table 1 gives an illustration of the maximum cooling effect for a heavy particle beam assuming no beam bunching after acceleration.

Table 1	
Maximum Cooling Effect	
Initial beam parameters	
Cathode temperature, eV	0.1
Beam radius at the cathode, cm	0.5
Parameters after acceleration	
Top energy after acceleration, MeV	100
Beam radius after acceleration, cm	0.02
Self-cooling effect	
Decreasing of beam emittance, times	100
Decreasing of transverse phase space volume, times	10^4

The adiabatic relation (Equation 1) and the conservation of Γ are valid under the condition

$$\lambda_d^{-1} \equiv 2 \left(\frac{\gamma v}{Ia}\right)' \cdot \frac{Ia}{\gamma v} << \lambda_{st}^{-1}. \tag{2}$$

where $'$ denotes change with respect to the longitudinal direction and λ_{st}^{-1} is the collision relaxation parameter defined from

$$(T_\perp - T_\parallel)'_{st} = -\lambda_{st}^{-1}(T_\perp - T_\parallel). \tag{3}$$

Near the equilibrium state $T_\perp = T_\parallel$, the parameter λ_{st}^{-1} is equal to

$$\lambda_{st}^{-1} = \lambda_{eq}^{-1} = g \cdot \frac{8\sqrt{\pi} n e^4 L}{5\sqrt{m} T^{3/2} \cdot \gamma^2 v} = g\frac{8\sqrt{\pi} L e^4}{5\sqrt{m}\gamma v} \cdot \frac{1}{\Gamma}, \tag{4}$$

where g is a numerical factor with an order of value of about 1, and L is the Coulomb parameter

$$L = \frac{1}{2}\ln\left(\frac{\gamma T^3}{4\pi n e^6}\right), \tag{5}$$

with an order of value of about 3–5.

We can see from formula (4) that it is very important to have as small a value of Γ as possible in order to get a maximum rate of temperature relaxation and satisfy the adiabatic condition (2).

SELF-COOLING OF AN ACTUAL BEAM

When accelerating an actual beam, the adiabatic condition cannot be satisfied in the region near the cathode, because the characteristic time of acceleration there is about equal to the inverse plasma parameter:

$$\omega_p^{-1} = \sqrt{m/4\pi n e^2},$$

which is small in comparison to the temperature relaxation time λ_{st}/v. With the acceleration, the longitudinal temperature goes down very fast, and one must take into account intra-beam scattering which can limit its decrease, i.e. the longitudinal entropy will increase with collisions between particles (see the Appendix). After a distance of about a_c from the cathode, we can equalize the transverse and longitudinal temperatures by having the transverse expansion of the beam and, if necessary, by deacceleration of the beam. In this state, we obtain an intermediate energy W_0 such that $T_c << W_0 << W_{max}$, with initial (maximum) radius a_0 and initial value of relaxation parameter λ_{st}^{-1}. With these parameters, we can start the adiabatic cooling process (1). In view of the presence of the non-adiabatic stage at the beginning of the beam evolution, the self-cooling effect will be less than potentially possible as was presented in Table 1 (see Table 2).

In addition, we should note the following conditions for the beam dynamics in focusing and accelerating:

1. A solenoidal magnetic field can be used in order to keep an intensive low energy beam from repulsion by the space charge effect.

2. The current distribution at the cathode and the accelerating electric field should have axial symmetry.

3. Electric and magnetic fields have to be matched in the region of the beam injection into the solenoid, in order to avoid radial beam excitation inside the solenoid, i.e., to reach the Brilluen's beam state.

<table>
<tr><th colspan="2">Table 2</th></tr>
<tr><td colspan="2">Self-cooling of an Actual Beam</td></tr>
<tr><td>Beam current I, A</td><td>1</td></tr>
<tr><td>Beam radius at cathode a_c, cm</td><td>0.5</td></tr>
<tr><td>Cathode temperature T_c, eV</td><td>0.1</td></tr>
<tr><td>Anode voltage V_A, kV</td><td>10</td></tr>
<tr><td>Longitudinal length of expansion section cm</td><td>6</td></tr>
<tr><td>Beam radius after expansion a_0, cm</td><td>3</td></tr>
<tr><td>Initial energy of the adiabatic process W_0, keV</td><td>10</td></tr>
<tr><td>Initial relaxation length λ_0, m</td><td>0.3</td></tr>
<tr><td>Final energy W_f, MeV</td><td>100</td></tr>
<tr><td>Maximum value of solenoidal field B_f, Tesla</td><td>10</td></tr>
<tr><td>Final beam radius in the solenoid a_f, cm</td><td>0.02</td></tr>
<tr><td>Final relaxation length λ_f, m</td><td>15</td></tr>
<tr><td>Cooling effect on beam emittance, times</td><td>25</td></tr>
<tr><td>Cooling effect on beam brightness, times</td><td>600</td></tr>
</table>

Appendix

To describe the evolution of a beam, we use the equations as follows:

$$\Gamma'_\perp = \pi a^2 (T'_\perp)_{st}, \quad \Gamma'_\parallel = \gamma v e (T'_\parallel)_{st}/2I\sqrt{T_\parallel} \quad (A1)$$

where $(T'_\perp)_{st}$, $(T'_\parallel)_{st}$ are the rates of temperature change under intrabeam collisions. With the definition (3) and taking into account heat energy conservation for collisions, we obtain:

$$(T'_\parallel)_{st} = -2(T'_\perp)_{st} = 2\lambda_{st}^{-1}(T_\perp - T_\parallel)/3 \quad (A2)$$

From $(A1)$ and $(A2)$, we get immediately:

$$\Gamma' = \Gamma'_\perp \Gamma_\parallel + \Gamma_\perp \Gamma'_\parallel = \lambda_{st}^{-1} \cdot (T_\perp - T_\parallel)^2 \Gamma/3T_\perp T_\parallel \quad (A3)$$

It is not difficult to calculate λ_{st}^{-1} in the cases of $T_\perp >>$ $T_\parallel$ and $T_\perp << T_\parallel$, using the Landau integral [1]:

$$\lambda_{st}^{-1} = \frac{2Ie^3}{\sqrt{m}v^2 a^2} \cdot \left(\begin{array}{c} \dfrac{\sqrt{\pi}}{T_\perp^{3/2}} L(T_\perp), \\ \dfrac{L^2(T_\parallel) - L^2(T_\perp)}{\sqrt{\pi}T_\parallel^{3/2}} \end{array} \right), \quad \begin{array}{l} \text{at } T_\perp >> T_\parallel \\ \\ \text{at } T_\perp << T_\parallel \end{array}$$

where $L(T_\parallel)$, $L(T_\perp)$ are functions like (5) with effective temperatures $T_\parallel$ and $T_\perp$, respectively. In order to calculate the numerical factor in (4) in the case when $T_\perp \approx T_\parallel$, we use the model with Gaussian distribution in temperatures $T_\perp$ and $T_\parallel$; then we get $g = 1$.

We can calculate the increase of Γ in the quasiadiabatic case. Taking into account that $\lambda_d >> \lambda_{st}$, we find from the equations $(A1)$ and $(A2)$ that

$$\Delta T \equiv T_\perp - T_\parallel \approx T\lambda_{eq}/\lambda_d;$$

then, after substituting ΔT in (A3), we get:

$$\Gamma' \approx (\lambda_{eq}/3\lambda_d^2)\Gamma$$

Solving this equation, one can establish the boundaries of stability of the quasiadiabatic process and calculate non-adiabatic effects.

The relaxation equations $(A1)$ and $(A2)$ can be used also to calculate collision effects when one of two temperatures, $T_\parallel$ or $T_\perp$, becomes small in relation to the other due to fast non-adiabatic space expansion of the beam. We consider two cases as follows:

1. $T_\parallel << T_\perp$; this case corresponds to non-adiabatic acceleration of a beam from the cathode. Then, we have from equations $(A1)$:

$$(v^2 T_\parallel)' = 2\sqrt{\frac{\pi}{T_\perp m}} \frac{Ie^3 L(T_\perp)}{a^2}.$$

We can integrate this equation from the cathode surface to the state $T_\parallel = T_\perp = T_0$ to logarithmic accuracy taking into account, that $T_\perp a^2 = $ const.

2. $T_\perp << T_\parallel$; we get this situation when the beam is expanded fast transversely after the adiabatic acceleration to maximum energy. Then $\Gamma_\perp$ increases with collisions as:

$$\Gamma'_\perp = 2\sqrt{\pi}Ie^3[L^2(T_f) - L^2(T_\perp)]/v_f^2\sqrt{T_f m},$$

with $T_\perp = \Gamma_\perp/\pi a^2$.

References

[1] L. D. Landau, Zh. Eskp. Teor. Fiz. **7**, 203 (1937).

Laser Cooling of Stored Beams in ASTRID

J.S. Hangst[1], K. Berg-Sørensen, P.S. Jessen, M. Kristensen, K. Mølmer, J.S. Nielsen, O. Poulsen, J.P. Schiffer[2] and P. Shi
Institute of Physics, University of Aarhus
DK-8000 Aarhus C, Denmark

Abstract

We report the results of laser cooling experiments on 100 keV Li+ beams in the storage ring ASTRID. The metastable fraction of the lithium beam has been laser cooled to a momentum spread dp/p~ 10^{-6}, corresponding to a rest frame temperature $T_\parallel = 1$ mK. Laser diagnostic methods have been employed to study the dynamics of intrabeam relaxation. A theoretical model of laser cooling has been used to interpret the experimental results. We also discuss Molecular Dynamics simulations of intrabeam interactions and the connection with crystalline beams.

I. INTRODUCTION

Laser cooling [1] is the result of the velocity-selective transfer of photon momentum from a laser beam to a moving atom or ion. In the most basic laser cooling scheme, particles having a closed transition between two energy levels are utilized. Those particles which are in resonance with a laser beam absorb photons. Each photon transfers momentum of magnitude hv/c to the absorbing particle, which recoils in the direction of the laser propagation. The photons are spontaneously reemitted. The average momentum transfer from the reemission vanishes due to symmetry. The net force is thus directed along the laser beam. The force is velocity selective because of the narrow width of the atomic transition, and velocity dependent through the Doppler effect. By tuning the frequencies of co- and counterpropagating lasers to accelerate slow particles and decelerate fast ones, one achieves cooling. The cooling limit, called the Doppler cooling limit, is given by $k_B T \sim h\Gamma/2$, where Γ is the homogeneous width of the cooling transition. Other schemes have been developed to cool particles to temperatures below the Doppler limit, but these will not concern us here. (See Reference [2] for an overview of lasercooling theory and experiment.)

The use of lasers for cooling ions or atoms stored in traps is well-documented [2]. More recently, it has become possible to cool energetic beams of ions circulating in storage rings. Unlike the trap experiments, which use up to six laser beams to cool all three spatial dimensions, storage ring laser cooling operates only on the longitudinal degree of freedom. In the first reported experiment, a beam of 7Li+ ions with kinetic energy of 13 MeV was laser cooled to a longitudinal temperature of less than 3K [3].

In addition to offering the potential for cooling to very low temperatures, the laser is a unique diagnostic tool. By measuring the laser-induced fluorescence (LIF) of a beam as a function of laser frequency, one can directly and nondestructively measure the velocity distribution of the circulating ions. Unlike electronic means to measure charge density fluctuations in the circulating beam (Schottky signals), the laser is insensitive to collective behaviour in the beam, which can render electronic signals very difficult to interpret.

In the current set of experiments, we have used the storage ring ASTRID to investigate laser cooling and diagnostics of 100 keV 7Li+ ions. We also discuss briefly recent experiments using 166Er+. Finally, we comment on the implications our experiments contain for the effort to achieve ordered or "crystalline" beams [4].

II. EXPERIMENTAL APPARATUS

The storage ring ASTRID and its associated optical experimental apparatus are shown schematically in Figure 1. ASTRID has been designed to function both as an ion storage ring and as a source of

synchrotron radiation [5]. For the current studies the machine has been operated as a fixed-energy storage ring for 100 KeV ions. The ring has four-fold symmetry, with four quadrupoles in each straight section. Bending is accomplished by 4 pairs of dipole magnets. Windows at the ends of two straight sections allow laser light to pass into the vacuum chamber. The ring circumference is 40 meters.

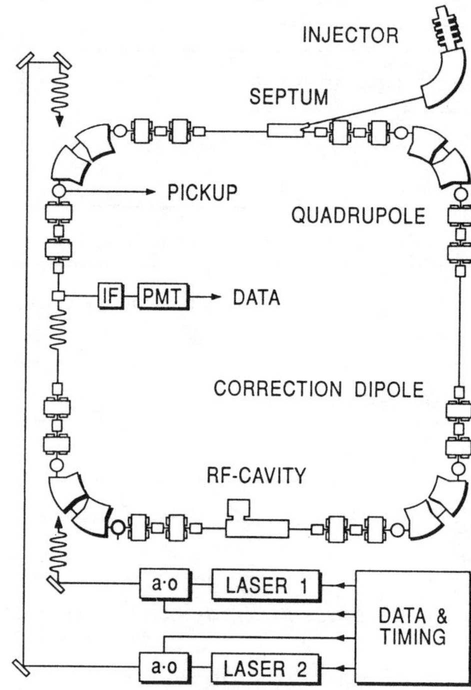

Figure 1. Schematic of the ASTRID storage ring and associated laser apparatus.

A variety of ion sources can be used with the electrostatic accelerator, which has a maximum energy of 200 keV. Ions travel from the accelerator through a mass-separating magnet, and are injected into the ring by a magnetic septum and a fast electrostatic kicker. Sixteen beam position monitors and sixteen correction dipole magnets provide orbit measurement and control. Additional beam diagnostics are provided by a longitudinal Schottky pickup. A radiofrequency cavity is available for bunching and accelerating the ions.

Laser light is provided by two ring-dye lasers, which can be intensity modulated using acousto-optic modulators. Fluorescent light from the ions is monitored through an additional window which looks transversely at the beam. A telescope collects light and directs it to a photomultiplier tube (PMT). An interference filter (IF) of bandwidth 0.6 nm (FWHM) discriminates between fluorescent light (which is Doppler-shifted relative to the direct laser light) and scattered light. The lithium experiments employed lasers and diagnostics in one straight section of the ring. In the erbium case, two straight sections have been used. A spectrometer replaces the interference filter in the second straight section.

III. EXPERIMENTAL PARAMETERS

The 100 kev lithium ion beam utilized is actually a two-component beam. A fraction of the beam (~10^{-4}) is produced in the meta-

---

[1]also FNAL, ANL, and Univ. of Chicago
[2]also ANL and Univ. of Chicago

1764

stable $1s2s^3S_1$ state; the remainder exists in the ground state. The metastable state is connected to the $1s2p^3P_2$ state by allowed optical transitions, and is subject to laser cooling by virtue of a closed hyperfine transition ($F''=5/2$ - $F'=7/2$). The rest frame wavelength of the transition is about 548.6 nm. The natural linewidth of the transition is $\Gamma=3.7$ MHz, corresponding to a lifetime of 43 ns. This implies a Doppler cooling limit of about 89 uK. The branching ratio to the ground state is less than 10^{-5}.

After acceleration, the beam has a flattened velocity distribution. The initial longitudinal energy spread is ~1 eV, coresponding to T~100mK or $\delta p/p \sim 10^{-5}$. The initial transverse temperature is about 1000K. Typical beam currents are ~10uA, corresponding to roughly 10^9 particles injected. The average beam radius, measured by scanning a slit across the LIF image of the beam, was $r_{rms} \sim$ 3mm. This yields a particle density of about $10^6/cm^3$. The average pressure in the ring during the experiments was 3×10^{-11} torr.

The erbium ion beam offers the potential for cooling the entire beam population. The cooling transition, of wavelength 541.6 nm, is between the ground state (J=13/2) and an excited state (J=15/2) at 18463 cm^{-1}. The lifetime of the excited state has not been accurately measured, but is of order 1 μs. The corresponding Doppler cooling limit is about 4 μK. Beam currents achieved from the ion source are typically a few microamps. The initial temperatures and densities of the beam from the accelerator are roughly the same as those of the lithium beam.

IV. LASER DIAGNOSTICS -BEAM RELAXATION

The laser has been used as a probe to monitor the longitudinal velocity distribution of the circulating beam. Figure 2 shows the time development of the longitudinal temperature of the metastable lithium ions. In 30 μs the beam heats from ~100 mK to ~2 K. At this point the velocity distribution is measured to be thermal with a FWHM of 100 m/s. The temperature continues to increase out to ~200 ms after injection, which was the latest time measurable with our experimental setup.

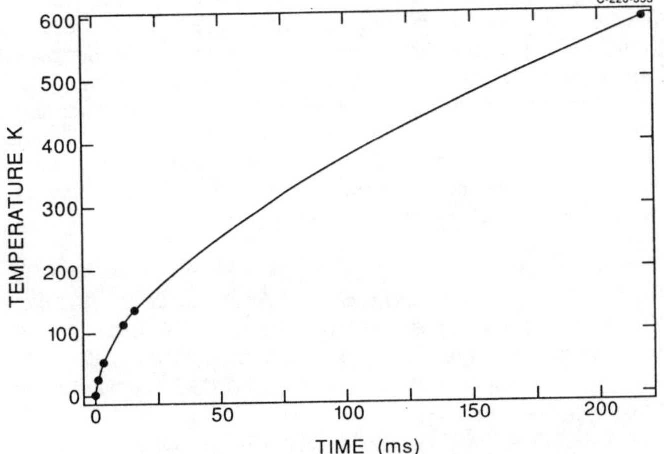

Figure 2. Longitudinal temperature vs. time after injection for $^7Li^+$ beam.

This observed rapid longitudinal heating may be due to intrabeam Coulomb interactions. Such interactions can transfer momentum from the transverse degrees of freedom to the longitudinal one. The high transverse kinetic energy of the beam particles is a likely source of heating for a beam of the relatively high charge density found in ASTRID. To estimate the importance of this process, we have employed a binary collision description of intrabeam scattering due to Sorensen [6]. The model calculates the energy transfer rate from transverse to longitudinal motion for a beam with an

initially anisotropic velocity distribution ($T_\perp \gg T_{\parallel}$). The measured beam size is first used to estimate the transverse energy of the beam. For ASTRID, the contribution of the space-charge force to the beam size is not negligible. The Sorensen description predicts an energy transfer rate of ~2 eV/s for the ASTRID beam at the time of injection. This is in good agreement with the initial slope of Figure 2, which gives 1.4 ± 1 eV/s. The shape of the curve in Figure 2 agrees qualitatively with the analytical predictions. A detailed comparison would require measurement of the transverse beam size as a function of time. This has not yet been undertaken.

The method of Molecular Dynamics [7] has also been applied to the problem of transverse to longitudinal energy transfer. The calculation starts with a system which has high kinetic energy in all three dimensions. When the system has equilibrated, the longitudinal motion is stopped. Following the reheating of the longitudinal degree of freedom is then analagous to the experiment represented in Figure 2. Computer size limitations prohibit simulating the actual particle density and transverse temperature corresponding to 10^9 particles in ASTRID. The calculations can, however, be employed to check the scaling relationships of the analytical model. For a system equivalent to 10^7 particles in ASTRID with transverse energies 0.3 to 2 meV, we obtain good agreement with the Sorensen method.

We have recently begun experiments with $^{166}Er^+$ ions. The laser measurement of the longitudinal heating of the erbium beam is shown in Figure 3. Improvements to our experimental set-up have permitted measurements to be carried out at later times after injection. The erbium curve is clearly richer in structure than that of lithium, with at least two different mechanisms providing time scales for heating. Detailed analysis of these data has just begun. In the erbium case, laser velocity distributions can be compared to measurements of the longitudinal Schottky spectrum. We note here only that the Schottky spectra show strong coherent behaviour of the injected beam for the time scale 0 – ~2 s after injection.

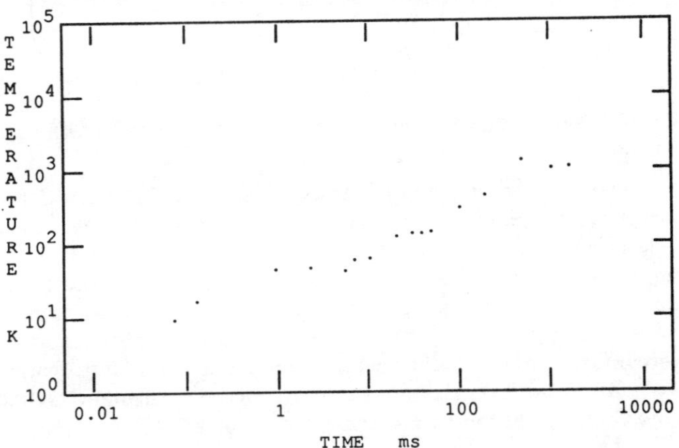

Figure 3. Longitudinal temperature vs. time after injection for $^{166}Er^+$ beam.

V. LASER COOLING OF METASTABLE LITHIUM

A. Fluorescence signal during the cooling process.

Laser cooling of the metastable fraction of the lithium beam is achieved using two lasers. The co- and counterpropagating laser beams are carefully aligned to achieve maximum overlap with the circulating ions. At the time of injection, both lasers are on, with the copropagating laser initially in resonance with the slower ions. The couterpropagating laser is likewise initially in resonance with the faster ions. After injection, the frequency of the counterpropagating laser is swept towards higher frequencies. The light pressure force from the swept laser decelerates the ions to the velocities in resonance with the fixed-frequency laser, thereby reducing the velocity spread in the beam. Figure 4 shows the fluorescent light produced in this process, plotted as a function of the frequency of the swept laser. This "cooling spectrum" is accumulated over many injection cycles of the machine in order to obtain good statistics.

The spectrum in Figure 4 exhibits some striking features. The sharp peak at the right side represents the fluorescence from the cooled velocity distribution as the swept laser decelerates the particles across the fixed laser. The detailed analysis of the cooling process, and determination of the temperature, appear below. The strong initial fluorescence, terminated by a sharp drop approximately 1 s after injection, is difficult to account for. The longitudinal heating of the beam due to Coulomb scattering contributes to the rapid onset of the fluorescence, as the tails of the velocity distribution come into resonance with the lasers after about 10 ms. However, the measured heating rate would predict a smoother rise in flourescence, and the rapid drop in signal after 1 s is not accounted for by this mechanism.

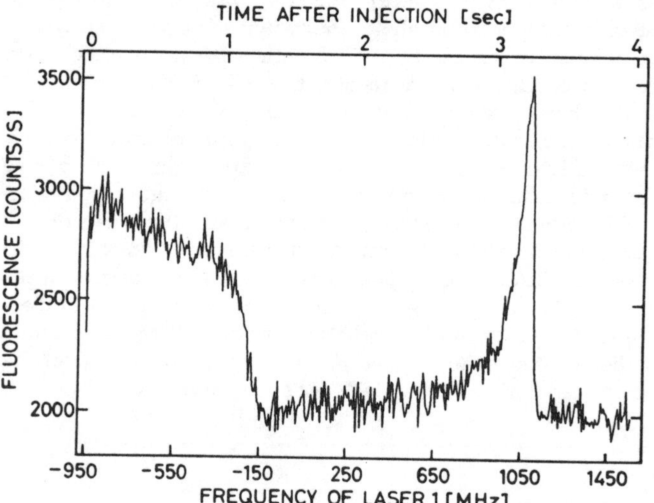

Figure 4. Fluorescence vs. laser frequency for the laser cooling process in ^{7}Li$^+$.

The position of the rapid drop in the signal has been determined to be independent of the intensity, scanning rate, and detunings of the lasers. In addition, it is impossible to produce the cold distribution, identified by the sharp fluorescence peak, before the initial fluorescence has dropped off. These observations appear to indicate that whatever forces are active in the beam at early times cannot be overcome by the laser cooling force. Without any competing heating mechanisms, the laser should be able to cool the metastable fraction to the Doppler limit in only 10 ms.

B. Measurement of the temperature

In order to measure the temperature of the cooled distribution, we use one laser as a probe. The intensity of the laser is lowered to minimize its perturbative effect on the velocity profile. Probe laser powers in the range of 10 to 100 times the saturation intensity of the transition were employed. We find that, for all powers used, the measured width of the distributuion is equivalent to that one would expect from power broadening alone. Using the systematic errors in measuring the laser intensity and laser beam profile and the statistical errors, we can put an upper limit on the residual width due to Doppler (i.e. thermal) broadening. We obtain $T_{\parallel} \sim 1$ mK.

C. Calculation of the temperature

The velocity distribution of the cooled ions can be calculated from the velocity-dependent laser-induced force and diffusion. In the following we neglect intrabeam and other forces not related to the light force. We assume that at any instant the velocity distribution is stationary and determined by the actual laser field parameters. This assumption is supported by the experimental observation that, after the initial drop in fluorescence, the effectiveness of the cooling process is independent of the speed with which the frequency of the cooling laser is swept.

The velocity-dependent force is calculated by the continued fraction method [7]. The fluctuations in the light-induced force are described by a diffusion coefficient, derived using a procedure due to Minogin [8]. The velocity distribution for a given field configuration is obtained as a stationary solution to the Fokker-Planck equation. For the laser intensities and polarizations used in the ASTRID experiments, we obtain a minimum temperature of T_{min}=1mK.

VI. CONCLUSIONS —
LASER COOLING AND CONDENSED BEAMS

At the low temperatures attainable with laser cooling, the potential to achieve spatial ordering of the beam [9] exists. For ASTRID, Molecular Dynamics simulations predict that 10^7 particles should condense into three coaxial shells [10]. Similarly, 10^9 particles would form about 30 shells. The threshold for the transition from a one dimensional ordered string to a structure having transverse extent is about 10^6 particles. Of course, to achieve order, the beam must be cold in all three degrees of freedom. The metastable fraction of the lithium beam amounts to about 10^5 particles, which is below the string threshold. The influence of the ground state ions, which are not subject to the laser cooling force, cannot be overlooked. It is difficult to imagine an ordered state coexisting with the warm ground state ions.

Also pertinent to the effort to obtain an ordered beam is our observation of strong forces which compete with the laser cooling process. The ability to laser cool the lithium ions is strictly correlated with the drop in fluorescence (Figure 4). This may indicate that a threshold phenomenon (e.g. a beam instability) is active in the dense ASTRID beam. The inability of the laser to overcome the intrabeam forces may be due to the short range of the cooling force. The laser interacts only with those particles whose velocities lie within the power-broadened linewidth of the cooling transition. The vast majority of the particles are free to diffuse under the influence of other forces. The two-component nature of the lithium beam exacerbates this situation, since the ground state ions do not respond to the laser. Our experience suggests that further experimental efforts are necessary in order to understand the mechanisms which limit the effectiveness of the laser cooling force.

REFERENCES
[1]   D.J. Wineland & H. Dehmelt, Bull.Am.Phys.Soc.,20 (1975) 637 & T. Hänch and A. Schawlow, Optics Comm.13 (1975)68
[2]   S. Chu and C. Weiman, in J.Opt.Soc.Am., B6(1989)2020
[3]   S. Schröder et.al., Phys.Rev.Lett.64,2901 (1990) & private communication
[4]   J.P. Schiffer and P. Kienle, Z.Phys.A321,181 (1985)
[5]   S.P. Møller, these proceedings
[6]   A.H. Sørensen in CERN 87-10(ed. Turner, S.) 135-154 (CERN,Geneva,1987)
[7]   V.G. Minogin and O.T. Serimaa, Opt.Comm.30, 373(1979)
[8]   V.G. Minogin, JETP53,1164(1981)
[9]   J.P. Schiffer & O. Poulsen, Europhys.Lett.1,55 (1986)
[10]   A. Rahman and J.P. Schiffer, Physica Scripta T22, 133 (1988)

Theoretical Studies of the Ultra Slow Extraction for the Cooler Synchrotron COSY-Jülich

K. Bongardt, D. Dinev, S. Martin, P.F. Meads, H. Meuth,
D. Prasuhn, H. Stockhorst and R. Wagner

Forschungszentrum Jülich
Institut für Kernphysik
Postfach 1913, D-5170 Jülich, FRG

Abstract

The Cooler synchrotron COSY[1] will accelerate protons from 40 MeV up to 2.5 GeV. It is the goal of this machine to use the beam of maximum brightness for internal target experiments, as well as for external experiments. The tool for the extraction of a low emittance beam is the ULTRA SLOW EXTRACTION scheme developed by W. Hardt for the LEAR[2]. The 3rd order extraction resonances at $Q_X = 10/3$, 11/3 resp. will be driven chromatically. A spill rate of longer than 5 seconds will be controlled by noise generators acting in the momentum phase space. The dynamic acceptance during extraction will be controlled by 11 families of sextupoles. It is expected to extract a beam with a horizontal emittance of less than 1 π and a momentum spread of less than 0.02%.

I. OVERVIEW OVER THE SYSTEM

The lattice of the Cooler Synchrotron COSY has been described in detail in ref. [3]. It comprises 6 DFFD cells, each of which incorporates 4 dipole magnets, 4 quadrupole magnets, and 1 or 2 sextupole magnets. Three cells form one 180 degree bending arc. The two arcs are connected by 40 m long straight sections of four quadrupole quadruplets each.

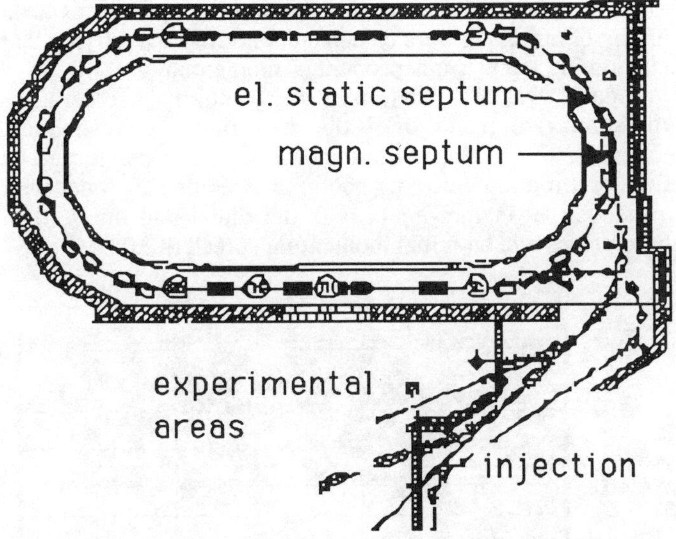

Fig. 1: Layout of the Cooler Synchrotron COSY

The straight sections normally provide an identity transformation with adjustable magnifications at their centers. Using four quadrupole families in the cell structure such that the 2nd and 5th cells are identical (two families) and the remaining four cells are identical (also two families), it is possible to make the telescopic insertions dispersion free and at the same time to meet one other condition.

The telescopes can be run in a triplet mode through most of the energy range. In this mode, there are common focal points between each quadruplet. A wide range of magnifications at the telescope centers is available. For the highest energies, the telescopes can be run in a doublet mode where the magnification at the telescope centers is unity. They can also be run in a π-2π mode where the magnification at the center is adjustable.

The electrostatic septum is located between the 4th and 5th cells. Here there is substantial dispersion of the order of 7 m. The magnitude and orientation of the dispersion vector can be varied slightly by adjusting the fourth quadrupole parameter mentioned above.

II. EXTRACTION WORKING POINTS

A wide range of tunes is available in the lattice. However, for efficient stochastic cooling, the phase advance between the pick-ups and the kickers should be an odd multiple of $\pi/2$. Because the pick-ups and kickers are separated by one half of the ring circumference for both the radial and the vertical planes, the preferred tunes are seen to be N+0.5 for some integer, N. For COSY, the best choices for extraction would be 10/3 or 11/3.

In order to minimize the angular spread in the extracted beam and thus minimize losses on the septa, we follow the LEAR method where the key unstable fixed point remains fixed as the momentum is varied. Going to the limiting momentum where the stable phase area vanishes completely, the equilibrium orbit must lie on top of this key unstable fixed point. Because the range in possible dispersion vectors is limited, this fixed point must lie around an angle of 70 degrees in the normalized phase space, thus fixing the outbound vertex of the triangular stability limit. There are now two choices for the outgoing separatrix; one of these is nearly vertical in the normalized phase space and thus provides no radial turn separation where the septum can be placed. This separatrix is obtained by approaching the resonance from above and clearly cannot be used. The other separatrix, obtained by approaching the resonance from below, is nearly parallel to the radial axis; it provides excellent radial separation with the resonant growth. Thus for either working point, we are constrained to approach the resonance from below.

The periodicity of the sextupoles in the dispersive cell structure is even. Using the 11/3 resonance for extraction has the advantage that only odd harmonics of the sextupole field drive the resonance. Thus the sextupoles in the arcs can be separately set to achieve the desired chromaticities, knowing that they will not otherwise influence the extraction. Then the sextupoles in the dispersion-free insertions can be set to obtain the desired extraction conditions, and they will not change the chromaticities.

III. NUMERICAL DESIGN METHOD

The linear *interactive* LATTICE code [4] has been modified to calculate the amplitude dependent tune shifts using the distortion function method of Collins and Ng [5]. It was further modified to calculate the second derivatives of the tune shifts with respect to $\Delta p/p$ and amplitude. The amplitudes and phases of the resultant sextupole fields with respect to the resonances $3Q_x=N$ and $Q_x+2Q_y=N$ are also calculated. All of these quantities can be fit simultaneously. The calculations are performed by the canonical method of linear transformations interspersed with nonlinear kicks at the centers of the sextupoles.

The method is to fit up to 11 sextupole families to yield the required radial chromaticity, the radial tune shift with radial amplitude and provide the correct phase of the resultant sextupole field in order that the unstable fixed point is superimposed on the dispersion vector. At the same time the coupling between the two planes, the spread in the vertical working point, and the magnitude of the driving term for the $Q_x+2Q_y=N$ resonance are minimized. Each sextupole strength is constrained to lie within limits that are both feasible and small enough that complicating higher-order fixed points do not upset the extraction procedure. This LATTICE does quite nicely within a few hours on an 80486 PC-AT computer. If the mixed derivatives are not specified, the solution speed is increased by an order of magnitude.

We show, as an example, a good solution for $Q_x=11/3-0.01$ and $Q_y=3.6$. The outgoing separatrix is nearly parallel to the x axis, and the dispersion vector is exactly superimposed upon the key unstable fixed point. Fig. 2 shows that an ion that just misses the electrostatic septum at 4.5 cm returns three turns later with an amplitude of 6 cm. The quadratic and higher order dependencies on $\Delta p/p$ are not calculated.

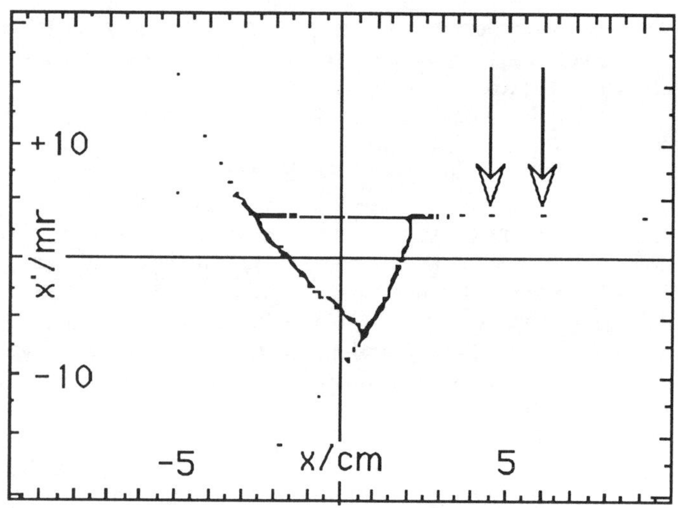

Fig. 2: Outgoing particles at the electrostatic septum

IV. NUMERICAL VERIFICATION

The radial and vertical tunes are similar and sextupoles driving the $3Q_x=N$ resonance can also drive the $Q_x+2Q_y=N$ resonance. That would result in a blow up of the beam in the vertical plane, possibly before it can be extracted radially. Therefore it is important to verify that coupling tune shifts and the driving term for the coupling resonance have been adequately limited. Fig. 3 shows the dynamic aperture close to the extraction resonance, calculated with LATTICE. The vertical physical aperture limitation of ± 2.9 cm lies well within the dynamic aperture.

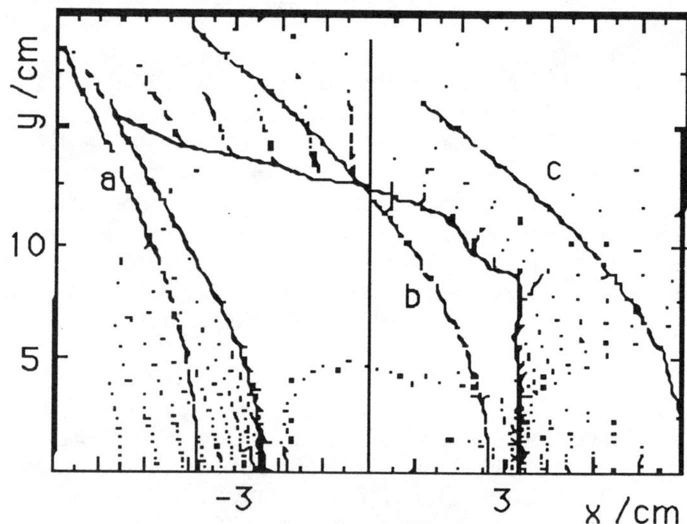

Fig. 3: Dynamic aperture near the extraction resonance (a); b) and c) correspond to coupling resonances

As a separate check on the method, the separatrices for a number of momenta were calculated with MAD, verifying that the outgoing leg was independent of momentum.

Finally DIMAD was used to gradually accelerate a representative group of 1000 ions into the resonance, increasing their relative momentum by 10^{-6} per turn. The initial distributions were parabolic in three phase planes with emittances of 5π mm-mr in both the radial and the vertical phase planes and an initial momentum spread of ±0.05%.

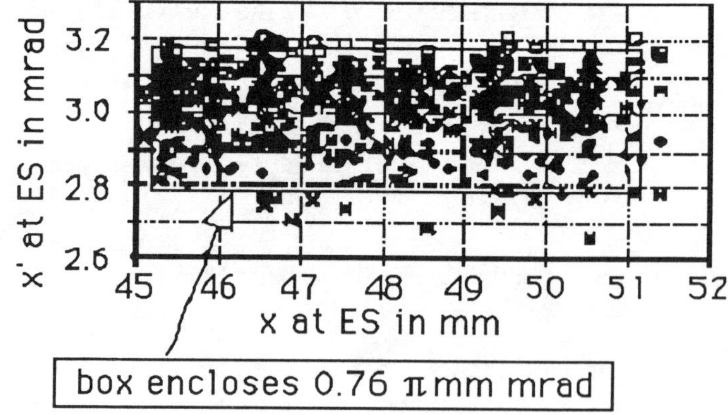

Fig. 4: Phase space of the extracted particles

a) resonance extraction

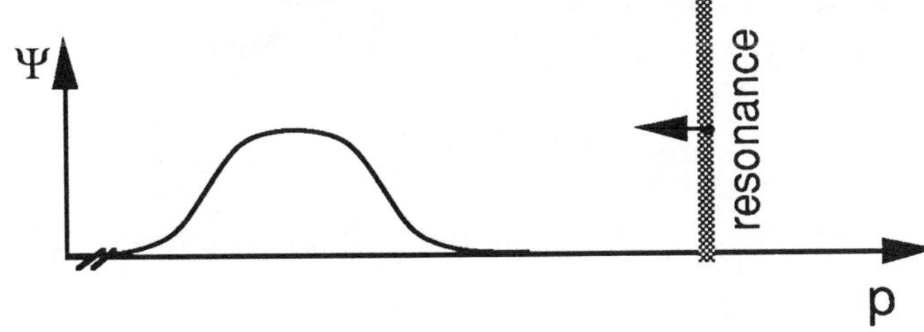

b) ultra slow extraction (USE)

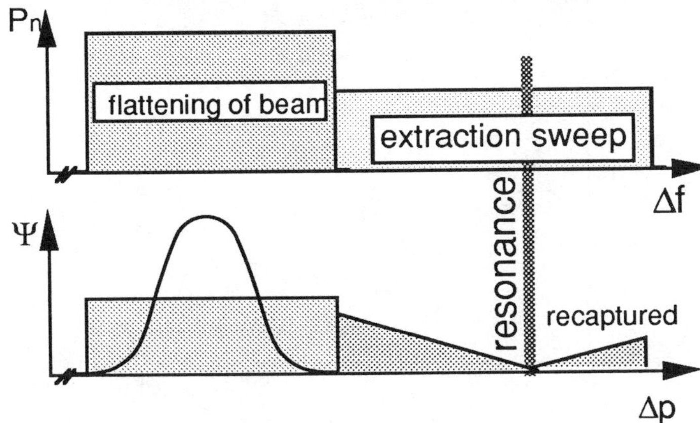

c) USE with "chimney"

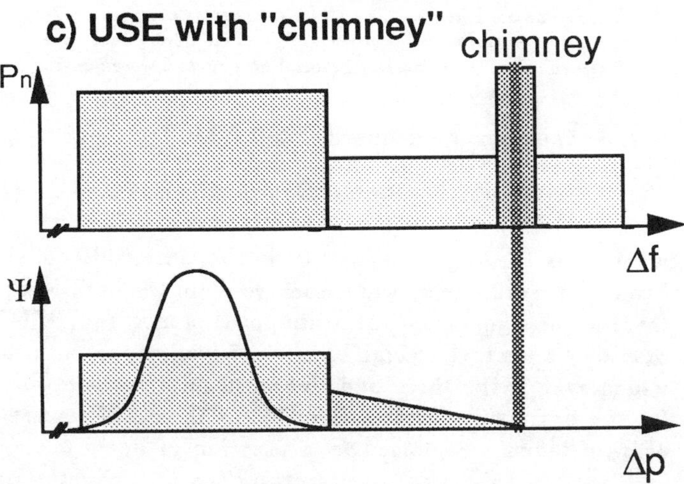

Fig. 5: Extraction schemes

About one half of these ions were extracted within a radial emittance of about 0.8π mm-mr as shown in Fig. 4. Correcting for terms in $(\Delta p/p)^2$ will reduce this emittance by an order of magnitude. The other half were recaptured as the separatrices reappeared as the momentum increased beyond the vanishing stable phase area. This is because the motion stagnates in the vicinity of the unstable fixed point, and the acceleration rate in the numerical simulation was unrealistically high. In the actual implementation, a strong noise signal, the chimney, is introduced such that ions are unable to move beyond the resonance momentum (see Fig. 5). In the simulation, these recaptured ions were accelerated into the septum.

A similar run where the acceleration rate was increased tenfold resulted in the extraction of a small fraction of the beam.

The DIMAD runs verify the lack of coupling between the vertical and the radial plane and small coupling between the initial momentum and the radial emittance of the extracted beam.

A separate DIMAD run was used to determine the locus of the equilibrium orbit and the key unstable fixed point as functions of momentum. This run revealed that the quadratic terms cause the unstable fixed point to move slightly and also caused the motion of the equilibrium orbit to depart from a straight line by a similar amount as it passes the unstable fixed point. The curvature increases noticeably for momenta above the maximum for the extracted beam. A desirable improvement would be to zero this quadratic term which LATTICE cannot calculate.

V. ACKNOWLEDGEMENTS

We are greatly indebted to our colleagues of CERN, TRIUMF and LBL Berkeley, especially Mr. M. Chanel and D. Möhl for giving us comprehensive information about the LEAR ultra slow extraction system, as well as R. Servranckx and M. Berz for their support with the programs DIMAD and COSY INFINITY.

VI. REFERENCES

[1] R. Maier and U. Pfister,"The COSY-Jülich project, May 1990 status", to appear in the proceedings of this conference

[2] W. Hardt, "Ultra slow extraction out of LEAR", PS/DL/LEAR Note 81-6, CERN, Geneva 21-April-1981

[3] K. Bongardt, D. Dinev, S. Martin, P.F. Meads, D. Prasuhn, H. Stockhorst and R. Wagner, "The COSY Lattice", Proc. EPAC 90, Nice, June 1990, p. 1491.

[4] P.F. Meads, Jr., "Application of Collins' Distortion Functions", Proceedings of the 1989 Particle Accelerator Conference, IEEE 89CH2669-0, p. 1382 (1989).

[5] T.L. Collins, "Distortion Functions", FERMILAB-84/114
N. Merminga and K.-Y. Ng, " Hamiltonian Approach to Distortion Functions", FN-493, FNAL, August 1988

Beam Property Measurements in the IUCF Cooler Ring

M. Ball, D.D. Caussyn, J. Collins, D. DuPlantis, V. Derenchuk, G. East
T. Ellison, D. Friesel, B. Hamilton, B. Jones, S.Y. Lee, M. G. Minty, and T. Sloan
Indiana University Cyclotron Facility*
2401 Milo Sampson Lane, Bloomington, Indiana, 47405 USA

Abstract

The high intensity stored beam stability limits of the IUCF Cooler-Storage Ring are being explored. Measurements of the equilibrium emittance and bunching factor for a 45 MeV electron cooled proton beam as a function of stored beam intensity are reported. In addition, calculations of space-charge tune shift based on these measurements are presented. The tune shift is found to be as large as about 0.3 in magnitude, depending on the degree to which the beam is cooled. The space-charge tune shift is discussed as a possible contributing factor to observed stored beam current limits.

I. Introduction

The Indiana University Cyclotron Facility Cooler Storage Ring and synchrotron accelerator ($B\rho = 3.6$ Tesla meters) was proposed in 1981, funded for construction in 1983, and completed in 1987. It was the first of the many similar accelerator storage rings designed specifically to employ electron cooling to produce and use high quality medium energy ion beams in equilibrium with thin internal targets for nuclear research. The status and development of the IUCF Cooler ring[1, 3], and cooling system [2, 4] have been reported in numerous accelerator conference proceedings.

We have accumulated 45 MeV proton beams in the Cooler using stripping injection of 90 MeV H_2^+ beams; further accumulation is possible using the electron cooling system to coalesce the injected beam into a stack such that the process can be repeated. We have observed coherent transverse instabilities for these low emittance, low momentum spread electron cooled beams near the maximum attainable beam currents, about $1 - 4$ mA for coasting beams. However, intensity limits are also reached at nearly the same current, but with no observable coherent transverse instabilities. Clearly, in addition to coherent transverse instabilities at least one other process limits the beam current. We hypothesize that one such additional limitation is due to the space-charge tune shift. In this work, we explore this hypothesis by using measurements of the bunching factor and the beam emittance to calculate a space-charge tune shift.

II. Procedure and Results

The bunching factor, B_f, is a measure of the peak-to-average beam current. Assuming that the beam distribu-

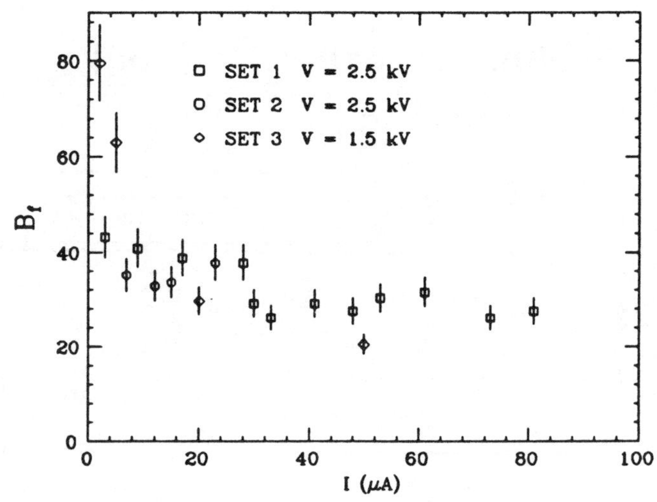

Figure 1:   Plot of typical values of the bunching factor versus beam current. The three different data sets correspond to data taken for three different machine conditions. In each case, the rf harmonic number is 6, and the rf cavity voltage, V, is listed.

tion is Gaussian, we define B_f (> 1) by

$$B_f = \frac{\tau}{\sqrt{2\pi}\sigma_t}, \tag{1}$$

where τ is the rf period, and σ_t is the time width of the beam. The value of σ_t was measured from the oscilloscope trace of the signal from the beam position monitor (BPM), or a low bandwidth or high bandwidth wall gap monitor[5], which ever of the three had the most suitable bandwidth for the beam pulse width. The B_f for three different sets of data taken are plotted as a function of beam current in Figure 1.   Each data set corresponds to a slightly different machine tune and, qualitatively, increasingly better cooling in going from the first to the third set. Generally, we observed that the B_f decreases with increasing beam current and does not appear to have a strong dependence on the quality of the electron cooling.

The horizontal transverse emittance, ϵ_x, is found from the beam profile. The profile was determined by horizontally sweeping the beam through a profile monitor, a vertically-mounted 10 μm diameter carbon fiber, and measuring the secondary emission current using a high input impedance amplifier. The profile monitor was mounted in a region of high momentum dispersion (4 m) where the horizontal motion of the beam was generated by ramping the

*This work supported by the National Science Foundation under Grant NSF PHY 90-15957.

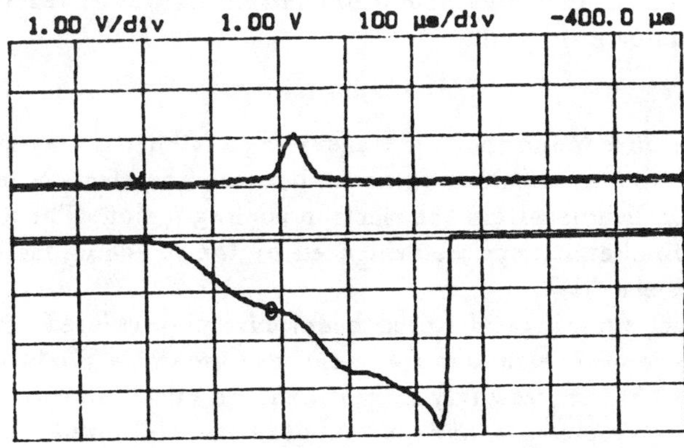

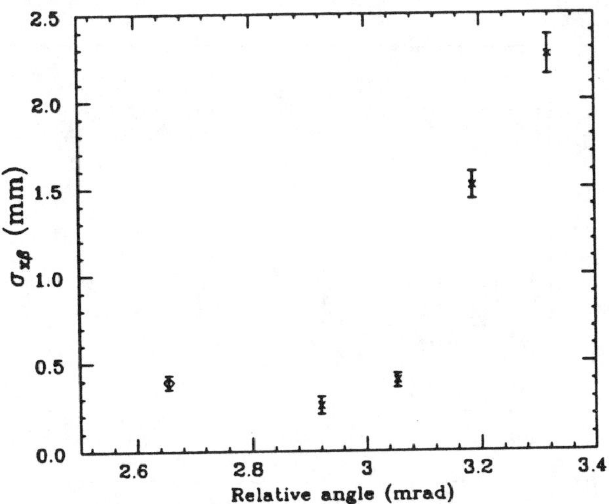

Figure 2: Typical oscilloscope traces of the beam profile monitor current (top trace) and the position signal from a Beam Position Monitor (bottom trace, delayed by 15 µsec). Horizontal scale is 100 µsec, and the BPM vertical scale is 2.1 mm/div. The slight oscillation in the BPM output is due to synchrotron oscillations.

Figure 3: Plot of the rms beam width due to betatron oscillations versus relative angular misalignment of the proton and electron beams in the cooler region. Note that the location of the origin for the angular misalignment is arbitrary.

rf cavity frequency. The transverse velocity of the beam while measuring the profile was as high as 19 m/sec in this region. Both position of the beam centroid, measured using a nearby BPM, and the secondary emission current from the profile monitor were recorded simultaneously as functions of time (see Figure 2). The width of the beam profile, σ_x can then be easily determined from these two plots.

The value of σ_x includes a contribution due to betatron oscillations, $\sigma_{x\beta}$, and another due to the beam momentum spread and the ring dispersion, σ_{xp}. The value of σ_{xp} is found from,

$$\sigma_{xp} = \eta \frac{\Delta p}{p}, \tag{2}$$

where η is the dispersion function. For small oscillations,

$$\frac{\Delta p}{p} = \frac{\nu_s}{\eta_0} \sigma_t \tag{3}$$

with

$$\eta_0 = \frac{1}{\gamma^2} - \frac{1}{\gamma_t^2}. \tag{4}$$

In these expressions, γ and γ_t are the relativistic factors associated with the beam energy and the transition energy respectively ($\gamma_t = 4.85$), and ν_s is the synchrotron frequency.

For an electron cooled proton beam, the Fokker-Planck equation gives basically a Gaussian distribution in the six dimensional phase space. In this case, the rms beam size is given by the Gaussian quadrature of the components from momentum and position space. So, the broadening due to the momentum dispersion can be removed,

$$\sigma_{x\beta} = \sqrt{\sigma_x^2 - \sigma_{xp}^2}. \tag{5}$$

Then, the transverse emittance is calculated using the relation, $\epsilon_x = \sigma_{x\beta}^2/\beta_x$, where β_x is the betatron amplitude function ($\beta_x = 2.0$m). Due to uncertainties in the momentum spread, this determination of ϵ_x is not as precise as could be accomplished with a flying wire profile monitor in a dispersion free region.

In Figure 3, $\sigma_{x\beta}$ is plotted versus the relative angular misalignment of the electron and proton beams in the electron cooling region. The misalignment was produced by "tilting" the magnetic field lines in the cooling region solenoid by varying the strength of a superimposed horizontal dipole magnetic field. The data plotted in this figure are for a proton beam current of about 60 µA and a B_f of about 30. All the data in this figure were taken with identical operating conditions with the exception of the point marked with a diamond for which there is an additional insignificant vertical misalignment. It is clear from this figure that ϵ_x is a strong function of the electron cooling. In fact, for a large part of the range covered by the data, the equilibrium proton beam rms divergence in the cooling region is to good approximation the angular misalignment between the electron and proton beams. This would be akin to the monochromatic instability[7]. It should also be noted that, based on this figure, we have determined that none of the three data sets used in the rest of this work were taken with minimum misalignment and, consequently, minimum emittance.

In Figure 4, values of ϵ_x are plotted as a function of beam current for each of the three sets of data taken. As in the discussion of the bunching factor, the electron cooling is improved in moving from the first to the third data set. With a possible exception for beam currents less than 20 µA, ϵ_x generally increases with beam current. There is also a systematic decrease in emittance in going from the first to the third data set corresponding to improved cooling.

Assuming that the beam is on-axis in a circular chamber with negligible effects due to image charges and currents,

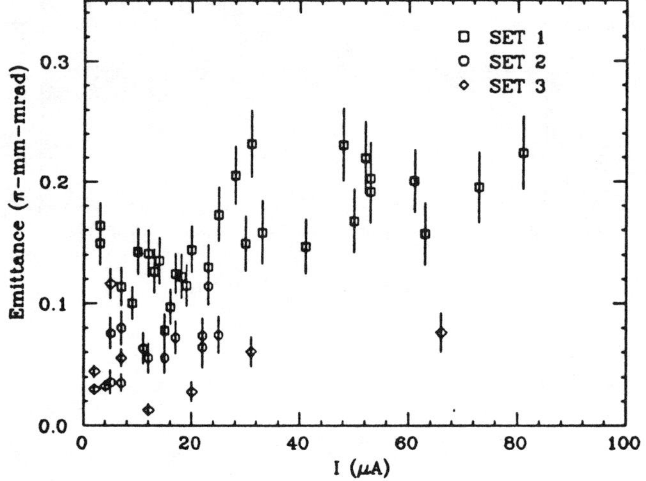

Figure 4: Plot of the radial transverse emittance versus beam current for data taken with three slightly different operating conditions.

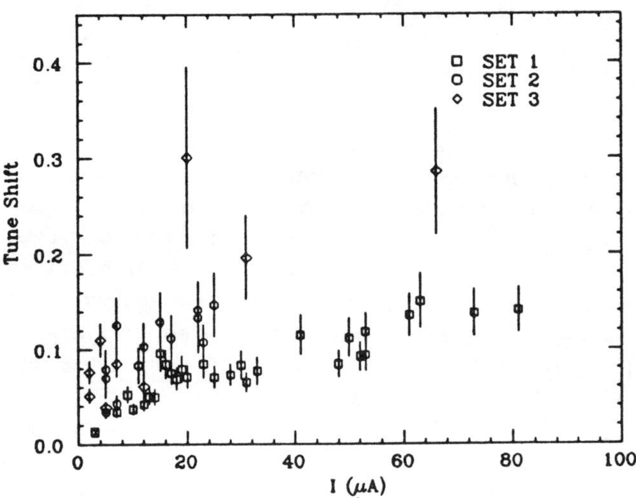

Figure 5: Plot of calculated space-charge tune shift versus beam current.

the space-charge tune shift is calculated from [6],

$$\nu_{sc} = -\frac{N R r_p}{\pi \nu_x \gamma^3 \beta^2} \frac{B_f \beta_x}{4\epsilon_x}. \tag{6}$$

In this expression, N is the number of particles in the ring, R is the ring radius, r_p is the classical proton radius, ν_x is the horizontal betatron tune of the cooler ring, and β and γ are the usual relativistic factors. The ν_x for the ring was measured using the ping-tune method [5], and was about 3.8.

Values calculated for the magnitude of $\Delta\nu_{sc}$ are plotted versus beam current in Figure 5 for each of the three data sets discussed previously. In this figure it is evident that $|\Delta\nu_{sc}|$ monotonically increases with beam current. It is also clear that as the cooling is improved, the values of $|\Delta\nu_{sc}|$ are greater at each current. It is also suggestive that $|\Delta\nu_{sc}|$ increases with beam current at greater rates as cooling is improved.

III. Conclusions

We have found that the transverse emittance is very sensitive to the angular alignment between the electron and proton beams within the electron cooling region. The longitudinal emittance, as evidenced by the bunching factor, is less sensitive.

Over the range of stored beam currents explored here, we have seen calculated space-charge tune shifts in a broad range, but increasingly large when the misalignment between the electron and proton beam is minimized. The magnitude of the observed tune shifts were as large as 0.3 with the best cooling. With higher currents and optimized cooling, larger space-charge tune shifts are expected. With tune shifts of this magnitude, it is evident that space charge is indeed a factor in stored beam current limitations observed in the IUCF Cooler Ring. Additional studies in which the beam lifetime corresponding to these large tuneshifts is measured, and more precise measurement of emittance is made, would be useful.

At IUCF our goal is to increase the luminosity by an additional two orders of magnitude to 10^{32} s^{-1}-cm^{-2}. Consequently, we will be adding active damping systems and other systems which will enable us to use the advantages of electron cooling to accumulate high intensity beams while avoiding the low momentum spreads and emittances which limit the intensity due to coherent transverse instabilities and the space charge tune shift.

References

[1] R.E. Pollock, "IUCF cooler ring status 1989", *Proc. 1989 IEEE P.A.C., Acc. Eng. and Tech.*, Chicago, IL, pp. 17–21.

[2] Timothy JP Ellison, Dennis L. Friesel, and Robert J. Brown, "Status and performance of the IUCF 270 keV electron cooling system", ibid, pp. 633–635.

[3] D.L. Friesel, T.J. Ellison and P. Schwandt, "Status report on the IUCF cooler-storage ring" *Nucl. Inst. Meth.* **B40/41**, 927–933 (1989).

[4] T. Ellison, "Electron Cooling", Indiana University, Bloomington, Indiana, Ph.D. Dissertation, 1990 (unpublished).

[5] Mark Ball, Timothy JP Ellison, and Brett Hamilton, "Beam diagnostic systems in the IUCF cooler ring", This conference.

[6] B. Zotter, "Betatron frequency shifts due to image and self fields", in *CERN Accelerator School, General Accelerator Physics*, eds. P. Bryant, S. Turner, CERN 85-19, p. 253

[7] Ya.S. Derbenev and A.N. Skrinsky, "The kinetics of electron cooling of beams in heavy particle storage rings", *Particle Accelerators*, **8**, pp. 1–20 (1977)

Intrabeam Scattering in the Fermilab Antiproton Accumulator

C.M. Bhat and J. Marriner

Fermi National Accelerator Laboratory *

P.O. Box 500, Batavia, Illinois 60510

Abstract

The growth rate of the horizontal and vertical emittance in the accumulator ring at Fermilab has been measured with 97.8×10^{10} antiprotons starting with completely cooled beam. We find that the emittance growth rate can not be fully explained by the existing intrabeam scattering calculations done including the detailed lattice of the storage ring.

I. INTRODUCTION

Recently a number of upgrade projects have been undertaken at Fermilab accelerator facility to improve operating conditions and be able to provide better quality beams. In this program, improving the performance of the antiproton storage ring, the accumulator, plays an important role both in the collider mode of operation of the Tevatron as well as the fixed target mode with antiprotons. This storage ring was originally built[1] for a final stack of 50mA (1mA $\simeq .9928 \times 10^{10}$ particles at about 8 Gev/c for the Fermilab accumulator ring) $\bar{p}$ with a momentum spread of $\pm 0.025\%$ and horizontal and vertical emittance of $2\pi - mm - mr$. But we have achieved $\bar{p}$ stacking in the accumulator ring up to 120mA, about a factor of two larger than the design intensity. We have also stored protons up to about 125mA during one of our runs dedicated to improve the storage ring performance. With the proposed Main injector we expect the maximum stored beam as high as 250-300mA. But, the limitation in the maximum stacking of antiprotons achievable in the accumulator could arise from intrabeam scattering. Hence it is highly interesting to investigate emittance growth of the stored beam in the accumulator.

The emittance growth in a storage ring for coasting as well as for the bunched beam due to intrabeam scattering is a rather well known subject. The initial theory of intrabeam scattering without including the detailed lattice of the ring was developed by Piwinski[2]. Further development lead to the inclusion of a general lattice[3,4]. A direct testing of the theory of intrabeam scattering was done at the CERN antiproton accumulator ring and a reasonable agreement

was obtained except for the vertical emittance growth rate[5].

In this paper we present results of our measurements on transverse emittance growth in the accumulator ring and comparison with the existing intrabeam scattering theory[3].

II. MEASUREMENTS AND DATA ANALYSES

During one of our $\bar{p}$ source study period we stacked about 100mA of $\bar{p}$ beam in the accumulator ring. Throughout stacking the core of the beam was cooled using high frequency momentum and transverse stochastic cooling systems[1]. After the stacking was stopped the cooling was left on for about 2-3 hours until a steady state beam temperature was reached (i.e. a balance between cooling and self heating was achieved). The initial beam parameters were,

Number of $\bar{p} = 98.5$mA ($\gamma = 9.48$ at kinetic energy

of 7.95GeV)

$$\epsilon_H = 1.50 \ \pi\text{-mm-mr}$$

$$\epsilon_V = 1.22 \ \pi\text{-mm-mr}$$

$$\left(\tfrac{\Delta p}{p}\right)_{99\%} = \pm 9.99 \times 10^{-4}$$

Vacuum $\simeq 4.0 \times 10^{-10}$ (Torr)

(life time about 800hr).

Where, ϵ_H and ϵ_V are horizontal and vertical emittances of the beam and Δp is momentum spread. Then, only the transverse cooling was turned off and the $\bar{p}$ stack was left to blow-up in the transverse plane. The emittance growth was monitored for about 3 hours. To monitor the emittance growth we used two different methods. In both cases the betatron side-band powers were measured which are directly proportional to the emittances of the beam, i.e. $\epsilon_{H,V} = \alpha P_{H,V}$. In the first method, the emittance measurements was done continuously by using a pair of schottky pick-up systems which were pre-tuned to the horizontal and vertical side-bands[6] and thereby measuring the powers. In the second method, spectrum analyzer measurements were taken of the schottky power once every five minutes and the power in each side-band is extracted from the spectra. In both cases the calibration constants α come from separate measurements with scrapers. For this we used slowly heated beam which gives a gaussian beam profile. The beam size measured using scraper is related to the beam emittance by[7],

$$\epsilon_{H,V} = \pi \frac{6\sigma^2_{H,V}}{\beta_{H,V}} \ .$$

Where σ the is size of the beam in the horizontal or vertical plane and β is the betatron function. The measured emittance by these methods have less than 5% errors. The Figures 1 and 2 show the time evolution of emittances ϵ_H and ϵ_V (95% of the particles

in phase space) respectively. The time t=0 corresponds to when the cooling was switched off. With in about three hours the emittances grew by a factor of four. The accumulator ring has horizontal and vertical acceptances of about $10\pi - mm - mr$ which are larger than the maximum emittances we reached.

To understand the emittance growth we carried out a number of intrabeam scattering calculations using a computer code obtained from CERN[3]. This uses the beam properties, betatron function of the accumulator ring as well as the dispersion functions and derivatives of the betatron functions as inputs. The results of calculations for horizontal emittance growth is shown in the figure 1 by a solid curve. The measured horizontal growth rate is larger than the intrabeam scattering predictions. This indicates that, there is some unknown heating mechanism that is larger than intrabeam scattering. We have calculated the contributions to the emittance growth arising from multiple Coulomb scattering on the residual gas and find the rate to be approximately $0.1\pi - mm - mr$/hr. As a check the calculated lifetime from nuclear and single Coulomb scattering is 1000hours compared to the observed lifetime of 800hours.

In the case of vertical emittance, the growth rate from intrabeam scattering is predicted to be negative. However, it is important to note that the model used does assume no coupling between horizontal and vertical betatron motions. But in reality the coupling in the accumulator is significant and we do not really expect to see damping of the vertical motion. In addition, the vertical motion may be subject to other heating mechanisms.

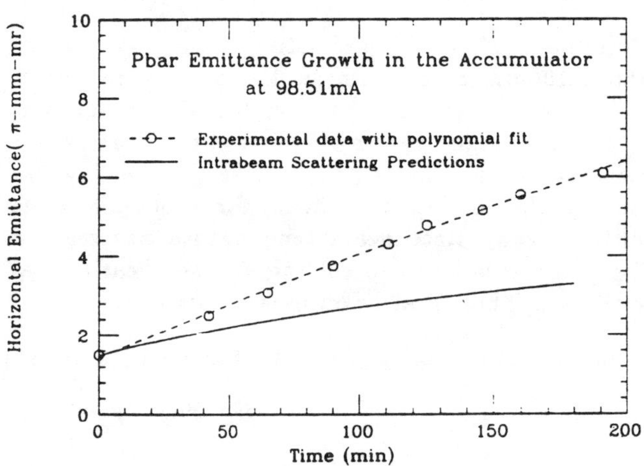

Figure 1. The Horizontal emittance versus time.

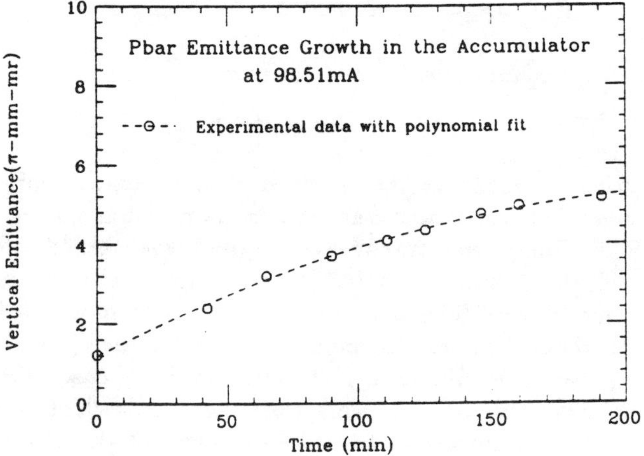

Figure 2. The vertical emittance versus time.

III. SUMMARY

We have measured the emittance growth in the Fermilab $\bar{p}$ Accumulator ring with a stack of about 98.5mA of antiprotons. The data have been compared with the intrabeam scattering model predictions. We find the observed growth rate is not explained by the intrabeam scattering model or by multiple Coulomb scattering.

REFERENCES

* Operated by the Universities Research Association, Inc., under contracts with U.S. Department of Energy.

1. Design Report Tevatron 1 Project, Fermi National Accelerator Laboratory, 1984.

2. A.Piwinski, Proc. 9th Int. Conf. on High Energy Accelerators, 1974,p405.

3. D. Mohl, Computer code INTBMS, CERN, Private Communications.

4. J.D. Bjorken and S.K. Mtingwa, Particle Accelerators, 1983, Vol.13, p115.

5. M. Conte and M. Martini, Particle Accelerators, 1985, Vol.17, p1.

6. D. Peterson, Fermilab, Private Communications (1991).

7. D.A. Edwards and M.J. Syphers, An introduction to the physics of particle accelerators, AIP Conference Proceedings Vol 184, (1989) page 7.

Ion Clearing by Cyclotron Resonance Shaking

P. Zhou and J.B. Rosenzweig*
Fermi National Accelerator Laboratory,†
P.O. Box 500, Batavia, IL 60510

Abstract

We discuss a new concept in ion clearing for storage rings, that of resonant removal of ions in dipoles by shaking the beam horizontally near the ion cyclotron frequency. This method of beam shaking is similar to the variations on ion bounce shaking developed by Orlov, Alves-Pires and coworkers at CERN, but has advantages in requiring a narrower bandwidth of shaking frequencies and in much higher achievable ion kinetic energies. The results of analytical theory, and computer simulations are discussed.

Introduction

Accumulation of ions has been a limiting factor in the performance of antiproton accumulators. Besides the direct method using clearing electrodes, beam shaking has been proven very effective in further clearing ions[1] and reducing the neutralization level in storage rings. In the so called "resonant shaking"[2] a driving voltage with frequency close to that of the ion oscillatory motion in the beam's electric field (bounce frequency) is applied to the beam, which responds by shaking, generating an oscillating transverse electric field which in turn drives transverse ion motion. Ions traversing different beam sizes, and thus bounce frequencies, through slow longitudinal motion will be locked-on to larger amplitudes; in this way the neutralization effects can be reduced[3]. The same or better results should be achievable through modulation of the driving frequency[4], which is called frequency modulation shaking. This way the process is controllable and does not rely on the ions' longitudinal motion. To distinguish from what we are going to describe here we call this kind of shaking bounce shaking because it is the ion bounce motion that is being driven.

An interesting and important experimental fact is that the bounce shaking described above doesn't work well in horizontal plane. This can be explained by noting that while ions in straight sections are easily removed by the clearing electrodes, ions inside dipole magnets cannot be easily cleared, since there are are often no clearing electrodes inside magnets due to tight space constraints, and thus the only clearing mechanisms are Coulomb heating and $\mathbf{E} \times \mathbf{B}$ longitudinal drift. The former is insignificant while the latter can be greatly reduced by the neutralization itself. As a result the neutralization level inside the magnets is much higher than the rest of the machine and deleterious ion effects are most likely caused mainly by the ions in dipoles. The strong magnetic field inside magnets leads to cyclotron motion of ions in the horizontal plane, which has, in general, a much higher oscillation frequency than that of the bounce motion, and therefore bounce shaking is not effective in the horizontal plane. This explanation leads us consider the possibility of horizontal shaking close to ion cyclotron frequency, which we call cyclotron shaking.

Theory of Cyclotron Shaking

The theoretical analysis for cyclotron shaking is parallel to that of bounce shaking[4], which uses the averaging technique originally developed by Krylov and Bogoliubov[5][6]. For completeness we reproduce analysis here.

Let x be the transverse coordinate of ion in the horizontal plane and z the longitudinal coordinate, then, ignoring the space charge of the ions themselves, the equations of motion for the ion are,

$$\frac{d^2x}{dt^2} = -\omega_c \frac{dz}{dt} + \frac{q}{m} E_x(x - b\cos\omega t)$$
$$\frac{d^2z}{dt^2} = -\omega_c \frac{dx}{dt}$$

where q and m are the charge and mass of ion respectively, ω_c is the cyclotron frequency of the ion, and b is the amplitude of the beam center oscillation driven by the applied voltage. For a round Gaussian beam with rms beam size σ the form of radial electric field is

$$E_r(r) = \frac{2\lambda}{r}(1 - e^{-\frac{r^2}{2\sigma^2}})$$

where λ is the beam line charge density.

*Present address: UCLA Dept. of Physics, 405 Hilgard Ave., Los Angeles, CA 90024

†Operated by the Universities Research Association under contract with the U. S. Department of Energy

The displacement of the ion from the beam center normalized by the beam rms size σ, $\tilde{x} = \frac{1}{\sigma}[x - b\cos(\omega t)]$, is then described by

$$\frac{d^2\tilde{x}}{dt^2} = A(\omega^2 - \omega_c^2)\cos(\omega t) - \omega_c^2(\tilde{x} - x_0) - \omega_b^2 f(\tilde{x})$$

where $\omega_b = \sqrt{|q\lambda|/m\sigma^2}$ is the maximum ion bounce frequency, $A = b/\sigma$ and $x_0\sigma$ is the horizontal position of the ion's guiding center. Unlike in the case of bounce shaking where the corresponding quantity is always zero, x_0 here is in general not. For simplification we take $f(\tilde{x}) = \frac{2}{\tilde{x}}[1 - \exp(-\frac{1}{2}\frac{\tilde{x}^2}{\sigma^2})]$ which implies that we are only considering ions close to the vertical center of the beam.

We look for the equilibrium solution of the form

$$\tilde{x} = a(t)\cos(\omega t + \theta(t)) + x_0$$
$$\frac{d\tilde{x}}{dt} = -\omega a(t)\sin(\omega t + \theta(t))$$

in which $a(t)$ and $\theta(t)$ are slow varying functions relative to the oscillation period. Averaging over a period yields:

$$\frac{da}{dt} = \frac{(\omega_c^2 - \omega^2)}{2\omega}A\sin\theta$$
$$\frac{d\theta}{dt} = \frac{(\omega_c^2 - \omega^2)}{2\omega}[1 - \frac{A}{a}\cos\theta] + \frac{\omega_b^2}{2\omega}G(a, x_0)$$

where $G(a, x_0) = \frac{1}{2\pi}\int_0^{2\pi} f(x_0 + a\cos\phi)\cos\phi d\phi$. The maximum amplitude satisfies $da/dt = d\theta/dt = 0$, therefore we have an implicit relation between the driving frequency and equilibrium ion amplitude:

$$\frac{\omega^2 - \omega_c^2}{\omega_b^2} = \frac{G(a, x_0)}{1 \pm \frac{A}{a}} \qquad (1)$$

Eq. 1 represents hysteresis which is well known for non-linear oscillators. For $x_0 = 0$, $G(a, 0) = \frac{4}{a^2}[1 - e^{(-\frac{a^2}{4})}I_0(\frac{a^2}{4})]$ and Eq. 1 is plotted in Fig. 1. The motion corresponding to the upper half of the left curve is unstable. The beam oscillation amplitude undergoes a jump at a particular driving frequency and it can reach much larger

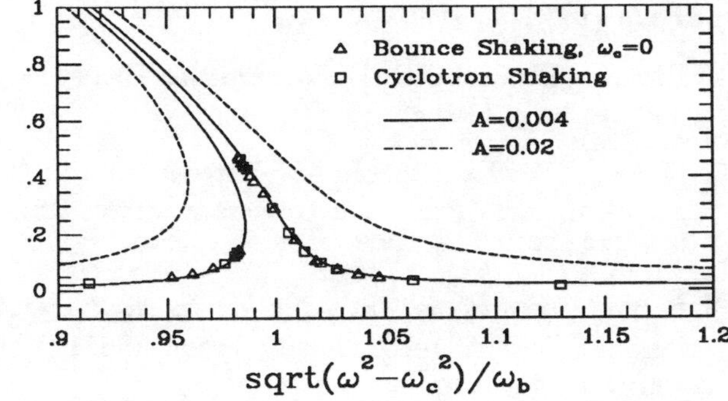

Figure 1: Equilibrium Ion Oscillation Amplitude vs. Shaking Frequency

values if the driving frequency is modulated from above that frequency downward, which is the concept frequency modulation shaking based upon.

In general function $G(a, x_0)$ cannot be expressed in closed form. Results by numerical methods are shown in Fig. 2. As x_0 increases G not only drops in magnitude but

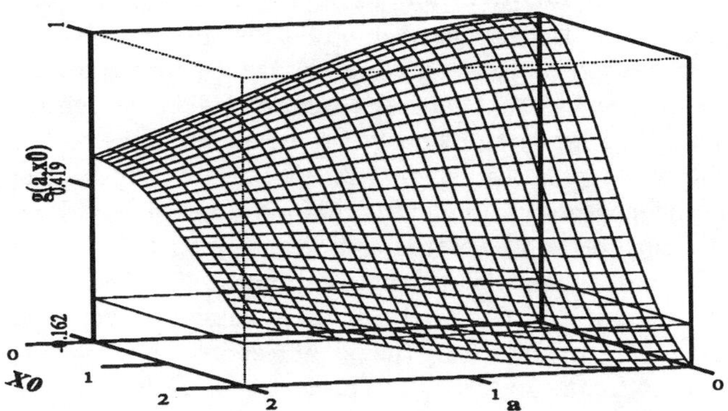

Figure 2: Function $G(a, x_0)$.

also reverses its trend as a function of a. This means not only the resonant frequency decreases, the direction of frequency modulation that increases the ion oscillation amplitude reverses also. Therefore the effect of cyclotron shaking with frequency modulation eventually vanishes. With this discussion in mind, it is clear that cyclotron shaking will only be effective to ions close to beam center. Fortunately, these ions are precisely the ones which pose the greatest threat to beam stability, because they slow the ions' $\mathbf{E} \times \mathbf{B}$ drift inside dipoles by neutralizing the electric field. The drift speed is $\sim 10^3 m/s$ for ions that are 1σ from the center of an unneutralized 100 mA beam in the Fermilab accumulator. This corresponds to a less than 0.2% equilibrium neutralization level inside dipoles, which is much better than the current operation conditions. If the cyclotron shaking can excite the ions created close to beam center to large amplitudes, the drifting motion of the other ions will provide sufficient clearing.

Discussion

Using the Fermilab antiproton accumulator parameters with 200mA beam current, we compared the theoretical predictions on the equilibrium amplitudes with the simulation results of both bounce and cyclotron shaking for protons. As shown in Fig. 1 the data points, which corresponds to $\sigma = 0.22cm$, $A\sigma = 0.01mm$ and no initial horizontal displacement agree remarkably well with the analytical theory. The modulation of frequency, in both shaking schemes, is also shown by simulation to be very effective, as predicted by theory.

As seen above cyclotron shaking and bounce shaking are very similar in many aspects. The equilibrium amplitudes can be described by the same plot and hence both have the same hysteresis "lock on" effect which is crucial in reducing neutralization effects[3]. However because of the very different frequencies of the cyclotron and bounce motion, the two ways of shaking do have different properties. The cyclotron frequency ω_c, at least in our case, is much higher than the bounce frequency ω_b. The corresponding frequency spread in cyclotron shaking is a factor of $\omega_c/\omega_b = 15$ smaller than that in bounce shaking. This difference affects the frequency modulation process because the beam response to the driving voltage depends strongly on the frequency and that has a big impact on the ion response (see Fig. 1). Obviously the beam response to the driving voltage varies rapidly around betatron sidebands. If in the process of frequency modulation the beam response changes too fast, that could cause ions to loose the lock-on and limit the effectiveness of modulated frequency bounce shaking. The small frequency range of cyclotron shaking helps in stabilizing the beam response in the whole process of frequency change. The factor of 15 in the case of Fermilab accumulator reduces the range to only a small fraction of revolution frequency. Note that if the cyclotron frequency falls very near a betatron side band, a beam-ion instability may be excited. This subject is analyzed in a separate paper[7].

The theory presented earlier only deals with equilibrium responses while frequency modulation inevitably introduces time varying effects. The simulation shows that cyclotron shaking with frequency modulation has a longer lasting transient ion motion which requires a longer modulation period. The long transient in the ion motion in the case of driving cyclotron resonant motion is easily understood by noting that there is a lot more kinetic energy in a cyclotron orbit than in a bounce orbit of the same horizontal amplitude. For the cases where cyclotron shaking are effective, the cyclotron kinetic energy is $E_k = (x_m\omega_c)^2 m/2 \simeq 500$ eV, and the bounce kinetic energy is smaller by a factor of $(\omega_b/\omega_c)^2$, or 2.2 eV. When a bounce-shaken ion has exited the end of the magnets through its $\mathbf{E} \times \mathbf{B}$ drift motion, its removal is still predicated on adequate clearing voltages, which may not always be provided. If the ion is cyclotron-shaken, however, it can by virtue of its large energy easily escape the beam potential well and clear completely. This is indeed verified by simulation. Shown in Fig. 3 are the normalized magnetic field and a sample of ion motion driven by frequency modulated cyclotron shaking. Notice that the tracks shown are actually the alias of ion's oscillation motion. The longitudinal positions that locked-on ions escape the beam and hit the vacuum wall are concentrated in a small range. This fact provides a potential diagnostic to directly measure the effect of the cyclotron shaking, by collecting these energetic ions, and measuring their energy spectra. In addition, inside the magnet, if the cyclotron-shaken ion has an elastic collision with an ion, a neutral, or a beam particle which redirects even 5% of its energy into the vertical plane, it will escape the beam entirely. Preliminary calculations indicate that this beneficial phenomenon may occur

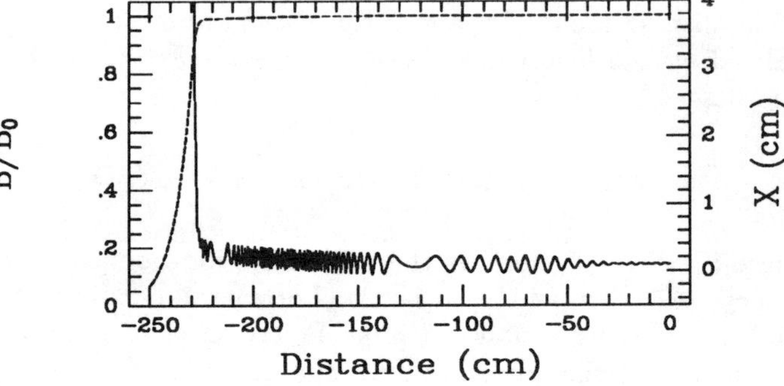

Figure 3: Dipole Field and Ion Motion

at a non-negligible rate in the Fermilab antiproton accumulator.

In short we conclude that the cyclotron shaking is less susceptible to the change of beam response to external driving voltage and therefore may be more effective in reducing ion neutralization effects. In addition, the large energy associated with exciting cyclotron orbits of radii on the order of the beam size may aid significantly in final removal of the ions from the beam. Experimental study is needed, and is presently being pursued on the Fermilab antiproton accumulator.

The authors would like to thank R. Alves Pires and A. Poncet for their valuable help.

References

[1] J. Marriner and A. Poncet, "Neutralization Experiments with Proton and Antiproton Stacks – Ion Shaking", Pbar-note 481, Fermilab Internal Note, March 1989.

[2] R. Alves Pires, "Beam Shaking for the Fermilab Antiproton Accumulator", Proceeding of the Fermilab III Instability Workshop, 1990

[3] Yuri Orlov, "The Suppression of Transverse Instabilities in the CERN AA by Shaking the $\bar{p}$ Beam", *CERN PS/89-01(AR)*, 1989.

[4] R. Alves Pires and R. Dilão, "On the Theory of Shaking", to be published

[5] N. Krylov and N. Bogoliubov, *Introduction to Nonlinear Mechanics* (Princeton University Press, Princeton, N.J., 1943)

[6] Ronald E. Mickens, *Introduction to Nonlinear Oscillations* (Cambridge University Press, 1981) (1989) p. 800

[7] J.B. Rosenzweig, "Beam-Ion Cyclotron Resonance Instability", Proceedings of Fermilab III Instability Workshop, July 1990

Envelope Instability in the Fermilab Booster

P. Zhou, J.B. Rosenzweig*and S. Stahl
Fermi National Accelerator Laboratory†
P.O. Box 500, Batavia, IL 60510

Abstract

The transverse emittance in the FNAL Booster grows at high beam intensity, in a time less than 100 turns. We examine the possible contributions to the fast emittance growth from coherent instability of the beam envelope. Theoretical analysis, through the use of envelope equations, is employed to predict the dependence of envelope instability on the peak current the Booster beam. These predictions are compared to the results of multi-particle tracking calculations.

Introduction

The intensity limit for operation of low energy proton synchrotrons is often determined by the space charge tune shift[1], which can drive individual particle tunes from the operating point onto a significant machine resonance. In this way a relatively small tune shift (usually a few percent) can cause the emittance of a synchrotron beam to grow significantly. Experimentally, this effect manifests itself by a linear asymptotic growth of the emittance with peak beam current. This type of behavior has been observed in experimental studies of the Fermilab Booster[2].

In contrast to scenario described above, linear transport allows much higher intensity beams to propagate without degradation of the beam emittance, as the beam travels much shorter distances in general, and does not encounter the periodic resonance-driving defects found in a circular machine. The limits of stable beam propagation in a linear focusing channel have been demonstrated experimentally by Tiefenback, et al. [3] at LBL. It was found that as long as the bare phase advance per cell σ_0 did not exceed approximately 90 degrees, the beam emittance was stable up to very high intensities. At phase advances above 90 degrees, however, it was verified that instability of the beam envelope caused emittance growth in the transmitted beam. This result is in agreement with the theoretical predictions of Hoffman, Laslett, Smith and Haber[4]. As

---

*Present address: UCLA Dept. of Physics, 405 Hilgard Ave., Los Angeles, CA 90024

†Operated by the Universities Research Association under contract with the U. S. Department of Energy

many synchrotrons are designed with bare phase advance per cell near 90 degrees to minimize the necessary aperture in the machine, it is reasonable to ask whether this effect is important for weaker intensity beams found in circular machines. In particular, this question was raised with regards to the Fermilab Booster, both in its present state and after the upgrade in injected energy from the linac[5].

We review below the standard derivation of rms transverse beam envelope equation, and extend it to include the effects of design orbit curvature, momentum spread and dispersion. The envelope equation obtained is then analyzed for its stability properties under perturbation. This technique is used to examine the Fermilab Booster to determine possible problems with envelope instability.

The Envelope Equation and Instability

The first envelope equations for describing beam propagation including the effects of space charge self-forces were derived by Kapchinskij and Vladimirskij[6] using K-V distribution function.

Sacherer generalized the treatment by concentrating on the evolution of the rms envelopes[7]. This approach is equivalent to following the second moments of the Vlasov equation for the four-dimensional transverse phase space of the beam. Assuming no x-y coupling, the second moments in the (x, p_x) plane, where $p_x \equiv x'$, satisfy the following equations:

$$\overline{x^2}' = 2\overline{xp_x} \tag{1}$$

$$\overline{xp_x}' = -K(z)\overline{x^2} + \overline{x\mathcal{F}_x} + \overline{p_x^2} \tag{2}$$

$$\overline{p^2}' = -2K(z)\overline{xp_x} + 2\overline{p_x\mathcal{F}_x} \tag{3}$$

where $\mathcal{F}_x$ is the beam's space charge self force divided by $\beta^2\gamma mc^2$. The term $\overline{p_x\mathcal{F}_x}$ is related to rms emittance by

$$\epsilon_{rms} = \sqrt{\overline{x^2\,p_x^2} - (\overline{xp_x})^2} \tag{4}$$

In addition, if the beam spatial density has elliptical symmetry, the term $\overline{x\mathcal{F}_x}$ depends only on the rms beam dimensions $\mathcal{X} \equiv \sqrt{\overline{x^2}}$ and $\mathcal{Y} \equiv \sqrt{\overline{y^2}}$,

$$\overline{x\mathcal{F}_x} = \frac{\mathcal{X}Q}{\mathcal{X}+\mathcal{Y}} \tag{5}$$

where $Q = I/(I_0 \beta^3 \gamma^3)$ with $I_0 \simeq 31$ MA for protons. and the rms envelope equations are

$$\mathcal{X}'' + K(z)\mathcal{X} = \frac{\epsilon_x^2}{\mathcal{X}^3} + \frac{Q}{\mathcal{X}+\mathcal{Y}} \tag{6}$$

$$\mathcal{Y}'' - K(z)\mathcal{Y} = \frac{\epsilon_y^2}{\mathcal{Y}^3} + \frac{Q}{\mathcal{X}+\mathcal{Y}} \tag{7}$$

The K-V envelope equations in rms form are contained in Eqs. 6 - 7 as a special case.

The eigenvalues obtained by perturbation analysis on Eqs. 6 - 7 tell the stability of envelope mismatches. Whenever eigenvalues move off of the unit circle on the complex plane, envelope instability occurs. The instabilities start when two eigenvalues collide on the unit circle. Necessary condition for envelope instability is that at least one of the bare phase advances per cell $\sigma_0 > 90$ degrees.

Circular Machines

The major changes to beam behavior in circular machines which have relevance to deriving envelope equations are due to curvature focusing, and to the presence of momentum dispersion[8].

Following Sacherer's analysis and ignoring the chromatic nature of the focusing we arrive at the equation for the rms horizontal beam size X in a circular machine

$$X'' + [K(z) + \frac{1}{\rho^2}]X - \frac{Q}{X+\mathcal{Y}} = \frac{\epsilon_x^2}{X^3} + \frac{1}{\rho X}\overline{x\frac{\Delta p}{p}}. \tag{8}$$

The self-force term takes the same form as before, implying the assumption that the overall beam density still has elliptical symmetry. We then follow the standard way of describing off-momentum orbits in accelerators, splitting the motion into a betatron component x_β (the quantity we have simply termed x to this point) and a dispersion component $\eta(\Delta p/p)$, i.e. $x = x_\beta + \eta(\Delta p/p)$. This approach is approximate in the presence of space charge forces, which introduce coupling that prevents the simple decomposition. In this approximation the rms beam size is $X = \sqrt{\mathcal{X}^2 + \eta^2\sigma_p^2}$ where $\sigma_p^2 = \overline{(\Delta p/p)^2}$, and the horizontal emittance takes the functional form

$$\epsilon_x^2 = \frac{X^2}{X^2 - \eta^2\sigma_p^2}[\epsilon_x^2 + \sigma_p^2(X\eta' - X'\eta)^2] \tag{9}$$

The other term on the right hand side of Eq. 8 is also easily obtained with the approximation. A new set of envelope equations then follow:

$$X'' + [K(z) + \frac{1}{\rho^2} - \frac{\eta}{\rho}(\frac{\sigma_p}{X})^2]X = \frac{\epsilon_x^2}{X^3} + \frac{Q}{X+\mathcal{Y}} \tag{10}$$

$$\mathcal{Y}'' - K(z)\mathcal{Y} = \frac{\epsilon_y^2}{\mathcal{Y}^3} + \frac{Q}{X+\mathcal{Y}} \tag{11}$$

where the vertical emittance is of course unchanged.

The perturbed envelope equations that are used to test for envelope instability are:

$$\delta X'' + [K(z) + \frac{1}{\rho^2} + \frac{\eta}{\rho}\frac{\sigma_p^2}{X^2} + \frac{3\epsilon_x^2}{X^4} + \frac{Q}{(X+\mathcal{Y})^2}]\delta X = \frac{-Q\delta\mathcal{Y}}{(X+\mathcal{Y})^2} \tag{12}$$

$$\delta\mathcal{Y}'' + [-K(z) + \frac{3\epsilon_y^2}{\mathcal{Y}^4} + \frac{Q}{(X+\mathcal{Y})^2}]\delta\mathcal{Y} = \frac{-Q\delta X}{(X+\mathcal{Y})^2} \tag{13}$$

Note that, in the spirit of Sacherer's treatment, we do not perturb the terms inside the emittance expression.

We now apply the analysis on Fermilab Booster[10] ($\sigma_{y0} = 102°$ and $\sigma_{x0} = 100.5°$) and check the predictions with the results of multiparticle tracking using computer code teapot[11]. Results from the envelope instability calculation are shown in Fig. 2 for cases with and without momentum spread. The rms emittance is taken to be

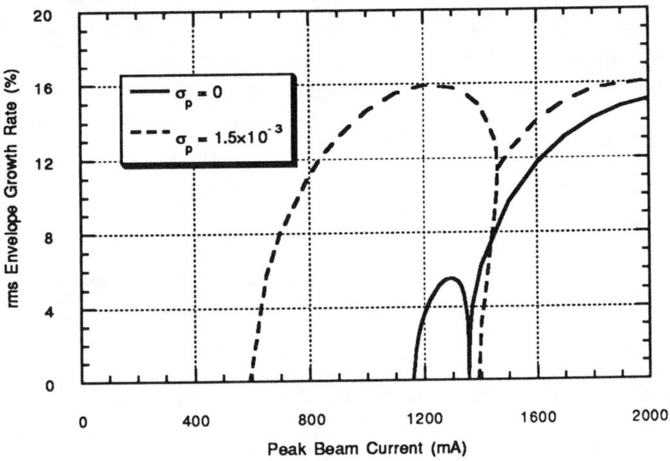

Figure 1: Emittance Growth vs. Beam Current

2.5π mm-mrad in both dimensions. The maximum operating peak current is 570 mA, assuming bunching factor $B = 0.35$ at 200 MeV injection energy. Instability onset occurs at approximately twice this current, assuming constant emittance. With momentum spread $\sigma_p = 1.5 \times 10^{-3}$ it is down to less than 600 mA, which is right on the assumed maximum operating current. This parametric resonance has a large growth rate (~ 15 percent/cell), which peaks at 1.2 A.

The tracking was done with 4000 macro particles with Gaussian density distribution and chromaticity of the ring was set to zero by tuning the sextupoles. DC beam was assumed and the number of space charge kicks used was 12. In Fig. 1, the rms emittance growth after about one turn is plotted as a function of beam current for both the cases of with and without momentum spread. We notice that first of all the emittance growth starts at much lower beam intensity (~ 500mA) compared to the predicted rms envelope instability threshold and increases gradually with beam current. Secondly, in both cases, emittance growth starts at much lower beam intensity than the envelope instability threshold. The amount of emittance growth pre-

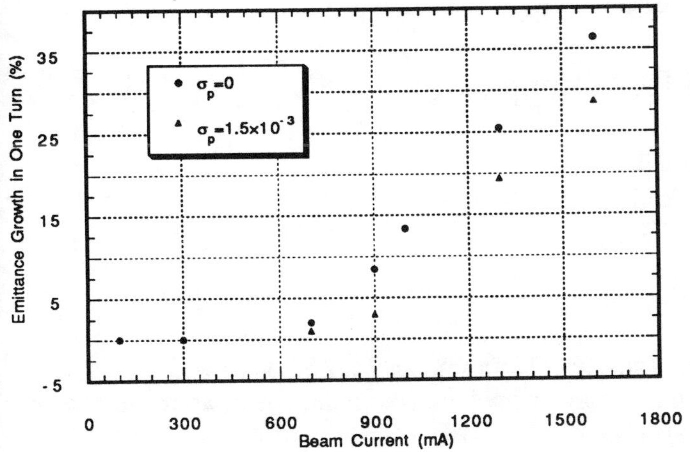

Figure 2: Instability Growth Rate vs Peak Beam Current

dicted by the non-linear field energy theory[9] is negligible compared to the growth observed in the simulation, thus not countable. However, the current at which emittance growth occurs corresponds to when the maximum tune shift pushes the individual particles' phase advance to below 90°. This seems to suggest that it is the fourth order resonance driven by the space charge force, which is exactly what drives the envelope instability. The fact that the emittance can change a noticeable amount in just one cell makes us believe that the simple perturbation method used in the calculation of envelope instability is not appropriate. Nonetheless, it does predict the point where a significant emittance growth occurs. The contribution of the above to the emittance growth is however very small for the Fermilab booster even at maximum operating current, thus cannot be the cause of the observed explosive emittance growth.

In the case including momentum spread, nothing significantly different was observed. Chromaticity was seen to play little role. We have to point out however that the simulation does not include any external impedance or field errors. The failure in this case again points to the fault of the way perturbation was done, i.e. keeping the emittance as a known function while perturbing the envelope equation. The effect of momentum spread, if any, is to introduce more incoherence and seen to reduce the emittance growth slightly.

Conclusions

The treatment of the envelope instability in circular machines is a limited success. While it does not predict the exact onset of emittance growth, it does point to the intensity of significant emittance growth. The former seems, as suggested by simulation, to be determined by when the particles with maximum tune shift, instead of the coherent moment, hit the fourth order resonance. In the case of Fermilab booster, the emittance growth due to this process is negligible even at the maximum operating beam intensity and could not have been responsible for the fast emittance growth observed. The treatment to include momentum spread failed to give reasonable description and simulation results suggest minimal effect of momentum spread.

References

[1] L.J. Laslett, "On Intensity Limitations Imposed by Transverse Space-Charge Effects in Circular Particle Accelerators", in *Proceedings of the 1963 Summer Study on Storage Rings, Accelerators, and Experimentation at Super-High Energies*, 324, Brookhaven National Laboratory, 1963 (BNL 7534).

[2] Y. Chao, J. Crisp, S. Holmes, J. Lackey, W. Merz, "Improving the Fermilab Booster Emittance", *Proceedings of the 1988 European Particle Accelerator Conference*, 663 (World Scientific, London, 1989).

[3] M.G. Tiefenback and D. Keefe, "Measurement of Stability Limits of a Space-Charge-Dominated Ion Beam in a Long A.G. Transport Channel," *IEEE Trans. Nucl. Sci.* **NS-32**, 2483 (1985). Also M.G. Tiefenback, Space-Charge Limits on the Transport of Ions Beam in a Long Alternating Gradient System," LBL-22465, PhD. Thesis, UC-Berkeley (1986).

[4] I. Hoffman, L.J. Laslett, L. Smith, and I. Haber, "Stability of the Kapchinskij-Vladimirskij (K-V) Distribution in a Long Periodic Transport Channel", *Particle Accelerators* **13**, 145 (1983).

[5] F. Mills and J. MacLachlan, private communication, and C. Celata and L.J. Laslett, "Envelope Instability Status Report", unpublished.

[6] I.M. Kapchinskij and V.V. Vladimirskij, "Limitations of Proton Beam Current in a Strong Focusing Linear Accelerator Associated with the Beam Space Charge", in *Proc. of the Second int. Conf. on High Energy Acc. and Inst.*", 274, CERN (1959).

[7] F.J. Sacherer, " RMS Envelope Equation With Space Charge", *IEEE Trans. Nucl. Sci.* **NS-18,** 1105 (1971).

[8] P. Zhou and J.B. Rosenzweig, "Envelope instability in low-energy proton synchrotrons", *Nucl. Instr. Meth.* **A297**, 24 (1990).

[9] M. Reiser, "Theory and Design of Particle Beams", 1989 U.S. Particle Accelerator Summer School notes (unpublished book draft).

[10] E.L. Hubbard, Ed. "Booster Synchrotron", Fermilab Publication FN-405 (Fermilab, 1973).

[11] L. Schachinger and R. Talman, "TEAPOT: A Thin-Element Accelerator Program for Optics and Tracking", *Particle Accelerators*, **22**, 35 (1987)

Multibunch Instability Investigation on a Cavity

equipped with a broad HOM suppressor

E. Karantzoulis
Sincrotrone Trieste, Padriciano 99
34012 Trieste, Italy

Abstract

An investigation of the longitudinal and transverse multibunch instabilities based on an assumed frequency shift behaviour during tuning of the remaining parasitic modes of an ELETTRA cavity prototype equipped with higher order mode (HOM) suppressor(s) is presented and compared with similar results from the undamped cavity. The goal of the analysis is to provide information about the importance of these HOMs and to define (if necessary) shunt impedance thresholds below which no multibunch instabilities occur.

I. INTRODUCTION

ELETTRA is the Synchrotron light source under construction in Trieste optimized for photon energies from ultraviolet to soft x-rays.The required performance is achieved by using an electron (positron) storage ring in the energy range from 1.5 to 2.0 GeV and a full energy linac as injector.

To satisfy the user requirements the storage ring will operate in two different filling modes; namely in a single(or few) bunch mode with high current per pulse (~ 9 mA) and in a multibunch operation with a large number of bunches (~ 400) each containing a relatively low current ($\leq$1 mA). However this high number of bunches may in general lead to multibunch instabilities due to HOMs of the cavities.

The main task of the RF-system is to replace the energy losses due to the radiation and to provide a large momentum acceptance to confine Touscheck scattered particles. The main parameters of the RF-system are listed in table 1.

Table 1
Main parameters for the synchrotron RF-system

Energy	2.0		1.5	GeV
Number of cavities	6		4	
Total rad. losses (+)	380		127.6	keV/turn
Peak effective voltage	1.8		1.7	MV
Synchronous phase	169.76		175.69	degree
Revolution freq.		1.1566		MHz
Harmonic number		432		
RF- frequency		499.654		MHz
Max. RF power (cw)		60		kW/cavity
Cavity radius		263		mm
Cavity axial length		302.6		mm

(+) includes radiation losses due to insertion devices.

The RF-system as well as the cavity design and construction was accomplished by the ELETTRA RF-group

[1]. The cavity has a smooth shape to prevent multipactoring and a few prototypes have been built. The measured resonance frequency is 500.1 MHz and the quality factor Q = 42000 which together with a R/Q of 166 Ω gives a shunt impedance of about 6.9 MΩ. The effective shunt impedance is 3.38 MΩ since the transit time factor of the accelerating mode is 0.7 and consequently the $R_{eff}/Q = 80.5$ Ω. In the following we shall use only R_{eff}/Q since this is the quantity relevant for the instability calculations.

The bandwidth of the accelerating mode $\Delta f = f/Q = 12$ kHz whereas when loading effects are considered it is estimated to be approximately 60 kHz so that the impedance curve will have an effective frequency width of some 200 kHz. To compensate for the beam loading the cavity has to be detuned below the RF-frequency by a maximum shift (for the maximum stored current) of 58 kHz and 82 kHz at 1.5 GeV and 2.0 GeV respectively (see table 1.).

The tuning is achieved by means of an external mechanical tuner acting on the length of the unit [1]. The tuning range is approximately 200 kHz with some 800 Hz /μm with a speed of 200 Hz/sec and a relative frequency stability of $\pm$ 4 10^{-7} within this range. A fast phase control will be also used (and has been tested) with response time of 5ms for 1° of phase variation. It is also worth mentioning that the frequency variation due to temperature changes is about 9 kHz/1°C.

II. MULTIBUNCH INSTABILITIES IN THE UNDAMPED ELETTRA CAVITY

That analysis was made using the well known theory for the calculation of the growth rates and then performing statistics [4]. The results of the investigation are summarized in tables 2 and 3.

Higher frequencies are not quoted here since they are above cut-off. The instability estimations were made for the first bunch shape modes and for the statistics 10,000 data sets were used. In the longitudinal case the instability growth is counteracted mainly by radiation damping (~4.5 ms at 2.0 GeV) whereas for the transverse case the Landau damping time resulting from the betatron frequency spread is dominant with a value comparable to that of the radiation damping.

From the measured HOM values may be observed that the bandwidths of the HOMs are of the order of ~20 kHz i.e. comparable with that of the fundamental. This means that by detuning the cavity to compensate for the beam loading also the HOMs will be detuned by a certain frequency shift, their induced growth rates being thus affected since the value of the impedance will drastically change due to the HOM narrow bandwidths.

Table 2
Longitudinal HOM and their instability probability

| Longitudinal (measured) HOMs | | | Instability probability | |
mode	frequency (MHz)	Q-factor	$R_{sh(eff)}$ (kΩ)	4 cavities (1.5 GeV)	6 cavities (2.0 GeV)
2	944	43500	1600	49%	99%
3	1060	57000	30	rad.damped	rad.damped
4	1421	49500	400	52%	74%
5	1509	53500	400	100%	100%
6	1614	53000	1600	80%	91%
7	1875	42000	400	100%	100%
8	1947	64000	200	88%	99%
9	2087	24000	40	rad.damped	rad.damped
10	2120	30000	2200	99%	100%

Table 3
Transverse (dipole) HOM and their instability probability

| Transverse (measured) HOMs | | | Instability probability | |
mode	frequency (MHz)	Q-factor	$R_{sh(eff)}$ (kΩ/m)	4 cavities (1.5 GeV)	6 cavities (2.0 GeV)
1	743	44000	320	Land.damped	Land.damped
2	749	40000	1240	>>	>>
3	1120	39500	1880	14%	17%
4	1221	89500	8	Land.damped	Land.damped
5	1248	37000	920	21%	26%
6	1307	58000	170	Land.damped	Land.damped
7	1561	38000	6	>>	>>
8	1638	32000	1440	8%	12%
9	1713	62000	144	Land.damped	Land.damped
10	1720	38000	340	>>	>>

However the cavities to be used are not (and can not be) identical due to manufacturing errors, temperature variations etc. (e.g. 800 Hz/μm, 9 kHz/1°C). Therefore although in one cavity might happen that a HOM is damped in another cavity the same mode can be unstable because its frequency has been shifted. This brings in the idea of performing statistics whereby it is assumed that the HOM frequencies are only known within a certain interval. This interval varies for every HOM and it has been measured [4] by shifting the frequency of the accelerating mode within its tuning range (i.e. ± 100 kHz).

By choosing randomly a number (equal to the number of the cavities e.g. 4 or 6) of frequency shifts within the tuning range of the fundamental, the corresponding shifts of the chosen HOM are extracted from the measured data and the growth rates are superimposed. This calculation is performed many times (arbitrary many frequency shifts in combinations of 4 or 6) and the probability of an instability that exceeds the damping rates is evaluated as % of probable instability shown in columns 5 and 6 of tables 2 and 3.

One may interpret this % of instability by imagining that in e.g. 100 days of machine operation t% of the time the mode will be unstable. According to this explanation the longitudinal mode number 4 for instance (see table 2) will be during 100 days of machine operation at 1.5 GeV 52 days unstable whereas mode number 5 or 7 will be always unstable!

It is clear that in the evidence of the above analysis most of the modes are unstable and therefore an adequate damping of the HOM spectrum of the RF cavity is necessary.

III. MULTIBUNCH INSTABILITIES IN A DAMPED CAVITY PROTOTYPE

The concept of broadband damping whereby one couples waveguides directly to the cavity in order to suppress, if possible totally, the HOM spectrum has been chosen and developed for the ELETTRA cavities [2,3].

Two different configurations were tested and measured in especially built prototypes. Until now measurements have been performed for the frequency identification of the remaining modes and their Q values. The first configuration involves only one waveguide coupled to the cavity which is enough to damp almost all longitudinal modes. However the dipole (first transverse) modes due to their polarization may partly be trapped into the cavity at an angle of 90°. Then the second configuration is to use two wave guides at an angle of 90° degrees. The results of these measurements [5] are summarized in tables 4 and 5.

Table 4
Longitudinal modes for 1 and 2 waveguides

| | 1-waveguide | | 2-waveguides | |
mode	frequency (MHz)	Q-factor	frequency (MHz)	Q-factor
1	482	30900	468	26000
2	935	690	935	90
9	2070	640	2070	320

Mode number 1 is the accelerating mode whereas modes 2 and 9 correspond to the ones shown at table 2

Table 5
Transverse(dipole) modes for 1 and 2 waveguides

| | 1-waveguide | | | 2-waveguides | |
mode	frequency (MHz)	Q-factor ϕ=0°	ϕ=90°	frequency (MHz)	Q-factor
1				732	300
2	747	10600	900	737	250
3	1110	2250	-----		
5	1248	1500	-----		
6	1305	700	-----		
7	1556	500	500		
8				1627	240

(compare also with table 3)

From the above tables it is obvious that the 2 waveguide solution is more effective in eliminating the HOMs, however it is interesting to see what kind of growth rates are introduced for the two schemes.

In doing this we must first assume a reasonable value for the HOMs effective shunt impedance. First of all we do not expect that the ratio R_{eff}/Q will exceed that of its undamped value and if one knows no better it seems reasonable to assume that this ratio will remain constant. Some preliminary measurements on the accelerating mode (where measurements are more precise since the resonance can be easier identified) show that R_{eff}/Q decreases about 20% but still we do not know the amount of the reduction for the higher modes.

Accordingly the investigation consists of firstly checking whether with the nominal R_{eff}/Q (taken from the undamped cavity data) the remaining modes are still unstable and if this is the case then by using the instability limits the maximum R_{eff}/Q that ensure the modes stability are found.

Table 6
Longitudinal HOMs and their growth rates

1-waveguide				stability	
mode	Q-value	R_{eff}/Q Ω	growth rate (Hz)	4 cavities (1.5 GeV)	6 cavities (2.0 GeV)
2	690	37_c	700	Unstable	Unstable
2	690	4	70	Rad.damped	Unstable
2	690	2.5	35	Rad.damped	Rad.damped
9	640	1.7_c	40	Rad.damped	Rad.damped
2-waveguides				stability	
mode	Q-value	R_{eff}/Q Ω	growth rate (Hz)	4 cavities (1.5 GeV)	6 cavities (2.0 GeV)
2	90	37_c	90	Unstable	Unstable
2	90	25	70	Rad.damped	Unstable
		15	35	Rad.damped	Rad.damped
9	320	1.7_c	20	Rad.damped	Rad.damped

Table 7
Transverse HOMs and their growth rates

1-waveguide $\phi=0°$				stability	
mode	Q-value	R_{eff}/Q Ω/m	growth rate (Hz)	4 cavities (1.5 GeV)	6 cavities (2.0 GeV)
2	10600	31_c	190	Land.damped	Land.damped
3	2250	48_c	60	Land.damped	Land.damped
5	1500	25_c	32	Land.damped	Land.damped
6	700	3_c	1.5	Land.damped	Land.damped
7	500	0.16_c	0.06	Land.damped	Land.damped
1-waveguide $\phi=90°$				stability	
mode	Q-value	R_{eff}/Q Ω/m	growth rate (Hz)	4 cavities (1.5 GeV)	6 cavities (2.0 GeV)
2	900	31_c	19	Land.damped	Land.damped
7	500	0.16_c	0.06	Land.damped	Land.damped
2-waveguides				stability	
mode	Q-value	R_{eff}/Q Ω/m	growth rate (Hz)	4 cavities (1.5 GeV)	6 cavities (2.0 GeV)
1	300	7_c	1.6	Land.damped	Land.damped
2	250	31_c	5.0	Land.damped	Land.damped
8	240	45_c	7	Land.damped	Land.damped

Next observation from the data in tables 4 and 5 is that the bandwidth of the most HOM is very large due to their low quality factor. For example the longitudinal mode number 2 in the 1-waveguide case has a bandwidth of 1.4 MHz whereas in the 2-waveguide configuration it becomes 10 MHz. This means that no substantial variations of their impedance is expected during the tuning of the cavity and therefore the

statistical analysis applied previously becomes marginal. However for modes with Q greater than 2000 statistical analysis was applied. For Q less than 2000 the norm of the instability frequency shifts (~growth rates) of the HOM is multiplied by the number of the cavities and the product is compared with the damping rates. In this case stability is ensured if the mode growth rate value is approximately N ($\equiv$number of cavities) times smaller than that of the corresponding radiation or Landau damping rate.

Concerning the behaviour of the HOM frequencies during tuning we assume that it is similar to that of the undamped case. However since the exact frequencies are not known, the frequency range between the undamped and damped modes were swept observing the variation of the growth rates. This variation was at the maximum 10% and therefore the calculations are not strongly dependent on the exact frequencies of the HOMs. The same order of variation was found between growth rates of 1.5 and 2.0 GeV and therefore the above quoted values are accordingly averaged. All this is shown in the tables 6 and 7 where also the safe maximum R_{eff}/Q is indicated. The subscript "c" in the values of R_{eff}/Q indicates that the ratio was taken to be the same as in the undamped case and therefore the shunt impedance will be scaled.

IV. CONCLUSIONS

The above analysis shows that in both 1- and 2- waveguide configurations, transverse instabilities induced by the remaining modes are damped. However the 2-waveguide solution appears better since it is more effective in reducing induced longitudinal instabilities. The very low Q of the second longitudinal HOM also suggests that probably this mode is or can be fully damped. It is then clear that the choice of the 2-waveguide HOM suppressor is justified leading towards an instability free cavity.

ACKNOWLEDGEMENTS

I would like to thank M. Svandrlik and the ELETTRA RF-group for many discussions and A. Wrulich for carefully reading the manuscript.

REFERENCES

[1] A. Massarotti et al., "500 MHz Cavities for the Trieste Synchrotron Light Source ELETTRA", in *Proceedings of the 2nd EUROPEAN PART. ACC. CONF.*, Nice, France, June 1990, pp. 919-921.
[2] A. Massarotti and M. Svandrlik, "Proposal for a broadband higher order modes suppressor for the RF cavity", Sincrotrone Trieste, Trieste, Italy, ST/M-90/5, 1990
[3] A. Massarotti et al., "Status report on the ELETTRA R.F. system", This conference
[4] E. Karantzoulis and A. Wrulich, "Multibunch Instabilities Investigation for the ELETTRA Cavities", in *Proceedings of the 2nd EUROPEAN PART. ACC. CONF.*, Nice, France, June 1990, pp. 1618-1620.
[5] M. Svandrlik, Private communication.

Beam Breakup in Recirculating Linacs*

B. C. Yunn

Continuous Electron Beam Accelerator Facility
12000 Jefferson Avenue
Newport News, VA. 23606

ABSTRACT

In general, a recirculation path length in a recirculating accelerator has to be an integer multiple of RF wavelength if recirculations are all in the same direction. However, it is not necessary to require such a relation to be satisfied with respect to the bunching frequency, when the bunch repetition rate is different from the RF frequency. An analytic model of multipass beam breakup(BBU) in recirculating linacs studied by Bisognano and Gluckstern[1] has been generalized to include the case of a subharmonic bunching scheme in the operation of such linacs.

INTRODUCTION

Currently at CEBAF, we are undertaking a feasibility study to accommodate possible two color FELs (IR and UV) within the existing CEBAF structure, taking advantage of the excellent beam quality expected with a superconducting RF accelerator.[2] The envisioned infra–red FEL will use once–recirculated electron beam of 85 MeV from the CEBAF injector. The repetition frequency of FEL bunches is chosen at 7.5 MHz, which produces the peak current required for high gain within the limiting RF–power available. As a result, the bunching is at the 200-th subharmonic of the operating cavity mode of 1500 MHz. The recirculation set–up, which will serve as a test bed for beam dynamics calculations in coming months, has a path length which is 307 RF–period long. Since 307 is obviously not an integer multiple of 200, it has been necessary to extend the work of reference [1].

BBU EQUATIONS

An analytic model of multipass beam breakup in recirculating linacs has previously been studied by Bisognano and Gluckstern, whose notation we will closely follow for the convenience of readers. In this note, a generalization of the model is presented to include the case of a subharmonic bunching scheme in the operation of such linacs.

Consider a bunch(say, the M-th one) entering the n-th cavity at its p-th pass through the linac with its motion represented by a two component column matrix, $U_p(n, M)$ of x and p_x. While traversing the cavity, the bunch will get a momentum kick due to transverse wakes excited in the cavity by preceded bunches. The bunch will then proceed to the next cavity. The equation of transverse bunch motion described in this physical picture is

---

*This work was supported by the U.S. Department of Energy under contract DE-AC05-84ER40150.

$$U_p(n + 1, M) = T_{n+1,n}^{p,p} U_p(n, M)$$

$$+ I T_{n+1,n}^{p,p} G \sum_{q=1}^{n_p} \sum_{L=1}^{M+S(p,q)} U_q(n, L) s_L^{pq}(\omega_n, \tau) , \qquad (1a)$$

and

$$U_p(1, M) = T_{1,n_c}^{p,p-1} U_{p-1}(n_c, M)$$

$$+ I T_{1,n_c}^{p,p-1} G \sum_{q=1}^{n_p} \sum_{L=1}^{M+S(p-1,q)} U_q(n_c, L) s_L^{p-1 q}(\omega_{n_c}, \tau) , \quad (1b)$$

where

$$s_L^{pq}(\omega_n, \tau) = Z_n e^{-\frac{\omega_n}{2Q_n}((M-L)\tau + \tau_p - \tau_q)}$$

$$\times \sin \omega_n((M - L)\tau + \tau_p - \tau_q).$$

$T_{n,m}^{p,q}$ is the transfer matrix from the m-th cavity site at the q-th pass to the n-th one at the p-th pass. G is a 2×2 matrix with all elements equal to zero except $G_{21} = 1$. n_p and n_c are the number of passes and cavities, respectively. τ_p is the summation over recirculation times up to the p-th pass, and $\tau_1 = 0$. In general, τ_p has to be an integer multiple of RF period, τ_{rf} if recirculations are all in the same direction. However, it is not necessary to be an integer multiple of the bunch spacing τ, when subharmonically bunched. To illustrate a complication which may arise, let us consider a two pass system with only one cavity,

$$\tau_2 = 25 \, \tau_{rf}$$
$$\tau = 3 \, \tau_{rf} \, .$$

When a bunch, say the 55-th one arrives at the cavity site on its first pass, the 46-th bunch just crossed the cavity $2\tau_{rf}$ ago. On the other hand, the same bunch on its second pass gets kicked by the 63-rd bunch crossed the cavity one τ_{rf} ahead.

A proper counting of bunches is assured by having $S(p, q)$ defined as

$$S(p, q) = \begin{cases} int.(\frac{\tau_p - \tau_q}{\tau}), & \text{if } p \geq q \\ int.(\frac{\tau_p - \tau_q}{\tau}) - 1, & \text{otherwise.} \end{cases} \qquad (2)$$

It is also necessary to introduce τ_{pq}, R_{pq}, and M_p

$$\tau_{pq} = \tau_p - \tau_q - S(p, q)\tau ,$$
$$R_{pq} = S(p, q) - M_p + M_q ,$$
$$M_p = int.(\frac{\tau_p}{\tau}) \, . \qquad (3)$$

We look for a steady state solution in the asymptotic region where $M \to \infty$. Assuming that such a solution exists, and we write

$$U_p(n, M) = e^{i\Omega M \tau} V_p(n) \qquad (4)$$

A negative imaginary part of Ω would mean an instability. One obtains from equations (1a) and (1b)

$$V_p(n+1) = T^{p,p}_{n+1,n} V_p(n)$$

$$+ IT^{p,p}_{n+1,n} G \sum_{q=1}^{n_p} e^{i(M_p - M_q)\Omega \tau} w^{pq}_n(\Omega) F_n(\Omega) V_q(n) , \qquad (5a)$$

and

$$V_p(1) = T^{p,p-1}_{1,n_c} V_{p-1}(n_c)$$

$$+ IT^{p,p-1}_{1,n_c} G \sum_{q=1}^{n_p} e^{i(M_{p-1} - M_q)\Omega \tau} w^{p-1\,q}_{n_c}(\Omega) F_{n_c}(\Omega) V_q(n_c) ,$$

$$(5b)$$

where

$$F_n(\Omega) = \frac{Z_n}{1 + H^2_n(\Omega) - 2H_n(\Omega)\cos\omega_n \tau} ,$$

$$H_n(\Omega) = e^{-\frac{\omega_n \tau}{2Q_n}} e^{-i\Omega \tau} .$$

All the complications due to subharmonic bunching are now isolated in one place with $w^{pq}_n(\Omega)$, where

$$w^{pq}_n(\Omega) = e^{-\frac{\omega_n \tau^{pq}}{2Q_n}} e^{i\Omega R_{pq}\tau} \qquad (6)$$

$$\times \left(\sin\omega_n \tau^{pq} + H_n(\Omega)\sin\omega_n(\tau - \tau^{pq}) \right) .$$

By iterating equations (5a) and (5b), an eigenvalue problem for the threshold current of multipass beam breakup emerges. In the most general circumstance where none of τ_{pr} is zero, we must deal with a $n_c \times n_p$-1 dimensional matrix:

$$\frac{1}{I} G \bar{V}_p(m) = \sum_{q=1}^{n_p} \sum_{n=1}^{m-1} w^{pq}_m(\Omega) F_n(\Omega) GT^{qq}_{m,n} G\bar{V}_q(n)$$

$$+ \sum_{q=2}^{n_p} \sum_{r=1}^{q-1} \sum_{n=1}^{n_c} e^{-i(M_q - M_r)\Omega\tau} w^{pq}_m(\Omega) F_n(\Omega) GT^{qr}_{m,n} G\bar{V}_r(n)$$

$$(7)$$

where

$$\bar{V}_p(m) = \sum_{q=1}^{n_p} e^{-iM_q\Omega\tau} w^{pq}_m(\Omega) V_q(m) .$$

The first term on the right hand side of equation (7) is understood to vanish in the case of a single cavity system. We also notice that $\bar{V}_p(m) = \bar{V}_r(m)$ if $\tau_{pr} = 0$. As a result, depending on the relation between recirculation lengths and the bunching frequency, the matrix dimension of this eigenvalue problem can be smaller. Therefore, caution is required in setting up the matrix for a numerical study.

When all $\tau_{pr} = 0$, equation (7) reduce to equation (4) of reference [1].

ANALYSIS

For the simplest example, let us consider a one cavity-two pass system. In this case, we have a one dimensional eigenvalue problem

$$\frac{1}{I} G\bar{V}_1(1) = e^{-iM_2\Omega\tau} w^{12}_1(\Omega) F_1(\Omega) GT^{21}_{1,1} G\bar{V}_1(1) . \qquad (8)$$

If $\tau_{21} = 0$, which will be the case when every bucket is filled, $w^{12}_1(\Omega)$ and $F_1(\Omega)$ combine to produce a function

$$\frac{Z_1 H_1(\Omega)\sin\omega_1\tau}{1 + H^2_1(\Omega) - 2H_1(\Omega)\cos\omega_1\tau} .$$

This is the function $Z_1 h_1(\Omega)$ defined in reference [1]. For a high Q mode such that $\frac{\omega_1\tau}{2Q} \ll 1$, a good approximation for the threshold current, when the mode is not near at a harmonic of the bunching, is obtained from

$$I_{th} = \frac{2\omega_1}{eZ''T^2 T^{21}_{1,1} \sin M_2(\omega_1\tau + K)}$$

where K is the solution of

$$K = \frac{\omega_1\tau}{2Q} \cot M_2(K + \omega_1\tau)$$

It is interesting to study how the threshold current depends on the bunch repetition frequency for a given higher order mode(HOM). We have looked at this problem and others by solving equation (8) numerically. The specific parameters chosen for the input are

$$\text{recirculation path length} = 300\ \tau_{rf}$$
$$\text{HOM frequency} = 1890.35\ MHz$$
$$\frac{Z''T^2}{Q} = 1.57 \times 10^5\ \Omega/m^3$$
$$\text{Q of the mode} = 32000$$

Figure 1 is the plot of the threshold current vs. the bunching subharmonic. For a given deflecting HOM, the BBU limit is quite sensitive to the relation between the three frequencies involved; the HOM frequency, the bunching frequency and the recirculation frequency determined by the path length. Let us take a 9-th subharmonic bunching for an example. We read from Figure 1 that BBU instability would not be seen for a beam current less than 38.5 mA. However, this will change if the mode frequency were shifted to a nearby value due to manufacturing tolerances of cavities or for many other reasons. In Figure 2, we see the effect of HOM frequency shifts on the threshold current. Minimum threshold is at $1.06 mA$.

We have also scanned in frequency a 31–dimensional matrix obtained by modeling CEBAF recirculating injector linac with 16 cavities. The acceleration takes place in two cryomodules, each module consisted of 8 superconducting 5–cell cavities. The dipole mode excited is the

1890 mode described in the previous paragraph, and the recirculation path length is adjusted to 307 τ_{rf}. It turns out that this mode produces the lowest threshold. Figure 3 is the result of such a scan showing the threshold current at 4.74 mA for the 200-th subharmonic bunching scheme envisioned. This has been compared with the prediction of a multipass beam breakup simulation code TDBBU developed at CEBAF[3]. Threshold displacement of bunches at the 16–th cavity site on their 2–nd pass is shown in Figure 4 when the current is 4.85 mA. The agreement between these two are quite encouraging.

CONCLUSION

Formulation of multipass BBU as a matrix eigenvalue problem has been properly extended to accommodate general subharmonic bunching schemes. Comparison of results with a bunch to bunch simulation based on equations (1a) and (1b) confirms also the consistency of our formulation, which is generally less time consuming in searching for a threshold and thus provides a useful complementary tool to the simulation.

ACKNOWLEDGEMENT

The author wishes to thank J. J. Bisognano and G. A. Krafft for their help in this work.

REFERENCES

[1] J. J. Bisognano and R. L. Gluckstern, *Proc. 1987 Particle Accelerator Conf.*, pp. 1078 (1987).

[2] G. R. Neil *et al.*, "FEL Design Using the CEBAF Linac," this conference.

[3] G. A. Krafft *et al.*, "Calculating Beam Breakup in Superconducting Linear Accelerators," to be published.

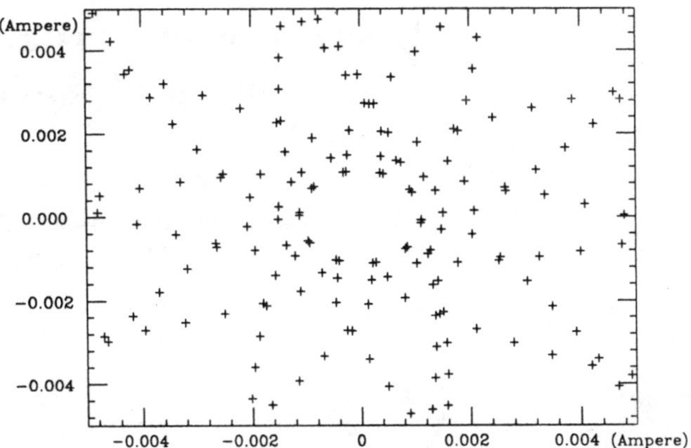

Figure 2. Plot of complex eigenvalues near origin for the case of the 9-th subharmonic bunching. The mode frequency is distributed uniformly with a spread of 2.9 MHz. Minimum threshold current is shown to occur at 1.06 mA.

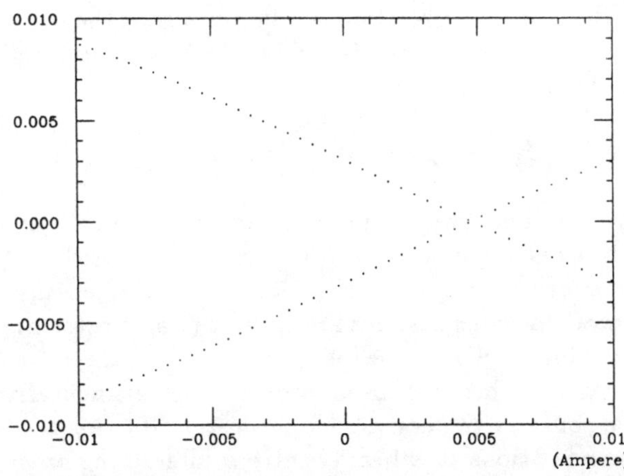

Figure 3. Scan of eigenvalues to locate the threshold current of 16–cavity model of recirculating CEBAF injector in a FEL operating mode.

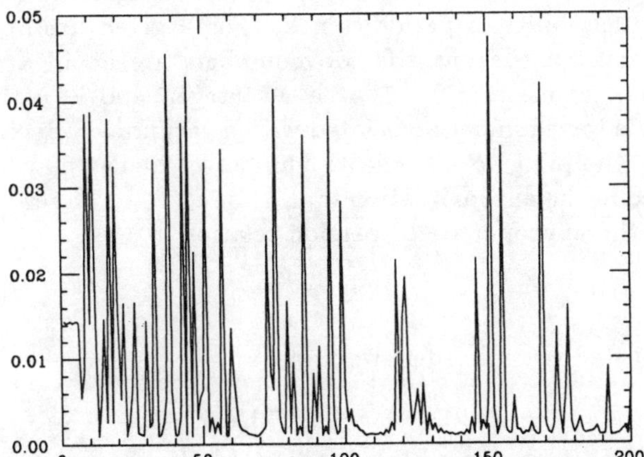

Figure 1. Threshold current of a single cavity–two pass system *vs.* bunching subharmonic. Sensitivity and degree of fluctuation decreases as the Q of offending mode reduces.

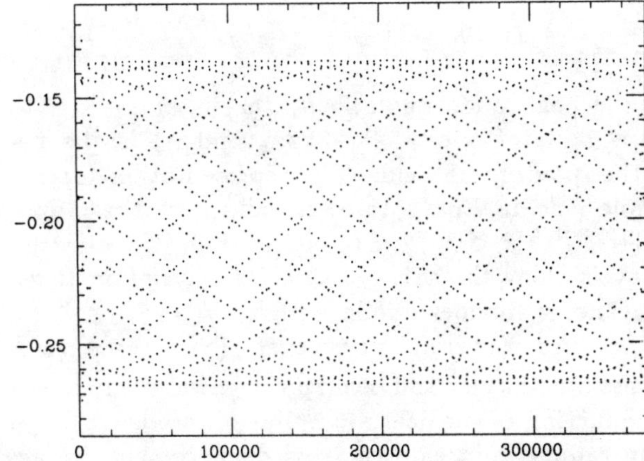

Figure 4. Transverse coordinates of successive bunches passing the last cavity. Bunch motion was tracked for 375,000 τ_{rf} long.

Electromagnetic Instability of an Intense Beam
in a Quadrupole Focusing System

Cha-Mei Tang and Jonathan Krall
Plasma Physics Division, Naval Research Laboratory
Washington, DC 20375-5000

Abstract

Discrete quadrupole focusing systems are subject to the electromagnetic instability previously found in helical quadrupole strong focusing systems. The dispersion relation is derived. Analytical growth rates are obtained. Instability boundaries are established. Numerical solutions of the dispersion relation results show that the overall stability properties are not favorable for long pulse, high-current electron beams.

I. Dispersion Relation

Discrete quadrupole focusing systems, also called FODO lattices, have been used to transport charged particle beams for a variety of applications. Helical quadrupole (stellarator) focusing systems are subject to an electromagnetic instability, which we referred to as the three-wave instability.[1-3] This has been observed experimentally.[4] Here, we consider the corresponding instability[5] for transverse perturbations of a beam centroid interacting with a FODO lattice field and a TE_{11} waveguide mode.

We consider an alternating gradient quadrupole field (B_{qx}, B_{qy}), where

$$B_{qx} = -B_q k_q f(z) y, \qquad B_{qy} = -B_q k_q f(z) x, \qquad (1a-b)$$

$B_q k_q$ is the peak quadrupole field, $f(z)$ is periodic with maximum value of one, $k_q = 2\pi/\lambda_q$ and λ_q is the period of the quadrupole field. The representation for the quadrupole field in Eqs. (1a-b) is valid near the z-axis, i.e., $(x^2 + y^2)^{1/2} << \lambda_q/2\pi$. In equilibrium, the electron beam travels along the axis of a circular waveguide at velocity v_o and is monoenergetic with $\gamma_o = (1 - \beta_o^2)^{-1/2}$, where $\beta_o = v_o/c$. For simplicity, we let $f(z) = \cos(nk_q z)$.

We assume the electron beam propagates within a perfectly conducting cylindrical waveguide of radius r_g. Both the beam radius and beam centroid displacement are assumed to be small in comparison to the waveguide radius. We expect TE_{11} mode to have the largest growth rate, because its electric field peaks on axis.

The dispersion relation for the TE_{11} mode in the presence of the beam is a function of the determinant of a tri-diagonal matrix of the form

$$det \begin{pmatrix} \ddots & \ddots & 0 & 0 & \cdots \\ \ddots & T_{-2} & \frac{K_d^4}{2}S_{-3} & 0 & \cdots \\ \cdots & \frac{K_d^4}{2}S_{+1} & T_0 & \frac{K_d^4}{2}S_{-1} & \cdots \\ \cdots & 0 & \frac{K_d^4}{2}S_{+3} & T_{+2} & \ddots \\ \cdots & 0 & 0 & \ddots & \ddots \end{pmatrix} = 0, \qquad (2)$$

where

$$S_m(\omega, k) = S(\omega, k + mk_q), \qquad (3)$$

$$S(\omega, k) = -(\frac{\omega}{v_o} - k)^2 + k_b^2 \frac{(\omega/v_o - k)^2}{(\omega^2/c^2 - k^2 - \mu_{11}^2)}, \qquad (4)$$

$$T_m(\omega, k) = S_m S_{m+1} S_{m-1} - \frac{1}{2} K_d^4 [S_{m-1} + S_{m+1}], \qquad (5)$$

and $K_d^2 = K_q k_q/\sqrt{2}$. Here, ω and k are the frequency and wavenumber of the TE_{11} mode, $K_q = \Omega_q/v_o$ is the wavenumber associated with the quadrupole field, where $\Omega_q = |e|B_q/\gamma_o m_o c$, $k_b^2 = 2\nu/\gamma_o I_{11}$, $\nu \simeq I_e/17\beta_o$, I_e is the electron beam current in units of kilo-Amperes, $I_{11} = \mu_{11}^{-2}(\mu_{11}^2 r_g^2 - 1)J_1^2(\mu_{11} r_g)$, and $\mu_{11} r_g$ is the the largest positive zero of Bessel function J_1'.

It can be shown from Eq. (2) that the growth rate of the instability is periodic in k. For a given unstable frequency ω_o, the unstable wavenumbers are at all $k = k_0 + nk_q$, where $n = 0, \pm 1, ...$ is an integer, and k_0 is the unstable wavenumber associated with a vacuum waveguide mode. Coupling to an infinite number of modes may be avoided in the approximation that $4K_d^2/k_q^2 \ll 1$. To zeroth order, the approximate dispersion relation is

$$T_0 = 0. \qquad (6)$$

To first and second order, we find

$$det \begin{pmatrix} T_{-2} & (K_d^4/2)S_{-3} \\ (K_d^4/2)S_{+1} & T_0 \end{pmatrix} = 0 \qquad (7)$$

and

$$det \begin{pmatrix} T_{-2} & (K_d^4/2)S_{-3} & 0 \\ (K_d^4/2)S_{+1} & T_0 & (K_d^4/2)S_{-1} \\ 0 & (K_d^4/2)S_{+3} & T_{+2} \end{pmatrix} = 0 \qquad (8)$$

respectively.

II. Analytical Results

Much insight can be obtained from analytical decomposition of the approximate dispersion relation, Eq. (6). This can give us analytical expressions for the growth rates and instability boundaries in parameter space.

The dispersion relation can be rewritten with the current coupling terms grouped together in a term $\bar{\sigma}$,

$$W_0 W_{+1} W_{-1} \Pi = \bar{\sigma}, \qquad (9)$$

where

$$W_m(\omega, k) = (\omega/c)^2 - (k + mk_q)^2 - \mu_{11}^2, \qquad (10)$$

$$\Pi(\omega, k) = \alpha_0^2 \alpha_{+1}^2 \alpha_{-1}^2 - \frac{K_d^4}{2}\left[\alpha_{+1}^2 + \alpha_{-1}^2\right], \qquad (11)$$

$$\alpha_m(\omega, k) = \omega/v_o - (k + mk_q), \qquad (12)$$

and

$$\bar{\sigma} = k_b^2 \alpha_0^2 \alpha_{+1}^2 \alpha_{-1}^2 \left[k_b^4 + W_{+1}W_{-1} + W_0 W_{-1} + W_0 W_{+1}\right]$$

$$-\frac{1}{2}k_b^2 K_d^4 W_0 \left[\alpha_{+1}^2 W_{-1} + \alpha_{-1}^2 W_{+1}\right]. \qquad (13)$$

The polynomial Π can be rewritten as

$$\Pi(\omega, k) = (\alpha_0 - k_q)^2 \Lambda - 2k_q K_d^4 \alpha_0, \qquad (14)$$

where

$$\Lambda(\omega, k) = \left[\left(\alpha_0 + \frac{k_q}{2}\right)^2 - \left(\frac{k_q^2}{4} - K_d^2\right)\right]$$

$$\times \left[\left(\alpha_0 + \frac{k_q}{2}\right)^2 - \left(\frac{k_q^2}{4} + K_d^2\right)\right]. \qquad (15)$$

Since K_d^4 is typically much less than 1, the last term in (14) can be dropped.

There are six waveguide modes, i.e., $W_0 W_{+1} W_{-1}$, and there are six beam modes. Four of the beam modes in (14) have the same characteristics as those in the helical quadrupole case.[2] As in the helical quadrupole case, the electron beam develops orbit instability, when

$$(k_q^2/4 - K_d^2) < 0, \qquad (16)$$

with or without electromagnetic modes in the waveguide. Electromagnetic instability exists at the intersection of the

$$W_{-1}(\omega, k) = 0 \qquad (17a)$$

waveguide mode and the

$$k + (-k_q/2 + \sqrt{k_q^2/4 - K_d^2}) - \omega/v_0 = 0 \qquad (17b)$$

beam mode.

Following the same procedure as outlined in Ref. 2, analytical expressions for the spatial or temporal growth rates can be obtained. The derivation for the spatial growth rate is presented here. Electromagnetic instability occurs at frequency ω which satisfies

$$\sqrt{(\omega/c)^2 - \mu_{11}^2} + k_q = (\omega/v_0) + k_q/2 - \sqrt{k_q^2/4 - K_d^2}. \qquad (18)$$

The dispersion relation (9) can be rewritten as

$$(k^2 - k_0^2)((k + k_q)^2 - k_0^2)((k - k_q)^2 - k_0^2)(k - k_2)^2$$

$$\times [(k - k_4)^2 - \Delta k_1^2][(k - k_4)^2 - \Delta k_2^2] \simeq -\bar{\sigma}, \qquad (19)$$

where $k_0 = \sqrt{(\omega/c)^2 - \mu_{11}^2}$, $k_1 = \omega/v_0$, $k_2 = (\omega/v_0) - k_q$, $k_3 = (\omega/v_0) + k_q$, $k_4 = (\omega/v_0) + k_q/2$, $\Delta k_1 = \sqrt{k_q^2/4 - K_d^2}$, and $\Delta k_2 = \sqrt{k_q^2/4 + K_d^2}$. Defining $k = k_0 + k_q + \delta k$, the imaginary part of δk is the spatial growth rate. The analytical growth rate expression can be simplified to

$$\delta k^2 = \sigma/D_a, \qquad (20)$$

where $\sigma = \bar{\sigma}|_{k=k_0+k_q}$,

$$\sigma \simeq k_b^2 (k - k_3)^2 (k^2 - k_0^2)((k + k_q)^2 - k_0^2)$$

$$[(k - k_1)^2 (k - k_2)^2 - \frac{K_d^4}{2}] \qquad (21)$$

and

$$D_a \simeq -4\Delta k_1 (\Delta k_2^2 - \Delta k_1^2) k_0 (k - k_2)^2$$

$$\times (k^2 - k_0^2)((k + k_q)^2 - k_0^2)|_{k=k_0+k_q}. \qquad (22)$$

When the more complete dispersion relation (8) is considered, growth rate at other intersections in (ω, k) of waveguide modes and other beam modes will collapse to an expression similar to Eq. (20) (with the wavenumber k shifted by $\pm nk_q$).

III. Examples

Here, the dispersion relations (8) and (9) are solved numerically. We verified that i) Eq. (9) is a fair approximation to (8), ii) the beam mode decomposition of (9) is accurate, iii) the boundary of orbit instability is as predicted by Eq. (16) and iv) analytical growth rate expressions are in good agreement with the results from (8).

Parameters for numerical examples are typical of a high-current, induction-accelerator beam, $I_e = 1\ kA$ and $\gamma_o = 7$. We chose a quadrupole gradient of $B_q k_q = 200$ G/cm, quadrupole wavenumber of $k_q = 0.5$ cm^{-1} and the waveguide radius of $r_g = 3\ cm$. Figure 1 plots the dispersion relation (9). The solid circle in Fig. 1 marks the instability denoted by Eqs. (17a,b). Figure 2 has plots of the spatial growth rate as a function of wavenumber, k, for $\omega > 0$ from the improved dispersion relation (8). The instability is periodic in k. The growth rates of all the instabilities are identical, except one. The instabilities at ($k \simeq k_0 - k_q$ and $k \simeq k_0 - 2k_q$) are from coupling to

T_{+2}. The instabilities from coupling to T_{-2} (not shown) are at ($k \simeq k_0 + 2k_q$ and $k \simeq k_0 + 3k_q$). The instability at ($k \simeq k_0 - 2k_q$), similar to the instability marked by the dashed circle in Fig. 1, is erroneous, and is modified to give the same growth rate as that at $k_0 + k_q$ as more terms of the full determinant (2) are included. The analytical result from Eq. (20) is indicated by the $\times$ on Fig. 2.

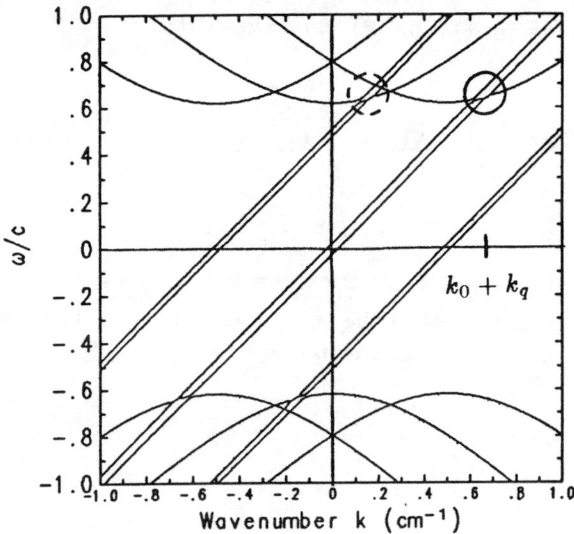

Fig. 1. Dispersion diagram of Eq. (9).

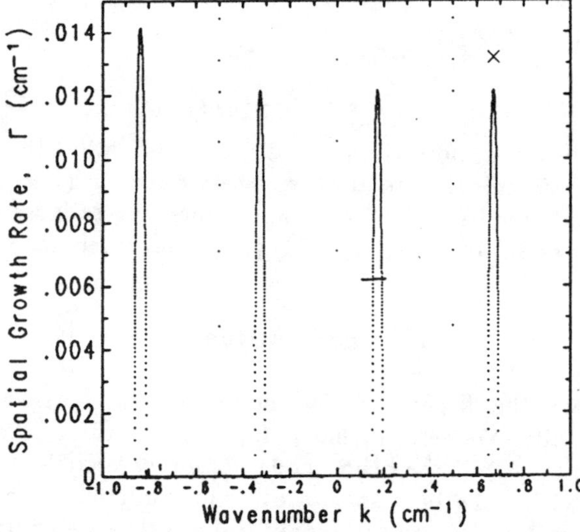

Fig. 2. Spatial growth rates from Eq. (8) for $\omega > 0$.

IV. Summary and Comments

Electromagnetic instability was found to grow on an intense beam in a FODO lattice. Exact and approximate dispersion relations were derived. Analytical growth rates were obtained and are in good agreement with numerical solutions of the dispersion relation. The boundary between electromagnetic and orbit instabilities was obtained and verified. The analysis will be extended to consider cases which include an additional nonzero axial (solenoidal) field.

Note that this instability has not been observed in rf linear accelerators because the electron beam pulse length is short. The instability group velocity is less than the speed of light and propagates out of the tail of the beam pulse before it can grow substantially.

This work was supported by the Defense Advanced Research Project Agency and the Office of Naval Research. We would also like to thank D. Chernin and T. P. Hughes for helpful discussions and T. Swyden for his assistance.

References

[1] T. P. Hughes and B. B. Godfrey, Phys. Fluids **29**, 1698 (1986).

[2] C. M. Tang, P. Sprangle, J. Krall P. Serafim and F. Mako, Part. Accel., **35**, 101 (1991).

[3] J. Krall, C. M. Tang, G. Joyce and P. Sprangle, Phys. Fluids B **3**, 204(1991).

[4] M. Tiefenback, private communication.

[5] T. P. Hughes and D. Chernin to be submitted for publication.

Instability Calculations for the MIT-Bates South Hall Ring [*]

K. D. Jacobs, P. T. Demos, J. B. Flanz, and A. Zolfaghari
MIT Bates Linear Accelerator Center
P.O. Box 846, Middleton, MA 01949

J. Wurtele and X. T. Yu
MIT Plasma Fusion Center
Cambridge, MA 02139

K. Balewski
Deutsches Elektronen-Synchrotron DESY
Notkestr. 85, 2000 Hamburg 52, Germany

Abstract

Instability growth rates and thresholds have been calculated for the MIT-Bates South Hall Ring. Both single bunch and coupled bunch instabilities have been investigated. For single bunch effects, a broad-band impedance budget has been developed. Numerical estimates of the impedances of ring components were made, and required to be within the budget. As part of this, the difficulty of fitting computed loss parameters to those derived from the usual broad band impedance model were studied. We conclude single bunch instabilities should not be a problem. However, coupled bunch instabilities are a serious concern, since all 1812 of the 2856 MHz RF buckets around the ring are filled. A variety of numerical and analytic methods have been used to calculate growth rates and thresholds, and thereby determine acceptable impedance limits. Ring components are being designed to meet these limits. Details of these calculations are presented.

I. Introduction

The MIT Bates Linear Accelerator Center is building an electron storage ring, known as the South Hall Ring (SHR), for medium energy nuclear physics experiments. The ring will have two modes of operation. In pulse stretcher mode, electrons from a 1 % duty factor linac will be injected into the ring, and then extracted uniformly over 1 ms to provide a CW beam to external target experiments. In internal target mode, the beam will be stored in the ring for a longer time (tens of milliseconds, or longer), providing high average currents for targets internal to the ring.

The SHR has a 190 m circumference, and is designed to store currents up to 80 mA at energies from 300 MeV to 1 GeV. The harmonic number is 1812, (ring RF frequency 2856 MHz), and every bucket will contain electrons. In order to achieve the desired beam storage times, with no significant degradation of beam quality, it is necessary to avoid instabilities of the electron beam. Both single bunch and coupled bunch instabilities are of concern. In this paper, we describe the analytic and numerical calculations which have been done regarding beam instabilities in the SHR.

II. Single Bunch Instabilities

A. Thresholds

The single bunch instabilities of greatest concern for the SHR are turbulent bunch lengthening in the longitudinal direction, and the transverse fast blowup instability. To avoid these instabilities, the broad band impedance of the ring must be kept below specified limits. Using analytic calculations, as well as the computer program ZAP [1], we have determined that the longitudinal broad band impedance limit is 25.7 Ω, and the transverse broad band impedance limit is 56 MΩ/m. As long as the total ring broad band impedances are below these limits, single bunch instabilities should not interfere with storage of a 300 MeV, 80 mA beam. (These parameters represent the most severe operating conditions with respect to instabilities.) These impedance limits are fairly large, since the charge in each bunch in the ring is small (1.75×10^8 electrons per bunch.)

B. Impedance determination

To determine the contribution of each ring component to the total ring broad band impedance, we used the computer programs TBCI [2] and MAFIA [3]. The procedure used was to numerically calculate the longitudinal loss parameter $k_\parallel$ for each component as a function of the bunch RMS length σ_l. This was then compared with the $k_\parallel$ vs. σ_l expected theoretically for a bunch traversing a device with some impedance, and from this the impedance was determined.

In order to find the theoretical loss parameter as a function of bunch length, it is necessary to assume some functional form for the impedance. The usual form taken is

$$Z_\parallel(\omega) = \frac{R}{1 + iQ(\omega_r/\omega - \omega/\omega_r)} \tag{1}$$

for the longitudinal case. Here R is the shunt impedance, ω_r is the characteristic resonant frequency, and Q is the usual quality factor.

For a Gaussian bunch of RMS length σ_l, the loss factor is given by

$$k_\parallel(\sigma_l) = \frac{1}{\pi} \int_0^\infty \mathrm{Re}[Z_\parallel(\omega)] e^{-(\omega \sigma_l / c)^2} d\omega. \tag{2}$$

[*] Work supported by the U.S. Department of Energy

Using the expression for $Z_\parallel(\omega)$ in Eq. (1), the above integral can be evaluated to give

$$k_\parallel(\sigma_l) = \frac{R\omega_r}{4QS}\left[(S-1)w\!\left(i\frac{\omega_1\sigma_l}{c}\right) + (S+1)w\!\left(i\frac{\omega_2\sigma_l}{c}\right)\right] \quad (3)$$

for $0 < Q < 1/2$,

$$k_\parallel(\sigma_l) = R\omega_r\left\{e^{\alpha^2}(1+2\alpha^2)[1-\mathrm{erf}(\alpha)] - \frac{2\alpha}{\sqrt{\pi}}\right\} \quad (4)$$

for $Q = 1/2$, and

$$k_\parallel(\sigma_l) = \frac{R\omega_r}{2Q\alpha}\left(1 - \frac{1}{4Q^2}\right)^{-1/2}\mathrm{Re}[zw(z)] \quad (5)$$

for $Q > 1/2$. In these equations, $S = \sqrt{1-4Q^2}$, $\omega_1 = -\omega_r/2Q(S-1)$, $\omega_2 = \omega_r/2Q(S+1)$, w is the exponentially scaled complex error function, erf is the error function, $\alpha = \omega_r\sigma_l/c$, and $z = (\sqrt{1 - 1/4Q^2} + i/2Q)\alpha$.

Figure 1 shows an example of fitting the loss parameter $k_\parallel$ determined by TBCI, as a function of RMS bunch length. Using the parameters of the fit, the broad band impedance can then be calculated.

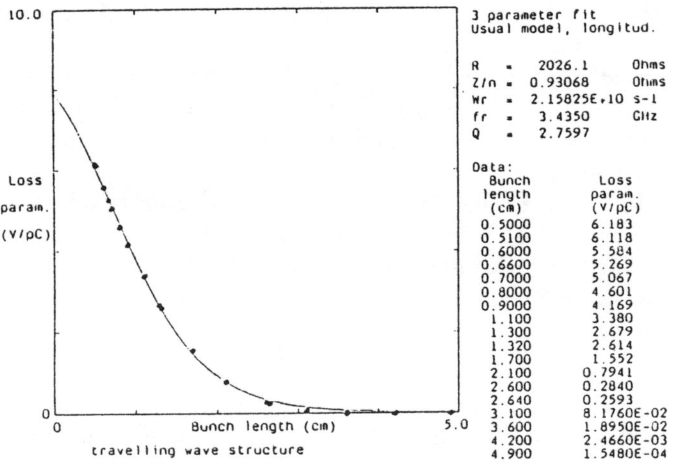

Fig. 1: Longitudinal loss parameter as a function of RMS bunch length for a travelling wave structure. The points are the result of TBCI calculations, and the solid line is a best fit of Eqs. 3–5 to the points.

C. Limitations

The fit of the loss parameter as a function of bunch length can produce erroneous results. In searching through parameter space to do the least squares fit, it is possible to become caught in some incorrect minimum. The incorrect minimum may have parameters (R, ω_r, and Q) significantly different from the true minimum, but nonetheless produce a fit of comparable quality. In order to understand this, a map of parameter space was made. The goodness of the fit was measured by the unnormalized χ^2, equal to the sum of the squares of the deviation of the fit curve from the data points generated by TBCI and MAFIA. All points were weighted equally. The parameters ω_r (the resonant frequency) and Q (the quality factor) were varied over a wide range. At each ω_r and Q, the shunt impedance R was adjusted to obtain the minimum χ^2. The χ^2 obtained in this fashion can then be plotted as a function of ω_r and Q. The results are shown in Fig. 2. From this we can see that it may be difficult to guarantee that the results of the fit are in the proper minimum.

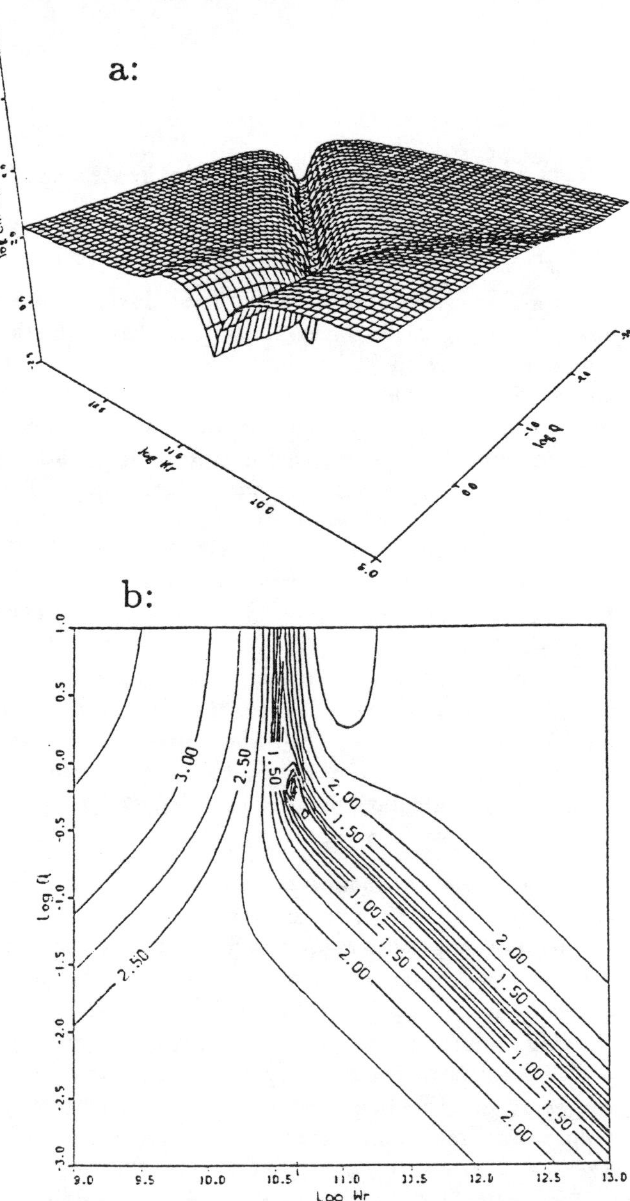

Fig. 2: a: Goodness of fit measured by χ^2, as a function of resonant frequency ω_r and quality factor Q. b: Contour plot showing the same results as the top three dimensional plot. Note that all scales are logarithmic.

In addition, the form of $Z(\omega)$ in Eq. (1) may not accurately describe the impedance of a ring component. Typically, in fact, any one component will have a more complicated dependence of the impedance $Z(\omega)$ on frequency.

Alternate forms of $Z(\omega)$ which may be more appropriate have been proposed [4], but we have not investigated them thoroughly.

D. Summary

Subject to the limitations described here, we have computed the broad band impedance of a variety of ring components. In general, it appears that it should not be too difficult to avoid single bunch instabilities, as long as reasonable care is taken in component designs. Because the 80 mA average current is spread among 1812 bunches in the ring, the charge per bunch is low, giving high impedance limits.

III. Coupled Bunch Instabilities

Although the large number of bunches around the ring helps reduce the problem of single bunch instabilities, there may be a serious problem with coupled bunch instabilities. Each bunch is separated from the next by 350 ps. As a bunch passes through a ring component with resonant modes, such as the RF cavity or an injection kicker tank, these modes may be excited. The fields of the excited modes will act on succeeding bunches, thus coupling the motion (both longitudinal and transverse) of all the bunches together. Instability may result.

The procedure used in analyzing the couple bunch problem was similar to that used for the single bunch case. A limit on the allowable impedances of the resonant modes was determined using analytic and numerical calculations. The impedances of all ring components were calculated using URMEL [5] and MAFIA [3], and compared to the impedance limits to see if they were below the instability thresholds.

The impedance limits found were $500\,\Omega$ for longitudinal modes, and $100\,\mathrm{k\Omega}$ for transverse modes. These limits are such that the growth rates of any unstable beam modes should be less than the damping arising from finite momentum spread and betatron tune spread. The impedance limits were determined using several different means. These include the computer program ZAP [1], analytic calculations and a program developed by Balewski [6], and a numerical tracking program by Yu [7]. The results of all methods were in good agreement.

As part or our investigations, we have developed a simulation code [7] to model the transverse and longitudinal coupled bunch instabilities for a partially filled ring. For the transverse motion in a ring with a single RF cavity, where the length of the cavity is much smaller than the betatron wavelength, we can neglect the betatron phase advance within the RF cavity. The transport of the bunch from the output of the RF cavity to the input of the RF cavity is then given by the ring transport matrix. The kick received by the bunch within the cavity is determined by the wakefields of all bunches (including itself) that have previously passed through the cavity.

The code allows for an arbitrary number of particles in each bunch and various ring elements such as rf cavities, kickers, and internal target structures. For a uniform filling of the ring, the growth rate calculated from our code agrees with ZAP [1] to within 3 %. The simulation extends the work of Thompson and Ruth [8] where bunches have identical charge, but are not uniformly distributed in the ring. By assigning a different charge to each bunch, the two turn injection used in the SHR, which results in a ~ 15 ns of half filled bunches, can be examined. Preliminary calculations show that, for this loading, there are no significant differences in the growth rate from that of a uniformly filled ring.

A frequency domain formalism, which will be presented elsewhere, [7] has been used to find dispersion relations which yield the growth rates of both transverse and longitudinal instabilities for any distribution of bunch charge. For the case of the SHR, with 1812 bunches, the direct simulation technique is probably superior to an analytical solution which requires the inversion of a large matrix.

Further extensions of the simulation to include the slow extraction of the beam during the millisecond storage (in pulse stretcher operation) and the inclusion of internal bunch structure are planned.

IV. Conclusions

We have investigated both single bunch and coupled bunch instabilities in the MIT-Bates South Hall Ring. By calculating the broad band impedances of the various ring components, and comparing these to the allowable broad band impedance limits, we conclude that single bunch instabilities in the SHR should not be a significant problem, as long as reasonable care is taken in component design. This is due to the low current per bunch in the ring.

However, the impedance limits on resonant modes are stringent. This is due to the fact that there are 1812 electron bunches around the ring, each separated by 350 ps. Thus, coupled bunch instabilities can easily arise. By using a variety of design tools and instability calculations, we are working to minimize the potential severity of the problem.

V. References

[1.] M. S. Zisman, S. Chattopadhyay, and J. J. Bisognano, LBL Report No. LBL-21270, Dec. 1986.

[2.] T. Weiland, *Nucl. Instr. and Meth.* **212,** 13 (1983).

[3.] T. Weiland, *Part. Accel.* **17,** 227 (1985).

[4.] A. Hofmann and B. Zotter, "Analytic Models for the Broad Band Impedance", *Proc. of the 1989 IEEE Part. Accel. Conf.*, p. 1041, Chicago, (1989).

[5.] T. Weiland, *Nucl. Instr. and Meth.* **216,** 329 (1983).

[6.] K. Balewski and R. D. Kohaupt, DESY Report No. DESY 90-152, Dec. 1990.

[7.] X. T. Yu, J. S. Wurtele, and K. D. Jacobs, to be published.

[8.] K. A. Thompson and R. D. Ruth, SLAC Pub. 4962, Sept. 1990.

Diagnosis and Cure of a Transverse Instability in the NSLS VUV Ring[*]

J.Rose[†], R.Biscardi, W.Broome, R.D'Alsace, J.Keane, J.M.Wang
National Synchrotron Light Source, Brookhaven National Laboratory

Abstract

A beam instability occurred as a function of cavity tuner position, restricting the operating range of the tuner and requiring the cavity to operate at an elevated temperature to compensate. The cavity mode responsible for the instability was diagnosed and subsequently damped with a tuned probe terminated in 50 Ω. The instability was cured, and has not been observed up to the design current of 1 ampere stored beam.

I. Introduction

The NSLS VUV ring is a 750 MeV electron storage ring with a design current of 1 ampere. The ring parameters are given in Table I.

Table I   VUV Ring Parameters

Nominal Energy	750 MeV
Nominal Current	1 amp
Circumference	51 meters
Rotation Frequency	5.87544 MHz
Nominal Tunes ν_x, ν_y	3.12, 1.2
R.F. Frequency	52.887 MHz
R.F. Peak Voltage	100 kV
Number of R.F. Buckets	9

The ring is in operations serving the UV user community with brief interruptions for maintenance and machine studies. During these operations, it was discovered that the beam would become unstable, undergoing violent oscillations, and above a nominal current threshold of 800 milliamps the beam would dump. The instability was correlated to a particular cavity tuner position. The tuner is required to compensate for the beam loading of the cavity to maintain the generator current in phase with the cavity voltage. In order to avoid the offending tuner position the cavity was operated at an elevated temperature to obtain the equivalent detuning. This study was undertaken to determine the cause of the instability, cure it and so remove the restriction on tuner movement.

* Work performed under the auspices of the U.S. Dept. of Energy
† Grumman Space Systems

II. Diagnosis

Machine studies were undertaken to discover the cause of the instability. Because of the correlation to the R.F. cavity tuner, a coherent bunch-bunch instability caused by a cavity higher order mode was suspected. After several beam crashes a current threshold was found, below which the beam would oscillate but not drop out. Under these conditions the instability could be studied in more detail, although the beam lifetime was severely curtailed. A spectrum analyzer was connected to an R.F. pickup in the cavity to study the cavity response to the beam excitation. Normal fills of seven bunches were used with rotation line separation of 5.8754 MHz. Sidebands at 800 kHz identified the instability as transverse (horizontal) since the fractional horizontal betatron tune times the rotation frequency equates to the sideband separation from the rotation lines. A deviation from the nominal horizontal tune is noted, which is due to operation with a slightly different tune and perhaps a tune shift associated with the instability.

Further studies identified sidebands around two rotation lines which were found to be of particularly large amplitude. The rotation line at 458 MHz had a lower sideband at 800 kHz separation, (horizontal betatron), whose amplitude was >20 dB above the rotation line, shown in Figure 1.

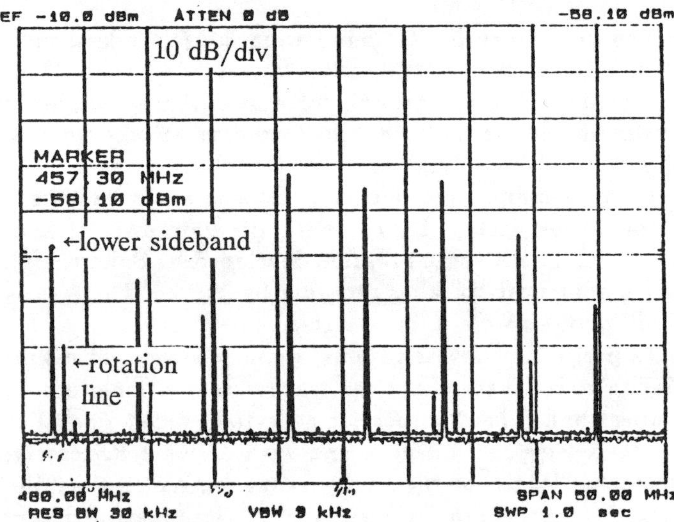

Figure 1. Lower betatron sideband at 458 MHz as measured in cavity with loop pickup, excited with 750 mA stored beam, 7 bunches, during instability.

A second rotation line at 352 MHz (not shown) had a lower betatron sideband whose amplitude was >10 dB above the rotation line. This indicated a high impedance parasitic mode to be slightly lower in frequency than the rotation line, such that the lower sideband can excite the mode. A narrower span with higher resolution clearly shows the upper and lower sideband response of the cavity to be skewed, again indicating that the cavity mode lies close to, but lower in frequency than the rotation line (Figure 2), which is the unstable (anti-damping) side.

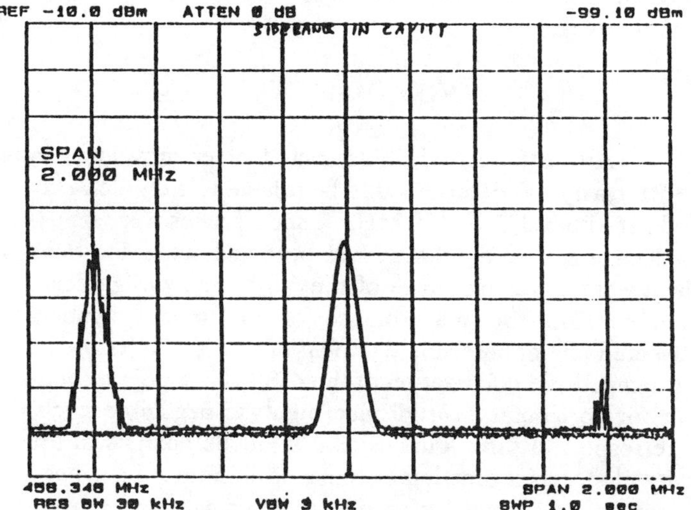

Figure 2. VUV Cavity response as excited by 700 ma 7 bunch.

These measurements were repeated for symmetric bunch fills with three and nine bunches. The beam instability still occurred for the three bunch fill, but could not be initiated with nine bunches (all buckets filled). The VUV ring has known ion trapping problems with nine bunch operation, and a possible explanation for the above result may be the increased Landau damping due to the bunch size increasing during ion trapping.

Cavity mode calculations were then performed with the code URMEL[1] identifying dipole modes near these frequencies. The URMEL cavity parameters for the two modes are listed in Table II.

Table II URMEL Data for dipole modes (mrot=2).

Frequency	Shunt Impedance	Q
353 MHz	4.6 MΩ/m	48139
449 MHz	1.69 MΩ/m	48392

Mode measurements on the cavity were conducted and confirmed cavity modes at these frequencies. Slight differences between calculated and measured mode frequencies are the result of various vacuum ports, slug tuners, and other penetrations on the cavity. Measured frequencies also vary with cavity temperature, and with the RF power on and the cavity actively cooled these modes can shift a MHz or more. A series of measurements made with an identical 52 MHz cavity showed that a dipole mode at 447 MHz had a strong coupling to the loop tuner which would split the dipole into two orthogonal modes with the higher frequency orientation changing frequency over 10 MHz from the retracted to fully inserted position of the tuner. This frequency shift is sufficient to guarantee the mode crossing the lower betatron sideband of a rotation line. The tuner on the VUV cavity is mounted in the horizontal plane, which would lead to the high frequency (magnetic field) perturbation of the dipole to be oriented with a vertical magnetic field in the gap. This corresponds to the observed horizontal instability. The 352 MHz dipole showed a similar behavior, except with the "traveling" or perturbed mode moving lower in frequency indicating electric field coupling to the tuner. The magnetic field in the gap due to the 352 MHz polarization of this dipole was measured to be parallel to the plane of the tuner, corresponding to a horizontal magnetic field in the VUV cavity. This orientation would give rise to vertical beam displacements, and so leads us to believe that it was not responsible for the instability.

III. CURE

Tuned probes were developed on a spare test cavity to damp both the 352 MHz and 458 MHz modes. These probes consist of an E-field probe which is inserted into the maximum electric field of the mode and terminated in a water cooled 50 Ω coaxial termination. There were only four available ports available for the damping antennas, limiting the access to the desired modes. In addition, all experiments had to be performed on a test cavity which was mirror symmetric with respect to the access ports and the tuner, causing differences in the fields between the test cavity and the operational cavity. This was particularly true of the modes we were interested in, which were highly coupled to the tuner.

The E-probe mode suppressors were installed during a maintenance period. The cavity response was measured via loop coupling and is shown in Figure 3 prior to damping and in Figure 4 after damping.

This measurement was made with a network analyzer in a transmission measurement using the main drive loop and a small pickup loop, in a plane at 45 degrees with respect to the tuner providing coupling to both vertical and horizontal polarizations of the dipole.

Significant differences in the performance of the mode suppressors on the test cavity and the VUV cavity were noted. This is assumed to be due to the difference in tuner location and the 8 tuned E-probe mode suppressors which were already in the VUV cavity. The dipole mode at 352 MHz was only damped by approximately 2 dB. The dipole at 458 MHz was damped by greater than 8 dB, however the Q of the mode was not measured before the damping probe was installed, and the response as measured in Figure 3 was made over too wide a span to accurately determine the peak, with too few samples taken for such a high Q mode. The Q was measured after the probes were

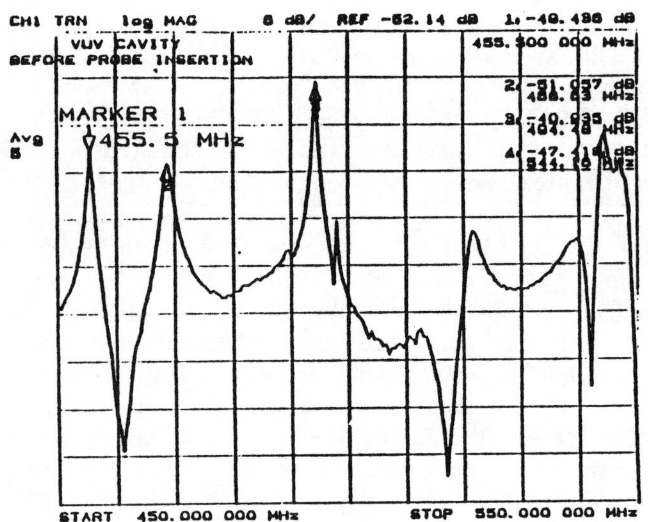

Figure 3. VUV cavity response before installation of tuned HOMS.

installed, and the lower frequency (451 MHz) polarization Q was 150. The higher frequency polarization, believed to cause the instability, response was in the noise, as shown by marker 2 in Figure 4.

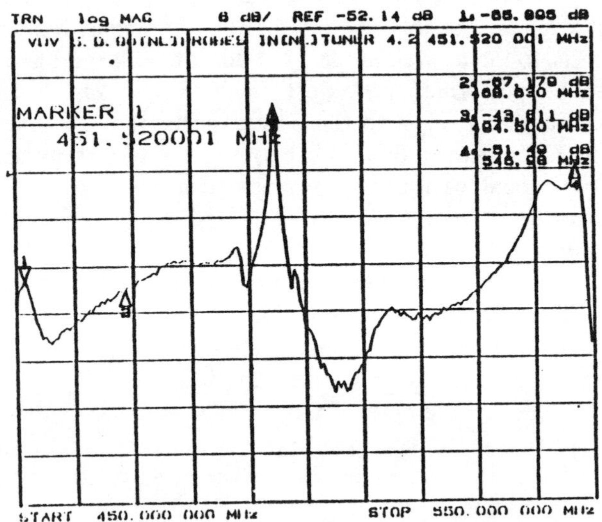

Figure 4. VUV Cavity response after HOMS installation.

After the installation of the mode suppressor probes the transverse instability could not be induced at any tuner position or beam current up to the design current of 1 ampere.

An unwelcome side effect of the probe installation was the shifting of an existing longitudinal instability which was also a function of tuner position to a different tuner position. The temperature setpoint of the cavity was changed to prevent operation about the new location. Prior to the study the tuner had to avoid positions of 130 mm insertion due to the transverse instability and 52 mm due to the

longitudinal. After the transverse modes were damped the longitudinal mode moved to a tuner position of 169 mm, close to the maximum insertion of 178 mm. The operating range was moved from the tuner restricted between 130 and 178 mm to values below 169 mm insertion. This allowed the cavity operating temperature, which also acts as a tuner, to be reduced from 67 degrees C to 59.9 degrees C at 800 ma beam current.

IV. SUMMARY

An instability which restricted tuner movement in the RF cavity was determined to be related to a dipole cavity mode at either 352 or 458 MHz. Tuned probes were developed on a test cavity which had port and tuner locations which were a mirror image of the VUV cavity. Significant differences in the performance of the mode dampers prevented any appreciable damping of the 352 MHz mode, but damped the polarization of the 458 MHz dipole responsible for horizontal beam displacements to negligible levels. In retrospect, a three dimensional RF code analysis of the cavity modes as a function of tuner position may have been warranted, and could have led to only a single probe being installed.

Based on the above experience, a broadband mode damper is being developed for the future synchrotron cavities at the NSLS.

1) T. Weiland, Nuclear Instruments and Methods 216, 329 (1983)

Modified Octupoles for Damping Coherent Instabilities[*]

M. Cornacchia
Stanford Synchrotron Radiation laboratory, Stanford, CA 94309

and

W. J. Corbett
Stanford Linear Accelerator Center, Stanford, CA 94309

and

K. Halbach
Lawrence Berkeley Laboratory, Berkeley, California 94720

Abstract: The introduction tune spread in circular e^+e^- accelerators with modified octupoles to reduce the loss of dynamic aperture is discussed. The new magnet design features an octupole field component on-axis and a tapered field structure off-axis to minimize loss of dynamic aperture. Tracking studies show that the modified octupoles can produce the desired tune spread in SPEAR without compromising confinement of the beam. The technique for designing such magnets is presented, together with an example of magnets that give the required field distribution.

I. INTRODUCTION

The use of octupoles to give amplitude dependent tune spread for Landau damping of transverse collective instabilities is a well known technique. This method is normally used in proton machines when the beam emittance is relatively large. In small emittance accelerators like synchrotron radiation sources, octupoles have found little application. This is because in order to provide appreciable tune spread within the core of the beam, a large octupole strength is required. At large amplitudes the strong octupole field can have a deleterious effect on the dynamic aperture.

In this paper we extend an idea already applied to chromaticity sextupoles [1]. The idea is to design a magnet (modified octupole) whose field behaves like an octupole close to the axis of the magnet (where the core of the beam circulates) and becomes progressively weaker at larger amplitudes in the plane of interest. In this way, the amplitude dependent tune shift required for Landau damping is obtained within the core of the beam, but large amplitude particles can be made to be more stable than with conventional octupoles.

We have carried out tracking studies with modified octupoles in the SPEAR synchrotron radiation source. In Section II, the tune spread required to damp transverse coupled bunch oscillations in SPEAR is estimated, and the field structure of a prototype modified octupole is discussed. In Section III, the results of the tracking simulations are documented. Section IV contains an example of how such magnets can be designed and built.

II. THEORY OF MODIFIED OCTUPOLES

A. Tune Spread Requirements: In order to test the idea on a real machine, we have considered SPEAR, which is now operating as a synchrotron radiation source. Multibunch operation is therefore common practice at a circulating beam current on the order of 100 mA, and the goal current of 200 mA is anticipated in the near future. We operate well below the expected transverse single bunch threshold, but coupled multibunch instabilities are a potential threat.

Although a discussion of the collective instabilities in SPEAR is not in the scope of this paper, we need to know the tune spread requirement to demonstrate the use of modified octupoles for Landau damping. We have computed (with the program ZAP [2], the growth times of transverse coupled bunch oscillations driven by parasitic modes in the rf cavities. The cavity modes have been computed with URMEL [3]. The ZAP results show that in order to stabilize the fastest growing (dipole) mode at 3 GeV and 200 mA beam current, an amplitude dependent tune spread of 0.0015 within $\sqrt{2}\sigma$ of the betatron amplitude distribution is required in whatever plane the instability develops.

B. Modified Octupoles: In what follows we will use complex notation for the magnetic field **B** with $z = x + iy$ and $\mathrm{B}^* = \mathrm{B}_x - i\mathrm{B}_y$. Consider now the following example of a magnetic field structure which behaves like a standard octupole field for small excursions from the beam axis, but is modified for larger off-axis values of z:

$$\mathbf{B}^\star(Z) = K_1 z^3 \bullet (1 + K_2 z^2)^6 \qquad (1)$$

The constants K_1 and K_2 are used to regulate the octupole field strength and spatial field modification, respectively.

The vertical component of the field (B_y) along the horizontal axis satisfies the condition of behaving like an octupole in the vicinity of the origin, but increases more slowly than a pure octupole field at larger amplitudes when $K_2 < 0$ (see Fig. 1). The opposite is true in the vertical plane (B_x increases with increasing y) as was stated more generally in Ref. [1]. Thus, the field cannot decrease simultaneously in both planes.

However, one can exploit the strong focusing aspect of the accelerator when selecting placement of the magnets, and choose opposite signs for the constant K_2 depending on whether a horizontal or vertical tune spread is required. Then, given sufficient decoupling of the beta functions at the locations of the F- and D-octupoles, the loss of dynamic aperture caused by the strong octupole field can be mitigated by the $(1 + K_2 z^2)^6$ term in Eq. (1).

The amplitude dependent tune shift created by the modified octupole field can be estimated by averaging the angular kick over all possible phases. A similar technique has been used to compute the beam-beam resonances . [4,5] For the magnetic field function given in Eq. (1), the horizontal tune shift is given as a function of the amplitude by the following integral:

$$\Delta\nu_x(a_x) = \frac{K_1 l \bullet \beta_x}{4\pi^2 B\rho \bullet a_x} \bullet \qquad (2)$$

$$\int_0^{2\pi} (a_x \cos\phi_x)^3 [1 + K_2(a_x \cos\phi_x)^2]^6 \cos\phi_x\, d\phi_x$$

---

[*] Work supported by Department of Energy contract DE–AC03–76SF00515 and DE-AC03-76SF00098.

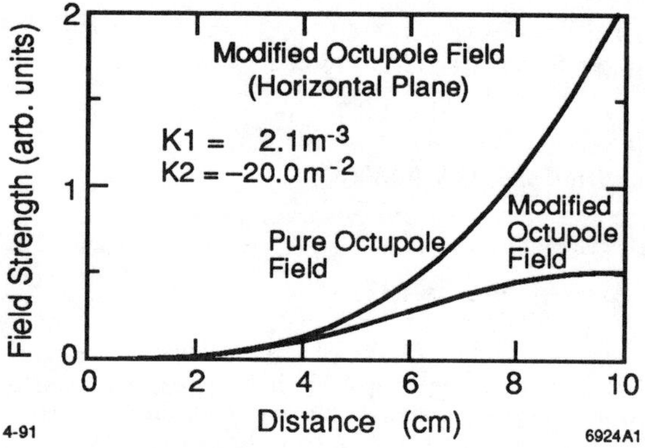

Figure 1. Magnetic field profile in the horizontal plane for a modified octupole.

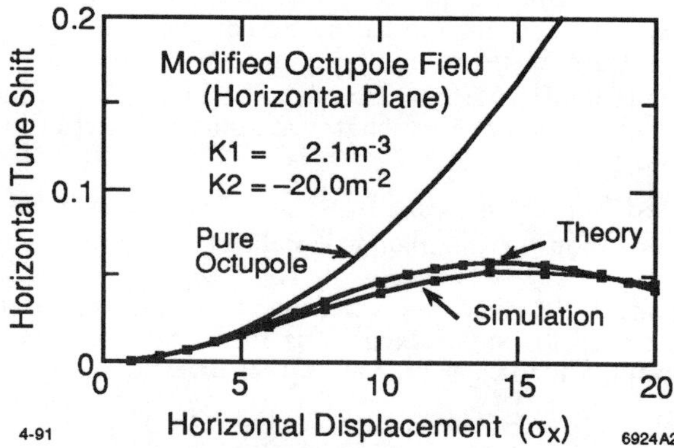

Figure 2. Tune Shift with amplitude in the horizontal plane. Theory and simulation curves refer to modified octupole studies for SPEAR.

where ℓ is the magnet length, β_x is the horizontal β-function at the octupole location (thin lens approximation), a_x is the maximum betatron amplitude at the same location, and the integral is calculated over all possible phases ϕ_x. $B\rho$ is the magnetic rigidity.

For example, Fig. 2 shows the tune shift as a function of amplitude computed with Eq. (2) and as calculated during tracking studies. Considering the sextupole tune shift is not included in Eq. (2), the two results agree quite well. Note that at large amplitudes in betatron space, the amplitude dependent tune shift for the modified octupoles is lower than with conventional octupoles. The tune spread is only required at the core of the beam in order to stabilize the beam against transverse collective instabilities.

III. Tracking Simulations in SPEAR

To simulate the effects of the modified octupoles in SPEAR, tracking studies were made using a collider optics configuration. In this configuration, locations with beta-function ratios β_x/β_y and β_y/β_x of $\approx 6:1$ were found in the insertion regions where $\eta_x = \eta_y = 0$ and sufficient space is available to install new magnets. By selecting locations where the betatron coupling is small, we try to confine the amplitude dependent tune shift to a single transverse plane. In terms of RMS beam size, the limiting geometric apertures for SPEAR are $15\sigma_x$ and $13\sigma_y$ for a horizontal emittance of 650 nm-rad with 20% emittance coupling into the vertical plane. In the same units, the dynamic aperture with sextupoles only is about $50\sigma_x$ by $100\sigma_y$ in the principle planes for betatron tunes of $\nu_x = 5.275$, $\nu_x = 5.110$.

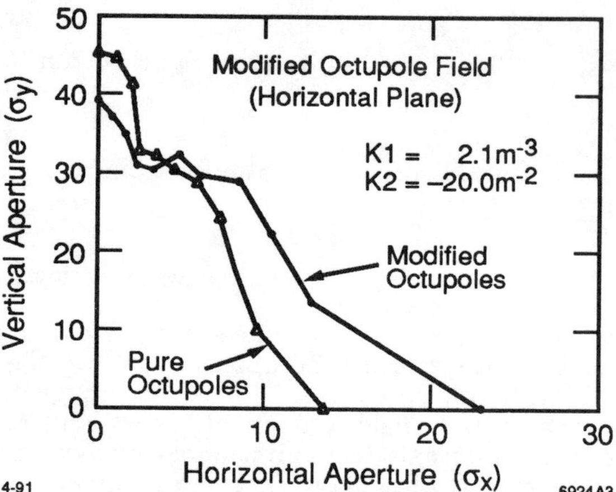

Figure 3. Dynamic aperture increases in the horizontal plane when modified octupoles are used.

The tune shift with amplitude and dynamic aperture are found by tracking with the lattice simulation code GEMINI/FUTAGO [6]. The modified octupole kicks are placed 30 cm from the nearest magnet pole face and represented by a multipole expansion of $B^\star(z) = K_1 z^3(1 + K_2 z^2)^6$. The simulation procedure is to first choose the octupole coefficient for a given transverse plane to produce a tune shift of $\Delta\nu = 0.0015$ at $\sqrt{2}\sigma$ in a given transverse plane. The dynamic aperture for the pure octupole is then found by tracking for up to 20000 turns. No synchrotron oscillations were included. Next, we search for the optimum value of K_2 to increase the aperture in the plane of interest. Generally, K_2 is restricted to values such that $(1 + K_2 z^2) > 0$ is not zero anywhere within the vacuum chamber.

The result of a simulation for a modified F-octupole in SPEAR is shown in Fig. 3. For this case, we have $K_1 = 2.1$ m⁻³, $K_2 = -20m^{-2}$ and the horizontal aperture is increased by almost a factor of two by modifying the field structure. Figure 2 shows the tune shift with amplitude superimposed on the theoretical curve.

In the vertical plane, the F-octupole has less effect because the beta functions are decoupled at the magnet location and the emittance is low. Low vertical emittance, however, complicates the design of a modified D-octupole. In this case, we had to increase the emittance coupling coefficient to value of 30% (limited by insertion device gaps) and reduce the tune shift to < 0.001 (at $\sqrt{2}\sigma_y$) to keep the horizontal dynamic aperture commensurate with the geometric limit.

IV. Design of a Modified Octupole

Figure 4 shows a cross section of the elliptical vacuum chamber (VC), the elliptical good field region (GFR), and a smaller circular GFR. Using the same procedure that was used for the design of modified sextupoles [7] [1,7], Fig. 5 shows the map of the VC and the GFR in the first quadrant of the $x - y$ coordinate system, using the map resulting from $dw/dz = z^3(1 + K_2 z^2)^6$ for $K_2 = -0.002$ cm⁻² $\bullet$ ($z = x + iy$).

It should be noted that the map of the y axis appears to fall onto the map of the x axis. However since one has advanced by 360 degrees on the Riemann surface in the w-plane by the time one has moved from the map of the x axis to the map of the y axis, these two maps are separated by a rather large distance in the w-plane, except very close to the origin. The symmetry of the desired field distributions

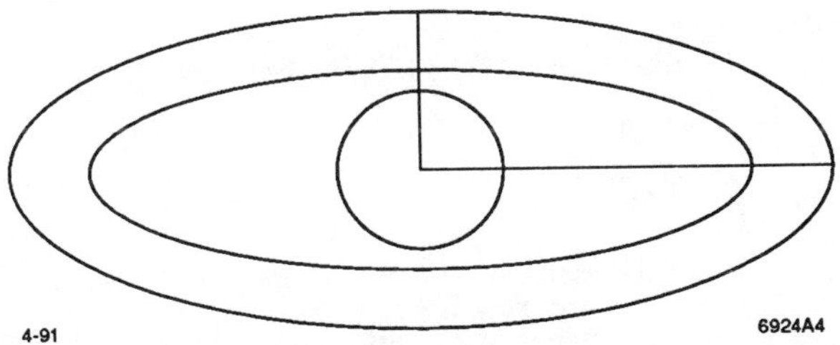

Figure 4. Cross section of a vacuum chamber with half axes 10 an 4 cm, and of good field regions.

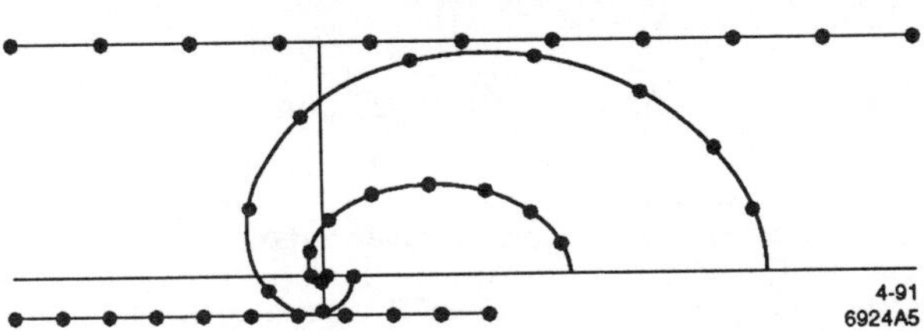

Figure 5. Conformal map of vacuum chamber and good field relations in the first quadrant of the X-Y plane, with flat poles giving a good dipole field in mapped geometry.

obviously demands that the fields on the axes have to be perpendicular to both of them, and their maps.

To get a feeling for the map, the maps of the VC and GFR show the maps of points that are equidistant on the circles in z-geometry before they are "squashed" into ellipses. The map of the circular GFR is so small that one does not see it, not surprising in view of the fact that close to the origin the map is close to $w = 0.25z^4$.

Also shown in Fig. 5 are the proposed flat poles in the w-geometry that should give a good dipole field in w-geometry, thus producing the desired field in z-geometry. Shims at the ends of the polefaces, the rest of the poles, coils, and other details would be incorporated into the design at a later stage.

Figure 6 shows, together with the VC and the GFR, the mapping of the two dipole-poles from the w-geometry to the z-geometry, indicating also the maps of the marked points on the dipole-poles in w-geometry. The excitation of the pole close to the y-axis is $-1/6$ times the excitation of the pole close to the x-axis, reflecting the fact that one pole is closer to the origin than the other, and the properties of the mapping function. This choice is dictated not so much by the desire to keep the excitation down as by the advisability to bring that pole as close as possible to the coordinate origin in order to reduce fringe fields at the ends of the magnet. Slight imperfections and fringe fields that do not break the overall symmetry of the magnet will probably cause a small quadrupole moment for the integrated fields, leading to a (sizeable) breakup of the ideal triple zero field points into three separate zero field points. Special magnetic field measurement techniques that take advantage of this fact to determine sub-harmonics, both in modified octupoles and modified sextupoles, will be described in a forthcoming paper. After measurement of the sub-harmonic strength, a correction is easily made by a small adjustment of the relative excitation of the poles close to the x and y axes.

Figure 7 shows a conceptual cross section of the whole magnet. To generate an integrated octupole field strength of $21\mathrm{Gcm}^{-2}$, the poles close to the x-axis need to be excited by about 478A-turns if one chooses an effective length of the magnet of about 20 cm.

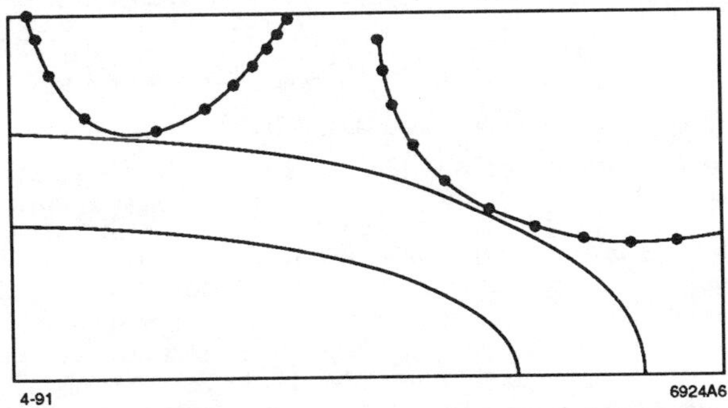

Figure 6. Flat pole faces in W-geometry mapped into Z-geometry.

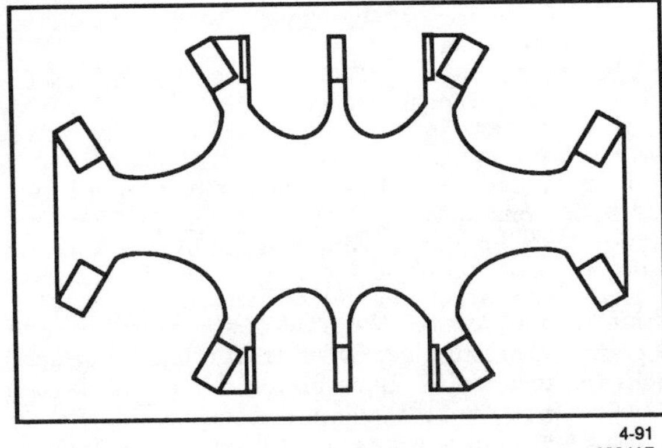

Figure 7. Conceptual design of complete modified octupole.

ACKNOWLEDGEMENTS

The authors would like to thank Albert Hofmann for suggesting the use of modified octupoles to suppress coherent instabilities as an extension of the idea where modified sextupoles are used for chromatic correction.

REFERENCES

[1] M. Cornacchia and K. Halbach, "Study of Modified Sextupoles and Dynamic Aperture Improvement in Synchrotron Radiation Sources," NIM A290, pp. 19–33, 1990.

[2] M. Zisman and S. Chattopadhyay "ZAP User's Manual," LBL–21270/UC–28, 1986.

[3] T. Weiland, NIM, Vol. 216, p. 329, 1983.

[4] L. R. Evans, CERN Internal Report, CERN SPS/DI–MST/NOTE 81/2, 1981.

[5] M. Cornacchia and L. R. Evans, "Nonlinear Resonance Strength Including Coupling," CERN Internal Report, CERN SPS/83/37 [DI/MST], 1983.

[6] E. Forest and H. Nishamura, "Vertically Integrated Simulation Tools for Self-Consistent Tracking and Analysis," in Proc. IEEE PAC, Chicago, IL, pp. 1304–1306, 1989.

[7] K. Halbach, "Understanding Modern Magnets Through Conformal Mapping," Int. J. of Mod. Phys. B, vol. 4, no. 6, pp. 1201-1222, 1990.

SIMULATION OF LONGITUDINAL PHASE SPACE IN THE SLC*

Karl L. F. Bane

Stanford Linear Accelerator Center, Stanford University, Stanford, CA 94309 USA

INTRODUCTION

Upon leaving the damping ring, the SLC beam passes through the Ring-to-Linac transfer line (RTL), the linac, and the arcs on its way to the collision point. In an earlier paper [1] calculations of the longitudinal distributions of the lengthened damping ring beam were presented, calculations that agreed remarkably well with measurements. In this paper we propagate the damping ring distributions through the rest of the machine. Among the effects that are included in the calculations are the curvature of the compressor rf and the limited energy aperture in the RTL, the wakefields in the linac, and the momentum compaction in the arcs. More details can be found in Ref. 2.

THE RTL

Bunch Compression

In a bunch compressor the beam first enters an rf section on the zero crossing of the rf wave (with particles in front gaining energy, those in back losing energy) and secondly traverses a dispersive beam line. The net result is a shorter beam with a larger energy spread. Note that bunch compression in the SLC is discussed in Refs. 3-5.

Let us describe bunch compression by a transformation of the initial beam coordinates (z, δ) to the final coordinates (u, v). Here as elsewhere in this paper a more negative value of position is more toward the front of the bunch; in addition, energy coordinates are given in units of relative energy deviation from the mean. Let us assume the beam position $z = 0$ enters the compressor on the zero crossing of the rf wave. For the moment let us, in addition, assume the compressor rf wave is linear. Then bunch compression is described by

$$v = -az + \delta \quad \text{and} \quad u = z + bv = \alpha z + b\delta \quad, \tag{1}$$

with $\alpha = 1 - ab$. The compressor rf strength factor a is given by $a = eV_c k_{rf}/E_0$, with V_c the peak amplitude of the compressor rf wave, k_{rf} the rf wave number, and E_0 the average beam energy, and b the momentum compaction.

The bunch distribution function $f(z, \delta)$ transforms to the new distribution function $g(u, v)$ as

$$f(z, \delta)\, dz\, d\delta = g(u, v)\, du\, dv = f(z, \delta) J(z, \delta; u, v)\, du\, dv \tag{2}$$

with J the Jacobian for the transformation. Our transformation is area preserving and $J(z, \delta; u, v) = 1$. Therefore,

$$g(u, v) = f(z, \delta) = f(u - bv, au + \alpha v) \quad. \tag{3}$$

The final position and energy distributions are

$$\lambda_u(u) = \int g(u, v)\, dv \quad \text{and} \quad \lambda_v(v) = \int g(u, v)\, du \quad. \tag{4}$$

We assume that the initial (*i.e.* damping ring) distribution function is separable as $f(z, \delta) = \lambda_z(z)\lambda_\delta(\delta)$, with λ_z and λ_δ representing respectively the initial position and energy distributions. Therefore the final distribution function is given by

$$g(u, v) = \lambda_z(u - bv)\lambda_\delta(au + \alpha v) \quad. \tag{5}$$

Several familiar properties of bunch compression can be derived by substituting Eq. (5) into Eq. (4). We see that at full compression, when $\alpha = 0$, $\lambda_u = a\lambda_\delta(au)$. If, in addition the beam energy spread is greatly increased in the process of compression (as is normal), then $\lambda_v = b\lambda_x(-bv)$. When $\alpha \neq 0$ and assuming that both λ_δ and λ_z are gaussian, with the respective lengths σ_δ and σ_z, we see that the final position and

* Work supported by Department of Energy contract DE-AC03–76SF00515.

energy distributions are also gaussian with lengths [4]

$$\sigma_u = \sqrt{\alpha^2 \sigma_z^2 + b^2 \sigma_\delta^2} \quad \text{and} \quad \sigma_v = \sqrt{a^2 \sigma_z^2 + \sigma_\delta^2} \quad. \tag{6}$$

We are mostly interested in the compressor behavior near full compression, i.e. where $\alpha\sigma_z \lesssim b\sigma_\delta$.

The damping ring bunch length increases with current; at higher currents the full length of the bunch approaches half the wavelength of the compressor rf. We will include the effects of the curvature of the compressor rf by replacing az in the transformation equations Eq. (1) by $(eV_c/E_0)\sin k_{rf}z$. Then

$$g(u, v) = \lambda_z(u - bv)\,\lambda_\delta(v + [eV_c/E_0]\sin k_{rf}[u - bv]) \quad. \tag{7}$$

Simulation Results

We begin our simulations with the damping ring distributions, λ_z and λ_δ. To obtain the bunch shape λ_z a modified version of potential well theory is used [1]. Since the damping ring impedance is inductive the bunch lengthens with current. As for the energy distribution λ_δ, it is taken to be gaussian. Below the threshold current for instability N_{th} the energy spread is taken to be constant, above threshold it varies as $N^{1/3}$. For a complete description of the damping ring calculations see Ref.1. Nominally the ring operates at a voltage $V_{ring} = 1$ MV, and in our calculations we will limit ourselves to this voltage. The nominal (low current) bunch length is 4.42 mm, the nominal energy spread is 0.07%. We will take $N_{th} = 1.9 \times 10^{10}$. To obtain the final distributions we substitute the ring distributions into Eq. (7), then into Eqs. (4) and then integrate numerically. The energy aperture limits of the RTL (at least on the North side) are $\pm 2.25\%$ [6]; we therefore set the limits of integration over dv to these values. Note that for the RTL $b = 0.603$ m, $E_0 = 1.15$ GeV, and $k_{rf} = 60$ m^{-1}.

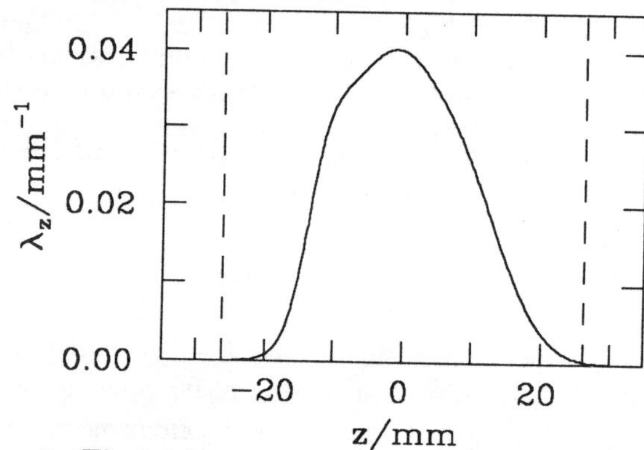

Figure 1. The bunch shape before compression, when $N = 5 \times 10^{10}$ and $V_{ring} = 1$ MV.

As an example, consider a beam with $N = 5 \times 10^{10}$. The initial (*i.e.* damping ring) energy distribution is gaussian with a spread $\sigma_\delta = 0.1\%$. The initial position distribution, with the beam centered on the zero crossing of the compressor rf, is shown in Fig. 1. The dashes locate the nearest extrema of the rf wave. The rms bunch length $z_{rms} = 8.7$ mm and the full-width(-at-half-maximum) $\Delta z_{FW} = 25.3$ mm. These values are respectively twice and 2.5 times the nominal values. Far from the beam core the tails of the distribution become gaussian. These tails, however, are centered on the nominal synchronous point in the ring, which is 4.2 mm to the right of beam center.

If we set $V_c = 30$ MV we obtain the results shown in Fig. 2. Fig. 2c shows contours of phase space; the dashed curve gives the ridge line of the distribution. The dotted lines mark the limits of the $\pm 2.25\%$ energy aperture; 4% of the beam lies outside these limits, and is lost. Fig. 2a shows the final energy distribution. This curve is similar to, though more rectangular

"

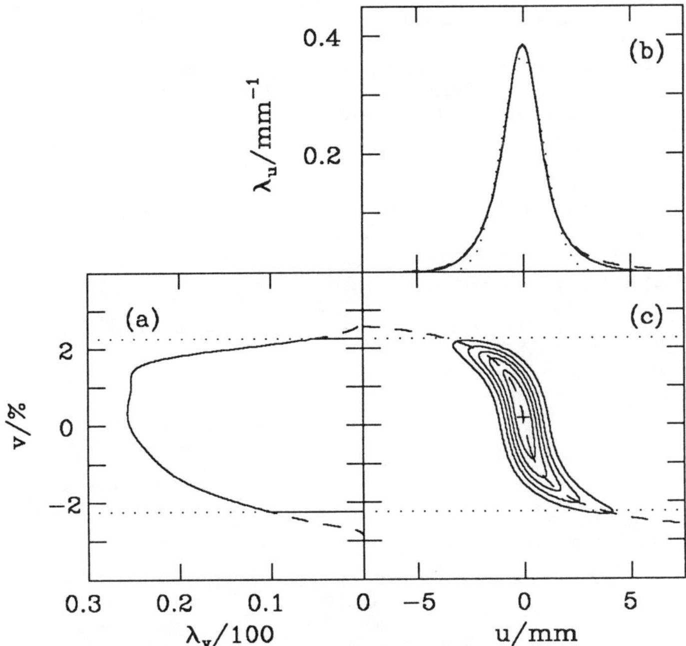

Figure 2. (a) The energy and (b) position distributions, and (c) contours of phase space, after bunch compression. For this example $N = 5 \times 10^{10}$, and $V_c = 30$ MV.

than, the mirror image of the initial bunch shape (see Fig.1). The distribution essentially fills the energy aperture. Its rms width $v_{rms} = 1.17\%$, its full-width $\Delta v_{FW} = 4.04\%$. In contrast, for a low current beam, the spectral shape is almost gaussian with $v_{rms} = 0.7\%$ [see Eq. (6)], and therefore $\Delta v_{FW} = 1.65\%$.

Fig. 2b displays the final bunch shape. As expected the shape is similar to that of the initial energy distribution, *i.e.* similar to a gaussian. The dotted curve gives a gaussian fit to the bunch shape, with an rms length of $\sigma_{ug} = 1.0$ mm. The dashed curve shows the position distribution when there is no limitation in the energy aperture. In Fig. 3 we plot again the same three curves of Fig. 2b, but this time showing in more detail the tails. In the bunch distribution (the solid curve) we see notches in both tails, a result of the energy collimation. The asymmetry in the distribution reflects the asymmetry of the tails of the initial distribution λ_z. The distribution when the energy aperture limitation is removed (the dashed curve) has tails that vary roughly as $e^{-c_1 u}$, with c_1 a constant, rather than as a gaussian (the dots), due to the rf curvature.

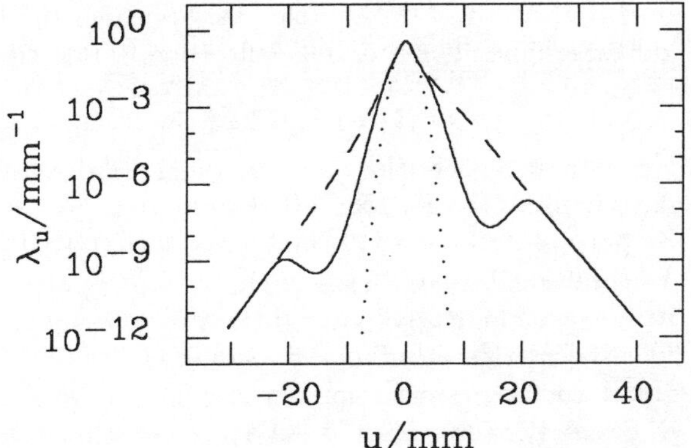

Figure 3. The bunch distributions of Fig. 2b, but with the axes expanded to show the behavior in the tails.

We have repeated the calculations for various currents and compressor settings. The length of the gaussian fit to the resulting bunch shapes σ_{ug} is plotted as function of V_c and as function of N in Fig. 4. The voltage that gives the minimum bunch length increases from 33 MV to 34 MV as we increase

the current from 0 to 6×10^{10}. In contrast the linear rf approximation has its minimum at 32 MV. The dotted curve in Fig. 4a gives the low current, linear rf result. Finally, recall that under the linear rf approximation and with $\alpha = 0$ the final bunch length is given by the initial energy spread, $\sigma_u = b\sigma_\delta$. The dotted curve in Fig. 4b shows this approximation.

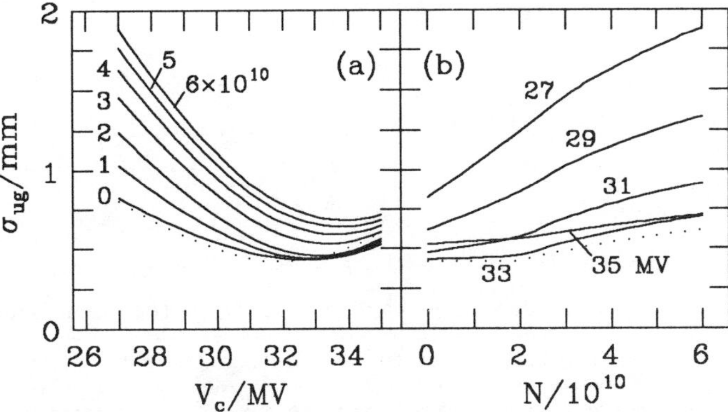

Figure 4. The bunch length after compression as function of (a) V_c and (b) N. The parameter σ_{ug} is the rms of the gaussian fit to the position distribution λ_u.

THE LINAC

The shape of the beam spectrum at the end of the SLC linac has been studied by many authors (*e.g.* Refs. 7-10). For a detailed description of the spectral properties of the bunch at the end of the linac as function of current, bunch length, and rf phase see Ref. 10.

The transformation to the final linac position and energy coordinates (u', v') is described by

$$u' = u \quad \text{and} \quad v' = \frac{E_0}{E_f} v + \bar{v}(u) \quad , \tag{8}$$

with E_f the final average energy. Normally the function

$$\bar{v}(u) = \frac{1}{E_f} [E_0 + E_a \cos(k_{rf} u + \phi) + eV_{ind}(u)] - 1 \quad , \tag{9}$$

with E_a the total peak rf energy gain, ϕ the beam phase with respect to the rf crest, and V_{ind} the induced voltage in the linac, gives essentially all the longitudinally correlated energy variation. The final position and energy distributions are

$$\lambda_{u'} = \lambda_u \quad \text{and} \quad \lambda_{v'} = \frac{E_f}{E_0} \int \lambda_u(u') \lambda_v \left(\frac{E_f}{E_0} [v' - \bar{v}(u')] \right) du' .$$

$$\tag{10}$$

Note that parentheses in the above equation surround the argument of a function.

If E_f/E_0 is large then the function λ_v in the integral of Eq. (10) is much more sharply peaked as function of u' than is λ_u. We can, therefore, approximate

$$\lambda_{v'} = \sum_w \lambda_u(w)/|\frac{d\bar{v}}{dw}(w)| \quad \text{with} \quad w = \bar{v}^{-1}(v') \quad . \tag{11}$$

The sum in this equation indicates that if the inverse function $\bar{v}^{-1}$ is multivalued at v' then all contributions must be summed. According to Eq. (11) wherever $d\bar{v}/du' = 0$ the energy distribution has an infinitely high spike. In reality, however, the distribution must be everywhere finite, and the minimum width of any spike is given by the width of the uncorrelated component of the energy variation.

Consider again the case $N = 5 \times 10^{10}$ and $V_c = 30$ MV. The total length of accelerating structure (needed for finding V_{ind}) is 2744 m. The acceleration gradient is chosen to give $E_f = 47$ GeV. The induced voltage, found by convolving the bunch shape with the wakefield for the SLAC structure [11], is given in Fig. 5. Then, since λ_u and λ_v are known from our earlier calculation, we obtain the final energy spectrum by performing the integration of Eq. (10).

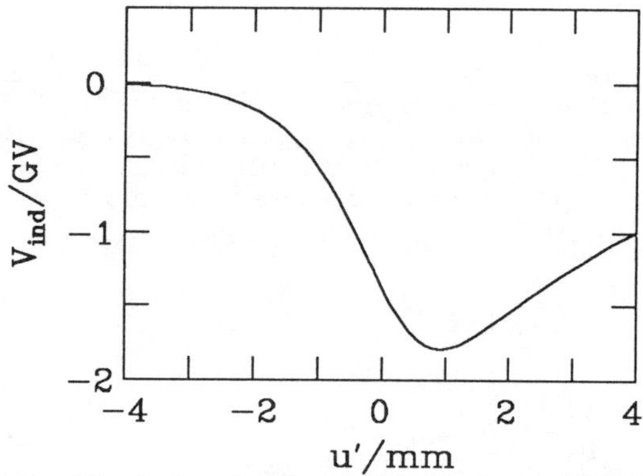

Figure 5. The induced voltage for the entire SLAC linac, when $N = 5 \times 10^{10}$ and $V_c = 30$ MV.

Fig. 6 displays the final spectrum when the beam is positioned at $5°$ and at $10°$ in front of the rf crest. The dashed curves, with their scales on the right, give $\bar{v}(u')$. At the top of the crest, the spectrum has two, widely-separated spikes. As the beam is moved forward the core of the spectrum narrows, and tails begin to grow. At its narrowest the spectrum full-width is $\Delta v'_{FW} = 0.06\%$. To optimize the luminosity we roughly want to maximize the number of particles within an energy window of $\pm 0.5\%$. (See Ref. 10 for more details.) For our examples 44% and 87% of the original beam are within this window; we find that $-10°$ is near the optimal phase.

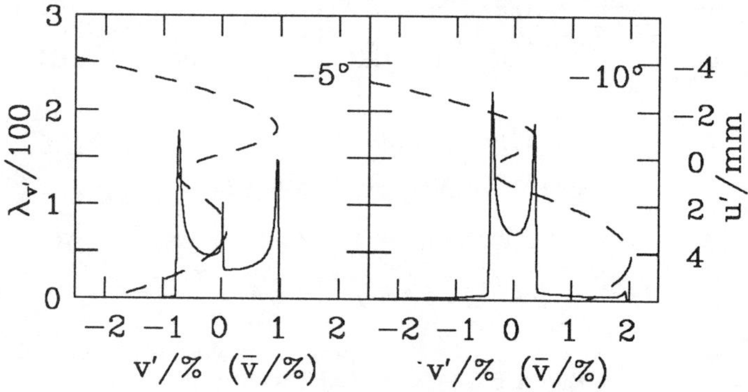

Figure 6. The bunch spectrum at the end of the linac $\lambda_{v'}$ for two values of rf phase. $N = 5 \times 10^{10}$ and $V_c = 30$ MV.

THE ARCS

Since the SLC arcs have non-zero momentum compaction the bunch shape will change in the arcs. Note that bunch compression in the arcs is discussed in Ref. 12. The transformation of a point at the end of the linac (u', v') to its position at the end of the arcs (u'', v'') is

$$u'' = u' + b_{arc}v' \quad \text{and} \quad v'' = v' \quad , \qquad (12)$$

with b_{arc} the arc compaction factor. The final distributions

$$\lambda_{u''} = \int \lambda_{u'}(u'' - b_{arc}v'')\lambda_{v'}(v'') \, dv'' \text{ and } \lambda_{v''} = \lambda_{v'} . \quad (13)$$

If E_f/E_0 is large then we can approximate

$$\lambda_{u''} = \sum_s \lambda_u(s)/\left|\frac{dh}{ds}(s)\right| \quad \text{with} \quad s = h^{-1}(u'') \quad , \qquad (14)$$

and $h(u) = u + b_{arc}\bar{v}(u)$. The bunch shape will have a spike wherever the denominator in Eq. (14) is zero. Again the spikes must be finite, with the minimum width of any spike given by b_{arc} times the width of the uncorrelated component of energy variation. Note also that wherever the slope of $\bar{v}(u)$ is negative the bunch will compress, wherever it is positive the bunch will expand. In our case the beam core will tend to compress and the tails expand. Note that if $\bar{v}(u) = c_1 u$, with c_1 a constant, then the final distribution is given by the initial distribution, but with its width scaled by the factor $(1 + b_{arc}c_1)$.

We numerically solve Eq. (13) to obtain the bunch shape after the arcs. The compaction parameter b_{arc} is 0.145 m. Synchrotron radiation in the arcs will smear the final bunch shape by an rms of 0.08 mm; we include this effect by convolving the bunch shape with a gaussian of this length. (Note the final spectrum is also smeared, by 0.08%.) At the top of the crest, the bunch distribution has two spikes that narrow as the beam is moved forward. Without synchrotron radiation the narrowest full-width would be 0.09 mm, with synchrotron radiation it is limited to 0.21 mm. For the same examples of Fig. 6 we show the final bunch shape in Fig. 7. The dashed curves give the final energy/position correlation, $\tilde{v}(u'') \equiv \bar{v}(u'' - b_{arc}v')$. For our two examples the full-widths of the bunch shape $\Delta u''_{FW}$ are 0.92 mm and 0.23 mm; the fraction of beam in the core $n_{u''c}$ is 75% and 50%.

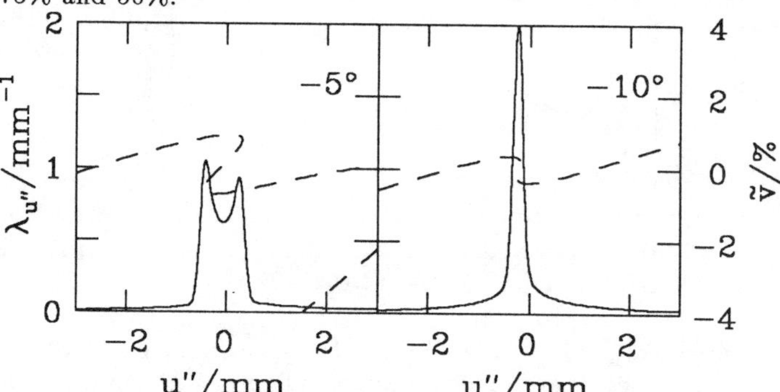

Figure 7. The bunch shape at the end of the arcs $\lambda_{u''}$ for two values of rf phase. $N = 5 \times 10^{10}$ and $V_c = 30$ MV.

Table 1 displays results for other currents, when $V_c = 30$ MV, for phases that yield near optimal luminosity. Included are the linac bunch length σ_{ug} and rf phase ϕ; the full-width and core population of the beam spectrum at the end of the linac, $\Delta v'_{FW}$ and $n_{v'c}$, and of the bunch shape at the end of the arcs, $\Delta u''_{FW}$ and $n_{u''c}$. We see that, in all these cases, the final bunch length $\Delta u''_{FW}$ is near the minimum possible value.

Table 1. Results for different currents. $V_c = 30$ MV.

N (10^{10})	σ_{ug} (mm)	$-\phi$ (deg)	$\Delta v'_{FW}$ (%)	$n_{v'c}$	$\Delta u''_{FW}$ (mm)	$n_{u''c}$
2	0.70	7	0.50	0.90	0.25	0.65
3	0.85	8	0.65	0.85	0.25	0.55
4	0.95	9	0.75	0.85	0.25	0.55
5	1.05	10	0.80	0.80	0.25	0.50

ACKNOWLEDGMENTS

The author thanks M. Ross, J. Seeman, W. Spence, R. Warnock, and members of the SLC Linac Group for having contributed to his understanding of this subject.

REFERENCES

[1] K. Bane and R. Ruth, Proc. of the 1989 IEEE Particle Acc. Conf., Chicago, 1989, p. 789.

[2] K. Bane, SLAC-AP-80 (1990).

[3] T. Fieguth, SLAC-CN-79 (1981).

[4] H. Wiedemann, SLAC-CN-117, (1981).

[5] The SLC Design Handbook p. 6-33 (1984).

[6] SLC Linac Log Book, Dec. 1987.

[7] R. Stiening, SLAC-CN-110 (1981).

[8] The SLC Design Handbook p. 2-21 (1984).

[9] J. Seeman and J. Sheppard, SLAC-PUB-3944 (1988).

[10] K. Bane, SLAC-AP-76 (1989).

[11] K. Bane and P.B. Wilson, Proc. of the 11th Int. Conf. on High-Energy Acc., CERN (1980), p. 592.

[12] A.W. Chao and Y. Kamiya, SLAC-CN-218 (1983).

Bench measurements of Coupling Impedance of AGS Booster Components*

A. Ratti and T. J. Shea
Brookhaven National Laboratory
Upton NY, 11973

Abstract

Quantifying instability thresholds for modern synchrotrons and storage rings requires some knowledge of the accelerator's coupling impedance. To this end, the wire technique has been implemented to measure the longitudinal coupling impedance of AGS Booster devices. The techniques are being refined to allow measurement of RHIC devices at higher frequencies. All the measurements are performed using an HP8753 Network Analyzer controlled via GPIB by a Macintosh computer. The computer provides an environment for automated data acquisition, data analysis, and report generation. Resistive matches between the 50Ω analyzer cables and the 300Ω pipe-and-wire structure allow the use of a simple response calibration in the measurement of S21 to 400MHz. Results from ferrite loaded rf cavities, position monitors and kickers are presented.

I. INTRODUCTION

The AGS Booster is an injector for the AGS and RHIC complex. It will run high intensity protons in order to increase the beam current available to AGS experimenters, as well as heavy ions both for fixed target experiments and eventually for injection in the RHIC. Since over 10^13 protons will be accelerated in each Booster pulse, resulting in high currents on the walls, some consideration has been given to the problem of coupling impedances and the related collective instabilities. This lead to the development of a lab setup capable of measuring the coupling impedance of various devices before installation in the Booster ring. Due to the time constraints of this effort, only the longitudinal effects have been measured; efforts for acquiring information on transverse impedances are planned for the future, both for spare Booster devices and RHIC prototypes.

II. MEASUREMENT TECHNIQUE

The wire method was adopted with resistive matches used to adapt the typical 50Ω measurement equipment to the 300Ω pipe-and-wire geometry. This simple geometry is detailed in Fig. 1. With non-reflective resistive matches, a full two port calibration of the network analyzer is not necessary and the simple response correction suffices. This type of resistive match shows acceptable performance for frequencies up to about 400 MHz.

All measurements were done in a frequency domain in order to take advantage of the large dynamic range available with modern network analyzers. The technique is based on measurements of the transmission parameter S21 of the

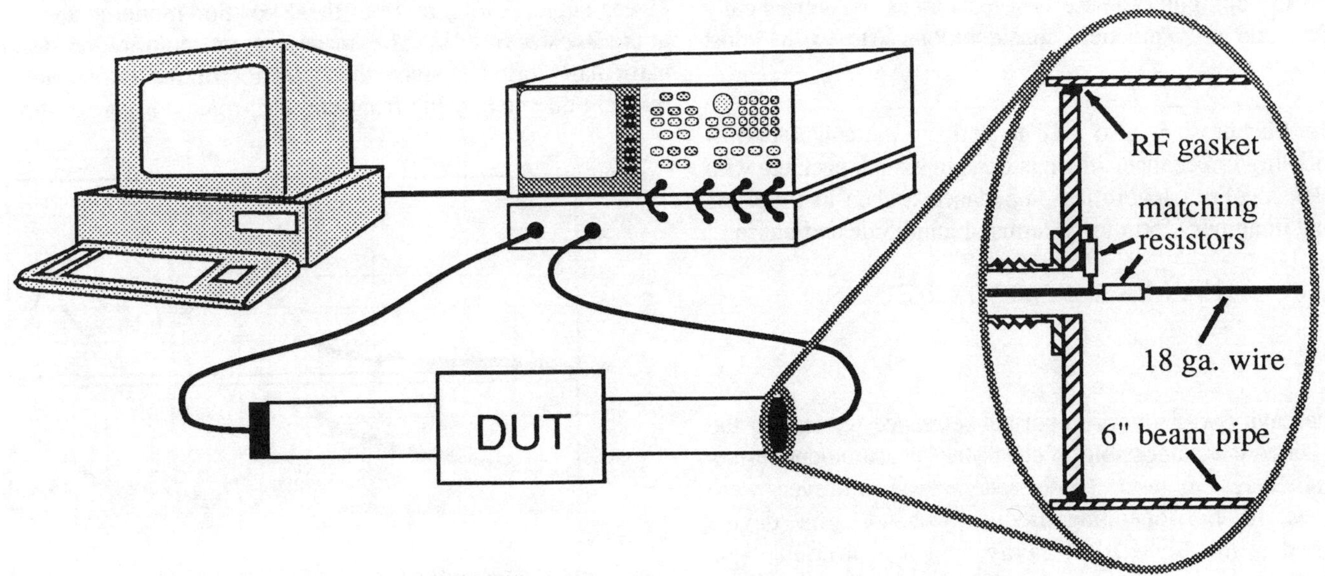

Figure 1. Setup for measuring longitudinal coupling impedance

*Work supported by the US Department of Energy

scattering matrix, from which the corresponding coupling impedance can be calculated. The characteristic impedance of the measuring apparatus was chosen to be 300Ω as set by practical limitations in the thickness of the inner conductor (wire) of the coaxial structure and by the 6 inch beam pipe diameter chosen for the Booster. From the measured scattering

$$Z = 2Z_0 \frac{1 - S_{21}}{S_{21}}$$

parameter, the longitudinal coupling impedance Z can be calculated using [1] :

Particular attention has been paid to keeping the measurement conditions as close as possible to those existing during the device's operation in the ring. To demonstrate the importance of this, the pick up electrode has been measured both with the front end electronics connected and with the two ports open, showing a dramatic change in the behavior of the impedance.

III. DESCRIPTION OF THE APPARATUS

The setup is shown in Fig. 1. A Macintosh computer, running LabVIEW routines interfaces via GPIB to an HP 8753 network analyzer. The calibration of this instrument is one of the most important tasks, and there are two possibilities available in this setup. The instrument can be calibrated on a straight pipe, assumed as a reference "through" to record the power dissipated on the matching resistors and the effect of the setup's electrical length. As an alternative, the analyzer may be calibrated on a standard 50 OHM "through", deferring the subtraction of the reference pipe to a later moment. In this case, the calibration is completed during the processing of the data in the computer. These two techniques are intrinsically identical and only practical considerations lead to the most convenient choice.

The Macintosh is also used to perform the calculation of the coupling impedance, which is done in a math package with complex algebra capabilities. Impedance is then available in real and imaginary form, or in terms of amplitude and phase.

IV. MEASUREMENTS RESULTS

Cavity

The cavity was a very important reference for testing the setup, since it was possible to check the measurement results with other system tests. These system tests, however, were restricted to the operating frequency band. The device measured is the Band III rf cavity, which is a double gap ferrite loaded coaxial resonator [2]. It is tunable over a frequency range from 2.5 to 4.2 MHz by changing the permeability of the ferrite with a biasing field. It is driven with a single ended power amplifier, and internally cross coupled with a "figure of eight" winding in order to guarantee its push

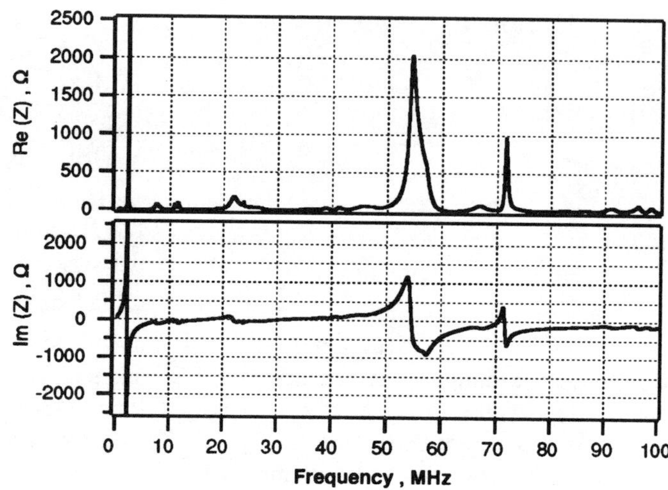

Figure 2. Rf cavity measurement

pull operation. Even though the power amplifier plays an important role in the loaded cavity, it was not practical to measure the system at full power. Nevertheless the results have been successfully compared to values obtained by extrapolation from the system calorimetry test data. The effects of the tuning currents have also been investigated and showed very little or no effect on the resonances found above the fundamental. The results of the measurements of one of the two Band III cavities used in the Booster are shown in Fig. 2.

PUE

The beam pick up electrode (PUE) is a cylinder split at a 45 deg angle. A total of 48 of these position monitors are used in the Booster ring [3]. The measurement setup in this case is particularly critical, since the behavior of the impedance is severely affected by the front end electronics. Figure 3 shows

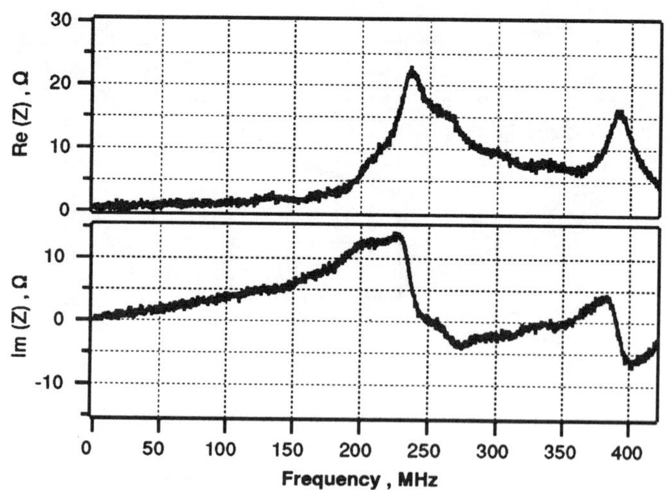

Figure 3. PUE with actual front end connected

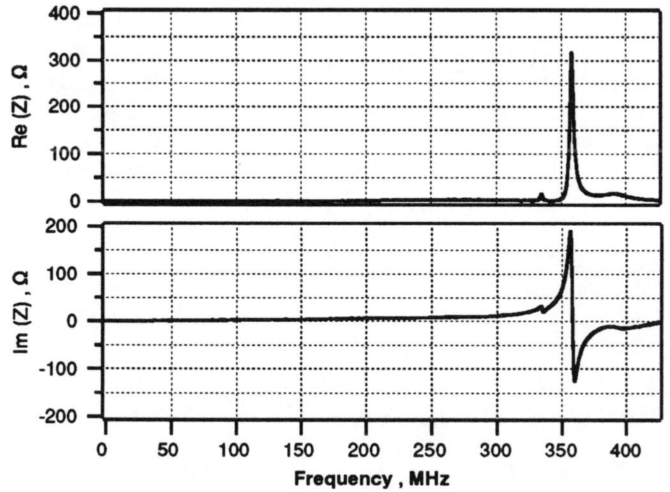

Figure 4. PUE with ports open

the result of a coupling impedance measurement with the position monitor setup that is being used in the Booster ring. In this case, the PUE is connected to a series resistor followed by 10 ft of matched cables leading to the front end electronics. However, if the PUE is left unterminated, the impedance behaves in a totally different manner, as shown in Figure 4.

Kicker

The Booster injection kicker measured here is a ferrite device with the azimuthal flux path split by two copper plates in order to reduce the beam induced effects [4]. The effectiveness of this design has been confirmed by the results of high power tests with and without the copper split. Figure 5 shows the measured longitudinal impedance of this device.

V. STATUS

In the future, most of the vacuum components for RHIC will be measured. For these devices, it will be important to measure at higher frequencies. An improved version of the resistive matches is under development and the time domain gating technique will also be implemented. In parallel with these experiments, some studies are underway with the available computer codes.

VI. ACKNOWLEDGMENTS

The authors would like to thank some of people who shared support and enlightenment: we are grateful to Peter Cameron and Ken Rogers for providing devices for each measurement, to Chris Degen for writing the data acquisition software, to F. Caspars, G. Lambertson and E. Raka for very useful discussions.

REFERENCES

[1] H. Hahn and F. Pedersen, "On coaxial wire measurements of the longitudinal coupling impedance", BNL report BNL-50870, April 1978

[2] R. T. Sanders, et al., "The AGS Booster high frequency rf system", these proceedings

[3] D. J. Ciardullo, et al., "The AGS Booster beam position monitor system" , these proceedings

[4] W. K. Van Asselt and Y. Y. Lee, "Beam induced heating of ferrite magnets", these proceedings

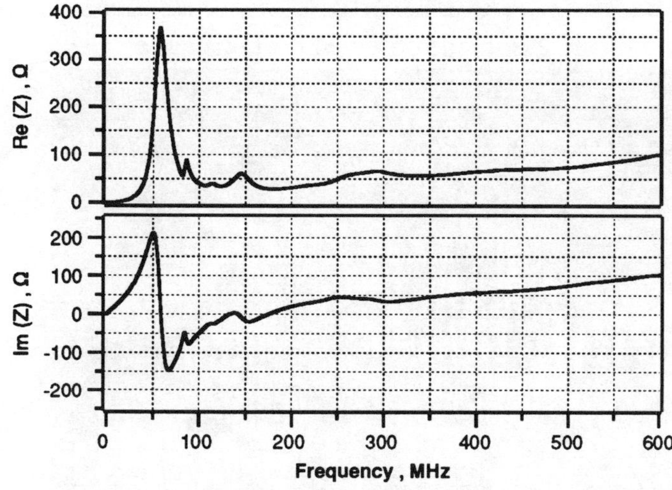

Figure 5. Kicker measurement

A Feedback for Longitudinal Instabilities in the
SLC Damping Rings

Y. Chao, P. Corredoura, T. Limberg, H. Schwarz, P. Wilson
Stanford Linear Accelerator Center, *
Stanford, California 94309

I. INTRODUCTION

Longitudinal coupled bunch instabilities which are being observed in the SLC Damping Rings present a serious limitation to the ultimate luminosity of the colliding beams. It has been observed at intensities above 1×10^{10} particles per bunch with growth rates of the order of 10^4 per second. At lower intensities we were able to control the growth by detuning the RF cavity through temperature or voltage changes. As the SLC operating intensity approaches the design value above 5×10^{10}, a more fundamental cure is definitely needed. In this report we sum up the studies done so far in characterizing and simulating this instability and the proposed remedies.

Tables 1 and 2 sum up the relevant machine and RF parameters of the Damping Rings. The longitudinal coupled bunch instability arises in the Damping Rings due to the coupling between the normal modes of the 2 bunch system and the cavity parasitic modes. A schematic of the Damping Ring is shown in Figure 1. The 2 cavities diametrically across from each other are powered by a single RF source which decouples any RF manipulation from the normal mode in which the 2 bunches execute synchrotron oscillations with opposite phases. If the latter mode,

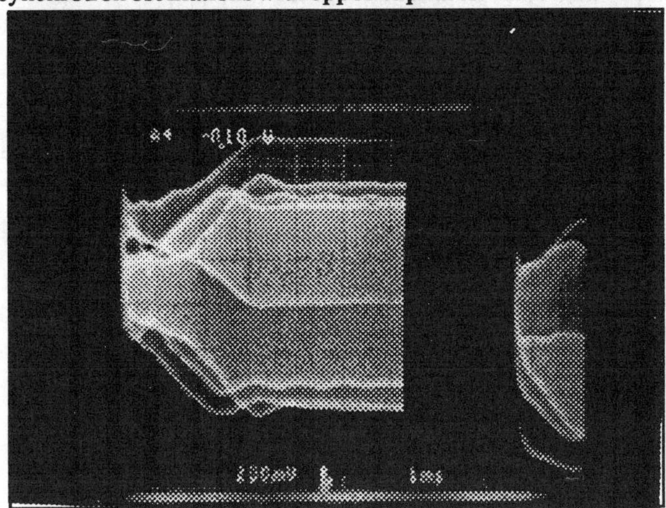

Figure 1. Damping Ring layout and π-mode

Momentum compaction (α)	0.01814	
Revolution frequency	8.5	MHz
Energy	1.21	GeV
Particles per bunch	5 & 5	10**10

Table 1. Damping Ring Parameters

Harmonic number	84	
Unloaded Q	24000	
Coupling factor (β)	2.5	
Synchronous angle	39.2	degree
Cavity detuning angle	68.3	degree
Shunt impedance	17.5	MΩ
Generator power/cavity	25.0	kW

Table 2. RF Parameters

Figure 2. π-mode observed over one cycle

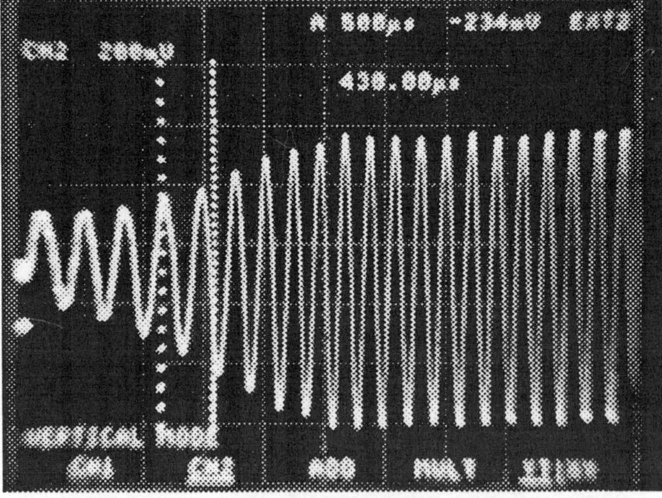

Figure 3. π-mode sampled at lower frequency

* Supported by DOE under contract # DE--AC03-76SF00515

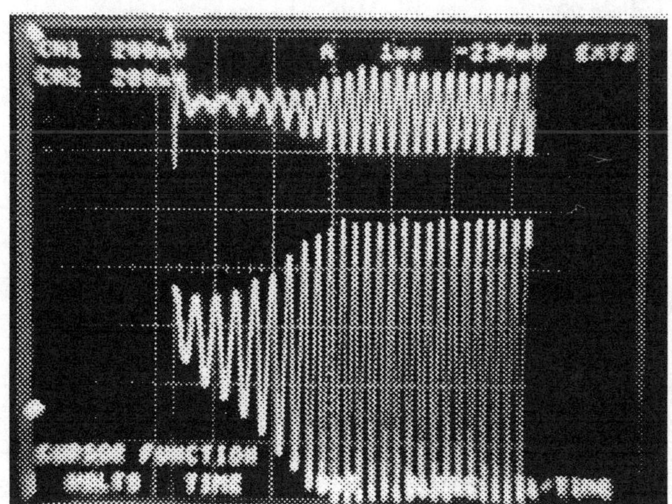

Figure 4. Observed 0 and π-modes

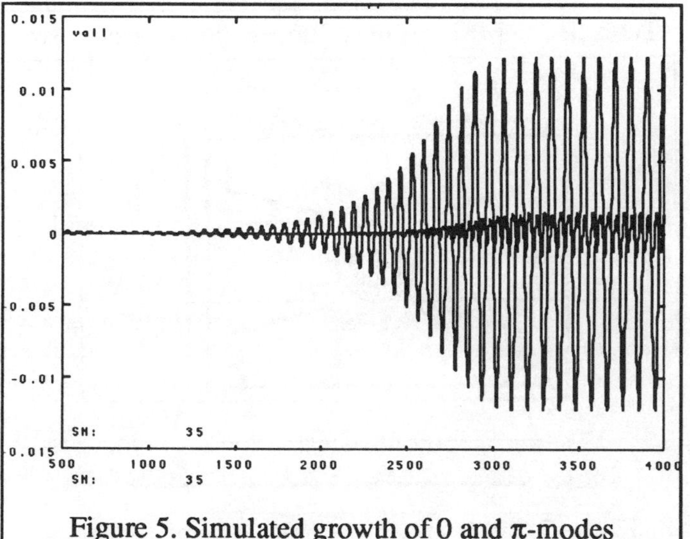

Figure 5. Simulated growth of 0 and π-modes

henceforth called the π-mode, falls within the resonance peak of some higher odd harmonic of the cavity, a resonant growth can result. Such instability was demonstrated in Figure 2, which displays a Damping Ring position monitor located at a high dispersion point picking up the growing amplitude in the energy oscillations throughout the cycle. Figure 3 shows a more clear pattern of this growth using a lower sampling frequency. Such signals from individual bunches can be added or subtracted to manifest the two normal modes, namely the zero mode and the π-mode. These are shown in Figure 4. It is seen that as the growth in the π-mode progresses, there is an accompanying growth in the

Mode no.	Δ freq.	R MΩ	Q	τ (rise time) μs
83	0	2.0	30000	17.4
127	0.4	2.64	30000	9.11
177	0.1	0.53	29000	23.1
209	0.3	0.82	27600	12.8
221	0.4	1.7	61800	8.22

Table 3. Odd parasitc modes calculated by URMEL

zero mode. Very similar phenomenon has been observed in the simulation program as shown in Figure 5 where the growth in the π-mode appears to influence the otherwise damped zero mode. In the latter case this is understood to be caused by the nonlinearity in the RF potential probed by the large amplitude oscillation, which deforms the limiting phase space contour and puts some growth into the zero mode, which in turn reaches a limiting value due to Robinson damping. It can not be determined presently whether the observed zero mode growth shares the same origin as the one in simulation. Calculation based on the RF cavity geometry was done with URMEL to look for the offending parasitic modes. Table 3 gives a partial list of the possibly existing odd harmonic modes which could drive the π-mode. Note that in calculating the respective growth rates, the frequencies of these modes have been shifted to the proper synchrotron sideband as indicated in the second column so as to give the worst possible growth. They are thus more benign in reality than is indicated in the table.

II. STUDY OF POSSIBLE SOLUTIONS

Several solutions have been studied in addressing the π-mode problem. Due to the special RF configuration in the Damping Ring, we can not contain the π-mode with any feedback based on RF manipulations using the direct π-mode signal. We give a brief account of these studies below, followed by a description of the passive damping cavity being designed and expected to be in service for the second 1991 SLC run.

Single bunch damping

A feedback mechanism has been experimented where the signal of a single bunch is picked up and used in the RF feedback which affects both bunches in the same way. This effectively damps the bunch being monitored and antidamps the other. The damped bunch is the one to be immediately extracted. This method was effective in controlling the final amplitude of the extracted bunch when the growth is mild in the first place.

Tune splitting

The idea of tune splitting is to superimpose the RF with an odd harmonic waveform so as to modify the RF potential differentially for the 2 bunches. The resulting tune splitting would then disrupt the coherence of the π-mode. Study based on computer simulation indicates that in order to control a π-mode

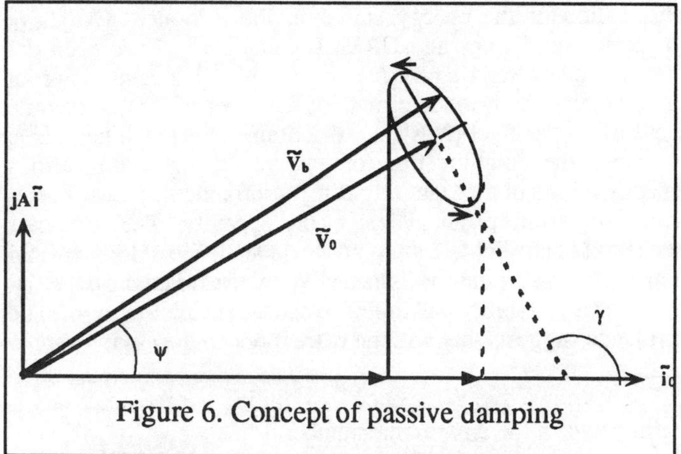

Figure 6. Concept of passive damping

with growth time of 81 μsec (mode number 83), a 15 kV RF modulation with mode number 85 is needed.

Active damping cavity

A feedback scheme using an extra odd harmonic cavity has also been studied. This cavity would couple exclusively to the π-mode and thus effectively damp it. It is found that a maximum gap voltage of 8 kV running at 253rd harmonic is needed to damp the 83rd harmonic π-mode with a growth time of 81 μsec.

Passive damping cavity

This method takes advantage of the finite filling time of any cavity which gives rise to either growth or damping of the normal modes of the circulating bunches, as manifested by Robinson damping of longitudinal oscillations by RF cavities. The damping mechanism is depicted in the phasor diagram of Figure 6[1], where it is shown that due to the finite filling time in the cavity, a phase oscillation in the bunches would induce an oscillation in both amplitude and phase of the induced voltage, which in turn leads to dissipation of the stored oscillation energy. The same concept applies to an odd harmonic cavity coupled to the π-mode. It has been worked out in [1] that the resulting damping rate is given by*

$$\frac{1}{\tau_d} = \frac{N_{pc}}{N_h} \cdot \frac{V_{br}\,\omega_s}{V_c \sin\phi} \cdot \frac{-\xi\,\eta}{[1+(\xi+\eta)^2]\,[1+(\xi-\eta)^2]}$$

where

$$\xi = -\tan\psi = (n\omega_0 - \omega_r)T_f \ , \quad \eta = \omega_s T_f$$

and

N_{pc} : harmonic number of passive cavity
N_h : harmonic number of main cavity
V_{br} : beam induced voltage
V_c : peak RF voltage

ϕ : synchronous angle
ω_s : synchrotron frequency
ψ : cavity tuning angle
T_f : $2\,Q_0/\omega_0(1+\beta)$
ω_0 : revolution frequency
ω_r : cavity resonant frequency
β : cavity coupling factor

Thus with an odd harmonic cavity in the Damping Ring passively coupled to the π-mode, one can expect to see some dissipation of the energy stored in the π-mode. Extensive tracking simulations and URMEL calculations have been devoted to characterizing the parameters of such a cavity in order to overcome the π-mode growth of the order of 100 μsec while conforming to other physical restrictions in the machine. Table 4 shows the damping time of passive cavities with various specifications in the Damping Ring environment. These correspond to various physical designs of the cavity. Table 5 shows the result of an URMEL study of the dependence of longitudinal impedance on the cavity diameter. We also studied the potential transverse instability that might be caused by this addition. The evidence suggests that with an extra tuner in the cavity we can

---

* We acknowledge W. Spence for pointing out the necessary ratio between the 2 harmonic numbers.

Mode no.	R M Ω	τ (damping time) μs
43	3	33.8
85	3	17.9
127	0.3	94.0
169	0.3	68.6
253	0.3	46.1
337	0.3	36.8
421	0.3	29.7
505	0.3	27.3
673	0.3	23.9
841	0.3	22.3

Table 4. Damping times for various passive cavities

Diameter (in)	R (MΩ)	τ (damping time) μs
0.59	4.8	12.6
1.0	3.9	14.5
2.0	2.1	22.0

Table 5. URMEL results of impedance vs. cavity size

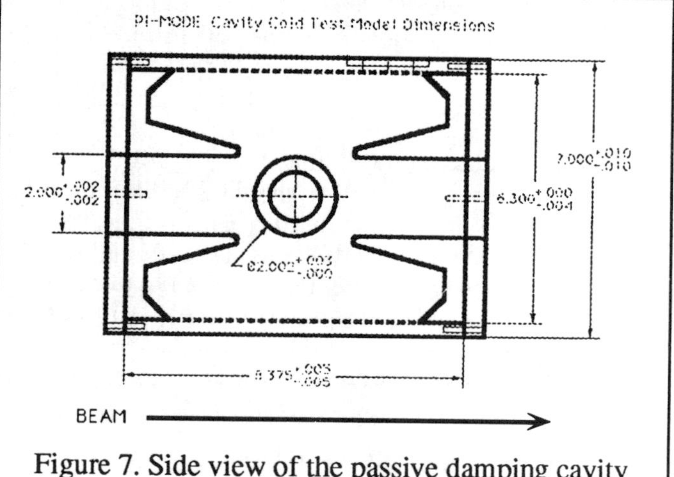

Figure 7. Side view of the passive damping cavity showing end caps and 2 tuners

safely steer the operation point clear of transverse resonances. A prototype design of this cavity is completed (see Figure 7) which incorporates the features dictated by the studies described above. It relies on two tuners to stay clear of any transverse resonances with the beam.

III. CONCLUSION

The longitudinal coupled bunch instability in the SLC Damping Rings has been thoroughly studied, as well as many possible cures thereof. We have decided to pursue the passive damping cavities based on power considerations. Extensive evaluation and design studies have been done to make sure this will live up to its expectation without causing adverse side effects. We expect to commission this cavity in December 1991.

REFERENCE

[1]. P.B. Wilson, SLAC-PUB 2884, 1982
   P.B. Wilson, Proc. 9th Int. Conf. on High Energy Accelerators, p.60

BEAM BREAKUP WITH FINITE BUNCH LENGTH

C. L. Bohn and J.R. Delayen

Argonne National Laboratory
Engineering Physics Division
9700 South Cass Avenue, Argonne, IL 60439

Abstract

Cumulative beam breakup in linear accelerators has been studied extensively for the case of infinitely short bunches. While this model is appropriate for high energy accelerators, in high-current ion accelerators the bunch length can occupy a significant fraction of an rf period of the deflecting mode. A semi-analytic model has been developed to study and predict beam breakup in the case of nonzero bunch length and arbitrary bunch shape. Simulations of steady-state bunch distortion are presented, and the role of focusing in controlling beam breakup with bunches of nonzero length is illustrated.

I. INTRODUCTION

A deflecting mode which induces cumulative beam breakup (BBU) can cause degradation of beam quality and possible beam loss to the cavity walls. In low-β accelerators, the bunches have nonzero length and can occupy a significant fraction of the rf period of the deflecting mode. In turn, they will drive wakefields which differ from those set up by the nearly delta-function bunches characteristic of high-β accelerators. The wakefields will also be affected by the distribution of current within each bunch. Therefore, the effects of cumulative BBU in low-β ion accelerators will generally be quantitatively different than in high-β electron accelerators.

This problem is of particular interest in connection with superconducting linacs for high-current ion beams.[1] In these linacs, the constituent cavities will be short and independently phased, and cumulative BBU is therefore expected to be the dominant transverse instability. Because superconducting linacs run cw, the steady-state properties of their beams are fundamentally important. Steady-state BBU predominates after times long compared to Q/ω, where Q and ω refer to the deflecting mode and represent the cavity quality factor and the angular frequency of this mode, respectively. The transient BBU which occurs earlier can be countered by slowly increasing the current during turn-on,[2] and the steady-state case is therefore of most interest in connection with linacs which run cw.

In this paper, we calculate cumulative BBU in a linac with smoothly varying parameters and arbitrary β. In particular, we calculate the effects of nonzero bunch length and arbitrary current distribution within each bunch on the steady-state BBU of a coasting beam. In choosing a coasting beam to model the problem, we are ignoring the effects of longitudinal synchrotron oscillations within each bunch which occur in the presence of acceleration.

II. EQUATION OF TRANSVERSE MOTION WITH PERIODIC CURRENT

We investigate cumulative beam breakup for the case of a periodic current composed of an infinite series of bunches of identical but arbitrary shape. The assumptions and notation of Ref. 3 are used. We will assume that $\beta\gamma$, κ, and ϵ are all independent of σ (coasting beam). As given in Ref. 3, the equation of transverse

motion is then

$$\left(\frac{d^2}{d\sigma^2} + \kappa^2\right)x(\sigma,\zeta) = \epsilon\int_{-\infty}^{\zeta}d\zeta'\, w(\zeta-\zeta')F(\zeta')x(\sigma,\zeta'),\tag{1}$$

with

$$F(\zeta) = \sum_{k=-\infty}^{+\infty} F_k e^{ik\frac{2\pi}{\omega\tau}\zeta}.\tag{2}$$

As also indicated in Ref. 3, after Fourier transformation eq. (1) becomes

$$\left(\frac{d^2}{d\sigma^2} + \kappa^2\right)\tilde{x}(\sigma,Z) = \epsilon\,\tilde{w}(Z)\sum_{k=-\infty}^{+\infty} F_k\,\tilde{x}\left(\sigma,Z-\frac{2\pi k}{\omega\tau}\right).\tag{3}$$

Equation (3) is a difference-differential equation for the Fourier transform of the displacement of periodic bunches of arbitrary shape. If the problem is completely periodic, i.e. steady state has been reached and the misalignment at the entrance to the linac has the same period as the bunches, then the displacement can be expanded in a Fourier series:

$$x(\sigma,\zeta) = \sum_{m=-\infty}^{+\infty} x_m(\sigma)e^{im\frac{2\pi}{\omega\tau}\zeta}.\tag{4}$$

In that case eq. (3) is replaced by

$$\left(\frac{d^2}{d\sigma^2} + \kappa^2\right)x_m(\sigma) = \epsilon\,\tilde{w}_m\sum_{k=-\infty}^{+\infty} F_k\,x_{m-k}(\sigma),\tag{5}$$

where

$$\tilde{w}_m = \int_0^{\infty}d\zeta\, w(\zeta)\, e^{-im\frac{2\pi}{\omega\tau}\zeta} = \frac{1}{1+\dfrac{1}{4Q^2}-\left(\dfrac{2\pi m}{\omega\tau}\right)^2+i\left(\dfrac{2\pi m}{Q\omega\tau}\right)}.\tag{6}$$

The BBU problem reduces to the calculation of the eigenvalues of the linear system associated with eq. (5).

Equation (5) can also be solved via series expansion of the Fourier components of the transverse displacement:

$$x_m(\sigma) = \sum_{j=0}^{\infty}\frac{x_{m,j}\,\sigma^j}{j!}.\tag{7}$$

The following recurrence relation is obtained:

$$x_{m,j+2} = \epsilon\,\tilde{w}_m\sum_{k=-\infty}^{+\infty} F_k x_{m-k,j} - \kappa^2 x_{m,j}.\tag{8}$$

For the case of a misaligned beam with initial conditions $x(0,\zeta)=x_0$, $x'(0,\zeta)=0$, the first terms of the expansion are

$$x_{0,2} = e\,\tilde{w}_0 F_0 x_0 - \kappa^2 x_0, \qquad x_{m,2} = e\,\tilde{w}_m F_m x_0, \tag{9}$$

and the beam displacement over the interval $0 \le \zeta \le \omega\tau$ is given by

$$x(\sigma,\zeta) \approx x_0 \left[1 + \frac{\sigma^2}{2}\left(e\sum_{m=-\infty}^{+\infty} \tilde{w}_m F_m\, e^{im\frac{2\pi}{\omega\tau}\zeta} - \kappa^2 \right) \right]. \tag{10}$$

The displacement of the center of the bunch ($\zeta = 0$) is

$$x(\sigma,0) = \sum_{m=-\infty}^{+\infty} x_m(\sigma) \approx x_0 \left[1 + \frac{\sigma^2}{2}(\Omega^2 - \kappa^2) \right], \tag{11}$$

with
$$\Omega^2 = e\sum_{m=-\infty}^{+\infty} \tilde{w}_m F_m = e\int_0^{+\infty} w(\zeta)\,F(-\zeta)\,d\zeta. \tag{12}$$

It is useful to define a "growth factor" $G(\sigma,\zeta;\kappa)$ as follows:

$$G^2(\sigma,\zeta;\kappa) \equiv \frac{1}{e\,x(\sigma,\zeta)} \frac{d^2 x(\sigma,\zeta)}{d\sigma^2}. \tag{13}$$

The growth factor of the bunch centroid at the entrance of the linac is

$$G^2(0,0;\kappa) = \frac{1}{e x_0}\frac{d^2 x(0,0)}{d\sigma^2} = \frac{1}{e}(\Omega^2 - \kappa^2). \tag{14}$$

The role of focusing is clear: $\kappa > \Omega$ generates an imaginary growth factor at the linac entrance which tends to stabilize the beam with respect to beam breakup.

III. EXAMPLE: UNIFORM BUNCHES

In the case where each bunch has the uniform current distribution

$$F(\zeta) = \begin{cases} \omega\tau/\alpha & \text{for } |\zeta| < \alpha/2, \\ 0 & \text{for } \alpha/2 < |\zeta| < \omega\tau/2, \end{cases} \tag{15}$$

the growth factor at the linac entrance with zero focusing is

$$\begin{aligned}
G^2(0,0;\kappa=0) = \;&\frac{\omega\tau}{\alpha}\frac{4Q^2}{4Q^2+1} \\
\times \Bigg\{ &\left[\left(2q+\frac{p}{Q}\right)\cos\left(\frac{\alpha}{2}\right)\sinh\left(\frac{\alpha}{4Q}\right) + \left(2p-\frac{q}{Q}\right)\sin\left(\frac{\alpha}{2}\right)\cosh\left(\frac{\alpha}{4Q}\right)\right] \\
&+ 1 - \left[\cos\left(\frac{\alpha}{2}\right) + \frac{1}{2Q}\sin\left(\frac{\alpha}{2}\right)\right]\exp\left(-\frac{\alpha}{4Q}\right) \Bigg\}
\end{aligned} \tag{16}$$

where p and q are the resonance functions defined in Ref. 3. In Fig. 1, $G^2(0,0;\kappa=0)$ is plotted as a function of "filling factor" $f \equiv \alpha/\omega\tau$ for various values of $\omega\tau$. The figure shows that the bunch length is an important factor in determining the stability of the beam with respect to BBU. This observation is accentuated by the plots in Fig. 2. There, the growth factor is plotted for various values of filling factor in the vicinity of the resonance $\omega\tau = 4\pi(1+1/2Q)$ for $Q = 1000$. As the bunch length increases, the stability of the beam with respect to BBU qualitatively reverses.

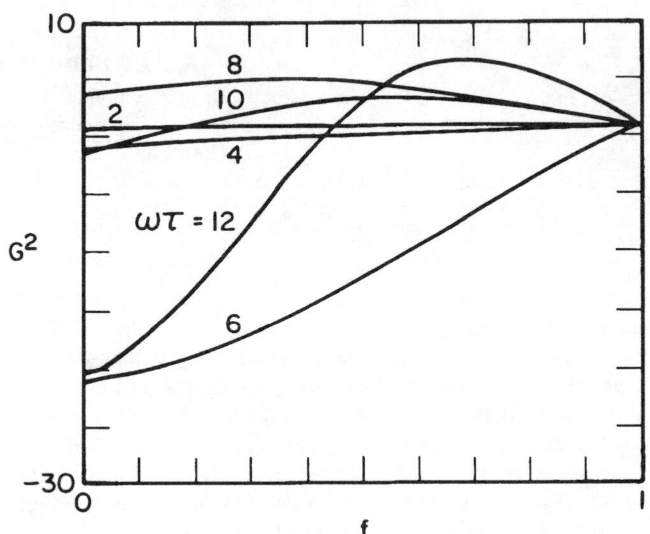

Fig.1. Growth factor $G^2(0,0;\kappa=0)$ vs. filling factor $f = \alpha/\omega\tau$ for $Q = 1000$ and for assorted values of $\omega\tau$. $f = 0$ corresponds to delta-function bunches, and $f = 1$ corresponds to the dc beam.

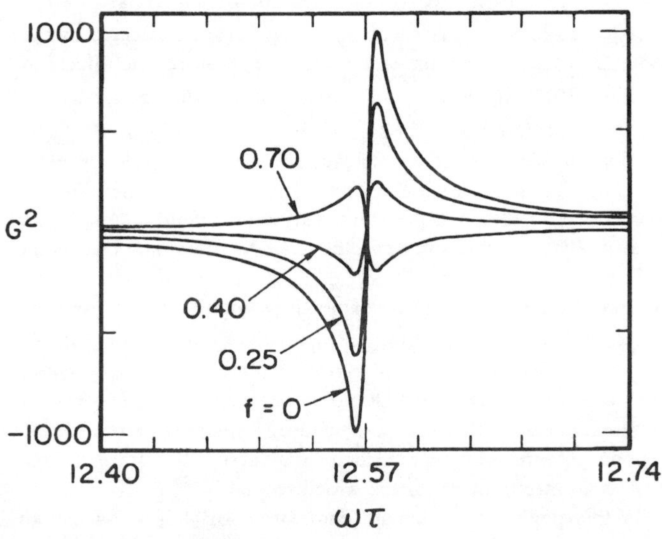

Fig. 2. Growth factor $G^2(0,0;\kappa=0)$ vs. $\omega\tau$ in the vicinity of the resonance $\omega\tau = 4\pi(1+1/2Q)$ for $Q = 1000$ and for assorted values of filling factor $f = \alpha/\omega\tau$.

In the case of finite but short bunches, we find

$$G^2(0,0;\kappa) \approx \frac{1}{e}\left[e\,\omega\tau\left(p+\frac{\alpha}{8}\right) - \kappa^2 \right]. \tag{17}$$

Thus, we find that a slight spread of the charge distribution within bunches exacerbates beam breakup, independently of the focusing strength κ. This is also true for arbitrary current distributions.

According to eqs. (12) and (14),

$$\Delta G^2 \equiv G^2(\alpha \neq 0) - G^2(\alpha = 0) = \sum_{m=-\infty}^{\infty} \tilde{w}_m (F_m - 1). \qquad (18)$$

For nearly delta-function bunches, the terms with large m predominate. From eq. (6), the corresponding $\tilde{w}_m$s are real and negative, so that

$$\Delta G^2 \approx 2 \sum_{m=1}^{\infty} \tilde{w}_m [Re(F_m) - 1]. \qquad (19)$$

We now see that $\Delta G^2 > 0$ because $Re(F_m)$ gradually decreases from unity as m increases. Accordingly, slightly spreading the bunches always exacerbates beam breakup. This conclusion is valid only for small deviations from the delta-function case, i.e. for $\alpha \ll \omega\tau$ and $\alpha \ll 2\pi$. In the limit $\alpha \to \omega\tau$, we recover the dc case from eq. (16).

Even when the beam is inherently stable, the deflecting modes can distort each bunch and thereby degrade beam quality. An example is presented in Fig. 3a for an unfocused beam with parameters which are plausible for a superconducting linac.[4] These results, which were generated numerically from eq. (8), reveal that portions of the bunch can be deflected to transverse displacements exceeding the initial displacement even though the bunch centroids are deflected toward the axis. However, as shown in Fig. 3b, focusing can be used to suppress bunch distortion.

IV. Conclusions

A formalism which enables the calculation of steady-state cumulative beam breakup with bunches of arbitrary shape and nonzero length comprising a coasting beam was developed. As an example, the formalism was applied to uniform bunches. Bunches of short, but nonzero, length always exacerbate BBU relative to delta-function bunches. Bunches of finite length may be severely distorted by the deflecting modes, even in circumstances which would be stable were the bunches to have zero length. In each case, however, focusing can be used as a cure.

V. Acknowledgements

This research was sponsored by the U.S. Department of Energy under contract W-31-109-ENG-38 and by the U.S. Army Strategic Defense Command.

VI. References

[1] J.R. Delayen, C.L. Bohn, W.L. Kennedy, C.T. Roche, and L. Sagalovsky, "Recent Developments in High-Current Superconducting Ion Linacs", these Proceedings.

[2] R.L. Gluckstern, F. Neri, and R.K. Cooper, "Cumulative Beam Breakup with Smoothly Varying Parameters", *Part. Accel.*, **23**, pp. 53-71, 1988.

[3] J.R. Delayen and C.L. Bohn, "Beam Breakup with Longitudinal Halo", these Proceedings.

[4] J.R. Delayen, C.L. Bohn, and C.T. Roche, "Application of rf Superconductivity to High-Brightness Ion Beam Accelerators", *Proc. 1990 Linear Accelerator Conference*, **LA-12004-C**, pp. 82-84, 1990.

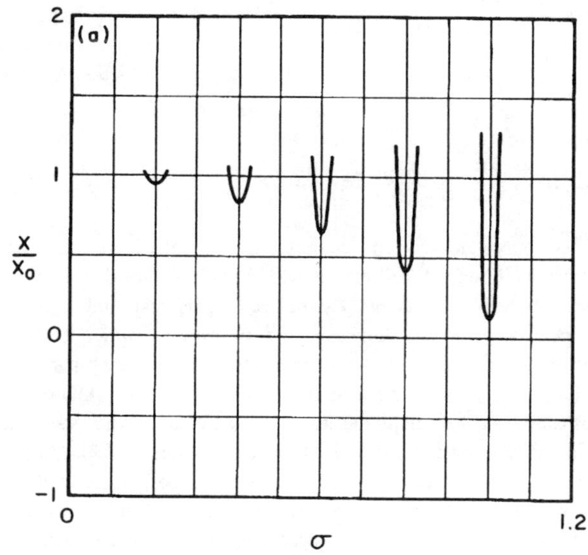

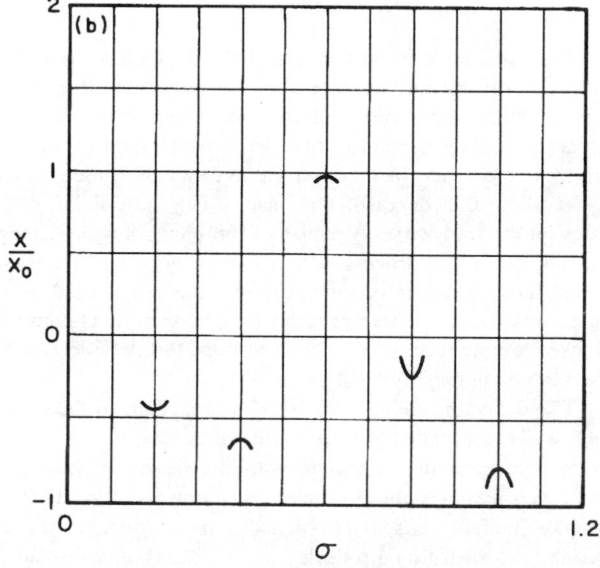

Fig. 3. Plots of bunch displacement and shape vs. position σ along the linac for $\omega\tau = 12$, $\epsilon = 0.2$, $Q = 10^6$, $f = 0.25$, and for (a) $\kappa = 0$ (no focusing) and (b) $\kappa = 10$. Transverse displacement, plotted as the ordinate, is normalized with respect to the initial displacement. $\sigma = 0$ denotes the entrance to the linac, and $\sigma = 1$ denotes the exit.

BEAM BREAKUP WITH LONGITUDINAL HALO

J.R. Delayen and C.L. Bohn
Argonne National Laboratory
Engineering Physics Division
9700 South Cass Avenue, Argonne, IL 60439

Abstract

We have developed an analytical model of cumulative beam breakup in linear accelerators that predicts the displacement of particles between bunches. Beam breakup is assumed to be caused by a periodic current consisting of an infinite bunch train. The particles in the halo do not contribute to the breakup but experience the deflecting fields and are displaced by them. Under certain circumstances, the displacement of particles in the halo can be considerably larger than that of the bunches. This may have important consequences for the design of high-current cw accelerators where even a small flux of particles striking components of the accelerator cannot be tolerated because of activation.

I. INTRODUCTION

The cumulative beam breakup (BBU) instability results when the beam is offset transversely and couples to a deflecting mode in the accelerating structures, thereby enhancing it.[1,2] The mode deflects the trailing portion of the beam, which then couples more strongly to the mode in the next cavity, and the process evolves likewise along the length of the linac. This description of BBU focuses on the transverse dynamics of the bunches which comprise the beam, yet in the process wakefields are generated between bunches. Any particles travelling between the bunches will respond to these wakefields. These particles may be viewed as constituting a diffuse "longitudinal halo". They respond to the deflecting fields but contribute negligibly to them.

The deflecting fields may drive the longitudinal halo into the cavity walls and thereby activate them. This is a particularly important concern in connection with the design of high-current linacs which are envisioned to run cw over long operational lifetimes. Examples include linacs envisioned for irradiation of fusion materials[3] and for tritium production.[4] The steady-state displacement of the halo is therefore of primary interest.

In this paper, we calculate the steady-state displacement of the longitudinal halo in the presence of cumulative BBU. We consider a linac with smoothly varying parameters and arbitrary β which drives a bunched beam. Focusing is included, and its role in mitigating activation caused by impingement of the longitudinal halo is quantified.

II. EQUATION OF TRANSVERSE MOTION WITH FOURIER TRANSFORMS

We calculate cumulative beam breakup within the framework of a model which incorporates a number of approximations. The cavities which comprise the linac are considered to have negligible length and to be the only source of deflecting fields. This approximation is, for example, justified for superconducting ion linacs which characteristically are comprised of short, independently phased cavities.[5] A single deflecting mode is assumed to be present, and its frequency is taken to be the same in every cavity. This corresponds to a worst-case analysis because a spread in deflecting-mode frequencies suppresses BBU,[6,7] but the worst case is of practical interest for the study of BBU control. A "continuum approximation" is used in which the discrete kicks in transverse momentum imparted by the cavities are considered to be smoothed along the linac. With this approximation, and allowing for arbitrary velocity, the equation of transverse motion of the beam in the presence of the single deflecting mode is:[8,9]

$$\left[\frac{1}{\beta\gamma}\frac{\partial}{\partial\sigma}(\beta\gamma\frac{\partial}{\partial\sigma}) + \kappa^2(\sigma)\right]x(\sigma,\zeta) = e(\sigma)\int_0^\zeta d\zeta'\, w(\zeta-\zeta')F(\zeta')x(\sigma,\zeta'). \quad (1)$$

Here, β and γ have their usual meanings; $\sigma = s/\mathcal{L}$ is a dimensionless spatial variable defined in terms of position along the linac, s, and total length of the linac, $\mathcal{L}$; $\zeta = \omega(t - \int ds/\beta c)$ is the time, made dimensionless by use of the angular frequency ω of the deflecting mode, measured after the arrival of the head of the beam at s; κ is the net transverse focusing wavenumber multiplied by $\mathcal{L}$; x is the transverse displacement of the beam centroid from the axis; $e(\sigma)$ is a dimensionless quantity which represents the strength of the BBU interaction; and $F(\zeta) = I(\zeta)/<I>$ is the form factor for the current defined in terms of the beam current $I(\zeta)$ and average beam current $<I>$. $w(\zeta)$ is a dimensionless "wake function" given by

$$w(\zeta) = e^{-\zeta/2Q}\sin\zeta\ u(\zeta), \quad (2)$$

where $u(\zeta)$ is the unit step or Heaviside function, and Q is the quality factor of the deflecting mode under consideration.

We now investigate beam breakup for the case of a periodic current composed of an infinite series of bunches. We will assume that $\beta\gamma$, κ, and e are independent of σ (coasting beam). When necessary, the results can be straightforwardly generalized using the WKBJ method.[10] The equation of transverse motion is then

$$\left(\frac{d^2}{d\sigma^2} + \kappa^2\right)x(\sigma,\zeta) = e\int_{-\infty}^\zeta d\zeta'\, w(\zeta-\zeta')F(\zeta')x(\sigma,\zeta'), \quad (3)$$

with

$$F(\zeta) = \sum_{k=-\infty}^{+\infty} F_k e^{ik\frac{2\pi}{\omega\tau}\zeta}. \quad (4)$$

Note that since the current is assumed to be periodic and started at $\zeta = -\infty$, the lower limit of the integral in eq. (3) is $-\infty$ and not 0 as it was in eq. (1). This does not restrict eq. (3) to the analysis of the steady state behavior since no assumption has been made about the initial conditions. For example, a beam which is turned on at $\zeta = 0$ can be modelled by assuming that $x(0,\zeta<0)=0$ since no deflecting field will be generated as long as the current travels on axis.

Introducing the Fourier transforms of $x(\sigma,\zeta)$ and $w(\zeta)$,

$$\tilde{x}(\sigma,Z) = \int_{-\infty}^{+\infty} e^{-iZ\zeta} x(\sigma,\zeta)\,d\zeta, \quad \tilde{w}(Z) = \int_{-\infty}^{+\infty} e^{-iZ\zeta} w(\zeta)\,d\zeta = \frac{1}{1-Z^2+\dfrac{1}{4Q^2}+\dfrac{iZ}{Q}}, \quad (5)$$

eq. (3) becomes

$$\left(\frac{d^2}{d\sigma^2} + \kappa^2\right)\tilde{x}(\sigma,Z) = e\,\tilde{w}(Z)\sum_{k=-\infty}^{+\infty} F_k\,\tilde{x}\left(\sigma,Z-\frac{2\pi k}{\omega\tau}\right). \quad (6)$$

Equation (6) is a difference-differential equation for the Fourier transform of the displacement of periodic bunches of arbitrary shape. It replaces and is equivalent to the integro-differential equation (3) which is more commonly used. Equation (6) can be used to study a variety of transient and steady-state BBU problems. In this paper we will address the problem of the transverse displacement of particles located between bunches. The problem of the displacement and distortion of bunches of finite length is treated in a companion paper.[11]

III. DELTA-FUNCTION BUNCHES

For the case of a beam composed of delta-function bunches, the Fourier components are $F_k = 1 \; \forall\, k$, so that eq. (6) becomes

$$\left(\frac{d^2}{d\sigma^2} + \kappa^2\right)\tilde{x}(\sigma,Z) = e\,\tilde{w}(Z)\sum_{k=-\infty}^{+\infty} \tilde{x}\left(\sigma,Z-\frac{2\pi k}{\omega\tau}\right). \quad (7)$$

If we define

$$\tilde{W}(Z) \equiv \sum_{k=-\infty}^{+\infty} \tilde{w}\left(Z-\frac{2\pi k}{\omega\tau}\right) = \frac{\omega\tau}{2}\,\frac{\sin\omega\tau}{\cosh[\omega\tau(1/2Q-iZ)] - \cos\omega\tau}, \quad (8)$$

and

$$\Lambda^2(Z) = e\,\tilde{W}(Z) - \kappa^2, \quad (9)$$

the solution of eq. (7) is

$$x(\sigma,\zeta) = x(0,\zeta)\cos(\kappa\sigma) + x'(0,\zeta)\,\frac{\sin(\kappa\sigma)}{\kappa}$$

$$+\frac{1}{2\pi}\int_{-\infty}^{+\infty} dZ\,e^{iZ\zeta}\,\frac{\tilde{w}(Z)}{\tilde{W}(Z)}\,A(Z)\left\{\cosh[\Lambda(Z)\sigma] - \cos(\kappa\sigma)\right\} \quad (10)$$

$$+\frac{1}{2\pi}\int_{-\infty}^{+\infty} dZ\,e^{iZ\zeta}\,\frac{\tilde{w}(Z)}{\tilde{W}(Z)}\,B(Z)\left\{\frac{\sinh[\Lambda(Z)\sigma]}{\Lambda(Z)} - \frac{\sin(\kappa\sigma)}{\kappa}\right\},$$

with $A(Z)$ and $B(Z)$ determined from the initial conditions at $\sigma = 0$:

$$A(Z) = \omega\tau\sum_{k=-\infty}^{+\infty} x(0,k\omega\tau)\,e^{-ik\omega\tau}; \quad B(Z) = \omega\tau\sum_{k=-\infty}^{+\infty} x'(0,k\omega\tau)\,e^{-ik\omega\tau}. \quad (11)$$

It is worth noting that the beam displacement is defined for all values of ζ, and not only for $\zeta = m\omega\tau$ which represents bunch m. In particular, eq. (10) can be used to calculate the transverse displacement of particles located outside the delta-function bunches, i.e. particles that do not contribute to the beam breakup but experience the transverse fields generated by the bunches. These particles comprise the "longitudinal halo". The transient displacement resulting from turn-on at $\zeta = 0$ of a misaligned beam may be obtained from the initial conditions $x(0,\zeta) = x_0\,u(\zeta)$ and $x'(0,\zeta) = 0$.

Steady-state BBU is of greater concern than transient BBU in connection with long-term cw operation. We now calculate the steady-state displacement induced by a misaligned beam for which $x(0,\zeta) = x_0$ and $x'(0,\zeta) = x_0'$. According to eq. (11), these initial conditions yield

$$A(Z) = 2\pi x_0\sum_{k=-\infty}^{+\infty}\delta\left(Z-\frac{2\pi k}{\omega\tau}\right); \quad B(Z) = 2\pi x_0'\sum_{k=-\infty}^{+\infty}\delta\left(Z-\frac{2\pi k}{\omega\tau}\right). \quad (12)$$

After integrating over Z, the beam displacement over the interval $0 \le \zeta \le \omega\tau$ is given by

$$x(\sigma,\zeta) = x_0\cos(\kappa\sigma) + x_0'\,\frac{\sin(\kappa\sigma)}{\kappa}$$

$$+ x_0\,e^{-\zeta/2Q}\left(\frac{1+q}{p}\sin\zeta + \cos\zeta\right)\left[\cosh[\Lambda(0)\sigma] - \cos(\kappa\sigma)\right] \quad (13)$$

$$+ x_0'\,e^{-\zeta/2Q}\left(\frac{1+q}{p}\sin\zeta + \cos\zeta\right)\left[\frac{\sinh[\Lambda(0)\sigma]}{\Lambda(0)} - \frac{\sin(\kappa\sigma)}{\kappa}\right]$$

with

$$\Lambda(0) = \sqrt{e\,\omega\tau\,p - \kappa^2}. \quad (14)$$

The functions p and q include the resonances between the frequencies $1/\tau$ of the accelerating mode and $\omega/2\pi$ of the deflecting mode:

$$p = p(\omega\tau,Q) = \sum_{k=1}^{\infty} e^{-\frac{k\omega\tau}{2Q}}\sin k\omega\tau = \frac{\sin\omega\tau}{4\left[\sinh^2\left(\dfrac{\omega\tau}{4Q}\right) + \sin^2\left(\dfrac{\omega\tau}{2}\right)\right]};$$

$$(15)$$

$$q = q(\omega\tau,Q) = \sum_{k=1}^{\infty} e^{-\frac{k\omega\tau}{2Q}}\cos k\omega\tau = \frac{\cos\omega\tau - e^{-\omega\tau/2Q}}{4\left[\sinh^2\left(\dfrac{\omega\tau}{4Q}\right) + \sin^2\left(\dfrac{\omega\tau}{2}\right)\right]}.$$

To find the displacement of the bunches themselves, we set $\zeta = 0$. The result is

$$x(\sigma,0) = x_0\cosh[\Lambda(0)\,\sigma] + x_0'\,\frac{\sinh[\Lambda(0)\,\sigma]}{\Lambda(0)}. \quad (16)$$

IV. TRANSVERSE DISPLACEMENT OF LONGITUDINAL HALO

Particles located outside the bunches can be deflected more than the bunches themselves. This is implied by Fig. 1, which provides example plots of the amplitude function $\chi = e^{-\zeta/2Q}\{[(1+q)/p]\sin\zeta + \cos\zeta\}$ with $Q = 1000$. As specific examples, we list three special cases where we assume $x_0' = 0$ for simplicity:

Case 1 -- $\omega\tau = 2n\pi(1 + 1/2Q)$ (resonance):
$$\frac{x(\sigma,\zeta)}{x_0} = \cos(\kappa\sigma) + (\sin\zeta + \cos\zeta)\left[\cosh\left(\sigma\sqrt{eQ - \kappa^2}\right) - \cos(\kappa\sigma)\right]. \quad (17)$$

Case 2 -- $\omega\tau = 2n\pi$:
$$\frac{x(\sigma,\zeta)}{x_0} = \cos(\kappa\sigma) + \frac{eQ}{\kappa}\,\sigma\,\sin(\kappa\sigma)\sin\zeta. \quad (18)$$

Case 3 -- $\omega\tau = (2n + 1)\pi$:
$$\frac{x(\sigma,\zeta)}{x_0} = \cos(\kappa\sigma) + \frac{e}{\kappa}\,\frac{(2n+1)\pi}{4}\,\sigma\,\sin(\kappa\sigma)\sin\zeta. \quad (19)$$

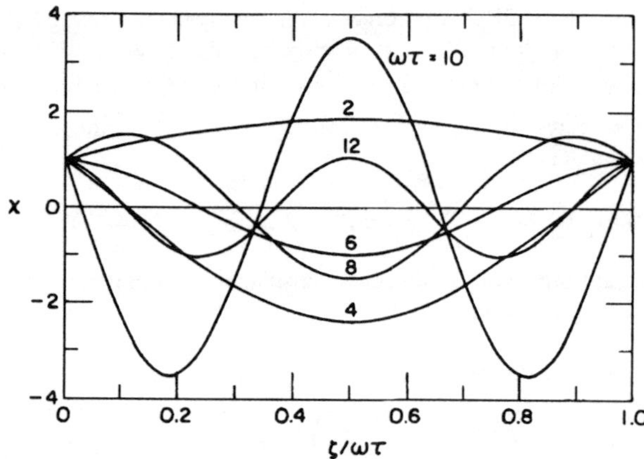

Figure 1. Plot of amplitude function χ vs. $\zeta/\omega\tau$ for delta-function bunches and for various values of $\omega\tau$. The bunches are located at $\zeta/\omega\tau = 0, 1$.

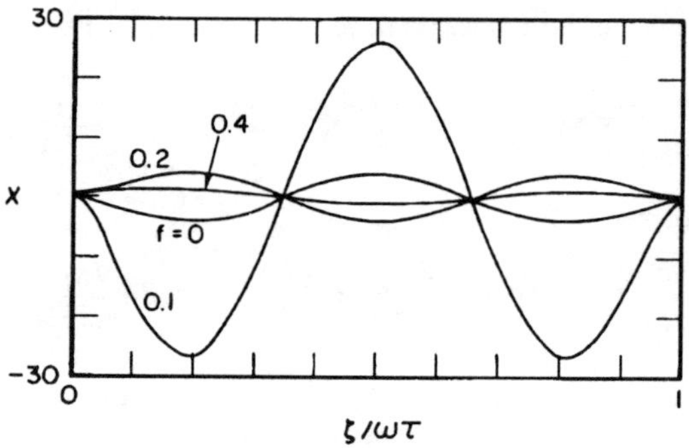

Fig.2. Generalized amplitude function χ vs. $\zeta/\omega\tau$ for $\omega\tau = 10$, $\epsilon = 0.2$, $Q = 1000$, $\kappa = 0$, $\sigma = 1$, and for assorted values of the filling factor f of the bunches (cf. Ref. 11). The bunch centers are located at $\zeta/\omega\tau = 0, 1$.

These expressions illustrate clearly the role of focusing in controlling the transverse displacement of the longitudinal halo. For example, Case 2 requires strong focusing, $\kappa > \epsilon Q$, to confine the halo. Considerations of focusing for halo control are potentially important in connection with the suppression of activation of the cavity walls due to particle impingement over the long-term cw operation of a high-current linac.

V. Uniform Bunches of Arbitrary Length

Lengthening the bunches can significantly alter the transverse dynamics of the longitudinal halo. To demonstrate this, it is useful to define a generalized amplitude function as follows, assuming again that $x_0' = 0$:

$$\chi \equiv \frac{x(\sigma,\zeta) - x_0 \cos\kappa\sigma}{x(\sigma,0) - x_0 \cos\kappa\sigma}. \tag{20}$$

For delta-function bunches in the steady state, we see from eq. (13) that χ depends only on ζ provided Q is large, and we plotted χ for this case in Fig. 1. Using the formalism of Ref. 11, we have numerically calculated χ for uniform bunches of arbitrary length. Results are given in Fig. 2 for the case $\omega\tau = 10$. This case exemplifies how a modest change of bunch length will sometimes cause a pronounced change in the amplitude function which, in turn, affects the dynamics of the longitudinal halo.

VI. Conclusions

The transverse displacement of particles comprising a diffuse longitudinal halo between bunches was calculated for the case of steady-state cumulative beam breakup. If the displacement were large enough to drive these particles into the cavity walls, then activation could become a problem due to the particulate flux accumulated over a long operational lifetime. It was seen that the longitudinal halo can sometimes be displaced much more than the bunches themselves. However, sufficiently strong focusing can be used as a cure.

VII. Acknowledgements

This research was supported by the U.S. Department of Energy under contract W-31-109-ENG-38 and by the U.S. Army Strategic Defense Command.

VIII. References

[1]   W.K.H. Panofsky and M. Bander, "Asymptotic Theory of Beam Break-Up in Linear Accelerators", *Rev. Sci. Instrum.*, **39**, 206, (1968).

[2]   R.H. Helm and G.A. Loew, "Beam Breakup", in <u>Linear Accelerators</u>, edited by P.M. Lapostolle and A.L. Septier, (North Holland, Amsterdam), 173, (1970).

[3]   G.P. Lawrence, T.P. Wangler, S.O. Schriber, E.L. Kemp, M.T. Wilson, T.S. Bhatia, G.H. Neuschaefer, F.W. Guy, and D.D. Armstrong, "A High-Performance D-Lithium Neutron Source for Fusion Technology Testing: Accelerator Driven Design", *Fusion Tech.*, **15**, 289, (1989).

[4]   T.P. Wangler, G.P. Lawrence, T.S. Bhatia, J.H. Billen, K.C.D. Chan, R.W. Garnett, F.W. Guy, D. Liska, S. Nath, G.H. Neuschaefer, and M. Shubaly, "Linear Accelerator for Production of Tritium: Physics Design Challenges", *Proc. 1990 Linear Accelerator Conference*, LA-12004-C, 548, (1990).

[5]   J.R. Delayen, C.L. Bohn, W.L. Kennedy, C.T. Roche, and L. Sagalovsky, "Recent Developments in High-Current Superconducting Ion Linacs", these Proceedings.

[6]   R.L. Gluckstern, F. Neri, and R.K. Cooper, "Cumulative Beam Breakup with Randomly Fluctuating Parameters", *Part. Accel.*, **23**, 37, (1988).

[7]   D.G. Colombant and Y.Y. Lau, "Effects of Frequency Spreads on Beam Breakup Instabilities in Linear Accelerators", *Appl. Phys. Lett.*, **55**, 27, (1989).

[8]   Y.Y. Lau, "Classification of Beam Breakup Instabilities in Linear Accelerators", *Phys. Rev. Lett.*, **63**, 1141, (1989).

[9]   A.W. Chao, B. Richter, and C.Y. Yao, "Beam Emittance Growth Caused by Transverse Deflecting Fields in a Linear Accelerator", *Nucl. Instrum. Methods*, **178**, 1, (1980).

[10] C.L. Bohn and J.R. Delayen, "Cumulative Beam Breakup in rf Linacs", *Proc. 1990 Linear Accelerator Conference*, LA-12004-C, 306, (1990).

[11] C. L. Bohn and J.R. Delayen, "Beam Breakup with Finite Bunch Length", these Proceedings.

Microwave Instability at Transition – Stability Diagram Approach

S.A. Bogacz

Accelerator Physics Department,
Fermi National Accelerator Laboratory[*]
P.O. Box 500, Batavia, Illinois 60510

Abstract

A simple model of a beam at transition driven by a storage ring impedance is formulated in the framework of the nonlinear Vlasov equation. This yields a set of coupled equations of motion describing time evolution of a single coherent mode and the overall equilibrium density distribution function. At transition, contour integration in the dispersion relation can be carried out analytically and a simple closed formula for the coherent frequency is obtained. From the resulting stability diagram further conclusions about the growth time of the microwave instability and the longitudinal emittance blowup at transition are derived.

I. INTRODUCTION

One would like to sort out purely kinematic contributions to the emittance blowup (due to the Johnsen and Umstätter effects)[1] from the intensity dependent one caused by the microwave instability, possibly building up at transition. The longitudinal phase space has a very peculiar structure; at transition the RF bucket does not exist in the usual sense and a beam can be considered a coasting beam to a good approximation. When the influence of the external restoring force disappears the beam is very susceptible to any fast growing instability which may in turn (through nonlinear driving terms) reshape the overall longitudinal phase space of the beam.

II. LONGITUDINAL BEAM DYNAMICS AT TRANSITION

Consider an initially uniform distribution of particles inside a storage ring modeled by a statistical density distribution function defined in a classical phase space as

$$f(\varepsilon,\theta,t) = f^0(\varepsilon,t) + \sum_{n \neq 0} h_n(\varepsilon,t)\, e^{i\theta n} \quad , \tag{1}$$

where θ is the azimuthal angle around the ring circumference and ε represents the energy deviation from its synchronous value, E_o. Here $f(\varepsilon,\theta,t)$ is normalized to the total number of particles in the storage ring, N. The phase space continuity equation which governs $f(\varepsilon,\theta,t)$ can be written as follows

$$\frac{\partial}{\partial t} f(\varepsilon,\theta,t) + \omega \frac{\partial}{\partial \theta} f(\varepsilon,\theta,t) + \dot{\varepsilon}\, \frac{\partial}{\partial \varepsilon} f(\varepsilon,\theta,t) = 0 \quad . \tag{2}$$

The revolution frequency, ω, of a given particle depends on its momentum offset, Δp, via the momentum compaction factor, α. The fractional frequency shift is given by

$$\frac{\Delta \omega}{\omega_o} = -\left(\alpha - \frac{1}{\gamma^2}\right)\delta \quad ,$$

where $\hspace{11cm}$ (3)

$$\delta = \frac{\Delta p}{p_o} = \frac{1}{\beta^2}\frac{\varepsilon}{E_o} \quad .$$

In principle the momentum compaction factor also exhibits some chromatic dispersion according to general expansion

$$\alpha = \frac{p}{C}\frac{dC}{dp} = \alpha_o + \alpha_p\, \delta + O(\delta^2) \quad . \tag{4}$$

Right at transition, $\gamma = \gamma_t$, the linear term in Eq.(3) disappears, since

$$\alpha_o - \frac{1}{\gamma^2} = 0 \tag{5}$$

and the leading term in $\omega(\varepsilon)$ happens to be quadratic in ε. One can summarize Eqs.(3)–(5) to the lowest leading term in ε as follows

$$\omega(\varepsilon) = \omega_o - K\, \varepsilon^2 \quad , \tag{6}$$

where the coefficient K is given by the following expression

$$K = \frac{\alpha_p \omega_o}{\beta^4 E_o^2} \quad . \tag{7}$$

Here α_p is a purely lattice dependent parameter, which is easy to express in terms of azimuthal averages of the lattice dispersion function[3].

The beam environment in Eq.(2) is modeled by the wakefield impedance of a storage ring represented in frequency domain by $Z(\omega)$. The energy of the beam changes by

$$\dot{\varepsilon} = -e\omega_o \sum_{n \neq 0} Z_n \phi_n(t)\, e^{i\theta n} \quad , \qquad Z_n = Z(n\omega_o). \tag{8}$$

where

---

[*] Operated by Universities Research Association Inc., under contract with the U.S. Department of Energy

$$\phi_n(t) = -e\omega_o \frac{1}{2\pi} \int_{-\infty}^{\infty} d\varepsilon\, h_n(\varepsilon,t) . \qquad (9)$$

Substituting Eqs.(1) and (8) into Eq.(2) and using orthogonality of azimuthal plane waves, one can rewrite Vlasov equation as a set of coupled equations of motion for individual azimuthal harmonics of the distribution function.

$$\frac{\partial}{\partial t} f^o(\varepsilon,t) - e\omega_o \sum_{n\neq 0} Z_n^* \phi_n^*(t) \frac{\partial}{\partial \varepsilon} h_n(\varepsilon,t) = 0 \quad , \qquad (10)$$

and

$$\frac{\partial}{\partial t} h_n(\varepsilon,t) + in\omega\, h_n(\varepsilon,t) - e\omega_o Z_n \phi_n(t) \frac{\partial}{\partial \varepsilon} f^o(\varepsilon,t)$$

$$- e\omega_o \sum_{m\neq 0} Z_{n-m} \phi_{n-m}(t) \frac{\partial}{\partial \varepsilon} h_m(\varepsilon,t) = 0 \quad . \qquad (11)$$

Here we introduce the instantaneous coherent frequency, $\Omega_n(t)$, describing evolution of the n-th mode within a small time interval (t, t') according to the formula

$$h_n(\varepsilon,t') = e^{-i\Omega_n(t)(t-t')} h_n(\varepsilon,t), \qquad t \approx t' . \qquad (12)$$

We also require that $f^o(\varepsilon,t)$ is a slowly varying function of time compared to rapidly oscillating coherent modes. Including both assumptions one can rewrite Eq.(11) as follows

$$h_n(\varepsilon) = (e\omega_o)^2 \frac{\partial}{\partial \varepsilon} f^o(\varepsilon,t) \frac{Z_n}{n\omega - \Omega_n(t)} \frac{1}{2\pi i} \int_{-\infty}^{\infty} d\varepsilon'\, h_n(\varepsilon'). \quad (13)$$

As was pointed out by Landau,[5] an appropriate integration of Eq.(13) leads to the following dispersion relationship defining the coherent frequency, $\Omega_n(t)$

$$1 = \left(\frac{e\omega_o}{2\pi}\right)^2 N Z_n \int_C d\varepsilon\, \frac{\frac{\partial}{\partial \varepsilon} \psi(\varepsilon,t)}{i[n\omega(\varepsilon) - \Omega_n]} \quad . \qquad (14)$$

Here, Ω_n and $\omega(\varepsilon)$, given explicitly by Eq.(6), define configuration of poles in the complex ε-plane.

III. STABILITY DIAGRAM

The dispersion relation, given by Eq.(14), can be solved with respect to the coherent frequency, Ω, then contours of constant growth rate, $\text{Im}(\Omega)$, can be composed in complex impedance plane. Here we assume a simple Gaussian distribution parametrized by $\sigma = \langle \varepsilon \rangle_{rms}$ as follows

$$\psi(\varepsilon) = \frac{1}{\sigma\sqrt{2\pi}} \exp\left(-\frac{\varepsilon^2}{2\sigma^2}\right) . \qquad (15)$$

Assuming a frequency dispersion given by Eq.(6), one can rewrite the dispersion equation in the following form

$$1 = \left(\frac{e\omega_o}{2\pi}\right)^2 N Z_n \frac{1}{2\sigma^2 niK} \int_C d\varepsilon\, \frac{2\varepsilon\psi(\varepsilon,t)}{(\varepsilon + \Xi)(\varepsilon - \Xi)} \quad . \qquad (16)$$

where

$$\Xi^2 = \frac{n\omega_o - \Omega}{nK} \quad .$$

As was shown in the Appendix the contour integral in Eq.(16) is given by Eq.(29) as follows

$$\int_C d\varepsilon\, \frac{2\varepsilon\psi(\varepsilon,t)}{(\varepsilon + \Xi)(\varepsilon - \Xi)} = 2\pi i\, \psi(\Xi,t) . \qquad (17)$$

Therefore, right at transition, the microwave stability is strictly governed by the Landau damping residuum term given by the right-hand side of Eq.(17). The last statement would translate to a global microwave stability at transition. Substituting Eq.(17) into Eq.(16) allows one to express the coherent frequency of a given mode in the following form

$$\Omega_n = n\omega_o - 2nK\sigma^2 \ln\left(\frac{(e\omega_o)^2}{(2\pi)^{3/2}} \frac{NZ_n}{2nK\sigma^3}\right) . \qquad (18)$$

Taking into account multivaluedness of the above expression one has to introduce appropriate cut-lines (connecting different Riemann sheets of the general solution) at the branch points. The growth rate, introduced as the imaginary part of the coherent frequency, is given by the following expression

$$\frac{1}{\tau_n} = -2\alpha_p n\omega_o \left(\frac{\sigma}{E}\right)^2 F_n(\omega)$$

with

$$F_n(\omega) = \arctan\left(\frac{Y_n}{X_n}\right), \quad Z_n = X_n + iY_n . \qquad (19)$$

For the purpose of this calculation we assume a resonant impedance centered at the frequency ω_r.

One can see from Figure 1 that the general solution for the phase factor, $F_n(x)$, spans an infinite family of curves joined by the vertical cut-lines at $x = \pm 1$. To select a physical solution we impose a following condition; the $x = 0$ point has to be equivalent to the $x \to \pm\infty$ asymptotics and it must be included in the stable region (zero impedance). This narrows down the allowed solutions to the one highlighted in Figure 1. A whole frequency spectrum is contained in the stable region (upper half-plane) with the resonant points, $x = \pm 1$, touching the stability curve.

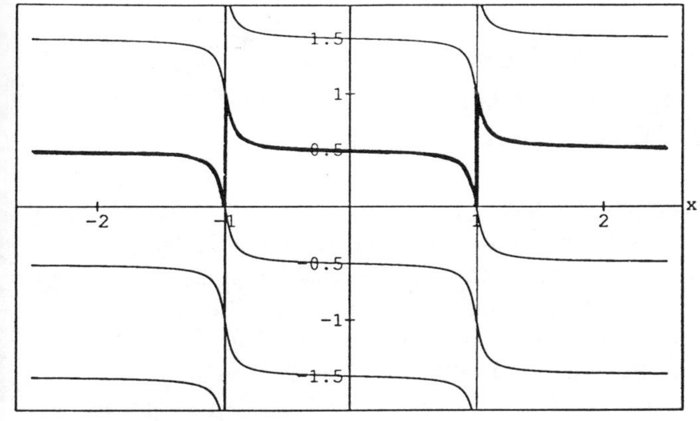

Figure 1   Microwave stability at transition, illustrated by a family of curves representing the phase factor, $F_n(\omega)$, in units of π, plotted versus $x = \omega/\omega_r$.

IV. CONCLUSIONS

We have shown that right at transition, the dispersion integral includes only the Landau damping term making a beam stable against the microwave instability. No instability develops, therefore the longitudinal emittance is not increased by the microwave instability.

V. APPENDIX

Here we evaluate the dispersion integral at transition, which is introduced as follows

$$I = \int_C d\varepsilon \, \frac{2\varepsilon\psi(\varepsilon)}{(\varepsilon + \Xi)(\varepsilon - \Xi)} \qquad (20)$$

where $\psi(\Xi,t)$ is a Gaussian parametrized by

$$\psi(\varepsilon) = \sqrt{\frac{\alpha}{\pi}} \, e^{-\alpha\varepsilon^2}. \qquad (21)$$

Using simple algebra one can rewrite our integral, I, in terms of the plasma dispersion integrals as follows

$$I = \sqrt{\frac{\alpha}{\pi}} \, [\, D(\sqrt{\alpha}\,\Xi) + D(-\sqrt{\alpha}\,\Xi) \,]. \qquad (22)$$

where the plasma dispersion integral is defined by the following formula

$$D(\xi) = \int_C d\varepsilon \, \frac{e^{-\alpha\varepsilon^2}}{(\varepsilon - \xi)} \qquad (23)$$

The contour C contains the real axis, an infinite semi-circle closed in the upper half-plane and a detour piece enclosing any singularity in the lower half-plane. Contributions along the first two pieces of the contour are given by the principle value integral, while the integral along the detour piece is equal to the residuum of the integrant at the singularity. Both contributions are summarized below

$$D(\xi) = W(\xi) + 2\pi i \theta_\xi \, e^{-\xi^2} \qquad (24)$$

where

$$W(\xi) = P \int_{-\infty}^{\infty} d\varepsilon \, \frac{e^{-\alpha\varepsilon^2}}{(\varepsilon - \xi)} \qquad (25)$$

and θ_ξ is defined as follows

$$\theta_\xi = \begin{cases} 0 & \text{if} & \text{Im}(\xi) > 0 \\ \dfrac{1}{2} & \text{if} & \text{Im}(\xi) = 0 \\ 1 & \text{if} & \text{Im}(\xi) < 0 \end{cases} \qquad (26)$$

Substituting a sequence of above equations into Eq.(22) one gets the following expression

$$I = \sqrt{\frac{\alpha}{\pi}} \, [W(\sqrt{\alpha}\,\Xi) + W(-\sqrt{\alpha}\,\Xi) + 2\pi i \, (\theta_\xi + \theta_{-\xi})e^{-\alpha\Xi^2}]. \qquad (27)$$

One can easily check from the definitions, Eqs.(25) and (26), that $W(\xi)$ is an odd function of ξ and the following simple identity for θ_ξ holds

$$\theta_{-\xi} = 1 - \theta_\xi \qquad (A28)$$

Applying these identities to Eq.(27) reduces it to the following final expression

$$I = 2\pi i \, \psi(\Xi). \qquad (29)$$

VI. REFERENCES

[1] S.A. Bogacz, "Transition crossing in the Main Injector - ESME simulation", Proceedings of the F III Instability Workshop, Fermilab, July 1990.

[2] S.A. Bogacz and K-Y Ng, "Nonlinear saturation of the longitudinal modes of the coasting beam in a storage ring", Phys. Rev. D, **36**, p 1538, (1987)

[3] S.A. Bogacz and S. Peggs, "Comments on the Behavior of α_1 in Main Injector γ_t Jump Schemes", This Proceedings

[4] S. Krinsky and J.M. Wang, "Longitudinal instabilities of bunched beams subject to a non-harmonic RF potential" Particle Accelerators, **17**, p 109, (1985)

[5] C.N. Lashmore-Davies, "Plasma Physics and Instabilities", Ch.3 and 5, CERN 81-13, Geneva 1981

Electron Beam Injector for Longitudinal Beam Physics Experiments*

J. G. Wang, D. X. Wang and M. Reiser
Laboratory for Plasma Research and Department of Electrical Engineering
University of Maryland, College Park, Maryland 20742

Abstract

Design parameters of the electron beam injector for longitudinal beam physics experiments are discussed. The performance characteristics of its components and the injector system is presented.

I. INTRODUCTION

An electron beam injector has been constructed to study problems of longitudinal beam in the University of Maryland electron beam transport experiment. These include studies of longitudinal pulse compression and resistive-wall instability. The injector consists of a variable-perveance gridded electron gun followed by three matching lenses and one induction module. In the compression experiment, it produces a 50 ns, 40 mA, 2-5 keV electron pulse with a time-dependent velocity spread. This beam will be injected into a 5-m long periodic transport channel with 36 short solenoid lenses. The pulse is expected to be compressed by a factor of 3 or greater when reaching the end of the channel. In the resistive-wall instability experiment, the injector produces a 5 ns, 100 mA and 2.5 keV beam pulse. This beam will be guided into a resistive-wall channel of a few meters length for the instability study. This paper reports on the design features and the performance of the injector components and system.

II. DESIGN STUDY OF THE LONGITUDINAL COMPRESSION AND INSTABILITY

2.1, Beam compression due to drift bunching

The design of the compression experiment applies to a beam with parabolic density distribution, that satisfies the longitudinal envelope equation [1]

$$Z_m'' - \frac{3gZ_i I_i}{\beta^3 \gamma^5 I_0} \frac{1}{Z_m^2} - \frac{\varepsilon_L^2}{\gamma^4} \frac{1}{Z_m^3} = 0 \tag{1}$$

where $2Z_m$ is the bunch length, g is a factor related to the beam radius, I_0 is the characteristic current of electrons, and ε_L is the normalized longitudinal emittance. Assuming a K-V distribution [2], one applies the transverse envelope equation

$$R'' + \kappa R - \frac{K}{R} - \frac{\varepsilon_T^2}{R^3} = 0 \tag{2}$$

where R is the transverse envelope, κ is the periodic focusing function of the lens, ε_T is the transverse emittance, and K is the generalized perveance. The two equations are coupled by the g factor and the product KZ_m which is a constant during the compression due to the conservation of the total number of electrons. Solving Eqs. (1) and (2) numerically yields the beam envelopes in both transverse and longitudinal directions. In our 5-m long channel, the solenoid lenses are spaced at intervals of s=13.6 cm. Fig. 1 shows these results for the initial parameters I_i=40 mA, T_i=50 ns, E_{head}=3.3 keV, E_{center}=5.0 keV. These are typical operational parameters expected for the new injector.

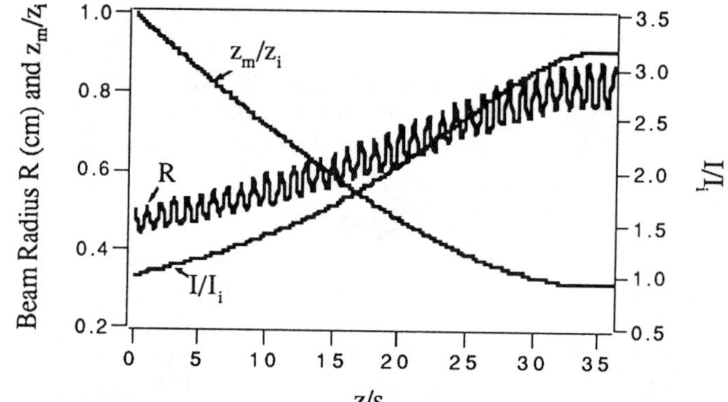

Fig. 1. Computer simulation of the longitudinal compression.

The experiment may have to compress a beam with an initial rectangular density distribution instead of a parabolic which is difficult to produce in practice. In this case the initial beam energy spread from head to tail will be smaller than the above results.

2.2, Resistive-wall instability

The resistive-wall instability growth rate is given by [3]

$$\omega_i \approx \frac{1}{2} v_0 R \left(\frac{4\pi\varepsilon_0 \eta\lambda_0}{g} \right)^{1/2}, \tag{3}$$

where λ_0, v_0 are the unperturbed line charge density and drift velocity, respectively, η=q/m denotes the ratio of the charge and mass of the charged particles, and R is the resistance of the wall per unit length. This equation yields the number of e-folds per meter:

$$N = \left(\frac{1}{2} \eta\pi^2\varepsilon_0^2 \right)^{1/4} \left(\frac{I}{g\, V^{1/2}} \right)^{1/2} R. \tag{4}$$

For the instability experiment, an electron beam pulse has been designed with a pulse width of 5 ns, a beam energy of 2.5 keV, and a peak current of 100 mA. Assuming a resistance of 1 kΩ/m and g=2, Eq. (4) yields N≅0.1. A transport channel of a few meters in length would produce observable perturbation growth. These parameters will be scaled in the experiment. This is also the requirement for the electron beam injector in the instability study.

* Research Supported by U.S. Department of Energy.

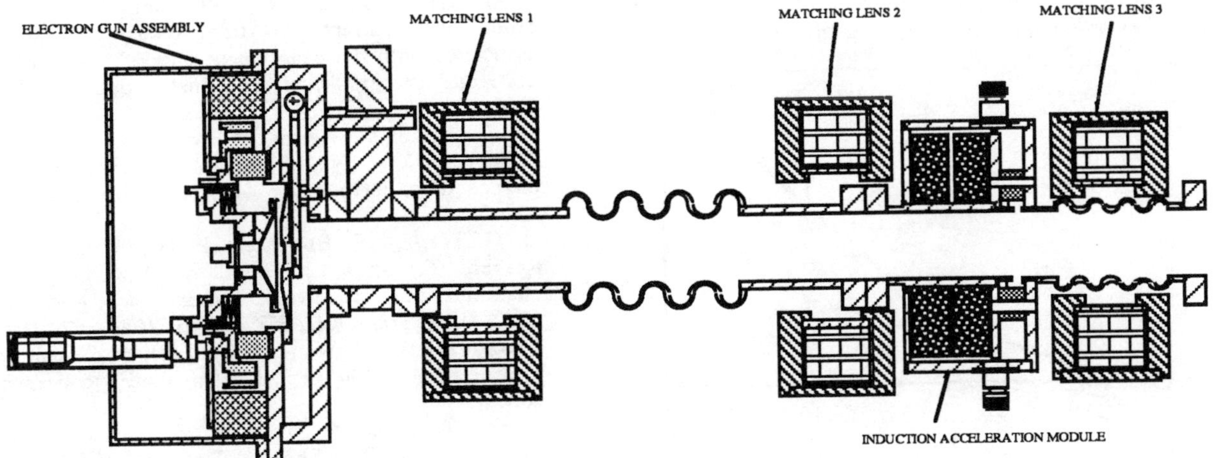

Fig. 2. Mechanical drawing of the electron beam injector system.

III. COMPONENTS OF ELECTRON BEAM INJECTOR

Fig. 2 shows the electron beam injector consisting of the following three major components:

3.1. Electron gun

The design of the variable-perveance gridded electron gun and its general performance characteristics were described in reference [4]. Since then, many improvements have been made to the gun, including replacement of the ML-EE55 oxide cathode by a Y646B dispenser cathode assembly. The A-K gap has been modified to vary from 0.93 cm to 2.3 cm, resulting in a perveance of 0.22 to 1.35 $\mu AV^{-3/2}$. This can be seen in Fig. 3, where the anode voltage is 2.5 kV.

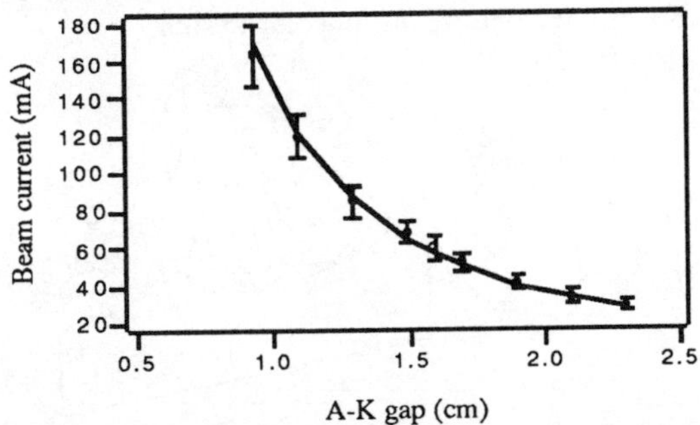

Fig. 3. Beam current vs. A-K gap where the smooth curve is from Child's law.

The beam emittance is measured by the pepper-pot method and calculated by [5]

$$\varepsilon_T = \frac{RD}{2z}\left(\frac{D_z}{D} - \frac{S_z}{S}\right)\left[1 - \left(\frac{S}{R}\right)^2\right]^{-1/2}, \qquad (5)$$

where R is the full beam radius, D and D_z are the sizes of the pepper-pot hole and its image, S and S_z are the radial distances from the axis of the centers of the pepper-pot hole and its image, respectively, z is the distance between the pepper-pot plate and the phosphor screen, The pepper-pot measurement yields a full beam emittance of 88 mm mrad with a standard deviation of 5 mm mrad.

3.2. Matching Lenses

There are three solenoid matching lenses in the injector system. The measured magnetic field $B_z(r=0)$ is plotted in Fig. 4. The experimental data can be well fitted by the following function

$$B_z(0, z) = B_0\left(1 + \frac{z^2}{a^2}\right)^{-1} \exp\left(-\frac{z^2}{d^2}\right), \qquad (6)$$

where the two constants d and a are obtained from least-square fitting. The magnetic field components $B_z(r,z)$ and $B_r(r,z)$ off the z axis can then be calculated from Eq. (6) by power series expansion. These data will be used to simulate beam dynamics in the compression and instability experiments.

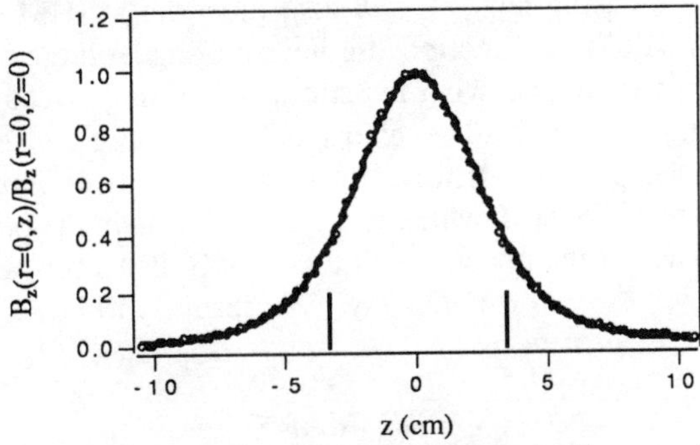

Fig. 4. The normalized magnetic field $B_z(r=0)$ vs. z. The width of the lenses is indicated by the two parallel lines.

3.3. Induction acceleration module

The design and performance characteristics of the induction acceleration module can be found in reference [6]. The induction module consists of a single-turn primary around 2 ferrite cores with the beam completing the single-turn secondary. The gap voltage is controlled by a PFN circuit. The measured gap voltage is shown in Fig. 5, which is approximately t^2 shaped. This waveform can be changed to linear by modifying the PFN circuit.

The induction linac employs a pseudospark switch to control the PFN operation. This component provides superior performance of the switch in the aspects of fast risetime, small jitter, current reversal capability, and long life time.

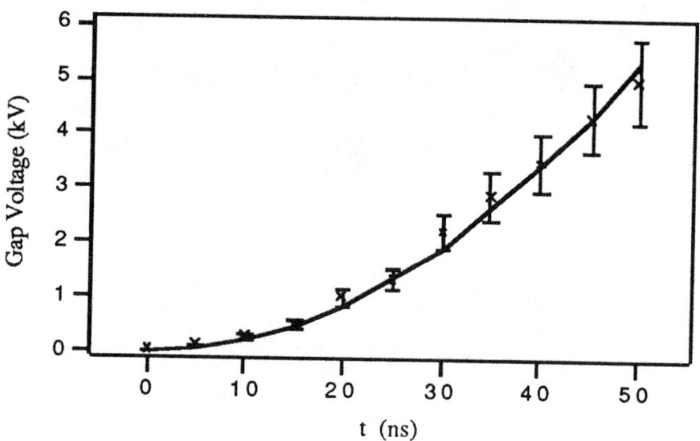

Fig. 5. Gap voltage vs. time in the induction module.

IV. BEAM CHARACTERISTICS IN THE INJECTOR

The beam transport in the injector and its interaction with the induction gap has been measured with a ML-EE55 cathode installed in the gun. Figure 6 shows the beam envelope obtained by a phosphor screen, where the two maximums are at the lens centers and the beam waist is focused at the induction gap. The beam image and profile is measured at 2 cm after the induction gap, and is shown in Fig. 7. The intensity of the beam after the induction gap as a function of the charging voltage V_0 of the induction module is shown in Fig. 8, indicating the energy transfer from the induction gap to the beam. This is a time integrated picture and needs to be calibrated. Besides the energy transfer, the induction gap also acts as a focusing lens. Fig. 8 also plots the FWHM of the beam profile as a function of the charging voltage of the induction module, showing the stronger focusing effect at the larger gap voltage. The average beam energy after the induction gap is proportional to the product of the intensity and square of FWHM, which is the straight line in the plot.

The test of the injector with the Y646B dispenser cathode assembly in the gun is underway. The results will be reported elsewhere in the near future.

V. SUMMARY

An electron beam injector has been designed for the longitudinal compression and resistive wall instability experiments at the University of Maryland. The injector has been constructed and preliminarily tested, showing satisfactory performance. The final test of the injector system is underway. The longitudinal physics experiments are expected in the near future and their results will be reported elsewhere.

VI. REFERENCES

[1] T. Shea, E. Boggasch, Y. Chen, and M. Reiser, "The University of Maryland electron pulse compression experiment", Proceedings of the IEEE Particle Accelerator Conference, Chicago Ill., March 20-23, 1989.

[2] I. M. Kapchinskij and V. V. Vladimirskij, "Limitations of proton beam current in a strong focusing linear accelerator associated with the beam space charge", Proc. of Int. Conf. on High Energy Accel. and Instr., pp. 274-288, CERN, Geneva, 1959.

[3] J. G. Wang, M. Reiser, W. M. Guo, and D. X. Wang, "Theoretical study of the longitudinal instability and proposed experiment", to be published in Particle Accelerators.

[4] J. G. Wang, E. Boggasch, P. Haldman, D. Kehne, M. Reiser, T. Shea, and D. X. Wang, "Performance characteristics of a variable-perveance gridded electron gun", IEEE Trans. ED, 37(12), pp. 2622-2628, Dec. 1990.

[5] J. G. Wang, D. X. Wang, and M. Reiser, "Beam emmitance measurement by the pepper-pot method", to be published in NIM, (A).

[6] J. G. Wang, D. X. Wang, E. Boggasch, D. Kehne, M. Reiser, and T. Shea, "Design and performance characteristics of a compact induction acceleration module for longitudinal compression experiments", NIM, (A)301, pp19-26, 1991.

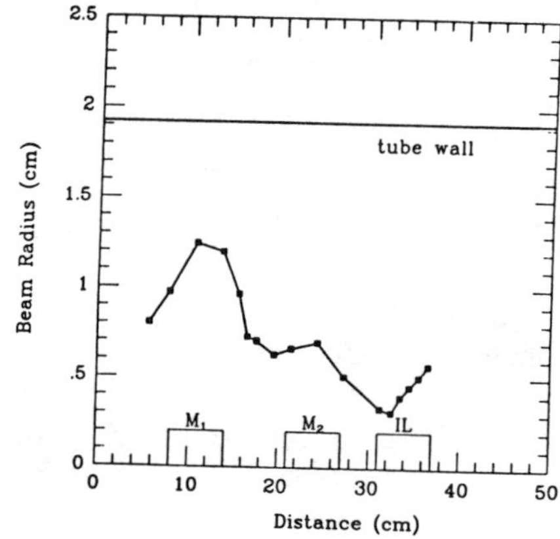

Fig. 6. Beam envelope in the injector system.

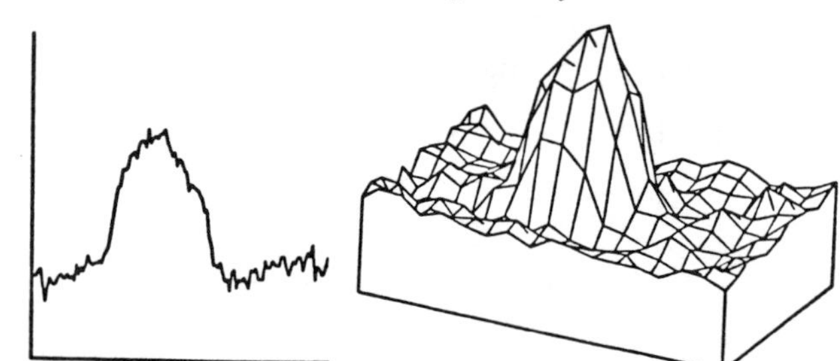

Fig. 7. Beam image and profile after the induction module.

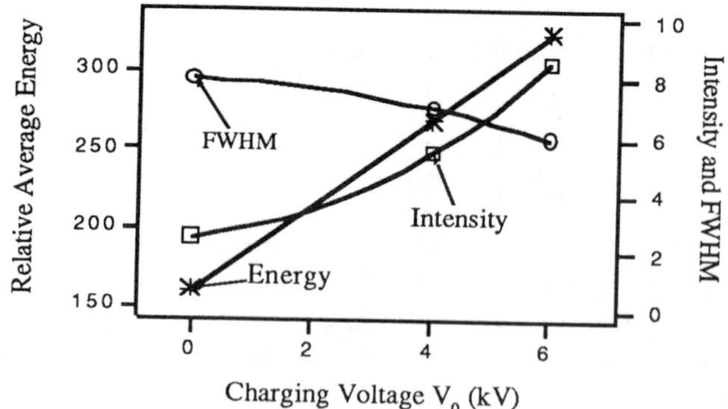

Fig. 8. Relative average beam energy after the induction gap vs. the charging voltage V_0 of the induction module.

Analysis of the Longitudinal Instability in an Induction Linac*

J. G. Wang and M. Reiser
Laboratory for Plasma Research and Department of Electrical Engineering
University of Maryland, College Park, Maryland 20742

Abstract

A theoretical model for the longitudinal instabilities of intense beams in a transport channel with complex wall impedances is presented. A dispersion equation is derived to relate the growth rates of the slow waves to all the relevant parameters including the space charge, the real and imaginary parts of the complex wall impedances. The growth rates for different wall structures are compared and discussed.

I. INTRODUCTION

The induction linac is a promising driver for Heavy Ion Inertial Fusion. An important issue in the design of such induction linacs is the longitudinal instability [1,2,3,4] which is predicted to develop and is detrimental to the beam. In contrast with the accelerators for high energy physics, the beam current in induction linacs as drivers for heavy ion inertial fusion is as high as kilo-amperes, while the beam particle velocity is only sub-relativistic. Thus, the space charge force plays an important role in the instability analysis. Another important factor is the interaction between charged particles and induction gaps which are usually modeled by discrete R, L, C circuits in the low frequency limit. Studies show that the impedance of induction gaps is a function of frequency and could change from inductive to capacitive as the frequencies shift. This requires different circuit models at different frequency ranges. Thus, it is desirable to have a theory which can analyse the instability in a transport channel with general wall impedances, rather than a specific circuit configuration. In a previous paper [5] we developed a theoretical model to relate the growth rates of the slow waves to the beam space charge and the complex impedances of the transport channel. This paper is an extension of the previous one. The derivation of the dispersion equations and the expressions for the general longitudinal field taking into account the real and imaginary parts of the wall impedance is given. The physical consequences of the resulting equations are discussed. The growth rates for the wall modeled by three different circuits are also compared. The results show explicitly how a capacitive wall can dramatically reduce the growth rate of the instability.

II. DERIVATION OF THE DISPERSION EQUATIONS

The longitudinal beam dynamics can be described by the linearized cold, one-dimensional fluid model which consists of the continuity equation and momentum transfer equation:

$$\begin{cases} \dfrac{\partial \lambda_1}{\partial t} + v_0 \dfrac{\partial \lambda_1}{\partial z} + \lambda_0 \dfrac{\partial v_1}{\partial z} = 0 \\ \dfrac{\partial v_1}{\partial t} + v_0 \dfrac{\partial v_1}{\partial z} \approx \eta E_z \end{cases} \tag{1}$$

where $\lambda(z,t)$ and $v(z,t)$ are the line charge density and the particle velocity, the subscripts 0 and 1 representing the unperturbed and perturbed quantities, respectively, $E_z(z,t)$ is the longitudinal electrical field induced by the a. c. component of the beam current, and $\eta = q/(m\gamma^3)$ denotes the ratio of the charge and the "longitudinal mass" of the charged particles. Under the long wavelength condition, the field $E_z(z,t)$ can be calculated as [6]

$$E_z(z, t) = - \frac{g}{4\pi \varepsilon_0 \gamma^2} \frac{\partial \lambda_1}{\partial z} + E_w(z, t) \tag{2}$$

where ε_0 is the permittivity of free space, g is a geometric factor of order unity, and $E_w(z,t)$ is the field induced on the pipe wall of the transport channel and depends on the perturbed beam current and the properties of the wall. In the complex frequency domain this field is related to the perturbed beam current and the wall complex impedance $Z_w(s)$ per unit length as

$$E_w(k, s) = \int_{-\infty}^{+\infty} dz \int_0^\infty dt E_w(z, t) e^{-(ikz + st)}$$
$$= - Z_w(s) \left[v_0 \lambda_1(k, s) + \lambda_0 v_1(k, s) \right]. \tag{3}$$

Solving Eqs. (1) to (3) by using Laplace transform in time and Fourier transform in space leads to the dispersion equation

$$\left(s + ikv_0 \right)^2 + \frac{k^2 \eta g \lambda_0}{4\pi \varepsilon_0 \gamma^2} + s\eta \lambda_0 Z_w(s) = 0 \tag{4}$$

Though this equation is exact in this theoretical model, its solution depends on the wall properties, namely $Z_w(s)$, and can not usually be solved analytically except in the simplest cases such as the pure resistive wall. Approximations have to be made for proper interpretation of the physical consequences of Eq. (4). In the real frequency domain the wall impedance per unit length has the general form

$$Z_w(\omega) = R(\omega) + iX(\omega), \tag{5}$$

where $R(\omega)$ and $X(\omega)$ are the real and imaginary parts of the complex impedance. It is usually supposed that the zeros of

---

* Research Supported by the U.S. Department of Energy.

"

Eq. (4) associated with the circuit $Z_w(s)$ are not of interest and that the approximation may be made by setting $s=-i\omega_0=-ikv_0$ in the last term of Eq. (4). Thus, one gets

$$\left(s + ikv_0\right)^2 - ik\eta\lambda_0 v_0 R\left(\omega_0\right) +$$

$$+ k^2\eta\left(\frac{g}{4\pi\varepsilon_0\gamma^2} - v_0^2\frac{X(\omega_0)}{\omega_0}\right)\lambda_0 = 0 \quad . \tag{6}$$

In the above approximation we actually rewrite Eq. (3) as

$$E_w(k, s) \approx - v_0\lambda_1(k, s)\left[R\left(\omega_0\right) - iX\left(\omega_0\right)\right]$$

$$= ikv_0^2\frac{X(\omega_0)}{\omega_0}\lambda_1(k, s) - R\left(\omega_0\right)v_0\lambda_1(k, s) \quad , \tag{7}$$

where the perturbed beam current component $\lambda_0 v_1(k, s)$ in Eq. (3), which is usually much smaller than the term $v_0\lambda_1(k, s)$ in most applications, is neglected. This equation shows that the real part of the wall impedance produces the field in the opposite phase with respect to the perturbed line charge density, while the imaginary part of the wall impedance produces the field that is 90° out of phase with respect to the perturbed line charge density. Inverse Laplace-Fourier transform of Eq. (7) yields

$$E_w(z, t) = v_0^2\frac{X(\omega_0)}{\omega_0}\frac{\partial\lambda_1}{\partial z} - R\left(\omega_0\right)v_0\lambda_1 \quad . \tag{8}$$

Therefore, the total longitudinal field acting on the beam can be expressed by

$$E_z(z, t) \approx - \left(\frac{g}{4\pi\varepsilon_0\gamma^2} - v_0^2\frac{X(\omega_0)}{\omega_0}\right)\frac{\partial\lambda_1}{\partial z} - \lambda_1 v_0 R\left(\omega_0\right) \quad . \tag{9}$$

It is interest to note that the inductive impedance which has a positive imaginary part can reduce the space charge field and the capacitive impedance just acts in the opposite way.

III. DISCUSSION OF THE GROWTH RATES

The dispersion equation (6) has the general solution for the perturbed wave frequency and the temporal growth rate:

$$\begin{cases} \omega_r = kc\left[\beta \pm \left[\frac{g}{2\gamma^5\beta}\frac{I}{I_0}\left(1 - \frac{2\beta\gamma^2}{g}\frac{X(\omega_0)\Lambda}{Z_0}\right)Y_+\right]^{1/2}\right] \\ \omega_i = \mp kc\left[\frac{g}{2\gamma^5\beta}\frac{I}{I_0}\left(1 - \frac{2\beta\gamma^2}{g}\frac{X(\omega_0)\Lambda}{Z_0}\right)Y_-\right]^{1/2} \end{cases} \tag{10}$$

where I is the beam current, $I_0=4\pi\varepsilon_0 mc^3/q$ is the characteristic current of the charged particles, Λ is the wavelength of the growing waves, $Z_0=1/(\varepsilon_0 c)=377$ ohms is the characteristic impedance of free space, and $Y_\pm=(1+y^2)^{1/2}\pm 1$ with y being determined by

$$y = \frac{2\beta\gamma^2}{g}\frac{R(\omega_0)\Lambda}{Z_0}\left(1 - \frac{2\beta\gamma^2}{g}\frac{X(\omega_0)\Lambda}{Z_0}\right)^1 \quad . \tag{11}$$

Equation (10) is valid for $y>0$. In the above equations the subscript 0 is dropped for β and γ for simplicity.

An interesting parameter range is $y<<1$, or

$$\frac{2\beta\gamma^2}{g}\frac{\left[R(\omega_0) + X(\omega_0)\right]\Lambda}{Z_0} << 1 \quad , \tag{12}$$

which requires that the algebraic sum of the real and imaginary parts of the wall impedance per perturbation wavelength must be much smaller than the characteristic impedance of free space. In this case Eq. (10) can be approximated as

$$\begin{cases} \omega_r \approx kc\left[\beta \pm \left(\frac{g}{\gamma^5\beta}\frac{I}{I_0}\right)^{1/2}\left(1 - \frac{\beta\gamma^2}{g}\frac{X(\omega_0)\Lambda}{Z_0}\right)\right] \\ \omega_i \approx \mp kc\beta\frac{R(\omega_0)\Lambda}{Z_0}\left(\frac{1}{g\gamma\beta}\frac{I}{I_0}\right)^{1/2}\left(1 + \frac{\beta\gamma^2}{g}\frac{X(\omega_0)\Lambda}{Z_0}\right) \end{cases} \tag{13}$$

For a pure resistive-wall where $R(\omega_0)=R$ and $X(\omega_0)=0$, the slow-wave growth rate is

$$\omega_i \approx kc\beta\frac{R\Lambda}{Z_0}\left(\frac{1}{g\gamma\beta}\frac{I}{I_0}\right)^{1/2} \quad . \tag{14}$$

Thus, the resistive instability growth rate is proportional to the resistance per unit length of the transport channel. This growth rate is also proportional to the wavelength of the perturbation, implying that the slow waves of perturbation with the fundamental frequency is more serious than the higher harmonics. Equation (14) also indicates that the resistive instability growth rate is higher for a beam with higher current and lower energy. This is the case in the induction linac as drivers for Heavy Ion Inertial Fusion. If the pipe wall is inductive, e.g. $Z=R+i\omega_0 L$, one gets

$$\omega_i \approx kc\beta\frac{R\Lambda}{Z_0}\left(\frac{1}{g\gamma\beta}\frac{I}{I_0}\right)^{1/2}\left(1 + \frac{\beta\gamma^2}{g}\frac{\omega_0 L\Lambda}{Z_0}\right) \quad . \tag{15}$$

Therefore, the addition of an inductive component increases the growth rate in comparison with Eq. (14). For a capacitive wall modeled by a circuit of R and C in parallel, Eq. (13) can be rewritten as

$$\omega_i \approx kc\beta \frac{R\Lambda}{Z_0}\left(\frac{1}{g\gamma\beta}\frac{I}{I_0}\right)^{1/2} \times$$

$$\times \frac{\left[1 + \left(\omega_0 RC\right)^2 - \frac{\beta\gamma^2}{g}\left(\omega_0 RC\right)\frac{R\Lambda}{Z_0}\right]}{\left[1 + \left(\omega_0 RC\right)^2\right]^2} \quad . \quad (16)$$

As expected, the growth rate of the slow wave is reduced when a capacitance is added. This result agrees with ref. [1] which states that in general the capacity reduces growth rates compared with the case of pure resistance by lowering the impedance as frequency increases.

Fig. 1 shows the relative growth rate versus the relative time constant of the wall impedance. In this example the growth rate for a pure resistive wall is set to unity. The lower branch of the plot is for the capacitive wall with $\tau/\tau_0=RC\omega_0$ while the upper branch is for the inductive wall with $\tau/\tau_0=\omega_0 L/R$. The curves A-1 and A-2 illustrate the results from Eqs. (14)-(16) where the parameters used are $\beta=0.1$, $R=100$ Ω/m and $\omega_0=2\pi\times10^8$ s^{-1} . The growth rate for the capacitive wall decreases rapidly when the capacitance is increased. By contrast, the growth rate for an inductive wall is slightly increased.

If the condition (12) is not satisfied, the results from Eqs. (13) to (16) are not valid any more. In another extreme case

$$\frac{2\beta\gamma^2\left[R(\omega_0) + X(\omega_0)\right]\Lambda}{g\quad Z_0} >> 1 \quad , \quad (17)$$

the perturbed frequency and growth rate can be expressed by

$$\left\{\begin{array}{l} \omega_r \approx kc\left\{\beta \pm \left[\frac{1}{\gamma^3}\frac{I}{I_0}\frac{R(\omega_0)\Lambda}{Z_0}\right]^{1/2}\right\} \\[2em] \omega_i \approx \mp kc\left[\frac{1}{\gamma^3}\frac{I}{I_0}\frac{R(\omega_0)\Lambda}{Z_0}\right]^{1/2} \end{array}\right. \quad , \quad (18)$$

which shows that the growth rate is determined by the real parts $R(\omega_0)$ of the complex impedance. In this case a R, C circuit in parallel will still reduce the growth rate, but a R, L circuit in series would not increase the growth rate in comparison with a pure resistive case. The curves B-1 and B-2 in Fig. 1 illustrate such an example where the parameters used are $\beta=0.3$, R=300 ohms/m and $\omega_0=2\pi10^6$ s^{-1}.

IV. SUMMARY

The cold one-dimensional fluid model has been used to analyze the longitudinal instability in a beam transport channel with general wall impedances. The dispersion equation derived explicitly relates the growth rate of the slow waves to all the relevant parameters including the space charge, the real and imaginary parts of the complex wall impedances. The temporal growth rate of the instability can be dramatically reduced by a capacitive component of the wall impedance along the transport channel. By contrast, the inductive wall component could enhance the instability.

V. REFERENCES

[1] E. P. Lee and L. Smith, "Asymptotic analysis of the longitudinal instability of a heavy ion induction linac", Proceedings of the 1990 Linear Accelerator Conference, Albuquerque, NM, September 10-14, 1990, pp. 716-718.

[2] E. P. Lee, "Resistance driven bunching mode of an accelerated ion pulse", Proceedings of the 1981 Linear Accelerator Conference, Santa Fe, NM, October 19-23, 1981, pp. 263-265.

[3] J. Bisognano, I. Haber, L. Smith, and A. Sternlieb, "Nonlinear and dispersive effects in the propagation and growth of longitudinal waves on a coasting beam", IEEE Trans. on Nuclear Science, Vol. NS-28, No. 3, pp. 2513-2515, June 1981.

[4] L. Smith, "Resistive instability growth rates", LBL HIFAR-Note-217, September 27, 1988.

[5] J. G. Wang, M. Reiser, W. M. Guo, and D. X. Wang, "Theoretical study of the longitudinal instability and proposed experiment", to be published in Particle Accelerators.

[6] A. Hofmann, "Single-beam collective phenomena-longitudinal", CERN 77-13, pp. 139-174 (CERN, Geneva, 1977).

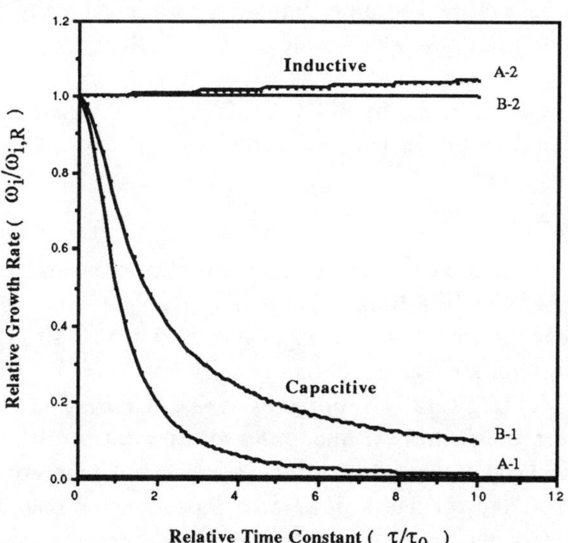

Fig. 1 . Relative growth rate $\omega_i/\omega_{i,R}$ vs. the relative time constant τ/τ_0 for a wall modeled by R, L in series or R, C in parallel, where ω_i is the growth rate for inductive or capacitive wall, $\omega_{i,R}$ is the growth rate for a pure resistive wall, $\tau=RC$ for the capacitive wall, or $\tau=L/R$ for the inductive wall, and $\tau_0=1/\omega_0$. The curves A-1 and A-2 illustrate the results from Eqs. (14)-(16) with the parameters $\beta=0.1$, R=100 Ω/m and $\omega_0=2\pi\times10^8$ s^{-1} . The curves B-1 and B-2 represent the results of Eq. (18) where the parameters are $\beta=0.3$, R=300 Ω/m and $\omega_0=2\pi\times10^6$ s^{-1}.

SHIELDED COHERENT SYNCHROTRON RADIATION AND ITS POSSIBLE EFFECT IN THE NEXT LINEAR COLLIDER*

Robert L. Warnock

Stanford Linear Accelerator Center, Stanford University, Stanford, CA 94309 USA

Abstract

Shielded coherent synchrotron radiation is discussed in two cases: (1) a beam following a curved path in a plane midway between two parallel, perfectly conducting plates, and (2) a beam circulating in a toroidal chamber with resistive walls. Wake fields and the radiated energy are computed with parameters for the high-energy bunch compressor of the Next Linear Collider (NLC).

I. INTRODUCTION

We consider a bunch of particles following a curved path. On first sight, one would expect the particles to radiate coherently at wavelengths comparable to the bunch size or larger. A closer examination reveals that the long-wavelength part of the synchrotron spectrum will be strongly suppressed in the presence of shielding, for instance the conducting surfaces of the vacuum chamber in an accelerator. This long-wavelength cutoff is not the usual wave-guide cutoff, comparable to the pipe diameter, but has a value that depends on the radius of curvature of the orbit as well as the pipe diameter, and is typically much smaller than the wave-guide cutoff. In fact, appreciable radiation under shielding is expected only in an interval of $\lambda = \lambda/2\pi$ given by $\lambda_0 \leq \lambda \leq \lambda_1$, where λ_0 is near the longitudinal bunch length σ, and

$$\lambda_1 \approx \frac{h}{\pi} \left(\frac{h}{R} \right)^{1/2} \quad , \tag{1}$$

where R is the orbit radius and h is the transverse size of the chamber (see Refs. [1] and [2]). It follows that shielding can be important in typical experimental situations. In electron storage rings one usually has $\lambda_1 < \lambda_0$, so that the shielding totally suppresses coherent radiation. Using a short bunch from a linac, and an apparatus with small R and large h, Nakazato et al. [3], observed coherent radiation. They verified a quadratic dependence on the number N of particles in the bunch, and checked coherence by an interference experiment. This observation is quantitatively consistent with the theory of shielding, as is the failure to see coherent radiation in earlier experiments at storage rings.

In this paper we give a brief review of the theory of shielding and an application to the NLC. Further details can be found in Refs. [4] and [2].

2. IMPEDANCE, ENERGY LOSS, AND WAKE FIELD

We work in cylindrical coordinates (r, θ, z) and treat a rigid bunch with centroid moving on a circular orbit in the plane $z = 0$. We are actually interested in more general orbits, for instance straight paths joined to arcs of circles, and a bunch that gradually changes shape. These situations will be treated in more detail in later work; for the present we use provisional approximations derived from the simpler theory. The charge density is assumed to have the form

$$\rho(r, \theta, z, t) = q\lambda(\theta - \omega_o t)f(r, z) \quad , \tag{2}$$

where ω_o is the revolution frequency and

$$\int_0^{2\pi} \lambda(\theta)d\theta = 1 \quad , \qquad \int rdr \int dz f(r, z) = 1 \quad . \tag{3}$$

The corresponding current density is $J_\theta = \omega_o r\rho$. The assumption of coherence enters at this point, since the atomic nature of the beam is ignored, and the continuously distributed charge radiates as a whole.

Since we are interested primarily in longitudinal effects, we make an average of the longitudinal electric field over r and z, weighted with the transverse charge distribution of the beam. We work with a Fourier analysis of the averaged field,

$$
\begin{aligned}
E_\theta(\theta, t) &= \int rdr \int dz \; f(r, z) \; E_\theta(r, \theta, z, t) \\
&= \int_{-\infty}^{\infty} d\omega \; e^{-i\omega t} \sum_{n=-\infty}^{\infty} e^{in\theta} \; E_\theta(n, \omega) \quad .
\end{aligned} \tag{4}
$$

The Fourier transform of the line density $\lambda(\theta - \omega_o t)$ is $\lambda_n \delta(\omega - \omega_o n)$, and the corresponding transform of the current is

$$
\begin{aligned}
I_\theta(n, \omega) &= \int dr \int dz \, J_\theta(r, n, z, \omega) \\
&= \omega_o q \, \lambda_n \, \delta(\omega - \omega_o n) \quad .
\end{aligned} \tag{5}
$$

If the environment of the beam has no longitudinal inhomogeneity, as in the models treated below, then there exists a complex function $Z(n, \omega)$, the longitudinal coupling impedance, such that

$$-2\pi R E_\theta(n, \omega) = Z(n, \omega) I_\theta(n, \omega) \quad . \tag{6}$$

This impedance accounts both for curvature of the orbit and for conductors in the environment of the beam.

If E_θ is the field due to the bunch, including effects of nearby conductors, then the radiated power is

$$\frac{dU}{dt} = -(q\omega_o)^2 \sum_{n=-\infty}^{\infty} |\lambda_n|^2 \, \text{Re} \, Z(n, n\omega_o) \quad . \tag{7}$$

This result is obtained by computing the work done per unit time on an infinitesimal element of charge by the field E_θ, then integrating over all elements [4].

★ Work supported by Department of Energy contract DE–AC03–76SF00515.

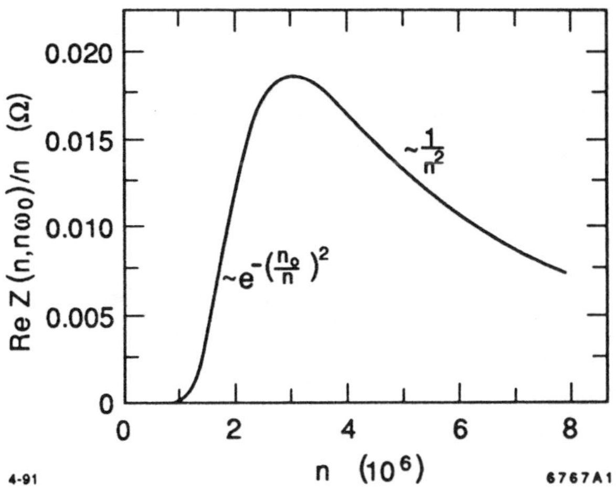

Figure 1. Re $Z(n, n\omega_0)/n$ versus n for the parallel-plate model, $R = 149.5$ m, $h = 2$ cm.

The wake field $E_\theta[\omega_o(t+\tau), t]$ is defined as the field on the trajectory at an angular distance $\omega_o\tau$ from the bunch center. We take τ to be positive for a point in front of the bunch. The wake field is expressed by

$$\mathcal{V}(\omega_o\tau) = -2\pi R\, E_\theta[\omega_o(t+\tau), t]$$

$$= \omega_o q \sum_{n=-\infty}^{\infty} \exp[in\omega_o\tau]\, \lambda_n\, Z(n, n\omega_o) \ .$$

$$(8)$$

The function $\mathcal{V}$ is the *wake voltage per turn*. A positive value of $\mathcal{V}(\omega_o\tau)$ means energy loss by particles at a distance $\omega_o\tau$ from the center of the bunch.

3. COUPLING IMPEDANCES FOR TWO MODELS

In order to state the impedance in a convenient analytic form, we choose a simple transverse beam profile,

$$f(r, z) = W(r)H(z) \ , \qquad W(r) = \delta(r - R)/R \ , \quad (9)$$

where $H(z)$ is a rectangular step function of width δh, symmetric about $z = 0$. More general beam profiles can be accomodated without much difficulty.

In our first model we have infinite, parallel, perfectly conducting plates, separated by a distance $h = 2g$. The beam circulates in the median plane between the plates. The impedance is computed by using a Fourier analysis in z to meet boundary conditions on the plates. The result is expressed in terms of Bessel functions of high order, that must be evaluated by appropriate asymptotic expansions [4,2]. Using the Olver expansions to lowest order, we find

$$\frac{\text{Re}\, Z(n, n\omega_o)}{n} = 2Z_o\left[\frac{\pi R}{hn}\right]^2 \exp\left[-\frac{2}{3n^2}\left(\frac{\pi R}{h}\right)^3\right] \ , \quad (10)$$

the value being in ohms with $Z_o = 120\pi$ Ω; the bending radius is R. Here only the dominant term (axial mode number $p = 1$) has been included, and the vertical size of the beam is much less than h. This simple formula, which seems not to have been noticed previously, gives an adequate representation of the exact result; a plot of the latter is shown in Fig. 1.

The maximum value of expression (10) is

$$\frac{720}{e}\frac{g}{R} \approx 265\,\frac{g}{R} \ , \quad (11)$$

in agreement with Faltens and Laslett [1]. The maximum occurs at $n = \pi 2^{1/2}(R/h)^{3/2}$, and the function cuts off exponentially below the threshold $n_o = \pi(R/h)^{3/2}$. Expressed in terms of wavelength, this threshold is λ_1 of Eq. (1).

Table 1. Energy losses for four versions of the NLC bunch compressor.

Version	1	2	3	4
R (m)	84.218	213.6	149.5	106.8
$\Delta\theta$ (degrees)	10.89	180	180	180
σ_i (10^{-6} m)	460	460	460	460
σ_f (10^{-6} m)	37	86	61	44
ΔU plates (MeV)	0.176	0.164	0.827	2.08
ΔU torus (MeV)	0.123	0.0121	0.337	1.30
ΔU incoherent (MeV)	2.19	14.2	20.4	28.5

In the second model the vacuum chamber is a torus of rectangular cross section. The cross section has height $h = 2g$ and width w. In this model we take the walls to be resistive. The expression for the impedance is given in Refs. [2] and [4]. Unlike the parallel plates, this closed chamber has resonances, and the impedance is negligible at frequencies below the lowest mode that is synchronous with the beam. This mode occurs at a value of n roughly equal to (actually, somewhat higher than) the threshold n_o for the parallel plate model.

4. NUMERICAL EXAMPLES OF WAKE FIELD AND LOSS

For numerical calculations, we take a Gaussian bunch of rms length σ; its spectral density, expressed as a function of λ, is

$$|\lambda_n|^2 = \frac{1}{(2\pi)^2} \exp\left[-\left(\frac{\sigma}{\lambda}\right)^2\right] \ . \quad (12)$$

For the parallel plate model, the spectral density of radiated power will be proportional to the product of Eq. (12) and Re Z from Eq. (10).

We illustrate with values of σ and R from four different conceptual designs for the NLC high-energy (16.2 GeV) bunch compressor [5]. Table I shows the compressor parameters, including the initial and final bunch lengths, σ_i and σ_f, and the total deflection angle $\Delta\theta$ of the compressor arc. We take $h = 2$ cm for the parallel plate model, and $h = w = 2$ cm for the toroidal model. The walls of the torus have the conductivity of aluminum.

To estimate the total energy loss in the arc, we assume that $dU/d\theta$, the instantaneous energy loss per unit angle of deflection at a particular bunch length, is the same as in our steady-state model running at the same (but fixed) bunch length. Repeating the steady-state calculation for many bunch lengths, we find a curve of $dU/d\theta$ versus σ. Since σ decreases in the compressor almost linearly with θ, this is equivalent to knowing $dU/d\theta$ as a function of θ. Integration with respect to θ produces the figures for total energy loss shown in Table 1. The values are in MeV per particle, supposing that the bunch contains $N = 2 \times 10^{10}$ electrons. For comparison, the values of the incoherent radiated energy are listed. For that, we assume that $dU/d\theta$ for each electron has the well-known value for steady state rotation on a full circle.

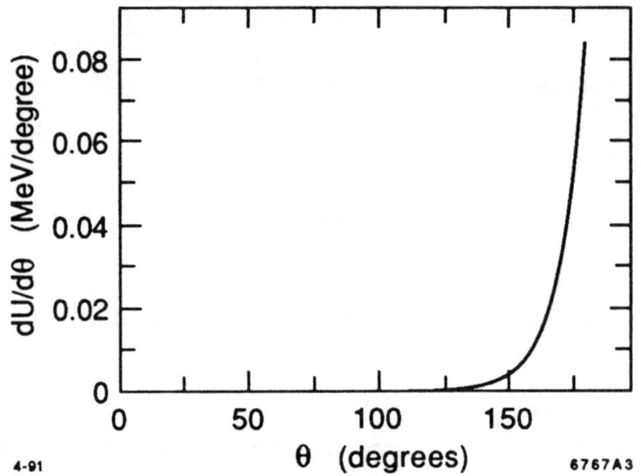

Figure 2. $dU/d\theta$ for the parallel-plate model in Version 3 of the bunch compressor..

The curve of $dU/d\theta$ for Version 3 of the compressor is shown in Fig. 2, for the parallel-plate model. In all four versions, the energy loss is sharply concentrated near the end of the arc, since elsewhere the impedance is too small at wavelengths within the bunch spectrum $|\lambda_n|^2$. The corresponding curve for the resistive toroidal chamber has a similar form, but is concentrated still closer to the end of the arc, due to the higher threshold of the impedance. The high-threshold effect is especially pronounced in Version 2 of the compressor, which has relatively large values of R and σ_f.

In Fig. 3 we show the wake voltage per turn, as defined by Eq. (8), for the parallel-plate problem and Version 3 of the compressor. Particles within one σ of the bunch center lose energy, whereas those around two σ on either side gain energy. The peak wake voltage per unit length is comparable to typical wakes in the SLAC linac structure, which amount to a few volts per picocoulomb over a 3.5 cm cell.

Corresponding results for the toroidal model are displayed in Fig. 4. Within two σ of the center the behavior

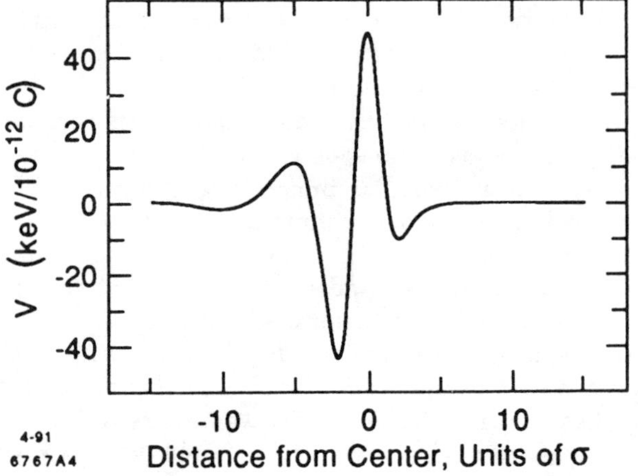

Figure 3. Wake voltage per turn for the parallel-plate model, $R = 149.5$ m, $h = 2$ cm, bunch length $\sigma = 61$ μm.

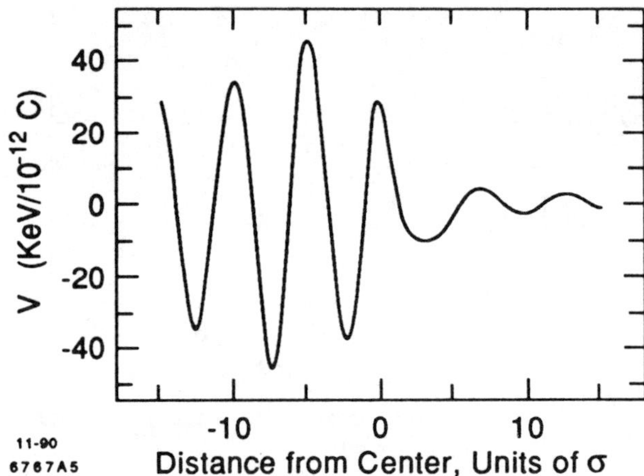

Figure 4. Wake voltage per turn for the toroidal chamber, $R = 149.5$ m, $h = w = 2$ cm, bunch length $\sigma = 61$ μm.

is similar, although the voltage is somewhat lower due to the higher threshold of the torus impedance. The persistent oscillations well beyond the bunch length can be understood if we refer to the limiting case of infinite conductivity. As that limit is approached the peak of Re $Z(n, n\omega_o)$ narrows to a delta function, and the wake field has the form $\exp[in\omega_o\tau]$ with essentially a single value of n. In fact, the oscillations in Fig. 4 have almost exactly the period of such a single exponential, if we choose n to have the value at the resonance peak of the impedance. For further remarks see Ref. [4].

5. OUTLOOK

There are a number of problems in this subject that remain to be investigated. For instance: (i) evolution of the phase-space distribution under the wake field; (ii) transverse impedances; (iii) possible microwave instabilities in rings due to curvature effects; (iv) polarization and angular distribution of coherent radiation; (v) accurate modelling of vacuum chambers for theory of bunch profile measurement through coherent radiation; etc.

REFERENCES

[1] A. Faltens and L.J. Laslett, Part. Accel. 4, 152 (1973).

[2] R.L. Warnock and P. Morton, Part. Accel. 25, 113 (1990).

[3] T. Nakazato, M. Oyamada, N. Niimura, S. Urasawa, O. Konno, A. Kagaya, R. Kato, T. Kamiyama, Y. Torizuka, T. Nanba, Y. Kondo, Y. Shibata, K. Ishi, T. Ohsaka, and M. Ikezawa, Phys. Rev. Lett. 63, 1245 (1989); *Proc. Fourth Advanced ICFA Beam Dynamics Workshop, 1990* (KEK, Tsukuba).

[4] R. L. Warnock, ibid.,pp. 151–160 and 30–35; available as SLAC-PUB-5375 and SLAC-PUB-5417.

[5] S. A. Kheifets, R. D. Ruth, and T. H. Fieguth, *Proc. Int. Conf. High Energy Accel.*, Tsukuba, Japan, 1990; available as SLAC–PUB–5034.

Experiments on the Beam Breakup Instability in Long-Pulse Electron Beam Transport Through RF Cavity Systems

R. M. Gilgenbach, P. R. Menge, R. A. Bosch, J. J. Choi, H. Ching, and T. A. Spencer
Intense Energy Beam Interaction Laboratory
Nuclear Engineering Department
The University of Michigan
Ann Arbor, MI 48109-2104

ABSTRACT

Experiments have been performed to investigate the beam-breakup-up (BBU) instability in high current electron beams transported through RF cavity systems. Experiments utilize long-pulse electron beam accelerators operating with parameters: energy = 0.3-0.8 MeV, current = 0.1-1 kA, and pulselength = 0.3-1.5 µs. The transport system consists of 10 RF cavities separated by tubes which are cut-off to the RF. Each cavity has a microwave probe to detect growth of e-beam emission in the TM_{110} mode at 2.5 GHz, corresponding to the BBU. Solenoidal magnetic fields of 0.8-5 kG are applied. Experiments show that 40% of the injected current was transported through the cavity system. The growth of the 2.5 GHz RF was found to be 4.4 dB per cavity; this compares well with the theoretical growth of 3.9 dB per cavity.

I. INTRODUCTION

The beam breakup instability is one of the most important instabilities in electron beam accelerators. This instability results in transverse beam deflection by the non-axisymmetric TM_{110} mode of the accelerating cavities. While the theory of the BBU instability has advanced considerably in recent years,[1-5] a great deal of experimental research is required to understand the scaling of the BBU over a wide range of electron beam parameters. Experiments at the University of Michigan utilize long pulse, high current electron beam accelerators to investigate the beam breakup instability in RF cavity structures.

II. EXPERIMENTAL CONFIGURATION

The accelerator for these experiments is the Michigan Electron Long Beam Accelerator (MELBA) [6], which operates with parameters: voltage = -0.8 to -1 MV, diode current = 1 - 10 kA, and pulselength = 0.3 - 5 µs, with flattop voltage provided by compensation over 1.5 µs. The experimental configuration is depicted in Figure 1. A planar velvet field emission cathode is utilized to generate an electron beam, part of which is extracted by a 2 cm diameter aperture in the graphite anode. A solenoidal magnetic field of 0.8-5 kG is applied to the 1 m transport region. A set of 10 microwave cavities are connected by cutoff sections of small diameter copper tubes. The cavities are designed so that the resonant frequency of the TM_{110} mode is 2.5 GHz; this frequency cannot propagate through the cutoff sections (dia.=3.9 cm, length=1 cm). This frequency is convenient for microwave priming of the first cavity by a magnetron. Each cavity has one or two small coupling probes oriented to detect the TM_{110} mode. The specifications of the cavities are given in Table 1. The growth of the BBU instability is detected by measuring the attenuation required to match the magnitudes of the TM_{110} mode RF signals at 2.5 GHz from the second and tenth cavity.

Extracted current from the diode is measured by a Rogowski coil in the diode flange. Current transported through the cavities is collected by a copper plate connected to a rod which is grounded through a Pearson current transformer.

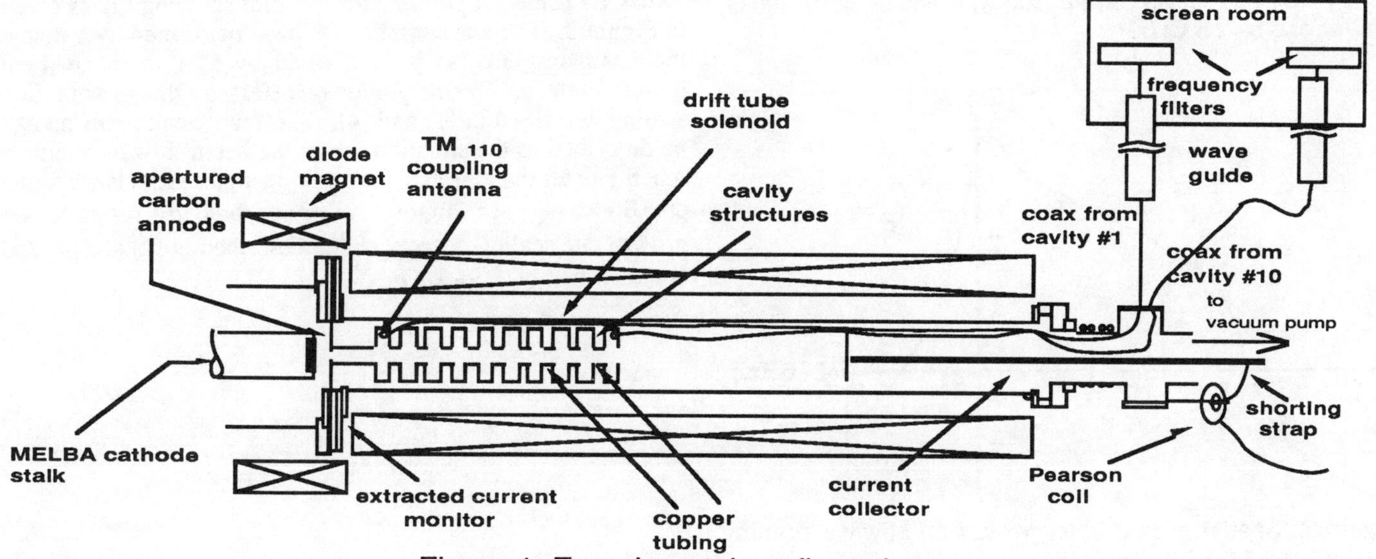

Figure 1. Experimental configuration

III. Theoretical Growth Rates for BBU Instability

The frequency dependence of the BBU growth rate is important in designing an experiment for which the TM_{110} mode is primed in the first cavity, because a mismatch between the priming frequency and the TM_{110} mode frequency could result in a substantial decrease in the initial growth rate. To examine this, the two-dimensional BBU dispersion relation is examined [5]. The 2-D model applies to the case of cylindrical symmetry where deflection in both transverse directions can couple and grow. The dispersion relation is given by eqn. (1).

Table 1
Experimental Parameters

e-beam voltage	0.6-1 MV
diode current	1-10 kA
extracted current	100-400 A
pulselength	0.5-1.5 μs
beam radius	1.0 cm
cavity radius	6.9 cm
TM_{110} frequency	2.5 GHz
average cavity Q	200
# of cavities	10
cavity length (ℓ)	2 cm
cavity spacing (L)	3 cm
magnetic field	800 G

Equation (1)

$$k(\omega) = \frac{2\omega \pm \sqrt{2}\left\{\omega_c^2 + \frac{NQ}{D}(\omega^2 - \omega_c^2) - i\frac{N\omega\omega_o}{D} \pm \left[\frac{-4NQ\varepsilon\omega_c^4}{D} + \left(-\omega_c^2 - \frac{NQ}{D}(\omega^2 - \omega_c^2) + i\frac{N\omega\omega_o}{D}\right)^2\right]^{\frac{1}{2}}\right\}^{\frac{1}{2}}}{2v}$$

where,

$$D = Q^2(\omega^2 - \omega_o^2)^2 - \omega^2\omega_o^2 + 2iQ(\omega\omega_o^3 - \omega^3\omega_o)$$

$$N = 4Q\varepsilon\omega_o^4$$

Here k is the wavenumber, ω is the frequency of the beam breakup wave, ω_o is the TM_{110} cavity frequency, ω_c is the relativistic cyclotron frequency, Q is the cavity quality factor, and ε is a dimensionless factor dependent on the beam current and energy.[3]

A plot of $Im[k(\omega)]$ versus $\omega/2\pi$ for the experimental parameters is shown in Figure 2. $Im(k)$ is the spatial growth rate at the frequency $\omega/2\pi$; $[Im(k)]^{-1}$ is the e-folding length. Here k has been multiplied by 1 m to make it dimensionless. Peak growth occurs for $f \approx f_o$. The parameters used are: beam energy=650 keV, beam current=300 A, magnetic field=800 G, Q=200, and f_o=2.5 GHz.

Figure 2. shows that if the priming source is mismatched above $\omega_o/2\pi$ by more than 25 MHz (1 %) then the initial growth rate is reduced by an order of magnitude. On the other hand, there is less critical fall off if the priming source is mismatched on the lower frequency side of the cavity TM_{110} frequency. Overall, however, it is important to accurately match the priming and cavity frequencies.

Further investigation of equation (1) using the above parameters reveals that our MELBA experiments fall under the "weak focusing" approximation. A plot showing this is given in Figure 3. The wavenumber has been multiplied by 1 m and the beam current has been divided by 1 A to make them dimensionless. *Weak focusing* refers to the growth rate scaling described in [1] and [4]. In effect, weak focusing can be described as the situation where the betatron wavelength is much greater than the BBU e-folding length. This leads to the condition $\omega_c^2 \ll 2\varepsilon Q\omega_o^2$. *Strong focusing* refers to the growth rate scaling approximation described in [2] and [4] and occurs when $\omega_c^2 \gg 2\varepsilon Q\omega_o^2$.

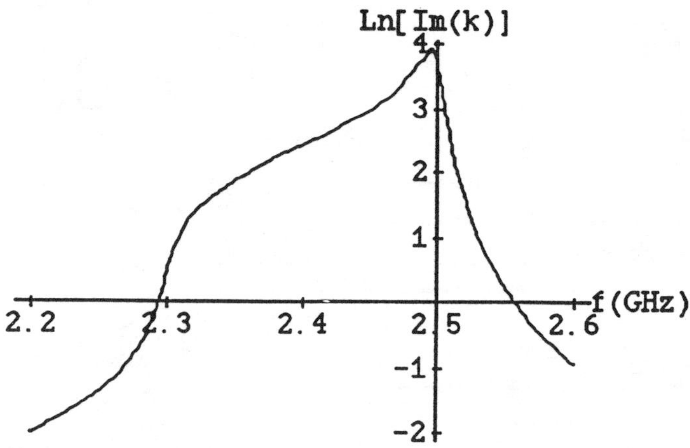

Figure 2. Spatial growth rates versus microwave priming frequency for MELBA parameters.

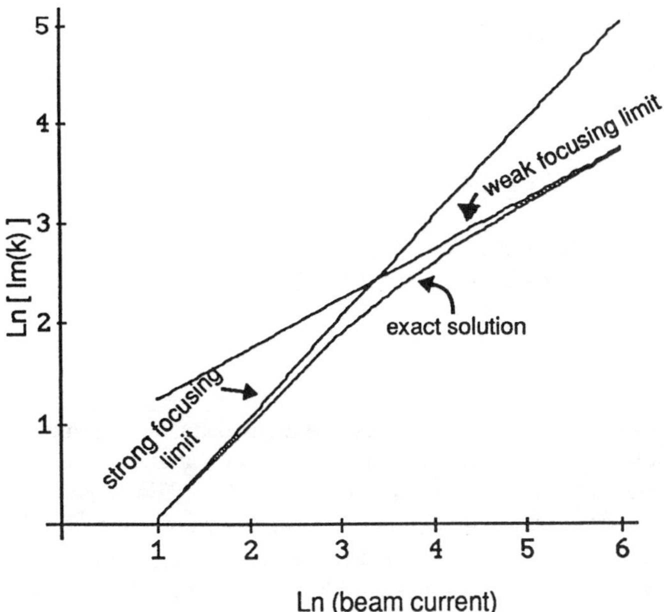

Figure 3. Spatial growth rate versus beam current for MELBA parameters

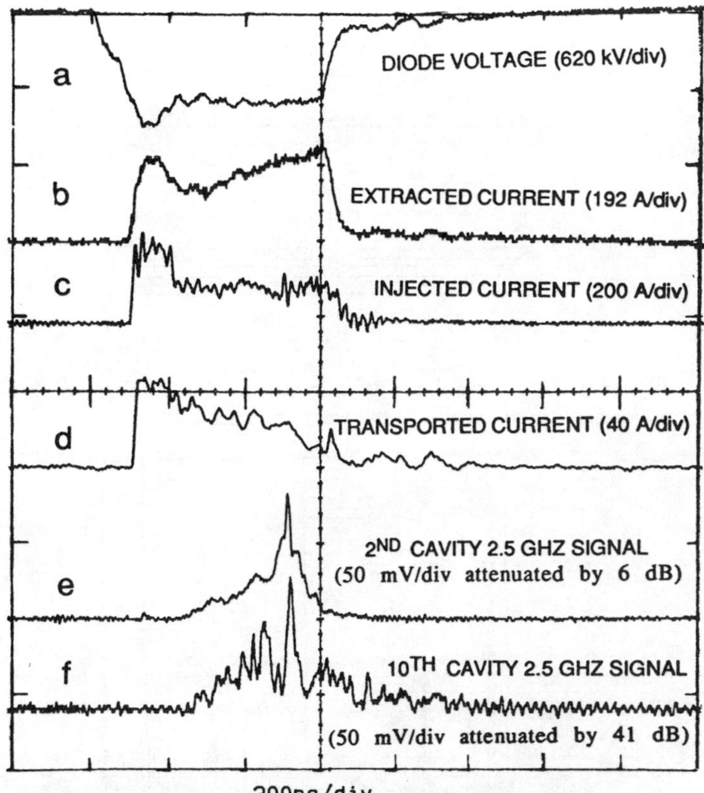

Figure 4. MELBA data: a) Diode voltage (Shot #M2578), b) Extracted current (current exiting 2.0 cm anode aperture) (Shot #M2578), c) 1st cavity exit current (Shot #M2592), d) 10th cavity exit current (Shot #M2578) e) 2nd cavity 2.5 GHz RF signal with 6 dB attenuation (Shot #M2581) f) 10th cavity 2.5 GHz RF signal with 41 dB attenuation (Shot #M2581). The time scale is 200 ns/div. Solenoidal magnetic field is 800 G.

IV. EXPERIMENTAL RESULTS

Experimental data from electron beam transport of the MELBA beam through a 10 cavity system are presented in Figure 4. The data shown are oscilloscope traces from three similar shots.

Extracted current varies from about 115 A to 230 A after the initial voltage overshoot. Injected current measured after the first cavity is about 100 A during the voltage flattop. The current transported through the 10 cavity system decreases from about 40 A to 10 A during the voltage flattop period. Radio frequency emission at 2.5 GHz was measured on the second and tenth cavities by similar detectors with adjustable attenuators. The second cavity RF signal increases over about 300 ns to a peak and then decays. The tenth cavity RF signal is 35 dB higher than the second cavity signal and exhibits a similar shape with more spiky structure.

This data shows that the beam current decreases substantially during its transit through the cavity system while the RF signal corresponding to the TM_{110} breakup mode increases with time and distance. This is indicative of BBU growth.

From equation (1) and Figure 3 which give amplitude growth, the calculated e-folding length is about 6.7 cm. The beam travels 24 cm between the second and tenth cavity giving a predicted growth of 3.9 dB per cavity which yields a total growth of 31 dB. This compares well with the observed 4.4 dB/cavity and 35 dB total growth.

V. ACKNOWLEDGEMENTS

We would like to thank Dr. Y. Y. Lau for valuable consultations. This research was partially supported by SDIO-IST through an ONR contract.

VI. REFERENCES

1. W.K.H. Panofsky and M. Bander, "Asymptotic Theory of Beam Breakup in Linear Accelerators", Rev. Sci. Inst. 39, 20 (1968)

2. V.K. Neil, L.S. Hall, and R. K. Cooper,"Further Theoretical Studies of the Beam Breakup Instability", Part. Accel. 9, 213 (1979)

3. Y.Y. Lau, "Asymptotic Growth of Cumulative and Regenerative Beam Breakup Instabilities in Accelerators", NRL Memo Report NO. 6237 (1988)

$$\varepsilon = 0.422 \frac{\ell}{L} \frac{I\ (kA)}{17} \frac{\beta}{\gamma}$$

where ℓ is the cavity length, L is the cavity spacing, I is the beam current, and β and γ are the usual relativistic velocity and mass factors.

4. Y.Y.Lau, "Classification of Beam Breakup in Linear Accelerators", Phys. Rev. Lett. 63 1141 (1989)

5. R.A. Bosch and R.M. Gilgenbach,"Effect of X-Y Coupling on the Beam Breakup Instability", Appl. Phys. Lett. 58, 699 (1991)

6. R.M. Gilgenbach, L.D. Horton, R.F. Lucey, Jr., S. Bidwell, M. Cuneo, J. Miller, and L. Smutek, "Microsecond Electron Beam Diode Closure Experiments", in *Proceedings of the 5th IEEE Pulsed Power Conference*, Arlington, VA, June 10-12, 1985 (IEEE, New York 1985), p. 126

Studies of Coupled-Bunch Modes in the Fermilab Booster

K. C. Harkay

Purdue University, Department of Physics, West Lafayette, IN 47907 *

V. K. Bharadwaj, P. L. Colestock

Fermi National Accelerator Laboratory, ** *P. O. Box 500, Batavia, IL 60510*

Abstract

Coupled-bunch instability and the resulting longitudinal emittance growth is a major limit to beam brightness in the Booster. The data show a strong correlation between various higher-order RF cavity modes and the growth. At present, both a beam feedback system [1] and higher-order RF mode dampers are being designed to control the longitudinal oscillations. A number of experimental studies have been done to characterize the instability; however, the relevant physical parameters prove difficult to measure. We are attempting, therefore, to approach the problem from both sides: experiment and theory. This paper describes modelling the coupled-bunch motion and calculating the instability growth rates using the measured longitudinal impedance. Also described are the dipole and higher-order form factors, which give the beam response as a function of frequency and as such, scale the effective electromagnetic "force" on the beam due to a given impedance. The results show a good correspondance with the data and will potentially impact feedback system and mode damper design.

I. Introduction

The Fermilab Booster [2] is an 8 GeV (kinetic), rapid-cycling proton synchrotron using 96 combined function magnets run from a 15 Hz resonant power supply system. Injection energy is 203 MeV. The RF system consists of 17 double-gap ferrite tuned cavities which sweep from 30 to 53 MHz during the 33 msec acceleration ramp. The total RF voltage is about 1 MV. Figure 1 shows a schematic of a Booster cavity.

At present, the maximum intensity in the Booster is 3.6E10 protons per bunch. While the Linac upgrade [3] should raise the space charge limit by a factor of three, beam brightness is expected to be limited by a long observed coupled-bunch instablity and resulting longitudinal emittance growth. Prior studies [4] have shown that the coupled-bunch motion is likely due to parasitic modes in the RF cavities.

Spectral measurements have indicated unstable growth of oscillations centered around modes 16 and 36, out of 42 possible coupled-bunch modes (h=84.) This is shown in Figure 2 and reported elsewhere in this conference [5]. The growth rate and coupling impedances, important in specifying gain and bandwidth for a beam feedback system, are very difficult to determine experimentally. These were calculated, therefore, using a beam model and measured single-gap impedances of the higher-order RF cavity modes. The model and growth rate calculations are described below.

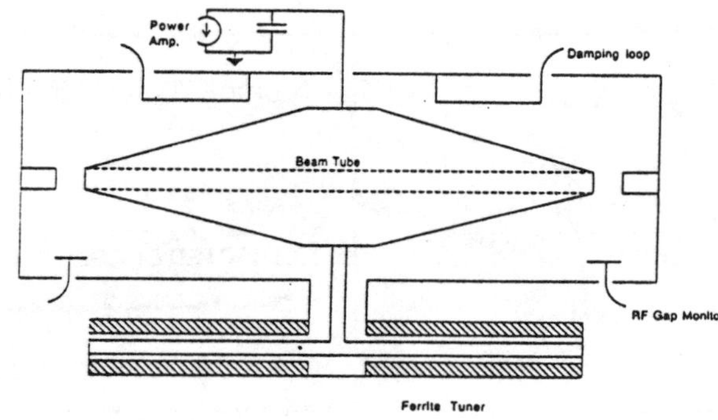

Figure 1. Booster RF cavity schematic. Not shown are two additional tuners located at the front and back.

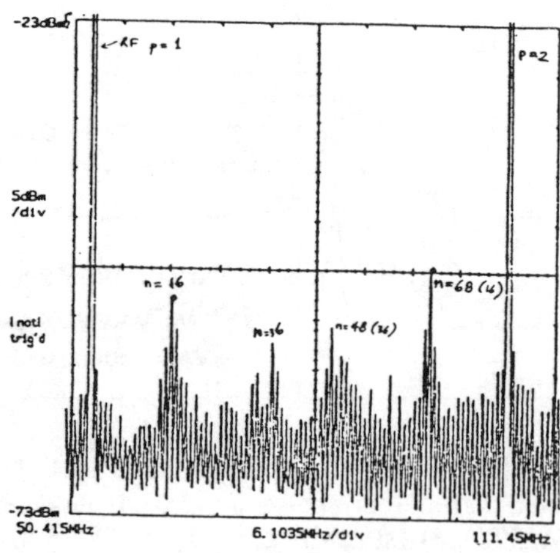

Figure 2. Booster beam spectrum showing coupled-bunch mode growth at n=16 and 36 (at 25 msec in the cycle.) The two sharp lines are the RF harmonics. The data were recorded on a 2 GHz digitizing oscilloscope using a broad band resistive wall pickup.

---

* Current address: Fermi National Accelerator Laboratory,** P.O. Box 500, MS 341, Batavia, IL 60510

** Operated by the Universities Research Association under contract with the U. S. Department of Energy

II. Beam model: coupled-bunch motion

Coupled-bunch instability may arise in a synchrotron if wake fields excited as the beam passes through structures in the ring feed back on the beam coherently. For low intensities and amplitudes, the characteristic frequencies of this coherent motion are harmonics of the synchrotron frequency. There are a number of references giving the complete treatment; we have used Laclare [6]. A few illustrative intermediate results are presented below. The intensity signal for a perturbed synchrotron beam undergoing coupled-bunch motion has a discrete spectrum with frequencies

$$\omega_n = (n+ph)\omega_0 + m\omega_s \qquad (1)$$

where n is the coupled-bunch mode number, p the RF harmonic, h the number of bunches, m the synchrotron mode number, ω_0 the revolution frequency, and ω_s the synchrotron frequency. The amplitudes, or rather envelopes, of the perturbed beam spectra for each m are given by form factors F_m. These give the effective response of the beam to an impedance at ω_n for coherent dipole (ω_s), quadrupole ($2\omega_s$) oscillation, etc. As such, the form factors also give the response of the beam to a feedback system.

Initially, the description is simplified by treating the beam as a single particle moving in (τ, $d\tau/dt$) phase space ($\tau=t-t_s$.) For coherent coupled-bunch motion, phase modulation occurs at the synchrotron phase such that $\tau=\tau_a\cos(\omega_s t+\psi_0)$, where τ_a is the amplitude of the motion and ψ_0 is an initial phase. For simplicity, we assume throughout that ω_s is constant, i.e. $\omega_s \neq \omega_s(\tau_a)$. The time-domain signal is Fourier analyzed, and we find that the amplitudes F_m of the single-particle spectra are Bessel functions $J_m = J_m(p\omega_0\tau_a)$.

This treatment is now extended to the case of multiple bunches. A parabolic amplitude longitudinal bunch distribution is assumed, which is consistent with measurements using a fast (2 GHz) oscilloscope. The amplitudes F_m are now integrals of the J_m and the bunch distribution over phase space. The stationary spectrum m=0 is solved simply and gives harmonics of the RF. The perturbed spectra are found by solving Vlasov's equation for the time evolution of the particle density function (with assumed longitudinal distribution and time dependence.) Laclare gives the result:

$$F_m(x) = \frac{2 \, (|m|+1)}{x^2}\int_0^x J_m^{\,2}(u)\, u\, du \; ; \quad x = \omega_n\frac{\tau_L}{2} \qquad (2)$$

Here τ_L is the bunch length. The form factors for m=0,1,2,3 corresponding to stationary, dipole, quadrupole, and sextapole coupled-bunch motion, respectively, are plotted in Fig. 3 as a function of x. Also shown in this figure are these form factors plotted as a function of frequency for various times in the Booster cycle. This result has immediate consequences for feedback system design. To feed back on dipole motion, for example, one might wish to operate the system at the peak of the dipole envelope for maximum beam signal. However, the RF signal here is large, yielding a poor signal-to-noise. If the RF signal is not filtered, one might choose instead to operate at the minimum of the RF envelope. As can be seen in Fig. 3, the bunch length varies through the Booster acceleration cycle as does, therefore, the optimum operating point. The final design is a compromise, considering also the observed onset in the cycle of the instability, which is around transition (19 msec.)

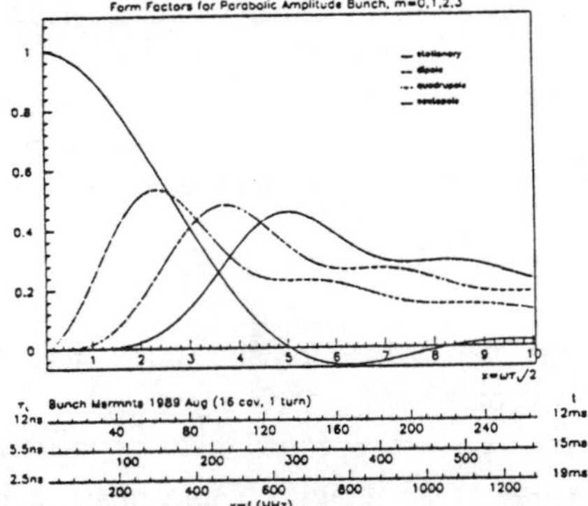

Figure 3. Form factors for stationary, dipole, quadrupole, and sextapole coupled-bunch motion in the Booster. Given at the bottom is the scaling with frequency for three different times in the cycle.

III. Higher-Order RF Mode Impedance Measurements

The higher-order RF cavity mode impedances have been measured at Fermilab using a variety of techniques. A description and critical survey of these impedance measurements is given by another paper in this conference [7]. For this paper, the impedance was measured at one gap on an RF cavity test stand for different tuned frequencies using a stretched wire. The single-gap stretched wire assembly, which fits into one end of the cavity beam pipe, is shown schematically in Fig. 4. Plotted in Fig 5 is the impedance when the cavity is tuned to a fundamental frequency of 52.2 MHz (corresponding to about 19 msec in the cycle.) The results for the single-gap impedances were multiplied by twice the total number of cavities (two gaps each) used during emittance growth measurements [4]. In this way, calculations may be compared to the data. The changing velocity of the beam through the cycle was not taken into

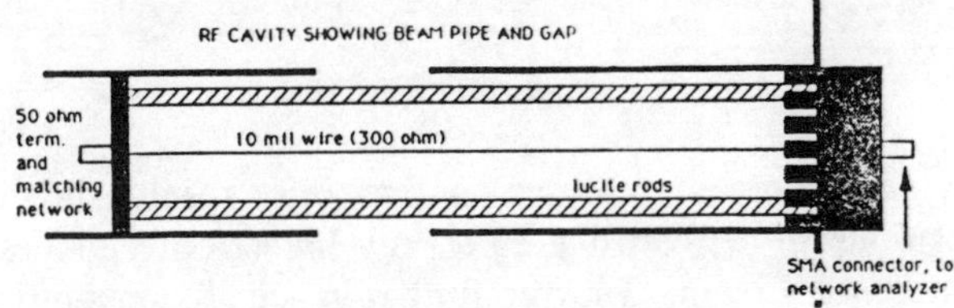

Figure 4. Schematic of single-gap stretched wire assembly used to measure gap impedance on an RF cavity test stand.

account. This will change the relative phase between the beam and the voltage at the two gaps; however, this is expected to be a small correction.

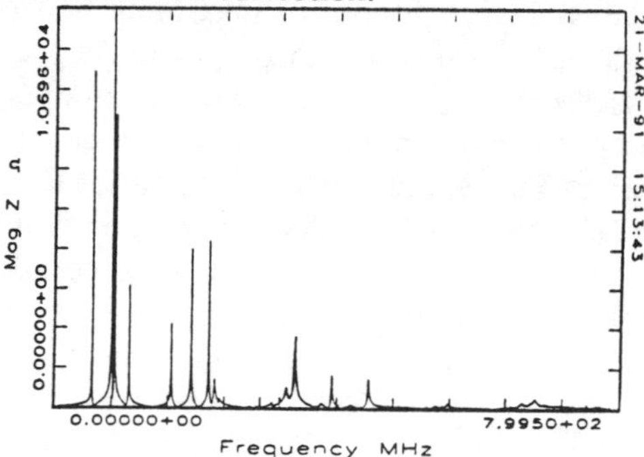

Figure 5. Single-gap impedance for a fundamental cavity tune of 52.25 MHz (bias current 1800 amps) using the probe shown in Fig. 4 above. This corresponds to about 19 msec in the cycle.

IV. Growth Rate

A program was written to calculate the form factors, read the impedance data, and calculate growth rates for coupled bunch modes. Only the impedances at $\omega = \omega_n$ contribute. The growth rate is the imaginary part of the coherent mode frequency shift $\Delta\omega_{cm} = \omega_{cm} - m\omega_s$ and is given by [6]

$$\mathrm{Im}\left(\Delta\omega_{cm}\right) = \frac{-\,N\,e\,m\,\omega_s}{f_0\,V\,\cos\phi_s\,(m+1)} \sum_{p=-\infty}^{\infty} \frac{\mathrm{Re}\,Z(p)}{p} F_m \qquad (4)$$

The results corresponding to 19 msec are shown in Fig.6. It is interesting to compare these results to the data in Fig.2, even though the times are different. The observed coupled-bunch mode 16(15) is predicted in the calculation. The calculation shows faster growth in modes 51(84-33) and 45(84-39.) In fact, experimentally, mode 16 saturates early in the cycle and we observe more pronounced growth around modes 32, 36. The data shows a narrow band at mode 16, but a broad band from 32-38. The latter could be driven by two cavity modes sweeping as the RF frequency sweeps.
The total effective impedance driving the instability is also calculated, as well as the dominant offending RF cavity mode. The maximum effective total coupling impedance due to the higher-order cavity modes was found to be 60 kohms early in the cycle and 15 kohms near extraction. This corresponds to an effective voltage on the beam of 18 and 4.5 kV, respectively, at maximum intensity. The offending RF modes appear to be 163 MHz for mode n=16 and 84, 79 MHz for n=32, 36.

V. Conclusions

These recent analyses confirm prior results showing a relationship between the RF cavity higher-order modes and emittance growth. They go further in actually predicting the excited coupled-bunch modes with associated growth rates comparing qualitatively well to the data. We plan to further develop the beam model to include in the form factors F_m the spread in synchrotron frequency with particle oscillation amplitude $\omega_s(\tau_s)$. Impedance measurements were performed with the cavity "shorted;" the short is a ferrite-loaded rod connecting the inner and out conductors. Also, impedance measurements were performed with prototype mode dampers installed on the RF cavity test stand. The effect of both the shorts and the dampers is to lower both the impedance and the Q of the higher-order modes. The theory assumes excitation by a high-Q resonator; it must now be further developed to bridge the gap between high-Q and broad band resonators. We plan to then analyze these data to compare calculated instability growth rates to emittance growth data taken with a few cavities shorted. The gap impedance measurements themselves are ongoing; we are currently assessing the various techniques employed for accuracy and ease of error elimination. Finally, we plan to do a dynamic study of coupled-bunch instability growth through the cycle using high-resolution impedance data.

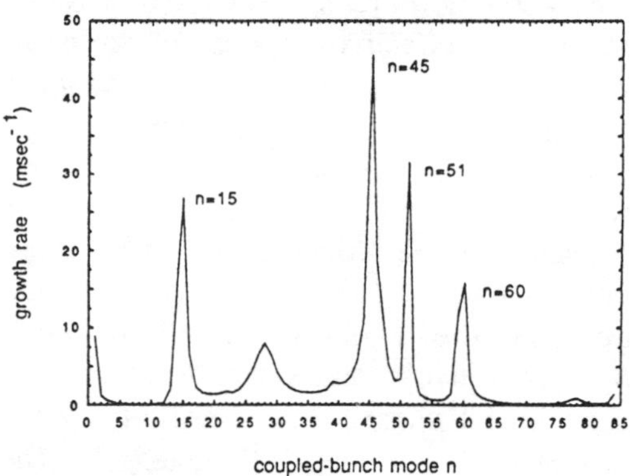

Figure 6. Calculated growth rates corresponding to 19 msec in the cycle. The spread in synchrotron frequencies has not been taken into account; this will reduce the absolute magnitudes. We expect growth times on the order of 1 msec in the Booster.

VI. References

[1] I. Haberman and I. Rypshtein, "Longitudinal Damping System for the Fermilab Booster," these proceedings.

[2] E. L. Hubbard, et.al., "Booster Synchrotron," FNAL, TM-405, Jan 1973.

[3] D. Young and R. Noble, "400 MeV Upgrade for the Fermilab Linac," in Proceedings of the 14th International Conference on High Energy Accelerators, Tsukuba, Japan, Aug 1989, pp. 205-10.

[4] V. K. Bharadwaj, "Coupled-Bunch Instability and Longitudinal Emittance Growth in the Fermilab Booster," in Proceedings of the 14th International Conference on High Energy Accelerators, Tsukuba, Japan, Aug 1989, pp.537-42.

[5] D. McGinnis, "Coupled Bunch Dipole Mode Measurements of Accelerating Beam in the Fermilab Booster," these proceedings.

[6] J. L. Laclare, "Bunched Beam Coherent Instabilities," CERN Accelerator School, CERN 87-03, Sept 1985.

[7] P. L. Colestock, "A Critical Survey of Stretched-Wire Impedance Measurements at Fermilab," these proceedings.

Simulation of Multibunch Instabilities in the Damping Ring of JLC

Kiyoshi KUBO

KEK, National Laboratory for High Energy Physics, Oho 1-1, Tsukuba, Ibaraki, 305 Japan

Abstract

Multibunch instabilities due to higher order modes of RF cavities in the damping ring of JLC (Japan Linear Collider) were studied by tracking simulations. Both transverse and longitudinal motions, with monopole and dipole modes in the RF cavities, were considered. Bunch motions were found stable in the condition of low Q-values of higher order modes of Q=100 and bunch to bunch betatron tune spread of $\Delta\nu=1\times10^{-3}$.

I. INTRODUCTION

The bunch configuration of the damping ring of JLC is shown in Fig.1 schematically. There are ten bunch trains in the ring and each bunch train contains ten bunches. The number of particles in a bunch is designed to be 2×10^{10}. The spacing between bunch trains are about 60 nsec and the spacing between bunches is 1.4 nsec in each train. There are two sections for RF cavities. Because of the large number of bunches and high current, multibunch instabilities due to higher order modes of RF cavities may be serious problems.

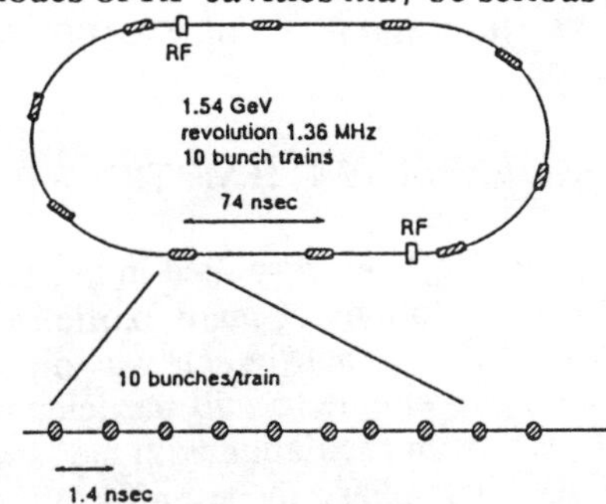

Fig. 1. Bunches in the JLC damping ring.

In section II and IV, each bunch was treated as a single charged particle (rigid bunch). Transverse instabilities (section II) and longitudinal instabilities (section IV) were simulated separately. In each case, only one higher order mode was considered. In section III, each bunch was assumed to consist of several macroparticles.

General parameters of the ring are as follows[1].

Beam energy	1.54 GeV
Revolution frequency	$\approx$1.4 MHz
RF frequency	1.428 GHz
Number of particle/bunch	2×10^{10}
Number of bunch/train	10
Number of train in a ring	10
Beam current	$\approx$0.4 A
Injection(extraction) rate	200 Hz
Radiation loss	0.47 MeV/turn
Radiation damping time	5 msec (transverse)
Betatron tune	$\approx$21.2
Betatron wave length β	5 m (RF sections)
dβ/ds	0 (RF sections)
Momentum compaction α	0.003

II. TRANSVERSE MOTION (RIGID BUNCH)

A. *Simulation steps and parameters*

The simulation consists of four steps as follows.
1) A higher order mode (dipole mode) in RF cavities are excited when bunches pass at the RF sections.
2) Each electron is kicked at RF sections.
3) The higher order mode oscillates and damps exponentially.
4) Each bunch takes place betatron oscillation and radiation damping through the ring.
The following parameters were used.

Frequency of the mode	$\approx$2.3 GHz
(R/Q) of the mode	10 KΩ/m/RF section
Injection error	0.1 mm

Note that one train (10 bunches) is injected (extracted) at a time. Injection rate is 200 Hz and number of train is 10, then each train stays in the ring for 50 msec. In our simulation, the first train was injected in an empty ring and the following trains were injected every 5 msec. After 50 msec, the first train was extracted and the next train was injected where the other trains were circulating. Motions of first injected train and those of trains which take the place of it (11th, 21th etc.) were observed. It was assumed that all bunches have the same transverse offset at the injection. Betatron tune was chosen not to be a simple fraction of any integers. The loaded Q-value of the dipole mode was varied to estimate the required value.

B. *Low Q-value case*

As examples, Fig. 2 shows offset of the last bunch of 10 bunches in a train vs. turn number for (a) Q=5 and (b) Q=10. It shows that Q=5 is low enough, but Q=10 is too high.

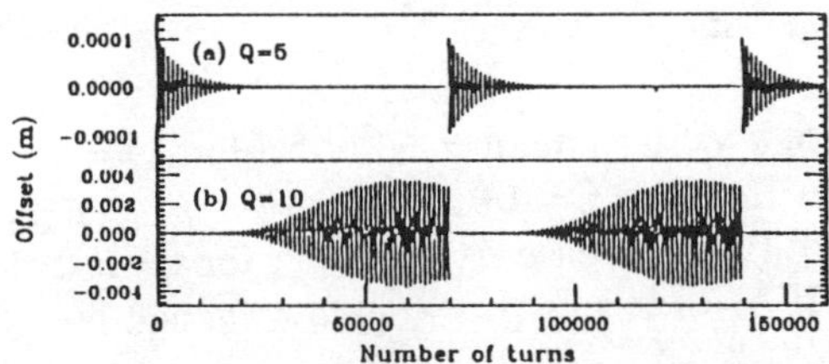

Fig. 2. Offset of 10th bunch in a train vs. turn number, (a) Q=5 (b) Q=10.

To see how bunches are stable (or unstable) for various conditions, we observed 'emittance of bunch center' defined as
$$\varepsilon=(x^2+v_x^2\beta^2)/\beta.$$
where x is transverse offset and v_x=dx/ds. For simplicity, we will observe only ε of the last(tail) bunch in a train because the last bunch will take place the largest oscillation in unstable cases. Fig. 3 shows ε vs. turn number for Q=5, 6, 7, 8, 10, 14, 50 and 100. In the case Q=50 and Q=100, oscillation is growing rapidly and never damped. For Q=8, 10, and 14, damping begins after growing but not enough. This result shows that Q=5 is low enough but Q=6 is not clearly good.

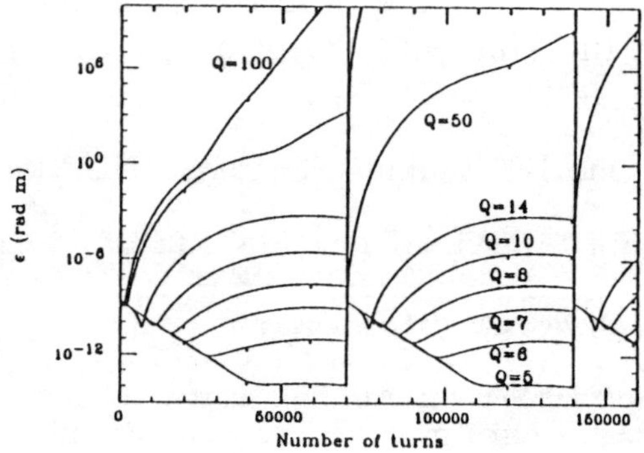

Fig.3. ε vs. turn number for Q=5, 6, 7, 8, 10, 14, 50 and 100.

To achieve such extremely low Q-value is very difficult and Q≈100 seems to be reasonable even applying 'damped cavity'.

Assuming that the damping is exponential, it was estimated how fast the damping should be for a certain Q value. It was found that the damping time should be smaller than 1 msec if Q=100. This damping rate can not be achieved by only radiation damping. The other damping mechanism were proposed[2] and these effects were simulated in section III.

C. Bunch to bunch tune distribution

Let's introduce another cure, 'bunch to bunch tune difference'. If bunches in a train have different betatron tune values, coherent oscillation between bunches is expected to be suppressed. Here, we introduce linear spread $\Delta\nu$ as shown in Fig. 4. The spread $\Delta\nu \approx 1\times10^{-3}$ is expected to be obtained in the JLC damping ring by RF quadrupole[3].

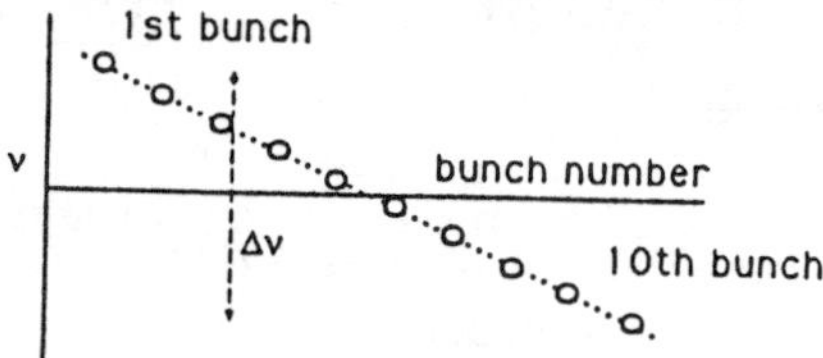

Fig.4. Definition of $\Delta\nu$.

In Fig. 5, ε vs. turn number is shown for $\Delta\nu = 3\times10^{-4}$ and 1×10^{-3} in the case Q=100. $\Delta\nu=1\times10^{-3}$ is enough. Fig. 6 shows results in the case $\Delta\nu=1\times10^{-3}$, for Q=100, 150, 200 and 300. If $\Delta\nu=1\times10^{-3}$, Q=150 is low enough but Q=300 is too high.

Some severer cases were studied. With higher R/Q of 20 KΩ/m/RF, it was simulated that Q=100 is low enough for $\Delta\nu=1\times10^{-3}$. It was also checked that longer damping time, about 10 msec, could be accepted with $\Delta\nu=1\times10^{-3}$, R/Q=10 KΩ/m/RF section and Q=100. In another simulation, the number of bunches/train was increased form 10 to 20, with the same charge/bunch, bunch to bunch spacing and train to train spacing (revolution frequency 1.14 MHz). If R/Q=10KΩ/m, Q=100 and $\Delta\nu=1\times10^{-3}$, it was seen that the bunches were stable.

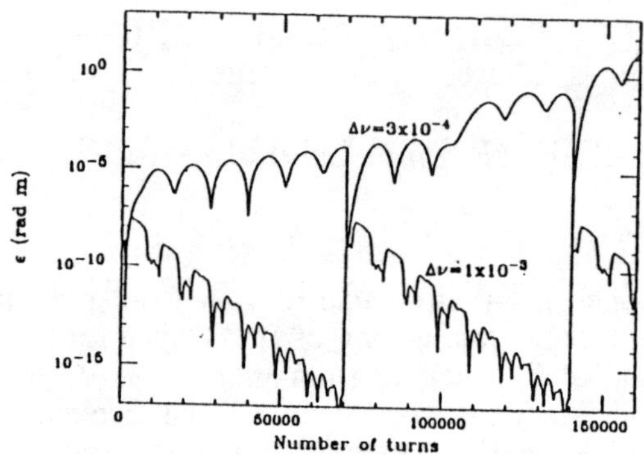

Fig. 5. ε vs. turn number for $\Delta\nu=3\times10^{-4}$ and 1×10^{-3}(Q=100).

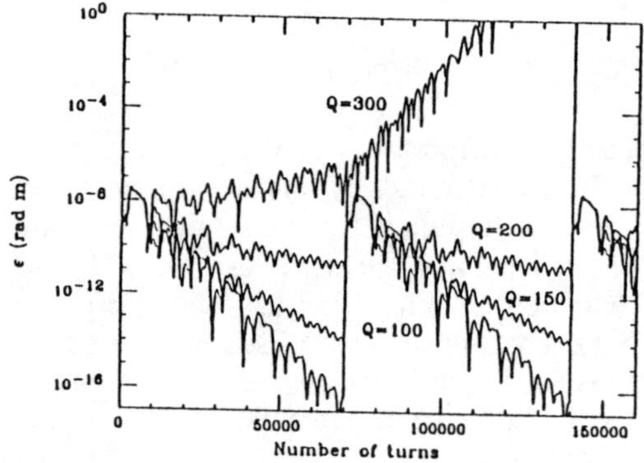

Fig. 6. ε vs. turn number for Q=100,150,200 and 300 ($\Delta\nu=1\times10^{-3}$)

III. TRANSVERSE (WITH MOTIONS IN A BUNCH)

Two damping effects were studied in this section, which can be expected from motions of many particles in a bunch. To simulate these effects, each bunch was divided into several macroparticles. For simplicity, all particles were assumed to take place synchrotron oscillation with the same amplitude and frequency, and their phase to be uniformly distributed. The following parameters were used.

Synchrotron frequency	25 KHz
Energy spread ($\Delta E/E$)	0.08% (equilibrium)
Energy spread ($\Delta E/E$)	0.6%(at injection)

$\Delta E/E$ was assumed to be damped to 0.08% exponentially after injection. Injection and extraction were treated in the same ways as section II. $\Delta\nu$ was set to be zero and Q-value of the dipole mode was fixed as Q=100. The other parameters were the same as those in section II.

If the chromaticity of the ring is not 0, the particles in a bunch have different betatron tunes corresponding to different energies. This tune spread will cause damping of the center of mass of the bunch. From simulations for chromaticity ξ=1, 5 and 10, it was observed that the effect would not suppress the transverse motion with reasonable value of chromaticity because of the rapid damping of the energy spread.

Another effect is due to short range wake field within a bunch. If chromaticity is positive, the head-tail effect can damp betatron oscillations of the bunch center.

To consider head-tail effect, wake function within each bunch was assumed to be constant, the strength (W_{short}) was changed as a parameter and the wake field was assumed to be localized at the RF sections. Chromaticity was fixed at $\xi=5$ and the number of macroparticles was 8/bunch. From analytic calculations with the hollow bunch model[4], the damping time of the zero mode oscillation is proportional to W_{short}^{-1} and estimated to be about 20 msec for $W_{short}=1$ V/pC/m^2 in our case. Fig. 7 shows the results (ϵ vs. turn number) for $W_{short}=0$, 1, 3, 6 ,10 and 30 V/pC/m^2. The head-tail effect could slow down the growth up to $W_{short}=6$ V/pC/m^2, but in the case of larger W_{short}, the rapid growth was observed. This growth can be regarded as a result of higher order motion in each bunch. This means also growth of the emittance of each bunch and it is not acceptable. It may be too optimistic to expect the effect damps transverse motion so rapidly compared with the radiation damping.

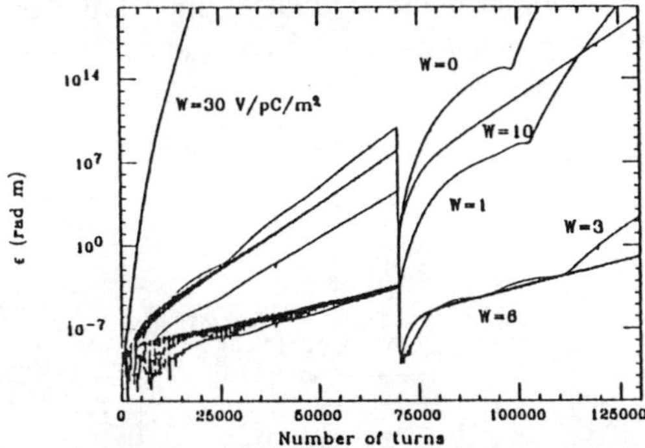

Fig. 7. ϵ vs. turn number for $W_{short}=0$, 1, 3, 6, 10 and 30 V/pC/m^2 (Q=100, $\xi=5$).

IV. LONGITUDINAL MOTION

A. Simulation steps and parameters

In our simulation, there are three field components in an RF cavity.

i. Field of the fundamental mode (accelerating mode), generated by power source.

ii. Wake field of the fundamental mode.

iii. Wake field of a higher order mode.

The strength and phase of the first field was fixed. The second one and the third one were simulated in the same way as the transverse wake field described in the previous section.

1) The fundamental mode and a higher order mode (monopole mode) in RF cavities are excited when bunches pass at the RF sections.

2) Each electron is accelerated at the RF sections.

3) The fields oscillate and damp exponentially.

4) Each bunch loses its energy through the ring (radiation loss) proportional to square of its energy.

5) Each bunch reaches to the next RF section with time delay depending on its energy.

The following parameters of the RF cavity were used.

Cavity voltage	0.6 MV/RF section
Impedance (fundamental)	10 MΩ/RF section
Q (fundamental)	≈ 5350
Frequency (fundamental)	1.428 GHz
(R/Q) (higher order)	100 Ω/RF section
Frequency (higher order)	2.4 GHz.

Note that, radiation damping time was about 2.5 msec corresponding to the radiation loss 0.47 MeV/turn. Initial state was chosen as 10 mm displacement for the first bunch in each train and 0 for other bunches and $E=E_0$ for all bunches. Q-value of the higher order mode was varied as a parameter.

B. Result

The results for Q= 100, 200, 300 and 500 are shown in Fig 8. Each figure shows the time delay of the tenth bunch in a bunch train vs. number of turns. The figure shows that Q=100 is low enough to suppress the longitudinal instabilities. Because of the low Q-values, the resonance frequency is not an important parameter. Even in the case R/Q=150 Ω/RF section, it was checked that Q=100 is low enough.

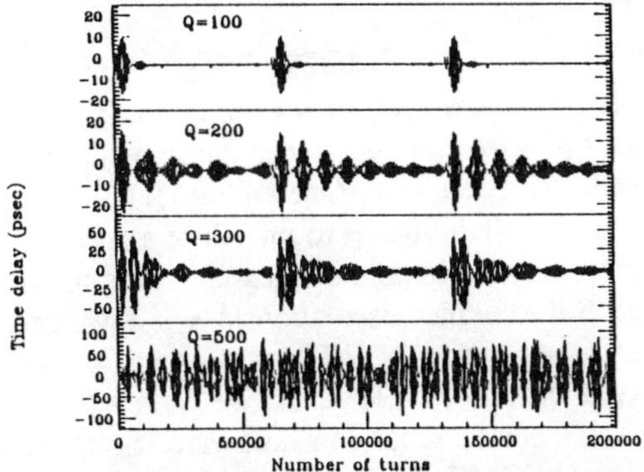

Fig.8. Time delay of 10th bunch, for Q= 100, 200, 300 and 500.

There may be higher modes with low R/Q and high Q-values. To see their effect, a simulation was performed for R/Q=20 Ω/RF section. Q=500 was checked to be low enough.

To study in the case of slower radiation damping, radiation loss/turn was changed as a parameter. Damping time of 4 msec (loss of 0.295 MeV/turn) is acceptable with R/Q=100 Ω/RF section and Q=100.

V. SUMMARY

Multibunch motion due to wake field of higher order mode in RF cavities for the JLC damping ring was simulated.

It was shown that with bunch to bunch tune spread $\Delta\nu=1\times10^{-3}$ and low Q-value of the dipole mode Q=100, transverse multibunch motion is stable. With low Q-value of the monopole higher order mode Q=100, the longitudinal motion is also stable. It was found that some severer design parameters are acceptable. The effect of bunch internal motion was considered preliminary.

VI. REFERENCES

[1] J. Urakawa et al. Linear Accelerator Conference, 1990, Albuquerque.
[2] K. A. Thompson and R.D. Ruth, SLAC-PUB-4962.
[3] S. Sakanaka, private communication.
[4] A.W. Chao, AIP Proc. 105(1983), p 353.

Construction of an RF Quadrupole Magnet for Suppressing

Transverse Coupled-Bunch Instabilities

Shogo Sakanaka and Toshiyuki Mitsuhashi

Photon Factory, National Laboratory for High Energy Physics (KEK)

1-1 Oho, Tsukuba-shi, Ibaraki-ken 305 Japan

Abstract

An RF Quadrupole Magnet (RFQM), which can produce a split in betatron tunes between individual bunches, was developed and effects on transverse coupled-bunch instabilities were investigated. The instability thresholds were raised by the use of the RFQM.

I. INTRODUCTION

Coupled-bunch instabilities arising from high coupling impedances of accelerating cavities cause a current limitation in multi-bunch machines. One of the cures for transverse coupled-bunch instabilities is to introduce a spread in betatron frequencies of individual bunches in order to destroy the coherence of the coupled oscillations[1-4]. It was reported that this cure was applied in DORIS, resulting in an increase of the instability threshold[1]. However, we cannot find any report on a detailed study on it nor on a hardware. So we developed a new device "RFQM" which can produce the bunch-to-bunch tune spread. We installed the RFQM in the Photon Factory 2.5 GeV positron storage ring (PF ring) at KEK, and made a detailed study. The hardware and the results are described in this paper.

II. HARDWARE

The RFQM system can provide a quadrupole magnetic-field oscillating at the revolution frequency of the PF ring (1.6 MHz). The principle parameters are given in Table 1. It consists of a magnet, a ceramic duct and a power supply.

A. RF Quadrupole Magnet

The RFQM was designed by the use of a computer code POISSON[5]. As shown in Figs. 1 and 2, the RFQM has a one-turn coil and a ferrite yoke. We adopted a flat shape for magnet poles to make the fabrication easy. The magnet can be devided into upper and lower pieces for the requirement of setting the ceramic duct in it.

A field distribution inside the RFQM was measured with a search coil of 3mm in diameter. An output signal from the coil was calibrated with a standard coil by the use of a dipole magnet. We obtained a field gradient of 0.0173 T/m at a peak current of 43 A; the measured field-gradient was 6 % larger than the calculated value.

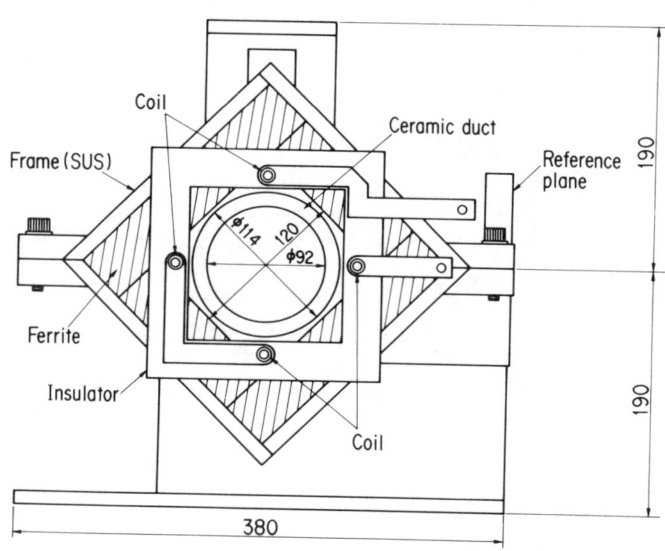

Fig. 1. Front view of the RF quadrupole magnet.

Fig. 2. Photograph of the RFQM installed in the PF ring.

Table 1. Principle parameters of the RF quadrupole magnet.

Maximum field gradient (peak value)	0.0173 T/m
Maximum peak current	43 A
Excitation frequency	1.6029 MHz
Bore diameter	120 mm
Core length	0.34 m
Effective magnetic length	0.38 m
Self inductance	2.1 μH
Maximum output power of RF amp	2 kW
Maximum horizontal tune shift (peak value)	1.23×10^{-3}
Maximum vertical tune shift (peak value)	6.8×10^{-4}

A quadrupole center of the RFQM was aligned to the beam orbit within ~0.5 mm. In addition, the RFQM was put on a movable horizontal-vertical stage which allowed us to adjust the position during ring operations.

B. Ceramic Duct

A ceramic duct, shown in Fig. 3, was adopted to avoid problems due to eddy currents. In order to reduce a beam-coupling impedance, inner and outer sides of the duct were coated with titanium stripes of $1.4\mu m$ thick; this configuration allowed us to prevent the coating from heating-up due to eddy currents. The inner-side stripes were connected with the left-side kovar flange (see Fig. 3), while the outer-side stripes the right-side kovar flange, in order to cut the currents induced by the RFQ-field. Then, high frequency components of the wall current flow through a capacitor between the inner and outer stripes. Low frequency components flow through a bypass conductor bridged over the RFQM. The duct caused no problems in both single and multi bunch operations. Temperature of the duct was raised up to 40°C when the maximum field was applied, and up to 70°C when a single bunch beam of 50 mA was stored.

An absorber was placed upstream the RFQM in order to prevent the synchrotron radiation from impinging on the duct.

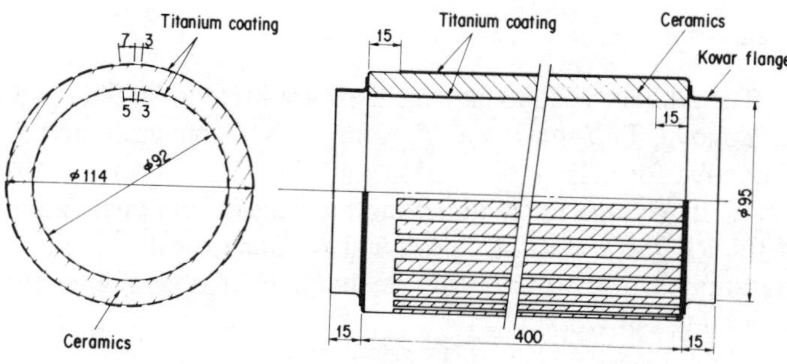

Fig. 3. Ceramic duct coated with titanium stripes.

C. Power Supply

A block diagram of the power supply is shown in Fig. 4. A sinusoidal signal synchronized to the beam revolution is input to a power amplifier. The output from the amplifier is fed to a parallel resonance circuit including the RFQM through an impedance matching circuit. The RFQM is driven by the resonance current.

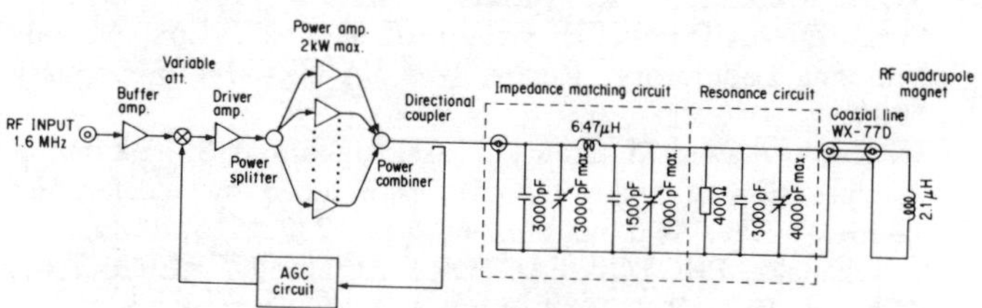

Fig. 4. Block diagram of the RFQM power supply.

III. STUDIES WITH BEAMS

A. Measurement of tune shifts produced by the RFQM

We measured tune shifts produced by the RFQM with a single bunch beam by changing the phase of the RFQ-field against the beam revolution signal. The result of the measurement is shown in Fig. 5. We obtained maximum tune shifts of 1.23×10^{-3} (hor.) and 6.8×10^{-4} (ver.).

Betatron tune distributions in a multi-bunch operation were also measured, and we observed splits in betatron tunes while the RFQM being excited.

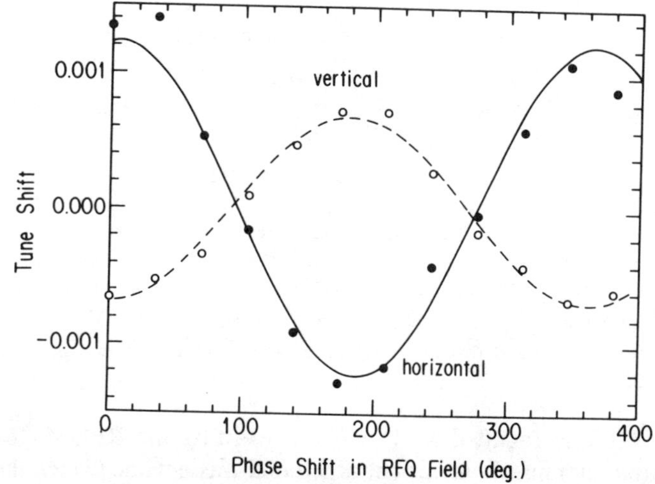

Fig. 5. Measured tune shifts produced by the RFQM in the single bunch operation. Solid and dashed lines denote fitted lines with sinusoidal curves. Peak current of the RFQM: 43 A.

B. Effects of the bunch-to-bunch tune spread on a horizontal coupled-bunch instability caused by TM111(H)-mode

In the PF ring, several coupled-bunch instabilities arising from higher-order-mode (HOM) resonances of the cavities have been observed. Though these instabilities are routinely avoided by a careful trimming of the HOM frequencies[6], they can be produced by tuning the HOM frequencies appropriately.

We first studied the effect of the RFQ-field on a horizontal instability caused by TM111(H)-mode. In order to induce this instability, a resonant frequency of the TM111(H)-mode of one of four cavities was tuned to the frequency of $(669-\delta\nu_x)f_r$, where $\delta\nu_x$ is a fractional part of the horizontal tune ($\delta\nu_x$ was 0.4518 in the experiment) and f_r is the revolution frequency; the tuning was made by adjusting a position of a tuning plunger of the cavity which was not powered.

We measured the threshold currents with three different excitation currents of the RFQM by slightly changing the resonant frequency; we intended to investigate the effects of the RFQM on the instability driven not only by a purely resistive impedance but also by an impedance having a large reactive part. The result is shown in Fig. 6.

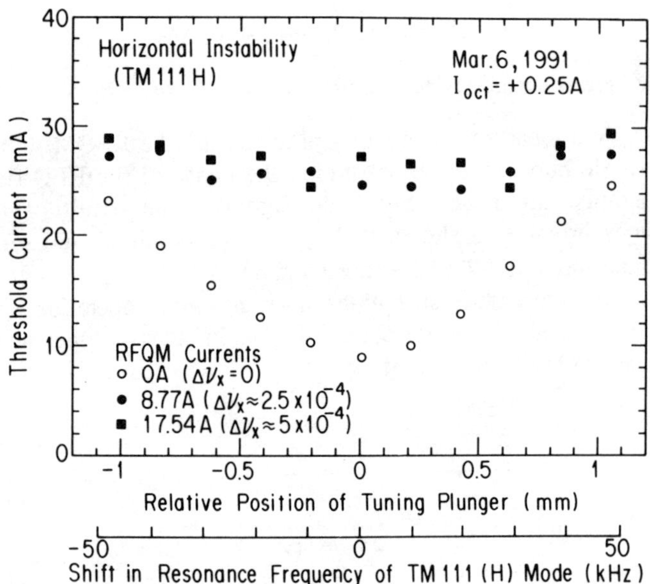

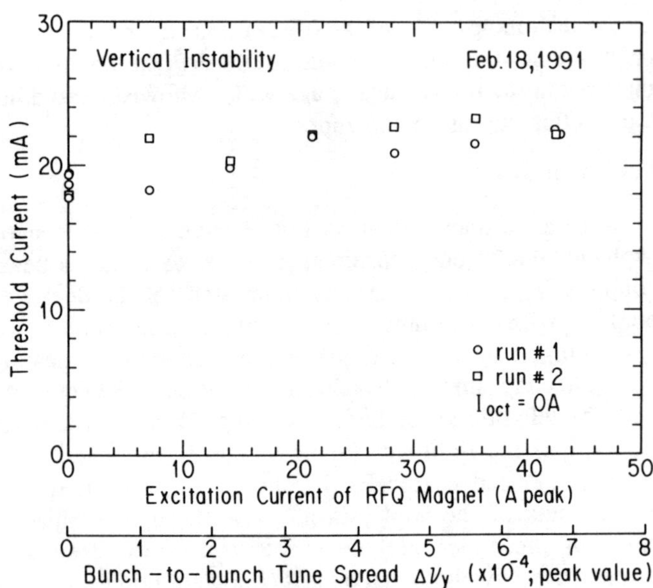

Fig. 6. Threshold currents of the TM111(H) instability with three different cases of the RFQM current. The data are shown as a function of the shift in the resonant frequency of the TM111(H) mode. The frequency at the center of abscissa, loaded-Q value and calculated R_T/Q of the TM111(H) mode were 1071.62 MHz ($\approx (669-\delta\nu_x)f_r$), 13,000, and 675 Ω/m, respectively. The transverse radiation damping time was 7.8 ms. The ring was uniformly filled.

It was found that: 1) without exciting the RFQM, the minimum threshold current of 9.0 mA was obtained when the resonanct frequency was tuned to $\approx (669-\delta\nu_x)f_r$; 2) at that frequency, the threshold current increased by a factor of 2.7 by introducing the tune spread of 2.5×10^{-4} (peak value); 3) increases in the threshold became small as the frequency was detuned from $(669-\delta\nu_x)f_r$. These results suggested that the RFQM was effective on the instability which was driven by a purely resistive impedance, while it was less effective when the instability was driven by an impedance which had a large reactive part. This result is qualitatively in agreement with the theory presented by Chin and Yokoya[4].

C. Effect on a vertical instability

In the PF ring, a vertical instability which leads to an increase of a beam size at high currents has been observed[7]. The principle features of this instability are: 1) the threshold current is about 18 mA with octupole magnets off; 2) observed only with positron beams; 3) it is accompanied with several betatron spectrum lines distributed over about 15 times the revolution frequency. The cause of this instability is not yet understood. We measured the threshold currents of the instability as a function of the RFQM current. The result is shown in Fig. 7. An increase in the threshold was only a factor of 1.2 with the maximum tune spread.

IV. CONCLUSION

Effects of the bunch-to-bunch tune spread on the coupled-bunch instabilities were investigated in detail. We observed

Fig. 7. Dependence of the threshold currents of the vertical instability on the excitation current of the RFQM. The ring was uniformly filled.

increases in the threshold currents in both TM111(H) instability and in the vertical instability. More quantitative analysis of the results is in progress.

Acknowledgement

The authors wish to express their gratitude to T. Shintake, K. Yokoya, T. Yamakawa, Y. Kamiya, M. Kobayashi and S. Ninomiya for helpful discussions and suggestions, to Y. Hori and Y. Takiyama for remodelling the ring for the installation of the RFQM and to M. Izawa and A. Ueda for their help in the studies. We also wish to thank H. Kobayakawa for promoting our work.

V. REFERENCES

[1] R.D. Kohaupt, "SINGLE BEAM INSTABILITIES IN DORIS", IEEE Trans. Nucl. Sci., NS-22, pp. 1456-1457, 1975.

[2] G. Saxon, "IMPLICATIONS OF TRANSVERSE INSTABILITY CRITERIA IN THE DESIGN OF HIGH CURRENT, MULTIBUNCH ELECTRON STORAGE RINGS", Daresbury Laboratory, Report No. DL/SCI/P 204A, October 1979.

[3] Y. Kamiya, "ON SOME PROPERTIES OF LONGITUDINAL AND TRANSVERSE COUPLED-BUNCH INSTABILITIES", KEK, Report No. KEK 82-15, February 1983.

[4] Y.H. Chin and K. Yokoya, "Landau Damping of a Multi-Bunch Instability due to Bunch-to-Bunch Tune Spread", DESY, Report No. DESY 86-097, August 1986.

[5] M.T. Menzel, H.K. Stokes, "User's Guide for the POISSON/SUPERFISH Group of Codes", Los Alamos National Laboratory, Report No. LA-UR-87-115, January 1987.

[6] H. Kobayakawa, M. Izawa, S. Sakanaka and S. Tokumoto, "Suppression of beam instabilities induced by accelerating cavities", Rev. Sci. Instrum., 60, pp. 1732-1735, 1989.

[7] S. Sakanaka, PHOTON FACTORY ACTIVITY REPORT #7, KEK, pp. R-7 - R-8, 1989.

ION TRAPPING IN THE CESR B–FACTORY*

David Sagan and Yuri Orlov
Laboratory of Nuclear Studies, Cornell University, Ithaca, NY 14853

Abstract

Analysis of the ion trapping expected for the Cornell B–factory shows that the presence of a 10% gap in the bunch train will be very effective in reducing the ultimate ion density. With a gap the average ion lifetime is seen from simulations to be under 10^{-5} seconds or about a few turns. While the average lifetime is short both simulation and analysis show that there exists near the center of the beam a small core of long lived ions. The resulting ion density distribution is then substantially different from the (gaussian) beam distribution.

Introduction

The phenomena of ion trapping by an electron beam has appeared in the majority of electron storage rings and there is no reason to suppose that the planned B–factory planned at Cornell, CESR–B, will be free of this problem. Indeed the large time averaged electron beam densities that will be present in CESR–B will produce unacceptably large ion densities unless the average ion lifetimes are kept to of order 10^{-5} seconds or less. The use of ion stripping electrodes in this case is not effective since the ions would not have enough time to migrate to the electrodes. In order to keep the ion density at a reasonable value it is necessary to introduce a gap in the bunch train in which a number of consecutive bunches are empty. The use of a gap has been used previously giving a significant enhancement in performance [1]. In CESR–B a gap is especially effective in clearing ions in consequence of the large velocities attained by the ions due to the large beam current.

Ion Stability and Density

Given a bunch train where a number of consecutive bunches are empty if the distance between bunches is small then the condition for linear stability in the horizontal (x) or vertical (y) plane is [4]

$$0 \le \theta_u \,(\mathrm{mod}\ \pi) \le \theta_{uc} \quad u = x \text{ or } y, \tag{1}$$

$$\theta_{uc} \equiv 2\tan^{-1}\left(\frac{2}{\omega_u t_{gap}}\right), \tag{2}$$

*Work supported by the National Science Foundation

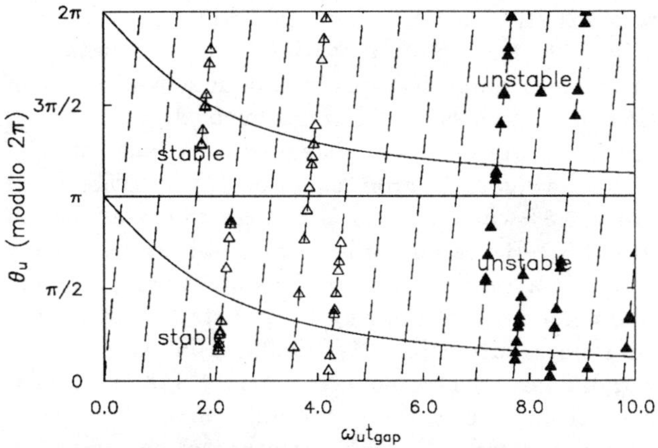

Figure 1: Stability region as a function of θ_u and $\omega_u t_{gap}$. The closed triangles (vertical) and open triangles (horizontal) correspond to the operating points at 50 'random' places in a CESR–B lattice.

where t_{gap} is the period of the gap, ω_u is the oscillation frequency given by

$$\omega_u = \left(\frac{2N\,c\,r_p}{A(t_0 - t_{gap})\sigma_u(\sigma_x + \sigma_y)}\right)^{1/2}, \tag{3}$$

$\theta_u = \omega_u(t_0 - t_{gap})$ is the total 'rotation angle' of the ion during the non-gap period, A is the atomic weight of the ion, N is the number of relativistic electrons, t_0 is the bunch revolution period, and σ_x and σ_y are the beam sigmas.

Ion stability as a function of θ_u and $\omega_u t_{gap}$ is shown in figure 1. With a 10% gap, an ion's 'operating point' will lie upon one of the dashed lines in figure 1. In the limit $\omega_u t_{gap} \to 0$ the ion is unconditionally stable while for $\omega_u t_{gap} \to \infty$ the width of a stability region decreases as $\delta\theta_u^{stable} \approx 4/\omega_u t_{gap}$. For $\omega_u t_{gap} \gtrsim 1$ the ratio of the width of the stable region $\delta\omega_u^{stable}$ to ω_u and the ratio of the width of the stable region to the distance between stable regions $\Delta\omega_u\,(= \pi)$ is given by [4]

$$\frac{\delta\omega_u^{stable}}{\omega_u} \approx \frac{4}{\omega_u(t_0 - t_{gap})\omega_u t_{gap}}, \tag{4}$$

$$\frac{\delta\omega_u^{stable}}{\Delta\omega_u} \approx \frac{4}{\pi\omega_u t_{gap}}. \tag{5}$$

Table 1 gives values for ω_u, $\omega_u t_{gap}$, and θ_u. The values shown in the table were obtained using CESR–B parameters with a revolution time of $2.54 \cdot 10^{-6}$ sec, a beam

u	ω_u (1/sec)	θ_u (rad)	$\omega_u t_{gap}$	% stable
x	$1.2 \cdot 10^7$	28	3.2	40%
y	$3.5 \cdot 10^7$	81	9.0	14%
x & y				6%

Table 1: Values for ω_u, θ_u, and $\omega_u t_{gap}$. The values were calculated for CO^+ using CESR–B parameters and a 10% gap

current of 2 Amps, a beam size of $\sigma_x \sim 1.6 \cdot 10^{-3}$ meters and $\sigma_y \sim 2.0 \cdot 10^{-4}$ meters, and a gap width of 10%. The values in table 1 are only approximate since variation in σ_x and σ_y from point to point in the ring will result in variation in ω_u and hence in variation of θ_u and $\omega_u t_{gap}$. Changes in ω_u will move the ion's operating point along a dashed line in figure 1. From equation (4) a 0.5% change in ω_y or a 4.5% change in ω_x is enough to destabilize an ion. In CESR–B the ions will be more or less uniformly distributed in a limited region of the stability diagram and thus the percentage of the ring that is stable can be calculated from equation (5). As tabulated in table 1, 40% of the ring is stable horizontally while 14% of the ring is stable vertically. Since the horizontal and vertical stable regions are uncorrelated, the percentage of the ring that is both horizontally and vertically stable is 6% (= 40% · 14%) with the length of a stable region being typically 0.4 meters. For CESR–B the above analysis is not effected by the dipole magnetic fields since the beam-ion force dominates the force due to the dipoles.

Besides the presence of the gap, a second major factor limiting the ion density is the nonlinearity of the transverse beam-ion force. This nonlinearity, in conjunction with the presence of a gap, produces destabilizing resonances. The widths of these resonances in frequency space are an increasing function of the ion's oscillation amplitude and for large enough amplitudes the ion motion will be unstable. Considering only cubic resonances the average amplitude beyond which the motion is unstable is given by [5]

$$A_x \simeq \sigma_x \sqrt{\frac{\delta\omega_x^{stable}}{\omega_x}}, \qquad (6)$$

$$A_y \simeq 2\sigma_y \sqrt{\frac{\delta\omega_y^{stable}}{\omega_y}}, \qquad (7)$$

Simulations show that for ions with amplitudes greater than the critical amplitude the resonance driven motion can quickly remove an ion from the vicinity of the beam.

The analysis of ion trapping using CESR–B conditions shows that one can naturally divide the ions into two classes depending upon their lifetimes. The ions with the shortest lifetimes, which will be referred to as 'background' ions, are the ions that are either born in a linearly unstable region of the ring or are born in a stable region with sufficient initial oscillation amplitude to have an unstable trajectory. Simulations have shown that for CO^+ the time that the background ions spend within 3σ of the beam

is on average about $5 \cdot 10^{-6}$ seconds or about $2t_0$. The second class of ions are those ions that are born in a linearly stable region and have a small enough amplitude to not be destabilized by the nonlinearities in the beam-ion force. These ions will be referred to as 'core' ions since they only inhabit the core of the beam. Simulations show that with CESR–B conditions this core has a characteristic width of order $\sigma/10$ either horizontally or vertically. The core ions can be further subdivided into two classes depending upon whether or not the ion is near a minimum in the longitudinal potential produced by the longitudinal component of the beam-ion force [3]. Core ions near a longitudinal potential minimum will be called '(longitudinally) stable core' ions while the other core ions will be denoted as '(longitudinally) unstable core' ions.

The lifetime for the unstable core ions is determined by the time it takes the ions to be pulled into an unstable region by the longitudinal component of the beam-ion force. With CESR–B conditions the longitudinal potential is of the order of 5 eV/meter which, for CO^+, gives a lifetime for the unstable core ions of typically $1.2 \cdot 10^{-4}$ seconds. The total percentage of the ring inhabited by the unstable core ions is the 6% of the ring that is linearly stable. The integrated density of the unstable core ions integrated over the entire ring is then comparable to the integrated density of the background ions but the transverse distribution of the unstable core ions is, as explained above, quite different.

The stable core ions, on the other hand, have a lifetime determined by the time for double ionization since the upward frequency shift experienced by the ion upon further ionization will almost certainly make the ion unstable. With CESR–B conditions this translates into a lifetime of about $6 \cdot 10^{-4}$ seconds. The total percentage of the ring inhabited by the stable core ions is determined by the number of longitudinal potential minimums in the ring that also correspond to a linearly stable region. A calculation[4] shows that this number is about 1% so that the integrated density of the stable core ions is similar to the integrated density of the unstable core ions.

With a design pressure of $5 \cdot 10^{-9}$ Torr, an ion tracking simulation program was used to calculate the expected density distribution. The resulting density profiles along the x and y axes are shown in figure 2 along with the gaussian beam density profile for comparison. As shown in the figure, the large ion density near the beam center is indicative of the presence of the core ions. The long tail of the ion distribution is due to the ions that are 'streaming' towards the vacuum chamber walls.

Tune Shift

From the calculated ion density profile and an electron beam energy of 3.5 GeV it is possible to calculate the amplitude dependent tune shift felt by electrons of the beam due to the ions. In this case the vertical tune shift is greater

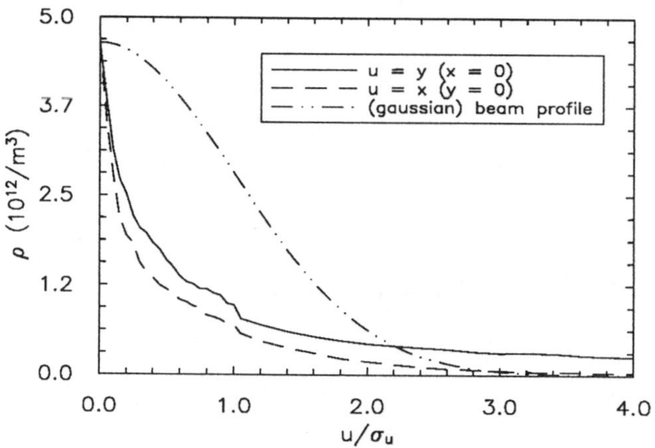

Figure 2: Ion density for a 10% gap. The gaussian beam profile shown for comparison has been normalized to be the ion density at $x = y = 0$

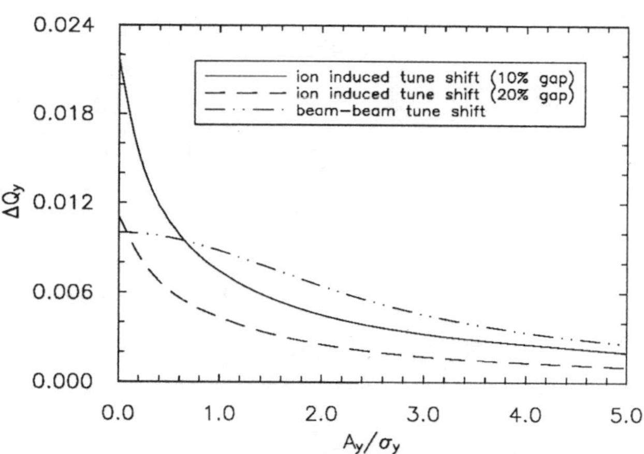

Figure 3: vertical tune shift vs. vertical amplitude

than the horizontal tune shift and so only the vertical tune shift needs to be considered. The vertical tune shift as a function of vertical amplitude is shown in figure 3. Also shown in the figure for comparison is the beam-beam tune shift scaled so that the zero amplitude tune shift is 0.01. As can be seen from the figure the ion induced tune shift falls off rapidly with increasing amplitude which is just a reflection of the large ion density near the center of the beam due to the core ions. This rapid fall off in tune shift is an advantage in terms of any incoherent instability since it is the electrons with amplitude of order the beam size or larger that will contribute to lifetime or beam blowup effects. Another consideration is that, unlike the beam-beam interaction, the ions are distributed around the ring so that the contribution of the ions to any beam resonances will be diminished due to phase averaging [2]. With the above considerations the ion induced tune shift is within tolerable levels in CESR–B.

The effect of increasing the gap to 20% is shown in figure 3. A 20% gap brings the tune shift well below the 0.01 beam-beam interaction tune shift for all amplitudes except at the lowest amplitudes. While the larger gap is effective in further suppressing the ions in terms of further increases in the gap size one is quickly reaching a region of diminishing returns since for a given current there is a minimum ion density that will accumulate regardless of the width of the gap. This can be seen by considering the fact that an ion will have stable oscillatory motion between the time of its creation and the time that the gap first passes by it. With CESR–B parameters this minimum density is about $3 \cdot 10^{11}/\mathrm{m}^3$ near the center of the beam.

Coherent Instabilities

Given a coherent normal mode of the beam-ion system, if any ions have a lifetime that is short compared to the growth time of the mode then the ions will not contribute to the growth of that mode. This is true since for mode growth there must be phase stability between ions and beam and the creation and destruction of ions will tend to randomize this phase. The background ions with their very short lifetimes will thus not contribute to coherent growth.

The analysis of the coherent interaction between the beam and the core ions is complicated by the time dependence in the beam-ion interaction introduced by the gap. There is also a further complication introduced by the differences in the density distributions between the core ions and the beam. An analysis of this situation[5] shows that for the dipole modes the effect of the gap is to increase the response of the ions to a given perturbation roughly in proportion to $\omega_u t_{gap}$. It can also be shown that for the dipole modes (the dipole modes are the most dangerous modes) the 'effective' ion density is the ion density that one would have if one took the core ions and spread them out over the whole beam. Taking the above into account the calculation of the effect of effect of Landau damping upon a mode can proceed upon standard lines [6]. With CESR–B parameters a calculation shows[5] that the dipole modes will be suppressed by Landau damping.

References

[1] M. Barton, Nuc. Inst. & Meth. A243, p. 278 (1983).

[2] S. Krishnagopal and R. Siemann, Phys. Rev. D, **41**, 2312 (1990).

[3] D. Sagan, Cornell CLNS 90/1040, (1990).

[4] D. Sagan, Cornell CBN 91-2 (1991).

[5] Y. Orlov, Cornell CBN 91-3 (1991).

[6] L. Laslett, A. Sessler, and D. Möhl, Nuc. Inst. & Meth. **121**, p. 517 (1974).

Longitudinal Beam Response Measurements at CESR*

John M. Byrd
Laboratory of Nuclear Studies, Cornell University
Ithaca, NY 14853 USA

Abstract

Measurements of the the beam transfer function (BTF) for longitudinal dipole motion of multiple electron bunches are presented. Quantitative agreement is found for low current measurements at various bunch lengths. Qualitative comparisons are made between the behavior of different multibunch modes at low and high current.

Introduction

Longitudinal beam stability historically has not been a problem at CESR but with the prospect of future upgrades and operation at higher current levels, we have begun a series of experiments to measure the effect of the longitudinal impedance on the longitudinal dipole motion of a single beam of single and multiple bunches of $e\pm$ and estimate the multibunch stability limit. Beam transfer function methods have been used successfully in the past to measure the impedance for coasting beams [1,2] and have been measured for bunched beams [3].

We excited the beam by modulating the RF phase and sweeping the excitation frequency through the synchrotron sidebands of revolution harmonics to drive particular longitudinal multibunch modes. The excitation and beam response were recorded by a LeCroy Transient digitizer operating as a network analyzer[4] and used to compute the longitudinal BTF, yielding both amplitude and phase response of the beam. The data acquisition system also allowed for measurement of motion of individual bunches in multibunch operation. Longitudinal beam motion was detected as transverse displacement due to coherent energy oscillations at a point of finite dispersion in the lattice. BTF measurements were made for all dipole multibunch modes for 3 and 7 bunches vs. current and vs. bunch length. Measurements of the transient response at low currents to impulse excitations were also used to measure Landau damping due to the spread in synchrotron frequencies resulting from the nonlinearity of the RF restoring force.

*Work supported by the National Science Foundation

Theory

Multibunch Mode Spectrum

For a beam of M equally spaced and equally populated bunches, the multibunch mode frequencies are given by

$$f_\ell = (n + \ell M)f_0 + m f_{s0} \tag{1}$$

where $n = -\infty, \ldots, \infty$, f_0 is the revolution frequency, f_{s0} is the synchrotron frequency, ℓ is the multibunch mode number and m specifies the mode of oscillation within the bunch. Each multibunch mode is characterized by an instantaneous bunch-bunch phase difference of $\Delta\phi = \frac{2\pi}{M}\ell$. This paper will be limited to discussion of the dipole (m=1) mode.

Multibunch Beam Transfer Function [5,6]

In the presence of an external driving term the longitudinal equation of motion of a single particle in a bunch, neglecting radiation damping, can be expressed as

$$\ddot{\tau} + \omega_{s0}^2 \tau = g_0 e^{j\omega t} \tag{2}$$

where τ is the time deviation of the particle from the synchronous particle, $\omega_{s0}^2 = \frac{\omega_0^2 h \alpha e V_{rf} cos\phi_s}{2\pi E}$, α is the momentum compaction, $\frac{\omega_0}{2\pi}$ is the revolution frequency, E is the particle energy, ϕ_s is the synchronous phase angle, V_{rf} is the RF voltage, and h $= \frac{\omega_{rf}}{\omega_0}$ is the harmonic number. g_0 and ω are the amplitude and frequency of the driving term due to the phase modulation.

For multiple bunches, the net coherent motion of a mode can be found using a linearized Vlasov equation approach. The coherent motion of mode ℓ is given as $\overline{\tau}_\ell(t) = |\overline{\tau}_\ell| e^{j\omega t}$ and the measured beam response is $\dot{\overline{\tau}}_\ell(t) = j\omega \overline{\tau}_\ell(t)$. The BTF is given as the ratio of the beam response to the external driving term in the frequency domain.

$$\frac{|\dot{\overline{\tau}}_\ell|}{g_0} = I \tag{3}$$

where I is the dispersion integral

$$I = \frac{\pi}{2} \int_0^\infty da \frac{\frac{\partial\psi_0}{\partial a}}{\omega - \omega_s(a)} \tag{4}$$

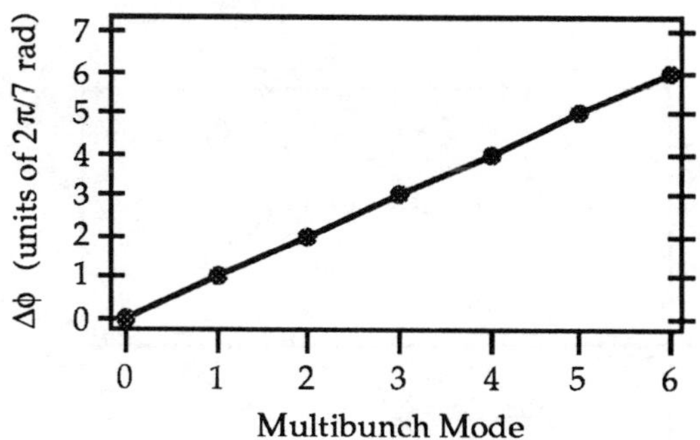

Figure 1: Measured Bunch-bunch phase difference vs. mode number. Error bars are too small to show.

a is the amplitude of longitudinal oscillations such that $\tau = a\cos\omega_s t$ and $\dot\tau = -\omega_s a\sin\omega_s t$, and $\psi_0(a) = \frac{1}{2\pi\sigma_\tau^2}exp(\frac{-a^2}{2\sigma_\tau^2})$ is the equilibrium phase space density.

The nonlinearity of the RF bucket is assumed to yield an amplitude-dependent synchrotron frequency of the form

$$\omega_s(a) = \omega_{s0} - \kappa\left(\frac{a}{\sigma_\tau}\right)^2 \tag{5}$$

where $c\sigma_\tau = \frac{c\alpha}{\omega_{s0}}\sigma_\epsilon$ is the RMS bunch length, σ_ϵ is the relative bunch energy spread, and κ is a weak function of ϕ_s.

The beam-induced voltage per turn due to the coherent motion is

$$V_{b\ell} = \frac{1}{2\pi}\int_{-\infty}^{+\infty}d\omega' Z_\parallel(\omega')\tilde{I}_\ell(\omega') \tag{6}$$

where $Z_\parallel(\omega)$ is the longitudinal impedance, and $\tilde{I}_\ell(\omega)$ is the Fourier transform of the coherent current signal for mode ℓ. The beam-induced voltage adds another effective driving term proportional to the coherent motion to the right hand side of Eq. (2). Jowett et. al. have found the self-consistent BTF for an inductive impedance to be[3]

$$\frac{|\dot{\tilde\tau}_\ell|}{g_0} = \frac{1}{\frac{1}{\bar{I}} + K} \tag{7}$$

where $K \propto j\bar{I}\left|\frac{Z_\parallel}{n}\right|$ and $\bar{I}$ is the DC current. The synchrotron frequency in Eq. (5) must also be modified to include effects due to potential well distortion.

Results

Relative Bunch Phase Measurements

Since the individual bunch motions are recorded, the relative phase of the bunches can be examined. Shown in figure 1 is the measured phase difference between bunches while the beam was being excited at the mode frequency given by Eq. (1). The phase difference has been corrected for the time delay between bunch passage at the detector.

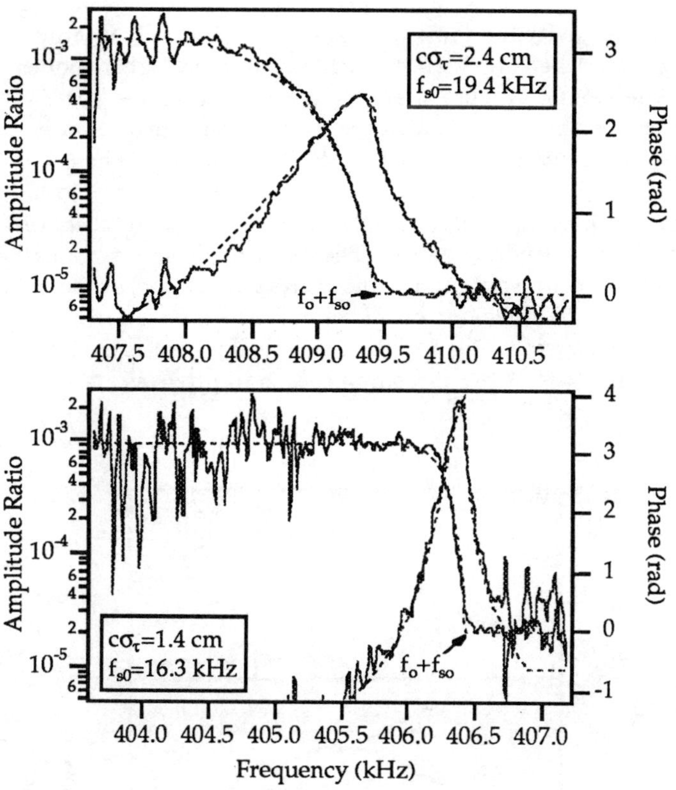

Figure 2: Low current transfer function measurements at 2 bunch lengths. Dashed curves are calculated from Eq. (4)

Low Current Measurements

In order to verify the model of the coherent response we made BTF measurements at several bunch lengths at low currents where the effect of the impedance can be ignored. Shown in figure 2 are typical BTF measurements for mode $\ell = 1$ measured with a bunch lengths of $c\sigma_\tau = 2.4cm$ and 1.4 cm and bunch currents of ~ 0.35 mA/bunch for 3 bunches. Dashed curves show the phase and amplitude response calculated from Eq. (4). Measurements of the FM sidebands in the RF cavity without beam were used to determine the amplitude of the driving term, g_0. Transfer function measurements of the RF cavity itself showed the various cavity feedback loops had no effect on the BTF other than a slow dependence of the amplitude of the effective phase modulation on frequency. Care was taken to minimize the phase modulation in order not to distort the equilibrium bunch distribution.

High Current Measurements

The dominant narrowband longitudinal impedance in CESR is the fundamental mode of the RF cavity. Since the revolution frequency is large compared to the bandwidth of the fundamental mode, only the $\ell = 0$ multibunch mode will interact with the cavity. This is commonly known as the Robinson effect. It is interesting to compare the BTF of the $\ell = 0$ mode with a multibunch mode not interacting with the cavity fundamental mode but potentially interacting with other impedances such as parasitic higher modes of the cavity. Figure 3 shows the BTF at three current levels for the $\ell = 0$ and figure 4 shows the corresponding measurements for the $\ell = 1$ mode for 3 bunches. The $\ell = 0$ mode is strongly affected by the cavity and shows significant broadening of its transfer function due to interaction with the cavity impedance. The $\ell = 1$ mode shows a small coherent tune shift and some distortion of the phase response. The behavior of all multibunch modes for 3 and 7 bunches, excluding $\ell = 0$, is practically identical. Complete analysis of the measurements is in progress.

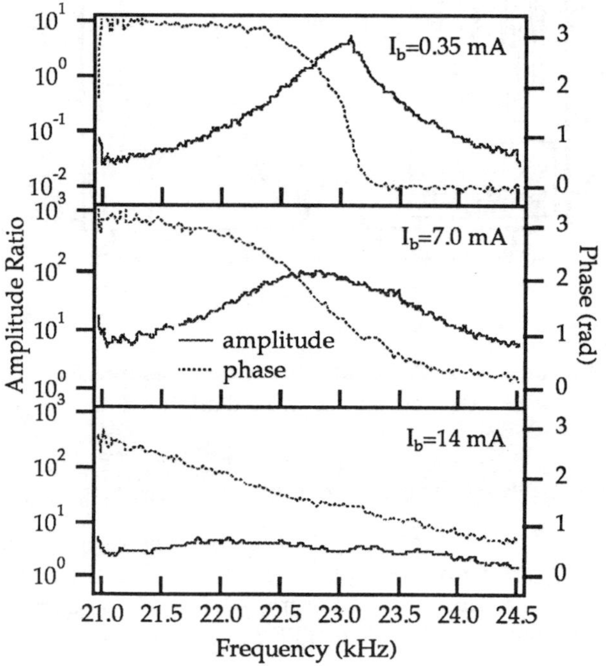

Figure 3: Beam transfer function vs. current for mode $\ell=0$.

Conclusions

Low current BTF measurements show good agreement with analytic calculation of the coherent response. High current measurements on various multibunch modes indicate very little impedance interaction except through the RF cavity fundamental mode. It may be possible to study single bunch effects such as potential well distortion using the BTF on multibunch modes which are not affected by the fundamental mode. Because the current dependence shown in fig. 4 is small, further analysis is necessary to obtain the effect of the impedance quantitatively. I would like to thank Richard Talman and Raphael Littauer for many thoughtful discussions.

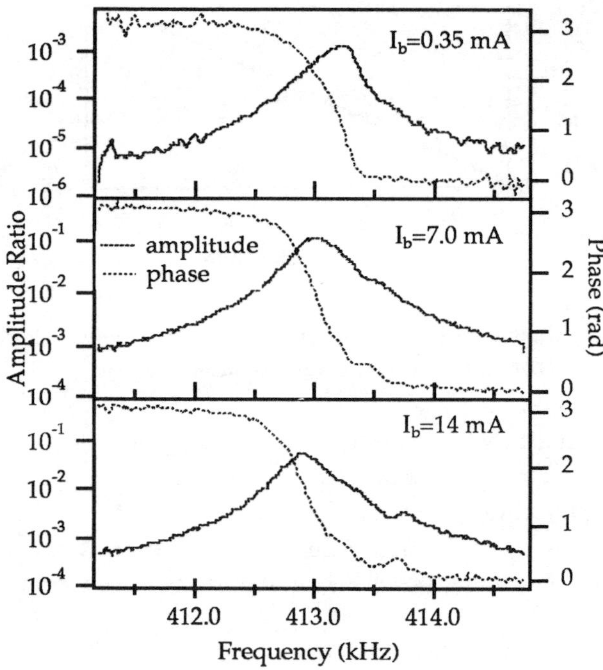

Figure 4: Beam transfer function vs. current for mode $\ell=1$.

References

[1] A. Hofmann and B. Zotter, *IEEE Trans. Nucl. Sci.* **NS 24-3** p1487 (1977).

[2] J. Borer et. al., *IEEE Trans. Nucl. Sci.* **NS-26**, No. 3, (1977).

[3] J. M. Jowett, A. Hofmann, K. L. F. Bane, P. L. Morton, W. Spence, and R. Stege, in *Proc. of European Particle Accelerator Conf.*, Nice, France (1990).

[4] S. Peggs, C. Saltmarsh, R. Talman, SSC-169 (1987).

[5] J. LeClare, CERN 87-03 (1987), LNS-043 and LNS-044 (1980).

[6] H. Hereward, in *Theoretical Aspects of the Behavior of Beams in Accelerators and Storage Rings*, Erice, (1977).

Nonlinear Effects in the SLC e^+ Transport Line*

H. Braun

Paul Scherrer Institut, 5232 Villigen, Switzerland

A. Kulikov, R. Pitthan and M. Woodley

Stanford Linear Accelerator Center, Stanford University, Stanford, California 94309

Abstract

The target area of the SLC positron system, 2 km down the LINAC, is connected back to the front end of the machine by a 2 km long transport line, consisting of 75 FODO cells, with a quadrupole aperture of 10 cm. The beam in the return line experiences a loss larger than predicted. Taking into account the relatively high dodecapole component in the FODO lattice quadrupoles of 2.5 % of the quadrupole component at a radius of 4.65 cm, it is found through particle tracking that nonlinear resonances occur for certain values of the phase advance per FODO cell, particularly at 90°, the standard value previously used. These resonances lead to particle loss in the outer layers of phase space due to the particles with a large amplitude being more affected by the dodecapole moment. These resonances are very similar to the ones found in circular machines. The influence of random variations of the quadrupole strength and position is also analysed. Initial measurements of the positron transmission as a function of the quadrupole focusing strength in the positron return line show a perceptible improvement for the values in phase advance favored by the tracking results.

Introduction

The positrons for SLC are produced by 33 GeV electrons extracted from the 3 km long main linear accelerator after two thirds of its total length. After acceleration to 220 MeV in a dedicated Linac the positrons are transported back to the front end of the main Linac in a 2 km long transport line called positron return line (PRL), mounted above the main Linac [1,2].

This transport line has, apart from bending and matching regions at both ends, a regular FODO structure with a period length of 25.42 m. Each F quadrupole is preceded by a horizontal beam position monitor and followed by a horizontal steering magnet. D quadrupoles are equipped with the corresponding vertical elements. The aperture radius of the PRL is 5 cm. The phase advance per FODO cell was chosen to 90°, to allow an easy beam steering. The whole system is designed to accept an energy spread of ±5%.

Since the linear acceptance of the PRL is higher than the calculated output emittance of $22\,\pi\,mm\,mrad$ of the 220 MeV

accelerator a 100% transmission is expected. However, during the SLC runs in 1989 and 1990 particle losses of the order of 15% distributed along this line were observed. These losses vanished when the beam were collimated at the upstream end of the return line. Thus effects due to beam interaction with the residual gas could be excluded. Wake field effects can be excluded as well, since no dependence of the relative losses on beam intensity could be observed.

Therefore nonlinear components of the PRL quadrupoles were taken into consideration as a possible source of the losses. The strongest nonlinear contribution in a quadrupole with fourfold symmetry is the dodecapole component [3]. This component has been measured with a rotating coil to be about 2.5% of the quadrupole component at a radius of 4.65 cm. The effects of this dodecapole component are studied with the methods of particle tracking.

The Tracking Code

The main goal of the tracking program was to estimate the reduction of the transport line acceptance due to nonlinear field components of the quadrupoles. At first a sample of particle coordinates in 4-dimensional phase space is generated, uniformly distributed in a 4-dim. hypersphere. The volume of this hypersphere V_i is chosen considerably larger than the expected linear acceptance. If N_i is the initial number of test particles and N_f the number of particles which passed the transport line, the 4-dim acceptance volume V_a is given by

$$V_a = \frac{N_f}{N_i} V_i$$

The tracking is done by the common method of representing linear elements with matrices and nonlinear elements with kicks, neglecting the finite length of the nonlinear elements. The aperture is represented by circular diaphragms in the middle of each quadrupole. A specific complication in the modelling of the PRL is the presence of the earth field, since no shielding is installed. The effect of the earth field can readily be observed by inspection of the horizontal PRL correctors settings. Therefore the effect of a drift space has to be modified to

$$\begin{pmatrix} x \\ x' \end{pmatrix} = \begin{pmatrix} 1 & L \\ 0 & 1 \end{pmatrix} \begin{pmatrix} x_0 \\ x_0' \end{pmatrix} + \begin{pmatrix} \frac{1}{2}\vartheta L^2 \\ \vartheta L \end{pmatrix}$$

where ϑ is the deflection angle due to the earth field per m. The value of ϑ in the case of the PRL is 0.095 mrad/m. It is

*Work supported by the U.S. Department of Energy under contract DE-AC03-76SF00515.

interesting to mention that a beam steered to zero displacement in all horizontal beam position monitors will have a systematic horizontal displacement in the D-quadrupoles of about 4.6 mm due to this effect.

The tracking program is also able to simulate random misalignments and coil excitation errors. The orbit errors due to the misalignments and earth field influence are compensated by appropriate changes of the steering magnets settings.

Tracking Results

The 4-dimensional acceptance of the PRL is calculated as a function of the phase advance per FODO cell. Fig. 1 shows the results for pure quadrupoles and for quadrupoles with the measured dodecapole component. In the first case one has

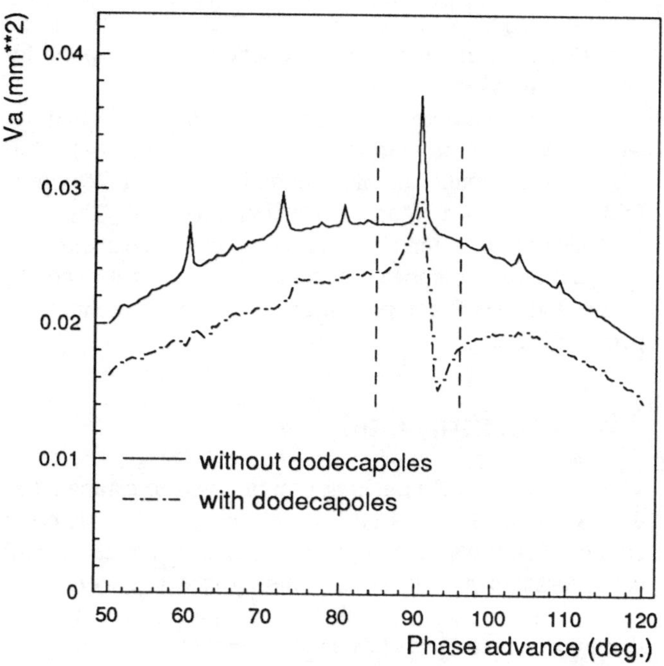

Figure 1: PRL acceptance as a function of phase advance per FODO cell

a smooth curve with a couple of sharp peaks superimposed. The sharp peaks correspond to phase advance values which give rational numbers when 360° is divided by them. They are particularly pronounced in the case of integer ratios. The peaks are explained by the fact that a diaphragm at the position of the β function maximum defines an acceptance area bounded by a polygon in the case of rational ratios, rather than an ellipse in the case of non rational ratios. In the case of integer ratios this polygon is regular. The width of these peaks will decrease with increasing number of FODO cells.

The smooth part of the curve has it is maximum close to the 76.5° value known from analytical calculations [4]. This is not obvious, since these calculations only deal with one plane in phase space, thus implying a rectangular aperture.

The acceptance curve calculated with dodecapole components shows a general depression of about 20% compared to

the other curve plus a resonant acceptance reduction with minimum value at 93° phase advance, due to a fourth order resonance. The nature and characteristic of this resonance is well known from synchrotron theory, but its astonishing that it can become effective within only 75 FODO cells.

The strong effect of the fourth order resonance is particular important for the PRL since 90° was the previous setting of this line. With this setting the phase advance covers a range indicated by the dashed vertical lines in fig. 1 due to the large energy spread of the positrons.

Introducing random misalignments of $\pm 2\,mm$ and quadrupolstrength errors of $\pm 1\%$ gives a reduction of acceptance of about 20% without changing the shape of the curves. The resonant behaviour is maintained. Disabling the earth field effects described in the preceding section gives an acceptance increase of about 7%. The shape of the curves is not much affected.

Due to a suggestion of K.L. Brown the calculations were redone with slightly different settings of F and D quadrupoles, each one 3% below respective beyond the value used for the calculation with equal quadrupole strengths. The results are

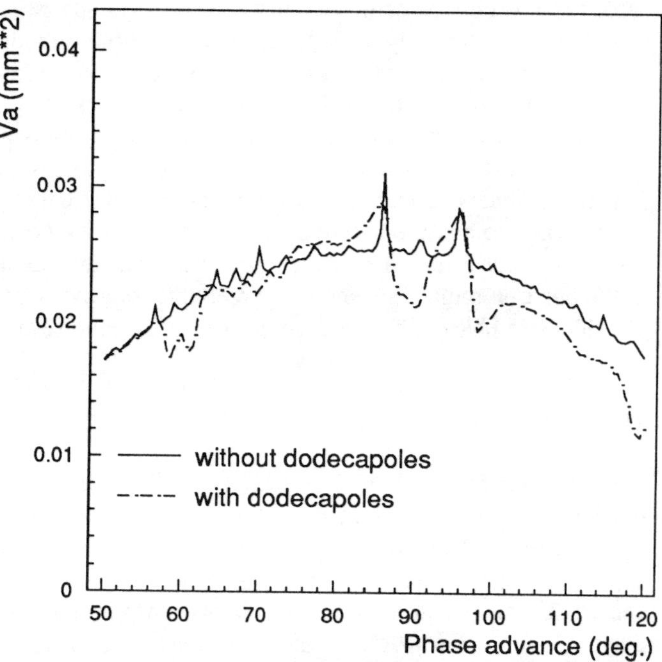

Figure 2: PRL acceptance, with 3% FD imbalance

shown in fig. 2. The abscissa gives the phase advance values which one would obtain for equal quadrupole strengths. The overall acceptance without dodecapoles is a little less than in the case with equal quadrupoles and no dodecapoles, but apart from regions with resonances there is no reduction of acceptance due to the dodecapoles for phase advance values lower than 90°. However the third and sixth order resonance become effective and due to the different phase advances in the horizontal and vertical plane a splitting of the resonances occurs.

It has to be mentioned that the calculated acceptance, even in the region of the fourth order resonance, is still 2.3 times larger than the positron emittance defined by the aperture of the positron capture accelerating section and the solenoid field strength in this section [2]. However, to obtain estimates for possible particle losses one has to take into account phase space filamentation due to nonlinearities and chromatic errors in the upstream systems, which was not done yet.

Measurements

During August 1990 some initial measurements were performed on the PRL, looking for effects of the phase advance on the transmission. In particular lattices with 75° and 55° phase advance per regular FODO cell and appropriate adjustment of the matching cells at both ends of the PRL were calculated and loaded. The transmission is also sensitive to the condition of

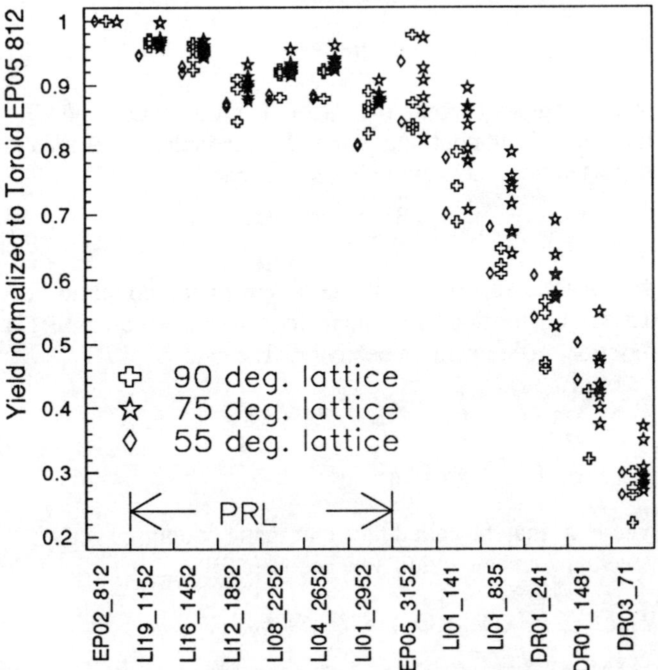

Figure 3: Toroid readouts in PRL and downstream systems. Values are normalized to the last toroid upstream of the PRL

the primary electron beam. To exclude effects due to variations of the primary electron beam, the new configurations were exchanged with the old one several times, each time recording the readouts of the beam current monitors. The raw results of these measurements are shown in fig. 3. In fig. 4 the average ratio of the toroid readouts for the 75° lattice to the 90° lattice is plotted. Apparently the 75° lattice gives on average the better transmission. The average gain in transmission at the injection of the SLC positron damping ring is 14%.

Most of the improvement in transmission with the 75° lattice is obtained in parts of the machine downstream of the PRL, namely during reinjection of the positrons into the SLC main Linac, while the reduction of losses in the PRL itself is only marginal (see fig. 4). A possible explanation for the larger losses with the 90° lattice in the systems downstream of the

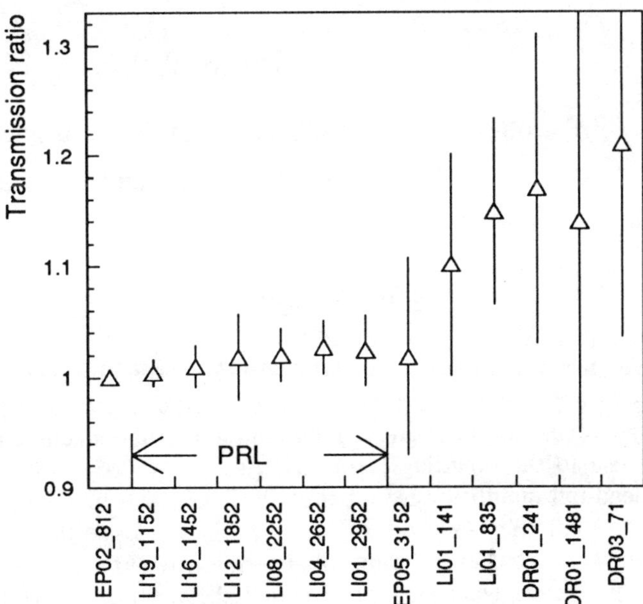

Figure 4: Transmission of the 75° lattice normalized to the transmission of the 90° lattice

PRL could be phase space filamentation due to the fourth order resonance.

Further measurements are needed to investigate the effect of a FD imbalance and to determine the real acceptance volume by measuring the transmission of a collimated beam as a function of upstream steering magnet strengths.

Conclusions

The tracking results show that resonance effects, well known from circular machines, can also become important in periodic transport lines, especially when large emittance beams have to be transported. To obtain a maximum acceptance in a transport line with a FODO structure phase advances per FODO cell between 70° and 80° should be chosen, with a slight difference between horizontal and vertical phase advance.

First measurements at the SLC positron return line indicate that the transmission can be improved by choosing phase advance values in this region. However further measurements are needed to get a more complete understanding.

References

[1] SLC Design Handbook, SLAC 1984

[2] F. Bulos et al., IEEE Trans. on. Nucl. Sci., Vol. NS-32, p.1832 (1985)

[3] G.E. Fischer, SLAC-PUB-3726 (1985)

[4] D.C. Carey, "The Optics of Charged Particle Beams", Harwood Academic Publishers, 1987

Head-Tail Stability and Linear Coupling in the Tevatron

G.P Goderre, S. Peggs, G. Annala, S.A. Bogacz, D. Herrup, S. Saritepe, T. Sullivan, and T. Williams

Fermi National Accelerator Laboratory[*], PO Box 500, Batavia, Illinois 60510

I. INTRODUCTION

During collider studies (1989-1990) it was observed that with large bunch intensities ($>60 \times 10^9$ particles per bunch) in the Tevatron, the beam would go unstable if the machine ran close to the coupling resonance ($\nu_+ - \nu_- < 0.005$). Simple head-tail stability only requires the horizontal and vertical chromaticities (χ_x and χ_y) to be positive. Since the beam went unstable even when this condition was met, a set of experiments were performed which showed that if there is significant linear coupling the head-tail stability criterion is modified. In fact, it was observed that the Tevatron could operate stably with either the horizontal or vertical chromaticity negative if the machine was strongly coupled. Also, when the machine was strongly coupled, it was observed that with both the horizontal and vertical chromaticities positive, the beam in the Tevatron can become unstable. The purpose of this paper is to present a formalism for calculating head-tail stability. Then the predictions of the formalism are compared to data taken with the Tevatron.

II. THEORY

A. *Head-Tail stability*

A naive model of head-tail stability says that a bunch of sufficient intensity goes unstable when one of the chromaticities becomes negative - when

$$\chi_{(x,y)} \equiv \frac{d\nu_{(x,y)}}{d\delta} < 0.0 \tag{1}$$

Here $\delta = \Delta p/p$ is the off-momentum parameter, while ν_x and ν_y are the horizontal and vertical betatron tunes. This is an approximation referring only to the lowest order head-tail mode. Nonetheless, it is a useful rule of thumb that is adequate for most practical situations.

Without proof, we conjecture that a similar expression holds when it is not possible to implicitly ignore the effects of residual linear coupling - due to solenoidal fields, rotated quadrupoles, and displaced sextupoles, et cetera. The uncoupled tunes are simply replaced by the coupled eigentunes,

$$\chi_{\pm} \equiv \frac{d\nu_{\pm}}{d\delta} < 0.0 \tag{2}$$

so that instability is now expected when one of the "eigenchromaticities", χ_+ or χ_-, goes negative.

[*]Operated by the Universities Research Association, Inc., under contract with the U.S Department of Energy.

B. *Off Momentum Coupling Parameterization*

The eigentunes of on-energy ($\delta = 0$) particles are conveniently parameterized by

$$\nu_{\pm} = \frac{1}{2}\{Q_x + Q_y \pm [(Q_x - Q_y)^2 + q^2]^{1/2}\} \tag{3}$$

where q is the two dimensional "coupling vector"[1,2]. In the limit of negligible coupling,

$$|Q_x - Q_y| \gg |q| \tag{4}$$

the eigentunes become the uncoupled tunes, Q_x and Q_y. In the limit of full coupling, when the uncoupled tunes are equal, the eigentunes have their closest approach,

$$\nu_{\pm} = Q_0 \pm \frac{1}{2}|q| \tag{5}$$

with a total separation of the length of the coupling vector. Three substitutions are made to generalize equation (3) to describe off-momentum behavior. They are

$$\begin{aligned} Q_x &\rightarrow Q_x + \chi_x \delta \\ Q_y &\rightarrow Q_y + \chi_y \delta \\ q &\rightarrow q + k \delta \end{aligned} \tag{6}$$

where k may be called the "chromatic coupling vector". The general parameterization now becomes [3]

$$\nu_{\pm}(\delta) = \frac{1}{2}\{(Q_x + Q_y) + (\chi_x + \chi_y)\delta$$
$$\pm [(\{Q_x - Q_y\} + \{\chi_x - \chi_y\}\delta)^2 + (q + k\,\delta)^2]^{1/2}\} \tag{7}$$

This incidentally suggests that it may not be possible for the eigentunes to become arbitrarily close, even if the unperturbed chromaticities are set to zero, the unperturbed tunes are equal, and the coupling vector q has been corrected to zero. The minimum difference is approximately

$$(\nu_+ - \nu_-)_{min} \approx |k|\frac{\sigma_p}{p} \tag{8}$$

proportional to σ_p/p, the beams finite momentum spread.

C. *Eigenchromaticities and stability*

In principle the problem is now solved, since it is formally possible to differentiate equation (7) with respect to δ, and then impose the constraints of equation (2). This is rather messy and unilluminating. Instead, consider the following practical problem: if the chromaticities are χ_x and χ_y when Q_x and Q_y are far apart, will the beams be stable when the tunes are brought together? When $Q_x = Q_y = Q_0$, equation (7) becomes

$$v_{\pm}(\delta) = Q_0 + \frac{1}{2}(\chi_x + \chi_y)\,\delta \qquad (9)$$
$$\pm \frac{1}{2}[\,\{\chi_x - \chi_y\}^2\delta^2 + (q + k\,\delta)^2\,]^{1/2}$$

Now consider two cases - when the on-energy lattice has been perfectly decoupled, so that $q = 0$, and when δ is small.

In the case of perfect decoupling the eigenchromaticities are

$$\chi_{\pm} = \frac{1}{2}(\chi_x + \chi_y) \pm \frac{1}{2}[(\chi_x - \chi_y)^2 + k^2]^{1/2} \qquad (10)$$

and equation (2) reduces to a single stability criterion

$$\chi_x\,\chi_y > \frac{k^2}{4} \qquad (11)$$

This shows that the area of stability in the upper right quadrant of (χ_x,χ_y) space shrinks significantly if k is large. Instability near the origin of this space has been observed in the Tevatron, when the unperturbed tunes are brought together.

In the second case, expanding (9) to first order in δ gives

$$\chi_{\pm}(\delta) = \frac{1}{2}[\,\chi_x + \chi_y \pm \frac{k \cdot q}{|q|}\,] + O(\delta) \qquad (12)$$

and the stability condition becomes

$$\chi_x + \chi_y > \frac{|k \cdot q|}{|q|} \qquad (13)$$

This is less restrictive than equation (11), since the boundary line is no longer a hyperbola, but a straight line entering three of four quadrants in (χ_x,χ_y) space. Instability has also been observed in Tevatron collider runs when the unperturbed tunes are <u>separated</u>, after the unperturbed chromaticities have been accidentally tuned so that one of them is large and positive, while the other is small and negative.

The apparent contradiction between equations (11) and (13) that arises when q is allowed to become small is resolved by properly considering the higher order term in equation (12). Note that both stability conditions (11) and (13) implicitly assume that it is the $\delta = 0$ eigenchromaticity that is important, in keeping with the spirit of equations (1) and (2).

III. EXPERIMENT

A. Setup

The Tevatron was operated with horizontal and vertical tunes of 20.408 and 20.419, respectively. The machine was decoupled using two somewhat orthogonal skew quadrupole circuits. The angel between the coupling vector associated with the two circuits was measured to be approximately 31°. The minimum tune split that could be achieved using these circuits was approximately 0.002, and 0.003 for uncoalesced and coalesced beam, respectively. High intensity beam was achieved by coalescing 11 bunches with approximately 8E9 particles/bunch into one bunch with intensity of

approximately 80E9. The tune split $(V_+ - V_-)$ is adjusted using the correction quadrupole circuit Q_x, which in principle changes only the horizontal tune (V_x). The chromaticity is adjusted using the chromaticity sextupole circuits (C_x,C_y). The chromaticity is measured by varying the RF frequency and measuring the change in tune. The chromaticity measurement is also made with $V_y - V_x > 0.01$. The tunes are measured by spectrum analysis of signals from horizontal and vertical Schottky plates[4].

B. Part I

A high intensity proton bunch (>60E9) was injected into the Tevatron. The chromaticity was set such that $\chi_x = \chi_y = \chi_0$. Then, the Q_x circuit was slowly varied to reduce the tune split from 0.010 to 0.003. The above procedure was repeated for several values of χ_0 (4,3,2, and 1 respectively). For χ_0 less than or equal to 2 units, the beam went unstable when Q_x was set to minimize the tune split (0.003). Substituting this result into equation 11 gives.

$$k \le 4.$$

Point A in figure 1 shows our result. Equation 13 implies that the cross hatched area for $\chi_x > 0$ and $\chi_y > 0$ will be unstable when the tune split is small (machine is coupled) and stable when the tune split is large (machine is decoupled).

Next, the Q_x circuit was used to place the tunes on top of one another (minimum tune split 0.003). χ_x and χ_y were set to 8 and -3 units, respectively, and the machine was operating stably at point B in figure 1. Then the Q_x circuit was used to separate the tunes. When the tune split was about 0.01 the vertical tune line on the spectrum analyzer went coherent (became very narrow) and total beam loss occurred in less than a second (i.e. beam went unstable).

Point B in figure 1 shows our result, and implies that the cross hatched region , excluding the first quadrant, will be stable when the beam is coupled and unstable when the beam is decoupled.

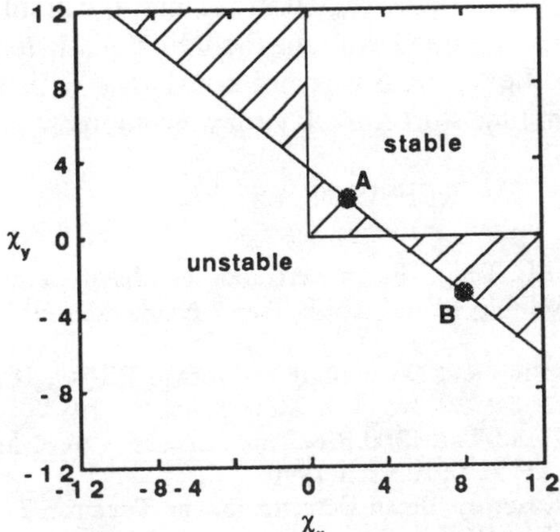

Figure 1: χ_x verses χ_y stability curve. The points A and B are either stable or unstable operating points as demonstrated in Part I. The cross hatched region is also believed to be either stable or unstable depending the amount of coupling.

C. Part II

The second part of the study used 20 uncoalesced bunches with approximately 8E9 particles/bunch in the Tevatron. Next the momentum (δ=dp/p) of the Tevatron was varied by changing the RF frequency. The eigentunes ($\nu_\pm$) were measured as a function of δ for three different settings of the Q_X circuit. The results of these scans are shown in figure 4. The sums and differences of the eigentunes (figures 2 and 3, respectively) were used to fit for the parameters in equation 7. The interesting results of the fits are that χ_x, χ_y, and q vary with Q_x settings (see table 1). Any attempt to hold these parameters constant caused the fit to become rather poor for $\delta>0$. Also, a fit was done excluding the k term in equation 7. The results of the fit again required χ_x, and χ_y to vary with Q_x while q remained constant at approximately 6.0E-3. One justification for including the k term in the fit was that it keeps the value of q consistent with the minimum tune split measurement (q<3.0E-3).

Table 1: Fitted Parameters

ΔQ_x circuit	.0000	.0030	.0060	Average value	σ
	fitted parameters				
Q_x	.4081	.4103	.4125		
Q_Y	.4183	.4189	.4197	.4190	.0007
χ_x	8.2	7.0	5.0	6.7	1.6
χ_y	-2.6	-1.8	-0.9	-1.8	1.0
q	.0041	.0030	.0025	.0032	0.0008
$\cos(k \cdot q)$	1	1	1	1.11	0.07
k	3.6	3.9	4.0	3.8	0.2

D. Conclusion

There are two measurements that require the k term in equation 7. First, the unstable region in the first quadrant of figure 1 cannot be explained without including the k term. Second, the fit to the $\nu_\pm$ vs δ was not consistent with the measured minimum tune shift if the k term was ignored.

IV. REFERENCES

[1] D. Edwards and L. Teng, "Parameterization of Linear Coupled Motion in Periodic Systems", IEEE Trans. Nucl. Sci., NS-20, No. 3, 1973.
[2] S. Peggs, "Coupling and Decoupling in Storage Rings", IEEE Trans. Nucl. Sci., NS-30, No. 4, p 2460, Santa Fe, 1983
[3] S. Peggs, "Head-Tail Stability and Linear Coupling", Fermilab Note AP 90-007, April 1990.
[4] D.Martin, "A Resonant Beam Detector for the Tevatron Tune Monitoring", FERMILAB-Conf-89/74.

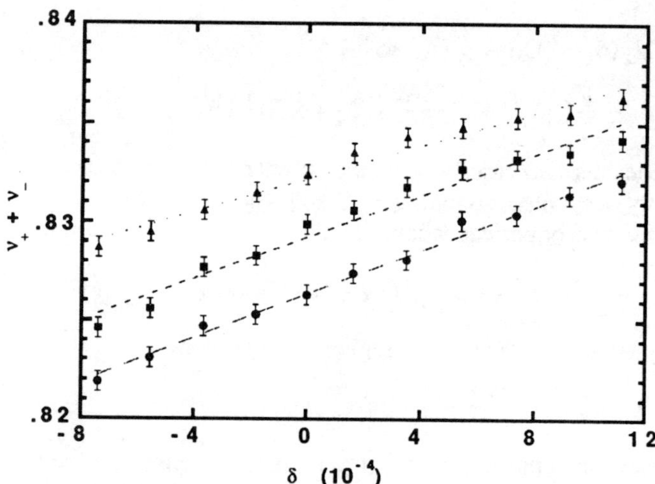

Figure 2: $\nu_+ + \nu_-$ verses δ - the curves are the best fit of the data to a straight line.

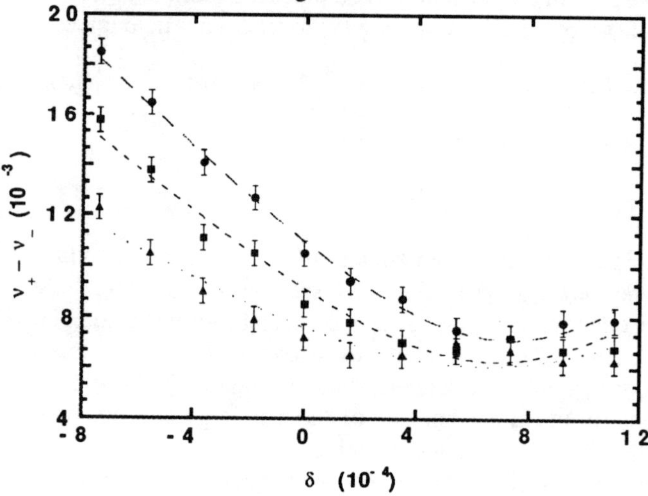

Figure 3: $\nu_+ - \nu_-$ verses δ - the curves are the best fit to the data. Table 1 gives the fit parameters.

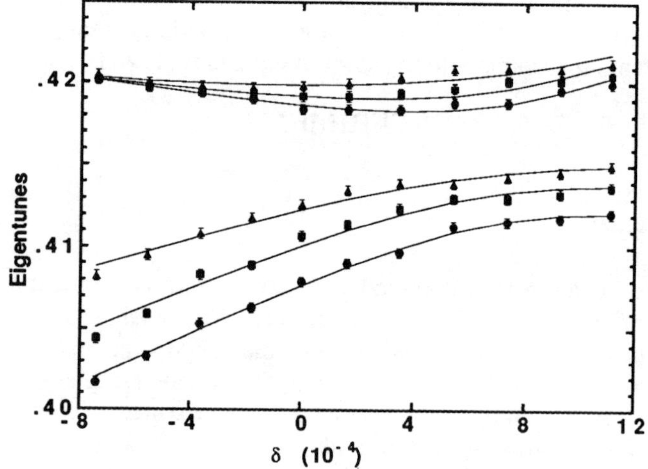

Figure 4: Eigentune verses δ - the points are the measurements and the curves are equation 7 with the parameters from table 1.

Low Loss Parameter for New CESR Electrostatic Separators [*]

James J. Welch, Laboratory of Nuclear Studies, Cornell University, Ithaca, NY 14853
Zhong Xiong Xu, Institute of HEP, Beijing

Abstract

The electrostatic separators in the e^+e^- storage ring CESR have been large contributors to the overall impedance. In addition, the RF fields generated by the beam passing through the separators have seriously degraded the high voltage performance. These problems are becoming more acute as the beam current and bunch current are increased to meet our luminosity goals. A new separator has been designed for which bench measurements on a full scale mockup indicate we can expect a loss parameter of only $0.123\ V/pC$ – about one quarter the value for the existing separators. A factor of two reduction came from tapering the ends. The other factor of two came from the unique ground electrode structure, which serves to keep electromagnetic energy close to the beam. At the ends of the separator, the taper and ground electrode are blended together in an unusual surface.

Separators and High Beam Currents

Electrostatic separators in e^+e^- storage rings are used to deflect electrons and positrons with opposite angles thereby providing different closed orbits. By carefully choosing the separator locations as well as controlling the betatron phase advance throughout the ring, closed orbits can be obtained with good separation between the beams at all unwanted collision points. This scheme allows multiple bunches of electrons and positrons to be stored, thereby increasing the luminosity.

When a bunch passes through a separator it leaves energy behind in the form of broad-band RF electromagnetic fields. This field energy is dissipated in the separator vacuum chamber, electrodes and especially any series resistors in the high voltage circuit. As the beam current is increased, the power absorbed from the beam by the separators also increases and several technical problems arise. The original CESR separators experienced increased breakdown (high voltage arcing) rates with beam current, burning of energy absorbing resistors, vapor lock due to boiling Freon coolant, and excessive Freon pressure due to Freon heating. These problems occurred at total stored beam currents below 150 mA where the estimated total power absorbed is about 3 kW per separator. Soon CESR will have the RF capacity to support 600 mA of average beam current. For a 14 bunch configuration, at this current the original separators would each dissipate about 17 kW.

Absorbing many kilowatts of broad band RF power in a separator is made quite difficult by the fact that the electrodes, feedthroughs, cables and power supplies must simultaneously operate around $\pm 100\ kV$ with a very low breakdown rate[1]. To obtain optimum performance at high beam current every effort must be made to reduce the overall amount of power absorbed by the separator.

New separators are being constructed whose design addresses the problems of higher beam currents without compromising a reliable DC high voltage design. In particular, electrode gaps and surface profiles were chosen so that the ratio of peak electric field on the electrodes to deflecting electric field is substantially lower than in the existing separators. Having more or less fixed the high voltage electrode design, it was somewhat surprising to find that there was still enough design freedom left to effectively reduce the loss parameter. Nevertheless, by concentrating on reducing the loss from the ends of the separator vacuum chamber, we were able to obtain a loss parameter of about one quarter that of the original CESR separators. At 600 mA of total stored beam current, the new separators are expected to dissipated only about 4 kW – almost the same power as the existing separators in CESR with only 150 mA of total beam current.

Design for Low Loss Parameter

There are two important ideas implemented in the design of the new separators which reduce the loss parameter:

- close proximity ground electrodes

- tapered ends

The ground electrode idea is to fill with metal some of the region which is effectively at ground potential midway between the plus and minus electrodes. (See Figure 1.) This

---

[*]Work supported by the National Science Foundation

[1] The RF fields themselves can add a voltage 'spike' of tens of kV to the DC voltage through the peak beam current and the characteristic transmission line impedance of the separator components.

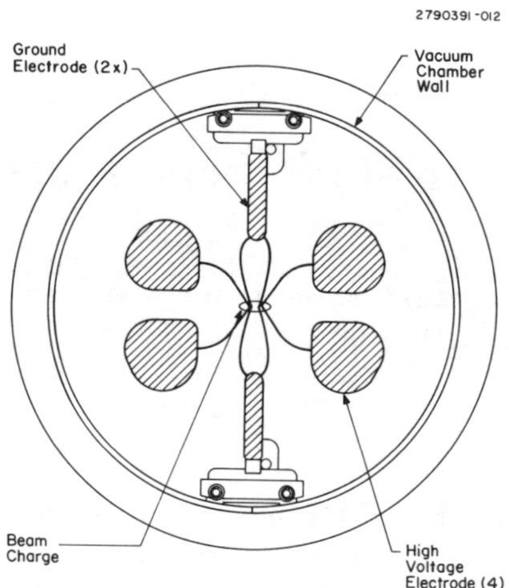

Figure 1: Electric field lines emanating from the charge in a bunch passing through the separator electrodes are shown schematically in this section taken through the middle of the separator. Very few field lines make it past the electrode structure where their energy would be lost, so most of the field energy is kept relatively close to the bunch.

metal helps to confine the electromagnetic fields that travel along with the bunch, keeping them from spreading into the larger part of the vacuum chamber where their energy would be lost. [1] Electromagnetic energy becomes 'lost' to the bunch if it is allowed to propagate far enough from the beam axis that it cannot reflect off the metal surfaces and catch up with the back of the bunch before the bunch leaves the separator structure. Additional electrodes at intermediate potentials would be even more effective at reducing the loss, but would also require additional feedthroughs thus reducing the high voltage reliability and raising the cost.

The other important idea, tapering the ends helps to reduce the amount of energy lost by effectively producing less perturbation of the bunch electromagnetic fields. (See Figure 2 on next page.) The surface profile of the vacuum chamber is chosen so that radii of curvature are large compared with the bunch length. Short radii would cause intensification of field lines and increase the amount of electromagnetic field energy at the surface.

These two ideas are merged in one unusual surface which smoothly guides the field energy from a normal almost elliptical vacuum chamber geometry to the multiple electrode geometry and then back to the normal vacuum chamber. (See Figure 2.)

Measurements

The bench measurement of the beam energy loss parameter k for the new CESR horizontal separators was performed on the Cornell University G machine [2]. A detailed

---

^{2}'G' is a currently less conventional symbol for the energy loss parameter. $k \ [V/pC] = G[m^{-1}] * 0.009$

Figure 3: This photo of the separator mockup shows the 30° taper, high voltage and ground electrodes, and the open wall configuration used in the measurements.

No.	Description	$k \ [V/pC]$
(1)	Empty Chamber, 30° tapers, no electrodes	0.206
(2)	Ground electrodes, no high voltage electrodes, straight ends	0.0917
(3)	Ground electrodes, no high voltage electrodes, blended ends	0.0730
(4)	All electrodes present, blended ends	0.123
(5)	High voltage electrodes, no ground electrodes	0.226

Table 1: The measured energy loss parameter k, is shown here for different configurations of the separator mockup. The effective bunch length was 2.1 cm.

description of this machine may be found in reference [2]. The method is based on the technique first described by Sands and Rees. [3]

The following is a physical description of the mockup designed for the G parameter measurement studies. A photo of the actual mockup is given in Figure 3. Considering the maximum measurement arm space available on the G machine is only 4 feet, a relatively short model was chosen. The tapers at both ends of mockup have the same size, both in longitudinal and transverse dimensions, as the actual separator to be built. We kept the gaps between the electrodes and chamber wall, the cross section of the ground electrode, and the transverse dimensions, close to the actual separator values. To smooth irregularities at the intersection of the ground electrode and taper surface, the ends of the ground electrode were spread using metal tape to form a concave surface.

Because the ground electrodes and the high voltage electrodes are so close to the central conductor they dominate

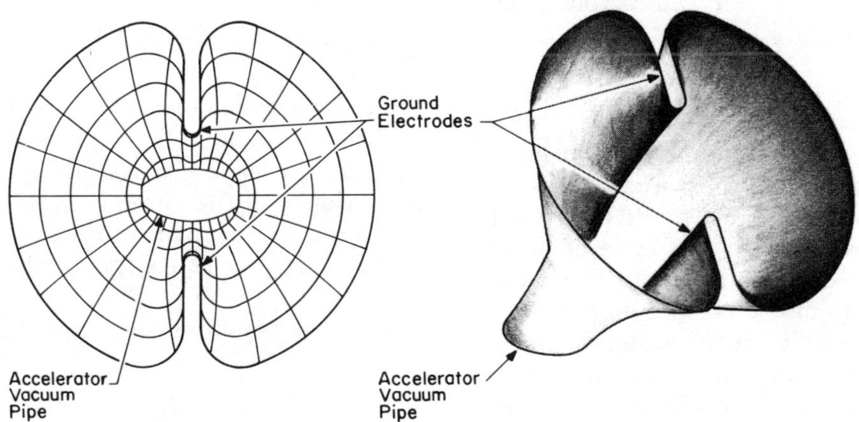

Figure 2: The surface connecting the beam tube to the main part of the separator vacuum chamber is shown in two views.

the pattern of the wake field. We chose to leave out the exterior wall (corresponding to the vacuum chamber wall) on the straight section; only four aluminum bars were used to fix the tapers together. A measurement of k with and without the outside wall in the straight section was made, and no significant difference was found.

Several measurements were made with various parts of the separator structure installed. The results presented in Table 1, show:

- Comparing (4) with (5), we can see that the addition of ground electrodes approximately halves the k value.

- A measurement was made for a mockup of existing CESR horizontal separators with the results [2]:

$$k \; [V/pC] = 1.05 \left(\frac{1}{\sigma[cm]} \right)^{1.176}$$

With $\sigma = 2.1 \; cm$, we have: $k = 0.440 \; V/pC$. Compared with (4), $k = 0.123 \; V/pC$, we find the new design has about one quarter the loss parameter of the existing separator.

- Blending the ends of the ground electrodes into the taper reduces k by about 20% compared with the straight ends' model.

The k values in Table 1 are the average of four measurements performed at different times. Typically, the variation of the measurements was $0.004 \; V/pC$. The uncertainty depended on many variables including the timing, the relative amounts of inductive to the resistive impedance of the mockup, the signal to noise ratio, the temperature drift, and uncertainty of baseline signal level, etc. A considerable source of variation came from bolting and unbolting the nulling piece and the mockup. Exchange of components can cause a change in the position of the

axis of the central wire and an error in the length of the transmission arms.

A systematic correction has been added to the calculation of k which takes into account the variation of the current pulse due to the secondary fields produced by the current pulse and the perturbations in the chamber geometry [3]. The correction is explained as follows: In the storage ring, the longitudinal electric fields due to the bunch will change the energy of the particles in the bunch, but will not have any significant effect on the charge and current distribution on the bunch. One other hand, in the G machine the fields induced by the simulated bunch interact with the chamber walls and return to the central conductor. The current pulse changes to match the boundary conditions of the conducting wire. The magnitude of the change is related with the dimension of central wire. In our case, about 10% correction was made to the calculation for G value.

References

[1] James J. Welch, *A Simple Analytic Estimate of the Loss Parameter of a Large Tapered Chamber*, these proceedings

[2] Michael G. Billing, Joseph L. Kirchgessner, Ronald M. Sunderlin. *Simulation Measurement of Bunch Excited Fields and Energy Loss in Vacuum Chamber Components and Cavities*. IEEE Transactions on Nuclear Science, Vol. NS-26, No. 3, June 1979.

[3] M. Sands, J. Rees. *A Bunch Measurement of the Energy Loss of A Stored Beam to A Cavity*. SLAC-report PEP-95 August 8,1974.

COUPLED BUNCH MOTION IN LARGE SIZE RINGS[†]

P.L. Morton, R.D. Ruth, K.A. Thompson

Stanford Linear Accelerator Center, Stanford, California 94309

I. INTRODUCTION

The growth of the quasi-steady-state motion of the coupled bunch oscillations in storage rings has been studied by means of a normal mode analysis to determine the beam stability. In this type of analysis, the initial amplitude displacements of the bunches are first written as a sum of the normal modes of the multiple bunch system, and then the stability of each mode is determined. If the amplitude of all modes decay then the amplitude of all of the individual bunches must eventually decay, and the motion is considered stable. However, if the beat frequency between the different modes is sufficiently high, compared to the decay rate of the modes, it is possible for the amplitude of some of the bunches to grow temporarily before eventually decaying. Thus, even if all normal modes are eventually damped it is possible during the transient phase for the amplitude of several individual bunch oscillations to grow and become lost. Mathematical complications also arise from a modal analysis when there is a gap in the bunch train and the wake fields from the last bunch in the train decays before arrival of the first bunch; for this case the coupled bunch motion more nearly represents that of beam breakup phenomena observed in linacs.

II. ANALYSIS

A. Simulations Results

In order to illustrate how the transient behavior can result in large amplitudes for the motion of individual bunches, even though the initial amplitudes of the individual bunches are small and the amplitude of each bunch is damped, we consider the special case of a daisy chain. The daisy chain has the property that the wake field left behind by a bunch only directly affects the next bunch. This is a special case of wake fields that decay in a time short compared to the revolution period of the ring. It is convenient to number the bunches by the integer p where $1 \leq p < P$ and to designate the leading bunch as number one. We denote the displacement of the pth bunch by x_p, the time derivative of the displacement by x_p', and assume that only the leading bunch has an initial displacement.

The model we use consists of a ring with a single cavity in which both the wake field and the damping field are produced. When a bunch passes through the cavity it receives both a wake field kick proportional to the previous bunch displacement and a damping kick, from a feedback system or radiation damping, proportional to the time derivative of the bunch motion. We use the turn number n in which the bunch passes through the cavity as the time variable, i.e. $t = nT$ with T the revolution period. Then passage through the cavity results in

$$\Delta x_1'(n) = H R x_P(n-1) - 2\alpha x_1'(n)$$
$$\Delta x_p'(n) = R x_{p-1}(n) - 2\alpha x_p'(n) \quad p \neq 1 \tag{1}$$

in which the wake field kick is given by R, and $H = 0$ when there is a gap in the beam train and $H = 1$ when the beam train is continuous . During the time between passages through the cavity the bunch oscillates about its equilibrium position $x = 0$. For the case of longitudinal motion where the oscillation frequency is much less than the revolution frequency it is not important whether the cavity is at one place in the ring or spread uniformly around the ring. Thus the difference equations of motion can be approximated by the differential equations

$$x_1''(n) + 2\alpha x_1'(n) + k^2 x_1(n) = H R x_P(n-1)$$

$$x_p''(n) + 2\alpha x_p'(n) + k^2 x_p(n) = R x_{p-1}(n) \quad p \neq 1 \tag{2}$$

in which $k = 2\pi v_s$, v_s is the number of oscillations per revolution, and α is the decay rate of the oscillation amplitude.

A simulation program has been used to solve the difference equation, Eqn. 1, for the oscillation amplitudes, a_p, as a function of turn number. The amplitudes as a function of turn number are shown for the first five bunches in Fig. 1 for three different damping rates. In Fig. 1(a) it is clear that damping rate is insufficient to prevent the transient buildup for the oscillation amplitudes for the bunches $p > 4$ before they decay. However, in Figs. 1(b) and 1(c) it is not as clear whether the decay rate is sufficient to limit the transient amplitude buildup for the bunches with numbers greater than five. The values for the coupling fields and the decay rate that have been used in these simulations are much larger than would be present in an actual storage ring. If one runs the simulation for realistic values of the coupling fields and

---

† Work supported by Department of Energy contract DE-AC03-76SF00515.

damping rates and for a sufficient number of bunches the maximum amplitude of each bunch can be determined, but the number of turns that we must run to find the maximum amplitudes of all P bunches goes like P/α which places a burden on the computer. We present below an analytic method that one can use to find the maximum amplitude and the turn number when it occurs. The circles shown on Fig. 1 are the maximum amplitude of each bunch and the turn number when it occurred as given by the analytic calculation below.

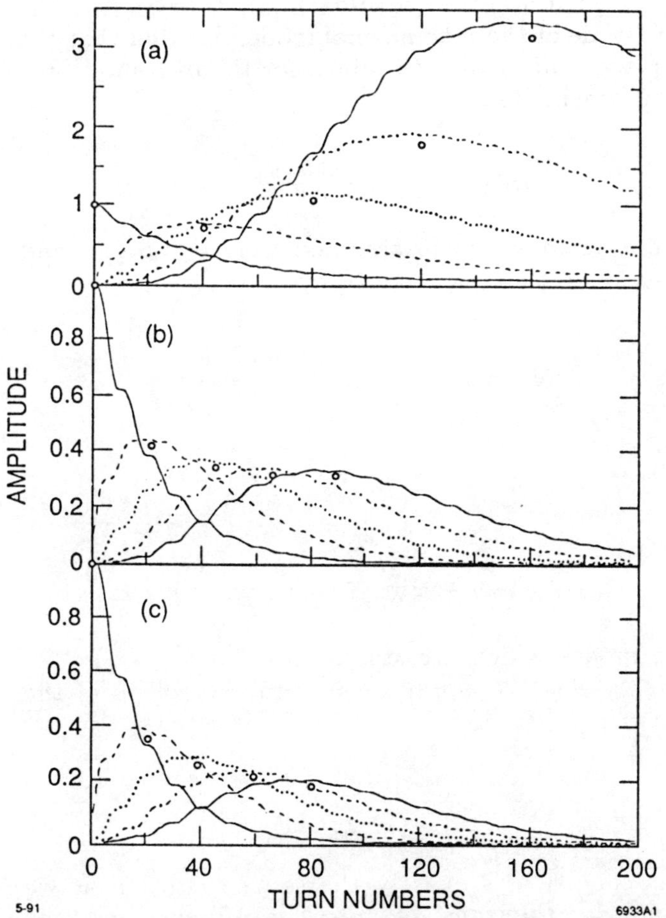

Figure 1. Oscillation amplitudes of the first five bunches versus turn number for $R/k = 0.1$ and $\nu = 0.05$. 1(a) $\alpha = 0.025$, 1(b) $\alpha = 0.045$, and 1(c) $\alpha = 0.05$. Circles represent analytic calculation of the maximum amplitudes.

B. Analytic Results

We write for the bunch displacement

$$x_p(n) = y_p(n)e^{-\alpha n}e^{i\bar{k}n} \qquad (3)$$

with $\bar{k}^2 = k^2 - \alpha^2$. We use the slowly varying amplitude and phase approximation in which we ignore y_p'' compared to ky_p' to obtain

$$y_{p+1}' = \frac{-iR}{2k}y_p \qquad (4)$$

where we also have assumed that $\alpha^2 << k^2$ so that $\bar{k} \approx k$.

At turn $n = 0$ we have the following initial conditions: $y_1 = a_1, y_1' = 0$ and $y_p = y_p' = 0$ for $p \neq 1$. The solution to Eqns. 3 and 4 is given by

$$x_{p+1} = \frac{a_1}{p!}\left(\frac{-iRn}{2k}\right)^p e^{-\alpha n}e^{ikn} \ . \qquad (5)$$

The amplitude of the oscillation a_p may be written as

$$a_p = \sqrt{x^2 + \left(x_p'/k\right)^2} = \sqrt{x_p x_p^*} \qquad (6)$$

or

$$a_{p+1} = \frac{a_1}{p!}\left(\frac{Rn}{2k}\right)^p e^{-\alpha n} \ . \qquad (7)$$

From Eqn. 7 we find that the maximum amplitude $\widehat{a}_{p+1}$ and the turn when the maximum occurs $\widehat{n}$ are given by

$$\widehat{a}_{p+1} = \frac{a_1}{p!}\left(\frac{Rp}{2k\alpha}\right)^p e^{-p}$$

and

$$\qquad (8)$$

$$\widehat{n} = \left(\frac{p}{\alpha}\right) \ .$$

The values of the coupling and the decay rate used in the simulations for Fig. 1 have been substituted into Eqn. 8 and the results are plotted in Figs. 1 and 2.

Finding the maximum oscillation amplitudes from a simulation program requires one to run the program for the number of turns in a damping period times the number of bunches in the ring which, for large multiple bunch colliders, is of the order of millions of turns.

We note that the maximum amplitude of $\widehat{a}_p$ is always bounded, but for large values of p the maximum amplitude can become quite large after many turns. Clearly, in a practical ring some restriction on the maximum amplitude for all bunches must be required. We specify that the coupled bunch motion will be acceptable if all oscillation amplitudes of the bunches are less than the initial amplitude of the first bunch, i.e. $\widehat{a}_p \leq a_1$. This requires that

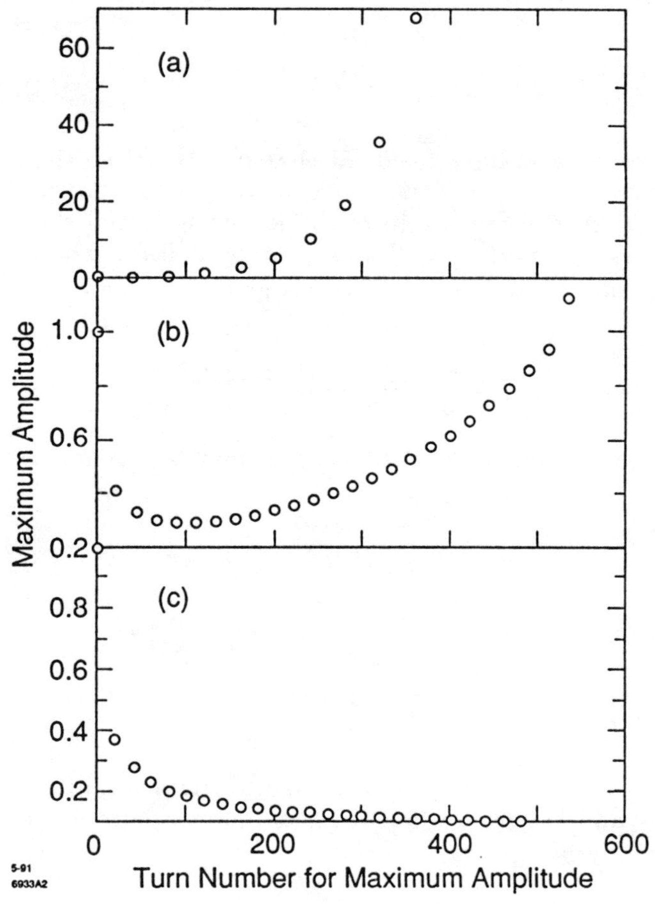

Figure 2. Maximum amplitudes of the bunches versus turn number when maximum occurs. 2(a) $\alpha = 0.025$, 2(b) $\alpha = 0.045$, and 2(c) $\alpha = 0.05$.

$$\left(\frac{Rp}{2k\alpha}\right)e^{-1} \leq \left(\frac{1}{p!}\right)^{\frac{1}{p}} . \tag{9}$$

If we use Stirlings formula for large p we have $\left(\frac{1}{p!}\right)^{\frac{1}{p}} \approx$

pe^{-1} which yields the following criteria for the required damping rate

$$\alpha \geq \frac{R}{2k} \tag{10}$$

for acceptable motion in the sense that $\hat{a}_p \leq a_1$ for all p.

C. Normal Mode Analysis

For the case where the bunch train is continuous ($H = 1$) it is easy to solve Eqn. 2 for $\xi_m(n)$, the amplitude of the mth normal mode. We find that the displacement of the pth bunch for the mth mode can be approximated by

$$x_p^{(m)}(n) = \xi_m(0)e^{\frac{i(2\pi m+k)p}{P}}e^{(\gamma_m-\alpha)n} \tag{11}$$

where again we have assumed that $k >> \gamma_m$ and $k >> \alpha$, so that γ_m is given by

$$\Re(\gamma_m) = \frac{R}{2k}\sin\left(\frac{2\pi m + k}{P}\right)$$

and

$$\tag{12}$$

$$\Im(\gamma_m) = -\frac{R}{2k}\cos\left(\frac{2\pi m + k}{P}\right) .$$

The modes which are stable have $\Re(\gamma_m - \alpha) \leq 0$, which yields the condition for stability of all of the modes

$$\alpha \geq \frac{R}{2k} . \tag{13}$$

This condition is exactly the same condition as we obtained in the transient case (Eqn. 10) when only the first bunch in the train had an initial displacement.

Simulation and Stability of a Crab Cavity *

Z. Greenwald, S. Greenwald, and D. H. Rice
Laboratory of Nuclear Studies, Cornell University, Ithaca, NY 14853

Abstract

A feasibility study of the stability of a superconducting crab cavity for a B Factory was carried out. The study was done for the two trapped modes: the longitudinal TM010 and transverse TE111. It was found that for both modes the voltage induced by the beam bunches will be less than ± 20 KV for the TM010 mode and $\pm$ 1KV for the TE111. Using the ZAP code it is shown that a stable frequency regime for the TM010 mode can be found in which the worse growth time is 8 times larger than the radiation damping time. For the TE111 mode the high energy ring is stable while the low energy ring will have to be stabilized by the feedback system.

1 INTRODUCTION

The Cornell B factory design employs a ± 12.5 mrad horizontal crossing angle at the interaction point. The undesirable side effects of this crossing angle will be compensated by 'crabbing' [1, 2] the beams with a pair of single cell superconducting RF cavities operating in a deflecting mode. The purpose of the crab cavity is to achieve a 2 MV transverse kick by the TM110 mode at the same frequency as the accelerating cavity *i.e.*, 500 MHz as described in references [3, 4]. The crab cell design allows all modes higher in frequency than the crabbing mode to propagate out the beam pipe and be damped outside the cryostat with Ferrite-50[†] absorbers [5] located on the beam pipe . Four unwanted modes remain trapped in the cell region: the fundamental TM010 mode, two polarizations of the TE111 mode, and one polarization of TM110 mode. A study of the first two unwanted modes is being carried out with this paper giving some of the initial results.

2 $TM010$ MODE

The TM010 mode in the crab cavity is an undesirable longitudinal mode. (since the purpose of the cavity is to create a transverse kick). Its resonant frequency is 367 MHz with $R/Q = 87$ Ω/cell.

---

*Work supported by the National Science Foundation
[†]Product of Trans-Tech, Inc., Adamstown, MD.

2.1 Induced Voltage

To study the induced voltage developed in the cavity by the beam bunches, the cavity is simulated by an impedance $Z_c(\omega)$ of a shunt RLC resonant circuit with the appropriate resonance frequency ω_r and quality factor Q obtained by the code URMEL.

$$Z_c(\omega) = \frac{R}{1 + jQ \left(\dfrac{\omega}{\omega_r} - \dfrac{\omega_r}{\omega} \right)} \tag{1}$$

For the sake of simplicity, the bunched current is simulated by a train of point charges (having no synchrotron motion) separated by ΔT_b seconds between two charges and circulating around the ring with a period of T.

$$I(t) = I_o \sum_{m=0}^{N_b-1} \left\{ \sum_{n=-\infty}^{\infty} \delta(t - m\Delta T_b - nT) \right\} \tag{2}$$

where N_b is the number of bunches. Taking a Fourier transform gives the current in the frequency domain:

$$I(\omega) = \frac{1 - e^{-j\omega \, \Delta T_b N_b}}{1 - e^{-j\omega \, \Delta T_b}} \omega_o$$

$$\cdot \left\{ \delta(\omega) + 2 \sum_{n=1}^{\infty} \delta(\omega - n\omega_o) \right\} \tag{3}$$

The induced voltage developed on the cavity will be:

$$V(\omega) = I(\omega) \cdot Z_c(\omega) \tag{4}$$

or by taking the inverse Fourier transform:

$$V(t) = \frac{\omega_o R}{\pi} N_b \frac{1}{1 - j2Q} \tag{5}$$

$$+ \frac{2\omega_o R}{\pi} \sum_{n=1}^{\infty} \left\{ \frac{e^{jn\omega_o t}}{1 + jQ \left(\frac{n\omega_o}{\omega_r} - \frac{\omega_r}{n\omega_o} \right)} \cdot \frac{1 - e^{-j\omega_o \, \Delta T_b N_b}}{1 - e^{-j\omega \, \Delta T_b}} \right\}$$

The envelope of the induced voltage of this TM010 mode with ΔT_b=5 RF buckets=10 nsec and revolution frequency f_o=391.97 KHz is plotted in Figure 1. Note that the voltage is plotted at the time of arrival of each bunch giving only the envelope and thus the 367 MHz mode resonance

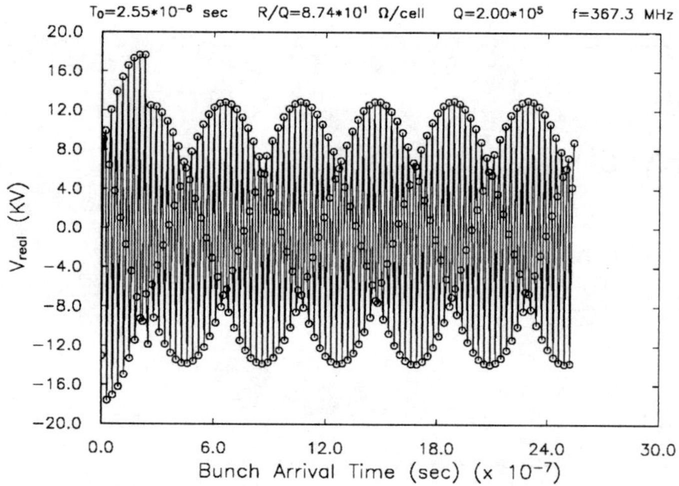

Figure 1: Voltage induced in the TM010 mode during one period

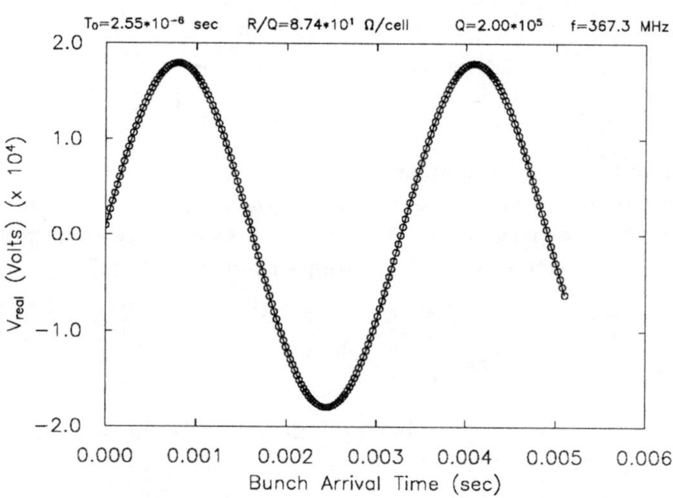

Figure 2: Induced Voltage seen at the gap during a thousand periods

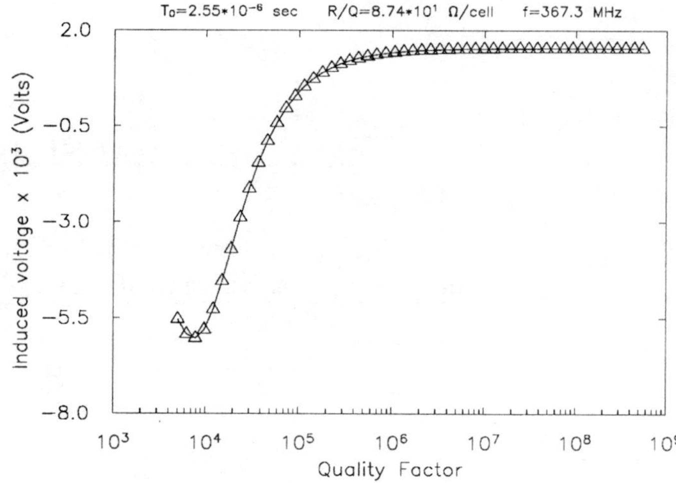

Figure 3: Induced voltage as function of quality factor

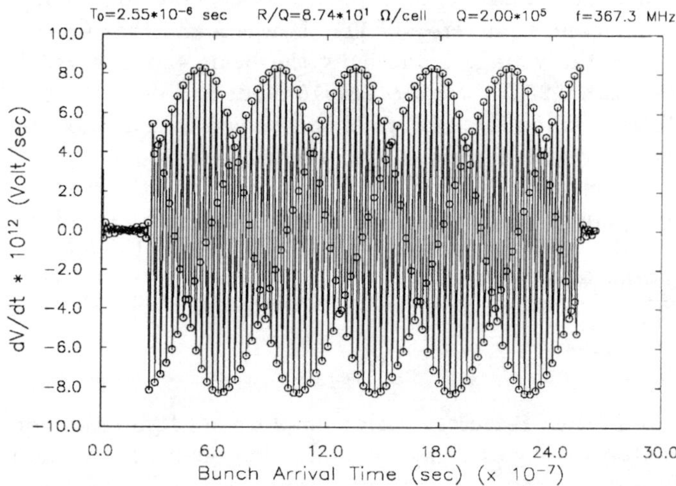

Figure 4: The derivative of the induced voltage as seen by the bunches in one period

frequency is not seen. Also note that since the spacing between the bunches is a multiple of the accelerating frequency *i.e.*, 500 MHz, and the longitudinal mode of the crab is oscillating at 367 MHz, the bunches will arrive at the cavity with different phases and as a result some of them will be accelerated and some decelerated causing to an irregular envelope of the built-up voltage. The amplitude of the induced voltage at small values of 't' is higher due to the gap of 151.2 nsec. at the end of the bunch train. Figure 2 shows the voltage seen at the gap during one thousand revolutions and Figure 3 shows the dependence of the induced voltage on the quality factor.

The amplitude of the induce voltage is smaller than ± 20 KV which is less than 2% of the driving voltage and can be neglected.

2.2 *Bunch Lengthening*

The other concern is a possible bunch lengthening due to the slope of the voltage developed in this TM010 mode. Figure 4 shows the slope of the voltage at the time of bunch arrival during one revolution obtained numerically from the simulation described in the previous section. This time of arrival would be different for each bunch if synchrotron motion had been in the bunch train. Since we are not interested in the slope of a specific individual bunch but in the range of all possible slopes, the simulation result is still valid. The slope obtained from tracking a bunch for 1000 revolutions is less than 10^{13} V/sec compared with a slope of 10^{16} V/sec from the main accelerating cavities so the perturbation in bunch length is negligible.

2.3 *Stability*

The longitudinal coupled bunch instability can be effectively controlled by tuning the resonant frequency of the

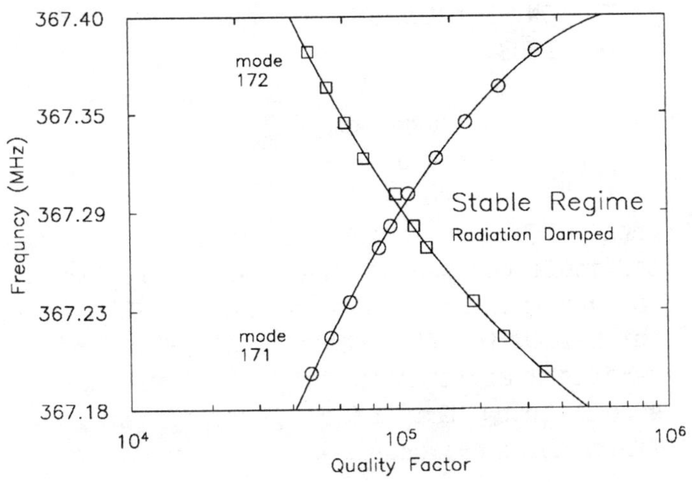

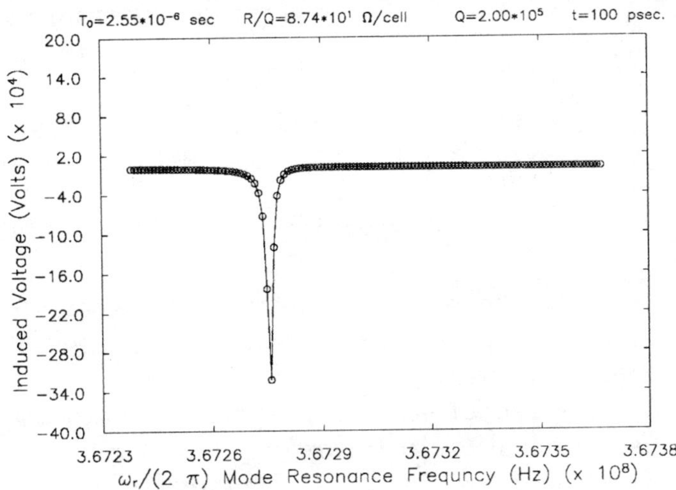

Figure 5: The stable frequency regime with growth time higher than 100 msec, in the TM010 mode as function of quality factor, for the low energy ring

Figure 6: Change of the induced voltage due to resonance mode tuning.

TM010 mode. A range of frequencies for which the beam will be either stable or Landau damped is searched for using the program ZAP [6]. The program calculates the frequency shift and growth time of all modes using the Wang formalism for electrons in gaussian bunches. For the low energy ring (3.5 GeV and $1.37 \cdot 10^{11}$ electrons/bunch) bunch modes 171 and 172 have the fastest growth time (few msec) and their synchrotron frequency shift is too large (1000 Hz at $Q=5 \cdot 10^3$) to use Landau damping. If we allow higher values of quality factor Q the bunch mode growth time can be increased by controlling if the mode frequency to allow damping by synchrotron radiation. Figure 5 shows the stable frequency regime as function of Q for which the worse growth time is higher than 100 msec which is 8 times the radiation damping time in the CESR-B low energy ring. It can be seen, for example, that for $Q=2 \cdot 10^5$ the stable frequency range is from 367.235 MHz to 367.347 MHz, while the frequency range is zero for $Q=1 \cdot 10^5$.

The change of the induced voltage during the tuning of the resonance frequency of the resonator using equation 5 can be seen in figure 6. The strong dip in the voltage occurs when the harmonic revolution frequency falls on the resonance frequency. The width of the voltage dip is very narrow due to the small frequency band of the mode resonance (1.8 KHz). In the high energy ring (8 GeV) the number of particles per bunch is smaller ($0.6 \cdot 10^{11}$) making the stability demands less severe. Bunch modes 85 and 86 have the shortest growth time but they will also be radiation damped for $Q=2 \cdot 10^5$.

3   *TE*111 MODE

A similar study done for the parasitic transverse mode with a resonance frequency of 484.7 MHz and transverse R_T/Q of 5.5 $\Omega/cell$ (obtained from the code URMEL) gives an in-duced voltage smaller than 1 KV. Assuming a tune spread of $2 \cdot 10^{-5}$, the beam in the high energy ring is stable with a frequency shift of less than 10 Hz and a growth time longer than 47 msec. For the low energy ring the growth time is greater than 20 msec but the beam is unstable and will have to be stabilized by the transverse feedback system.

4   *Conclusions*

By making the quality factor in the TM010 mode higher than $2 \cdot 10^5$, and by tuning its resonant frequency it is possible to make the growth time of the unstable modes to be greater than the radiation damping time. By avoiding the resonant frequency of 367.277 MHz (where it falls on the harmonic revolution frequency) the induced voltage is less than 20 KV and the bunch length distortion is negligible. In the transverse mode TE111 the high energy ring is stable while the low energy ring must be stabilized by the transverse feedback system.

References

[1] R. Palmer, SLAC-PUB-4707 (1988).

[2] K. Oide, K. Yokoya, SLAC-PUB-4832 (1989).

[3] H. Padamsee, et al, CLNS 90-1039.

[4] H. Padamsee, et al, This Conference.

[5] D. Moffat, et al, This Conference.

[6] M.S. Zisman, S. Chattopadhyay and J. Bibognano, LBL-21270 ESG-15.

EFFECTS OF QUADRUPOLE WAKE FIELD ON RF FOCUSING
IN LINEAR COLLIDERS.

A.N. Didenko, V.N. Gusarov, G.A. Kuzmenko,
Moscow Physical Engineering Institute.

1. INTRODUCTION.

It's known that transverse wakefield effects and the beam 'heating up' prevent small emittance preservation.

The use of an energy spread to damp the oscillation of the core of the bunch was first described by Balakin [1] (BNS damping). But efficiency of this method decreases at the final focus where the energy spread is to be reduced.

Another way to meet the damping criterion is to use a focusing field which has a variation along the bunch. For this purpose part of the accelerating sections are given nonsymmetric apertures (slots instead of circular holes) which are placed alternately vertically and horizontally at a proper period length. The slot apertures produce RF quadrupoles of considerable strength without appreciable loss of shunt impedance for acceleration. In this case the spread in betatron oscillation, necessary for BNS damping, can be introduced without any energy spread.

Usually, the damping criterion takes into account only dipole wakefield which is proportional to the distance between the path of the exciting charge and the axis of the structure, but independent of the transverse position of the particle on which it acts. The reason is that in symmetric structures there are other higher order wakefields which depend on higher powers of the distance between the path of the exciting and following particles and the axis of the RF structure, but whose influence on transverse dynamics is negligible.

But in azimuthally nonsymmetric structures (ANS) there is an accelerating field with strong quadrupole component used for RF focusing. So we can expect a high enough quadrupole component of wakefield.

In this paper the use of ANS in cases when BNS criterion fulfilling becomes difficult because of energy spread limitation is investigated. Two problems are pointed out below, the first being that of quadrupole wakefield utilization for beam focusing, the second being that of RF generator and wake fields interaction.

2. ELECTROMAGNETIC FIELDS.

Electromagnetic fields in the ANS can be analyzed in detail in case of the slotted rectangular waveguide. In such structures slow LE mode waves and a fast LM mode waves propagate [2]. In the lower band in-phase waves include both longitudinal (accelerating) and transverse quadrupole components, while anti-phase waves include a dipole component thus being useless for acceleration.

The same considerations are preserved in a slotted circular waveguide, in a waveguide with elliptical cross-section and hyperbolically-shaped irises.

Electromagnetic properties of such structures have been studied rather in detail.

3. WAKEFIELDS IN THE ANS.

If particles travel with velocity $v=c$, focusing gradient can be calculated based upon Panofsky-Ventzel theorem and expression for the longitudinal field. In the region of high frequencies so-called optical resonator model can be applied. A. Sessler developed the formula of radiation losses based on analogy between a set of finite plates with circular holes and a pair of circular mirrors. To estimate radiation fields in this frequency region one can make use of equivalent optical resonator for iris-loaded structure, suggested by Weinstein [3].

4. BETATRON WAVELENGTH SPREAD IN THE ANS.

Expression for wake fields and external electromagnetic ones achieved allow us to investigate the bunch motion in each particular azimuthally asymmetrical structure.

Calculations show that dipole wakefield may amplify the influence of accidental misalignments of both accelerating and focusing systems, that can lead to large emittance growth and even beam loss.

Quadrupole wakefield can make nonlinear spread of betatron wavenumbers along the bunch. Note, that it can 'disturb' BNS criterion for bunch coherent oscillations.

Large enough quadrupole wakefield can result from the structure asymmetry variation. This can be used for particles focusing. Provided that quadrupole wakefield is linear vs longitudinal coordinate inside the bunch, its influence on transverse dynamics is similar to that of quadrupole magnets. Wakefield nonlinearity along the bunch leads to wavenumber spread of this new 'magnetic-RF' focusing system even in case of a monoenergetic bunch.

Taking into account all the above said, charged particle motion equation will be:

$$m\,c^2\,\frac{\mathrm{d}}{\mathrm{d}z}\left[\gamma(s,z)\,\frac{\mathrm{d}x(s,z)}{\mathrm{d}z}\right] = qx(s,z) \times$$

$$\left\{ cG_{rf}\sin\phi(s,z) + \int_{0}^{s}\mathrm{d}s'\,\rho(s')W_{\perp}^{q}(s-s') + cB' \right\} +$$

$$q\int_{0}^{s}\mathrm{d}s'\,\rho(s')W_{\perp}^{d}(s-s')x(s',z) \qquad (1)$$

with $\quad G_{rf} = \dfrac{1}{2\pi f}\dfrac{\partial^2 E_{zo}(x,0)}{\partial x^2}\quad$ and $B' = \dfrac{\mathrm{d}B_y}{\mathrm{d}x}$

gradients of RF and magnetic field respectively, m and q the particle mass and charge, γ the normalized energy, $W_{\perp}^{d}$ and $W_{\perp}^{q}$ the transverse dipole and quadrupole wake potentials. z is the coordinate along the linac, s is the longitudinal position in the bunch (from head to tail), x is the particle transverse offset.

5. WAKEFIELD QUADRUPOLE (WFQ).

Quite a number of such accelerating - focusing structures is possible:

A. <u>Both accelerating and wakefield focusing exist at the lowest mode.</u>

This version seeming to be simple in technology, has some disadvantages:

- Quadrupole component growth can be accompanied by accelerating properties degradation. Thus, varying of slotted rectangular waveguide sizes results in a significant growth of quadrupole component (that is the best) and unfortunately falling of group velocity value more than 10-fold and corresponding filling time growth.

- Since quadrupole component of accelerating field and first mode of wakefield are oppositely directed and, therefore, are subtracted from each other, betatron wavenumber spread along the bunch resulting from the WFQ diminishes.

B. <u>WFQ on higher modes.</u>

In some accelerating structures a quadrupole component corresponding not to the mode used for acceleration but to the higher modes, for instance, to the second mode, is excited. Example of such a structure is elliptical waveguide having elliptical irises. RF source excites structure on the first mode, no RFQ effect appearing in it. WFQ exists but on the second mode. Thus, accelerating and WFQ become separated.

C. <u>Accelerating and WFQ are separated in space.</u>

The main structure of this construction being a conventional iris-loaded waveguide, some passive high gradient ANS are inserted between sections. Unfortunately, this result in accelerating gradient reducing. It may be useful to combine such passive cavities with quadrupole magnets.

6. CONCLUSION.

Influence of wakefields on transverse dynamics of charged particles in linear colliders has been analyzed. Problem of betatron wave length spread producing has been discussed, some ways of its realization having been proposed. Interaction between quadrupole components of external and wake field has been considered. The new focusing element referred to as wakefield quadrupole (WFQ) has been proposed. All mentioned above allows us to conclude that application of the ANS in high energy collider designing seems to be promising.

7. ACKNOWLEDGEMENT

The authors wish to thank A.V. Izmailov for his participance in discussions on this paper and assistance in its preparation.

8. REFERENCES.

[1] V.E. Balakin , "Podavlenie stokhasti-tcheskogo razogreva puchka v lineynom kollaidere." Preprint INP 88-100, Novosibirsk, 1988.

[2] E.S. Kovalenko, V.I. Shimansky,"Sinfaznye volny v diafragmirovannom volnovode pryamougolnogo setcheniya" // Izvestia Vuzov SSSR, Radiotekhnika, N2, 1960, pp. 153-167.

[3] L.A. Weinstein, Electromagnitnye volny, Moscow, 'Sovietskoe Radio', 1957.

[4] J.T.Seeman, N. Merminga "Mutual compensation of wakefield and chromatic effects of intense linac bunches" // SLAC-PUB-5220, May 1990 (A).

Realistic Modeling of Microwave Instability Effects on the Evolution of the Beam Energy-Phase Distribution in Proton Synchrotrons

J. A. MacLachlan

*Fermi National Accelerator Laboratory, Box 500, Batavia IL   60510**

Introduction

Either bunched or coasting beam in a synchrotron may exhibit microwave instability if the momentum spread is less than[1]

$$\left(\frac{\Delta p}{p}\right)^{(\text{th})}_{\text{FWHM}} = \sqrt{\frac{e\hat{I}(Z/n)}{F\beta^2 E|\eta|}} \ , \qquad (1)$$

where $\hat{I}$ is the peak current, (Z/n) is the longitudinal coupling impedance at the n-th harmonic of the beam circulation frequency divided by the harmonic number n, F is a form factor of order 1 accounting for the particle distribution (including whether bunched or not), β is v/c, E is the total energy of the beam particles, η is the time dispersion $\gamma_T^{-2} - \gamma^{-2}$; γ's are $E/m_o c^2$ and T subscript labels transition energy. A useful physical picture is that beam particles are captured in buckets generated by the beam image current flowing in the longitudinal coupling impedance. Qualitatively, trapping and auto-deceleration occur when the height of the buckets exceed the FWHM energy spread of the beam. "Microwave instability" implies in addition that the coupling impedance is largest at several times the rf frequency and that the decay of the wakefield is fast enough that bunches do not affect each other. This high frequency, low Q impedance is is represented in the reported work by a single resonance at 1.7 GHz (approximately microwave cutoff of vacuum chamber), Q=1, and $R_{\text{shunt}} = Z$. The parameters used in this paper are influenced by the Fermilab Main Ring and design of the Main Injector.[2]

The numerical modeling uses standard features of the code ESME.[3],[4] In most of the reported simulations $2 \cdot 10^4$ macroparticles and 32 values of n separated by 1113 provide the current spectrum; the justification for these choices is given in ref. [5] which gives some detail on methods. Microwave instability may be an intensity limitation during parts of the acceleration cycle where the beam is debunched or loosely bunched, perhaps at injection or high duty factor extraction. Probably of more general importance is the time near transition when the spread in circulation frequency is sharply reduced, *i.e.*, when $\eta \approx 0$. Concrete examples are given below.

*Operated by the Universities Research Association under contract with the U. S. Department of Energy

Coasting Beam

The continuous coasting beam case involves a single time scale, the growth time. It has therefore reasonably simple evolution and is tractable analytically. However, even in this case the early decay of the highest frequency modes and the absence of a dominant mode or band of modes is difficult to calculate in detail. Because no one amplitude grows steadily, the growth time is defined here as the time required for the momentum spread to grow to the threshold value. This time is plotted against $Z/Z_{\text{th}} - 1$ for a 0.5 A beam with $\Delta E/E$ (FWHM) = ± 13.28 MeV in a parabolic energy distribution at 150 GeV with $\eta = 0.0028$ in fig. 1. The fit is second order; both linear and quadratic terms are important. The instability saturates short of the so-called "overshoot" value

$$(\Delta p/p)_{\text{in}}(\Delta p/p)_{\text{fin}} = (\Delta p/p)^2_{\text{th}} \ . \qquad (2)$$

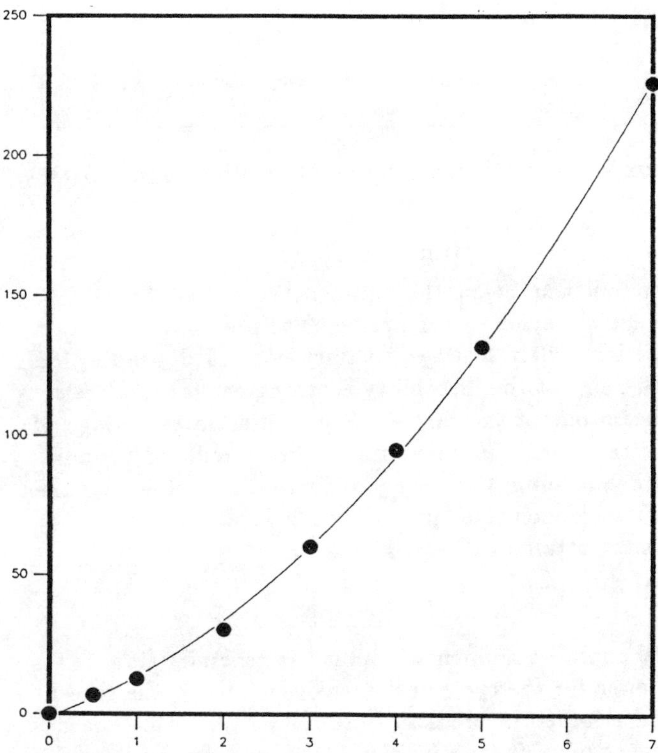

Figure 1: Rate of growth of coasting beam instability to threshold [s^{-1}] vs. $Z/Z_{\text{th}} - 1$

Debunching with RF Off

If the rf voltage is removed at fixed momentum, instability appears when the bunches shear to the point that local momentum spread drops below threshold. Note from the phasespace plot in fig. 2 that the total Δp of the bunch is not what determines stability. Note also that the instability proceeds independently for each bunch even after overlap, a result predicted by Ng.[6] There is a growth time and a debunching time dependence in this example; therefore, the dependence of the blowup on parameters is more complex.

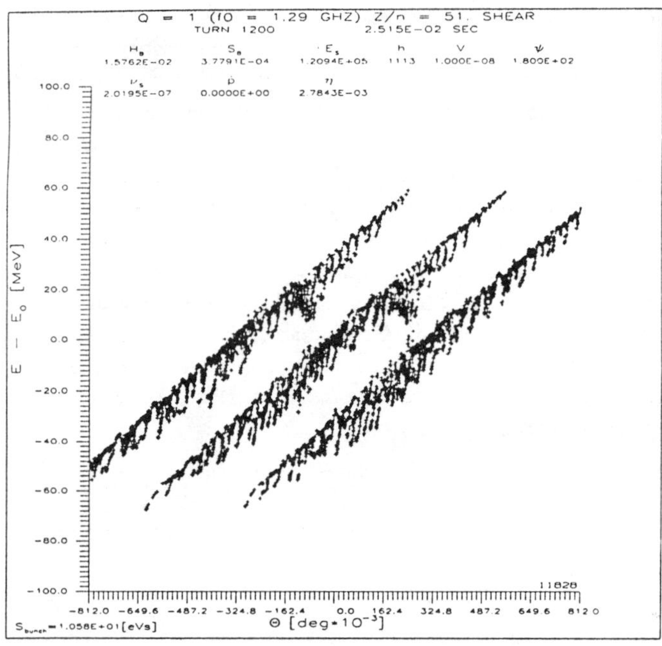

Figure 2: RF off debunching — Main Ring, $Z/n = 51\Omega$

Bunched Beam

For bunched beam the time scales are set by growth time and the synchrotron period. The long term evolution of the instability has two distinct courses depending on whether or not the instability is strong enough to decelerate beam out of the bucket. Fig. 3 illustrates a stage of the latter course. If the voltage is being reduced for adiabatic debunching, the voltage reduction is so slow that the threshold momentum spread gives a good estimate of the minimum attainable.

Transition

Although the momentum spread is generally close to its maximum for the cycle near transition energy, the threshold criterion eq. 1 predicts instability as $\eta \to 0$. The dependence of orbit length on momentum difference may be written

$$\Delta C/C = (\alpha_0 + \alpha_1\delta + \cdots)\delta \ , \qquad (3)$$

where δ is $\Delta p/p$ and $\alpha_0 = \gamma_T^{-2}$. To first order the momentum dependence of η is

$$\eta = \eta_0 + \eta_1\delta \approx \alpha_0 - \gamma^{-2} + (\alpha_1 + 3\alpha_0/2)\delta \ . \qquad (4)$$

With parameters similar to the Main Ring ($\gamma_T = 18.75$, $\alpha_1 = .005689$, $Z/n = 9\Omega$, $\delta = 0.34\%$) the coasting beam threshold is ~ 0.75 A, and emittance growth about one percent is evident in 10 ms at 1.5 times threshold.

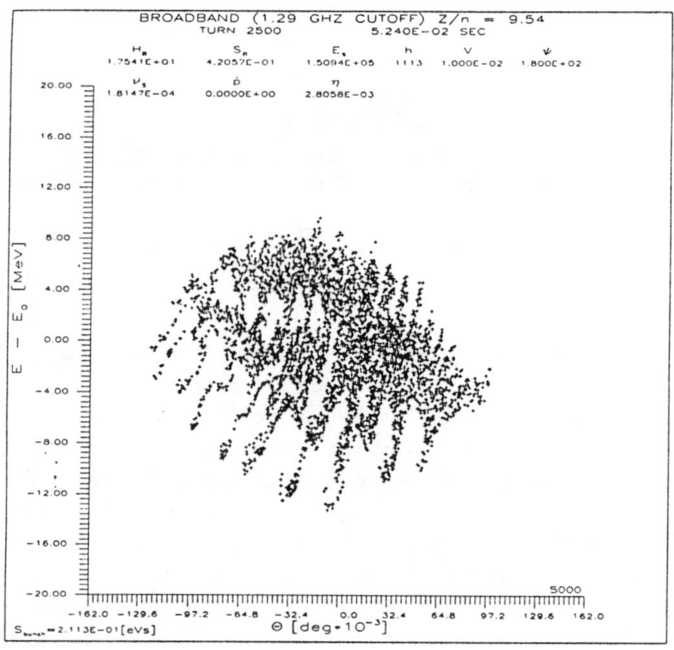

Figure 3: 53 MHz bunch in Main Ring — $2 \cdot 10^{10}$ proton, $Z/n = 9.54\Omega$, $f_{co} = 1.3$ GHz, $V_{rf} = 10$ kV

By reversing the sign of α_1 one meets the condition $\alpha_1 = -2\alpha_0$ which makes the lattice isochronous through second order at transition. Figure 4 shows the growth of $\Delta\varepsilon_l/\varepsilon_l$ for coasting beam in 0.0119 s at Main Ring transition with α_1 set for isochronicity. With the sign of α_1 reversed growth is not visible on this scale below 2 A.

The practical questions for transition crossing are what is the emittance blowup and what parameter choices minimize the blowup or beam loss for a given beam current. Figure 5 shows one cut through parameter space for a so-called duck-under scheme specifying the rf voltage in the transition region. The plot shows growth of an initial 0.1 eVs bunch during transition crossing vs. $\hat{I}$ in the range 0. to 4.57 A for an accelerator like the Main Injector ($\gamma_T = 20.4$, $\alpha_1 = 0.$) except that the ramp is slower ($\dot{\gamma} = 107$) and Z/n is higher (10 Ω). At 7.1 A there is 6% beam loss making emittance incomparable.

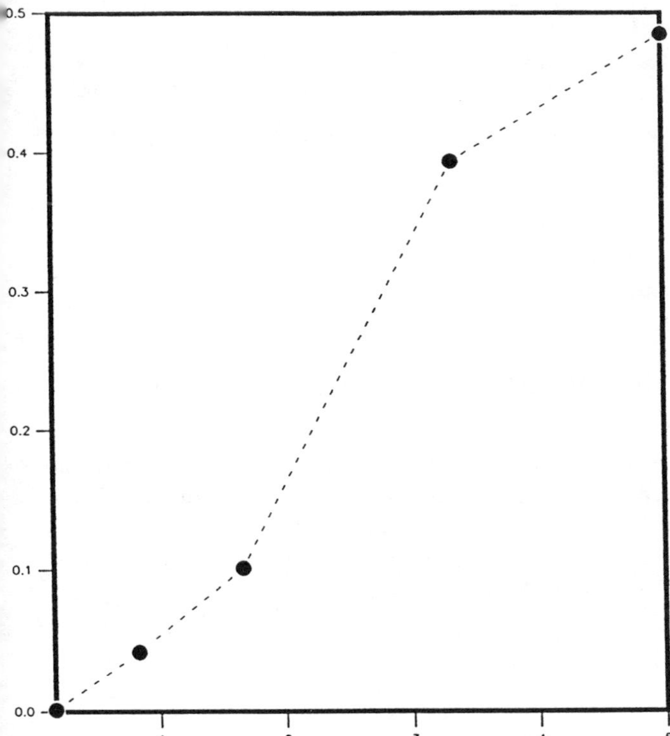

Figure 4: Growth of $\Delta\varepsilon_l/\varepsilon_l$ for Main Ring coasting beam at transition with $\alpha_1 = -2\alpha_0$ for isochronous orbit vs. number of protons per bunch $[10^{10}]$

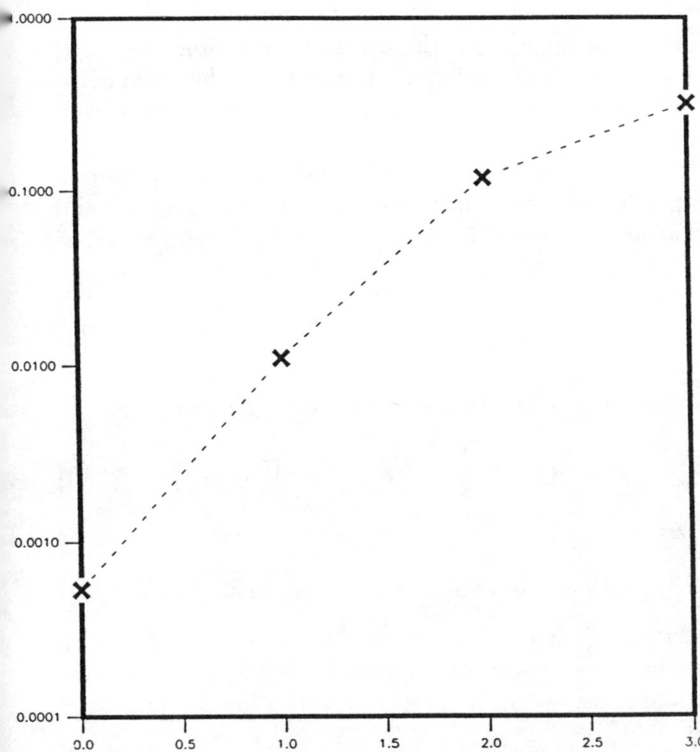

Figure 5: $\Delta\varepsilon_l/\varepsilon_l$ [eVs] for "duck-under" transition crossing of 0.1 eVs bunch in modified Main Injector vs. number of protons per bunch

<u>Remarks</u>

Some situations in which microwave instability can set the effective limit to accelerator performance have been examined by particle tracking simulation. The emphasis has been on the practicability of testing alternative parameters by numerical experiment for processes with multiple time scales which would not be tractable analytically. It is not possible to exhibit here the evolution of the instability to lend intuitive support for the interpretations offered nor to make thorough comparison with analytic results. Little space has been devoted either to fundamental questions of dynamics or details of numerical method. These matters are treated more fully in ref. [5] by MacLachlan, Kourbanis, Ng, and Wei. The modeling approach may have practical benefit if it leads designers to evaluate the overall growth of emittance in a process instead of considering only the instability threshold.

References

[1] A. Hofmann, "Longitudinal Instabilities" in "Theoretical Aspects of the Behavior of Beams in Accelerators and Storage Rings", CERN 77-13 (19 July 77) p139ff

[2] "Fermilab Upgrade: Main Injector Conceptual Design Report", Rev. 2.3, Fermilab (April 90)

[3] J. A. MacLachlan, "Fundamentals of Particle Tracking for the Longitudinal Projection of Beam Phasespace in Synchrotrons", Fermilab note FN-481 (15 April 90)

[4] S. Stahl and J. MacLachlan, "User's Guide to ESME v. 7.1", Fermilab internal note TM-1650 (26 Feb. 90)

[5] Proceedings of the Fermilab III Instabilities Workshop", Fermilab internal note TM-1696 (Dec. 90)

[6] K.-Y. Ng, "Microwave Instability for Overlapped Bunches", Fermilab note FN-432 (May 86)

COMPARISON BETWEEN COASTING AND BUNCHED BEAMS ON OPTIMUM STOCHASTIC COOLING AND SIGNAL SUPPRESSION*

J. Wei

Brookhaven National Laboratory
Upton, NY 11973, USA

Abstract

A comparison has been performed between coasting and bunched particle beams pertaining to the mechanism of stochastic cooling. In the case that particles occupy the entire sinusoidal rf bucket, the optimum cooling rate for the bunched beam is shown to be the same as that predicted from the coasting-beam theory using local particle density. However, in the case that particles occupy only the center of the bucket, the optimum rate decreases in proportion to the ratio of the bunch area to the bucket area. Furthermore, it has been shown for both coasting and bunched beams that particle motion is stable upon signal suppression if the amplitude of the gain is less than twice the optimum value over the entire frequency bandwidth of the cooling system.

I. INTRODUCTION

Stochastic cooling for both coasting and bunched beams[1–7] has been successfully applied to many accelerators. Theories, mostly using Fokker–Planck approach, have been developed to investigate the particle motion.

This paper provides an analytic comparison between coasting and bunched beams on the cooling mechanism. In section II, the optimum cooling rates for coasting and bunched beams are derived from the Fokker-Planck equations. In the case that particles occupy the entire sinusoidal rf bucket, the optimum rate for the bunched beam is shown to be the same as that predicted from the coasting-beam theory using local particle density. However, in the case that particles occupy only the center of the bucket, the optimum rate decreases in proportion to the ratio of the bunch area to the bucket area, which contradicts the coasting-beam prediction. In section III, the effect of signal suppression is evaluated based on the equations of motion. It is shown that the particle motion under cooling is stable if the amplitude of the gain of the system is less than twice the optimum value over the entire frequency bandwidth.

Although the discussion is restricted to transverse stochastic cooling, it has been found that the conclusion is also true for longitudinal cooling, where the analysis is complicated by the mixing factor that depends on longitudinal particle distribution.

*Work performed under the auspices of the U.S. Department of Energy.

II. COMPARISON ON OPTIMUM COOLING

A. Coasting Beam

Consider a beam of N charged particles azimuthally distributed along the accelerator. The increment $U_{x'}$ in $x' = dx/ds$ that is experienced by the ith particle per unit time at the kicker, is proportional to the displacement x^P of all the particles at the pick-up,

$$U_{x',i} = \sum_{j=1}^{N} U_{ij} x_j^P,   \tag{1}$$

where

$$U_{ij} = \frac{\omega_0}{2\pi\sqrt{\beta_x^P \beta_x^K}} \sum_{m=-\infty}^{\infty} G\left(m^{\pm}\omega_j\right) e^{im(\omega_j - \omega_i)t}.   \tag{2}$$

Here, $m^{\pm} = m \pm \nu_x$, ν_x is the transverse tune, $G(\omega)$ is the gain of the cooling system, ω_i is the revolution frequency of the ith particle, ω_0 is the average revolution frequency, and β_x is the Courant-Snyder parameter. The superscript P and K denote values at the pick-up and the kicker, respectively.

It is convenient to describe the transverse motion of the particles in terms of the angle- action variables φ and I that are generated from the original variables x and x' by a generating function

$$F_1\left(x, \varphi; s\right) = -\frac{x^2}{2\beta_x}\left(\tan\varphi - \frac{\beta_x'}{2}\right).   \tag{3}$$

The equations of motion thus become

$$\varphi' = \frac{1}{\beta_x} + U_\varphi, \quad I' = U_I,   \tag{4}$$

where

$$U_I = -\sqrt{2\beta_x I}\sin\varphi U_{x'}, \quad U_\varphi = -\sqrt{\beta_x/2I}\sin\varphi U_{x'}.   \tag{5}$$

Evolution of the transverse distribution function Ψ can be described by a transport equation, which is obtained by averaging the two-dimensional Fokker-Planck equation[6] over φ,

$$\frac{\partial\Psi}{\partial t} = -\frac{\partial}{\partial I}\left(F\Psi\right) + \frac{1}{2}\frac{\partial}{\partial I}\left(D\frac{\partial\Psi}{\partial I}\right).   \tag{6}$$

Neglecting the thermal noise of the cooling system, the coefficients of coherent correction F and diffusion D can be evaluated

$$F(I) = F^0 I, \quad D = D^0 I \langle I\rangle,   \tag{7}$$

where $\langle\,\rangle$ denotes the average, and

$$F^0 = -\frac{\omega_0 \sin \nu_x \Delta\theta^{PK}}{2\pi} \sum_{m=-\infty}^{\infty} G\left(m^\pm\omega_i\right) e^{-im\Delta\omega_i\Delta\theta^{PK}/\omega_0}$$

$$D^0 = \frac{\omega_0^2 \rho\left(\omega_i\right)}{4\pi} \sum_{m=-\infty}^{\infty} \frac{\left|G\left(m^\pm\omega_i\right)\right|^2}{|m|}.$$

$$(8)$$

Here, $\rho(\omega_i)$ is the density in frequency seen by the test particle i, and $\Delta\theta^{PK}$ is the azimuthal distance in radian between the kicker and the pick-up. The factor containing the difference $\Delta\omega_i$ in revolution frequency represents "bad mixing". Because F and D are both independent of Ψ, the cooling rate can be obtained by integrating Eq. (6) using[6] relevant boundary conditions,

$$\tau^{-1} = \frac{1}{\langle I\rangle}\frac{d\langle I\rangle}{dt} = F^0 + \frac{D^0}{2}. \qquad (9)$$

The average gain G_{opt} to achieve the optimum cooling rate τ_{opt}^{-1} is then

$$G_{opt}^{-1} = \frac{\omega_0\langle\rho\left(\omega_i\right)\rangle}{2\langle n\rangle}, \quad \tau_{opt}^{-1} = \frac{\Delta n\langle n\rangle}{\pi\langle\rho\left(\omega_i\right)\rangle}, \qquad (10)$$

where $\Delta n f_0$ is the frequency bandwidth, and $\langle n\rangle$ is the average harmonic of the cooling system.

B. Bunched Beam

Consider a bunch of N_0 particles performing synchrotron oscillation with frequency Ω_i and amplitude τ_i. U_{ij} can now be expressed

$$U_{ij} = \frac{\omega_0}{2\pi} \sum_{m=-\infty}^{\infty} \sum_{l=-\infty}^{\infty} \frac{(-i)^l\, e^{il\phi_j^0}}{\sqrt{\beta_x^P \beta_x^K}} J_l\left(m\omega_0\tau_j\right) G\left(m^\pm\omega_0 - l\Omega_j\right)$$

$$\times \sum_{n=-\infty}^{\infty}\sum_{k=-\infty}^{\infty} (-i)^k\, J_k\left(n\omega_0\tau_i\right) e^{ik\phi_i^0} e^{it(m\omega_0 - l\Omega_j + n\omega_0 - k\Omega_i)},$$

$$(11)$$

where J_l is the Bessel function of lth order, and ϕ^0 is the initial phase. The coefficients of the transport equation (Eq. 6) in terms of the action variable can be evaluated

$$F^0 = -\frac{\omega_0 \sin \nu_x \Delta\theta^{PK}}{2\pi} \sum_{m=-\infty}^{\infty}\sum_{l=-\infty}^{\infty} G\left(m^\pm\omega_0 - l\Omega_i\right)$$

$$\times\, e^{il\Omega_i\Delta\theta^{PK}/\omega_0} J_l^2\left(m\omega_0\tau_i\right)$$

$$(12)$$

$$D^0 = \frac{\omega_0^2}{4\pi} \sum_{l=-\infty}^{\infty}\sum_{k=-\infty}^{\infty} \frac{\rho\left(J'\right)}{|l|\left|\frac{d\Omega(J')}{dJ'}\right|}\Bigg|_{\Omega(J')=k\Omega(J)/l}$$

$$\times\left\{\left|G_D\left(-\right)\right|^2 + \left|G_D\left(+\right)\right|^2 + 2Re\left[G_D\left(-\right)G_D\left(-\right)\right]\right\},$$

$$(13)$$

where $e^{il\Omega_i\Delta\theta^{PK}/\omega_0}$ represents "bad mixing", and

$$G_D\left(\pm\right) = \sum_{m=1}^{\infty} G\left(m^\pm\omega_0 \pm l\Omega_j\right) J_{\mp l}\left(m\omega_0\tau_j\right) J_{\mp k}\left(m\omega_0\tau_i\right).$$

$$(14)$$

Here, $\rho(J)$ is the particle density, and J represents the longitudinal phase-space area enclosed by the particle trajectory. The rate of change in synchrotron-oscillation frequency under a sinusoidal rf voltage is[6]

$$\frac{d\Omega\left(J\right)}{dJ} \approx -\frac{C_W}{8\pi}, \quad C_W = \frac{h^2\omega_0^2\eta}{2E\beta^2}, \qquad (15)$$

where h is the rf harmonic number.

The average gain to achieve the optimum cooling rate τ_{opt}^{-1} can be similarly obtained. Because the number of significant synchrotron sideband $\langle k\rangle$ ($\langle k\rangle = \langle n\rangle\omega_0\langle\tau\rangle$) is typically much larger than 1, $J_{\mp k}$ in Eq. (14) may be replaced by their asymptotic forms. Employing the identity

$$J_0^2\left(x\right) + 2\sum_{k=1}^{\infty} J_k^2\left(x\right) = 1, \qquad (16)$$

the optimum gain can be derived

$$G_{opt}^{-1} = \frac{\Delta n\omega_0\langle\rho\left(\Omega_i\right)\rangle}{\pi\langle k\rangle^2}, \quad \tau_{opt}^{-1} = \frac{\langle k\rangle^2}{2\langle\rho\left(J\right)\rangle}\left|\frac{d\Omega}{dJ}\right|. \qquad (17)$$

C. Comparison

In the case that particles in the bunch occupy the entire rf bucket,

$$\rho\left(J\right) \approx \frac{N_0}{8}\sqrt{\frac{C_W}{C_\phi}}, \quad C_\phi = \frac{qe\hat{V}}{\pi h}. \qquad (18)$$

The spread $\Delta\Omega$ in synchrotron-oscillation frequency is comparable to the zero- amplitude frequency Ω_s. Expressing the oscillation amplitude

$$\langle\tau\rangle = \frac{2}{h\omega_0}\sqrt{\frac{C_W}{C_\phi}}\langle W\rangle \approx \frac{1}{h\omega_0}, \qquad (19)$$

Eq. (17) becomes

$$\tau_{opt}^{-1} \approx \frac{\langle n\rangle^2\langle\Delta\omega_i\rangle}{2\pi h N_0}, \qquad (20)$$

where $\langle\Delta\omega_i\rangle$ is the mean spread in revolution frequency. Compared with Eq. (10), the optimum cooling rate for the bunched beam is the same as that predicted from the coasting-beam theory if $4hN_0$ is considered the effective number of particles in the ring.

In the case that particles in the bunch occupy only the center of the bucket, the spread $\Delta\Omega$ is small compared with Ω_s. Consequently, coasting-beam prediction is no longer applicable. In order to demonstrate the difference, consider the situation when the peak rf voltage is increased while the bunch area is kept constant. According to the coasting-beam theory (Eq. 10), the optimum cooling rate is unchanged because the effective density $\langle\rho(\omega_i)\rangle$ is unchanged. However, according to the bunched-beam theory (Eq. 17), the rate decreases in proportion to $\hat{V}^{-1/2}$. Cooling becomes difficult because of the effectively higher particle density in synchrotron sideband. Define the bucket filling ratio as the ratio of the bunch area to the bucket area, Fig. 1 shows the change of optimum cooling rate calculated by assuming constant bunch area, and by including synchrotron sideband overlapping.

III. CONDITION FOR BEAM INSTABILITY

A. Coasting Beam

In the previous section, the undisturbed value x^{P0} has been used (Eq. 1) as an approximation for the transverse displacement x^P. In reality, the solution x^P to the equations of motion valid to the first order in $U_{x'}$ is

$$x_i^P = x_i^{P0} - \sqrt{\beta_x^P \beta_x^K} \int_0^t dt' \cos \nu_x \omega_i (t - t') \sum_{j=1}^N x_j^P U_{ij}(t').$$

$$(21)$$

The center-of-mass displacement can be solved[3] by using the Laplace-Fourier transformation,

$$X_m(t) = \sum_{j=1}^N e^{-im\omega_j t} x_j^P = \sum_{j=1}^N \frac{e^{im\omega_j t} x_j^{P0}}{1 + \tilde{F}_m(m^\pm \omega_j)}, \quad (22)$$

where $\tilde{F}_m$ is transformed from

$$F_m(t) = \frac{\omega_0 G(m^\pm \omega_0)}{2\pi} \sum_{j=1}^N e^{im\omega_j t} \cos \nu_x \omega_j t. \quad (23)$$

Instability occurs if the denominator vanishes, i.e.

$$1 = \frac{\omega_0 G(m^\pm \omega_0) \rho(\omega_j)}{4m^\pm} + \frac{i\omega_0 G(m^\pm \omega_0)}{4\pi m^\pm} \mathcal{P} \int \frac{\rho(\omega)\, d\omega}{\omega + \omega_j},$$

$$(24)$$

where $\mathcal{P}$ denotes the Cauchy principle value. Compared with Eq. (10), Eq. (24) implies that particle motion is stable upon signal suppression if the magnitude of the gain is less than twice the optimum value G_{opt} over the entire frequency bandwidth of the cooling system.

B. Bunched Beam

Signal suppression for the bunched beam is complicated due to the coupling between different harmonics. Define the center-of-mass displacement

$$X_n(t) = \sum_{k=-\infty}^{\infty} \sum_{j=1}^N (-i)^k J_k(n\omega_0 \tau_j) e^{ik\phi_j^0} e^{-ik\Omega_j t} x_j^P. \quad (25)$$

The Laplace-Fourier transform satisfies

$$\tilde{X}_n(\nu) = \tilde{X}_n^0(\nu) - \sum_{m=-\infty}^{\infty} \tilde{X}_m(\nu) \tilde{F}_{mn}(\nu), \quad (26)$$

where $\tilde{F}_{mn}(\nu)$ is transformed from

$$F_{mn}(t) = \frac{\omega_0 G(m^\pm \omega_0)}{2\pi} \sum_{l=-\infty}^{\infty} \sum_{j=1}^N J_l(n\omega_0 \tau_j) J_l(m\omega_0 \tau_j)$$

$$\times e^{-il\Omega_j t} \cos \nu_x \omega_0 t.$$

$$(27)$$

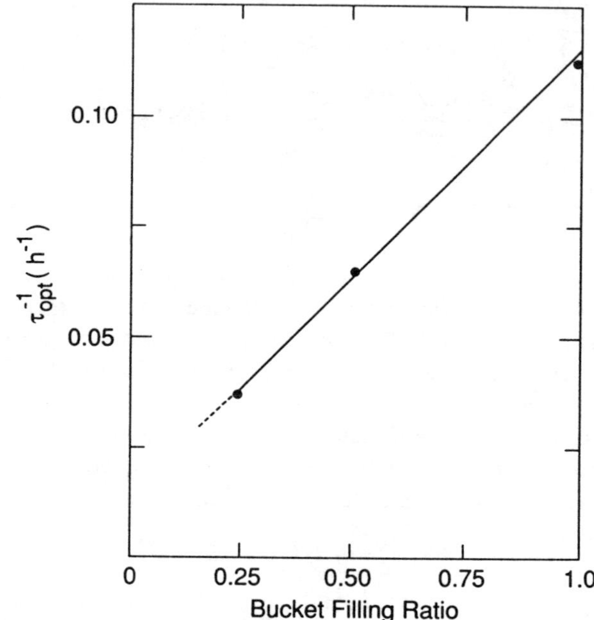

Figure 1: Optimum cooling rate for the bunched beam as a function of the bucket filling ratio.

Solving Eq. (26) is equivalent to obtaining the eigenvalues of a matrix $\mathbf{F}$, which is defined with $\tilde{F}_{mn}$ as its elements. The instability that most likely occurs corresponds to the largest eigen-value being equal to 1. Using Eq. (16), this criterion is

$$1 \approx \frac{\Delta n G(m^\pm \omega_0) \rho(\Omega_j)}{2\pi k m^\pm \tau_j} + i\frac{\Delta n G(m^\pm \omega_0)}{2\pi^2 k m^\pm \tau_j} \mathcal{P} \int \frac{d\Omega \rho(\Omega)}{\Omega - \Omega_j}.$$

$$(28)$$

Compared with Eq. (17), this again implies that particle motion is stable if the magnitude of the gain is less than twice the optimum value over the entire frequency bandwidth.

IV. ACKNOWLEDGMENT

The author would like to thank Dr. J.Marriner for helpful discussion, and Dr. A.G. Ruggiero for comments.

V. REFERENCES

[1] F. Sacherer, CERN-ISR-TH/78-11.

[2] D. Möhl, G. Petrucci, L. Thorndahl and S. van de Meer, CERN-PS/AA 79-23.

[3] Y.S. Debenev and S.A. Kheifhets, Particle Accelerators 9, 237 (1979).

[4] J. Bisognano, LBL Internal Report, Becon-21, 1981.

[5] S. Chattopadhyay, LBL-14826, 1982.

[6] J.Wei and A.G.Ruggiero, AD/RHIC-71, AD/RHIC-81 (1990), Brookhaven National Laboratory.

[7] J. Marriner, G. Jackson, D. McGinnis, R. Pasquinelli, and D. Paterson, to be published (1990).

BEAM LIFE–TIME WITH INTRABEAM SCATTERING AND STOCHASTIC COOLING[*]

J. Wei and A.G. Ruggiero

Brookhaven National Laboratory
Upton, NY 11973, USA

Abstract

A transport equation has been derived in terms of the longitudinal action variable to describe the time evolution of the longitudinal density distribution of a bunched hadron beam in the presence of intrabeam scattering and stochastic cooling. A computer program has been developed to numerically solve this equation. Both beam loss and bunch–shape evolution have been investigated for the $^{197}\mathrm{Au}^{79+}$ beams during the 10–hour storage in the Relativistic Heavy Ion Collider currently under construction at the Brookhaven National Laboratory.

I. INTRODUCTION

During the 10–hour storage of the heavy–ion beams in the RHIC, one of the major concern is the particle loss and luminosity decrease caused by Coulomb interaction between the particles in the bunch. Most of the theories[1–4] developed on the subject of intrabeam scattering (IBS) concentrate on the growth of rms beam dimensions under the assumption that particle distribution remains Gaussian in both transverse and longitudinal phase space, thus disregarding boundary limitations and particle loss. Previous studies[4, 5] on RHIC using those theories indicate that the bunch area is in most cases comparable to the rf–bucket area during the storage. It is thus expected that particle loss through the edge of the rf buckets is appreciable. Under this circumstance, intrabeam–scattering calculation without taking into account the beam loss is inadequate to describe the particle motion.

This paper presents a new approach to the problem of beam life–time based on the Fokker–Planck equation for the density distribution function Ψ of the particles in the presence of intrabeam scattering and longitudinal stochastic cooling. Part A of section II introduces the transport equation in terms of the action variable J which describes the time evolution of Ψ. The coefficients of dynamic friction and diffusion due to IBS are obtained in part B, while the coefficients of coherent correction and diffusion due to stochastic cooling are evaluated in part C. Numerical method is addressed in section III to solve the transport equation for given initial distribution and boundary conditions. Loss and instantaneous distribution of the fully–stripped gold ions during the storage in the RHIC are investigated in section IV.

___
[*]Work performed under the auspices of the U.S. Department of Energy.

II. THEORETICAL APPROACHES

A. The Transport Equation

A transport equation in terms of angle–action variables Q and J describing the evolution of longitudinal particle distribution, can be obtained by averaging the 6–dimensional Fokker–Planck equation over all transverse variables. Because the time for intrabeam scattering and stochastic cooling to produce appreciable effect is typically much longer than the synchrotron–oscillation period, which is again much longer than the correlation time of the collision process, this equation can be further reduced by averaging over the angle variable Q for one synchrotron–oscillation period

$$\frac{\partial \Psi}{\partial t} = -\frac{\partial}{\partial J}\left(F\Psi\right) + \frac{1}{2}\frac{\partial}{\partial J}\left(D\frac{\partial \Psi}{\partial J}\right),\tag{1}$$

where

$$F\left(J;t\right) = \int_0^1 dQ \left.\frac{\langle\Delta J\rangle_{C,T}}{\Delta t}\right|_0 ,\; D\left(J;t\right) = \int_0^1 dQ \left.\frac{\langle(\Delta J)^2\rangle_{C,T}}{\Delta t}\right|_0 .\tag{2}$$

Here, $\langle\;\rangle_{C,T}$ denotes the average over both collision events and transverse variables. The subscript 0 in Eq. (2) implies that the integration over Q is performed along contours of particle motion in the absence of IBS and cooling. The boundary condition to this equation is

$$\begin{cases} J = 0: & -F\Psi + \dfrac{D}{2}\dfrac{\partial \Psi}{\partial J} = 0\,, \\ J = \hat{J}: & \Psi = 0 \end{cases}\tag{3}$$

B. Intrabeam Scattering

In terms of ϕ and W, the longitudinal equations of motion at storage are

$$\dot{W} = -\frac{C_\theta}{2}\sin\phi + U_W\,,\quad \dot{\phi} = 2C_W W,\tag{4}$$

where

$$C_W = \frac{h^2\omega_0^2\eta}{2E\beta^2}\;\text{and}\;C_\phi = \frac{qe\hat{V}}{\pi h}.$$

$\hat{V}$ is the peak voltage, $\eta = 1/\gamma_T^2 - 1/\gamma^2$, γ_T is the transition energy, $E = Am_0c^2\gamma$, βc is the synchronous velocity, ω_0 is the revolution frequency, and q and A are the charge and atomic number, respectively. Regarding the rate of energy increment U_W due to IBS and cooling as a perturbation, the unperturbed particle motion is

Hamiltonian. The constant of motion J can be obtained through a canonical transformation,

$$J = 8\sqrt{\frac{C_\phi}{C_W}} \left[\left(k^2 - 1\right) \mathrm{K}\left(k\right) + \mathrm{E}\left(k\right) \right], \quad k \le 1, \qquad (5)$$

where k is[8] the normalized Hamiltonian, K and E are complete elliptical integrals of first and second kind. With the perturbation, the rate of increment in J can be expressed in terms of U_W

$$\frac{\Delta J}{\Delta t} = \left.\frac{\partial W}{\partial J}\right|_\phi^{-1} U_W, \qquad (6)$$

where

$$\left.\frac{\partial W}{\partial J}\right|_\phi^{-1} = 8k\,\mathrm{K}\left(k\right)\cos 2\pi Q\,\left[1 - 4\xi \sin^2 2\pi Q + O\left(\xi^2\right)\right]. \quad (7)$$

Here, ξ becomes significant[8] only near the separatrix.

U_W may be evaluated in the rest frame of the synchronous particle where the motion of the particles is non–relativistic. Let $|\mathbf{u}| = |\bar{\mathbf{v}}_1 - \bar{\mathbf{v}}_2|$ denote the relative velocity between the test particle 1 and a media particle 2 in the rest frame. Rutherford's formula for the cross section of the Coulomb scattering is

$$\sigma\left(u, \theta\right) = \frac{q^4 e^4}{A^2 m_0^2 u^4 \sin^4 \theta/2}, \qquad (8)$$

where θ is the angle through which the velocity vector $\mathbf{u}$ undergoes a rotation during the collision. Integrating over both θ and the azimuthal scattering angle, the change in the longitudinal component of the velocity of the test particle and its square per unit rest–frame time[6–8] can be shown as

$$\langle \Delta \bar{v}_{z1} \rangle_\Omega = -2\Gamma\,\frac{u_z}{u^3}, \quad \text{and} \quad \langle (\Delta \bar{v}_z)_1^2 \rangle_\Omega = \Gamma\,\frac{u_x^2 + u_y^2}{u^3}, \quad (9)$$

where

$$\Gamma \equiv \frac{4\pi q^4 e^4 \mathrm{Log}}{A^2 m_0^2}, \quad \mathrm{Log} \equiv -\ln \sin \frac{\theta_{min}}{2},$$

and θ_{min} is the minimum scattering angle. The Coulomb logarithm Log can be verified to be much larger than 1, which implies that the Fokker–Planck equation is a good approximation to describe the particle motion. A value of 20 is designated[3] to Log for simplicity.

Based on Eq. (9), the rate of average energy increment of the test particle in the laboratory frame is evaluated by integrating over all the velocity components of the media particles involved in the collision. If the distributions in horizontal and vertical phase space are assumed as Gaussians, F and D can be obtained by integrating over all the transverse components of the test particle,

$$F\left(J\right) = \int \frac{2dz}{\pi R} \int_0^{\frac{1}{4}} dQ \left.\frac{\partial W}{\partial J}\right|_\phi^{-1}\left(Q, J\right) \int_{J_{min}}^{\hat{J}} \left.\frac{\partial W}{\partial J}\right|_\phi\left(Q', J'\right)$$
$$\times \left[A_F\left(\lambda_-\right) + A_F\left(\lambda_+\right)\right]\Psi\left(J'\right) dJ', \qquad (10)$$

$$D\left(J\right) = \int \frac{2dz}{\pi R} \int_0^{\frac{1}{4}} dQ \left[\left.\frac{\partial W}{\partial J}\right|_\phi^{-1}\left(Q, J\right)\right]^2 \int_{J_{min}}^{\hat{J}} \left.\frac{\partial W}{\partial J}\right|_\phi\left(Q', J'\right)$$
$$\times \left[A_D\left(\lambda_-\right) + A_D\left(\lambda_+\right)\right]\Psi\left(J'\right) dJ', \qquad (11)$$

where $\lambda_\mp = \dfrac{h\omega_0 a}{\gamma\beta^2 E}\left(W \mp W'\right)$, $a = \frac{1}{2}\sqrt{\dfrac{6\beta\gamma\beta_{x,y}}{\epsilon_{Nx,y}}}$,

$$A_F\left(\lambda\right) = -\frac{2\Gamma E}{\beta^2\gamma^4 c^4}\,\frac{I_F\left(\lambda\right)}{4\pi\sigma_{x\beta}\sigma_y}, \quad A_D\left(\lambda\right) = \frac{\Gamma E^2 R}{\gamma^3\beta c^5 h}\,\frac{I_D\left(\lambda\right)}{4\pi\sigma_{x\beta}\sigma_y}. \qquad (12)$$

Here, $\sigma_{x\beta,y} = \sqrt{\beta_{x,y}\epsilon_{Nx,y}/6\beta\gamma}$ are the rms bunch widths that are time dependent. $\beta_{x,y}$ and $\alpha_{x,y}$ are the Courant–Snyder parameters. $\epsilon_{Nx,y}$ are the normalized emittances. The first integral in Eqs. (10&11) represents the average over the machine lattice; the second integral represents the average over synchrotron–oscillation period; while the third integral describes particles of different action J' involved in the collision. The integration over J' is performed such that $k(J')\sin 2\pi Q' \approx \sin[\phi(Q, J)/2]$, extending from J_{min} to the bunch edge $\hat{J}$, with $k(J_{min}) \approx [\sin \phi(Q, J)/2]$. For a round beam with $\beta_x x_p' + \alpha_x x_p \sim 0$,

$$I_F\left(\lambda\right) = 2a^2 \mathrm{sgn}\left(\lambda\right)\chi\left\{1 - \sqrt{\pi}|\lambda|e^{\lambda^2}\left[1 - \Phi\left(\lambda\right)\right]\right\}$$
$$I_D\left(\lambda\right) = a\chi\left\{\sqrt{\pi}\left(1 + 2\lambda^2\right)e^{\lambda^2}\left[1 - \Phi\left(\lambda\right)\right] - 2|\lambda|\right\}, \qquad (13)$$

where Φ is the error function. Note that $\chi = e^{-\left(x_p\gamma\lambda/2a\sigma_{x\beta}\right)^2}$ represents the damping of the dispersion x_p on the diffusion process.

C. Longitudinal Stochastic Cooling

The corresponding coefficients of Eq. (1) due to longitudinal stochastic cooling can be obtained by similar averaging procedures. The coherent correction F provided by the cooling system is[8–9]

$$F\left(J\right) = \frac{q^2 e^2 \omega_0}{\pi} \sum_{m=1}^{\infty} \sum_{l=1}^{\infty} \frac{l}{m}\mathrm{J}_l^2\left[m\omega_0\tau\left(J\right)\right]$$
$$\times Re\left[G_F\left(m, +l\right) - G_F\left(m, -l\right)\right], \qquad (14)$$

where

$$G_F\left(m, \pm l\right) = G\left[m\omega_0 \pm l\Omega\left(J\right)\right]e^{\pm il\Omega(J)\Delta\theta^{PK}/\omega_0}. \qquad (15)$$

Here, $\tau(J) = \arccos\left(1 - 2k^2\right)/h\omega_0$ is the oscillation amplitude in time, $\Omega(J)$ is the synchrotron–oscillation frequency, and $\Delta\theta^{PK}$ is the azimuthal distance between the pick–up and the kicker. The summation on the revolution bands in Eq. (13) is performed over the system bandwidth, while the summation on the synchrotron sidebands is actually performed from $l = 1$ to $m\omega_0\hat{\tau}$. The factor $e^{il\Omega(J)\Delta\theta^{PK}/\omega_0}$ represents the phase slip that non– synchronous particles experience during their passage from the pick–up to the kicker. In order to minimize this "bad mixing", the distance between the pick–up and the kicker should be for any harmonic n in the bandwidth

$$\Delta\theta^{PK} n\hat{\tau}\Omega(0) \ll 1, \qquad \hat{\tau} \equiv \tau\left(\hat{J}\right). \qquad (16)$$

The diffusion is contributed from both the particle Schottky noise and the system thermal noise. If the revolution bands are not overlapping within the system bandwidth, the Schottky coefficient D_{SH} is

$$D_{SH}(J) = \frac{Nq^4e^4\omega_0^2}{\pi}\sum_{l=1}^{\infty}\sum_{k=1}^{\infty}\frac{k^2\Psi(J')}{l\left|\frac{d\Omega(J')}{dJ'}\right|}\Bigg|_{\Omega(J')=k\Omega(J)/l}$$
$$\times\left\{|G(k,l)|^2 + |G(-k,-l)|^2 - 2Re[G(k,l)G(-k,-l)]\right\}, \qquad (17)$$

where the double summation on l and k represents synchrotron sideband overlapping, and

$$G(k,l) = \sum_{m=1}^{\infty}\frac{G[m\omega_0 + l\Omega(J')]}{m}J_k[m\omega_0\tau(J)]J_l[m\omega_0\tau(J')]. \qquad (18)$$

The thermal coefficient D_T is expressed in terms of the thermal temperature T^P at the pick–up,

$$D_T(J) = 2k_B T^P q^2 e^2 \sum_{m=1}^{\infty}\sum_{l=1}^{\infty}\frac{l^2}{m^2}J_l^2[m\omega_0\tau(J)]$$
$$\times\left[|G_T(m,+l)|^2 + |G_T(m,-l)|^2\right], \qquad (19)$$

where G_T is the gain excluding the pick–up. The thermal contribution often appears negligible for the cooling of beams of high charge–state ions.

III. COMPUTER TECHNIQUES

A computer program has been developed to numerically solve Eq. (1) with given initial distribution and boundary conditions. The instantaneous rate of change of the transverse emittance is provided by standard IBS calculations.[4] The accuracy of this numerical approach is obtained by evaluating the zeroth moment of J and then by comparing to the expected value, which is equal to 1 in the case of proper cooling without particle loss.

IV. BEAM LIFE–TIME IN RHIC

Consider the beam of one of the highest charge–state ions $^{197}Au^{79+}$ in the RHIC where intrabeam scattering is expected to be the severest. Each bunch contains 10^9 particles. The initial bunch area is 0.3eV·s/u. In order to minimize beam growth and luminosity variation, the transverse emittance is initially blown–up to a normalized value of 60πmm–mrad. Thereafter, the growth in transverse emittance is small.

In the absence of cooling, particle loss can be minimized by keeping the peak rf voltage at a constant, maximally achievable value of 4.5MV. Figure 1 shows the time evolution of Ψ during the 10–hour period of operation. The beam loss becomes significant after about 3 hours when the bunch area is comparable to the bucket area. The total beam loss is about 20%. The asymptotic distribution in longitudinal phase space is found to be Gaussian–like, independent of initial conditions.

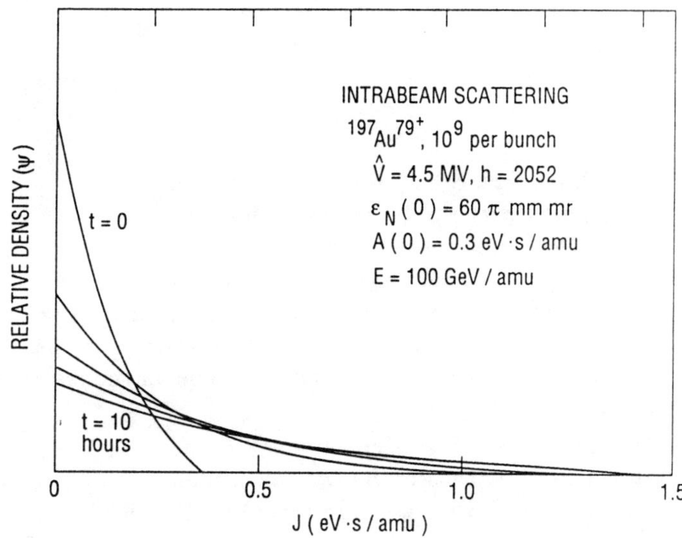

Figure 1: Evolution of the longitudinal distribution function Ψ as a function of longitudinal phase space area J. (t=0, 2.5, 5, 7.5, and 10 hour).

With a longitudinal stochastic cooling system of 4-8 GHz bandwidth, the bunch will reach a stationary state after about 3 hours. Beam loss is essentially eliminated. Because the mixing factor is much larger than 1, full–turn delay between pick–up and kicker is plausible. A factor of 3 increase in average luminosity can be achieved by applying the cooling without initially blowing up the transverse emittance.

V. ACKNOWLEDGMENT

The authors would like to thank Dr. G.Parzen for the valuable information on transverse beam growth, and Dr. J.Marriner for helpful discussion.

VI. REFERENCES

[1] A.Piwinski, CERN Accel. Sch. on General Accel. Phys., Gifsur–Yvette, Paris, 1984, p.405.
[2] M.Martini, CERN PS/84-9 (AA) (1984).
[3] J. Bjorken and S.Mtingwa, Part. Accel. **13**, 115 (1983).
[4] G.Parzen, Nucl. Instr. Meth. **A256**, 231 (1987), etc.
[5] Conceptual Design of the RHIC, BNL-52195 (1989).
[6] R.Cohen, L.Spitzer, and P.McRoutly, Phys. Rev. **80**, 230 (1950).
[7] A.H. Sorenssen, CERN Accel. Sch. Second General Accel. Phys. Course (Geneva, 1987), p.135.
[8] J.Wei and A.G.Ruggiero, AD/RHIC-71, AD/RHIC-81, Brookhaven National Laboratory (1990).
[9] S. Chattopadhyay, LBL-14826, 1982.

BETA FUNCTIONS IN THE PRESENCE OF LINEAR COUPLING*

G. Parzen

Brookhaven National Laboratory
Upton, NY 11973, USA

I. INTRODUCTION

One effect of random skew quadrupole field errors is to perturb the β–functions. This effect may be large in proton accelerators using superconducting magnets, because of the relatively large random skew quadrupole field errors that are expected in these magnets. The effect is also increased by the required insertions in proton colliders which generate large β–functions in the insertion region.

This effect has been studied in the RHIC accelerator (the Relativistic Heavy Ion Collider proposed at Brookhaven National Laboratory). For RHIC, large changes in the β–functions were found, as large as 100% increase in the β–function in one worse case.

An analytic result has been found for the changes in the beta functions caused by the random a_1. This result indicates that the important harmonics of a_1 that need to be controlled are the harmonics near $\nu_x + \nu_y$. This has been verified in computer studies. The random a_1 will also generate a large tune splitting[1,2] in RHIC. The analytic result indicates that the correction of the tune splitting will also correct a large part of the errors in the beta functions. This has been shown in computer studies. The a_1 correction system that has been developed to correct the tune splitting in RHIC appears able to correct most of the error in the beta functions.

II. RESULTS FOR THE β–FUNCTIONS

The random quadrupole field errors expected in RHIC[2] are used in this study. The effect of the random a_1, the skew quadrupole error, is to couple the x and y motions. The x motion and the y motion can each be written as the sum of 2 normal modes which have the ν–values ν_1 and ν_2. Each of the normal modes also have β–functions denoted by β_1 and β_2. For a given distribution of field errors, β_1 and β_2 can be computed using the results of Edwards and Teng.[3]

An analytic result has been found for the change in β_1 and β_2 which is valid close to the resonance $\nu_x - \nu_y = p$, p being some integer. Let β_1 be the beta function that approaches β_x, the unperturbed beta function, when $a_1 \to 0$, and similarly for β_2 and β_y. Then $(\beta_1 - \beta_x)/\beta_x$

and $(\beta_2 - \beta_y)/\beta_y$ are given by

$$(\beta_1 - \beta_x)/\beta_x = \sum_n \left[\left(\frac{\nu_1 - \nu_x}{\Delta\nu\,(\nu_1, \nu_1 - p)} \right) b_n \frac{\exp\left[i\,(n-p)\,\theta_x\right]}{n + \nu_x + \nu_y} \right] + c.c.$$

$$(\beta_2 - \beta_y)/\beta_y = \sum_n \left[\left(\frac{\nu_2 - \nu_y}{\Delta\nu\,(\nu_2 + p, \nu_2)} \right) c_n \frac{\exp\left[i\,(n+p)\,\theta_y\right]}{n + \nu_x + \nu_y} \right] + c.c.$$

$$b_n = (1/4\pi\rho) \int ds\; a_1\,(\beta_x\beta_y)^{\frac{1}{2}} \exp\left[i\,(-\,(n + \nu_y)\,\theta_x + \nu_y\theta_y)\right]$$

$$c_n = (1/4\pi\rho) \int ds\; a_1\,(\beta_x\beta_y)^{\frac{1}{2}} \exp\left[i\,(-\,(n + \nu_x)\,\theta_y + \nu_x\theta_x)\right]$$

$$\Delta\nu\,(\nu_x, \nu_y) = (1/4\pi\rho) \int ds\; a_1\,(\beta_x\beta_y)^{\frac{1}{2}} \exp\left[i\,(-\nu_x\theta_x + \nu_y\theta_y)\right]$$

$$\theta_x = \psi_x/\nu_x,\; \theta_y = \psi_y/\nu_y$$

$$\tag{1}$$

ν_x and ν_y are the unperturbed tunes. The above results will be derived in a future paper. Eqs. (1) can be written in integral form as

$$(\beta_1 - \beta_x)/\beta_x = -\frac{\nu_1 - \nu_x}{|\Delta\nu\,(\nu_1, \nu_1 - p)|}\frac{1}{2\rho\sin\pi\,(\nu_x + \nu_y)}$$

$$\times \int ds'a_1\left(s'\right)\left(\beta_x\left(s'\right)\beta_y\left(s'\right)\right)^{\frac{1}{2}}$$

$$\cos\left[\pm\pi\,(\nu_x + \nu_y) - (\nu_x + \nu_y)\,(\theta - \theta') + \nu_y\left(\theta_y' - \theta_x'\right) - \delta_1\right]$$

$$\tag{2}$$

The integral form for $(\beta_2 - \beta_y)/\beta_y$ is obtained from Eq. (2) through the substitutions in Eq. (2) of $\beta_1 \to \beta_2$, $\beta_x \to \beta_y$, $\nu_x \to \nu_y$, $\nu_y \to \nu_x$, $\delta_1 \to \delta_2$, $\Delta\nu\,(\nu_1, \nu_1 - p) \to \Delta\nu\,(\nu_2 + p, \nu_2)$, $\nu_1 \to \nu_2$, $\delta_1 = $ phase $[\Delta\nu\,(\nu_1, \nu_1 - p)]$ and $\delta_2 = $ phase $[\Delta\nu^*\,(\nu_2 + p, \nu_2)]$. For the $\pm$, the top sign is used for $\theta > \theta'$ and the bottom sign for $\theta < \theta'$.

Eq. (1) shows that the important harmonics in a_1 are the harmonics near $\nu_x + \nu_y$, and a correction system for β_1, β_2 should probably control the harmonics of a_1 near $\nu_x + \nu_y$. However, Eq. (1) shows that the dominant harmonic in $(\beta_1 - \beta_x)/\beta_x$ is the $2\nu_x$ harmonic and in $(\beta_2 - \beta_y)/\beta_y$ the $2\nu_y$ harmonic. Near the $\nu_x - \nu_y = 0$ resonance, the $2\nu_x$, $2\nu_y$ and $\nu_x + \nu_y$ harmonics are about the same. However they are different near the $\nu_x - \nu_y = p$ resonance. The above formulae appear to suggest the effects of a $\nu_x + \nu_y = $ integer sum resonance. However it may be more proper to label it as a half–integer resonance effect in the coordinate system of the normal modes.

One may note the factor $(\nu_1 - \nu_2)/\Delta\nu$. For large $\Delta\nu$, $\Delta\nu >> (\nu_x - \nu_y - p)$, then this function approaches 1. This may be seen from the result for ν_1 and ν_2 near

*Work performed under the auspices of the U.S. Department of Energy.

the $\nu_x - \nu_y = 0$ resonance.

$$\nu_1 = \bar{\nu} \pm \left[(\nu_x - \nu_y)^2 / 4 + |\Delta\nu|^2 \right]^{\frac{1}{2}},$$

$$\nu_2 = \bar{\nu} \mp \left[(\nu_x - \nu_y)^2 / 4 + |\Delta\nu|^2 \right]^{\frac{1}{2}}, \quad (3)$$

$$\bar{\nu} = (\nu_x + \nu_y)/2, \quad \Delta\nu = \Delta\nu(\bar{\nu}, \bar{\nu}).$$

For the $\pm$, the top sign is used for $\nu_x > \nu_y$ and the bottom sign for $\nu_x < \nu_y$. According to Eq. (3) $(\nu_1 - \nu_x)/\Delta\nu \to 1$ for large $\Delta\nu$ and $(\nu_1 - \nu_x)/\Delta\nu \simeq 2\Delta\nu/|(\nu_x - \nu_y)|$ for small $\Delta\nu$. For very small $\Delta\nu$, the result is not easy to compute correctly because of higher order term in a_1 that are not given in Eqs. (3). Computer studies show that correction of ν_1 and ν_2 to make them closer to ν_x, ν_y tends to decrease $(\beta_1 - \beta_x)/\beta_x$ and $(\beta_2 - \beta_y)/\beta_y$ which may be due in part to this $(\nu_1 - \nu_x)/\Delta\nu$ factor.

Table 1: Maximum β_1, β_2 for 10 distributions of random a_1 fields for the RHIC $\beta^* = 2$ lattice. No corrections are present.

Error Field	$\beta_{1,max}$ at QF (m)	$\beta_{2,max}$ at QD (m)
1	68	75
2	138	95
3	89	78
4	83	74
5	69	65
6	65	67
7	76	82
8	78	79
9	80	78
10	77	87

One may also note that $(\beta_1 - \beta_x)/\beta_x$ is linear in a_1 for large $\Delta\nu$, and quadratic in a_1 for small $\Delta\nu$.

A result for the rms value of $(\beta_1 - \beta_x)/\beta_x$ due to random distribution of a_1 errors may be obtained from the integral form Eq. (2), for the case when $|\Delta\nu| >> |\nu_x - \nu_y - p|$. In this case $|\nu_1 - \nu_x|/|\Delta\nu| \simeq 1$ and

$$\left(\frac{\beta_1 - \beta_x}{\beta_x} \right)^2_{rms} = \sum_k \left(\frac{\beta_1 - \beta_x}{\beta_x} \right)^2_{k,rms}$$

$$\left(\frac{\beta_1 - \beta_x}{\beta_x} \right)_{k,rms} = N_k^{1/2} \frac{\left((\beta_x \beta_y)^{1/2} a_{1,rms} \right)_k}{2.8\rho \sin \pi (\nu_x + \nu_y)} \quad (4)$$

where the index k indicates the different types of magnets. N_k is the number of magnets of a certain type. Eq. (4) also gives the result for $((\beta_2 - \beta_y)/\beta_y)_{rms}$. One also sees that

$$((\beta_1 - \beta_x)/\beta_x)_{rms} = [4\pi/(2.8 \sin \pi (\nu_x + \nu_y))] \Delta\nu_{rms} \quad (5)$$

where $\Delta\nu_{rms}$ is the rms value of $\Delta\nu$.

Table 2: Maximum β_1 at QF and the maximum β_2 at QD for 10 distributions of random a_1 fields, for the RHIC $\beta^* = 2$ lattice, showing the effect of correcting the tune splitting on β_1, β_2.

Error Field	2 Family Correction System		Enlarged a_1 Correction System	
	$\beta_{1,max}$	$\beta_{2,max}$	$\beta_{1,max}$	$\beta_{2,max}$
1	58	58	57	56
2	57	54	59	74
3	56	59	56	59
4	63	57	61	56
5	63	60	63	60
6	55	58	58	62
7	102	84	61	59
8	60	57	64	60
9	58	59	58	58
10	60	78	57	56

Table 1 lists the beta functions, before any correction of the random a_1 fields for the RHIC $\beta^* = 2$ lattice for ten distributions of the random a_1. The table lists the maximum β_1 found at the normal focusing quadrupoles, QF, and the maximum β_2 at the normal defocusing quadrupoles, QD. The unperturbed value of the maximum β_x and β_y is 50 m. One sees that change in the beta functions of the order of 100% are computed. β_1 and β_2 were computed using the exact results of Edwards and Teng.[3]

Using Eq. (4), one can compute the expected rms value of $(\beta_1 - \beta_x)/\beta_x$ which gives for RHIC $((\beta_1 - \beta_x)/\beta_x)_{rms} = 0.29$. For 72 QF magnets, the 90% probability result is $\beta_{1,max} = \beta_{2,max} = 95$. This is in fairly good agreement with the results in Table 1. The exceptionally large $\beta_{1,max} = 138$ for seed 2 is due to ν_1 being shifted close to the $\nu_1 = 29$ resonance where β_1 would become infinite. This effect of the $\nu_1 = 29$ resonance is not included in our results for $(\beta_1 - \beta_x)/\beta_x$.

III. CORRECTION OF β_1, β_2

Computer studies indicate that when the tune shifts $\nu_1 - \nu_x$ and $\nu_2 - \nu_y$ are corrected, then $\beta_1 - \beta_x$ and $\beta_2 - \beta_y$ are also reduced. This is indicated by the analytical result Eqs. (1). It is partly due to the $(\nu_1 - \nu_x)/\Delta\nu$, $(\nu_2 - \nu_y)/\Delta\nu$ factors in Eqs. (1). It is also partly due to the driving terms b_n, c_n in Eq. (1), which are also important driving terms for the tune splitting.[4] In RHIC the tune splitting correction includes a 2 family a_1 correction system, that corrects the different resonance, $\nu_x - \nu_y = 0$, and most of the tune splitting, and an enlarged a_1 correction system[4] to correct the residual tune splitting by controlling the harmonics of a_1 near $\nu_x + \nu_y$.

Table 2 shows what happens to β_1, β_2 when the tune splitting is corrected by the above 2 family tune splitting correction systems. One sees that the $\beta_1 - \beta_x$, $\beta_2 - \beta_y$ errors are considerably reduced. The results in Table 2, include the effects of random b_1 field which can generate a 20% error in β_1 and β_2. However, a considerable error in β_1, β_2 remains for some distributions.

Table 2 also shows what happens to β_1, β_2 when the tune splitting is further reduced using the enlarged a_1 correction system to control harmonics near $\nu_x + \nu_y$. One sees a considerable improvement in β_1 and β_2.

The results for β_1, β_2 can be further improved by adjusting the harmonics of a_1 near $\nu_x + \nu_y$, or by moving the ν-values slightly closer to the resonance line $\nu_x - \nu_y = 0$. At this level, the correction of the tune splitting and the β_1, β_2 correction may be competing with each other.

Using these procedures the largest error in β_1, β_2 was reduced to 28%, which includes the 20% effect due to the random b_1 which has not been corrected.

IV. REFERENCES

[1] G. Parzen, BNL Report AD/RHIC–AP–72 (1988).

[2] G. Parzen, BNL Report AD/RHIC–82 (1990).

[3] D. Edwards and L. Teng, IEEE 1973 PAC, p. 885.

[4] G. Parzen, Tune splitting in the presence of linear coupling, these proceedings.

DYNAMIC APERTURE EFFECTS DUE TO LINEAR COUPLING*

G. Parzen

Brookhaven National Laboratory
Upton, NY 11973, USA

I. INTRODUCTION

The coupling introduced by the random a_1 can produce considerable distortion of the betatron motion. For a given initial x_o, x'_o, y_o, y'_o, the maximum x and the maximum y for the subsequent motion can be considerably larger when coupling is present. The maximum x and y can be used as a measure of the betatron distortion. One effect of this betatron distortion shows up in the dynamic aperture, in a loss in the stability limit, A_{SL}, found by tracking. The betatron distortion causes the particle to move further out in the magnets, where it sees stronger non–linear fields. Previous tracking[1,2] showed a loss in A_{SL} due to random a_1 and b_1. It was noticed then that the loss in A_{SL} was associated with a betatron motion distortion which was primarily a linear effect. For a given initial x, x', y, y' the x_{max} and y_{max} in the high–β magnets were considerably larger (about 30% larger) for those random a_1 distributions which produced the smaller A_{SL}. The studies described below show that the stability limit, A_{SL}, depends on the starting location around the ring. This variation in A_{SL} around the ring can be correlated with the variation in the betatron distortion for particles starting at different places around the ring. It is proposed that the average of the A_{SL} found by starting at each of the QF, the focusing quadrupoles in the arcs, can be taken as a measure of the dynamic aperture. It is found that the average A_{SL} is reduced by about 15% by the random a_1 multipoles expected in RHIC.

Computing the stability limit A_{SL} is made more difficult by the dependence of A_{SL} on the starting location around the ring. In order to avoid having to compute A_{SL} for all possible starting locations at a QF, one can make use of an observed correlation between A_{SL} and a linear parameter, CDF, which is called the coupling distortion function, and which is defined below. Roughly, the CDF is a measure of how large x or y may become because of the presence of coupling for a given set of starting conditions x_o, x'_o, y_o, y'_o. The CDF being a linear parameter can be computed quickly starting at each QF in the ring. The QF with the largest CDF is found to have the smallest A_{SL}, and the smallest CDF corresponds to the largest A_{SL}. After finding the QF with the maximum and minimum CDF, one can compute the A_{SL} for these two starting locations with tracking runs. It is proposed that

the average A_{SL} around the ring be approximated by taking the average of these two values of A_{SL} found at the locations where the CDF has its maximum and minimum.

II. CORRELATION OF CDF AND A_{SL}

The coupling distortion function, CDF, is defined by

$$\text{CDF} = \frac{x_{max}(s)}{x_{max,nc}(s)} \tag{1}$$

for a given initial x, x', y, y' starting at $s = s_o$, and in the absence of non–linear fields. $x_{max,nc}$ is the x_{max} in the absence of coupling. A similar CDF exists also for the y–motion. The CDF depends on s and also on the starting conditions of x, x', y, y'. Since we are interested in the correlation between the CDF and the A_{SL} found in tracking studies, we will compute the CDF for the starting conditions $x' = y' = 0$, $\epsilon_x = \epsilon_y$, and at the s which corresponds to the location of the high–β quadrupoles in the insertions.

To compute CDF, one has to compute $x_{max}(s)$ in the presence of coupling. An analytical result was found for $x_{max}(s)$ for a given starting x, x', y, y'. This result is given in Section III.

Table 1: CDF at High β Quads for a particle starting at the focusing quadrupole at the middle of the first arc in RHIC for the RHIC $\beta^* = 2$ Lattice.

Field Error No.	CDF	β_1 (m)	β_2 (m)	A_{SL} (mm)
1	1.32	695	684	8.5
2	1.46	650	644	6.5
3	1.30	682	744	8.5
4	1.55	723	797	7.5
5	1.10	766	752	8.5
6	1.38	625	654	8.5
7	2.10	956	970	4.5
8	1.63	816	764	6.5
9	1.30	687	655	8.5
10	1.35	797	670	7.5

Table 1 lists the coupling distortion function, CDF, and the β–functions of the normal modes, β_1 and β_2 for ten different distributions of the random field errors expected in RHIC.[3] The CDF is the largest CDF found at any of the high β quadrupoles in the insertions for

*Work performed under the auspices of the U.S. Department of Energy.

Table 2: The coupling distortion function, CDF, starting at different QF for a RHIC lattice with $\beta^* = 6$ and for a_1 errors corresponding to Seed 8.

QF Element No.	CDF	QF Element No.	CDF
1	1.86	113	1.67
15	1.48	127	1.67
29	1.81	141	1.62
43	1.52	145	1.67
57	1.81	169	1.62
71	1.57	1299	1.28
85	1.71	1313	2.00
99	1.62		

Table 3: The coupling distortion function, CDF, and the A_{SL} found starting at different Q for a RHIC lattice with $\beta^* = 6$ and Seed 8 a_1 errors.

QF Element No.	CDF	A_{SL} (mm)
1313	2.00	10.5
1	1.86	11.5
85	1.71	12.5
15	1.48	13.5
1299	1.28	14.5

either x or y motion for a particle starting at the focusing quadrupole in the middle of the first arc in RHIC. The β_1, β_2 are the largest β–functions of the normal modes found around the accelerator. For $a_1 = b_1 = 0$, the largest β functions are $\beta = 220$ m for $\beta^* = 6$ and $\beta^* = 630$ for $\beta^* = 2$. A CDF as large as CDF $= 2.1$ is found for one field error distribution. A simple picture of coupling would suggest CDF $\simeq 1.4$ resulting from a complete exchange of the emittance between the x and y motions. The betatron distortion is about 40% larger for the worst case than what one would expect from the simple picture of a complete exchange of the emittances.

There is some correlation between CDF and β_1 and β_2. However, the CDF depends not only on β_1, β_2 but also on the mismatch in the emittances. For the given initial ϵ_x, ϵ_y, the corresponding ϵ_1, ϵ_2 may be larger than ϵ_x, ϵ_y where ϵ_1, ϵ_2 are the emittances for the normal modes.

Table 1 also shows the correlation between the CDF and the dynamic aperture which is measured by A_{SL}. A_{SL} is found from tracking studies with the random field errors, including higher multipoles, present. A_{SL} is the largest stable initial x with the starting conditions $\epsilon_x = \epsilon_y$, $x' = y' = 0$, and starting at the focusing quadrupole at the middle of the first arc in RHIC.

The correlation between the coupling distortion function, CDF, and the dynamic aperture A_{SL} is good. Note that the A_{SL} are computed after the coupling has been corrected[3] using the a_1 correctors in the insertions.

In the absence of the random a_1, b_1, the stability limit has been found to be $A_{SL} = 7.5$ mm for $\beta^* = 2$. The results in Table 1 indicate that for this particular starting place around the ring, A_{SL} has been reduced to $A_{SL} = 4.5$ mm for $\beta^* = 2$ m.

The above result for A_{SL} is misleading. It was found at a QF where the coupling distortion function, CDF, has a large value. If the tracking study to find A_{SL} is done starting on a different QF, a different value of A_{SL} will be found which is inversely correlated with the CDF at that QF.

If the particle is started at different QF, the value of CDF and A_{SL} will change at each QF. Table 2 shows this variation in CDF at 14 different QF. The CDF listed is the largest CDF found at high–β quadrupoles in the insertions, when the particle is started at different QF.

A_{SL} will also vary around the ring, since A_{SL} and CDF are correlated. In Table 3 A_{SL} and CDF are listed for a particle starting at 5 different QF in the ring. These 5 points include the largest and smallest CDF found in the ring. A_{SL} is computed at the QF at which the particle is started.

Since A_{SL} now depends on which QF the particle is started at, one has the problem of which value of A_{SL} is to be used for the dynamic aperture. The procedure proposed here is to use the average A_{SL} found from the A_{SL} computed at all the QF in the ring.

To compute $A_{SL,av}$ the average A_{SL}, one should in principle compute A_{SL} at all the QF in the ring and then compute the average A_{SL}. This would require a good deal of effort. Instead, the following procedure was used. The coupling distortion function at the high β quadrupole is computed starting at each QF in the ring. One then locates the QF that give the largest and smallest CDF, CDF_{max} and CDF_{min}. The stability limit A_{SL} is computed starting at these two QF. It is assumed that this gives the largest and smallest A_{SL}, $A_{SL,max}$ and $A_{SL,min}$. The average A_{SL} is then computed as

$$A_{SL,av} = \frac{1}{2} \left(A_{SL,max} + A_{SL,min} \right)$$

The computation of $A_{SL,av}$ for the two worst field errors, errors 7 and 8, are shown in Table 4 for two RHIC lattices having $\beta^* = 6$ and $\beta^* = 2$.

The loss in A_{SL} is now 15%. A_{SL} has been decreased from 15.5 mm to 13 mm, for $\beta^* = 6$ from 7.5 mm to 6.5 mm for $\beta^* = 2$.

Table 4: Computation of $A_{SL,av}$.

β^*	Error No.	A_{SL} max	A_{SL} min	A_{SL} av	CDF Max	CDF Min
6	7	14.5	11.5	13	1.67	1.29
6	8	16.5	10.5	13.5	2.0	1.24
2	7	8.5	4.5	6.5	2.1	1.10
2	8	8.5	6.5	7	1.8	1.21

III. ANALYTICAL RESULTS FOR X_{max} and Y_{max}

In this section, analytical results are given for the maximum x and y, x_{max} and y_{max}, that will be reached for a particle with a given initial x, x', y, y' in the presence of coupling. These results are needed in order to compute the coupling distortion function, CDF. Derivations of these results will be given in a future paper.

Edwards and Teng showed how to transfer to a new set of coordinates v, v', u, u' which are uncoupled.[5] These new normal coordinates are related to x, x', y, y' by a 4×4 matrix R

$$x = R\,v\,.$$

The normal coordinates have emittances ϵ_1 and ϵ_2 and β–functions β_1 and β_2. ϵ_1 and ϵ_2 are invariants. The x and y motion can be written as the sum of these two normal modes.

β_1 and β_2 and the R matrix can be computed from the one turn transfer matrix.[5] For RHIC, β_1 and β_2 can be considerably larger than β_x, β_y by as much as 100% in the worse case found. However, the x_{max}, y_{max} for a given initial x, x', y, y' are not simply related to β_1, β_2.

For a given set of initial x, x', y, y', x_{max} and y_{max} are given by

$$
\begin{aligned}
x_{max} &= (\beta_1 \epsilon_1)^{\frac{1}{2}} \cos \phi + (\beta_{x,2}\epsilon_x)^{\frac{1}{2}} \sin \phi, \\
\beta_{x,2} &= \overline{D}_{11}^2 \beta_2 + \overline{D}_{12}^2 \gamma_2 + 2\overline{D}_{11}\overline{D}_{12}\alpha_2, \\
y_{max} &= (\beta_2 \epsilon_2)^{\frac{1}{2}} \cos \phi + (\beta_{y,1}\epsilon_1)^{\frac{1}{2}} \sin \phi, \\
\beta_{y,1} &= D_{11}^2 \beta_1 + D_{12}^2 \gamma_1 + 2D_{11}D_{12}\alpha_1\,.
\end{aligned}
\tag{2}
$$

D and $\overline{D}$ are 2×2 matrices, which together with ϕ define the 4×4 R matrix R is given by

$$
R = \begin{pmatrix} I \cos \phi & D \sin \phi \\ -\overline{D} \sin \phi & I \cos \phi \end{pmatrix}
\tag{3}
$$

$\overline{D} = D^{-1}$, $|D| = 1$ and I is the 2×2 identity matrix.

β_1, α_1, γ_1 and β_2, α_2, γ_2 are the orbit parameters of the normal modes, ϵ_1 and ϵ_2 are the emittances of the normal modes that correspond to the initial x, x', y, y'.

One can also find expressions for x'_{max} and y'_{max}. These are given by

$$
\begin{aligned}
x'_{max} &= (\gamma_1 \epsilon_1)^{\frac{1}{2}} \cos \varphi + (\gamma_{x,2}\epsilon_2)^{\frac{1}{2}} \sin \phi, \\
\gamma_{x,2} &= \overline{D}_{21}^2 \beta_2 + \overline{D}_{22}^2 \gamma_2 + 2\overline{D}_{21}\overline{D}_{22}\alpha_2, \\
y'_{max} &= (\gamma_2 \epsilon_2)^{\frac{1}{2}} \cos \varphi + (\gamma_{y,1}\epsilon_1)^{\frac{1}{2}} \sin \phi, \\
\gamma_{y,1} &= D_{21}^2 \beta_1 + D_{22}^2 \gamma_1 + 2D_{21}D_{11}\alpha_1\,.
\end{aligned}
\tag{4}
$$

Edwards and Teng[5] describe how to compute the R matrix and β_1, α_1, γ_1 and β_2, α_2, γ_2 from the one turn transfer matrix. This then allows one to compute ϵ_1 and ϵ_2 from the initial x, x', y, y'. Equation (2) can then be used to compute x_{max} and y_{max} for the given initial x, x', y, y', which is needed to compute the coupling distortion function, CDF (see Eq. (1)).

IV. REFERENCES

[1] G. Parzen, AD/RHIC–AP–80 (1989).
[2] G. Parzen, AD/RHIC–72 (1990).
[3] G. Parzen, AD/RHIC–AP–72 (1988).
[4] G. Parzen, AD/RHIC–82 (1990).
[5] D. Edwards and L. Teng, IEEE 1973 PAC, p. 885.

BROWN'S TRANSPORT UP TO THIRD ORDER ABERRATION BY ARTIFICIAL INTELLIGENCE

Xia Jiawen , Xie Xi , Qiao Qingwen

Institute of Modern Physics , Academia Sinica

P.O. Box 31 , Lanzhou , China

Abstract

Brown's TRANSPORT [1] is a first and second order matrix multiplication computer program intended for the design of accelerator beam transport systems, neglecting the third order aberration. Recently a new method was developed [3] to derive analytically any order aberration coefficients of general charged particle optic system, applicable to any practical systems, such as accelerators, electron microscopes, lithographs, etc., including those unknown systems yet to be invented. An artificial intelligence program in Turbo Prolog was implemented on IBM-PC 286 or 386 machine to generate automatically the analytical expression of any order aberration coefficients of general charged particle optic system. Based on this new method and technique, Brown's TRANSPORT is extended beyond the second order aberration effects by artificial intelligence, outputing automatically all the analytical expressions up to the third order aberration coefficients.

INTRODUCTION

Among all charged particle optic theories, K.L. Brown's first and second order matrix theory [1] is the most successful one, and the TRANSPORT program [2] based on Brown's theory has already been applied to most areas for beam transport calculation. But with the development of accelerator and other physical experiment technique, the beam quality will be needed to be much finer. Thus in beam transport calculation, the third and more high order aberration would be considered besides the first and second order approximation.

Rencently a new method was developed[3] to derive analytically any order aberration coefficients of general charged particle system by artificial intelligence, which is realized in IBM-PC 286 or 386 computer. Based on the new method and technique, this paper extends Brown's TRANSPORT up to third order aberration effects, and gives out the analytical expressions of third order aberration coefficients for charged particle in a static magnetic field. As the method in this paper is general, according to practical requirement, we can extend Brown's TRANSPORT up to any order aberration effects by artificial intelligence. In the future, if a numerical calculating program is written based on these any order aberration coefficient expressions, then, this new methed would have enormous applied area. Especially, with the application of artificial intelligence, pepole not only can calculate any order aberration effects, but also can be further saved out from the chore of calculations.

DERIVING THE THIRD ORDER EQUATION OF MOTION

In K.L. Brown's theory[1], charged particle is moving in the static magnetic field with midplane symmetry, and in a curvilinear coordinate system (x,y,t), the component differential equations of motion are derived[1]:

$$
\begin{cases}
x'' - h(1 + hx) - \frac{x'}{(T')^2}[x'x'' + y'y'' + (1 + hx) \\
\quad \cdot (hx' + h'x)] = \frac{e}{p}T'[y'B_t - (1 + hx)B_y], \\
y'' - \frac{y'}{(T')^2}[x'x'' + y'y'' + (1 + hx)(hx' + h'x)] \\
\quad = \frac{e}{p}T'[(1 + hx)B_x - x'B_t],
\end{cases}
\tag{1}
$$

where

$$
(T')^2 = x'^2 + y'^2 + (1 + hx)^2.
\tag{2}
$$

Note that in this form, no approximation is made, and the equations of motion are still valid to all orders in the variables x,y and their derivatives. The magnetic scalar potential is:

$$
\varphi(x, y, t) = \sum_{m=0}^{\infty} \sum_{n=0}^{\infty} A_{2m+1,n} \frac{x^n}{n!} \frac{y^{2m+1}}{(2m + 1)!},
\tag{3}
$$

giving the magnetic field:

$$
\begin{cases}
B_x = \frac{\partial \varphi}{\partial x} = \sum_{m=0}^{\infty} \sum_{n=0}^{\infty} A_{2m+1,n+1} \frac{x^n}{n!} \frac{y^{2m+1}}{(2m+1)!}, \\
B_y = \frac{\partial \varphi}{\partial y} = \sum_{m=0}^{\infty} \sum_{n=0}^{\infty} A_{2m+1,n} \frac{x^n}{n!} \frac{y^{2m}}{(2m)!}, \\
B_t = \frac{1}{(1+hx)} \frac{\partial \varphi}{\partial t} = \frac{1}{(1+hx)} \\
\quad \cdot \sum_{m=0}^{\infty} \sum_{n=0}^{\infty} A'_{2m+1,n} \frac{x^n}{n!} \frac{y^{2m+1}}{(2m+1)!}.
\end{cases}
\tag{4}
$$

In K.L. Brown's transport theory, the equation of motion (1) is approximated up to second order. Now it is approximated up to third order as follows:

$$\begin{cases}
x'' - h(1 + hx) - h'xx' - hx'^2 + hh'x^2x' \\
\quad - h^2(n-2)xx'^2 + h^2nx'yy' - hx'^2\delta \\
\quad = \frac{c}{p}T'[y'B_t - (1 + hx)B_y], \\
y'' - h'xy' - hx'y' + hh'x^2y' \\
\quad - h^2(n-2)xx'y' + h^2nyy'^2 - hx'y'\delta \\
\quad = \frac{c}{p}T'[(1 + hx)B_x - x'B_t],
\end{cases} \tag{5}$$

where it is enough for B_x and B_y to be approximated through third order in x and y and their derivatives, and for B_t through second order by equations (4) to get the following expressions.

$$\begin{cases}
B_x = A_{11}y + A_{12}xy + \frac{1}{2}A_{13}x^2y + \frac{1}{6}A_{31}y^3, \\
B_y = A_{10} + A_{11}x + \frac{1}{2}A_{12}x^2 + \frac{1}{2}A_{30}y^2 \\
\quad + \frac{1}{6}A_{13}x^3 + \frac{1}{2}A_{31}xy^2, \\
B_t = A'_{10}y + (A'_{11} - hA'_{10})xy.
\end{cases} \tag{6}$$

After some algebra, we finaly obtain the equation of motion approximated up to third order in x, y and their derivatives:

$$\begin{cases}
\begin{aligned}
x'' &= (n-1)h^2x + h\delta + (2n-1-\beta)h^3x^2 \\
&+ h'xx' + (2-n)h^2x\delta + \frac{1}{2}hx'^2 - \frac{1}{2}a_{30}y^2 + h'yy' \\
&- \frac{1}{2}hy'^2 - h\delta^2 + (n-2\beta-\gamma)h^4x^3 - hh'x^2x' \\
&+ (1-2n+\beta)h^3x^2\delta + (\frac{3}{2}n-2)h^2xx'^2 \\
&- \frac{1}{2}(a_{31} + 2ha_{30})xy^2 + \frac{1}{2}nh^2xy'^2 - (2nhh' \\
&+ n'h^2)xyy' + (n-2)h^2x\delta^2 + \frac{3}{2}hx'^2\delta \\
&- nh^2x'yy' + \frac{1}{2}a_{30}y^2\delta - h'yy'\delta + \frac{1}{2}hy'^2\delta + h\delta^3,
\end{aligned} \\[2ex]
\begin{aligned}
y'' &= -nh^2y - 2(n-\beta)h^3xy - h'x'y + h'xy' \\
&+ hx'y' + nh^2y\delta + (3\gamma+4\beta-n)h^4x^2y - hh'x^2y' \\
&+ (2nhh' + n'h^2)xx'y + (n-2)h^2xx'y' \\
&+ 2(n-\beta)h^3xy\delta - \frac{1}{2}nh^2x'^2y + h'x'y\delta + hx'y'\delta \\
&+ \frac{1}{6}a_{31}y^3 - \frac{3}{2}nh^2yy'^2 - nh^2y\delta^2.
\end{aligned}
\end{cases} \tag{7}$$

where

$$a_{30} = h'' + nh^3 - 2\beta h^3,$$

$$a_{31} = 4n'hh' + 2nh'^2 + 2nhh'' + n''h^2 + 2hh''$$
$$+ h'^2 - 6\gamma h^4 - 2\beta h^4 - nh^4. \tag{8}$$

OBTAINING THE THIRD ORDER ABERRARION EXPRESSIONS

By the new developed method [3], for general charged particle system with general third order equation of motion:

$$\ddot{x}^i(t) = \alpha^i_{j_1}(t)x^{j_1}(t) + \alpha^i_{j_1j_2}(t)x^{j_1}(t)x^{j_2}(t)$$
$$+ \alpha^i_{j_1j_2j_3}(t)x^{j_1}(t)x^{j_2}(t)x^{j_3}(t), \tag{9}$$

we have the general solution:

$$x^i(t) = R^i_{j_1}(t)x_0^{j_1} + T^i_{j_1j_2}(t)x_0^{j_1}x_0^{j_2} + T^i_{j_1j_2j_3}(t)x_0^{j_1}x_0^{j_2}x_0^{j_3}, \tag{10}$$

where the first order aberration coefficient is the wellknown transfer matrix $R^i_{j_1}(t)$, and the second order aberration coefficient is the tensor:

$$T^i_{j_1j_2}(t) = R^i_l(t)\int_0^t [R^{-1}(\tau)]^l_j \alpha^j_{k_1k_2}(\tau)R^{k_1}_{j_1}(\tau)R^{k_2}_{j_2}(\tau)d\tau, \tag{11}$$

and the third order aberration coefficient is the tensor:

$$T^i_{j_1j_2j_3}(t) = R^i_l(t)\int_0^t [R^{-1}(\tau)]^l_j \{2\alpha^j_{k_1k_2}(\tau)R^{k_1}_{j_1}(\tau)T^{k_2}_{j_2j_3}(\tau)$$
$$+ \alpha^j_{k_1k_2k_3}(\tau)R^{k_1}_{j_1}(\tau)R^{k_2}_{j_2}(\tau)R^{k_3}_{j_3}(\tau)\}d\tau. \tag{12}$$

For Brown's TANSPORT up to third order aberration as discussed above, we have the following parameters from the equation of motion (7):

$$R(t) = \begin{pmatrix}
C_x(t) & S_x(t) & 0 & 0 & D_x(t) \\
C'_x(t) & S'_x(t) & 0 & 0 & D'_x(t) \\
0 & 0 & C_y(t) & S_y(t) & 0 \\
0 & 0 & C'_y(t) & S'_y(t) & 0 \\
0 & 0 & 0 & 0 & 1
\end{pmatrix},$$

$$R^{-1}(t) = \begin{pmatrix}
S'_x(t) & -S_x(t) & 0 & 0 & S_x(t)D'_x(t) - D_x(t)S'_x(t) \\
-C'_x(t) & C_x(t) & 0 & 0 & -C_x(t)D'_x(t) + D_x(t)C'_x(t) \\
0 & 0 & S'_y(t) & -S_y(t) & 0 \\
0 & 0 & -C'_y(t) & C_y(t) & 0 \\
0 & 0 & 0 & 0 & 1
\end{pmatrix}, \tag{13}$$

$$\alpha^2_1 = (n-1)h^2, \ \alpha^2_5 = h, \ \alpha^2_{11} = (2n-1-\beta)h^3, \ \alpha^2_{12} = h',$$
$$\alpha^2_{15} = (2-n)h^2, \ \alpha^2_{22} = \frac{1}{2}h, \ \alpha^2_{33} = \frac{-1}{2}a_{30}, \ \alpha^2_{34} = h',$$
$$\alpha^2_{44} = \frac{-1}{2}h, \ \alpha^2_{55} = -h, \ \alpha^2_{111} = (n-2\beta-\gamma)h^4,$$
$$\alpha^2_{112} = -hh', \ \alpha^2_{115} = (1-2n+\beta)h^3, \ \alpha^2_{122} = (\frac{3}{2}n-2)h^2,$$
$$\alpha^2_{133} = -\frac{1}{2}(a_{31} + 2ha_{30}), \ \alpha^2_{134} = -(2nhh' + n'h^2),$$
$$\alpha^2_{144} = \frac{1}{2}nh^2, \ \alpha^2_{155} = (n-2)h^2, \ \alpha^2_{225} = \frac{3}{2}h, \ \alpha^2_{234} = -nh^2,$$
$$\alpha^2_{335} = \frac{1}{2}a_{30}, \ \alpha^2_{345} = -h', \ \alpha^2_{445} = \frac{1}{2}h, \ \alpha^2_{555} = h,$$
$$\alpha^4_3 = -nh^2, \ \alpha^4_{13} = -2(n-\beta)h^3, \ \alpha^4_{23} = -h', \ \alpha^4_{14} = h',$$
$$\alpha^4_{24} = h, \ \alpha^4_{35} = nh^2, \ \alpha^4_{113} = (3\gamma+4\beta-h)h^4,$$
$$\alpha^4_{123} = (2nhh' + n'h^2), \ \alpha^4_{135} = 2(n-\beta)h^3,$$
$$\alpha^4_{235} = h', \ \alpha^4_{223} = \frac{-1}{2}nh^2, \ \alpha^4_{344} = \frac{-3}{2}nh^2, \ \alpha^4_{114} = -hh',$$
$$\alpha^4_{124} = (n-2)h^2, \ \alpha^4_{245} = h, \ \alpha^4_{355} = -nh^2, \ \alpha^4_{333} = \frac{1}{6}a_{31}. \tag{14}$$

Inserting the parameters in (13) and (14) into expressions (11) and (12), we get all Brown's transport aberration coefficients up to third order approximation. For lack of space, only some typical third order aberration coefficients are typed out by Artificial Intelligence in the following:

$$
\begin{aligned}
T(1;1,1,1) =\ & Cx(t)\int\{(-Sx(t))[(n-2\beta-\gamma)h^4 Cx(t) \\
& \cdot Cx(t)Cx(t)+(1/3)(-hh')(Cx(t)Cx(t)C'x(t) \\
& +Cx(t)C'x(t)Cx(t)+C'x(t)Cx(t)Cx(t)) \\
& +(1/3)((3/2)n-2)h^2(Cx(t))C'x(t)C'x(t) \\
& +C'x(t)Cx(t)C'x(t)+C'x(t)C'x(t)Cx(t))]\} \\
& +Sx(t)\int\{Cx(t)[(n-2\beta-\gamma)h^4 Cx(t)Cx(t)Cx(t) \\
& +(1/3)(-hh')(Cx(t)Cx(t)C'x(t)+Cx(t)C'x(t) \\
& \cdot Cx(t)+C'x(t)Cx(t)Cx(t)) \\
& +(1/3)((3/2)n-2)h^2(Cx(t)C'x(t)C'x(t) \\
& +C'x(t)Cx(t)C'x(t)+C'x(t)C'x(t)Cx(t))]\},
\end{aligned}
\tag{15}
$$

$$
\begin{aligned}
T(4;1,2,3) =\ & C'y(t)\int\{(-Sy(t))6[(1/3)(3\gamma+4\beta-n)h^4 \\
& \cdot(Cx(t)Sx(t)Cy(t)) \\
& +(1/3)(-hh')(Cx(t)Sx(t)C'y(t))+(1/6)(2nhh' \\
& +n'h^2)(Cx(t)S'x(t)Cy(t)+C'x(t)Sx(t)Cy(t)) \\
& +(1/6)(n-2)h^2(Cx(t)S'x(t)C'y(t)+C'x(t)Sx(t) \\
& \cdot C'y(t))+(1/3)(-1/2)nh^2(C'x(t)S'x(t)Cy(t))]\} \\
& +S'y(t)\int\{Cy(t)6[(1/3)(3\gamma+4\beta-n)h^4(Cx(t)Sx(t) \\
& \cdot Cy(t))+(1/3)(-hh')(Cx(t)Sx(t)C'y(t)) \\
& +(1/6)(2nhh'+n'h^2)(Cx(t)S'x(t)Cy(t) \\
& +C'x(t)Sx(t)Cy(t)) \\
& +(1/6)(n-2)h^2(Cx(t)S'x(t)C'y(t)+C'x(t)Sx(t) \\
& \cdot C'y(t))+(1/3)(-1/2)nh^2(C'x(t)S'x(t)Cy(t))]\},
\end{aligned}
\tag{16}
$$

with machine notation:

$$T(1;1,1,1)=T^1_{111}\ ,$$

$$T(4;1,2,3)=T^4_{123}\ ,$$

$$\int=\int_0^t d\tau.$$

References

[1] Karl L. Brown, A first and second order matrix theory for the design of beam transport systems and charged particle spectromenters, SLAC-75, 1971.

[2] K. L. Brown and F. Rothacker, A computer program for designing charged particle beam transport systems, CERN 80-04, 1980.

[3] Xie Xi and Liu ChunLing, Any order approximate analytical solution of the accelerator nonlinear dynamic system equation, Chinese Journal of Nuclear Physics, No. 3, 1990.

Resonance Seeding of Stability Boundaries in Two and Four Dimensions

LEO MICHELOTTI

Fermilab*, P.O.Box 500, Batavia, IL 60510

1 Introduction.

"Resonance seeding" refers to the hypothesis that the stochastic layer delineating the dynamic aperture of a Hamiltonian system grows out of separatrices generated by very low order resonances.[2, 3] This is a *physics* hypothesis and should not be interpreted as arising from any particular technique for writing perturbative expansions, such as the ones developed by Deprit, Dragt, or Forest. Although analytic representations of the resonances are indeed obtained via perturbation theory, existence of the separatrices and the validity (or otherwise) of resonance seeding are separate from it. We shall describe some of the evidence supporting this idea in two and four dimensions.

2 Two-dimensional maps

With suitable definitions of transverse position and "momentum," the Poincarè map representing insertion of a thin sextupole into a storage ring can be written,

$$\begin{pmatrix} x \\ p \end{pmatrix} \Leftarrow \begin{pmatrix} \cos 2\pi\nu & \sin 2\pi\nu \\ -\sin 2\pi\nu & \cos 2\pi\nu \end{pmatrix} \begin{pmatrix} x \\ p - \lambda x^2 \end{pmatrix} \ , \quad (1)$$

where $\lambda = -e\beta B''l/2p_3$, e being the charge on a proton (the assumed probe), p_3 its longitudinal momentum, $B''l$ the integrated second derivative of the sextupole field, and β the envelope function evaluated at the position of the sextupole. We can set $\lambda \equiv 1$ without loss of generality by rescaling, $x \to x/\lambda$ and $p \to p/\lambda$. (In nonlinear circles, this is called the Hénon map. See Henon,M. Quart. App. Math. **27**,291(1969).)

Comparison of this map with perturbation theory calculations was reported in [2]. We have studied this mapping in the tune range $0 < \nu < \frac{1}{2}$; Figures 1a and 2a illustrate a few orbits at the tunes $\nu = 0.1$, $\nu = 0.29$ respectively. The general features in these drawings are not surprising: (i) near the origin there are smooth (on the scale of the observations) KAM tori; (ii) as one gets farther in phase space a structure of islands and sub-islands develops; (iii) which

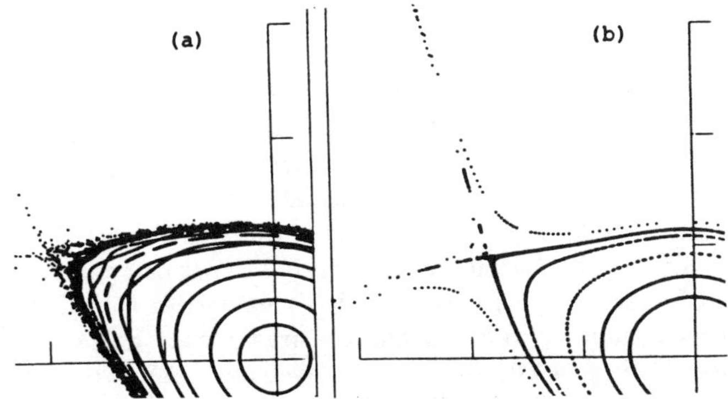

Figure 1: (a) Orbits of the sextupole mapping for $\nu = 0.15$. (b) Second order perturbation theoretic calculation of the stability boundary. The tic marks on the axes are separated by 0.5.

finally breaks into a chaotic sea, the "stability boundary" marking the system's dynamic aperture.

Figures 1b and 2b illustrate calculations done by applying second order $(O(\lambda^2))$ perturbation theory to the map Eq.(1). The dynamics in Figure 1 are dominated by a first order integer resonance, which is put explicitly into a resonant normal form Hamiltonian. With the appropriate distortion, also given by the perturbation expansion, *the separatrix of the resonance then can be associated with the stability boundary of the exact mapping.* By making this identification, we can estimate the location of the latter reasonably accurately.

Figure 2 is a remarkable case. Its most dramatic feature is the very large 2/7 resonance which produces a system of seven islands. Seventh "order" resonances (i.e., resonances with winding number seven) should not appear until fifth order in the perturbation expansion, while the island chain is certainly more than a fifth order effect. In fact it is due to an *interference* between the 1/3 resonance, which appears at first order in the perturbation expansion, and the 1/4 resonance, which appears at second order. This is confirmed in Figure 2b which shows the perturbation theoretic prediction when those two resonances are explicitly taken into account.

Carrying out similar comparisons at other values of the

---

*Operated by the Universities Research Association, Inc. under contract with the U.S. Department of Energy.

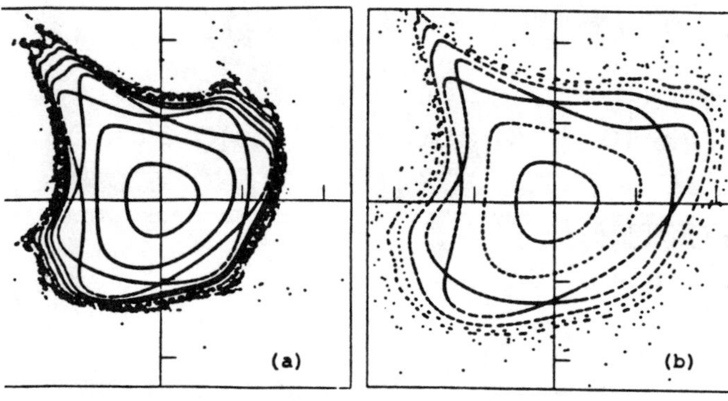

Figure 2: Same as Figure 1, but with $\nu = 0.29$.

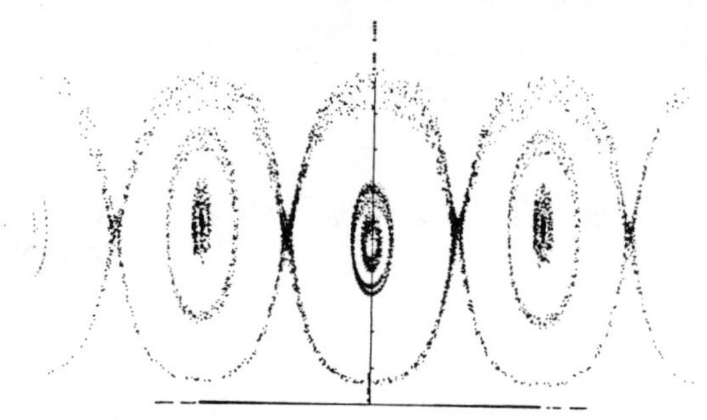

Figure 3: A slice through the separatrix for closed helix.

tune we have found that separatrices associated with first and second order resonant normal form Hamiltonians can *usually* predict the stability boundary surprisingly accurately. Further studies were carried out on Hamiltonians including octupole kicks with similar results.[1] The large amplitude chaotic layers marking dynamic aperture could frequently be associated with separatrices from low order resonances. The conclusion drawn was that resonance seeding in two-dimensional systems worked better than we had a right to expect.

3 Four-dimensional maps

In order to control the beam-beam tune shift (and spread) and allow Fermilab to increase luminosity in the Tevatron, electrostatic separators will be used to place the fiducial (design) p and $\bar{p}$ orbits on the branches of a double helix. Head-on collisions will occur only at two interaction regions, B0 and D0, where the helix will be pinched. Theoretical and experimental studies were conducted to study orbit stability under the double-helix scenario. The problem represents an interesting physical situation: even when individual kicks are small the combined beam-beam tune shift can be extremely large. Thus, the effects are distributed, much like those of small nonlinear fields distributed around the ring. In this type of situation, we can expect the dominant sources of instability to be driven by resonances. The principal observations we report here involve (a) bifurcations of chaotic separatrices leading to (b) the existence of very low entropy chaotic orbits. The control parameter for these bifurcations was the separator strength, or the size of the double helix. What follows is necessarily a cursory discussion of observations; a more detailed paper is being written.[4]

The non-trivial task of following chaotic, multidimensional separatrices through bifurcations was accomplished using AESOP, an "Exploratory Orbit Analysis" graphics shell for studying four-dimensional phase space. (See *C++ Objects for Beam Physics*, this Proceedings.) The low entropy orbits referred to above, previously called "tangled," were observed a few years ago using

AESOP.[3, 5] What we were not able to do at that time was trace their evolution. Since then, a four-dimensional cursor was introduced into AESOP, making it easier to find and to follow separatrices.

At zero separation, we see a slice through the separatrix which exhibits the familiar four-lobe structure associated with a $2\nu_1 - 2\nu_2$ resonance, one excited strongly by the beam-beam interaction. (See Figure 3.) The coordinates for this three dimensional projection are the horizontal and vertical "angle" variables and an "action" variable. In AESOP's standard mode, the action variable displayed is simply proportional to the horizontal emittance, not a true action variable with nonlinearities present. This is basically the "angle-angle-action" plot which has been used before by the author, by Ruth *et al* [6], and by others. Figure 3 shows an orthographic projection of phase space along the $\delta_1 = \delta_2$ diagonal, with a number of orbits plotted. Regular, stable resonant orbits lie in manifolds which are near the regions $S_1 : \delta_1 - \delta_2 \simeq 0$ or $S_2 : \delta_1 - \delta_2 \simeq \pi$, while the unstable resonant orbits are near $U_1 : \delta_1 - \delta_2 \simeq \pi/2$ or $U_1 : \delta_1 - \delta_2 \simeq 3\pi/2$. Some care must be taken in interpretation. The apparently two stable regions just left and right of center are actually the *same* region, and the ones at the extreme ends coincide with the on in the center. We are viewing a projection of angle data which gets wrapped with 2π periodicity. The "four" lobes of this picture actually represent only two stable regions.

Seen from this projection, the orbits in regions S_1 and S_2 appear similar. In fact, they are very different. Tori surrounding S_1 are banded, as can be seen in the slightly rotated view of Figure 4. This indicates the existence of more complicated, "secondary" resonances winding around the "principal" tori, which envelop the "principal" resonant orbit. In addition, if one observes closely, these bands are themselves striated, pointing to a yet more complicated resonance structure. In contrast, the resonant orbit in region S_2 is actually a family of period eight orbits strung together. The small tori appearing in the side view of Figure 4 surround one set of orbits from this family.

When the electrostatic separators are powered and the helix begins to open, the first global bifurcation takes place, as illustrated in Figure 5. The separatrix no longer

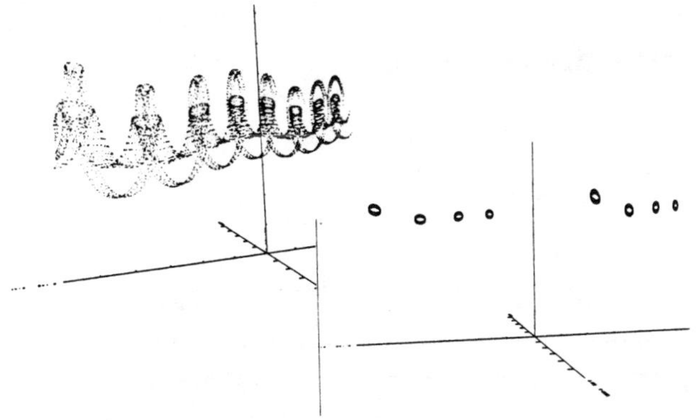

Figure 4: Side view of representative orbits from stable regions S_1 and S_2.

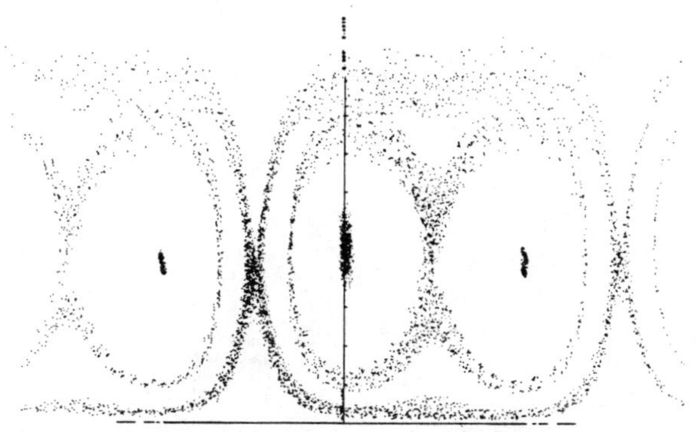

Figure 5: Appearance of the separatrix for small helix.

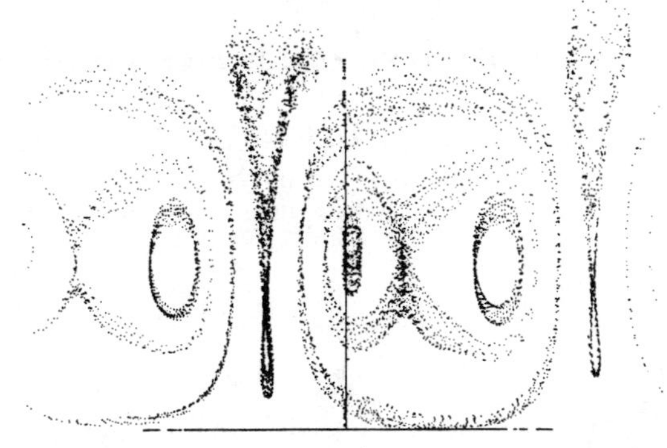

Figure 6: Appearance of the separatrix for larger helix.

ter five iterates. Correspondingly, the lengthening of the "thread" is slow; hundreds of iterates may be required to form a single "loop." From the standpoint of the theory of chaos, we must interpret such orbits as chaotic with exceedingly small KS entropy. That is, the tangled orbits are *almost but not quite* regular. They appear, at least superficially, as one-dimensional objects. A great many iterates would be needed before the fractional parts could be determined with any accuracy.

References

[1] Enrique Henestroza. Perturbative analysis of nonlinear betatron oscillations. Master's thesis, Massachusetts Institute of Technology, 1988.

[2] Leo Michelotti. Perturbation theory and the single sextupole. In Y. S. Kim and W. W. Zachary, editors, *The Physics of Phase Space*. Springer-Verlag, 1987.

[3] Leo Michelotti. Phase space concepts. In Melvin Month and Margaret Dienes, editors, *Physics of Particle Accelerators*. American Institute of Physics, 1989. A.I.P. Conference Proceedings No. 184.

[4] Leo Michelotti and Selcuk Saritepe. Observations from exploratory orbit analysis of the distributed, displaced beam-beam interaction. In preparation.

[5] Leo Michelotti and Selcuk Saritepe. Orbital dynamics in the tevatron double helix. In Floyd Bennett and Joyce Kopta, editors, *Proceedings of the 1989 IEEE Particle Accelerator Conference, Chicago, IL*. IEEE Press, March 20-23, 1989. Vol.2, pp.1391-1393.

[6] R. D. Ruth, T. Raubenheimer, and R. L. Warnock. Superconvergent tracking and invariant surfaces in phase space. *IEEE Transactions on Nuclear Science*, NS-32(5):2206–2208, 1985.

connects the two unstable resonant orbits. Rather, it has split into two branches, each of which is attached to one of them: the central "rotated figure eight" and its "cocoon."[1] As the separation increases, a second bifurcation occurs in the vicinity of the unstable resonant orbit. (See Figure 6) Rather then forming the "cocoon" of Figure 5 the branch closes quickly, forming a new figure eight. Unlike the previous bifurcation, this is not one which takes a heteroclinic branch and changes it into a homoclinic one. It is homoclinic both before and after the bifurcation; what has changed is how it closes back on itself at large amplitudes.

Even though the separatrix is chaotic, its shape and *behavior under bifurcation* is clearly determined by an underlying, low order, integrably resonant system, supporting the reality of resonance seeding in four dimensions. (I suspect, but am not certain, that this is the first attempt to follow in detail the bifurcations of four-dimensional separatrices.)

The "tangled orbits" mentioned above (see the references for figures; there is no more space available here) appeared as the helix continued to open. They are almost periodic, with period five: watching one develop on the screen reveals that it returns on itself almost exactly af-

---

[1]The situation may be even more complicated. There is some suggestion that one of these is actually two resonant orbits very close together, but this was not resolved.

Beam Dynamics Design of an RFQ for the SSC Laboratory

T. S. Bhatia, J. H. Billen, A. Cucchetti, F. W. Guy*, G. Neuschaefer, L. M. Young
Los Alamos National Laboratory, Los Alamos, NM 87545
*Superconducting Super Collider Dallas, TX 76237

Abstract

The design of the Radio Frequency Quadrupole (RFQ) accelerator for the Superconducting Super Collider (SSC) is presented. The RFQ, which accepts the beam from the ion source through the low energy beam transport (LEBT) line, is designed to accelerate H^- beam from 35 keV to 2.5 MeV. Key design considerations and the final design parameters for the RFQ are presented. Results of simulation studies with and without misaligned input beam are also discussed.

I. INTRODUCTION

The Radio Frequency Quadrupole (RFQ) accelerator section of the SSC linac is designed to provide ~25 mA of accelerated H^- beam at 2.5 MeV with a beam bunch frequency of 427.617 MHz. The ion source[1] followed by a suitable LEBT section is expected to provide up to 70 mA of H^- beam at a normalized rms transverse emittance, ε, of 0.020 π cm mrad. In keeping with the design philosophy of operational flexibility and reliability as stated in Ref. 1, the RFQ should be able to provide output beam varying from 5 to 50 mA. The maximum peak surface field should not exceed ~36 MV/m (1.8 Kilpatrick). The low energy booster (LEB) ring filling requirement dictates the linac to operate at 10 Hz with a pulse length of ~37 μs, limiting the duty factor to less than ~0.05%. The transmission through the RFQ should also be high, preferably around 90%. The RFQ beam dynamics design, described in the following sections, meets these goals.

BEAM DYNAMICS DESIGN

The fixed starting parameters of the RFQ are contained in Table I.

TABLE 1

Ions		H^-
Frequency	f	427.617 MHz
Injection Energy	w_i	0.035 MeV
Output Energy	w_f	2.500 MeV
Injection Current (nominal)	c_i	30 mA
Input Emittance	ε_i^t(n, rms)	0.02 π cm mrad
Maximum Peak Surface Field		~1.8 E_k (36 MV/m)

A schematic of the RFQ vane geometry is shown in Figure 1. In the radial matching section, the beam changes from a round time independent shape to a transverse time dependent quadrupole shape. Next the shaper section, accelerates the beam to 75 keV. The acceleration efficiency A and synchronous phase ϕ_s vary linearly with longitudinal position.

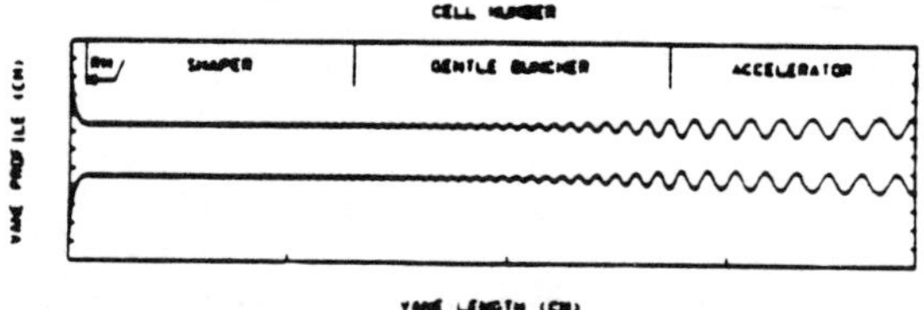

Figure 1. Schematic of the RFQ Geometry

The potential used in the design has eight terms. The first four terms of the radial electric field

$$E_r = \frac{-V}{r_o} \left[A_{01} \left(\frac{r}{r_o} \right) \cos 2\Theta + A_{03} \left(\frac{r}{r_o} \right)^5 \cos 6\Theta \right.$$
$$\left. + \left[A_{10} \frac{kr_o}{2} I_0'(kr) + A_{12} \frac{kr_o}{2} I_4'(kr) \cos 4\Theta \right] \cos kz \right]$$

$$\text{where} \quad k = \frac{2\pi}{\beta\lambda},$$

correspond to quadrupole, duodecapole, rf defocusing and the octupole field components respectively. Alternate focusing and defocusing is not provided by the 'rf defocusing' and 'octupole' terms because the voltage (cos ωt) and cos kz factors both change sign with the same periodicity. Therefore, the defocusing fields can be very important even though these fields are an order of magnitude weaker than the quadrupole fields which focus with alternately focusing and defocusing fields. Figure 2 shows the relative contribution of the different radial electric components (the quadrupole term is ~1 and is not shown) along the entire RFQ length.

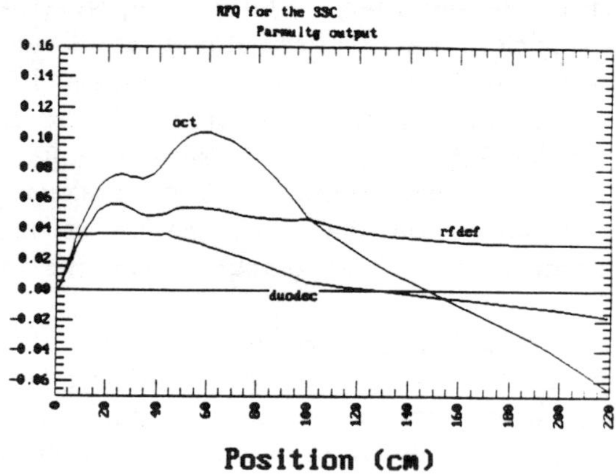

Figure 2. Higher order multipole components and the rf defocusing field along the length of the RFQ. evaluated at r = r_o.

*Work supported by the U.S. Department of Energy, Office of the Superconducting Super Collider.

Table 2 lists the relevant design final parameters. The gentle buncher section follows the shaper. The accelerator and the exit fringe field sections follow the gentle buncher.

Table 2

Key Final Design Parameters

Injection energy	w_i	0.035 MeV
Energy at end of Shaper	w_s	0.075 MeV
Energy at end of GB (Gentle Buncher)	w_g	0.115 MeV
Final energy	w_f	2.500 MeV
Synchronous phase at end of GB	ϕ_g	−55°
Final synchronous phase	ϕ_f	−30°
Modulation at end of GB	m_g	1.23
Final modulation	m_f	1.93
Aperture radius, minimum	a_{min}	0.178 cm
Aperture radius, minimum (pt. of symmetry)	$r_{o,min}$	0.198 cm
Final aperture radius	a_f	0.240 cm
Final aperture radius (pt. of symmetry)	$r_{o,f}$	0.366 cm
Intervane Voltage	V	54.8 to 88.5 kV
Power (copper)		278 kW
Maximum peak surface field		36.24 MV/m = 1.81 E_s
RFQ Length		219.64 cm
Accelerating efficiency at end of GB	A_g	0.133

The variation of several of the parameters along the length of the RFQ is shown in Figure 3, as a function of position. Vane voltage and peak surface field along the length are plotted in Figure 4.

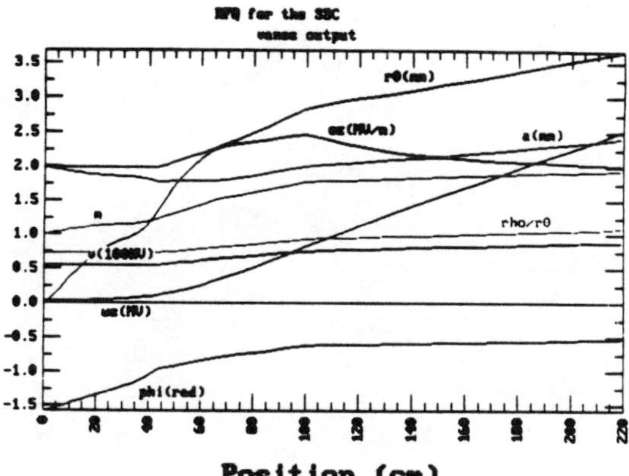

Figure 3. Variation of key design parameters along the length of the RFQ.

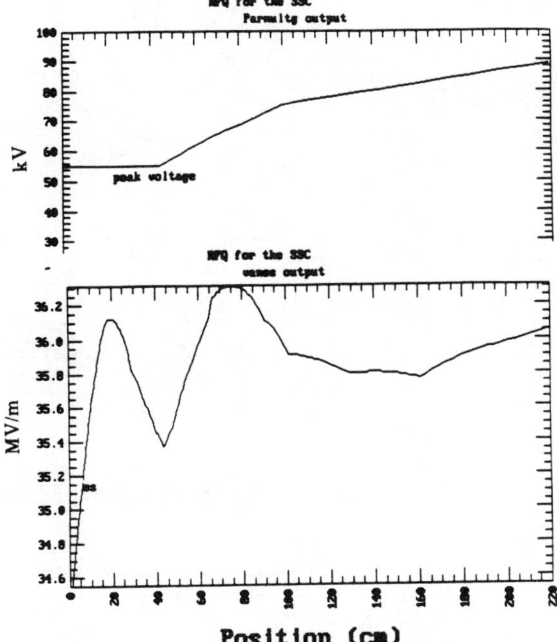

Figure 4. Vane voltage and the peak surface field along the length of the RFQ.

BEAM DYNAMICS SIMULATIONS

Particle simulation runs were made using several versions of PARMTEQ. One version [2] includes the eight term potential to incorporate the higher order multipoles discussed earlier. Another version [3] calculates 3D space charge and image charge effects. Figure 5 shows the transverse and longitudinal emittances and the transmission along the length of the RFQ. Figure 6 shows the profile plots for the standard 30 mA input run. We have also considered the case when 30 mA and 70 mA misaligned beams are injected 0.25 mm off-axis and 5 mrad off-angle in the x-plane. The beam profile plots for such a misaligned input beam of 30 mA are shown in Figure 7. In Table 3, transmission and emittance growths are presented for the case of an aligned and misaligned 30 mA and 70 mA input beam.

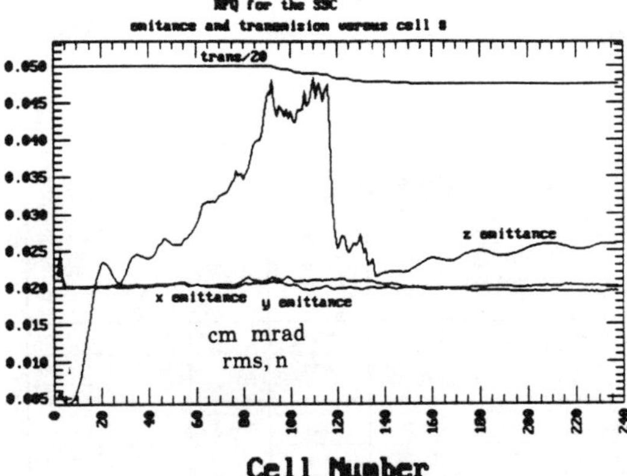

Figure 5. Transverse and longitudinal emittances, and the transmission along the length of the RFQ as a function of the cell number, for an input beam of 30 mA.

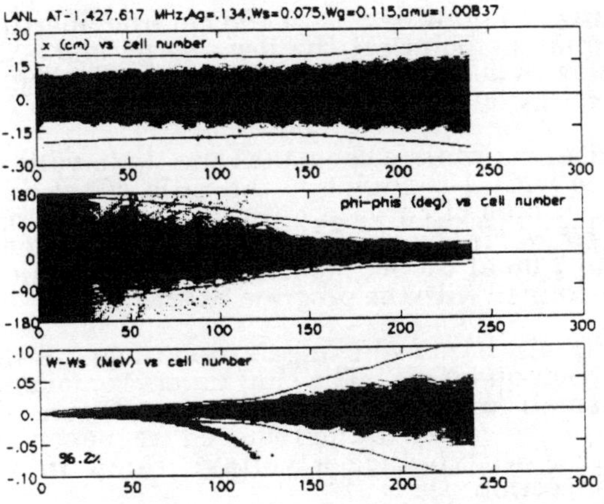

Figure 6. Profile plots for a matched beam of 30 mA. Multipole code.

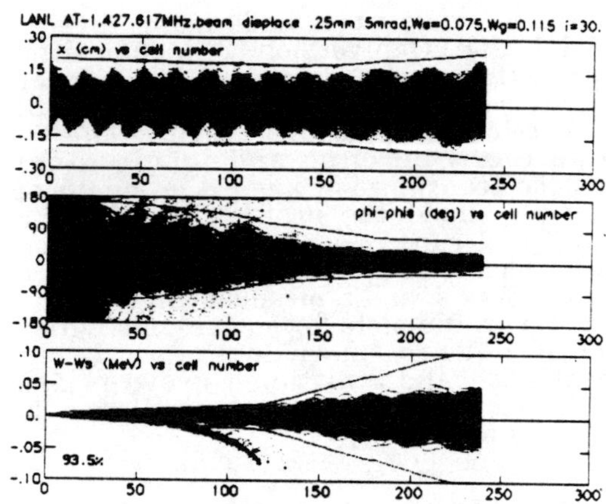

Figure 7. Profile plots for a misaligned (Δx = 0.25 mm, $\Delta x'$ = 5 mrad) input beam of 30 mA. Transmission is 93.5%

CONCLUSION

An RFQ for the SSC has been designed which exceeds the baseline 25 mA output current (30 mA input) and the upgrade 50 mA output current (70 mA input) requirements. The peak surface field, 1.81 E_k, (target value = 1.80 E_k) is conservative for the low duty factor of <0.05%. There is essentially no growth in the transverse emittances. Simulations with 2D and 3D space charge, multipoles and images charges predict high (~95%) transmission through the RFQ for an input beam of 30 mA. For 70 mA input beam the transmission value is also fairly high, i.e., 79%.

REFERENCES

1. J. Watson, et.al., contribution to this conference.
2. K. Crandall, private communication, AccSys Technology
3. F. W. Guy, private communication, SSC Laboratory

TABLE 3

Code	Input Beam Displacement mm mrad		Input Beam Current mA	Percent Transmission	$\mathcal{E}_x$n, rms mm mrad	$\mathcal{E}_y$n, rms mm mrad	$\mathcal{E}$,rms MeV deg
2D Space Charge	0	0	70	83.9	.208	.224	.130
3D Space Charge	0	0	70	85.1	.199	.208	.098
3D & Image Charge	0	0	70	79.7	.206	.198	.101
2D & Multipoles	0	0	70	82.8	.228	.192	.124
2D & Multipoles	0.25	5	70	80.0	.204	.212	.123
3D & Image Charge	0.25	5	70	73.6	.195	.202	.100
3D Space Charge	0.25	5	70	80.3	.207	.200	.093
2D & Multipole	0	0	30	95.4	.197	.203	.127
2D & Multipole	0.25	5	30	93.5	.200	.197	.125
3D & Image Charge	0.25	5	30	94.0	.205	.198	.102
3D Space Charge	0.25	5	30	95.3	.207	.191	.100
2D & Multipole	0	0	5	98.2	.205	.199	.166

RF STRUCTURE

This RFQ design uses r_o = 0.198 cm (the aperture, at the symmetry point) at the low energy end. The beam dynamics design then increases r_o to 0.366 cm at the high energy end.

In order to maintain a constant vane tip to vane tip capacitance with a changing r_o the ratio of the vane transverse radius of curvature (ρ) to r_o must change. The ratio ρ/r_o varies from 0.75 at the low energy end of the RFQ to 1.06 at the high energy end. The ratio of ρ/r_o is determined with the program SUPERFISH.

The RFQ will be tuned using 32 slug tuners, 8 slug tuners in each quadrant. The slug tuners remove the effects of small misalignments of the vanes, tune the frequency of the cavity to the required frequency and in this RFQ will induce a voltage ramp in the accelerator section.

Dipole tuners that are placed on the endwalls will be used to adjust the frequency of dipole modes away from the quadrupole mode. These dipole tunes have virtually no effect on the quadrupole mode.

On Dynamic Aperture

Zohreh Parsa

Brookhaven National Laboratory

Upton, NY 11973

Abstract

Nonlinear perturbation can alter the beam behavior in dynamical systems. For example, nonlinear fields due to sextupoles or octupoles can perturb the motion of the beam of particles in accelerators. If the perturbation is increased, at some point, the motion of the beam become unstable and chaotic. In the Superconducting Super Collider (SSC) a proton would interact 10^{12} times (in $\sim 10^9$ revolutions) with the local nonlinear magnetic fields which can cause the deviations, of the beams from the designed orbit, and lead to loss of particles. We present an analytical method to study the beam stability in large accelerators such as SSC, (and an alternate to using "kick" approximation (particle tracking) over large time intervals). This includes stability factor "f" which determines the convergence of the superconvergent perturbation theory, and the behavior of the system, e.g., stable if $f > 1$, and unstable if $f < 1$. The critical point where the system makes the transition from stable ($F > 1$) to chaotic ($f < 1$) behavior corresponds to $f = 1$, the point at which the size of the beam is defined as the dynamic aperture in accelerators.

I. INTRODUCTION

A method for determining the dynamic aperture in accelerators is discussed, which corresponds to finding the transition point from a stable to chaotic behavior. When the stability factor f, which determines the convergence of the superconvergent perturbation theory, (a method used to study the stability of dynamical systems, e.g. used in proof of K.A.M. Theorem), is equal to 1.

II. FORMALISM

According to K.A.M. Theorem [1], if a dynamical system described by the Hamiltonian

$$H = T + V_N \tag{1}$$

has no degenerecies in its frequencies, (there exist an $m > 0$ such that $|V_N| < m$), then for a small perturbation (V_N), the motion is not stochastic and can be approximated by analytic functions. That is, under a small analytic perturbation the majority of the invariant tori do not collapse but are only slightly deformed [1]. Our aim is to find m, the upper limit on the strength of this perturbation, such that K.A.M. theorem still holds.

In a case where the KAM boundary between stable and stochastic motion might be at a finite amplitude but the perturbation series does not converge to analytic functions for the constants of motion even below the limit (e.g. where there is a stability bound beyond which the phase plot exhibits island structure, although the motion is not stochastic), another test (e.g. overlap criteria [2–9]), should be used for confirmation.

Consider a Hamiltonian:

$$H_l = T_l\left(\overrightarrow{J}_l, s\right) + V_l\left(\overrightarrow{J}_l, \overrightarrow{\Phi}_l, s\right) \tag{2}$$

where T is the kinetic energy, V is the perturbed potential, $l =$ integer, J_l and Φ_l are l dimensional (action and angle conjugate) vectors, such that $H_0 = H, T_0(\overrightarrow{J}_0, s) = T(\overrightarrow{J}, s)$, $V_0(\overrightarrow{J}_0, \overrightarrow{\Phi}_0, s) = V_N(\overrightarrow{J}, \overrightarrow{\Phi}, s)$, $\overrightarrow{J}_0 = \overrightarrow{J}$ and using the generating function

$$F_l = \overrightarrow{\Phi}_l \bullet \overrightarrow{J}_{l+1} + G_l\left(\overrightarrow{J}_{l+1}, \overrightarrow{\Phi}_l, s\right) \tag{3}$$

from H_l, $\overrightarrow{\Phi}_0 = \overrightarrow{\Phi}$. H_{l+1} is found and we obtain:

$$H_{l+1} = T_{l+1}\left(\overrightarrow{J}_{l+1}, s\right) + V_{l+1}\left(\overrightarrow{J}_{l+1}, \overrightarrow{\Phi}_{l+1}, s\right) \tag{4}$$

where

$$T_{l+1} = T_l\left(\overrightarrow{J}_{l+1}, s\right) + A_l\left(\overrightarrow{J}_{l+1}, s\right) \tag{5}$$

and

$$
\begin{aligned}
V_{l+1} = &\left[T_l\left(\overrightarrow{J}_{l+1} + \partial_{\overrightarrow{\Phi}_l} G_l, s\right) - T_l\left(\overrightarrow{J}_{l+1}, s\right)\right. \\
&\left. - \partial_{\overrightarrow{J}_{l+1}} T_l\left(\overrightarrow{J}_{l+1}, s\right) \bullet \partial_{\overrightarrow{\Phi}_l} G_l\right] \\
&+ \left[V_l\left(\overrightarrow{J}_{l+1} + \partial_{\overrightarrow{\Phi}_l} G_l, \overrightarrow{\Phi}_l, s\right) - V_l\left(\overrightarrow{J}_{l+1}, \overrightarrow{\Phi}_l, s\right)\right] \\
&+ \left[V_l\left(\overrightarrow{J}_{l+1}, \overrightarrow{\Phi}_l, s\right) - \left\{V_l\left(\overrightarrow{J}_{l+1}, \overrightarrow{\Phi}_l, s\right)\right\}_{2^l N}\right] \\
&- A_l\left(\overrightarrow{J}_{l+1}, s\right)
\end{aligned}
\tag{6}
$$

The term $A_l(\overrightarrow{J}_{l+1}, s)$ is included in the above two equations to eliminate the nonresonant term from V_{l+1} in Eq. (6). The Fourier expansion of V_l about $\overrightarrow{\Phi}_l$ is

$$\left\{ V_l\left(\overrightarrow{J}_{l+1}, \overrightarrow{\Phi}_l, s\right) \right\}_{2^l N}$$
$$= \sum_{0 < |\overrightarrow{m}| \le 2^l N} v_{l\overrightarrow{m}}\left(\overrightarrow{J}_{l+1}, 2\right) e^{i\overrightarrow{m}\cdot\overrightarrow{\Phi}_l} \tag{7}$$

where $\overrightarrow{m}$ is an n-dimensional vector with integer components and $v_{l\overrightarrow{m}}$ are the Fourier amplitude of V_l; and we expect that the potential term V_l goes to zero in the limit $l \to \infty$ if there exist "n" invariants i.e.,

$$\lim_{l\to\infty} V_l\left(\overrightarrow{J}_l, \overrightarrow{\Phi}_l, s\right) = 0. \tag{8}$$

Next we express the G_l term in the generating function as:

$$G_l = \sum_{\overrightarrow{m}} g_{l\overrightarrow{m}}\left(\overrightarrow{J}_{l+1}, s\right) e^{i\overrightarrow{m}\cdot\overrightarrow{\Phi}_l} \tag{9}$$

and define the phase advance $\overrightarrow{\Psi}_l$ and tune $\overrightarrow{\nu}_l$ as:

$$\overrightarrow{\Psi}_l\left(\overrightarrow{J}_{l+1}, s\right) \equiv \partial_{\overrightarrow{J}_{l+1}} T_l\left(\overrightarrow{J}_{l+1}, s\right) \tag{10a}$$

$$\overrightarrow{\nu}_l\left(\overrightarrow{J}_{l+1}\right) \equiv \frac{1}{2\pi}\int_0^L \overrightarrow{\Psi}_l\left(\overrightarrow{J}_{l+1}, s\right) ds \tag{10b}$$

where Ψ is periodic in s, with L the length of the period (e.g. a circumference of an accelerator) i.e. $\overrightarrow{\Psi}_l(\overrightarrow{J}_{l+1}, s) = \overrightarrow{\Psi}_l(\overrightarrow{J}_{l+1}, s + L)$, and obtain

$$G_l\left(\overrightarrow{J}_{l+1}, \overrightarrow{\Phi}_l, s\right) = \sum_{\overrightarrow{m}} \frac{1}{2\sin\pi\left(\overrightarrow{m}\cdot\overrightarrow{\nu}_l\left(\overrightarrow{J}_{l+1}\right)\right)}$$
$$\int_s^{s+L} v_{l\overrightarrow{m}}\left(\overrightarrow{J}_{l+1}, t\right) \exp\left\{ i\overrightarrow{m}\cdot\left[\overrightarrow{\Phi}_l + \overrightarrow{\Psi}_l\left(\overrightarrow{J}_{l+1}, t\right)\right.\right.$$
$$\left.\left. - \overrightarrow{\Psi}_l\left(\overrightarrow{J}_{l+1}, s\right) - \pi\overrightarrow{\nu}_l\left(\overrightarrow{J}_{l+1}\right)\right] \right\} dt \tag{11}$$

from this we get a bound on V_{l+1} given the magnitude of v_l using Eq. (6) and expanding the terms in V_{l+1} (which depends on $\overrightarrow{J}_{l+1} + \partial_{\overrightarrow{\Phi}_l} G_l$) in the Taylor series about $\overrightarrow{\Phi}_l$

$\overrightarrow{J}_{l+1}$:

$$v_{l+1}\left(\overrightarrow{J}_{l+1}, \overrightarrow{\Phi}_l, s\right) \equiv \tfrac{1}{2}\partial_{\overrightarrow{\Phi}_l} G_l \bullet \partial_{\overrightarrow{J}_{l+1}} \overrightarrow{\Psi}_l\left(\overrightarrow{J}_{l+1}, s\right)$$
$$\bullet\, \partial_{\overrightarrow{\Phi}_l} G_l + \partial_{\overrightarrow{J}_{l+1}} V_l\left(\overrightarrow{J}_{l+1}, \overrightarrow{\Phi}_l, s\right)$$
$$\bullet\, \partial_{\overrightarrow{\Phi}_l} G_l + 0\,(v^3) + \left[V_l\left(\overrightarrow{J}_{l+1}, \overrightarrow{\Phi}_l, s\right)\right.$$
$$\left. - \left\{ V_l\left(\overrightarrow{J}_{l+1}, \overrightarrow{\Phi}_l, s\right) \right\}_{2^l N} \right] \tag{12}$$

Substituting the generating function G_l into (12) and after changing the angle variables from Q_l to Q_{l+1}; (for $|\overrightarrow{m}| < 2^l N$), and using triangle inequality, defining

$$D_{l\overrightarrow{m}}\left(\overrightarrow{J}_{l+1}\right) \equiv \sup_{s\,\to\,s+L} |v_{l\overrightarrow{m}}\left(\overrightarrow{J}_{l+1}, s\right)| \tag{13}$$

and the $M_l(\overrightarrow{J})$ as the upper bound of the potential

$$M_l\left(\overrightarrow{J}_l\right) \equiv \sum_{\overrightarrow{m}} D_{l\overrightarrow{m}}\left(\overrightarrow{J}_l\right) \tag{14}$$

we find the relation between $M_l(\overrightarrow{J}_l)$ and $M_{l+1}(\overrightarrow{J}_{l+1})$:

$$M_{l+1} \le \frac{n^2}{4} L\left(\frac{A_l L}{2}\mu_l\xi_l + \frac{1}{|J_{l+1}|}\lambda_l\zeta_l\right) M_l^2 \tag{15}$$

with

$$\xi_l\left(J_{l+1}\right) \equiv \sum_{\overrightarrow{m}_1} \frac{|\overrightarrow{m}_1|}{|\sin\pi\overrightarrow{m}_1\bullet\overrightarrow{\nu}_l|} \quad \text{and} \quad \zeta_l \equiv \sum_{|m_1|} |\overrightarrow{m}_1| \tag{16}$$

where $\mu_l(\overrightarrow{J}_{l+1})$ and $\lambda_l(\overrightarrow{J}_{l+1})$ are $\approx 1/$the number of terms in each sum respectively. Raising Eq. (15) to (2^{-l-1}) power and defining $F_l \equiv m_l^{2^{-l}}$ (i.e. translating Eq. (15) to a linear difference equation) we obtain

$$F_{l+1} \le \left[\frac{n^2}{4}L\mu_l\xi_l\left(\frac{A_l L}{2}\mu_l\xi_l + \frac{1}{|\overrightarrow{J}_{l+1}|}\lambda_l\zeta_l\right)\right]^{2^{-l-1}} F_l \tag{17}$$

defining

$$F_{l+1} \equiv b\,(l, J_{l+1})\, F_l \tag{18}$$

where

$$b\left(l, \overrightarrow{J}_{l+1}\right) \le \left[\frac{n^2}{4}L\mu_l\xi_l\left(\frac{A_l L}{2}\mu_l\xi_l + \frac{1}{|\overrightarrow{J}_{l+1}|}\lambda_l\zeta_l\right)\right]^{2^{-l-1}} \tag{19}$$

we obtain F_l as the solution to Eq. (18) to be:

$$F = \left[\prod_{k=0}^{l} b\left(k, \overrightarrow{J}_{k+1}\right)\right] F_0 \tag{20}$$

From this equation we determine the stability of the dynamical system, e.g. when the $f \equiv \lim_{l\to\infty} F_l = 1$ (which corresponds to $\lim_{l\to\infty} M_l = 0$), the dynamical ystem remains stable, (this is similar to the "root test" for an infinite series) [3]. Thus we can define a stability factor

$$f \equiv \lim_{l \to \infty} F_l = \left[\prod_{k=0} b\left(k, \vec{J}_{k+1}\right) \right] M_0 \qquad (21)$$

such that as long as $f < 1$ the dynamic system remains stable, becomes chaotic when $f > 1$ (e.g. for large perturbation), and makes the transition from stable to unstable behavior for $f = 1$ (or could occur at $\sim f = 1^-$ or $f = 1^+$). The dynamical aperture in accelerators, is defined as the size of the beam at which the particles become unstable. This corresponds to $f = 1$, which is the transition point from stable to chaotic behavior. Using Eq. (19) one can estimate the dynamic aperture for accelerators. For example the presence of the large β functions, at the SSC (Superconducting Super Collider) quadrupole triplets in the low β insertions, leads to a large perturbing potential (V_N, due to multipoles, etc.) that adds a large contribution to M_0 in Eq. (21), and greatly reduces the dynamic aperture in that accelerator. Thus requiring a smaller size beam in order to avoid instability (for $f = 1$). This suggests, that by reducing the multipole contributions in the SSC triplet quadrupole magnets, the dynamic aperture may be improved.

III. SUMMARY

Using a full superconvergent perturbation theory, one can calculate ($b(l, J_{l+1})$ exactly resulting in) a more precise estimate of the dynamic aperture (the transition point from stable to stochastic motions with $f = 1$), which is a useful analytical tool in accelerator design.

IV. REFERENCES

1. V.I. Arnold, Russian Math. Surveys, vol. 18, 1963.

2. Z. Parsa, "Accelerator dynamics and beam aperture", BNL 29194, 1986.

3. Z. Parsa, S. Teikian and E. Courant, "Second order perturbation for accelerators", Particle Acc., vol. 22, pp. 205–230, 1987.

4. Z. Parsa, S. Tepikian, (unpublished).

5. Z. Parsa, "Analytical method for obtaining the variations of the beam emittance, particle action and linear aperture in accelerators", *Proc. of the 1986 Summer Study on the Physics of the SSC*, Snowmass, Co., Edited by R. Donaldson and J. Marx, June 23–July 11, 1986, p. 573, and BNL Report #38735.

6. SSC Central Design Group, "SSC conceptual design", SSC-SR-2020, March 1986.

7. R.C. Buck, *Advanced Calculus*, McGraw-Hill Inc. 1978.

8. B.V. Chirikov, Phys. Rept. 52, 1979.

9. Z. Parsa, "Analytical method for treatment of non-linear resonances in accelerators", *Proc. of the 1986 Summer Study on the Physics of SSC*, Snowmass, Co., Edited by R. Donaldson and J. Marx, June 23–July 11, 1986, p. 579, and BNL Report #38734.

10. J.M. Greene, J. Math. Phys., vol. 20, pp. 1183–1201, 1979.

* We thank Ms. Fern Simes for assisting with the typing of the manuscript.

Modelling of Space Charge Effects in the CERN Proton Synchrotron

Michel Martini and Oleg Ponomarev*
CERN, CH-1211 Geneva 23

Abstract

The performance of the CERN Proton Synchrotron (PS) at injection is limited by resonance effects caused by space charge tune spreads.

This paper describes a new computer code for calculating the incoherent tune spreads and the beam envelopes in linear coupled synchrotron lattices in the presence of transverse and longitudinal space charge fields. This work is based on an existing theory devised at DESY using the six-dimensional phase space formalism. It has been extended to deal with the nonuniform charge density within the bunches. The application to high intensity beams at present in operation in the PS is discussed.

I.   INTRODUCTION

The understanding of the space charge effects is of primary importance for low energy circular accelerators and storage rings. Theoretical models which permit reliable numerical simulation are desirable to analyze the beam behaviour.

In this paper the original formalism developed at DESY [1, 2] is extended by taking into account nonuniform charge density within bunches. Throughout this text the same variables will be used as those in [2], i.e. $x, z, \sigma = s - \beta ct, \eta = \Delta E / E$, where x, z describe the transverse betatron oscillations, σ, η the synchrotron oscillations in the longitudinal plane, s the arc length of the reference orbit, and β, E the velocity in units of c and the total energy of the reference particle respectively.

II.   EQUATIONS OF MOTION

In a matrix form the equation of motion of a particle of rest mass m_0 and charge e in a linear lattice in presence of space charge effects and coupling writes

$$\vec{y}\,' = \underline{A}(s, \vec{y})\, \vec{y} \tag{1}$$

with $\vec{y}\,^t = (x, p_x, z, p_z, \sigma, \eta)$ where

$$p_x = \beta^2\, x' - H(s)z \tag{2}$$

$$p_z = \beta^2\, z' + H(s)x \tag{3}$$

and

$$A_{1,2} = A_{3,4} = \beta^{-2}$$

*On leave of absence from Academy of Sciences of the USSR INR, Moscow

$$A_{1,3} = A_{2,4} = H$$
$$A_{2,1} = -\beta^2 \left(K_x^2 + g + H^2 - F_{xx}(\vec{y}) \right)$$
$$A_{2,3} = A_{4,1} = \beta^2 \left(N + F_{xz}(\vec{y}) \right)$$
$$A_{2,6} = -A_{5,1} = K_x$$
$$A_{3,1} = A_{4,2} = -H$$
$$A_{4,3} = -\beta^2 \left(K_z^2 - g + H^2 - F_{zz}(\vec{y}) \right)$$
$$A_{4,6} = -A_{5,3} = K_z$$
$$A_{5,6} = \beta^{-2}\gamma^{-2}$$
$$A_{6,5} = \frac{2\pi h e \hat{V}}{E_0 L} \cos\varphi \sum_{\nu=1}^{m} \delta(s - s_\nu) + F_\sigma(\vec{y})$$
$$A_{i,j} = 0 \;\; otherwise \tag{4}$$

with $\gamma = 1/\sqrt{1 - \beta^2}$. Here $g(s)$ is the quadrupole strength, $N(s), H(s)$ are the skew quadrupole components and solenoid fields, $K_x(s), K_z(s)$ are the curvatures in the x and y directions, and m is the number of cavities (assumed point like at $s = s_\nu$). The peak electric field in a cavity is $\hat{V}$, $\varphi = 0, \pi$ (no acceleration), h is the harmonic number and L is the length of the reference orbit [2].

The space charge forces depend on $\vec{y}$ for nonuniform charge distribution. The terms $F_{xx}(\vec{y})$, $F_{xz}(\vec{y})$, $F_{zz}(\vec{y})$ and $F_\sigma(\vec{y})$ describe the self field space charge forces and will be defined later.

III.   SPACE CHARGE FORCES

Bunched beams of ellipsoidal shape with nonuniform charge density in the ellipsoid are considered. The three-dimensional model of the bunch is a nest of N concentric ellipsoids (numbered from the inner towards the outer) whose shells (each of sufficiently small thickness) are assumed to have a uniform charge density.

Synchro-betatron coupling other than by space charge forces will be neglected. Thus the twist angles $\theta_{x\sigma}$ and $\theta_{z\sigma}$ of the bunch with respect to the σ axis in the x–σ and z–σ planes can be ignored.

The force acting on a particle in the nth layer can be obtained from the expressions for the potential of homogeneous ellipsoids. Assuming that the longitudinal dimension is much greater than the transverse dimensions ($\gamma E_\sigma \gg E_1, E_2$), the components of the transverse space charge force in the nth layer in a rotated coordinate system $(\tilde{x}, \tilde{z}, \sigma)$ parallel to the half axes $E_{1,n}, E_{2,n}, E_{\sigma,n}$ of the nested ellipsoids are (in the laboratory system)

$$F_{\tilde{x}}^n = \frac{e}{\gamma \epsilon_0} I_1^n(\tilde{x}, \tilde{z}, \sigma)\tilde{x} \tag{5}$$

$$F_\sigma^n = \frac{e\gamma}{\epsilon_0} I_3^n(\tilde{x}, \tilde{z}, \sigma)\sigma \qquad (6)$$

with the approximation

$$I_1^n(\tilde{x}, \tilde{z}, \sigma) =$$
$$\sum_{i=n+1}^{N} \rho_i \left(\frac{E_{2,i}}{E_{1,i} + E_{2,i}} - \frac{E_{2,i-1}}{E_{1,i-1} + E_{2,i-1}} \right) +$$
$$\sum_{i=1}^{n-1} (\rho_i - \rho_{i+1}) \frac{E_{1,i} E_{2,i}}{E_{1,i}^2 - E_{2,i}^2} \left(1 - \sqrt{\frac{k_{i,n} + E_{2,i}^2}{k_{i,n} + E_{1,i}^2}} \right) +$$
$$\rho_n \frac{E_{2,n}}{E_{1,n} + E_{2,n}} \qquad (7)$$

Similar expression hold for $F_{\tilde{z}}^n$ and I_2^n provided that $\tilde{z}$ replaces $\tilde{x}$ in Eq. 5 and $E_{1,i}$ is interchanged with $E_{2,i}$ in Eq. 7. Further

$$I_3^n(\tilde{x}, \tilde{z}, \sigma) =$$
$$\sum_{i=n+1}^{N} \rho_i \left(\frac{E_{1,i} E_{2,i}}{E_{\sigma,i}^2} P_i - \frac{E_{1,i-1} E_{2,i-1}}{E_{\sigma,i-1}^2} P_{i-1} \right) +$$
$$\sum_{i=1}^{n-1} (\rho_i - \rho_{i+1}) \frac{E_{1,i} E_{2,i}}{E_{\sigma,i}^2} (P_i + Q_{i,n}) +$$
$$\rho_n \frac{E_{1,n} E_{2,n}}{E_{\sigma,n}^2} P_n \qquad (8)$$

with

$$P_i = \ln \left(\frac{E_{\sigma,i} + \sqrt{E_{1,i}^2 - E_{2,i}^2 + E_{\sigma,i}^2}}{E_{1,i} + E_{2,i}} \right) \qquad (9)$$

$$Q_{i,n} = \ln \left(\frac{\sqrt{k_{i,n} + E_{1,i}^2} - \sqrt{k_{i,n} + E_{2,i}^2}}{E_{1,i} - E_{2,i}} \right) \qquad (10)$$

where ρ_i is the constant charge density inside the ith layer and $k_{i,n}$ is the largest root of the equation

$$\frac{\tilde{x}^2}{E_{1,i}^2 + k_{i,n}} + \frac{\tilde{z}^2}{E_{2,i}^2 + k_{i,n}} + \frac{\sigma^2}{E_{\sigma,i}^2 + k_{i,n}} = 1 \qquad (11)$$

which depends on the location $(\tilde{x}, \tilde{z}, \sigma)$ in the layer. An approximate mean value for $k_{i,n}$ is obtained by taking the average of Eq. 11 over the volume of the shell. This approximation means that I_1^n, I_2^n, I_3^n keep constant values within the nth shell. Thus the nonlinear character of space charge forces will be approximated by piecewise linear functions in the N ellipsoidal level layers.

The space charge force components with respect to the (x, z) axes become, using Eqs. 5–6

$$F_x^n = m_0 c^2 \beta^2 \gamma \left(F_{xx}^n x + F_{xz}^n z \right) \qquad (12)$$

$$F_z^n = m_0 c^2 \beta^2 \gamma \left(F_{xz}^n x + F_{zz}^n z \right) \qquad (13)$$

with

$$F_{xx}^n = \frac{e}{\epsilon_0 m_0 c^2 \beta^2 \gamma^2} \left(I_1^n \cos^2 \theta_{xz}^n + I_2^n \sin^2 \theta_{xz}^n \right) \qquad (14)$$

$$F_{zz}^n = \frac{e}{\epsilon_0 m_0 c^2 \beta^2 \gamma^2} \left(I_1^n \sin^2 \theta_{xz}^n + I_2^n \cos^2 \theta_{xz}^n \right) \qquad (15)$$

$$F_{xz}^n = \frac{e}{\epsilon_0 m_0 c^2 \beta^2 \gamma^2} \left(I_1^n - I_2^n \right) \sin \theta_{xz}^n \cos \theta_{xz}^n \qquad (16)$$

where θ_{xz}^n is the twist angle of the nth bunch layer with respect to the x axis in the x-z plane.

Hence, the space charge terms $F_{xx}(\vec{y})$, $F_{xz}(\vec{y})$, $F_{zz}(\vec{y})$ and $F_\sigma(\vec{y})$ in Eq. 4 are constituted of the piecewise approximations F_{xx}^n, F_{xz}^n, F_{zz}^n and F_σ^n, for $n = 1 \ldots N$.

IV. SOLUTION OF THE EQUATIONS

For a particle with coordinates (x, z, σ) lying in the nth ellipsoidal shell, Eq. 1 can be linearized so that its solution can be written in the form

$$\vec{y}^n(s) = \underline{M}^n(s, s_0) \vec{y}^n(s_0) \qquad (17)$$

where $\underline{M}^n(s, s_0)$ is the transfer matrix and $\vec{y}^n(s_0)$ is an initial vector.

Although a rigorous symplectic thin lens approximation for the transfer matrix has been established [1, 2], the simpler commonly used thin lens linear approximation (non symplectic) is considered here because it yields in practice similar results

$$\underline{M}^n(s + \Delta s, s) = \underline{I} + \Delta s\, \underline{A}^n(s) \qquad (18)$$

Here $\underline{A}^n(s)$ is the matrix $\underline{A}(s)$ in which the space charge terms are replaced by their piecewise approximation.

Let $\vec{y}_k^n(s_0)$ $(k = 1 \ldots 6)$ be linearly independent vectors. Then the following vector spans an hyperellipsoid in the six dimensional phase space x-p_x-z-p_z-σ-η by varying the angles $\varphi, \chi, \delta_1, \delta_2, \delta_3$ [2]

$$\vec{y}^n(s_0; \varphi, \chi, \delta_1, \delta_2, \delta_3) =$$
$$\cos\varphi \cos\chi \left(\vec{y}_1^n(s_0) \cos\delta_1 + \vec{y}_2^n(s_0) \sin\delta_1 \right) +$$
$$\cos\varphi \cos\chi \left(\vec{y}_3^n(s_0) \cos\delta_2 + \vec{y}_4^n(s_0) \sin\delta_2 \right) +$$
$$\sin\varphi \left(\vec{y}_5^n(s_0) \cos\delta_3 + \vec{y}_6^n(s_0) \sin\delta_3 \right) \qquad (19)$$

The projections of the hyperellipsoid onto the x-z, x-σ, z-σ and x-p_x, z-p_z, σ-η planes yield the bunch cross sections for the nth layer (i.e. the half axes of the ellipses obtained by projection of the hyperellipsoid) [2]. Hence the bunch envelope comprising, say, 95% of particles, will be given by the cross sections of the $n_{0.95}$th layer.

V. INITIAL CONDITIONS AND CHARGE DENSITY FOR THE HYPERELLIPSOID

The transfer matrix over one machine turn must be periodic. This is the case if the hyperellipsoid spanned by (19) recovers its original shape after one revolution, i.e. if $\vec{y}^n(s_0; \varphi, \chi, \delta_1, \delta_2, \delta_3)$ transforms after one turn into $\vec{y}^n(s_0; \varphi, \chi, \delta_1 - 2\pi Q_1^n, \delta_2 - 2\pi Q_2^n, \delta_3 - 2\pi Q_3^n)$. Here $Q_{1,2}^n$ and Q_3^n characterize oscillation modes which no longer correspond to pure horizontal, vertical and longitudinal motion. However they are still identified with the transverse machine tunes and the synchrotron tune respectively. The periodic condition involves the following computational steps ($n = 1 \ldots N$, $j = 1, 2, 3$):

- Find the spectrum of the revolution matrix (eigenvalues must be complex of module unity)

$$\underline{M}^n(s_0 + L, s_0)\, \vec{v}_j^n(s_0) = e^{-2i\pi Q_j^n}\, \vec{v}_j^n(s_0) \qquad (20)$$

- Normalize the vectors $\vec{v}_j^n(s_0)$ and form

$$(\vec{v}_j^n)^+(s_0)\, \underline{S}\, \vec{v}_j^n(s_0) = 2i \qquad (21)$$

$$\sqrt{\epsilon_j^n(s_0)}\, \vec{v}_j^n(s_0) = \vec{y}_{2j-1}^n(s_0) - i\vec{y}_{2j}^n(s_0) \qquad (22)$$

where $\underline{S}$ stands for the *unit symplectic* matrix.

The quantities $\vec{v}_j^n(s_0)$, Q_j^n and ϵ_j^n must be computed iteratively because the revolution matrix depends on its eigenvalues and eigenvectors. In the first iteration the space charge forces are ignored and the values $\epsilon_{1,2,3}^n$ are derived from a system of three nonlinear algebraic equations defining the boundary of the beam cross sections at injection

$$\pi\epsilon_{x,z,\sigma} = f_{x,z,\sigma}(\vec{y}_1^N(s_0), \ldots, \vec{y}_6^N(s_0), \epsilon_1^N, \epsilon_2^N, \epsilon_3^N) \qquad (23)$$

and for the subsequent ellipsoidal level layers

$$\epsilon_j^n = r_n^2\, \epsilon_j^N \qquad (24)$$

where $\pi\epsilon_x$, $\pi\epsilon_z$ and $\pi\epsilon_\sigma$ are the areas of the elliptical projection of (19) onto the corresponding phase planes, and r_n is obtained from the initial charge distribution. Eqs. 23 are solved for the case of beams injected with no twist.

In the limiting case of negligible space charge forces and no coupling between the synchrotron and betatron oscillations and between the betatron oscillations themselves, $\epsilon_{1,2}^N$ and ϵ_3^N are approximately equal to the transverse and the longitudinal emittances respectively.

The initial charge distribution $\rho(x, p_x, z, p_z, \sigma, \eta)$ occupies the six dimensional ellipsoid spanned by (19). Assuming a charge density with ellipsoidal symmetry, this hyperellipsoid can be transformed into an hypersphere so that charge distribution $\rho(\tilde{r})$ only depends on the radius $\tilde{r} = \sqrt{r^2 + r'^2}$, with $r^2 = \bar{x}^2 + \bar{z}^2 + \bar{\sigma}^2$ and $r'^2 = \bar{p_x}^2 + \bar{p_z}^2 + \bar{\eta}^2$ (the *bar* means that the values are normalized) [3]. Hence, since the charge density is constant on any hypersphere of radius $\tilde{r}$, the distribution $\rho(r)$ in the real space is obtained by integration of $\rho(r, r')$ over r'.

Thus a nest of N concentric spheres in the $\bar{x}$–$\bar{z}$–$\bar{\sigma}$ space (of density $\rho(r)$) can be derived from a family of N circles of radius $\tilde{r}_n$ (of density $\rho(\tilde{r})$) by projection onto the r axis, yielding a family of segments of length r_n (the outer circle has a radius $\tilde{r}_N = 1$). If two consecutive circles have radii $\tilde{r}_n$ and $\tilde{r}_{n+1}$ close to each other, the local charge density $\rho_n(\tilde{r})$, and so $\rho_n(r)$, can be assumed uniform.

Denormalizing the six dimensional phase space variables yields a nest of concentric ellipsoids which can be

used as a model for the bunch. A Gaussian distribution has been chosen to represent the charge distribution $\rho(x, p_x, z, p_z, \sigma, \eta)$, yielding by projection a Gaussian distribution in the real space. Uniform charge density in the real space is also considered as a limiting case [2].

VI.   CONCLUSION

A simulation code has been written and was used to study the effects of space charge self fields at 1 GeV (kinetic energy) injection into the PS for the present high intensity beam delivered to the SPS (20 bunches, each of 10^{12} protons, 55 ns long (at 4σ), 1.3 MeV half energy spread and normalized emittances $\epsilon_x^* = 50\mu m$, $\epsilon_z^* = 25\mu m$ (at 2σ)).

Both Gaussian and uniform charge distributions have been used for the simulation. The particle bunch was sliced into 8 ellipsoidal layers. The tune values (without space charge) are $Q_{0,1} = 6.217$ and $Q_{0,2} = 6.361$. Fig. 1 shows the tune spread in the PS for a Gaussian charge density within the bunch. When a uniform charge density is considered the calculated tune shifts are $\Delta Q_1 = -0.147$ and $\Delta Q_2 = -0.214$. The calculated tune spread has no practical effect on beam performance as no significant beam losses or beam blow-up have ever been observed for this intensity.

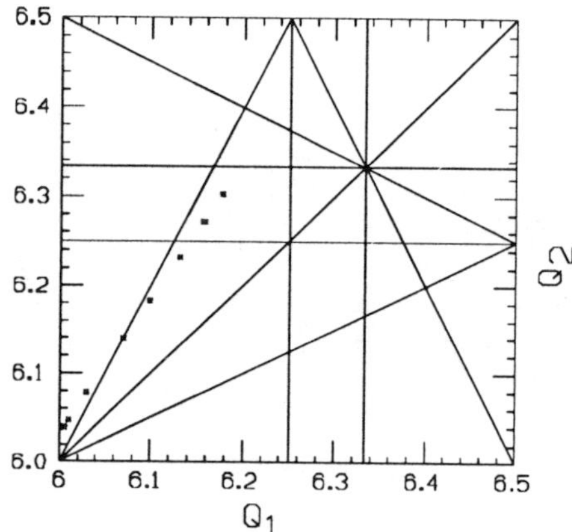

Figure 1: Distribution of incoherent tune at 1 GeV in the PS

REFERENCES

[1] I. Borchardt, E. Karantzoulis, H. Mais, G. Ripken, "Calculation of beam envelopes in storage rings and transport systems in the presence of transverse space charge effects and coupling", DESY internal report 87-161, December 1987.

[2] I. Borchardt, E. Karantzoulis, H. Mais, G. Ripken, "Calculation of transverse and longitudinal space charge effects within the framework of the fully six-dimensional formalism", DESY internal report 88-013, February 1988.

[3] J. Struckmeier, J. Klabunde, M. Reiser, "On the stability and emittance growth of different particle phase space distributions in a long magnetic quadrupole channel", GSI internal report 83-33, November 1983.

Observation of Space-Charge Effects
in the Los Alamos Proton Storage Ring*

D. Neuffer, D. Fitzgerald, T. Hardek, R. Hutson
R. Macek, M. Plum, H. Thiessen and T.-S. Wang
Los Alamos National Laboratory, Los Alamos, NM 87545

Abstract

In recent operation of the Los Alamos Proton Storage
Ring (PSR), the vertical and horizontal tunes have been
moved closer to the integers ($\nu_y = 2.12$, $\nu_x = 3.17$) to
enlarge the low-loss working region. In this region, the
beam can be significantly affected by space charge. The
first observed effects are a nondestructive distortion of the
beam profile and vertical growth of beam size sufficient
to keep the shifted tunes from crossing the integer, but
without large beam loss. At higher intensities, or with
tunes closer to the integer, beam blow-up, accompanied
by beam losses, can occur. In this paper, we report recent
observations of this intensity-dependent effect and discuss
implications for future PSR operation.

I. Introduction

The Los Alamos Proton Storage Ring (PSR) is a fast-
cycling storage ring designed to accumulate beam over a
macropulse of the LAMPF linac ($\sim$1 ms) by multiturn
injection through a stripper foil, and compress that beam
into a short, single-turn extracted pulse (0.25 μs), that
drives a spallation neutron source. The design intensity
is 100-μA on target with a 12-Hz repetition rate, which
implies $N = 5.2 \times 10^{13}$ protons/pulse. Currently, the PSR
achieves 80 μA at 20 Hz ($N = 2.5 \times 10^{13}$), and is limited
by beam losses, as well as an instability.[1] Other key PSR
parameters include proton kinetic energy ($T = 797$ MeV)
with kinetic factors ($\beta, \gamma = 0.842, 1.85$); ring circumference
($C = 2\pi R = 90.1$m); mean beam-pipe radius ($r = 0.05$m);
revolution frequency ($f = 2.795$MHz); and a single-bunch
harmonic ($h = 1, V = 10$ kV) rf system.

In recent operation, the horizontal and vertical tunes
have been moved closer to the integer ($\nu_x = 3.17$, $\nu_y =
2.12$) to obtain more stable operation within a larger low-
loss working region. The closer proximity to the integer
enhances space-charge effects, and in this paper we report
some observations of these effects.

II. Space-Charge Tune Shifts

The PSR is a high-intensity, low-energy ring, and
therefore is expected to have relatively large space-charge

effects. Following previous analyses, we expect space-
charge forces, as well as tune shifts and spreads from
electromagnetic self-fields and from electric and magnetic
image forces. At PSR parameters, the largest tune shift is
due to the electromagnetic self-field and is given by

$$\Delta\nu_y = -\frac{r_p\,R\,\overline{\beta}_y}{2\beta^2\gamma^3\,\sigma_y(\sigma_x + \sigma_y)}\frac{dN}{dz}\ , \qquad (1)$$

where $r_p = 1.536 \times 10^{-18}$ m, (σ_x, σ_y) are the rms beam
width and height, $\overline{\beta}_y \cong R/\nu_y$ is an averaged betatron
function, and dN/dz is the longitudinal density. $\Delta\nu_x$ is
obtained from the same formula with x and y exchanged.
The formula assumes Gaussian density profiles and should
be modified by a shape factor F for different distributions.
The tune shifts apply to small-amplitude motion; larger-
amplitude particles have smaller tune shifts, and the $\Delta\nu_{x,y}$
represent maximal "incoherent" tune spreads for the entire
beam.

There are also electric- and magnetic-image-field
tune shifts, which include both coherent and incoherent
effects (see references 2,3,4). At PSR parameters, the
image terms are almost an order of magnitude smaller
than the direct terms [Eq. (1)] and will not be explicitly
discussed here.

The size parameters σ_x, σ_y are given in terms of
emittances and betatron function by

$$\sigma_y = \sqrt{\frac{\epsilon_y\overline{\beta}_y}{4\pi}}\ ,$$
$$\sigma_x = \sqrt{\frac{\epsilon_x\overline{\beta}_x}{4\pi} + \left(\overline{\eta}\frac{\Delta p}{p}\right)^2}\ , \qquad (2)$$

where the betatron and dispersion functions $(\overline{\beta}_x, \overline{\beta}_y, \overline{\eta})$
are averaged around the ring and $(\Delta p/p)$ is the rms
momentum spread.

The averaging and other approximations introduce
several possible inaccuracies in applying the tune-shift
formulae to the PSR. The momentum and amplitude
distributions are not Gaussian. Neutralization of the beam
by stray electrons may occur and would reduce the tune
shift. And the dependence on longitudinal density dN/dz
can vary substantially. In the calculations below, we have
assumed that the longitudinal density profile is a Gaussian
distribution with standard deviation approximately equal
to the ring radius, so the maximum density is given by

$$\frac{dN}{dz}\bigg|_{\max} = \frac{N}{\sqrt{2\pi}R}\ , \qquad (3)$$

* Work supported by Los Alamos National Laboratory
Institutional Supporting Research, under the auspices of
the United States Department of Energy.

which implies a bunching factor of 2.5. The density profiles were not measured directly in this case and could be somewhat different. However, consistent treatment of similar cases will allow accurate exploration of the changes in the beam.

At typical PSR parameters ($N = 2 \times 10^{13}$, $\Delta p/p = 0.0025$, $\epsilon_x = \epsilon_y = 25\pi$ mm-mR, where $\epsilon \sim 4\sigma^2/\beta^*$), Eq. (1) obtains $\Delta\nu_x = 0.07$, and $\Delta\nu_y = 0.10$, comparable to the distance-to-integer tune shift. The measurements below explore the effects of this proximity.

III. MEASUREMENTS OF EFFECTS ON BEAM

The space-charge effects are greatly dependent on beam size, and in usual operation the beam is injected off-center vertically. This leads to an enlarged beam with a non-Gaussian density and reduced $\Delta\nu_y$. It was noticed that when injected on-axis the beam had a final vertical size that was dependent on beam current. Subsequent experiments and calculations showed that this was a space-charge effect.

In the experiments, a constant total injection width of 625 μs was used with a chopped micropulse width of 250 ns, centered within the 360-ns ring circumference. Intensity was varied by changing the "countdown," (CD) which sets the fraction of linac buckets that contain beam. (CDs of 1, 2, and 4 imply full, half, and one-fourth intensities, respectively.) In this experiment, full intensity was $N \cong 2.3 \times 10^{13}$ protons, although calibration was not precise. The horizontal tune was fixed at $\nu_x = 3.155$, while the vertical tune was varied to test dependences. The beam was extracted immediately at the end of injection, and beam profiles were measured by wire scans in the extraction lines. The results of one vertical and two horizontal wire scans at extraction are used, in conjunction with the TRANSPORT betatron functions at the wire scanner locations, to obtain values of $\Delta p/p$, ϵ_x, and ϵ_y, using Eq. (2). Figure 1 displays vertical wire-scan data at $\nu_y = 2.100$ for countdowns of 4, 2, 1 and shows beam-size increases with intensity. The direct tune shifts were calculated using Eq. (1); image terms were not included but would move the vertical tunes a bit closer to the integer (by $\sim 0.01 - 0.03$) and horizontal tunes farther away.

Complete data for $\nu_y = 2.193$, 2.142, 2.100, and 2.059 are displayed in Table I. Beam-size increase with intensity is shown, particularly for tunes near the integer. For $\nu_y = 2.142$, 2.100, and 2.059, the beam size increases until the tune shift saturates at values such that $\nu_y - |\Delta\nu|$ remains above the integer.

The beam blow-up is not accompanied by losses until the beam becomes too large (at $\nu_y = 2.100$, CD = 1, and at $\nu_y = 2.059$, CD $\leq$2). Also, significant horizontal beam-size increase occurs only in extreme cases ($\nu_y = 2.142$, 2.100 at CD = 1) with large coupling or large losses.

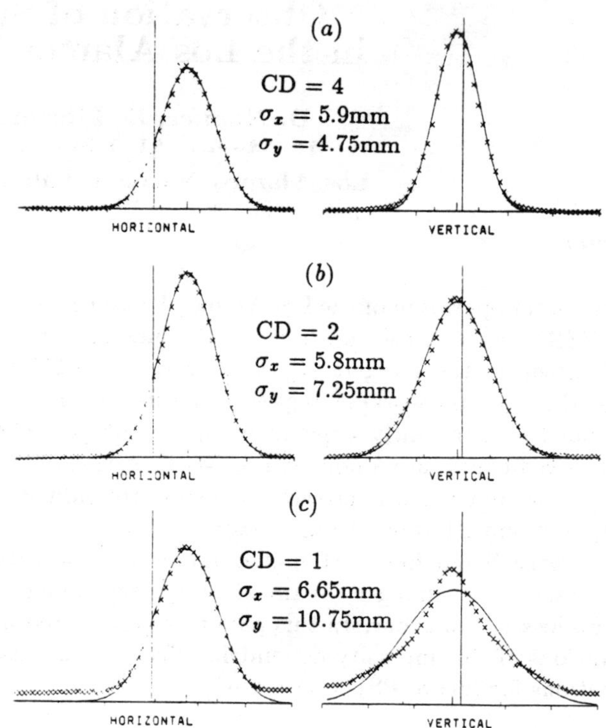

Figure 1. Beam profiles measured by wire scans for (a) Countdown = 4, (b) Countdown = 2, and (c) Countdown = 1.

Table I. Complete Data for ν_y.

ν_y	Count-down	N (10^{13})	$2\sigma_y$ (mm)	$2\Delta p/p$ (%)	ϵ_x (mm-mR)	ϵ_y (mm-mR)	$\Delta\nu_x$ (calculated)	$\Delta\nu_y$ (calculated)
2.193	4	0.6	8.5	0.38	20.1	8.6	0.035	0.075
	2	1.18	9.9	0.41	22.4	11.6	0.055	0.115
	1	2.3	13.3	0.38	26.6	20.0	0.090	0.148
2.142	4	0.6	8.4	0.41	19.7	8.4	0.032	0.075
	2	1.18	11.5	0.45	16.7	15.7	0.056	0.095
	1	2.3	15.5	0.41	25.4	28.6	0.082	0.113
2.100	4	0.6	9.5	0.47	15.5	10.7	0.031	0.061
	2	1.18	14.5	0.45	16.4	25.0	0.050	0.068
	1	2.3	21.5	0.31	44.0	55.0	0.055	0.064
2.059	4	0.6	12.6	0.46	15.0	18.9	0.028	0.043
	2	1.18	20.6	0.45	16.0	50.5	0.041	0.039
	1	beam loss prevented operation						

In the calculation, we used $\overline{\beta}_x = 4.6m$, $\overline{\beta}_y = 6.8$, $\overline{\eta} = 1.8m$ and $B = 2.507$.

IV. OPERATIONAL OBSERVATIONS

In usual operation the beam is injected off-axis vertically, so that the beam fills the aperture. This minimizes $\Delta\nu_{x,y}$. The injected beam is far from Gaussian and is actually relatively "hollow" vertically. Figure 2 shows vertical beam profiles at extraction from production conditions ($N \cong 2.25 \times 10^{13}$ at a CD of 1). Figure 2 shows profiles at CDs of 4, 2, 1 (1/4, 1/2, and full intensity). At lower intensities, the beam has a pronounced "hollow" shape vertically with an intensity minimum at $y = 0$. With higher intensity, the distribution changes becoming smoother, with a flat density near $y = 0$. This change reduces the space-charge tune spread. Beam losses are not greatly increased by the distribution shifts; in changing from CD = 2 to CD = 1, loss-monitor readings increased from 24 to 52 μV-s, only slightly more than linear with current.

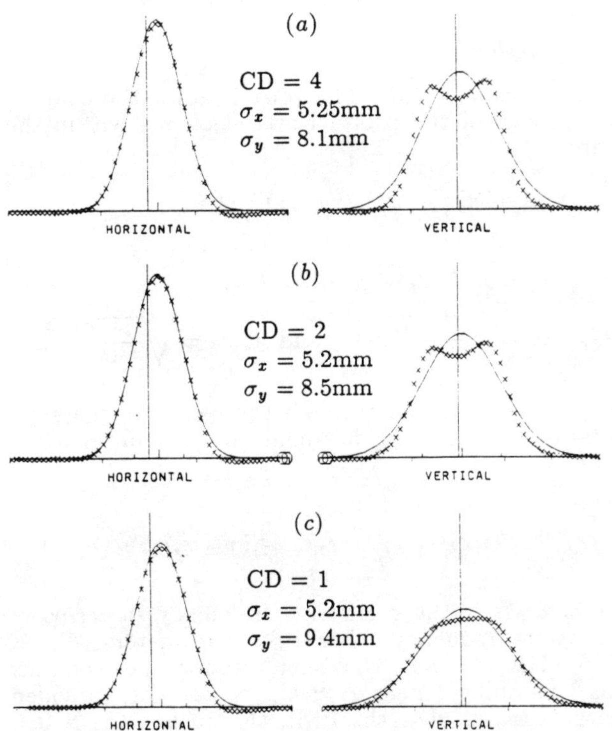

Figure 2. Beam profiles measured by wire scans for (a) Countdown = 4, (b) Countdown = 2, and (c) Countdown = 1. The beam was injected off-axis vertically.

At these parameters ($N = 2.25 \times 10^{13}$, $\epsilon_x \cong 17\pi$mm-mR, $\Delta p/p \cong 0.0018$, $\epsilon_y \cong 42\pi$, $B = 2.507$) and Gaussian formulae, we obtain $\Delta\nu_x = 0.092$, and $\Delta\nu_y = 0.089$. Because $\nu_x = 3.17$ and $\nu_y = 2.12$, the shifted vertical tune is close to the integer. At CD = 1, the proximity to the integer has been reduced by the flattening of the distribution, but without much increase in beam loss.

V. SUMMARY AND DISCUSSION

Under both production and development conditions, space-charge effects on the PSR beam have been observed. The first effect is a nondestructive distortion of the vertical beam profile and/or a size increase sufficient to keep shifted tunes from crossing the integer, but without large beam losses. At higher intensities, or ν_y nearer the integer, beam blow-up with losses occurs. This general behavior should also occur in other high-intensity synchrotrons with large, direct space-charge forces (e.g., BNL AGS and Booster, Fermilab Booster, CERN PS and Booster, SSC Low-Energy Booster) and should be considered in their design and operation. We note that in some of these cases (unlike the PSR), nondestructive emittance increase is undesirable. Our observations apply to situations in which the tunes are in proximity to an integer resonance and not to other resonances. When the tunes are in proximity to other resonances ($N/3$, $N/2$) the results could be somewhat different and should be explored.

Current operational parameters ($\nu_x \cong 2.17$, $\nu_y \cong 2.12$) are close to vertical space-charge limits. Significant increases in intensity (from $N \cong 2.5 \times 10^{13}$) will require increasing the vertical tunes.

VI. REFERENCES

1. D. Neuffer et al., contributed paper, 1991 Particle Accelerator Conference, San Francisco, California, May 1991.
2. L. J. Laslett, in *Proceedings of the 1963 Summer Study on Storage Rings*, Brookhaven National Laboratory report BNL-7534 (1963), p. 324.
3. P. Bryant, in *Proc. CERN Accelerator School— Second General Accelerator Physics Course*, CERN 87-10 (1987), p. 62.
4. G. Guignard, "Selection of Formulae Concerning Proton Storage Rings," CERN Report, CERN 77-10, 1977.

Observations of the PSR Transverse Instability*

E. Colton
EK-92, GTW, US Department of Energy, Washington, DC 20545
D. Fitzgerald, T. Hardek, R. Macek, M. Plum, H. Thiessen and T. Wang
Los Alamos National Laboratory, Los Alamos, NM 87545
D. Neuffer
CEBAF, Newport News, VA 23606

Abstract

A fast instability with beam loss is observed in the Los Alamos Proton Storage Ring (PSR) when the injected beam current exceeds thresholds, with both bunched and unbunched beams. Large coherent transverse oscillations occur before and during beam loss. Recent observations of the instability indicate that it is an "e-p"-type instability, driven by coupled oscillations due to electrons trapped within the proton beam.

I. INTRODUCTION

The PSR is a fast-cycling, high-current storage ring designed to accumulate beam over a macropulse of the LAMPF linac ($\sim$ 1 ms) by multiturn injection through a stripper foil, and compress that beam into a short, single-turn extracted pulse ($\sim 0.25\mu$s). Key PSR parameters include beam kinetic energy $T = 797$ Mev; circumference $2\pi R = 90.1$ m; revolution frequency $\Omega/2\pi = 2.795$ MHz; betatron tunes $Q_x, Q_y \approx 3.17, 2.13$; and current operating intensity $N \approx 2.35 \times 10^{13}$ particles. The design intensity is 100 μA on target at 12 Hz, which implies $5.2 \cdot 10^{13}$ protons/pulse. Average and peak intensities have been somewhat less (80 μA at 20 Hz in 1990, and $4 \cdot 10^{13}$ maximum pulse size). The average current has been limited by slow beam losses, and individual pulse intensities are limited by the fast instability.

The instability appears when more than $\sim 1.5 \times 10^{13}$ protons are stored in bunched mode (rf on) and when more than $\sim 0.5 \cdot 10^{13}$ are stored in unbunched mode. Transverse oscillations at ~ 100 MHz are seen, and grow exponentially at time scales of 10-100 μs, causing beam losses (see Fig. 1). Initial experiments and observations of the instability reported by Neuffer et al. [1] showed dependences of the thresholds on rf voltage, beam size, momentum spread, and sextupole and octupole strengths. Impedance couplings were suspected to be the cause of the instability. Searches for a possible impedance source were unsuccessful, and recent observations (both with unbunched and with bunched beam) are not consistent with a hardware impedance source. The observations are consistent with the possibility that the instability is an e-p instability, and supporting calculations have shown that conditions for e-p instability may occur with bunched beam in the PSR [2,3].

In an e-p instability, background low-energy electrons are trapped within the space-charge potential of the circulating proton beam. Coupled transverse oscillations of the beam due to the trapped electrons develop, leading to beam losses. The instability has been seen in the Bevatron and CERN ISR, and simple, linearized theoretical models

* Work supported by Los Alamos National Laboratory Institutional Supporting Research, under the auspices of the United States Department of Energy.

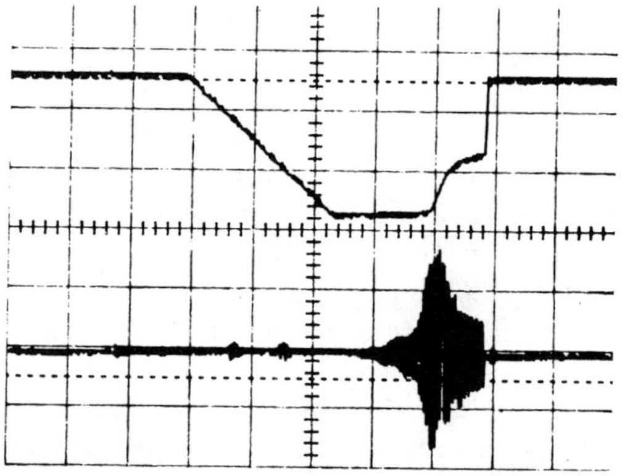

Fig. 1. Beam current (upper trace) and vertical difference signals (lower trace) under unstable conditions.

have been developed [4,5]. The equations for the coupled vertical motions of the protons and electrons within the beams are:

$$\ddot{y}_p + (Q_y^2 + Q_p^2)\Omega^2 y_p = Q_p^2\Omega^2 \bar{y}_e \quad ,$$

and

$$\ddot{y}_e + Q_e^2\Omega^2 y_e = Q_e^2\Omega^2 \bar{y}_p \quad ,$$

where $Q_e\Omega \cong \sqrt{\frac{2Nr_ec^2(1-\eta_e)}{\pi b(a+b)R}}$, and $Q_p\Omega = \sqrt{\frac{2\eta_e Nr_pc^2}{\pi b(a+b)\gamma R}}$.

The motions are coupled through the center of mass $\bar{y}_e$, $\bar{y}_p$ oscillations. Assuming harmonic motion obtains the dispersion relation.

$$(Q_e^2 - x^2)(Q_y^2 + Q_p^2 - (n - x)^2) = Q_e^2 Q_p^2 \quad ,$$

where $x = \omega/\Omega$ is the oscillation frequency in terms of the revolution frequency. For PSR parameters, $Q_e \cong 40$ (~ 100 MHz). The dispersion relation has complex solutions (instability) near $x \approx Q_e \approx n - Q_y$, provided Q_p is large enough. For the PSR, this means $Q_p > 0.1$, which implies that a neutralization of $\eta_e > 0.01$ (1%) can lead to instability at a relatively low electron density. Some stabilization by Landau damping (frequency spread) is possible; the stabilization effects seen in the PSR are qualitatively in agreement with the e-p model.

For an e-p instability to exist, stable trapping of electrons must occur within the space-charge potential of the circulating protons. With unbunched beam, the space charge is quite strong and should easily trap electrons. With bunched beam, a beam-free interbunch gap of ~ 100 ns should pass through the electrons every turn, freeing rather than trapping electrons. However, recent

calculations [2] have shown that if a low-density beam
having a smooth overall density distribution, leaks into
the gap, electron trapping can occur with bunched beam
at PSR parameters. Recent experiments (see Section
II) do show that the instability is associated with beam
leakage, and that such leakage is a plausible result of the
PSR longitudinal dynamics, involving rf voltage (relatively
small effect), longitudinal space charge (large effect), and
injection phase-space mismatch (large effect).

II. RECENT PSR EXPERIMENTS

The results of these experiments are reported in
detail in a forthcoming article [3]; we summarize some
critical observations below.

A. Background Charge Experiments

An e-p instability depends on a source of electrons
that can be trapped within the circulating proton
beam. Possible sources include secondaries from beam-
foil interactions, beam-gas interactions, and beam losses
on the walls. Both beam-foil interactions and beam losses
are relatively large in the PSR. Also, there are no clearing
electrodes to remove charges. In some recent experiments,
PSR background charge conditions have been modified,
leading to changes in instability thresholds. Such changes
are not consistent with a $Z_\perp$-instability, which should be
independent of background.

In one experiment, sufficient voltage was placed on
the foil to clear electrons in the vicinity. With unbunched
beam, an increase to 300 V (the expected space-charge
potential) increased thresholds by $\sim$10%; but further
increases (to 2000 V) showed no further improvement. In
several experiments, the vacuum was degraded from $\sim$2-5
10^{-8} up to 10^{-6} Torr. The beam, bunched or unbunched,
became more unstable; thresholds were changed by $\sim$10%.
The further instability could be caused by increased
e^--density from the beam-gas scattering. In another
experiment, beam losses were increased by moving the halo
scrapers toward the beam. The beam again became more
unstable, even though intensity was decreased. The model
is that losses increased secondary e^- production, leading
to increased instability. That changes in e^- density does
change the stability of the proton beam is consistent with
the e-p instability hypothesis; however, a dominant e^-
source has not yet been identified.

B. Gap Filling and Instability

Calculations show that bunched beam e-p instability
should not occur in the PSR, unless the interbunch gap
has filled in. Recent experiments do indeed show that
instability only occurs when the interbunch gap has filled
in, providing a continuous trapping force for low-energy
electrons. In one experiment, beam was injected with rf on,
forming a stable bunch. During storage the rf was turned
off, and it was observed that as gap-filling proceeded, the
beam became unstable.

With bunched beam (rf on), it has generally been
observed that instability occurs *only* when the interbunch
gap has filled in to some extent. Figure 2 shows
bunch shapes in cases slightly below and slightly above
threshold, at $V_{rf} = 10\ kV$. In the unstable case, the
interbunch gap has filled in, forming a smooth, sinusoidal
density variation. Figure 3 shows some longitudinal beam

profiles at end of injection. All cases with beam leakage
showed strong instability, and all cases with a beam-free
interbunch gap were stable.

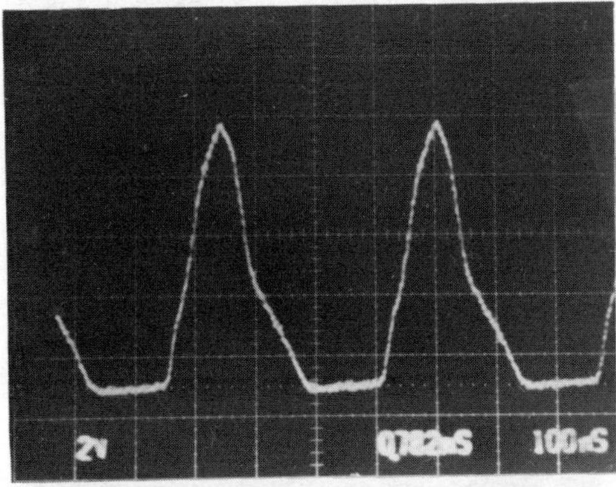

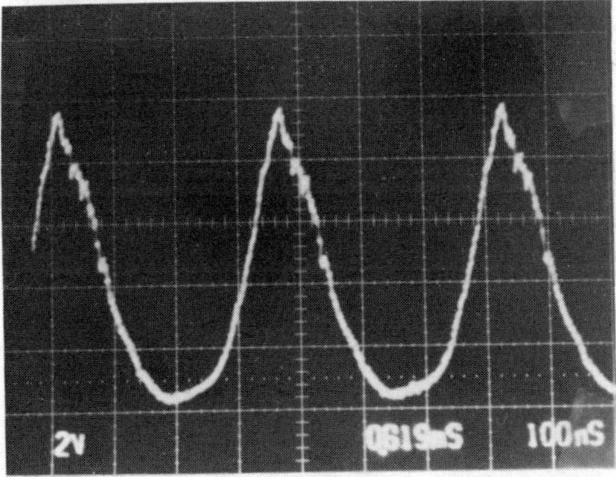

Fig. 2. Beam profiles below and above instability
threshold. Note smoothly-filled gap in unstable case.

Measurements under various conditions indicate that
gap filling occurs before or simultaneously with the
beginning of exponentially growing oscillations, well before
beam loss. This indicates that gap-filling is a cause, not a
result, of the instability.

In addition, experiments were performed in which a
small amount of beam was deliberately injected into the
interbunch gap (by degrading injection chopping). The
beam became more unstable; instability thresholds were
reduced by $\sim$20%. In a complementary experiment, beam
was kicked out of the interbunch gap during storage (by a
gated transverse kick); the beam was stabilized. Injection
can also be modified to make leakage more difficult, by
injecting with a shorter width. The beam is then injected
deeper into the confining rf bucket. Results show that
much more stable beam can be stored by this method.

C. Frequency Spectra Observations

The frequency spectra of the unstable vertical
oscillations have been measured by an HP spectrum
analyzer and by Fourier transforms of digitized, position-
monitor data. Oscillation peaks near 100 MHz with 10-
50-MHz widths occur with instability; however, the peak
location can vary between 40 and 200 MHz, depending on
beam conditions. This is inconsistent with a $Z_\perp$ instability,
in which the oscillation frequencies should remain

unchanged. However, in an e-p instability, the frequencies should be near $Q_e f_0$, and should vary with beam density $\{Q_e f_0 \propto \sqrt{(N'/b(a+b))}\}$. The variations in peak location and width that we observed are consistent with the measured and expected variations in three-dimensional beam density.

III. Summary and Discussion

Recent experiments and calculations indicate that the PSR transverse instability is an e-p instability. Because electron trapping, which triggers the instability, occurs only when there is some leakage of beam into the gap, maintaining a beam-free interbunch gap is desirable in PSR operations. Leakage can be avoided or delayed by manipulating PSR parameters, such as injection width, rf voltage, and phase; such measures have empirically improved operation and assisted in increasing intensity to current levels.

Future experiments will search for more definite proof of e-p instability. A critical experiment would be to investigate more fully the use of a transverse kicker to remove beam from the interbunch gap, to determine whether gap clearing consistently stabilizes the beam. Other experiments should try to identify a dominant e$^-$ source (possibly stripper foil or beam losses); this could point the way for installation of clearing electrodes.

IV. References

[1] D. Neuffer, et al., "Transverse Collective Instability in the PSR," *Particle Accelerators* **23**, 133 (1988).
[2] D. Neuffer, "Calculations of the Conditions for Bunched-Beam e-p Instability in the Los Alamos Proton Storage Ring," 1991 Particle Accelerator Conference, San Francisco, CA (May 1991).
[3] D. Neuffer et al., "Observations of a Fast Transverse Instability in the PSR," submitted to *Particle Accelerators*.
[4] H. Grunder and G. Lambertson, "Transverse Beam Instabilities at the Bevatron," Proc. 8th Int. Conf. on High-Energy Accelerators, CERN (1971), p. 308.
[5] E. Keil and B. Zotter, "Landau-Damping of Coupled Electron-Proton Oscilltions," CERN Report CERN/ISR-TH/71-58 (1971).

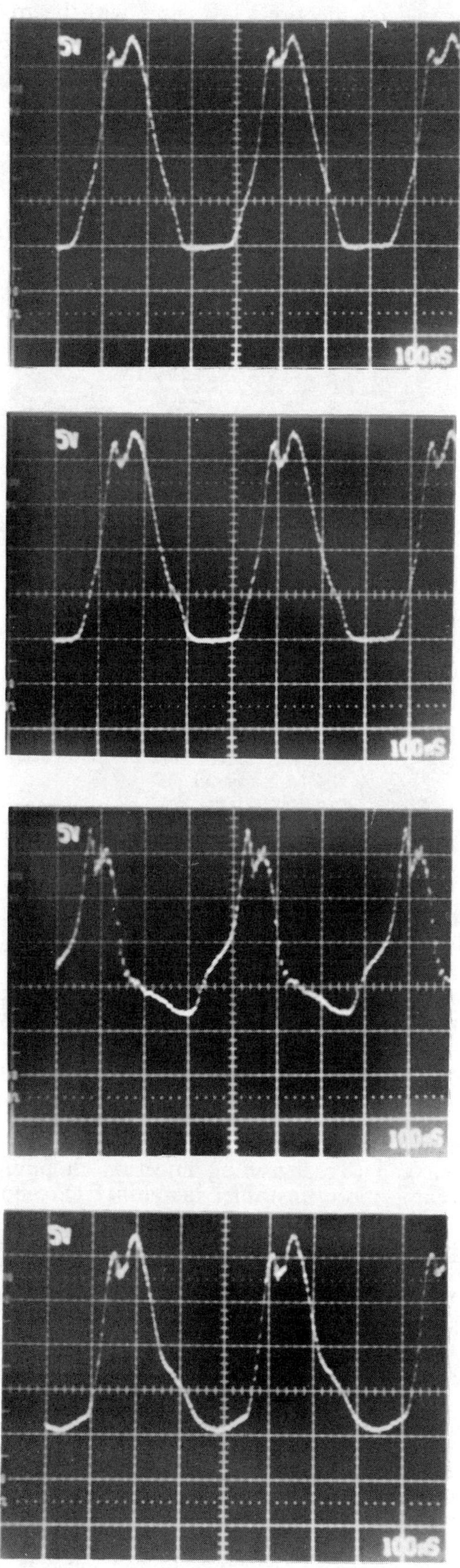

Fig. 3. Beam profiles at end of injection under stable (upper two) and under unstable (lower two) conditions.

NONLINEAR DYNAMICS IN THE BOOSTER OF THE MOSCOW KAON FACTORY

N. I. Golubeva, A. I. Iliev and Yu. V. Senichev, INR, Moscow, USSR

I.   INTRODUCTION

The proposed lattice for the MKF Booster low-energy ($7.47\,GeV$) synchrotron [1] has a racetrack shape with two $180°$ arcs and two long dispersion-free straight sections. To obtain the desirable high transition energy $\gamma_t \simeq 20$ each arc has a 'missing magnet' $FODO$ structure with relatively small modulation of the β-functions [2]. The horizontal betatron tune of each arc Q_x equals to 3 to provide zero dispersion in long straight sections. In order to reduce the tune spread 32 sextupoles are placed in the arcs for chromaticity correction. In contrary to sextupoles introduced intentionally, in the Booster lattice there are casual and, generally speaking, unavoidable sources of nonlinearities like multipolar field imperfections in the magnets. Since, the Booster is high-intensity synchrotron (average current $= 250\,\mu A$) with large transverse beam size, the nonlinear dynamics of the particle motion is a matter of serious interest. In order to be sure that the lattice is satisfied the design performance in presence of chromaticity-correcting sextupoles we have to estimate the phase-space distortion, the growth of the effective emittances[1], nonlinear tune dependence on the amplitudes, etc. We should also determine the acceptable multipole contents of the Booster magnets.

In general, the study of nonlinear effects is very complicated problem and it has various aspects. And, we are not stupid enough to have a hope to solve the problem completely in one 'jump'. The present note is an attempt to understand the phase-space topology and rather the plan for future work should be done than a full description of nonlinear properties of the lattice.

As a first step, to be sure that the lattice allows for chromaticity correction, we have studied the nonlinear actions on particle motion provided by sextupoles for the perfect machine, i.e. without misalignments and multipolar imperfections in the magnets.

II.   NONLINEARITY DUE TO SEXTUPOLES

The oscillation of the dispersion on the arcs, corresponding to high γ_t, significantly reduces a number of places in which the sextupoles can be installed. Consequently, the chromaticity correction requires relatively strong sextupoles. This is the general disadvantage of any racetrack high transition energy lattice with relatively small circumference.

A.   The principles of analysis

Let us give some remarks concerned the principles of analysis of nonlinear dynamics. In this study we have attempted to obtain a clear view of the following regions of different particle's behavior:

- a region of quasi-linear motion that is often called "linear aperture";

- a region of chaos;

- a boundary region beyond which motion becomes unbounded — "dynamic aperture".

Linear aperture

Since we suppose the particle motion would be in the range of linear aperture we try to estimate the size of this region. To do this it is essential to have a criterion of "linearity". The good candidates for such figure of merit are

— amplitude depended tune shifts from the nominal values Q_{x0}, Q_{y0};

— some kind of "SMEAR";

In analysis of tracking data the set of particle positions (x, p_x, y, p_y) obtained after each turn is used to compute the quantities $\varepsilon_x, \varepsilon_y$ and w, where ε_z, $z = \{x, y\}$ is invariant of linear betatron theory ('emittance'): $\varepsilon_z = \gamma_z z^2 + 2\alpha_z p_z z + \beta_z p_z^2$. And $w = (\varepsilon_x^2 + \varepsilon_y^2)^{1/2}$. Then we have computed the average and and rms values of them: $\overline{\varepsilon}_x, \overline{\varepsilon}_y, \overline{w}, \sigma_x, \sigma_y, \sigma_w$. The three kinds of smear S_x, S_y, S_w are defined as

$$S_x = \frac{\sigma_x}{\overline{\varepsilon}_x}, \quad S_y = \frac{\sigma_y}{\overline{\varepsilon}_y}, \quad S_w = \frac{\sigma_w}{\overline{w}}. \tag{1}$$

At this stage of the study the phase-space region in which the following criteria (both or one) are fulfilled:

- $|\Delta Q| \leq 0.005$;

- $S_x, S_y, S_w \leq 15\%$,

we call "linear aperture".

Dynamic aperture

The most general definition of dynamic aperture (DA) is

Def. 1 *The DA is the phase-space volume (set of initial conditions $\{x_0, p_{x0}, y_0, p_{y0}, \tau_0, \delta\}$) for which the particle motion is stable (remains finite) over a large enough number of turns N_t i.e. $|z(n)| \leq A_z$, $z = \{x, y\}$, $n = 1 \ldots N_t$, A_z is a boundary.*

Thus to obtain the DA one should test many particles with initial conditions within a range of the phase-space in which the particles are injected. In practice, since a number of the testing particles and number of turns for tracking are limited by computer power and 'staying power' of scientist, one should accept some different and less general definition of the dynamic aperture:

Def. 2 *The DA is a set of initial conditions of actions $\{J_{x0}, J_{y0}\}$ ($J_z = \varepsilon_z/2$) and δ at fixed initial betatron phases ϕ_{x0}, ϕ_{y0} for which the particle motion remains finite over a fixed number of turns N_t i.e. $|z(n)| \leq A_z$, $z = \{x, y\}$, $n = 1 \ldots N_t$, A_z is a boundary.*

Obviously, the finite number of testing particles and a particular choice in the initial betatron phases ϕ_{x0}, ϕ_{y0} leads to that DA defined by Def. 2 becomes

- in dependence on the azimuthal position of the inspection point IP (i.e. starting point for tracking) and

---

[1]It is important to provide some safety factor in the aperture definition.

- in dependence on the particular choice in the initial phases.

This dependencies require to analyze larger than one point in the lattice. In section B. you will see why we have chosen two inspection points: the first point is localized at the entrance of the arc, the second one is placed in the middle of same arc.

At this stage of the research we have found by tracking over $N_t = 1000$ turns only cuts of the stable volume corresponding to on-momentum particles $\delta = 0$ with full coupling, i.e. $\varepsilon_x = \varepsilon_y = \varepsilon$. We have tested only one particle with particular choice $p_{x0} = p_{y0} = 0$ ($\phi_x = \phi_y = 0$) for each value of ε. The boundary A_z was chosen equals to $1\,m$ for both transverse planes.

B.   Linear arc's optic and geometric aberrations

In proposed lattice design for the Booster the conditions for the arcs to be a second-order pseudo-achromat are fulfilled [3]:

- the horizontal betatron tune of each arc Q_x equals to the integer ($Q_x = 3$), the vertical tune of the arc Q_y is only slightly differed from integer number $Q_y = 3.125$ (it is needed for acceleration of polarized proton beam);

- the tunes of the superperiod ($\nu_x = 0.75\,, \nu_y = 0.78125,$) are not satisfied the resonance conditions: $m\nu_x + n\nu_y \neq integer$, $(m, n) = \{(1, 0), (3, 0), (1, 2), (1, -2)\}$.

In this case all low order resonances introduced by sextupoles are compensated after passing of each arc. But, in any second-order achromat there is a point $\pi \times$ *an odd integer* distant from the entrance of the achromat (in our case this point is the middle of the arc), in which the influences of these resonances are still great. Really, in the first order in sextupole's strength the perturbation of the linear invariant of motion (or action) J after passing N superperiods having identical sextupole's arrangement is expressed by

$$|\Delta J| \simeq |\Delta J_{sup}| \cdot \frac{\sqrt{1 - \cos(2\pi N(m \cdot \nu_x + n \cdot \nu_y))}}{\sqrt{1 - \cos(2\pi(m \cdot \nu_x + n \cdot \nu_y))}}, \qquad (2)$$

where ΔJ_{sup} is an integral over one superperiod. It is clear from (2) that at the ends of the arc ($N = 4$) there are not the phase-space distortions provided by low order sexdtupole's resonances, but near the middle of the arc ($N = 2$, $N \cdot \nu_{x,y} \simeq$ 1.5 the influences of that resonances are still great. Just because we have carried out the analysis for two different points in the lattice: the first inspection point (IP1) is the entrance of the arc, the second one (IP2) is the middle of the same arc. In Fig. 1,2 the phase-space plots $p_{xn} = -1/\sqrt{\beta}(\alpha x + p_x \beta)$ versus $x_n = x/\sqrt{\beta}$ for horizontal betatron motion are shown for these points of the arc. One can see the large phase-space distortion in the middle of the arc corresponding to the strong 1/3 resonance.

C.   Results

Following the principles described above we have used the program DIMAD [4] for tracking particles. The phase-space coordinates of the particle, obtained for each turn in points of the arc IP1 and IP2, are used to calculate all kinds of smear defined by (1), the average and *rms* values of the betatron tunes. The results are summarized in Fig. 3-7. In Figures 3-5 the smears S_x, S_y and S_w versus initial amplitude of the testing particles ε are shown. As you can see, in all cases the

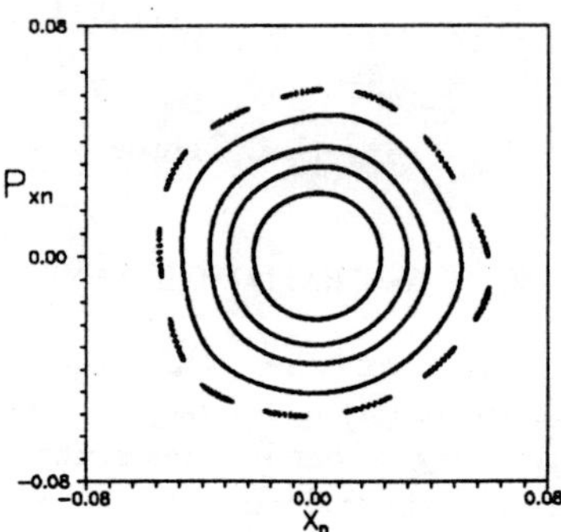

Figure 1: The turn-by-turn tracking data displayed in normalized phase space. Plot corresponds to the entrance of the arc.

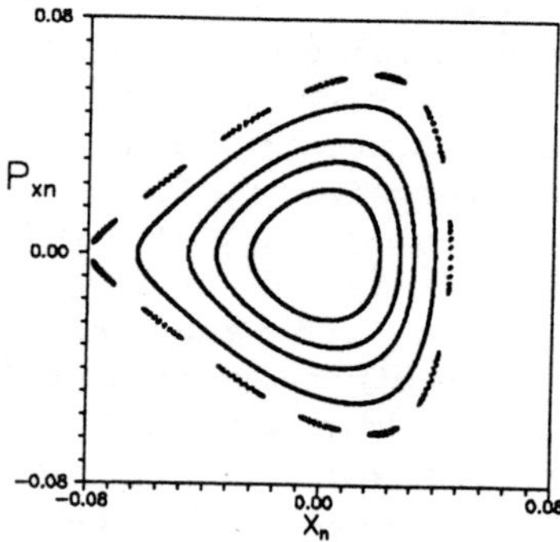

Figure 2: The same phase-space plot as in Fig. 1, but the indicated particle positions correspond to the middle of the arc.

smear computed for inspection point IP2 considerably larger than smear for IP1. The same is true for *rms* values of the tunes (Fig. 6). Based on these results we derive that the linear aperture is about $250\,\pi\,mm\,mrad$. The dynamic aperture in sense described above, is approximately $2300\,\pi\,mm\,mrad$. Near the DA the smear becomes $\simeq 100\%$. As it is seen from Fig. 3,7 at the amplitudes about $1000\,\pi\,mm\,mrad$ there is a strong coupling resonance.

III.   THE EFFECT OF HIGH-ORDER MULTIPOLE IMPERFECTIONS ON DA

In order to estimate the effect of multipole imperfections of magnets on DA, the systematic multipole components have been taken into consideration. Two thin-multipoles have been used to simulate the field errors in each dipole magnet. They have been placed in the both ends of the dipoles. At present stage, we use the following integrated strengths of the multipoles: $K_2 = \beta = 102.81$, $K_4 = 7.68$, $K_6 = 2495.5$, $K_8 = -6235.5$ [2]. To estimate the importance of each multipole, we plot DA versus number of the field harmonic (Fig. 8). We have found that the systematic multipole field harmonics in the bending magnets K_4 and K_6 could significantly reduce the dynamic aperture (Fig. 8). Further study should be done to obtain the upper limits of the tolerable multipole field errors.

---

[2] We use the DIMAD's definition of K_n.

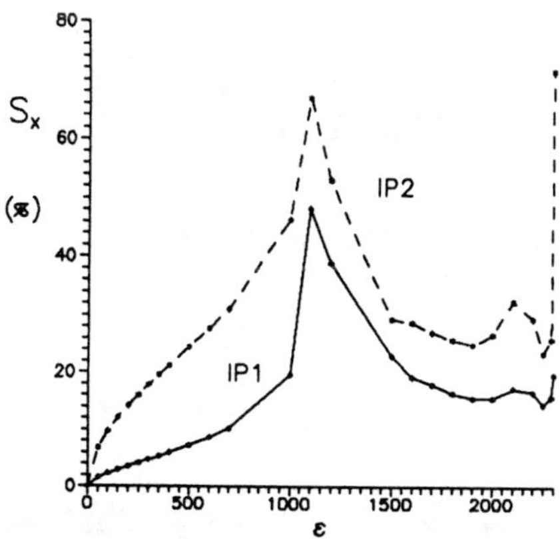

Figure 3: Smear S_x vs. initial particle's amplitude $\varepsilon\,[\pi\,mm\,mrad]$

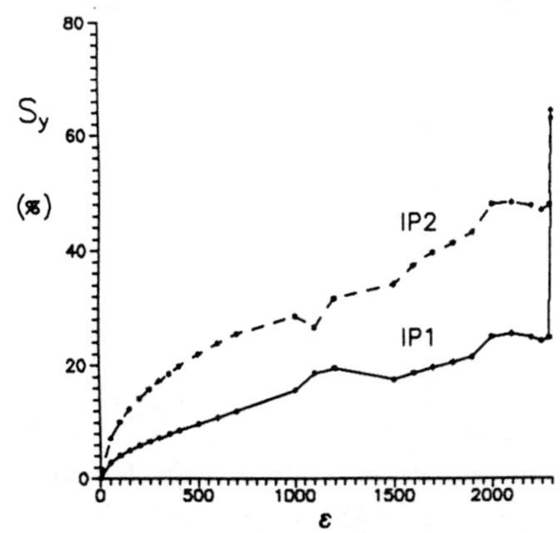

Figure 4: Smear S_y vs. initial particle's amplitude $\varepsilon\,[\pi\,mm\,mrad]$

IV.   CONCLUSION

It seems from the obtained results that proposed lattice for MKF Booster is allowed for chromaticity correction. Much more systematic work should be done to study the effects on the beam dynamics of nonlinear field imperfections.

V.   REFERENCES

[1] Yu. V. Senichev et al.," The accelerator complex of the Moscow Kaon Factory," in this proceedings

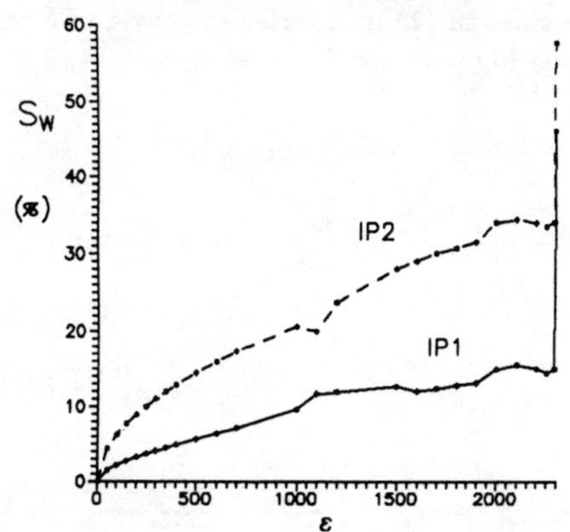

Figure 5: Smear S_w vs. initial particle's amplitude $\varepsilon\,[\pi\,mm\,mrad]$

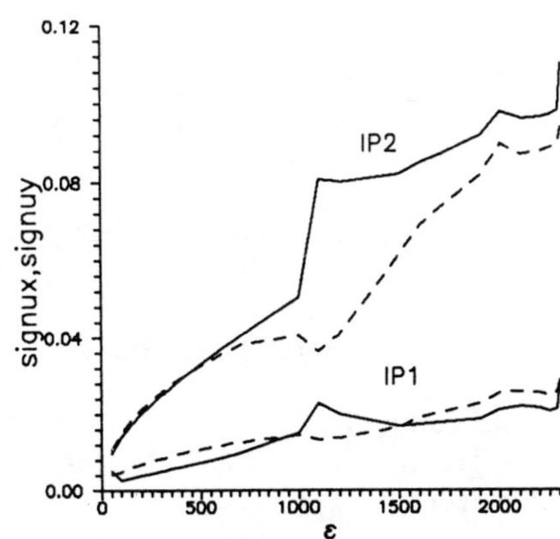

Figure 6: The *rms* values of betatron tunes vs. initial amplitude of particles

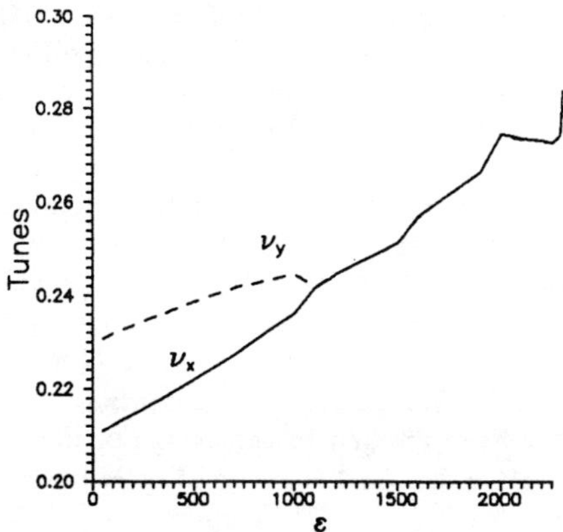

Figure 7: The average betatron tunes vs. ε

[2] A. I. Iliev and Yu. V. Senichev, "Racetrack lattices for low-medium-energy synchrotrons," in this proceedings

[3] S. A. Kheifets, T. H. Fieguth and R. D. Ruth, "Canonical description of a second-order achromat," in proc. of Second Advanced ICFA Beam Dynamics Workshop, CERN 88-04, pp. 52–61, Geneva, 29 July 1988

[4] R. V. Servranckx et al., "User guide to the program DI-MAD," SLAC Report 285 UC-28 (A), May 1985

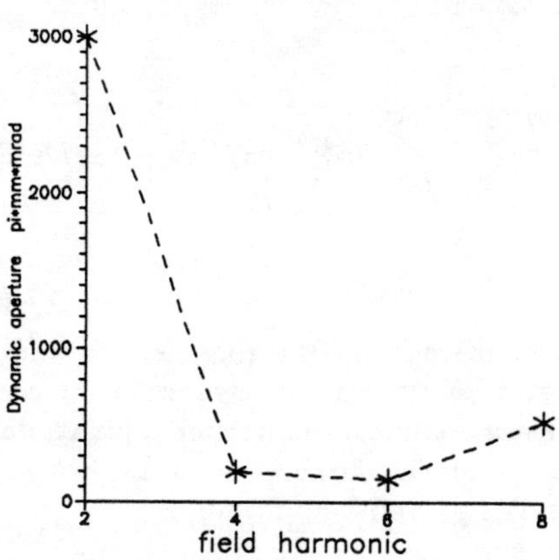

Figure 8: The dynamic aperture versus systematic multipole field harmonic of the dipoles

HOW TO GET A SEPARATRIX BRANCH WITH LOW DIVERGENCE AT A 1/3-INTEGER RESONANT BEAM SLOW EXTRACTION

S. P. Volin,
Institute for Nuclear Research of the Academy of Sciences of the USSR Moscow, USSR

I.   INTRODUCTION

After the invention by H.Hereward of the resonant slow extraction scheme [1], this method is widely used to almost all existing synchrotrons. An analytical treatment of the resonant extraction based upon 1/2 or 1/3-integer resonances are proposed in numerous works [2,3,4]. This article deals with the particular problem concerned with the separatrix dependent on chromaticity in 1/3-integer resonance. It means that the outgoing separatrices do not overlap for different momentum in other words they occur to be smeared. It is desired to get a separatrix branch with low divergence at the septum position in order to minimize the particle losses on it. Hardt [5] proposed the method of compensation of this effect by introducing nonzero dispersions in the beam extraction straight section. But in his formula only global chromaticity is included and not local. In this article some handy formulas are proposed using analytic techniques for solving this problem including both global and local chromaticity effects. These formulas have been applied to the design of the resonant slow extraction system for the Extender Ring of the Moscow Kaon Factory [6].

II.   DERIVATION OF THE FORMULAS

In this Section we will use the formalism of the resonance canonical perturbation theory expounded in [7]. We start with the Hamiltonian describing the particle dynamics near 1/3-integer resonance:

$$H = \Delta e J + \varepsilon J^{3/2} \cos 3\varphi . \tag{1}$$

Here $\Delta e = Q_x - p/3$ is the frequency difference, $Q_x = Q_x^{(0)} + Q_x' \delta$ is the horizontal betatron tune, Q_x' is the horizontal chromaticity, $\delta = \Delta P/P_0$ is the fractional difference of the particle momentum P from that of the reference momentum P_0, ε is the resonance force which is determined by the strength S and location of the sextupoles as follows:

$$\varepsilon = (A^2 + B^2)^{1/2}$$

$$A = \frac{1}{24\pi} \sqrt{\frac{2}{1+\delta}} \int_0^C S\beta_x^{3/2} \cos 3(\mu_x - \Delta es/R) \, ds ,$$

$$B = -\frac{1}{24\pi} \sqrt{\frac{2}{1+\delta}} \int_0^C S\beta_x^{3/2} \sin 3(\mu_x - \Delta es/R) \, ds . \tag{2}$$

Here C is the circumference of the ring, α_x, β_x and μ_x are the Twiss parameters and the phase advance in the horizontal space along the closed nonlinear equilibrium orbit at s location. The action J and angle φ variables are transformed to the Cartesian coordinates p_x and x by formulas:

$$x = \delta D_x + \sqrt{\frac{2\beta_x J}{1+\delta}} \cos \Phi$$

$$p_x = \delta D_x' - \sqrt{\frac{2(1+\delta)J}{\beta_x}} \cos \Phi \left[\tan \Phi + \alpha_x\right], \tag{3}$$

where

$$\Phi = \varphi + \mu_x - \Delta es/R - \chi/3$$

$$\chi = -\arctan(B/A) . \tag{4}$$

The phase portrait looks very simply in the coordinates $(2J)^{1/2} \cos \varphi$ and $(2J)^{1/2} \sin \varphi$ — it is a stable triangle limited by straight separatrises. The fixed points of the hamiltonian system with the Hamiltonian (1) for $\Delta e < 0$ are:

$$J_i^u = \left(\frac{2\Delta e}{3\varepsilon}\right)^2$$

$$\varphi_i = i\frac{2\pi}{3}, \quad i = 0, 1, 2 . \tag{5}$$

and for $\Delta e > 0$ are:

$$J_i^u = \left(\frac{2\Delta e}{3\varepsilon}\right)^2$$

$$\varphi_i = \frac{\pi}{3} + i\frac{2\pi}{3}, \quad i = 0, 1, 2 . \tag{6}$$

Henceforward for definition without violation generality we suppose that $\Delta e < 0$. As far as we are interested in functions at the septum location so also we will put the origin of the beam axis at this location. Then in coordinates x and p_x the formula (5) looks as:

$$x_i = \delta D_x + \sqrt{\frac{2\beta_x}{1+\delta}} \frac{2|\Delta e|}{3\varepsilon} \cos\left(i\frac{2\pi}{3} - \frac{\chi}{3}\right)$$

$$p_{xi} = \delta D_x' - \sqrt{\frac{2(1+\delta)}{\beta_x}} \frac{2|\Delta e|}{3\varepsilon} \cos\left(i\frac{2\pi}{3} - \frac{\chi}{3}\right)$$

$$\times \left[\tan\left(i\frac{2\pi}{3} - \frac{\chi}{3}\right) + \alpha_x\right], \quad i = 0, 1, 2 . \tag{7}$$

As seen from (7) the linear dependence of the stable triangle position on δ is determined by the dispersions D_x and D_x' and the horizontal tune chromaticity Q_x'. Meanwhile the nonlinear dependence is determined by $\beta_x(s,\delta)$, $\alpha_x(s,\delta)$, $\mu_x(\delta)$, $\varepsilon(\delta)$, $\chi(\delta)$ and impede the further analytical treatment. We simplify the problem by consideration only the linear theory in δ. We expand the right sides in (7) in a series in power of δ and omit the terms of power higher than 1. It leads to

$$x_i = \delta D_x + \frac{2\sqrt{2\beta_x^{(0)}}}{3} \frac{|\Delta e_0|}{\varepsilon_0} \cos \chi_i^{(0)}$$

$$\times \left\{1 + \left[\frac{\beta_x^{(1)} - 1}{2} + \frac{Q_x'}{\Delta e_0} - \varepsilon_1 + \frac{\chi_1}{3} \tan \chi_i^{(0)}\right]\delta\right\}$$

$$p_{xi} = \delta D_x' - \frac{2}{3} \sqrt{\frac{2}{\beta_x^{(0)}}} \frac{|\Delta e_0|}{\varepsilon_0} \cos \chi_i^{(0)} \left\{\tan \chi_i^{(0)} + \alpha_x^{(0)}\right.$$

$$+ \left[\left(\tan \chi_i^{(0)} + \alpha_x^{(0)}\right)\left(\frac{Q_x'}{\Delta e_0} - \frac{\beta_x^{(1)} - 1}{2} - \varepsilon_1\right) + \alpha_x^{(1)}\right.$$

$$\left. - \frac{\chi_1}{3}\left(1 - \alpha_x^{(0)} \tan \chi_i^{(0)}\right)\right]\delta\right\}, \quad i = 0, 1, 2 \tag{8}$$

where $\chi_i^{(0)} = i(2\pi - \chi_0)/3$ or in short form:

$$
\begin{aligned}
x_i &= x_i^{(0)} + (D_x + x_i^{(1)})\delta \\
p_{xi} &= p_{xi}^{(0)} + (D'_x + p_{xi}^{(1)})\delta
\end{aligned}
\tag{9}
$$

Here we used the next notations:

$$
\begin{aligned}
\beta_x(\delta) &= \beta_x^{(0)}(1 + \beta_x^{(1)}\delta + \dots) \\
\alpha_x(\delta) &= \alpha_x^{(0)} + \alpha_x^{(1)}\delta + \dots \\
\varepsilon(\delta) &= \varepsilon_0(1 + \varepsilon_1\delta + \dots) \\
\chi(\delta) &= \chi_0 + \chi_1\delta + \dots
\end{aligned}
\tag{10}
$$

Let x_s is a septum location and p_{xs} is a value of straight separatrix crossing two fixed points, e.g. M_0 and M_2 at $x = x_s$. We get for p_{xs}:

$$
p_{xs} = \frac{(p_{x2} - p_{x0})}{(x_2 - x_0)}(x_s - x_0) + p_{x0}
\tag{11}
$$

Certainly p_{xs} is a function of δ and is of our interest. We demand of this function $p_{xs}(\delta)$ that its first derivative with respect to δ equals to zero at $\delta = 0$:

$$
\left.\frac{dp_{xs}}{d\delta}\right|_{\delta=0} = 0
\tag{12}
$$

For small δ which is a common case for large synchrotrons this demand provides approximately constant value of the function $p_{xs}(\delta)$ and what is actually need. So from (9), (12) it follows that

$$
\left.\frac{dp_{xs}}{d\delta}\right|_{\delta=0} = F_0 D_x + D'_x + F_1 = 0,
\tag{13}
$$

where

$$
F_0 = -\frac{p_{x2}^{(0)} - p_{x0}^{(0)}}{x_2^{(0)} - x_0^{(0)}}
$$

$$
\begin{aligned}
F_1 = {}& p_{x0}^{(1)} - \frac{(p_{x2}^{(0)} - p_{x0}^{(0)})}{(x_2^{(0)} - x_0^{(0)})}x_0^{(1)} + (x_s - x_0^{(0)}) \\
&\times \frac{(p_{x2}^{(1)} - p_{x0}^{(1)})(x_2^{(0)} - x_0^{(0)}) - (p_{x2}^{(0)} - p_{x0}^{(0)})(x_2^{(1)} - x_0^{(1)})}{(x_2^{(0)} - x_0^{(0)})^2}.
\end{aligned}
\tag{14}
$$

From the mathematical point of view we have one equality which binds the values of 11 functions D_x, D'_x, Q'_x, $\beta_x^{(0)}$, $\beta_x^{(1)}$, $\alpha_x^{(0)}$, $\alpha_x^{(1)}$, ε_0, ε_1, χ_0, χ_1 at the septum location and so have 10 independent variables. But the practical design of a synchrotron may impose additional demands. For example, the Extender ring of the Moscow Kaon Factory has a racetrack lattices structure with two long straight sections one of which provides beam extraction and has fixed lattices excluding matching lines. To get the extraction section with the periodic dispersion functions a multiple of 2π horizontal phase advance is required. So if μ_x is a phase advance on the line matching with the arcs then the dispersion functions at the septum location must satisfy the following equation:

$$
D'_x = -\tan(\mu_x)\frac{D_x}{\beta_x^{(0)}}
\tag{15}
$$

The procedure of lattices design for the Extender ring of the MKF allows to vary parameters of the optic elements only in the dispersion supressors, in the matching lines of the extraction straight section and in the second straight section. As far as properly the lattices with the extraction straight section are fixed so the Twiss parameters — $\beta_x^{(0)}$, $\beta_x^{(1)}$, $\alpha_x^{(0)}$, $\alpha_x^{(1)}$ — at the septum location are fixed too. Moreover such procedure disturbs the other parameters — ε_0, ε_1, χ_0, χ_1 — only slightly. So two or three iterations of such procedure with the successive applying of the given formulas lead to the success.

Thus only one independent variable — Q'_x — is left and it is allowed to introduce any pure mathematical requirement for simplification of the above equations. Such requirement is dictated only by the user's needs and taste and finally is not obliged.

III. RESULTS

The actual divergence slightly differs from those which Hardt's and the present theory supply because they do not consider the high order perturbations at all. Consequently one should take care of the minimizing these perturbations before using these theories. From the other side the more absolute value of the chromaticity Q'_x as well as δ correspond to the more difference between the results due to Hardt and the present theory. In the present time the designed lattice structure for the Extender Ring of the Moscow Kaon Factory can support only achromatic mode when $Q'_x = 0$. In that case Hardt theory proposes $D_x = D'_x = 0$ and as a result zero beam divergence at the septum position. The present theory with the same dispersion functions gives non zero magnitude of the beam divergence equal to $0.002\,mrad$ at the septum position for $|\delta| \leq 0.00035$. The actual divergence has been executed by means of the DIMAD code [8] which provides the exact values of the fixed points at a $1/3$ - integer resonance. The value of the divergence has been found at $0.005\,mrad$ for the same rang of δ. The advantage of the present approach over the other is hardly visible in this example but expected more considerable in the chromatic mode of the resonant slow extraction.

IV. REFERENCES

[1] H. G. Hereward, "The Possibility of Resonant Extraction from the CPS," CERN AR/Int.GS/61-5, 1961.

[2] K. R. Symon, "Beam Extraction at a Third Integral Resonance," NAL-FN-130, 134, 140, 144. 1968.

[3] T. Suzuki and S. Kamada, "Theory of Half-Integral Resonant Extraction," KEK-76-7.

[4] L. C. Teng, "Half Integral Resonant Extraction from the Main Ring," NAL-TM-375, 1972.

[5] W. Hardt. PS/DL/LEAR Note 81-6, CERN, 1981.

[6] A. Chursin et al., "The Accelerator Complex of the Moscow Kaon Factory," in this proceedings.

[7] S. P. Volin, "Resonance Canonical Perturbation Theory for Circular Accelerators with Sextupoles," INR, P-0670, Moscow, 1990, (in Russian).

[8] R. V. Servranckx, K. L. Brown, L. Schachinger and D. Duglas, "Users Guide to the Program DIMAD," Report 285 UC-28 (A), SLAC, 1985.

RACETRACK LATTICES FOR
LOW-MEDIUM-ENERGY SYNCHROTRONS

A. I. Iliev and Yu. V. Senichev, INR, Moscow, USSR

I. INTRODUCTION

In design of magnet lattice for an accelerator the various requirements depending on machine specifications can be met. In this article we discuss the possible lattice design for low-energy ($\gamma_{max} \simeq 3 \div 10$) and medium-energy ($\gamma_{max} \simeq 30 \div 50$) high-intensity synchrotrons. We will be restricted and concentrate on those lattices in which the following requirements are fulfilled:

- the transition energy γ_t is kept high (or low) enough to satisfy many conditions of beam stability;

- the accelerator ring has a racetrack shape with two 180° arcs and two long dispersion-free straight sections desired for rf cavities, injection and extraction elements, Siberian Snakes, etc.;

- the dynamic aperture in presence of chromaticity-correcting sextupoles is sufficiently large.

A careful analysis shows that these requirements are not really contradictory ones[1], but often it is hard to fulfil them together.

II. MAIN TYPES OF MAGNET LATTICES

Since a choice of lattice for arc depends to a large extent on the requirement to avoid the transition crossing, let us remind the methods of creating transitionless lattices. All lattices can be divided in two groups

- lattices with regular arc;

- lattices, in which arc superperiodicity is introduced.

A. Regular lattice

The term — "regular" is concerned to such lattices, in which the arc periodicity coincides with a number of the basic optical building blocks — cells. In such lattices the transition energy $\gamma_t \approx Q_x$ [1], where Q_x is the horizontal betatron tune. The dispersion in long straight sections is canceled by using dispersion suppressors placed at the ends of arcs. As a rule, the dynamic aperture of such rings with full chromaticity correction is a sufficiently large. If it is allowed to keep γ_t not far from the range $[\gamma_{min}, \gamma_{max}]$, then a regular lattice with a high tune Q_x can be adopted for low-energy rings and, with a low tune, for medium-energy (with $\gamma_{min} \simeq 7 \div 10$) synchrotrons. This method has been used in the lattice design of the MKF Main Ring [2].

However, for increasing of the threshold of microwave instability and for the longitudinal matching it is often important to obtain very high or even imaginary γ_t. In the next section we will consider three different methods, by which the γ_t can be altered over a broad range of values.

B. Superperiodic lattice

The basic principle is well known [1,3] — it is necessary to create a perturbation of the dispersion function, that requires to break the natural arc periodicity and to introduce a new periodicity — superperiodicity S. Then, if the horizontal tune of the arc $Q_x \simeq pS$, where p is an integer, a modulation of the dispersion can be achieved, and (for $Q_x < pS$) the transition energy may become high. In the case of a superperiodic lattice the use of special dispersion suppressors is unsuitable, and the dispersion in the long straight sections is canceled by tuning the horizontal tune of each arc to an integer number (the first-order achromat) [4].

The oscillation of the dispersion function corresponding to a high γ_t significantly reduces a number of places in which the chromaticity-correcting sextupoles can be installed. In general, the sextupoles can not be organized in the non-interleaving pairs with phase advance of π in both planes between elements of each pair to cancel the nonlinear effects of the sextupoles. This may result in serious problem of obtaining of a large enough dynamic aperture. The possible way to improve the dynamic aperture is to fulfil the conditions for the arc to be a second-order achromat[2][5]:

- the chromaticities $\xi_x = 0$ and $\xi_y = 0$;

- the tunes of each arc $Q_x = S\nu_x$, $Q_y = S\nu_y$ are equal to integers;

- the tunes of the superperiod are not in resonances $m\nu_x + n\nu_y \neq integer$, where $(m,n) = \{(1,0),(3,0),(1,2),(1,-2)\}$.

In this case all low order resonances introduced by sextupoles are compensated at the ends of each arc. Nevertheless, the compensation has the local nature and the influences of these resonances are still great near the middle of the arcs. Just because one should carry out the analysis of a phase-space distortion and growth of the effective emittances for two different points in the lattice: the first inspection point is located at the entrance of the arc, the second one is located near the middle of the arc. In Fig. 1,2 the phase-space plots $p_{xn} = -1/\sqrt{\beta}(\alpha x + p_x\beta)$ versus $x_n = x/\sqrt{\beta}$ for horizontal betatron motion are shown for these points of the arc. One can see the large phase-space distortion in the middle of the arc corresponding to the strong 1/3 resonance.

Thus the fact that arcs form the second-order achromats does not work out completely the problem of the improvement of the dynamic aperture. The only reliable method to minimize the effects of the nonlinear aberrations is to reduce them — keeping the betatron amplitudes as low as possible and decreasing strength of sextupoles. And a work should be done along this line.

There are following different methods to create a superperiodicity S [3]:

- to change the field gradients in arc's quadrupoles (β-function modulation) or/and

---

[1]Based on these requirements we have proposed the lattice designs for MKF and TRIUMF low-energy Boosters.

[2]We suppose that the periodicity in the sextupole's distribution in the arcs coincides with the number of arc's superperiods.

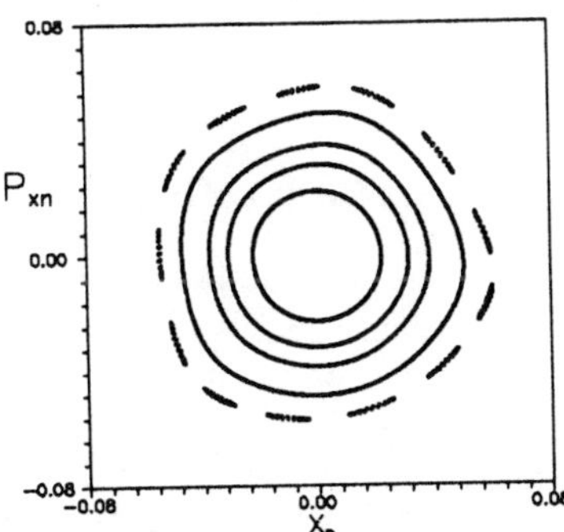

Figure 1: The MKF Booster turn-by-turn tracking data displayed in normalized phase space. Plot corresponds to the entrance of the arc.

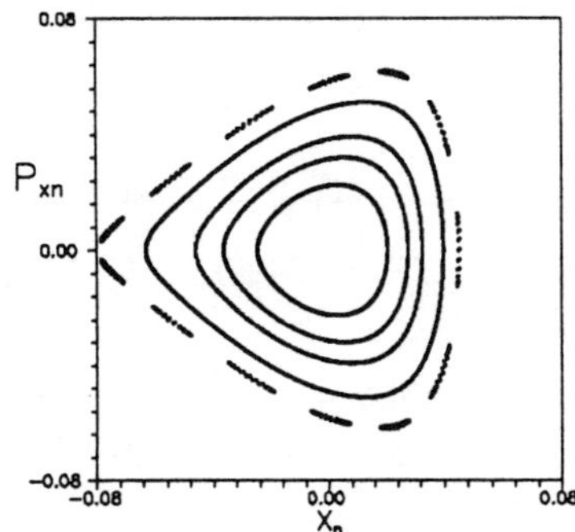

Figure 2: The same phase-space plot as in Fig. 1, but the indicated particle positions correspond to the middle of the arc.

- to violate a regular dipole arrangement (modulation of curvature radius ρ).

Let us consider the first method.

Perturbations in quadrupole strengths

In the past this method had been used only for 'circular' lattice design. For the fist time the racetrack lattice with perturbations in quadrupoles was realized for design of TRIUMF KAON Factory medium-energy rings [6]. Some later the same idea was used in the first proposal of the racetrack lattice for MKF 45 *GeV* Main Ring [7].

Starting with a regular lattice, the arc superperiodicity S is created by introducing perturbations in field gradients of some quadrupoles. These perturbations result in a modulation of the β-function, and the dispersion becomes negative in some regions of the arc so that the transition energy γ_t is increased to a high (or imaginary) value. In Fig. 3 the maximum fractional changes in β-function and dispersion computed by using analytic formulas [8], for case $\gamma_t \simeq \infty$, versus tune of a superperiod ν_x are shown. We can see to avoid high peaks of the β-function one should put ν_x near 1. But to prevent a high dispersion, ν_x should be chosen far from 1. Since a high peak value of the β-function is more dangerous, the optimal values for ν_x lie in the range $0.8 \div 0.875$. Thus the large undesirable changes in β-function can be reduced by making the horizontal tune of each arc equals to $S-1$ so that $\nu_x = 1-1/S$. For large rings, in particular for medium-energy synchrotrons, consisted of a large number of cells, it is possible to create a high arc periodicity $S = 5 \div 8$ ($\nu_x = 0.8 \div 0.875$) and, as consequence, to obtain a high γ_t using relatively small modulation

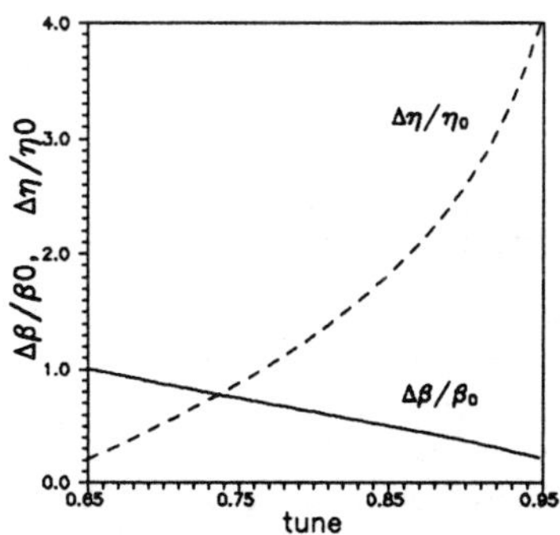

Figure 3: Maximum fractional changes in β-function and dispersion for $\gamma_t \simeq \infty$.

of the β-function. The analysis of resonance properties of such lattices shows that the dynamic aperture is large enough.

After successful progress of lattices for medium-energy synchrotrons the same idea was realized for lattice designs of low-energy boosters of TRIUMF and Moscow KAON Factories [7,9]. Since the typical value of arc superperiodicity is $3 \div 4$, then $\nu_x \simeq 0.67 \div 0.75$. Then, as we can see from Fig. 3, for obtaining a high γ_t it is required more modulation of the β-function. And since the resonance strengths excited by sextupoles are proportional to $\beta^{3/2}$, the dynamic aperture should be more sensitive to chromaticity correction. As it was to be expected, dynamic aperture for such lattices with full chromaticity correction is unacceptable reduced [9,10]. Besides in this lattice there are not enough room in arcs to accommodate desirable diagnostic devices, collimators, correctors etc.

Thus this method seems to be unsuitable for low-energy rings and one should consider some different approach.

Scheme with modulated orbit curvature

The first proposal for eliminating of transition energy [11] have consisted in using of dipoles with reversed curvature. The evident defect of the method — unacceptable large increase of ring's length. A less extreme variant is the use of a regular focussing lattice, in which some of cells have not dipoles — 'missing magnet' scheme. The method was used for design of many lattices having a 'circular' shape. In such lattice the empty cells have a double meanings: they are needed for high γ_t and they form the straight sections used for injection,extraction, etc. The application of this method for a racetrack design has essential difficulties. As it can be shown [3,12] to obtain a high γ_t one should have the horizontal tune of a superperiod very close to 1: $\nu_x \simeq 0.9$. Obviously, this requirement can not be satisfied in a racetrack low-energy rings with $Q_x = S - 1$. And one should choose the horizontal tune of a superperiod ν_x so close to 1 as it is required. In this case, evidently, the condition of the first-order achromat can not be fulfilled, and we have to renounce our requirement of dispersion cancelation in both straight sections. Thus in design of a racetrack low-energy synchrotron we have met with two oppositions: high γ_t and zero dispersion in straight sections. Keeping within the bounds of *only* 'missing magnet' scheme we should

1. either to require the dispersion cancelation in only one straight section. This condition can be written as [10,13] $\nu_x{}_{str1} = \frac{2m+1}{2} - \frac{Q_x}{2}$, where $\nu_x{}_{str1}$ — horizontal tune of dispersion-free straight section, $Q_x/2 \neq$ an integer, m

– an integer. However, in such lattice it is so hard to accommodate all rf cavities in only one dispersion-free straight section. The one superperiodicity of whole ring is not so friendly for spin control and should result in exciting of strong structure resonances;

2. or finally say 'good-bye' to hope of obtaining zero dispersion in straight sections and put $\nu_{x\,str1} = \nu_{x\,str2}$. In this case the dispersion would have nonzero although small value in the straight sections, and a careful analysis of synchro-betatron coupling in cavities should be done.

Now we are in the position to consider a scheme that seems to be the most convenient for a racetrack design of low-energy synchrotrons.

Scheme with both modulated ρ and β

As far as we know the fist proposal of using a scheme with two modulations for 'circular' lattice design was done in [3]. In that paper a superperiodicity had been created by modulating drift spaces between quadrupoles and dipoles. It was shown that by increasing drift length, for example, near the ends of each superperiod and by increasing magnet density around center of the one, the variations in both β and ρ, and as a result a high γ_t may be achieved. But it is not difficult to see that the scheme is hardly adopted for racetrack designs. Since to obtain a high value of γ_t it is needed a large difference between long and short drifts. It leads to considerable increase of arc length and large beatings of the lattice functions.

The alternative approach is the use of 'missing magnet' and β-modulation schemes simultaneously. As an example of the method let us consider the low-energy ($7.5\ GeV$) booster of MKF [10,13]. The lattice has a racetrack shape. Each arc has superperiodicity $S = 4$. The superperiod contains four $FODO$ cells, two central halve-cells have not dipoles ('missing magnets'). The horizontal tune of arc equals to 3, that gives zero dispersion in long straight sections. It is clear that γ_t is not so high, but using small perturbations in quadrupoles it is possible to obtain any γ_t without significant modulation of the β-function (see Fig. 4). As long as,

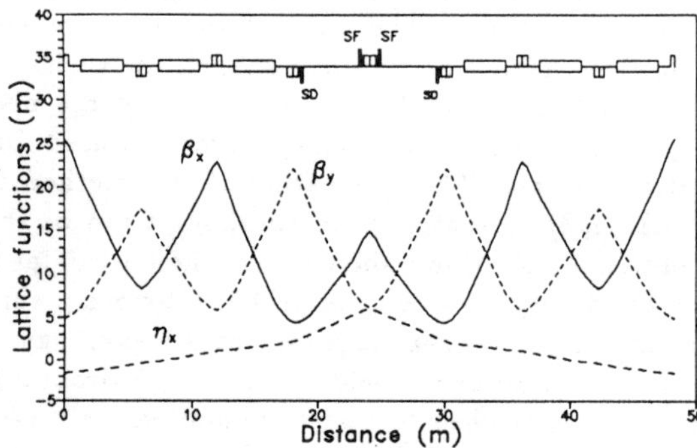

Figure 4: Lattice functions for a superperiod of MKF booster (a scheme with both β and ρ modulation).

- the analysis shows the dynamic aperture is sufficiently large when the full chromaticity correction is made;

- empty cells do not considerable increase the length of the ring (the superperiodicity of arc is not so great), moreover the short straight sections can be useful for accommodation of collimators, diagnostic devices, correctors and injection systems,

it seems that proposed lattice is the best solution for low-energy synchrotrons.

III. CONCLUSION

We have described an alternative transitionless lattices for low-medium-energy synchrotrons, which have a racetrack shape and a large enough dynamic aperture. We have shown that a lattice based on β-modulation is a good solution for medium-energy synchrotrons, and a lattice with modulation of β-function together with 'missing magnet' scheme is most suitable for low-energy rings.

IV. REFERENCES

[1] E. D. Courant, H. S. Snyder, "Theory of the alternating-gradient synchrotron," Annals of Phys., vol. 3, pp. 1–48, 1958

[2] A. G. Chursin et al., "The accelerators complex of the Moscow Kaon Factory,"in proc. of Intern. Seminar on Intermediate Energy Phys., INR, Moscow, November 1990

[3] R. C. Gupta, J. I. M. Botman and M. K. Craddock, "High transition energy magnet lattices," IEEE Trans. Nucl. Sci., vol. NS-32, pp. 2308–2310, October 1985

[4] Karl L. Brown and Roger V. Servranckx, "Optics modules for circular accelerator design," SLAC–PUB–3957, May 1986

[5] S. A. Kheifets, T. H. Fieguth and R. D. Ruth,"Canonical description of a second-order achromat," in proc. of Second Advanced ICFA Beam Dynamics Workshop, CERN 88-04, pp. 52–61, Geneva, 29 July 1988

[6] R. Servranckx, "New lattice for C, D and E rings, Design Note, TRI–DN–88-3, TRIUMF, January 1988

[7] A. G. Chursin et al., "The proposal of the accelerator complex of the Moscow Kaon Factory,"in proc. of AHF Accel. Design Workshop, LA–11684–C, vol. 1, pp. 207–219, February 1989

[8] A. I. Iliev, "Analytic approach to design of high transition energy lattice with modulated β-function," in this proceedings

[9] R. V. Servranckx, et al. "Racetrack lattices for the TRIUMF KAON Factory," in proc. of AHF Accel. Design Workshop, LA–11684–C, vol. 1, 1989, pp. 182–188, February 1989

[10] N. I. Golubeva, A. I. Iliev and Yu. V. Senichev, "The new lattices for the booster of Moscow Kaon Factory," in proc. of Intern. Seminar on Intermediate Energy Phys., INR, Moscow, November 1990

[11] V. V. Vladimirscij, E. K. Tarasov, "Problems of cyclic accelerators," USSR, Academy of Sciences, Moscow, 1955

[12] A. I. Iliev, "Analytic approach to design of high transition energy lattice using 'missing magnet' scheme," unpublished

[13] N. I. Golubeva, A. I. Iliev and Yu. V. Senichev, "The booster lattice for the Moscow Kaon Factory," in proc. of XII All Union Conf. on Charge Particle Accel., Moscow, 1990

ANALYTIC APPROACH TO DESIGN OF HIGH TRANSITION ENERGY LATTICE WITH MODULATED β-FUNCTION

A. I. Iliev, INR, Moscow, USSR

I. INTRODUCTION

In high intensity synchrotrons the choice of the transition energy γ_t has the essential significance. Since just γ_t defines the threshold of microwave instability, a required sign of chromaticity and to a large extent the magnetic lattice of the ring. Lattices with a high γ_t are often used for design of low-medium-energy accelerators. The review of methods of creating a high transition energy lattice is given in [1,2]. But these papers contain only qualitative analysis of the methods. However, in the initial stage of a lattice design it is extremely essential to have a prior view of lattice properties: the peaks and behaviors of the β and dispersion functions, the potentiality for chromaticity correction, etc. From the other hand, the existent computer codes like MAD [3] are very powerful in beam calculations with fixed lattice parameters. But often it is required a much work to optimize a lattice. So, in general, before using a powerful code it is desirable to have some approximate solution.

In this report an analytic approach to design of high transition energy lattice with modulated β-function is proposed. An application to the racetrack lattice design of MKF medium energy synchrotron [4] is given and the agreement between the results predicted by analytic formulas and the results obtained by the computer code MAD is found to be good.

II. THEORY

A. General

The transition energy can be expressed as [5]

$$\gamma_t^{-2} = \frac{Q_x^3}{R} \sum_{n=-\infty}^{+\infty} \frac{|a_{nS}|^2}{Q_x^2 - n^2 S^2}, \qquad (1)$$

where R is machine's average radius, Q_x is the horizontal tune, S is the periodicity of lattice and a_{nS} is defined by

$$a_{nS} = \frac{1}{2\pi} \int_0^{2\pi} \frac{\beta_x^{3/2}(\phi)}{\rho} e^{inS\phi} d\phi, \qquad (2)$$

where β_x is the horizontal β-function, ρ is the curvature radius of the equilibrium orbit, and $\phi = \mu_x/Q_x$ is the normalized horizontal phase. If an accelerator has a regular lattice ($S = M$, M is a number of magnetic cells) with $Q_x \ll M$, then $\gamma_t \approx Q_x$. But, as it is clear from (1), if $Q_x \simeq pS$ and $Q_x < pS$, where p is an integer, a harmonic component a_{pS} may result in a large γ_t. As a rule, a machine tune and a number of magnetic cells M can not be varied in wide ranges, so to produce a high γ_t one should create a new lattice periodicity — superperiodicity S. A superperiodicity can be obtained either by violating a regular magnet arrangement (a modulation of bending radius ρ) or/and by changing field gradients in some quadrupoles (a modulation of the β-function).

Let me discuss the last method — introducing a periodic modulation of the β-function.

B. Lattice with modulated β-function

Starting with a regular lattice a superperiodicity is created by a periodic perturbations $k(s)$ in the quadrupole strengths. The perturbations give changes in the lattice functions and cause the tune shifts. Let a perturbation be a small parameter $\epsilon \cdot k$. Then according to perturbation theory, the perturbed β-function is expanded in a power series in ϵ: $\beta(\phi_0) = \beta_0 \cdot (1 + \Delta_1\beta\epsilon + \Delta_2\beta\epsilon^2 + \cdots)$. The β-function satisfies the second-order differential equation

$$\frac{d^2}{ds^2}\sqrt{\beta(s)} + K(s)\sqrt{\beta(s)} - \beta^{-3/2}(s) = 0, \qquad (3)$$

where $K(s)$ is a quadrupole strength, s is a distance along the equilibrium orbit. The unperturbed β-function β_0 is a solution of equation (3) with $K(s) = K_0(s)$. Using in (3) the phase variable ϕ_0 rather than a geometrical variable s, the following equations for $\Delta_1\beta$, $\Delta_2\beta$ can be found

$$\ddot{\Delta}_1\beta + 4Q_0^2 \cdot \Delta_1\beta = -2\beta_0^2 Q_0^2 k, \qquad (4)$$

$$\ddot{\Delta}_2\beta + 4Q_0^2 \cdot \Delta_2\beta = -2\beta_0^2 Q_0^2 k \cdot \Delta_1\beta + \frac{1}{2}(\dot{\Delta}_1\beta)^2$$
$$+ 2Q_0^2(\Delta_1\beta)^2, \qquad (5)$$

where $k = K - K_0$. The equation (4) has the periodic solution

$$\Delta_1\beta = -\frac{\nu_0}{\pi} \sum_{n=-\infty}^{+\infty} \frac{J_n e^{inS\phi_0}}{4\nu_0^2 - n^2}, \qquad (6)$$

$$J_n = \nu_0 \int_0^{2\pi} \beta_0^2(\phi_0) k(\phi_0) e^{-inS\phi_0} d\phi_0, \qquad (7)$$

where $\nu_0 = Q_0/S$. That agrees with an expression of Courant and Snyder [5] found by using somewhat different approach. It is easy to solve (5) and, taking into consideration (6), one can compute the tune shift $\nu = \nu_0 \cdot (1 + \Delta_1\nu\epsilon + \Delta_2\nu\epsilon^2 + \cdots)$ up to second order

$$\Delta_1\nu = \frac{J_0}{4\pi\nu_0}, \quad \Delta_2\nu = -\frac{1}{8\pi^2} \sum_{k=-\infty}^{+\infty} \frac{|J_k|^2}{4\nu_0^2 - k^2}, \qquad (8)$$

hence it appears that a large J_0 may produce a large tune shift. From (6,8) one can see also, that for $\nu \approx p$, where p is an integer, the harmonic J_{2p} would give large changes in the β-functions and the tune. Therefore, the harmonics J_0, J_{2p} should be kept small.

Further note that a_n is the integral over new phase ϕ (see (2)), but the derived expression of the β-function depends on 'old' phase ϕ_0. But $\beta_0(\phi)$ can be expanded in power series in ϵ too: $\beta_0(\phi) = \beta \cdot (1 + \tilde{\Delta}_1\beta\epsilon + \tilde{\Delta}_2\beta\epsilon^2 + \cdots)$. It is easy to see, the equations for $\tilde{\Delta}_1\beta$ and $\tilde{\Delta}_2\beta$ should be the same as (4,5), and can be obtained from (4,5) by the following substitutions: $\phi_0 \to \phi, \beta_0 \to \beta, \Delta \to \tilde{\Delta}, k \to -k, Q_0 \to Q$. Then $\tilde{\Delta}_1\beta$ is

$$\tilde{\Delta}_1\beta = -\frac{\nu}{\pi} \sum_{n=-\infty}^{+\infty} \frac{\tilde{J}_n e^{in\phi}}{4\nu^2 - n^2}, \qquad (9)$$

$$\tilde{J}_n = -\nu \int_0^{2\pi} \beta^2(\phi) k(\phi) e^{-inS\phi} d\phi. \qquad (10)$$

Expanding $\beta^{3/2}(\phi)$ in terms of ϵ up to first order[1] one can derive the rough expression for a_n. Then this expression are substituted into (1) and one obtains approximately[2]

$$\gamma_t^{-2} \approx \gamma_{t0}^{-2}\left[1 + \frac{9\nu_x^4}{2\pi^2}\sum_{n>0}^{\infty}\frac{|J_n|^2}{(4\nu_x^2 - n^2)^2(\nu_x^2 - n^2)}\right], \qquad (11)$$

where it has been supposed that $\nu = \nu_0$, the contributions from J_0 have been neglected and the fact that $J_k = -\tilde{J}_k$ in the first order in ϵ have been used. It is clear from (6,11) that if ν_x is closed to an half-integer $\nu_x \approx (2k+1)/2$ then the only component J_{2k+1} is a favorable for obtaining a high γ_t but the same one may give a large modulation of the β-function. But if by chance ν_x is closed to an integer p, there are two effective components for changing γ_t : J_p and J_{2p}, and the second one can be required be small. In such case one can obtain a high γ_t and control beating of the β-function simultaneously.

Thus, let ν_x be closed to an integer. Further, let us assume that.

- components J_n are real: $J_n = J_n^* = J_{-n}$;

- if $\nu_x \approx p$, p is an odd integer, we can suppose that

    - even harmonics J_{2k} , $k = 0, 1, 2, \ldots$ are small and can be neglected;

    - all odd harmonics J_{2k+1} are identical with each other: $J_1 = J_3 = J_5 = \ldots$

Then from (11) one obtains

$$|J_1| = \frac{\pi}{3\nu_x^2}\sqrt{\frac{2}{\sigma}\left(\frac{\gamma_{t0}^2}{\gamma_t^2} - 1\right)},$$

$$\sigma = \sum_{k=0}^{\infty}\frac{1}{\left(\nu_x^2 - (2k+1)^2\right)\left(4\nu_x^2 - (2k+1)^2\right)^2}. \qquad (12)$$

The sum σ can be evaluated explicitly:

$$\sigma = \frac{\pi}{12\nu_x^5}\left[\frac{11}{48}\tan\pi\nu_x - \frac{1}{3}\tan\frac{\pi\nu_x}{2} - \frac{\pi\nu_x}{16}\frac{1}{\cos^2\pi\nu_x}\right].$$

Knowing J_1 the perturbed β-function can be computed

$$\beta(\phi_s) \approx \beta_0(\phi_s)\cdot\left(1 + \frac{J_1}{4\cos\pi\nu}\cdot f(\phi_s)\right) ,$$

$$f(\phi) = \begin{cases} \sin(\nu(\pi - 2\phi)), & \text{if } 0 \le \phi \le \pi; \\ -\sin(\nu(3\pi - 2\phi)), & \text{if } \pi \le \phi \le 2\pi, \end{cases} \qquad (13)$$

where the phase variable ϕ_s, with ranges from 0 to 2π in the superperiod, is given by

$$\phi_s = \phi_{s0} + \frac{J_1}{8\nu\cos\pi\nu}$$

$$\times \begin{cases} -\cos(\nu(2\phi - \pi))\Big|_0^{\phi_{s0}}, & \text{if } 0 \le \phi_{s0} \le \pi; \\ \cos(\nu(2\phi - 3\pi))\Big|_\pi^{\phi_{s0}}, & \text{if } \pi \le \phi_{s0} \le 2\pi. \end{cases} \qquad (14)$$

And the dispersion is

$$\eta_x(\phi_s) \approx \eta_{x0}\cdot\left(1 + \frac{J_1}{8\cos\pi\nu}\cdot f(\phi_s)\right)$$

$$\times\left(1 - \frac{3\nu^3}{\pi}J_1\sigma_1(\phi_s)\right), \qquad (15)$$

---

[1] It is not difficult to convince that the second order terms affects on only zero harmonic of the expansion of γ_t. In present report this effect is not taken into account.

[2] The same formula is adduced in [1] with reference to [6].

where

$$\sigma_1(\phi) = \sum_{k=0}^{\infty}\frac{\cos((2k+1)\phi)}{\left(\nu^2 - (2k+1)^2\right)\left(4\nu^2 - (2k+1)^2\right)}.$$

It can be convinced that

$$\sigma_1(\phi) = \frac{\pi}{24\nu^3}$$

$$\times \begin{cases} -\left[\frac{2\sin(\nu(\pi/2-\phi))}{\cos\pi\nu/2} - \frac{\sin(\nu(\pi-2\phi))}{\cos\pi\nu}\right], & \text{if } 0 \le \phi \le \pi; \\ \left[\frac{2\sin(\nu(3\pi/2-\phi))}{\cos\pi\nu/2} - \frac{\sin(\nu(3\pi-2\phi))}{\cos\pi\nu}\right], & \text{if } \pi \le \phi \le 2\pi. \end{cases} \qquad (16)$$

In the case of *FODO* lattice in the thin-lens approximation there are the following analytic formulas for the unperturbed β-functions, dispersion and γ_{t0}

$$\beta_{Fx,y} = \frac{L}{\sin\mu_{x,y}}\left(1 \pm \frac{L\delta_D}{4}\right), \quad \beta_{Dx,y} = \frac{L}{\sin\mu_{x,y}}\left(1 \mp \frac{L\delta_F}{4}\right),$$

$$\eta_F = \frac{L\Phi(\delta_D L + 8)}{8(1 - \cos\mu_x)}, \quad \eta_D = \frac{L\Phi(\delta_D + \delta_F)}{2\delta_D(1 - \cos\mu_x)} - \frac{\Phi}{\delta_D},$$

$$\gamma_{t0}^{-2} = \frac{\Phi^2}{24}\frac{5\cos\mu_x - 3\cos\mu_y + 46}{1 - \cos\mu_x},$$

where L is a cell length, μ is a phase advance per cell, Φ is a bending angle of dipole magnets and labels F,D denote the focussing and defocussing quadrupoles correspondingly. The integrated quadrupole strengths δ_F, δ_D are given by

$$\delta_F \cdot \delta_D = -\frac{4(\cos\mu_x + \cos\mu_y - 2)}{L^2},$$

$$\delta_F - \delta_D = \frac{\cos\mu_y - \cos\mu_y}{L}.$$

Thus, the closed analytic formulas for lattice functions in terms of the main machine parameters (γ_t, betatron tunes, etc.) have been obtained. One notes, however, that at deriving the formulas some important assumptions have been accepted, and the question should arise: are there lattices in which the foregoing constrains on J_n are satisfied? The answer is yes and below we discuss how to design such lattice.

C. Suitable lattice

Let us consider a lattice, in which a superperiod has an even number of *FODO* cells and, basically, two quadrupoles, that are placed by π in the normalized phase, are perturbed. Then, evidently,

- $J_n = J_n^* = J_{-n}$;

- if the perturbations in the quadrupoles have different signs, but the same absolute values, then

    - $J_{2n} \approx 0$, $n = 0, 1, 2, \ldots$ and J_{2n+1} are equal to each other for both planes simultaneously;

    - there is a simple relation between the component J_{1x} for horizontal plane and the component J_{1y} for vertical plane: $J_{1x} = -(\beta_{Fx}/\beta_{Fy})\cdot J_{1y}$.

Thereby, in cases such that $\nu_x \approx p$, p is an odd integer, the undesirable component J_{2p} can be kept small and all our assumptions are valid. But, if a superperiod contains an odd number of cells, one can not satisfy the condition $J_{2n} \approx 0$ for both planes simultaneously using the same perturbations. In this case a special analysis is required.

III.   AN EXAMPLE

To illustrate the results of previous section let us consider the lattice which has been adopted for 45 GeV main synchrotron of Moscow Kaon Factory. The first proposed lattice [4] has a racetrack shape. In each arc included 32 $FODO$ cells the superperiodicity $S = 8$ is created. The strength of the focussing quadrupole at the origin of the superperiod and the strength of the focussing one at the middle of the superperiod are tuned to produce a desirable value of γ_t. The parameters of the superperiod are given in Table 1. In this lattice the β-functions and

Table 1: Superperiod parameters

Length [m]	65.72367
Number of cells	4
Horizontal tune ν_x	0.875
Vertical tune ν_y	0.875
Length of dipole [m]	6.015459
Bending radius [m]	122.5582
Short drifts [m]	0.6
Half-length of quadrupoles [m]	0.5

dispersion have been computed by analytic formulas in wide range of transition energy and they have been compared with the results obtained by computer code MAD [3]. The behaviors of peak values of the β-function and dispersion are shown in Fig. 1, 2 (sign "$-$" corresponds to an imaginary γ_t). It is

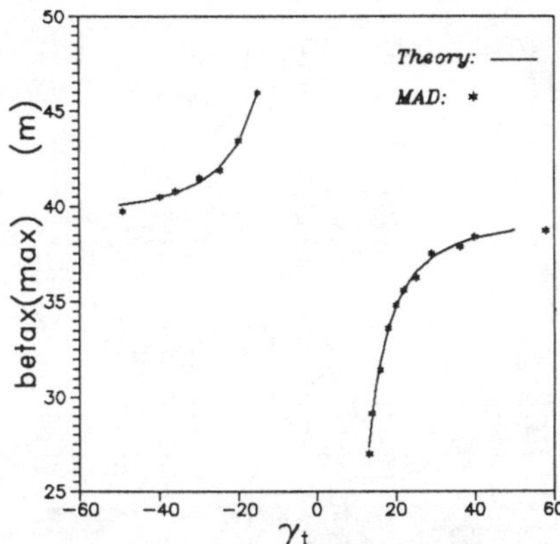

Figure 1: Peak β_x-function as a function of γ_t, using the analytic approach and the computer code MAD.

seen that peaks of the lattice functions predicted by analytic formulas are in good agreement with the values obtained by numerical computations. Moreover, as you can see from Fig. 3, the approximations of the lattice functions at any point in the superperiod is quite well too. Consequently, some important information about the lattice can be extracted, e.g. one can estimate the beam size, the strengths of sextupoles adjusted the chromaticities and, using either an analytic method or by tracking, define the limit of stable motion, etc.

IV.   CONCLUSION

The analytic approach, which allows to compute the lattice functions of a high transition energy lattice have been developed. This method seems to be useful for pre-design of an accelerator and in the optimization of lattice parameters.

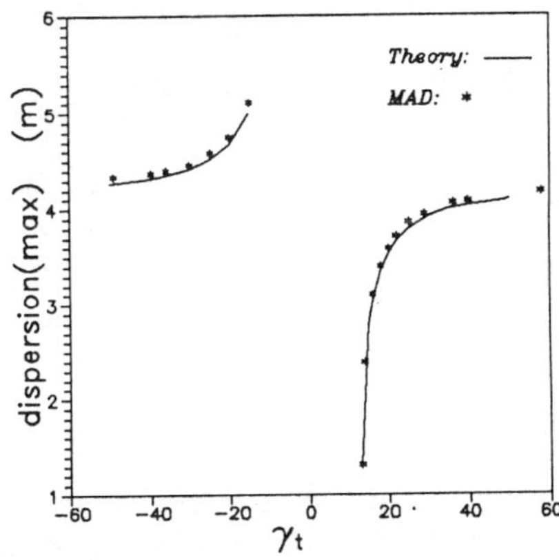

Figure 2: Peak dispersion as a function of γ_t, using the analytic approach and the computer code MAD.

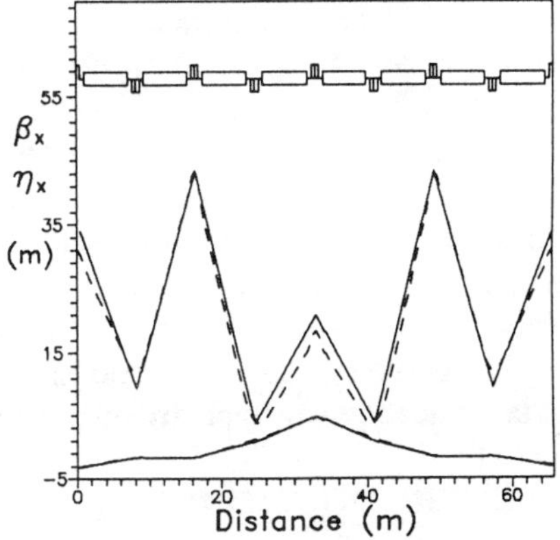

Figure 3: β_x-function and dispersion for the superperiod ($\gamma_t = 19.9\,i$): theory — solid line, MAD — dashed line.

V.   ACKNOWLEDGMENTS

I would like to thank Yu. Senichev for suggesting this study and for careful reading of the manuscript.

VI.   REFERENCES

[1] R. C. Gupta, "Methods of designing a synchrotron lattice with high energy transition energy," Design Note, TRI–DN–83–51, TRIUMF, December 1983

[2] R. C. Gupta, J. I. M. Botman and M. K. Craddock, "High transition energy magnet lattices," IEEE Trans. Nucl. Sci., vol. NS–32,pp. 2308–2310, October 1985

[3] F. Christoph Iselin and James Nieder, "The MAD program — version 6 User's reference manual," CERN/LEP–TH/87–33, 1987

[4] A. G. Chursin et al., "The proposal of the accelerators complex of the Moscow Kaon Factory," in proc. of AHF Accel. Design Workshop, LA-11684-C, vol. 1, 1989,pp. 207–219, February 1989

[5] E. D. Courant, H. S. Snyder, "Theory of the alternating-gradient synchrotron," Annals of Phys., vol. 3, pp. 1–48, 1958

[6] S. Ohnuma, Fermilab $\overline{P}$–Note 105, 1980
L. C. Teng, NAL, FN–276, 1970

A Compact RF Driven H⁻ Ion Source for Linac Injection[+]

J. Patrick Rymer, G. A. Engeman, R. W. Hamm and J. M. Potter
AccSys Technology, Inc.
1177 Quarry Lane
Pleasanton, CA 94566

Abstract

A compact rf driven H⁻ ion source has been developed for use as an injector for the AccSys radio frequency quadrupole (RFQ) linacs. A multicusp magnetic bucket geometry developed at Lawrence Berkeley Laboratory confines the plasma created by an antenna driven by 35 kW (peak) of pulsed rf power at 1.8 MHz. A three electrode system is used to extract and accelerate the H⁻ beam, which is then focused into the RFQ by an einzel lens. Permanent magnets in the extraction region sweep electrons onto the second electrode at energies up to half of the full acceleration voltage. A fast pulsed valve allows the hydrogen gas supply to be pulsed, thus minimizing the average gas flow rate into the system. The design features and performance data from the prototype are discussed.

I. INTRODUCTION

A compact H⁻ ion injector has been developed for the RFQ linac system that will be used as a calibration source for the L3 experiment at CERN. The injector consists of an rf driven ion source, extraction and focusing optics, and associated vacuum equipment. The 30 keV H⁻ beam from the injector is accelerated by an RFQ to 1.85 MeV, and then stripped to H⁰ in a gas neutralizer, so that it can drift through the magnetic field of the L3 magnet to a lithium target inside the electromagnetic calorimeter. The radiative capture gammas from the $Li(p,\gamma)$ reaction are used to calibrate the BGO detectors in the calorimeter. This application of an RFQ has been previously described[1].

The injector parameters required for operation with the standard AccSys PL-2 RFQ used in the L3 system are listed in Table I.

Table I

Injection energy	30	keV
H⁻ current (peak)	20	mA
Maximum e⁻/H⁻	50	
Minimum pulse width	50	μsec
Max. pulse repetition rate	150	Hz
Emittance	<0.5	πmm-mrad

[+] This work was funded by the US Department of Energy, SBIR Contract No. DE-AC03-88ER80674.

II. ION SOURCE DESCRIPTION

The rf driven volume ion source used in the AccSys H⁻ injector is based on the one developed by Leung, et. al. at Lawrence Berkeley Laboratory (LBL) under subcontract to AccSys[2]. A multicusp magnetic field geometry using samarium cobalt magnets provides plasma confinement inside the 10 cm diameter bucket. Hydrogen plasma is produced by driving a helical antenna with up to 35 kW of peak rf power at a frequency of about 1.8 MHz. The water cooled antenna is made of copper tubing which is coated by a thin glass layer applied using the technique developed at LBL. The glass insulates the antenna from the plasma and protects the copper tubing from plasma bombardment. A small heated tungsten filament provides electrons to allow plasma ignition at lower pressures than possible with only the rf drive. The source geometry and its mounting arrangement in the injector are shown in Figure 1.

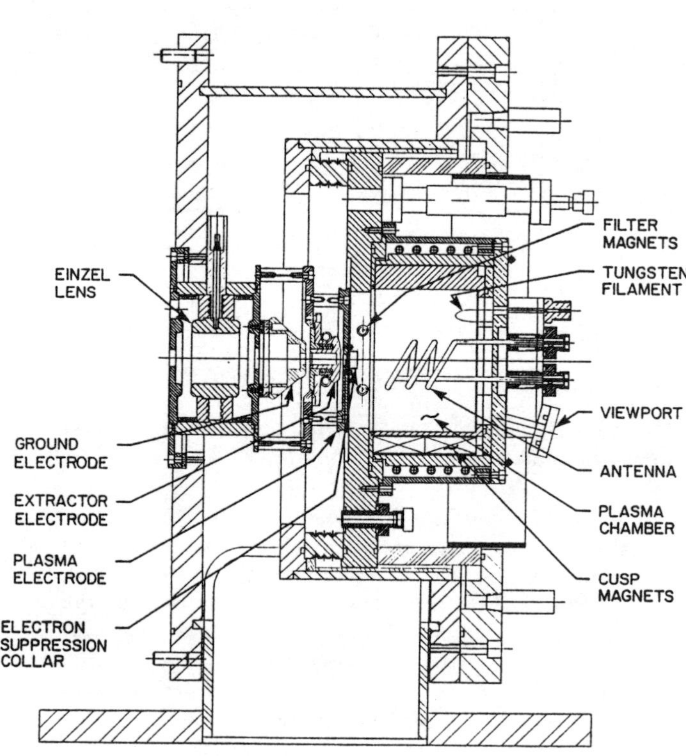

Figure 1. Cross-sectional view of AccSys H⁻ injector.

The ion source is divided into two plasma regions by water cooled tubes containing NdFe permanent magnets which produce a maximum dipole field of about 120 G. A tainless steel insert containing the extraction aperture is mounted on the plasma electrode along with a stainless steel collar which projects back into the source to reduce the extracted electron current[3]. Hydrogen gas is injected by a modified automotive fuel injector valve that is pulsed at the same repetition frequency as the plasma[4]. The pulse width applied is variable, as is the delay between the valve pulse and the plasma pulse. Average source pressure is monitored by a thermocouple gauge.

The antenna is driven by a pulsed amplifier which uses a 3CPX1500A7 tube driven in Class D operation by power MOSFETs. Isolation from the ion source high voltage is provided by a balanced rf transformer. The antenna is resonated by two external inductors and a capacitor located at the back of the source. Impedance matching is provided by connecting the isolation transformer output to various tap points on the inductors. A simplified schematic of the rf and high voltage circuits is shown in Figure 2.

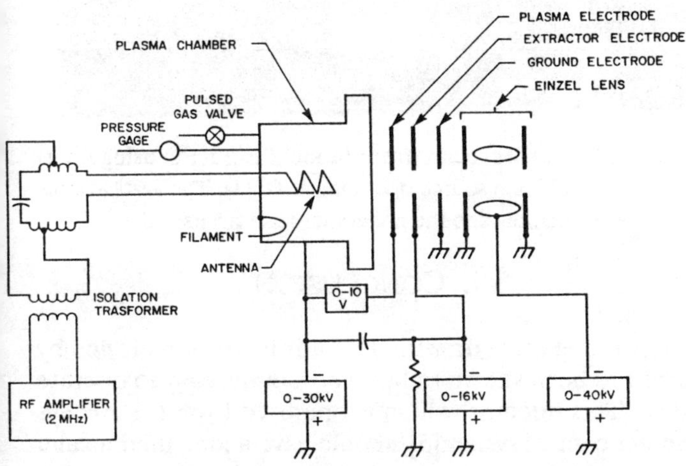

Figure 2. Schematic of the H$^-$ ion injector circuits.

III. DESCRIPTION OF THE INJECTOR

As shown in Figure 1, the ion source is mounted into a re-entrant vacuum chamber, with the extraction and focusing optics mounted on the opposite flange. A vacuum valve and cryopump are mounted on a flange which is attached to a rectangular opening in the chamber. A high voltage enclosure is also mounted on the vacuum chamber to provide high voltage protection of the ion source and its connections. Hence, there is no exposed high voltage during operation of the injector. In addition, the ion source and beam electrodes are mounted precisely with respect to two pins which align the injector beam axis to the axis of the RFQ when the ion source vacuum chamber is attached to it.

IV. EXTRACTION AND FOCUSING OPTICS

A three electrode system is used to extract the ion beam and accelerate it to 30 keV. This acceleration column is followed by an einzel lens to focus the beam into the RFQ.

The high voltage circuit used to implement these voltages is shown in Figure 2. The plasma electrode is isolated from the source so that it can be biased up to +10 V with respect to the source body. Injection energy and extraction voltage are independently adjustable. A potential difference of 15 kV is nominally applied between the plasma electrode and the extractor electrode. Two permanent magnets in the extractor electrode create a dipole field which bends extracted electrons into a cavity in this electrode, while deflecting the ions only slightly. The ions are then accelerated across another 15 kV gap between the extractor and the ground electrode. A second magnetic dipole in the ground electrode corrects the first angular deflection of the ion beam. A current transformer mounted in the ground electrode measures current leaving the accelerating column.

The accelerating electrode geometry is made as open as possible for vacuum pumping, to minimize H$^-$ stripping. The injector chamber ion gauge pressure is typically 1 to 2×10^{-5} Torr during operation, while the source chamber pressure is typically 15 to 20 mT. Differential pumping of the RFQ vacuum chamber is provided by the restricted openings of the ground electrode and einzel lens mounted on the flange separating the ion injector and RFQ.

V. DATA

Extracted H$^-$ and electron current from the rf driven ion source were first measured by temporarily replacing the ground electrode with a simple spectrometer as shown schematically in Figure 3. Two permanent magnets were mounted in the extraction electrode to bend the extracted electrons into a recess in the spectrometer body. The slightly deflected H$^-$ ions were collected on an isolated electrode at ground potential and measured through a resistor to ground. A guard ring biased at -300 V was used to suppress secondary electrons from the ion collector. Total power supply current was monitored to determine the extracted electron current.

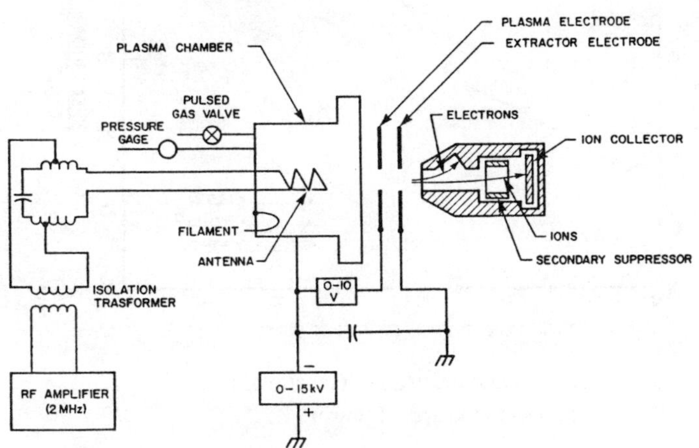

Figure 3. Schematic of spectrometer setup for H$^-$ measurements.

Figure 4 shows the H⁻ current and the ratio of electron current to H⁻ current measured by the spectrometer as the gas pressure in the ion source chamber was varied using a plasma aperture of 0.48 cm diameter. Both quantities change slowly with pressure. Figure 5 shows the same two quantities measured as a function of antenna rf power. The H⁻ current was still increasing at the maximum rf power available from the amplifier, but the current achieved was sufficient for use with the L3 calibration system.

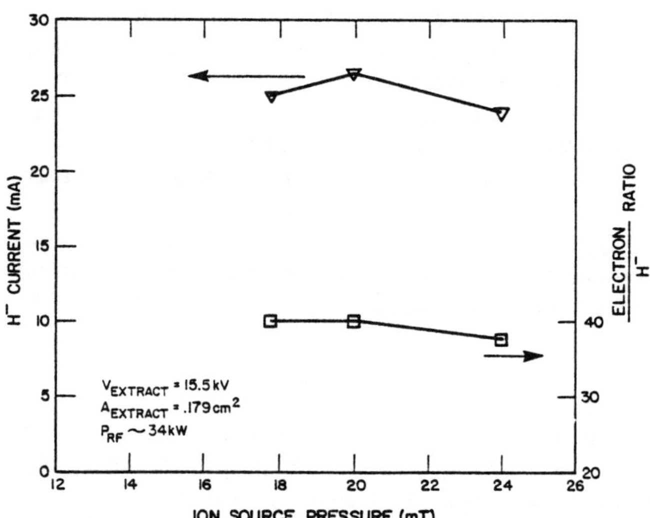

Figure 4. H⁻ current and electron to H⁻ ratio as a function of gas pressure.

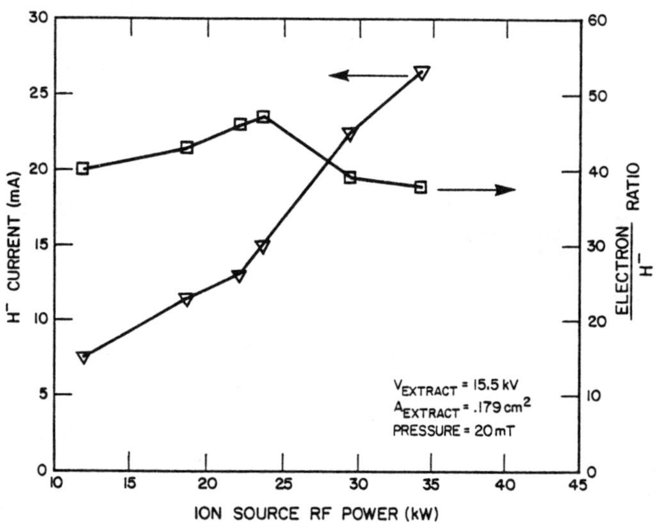

Figure 5. H⁻ current and electron to H⁻ ratio as a function of ion source rf power.

The ion injector has been assembled with all the electrodes present and H⁻ ions have been accelerated to 1.85 MeV in the RFQ. An output current of 5 mA was obtained from the RFQ in the initial tests with an input current of 8 mA through the injector einzel lens. The accelerated beam that was transported to the gas stripper is shown in Figure 6. Further tests are in progress to improve the beam transmission through the ion injector beam optics and to match the beam into the RFQ linac.

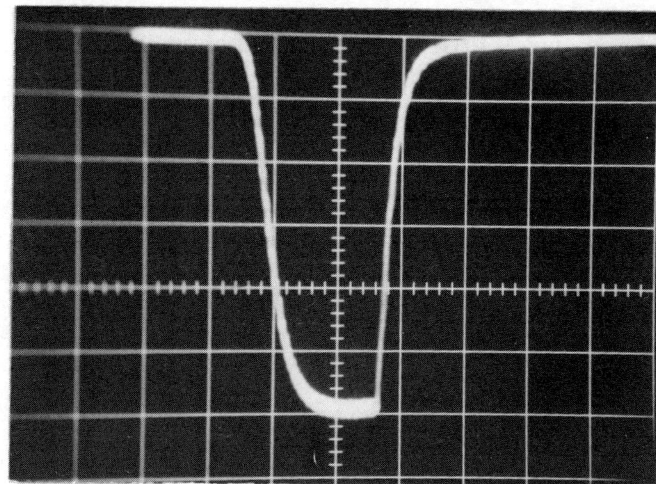

Figure 6. 1.85 MeV H⁻ beam from Model PL-2 RFQ using the rf driven H⁻ ion source operating at 60 Hz. The vertical scale is 0.5 mA/div and the horizontal scale is 5 μsec/div.

VI. CONCLUSION

The compact rf driven H⁻ ion injector developed by AccSys has been shown to be a very stable, easy to operate system. This injector, which is operated from the AccSys computer control system[5], should have a long lifetime and produce a high current beam for injection into an RFQ linac. Installation of this first system will occur later this year at CERN.

REFERENCES

[1] H. Ma, et. al., "Calibration of the L3 BGO Electromagnetic Calorimeter with a Radio Frequency Quadrupole Accelerator," Nucl. Inst. Meth. Phys. Res. A274, pp. 113-128, 1989.

[2] K.N. Leung, et. al., "An rf driven multicusp H⁻ ion source," Rev. Sci. Instr. 62, No. 1, pp. 100-104, January 1991.

[3] K.N. Leung, et. al., "Electron Suppression Experiments in a Small Multicusp H⁻ Source," Rev. Sci. Instr. 61, No. 3, pp. 1110-1116, March 1990.

[4] J.L. Dobson and B.R.F. Kendall, "Controllable Leaks Using Electrically Pulsed Valves", J. Vac. Sci. Technol. A8(3), pp. 2790-2799, May/June 1990.

[5] M.E. Hamm and J.M. Potter, "An Inexpensive PC-Based Ion Linac Control System," Proceedings of this conference.

THE BNL TOROIDAL VOLUME H⁻ SOURCE*

J.G. Alessi and K. Prelec
AGS Department, Brookhaven National Laboratory
Upton, New York 11973

Abstract

The BNL toroidal volume H⁻ ion source, in pulsed operation, is now producing up to 35 mA with an electron to H⁻ ratio of less than 5, and a ratio of less than 3 for currents up to 20 mA. This improvement came about by increasing the strength of the conical filter field. The source has also been operated steady state at low arc currents, where up to 6 mA of H⁻ was extracted. The electron to H⁻ ratio is 2-3 times larger for dc operation. For dc currents up to 5 mA, the arc power efficiency was 5 mA/kW. Pulsed performance with Ta and W filaments were very similar, except for the large gas pumping observed with the Ta filament. In dc operation, the Ta filament performed somewhat better than W. Extraction from 7 apertures having a total area of 1 cm^2 produced the same results as a single 1 cm^2 aperture.

Introduction

Studies have been in progress at BNL since 1988 on a volume H⁻ ion source of unique design. While surface production H⁻ sources are still used on most high energy accelerators, volume sources have the attractive feature that cesium is not required for H⁻ production. Volume H⁻ sources have two plasma regions, separated by a magnetic "filter field". In this type source, the hydrogen discharge results in the production of highly vibrationally excited H_2 molecules. H⁻ ions are then produced by dissociative attachment of the excited H_2 with slow electrons. The filter field separates the plasma generation region, which requires fast electrons, from the H⁻ production region, where slow electrons are required for the dissociative attachment process and fast electrons are detrimental in that they result in destruction of H⁻. The novel feature in the BNL source is the conical shape of this magnetic filter field.

Previously reported parametric studies of the source[1,2] included the effect of filament position and size, gas pressure, plasma electrode bias, anode aperture diameter, and filter field strength. We have also reported on emittance measurements,[3] operation with deuterium[4] (D⁻ production), and the effect of cesium on source performance.[4] In this paper, new results will be

*Work performed under the auspices of the U.S. Department of Energy.

presented on filter field effects, a comparison of operation with tantalum and tungsten filaments, operation with a multiaperture extraction system, and steady state operation of the source.

Source Geometry and Measurement Setup

Figure 1 shows the source schematically. The source has a cylindrical discharge chamber, 6 cm long and 20 cm i.d. SmCo magnets are arranged around the outside of the source to produce circular cusp magnetic fields as shown in the figure. A SmCo disk magnet opposite the extraction aperture produces the conical filter field. The filament is a single loop of wire located outside the filter region. The cathode voltage is applied to the filament. All results given below are with the plasma electrode grounded. Floating the plasma electrode generally resulted in a slight (<10%) increase in H⁻ current, accompanied by a large increase in the extracted electron current.

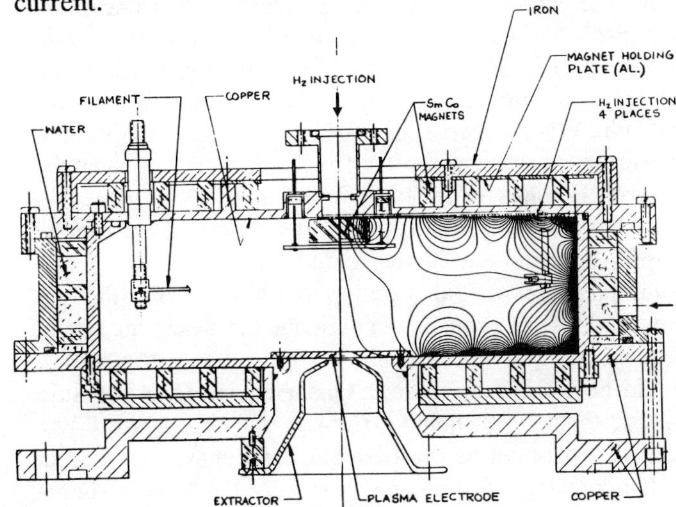

Fig. 1. Cross section of the BNL volume H⁻ source. The calculated magnetic field is also shown.

For all but the dc operation discussed below, the source was operated with a pulsed discharge of ≈1.2 ms, at a repetition rate of 0.5-1.3 Hz. The gas was pulsed, and the extraction voltage was dc. Beam current was measured on a Faraday cup located 10 cm from the source. A strong dipole field between the extractor and Faraday cup prevented any electrons from hitting the cup. A current transformer on the output of the extractor supply measured the total supply drain current. The difference between the H⁻ current measured on the Faraday cup and the total supply current is assumed to be electrons.

Filament Studies

In previous studies with a tungsten filament, it was found that a smaller (9 cm) filament loop diameter (filament closer to the center of the source) gave a better H⁻ output than a 16 cm loop.[2] However, with the small loop diameter, the arc current was limited to 150 A due to the small emitting area. A filament (1.25 mm wire diameter) was therefore tried with the 9 cm loop diameter, but with waves bent into the filament parallel to the cylindrical chamber axis. In this way, we were able to increase the total length of the filament by a factor of 2-3 while keeping the filament at the optimum location with respect to the axis of the source. The H⁻ output with this filament was the same as with the loop without waves, but we were able to push the arc current to > 350 A without the filament burning out, and the filament lifetime for operation at lower arc currents should be improved. (Life tests have not yet been done).

The source was also tested with a tantalum filament with a shape identical to the W filament described above. While the evaporation rate of Ta is higher than that of W at a given temperature, the electron emission rate is also higher for Ta. The net effect, for operation at the same electron emission rate, is very similar evaporation rates for the two. When compared, the H⁻ output under optimum conditions was very similar for Ta and W, as was the electron/H⁻ ratio for the two. With Ta, however, there was more conditioning required when first starting the source, before the optimum current could be reached. A strong pumping effect was also observed when the Ta filament was used. As the source ran with Ta, the discharge would become gas starved, and the gas pulsed into the source would have to be increased. Observation of the pressure outside the source chamber showed that the pressure was actually dropping as the filament current was increased, and increasing the pulsed gas restored it to its original pressure. It is known that Ta pumps hydrogen very well. (It has a sticking probability for hydrogen of 0.48 at 330 K).[5] From the observed pressure changes, we estimate that the Ta depositing on the source walls from the filament gave 30-40 l/s of pumping for hydrogen during the gas pulse. Some slight pumping was also observed with the tungsten filament.

Filter Field Studies

The dependence of the H⁻ and electron outputs on the strength of the conical filter field had been studied previously by placing a small pulsed coil in the source in place of the SmCo disk magnet.[2] It was found that the electron current was very sensitive to the strength of this field, while the H⁻ current had a fairly broad optimum. This experiment had presented practical difficulties, so normal operation continued to be with the disk magnet. We are able, however, to increase the strength of the filter field slightly by appropriate placement of some small SmCo magnets outside the source. Figure 2 shows the H⁻ current and e⁻/H⁻ ratio as a function of arc current, with and without the external magnets. The addition of these magnets (20-30% increase in filter field strength) has allowed us to now run routinely with e⁻/H⁻ ratios lower than we had previously reported. Figure 3 is a plot of the e⁻/H⁻ ratio as a function of H⁻ current, showing the source performance under many different conditions (W and Ta filaments, 1 and 2 cm² apertures). This electron-to-H⁻ ratio is now at least an order of magnitude smaller than what others typically obtain from Cs-free volume H⁻ sources.

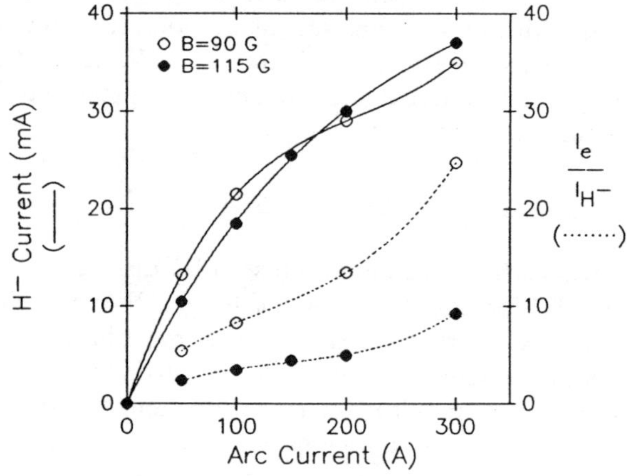

Fig. 2. H⁻ current and electron-to-H⁻ ratio vs. arc current for filter field strengths of ~ 90 G and ~ 115 G (W filament, 2 cm² aperture).

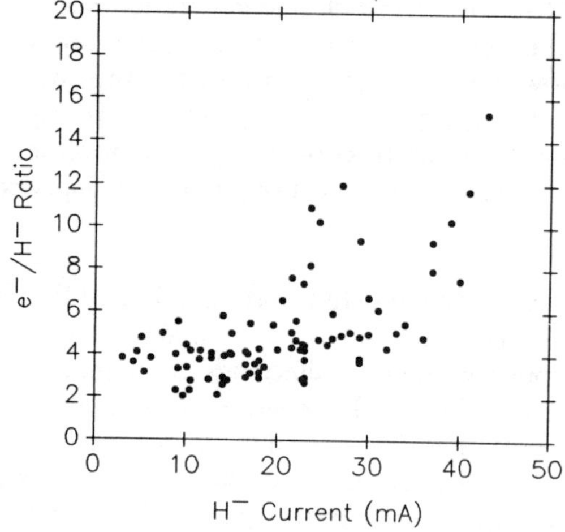

Fig. 3. Electron-to-H⁻ ratio vs. H⁻ current for various pulsed operating conditions and source geometries (W and Ta filaments, 1 cm² and 2 cm² apertures).

Single and Multiaperture Extraction

Figure 4 shows the H$^-$ current vs. arc current obtained for 1 cm^2 and 2 cm^2 apertures. A saturation in output as arc current increases is typically seen, and the H$^-$ output does not scale well with increasing extraction aperture. A multiaperture extraction system was tried and is also shown in Figure 4. Both the anode and extraction electrode had 7 apertures of 0.44 cm diameter each (total extraction area of 1.06 cm^2), spaced within a 1.54 cm total diameter. The extraction gap was 0.76 cm. The H$^-$ current and e$^-$/H$^-$ from this extractor was the same as that from the single 1 cm^2 extractor. We have not yet tested to see how the H$^-$ current scales with total extraction area for multiaperture extraction.

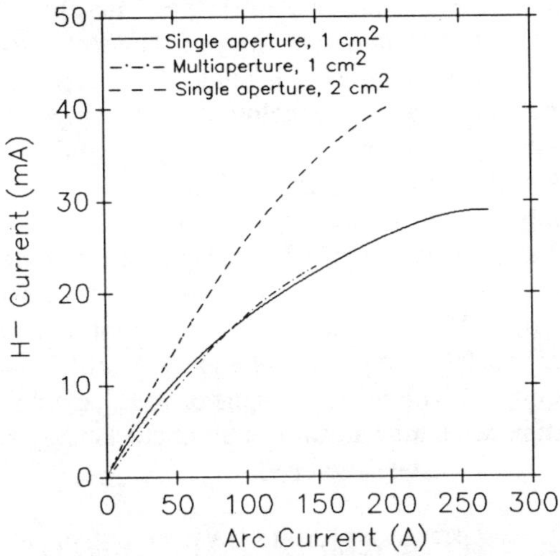

Fig. 4. H$^-$ current vs. arc current for single apertures of 1 cm^2 and 2 cm^2, and a seven-aperture extractor of 1 cm^2 total area (Ta filament).

Steady State Operation

Although the source was initially designed for pulsed operation, the addition of one cooling loop and some modifications to the filament feedthroughs allowed us to operate it dc at reduced arc powers. For these tests, the arc and gas were steady state, but the extraction voltage was pulsed (1.2 ms) in order to reduce the power on the extraction electrode and Faraday cup. The anode and extraction electrode were 1.13 cm diameter (1 cm^2 area).

Running steady state, we were able to measure up to 6 mA of H$^-$ at the maximum arc current we attempted of 20 A and 150 V. This current was essentially the same as that obtained at 20 A arc for pulsed operation. However, the e$^-$/H$^-$ ratio for dc operation was 2-3 times that obtained when running pulsed. (A test with a pulsed arc but dc gas flow gave the same performance as normal pulsed operation). In pulsed operation, as the source

pressure was varied, the minimum e$^-$/H$^-$ ratio occurred at about the same pressure as the maximum H$^-$ current. When running dc, the minimum in e$^-$/H$^-$ occurred at about half the pressure of that giving the maximum H$^-$. Also unlike pulsed operation, the dc performance was somewhat better with the Ta filament when compared with W. Measured gas flows for dc operation were typically in the range of 15-30 sccm, which is consistent with estimates based on previously measured pressures of 5-15 mT in the source for pulsed operation.

In pulsed operation, the source is typically operated with arc voltages of 200-350 V for arc currents of 50-300 A. There, the arc voltage is used to optimize H$^-$ current without regard for power efficiency. When running dc, however, we did not push the discharge voltage above 150 V. For H$^-$ currents up to 5 mA, the arc power efficiency was 5 mA/kW. The combination of the H$^-$ current not scaling linearly with arc current, and higher arc voltages being required at higher arc currents, leads to a much worse (5-10 times lower) power efficiency when producing 10's of mA's in pulsed operation.

Acknowledgements

We would like to thank D. McCafferty for the excellent job he has done in carrying out the many modifications to the source required for these studies, and K. Welch, for helpful discussions on the pumping of hydrogen by tantalum. One of us (JA) would like to thank R. York of LANL for advice on filament shapes and fabrication, and multiaperture extraction.

References

[1]   K. Prelec, Proc. 1989 IEEE Part. Accel. Conf. (Catalog No. 89CH2669-0), 340.

[2]   K. Prelec, Proc. Fifth International Symp. on the Production and Neutralization of Negative Ions and Beams, AIP Conf. Proc. No. 210, 304 (1990).

[3]   J.G. Alessi, Proc. Fifth International Symp. on the Production and Neutralization of Negative Ions and Beams, AIP Conf. Proc. No. 210, 526 (1990).

[4]   J.G. Alessi and K. Prelec, Proc. Second European Part. Accel. Conf., Nice, France, 656 (1990).

[5]   D.O. Hayward and N. Taylor, Proc. Fourth International Vac. Cong., 115 (1968).

LASER DIAGNOSTICS OF H⁻ FORMATION IN

A MAGNETIC MULTICUSP ION SOURCE

A. T. Young, P. Chen, W. B Kunkel, K. N. Leung, C. Y. Li, and G. C. Stutzin
Lawrence Berkeley Laboratory
Berkeley, CA 94720

Abstract

The populations of ground electronic state atomic hydrogen and ground electronic state, vibrationally-rotationally excited hydrogen molecules in an negative hydrogen ion source discharge have been measured using vacuum ultraviolet (VUV) laser absorption spectroscopy. Vibrational states up to v=8 and rotational levels as high as J=15 have been measured. The measurements have been made under a range of discharge conditions. The complete vibrational population distribution for v=1-8, J=1 has been obtained. The vibrational distribution appears to be thermalized and does not exhibit a "plateau" at the higher vibrational levels, in contrast to most models of this system. In contrast, the high rotational states are populated suprathermally. These determinations indicate that rotationally excited molecules may play an important role in the production of H⁻ in these sources.

I. INTRODUCTION

The development of intense sources of high energy neutral beams is of critical importance to the magnetic fusion community [1]. Central to this development are efficient sources of multi-ampere beams of H⁻ (D⁻). Several types of sources have been proposed to produce these ions. One of

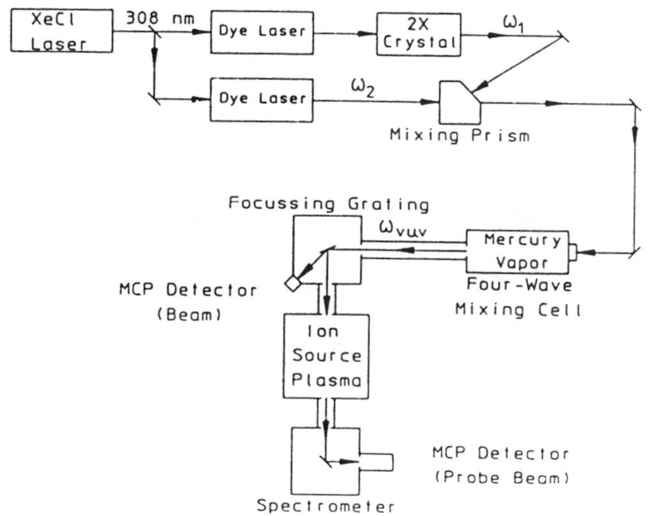

Fig. 1: Schematic diagram of VUV laser absorption spectroscopy system.

these is the so-called magnetic-multicusp volume production ion source. This source has been under development in many laboratories around the world. Yet, in spite of this importance, a precise understanding of the mechanism of H⁻ production in these sources is lacking. This paper describes experiments to measure the absolute density, temperature, and internal energy distribution of the hydrogen atoms and molecules in this type of source. These species, in particular vibrationally excited molecules, are thought to play critical roles in the production of H⁻. These measurements are obtained using vacuum ultraviolet laser absorption spectroscopy. This technique is capable of making non-perturbative measurements of these species directly in the the plasma. Coupled with measurements of the charged particles, these measurements may allow a better understanding of the H⁻ formation kinetics to be developed.

II. EXPERIMENTAL APPARATUS

The VUV laser absorption system has been previously described in detail [2]. A schematic of the system is shown in Figure 1. Using the technique of four-wave sum mixing (FWSM) in mercury vapor, narrow bandwidth pulses of conerent VUV are produced. With FWSM the energies of three photons (waves) of energies ω_1, ω_1, and ω_2 are combined to produce a fourth photon with energy ω_{vuv}. VUV with wavelengths from 97 nm to 128 nm has been produced in pulses approximately 20 ns long. These wavelengths encompass both the atomic Lyman series and the (B-X) and (C-X) molecular absorption bands.

The VUV is directed through the ion source where it is detected using microchannel plates. Absorption spectra are obtained by tuning the VUV wavelength through a molecular or atomic transition. The intensity of the transmitted VUV is measured as a function of the wavelength, producing the absorption lineshape. The area under the absorption lineshape is proportional to the density of the absorbing species, while the width of the lineshape is proportional to its translation temperature.

1916

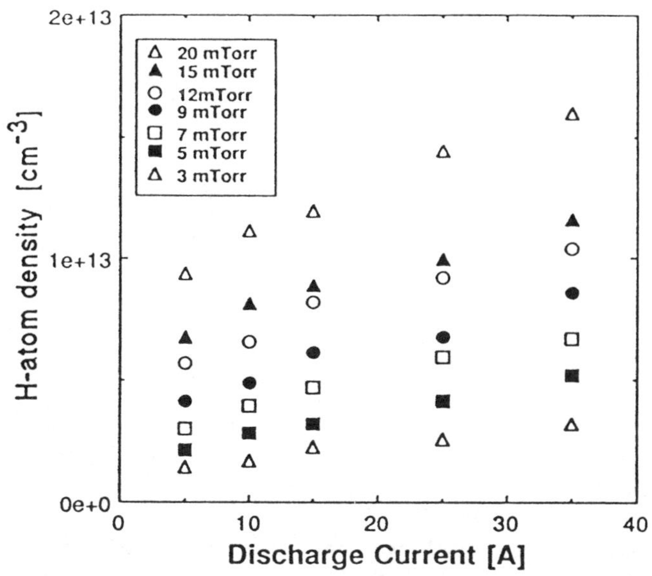

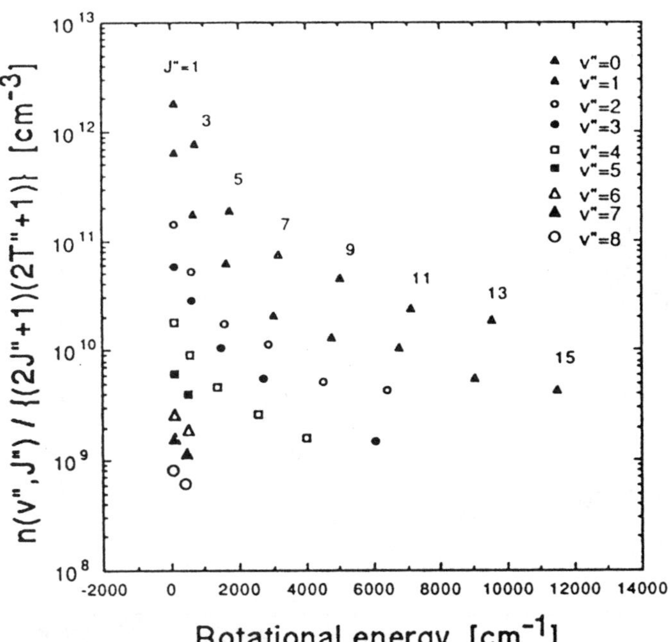

Fig. 3   The measured rotation-vibration populations of hydrogen. Discharge conditions are 25A, 120 V, and 8 mTorr.

Fig. 2   Atom density as a function of discharge current for several pressures. Discharge voltage is 120 V.

The ion source studied is the Lawrence Berkeley Laboratory 25 cm diameter volume source.   A magnetic multicusp ion source, it is cylindrical and has rows of magnets to provide a confining magnetic field for the charged particles of the plasma. The cathodes are tungsten filaments. Discharge was obtained with 25 A discharge current and 120 V discharge voltage. Typical hydrogen pressure is 8 mTorr.

III. RESULTS AND DISCUSSION

The results of typical measurements of H-atom are shown in Figure 2.   Here, the atom density was measured as a function of discharge current for several hydrogen pressures. The discharge voltage was a constant 120 V for all data shown. As can be seen, the atom density rises monotonically with increasing current, although with only a small slope. The degree of dissociation of the hydrogen is small, less than 3% of the filling pressure at even the highest current.   The translational temperature of the atoms is also observed to increase with discharge current. The absorption lineshapes are well characterized by a single temperature, ranging from 0.06 to 0.17 eV at 5 A and 35 A of discharge current, respectively.

The absolute population in many vibrational-rotational states of molecular hydrogen is also measured.   Figure 3 displays the results obtained for many of these states. The population in each vibrational level is subdivided among the different rotational states (labeled by J in the figure.) In this type of representation, a thermalized rotational population distribution would be characterized by a straight line. As can be seen, although the population of the lower J states (J=1-4) fall on a line, the higher J state populations do not. The linear portion of the population is described by a rotational temperature of 500K. The high J states are suprathermally

populated and cannot be assigned a temperature. The excess population in these states may have important implications for the production of H⁻.

If the rotational distribution is the same for all vibrational levels, the correct shape of the vibrational state population distribution will be obtained even if only the populations in the v=n, J=1 states are plotted against vibrational energy. This analysis has been performed and the results are shown in Figure 4. As can be seen, the population distribution is linear and is described by a vibrational temperature of 4900 K. However, a more accurate vibrational population distribution for the hydrogen molecules would be obtained by summing the population in all J states for each vibrational level. As was shown in Figure 3, high J levels can have a significant population.   Based on the trends exhibited by the states that were measured, estimates of the population were made for J states all the way out to the dissociation limit.   Summing these and plotting the resulting populations yielded a distribution which, although different than the original, could still be well fit by a temperature near 6000 K. This indicates that the high rotational levels can contribute substantial populations to vibrational levels thought to be important in forming H⁻.

The effect of increasing the discharge power of the source on rotational-vibrational populations is shown in Figure 5. Here the populations in v=4, J=1 and 7 are shown as a function of discharge current. As can be seen, the population in J=1 peaks at only 12 A and decreases slightly as the current is increased.   On the other hand, the population in J=7 increases over the range of currents used. This increase in the

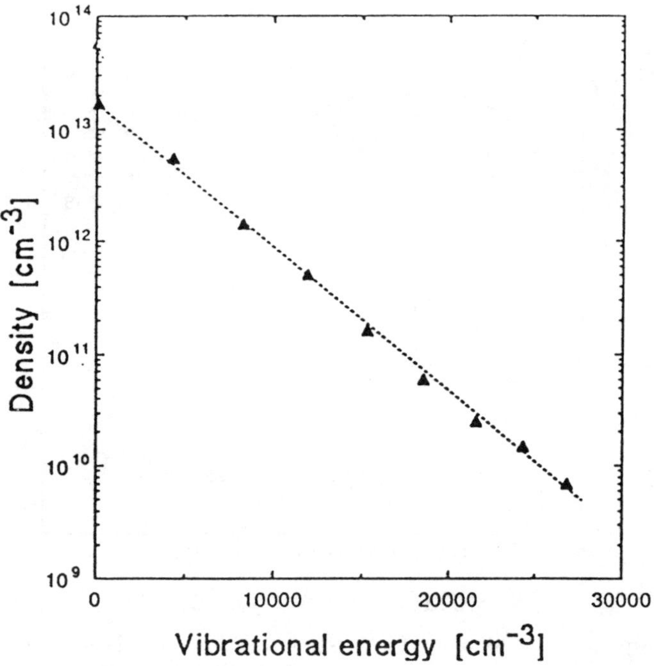

Fig. 4 Vibrational population distribution using J=1 states.

population of the higher J states with increases in discharge current is also seen in other vibrational levels.

The increase of high J state populations with discharge current has important implications for H^- formation. Calculations [3] indicate that both vibrational and rotational excitation will give enhanced H^- production. Therefore, increasing the population in v=4, J=7 relative to J=1 will increase the total rate of H^- production With sufficient population in the high rotational levels of even moderate vibrational levels($4 \leq v \leq 7$), large populations in the high vibrational levels (v>7) would not be necessary. In particular, a "plateau" in the vibrational distribution, a feature common to many models [4], might not be necessary.

Because the density of molecules in these high J states has not been measured, it is not known how much they contribute to H^- production. However, it is known that for this source increases in discharge current in this parameter range lead to enhanced H^- emission current [5]. Since increasing the discharge current also enhances the population in states which are more efficient in producing H^-, i.e. high J states, the increasein the emission current with discharge current may be due in part to populations in high rotational levels rather in high vibrational levels.

SUMMARY

VUV laser absorption spectroscopy has been used to measure the density and temperature of hydrogen atoms and molecules in an H^- ion source discharge. The translational temperature of the atoms ranges from 0.06 to 0.17 eV,

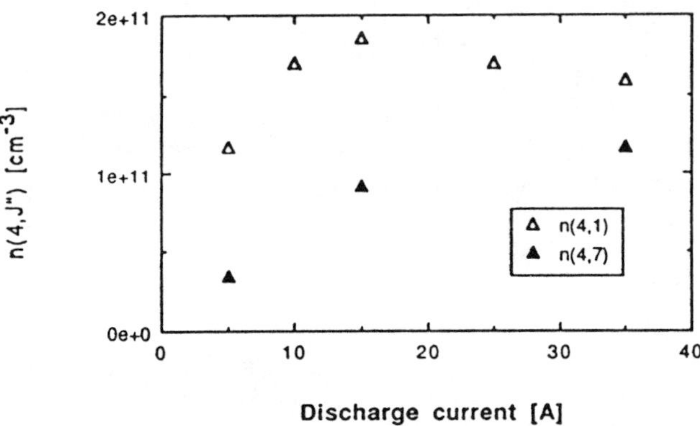

Fig. 5 Variation of rotational population with discharge current. States shown are J=1 and J=7 of v=4. Note the enhancement of J=7 with increasing current.

depending on the discharge conditions. The density of the atoms increases mildly with discharge current, but remains below 3% of the molecular density. The molecules exhibit a large degree of vibrational and rotational excitation. The rotational distribution is not thermalized, with the high J states being overpopulated. The vibrational distribution appears to be thermalized, but the temperature of the distribution depends on the assumptions made about the population in the highest J states.

The population in the high J states is of crucial importance in understanding the H^- formation mechanism. The scaling of the H^- current and the high J state populations suggest that these states play a significant role in negative ion formation. Experiments are now underway to measure these states and elucidate their role in H^- production.

ACKNOWLEDGEMENTS

This work has been supported by the Air Force Office of Scientific Research and the Department of Energy under contract number DE-AC03-76SF00098.

REFERENCES

[1] W. B. Kunkel "Giant Ion Sources of Neutral-Beam Injectors for Fusion ", Rev. Sci. Inst. 61, pp. 354-359, 1990
[2] G.C. Stutzin, A.T. Young, A.S. Schlachter, J.W. Stearns, K.N. Leung, W. B. Kunkel, G.T. Worth, and R.R. Stevens, "VUV Laser Absorption Spectrometer System for Measurement of H Density and Temperature in a Plasma", Rev. Sci. Inst. 59, pp. 1363-68, 1988
[3] J.M. Wadehera, "Dissociative Attachment to Rovibrationally excited Hydrogen" Phys. Rev. A, 29, pp. 106-10, 1984
[4] J.R Hiskes and A.M.Karo, "Analysis of the H_2 Vibrational Distribution in a Hydrogen Discharge",Appl. Phys. Lett. 54, pp 508-11, 1989
[5] G.C. Stutzin, to be published

OPTIMIZATION OF AN RF DRIVEN H⁻ ION SOURCE

K. N. Leung, W. F. DiVergilio, C. A. Hauck, W. B. Kunkel and D. S. McDonald
Lawrence Berkeley Laboratory
University of California
Berkeley, CA 94720

Abstract

A radio-frequency driven multicusp source has recently been developed to generate volume-produced H⁻ ion beams with extracted current density higher than 200 mA/cm^2. We have improved the output power of the rf generator and the insulation coating of the antenna coil. We have also optimized the antenna position and geometry and the filter magnetic field for high power pulsed operation. A total H⁻ current of 30 mA can be obtained with a 5.4-mm-diam extraction aperture and with an rf input power of 50 kW.

I. INTRODUCTION

Multicusp plasma generators have been operated successfully both as volume production or surface conversion H⁻ sources.[1] The H⁻ ions generated by volume-production processes have lower beam emittance and therefore are useful for the generation of high-brightness beams. In order to achieve high current densities, volume H⁻ sources require high discharge power. For this reason, the lifetime of the ordinary filament cathodes is short for steady-state or high repetition rate pulse operations.

A new radio-frequency (rf) driven H⁻ source[2] has recently been developed at Lawrence Berkeley Laboratory (LBL) for use in a calibration beam system and possibly in the injector unit of the Superconducting Super Collider (SSC). Initial study showed that rf power as high as 25 kW could be coupled inductively to the plasma via a glass-coated copper-coil antenna. It has also been demonstrated that the source is capable of generating 1-ms H⁻ beam pulses with a repetition rate as high as 150 Hz.

Experiments have been conducted at LBL to improve the efficiency and reliability of the source, and the H⁻ to electron ratio in the extracted beam by optimizing the filter magnetic field, the position and geometry of the antenna and the extraction aperture configuration. To date, it has been demonstrated that rf power higher than 50 kW could be coupled to the plasma via an improved porcelain-coated antenna. The extracted H⁻ current achieved is higher than 30 mA.

II. EXPERIMENTAL SETUP

Figure 1 is a schematic diagram of the ion source. The source chamber is a thin-walled copper cylinder (10 cm diam by 10 cm long) surrounded by 20 columns of samarium-cobalt magnets that form a longitudinal linecusp configuration. The magnets, in turn, are enclosed by an outer anodized aluminum cylinder, with the cooling water circulating around the source between the magnets and the inner housing. The back flange has four rows of magnets cooled by drilled water passages in the copper. In order to enhance the H⁻ yield, a pair of water-cooled permanent magnet filter rods is installed near the extraction region.

The open end of the source chamber is enclosed by a two-electrode extraction system. H⁻ beams are normally extracted from the source through a 2-mm-diam aperture. a permanent-magnet mass analyzer is used with a Faraday cup to measure the electron and the H⁻ currents in the accelerated beam. In order to reduce the electron contents in the beam, a stainless-steel cylindrical collar[3] is installed at the exit aperture (Fig. 1). This collar is capable of reducing the electron current by a factor of 2 without any degradation in the H⁻ output current.

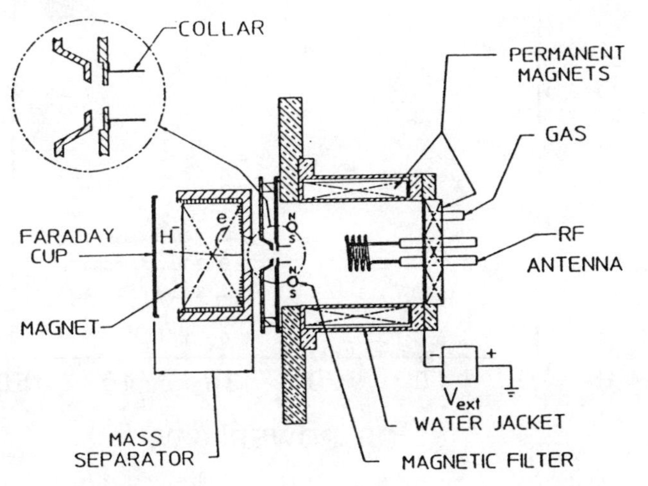

Fig. 1 Schematic diagram of the rf multicusp H⁻ ion source.

The rf antenna is fabricated from 4.7-mm-diam copper tubing and is coated with a thin layer of hard porcelain material. This new porcelain- coating performs better than the glass-coating which we used in the previous experiments. The thin coating is slightly flexible and is therefore resistant to cracking. It can maintain a clean plasma and can last for a long period of operation.

A sine-wave oscillator drives a gated 400 W solid state amplifier at a nominal operating frequency of 1.8 MHz. The resulting rf pulses then drive a Class C tube amplifier

1919

with a maximum pulse output power of 50 kW. In order to couple the rf power efficiently into the source plasma, a matching network (which is a tunable resonant parallel LC circuit) is employed. In this experiment, a small hairpin tungsten filament is used as a starter for the rf induction discharge.

III. EXPERIMENTAL RESULTS

The multicusp source is equipped with a pair of neodymium-iron magnet filter rods which are placed 5 cm apart. The maximum field at the center of the filter plane is about 125 G. As we reduce the filter separation to 4 cm, the magnetic field is increased to about 190 G. With this new filter arrangement, we are able to obtain higher H⁻ output and lower electron current in the extracted beam. All the source optimization data reported in this paper are obtained with this filter geometry.

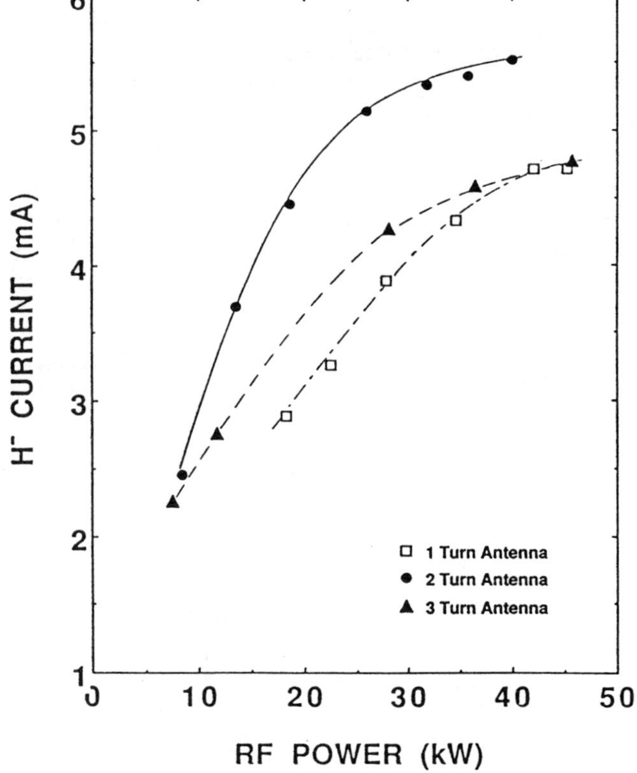

Fig 2. H⁻ output as a function of rf power for three different antenna geometries.

Figure 2 shows the H⁻ output current as a function of rf input power for 1, 2 or 3 turn antenna coil. Within the range of rf power considered, the 2-turn antenna provides more H⁻ current than the 3 or 1-turn antenna. The diameter of the antenna coil can also affect the H⁻ yield. We have tested two different diameter antenna coils (5.3 cm diam and 7 cm diam), both having the same number of turns. Figure 3 is a plot of the extracted H⁻ current versus rf input power. The 7-cm-diam coil improves the H⁻ output by approximately 18%.

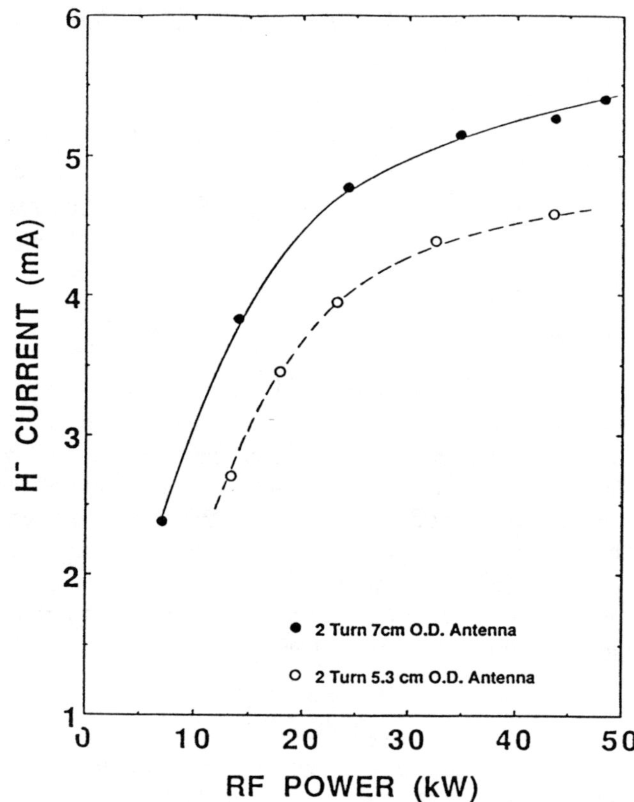

Fig. 3 H⁻ output as a function of rf power for two different diameter antenna coils.

The dependence of the H⁻ output current on the axial position of the antenna has been investigated. Since the antenna is supported through two sliding seals on the back flange, the position of the antenna can be easily varied without opening the source chamber. Figure 4 shows the H⁻ current as a function of rf power for three different axial positions. (The distance is measured from the back flange surface to the end of the antenna). It can be seen that the extractable H⁻ current increases as the antenna approaches the filter. The optimum position occurs at about 8.35 cm. Further increase in the distance from the back flange results in a reduction of the H⁻ current.

The scaling of H⁻ current as a function of extraction area is of great interest in H⁻ source development. By using the optimized filter geometry and position, we have operated the source with a 5-mm-diam and a 5.4-mm-diam extraction aperture. Figure 5 shows the dependence of the extracted H⁻ current as a function of rf power. With a 5-mm-diam aperture, the highest H⁻ current achieved is about 26 mA. By enlarging the aperture diameter to 5.4 mm, the H⁻ output current can exceed 30 mA for an rf input power of 50 kW. The required source operating pressure is approximately 17 mTorr.

The present result as well as the data from the previous investigation have demonstrated that the rf driven volume-production source can provide substantial H⁻ current. Recent experimental measurement also shows that the beam emittance is nearly the same for both the rf and filament dc discharge for a given input power.[4] For this reason, the rf induction driven source is extremely useful for the production of high brightness H⁻ beams.

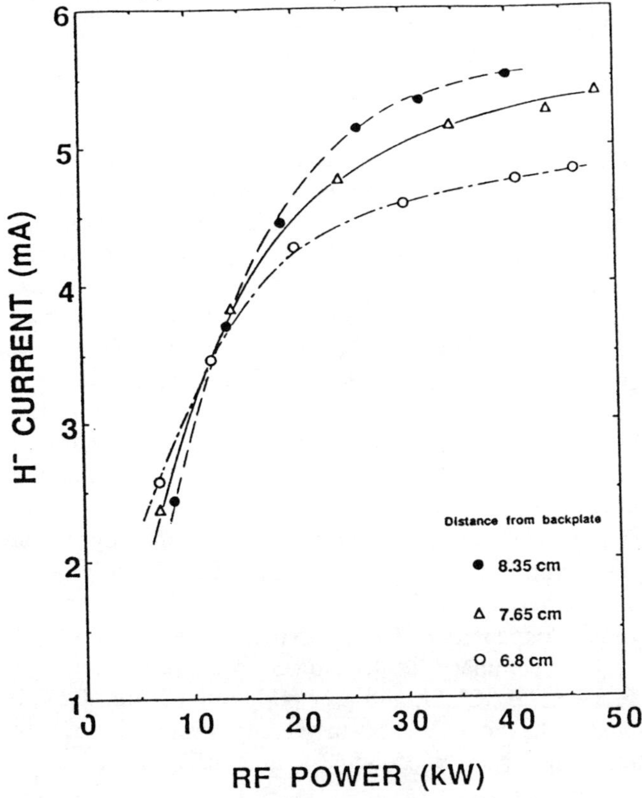

Fig. 4 H⁻ output as a function of rf power for three different antenna positions.

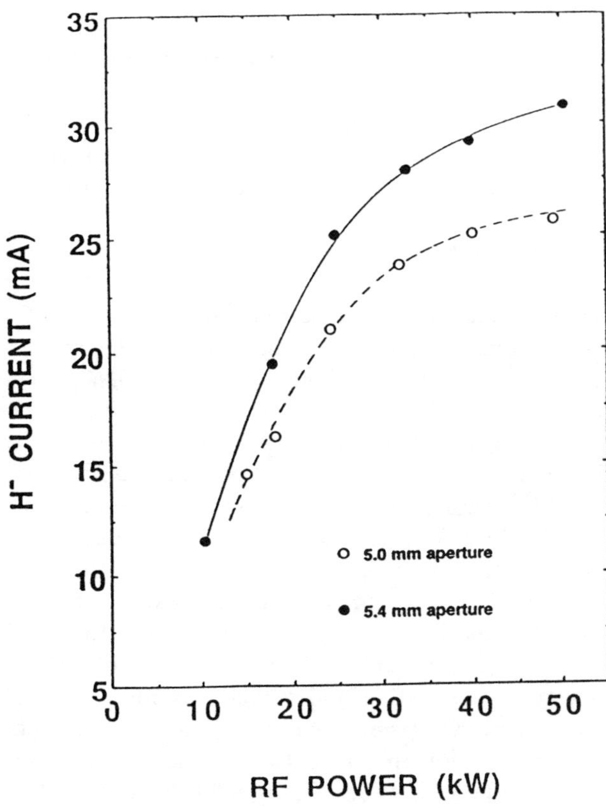

Fig. 5 H⁻ current versus rf input power for two different extraction aperture sizes.

We would like to thank D. Moussa, S. Wilde, P. Rosado, G. DeVries , G. Koehler, R. Wells and M. Williams for technical assistance. This work is supported by the Superconducting Super Collider Laboratory and the Director, Office of Energy Research, Office of Fusion Energy, Development and Technology Division, of the U.S. Department of Energy under Contract No. DE-AC03-76SF00098.

REFERENCES

1. K. N. Leung, Nucl. Instrum. Method B40/41, 1028 (1989).
2. K. N. Leung, G. J. DeVries, W. F. DiVergilio, R. W. Hamm, C. A. Hauck, W. B. Kunkel, D. S. McDonald, and M. D. Williams, Rev. Sci. Instrum., 62, 100 (1991).
3. K. N. Leung, C. A. Hauck, W. B. Kunkel, and S. R. Walther, Rev. Sci. Instrum., 61, 1110 (1990).
4. G. Gammel, Proc. of the 1991 IEEE Particle Accelerator Conference, San Francisco, CA. (May, 1991).

IUCF HIGH INTENSITY POLARIZED ION SOURCE*

M. Wedekind, R. Brown, J. Collins, V. Derenchuk, D. Dale, D. DuPlantis, T. Ellison, D. Friesel, J. Hicks,
D. Jenner, A. Pei, H. Petri, P. Schwandt, J. Sowinski
Indiana University Cyclotron Facility, Bloomington, IN 47408

I. INTRODUCTION

A major fraction of the experimental program at IUCF has always concentrated on studies of spin degrees of freedom. The newly commissioned electron cooled storage ring/synchrotron will provide unique opportunities in spin physics research which require circulating beam intensities near the limit of ring operation, 10^{16} particles/sec, because target densities and/or reaction cross sections are very low.

In order to meet these experimental requirements we are replacing the existing ANAC source built in 1976 with a modern high intensity source expected to yield in excess of 100μA DC $\vec{H}^+$ and $\vec{D}^+$ ion beams and/or at least 10μA DC $\vec{H}^-$ and $\vec{D}^-$ ion beams. Such a source, coupled with a high-efficiency bunching system and a high-transmission beam line into the first cyclotron, should allow 10^{10} protons to be stored in the Cooler ring in a few seconds. Besides the options of intense positive or negative ion beams and CW or pulsed mode of operation, other source requirements are: low transverse phase space emittance ($< 15\pi$ mm mrad $\text{MeV}^{1/2}$) to match the cyclotron acceptance, low longitudinal phase space emittance (i.e. energy spread < 10eV) to permit efficient bunching, and proton polarization of 75% or higher, with equivalently high deuteron polarization in both vector and tensor states separately.

Of the various proven state-of-the-art polarized ion source technologies, the system recently built and brought into successful operation at TUNL by T.B. Clegg and associates [1] best meets our needs. In 1989 we began acquiring design specifications from TUNL to build HIPIOS (High Intensity Polarized Ion Source). Based on TUNL's experience, HIPIOS should produce in excess of 100 μA DC $\vec{H}^+$ and $\vec{D}^+$ ion beams. It employs cold (~30 K) atomic beam technology, a proven method for generating intense neutral atomic beams. The ionizer is an electron cyclotron resonance (ECR) device similar to those routinely employed for many years for production of intense beams of heavy ions in high charge states.

II. INITIAL DESIGN AND TESTING

Throughout the design process we have been greatly assisted by the TUNL staff who provided us with all their construction drawings and many useful recommendations based on their operating experience.

---

* Work supported by NSF Grant PHY-8914400

A. Dissociator

The dissociator geometry is a close copy of the TUNL design. The discharge is excited by an ENI OEM-6A, 15 MHz, 600 watt supply and is contained inside a 10 mm OD Pyrex tube. About 35 SCCM of hydrogen gas flows into the discharge tube which has a 20 mm OD Pyrex water jacket. The copper cold nozzle is slipped over a MACOR thermal adapter and is cooled by a CTI 1020 cold head that has a capacity of 10 W at 20 K. The cold nozzle temperature is controlled with a heater and sensor attached directly to the nozzle clamp. As in the TUNL design, N_2 is bled into the nozzle at a flow rate of 0.02 SCCM to prevent recombination on the copper surface.

The dissociator has recently been put into operation without sextupoles, and measurements of the atomic beam velocity have been made using an atomic beam chopper generously loaned to IUCF by Professor W. Gruebler of ETH [2]. Preliminary results indicate that with an H_2 flow rate of about 20 SCCM, and a nozzle temperature of 25 K, 30 K and 40 K, the atomic beam velocity was consistent with a beam temperature of 35 K, 42 K and 54 K respectively.

B. Sextupoles

Conventional sextupole magnets are used to separate the hydrogen atomic spin states hence generating atomic polarization. Two independent magnets with a 26 cm drift space between are used for improved focusing of the atoms into the ionization region. The optics of the first magnet is optimized by using a tapered (14 mm - 28 mm) aperture resulting in a ramped sextupole strength. We have increased the length of the first sextupole by 50% from previous designs to provide the increased integrated sextupole strength recommended. Ray trace calculations predict that this modification should have an effect similar to that obtained by increasing the strength of the shorter magnet - a much more difficult task. We have measured a maximum pole tip field for the tapered sextupole of 7600 G at the design current of 250 Amps, and 5600 G for the second sextupole at the design current of 200 Amps.

C. Rf Transitions

The rf transitions, which convert the atomic polarization to nuclear polarization, utilize the conventional adiabatic fast passage principle causing transitions between particular magnetic substates. One strong field and weak field transition pair is located between the sextupoles and another pair between the second sextupole and the ionizer.

Each strong field transition consists of a magnetic field transverse to the beam direction (85 G for deuterons, 150 G for protons) and an rf magnetic field (330MHz, 400MHz, 460MHz for deuterons and 1485MHz for protons) generated by a cavity. Each weak field transition consists of a magnetic field of 10 G transverse to the beam direction and an rf magnetic field of 15 MHz generated by a small coil with its axis along the beam. The strong field transition is shielded from the fields of the sextupole magnet by a 1.6 mm thick iron field clamp. The weak field transitions are located inside a cylinder of magnetic shielding to prevent interference from other magnetic fields.

The mechanical design of the cavities follows the design of Robinson et al. [3]. For driving the cavities, we opted to change from a medium power signal generator to a self-oscillating loop utilizing the cavities self-resonance and a high gain amplifier. After phase adjusting the output of the cavity with the input we were able to operate from 500 milliwatts to over 5 watts.

Assembly of the 330 MHz and the 400 - 460 MHz dual unit has been completed and development is continuing to enhance their operating characteristics. Data taken thus far show we have approximately 10 G per square root of rf watt.

D. ECR Ionizer

The ECR ionizer design follows the now standard configuration employed in several recently commissioned polarized ion source systems [4,5]. A polarized atomic beam is ionized while passing through a resonantly excited electron plasma confined radially by a permanent magnet hexapole and longitudinally by solenoidal mirror coils. Our ionizer closely follows the TUNL design which has the hexapole magnet mounted within a vacuum chamber on which the external mirror solenoids and return yoke are assembled. We have, however, made several modifications to the basic configuration to address special requirements for our application, accomodate competing technical requirements, and incorporate improvements resulting from source commissioning work at TUNL.

A primary technical consideration involves the design and mounting of the permanent magnet (SmCo) hexapole, which can be damaged by plasma or beam heating above 300 °C or by exposure to the hydrogen gas in the ionizer vacuum chamber. An ECR chamber pressure less than 0.1 uTorr is required. A further complication is that the internal structure (hexapole magnet, pre-buncher, and extraction electrodes) of the ECR ionizer is isolated from and operated at 20 kV above platform ground to provide an extraction gradient. This allows the majority of the ion source apparatus to remain at ground potential, and also simplifies the pre-bunching of the low energy dc beam from the ionizer for later injection into the IUCF cyclotrons.

To meet the competing requirements of vacuum, temperature, rf and high voltage constraints on the mounting of the hexapole structure, it was decided to vacuum encapsulate the permanent magnets in thin walled (0.5 mm) stainless steel cans welded onto thicker ring brackets to form the hexapole field configuration. The magnets are inserted into the welded structure following fabrication, mapped and shimmed to minimize other multipole field components, and then epoxy impregnated. This structure is supported from one end of the ECR vacuum chamber by six 2.5 cm diameter ceramic insulators. A separate freon cooled copper heat shield is mounted inside the hexapole structure to conduct the heat generated by the plasma and rf ($\leq$ 150 watts) away from the magnets. This overall design provides for an open geometry permitting good radial and axial pumping in the region of the ECR plasma.

The rf generator for the ECR plasma will operate over a range of 2.6 to 4.7 GHz at 150 watts. The system will consist of a voltage controlled oscillator based on an Avantec HTO-2600 oscillator and AGT-8235 variable gain preamplifier. Final amplification of the microwave system is provided by a Keltec TWTA 200 watt amplifier. Power will be fed into the ECR chamber via Heliax coaxial cable to an internally mounted horn using a waveguide adapter at the chamber. We expect to achieve higher beam intensities and polarizations at the higher frequencies, as reported by the TUNL group.

E. ECR Extraction and Acceleration

Extraction of positive ions from the ECR plasma region will take place in an axial (solenoidal) magnetic field of about 1 kG, using a computer designed electrostatic electrode system. The ECR extraction beam energy will be relatively low, 1-2 keV, to permit efficient prebunching using a low-voltage ramp waveform. The low-energy beam will drift in the uniform solenoid field for about 30 cm to achieve bunch formation at the second acceleration stage (to 20 keV beam energy) at the termination of the solenoid field. The solenoid field provides preferential focusing of the proton beam relative to the underfocused nitrogen ions, allowing for early rejection of much of the large nitrogen component which originates from the nitrogen buffer gas used to create the ECR plasma.

The ray trace code BEAM3D specifically designed for ECR extraction at Michigan State University was installed on the IUCF VAX computer system with the invaluable help of T. Antaya of MSU. For a given axially symmetric electric and magnetic field configuration (calculated from any electrode and coil geometry using the POISSON group of codes), the BEAM3D code performs exact 3D ray tracing for any multi-species ion beam, includingspace charge and finite ion temperature effects. Optimization of the beam extraction, focusing and acceleration geometry for minimum beam phase space and maximum transmission of protons has been performed for HIPIOS.

F. Buncher

A prototype of the wideband, ramp-waveform pre-buncher which follows the ECR was built and tested at low power by Lars Hermansson, a visiting engineer from The Svedberg Laboratory in Uppsala, Sweden. In order to compress 90% of the beam from the source into the $\pm 3°$ cyclotron phase acceptance, this buncher should have a linearity of $\pm (30/n)\%$ of the voltage amplitude over 90% of the fundamental rf period, where $n = 1,2,3$ is the pulse-selection ratio (subharmonic). A buncher with this linearity specification should ideally produce a beam phase width of less than $\pm 30°$ at the entrance to the resonant buncher in the 600 kV beam line to the cyclotron and the nonlinearity resulting from the sinusoidal waveform of this second buncher should result in less than $\pm 1°$ beam phase width at the cyclotron (neglecting all other effects). Test results show that the design specifications for the most part have been met or exceeded.

G. Control System

We made the decision not to use the existing cyclotron control systems for HIPIOS since their limited functionality and numerous obsolete components placed limits on the proposed expansion. Instead, we are implementing a new system that will operate in parallel with the present controls and provide a platform for future facility expansion.

Hardware controls are based on the use of Allen Bradley programmable logic controllers (PLC's) for vacuum and interlock systems, and VME modules for analog readout and control of temperatures, voltages, currents, and vacuum levels. Direct control of the VME systems is provided by a VAX RT300 controller supplied by AEON.

Because the HIPIOS terminal operates at 600 KV potential, special care will be been taken to minimize the possibility of damage during sparking. These efforts include placing intelligent modules at ground potential, ensuring that all I/O racks are EMI sealed, and that all cabling uses EMI connectors with Tanszorb protectors. VME optical links will provide the necessary control isolation up to the terminal.

To drive the VME hardware we will use the graphically oriented workstation based software system marketed by Vista Control Systems, Inc. [6], chosen because of our familiarity with their VAX/VMS product line, the ease of accessing the existing control system PDP-11s running RSX via Decnet, their support of VME hardware and our judgment that the system is well suited to the comparatively small nature of our project. At present we are learning how to use the Vista system, writing low level handlers for specific VME modules and creating simple applications.

X-Windows terminals will be used to provide multiple access points to the control system for operation and maintenance - directly in the terminal, for example. We will also provide services similar to those now available to the cyclotron operator, such as DAC setup and scaling, DAC and ADC archiving and limits checking.

H. Power Supply Modifications

To reduce the size, weight, and power consumption of power supplies for the source, a commercial line of 5 and 6 kW switching mode power supplies for powering the magnet loads has been evaluated for performance. While stability was found to be acceptable, the high harmonic content of the three phase line currents (especially at light loading) was excessive (in excess of 110% Total Harmonic Distortion, THD). Since the AC system is alternator driven and presents higher than average source impedance (as opposed to a normal AC distribution network), AC line voltage distortion could be well in excess of acceptable limits (3-5% THD). The manufacturer agreed to a modification of the units by adding a filter choke to the DC output of the rectifier bridge. This lowered current distortion to about the 35% THD level while AC RMS line current dropped 30%.

Because of the high stability requirement for the ECR electrostatic extraction elements (better than 100PPM), the extraction high voltage supplies have had remote sensing modifications designed, installed, and tested. These modifications compensate for the voltage variation that would normally be seen at the load due to changes in beam loading as a consequence of having a 180 KΩ series surge resistor in the high voltage load cable. Worst case voltage variations at the element of 1% have been reduced to .02% by use of this remote sensing technique.

III. ACKNOWLEDGMENTS

The authors express their appreciation to Tom Clegg and to the entire technical staff at IUCF.

IV. REFERENCES

[1] T.B. Clegg in Proc. of the High Energy Spin Physics 8th Intl. Symp., Minneapolis, MN, ed. K.J. Heller, A.I.P., New York, 1989, Vol. 187, 1227.

[2] D. Singy, P.A. Schmelzbach, W. Gruebler, and W.Z. Zhang, N.I.M., A278, 349 (1989).

[3] H.G. Robinson et al., N.I.M., A278, 655 (1989).

[4] T.B. Clegg, U. Konig, P.A. Schmelzbach and W. Gruebler, N.I.M., A238, 195 (1985).

[5] L. Friedrich, E. Huttel and P.A. Schmelzbach, N.I.M., A272, 906 (1988).

[6] P. Clout, "Past, Present and Future of a Commercial Graphically Oriented Control System", International Conference on Accelerator and Large Experimental Physics Control Systems, Vancouver, B.C., Oct. 30 - Nov. 3, 1989.

Operational Experience with the TRIUMF Optically Pumped Polarized H⁻ Ion Source

P.W. Schmor, L. Buchmann, K. Jayamanna, C.D.P. Levy, M. McDonald, and R. Ruegg
TRIUMF, 4004 Wesbrook Mall, Vancouver, B.C., Canada V6T 2A3

Abstract

The initial goal of a polarized proton beam extracted from the TRIUMF cyclotron, having a current of 5 μA with 60% polarization, has been achieved with the development of the optically pumped polarized H⁻ ion source. This beam is now being used to produce an intense secondary beam of polarized neutrons for the TRIUMF experimental program. Much of the recent development effort has addressed the reliability requirements for routine operation. This paper describes the results with emphasis on the laser stabilization subsystem, the modifications to the electron cyclotron resonance proton ion source (ECRIS), the sodium charge exchange cells and the development of a low energy polarimeter. Also discussed are the developments which should lead to a higher polarization.

I. INTRODUCTION

TRIUMF has developed a 100% duty cycle, high intensity, optically-pumped, polarized ion source which now produces 5 μA of 60% polarized beam after acceleration to 200 MeV in the cyclotron [1]. The polarization is slightly less at 500 MeV due to two depolarizing resonances which are crossed during acceleration.

The polarized H⁻ beam is transported at 300 keV, approximately 50 m from the source to the cyclotron, through a beam line which uses only electrostatic focusing elements. The H⁻ beam is injected into the median plane of the cyclotron with an electrostatic spiral inflector. Protons are extracted from the cyclotron, by stripping the two electrons from the accelerated H⁻ ions, at energies which can be varied from 180 to 500 MeV, independently, into two beam lines. A Wien filter in the 300 keV beam transport line is used to rotate the spin direction from the horizontal into the vertical; i.e., aligned with the cyclotron magnetic field in order to preserve the polarization during acceleration.

The optically pumped polarized ion source (OPPIS) is based on a proposal by Anderson [2]. A hydrogen plasma is created within an electron-cyclotron-resonance (ECR) cavity by 28 GHz microwave ionization in an axial magnetic field. The typical absorbed power is 850 W. Protons are extracted from the ECR cavity plasma and are directed, at 5 keV, through a polarized sodium vapour where a fraction of the protons are neutralized by picking up a polarized electron. The polarization is induced by optical pumping of the sodium vapour with laser light tuned to the sodium D_1 transition at 590 nm. Most of the neutralized hydrogen is created in an excited atomic state and it is necessary to use the high magnetic field of a superconducting solenoid to preserve the polarization as the hydrogen atom decays to the ground state. An electrostatic deflector immediately downstream of the neutralizer removes all charged species from the beam which have passed through the sodium. The axial magnetic field reverses sign, between the neutralizer and a subsequent negative ionizer, in order to enhance the nuclear polarization through a Sona type transition. The sodium vapour of the ionizer produces an equilibrium fraction converting about 7% of the atomic hydrogen beam into an H⁻ ion beam. This negative ion beam is accelerated to 300 keV and transported to the cyclotron.

II. DEVELOPMENTS

A. Extraction Electrodes

Several extraction electrode systems for the ECRIS were investigated. The best results were obtained from a system of three electrodes, consisting of 1 mm thick molybdenum disks spaced 1 mm apart, with an hexagonal array of 1 mm diameter holes. The electrodes are powered in an accel-accel mode with the first electrode at the cavity potential, the second electrode about 1 kV below the cavity potential and the final electrode at ground (or slightly negative) potential.

B. Neutralizer

The sodium neutralizer cell consists of three thermally isolated components; i.e., an entrance snout, a vapour canal and a sodium reservoir. The canal and reservoir are independently heated. The sodium thickness (usually about 3 to $5\cdot10^{13}$ atoms/cm²) can be maintained reliably stable over periods of a week, with or without beam. While the optically pumped region of the canal is roughly only 1 cm in diameter, the actual canal diameter is, in fact, 3 cm in order to improve vacuum pumping of hydrogen from the canal and thereby reduce the probability of proton neutralization from the unpolarized hydrogen (and consequently increase proton polarization).

C. Ionizer

The ionizer cell, which also uses sodium vapour, is about 8 cm long and has 1.6 cm diameter condensation canals, both at the entrance and exit to the cell. With the higher sodium thickness (typically 40 to $80\cdot10^{13}$ atom/cm²) in the ionizer, it is necessary to recirculate the sodium in order to achieve a reasonable operating period for each sodium charge and to avoid frequent maintenance of the electrostatic elements.

C. Laser System

Two Coherent CR-599 dye lasers, which have been modified with 0.5 mm thick solid etalons to reduce the bandwidth to 3 GHz (FWHM), now each provide approximately 3 W in stable operation. Dye laser power up to 5 W each has been attained with argon pump powers of 25 W, although the stability was poor. The modifications of the dye lasers to increase the output power included: the installation of higher output transmission mirrors, the addition of 20% water to the ethylene glycol solvent and continuous charcoal filtering, improved cooling of the dye solution, the addition of water-cooled beam blocks inside the laser cavities, and higher dye flow rates.

The divergent beams from the dye lasers are brought by a relay lens assembly to a 1 mm diameter waist near a mirror in the beam line. A small diameter beam at that location is important as it allows the laser beams to be brought as close as possible to the source axis without obstructing the ion beam and enter the source at an angle only 3 mrad from parallel to the source axis. Each beam expands to approximately fill the central 1 cm of the neutralizer canal. To prevent damage to the mirror by the 5 keV neutrals and sodium escaping from the ionizer, it is protected by a movable microscope slide.

The laser frequency is in resonance with the Zeeman shifted D_1 absorption line in the neutralizer. The proton polarization is reversed by changing the helicity of the light and simultaneously changing the laser frequency, by tilting the etalon, to account for the Zeeman shift. The laser frequency is stabilized with an analog feedback signal derived from a scanning spectrum analyzer that samples a small fraction of the laser output. The correction signal is applied to the etalon galvanometer. The spectrum analyzers have a reference drift of about 0.3 GHz over a 24 hour period, when hermetically sealed. The laser power must also be stabilized in order to maintain the laser frequency over long periods. As the power drops, due to drifts in the alignment of the birefringent filter (BRF), the laser will begin to oscillate on an adjacent etalon peak which is separated in frequency by 200 GHz. The laser power is stabilized by a computer system which samples the laser power and makes adjustments to the BRF. The system is described in a separate paper in these proceedings [3].

D. Polarimeter

A nuclear polarimeter based on the low energy analyzing power of the $^6Li(p,^3He)\alpha$ reaction has been developed. Although the polarimeter is used at 300 keV with current up to 2 μA, it has, in fact, been shown to work at energies as low as 200 keV [4]. The analyzing power for detecting ^3He at 130° is approximately 0.21 at 300 keV. With a current of 2 μA of H^-, 1% statistics are achieved in about 3 minutes. A second polarimeter based on selective quenching of the metastable 2S state of hydrogen [5] has been used for source polarization studies but this device can not easily be adapted to co-exist within the operational demands of polarized beam production.

III. Operational Experience

A. Source Characteristics

The H^- current from the source depends on a number of parameters such as the thicknesses of vapour in both the ionizer and the neutralizer as well as the proton current drawn from the ECR plasma. The current can easily be changed, albeit at the expense of the polarization, by changing the thickness of the neutralizer vapour. There is a substantial unpolarized component in the beam which originates from proton neutralization by hydrogen upstream of the charge deflection plates. This unpolarized background is nearly independent of the sodium thickness and its relative contribution to the total H^- becomes less as the sodium neutralizer vapour thickness is increased. However, the polarization of the sodium vapour drops as the thickness is increased, due to radiation trapping and to the limited number photons from the dye lasers. As a result, there is an optimal sodium thickness of 3 to $5 \cdot 10^{13}$ atoms/cm² which gives the maximum proton polarization.

With a 31 hole extraction electrode system, a proton current of 22 mA is extracted from the plasma. With the neutralizer at $5.1 \cdot 10^{13}$ atom/cm² and the ionizer at $55 \cdot 10^{13}$ atoms/cm², there is about 35 μA on the first beam stop in the 300 keV beam line and 9 μA extracted from the cyclotron. By reducing the hydrogen fed to the ECRIS, to reduce the extracted current from the cyclotron to 5 μA and consequently improving the polarization slightly because of the reduced unpolarized background, the polarization (at 200 MeV) as measured on a polarimeter in an external beam line is 60.9%.

B. Source Emittance

The effective emittance of the H^- beam leaving the source is primarily determined by the ionizer [6]. The total change in emittance, in the transverse plane, as the beam passes through the ionizer is given by, $\delta\epsilon = \pi r^2/2r_c$, where r_c is the radius of the ion orbit in the peak magnetic field of the ionizer and r is the radius of the beam envelope which in this case is equal to the radius of the ionizer canal condenser. Beam transmission between the source and the cyclotron inflector is about 35% for a 61 hole extraction electrode to 45% for a 31 hole array which is consistent with a normalized source emittance of 0.7 $\pi \cdot$mm$\cdot$mrad for the 61 hole system.

C. Stability

A typical polarized proton user requests a few percent stability in the current and polarization over periods of weeks. At the source, this implies stability criteria to: a) the proton current from the ECRIS, b) the sodium neutralizer thickness, c) the sodium ionizer thickness, d) the laser power into the neutralizer, and e) the laser frequency.

The proton current, from the ECRIS, remains remarkably stable for periods of 24 hours and then only requires small changes to the hydrogen flow and to the voltage on the second electrode; processes which require only a few minutes to complete and are transparent to the user. After the first few hours, to reach thermal equilibrium, the sodium thicknesses in the neutralizer and the ionizer remain stable without further adjustment for periods of one week. The laser power into the neutralizer gradually decreases due to deposits on the microscope slide which is upstream of the beam-line mirror. The slide must regularly be moved to insert clean segments and ensure high laser transmission. The rate of slide movement (typically about once per day) depends on the quality of the vacuum around the slide. This procedure requires a negligible 5 minute beam interruption. In addition, approximately once per week the slide must be replaced; a procedure which shuts the beam off for a 2 to 3 hour period.

Since most experiments require that the proton spin regularly be flipped, it is necessary to quickly restabilize the lasers after each spin flip (84 GHz frequency shift). To achieve this rapid restabilization, the BRF is kept in a position where its transmission peak is midway between the optimum positions of the two frequencies. The resulting 10% loss in laser power is acceptable. The laser power stabilization system is programmed to keep the laser power roughly equal in both states. This technique works for spin flip rates up to once per minute. Gating the power measurements should allow the rates to be increased. Spin flip rates are limited to 1 Hz if only the laser frequency is stabilized and to about 10 Hz with no stabilization. The dye lasers are designed to permit a more rapid frequency switching, eventually up to 100 Hz. The etalon has a measured mechanical response time of 1 ms. The laser system is still not fully automated and requires periodic manual intervention. Nevertheless, the ion source has, recently, operated flawlessly for a continuous 290 hour experimental run.

IV. Future Planned Developments

A. Optical Pumping

The optical pumping needs to be improved to yield a proton polarization beyond the present 60% maximum. To achieve this goal, the dye lasers are now being replaced with inherently more stable titanium sapphire lasers. In addition the sodium neutralizer is being replaced with a rubidium neutralizer. This change should simplify and improve the polarization stability task. In addition, a higher nuclear polarization is expected for several reasons. The maximum output power of each titanium sapphire laser exceeds that from the modified dye lasers. Moreover since rubidium is pumped at 795 nm (compared to 590 nm for sodium), the optical pumping rates are higher as a result of the 33% more photons per watt. The major loss of polarization in the neutralizer is due to the wall depolarization

rate and this rate will be less with the slower velocity (lower temperature, higher mass) rubidium atoms. Finally, rubidium has a wider absorption bandwidth than sodium and as a result there is less radiation trapping (therefore higher ultimate polarization) at a given neutralizer thickness. These factors should allow thicker targets to be more highly polarized, leading to relatively less unpolarized background and higher nuclear polarization. The charge exchange cross sections are similar for both rubidium and sodium. These changes to OPPIS are expected to be completed prior to an experimental run in September 1991.

B. Polarimeter

A more compact version of the 300 keV polarimeter is being installed. It incorporates somewhat larger solid angles and includes two (angular) channels of observation (at about 100° and 140°). The target ladder and the associated beam collimator are retractable so that the proton beam can be transmitted, without additional losses, into the TRIUMF cyclotron. A cryopump is mounted close to the polarimeter target, to ensure a good clean vacuum in the region and thus reduce the amount of (lifetime-limiting) deposits on the target.

C. Spin Exchange

The source conversion from sodium to rubidium is the first step of a planned study to investigate the feasibility of using spin exchange rather than charge exchange to establish the polarized atomic hydrogen beam [7]. The higher current from this type of ion source could substantially increase the beam availability to TRIUMF polarized users.

V. References

[1] L. Buchmann, K. Jayamanna, C.D.P. Levy, M. McDonald, R. Ruegg, P.W. Schmor, A. Belov, V.G. Polushkin, A.N. Zelenskii, "A DC Optically Pumped, Polarized Ion Source", Nucl. Inst. Meth., in press, 1991.

[2] L.W. Anderson, "Optically Pumped Electron Spin Polarized Targets for use in the Production of Polarized Ion Beams", Nucl. Inst. Meth., 167, pp. 363-370, 1979.

[3] S. Sarkar, C.D.P. Levy, "Laser Power Stabilization in the TRIUMF Optically Pumped Polarized H⁻ Ion Source", these proceedings.

[4] L. Buchmann, "A Proton Polarimeter for Beam Energies Below 300 keV", Nucl. Inst. Meth., in press, 1991.

[5] A.N. Zelenskii, S.A. Kokhanovskii, V.M. Lobashev, V.G. Polushkin, "A Laser Source of Polarized Protons and H⁻ Ions", Nucl. Inst. Meth., A245, pp. 223-229, 1986.

[6] W.M. Law, C.D.P. Levy, P.W. Schmor, J. Uegaki, "Emittance Growth due to Hydrogen Ionisation by Charge Exchange in a Solenoid", Nucl. Inst. Meth., A263, pp. 537-539, 1988.

[7] A.N. Zelenskii, S.A. Kokhanovskii, V.G. Polushkin, K.N. Vishnevskii, "Investigation of Spin-Exchange Polarized Ion Source", in *Proceedings of the International Workshop on Polarized Ion Sources and Polarized Gas Jets*, KEK, Tsukuba, Japan, February 1990, pp.310-316.

OPERATION OF THE OPTICALLY PUMPED POLARIZED H⁻ ION SOURCE AT LAMPF

R. L. York, D. Tupa, D. R. Swenson, and O. B. van Dyck

Los Alamos National Laboratory, Mail Stop H838, Los Alamos, New Mexico 87545

Abstract

We report on the first five months of operation of the Optically Pumped Polarized Ion Source (OPPIS) for the nuclear physics research program at LAMPF. The LAMPF OPPIS is unique in using Ti:Sapphire lasers to polarize the potassium charge-exchange medium, and until recently was unique in using a superconducting magnet in the ECR source and polarizer regions. The ECR extraction electrode biasing arrangement is also unique. Typical performance was 25 microamps of peak current (measured at 750 keV) with 55% beam polarization or 15 microamps at 62%. Ion source availability was greater than 90%. We also report our planned improvements in preparation for research operation in May of 1991.

1 Introduction

In its first year of operation, OPPIS exceeded its design goal by producing a polarized beam whose value of P^2I was more than a factor of 10 greater than that of the Lamb-shift source. The source was operated for up to 500 hrs between maintenance periods. The 25-μA peak current results in a 600-nA average beam current in the experimental area at 6% duty with 100-nS chopping for neutron-spin interaction experiments.

Full descriptions of optically pumped ion sources can be found in the literature.[1] A drawing of OPPIS at LAMPF is shown in Figure 1. This ion source has several unique features, both in its H^- source and its optical pumping system.

2 Features of LAMPF OPPIS

The OPPIS ECR source operates at a frequency of 18 GHz and a power level of 1 kW. The axial magnetic field of the ECR and the polarizer cell is provided by a superconducting magnet with a 14 cm bore. This large bore allows for vacuum pumping between the ECR extraction lenses and the polarizer cell, minimizing the unpolarized H^0 beam formed from residual hydrogen gas.

The ECR extraction system consists of three multiaperture lenses. The electrodes are biased in the "accel-accel" configuration [2] to produce a low divergence H^+ beam.

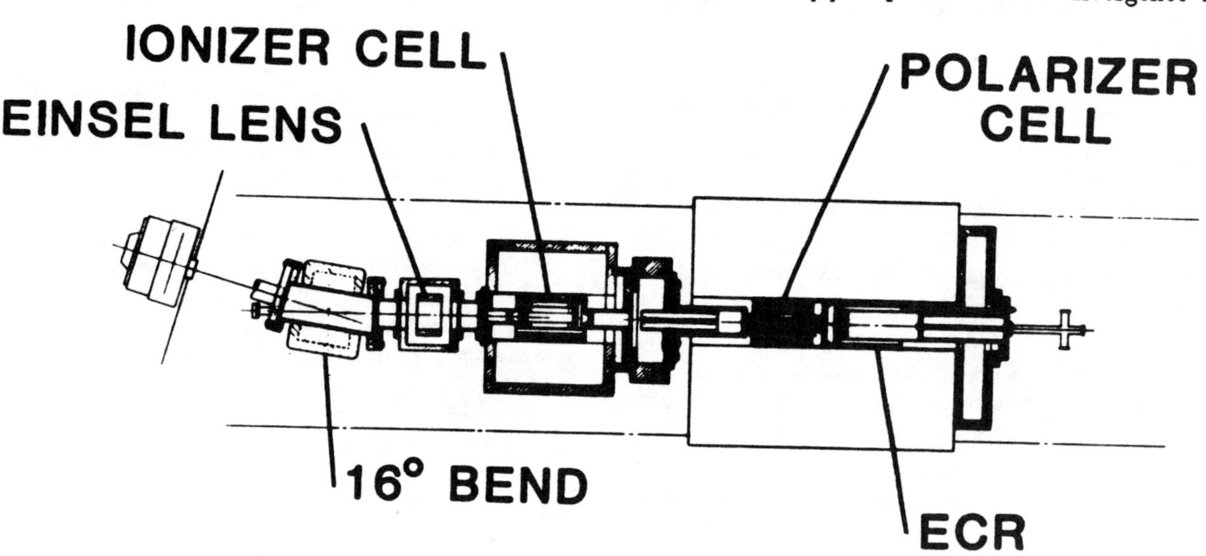

Figure 1: A diagram of OPPIS. A profile of the axial magnetic field in the source overlays the diagram.

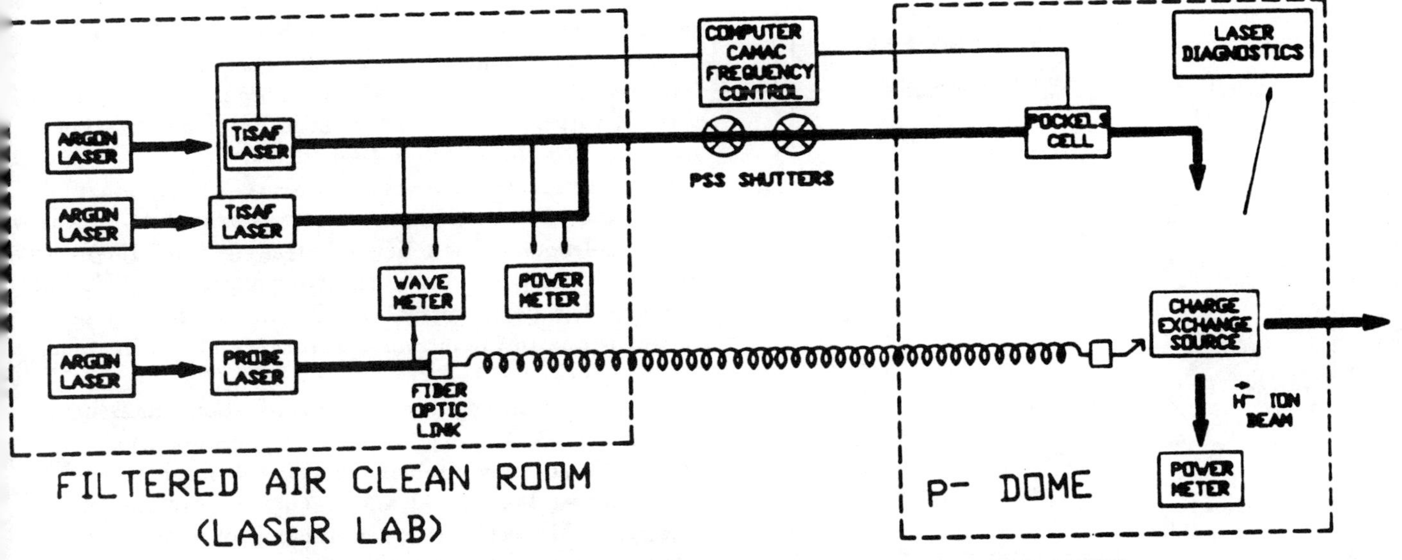

Figure 2: A schematic of the optical pumping system of LAMPF OPPIS.

For the 1990 operating period, lenses of 1-mm thickness and 1-mm spacing and a hexagonal close-packed array of 37 or 19 1-mm diameter holes with 0.25-mm web between holes were used. The lens configuration has since been further optimized; details of these experiments are given later in this paper.

A schematic of the optical pumping system of OPPIS is given in Figure 2. LAMPF was the first facility to use K rather than Na in the polarizer cell. Potassium can be optically pumped by the solid-state Ti::sapphire laser. These lasers can produce over 3 W of light with a bandwidth of ≈ 500 MHz, which efficiently pumps the 1.3 GHz bandwidth of K. This large laser power makes it possible to maintain high polarization while dedicating a single laser to a particular spin state, simplifying the spin-flip procedure greatly and allowing rapid spin flip. The Ti::sapphire lasers have stable frequency and power. Their reliability contributed to the high availability of polarized beam.

As shown in Fig. 1, the OPPIS 4-keV beamline has a $16°$ bend before the beam is injected into the 750-kV high voltage column. This feature allows for efficient insertion of the laser beam and avoids depositing the high-intensity neutral hydrogen beam and alkali vapor in the high voltage column. This bend is accomplished by combined electric and magnetic fields balanced to give no net spin precession with respect to the particle direction of motion. This allows a longitudinally polarized beam to be injected into the 750-keV spin precession system, which gives the experimental areas complete freedom of spin orientation. Since the spin precession is proportional to the anomalous part of the particle gyromagnetic ratio, which for H^- ions is -3.78, the spin correction device has a magnetic field to give a $4.2°$ left bend and electric field to give approximately a $20.2°$ right bend for net bend of $16°$. The effective field length on axis is 194 mm. The fields have the typical $E \times B$ or Wien filter geometry in a very compact space, so the device is termed the "Teeny Wieny filter." The Teeny Wieny filter accomplishes the $16°$ bend and spin precession correction with no measurable loss of beam intensity.

3 Development Experiments

One of the reasons for the successful performance level of OPPIS is the change from a accel-decel to an accel-accel extraction system. The accel-accel mode produces an H^- current density that is seven times that of the accel-decel mode [2]. Since the end of the LAMPF experiment run cycle in October 1990, experiments to further optimize the ECR extraction lenses have been performed. When the extraction lenses are operated in the accel-accel mode and the net extraction voltage is lowered, we believe the plasma sheath becomes less concave, resulting in a more parallel H^+ beam. This parallel H^+ beam produces a less divergent H^0 beam which better matches the small acceptance angle of the ionizer cell (10 mrad).

In addition to the extraction voltage, an important parameter determining the divergence angle of the H^+ beam is the spacing between the first two lenses. Data exploring the effect of this spacing as a function of extraction voltage is shown in Fig. 3. The data shows that for 2-mm spacing the accel-accel effect is not evident. For the 0.5-mm spacing, significant beam heating of the K cell was observed, so this data was taken with a 0.5-cm diameter beam for comparison. The optimum spacing for 1-mm holes and 1-mm thickness seems to be 1 mm.

Another important parameter for the H^+ beam quality is aspect ratio, where aspect ratio is defined as the ratio of a single extraction hole diameter to lens thickness. Studies indicate that the accel-accel mode does not work well for aspect ratio greater than 1.0. The data shown in Fig. 4 indicates that an aspect ratio of 0.75 is 20% better than that of 1.0.

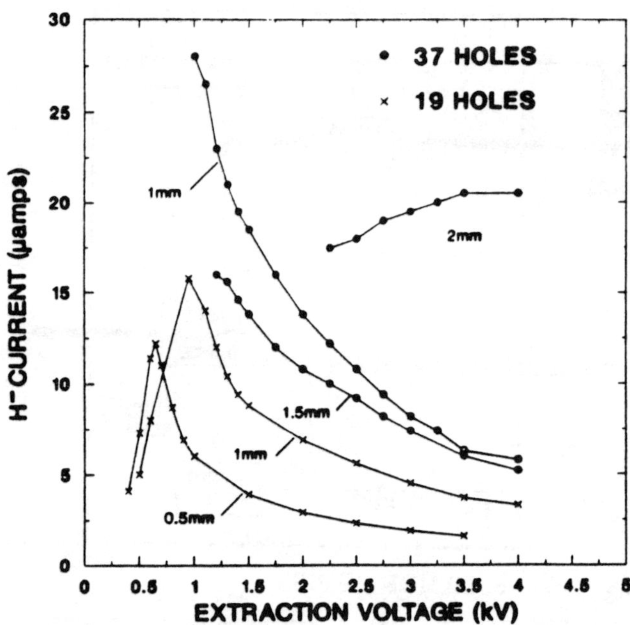

Figure 3: H⁻ current versus the extraction voltage for different spacings between the first two lenses.

Because some of the nuclear physics experiments at LAMPF have limited current tolerance, which enhances the beam polarization requirement, considerable effort has been concentrated on improving polarization. The K cell diameter is 0.8 cm and the H^+ beam, with 37 holes in a close-packed hexagonal pattern, has a flat-to-flat diameter of 0.7 cm. When the H^+ extraction geometry is changed to a 19-hole pattern with a beam diameter of 0.5 cm, and half the extraction area, the H^- beam polarization increases from 55% to 62%. Two effects contribute to this improvement. Lowering the extraction area reduces the required hydrogen flow, thus limiting the unpolarized beam formed from residual gas. Also, there is a radial dependence of K polarization in the polarizer target due to insufficient laser coverage and depolarizing wall collisions. Thus, a smaller diameter H^+ beam selects only the more polarized portion of the K target. The improvement in beam polarization is probably a combination of these effects.

The dependence of beam polarization on the magnetic field strength in the Na ionizer cell was explored. Increasing the solenoidal magnetic field strength from 1500 g to 2000 g increases the beam relative polarization 3%, but decreases the beam current 25%. The decrease in beam current is due to the increased emittance of the H^- beam [3].

Another enhancement of beam polarization was achieved by tuning both Ti::sapphire lasers to pump a single spin state. This not only doubles the available laser power but also improves the spacial and frequency coverage of the K vapor. This results in a relative improvement in beam polarization of 8%. Using this effect, increasing the magnetic field in the ionizer solenoid to 2000 g, and using a 0.5 cm H^+ beam yields 13 μA with beam polarization of 70%. Work is currently in progress to develop an automated laser frequency adjustment, to allow spin flipping with both lasers tuned to each spin state. Progress in this endeavor is reported by Swenson, *et al.* at this conference.

Conclusion

The performance of OPPIS has enhanced the LAMPF neutron-spin interaction program. Development efforts are focussed to improve the beam polarization to better satisfy the full range of nuclear spin experiments. The performance level goal for 1991 is 20 μA at 60% polarization.

References

[1] R. L. York, in *Proceedings of the International Workshop on Polarized Ion Sources and Polarized Gas Jets*, edited by Y. Mori (KEK, Tsukuba, Japan, 1990, 142.

[2] R. L. York and D. Tupa, in *Proceedings of the 10th International Workshop on ECR Ion Sources*, edited by F. W. Meyer and M. I. Kirkpatrick (ORNL, Oak Ridge, TN, 305 (1991).

[3] G. G. Ohlsen, J. L. McKibben, R. R. Stevens, Jr., and G. P. Lawrence, *Nucl. Instr. and Meth.*, **73**, 45 (1969).

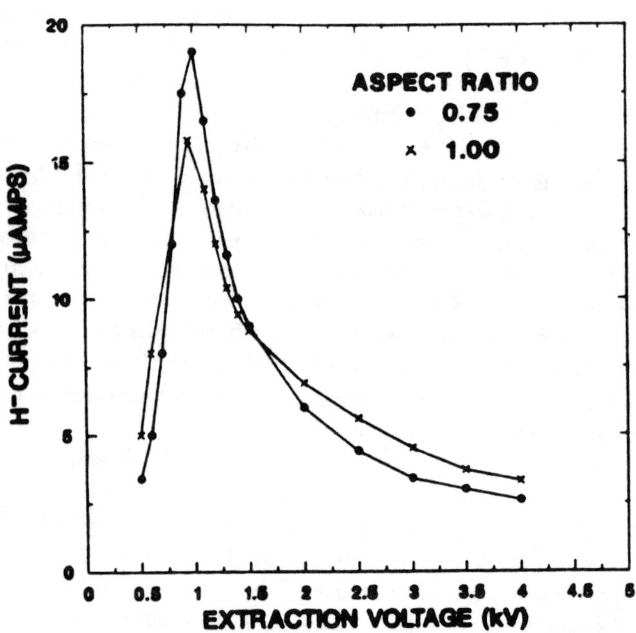

Figure 4: H⁻ current versus extraction voltage for extraction lens aspect ratios of 1.0 and 0.75.

OPTICAL PUMPING OF THE POLARIZED H⁻ ION SOURCE AT LAMPF

D. R. Swenson, D. Tupa, O. B. van Dyck, and R. L. York

Los Alamos National Laboratory, Mail Stop H838, Los Alamos, New Mexico 87545

Abstract

We report experiments to understand the laser optical pumping efficiency for the Optically Pumped Polarized Ion Source (OPPIS) at LAMPF. We have measured ion beam polarization and current vs. laser power and potassium vapor thickness in order to understand the dependence of source performance on laser power and other parameters. Attempts to fit the data with simple scaling models to evaluate projected performance improvements are shown. Development of economical ways to make more efficient use of available laser power are described.

1 Introduction

The H⁻ current from OPPIS [1] is an increasing function of the optical pumping target thickness. However, thicker targets require more laser power to achieve the same degree of polarization. Thus, optimizing source performance (in terms of some figure of merit combining current and polarization) requires determining the optimum thickness for the available laser power, or stated differently, determining the laser power necessary to reach some performance level. Section 2 shows measurements from the LAMPF OPPIS addressing these issues.

The target polarization is determined by equilibrium between the rate of polarization by optical pumping, and the rate of depolarization by wall collisions. These factors alone give a simple model of optical pumping which can be useful for understanding qualitative features of the measurements; this is discussed in Section 3. Section 4 describes modifications of our Ti:sapphire laser system to realize the benefits of higher laser power.

2 Nuclear polarization vs. laser power and target thickness

For high-duty-factor sources, available laser power is critical to source performance. Fig. 1 shows H⁻ nuclear polarization, measured at the 800-MeV polarimeter, as a function of laser power for different target temperatures. Data for laser powers less than 5 W was taken with one laser and

a variable attenuator, while powers greater than 5 W were obtained with two lasers. The laser powers quoted were measured in the laser lab before the beam was transported to the cell; we estimate that 75±10% of the laser light reaches the cell. Although laser powers above 4 W afford only a modest gain in polarization, a thicker target can be polarized; the H⁻ current increases 30% by changing the target temperature from 141°C to 150°C.

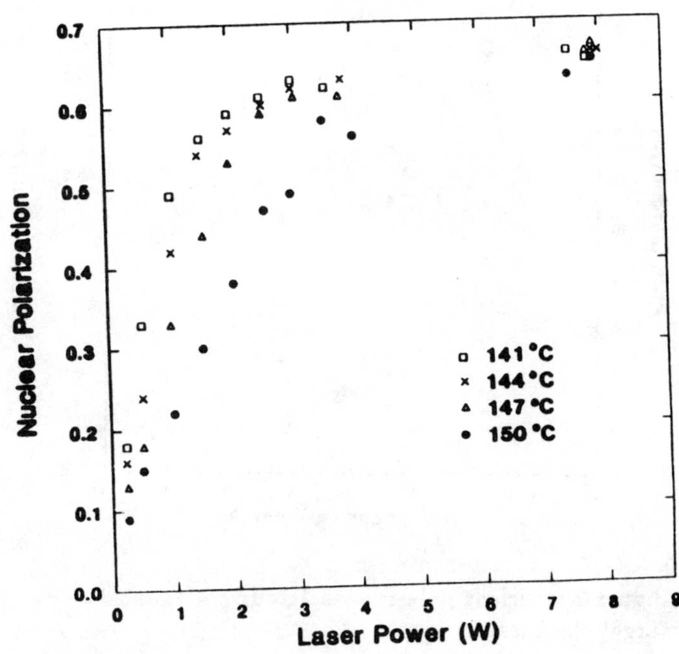

Figure 1: Nuclear polarization vs. laser power.

At laser power above 4 W, it appears that the H⁻ polarization saturates with limiting value near 70%. However, it is not conclusively known that the corresponding values of vapor polarization are near 100% because of the questions of spectral overlap with the Doppler-broadened K absorption profile and spatial overlap of the laser beam with that portion of the vapor which contributes to the final ion beam. Exploration of these questions will be facilitated by two-laser operation this year.

Fig. 2 shows data for optimizing target thickness when polarized by one 4-W laser. The beam polarization and current were measured for different target temperatures. The quoted target thicknesses are accurate to $\pm 20\%$ and were determined from the temperature using a vapor pressure curve and a fit to Faraday rotation measurements. The H^- beam currents were measured at 750 keV and were limited to less than our typical values (25 μA peak at 5×10^{13} cm^{-2}) because all parameters were not optimized. The current measurements above 5×10^{13} cm^2 show an anomalous laser-induced effect [2] and are not included. The maximum polarization was obtained for thicknesses less than 4.0×10^{13} cm^{-2} ($<137°$C) while the best P^2I was achieved at approximately 5×10^{13} cm^{-2} (140°C). In subsequent experiments, finer adjustments of the laser allowed us to achieve 57% polarization at 5.3×10^{13} (141°C). It should be noted that the measurements in Fig. 2 were made using a 7-mm-diameter ion beam and a 10-cm-long vapor cell, while those of Fig. 1 were with a 5-mm beam and a 16-cm cell. The higher polarization shown in the earlier figures is related to the smaller ion beam size as explained in the companion paper [3].

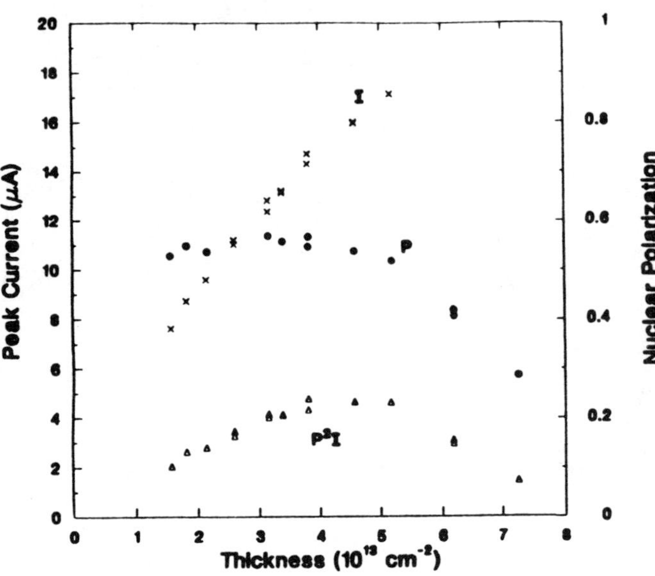

Figure 2: Nuclear polarization P, current I, and P^2I vs. target thickness.

3 Two level optical pumping model

The saturation of polarization in the optical pumping of thick alkali vapors can be illustrated with a simple two-level model [4] which yields analytical equations that can be fit to the data. Briefly, two rate equations describe the evolution of the populations of the two spin-orientation states of the ground level of the atom under the process of optical pumping and wall relaxation. Solving these equations for steady-state conditions gives expressions which relate the average vapor polarization to the incident and transmitted laser power:

$$\frac{I_O - I_L}{I_S} + \log\left(\frac{I_O}{I_L}\right) = \frac{\sigma_e \rho L}{2} \tag{1}$$

$$P_N = \alpha \bar{P} = \frac{2\alpha}{\sigma_e \rho L I_S}(I_O - I_L) \tag{2}$$

$$I_S = \frac{3Ah\nu}{\sigma_e \tau} \tag{3}$$

where I_0 is the incident laser power, I_L is the laser power after passing through the vapor, I_s is the "saturation power," ρL is the thickness of the vapor, σ_e is the effective cross section for laser absorption, P_N is the H^- nuclear polarization, α is the polarization transfer efficiency, $\bar{P}$ is the average polarization in the vapor, **A** is the area of the beam, and τ is the relaxation time of the polarization.

Fig. 3 shows a fit of the polarization and laser transmission data from Fig. 1 at 141°C to the two-level model of the optical pumping. For these data the fit is quite good. The values for the fit are $I_S = 170$ mW, $\sigma_e \rho L/2 = 8.8$ and $\alpha = 0.67$. Fitting the data for the higher target temperatures gave erratic results and it is not clear whether the problem was in the accuracy of the data or a breakdown in the assumptions of the model; this remains to be explored further.

4 Spin flip with two lasers

To realize the benefits of higher laser power, we are developing a system to use both lasers to pump both spin states.

In 1990 we used one laser (Spectra-Physics 3900) per spin state. In order to use both lasers for each spin

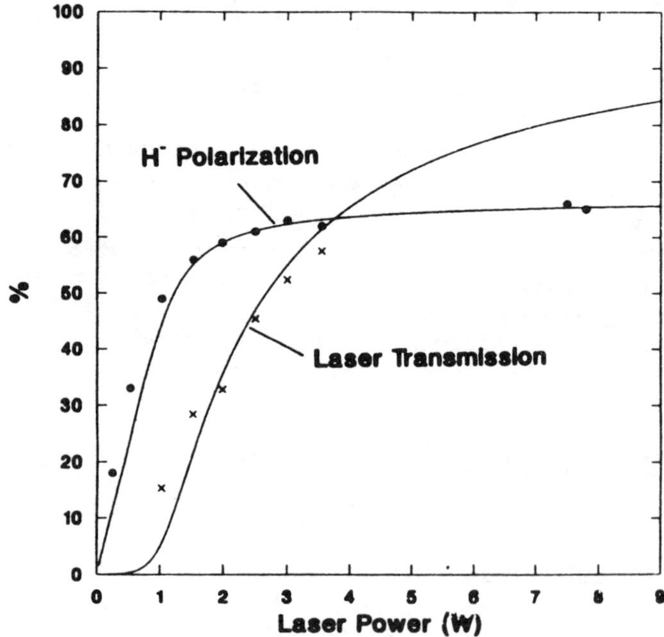

Figure 3: H^- polarization and laser transmission vs. laser power.

state, is necessary to change the helicity of the laser beams from σ^+ to σ^- and tune laser frequencies to both σ^+ and σ^- transitions. The transitions are separated by $2.00\,\mathrm{cm}^{-1}$ in the 16-kG magnetic field of the polarizer cell.

The frequency of the lasers is determined by their tuning elements. Each laser has a birefringent filter (BRF), a thin etalon with a 7-cm^{-1} free spectral range (FSR), and a temperature-stabilized thick etalon with a $0.67\,\mathrm{cm}^{-1}$ FSR. The etalon stack is chosen to reduce the laser bandwidth to about 500 MHz to better match the K-absorption linewidth. Tilting or changing the temperature of the thick etalon gives essentially continuous tuning in a 0.67-cm^{-1} range. Tilting the thin etalon causes the frequency to jump in quanta of 0.67 cm^{-1}, one thick etalon mode. Thus, tuning the lasers from σ^+ to σ^- requires tilting the thin etalon to hop three thick-etalon modes, or 2.00 cm^{-1}.

The micrometer positioners for the BRF and thin etalon have been replaced with computer-controlled stepping-motor micrometers (Oriel models 18500 and 18512). When a spin flip is desired, the micrometers move the approximate distance to hop three thick-etalon modes. A feedback loop then peaks the laser power with the BRF and thin etalon. The frequency of the laser is confirmed with a wavemeter and corrected if necessary. The final system will flip spin in under 10s.

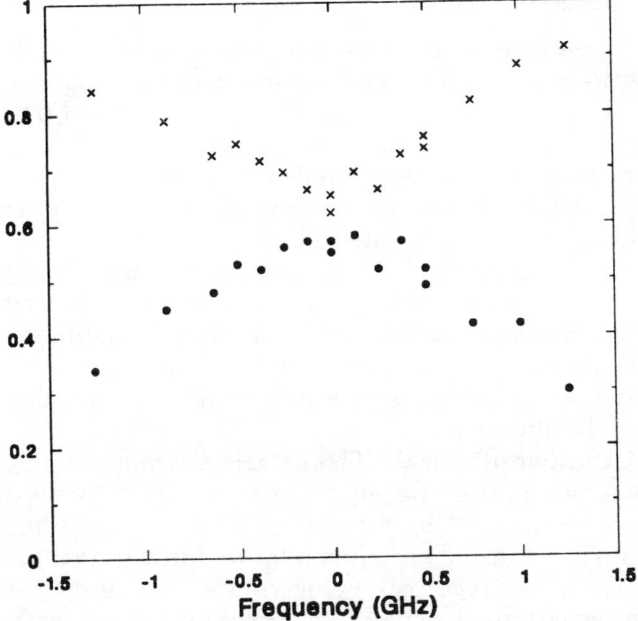

Figure 4: H$^-$ polarization (circles) as a function of pump-laser detuning from K line center. The polarization remains high for detunings within $\pm\,0.01$ cm^{-1} (± 300 MHz). The transmission of the laser beam through the target (crosses) is an operationally useful relative indication of polarization.

The temperature-controlled stabilization of the thick etalons limits the frequency drift to within $\pm\,0.01$ cm^{-1}; this drift has minimal effect on polarization as illustrated in Fig. 4. For this method to succeed, the FSR of the thick etalons in the two lasers must be closely matched to each other and to the σ^+/σ^- splitting determined by the magnetic field. Our two etalons have a frequency change of 1.980 and 1.965 cm^{-1} per three mode hops. The superconducting magnet has been adjusted slightly to give a 1.97 cm^{-1} splitting between the σ^+ and σ^- transitions in K.

The two laser beams are combined with little power loss and with parallel polarizations by using a sharp-edged mirror. The beams are about 3 mm apart at the mirror and essentially overlapping when they reach the source 8 m away. A Pockels cell (Quantum Technology QK-24) transforms the linearly polarized beams into σ^+ or σ^- helicity, depending upon the applied voltage.

References

[1] R. L. York, in *Proceedings of the International Workshop on Polarized Ion Sources and Polarized Gas Jets*, edited by Y. Mori (KEK, Tsukuba, Japan, 1990), p. 142.

[2] A. N. Zelenskii, et al. (above proceedings) p. 154.

[3] R. L. York et al., Operation of the Optically Pumped Polarized H$^-$ Ion Source at LAMPF, in *Proceedings of the IEEE Particle Accelerator Conference* (to be published).

[4] O. B. van Dyck, in *Proceedings of the International Seminar on Intermediate Energy Physics*, (INR, Moscow, USSR, 1990), p. 65.

Installation of the Legnaro ECR Ion Source

M.Cavenago, G.Bisoffi, G.Carugno, F.Cervellera,
G.Fortuna, M.F.Moisio, V. Palmieri, K.Rudolph
Laboratori Nazionali di Legnaro, INFN
via Romea n. 4, I-35020 Legnaro (PD), Italy

Abstract

The mechanical parts of the 14.4 Ghz Legnaro Electron Cyclotron Ion Source "Alice" were built, in particular the 24 bars NdFeB Halbach hexapole. A newly designed extractor was implemented, with a tapered focus electrode and a reentrant puller. Several delays forced us to postpone the first tests to late June. The source is expected to inject a superconductive linac, which is being completed, via an RFQ.

I. INTRODUCTION

Electron Cyclotron Resonance (ECR) ion sources now routinely used as first elements to inject cyclotron [1] or heavy ion linacs [2] as an interesting alternative to tandem accelerators, require an intense magnetic field of a prescribed topology to confine the ions. In order to limit the dissipated power in the coil to about $30\,kW$ (total power $60\,kW$), with a confining magnetic field of $B_w = 7300\,G$, the design of our source "Alice" (see Fig. 1), envisions a small main plasma chamber (radius= 3 cm, length= 19 cm), and a compact packing of the coils, their yoke and the

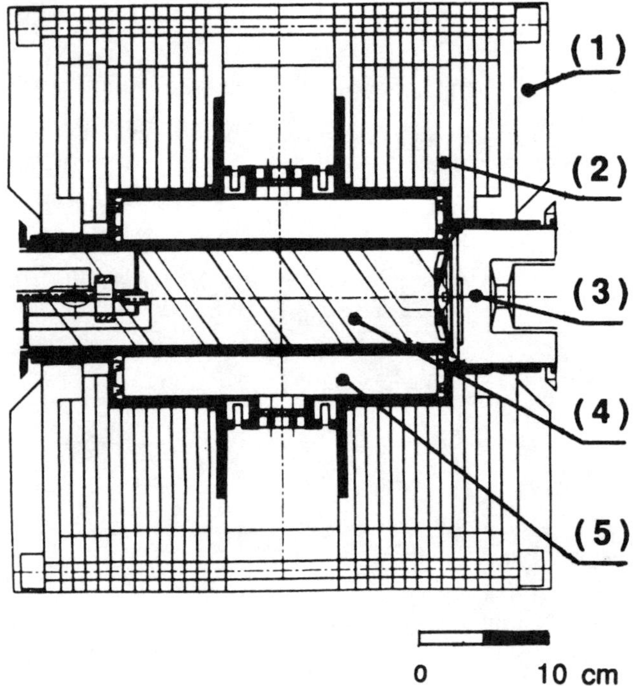

Figure 1 . Section of Alice source. The main components of the source are indicated: (1) iron (2) solenoid panacakes (3) extraction system (4) plasma chamber (5) NdFeB hexapole.

hexapole, the whole occupying the region $r > 3.6cm$ and $|z| < 16.8cm$; here the origin of the cylindrical coordinates r, ϑ, z is the middle of the main plasma. In this context, the implementation of a first stage, a differential pumping on one side of the main plasma ($z < -9.5cm$) and the construction of a good beam quality extractor system are subjected to severe constraints [3], and few significant but overcome difficulties will be discussed here.

The optimal performances of the superconductive linac Alpi [4] and the superconductive RFQ (under development in our laboratory [5] in collaboration with State Univ. New York) are obtainable with a mass to charge ratio $A/Q \cong 7.3$, which implies a rather demanding charge state for uranium (32+): no stripping is then needed downstream to provide $6MeV/u$. This motivated us to an earlier installation of Alice, with the purpose of a long tuning up period.

II. MAGNETIC CONFIGURATION

The disposition of the hexapole and the coils is given in Fig. 1 ; it is well known that the hexapole bars exert forces up to 200 Kg one onto the other [6]; each bar is glued to its steel cage. From Fig. 1 we see that central rings, needed to keep the bars together, can be bolted by screws, since enough space is available there; yet lateral binding rings are necessary too, and it is possible to lock them in position by pressure. The assembling is done thanks to a 24 arm manual machine, so that the hexapole is demountable in its 24 constituent bars, whenever this may prove necessary. A stronger vice has been constructed to precisely place the bars.

Another practical difficulty arising from our magnets is the limited space ($< 4\,mm$) available for insulation, which must be cut in two parts for assembling; again the solution was an insulation built in two layers, the outer layer cut being rotated by 90 degrees with respect to the inner cut, so that discharge paths are at least 5 cm long.

Coils are standard double pancakes.

III. VACUUM

A classical differentially pumped two stage plasma was implemented for the configuration of the Alice first tests with gases, leaving further options open. There are two turbopumps (using Fomblin oil) for the two stages; a significant contribution to the main stage plasma pumping comes from holes (which can be closed) towards the extractor region, pumped by a third turbopump.

The vacuum chambers are made out of steel, the seals are metallic (except for the Pirani ones), so that good control of the plasma composition and clean current spectra are expected. The three insulation breaks (two 90 mm inner diameter and one 138 mm inner diameter) are made of brazed alumina, while the einzel lens is sustained by "stesalite" bars.

A removable and water cooled subassembly, called the "plasma chamber", is inserted inside the vacuum chamber. The plasma chamber houses the main plasma chamber, an iron ring so as to create the magnetic field depression of the first stage and a backbone which separates the two pumping manifolds and holds two waveguides and two gas-feed capillary tubes.

The purpose of the plasma chamber was to make tailoring to several metallic elements possible; for example, by building a new plasma chamber without the iron ring and with the backbone replaced by a suitable oven, the source may be optimized for fairly intense Bismuth beams. During the construction, the opportunity of not welding the backbone to the plasma chamber was evident.

Gate valves have not been purchased yet.

IV. BEAM EXTRACTION AND TRANSPORT

It can be argued that not only will a good extraction optics be of direct benefit to the extracted beam, but it will also allow for a better understanding of the source working regime. The Alice optics indeed envisioned the extraction of a slightly divergent beam, entering — after a 70 cm path — into a 50 cm-bending radius, 8 cm-gap analysis magnet, which is astigmatic and symmetric: an einzel lens of limited focal power $1/f_e < 1.m^{-1}$ matches the ion beam to the object point of the magnet, 115 cm far from the magnet entrance border. Beam diameter at magnet entrance is about $4 \div 5\,cm$.

This configuration minimizes the space charge effects, since the full extracted current is never focused to a waist, but a compact assembling is then required; between the source and the dipole magnet, we have: the mounting of the extractor, with a bended arm, and the suspended einzel lens (see Fig. 2), the $x - y$ steerers where the beam pipe diameter shrinks in order to enter into the magnet, a tentative concept of a x slit (see Fig. 2), and the field clamps of the magnet, closing a slanted CF100 flange.

The (removable) x-slit is determined by a prolate ellipsoidal surface, which is rotatable by 90 degrees around its axis. Since two opposite 90 degree sectors of this surface are removed, an elliptic slit is projected, with the minor axis $0 < a < 5.0\,cm$ and the major axis $b = 5.0\,cm$.

Due to the limited space available, interaction of the fringe fields of the beam transport elements — namely the einzel lens, the steerers and the magnet

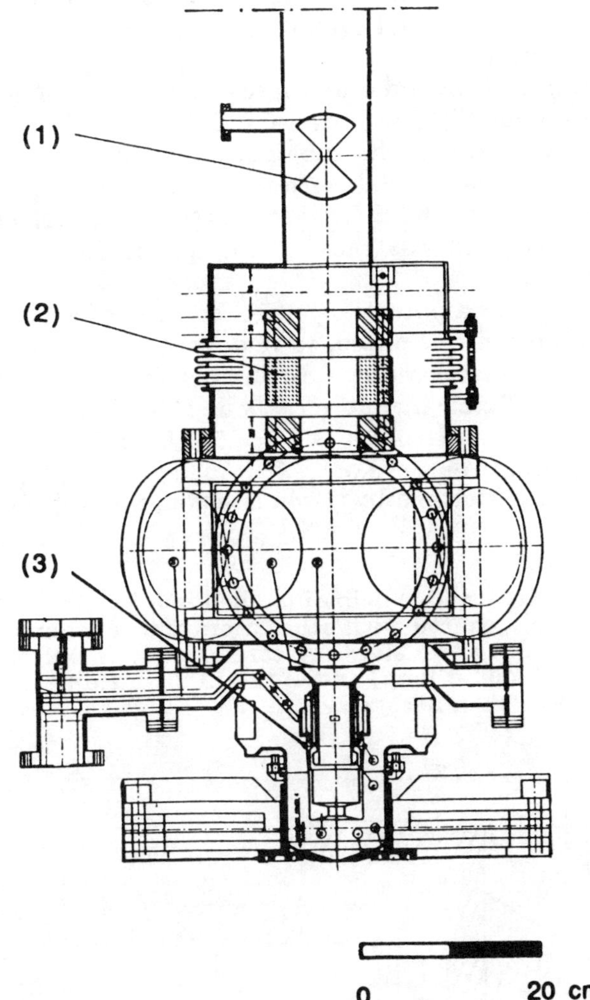

Figure 2 . Transport line to the magnet: the movable extractor (3), the einzel lens (2) and the x-slit (1) are shown.

— can not be ruled out. Studies of the magnet section showed that the C-shape solution was still possible.

The newly designed extractor is treated elsewhere [6]; we implemented a tapered focusing electrode [7] and a reentrant puller. The former, going from a 67.5^0 angle (with the z-axis) near the beam to a 74.16^0 angle when $r \to \infty$ is precisely given by analytic formulas and extensive computer simulations show its convenience with respect to the simple 67.5^0 semiangle cone [7] . The reentrant puller is proposed as a correction of the perturbing effect of the vacuum pipe.

We avoided to braze insulating and metal pieces in the puller assembly, the whole package being secured by screws.

V. SETTING THE TEST STAND

1. Electronic control

The ECR source requires some control at three different potentials, namely the plasma potential, the platform potential and the whole linac potential (clearly connect to the ground); fiber optics link them.

The basic control of the source is at platform potential, setting and reading the klystron amplifier, the power supplies, and several safety interlocks (e.g. boiling water). For its inherent simplicity and the wealth of boards available from several suppliers, the G-64 bus with the 6809 CPU results to be attractively unexpensive and was chosen. Equipment interfaces are RS-232C.

The plasma potential control consists of a gauge reading unit in this phase, so that only a 50 kV insulation transformer and a RS-232 fiber optic links are needed (negligible manpower, about 1500 ECU worth of components; usually it is $1\ \$ = 0.9 \pm 0.1 ECU$).

The final supervision and software development of Alice is performed by a workstation, not yet purchased.

B. Logistic

The test stand is being installed in a 8.5 times 23 meters room, shared with other users and surrounded by offices along two sides. A careful planning of our section (7.5 meter long) and of the ECR shielding, see Fig. 3 , was then necessary.

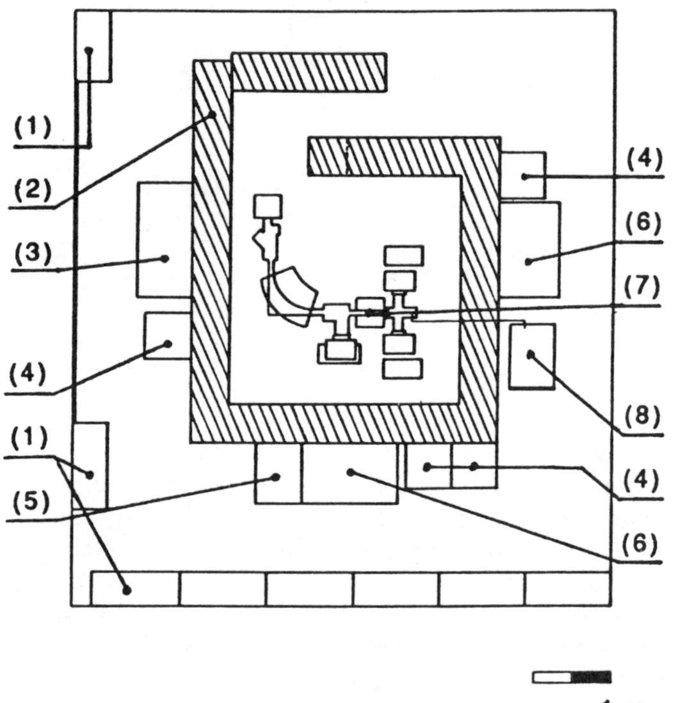

Figure 3 . Test stand of Alice. We see (1) shelves (2) concrete walls (3) table (4) racks (5) dipole magnet power supply (6) solenoid power supplies (7) the source itself (8) klystron amplifier

C. Major costs

The costs were generally kept at the lower expectation value, while several delays occurred. Mechanical construction was performed by an external contractor. In table 1, we quote the construction costs and manpower (not the installation work) of only what we have at the moment, excluding therefore any necessary equipment for metallic material, any plasma diagnostics and flow control equipment (a significant part of the whole).

Table 1: Construction costs

Item	Cost (no taxes)	Manpower (our workshop)
Basic mechanical parts	110000 ECU	none
Klystron & waveguides	80000 ECU	36 h.
Power supplies	70000 ECU	
Pumps	60000 ECU	
Analisys magnet	26000 ECU	
NdFeB magnet	16000 ECU	
Gauges, gas, etc.	15000 ECU	36 h.
Electronics	8000 ECU	72 h.
Magnetometry	6000 ECU	
Flanges, coil spares, etc.	15000 ECU	

Next year budget must include also the HV platform construction; the major technical problems to be addressed are the insulation transformer and its voltage compensation at a variable load.

VI. ACKNOWLEDGMENTS

We appreciated fruitful discussions with our collegues L. Badan, G. Bezzon, G. Binelle, F.Scarpa, A. Zanon and L. Ziomi.

VII. REFERENCES

[1] R.Geller, "Highly charged ECR ion sources: Summary and comments", Rev. Sci. Instr., vol 61, pp. 659-661, Jan. 1990

[2] R.C.Pardo, "Operating experience of an ECR on a high voltage platform", Nucl. Instr. Meth., vol. B40/41, pp. 1014-1019, 1989.

[3] M.Cavenago, G.Bisoffi "Optimization studies of a 14.4 GHz ECR ion source for the superconductive linac at Legnaro" Nucl. Meth. Instr., vol. A301, pp. 9-18, Feb. 1991.

[4] G.Fortuna et al., "The ALPI Project at the Laboratori Nazionali di Legnaro", Nucl. Inst. Meth, vol. A287, 253-256, 1990.

[5] A.Jain,I. Ben-Zvi, P.Paul, H.Wang, A.Lombardi, "Status of the SUNY superconducting RFQ", in Conference Record of the IEEE 1991 Particle Accelerator Conference (these proceedings), S.Francisco, June '91.

[6] M.Cavenago, G.Bisoffi "Status report of the 14.4 GHz ECR in Legnaro", in "Proceedings of the 10th Int. Conf. ECR Ion Source", Oak Ridge National Laboratory Report, March 1991.

[7] M.Cavenago, G.Bisoffi "Cylindrically symmetric extractors with space charge dominated flow", submitted to Rev. Sci. Instr.

DESIGN ASPECTS FOR A PULSED-MODE, HIGH-INTENSITY, HEAVY NEGATIVE ION SOURCE

G. D. ALTON

Oak Ridge National Laboratory*
Oak Ridge, TN 37831-6368

ABSTRACT

A high-intensity, plasma-sputter, negative ion source, which utilizes multi-cusp, magnetic-field, plasma-confinement techniques, has been designed at the Oak Ridge National Laboratory (ORNL). The source is an axial-geometry version of the radial-geometry source which has demonstrated pulsed-mode peak intensity levels of several mA for a wide spectrum of heavy negative ion species. The mechanical design features include provisions for fast interchange of sputter samples, ease of maintenance, direct cooling of the discharge chamber, and the use of easily replaced coaxial LaB_6 cathodes.

INTRODUCTION

The radial-geometry, plasma-sputter, negative ion source, described in Refs. 1-5, has demonstrated the capability of producing high-intensity (several mA) pulsed beams of a wide spectrum of atomic and molecular negative ion species and also has shown promise for operation in dc mode.[5] The pulsed-mode performance characteristics of this source type are particularly well suited for use in conjunction with the tandem electrostatic accelerator when used as an injector for a heavy ion synchrotron, while the dc mode of operation is commensurate with stand-alone tandem accelerator operation. For heavy ion synchrotron injection applications, high-intensity, pulsed beams of widths 50-300 μs at repetition rates of 1-50 Hz for a wide spectrum of atomic and molecular species are typically required.

SOURCE DESIGN

The high-intensity, radial-geometry, plasma-sputter, negative ion source, described in Refs. 1-5, has proved to be a reliable, stably operating source with a long lifetime for pulsed-mode operation. The intensity levels obtained are often higher, by factors of 30-100, than those which can be generated in cesium sputter-type sources such as described in Refs. 6-8 and yet the emittances of beams provided by the source are comparable for pulsed-mode operation to those provided by conventional sputter-type sources.[9] The emittances of these beams also match the calculated acceptances of large tandem accelerators such as the 25URC tandem accelerator at ORNL.[10]

The operational characteristics of the axial-geometry source, shown in Fig. 1, are expected to be identical in almost

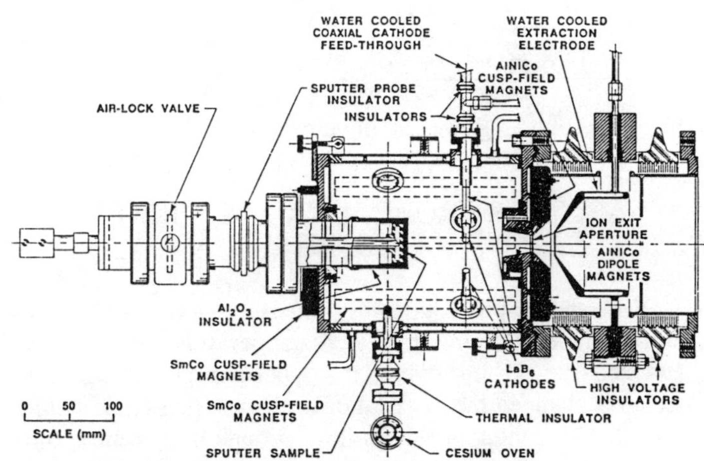

Fig. 1. Axial-geometry, plasma-sputter,
negative ion source (top view).

every detail to the radial-geometry source.[1-5] The source is constructed primarily of stainless steel and utilizes metal-to-ceramic bonded high-voltage insulators and low-voltage feedthroughs. Design emphasis has been placed on the ability to rapidly change the source itself and all degradable components. The cesium oven is mounted externally, permitting easy access for servicing, while providing good thermal isolation between the discharge chamber and the oven itself. The main source can be quickly and easily disassembled for cleaning and other maintenance operations. The source assembly is composed of four major independent assemblies: (1) the sputter probe vacuum airlock assembly; (2) a freon-cooled, stainless steel discharge chamber onto

* Research sponsored by the U.S. Department of Energy under contract DE-AC05-84OR21400 with Martin Marietta Energy Systems, Inc.

which is attached the cesium oven, discharge gas support system, coaxial geometry LaB_6 cathodes, and SmCo and AlNiCo plasma confinement magnets, (3) a ceramic-to-metal bonded alumina (Al_2O_3) insulator to which is attached the high-voltage extraction electrode system, and (4) the coaxial LaB_6 cathode assembly. The power supply arrangement required for operation of the source is shown in Fig. 2.

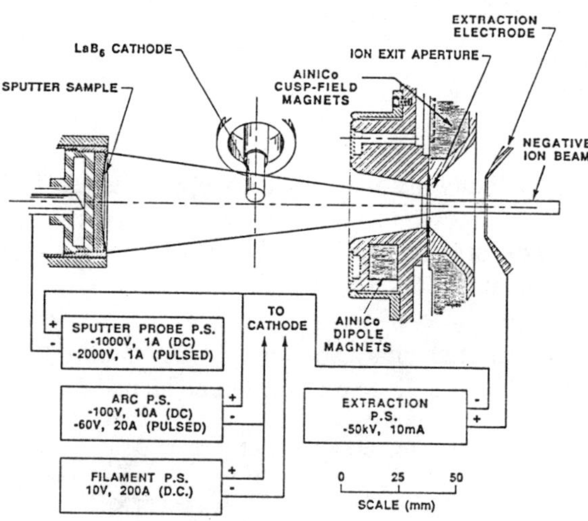

Fig. 2. Power supply arrangement for the axial-geometry, plasma-sputter, negative ion source.

SPUTTER PROBE VACUUM AIRLOCK ASSEMBLY

The sputter probe assembly consists of a thin wall, 12.7-mm-diameter chromium plated copper tube to which is attached the 50-mm-diameter copper sample holder onto which is clamped the material of interest. The sample holder is cooled by continuous freon flow through a concentric tube arrangement. The probe can be inserted into and withdrawn from the source through the airlock valve which is sealed against atmospheric pressure by a conventional elastomer gasket. The vacuum interlock assembly is attached to an externally mounted ceramic-to-metal insulator which in turn is fastened to the back flange of the source. The insulator is used to isolate the probe-vacuum airlock assembly from the source body. Based on experience with sources equipped with similar provisions, the total time required for withdrawing, replacement, and reinsertion of the sputter probe sample material is expected to be the order of 10 minutes.

The sputter probe/airlock assembly is insulated from the source housing with an insulator designed to withstand potentials up to ~–5 kV. When fully inserted into the discharge chamber, the sputter probe is surrounded by an Al_2O_3 insulating sleeve which prevents all internal negatively biased components other than the sample material from being bombarded by positive ions extracted from the surrounding plasma. Two geometries of sputter samples will be utilized. In cases where malleable metal sheet material containing the species of interest is readily available, samples 1-1.5 mm in thickness will be pressed by means of a die fixture into a 50-mm-diameter spherical sector probe with radius of curvature $\rho \simeq 210$ mm for focusing the negative ion beam generated in the sputtering process through the exit aperture of the source. Samples which are brittle must be formed from solid materials. Composite sintered compounds or mixtures of compounds will be typically 5 mm in thickness with a spherical radius of 210 mm machined into the face of the material. These samples will be indirectly cooled by clamping the sample to a spherical or flat geometry copper heat sink appropriately contoured to the respective sample geometry. The sputter probe assembly will be cooled by a freon heat exchange unit maintained at 15°C.

VACUUM DISCHARGE CHAMBER AND PLASMA CONFINEMENT ARRANGEMENT

The vacuum/discharge chamber is made of stainless steel equipped with freon coolant passages to protect the ten sets of equally spaced SmCo and AlNiCo plasma discharge confinement magnets from thermal degradation by the radiant power incident on the walls of the chamber arising from the plasma discharge and high temperature LaB_6 cathodes. Plasma confinement is effected by the use of ten rows of SmCo permanent magnets, equally spaced circumferentially around the diameter of the cylindrical chamber. The external and internal flanges of the chamber are equipped with ten equally spaced, azimuthally oriented SmCo and AlNiCo magnets, respectively. The internal flange is equipped with a set of dipole AlNiCo magnets to inhibit electron extraction from the source and thereby reduce loading to the extraction power supply and a replaceable tantalum aperture. Initially, the exit aperture will be 10 mm. The chamber is attached to the rear and front flanges by means of thumb screws for ease in disassembly for cleaning and other maintenance operations.

COAXIAL LaB_6 ASSEMBLY

The source will utilize directly heated coaxial-geometry LaB_6 cathodes to initiate and sustain the plasma discharge. The cathode structure is very similar to the design described by Leung et al. in Ref. 11. The high melting point, chemical inertness, low work function, and low sputter ratio properties make LaB_6 especially attractive for such applications. The coaxial geometry is desirable because it minimizes the magnetic field surrounding the cathode and thus permits emission and escape of low-energy electrons from the surface. More importantly, the design allows easy and fast interchange of cathodes through a single metal-to-metal vacuum seal feedthrough.

THE ION EXTRACTION ELECTRODE SYSTEM AND EXTRACTION OPTICS

One of the advantages of the plasma-type sputter negative ion source lies in the fact that, when operated in a high-density plasma mode, the negatively biased sputter probe containing the material of interest is uniformly sputtered. Negative ions created in the sputter process are accelerated and focused through the plasma to a common focal point, usually chosen as the ion exit aperture of the source, and then pass into the field region of the extraction electrode system. Within the plasma, the ion beam is free of space charge effects which only become important upon exit from the plasma region of the source. Since the beam energy is 500-1000 eV upon exit from the plasma region of the source, space charge influences on the beam are reduced. After exiting the source, the beam is further accelerated by a two-stage electrode system insulated by high-quality Al_2O_3 insulators. Typically, the ion beam will be accelerated through a potential difference of 20 kV in the first stage and by an additional 30 kV in the second stage.

The ion optics of the ion generation and extraction regions of the source have been studied computationally through use of the code described in Ref. 12. Examples of such calculations which display ion trajectories for a 3.5-mA O^- or a 1-mA Au^- ion beam are shown in Fig. 3. Negative ions,

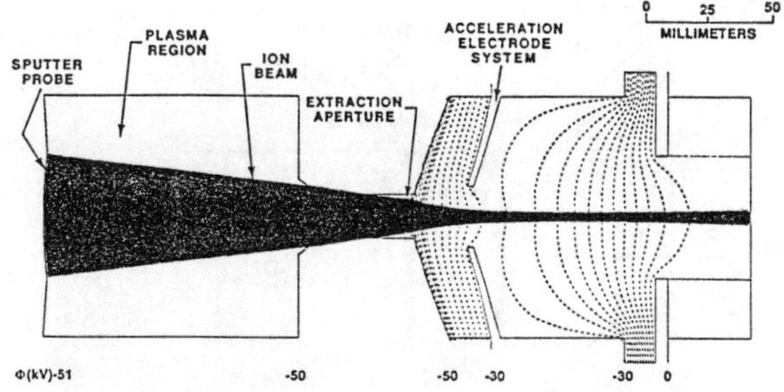

Fig. 3. Negative ion optics of the axial-geometry, sputter heavy negative ion source equipped with a two-stage extraction electrode system shown in Fig. 1. Ion beam: 3.5 mA O^- or 1 mA Au^-; ion energy: 51 keV.

generated at the sputter probe surface and accelerated across the plasma sheath, drift through the field-free region of the plasma and exit into the extraction lens system where they are accelerated further to energies up to 51 keV. The characteristics of mA-intensity-level ion beams in the extraction region of the source are found to be dominated by space charge effects.

EXPECTED PERFORMANCE OF THE SOURCE

The operational parameters and intensity capabilities of the subject source are expected to be similar to those of the radial-geometry source and therefore, the reader is referred to Refs. 1-5 for specific information concerning the species and intensity capabilities, as well as the qualities (emittances) of beams produced in this type of source. The anticipated operational parameters for pulsed-mode operation of the source are shown in Table 1. For example, the source is expected to generate peak beam intensities close to those reported in Ref. 4 for the radial-geometry source, which includes a list of more than 20 negative ion species, including 6 mA C^-, 10 mA Cu^-; 8 mA Pt^-; and 10 mA Au^-. The emittances of beams extracted from the source are also expected to be close to those of the radial-geometry source (typically, 11-17 π mm.mrad $(MeV)^{1/2}$ at the 80% contour level, depending on the beam intensity).

ACKNOWLEDGEMENTS

The author is indebted to Dr. K. N. Leung for supplying information concerning the coaxial LaB_6 cathode assembly design used at Lawrence Berkeley Laboratory.

Table 1. Expected pulsed-mode source operating parameters.

Arc current	15-20 A
Arc voltage	30-60 V
LaB_6 cathode current	130 A
LaB_6 cathode temperature	1450°C
Xe gas pressure	2.3 x 10^{-4} Torr
Sputter probe voltage	500 V
Beam extraction voltage	50 kV
Cesium oven temperature	190-205°C
Beam pulse width	1-5 Hz
Repetition rate	1-50 Hz

REFERENCES

1. G. D. Alton, Y. Mori, A. Takagi, A. Ueno, and S. Fukumoto, Nucl. Instrum. and Meth. A270 (1988) 194.
2. Y. Mori, G. D. Alton, A. Takagi, A. Ueno, and S. Fukumoto, Nucl. Instrum. and Meth. A273 (1988) 5.
3. G. D. Alton, Y. Mori, A. Takagi, A. Ueno, and S. Fukumoto, Nucl. Instrum. and Meth. B40/41 (1989) 1008.
4. G. D. Alton, Y. Mori, A. Takagi, A. Ueno, and S. Fukumoto, Rev. Sci. Instrum. 61 (1990) 372.
5. Y. Mori, A. Takagi, A. Ueno, K. Ikegami, and S. Fukumoto, Nucl. Instrum. and Meth. A278 (1989) 605.
6. G. D. Alton, Nucl. Instrum. and Meth. B37/38 (1989) 45.
7. G.D. Alton, IEEE Cat. No. 89CH2669-0, Vol. 2 (1989) 1112.
8. G. D. Alton, Nucl. Instrum. and Meth. A287 (1990) 139.
9. G. D. Alton, Phys. Div. Progress Report, ORNL-6420 (1987) p. 222.
10. J. D. Larson and C. M. Jones, Nucl. Instrum. and Meth. 140 (1977) 489.
11. K. N. Leung, D. Moussa, and S. B. Wilde, Rev. Sci. Instrum. 57 (1986) 1274.
12. J. H. Whealton, Nucl. Instrum. and Meth. 189 (1981) 55.

Current Density Calculation of a High Frequency Ion Source

H. Heydari
Technical University, EN-2
D-1000 Berlin 10

Abstract

The current density of a high frequency charged particle
source is calculated using the Child-Langmuir formula and
the quantummechanical transition probability for ionization processes at fixed gas pressure and temperature. The
dependence of the currrent density on the differential ionization coefficient and various geometrical and electromagnetic quantites is taken into account. The fundamental mechanism of the ion source considered consist in the
initial natural ionization and in the application of high
frequency fields in order to accelerate the electrons and
ions against the gas molecules thus yielding considerable
ionization. Plasma physical phenomena are not relevant
and the calculation agrees well with experimental data for
the range considered.

I. Introduction

We present a theoretical approach concerning the determination of the current density of an ion source, for which
the construction scheme is illustrated in Fig. 1.

While applying a high frequency field on the mid-section
of the piston neutral atoms will be ionized. In the beginning the ionization is caused mainly through accelerated
free electrons. Later, however, also secondary electrons
originating from shells will play a stronger role for the
ionization. Inside the mid-section of the piston one can
consider a highly conductive plasma with only small variations of the potential. Near cathode and anode, however,
one expects an ensemble of positiv ions and electrons respectively. As approximation the Child-Langmuir equation
[1] describes the correlation between current and voltage
inside the high vacuum tube.

$$I_- = \sqrt{\frac{2e}{m_e}} \frac{U_-^{3/2}}{9\pi \, ds^2} \tag{1}$$

The indirect heating of the cathode through collisions with
ions as well as the tunneleffect of the neutral gas atom
are subject to no considerable contribution of the ionization rate in this apparatus. Also statistical physical effects (i.e. Boltzmann-distribution) are at temperatures
around $300^o K$ irrelevant. Only temperatures greater than

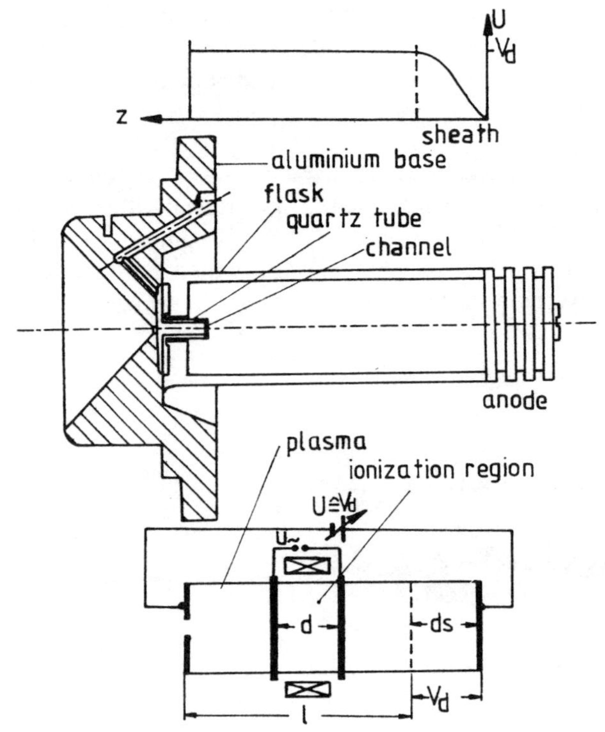

Figure 1: The high frequency ion source plane section

$16 \cdot 10^4 K$ would offer the necessary thermic energy for ionization.

II. Calculation of the ionization probability

To calculate the number of produced ions during one second and inside one cm^2 we first determine the ionization
probability for a collision of a fast electron and a neutral
atom (see Fig. 2). In a first approximation the colliding
electron will move with a velocity v_e along a straight line,
having in D the shortest distance to the nucleus. The transition probability for the transition from state n to state
m can be calculated by taking the timedependent perturbation theory into account. With the "Fermi golden rule"

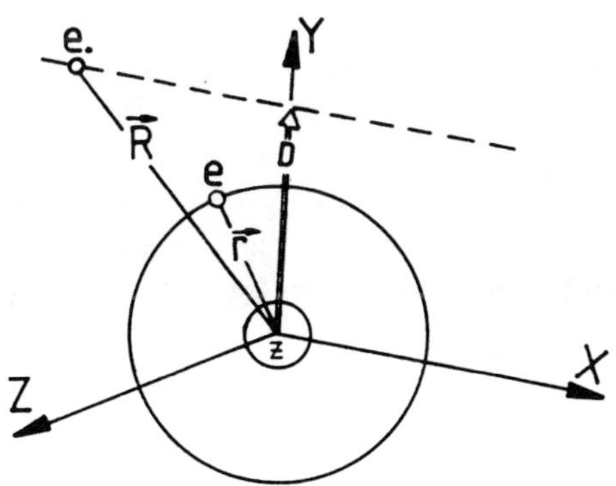

Figure 2: Simplified model to determine the ionization-probability

and the Hamilton operator [2] for the discussed system of atoms colliding with electrons we receive the following expression for the transition probability

$$
U_{nm}(\tau) = -\frac{1}{\hbar^2} \left| \int_\tau \langle n|w(t)|m\rangle \, e^{i\omega_{nm}t} \, dt \right|^2 \tag{2}
$$

with the perturbation operator

$$
w(t) = (-1)\frac{Ze^2}{4\pi\varepsilon_0|\vec{R}-\vec{r}|} \tag{3}
$$

depending on the transition energy

$$
\hbar\omega_{nm} = E_n^0 - E_m^0 \tag{4}
$$

After introducing an effective collision time $t_{st} = D/v_e$ one

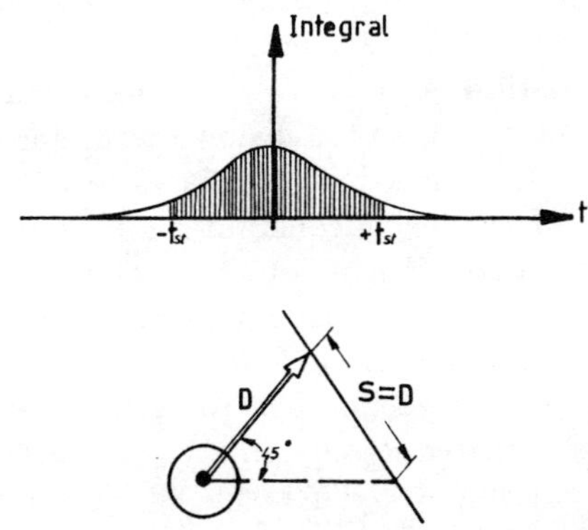

Figure 3: For determination of the distance D and time t_{st}

receives the following equation

$$
U_{nm}(-\infty, +\infty) \approx \frac{1}{\hbar^2}Z^2\left(\frac{e^2}{4\pi\varepsilon_0}\right)^2 \cdot \frac{4|Y_{nm}|^2}{v_e^2 D^2} \tag{5}
$$

$$
|\vec{D}| \gg |\vec{r}| = a
$$

for an ionization process caused by Coulomb interaction between the colliding electron and electron shell of the investigated atom. $Y_{nm} = \langle n|y|m\rangle$ is the matrix-element of the y-component of the position operator.

Here we describe a non-adiabatic case, since otherwise the passing electron would have not excited the atom. To calculate the probability dP_{nm} to exite an atom from state n to m caused by interaction of many colliding electrons with a current density of

$$
j = \frac{dN}{dt\, dA} \tag{6}
$$

one multiply the transition probability of (5) with the current density of (6) and integrate the result over the area $dt\, 2\pi D\, dD$ between $D = 0$ and $D = v_e/\omega_{nm}$. We receive the expression

$$
dP_{nm} = \int_0^{v_e/\omega_{nm}} \underbrace{dt}_{time}\, \underbrace{2\pi D\, dD}_{area}\, j U_{nm}^D(-\infty, +\infty) \quad . \tag{7}
$$

The upper integration border v_e/ω_{nm} characterizes a separation line between the non-adiabatic case ($D < v_e/\omega_{nm}$) and the adiabatic case ($D > v_e/\omega_{nm}$). Since $j\, dt2\pi D\, dD = dN$ describes the number of collisions inside a ring around the nucleus of a radius D and thickness dD one receives for the ionization probability per unit time

$$
\frac{dP_{nm}}{dt} =
$$
$$
= 8\pi j Z^2\alpha^2\left(\frac{c}{v_e}\right)^2 |Y_{nm}|^2 \ln\left(\frac{v_e}{\omega_{nm}a}\right)
$$
$$
= 8\pi j \frac{Z^2\alpha^2}{2}\frac{c^2}{W_e}\frac{m_e}{2}|Y_{nm}|^2
$$
$$
\times \left(\ln W_e - 2\ln(eV_i) + 2\ln\hbar - 2\ln a - \ln\frac{m_e}{2}\right) \quad . \tag{8}
$$

W_e symbolizes the collision energy and eV_i the ionization energy.

III. Calculation of the number of produced ions

If one divides the ionization probability per unit time with the area perpendicular to the particle current and considers the number of neutral gasatoms

$$
N_{n_A} = \frac{pV}{kT} \tag{9}
$$

following equation describes the number of ionized atoms per unit time and area

$$
\dot{n}_+ = \frac{p\,l}{kT}4\pi\frac{I_-}{e}\alpha^2\frac{2}{W_e}\frac{m_e}{2}
$$
$$
\times |Y_{gas\to ion}|^2 \left\{\frac{W_e\hbar^2}{(eV_i)^2 a^2 \frac{m_e}{2}}\right\} \quad . \tag{10}
$$

With the velocity of light c, finestructure constant α, length l, ionization energy eV_i, collision energy W_e of electrons with the charge e and the abbreviations

$$C_1 = \frac{4\pi}{kT}\alpha^2 \frac{m_e}{2}|Y_{gas\rightarrow ion}|^2 eV_i$$

$$C_2 = \hbar\left(1/a^2\frac{m_e}{2}eV_i\right) \tag{11}$$

plus the so called differential ionization coefficient S_e

$$(C_1/W_e\,eV_i)\ln\{C_2(W_e/e\,V_i)\} = S_e \tag{12}$$

expression (10) simplifies to

$$\dot{n}_+ = (I_-\,p\,l/e)S_e \quad . \tag{13}$$

Presently one can find also other expressions for the ionization coefficient $S_e = S_e(W_e)$ [3]. In our case, however, only equation (12) is relevant, since the collision energy is far greater than the ionization energy eV_i and quantummechanical treatment of the ionization process is preferred. Still missing is the determination of the collision energy W_e of the electrons, which extract their kinetic energy from the high frequency field:

$$E_\sim = \frac{U_\sim}{d} = \frac{U_0}{d}\sin(\omega t)$$

The motion of an accelerated electron is described by

$$m_e\ddot{x} = e\frac{U_0}{d}\sin(\omega t) \tag{14}$$

with the start conditions $x(0) = 0$, $\dot{x}(0) = v_e$, since the electrons posses an initial velocity v_e due to the constant electrical field (highvoltage U_-). Integration of the equation of motion (14) delivers the velocity $\dot{x}$ of the electrons and therefore the collision energy $W_e = (m_e/2)\dot{x}^2$:

$$\int_0^t dt\,\ddot{x}(t) = \int_0^t \frac{e}{m_e}\frac{U_0}{d}\sin(\omega t)\,dt \tag{15}$$

The time average of the collision energy results to

$$\overline{w}_e = \frac{1}{T}\int_0^T W_e\,dt =$$

$$= \frac{m_e}{2}\left\{v_e^2 + 2\,v_e\frac{eU_0}{m_e\omega d} + \frac{3}{2}\left(\frac{eU_0}{m_e\omega d}\right)^2\right\} \tag{16}$$

$$T = \frac{2\pi}{\omega}$$

IV. Results, comparision with available data

One receives a good approximation for the number of produced ions per unit time and area, if the calculation for the ionization coefficient (12) is done with the time average of the collision energy $\overline{w}_e$ (16) instead of the individual collision energy W_e itself. The result for (13) depends on the voltage U_- (see Fig. 1), frequency $f = \omega/2\pi$ of the high frequency source and geometry of the ion source, so that the following comparision for S_e can be presented (Fig. 4).

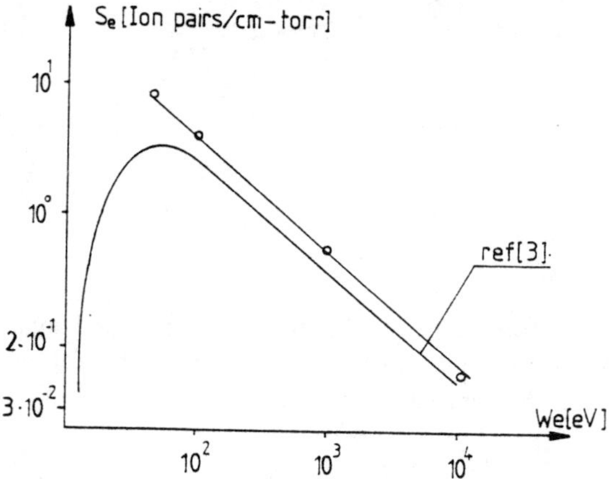

Figure 4: Results from eq.(12) compared with [3]

V. Conclusion

Introducing an approach based on a very simple theoretical model for collisional processes of an accelerated electron with a neutral gas atom we presented here an expression describing the number of produced ions per unit time and area in a high frequency ion source. As important parameters the result depends on geometry dimensions, frequency and amplitude of the high frequency source, U_--voltage and the gas characteristics. As a further simplification we also employed the time average of collision energy for our calculations. This approach is justified as experimental results demonstrate, where only a constant ionization current density was registered. We believe that plasmawave effects have only a minor effect in the time average. The in time and position constant magnetic field (see Fig.1), not discussed in this paper, has no influence on the resulting current density, since it proves to only enlarge the effective path length of the colliding electron inside the ionization area.

References

[1] R.G. Wilson and G.R. Brewer, "Ion Beams with applications to ion implantation", J. Wiley & Sons, New York, 1973

[2] A. S. Dawydow, "Quantenmechanik", VEB Deutscher Verlag der Wissenschaften, Berlin, 1978

[3] A.V. Eengel and M. Steenbeck, "Elektrische Gasentladungen", Band 2, Berlin, Heidelberg, New York: Springer, 1934

REVIEW OF MEVVA ION SOURCE PERFORMANCE FOR ACCELERATOR INJECTION

I. G. Brown and X. Godechot*
Lawrence Berkeley Laboratory
Berkeley, CA 94720

and

P. Spädtke, H. Emig, D. M. Rück and B. H. Wolf
Gesellschaft für Schwerionenforschung (GSI)
Postfach 110552, 6100 Darmstadt, Germany

Abstract

The Mevva (metal vapor vacuum arc) ion source provides high current beams of multiply-charged metal ions suitable for use in heavy ion synchrotrons as well as for metallurgical ion implantation. Pulsed beam currents of up to several amperes can be produced at ion energies of up to several hundred keV. Operation has been demonstrated for 48 metallic ion species: Li, C, Mg, Al, Si, Ca, Sc, Ti, V, Cr, Mn, Fe, Co, Ni, Cu, Zn, Ge, Sr, Y, Zr, Nb, Mo, Pd, Ag, Cd, In, Sn, Ba, La, Ce, Pr, Nd, Sm, Gd, Dy, Ho, Er, Yb, Hf, Ta, W, Ir, Pt, Au, Pb, Bi, Th and U. When the source is operated optimally the rms fractional beam noise can be as low as 7% of the mean beam current; and when properly triggered the source operates reliably and reproducibly for many tens of thousands of pulses without failure. In this paper we review the source performance referred specifically to its use for synchrotron injection.

I. INTRODUCTION

Beam transport at the low energy injection end of a heavy ion synchrotron is a concern which becomes particularly important for high current beams because of the increasingly important role of space charge forces within the beam [1]. When the beam is transported through focusing or mass selection elements in which there are transverse magnetic field components, the maintenance of good space charge neutralization may be at risk if the beam suffers from too great a level of high frequency fluctuation in beam current. Another concern has to do with possible accelerator rf regulation problems that might arise due to fast changes in beam loading during the pulse. Beam noise is also a detriment to carrying out a wide range of experiments, quite apart from the concern of beam transport. Reproducibility of the beam pulse shape is important to the accelerator user and has bearing on the kinds of experimental techniques necessary to collect data. The operational lifetime of the ion source between scheduled maintenance periods and the downtime needed to change ion species are important to the accelerator operations and to the experimenter. The Mevva ion source is a powerful method for the production of pulsed, high current beams of metal ions. Here we summarize a range of observations that we have made about its beam noise, pulse shape reproducibility, triggering reliability, lifetime and charge state distribution.

II. EXPERIMENTAL BACKGROUND

The Mevva ion source has been described in detail elsewhere [2-6]. The experiments reported here were carried out both at LBL and GSI. At LBL the Mevva IV ion source embodiment was used; this is a multi-cathode source incorporating 16 separate cathodes in which the operational cathode can be changed by rotation of an external control knob. The source was mounted on a test stand which incorporates magnetically-suppressed Faraday cups for measurement of beam current and a time-of-flight diagnostic for measurement of the ion charge state distribution [7]. Beam extraction voltage was up to 100 kV and the total ion beam current was up to 570 mA. At GSI we used a source configuration in which the Mevva cathode stem was attached to the anode chamber of a CORDIS [8,9] ion source. Beam extraction voltage was up to 35 kV and the beam current measured 50 cm downstream was up to 40 mA. Experiments have been carried out using both an ion source test stand facility, and the injector terminal of the UNILAC heavy ion linear accelerator [10].

III. RESULTS

A. Beam Noise

The beam current fluctuation level varies according to the arc current at which the source is operated. There is an optimum operating point at which the rms fluctuation level reaches a minimum, typically about 7%. This operating point corresponds to the perveance match condition, at which the plasma density (which varies with arc current) is optimally matched to the extraction optics [11]. This optimum occurs for an arc current of about 100 A for the ion source used, and is only weakly dependent on metal species employed. Figure 1 shows a typical beam pulse for Ti at the optimum current level. Figure 2 shows the variation of beam noise as a function of arc current for a titanium beam. The GSI experiments found no difference in beam noise whether or not magnets were installed in the CORDIS multipole structure [12].

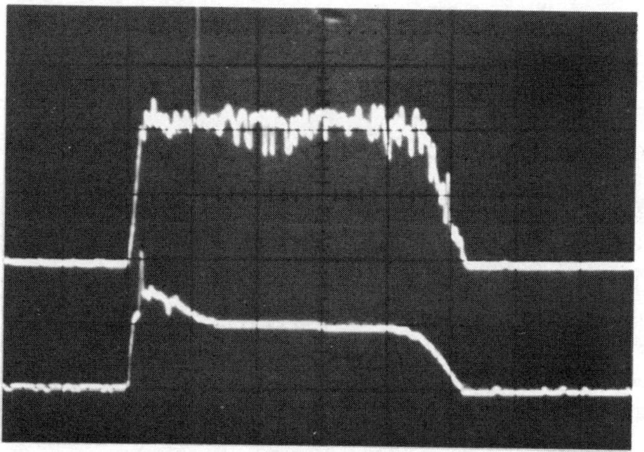

Fig. 1 Beam current pulse at perveance match. Upper trace: I_{beam}, 100 mA/cm; Lower trace: I_{arc}, 100 A/cm. Sweep speed is 50 μs/cm. Titanium beam. (XBB 890-9887)

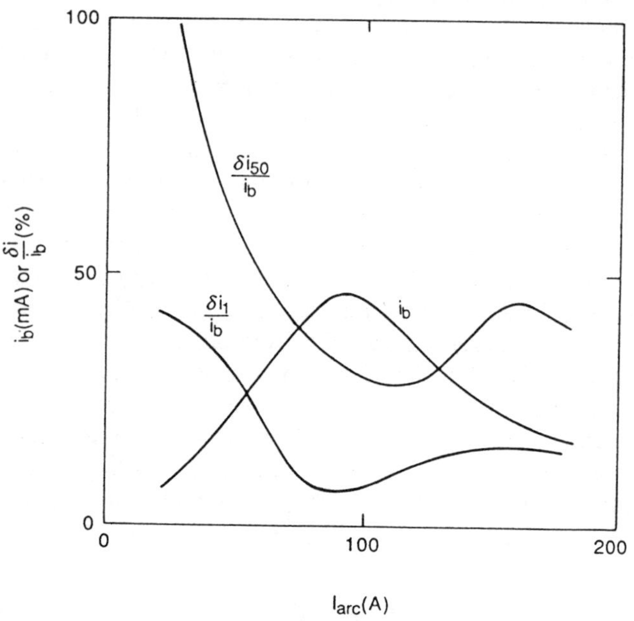

Fig. 2 Beam current i_b, fractional rms noise measured on a single-shot basis, $\delta i_1/i_b$, and the fractional rms noise measured for a sequence of 50 consecutive shots, $\delta i_{50}/i_b$, as a function of arc current. Titanium beam. Current measured 75 cm downstream from ion source by a 5-cm diameter Faraday cup. (XBL 8911-7325)

B. Shot-to-Shot Reproducibility

Variation in pulse shape from one pulse to the next is another performance characteristic that is important. Typical Mevva performance in this respect is shown in Figure 3. Here the oscilloscope was in "envelope mode" and thus recorded the envelope of pulse shape extrema for a succession of 50 beam pulses. The pulse amplitude varies from one pulse to the next by the extent indicated. Note that there were no missing pulses.

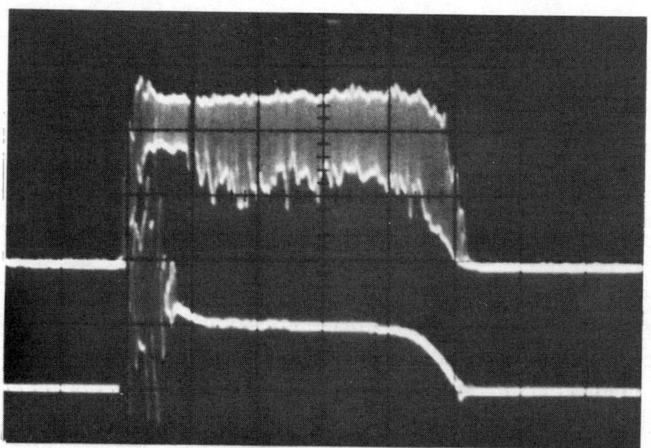

Fig. 3 Shot-to-shot variation in beam current at perveance match. Upper trace: I_{beam}, 100 mA/cm, this is the trace obtained when the oscilloscope is operated in envelope mode for 50 consecutive shots. Lower trace: I_{arc}, 100 A/cm. Sweep speed is 50 μs/cm. Titanium beam. (XBB 890-9887)

C. Triggering Reliability

It is important that the source trigger reliably for a large number of pulses before it must be removed for maintenance. The triggering failure rate should be small. We have found that when the trigger pulse is electrically adequate (peak voltage and total energy are the important parameters) and the trigger itself (annular alumina insulator) in good shape, then the triggering reliability can be excellent. As examples, the following triggering statistics were recorded for recent runs:

Cathode	Total pulses	Misfires
Pt	10,000	4 per 1000
Pd	10,000	0 per 1000
Ti	10,000	1 per 1000
Ti	100,000	10 per 1000
Ti	200,000	6 per 1000

Here "misfires" means the fraction of missing pulses (failure to produce a beam pulse when trigger pulse is applied), and "total pulses" means the number of pulses accumulated prior to making the misfires measurement. Some cathode materials (e.g., Mo) are difficult to make trigger reliably at arc current less than several hundred amperes.

D. *Lifetime*

The time for which the source can run between necessary maintenance periods is determined by the need for cathode replacement. Under typical operating conditions we obtain up to several hundred thousand pulses before the cathode is eroded away enough to require change. For the multi-cathode source embodiments (Mevva IV has 16 cathodes and Mevva V has 18) one can simply switch to another cathode at will, and in this way the total number of shots can be extended up to several million before needing to replace all the cathodes in the cathode assembly. At a pulse rate that might suit synchrotron application, say 4 pulses per second, the source can thus be run steadily 24 hours per day for a duration of order a week between changes. Cathode assemblies can be changed in an up-to-air time of less than a minute.

E. *Charge State Distribution*

The ions generated are in general multiply-stripped with a mean charge state of from 1 to 3, depending on the particular metal species, and the charge state distribution can have components from $Q = 1+$ to $6+$. This means that the ion energy is greater than the extraction voltage by a factor equal to the charge state; for 100 kV extraction voltage the beam can thus be of mean energy up to 300 keV and can have components with energy up to 600 keV. Lower Z and lower boiling point metals tend to have lower mean charge state. An empirical expression that we have found to provide a reasonable predictor for the mean charge state (referred to the distribution in particle current) is

$$\overline{Q}_p = 0.38(T_{BP} / 1000 + 1)$$

where T_{BP} is the boiling point of the metal in °C. We have studied the charge state distributions in detail and the results have been reported in the literature [13-15].

IV. CONCLUSION

The Mevva ion source provides a means for the production of high current beams of metal ions that can be useful for accelerator injection, particularly for heavy ion synchrotrons. When the source is operated appropriately ("tuned" correctly) and triggered properly, characteristics such as beam noise, shot-to-shot pulse reproducibility, triggering reliability and lifetime can be excellent. The multiply-charged ions produced offer an advantage in beam energy for given extraction voltage.

ACKNOWLEDGEMENTS

This work was undertaken as part of an ongoing collaboration between the Lawrence Berkeley Laboratory and the Gesellschaft für Schwerionenforschung. That part of the work that was carried out at LBL was supported in part by the U.S. Army Research Office under Contract No. ARO 116-89, the Office of Naval Research under Contract No. 88-F-0093, and the U.S. Department of Energy under Contract No. DE-AC03-76SF00098.

*Present address: SODERN, 1 Ave. Descartes, 94451 Limeil-Brevannes, France. Supported at LBL by a grant from the French Ministère des Affaires Etrangères, Bourse Lavoisier, and a grant from SODERN.

REFERENCES

1. See for instance A. J. T. Holmes, in "The Physics and Technology of Ion Sources", edited by I. G. Brown (Wiley, N.Y., 1989), p. 53.
2. I. G. Brown, in "The Physics and Technology of Ion Sources", edited by I. G. Brown (Wiley, N.Y., 1989), p. 331.
3. I. G. Brown, J. E. Galvin and R. A. MacGill, Appl. Phys. Lett. 47, 358 (1985).
4. I. G. Brown, J. E. Galvin, B. F. Gavin and R. A. MacGill, Rev. Sci. Instrum. 57, 1069 (1986).
5. I. G. Brown, R. A. MacGill, J. E. Galvin, and R. T. Wright, 1987 Particle Accelerator Conference, Washington, D.C., March 16-19, 1987; Proceedings pub. by IEEE, New York, Catalog No. 87CH2387-9, p. 343 (1987).
6. See, for instance, Rev. Sci. Instrum. 61, 577-591, (1990).
7. I. G. Brown, J. E. Galvin, R. A. MacGill and R. T. Wright, Rev. Sci. Instrum. 58, 1589 (1987).
8. R. Keller, Proceedings of the 1986 Linear Accelerator Conference, Stanford, CA; Report SLAC-303, Stanford, (1986), p. 232.
9. R. Keller, GSI Scientific Report GSI-81-2 (1981), p. 263.
10. P. Spädtke, H. Emig, J. Klabunde, D. M. Rück, B. H. Wolf and I. G. Brown, Nucl. Instr. and Meth. A279, 643 (1989).
11. T. S. Green, Rep. Prog. Phys. 37, 1257 (1974).
12. I. G. Brown, P. Spädtke, H. Emig, D. M. Rück and B. H. Wolf, Nucl. Instr. Meth. A295, 12 (1990).
13. I. G. Brown, B. Feinberg and J. E. Galvin, J. Appl. Phys. 63, 4889 (1988).
14. I. G. Brown and J. E. Galvin, IEEE Trans. Plasma Sci. PS-17, 679 (1989).
15. I. G. Brown and X. Godechot, paper presented at the 14th Intl. Symp. on Discharges and Electrical Insulation in Vacuum, Santa Fe, NM, September 16-20, 1990; to be published in IEEE Trans. Plasma Sci. (1991).

An Antiproton Target Design for Increased Beam Intensity

K. Anderson, C.M. Bhat, J. Marriner, and Z. Tang
Fermi National Accelerator Laboratory *
P.O. Box 500
Batavia, Illinois 60510

Abstract

Upon the commissioning of the higher energy LINAC and Main Injector at FNAL, increased proton beam intensities will be achieved. Despite these upgrades, the target used in the production of antiprotons must maintain mechanical integrity when exposed to the more intense primary proton beam. This paper describes the design parameters and thermal bench testing of an efficient air cooled target designed for effective bulk cooling and dynamic stress wave amplitude control when the target is exposed to a 120 GeV primary proton beam of 1.6μsec pulse duration, intensity of 5E12 protons/pulse, and beam size of σ=0.1mm.

I. INTRODUCTION

Increased primary proton beam intensity and small beam spot size at the antiproton target is desirable at FNAL to maximize the phase space density of antiprotons collected. However, targeting a more intense proton beam for antiproton production exacerbates problems associated with cooling as well as with mechanical wave propagation. Previous collider runs have utilized Cu, Ta, W-Re, and Heavy Met (90% W, 6% Cu, 4% Ni) alloy disks as target materials. Of these materials, Cu exhibits the most desirable combination of thermophysical properties and is able to withstand beam intensities on the order of 2E12 protons/pulse at a repetition rate of 2 seconds and a beam size of σ=0.15mm without exhibiting evidence of target material melting or mechanical fracture.

One must consider three primary aspects when designing a target for antiproton production: selection of appropriate materials, design for thermal control (i.e., both bulk and localized beam region cooling), and optimized geometry for controlling mechanical wave propagation.

II. DESIGN CONSIDERATIONS

A. Energy Deposition Calculation

A knowledge of the energy deposition due to high energy particle beam interaction with a thick target is essential to properly design a target for antiproton production. In the present case, the energy deposition per Main Ring proton pulse has been calculated using the Monte Carlo computer code MARS10[1] which simulates the three dimensional hadron and electromagnetic cascades. The analysis assumes radial symmetry and the energy deposition per grid zone as a result of beam interaction is calculated. The result is that for a 120 GeV proton beam consisting of 5E12 protons/pulse, the total energy deposited is about 770 Joules per pulse. Assuming a repetition rate of 1.5 sec, the average power is approximately 514W.

B. Material Considerations

Material subject to a sudden deposition of beam energy experiences an instantaneous increase in pressure. The change in energy density dE and the corresponding change in hydrostatic pressure dP are related by the Mie-Grüneisen equation of state[2]:

$$dP = \Gamma \rho \, dE \qquad (1)$$

The quantity Γ is known as the Grüneisen parameter. This parameter is useful in determining the response of materials to rapid heating at constant volume and can be calculated from material thermoelastic constants.

$$\Gamma = \frac{B_t (3\alpha)}{\rho \, C_v} \qquad (2)$$

B_t is the bulk modulus, α is the linear thermal expansion coefficient, ρ is the material density, and C_v is the constant volume specific heat. The rapid heating and subsequent sudden pressure variations developed in the material due to a single beam pulse result in mechanical wave propagation. Knowledge of the quantity $\rho\Gamma$ for various materials allows one to estimate beam zone pressures as a function of energy deposition. It may be readily shown that for the same amount of energy deposition, heavy metals (e.g., tungsten) experience about twice the pressure as that experienced in copper.

The atomic structure also must be considered when selecting a suitable target material. Body-centered cubic crystals such as iron and tungsten exhibit high sensitivity to notch embrittlement and relatively low impact strength. Face centered cubic crystals, such as copper, exhibit better impact resistance. Additionally, one of the effects of radiation on structural materials is the loss of ductility. However, the face-centered cubic materials exhibit less of a tendency to become brittle when irradiated.

*Operated by the Universities Research Association Under Contract with the U.S. Department of Energy

In addition, the target material should have high thermal conductivity to minimize temperature gradients and allow for quick heat dissipation. For these noted reasons, copper has been selected as a suitable starting point for target material selection.

C. Prototype Target Discussion - Design for Thermal and Mechanical Stress Wave Control

Practical design consideration for bulk cooling focuses on the application of a simple and reliable cooling scheme. Because of the inconvenience involved in the handling and maintenance of systems incorporating water cooling in which the water becomes tritium laden after prolonged exposure to beam environments, an air cooled target design was selected. The cooling system must dissipate the 514W generated by the beam interaction while maintaining reasonable temperature levels and gradients. Mass airflow rate and cooling channel geometry was specified to accommodate a design goal of maintaining bulk target temperatures below 175°C during service.

In addition to the bulk cooling requirements, the geometry of the target must also be designed to control the amplitude of stress waves resulting from the sudden deposition of beam energy. The basic mechanism of stress wave propagation was investigated using the finite element code ANSYS®. Mechanical analyses were conducted using the energy deposition calculations assuming a copper target subjected to a 120 GeV primary proton beam comprised of 5E12 protons/pulse with a beam size of σ=0.1mm. Initial models analyzed a simple cylindrical copper geometry of length 7cm and radius 25mm. Material in the beam region was modeled as a hydrodynamic solid using the Mie-Grüneisen equation of state to determine initial pressures. The remainder of the cylindrical target was modeled as a thermal elastic-plastic solid obeying von Mises yield criterion with kinematic hardening.

The mechanical analysis results reveal a radially outgoing compressive wave immediately after beam spill. Upon reaching the free surface, the compressive wave is reflected as an ingoing tensile wave. Due to focusing effects, the wave intensity increases with decreasing radius and a large tensile peak occurs in the region surrounding the core of maximum energy deposition. Such a tensile peak is potentially destructive. Figure 1 presents an ANSYS® plot of circumferential and radial stress vs. time, displaying such phenomena at a location 3mm from the beam centerline. The curve indicates a maximum circumferential stress of about .38 GPa (55000 psi) occuring 12 µsec after beam spill. The initial compressive and half-wavelength rarefaction peak amplitudes located near the beam immediately after beam exposure are difficult to control through geometry. However, one may control the magnitude of the reflected tensile peak

---

®ANSYS is a registered trademark of Swanson Analysis Systems, Inc.

through selection of appropriate geometry.

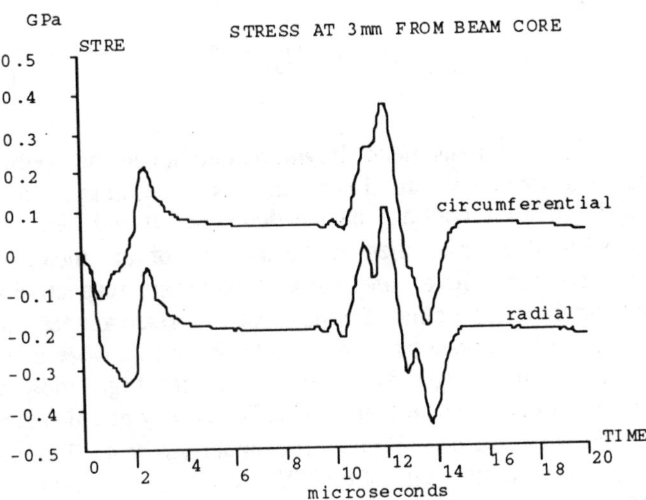

Figure 1. ANSYS® plot of wave propagation in Cu cylinder

Figure 2 outlines the basic features of the prototype antiproton target which awaits the first stages of beamline testing. The target length is 7cm and the effective target diameter is 12.7mm. Also present are 81 cooling channels each 4.8mm in diameter. Such geometry allows sufficient area for convective heat transfer to occur. In addition, the array of holes is effective in dispersing the outgoing compressive wave which dramatically decreases the magnitude of the reflected ingoing tensile wave. Simplified one-dimensional analytical models indicate that a single layer of holes may disperse 30 to 50% (i.e., depending on spacing) of the outward travelling wave energy.

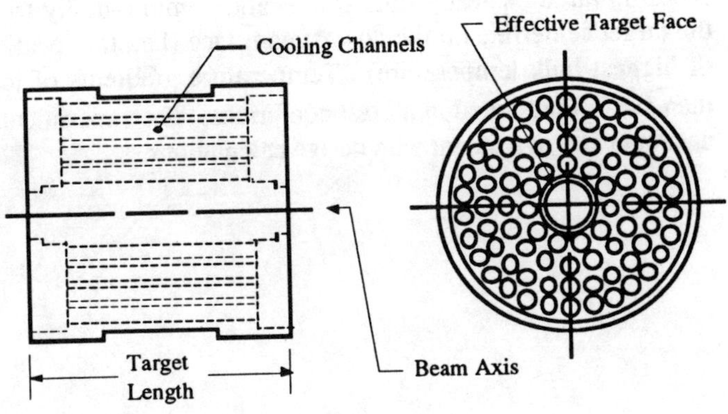

Figure 2. Prototype antiproton target core schematic.

The efficiency of the design relative to bulk cooling was verified using a simple bench test. The target core cooling channels present 616 cm² of surface area for convection cooling. Required air flow rates and heat transfer coefficients were calculated using a correlation from Seider and Tate[3] to determine the Nusselt number for combined entry length flow (i.e., the case in which the temperature and velocity profiles develop simultaneously).

$$\overline{Nu}_D = 1.86 \left(\frac{Re_D Pr}{L/D}\right)^{1/3} \left(\frac{\mu}{\mu_s}\right)^{0.14} \tag{3}$$

Re is the dimensionless Reynolds number in the cooling channels, Pr is the Prandtl number, L is the cooling channel length, D is the cooling channel diameter, μ and μ_s is the viscosity of air evaluated at the average of the mean air temperature and the channel surface conditions, respectively. The correlation is accurate for the constant surface temperature condition (determined by experiment to be the case when using copper as the target material due to its high thermal conductivity). The heat transfer coefficient may be calculated directly using equation (3) and was estimated to be 45 W/m^2/K for an air mass flow rate of 6.577E-3 kg/sec (i.e., volumetric airflow of 12 cfm at STP). At such flow, the steady state temperature was estimated to be about 200°C. Further calculations indicated an air mass flow rate of 8.22E-3 kg/sec (i.e., volumetric airflow of 15 cfm at STP) would result in steady state temperatures of approximately 165°C.

Bulk cooling tests were conducted by machining a hole along the axis of the target core to accept a 9.5mm diameter, 7 cm length cartridge rod heater for the purpose of simulating the thermal energy deposited via beam interaction. The assembly was instrumented with thermocouples and wrapped with ceramic cloth such that convective heat transfer in the cooling channels was the only means by which input energy would be dissipated. Testing was conducted at volumetric airflow rates of 12 and 15 cfm for cartridge heater powers up to 600W. Figure 3 presents a portion of experimental data showing temperature as a function of time immediately after heater powerup through steady state at a location 8mm radially from the target centerline on the downstream face (i.e., the location of highest bulk temperature). Temperature gradients of less than 10°C were noted in all test conditions. The experimental data indicates agreement with design calculations.

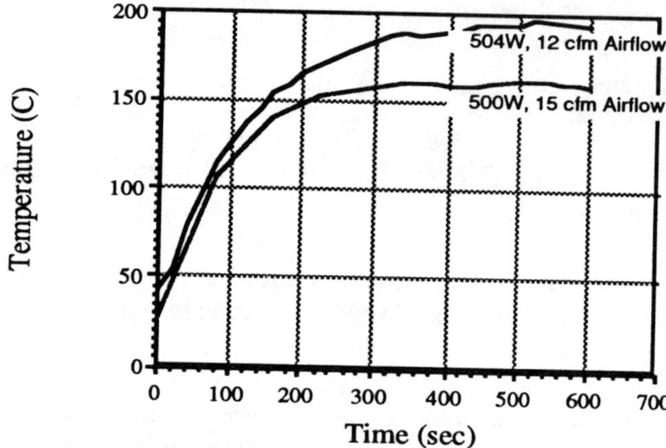

Figure 3. Thermal bench test results for 12 and 15 cfm volumetric airflow through target cooling channels

For initial beamline testing, an air delivery system consisting of a roots type blower will be incorporated. Such a system is capable of supplying continuous volumetric airflow rates of 25 cfm. An increased airflow rate over that used in the bench test coupled with natural convection occuring at the target surface and conduction heat transfer through the target mounting mechanism should result in substantially lower bulk target temperatures than those measured in the bench test.

Forthcoming beamline testing is scheduled to evaluate the design on a per pulse basis and determine if localized melting in the beam region is to occur. Should localized melting be apparent, endcaps that provide a degree of axial preload could be utilized to prevent material from spalling at the target face ends. Further testing will focus on long term fatigue effects and antiproton yield vs. the number of cumulative beam pulses.

III. REFERENCES

(1) N.V. Mokhov, Private Communication & *MARS10 Manual*, Fermilab FN-509, 1989.
(2) G.H. Bloom, "Grüneisen Parameter Measurements for High Explosives", in *Shock Waves in Condensed Matter*, New York: American Institute of Physics, 1982, pp. 588-590.
(3) F.P. Incropera and D.P. DeWitt, *Fundamentals of Heat and Mass Transfer*, New York: John Wiley & Sons, 1985, p. 393.

PROJECT OF COMPLEX TANDEM – RF LINEAR POSTACCELERATOR

O.A. Valdner, V.Ph. Gass, A.D. Koljaskin,
A.N. Pronin, P.B. Shurupov

Moscow Physical Engineering Institute
Moscow, 115409, USSR

Construction of tandem-RF linac facilities is a dynamically developing trend in acceleration technique [1]. Thre are several tandems in the USSR [2] which can be provided by the postaccelerators. This paper presents the first results of development of this type of accelerators. The main parameters of the combination are the following:

Output energy, W	10 MeV/n
Atomic mass, A	12–35
Tandem terminal potential, V	5–6 MV
Operation regime	continuous

Traditional scheme of the postacceleration is accepted (fig.1). It includes stripping target, multiharmonics buncher, many gaps aceleration resonators, focusing elements and operation system.

The acceleration channel parameters were chosen from the beam dynamics simulation using the results of the scaling model investigations. For acceleration of different particles the followings scaling laws are used:

$$\frac{\sqrt{P\,q}}{W_0} = \text{const}, \quad \beta = \text{const}, \qquad (1)$$

where P –power stored in resonator, W_0 – rest energy, q –charge, β – reduced velocity. Following these conditions the trajectories of different particles are identical, whichgives the possibility to optimize accelerati- on system.

Initial velocity of the particles was obtained from the criterion of highest intensity of the beam under the highest charge condition. As a result the value $\beta=0.04936$ was chosen. Possible parameters of the accelerated particles are presented in table 1, where the coefficient of beam using k means the intensity of the beam with charge q_2. It was assumed that Cl ions are injected to the first resonator, S, Si ions – to the second and F, O ions –to the third resonator. The ratio A/q_2 is in the range of 2.3 – 2.9. The maximum effective RF voltage for $^{35}Cl^{12}$ is 26 MV.

Table 1. Parameter of particles under injection.

Elements	V, MV	q_1	q_2	W, MeV	k
^{35}Cl	5.0	7	12	40.0	0.1
^{32}S	5.94	7	12	47.5	0.12
^{28}Si	5.94	6	11	41.6	0.16
^{19}F	5.79	5	8	34.7	0.22
^{16}O	4.87	5	7	29.2	0.24

For chosen β the value of the RF frequency f = 152.5MHz was adjusted. Starting

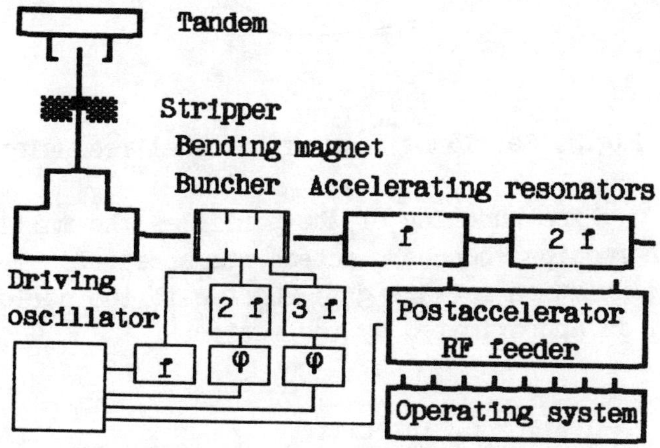

Fig.1. The scheme of the postaccelerator.

with β=0.1 it seems reasonable to increase the value of RF frequency by a factor of 2 to reduce the longitudinal size of the postaccelerator. The transverse stability of particles is provided by doublets of magnetic quadrupole lenses placed between resonators. The main parameters of the acceleration channel are presented in table 2.

Table 2. Postaccelerator parameters.

RF frequency 1 stage, MHz	152.5
RF frequency 2 stage, MHz	305.0
Number of resonators 1 stage	9
Number of resonators 2 stage	15
Resonator length, m	0.75-1.5
Total length, m	25
Synchronous phase, deg.	-20
Number of gaps in resonator	15
RF gap voltage, kV	90
Gap width, mm	15
Gap efficiency	0.82-0.94
Resonator effective voltage, kV	1.1-1.2
Drift tubes aperture, mm	15
Lens aperture, mm	20
Max gradient, Tl/m	32
Transversal acceptance, π mm mrad	1.3
Max RF power of oscillator, kW	6
Total RF power consumption, kW	100-120

RF system consists of 24 resonators. A voltage of 1.1-1.2 MV is applied to every resonator, which gives the possibility to change the output energy both sharply (switching off the power in resonator) and smoothly (changing the phase and amplitude of the field in the last resonator). Maximum total RF power consumption is 100-120 kW.

Two variants of RF structure are under consideration. The first is an interdigital H-resonator loaded by the rods (fig.2). It consists of the tank body (1) with the plates (2), where the rods with the drift tubes (3) are mounted, and the cooling system (4). The necessary distribution of electric field is achieved by the tuning elements (5). The second variant of RF structure (fig.3) is a toroidal resonator with the drift tubes mounted at the plates. Experimental parame-

ters of both scale models are presented in table 3. The shunt impedance of the first structure at the frequency f=152.5 MHz is a bit higher than the second one, but the both variants should be optimized. The final choice will be made later.

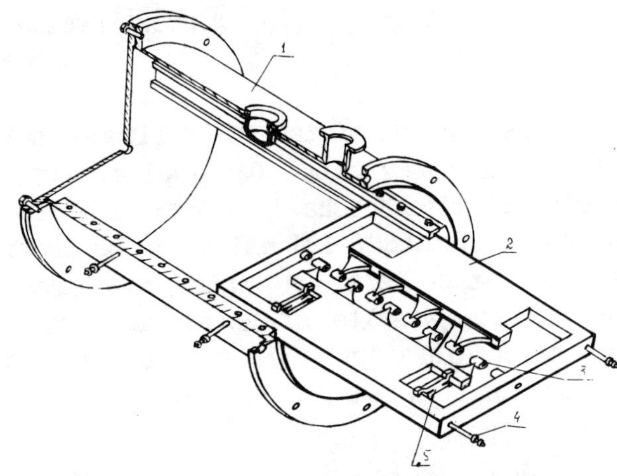

Fig.2. Sectional view of H-resonator.

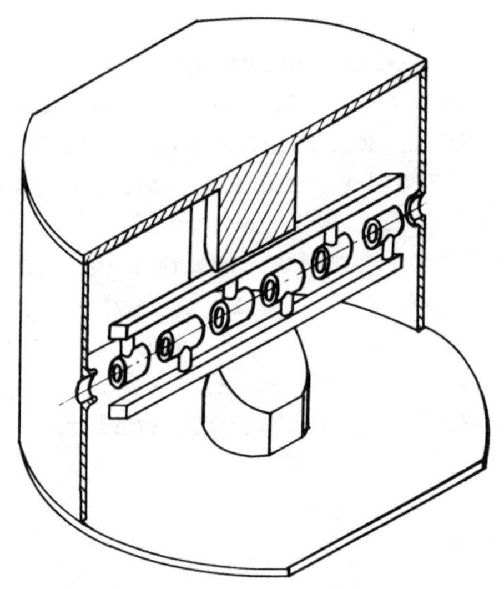

Fig.3. Sectional view of toroidal resonator.

For bunching of the particles the multi-harmonics buncher scheme was accepted. The modulation voltage $g(t)=2U_m ft$ with the period T is approximated by the series:

$$u(t)=\sum_{i=1}^{\infty} b_i \sin(2\pi i ft), \qquad (2)$$

Table 3. Parameters of resonators.

Parameter	1 model	2 model
Diameter, mm	366	420
Length, mm	537	282
Operation frequency, MHz	187	171
Quality factor	10000	9500
Drift tube outside diameter, mm	20	16
Drift tube inner diameter, mm	14	12
Drift tube length, mm	20	20
Gap width, mm	20	12
Number of drift tubes	13	12
Shunt impedance, MΩ/m	203	240

where the frequency f is a frequency of the first stage of postaccelerator. The voltage amplitudes b_1 are defined from the linear matrix system:

$$\frac{\partial}{\partial b_1} \left(\int_{-\tau/2}^{\tau/2} (g(t)-u(t))^2 dt \right) = 0, \qquad (3)$$

where τ is the time interval of the linearized part of the function $g(t)$. The dependences $|b_1|(\tau/T)$ normalized on $(2U_m/\pi)$ are shown in fig.4. For obtaining the short bunches it is reasonable to exclude the sharp voltage peculiarities and choose $\tau/T=0.7-0.8$.

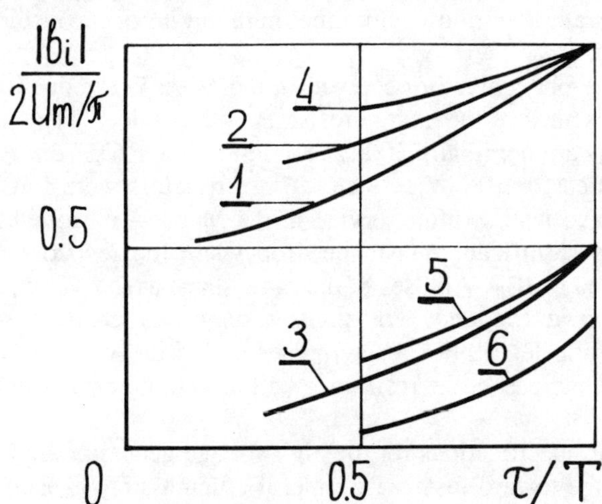

Fig.4. The normalized dependences $|b_1|(\tau/T)$:
    1   - one harmonic buncher,
   2-3 - two harmonics buncher,
   4-6 - thre harmonics buncher.

The value U_m is connected with the distance L between buncher and postaccelerator by the expression:

$$\frac{U_m}{(W_0/q_2)} = \frac{\beta^3}{2(L/\lambda) - \beta} \qquad (4)$$

For considering project we have L = 5 m, $\tau/T=0.75$. For $^{35}Cl^{12}$ $U_m=65.4$kV, $b_1=37.6$kV, $b_2=13.7$kV, $b_3=4.9$kV. It is assumed to use three-gaps resonator with the voltage distribution among the gaps 0.5:1:0.5. The bunch duration at the entrance of postaccelerator is less than 0.2 ns.

At the moment the RF power system is under construction. The driving three-harmonics oscillator with transistor amplifiers for the power 5W, 40W, 20W and the frequencies f, 2f, 3f has been constructed. The 2kW amplifier with the tetrod feeding system which is the model for 5-6kW amplifier has been constructed also.

REFERENCES

[1] I. Ben-Zvi, "Linac boosters for electrostatics accelerators machines," Nucl. Instr. & Meth., A287 (1990), pp.216-223.

[2] B.M. Hochberg, V.D. Michailov, V.A. Romanov, "Some tandem accelerators in the Soviet Union," Nucl. Instr. & Meth., 122 (1974), pp. 119-128.

[3] U. Ratzinger, R. Geier, W. Schollmeier, S. Gustavsson and E. Nolte, " The three-harmonics double-drift buncher at the Munich heavy ion postaccelerator," Nucl. Instr. & Meth., 205 (1983), pp. 381-386.

An Induction Linac Injector for Scaled Experiments*

H.L. Rutkowski, A. Faltens, C. Pike, D. Brodzik, R.M. Johnson, D. Vanecek, and D.W. Hewett**

Lawrence Berkeley Laboratory,
University of California
Berkeley, California 94720

ABSTRACT

An injector is being developed at LBL that would serve as the front end of a scaled induction linac accelerator technology experiment for heavy ion fusion. The ion mass being used is in the range 10-18. It is a multi-beam device intended to accelerate up to 2 MeV with 500 mA in each beam. The first half of the accelerating column has been built and experiments with one carbon beam are underway at the 1 MeV level.

I. INTRODUCTION

The Heavy Ion Fusion Accelerator Research program at LBL is planning to construct an experimental induction linac called ILSE [1] in order to study such phenomena as beam combining, bending of bunches with velocity tilt and pulse compression. In order to build this machine, a multi-beam injector providing ions in the mass range from carbon to potassium and with particle energy for a singly charged ion of about 2 MeV is required. An experimental injector is under construction, however as the ILSE design evolves, its design goals will also evolve. The original design called for sixteen two inch diameter beams providing up to 500 mA of C^+ ions per beam. The design normalized emittance per beam was $5 \times 10^{-7} \pi$ m-rad. The high voltage generator, ion source, control system, and the first half of the accelerating column have been constructed and beam experiments at the 1 MeV level are now occurring. We now discuss the existing system and future plans in terms of the three major components:

a) High voltage pulse generator
b) Ion source-extraction system
c) Accelerating column.

II. HIGH VOLTAGE PULSE GENERATOR

This system is an unusual inductively loaded, gas insulated system designed for slow rise long pulse operation. The slow rise time is mandated by the fact that stray capacitances associated with the electrode structure in the column can cause voltage overshoot between gaps when a fast rise pulse is applied. The Marx circuit is loaded by placing one inductor coil in series with the Marx discharge circuit at every other capacitor-spark gap stage. The self inductance of one coil is 18 mh. The full system consists of eighteen plastic trays each containing two capacitors and two spark gaps along with associated charging resistors. The trays hang from a pair of plastic impregnated wood beams inside the pressure vessel which can contain up to 80 psig. of SF_6 gas for insulation. Thus, the structure is modularized and individual trays can be removed for servicing. One coil is mounted around each tray. At the present time, a twelve tray system sufficient for use with the first half column is being used. The same system was used to fire into an open circuit in order to ring the discharge up to 2 MV for a dome-pressure vessel breakdown test. The proper loading of the inductive Marx circuit, to get the desired waveform, is provided by the resistor grading structure on the column itself. This resistor is a double-helix of Tygon tubing filled with conductive water. One leg of the helix runs from ground to the high voltage terminal. The other leg returns back to ground. The conductivity of the $NaSO_4$-water solution in the resistor is controlled by a commercial unit at a level of 27,000 μS for the present configuration. According to circuit simulations, the present system is suited for operation as fast as one shot/5 sec. Normally we use it at 1 shot/10-12 sec for high voltage column conditioning and beam experiments. This rate matches the initial ILSE specification. The purpose of the slower rate is to allow additional pumping to occur inside the column.

One phenomenon observed in the Marx is a reduction of the resistance of the charge resistors with service. The charge resistors are nominally 40 kΩ carbon resistor 1.675" dia. x 12" long rated for 100 W service. These resistors were found to drop in resistance after service in the Marx. Controlled tests showed significant resistance drop when the resistors were subjected to 10 KV pulses of the same shape as the Marx pulse. There is a tendency for the resistance to drop to some asymptotic level after a few pulses and if the voltage of the pulse is increased, the resistance will drop to a new asymptotic level.

The specifications for the high voltage generator are likely to change toward faster rise time, if column stray capacitance will permit, and toward a longer flat voltage duration. Calculations are currently being done to determine if it is possible to reduce the rise time of the pulse applied to the column from the present 30 μsec to ~5 μsec. Finally, we are seeking to increase the rep rate capability of the system to

* Work supported by the Office of Energy Research, Office of Basic Energy Sciences, U.S. Department of Energy. Contract DE-AC03-76SF00098.

** Lawrence Livermore National Laboratory, P.O. Box 808, Livermore, CA 94550

1 Hz. Of course other factors affect the useable rep rate notably the pumping speed of the accelerating column.

III. ION SOURCE AND EXTRACTION SYSTEM

The ions source being used at present is a carbon vacuum arc in conjunction with an electrostatic plasma confinement device (plasma switch) to control emission optics during fast pulse extraction [2]. The carbon arc provides sufficient ion flux for our purposes (>25 mA/cm^2) and in a single charge state [3]. Unfortunately the trigger that initiates the arc discharge does not last more than a few tens of thousands of shots. This is due to the fact that the trigger becomes coated with carbon over time and no remote cleaning technique has proved satisfactory for the long term. It is possible to change from a surface flashover trigger to a gas trigger [4] thus avoiding the carbon coating problem. However, there are other disadvantages to the source, mainly the noisy plasma from the cathode spot.

The ion source has been used together with the plasma switch in a pulsed extraction mode both in the injector itself and in a test stand. The plasma switch voltage is constant for these tests. The plasma switch allows one to extract quiescent beams from the source by electrostatically creating a planar virtual anode layer from which to extract ions. This means that the extraction system does not see the noisy streaming plasma from the cathode spot but rather a layer of ions captured by space charge forces near the plasma switch grid. This advantage of quiescent extraction is affected by the fact that the electric fields around the wires give transverse energy to the ions which ultimately becomes added emittance. Attempts have been made to circumvent this problem by reducing the plasma switch voltage during extraction as suggested by simulation results. The experimental result was that when the plasma switch voltage was reduced, oscillatory pulses were observed in the emittance measuring system and these pulses were not reproducible. Computer simulations were performed to determine whether or not the plasma drift velocity which was known to exceed the Bohm velocity by ~2 was responsible for this behavior. The simulations showed that the equilibrium near the switch changed but was not unstable The most probable explanation for the beam noise is that making the plasma switch transparent by lowering its voltage causes the extractor to use the noisy streaming plasma emitted directly from the arc because the virtual anode layer is quickly destroyed when the plasma switch voltage is dropped. This technique for reducing grid induced emittance may still work if a quiescent plasma is available for the plasma switch.

If a quiescent plasma is available the switch would be used for D.C. shutoff of the plasma from the extraction gap. When the voltage is reduced extraction would occur from the plasma behind the switch. Simulation indicates that the extraction surface should remain very close to the switch grid thus preserving the geometry of the emission surface during pulsed extraction. We are presently constructing a 25 cm dia. cusp field source which would use RF power to generate neon ions. This source is capable of generating up to 10 cm dia. beams which are of interest to the ILSE designers so that higher linear charge density can be generated in the beam. The gas will be admitted to the cusp chamber using a piezoelectric puff valve to reduce the acceleration column gas load. If, for some reason, the concept of reducing emittance generation by modulating the voltage on the plasma switch fails, the cusp field source would be usable for direct extraction of long pulses. If this approach is used, some way of selectively switching the output pulse of the injector must be found.

Meanwhile we continue to use the arc source-D.C. plasma switch combination for tests of the injector first half column. The inherent source emittance available is 2-2.5 x 10^{-6} π m-rad. No further development of the arc source is anticipated because of the perceived advantages of the cusp field source in extending lifetime, use of inert gases, and plasma quiescence.

IV. ACCELERATING COLUMN

The first half of the accelerating column is designed for service at 944 KV. A drawing of the whole assembly complete with ion source and plasma switch is shown in Fig. 1. The cross hatched area represents the beam envelope as calculated with the EGUN [5] computer code for the case of 500 mA C$^+$ ions and full column voltage. The electrodes in the column are constructed of 6A14V titanium. The ion source is located on the left. The plasma drifts from the arc cathode to the plasma switch which is operated at -55 V D.C. The 1 cm gap to the right is called the current valve. It is connected to a pulse line which pulses the source positively to inject a 1 μsec pulse into the column at the peak of the Marx voltage pulse. At the output end of the column a long tube (3") is biased at -3 kv with respect to ground to keep electrons generated by background gas ionization from entering the exit of the column and being accelerated toward the source.

The column has been operated to 10% above full design voltage without beam. In the beam extraction mode the column was first used for long pulse extraction directly from the source. For these experiments, the current valve was removed and the plasma switch grid was moved forward to the normal position of the current valve grid. The rest of the source assembly was moved forward by the same amount. The geometry makes it impossible to extract full current (500 mA) into the column and the EGUN predicted value is ~300 mA at full voltage. The measured current with a 4" diameter deep Faraday cup was 210-230 mA. Emittance measurements to study the optics of the beam were undertaken and the results were in good agreement with EGUN. The fact that such long current pulses at full column voltage could be transported reliably is very encouraging.

Experiments are now underway to use the current valve pulser to inject into the column. The voltage pulse is variable from 5-14 KV depending on the charge voltage of the pulse line. The voltage pulse remains unloaded by the presence of the source discharge and shows no pulse distortion. This behavior remains consistent for a full range of delays between the arc discharge and the current valve pulse. Control of the delay is maintained when the injector is fired as is evidenced by fibre optic monitors of the in-dome pulsers.

Initial experiments with the large aperture Faraday cup revealed a slow leakage current following the Marx voltage pulse and a noisy pulse at the time of the current valve discharge. This was verified to not be electromagnetic noise. After trying many Faraday cup variations a fast response current transformer was installed at the beam exit which verified the final Faraday cup results that net electron current was exiting the column.

Experiments using a current transformer installed inside the column at the current valve exit verified that a noisy ion current pulse was entering the column. Increasing the rise time of the pulse from the initial 300 ns has showed an improvement trend. The current valve diode in 9.75 mm wide. Therefore a 300 ns rise time corresponds to several ion transit times across the gap at the design voltage of 13.6 KV for 500 mA of C$^+$ ions. The

rise time of the pulser had to be increased to 1.5 µsec before a reasonably clean current pulse could be obtained. Tests using this new pulse shape are underway.

REFERENCES

[1] Induction Linac Systems Experiments, Conceptual Engineering Design Study, Lawrence Berkeley Laboratory, March 1989, PUB-5219.

[2] H.L. Rutkowski, R.M. Johnson, W.G. Greenway, M.A. Gross, D.W. Hewett, S. Humphries Jr., Review of Sci. Inst. 61, p. 553, January 1990.

[3] S. Humphries Jr. and H. Rutkowski, J. Appl. Phys. 67, p. 3223, April 1, 1990.

[4] S. Humphries Jr., private communication.

[5] W.B. Herrmannsfeldt, Electron Trajectory Program, SLAC-226, November 1979.

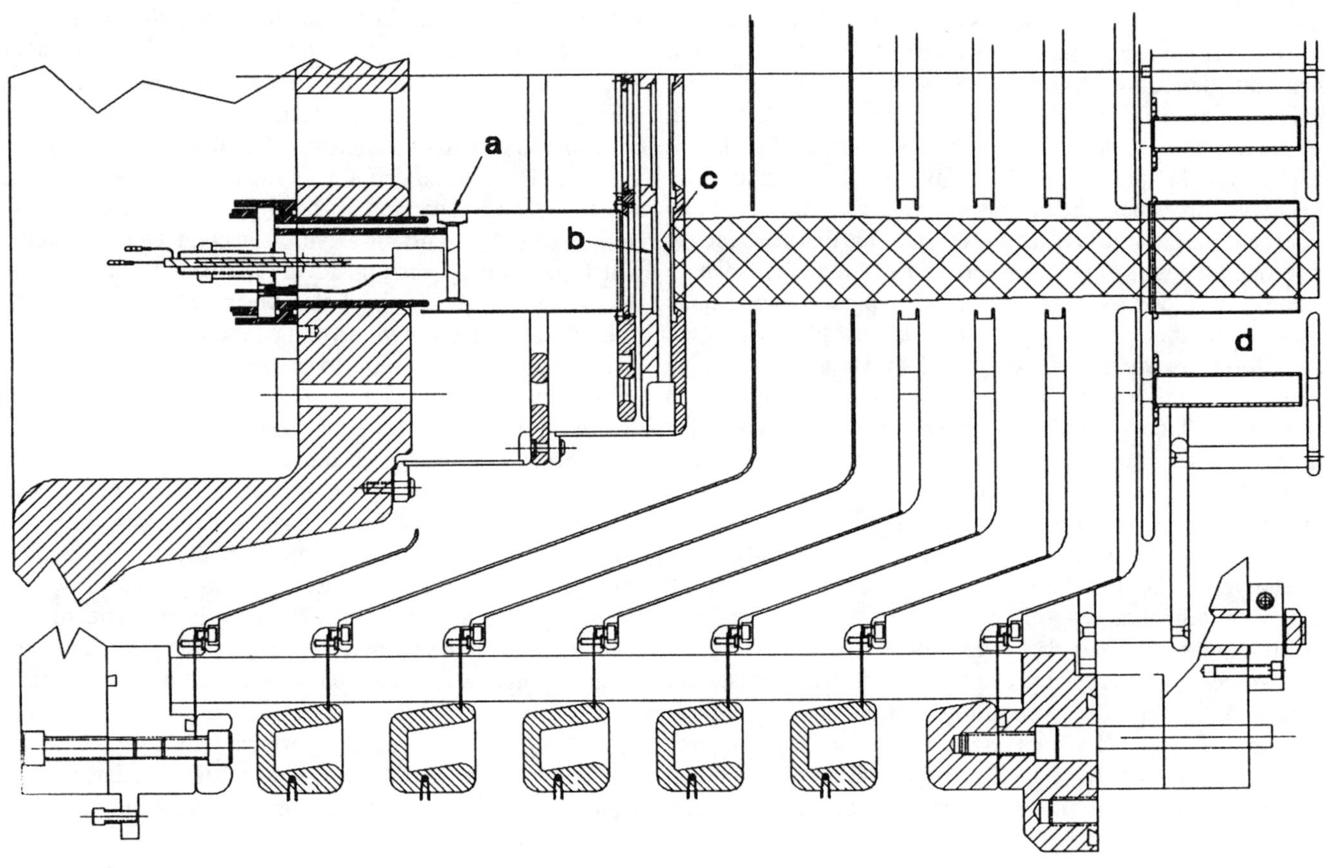

Fig. 1. First half of the full 2 MV injector accelerating column showing a) ion source, b) plasma switch grid, c) current valve grid, d) electron trap. Crosshatched area is calculated beam envelope for 500 mA, C$^+$ ions at full column voltage.

Testing of a High Current DC ESQ Accelerator

J. W. Kwan, G. D. Ackerman, O.A. Anderson, C.F. Chan, W.S. Cooper,
G. J. deVries, W. B. Kunkel, L. Soroka, W. F. Steele, R. P. Wells

Lawrence Berkeley Laboratory, University of California,
Berkeley, CA 94720

A high current dc electrostatic quadruple (ESQ) accelerator is being developed for negative-ion-based neutral beam heating and current drive on the next generation tokomak. Beam energy and current will eventually be in the MeV and multi-ampere range. This CCVV (constant-current variable-voltage) accelerator uses a series of identical ESQ modules. We have successfully tested a prototype CCVV accelerator up to 200 keV with a 100 mA He$^+$ beam (with space charge equivalence of 140 mA of D$^-$) for a pulse length of 1 s. Testing was also done with a 42 mA H$^-$ beam (H$^-$ beam current was limited by source performance). There was almost no beam loss in the ESQ accelerator. No emittance growth was found if the beam injected from the preaccelerator into the ESQ accelerator had low aberration. We are presently designing a proof-of-principle one-channel CCVV accelerator that would accelerate 1.0 A of D$^-$ to 1.3 MeV energy.

I. Introduction

Next generation tokomak fusion reactors, e.g. the International Thermonuclear Experimental Reactor (ITER), require D$^-$ neutral beams with energy as high as 1.3 MeV and with pulse lengths as long as two weeks. To meet these requirements, we are developing CCVV accelerators that can carry high current in a single channel. The ESQ sections provide strong focusing and quickly remove secondary electrons before they gain significant energy. Concept and design details of the CCVV accelerator and the neutral beam line were reported earlier by Anderson et al.[1,2]

One important characteristic of this design is that the accelerator is modular. A prototype, which has a matching/pumping module plus one 100-keV acceleration module, was built for testing. It is capable of accelerating 0.14 A of D$^-$ beam up to 200 keV. This beam has the same equivalent current (i.e. the same space charge) as 0.20 A of H$^-$ ions or 0.10 A of He$^+$ ions. Earlier we tested the prototype with 42 mA of H$^-$ beam. At present we lack an ion source that can produce 0.2 A of H$^-$ ions (or 0.14 A of D$^-$) in a single channel operating with long pulses. In this paper, we report the result of testing the prototype with 0.10 A of He$^+$ beams.

In testing the CCVV accelerator prototype, we measured the beam current and the beam emittance before and after the ESQ modules to determine the amount of beam loss, aberration and emittance growth. Measured envelope parameters were compared with calculated envelope parameters to evaluate the usefulness of the envelope code.

II. Apparatus

A schematic diagram of the CCVV prototype is shown in Fig. 1. There are two ESQ modules; the first has three quadrupole units and the second has two. The quadrupole electrodes are round rods, supported at one end by flat plates. Beam acceleration occurs at the free rod ends and in the gaps between plates. The prototype is designed to transport the beam at 100 keV through the first ESQ module and then accelerate the beam to 200 keV energy in the second ESQ module.

Between the ion source and the ESQ modules is a Pierce type preaccelerator designed to inject 0.14 A of D$^-$ at 100 keV energy into the first ESQ module. Permanent magnets are

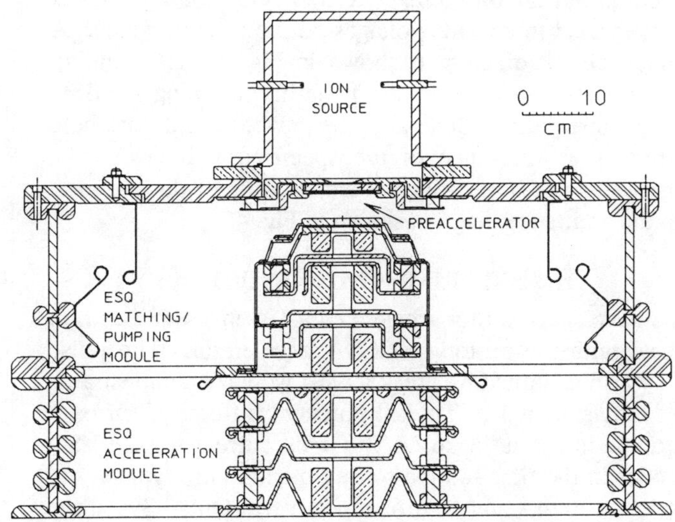

Fig. 1. Schematic Diagram of the prototype ESQ Accelerator

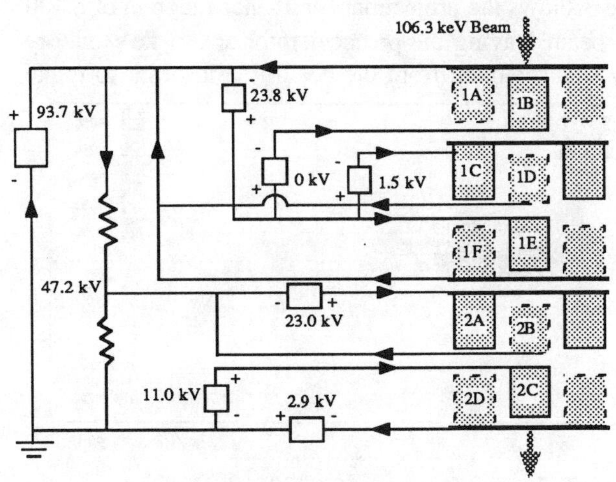

Fig. 2. Circuit diagram of the ESQ accelerator.

Supported by U.S. DOE Contract No. DE-AC03-76SF00098

embedded inside the second electrode of the preaccelerator to trap electrons only when the apparatus is used to produce negative ion beams. These permanent magnets are removed for positive ion beams.

Stripping of negative ions produces electrons which in turn produce X-rays. In order to minimize the stripping loss, thus optimize the power efficiency of the acceleration system, the apparatus is designed for high conductance pumping to quickly remove the gas from the beam axis.

The ion source is a 20-cm-diam, 23-cm-deep multicusp dc source developed for volume-production of H$^-$ (or D$^-$) ions.[3] The same ion source, operating without the "filter rods", can be used to produce He$^+$ ions . With a 3-cm-diam aperture, the ion source can produce up to 100 mA of H$^-$ beam in 10 ms pulses. Nevertheless, a 2-cm-diam aperture was used when we tested the ESQ with a 42 mA H$^-$ beam because the smaller aperture produced a H$^-$ beam with less aberration and lower effective beam temperature. In the case of testing the ESQ with a He$^+$ beam, a 2.6-cm-diam aperture was used to produce more than 100 mA of steady He$^+$ beam current.

A combination of voltage divider and floating power supplies is used to provide voltages on the ESQ modules. A schematic circuit diagram is shown in Fig. 2. The floating power supplies allow maximum flexibility in tuning the ESQ during the test phase. Generally, the power supplies are held to within 1 % tolerance of the specified voltages. The voltages indicated in the diagram are the actual values used to obtain the emittance diagram shown in Fig. 4.

III. EXPERIMENTAL RESULTS

We have successfully accelerated a 105 mA He$^+$ beam to 202 keV with the prototype CCVV accelerator. The pulse length was not limited by breakdowns; we have demonstrated pulses longer than 1 s. Typical potential difference across a quadrupole is about 25 keV. The drain current for the ESQ electrodes in the first and the second modules are typically in the order of 20 mA and 10 mA respectively. Only ≈3% of the He$^+$ beam was lost in the ESQ; the uncertainty of the calorimetric measurements can be as high as 2-3%.

Figure 3 shows the projectional emittance diagram of a 100 mA He$^+$ beam leaving the preaccelerator at 101 keV energy (the ESQ was removed from the beam line in order to make

this measurement). The 63% intensity contour on the diagram gives a normalized emittance of 0.015 π-cm-mrad. With a 2.6-cm-diam beam-forming aperture, this emittance value corresponds to that of a 0.5 eV Gaussian beam. Aberrations in the beam optics are preserved as the beam propagates through the ESQ modules. Figure 4 shows the emittance diagram as the beam emerges from the ESQ at 200 keV energy (the emittance diagram on the other transverse axis has a much wider beam envelope thus confirming an elliptical beam spot). In propagating through the two ESQ modules, the projectional emittance increases by 13% from 0.015 to 0.017 π-cm-mrad ±5% of experimental errors.

When the beam current was reduced to 84 mA, the aberrations introduced in the preaccelerator disappeared and we found no emittance growth in the ESQ. Figures 5a and 5b are the x-x' and y-y' emittance diagrams of a 84 mA He$^+$ beam leaving the ESQ at 200 keV.

Figure 6 shows the result from the envelope code[1] using input data from the actual potentials applied to the ESQ electrodes and from the measured beam parameters leaving the preaccelerator (in absence of the ESQ modules). Comparing Fig. 5a and 5b with Fig. 6, we found a small discrepancy between the measured and the calculated beam envelopes at the ESQ exit which is corrected by assuming that the addition of the ESQ module at the end of the preaccelerator increases the divergence of the beam entering the ESQ by 20 mrad. We shall check this assumption in future experiments. Nevertheless, the envelope code was proven to be the most valuable tool in assisting tuning of the ESQ accelerator. We are improving the code to account for the gap focusing between units of quadrupole electrodes.

Similar results were obtained when we operated the ESQ accelerator with H$^-$ beams. Using the ESQ accelerator, we have accelerated 42 mA of H$^-$ to 200 keV. The beam loss within the ESQ modules was estimated to be 9%, presumably due to stripping of the H$^-$ ions by the H$_2$ gas. We measured no emittance growth within the 5% experimental uncertainty. Most of the current drain from the quadrupole electrodes in the matching/pumping ESQ module (i.e. the first module) were related to the electrons from beam stripping and their secondary electrons. The electrical power consumed by the ESQ module was less than 2% of the beam power.

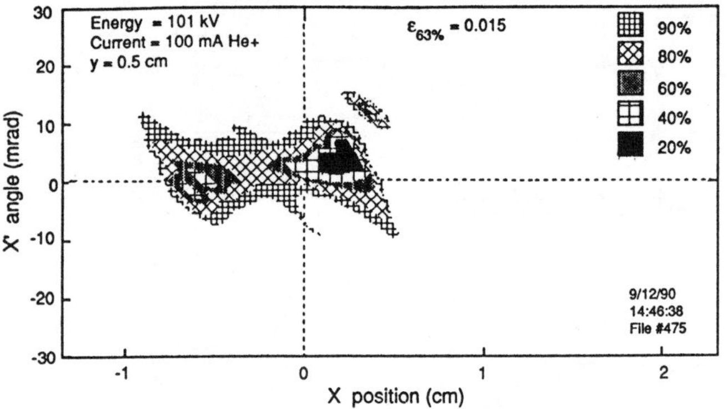

Fig. 3. Emittance Diagram of a 100 mA He+ beam leaving the preaccelerator at 101 keV energy.

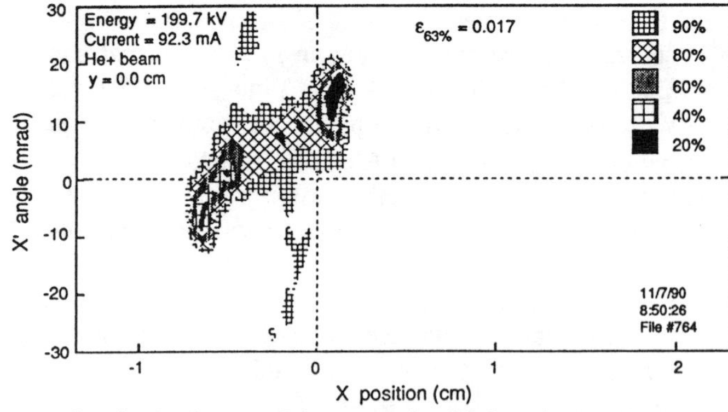

Fig. 4. Emittance Diagram of a 92.3 mA He+ beam leaving the ESQ accelerator at 200 keV energy.

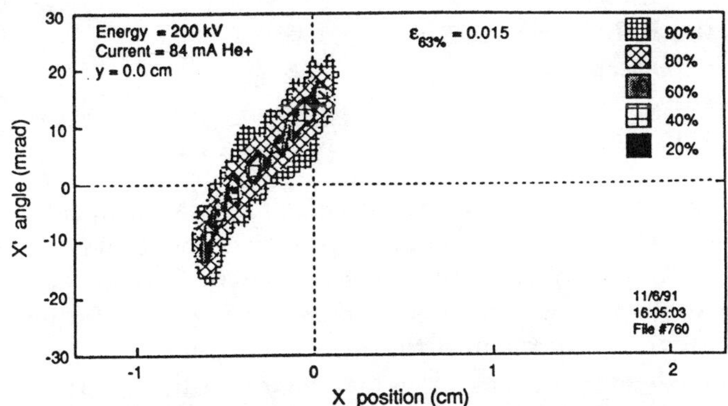

Fig. 5a. Emittance (x-x') of a 84 mA, 200 keV He+ beam.

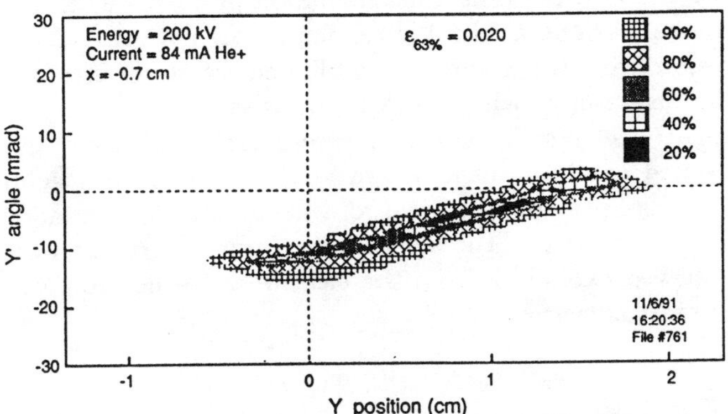

Fig. 5b. Emittance (y-y') of a 84 mA, 200 keV He+ beam.

IV. DISCUSSION

In summary, we have successfully accelerated 0.10 A of He$^+$ beam (equivalent to 0.14 A of D$^-$) to 200 keV using the prototype CCVV accelerator. Beam loss and emittance growth are both small; the latter can be improved by providing beams with better optics at the preaccelerator. Tuning the accelerator can be done with the assistance of the envelope code. Eventually, when the performance of the CCVV accelerator is well characterized, frequent tuning would not be necessary and the floating power supplies can be replaced by voltage dividers.

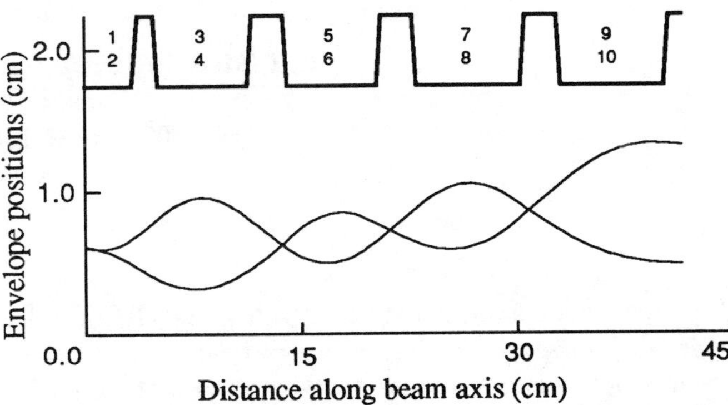

Fig. 6. Beam envelopes of a 84 mA He+ beam in the ESQ.

We are proposing to build a proof-of-principle 1.3 MeV, 1.0 A (D$^-$) CCVV accelerator (Fig. 7)[4]. This new accelerator will have quadrupole units approximately 2.5 times larger in length and in diameter than the present prototype. By keeping the same electric field strength, the new ESQ accelerator will accelerate ion beams at 250 keV per module; thus only 4 ESQ modules are required to reach the 1.3 MeV final beam energy (here the preacceleration is 300 keV). The accelerator will also produce beams of half or three-quarter energy with no change in current.

V. REFERENCES

[1] O.A. Anderson, L. Soroka, C.H. Kim, R.P. Wells, C.A. Matuk, P. Purgalis, W.S. Cooper, and W.B. Kunkel, Proc. 1988 European Particle Accelerator Conf., Rome, Italy; S. Tazzari, Ed.; World Sci. Pub. Co, Singapore, p.470 (1989).

[2] O.A. Anderson, L. Soroka, C.H. Kim, R.P. Wells, C.A. Matuk, P. Purgalis, J.W. Kwan, M.C. Vella, W.S. Cooper, and W.B. Kunkel, Nucl. Instrum. and Meth. B40/41, 877 (1989).

[3] J.W. Kwan, G.D. Ackerman, O.A. Anderson, C.F. Chan, W.S. Cooper, G.J. deVries, K.N. Leung, A.F. Lietzke, and W.F. Steele, Rev. Sci. Instrum. 61, 369 (1990).

[4] O.A. Anderson, et al., IAEA Thirteenth International Conference on Plasma Physics and Controlled Nuclear Fusion Research, Washington, DC, 1990 (to be published).

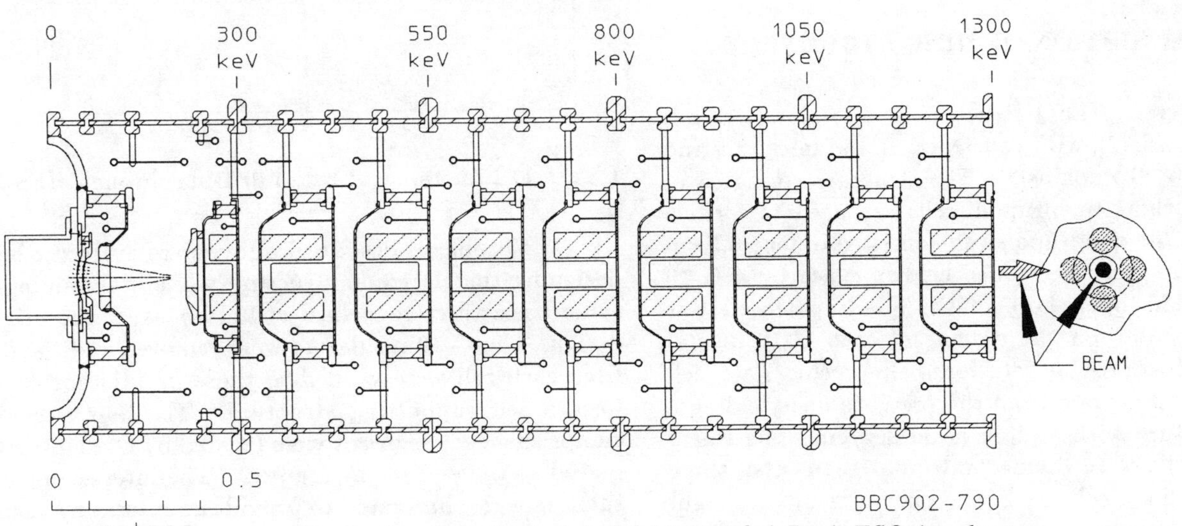

Fig. 7. A proof-of-principle 1.3 MeV, 1.0 A D$^-$ dc ESQ Accelerator.

Transport Properties of a Discrete Helical Electrostatic Quadrupole

C.R. Meitzler, K. Antes, P. Datte, F.R. Huson, L. Xiu
Texas Accelerator Center†
4802 Research Forest Drive, Bldg. 2
The Woodlands, Texas 77381

Abstract

The helical electrostatic quadrupole (HESQ) lens has been proposed as a low energy beam transport system which permits intense H$^-$ beams to be focused into an RFQ without seriously increasing the beam's emittance. A stepwise continuous HESQ lens has been constructed[1], and preliminary tests have shown that the structure does provide focusing. In order to understand the transport properties of this device, further detailed studies have been performed. Emittances were measured 3.5 cm from the end of the HESQ at two different voltages on the HESQ electrods. A comparison of these experimental results with a linear model of the HESQ beam transport is made.

I. INTRODUCTION

Modern accelerators usually require intense charged particle beams to be produced by the ion source and efficiently injected into the first stage of acceleration, typically an RFQ. The problem of capturing the beam from the ion source and focusing it into the acceptance of the RFQ has been the subject of considerable recent development. During transport, space charge forces cause the beam emittance to increase while the build up of neutralization in the back ground gas causes emittance to rotate. Both of these problems can be solved by using a purely electrostatic transport system. The following paper describes a measurement of the emittance of a Helical Electrostatic Quadrupole (HESQ) lens and compares it to a simulation of the transport properties of the lens.

II. DESCRIPTION OF HESQ STRUCTURE

The discrete HESQ structure was originally developed by Raparia[2], with construction and initial testing being done by Tompkins[3]. The structure consisted of thirty eight short quadrupole cells arranged to form a helix. The basic quadrupole cell consisted of four 0.5 cm thick stainless steel electrodes held in place by a G-10 insulator. The open bore of the quadrupole cell was 26 mm. A notch on the outside of each G-10 insulator was used to provide rotational alignment. Each cell was rotated 18 degrees from the previous one yielding a helical structure with a pitch of 36 deg/cm. The thirty eight unit cells were divided into four segments: three segments with ten cells each, and one segment with eight cells. The segments were separated by thin G-10 disks.

The positive and negative voltages of each segment could be independently varied for flexibility during tuning.

A tandem faraday cup was used to initially show that the lens was capable of transmitting 55% of a low current H$^-$ beam into the acceptance of an RFQ, although the emittance was not determined[1]. Furthermore, operational experience showed that the presence of the G-10 posed some voltage holdoff problems when the source was operated for extended periods of time. First, cesium from the ion source would deposit on the inside of the insulator causing sparking between the positive and negative electrodes along the inner surface. Second, gas seemed to accumulate in the voids between the insulator disks providing a break down path to the grounded outer supports of the structure. Numerous tracks were observed extending from the electrodes to the grounded support structure.

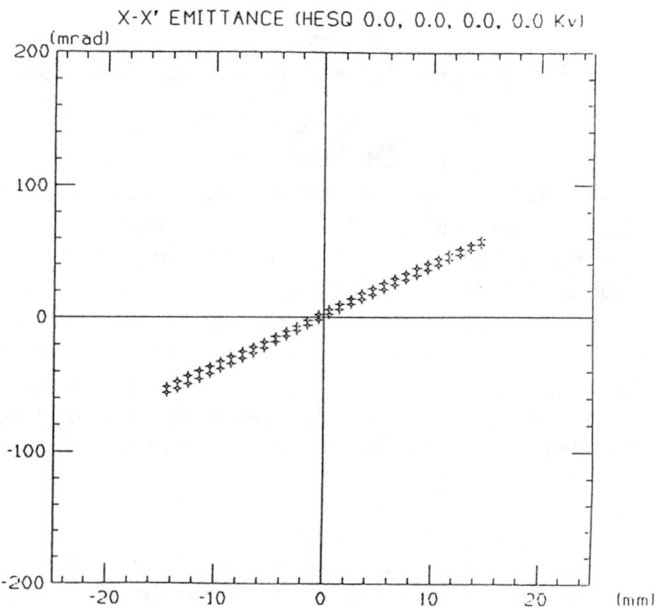

Figure 1. Emittance of beam drifting through HESQ.

The problems described above were overcome by redesigning the electrode structure while maintaining the same basic discrete helical structure as in the original design. The G-10 insulators were removed and the disks from each cell were welded to those in adjacent cells to form a self supporting structure. The four electrodes comprising each segment were located by G-10 insulators placed at the end of the segment. The inter-segment insulators were eliminated to provide an insulating vacuum between segments. This permitted better radial pumping than the original structure; however, the placement accuracy for each individual electrode disk decreased. This modification did not completely eliminate sparking between electrodes when the pressure in the vacuum chamber rose into the mid 10^{-6} Torr range.

III. EMITTANCE MEASUREMENTS

The emittance was measured with an electrostatic emittance scanner[4] located 3.5 cm from the output end of the HESQ. The plate voltages could be swept in one of two different modes: the plate voltage would be held constant during the beam pulse, or the voltage could be swept by a 2.5 kV high voltage amplifier. The current incident upon the front face of the scanners could be measured to provide pulse-to-pulse normalization of the EES signal. The analyzed signal was recorded for later off-line analysis. In the current set of measurements, the voltage on the deflection plates was held constant during the pulse.

IV. RESULTS AND DISCUSSION

In order to estimate the transport properties of the lens and show focussing, emittance data were collected for two HESQ conditions. First, the voltage on all of the electrodes was set to zero to allow the beam to coast through the lens. A second set of measurements was made with 8.5 kV on all of the electrodes which avoided complications in the boundary regions between different segments. The beam was provided by a magnetron H$^-$ ion source whose current was 6 mA, the normalized horizontal and vertical emittances were 0.29 π mm–mrad and 0.36 π mm–mrad, respectively[1,3].

Figure 1 shows the horizontal 100% emittance contour when the beam is allowed to drift through the lens. The apparent cut-off of the contour at large radii is due to the 26 mm diameter aperture of the HESQ. The vertical emittance has a similar shape and is not provided here.

Figure 2 shows the horizontal 100% emittance contour when the electrode power supplies are all set to 8.5 kV, while Figure 3 shows the vertical emittance. The emittances shown in Figures 2 and 3 have very different shapes and possibly arise from non-linearities introduced by the fraction of the beam which goes out to large radii during transport.

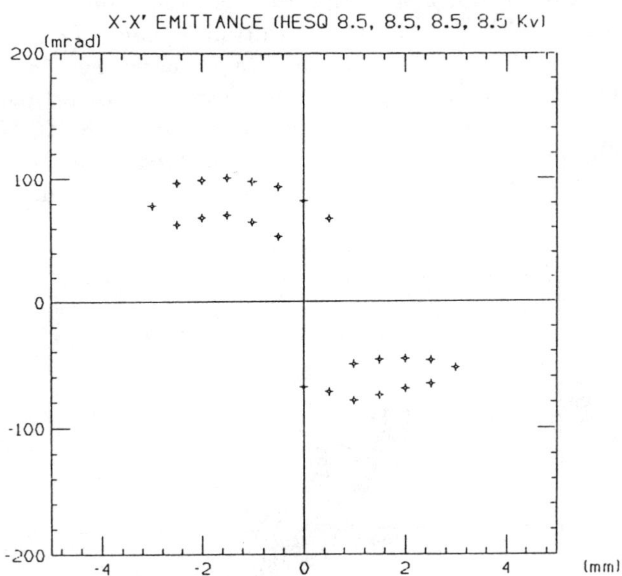

Figure 2. Measured horizontal emittance when the HESQ electrodes are set at 8.5 kV

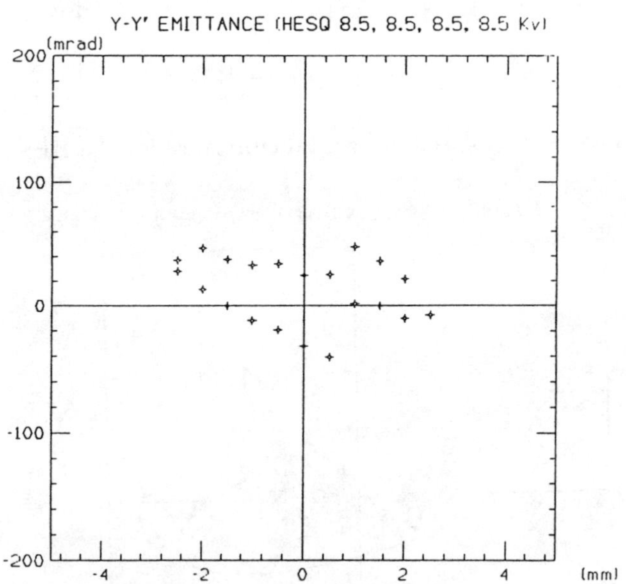

Figure 3. Measured vertical emittance when the HESQ electrodes are set to 8.5 kV.

The performance of the HESQ was simulated using the program HRK developed at TAC. The input emittance for this code was the measured ion source emittance obtained by Tompkins[3]. This tracking code analytically tracks the particles through a continuous with linear space charge forces and a linear fringe field at the ends. A calculation of the fringe field with RELAX-3D showed that the fringe field has a tilt as one gets closer to the first quadrupole cell; therefore, the first quadrupole in this model is skewed 45 degrees to simulate this behavior. Figure 4 shows the simulated horizontal emittance when all of the electrode voltages are set to 8.5 kV. Figure 5 shows the simulated vertical emittance under the same conditions. A comparison of these simulated emittances with the measured emittances above (Figures 2 and 3) shows reasonable qualitative agreement in both beam size and orientation; however, the distortions in the measured emittances are not reproduced by the simulations. This is because the model assumes that the electrodes create a pure quadrupolar electric field.

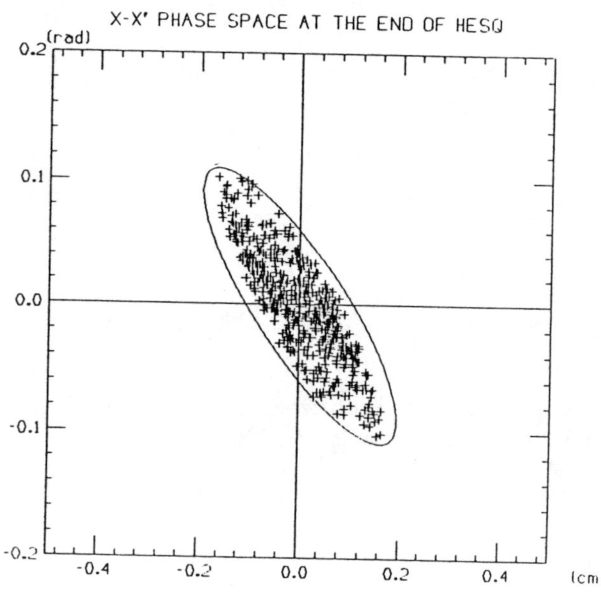

Figure 4. Simulated horizontal emittance for the HESQ.

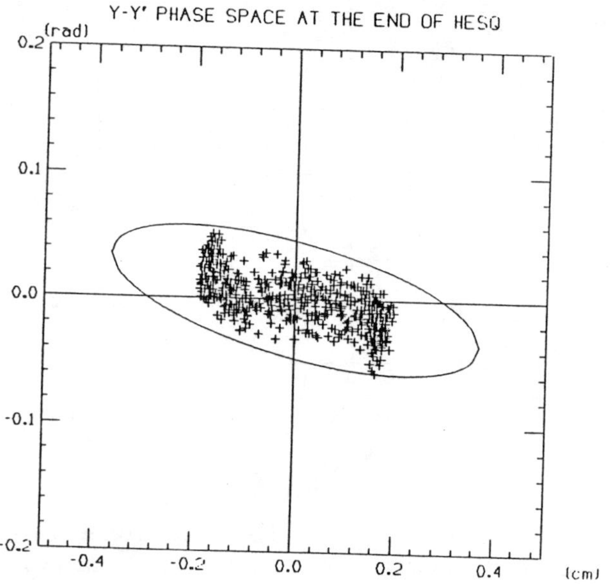

Figure 5. Simulated vertical emittance after HESQ.

IV. CONCLUSIONS

The HESQ is a promising structure for transporting intense low energy ion beams from an ion source to an accelerator. The discrete structure seems to introduce aberrations into the beam, presumably from the beam filling a large part of the open aperture of the lens. The general agreement between the measured and simulated emittances show that the fringe field has a slight tilt which must be included in any simulations. Finally, the first HESQ lens has been shown to provide focussing although there is still considerable work to be done to bring this new technology to a mature state.

V. ACKNOWLEDGEMENTS

The authors would like to thank Deepak Raparia and Perry Tompkins for many useful discussions. We would also like to thank Larry Crowe for his diligent and capable work as the technician/machinist of the project.

V. REFERENCES

† The Texas Accelerator Center at HARC is a consortium of Rice University, Texas A&M University, the University of Houston, the University of Texas, Prarie View A&M University, Sam Houston State University, and the Baylor College of Medicine MR Center.

1. C.R. Meitzler, P. Datte, F.R. Huson, P. Tompkins and D. Raparia, Proceedings of the 1990 Linear Accelerator Conference Proceedings, Albuquerque, New Mexico, 1990.

2. D. Raparia, "Beam Dynamics of the Low Energy Beam Transport and Radio Frequency Quadrupole," University of Houston, Houston, Texas, Ph.D. Dissertation, 1990.

3. P.A. Tompkins, "The TAC H$^-$ Ion Source and Low Energy Beam Transport," Texas A&M University, College Station, Texas, Ph.D. dissertation, 1990.

4. P.W. Allison, J.D. Sherman, and D.B. Holtkamp, IEEE Trans. Nucl. Sci. NS-30, 2204 (1983).

Low Energy H$^-$ Beam Transport Using an Electrostatic Quadrupole Focusing System*

S.K. Guharay, C.K. Allen, M. Reiser, V. Yun
Laboratory for Plasma Research
University of Maryland
College Park, MD. 20742

Abstract

An experiment is designed to transport a round and diverging H$^-$ beam, with beam current of 30 mA and voltage of 35 kV, over a length of about 30 cm with the aim to match it to an RFQ with a modest emittance growth. The low energy beam transport system consists of six electrostatic quadrupole (ESQ) lenses. A linear KV code, in conjunction with a 3-D Laplace solver, is used to configure the electrode geometry. Fringe field integrals from the Laplace solver are given as input to a modified PARMILA code to evaluate the influence of the nonlinear forces and calculate the emittance growth. The ESQ system is designed with a mechanical tolerance of < 1 mil; the critical issue of alignment of the apparatus with respect to the beam is addressed. Details on the design and construction of the ESQ system and some preliminary experimental results are presented.

I. Introduction

In accelerator research, designing an efficient low energy beam transport (LEBT) section is of great relevance even today. The electrostatic quadrupole (ESQ) lens system presents a good prospect in this regard in handling beam current up to $\lesssim 30$ mA at acceleration voltage of 35 kV. In this context, several experimental efforts are currently in progress, e.g., at the SSCL using helical quadrupole lenses[1], in this laboratory with conventional configuration of the ESQ lenses[2].

The beam transport section is designed here to transport a 30 mA, 35 kV H$^-$ beam over a length of about 30 cm, and eventually to match this beam into a radio frequency quadrupole (RFQ). This experiment is planned following the beam parameters of a modified Penning-Dudnikov source[3] on the AT-1 test stand at Los Alamos National Laboratory. The design of the LEBT section is developed through computer code simulations. Results from a linear beam optics code solving the well-known K-V envelope equation, and an analysis of the focusing function and fringe-field matrices from a 3-D Laplace solver form the basis of the lens design. A modified PARMILA code[4] calculates the emittance growth in the system.

The key features of the present system are: (1) The lens assembly consists of six quadrupole lenses in a symmetric triplet configuration. (2) The entire unit is adjustment-free mechanically. This points to ruggedness of the system. (3) Estimated emittance growth is about 78%.

In an earlier paper, design procedure of the LEBT section is described in detail[2]. The emphasis of this article is on issues related to fabrication of components, interfacing the LEBT section with the ion source and beam diagnostics. Some characteristic test results of a prototype lens system are given.

II. The LEBT System

A. Essential Design Issues

The LEBT system is designed to transport a high-brightness (normalized brightness $= 7.98 \times 10^{10}$ A/(m-rad)^2) H$^-$ beam from the aforementioned ion source at Los Alamos. The input beam, carrying a current of 30 mA, is taken as round and diverging with beam radius $= 1$ mm and divergence of the beam envelope $= 20$ mrad. It is aimed to generate a round, converging beam (radius ~ 1 mm, convergence angle ~ 20 mrad) at the output of the transport channel. Details of the design are described earlier[2]. The key points in this regard are mentioned here and some important modifications of the previous design are highlighted.

A linear beam optics code, solving the K-V equation and coupled with a 3-D Laplace solver, generates parameters of the ESQ lenses. In order that the influence of spherical and chromatic aberrations as well as nonlinear image fields may be minimized, some design constraints are imposed on the maximum radius of the beam envelope in various sections of the transport channel, dimensions of the quadrupole lenses, and applied voltage on the quadrupoles[2]. Table 1 presents characteristic lens parameters.

The symbols carry their usual significance. Effective length, l_{eff}, of the quadrupoles is calculated from the 3-D Laplace solver following the relationship $l_{\text{eff}} = 1/\kappa_0 \left(\int_{z_1}^{z_2} \kappa(z) dz \right)$. Here z_1 and z_2 correspond to the zero-crossing points of the focusing function $\kappa(z)$ at the entrance and exit ends of the lens respectively; the maximum value of κ is κ_0. V_q is obtained from the K-V code using the

*Research supported by ONR/SDIO.

Table 1: Lens parameters from linear code analysis and 3-D Laplace solver.

Quad number		1/6	2/5	3/4
l	(mm)	15.00	59.00	47.00
L^*	(mm)	6.00	6.00	6.00
R_q	(mm)	8.00	12.00	12.00
l_{eff}	(mm)	14.50	55.40	42.90
V_q	(kV)	8.08	3.98	3.92

*The drift space between lenses, L, is longer here compared to the previous design[2].

Table 2: RMS normalized emittance values and emittance growth .

$\tilde{\epsilon}^{(i)}_{nx,ny}$	$\tilde{\epsilon}^{(f)}_{nx}$	$\tilde{\epsilon}^{(f)}_{ny}$	$\tilde{\epsilon}^{(f)}_{n}$	$\tilde{\epsilon}^{(f)}_{n}/\tilde{\epsilon}^{(i)}_{n}$
0.069	0.107	0.138	0.123	1.78

quadrupole length as l_{eff}, when an almost round, converging output beam envelope ($X_f = 1.04$ mm and $Y_f = 0.98$ mm; $X'_f = -26.4$ mrad and $Y'_f = -25.6$ mrad) is generated at a distance of 15 mm downstream from the last lens. The beam envelope and the focusing function are shown in Fig. 1. The beam convergence gets appreciably high moving the end point very little towards the last lens– at $z = 29.8$ cm, i.e., moving the end point 4 mm closer to the last lens, $X'_f = -42.4$ mrad and $Y'_f = -42.2$ mrad, while $X_f = 1.19$ mm and $Y_f = 1.13$ mm. This feature may be exploited in coupling the LEBT system to an RFQ, when a highly convergent beam (e.g., about -76 mrad for BEAR RFQ) is usually required. However, the aim of our initial effort is to study effectiveness of transport of a high-brightness beam through the ESQ lens system and identify the various control parameters.

A modified PARMILA code is used to simulate the beam envelope in the presence of nonlinear fringe-field forces; this code estimates the emittance growth in the LEBT system as well. Initially, the code is run with input parameters corresponding to Table 1. Next, voltage on the quadrupoles is adjusted to establish a round, converging output beam. Figure 2 shows phase space plots of particle distribution at $z = 30.2$ cm in the most optimized attempt made here so far. Here, $X_f = 1.25$ mm, $Y_f = 1.31$ mm, and $X'_f = -30.6$ mrad, $Y'_f = -19.2$ mrad. Shifting the endpoint 4 mm closer to the last lens, i.e., at $z = 29.8$ cm, the beam envelope shows $X_f = 1.4$ mm, $Y_f = 1.4$ mm, and $X'_f = -44.4$ mrad, $Y'_f = -34.5$ mrad. The maximum deviation of the new set of voltages from the respective $V_q s$ in Table 1 is about -14%. The code analysis thus appears to yield sufficient guideline to the possible range of voltage setting on the quadrupoles to generate a round, converging output beam in the beam transport experiment. Results on beam emittance in mm-mrad are given in Table 2. Emittance growth appears to be quite modest.

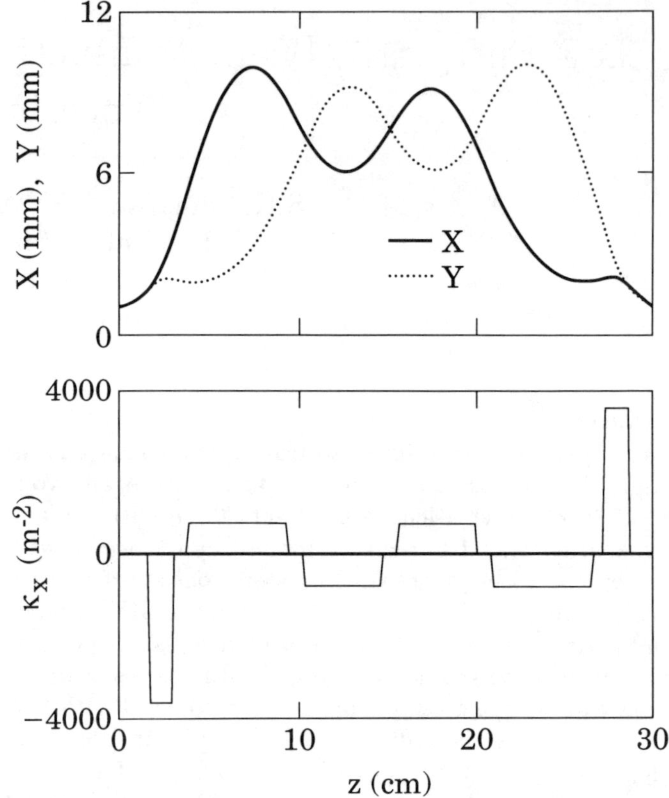

Figure 1: K-V beam envelope solution (top), hard-edge focusing function (bottom).

We have also studied the influence of the boundary of the cylindrical grounded body containing the lens assembly. It appears that the radial dimension of this housing plays an important role in determining focusing function in the transport channel. In context of our particular lens geometry and beam parameters, significant mutual influence of neighboring lenses is noted for radius of the housing less than about 8 cm, and effectiveness of the ground plates between lenses is reduced. This weakens the focusing function within the lens and may cause a deleterious effect on the beam envelope in terms of emittance dilution. Further studies on this issue are warranted.

B. Construction of ESQ Lenses and Hardware Assembly

The ESQ lens system in the present experiment is shown in Fig. 3. The three principal elements of the lens system are: (1) six lenses, each consisting of four quadrupoles, (2) ground plates between the lenses, and (3) precision ceramic insulator balls. The four quadrupoles of each lens are machined out of a single piece of finished cylindrical aluminium rod. The dimensional tolerance of each electrode is held within 0.2 mil, which is checked by a precision micrometer. The electrodes are contoured in a manner to reduce field nonlinearities[2]. The aluminium ground plates are 2 mm thick. Ceramic balls are used here to develop an adjustment- free, self-aligned lens assembly. Two

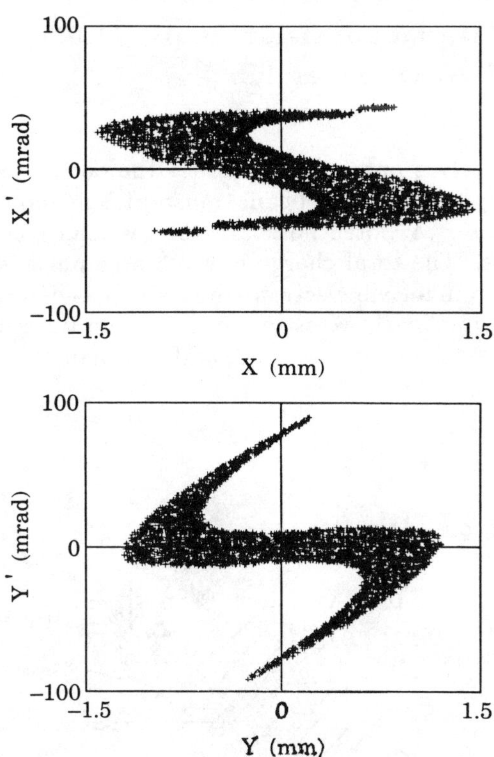

Figure 2: Particle distribution in phase space at $z = 30.2$ cm from modified PARMILA code.

sets of precision balls, having tolerance in sphericity within ± 0.025 mil, are used. Larger balls of 3/8 inch diameter are used between two opposing quadrupoles of adjacent lenses; smaller ones of 1/8 inch diameter are placed between a quadrupole and its neighboring ground plate. Such an arrangement delivers much better alignment than the earlier one[2]. The entire lens assembly is mounted in a cylindrical housing, which is installed in a vacuum vessel.

Currently, efforts are being made to build up mechanical hardwares to interface the front end of the vacuum vessel with the ion source. The back end will be connected to a diagnostic box equipped with emittance scanners.

III. TEST RESULTS AND EXPERIMENTAL PLANS

A prototype lens system has been constructed and voltage hold-off tests were done. At a hydrogen gas pressure of about 3×10^{-4} Torr, a voltage level of ± 12.5 kV can be maintained between quadrupoles of opposite polarities in a lens. In this effort, corona discharges occurred quite frequently at lower voltage initially. With gradual conditioning of the surface of the electrodes, the lens system could be driven at higher voltage without any noticeable voltage breakdown. This insures practicability of the present compact ESQ lens design.

Preparations are being made to characterize the H^- beam from the ion source using Faraday cup and emittance scanners. Afterwards, the LEBT system will be coupled to

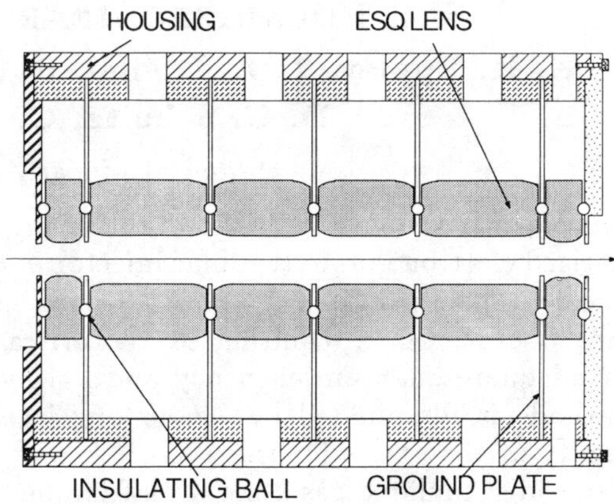

Figure 3: ESQ lens assembly.

the ion source and efficiency of beam transport with ESQ lenses will be studied.

IV. REFERENCES

[1] C.R. Meitzler, P. Datte, F.R. Huson, P. Tompkins, D. Raparia, "Progress on the TAC ion source and LEBT," Proc. LINAC Conf., Albuquerque, N.M., September 1990, pp.710-712.

[2] S.K. Guharay, C.K. Allen, and M. Reiser, "Electrostatic focusing and RFQ matching system for a low energy H^- beam," Proc. SPIE Conf. on Intense Microwave and Particle Beams, Los Angeles, CA, January 1991, vol.1407, pp.610-619.

[3] P.W. Allison and J.D. Sherman, "Operating experience with a 100 keV, 100 mA H^- injector", Proc. 3rd Symp. on Production and Neutralization of Negative Ions and Beams, AIP Proc., 1983, vol. 111, p. 511.

[4] C.R. Chang, E. Horowitz, and M. Reiser, "Conceptual design of an electrostatic quadrupole transport system for high-brightness H^- beams", Proc. SPIE Conf. on Intense Microwave and Particle Beams, Los Angeles, CA, January 1990, vol. 1226, pp. 483-498.

SLC POLARIZED BEAM SOURCE ELECTRON OPTICS DESIGN[*]

K. R. Eppley, T. L. Lavine, R. A. Early, W. B. Herrmannsfeldt, R. H. Miller,
D. C. Schultz, C. M. Spencer, and A. D. Yeremian

Stanford Linear Accelerator Center, Stanford, CA 94309

This paper describes the design of the beam-line from the polarized electron gun to the linac injector in the Stanford Linear Collider (SLC). The polarized electron source[1] is a GaAs photocathode, requiring 10^{-11}-Torr-range pressure for adequate quantum efficiency and longevity. The photocathode is illuminated by 3-nsec-long laser pulses. The quality of the optics for the 160-kV beam is crucial since electron-stimulated gas desorption from beam loss in excess of 0.1% of the 20-nC pulses may poison the photocathode. Our design for the transport line consists of a differential pumping region isolated by a pair of valves. Focusing is provided by a pair of Helmholtz coils and by several iron-encased solenoidal lenses. Our optics design is based on beam transport simulations using $2\frac{1}{2}$-D particle-in-cell codes to model the gun and to solve the fully-relativistic time-dependent equations of motion in three dimensions for electrons in the presence of azimuthally symmetric electromagnetic fields.

The design of the polarized beam source transport system is nearly complete. We are beginning to fabricate two identical systems. One system is to be installed in the SLC this summer for operation this autumn. A copy is to be installed in a polarized gun test laboratory this summer, in order to test the SLC polarized source prior to operating the source in the SLC, and for future tests of new photocathode guns.

In the present configuration, shown in Figure 1, two guns can be mounted on the SLC injector. One gun has a GaAs photocathode suitable for producing a polarized electron beam. The other gun, which has a thermionic cathode, has produced the unpolarized beams used in previous SLC operations. Both of these guns aim into a "Y"-shaped vacuum chamber inside a DC bend magnet that can bend the beam from either gun into the SLC injector.

Isolation of the photocathode gun vacuum from the "Y" region is necessary for two reasons: (1) It will be necessary occasionally to remove the photocathode gun from the accelerator and to replace it with a spare. When swapping guns, it is desirable to maintain vacuum in the guns and in the "Y" chamber. (2) The photocathode demands pressure below 10^{-10} torr for adequate quantum efficiency and lifetime. The best pressure attained in the "Y" chamber is of order 10^{-9} torr.

The vacuum system design is described elsewhere in these proceedings.[2] The goal of photocathode vacuum isolation is met by a pair of all-metal straight-through valves with low magnetic permeability.[3] The valves have 2.9-cm-diameter aperture and are 18-cm long. One valve seals-off the gun. The other valve seals-off the "Y" chamber.

The valves pose a challenge for the optics design because they lengthen the beam transport line and constrict its aperture. A potential source of gas load comes from beam loss. The total charge in the 3-nsec pulse is 20 nC. In order to maintain electron-stimulated gas desorption at an acceptable level, we estimate that beam losses must be less than 0.1% before the bend, and less than 1% after the bend.

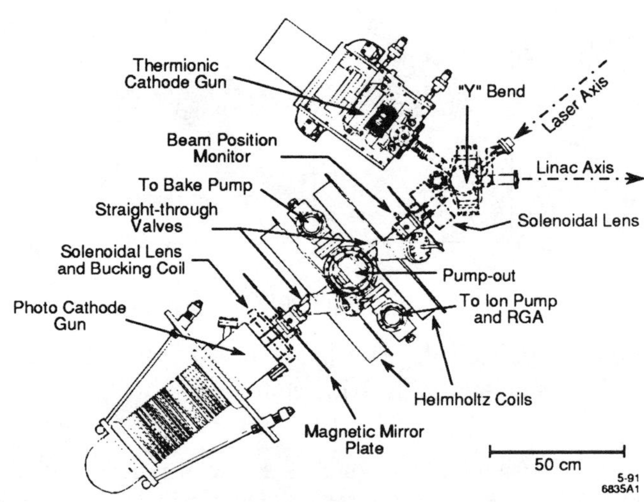

Figure 1: Plan view of the SLC gun region. The polarized electron beam emerges from the photocathode gun (at lower left), passes through a pair of straight-through valves and Helmholtz coils, and is bent through the "Y" chamber into the linac injector. The upper beam-line is the (unpolarized) thermionic gun.

The optics design is further complicated by the fact that the photocurrent is not space-charge limited at the head, tail, and outer radial edges of the beam pulse. This situation arises because the chopped 3-nsec-long laser pulse has finite rise- and fall-times, and because the spatial distribution of the laser spot on the photocathode has edges of diminishing intensity. The temporal distribution used in the simulations is shown in Figure 2.

The basic optics design was calculated using Herrmannsfeldt's electron optics and gun design code[4] to model the gun electrode structure, and the CONDOR electromagnetic simulation code to model the transport line.[5, 6]

The magnetic field configuration for the optics design is shown in Figures 3 and 4, which were calculated by the POISSON computer program. The focusing field necessary to transport the beam efficiently through the pair of low-permeability straight-through valves[3] is provided by a pair of air-core Helmholtz coils. The coils, which are

---

* Work supported by the U.S. Department of Energy under contract DE-AC03-76SF00515.

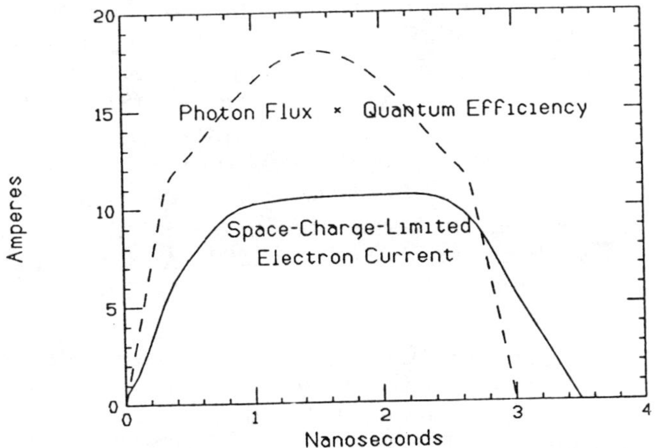

Figure 2: Temporal profile of the laser pulse and the resulting photoemission pulse.

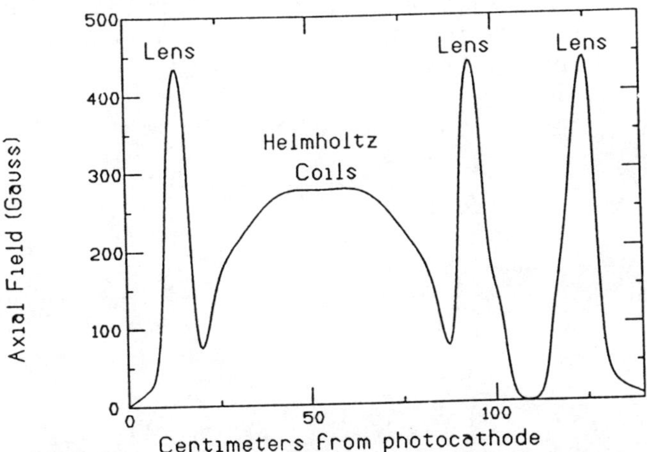

Figure 3: Axial focusing field profile from the photocathode to beyond the "Y" bend chamber, computed by POISSON.

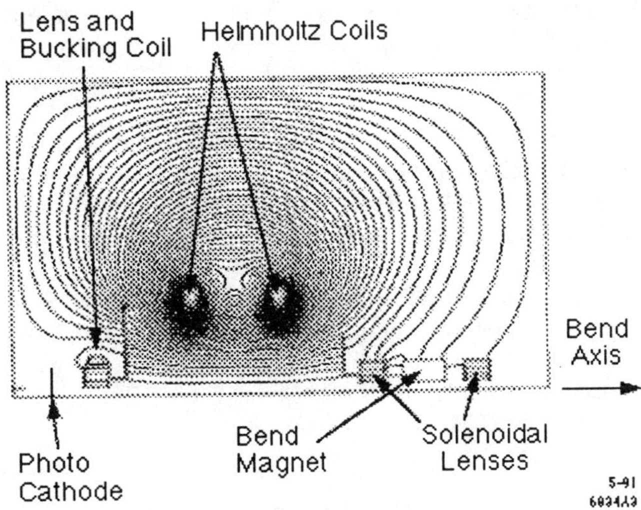

Figure 4: Magnetic flux plot from the photocathode to beyond the "Y" bend chamber, computed by POISSON.

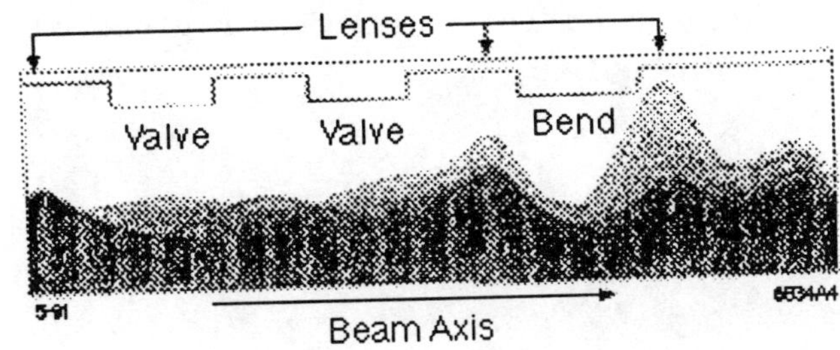

Figure 5: The beam envelope from the photocathode to beyond the bend chamber, computed by CONDOR.

wound from hollow water-cooled conductor, are to be run at half of their 19000-amp-turn peak capability. The coils, which encircle the valves, have 40-cm inner-diameter, 63-cm outer-diameter, and are separated by 26 cm.

Additional focusing is provided by 2500-amp-turn iron-encased solenoidal magnetic lenses. Lenses are located 40 cm on either side of the center of the Helmholtz pair. One of these lenses is mounted on the gun. Another is mounted on the entrance to the "Y" chamber. A third lens is mounted on the exit of the "Y" chamber. The lenses are bakeable to 250°C.

A 1100-amp-turn bucking coil is wound around the first lens. The coil is made of solid wire, is bakeable to 250°C, and has an external cooling water circuit. The purpose of the bucking coil is to zero the DC magnetic flux through the cathode, which is important for producing a low-emittance beam. The normalized beam emittance due to axial field B_c through the cathode is $(e/2mc)R^2B_c$, where R is the beam radius at the cathode.

The beam envelope, displayed as a superposition of CONDOR's step-by-step snapshots, is displayed in Figure 5.

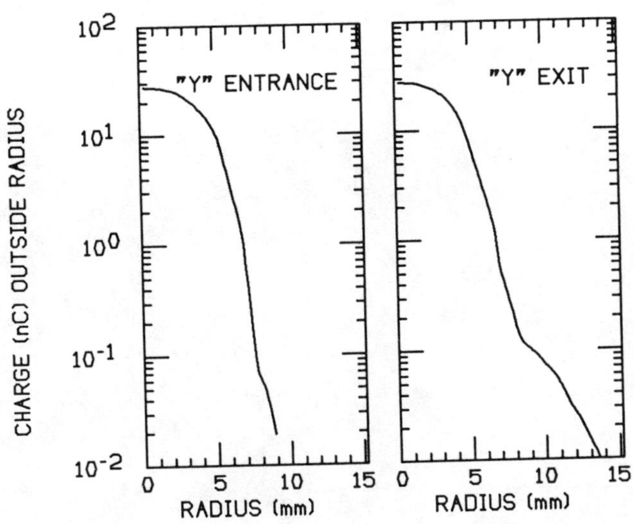

Figure 6: The beam profile before and after the bend, computed by CONDOR. The horizontal axis in both plots extends to the beam pipe wall. The plot indicates that the desired limits on interception (0.02 nC before the bend and 0.2 nC after the bend) have been met.

Not all dots in this figure represent the same amount of charge. A quantitatively precise plot of the beam profile in the CONDOR simulation is shown in Figure 6, both before and after the bend.

We are extending our beam dynamics simulation studies through the buncher region of the SLC injector in order to study effect of beam-size and scalloping on capture and bunching, and on depolarization induced by the intense beam's self-fields.

We thank SLAC's Accelerator Theory and Special Projects Department and Mechanical Fabrication Department for technical assistance. In particular, we wish to thank J. E. Clendenin, N. R. Dean, L. A. Klaisner, J. A. Nuttall, and B. Woo for assistance.

References

1. J. E. Clendenin, "The SLC Polarized Electron Source" (SLAC-PUB-5368), presented at the Workshop on Polarized Electron Sources and Electron Spin Polarimeters, Bonn, Federal Republic of Germany, September 6–7, 1990.

2. T. L. Lavine, J. E. Clendenin, E. L. Garwin, E. W. Hoyt, M. W. Hoyt, R. H. Miller, J. A. Nuttall, D. C. Schultz, and D. Wright "SLC Polarized Beam Source Ultra-High Vacuum Design" (SLAC-PUB-5454, May 1991), published in these proceedings.

3. The straight-through valves are Vacuum Generators' models CSD-32 being fabricated with special low magnetic permeability movement mechanisms.

4. W. B. Herrmannsfeldt, "EGUN—An Electron Optics and Gun Design Program," SLAC Report No. 331,

5. B. Aiminetti et al., "CONDOR User's Manual," Livermore Computing Systems Document, Lawrence Livermore National Laboratory, Livermore, California, April, 1988.

6. A. Palevsky and A. T. Drobot, "Application of E-M PIC Codes to Microwave Devices," Proceedings of the Ninth Conf. on Numerical Simulation of Plasmas (Northwestern University, Evanston, Illinois, June 30–July 2, 1980), Paper PA-2.

OBSERVATIONS ON FIELD-EMISSION ELECTRONS FROM THE LOS ALAMOS FEL PHOTOINJECTOR*

Alex H. Lumpkin
Physics Division
Los Alamos National Laboratory
Los Alamos, NM 87545 USA

Abstract

A background source of electrons from the photoelectric injector (PEI) of the Los Alamos FEL experiment has been identified. This source is present without the drive laser irradiation and when the rf power is applied to the injector accelerator. Using intensified cameras and a synchroscan streak camera, these electrons have been imaged via optical transition radiation and Cherenkov radiation and characterized. The basic questions of location (photocathode), timing (~40 to 90° of the RF cycle), magnitude (2.2 nC per μs of rf power at 26 MV/m at the photocathode), and parameter sensitivity (accelerator A field's duration and magnitude) have been answered. The properties are consistent with a field-emission mechanism.

I. INTRODUCTION

The Los Alamos FEL facility incorporates a photoelectric injector (PEI) as a source of low-emittance electrons [1]. During the initial accelerator commissioning phase for beam energies of 14-17 MeV, electrons were accelerated, transported, and detected even when the PEI drive laser beam was blocked and only the RF power was on. This low power beam was initially detected with x-ray detectors and intensified television cameras that viewed intercepting beam-profile screens [2]. Using optical transition radiation (OTR) and Cherenkov radiation (CR) conversion mechanisms, the electron beam information was converted into visible light that was recorded by the intensified cameras and synchroscan streak camera, respectively. Several transport conditions were then used to image the background source and determine its location, spatial extent, timing, and parameter sensitivity. The characterizations led to the identification of the CsK_2Sb photocathode material as the source and field emission as the most probable mechanism. These results are some of the first to

be reported from a high quantum efficiency (QE) photocathode/photoinjector.

II. EXPERIMENTAL PROCEDURES

The commissioning experiments began in the summer of 1989 and ended in May 1990. The beamline diagnostics used for the majority of this time are schematically shown in Fig. 1. No intercepting diagnostics or even wall current monitors (WCM) were allowed by the designers in the beamline until after the second accelerator tank. The electrons were transported about 7 m from the photocathode to the front-surface, aluminized fused-silica screen at position #3. The screen was oriented at 45° to the beam direction so that OTR was viewed at 90° to the beam direction and CR was viewed at 46° to the downstream beam direction. The beam position and profile were determined from the front surface (OTR) source and the CR from the fused silica substrate was optically relayed to a synchroscan streak camera several meters away from this screen. The Hamamatsu C1587 streak camera was phase-locked to the 108.3 MHz reference frequency, which is a subharmonic of the master 1300 MHz accelerator frequency. This technique allows the synchronous summing or integration of signals from many micropulses with relatively low jitter, 4 ps (FWHM), and reasonable temporal resolution, 5 ps (FWHM). In the case of this low current source we integrated the whole rf macropulse time at the 108.3 MHz repetition frequency [3].

III. EXPERIMENTAL OBSERVATIONS

As reported earlier [3] there was a strong dependence of the background source intensity on the photoinjector accelerator (A) field amplitude and duration. Starting at 26 MV/m, a 20% reduction in field resulted in a factor of 5 decrease in the box-average peak intensity of the field-emission image. Under some conditions a single laser-generated micropulse was comparable to the "field emission" electrons from about 50 μs of rf power. A synchroscan streak image exhibited a triplet temporal structure, which was attributed to a

---

*Work supported by the US Department of Defense under the auspices of the US Department of Energy.

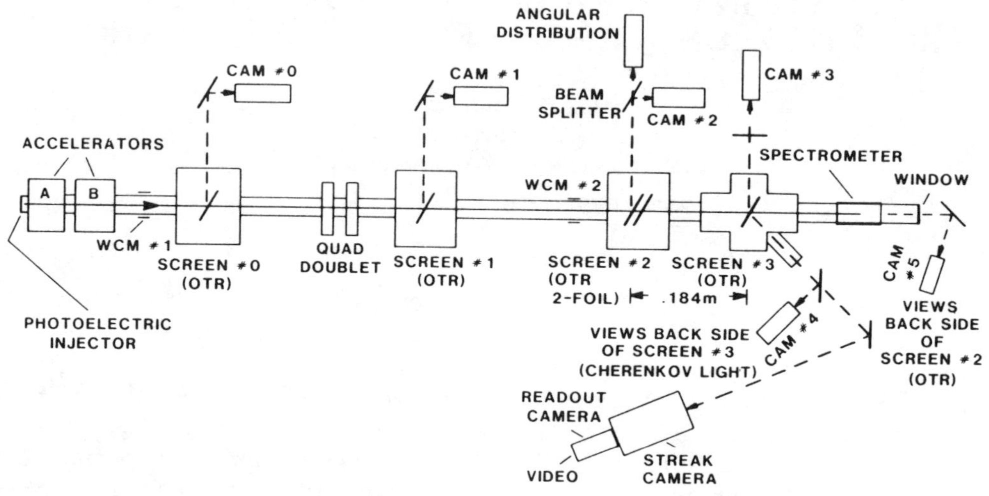

Fig. 1. Schematic of the accelerator beamline and diagnostics.

successful transport of electrons on only part of the rf phase, the partial sampling by the streak camera's entrance slit of the Cherenkov image, and the overlap of two pairs of time doublets via the display of deflections from both sides of the 108-MHz rf deflection in the streak camera. The latter effect was further proven when a change of 42 ps on the Narda phase shifter in the line between the rf source and the streak's synchroscan input moved the doublets in opposite "time directions."

Figure 2 shows the simultaneous imaging of PEI and field-emission injector (FEI) electrons with the synchroscan-streak camera. In this case, the drive-laser phase was at 20° to zero rf phase, and the FE electrons appear about 30 ps later and

extend for ~30 to 40 ps (there is, thus, a several percent-energy spread). Due to transit-time effects in the first cell, the FE electrons can be generated from 40 to 90° of phase and arrive in this temporal window. In this case, beam transport seems to have included more of the differently phased electrons than in the March 20 data with the "doublet" structure.

Verification of the importance of the photocathode material's presence is illustrated in Figs. 3 and 4. In Fig. 3a, the quadrupole focus was adjusted to preserve the PEI electron beamlet pattern from the photocathode on the downstream screen. The same transport then allows the "imaging" of the background emission source distribution in Fig. 3b when the drive laser is blocked and higher camera gain is used. The swirling scene on the source (Fig. 3b) may be due partly to machining grooves on the MOLY substrate of the photocathode. Such edges could lead to enhanced local field gradients and, hence, emission of electrons from the lower work-function photocathode. Figure 4 shows an even more graphic result of the source before (upper) and after (lower) the photocathode material was baked off the plug. The halo remaining is larger than the plug diameter, and the camera gain had to be increased to see it. The absence of electrons from the central region is evidently because the photocathode material had been removed. Some photocathode material may be on the accelerator cell wall around the plug. The next day we also pulled the PC back about 2 mm in its slot, and the FE source strength decreased dramatically (10-20) with the reduction in rf field.

Subsequently, we performed a cross-comparison of integrated intensities under field-emission

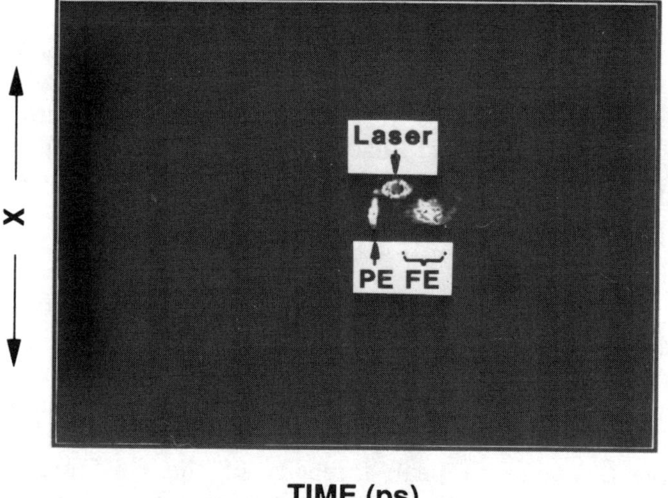

Fig. 2. "Simultaneous" synchroscan image of the drive laser, (PE) and FE sources.

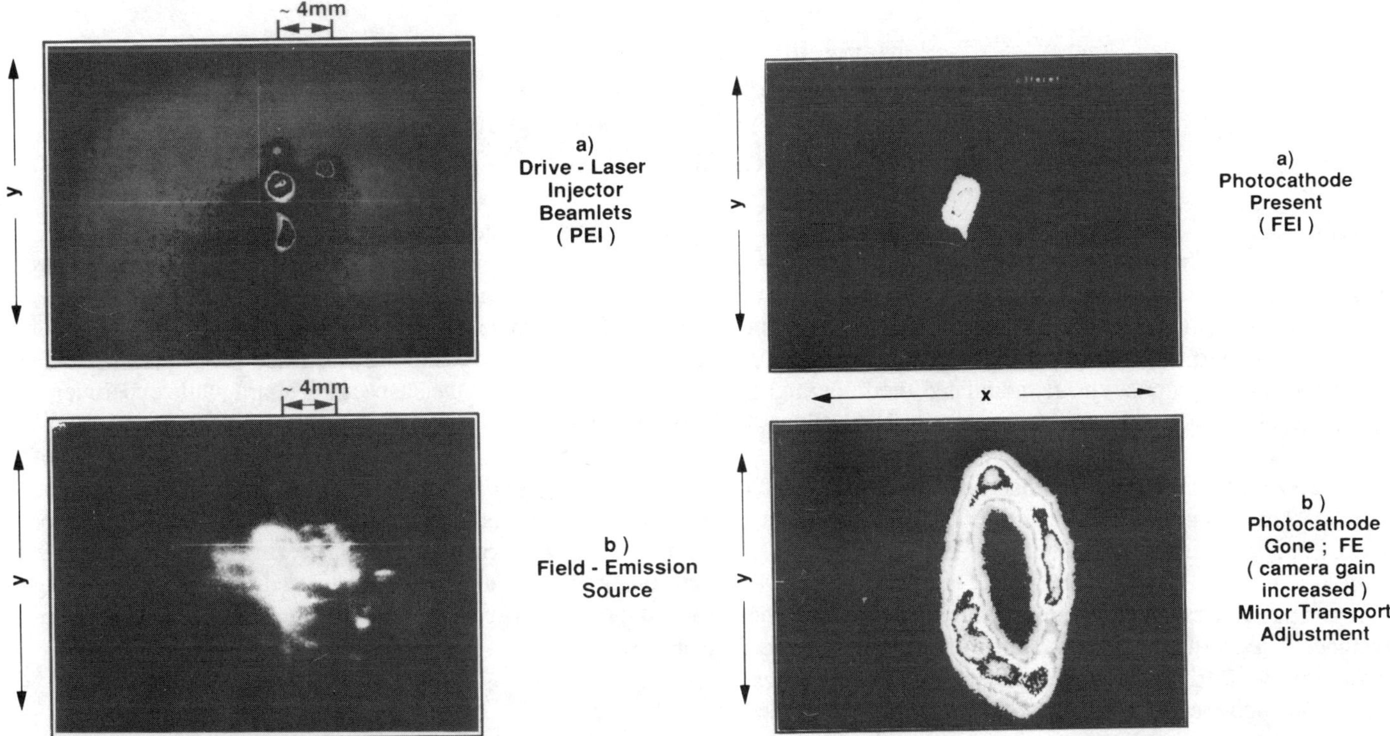

Fig. 3.   Beam transport adjusted to preserve (a) PE beamlet pattern and (b) FE source distribution.

electron images and PEI electron-beam spot images and referenced the charge via a WCM. This comparison implied the FEI emission source strength of about 2.2 nC per microsecond of rf power for a field gradient of 26 MV/m. A plot of FE-electron intensity versus accelerator A field gradient showed an almost exponential dependence.   It should be noted that one was only measuring those electrons transported through both accelerators and to this particular diagnostic station.

IV. SUMMARY

In summary, the background source of electrons in our photoinjector has been characterized. These electrons are generated from the photocathode material, depend nonlinearly (exponentially) on the injector accelerator field, depend linearly on the injector accelerator field duration, depend on the mechanical roughness of the photocathode substrate, and had a temporal extent of tens of picoseconds in our tests. Further experiments are planned at 6 MeV and ~40 MeV on our facility. This phenomenon is sufficient in strength to interfere with single micropulse beam parameter studies but not single macropulse studies. It would

Fig. 4.   Images (via OTR) of the photocathode source (a) before and (b) after bake off of the CsK_2Sb.

need to be addressed in PEI applications with high-duty factor rf power.

V. ACKNOWLEDGMENTS

The author acknowledges the assistance of Los Alamos National Laboratory staff Renee Feldman, Don Feldman, Pat O'Shea, and Scott Apgar in various aspects of this preliminary work.

VI. REFERENCES

[1] J. S. Fraser and R. L. Sheffield, IEEE J. of Quant. Elect., Vol. QE-23, p. 1489, September 1987.

[2] A. H. Lumpkin, et al., NIM. A 296, p. 769, 1990.

[3] Alex H. Lumpkin, Bruce E. Carlsten, and Renee B. Feldman, "First Measurements of Electron-Beam Transit Time and Micropulse Elongation in a Photoelectric Injector," in *Proceedings of the 12th International FEL Conference*, Paris, France, September 17-21, 1990.

An Rf Modulated Electron Gun Pulser For Linacs

Robert Legg and Robert Hartline

Maxwell Laboratories, Inc.

Brobeck Division

4905 Central Avenue

Richmond, California 94804

Abstract

Present linac injector designs often make use of sub-harmonic pre-buncher cavities to properly bunch the electron beam before injection into a buncher and subsequent accelerating cavities. This paper proposes an rf modulated thermionic gun which would allow the sub-harmonic buncher to be eliminated from the injector. The performance parameters for the proposed gun are 120 kV operating voltage, macropulse duration-single pulse mode 2 nsec, multiple pulse mode 100 nsec, rf modulating frequency 500 MHz, charge per micropulse 0.4 nC, macropulse repetition frequency 10 Hz (max).

The gun pulser uses a grid modulated planar triode to drive the gun cathode. The grid driver takes advantage of recently developed modular CATV rf drivers, high performance solid state pulser devices, and high-frequency fiber optic transmitters for telecommunications. Design details are presented with associated SPICE runs simulating operation of the gun.

I. Introduction

The proposed electron gun pulser, shown in figure 1, is designed around the Eimac Y796 gun cathode used by SLAC. It is designed to allow high performance gun operations without inordinate cost or maintenance requirements. It is particularly suited to those resquirements of an electron storage ring injector. To accomplish these goals performance of the gun itself is improved by the superposition of rf modulation on the gun cathode pulse to shorten the micropulse pulse width before entrance into the accelerator. This improves bunching and results in less lost charge in the injector.

To reduce maintenance time and expense, off-the-shelf, modular CATV amplifiers are used. The design philosophy has been to shorten down time by using inexpensive modular parts instead of custom or semi-custom PC boards. Modules are replaced in the case of a failure and fixed off-line, rather than trying to trouble-shoot the suspect PC board while experiments wait for beam.

The use of modular components also allows the upgrade of the pulser as higher performance pulsers and amplifiers become available from component manufacturers without the need for extensive reengineering of the pulser. This makes the upgrade path much cheaper and prevents the end user from becoming locked into one pulse format dictated by custom pulser components.

II. Pulser Design

The proposed electron gun is of the standard thermionic variety. It uses a dispenser type, 2 cm^2 grounded grid cathode. The cathode and grid are biased at -120 kV with respect to a flow-through anode. The cathode is biased positively with respect to the grid to control emissions. The cathode is then pulsed with a negative voltage of approxiamately 150 volts to accelerate electrons toward the more positive grid. Most of the electrons pass through the grid and accelerate toward the much more positive anode and into the accelerator.

The pulser which drives the cathode is a high-frequency planar triode operating in a switched grid mode. The cathode of the triode is biased at -1 to -2 kV, with the grid biased at -75 volts with respect to the cathode to cut off emission. The grid is driven by three sources; a 500 MHz amplifier, a 100 volt, 100 nsec wide pulser, and a 100 volt, 1 nsec wide pulser. This second pulser allows single micropulses to be accelerated which allows a ring to stack beam into a single rf bucket.

The rf drive is superimposed on the grid bias but is of insufficient amplitude to put the triode into forward conduction. When the long and short pulsers are triggered the tube is put into conduction with the rf signal superimposed on the grid voltage acting to "pre-bunch" the beam. The rf reference and the two pulser triggers are brought up to the deck on fiber optic cables and provide direct drive for the pulsers and rf amplifier, as shown in figure 2.

The feed from the planar triode to the gun structure is a tapered impedance transmission line. This allows the output impedance of the planar triode to couple efficiently to the lower impedance of the gun. The bias for the cathode is provided by several coaxial resistors with bypass capacitors.

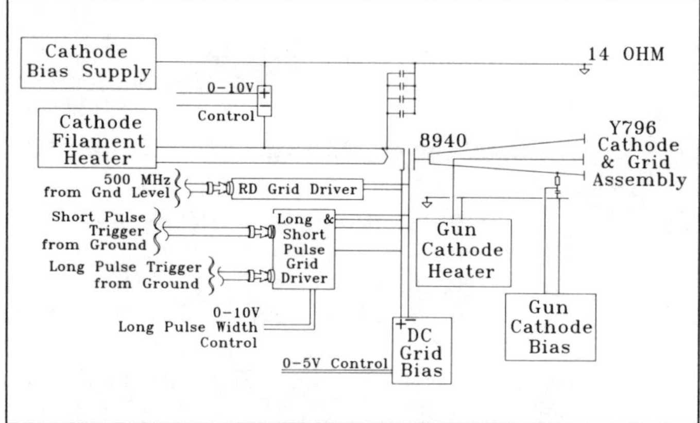

Figure 1. Electron gun pulser.

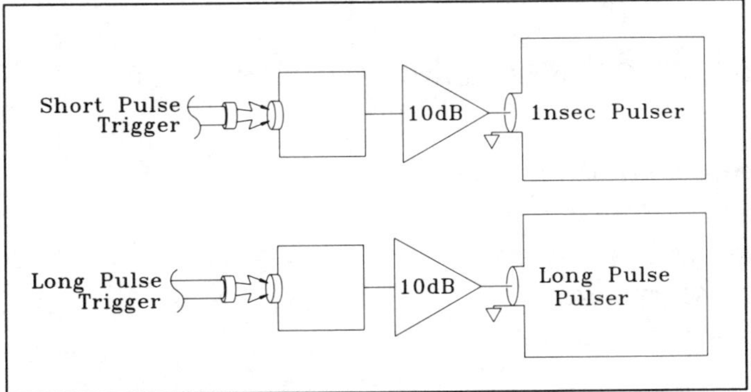

Figure 2. Fiber optic triggers.

Performance for the device is predicted from manufacturers data and Spice modeling of the system. The triode grid voltage, figure 3, demonstrates the typical pulse format to the gun for a long pulse. The output current for the planar triode with this amount of grid drive is about 3 amps according to Eimac. This would equate to about 3 nC per micropulse output from the gun. The nominal operation level for the gun is 0.3 nC per micropulse, so the triode grid and gun cathode are biased to optimize micropulse shape at the expense of charge. In this way, micropulses as short as 1 nsec FWHM can be produced. The output for the single micropulse mode is shown in figure 4.

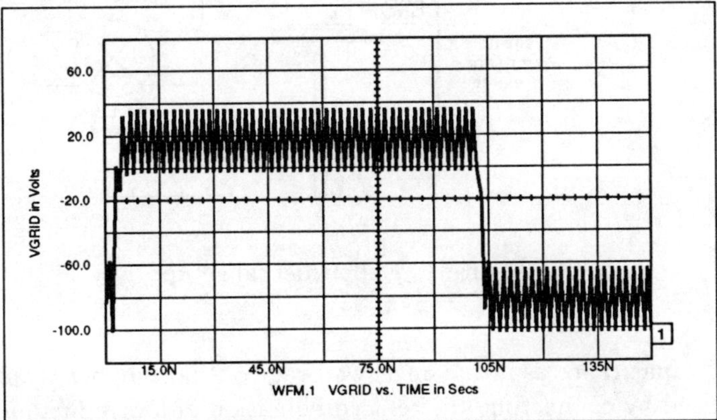

Figure 3. Triode grid voltage.

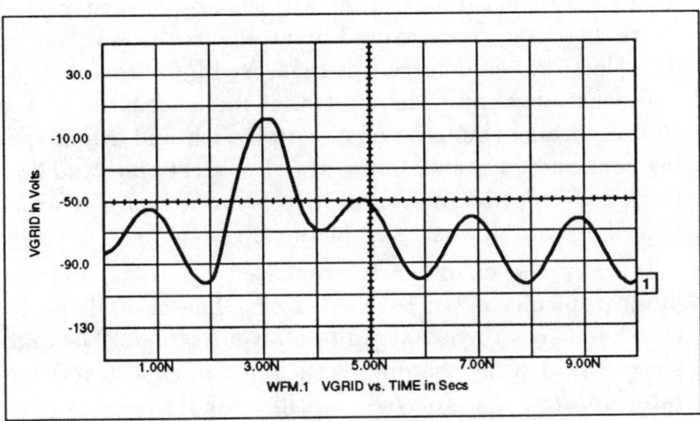

Figure 4. Output in single micropulse mode.

The gun has a second mode of operation, which sacrifices beam quality for quantity. In this mode, output charge is optimized at the expense of micropulse pulse width. The gun deck voltage can be increased to -150 kV to stiffen the beam and reduce space charge effects. Beam quality is severely degraded because of the looser "bunching" from the gun, which translates into higher emittance in the injector output. This mode is for operations where the beam is used to strike a target and the secondary emissions are of interest — for example, as a gamma source for radiation effects testing or a positron source for a medical linac.

II. CONCERNS

The major concerns with the design are two fold. One is the concern over insufficient transient protection of the sensitive pulser components. Electron guns do arc as much as designers wish they wouldn't and the destruction of components when they do must be minimized. The second concern is the uncertainty in modeling very high frequency structures using a discrete element model. There is always the fear that a stray or distributed element will interfere with operation of the circuit.

To minimize the effect of gun arcs it is necessary to control the current path taken by the arc. Almost all gun arcs initiate on the gun grid and progress to the gun cathode. To control the initial arc the return for the gun grid is made slightly inductive to slow down transients which occur during the arc. These inductances do not effect normal operation of the gun as very little grid current flows during normal operation, but they slow down the transient enough to allow overvoltage gaps to breakdown and protect the pulsers from the transients. Chokes to isolate common mode signals are also installed on the pulser module power feeds.

An example of an effect which is not accounted for in discrete element models is dynamic input capacitance between the grid and cathode of the planar triode. This capacitance can spoil the risetime of the pulse train. It is determined by the tube gain, the load impedance, and the anode to grid capacitance in the triode. The effect is overcome by driving the grid with a low impedance source which can sink as well as source current.

III. CONCLUSIONS

An electron gun pulser using modular components is described. The design is easy to maintain and upgrade because it is not dependent on a single custom gun assembly. The pulser does take advantage of recent advances in fiber optic telecommunication hardware, cable television amplifiers, and solid state pulsers to improve overall performance.

Scaling Study of Pseudospark Produced Electron Beam*

K. K. Jain,[†]B. N. Ding,[‡]and M. J. Rhee
Laboratory for Plasma Research and
Electrical Engineering Department
University of Maryland
College Park, Maryland 20742

Abstract

The characteristics of the breakdown voltage of a pseudospark device, the current, and the rms emittance of electron beam produced in the device are determined. It is found that the breakdown voltage is a function of the product of the gas pressure squared and the anode-cathode gap distance; the electron beam current increases with the breakdown voltage up to 35 kV and then decreases; the rms emittance appears to be nearly constant up to the breakdown voltage 25 kV and then increases.

I. Introduction

Recently, the pseudospark discharge[1-8] has gained considerable attention because of its interesting discharge characteristic[2-4] as well as its capability of producing a high-quality electron beam.[5-8] It is imperative to determine the characteristics of the breakdown voltage, the electron beam current, and the beam quality of the pseudospark device not only for a better understanding of the discharge but also for the scaling of further applications.

In this work, we report least-squares-fit analyses of breakdown voltages of a multigap pseudospark device, peak currents and rms emittances of the electron beam produced in the device. The results show that the breakdown voltage is a function of p^2d; the electron beam current increases up to 35 kV and then decreases and it increases linearly with the capacitance; the rms emittance appears to be nearly constant up to 25 kV and then increases.

II. Experiment

The experimental setup is shown in Fig. 1. The discharge chamber consists of a planar cathode with a hollow cavity, a number of sets of intermediate electrodes and insulators, and a planar anode. A center hole is present through the entire electrode system. The anode-cathode gap distance is varied by employing different number of intermediate electrode sets (each set is 6.4 mm thick). The

*This work was supported by the U.S. Department of Energy.

†Present address: Institute for Plasma Research, Bhat, Gandhinagar-382 424, India

‡Permanent address: China Academy of Engineering Physics, P.O. Box 523-56, Chengdy, Sichuan, China

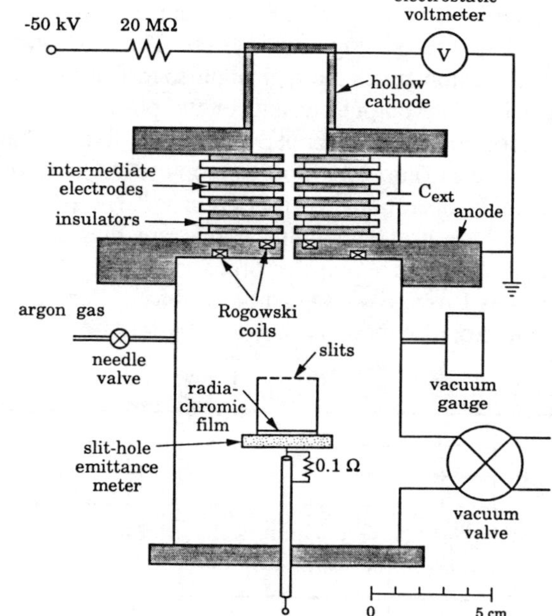

Figure 1: Experimental setup.

capacitance as initial energy storage of the chamber is varied by connecting different combination of low inductance type capacitors. Argon gas is used in this experiment and the pressure is measured by a capacitance-manometer type vacuum gauge. The cathode side of the chamber is charged up to -50 kV through a 20 MΩ charging resistor. The charging voltage is measured by an electrostatic voltmeter. The electron beam current is measured by a Rogowski coil and a Faraday cup. The emittance meter consists of a series of parallel thin slits of 200 μm width and 2 mm spacing constructed from 0.5 mm stainless steel plate and 2 mil thick radiachromic film used as a beam detector, which is placed 10 mm downstream of the slit plane.

Initially the chamber is evacuated by an oil diffusion pump typically down to 10^{-5} Torr. The cathode side of the chamber is then charged to a given voltage. The argon gas pressure of the chamber is then increased at a very slow rate, $dp/dt \approx 1$ mTorr/sec, until breakdown takes place. For the given voltage, the pressures at which the breakdown takes place and the waveforms of the electron beam current produced are recorded for a wide range of experimental parameters: voltage, anode-cathode gap distance, and capacitance.

III. THE BREAKDOWN VOLTAGE

In order to determine the functional dependence, the measured pressures are least-squares fitted to a regression plane given by a simple two variable function

$$p(d, v) = a v^b d^c \qquad (1)$$

where p, v, and d are the pressure in Torr, voltage in kV, and the anode-cathode gap distance in mm, respectively, and the coefficients a, b, and c are to be determined.

The least-squares fitting is done by employing a numerical method, called the grid search, described in Ref. 9. The algorithm of the method includes the following steps: Starting with the initial values of the coefficients a, b, and c, chi-square is evaluated as $\chi^2 = \sum [p_i - a v_i^b d_i^c)]^2 / (N - n - 1)$ for the entire data set, where $N = 250$ is the total number of data points and $n = 3$ is the number of coefficients. The value of a is then increased (or decreased) by a small amount and χ^2 is again evaluated to compare with the previous value. This step is repeated until χ^2 reaches its minimum. The same procedure is repeated for the coefficients b and c. All of the above steps are iterated many times until χ^2 converges to a minimum value at which point the values of a, b, and c for the best fit are obtained.

The results of this numerical analysis are summarized in the following: $a \pm \sigma_a = 0.6851 \pm 0.0003$, $b \pm \sigma_b = -0.2252 \pm 0.0014$, $c \pm \sigma_c = -0.5012 \pm 0.0016$. The regression plane with the least-squares-fit coefficients is depicted by the solid lines and plotted together with the data points in Fig. 2. It should be noted that the choice of p as the dependent variable in Eq. (1) instead of v (or d) is because of the convergence speed in the numerical calculation. Equation (1) is rewritten in a more traditional form,

$$v(p, d) = \alpha / (p^\beta d)^\delta \qquad (2)$$

where $\alpha = 0.1865 \pm 0.0019$, $\beta = 1.9952 \pm 0.0064$, and $\delta = 2.226 \pm 0.016$ are found from the values of coefficients a, b, and c. It is interesting to note that the value of β is very close to 2 which is well within the error range. This suggests that the breakdown voltage is a function only of the product, $p^2 d$. Other results[10] show a dependence of $p d^{0.58}$, which seems to closely support this $p^2 d$ dependence.

IV. THE ELECTRON BEAM CURRENT

The peak current of the electron beam measured with the anode-cathode gap distance of 38.4 mm is least-squares fitted to a function

$$I_p = (a_1 + a_2 C) \exp[-(V - a_3)^2 / 2 a_4^2], \qquad (3)$$

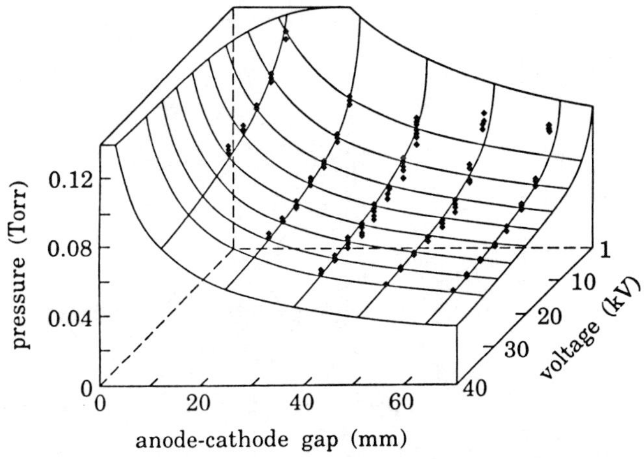

Figure 2: Measured pressures and the least-squares-fit regression plane.

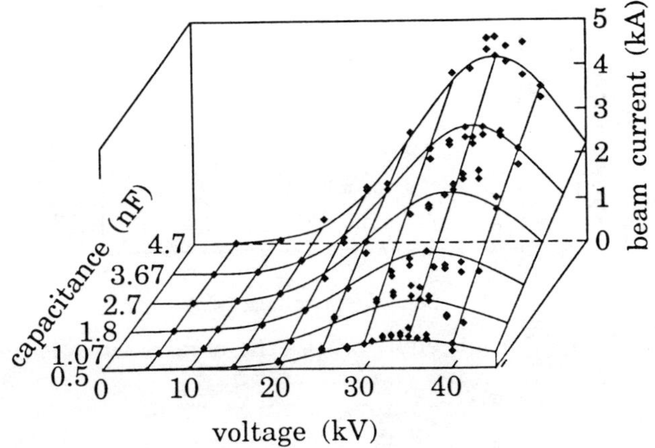

Figure 3: Measured peak currents and the least-squares-fit regression plane.

where C is the storage capacitance (nF), V is the breakdown voltage (kV), a_1, a_2, a_3, and a_4 are the coefficients to be determined. The same numerical method described above is used. The coefficients for the best fit are found to be $a_1 = 213.6$ (A), $a_2 = 842.4$ (A/nF), $a_3 = 34.76$ (kV) and $a_4 = 9.05$ (kV). The resultant regression plane and data points are shown in Fig. 3.

V. THE RMS EMITTANCE

For the emittance measurement, exposure of approximately ten consecutive beam pulses to a radiochromic film through the slits in the emittance meter is made. An attempt is made to scale the emittance with the breakdown voltage by measuring the emittance at several different breakdown voltages. Typical optical density profile of the radiochromic film after exposure is shown in Fig. 4. It is very reasonable to assume that as in Ref. 11, the beam produced in this experiment is axisymmetric and of Maxwellian transverse velocity distribution. This allows us to use the simple slit-hole type emittance meter, whose results can be easily analyzed.[11] The emittances vs the breakdown voltages are plotted in Fig. 5.

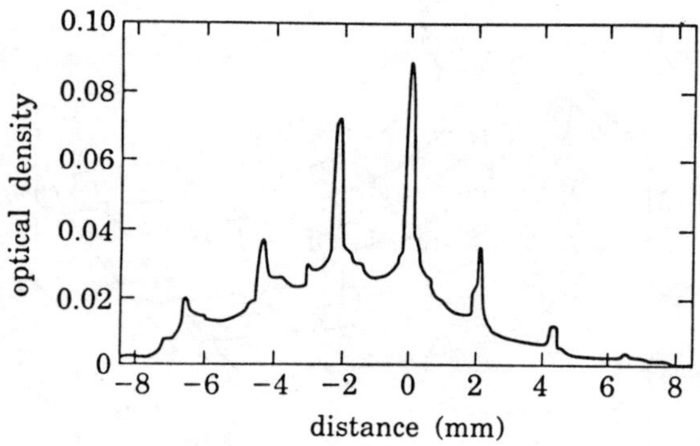

Figure 4: Typical optical density profile of the radiachromic film.

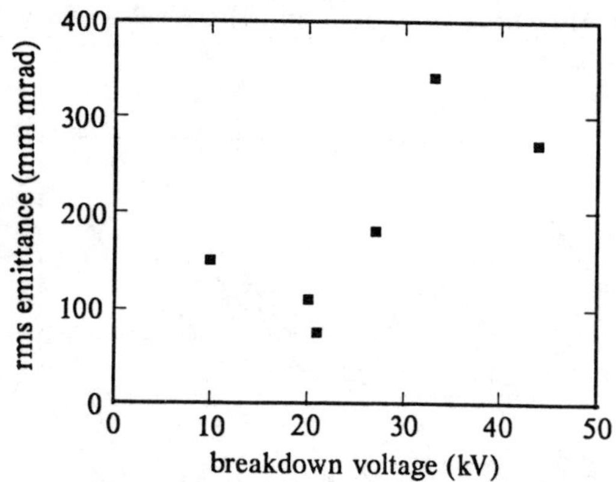

Figure 5: The rms emittance vs the breakdown voltage.

V. CONCLUSIONS

In conclusion, the breakdown voltage of a pseudospark device, the peak current and rms emittance of the electron beam produced in the device have been measured for a wide range of gas pressure, anode-cathode gap distance, and storage capacitance. The measured data are analyzed by a least-squares-fit method. It is found that the breakdown voltage is a function of the product of the gas pressure squared and the anode-cathode gap distance; the electron beam current increases linearly with the capacitance and increases with the breakdown voltage up to 35 kV and then decreases; the the rms emittance is nearly constant up to 25 kV and then increases.

V. REFERENCES

[1] J. Christiansen and C. Schultheiss, Z. Physik A **290**, 35 (1979).

[2] D. Bloess, I. Kamber, H. Riege, G. Bittner, V. Bruckner, J. Christiansen, K. Frank, W. Hartmann, N. Lieser, Ch. Schultheiss, R. Seebock, and W. Steudtner, Nucl. Instrum. Methods **205**, 173 (1983).

[3] G. F. Kirkman and M. A. Gundersen, Appl. Phys. Lett. **49**, 494 (1986).

[4] H. Riege and E. Boggasch, IEEE Trans. Plasma Sci. **PS-17**, 775 (1989).

[5] P. Choi, H. H. Chuaqui, M. Favre, and E. S. Wyndham, IEEE Trans. Plasma Sci. **PS-15**, 428 (1987).

[6] M. Hobel, J. Geerk, G. Linker, and C. Schultheiss, Appl. Phys. Lett. **56**, 973 (1990).

[7] E. Boggasch and M. J. Rhee, Appl. Phys. Lett. **56**, 1746 (1990).

[8] K. K. Jain, E. Boggasch, M. Reiser, and M. J. Rhee, Phys. Fluids B **2**, 2487 (1990).

[10] K. Frank, Physics and Applications of Pseudosparks, edited by M. A. Gundersen and G. Schaefer, (Plenum, New York, 1990), p. 15.

[9] P. R. Bevington, Data Reduction and Error Analysis for the Physical Sciences, (McGraw-Hill, New york, 1969), p. 180.

[11] M. J. Rhee and R. F. Schneider, Part. Accel. **20**, 133 (1986).

Low-Emittance Uniform-Density Cs$^+$ Sources for Heavy Ion Fusion Accelerator Studies*

S. Eylon, E. Henestroza, T. Garvey, R. Johnson and W. Chupp

Lawrence Berkeley Laboratory,
University of California
Berkeley, California 94720

ABSTRACT

Low-emittance (high-brightness) Cs$^+$ thermionic sources were developed for the heavy ion induction linac experiment MBE-4 at LBL. The MBE-4 linac accelerates four 10 mA beams from 200 keV to 900 keV while amplifying the current up to a factor of nine. Recent studies of the transverse beam dynamics suggested that characteristics of the injector geometry were contributing to the normalized transverse emittance growth. Phase-space and current density distribution measurements of the beam extracted from the injector revealed overfocusing of the outermost rays causing a hollow density profile. We shall report on the performance of a 5 mA scraped beam source (which eliminates the outermost beam rays in the diode) and on the design of an improved 10 mA source. The new source is based on EGUN calculations which indicated that a beam with good emittance and uniform current density could be obtained by modifying the cathode Pierce electrodes and using a spherical emitting surface. The measurements of the beam current density profile on a test stand were found to be in agreement with the numerical simulations.

I. INJECTOR DESCRIPTION AND CHARACTERISTICS

Four 10 mA Cs$^+$ ion beams are emitted thermionically into the diode gap from alumino-silicate layers coated on molybdenum cups[1]. The cups are mounted on a Pierce shaped graphite plate anode connected to a -200 kV Marx pulse generator. The beams, focused along the diode, pass through four holes in the cathode ground plate into the MBE-4 matching section. The ion source phase-space and current density distributions were determined experimentally using a pinhole and slit combination coupled to a Faraday cup. The pinhole was placed as close as possible to the ion source exit at a position where the beam is still cylindrically symmetric. Figure 1 (lower) shows the agreement between the measured beam phase-space distribution and the result of an EGUN[2] simulation for a zero emittance beam. One can see that the outermost rays of the beam are turned toward the diode axis due to overfocusing as a result of the diode field aberrations. This leads to current accumulation at the beam edges resulting in a hollow beam (figure 2, lower). Recent studies[3] have suggested that the source properties may contribute to the measured emittance growth for the 10 mA accelerated beam. A quick solution to improve the source beam dynamics by scraping

the beam current to 5mA was temporarily adopted. Along with this solution we have developed a new improved 10 mA source.

II. THE 5 mA SOURCE

A simple way of eliminating the outermost beam rays is to use a circular aperture ring (scraper) placed at the cathode plate, thus stopping down some of the beam current. Faraday cup measurements of the current taken at the entrance to the MBE-4 matching section initially showed a deteriorated current waveform. This appeared as an increase in the current rise-time and the total beam current. This current increase may be due to additional secondary electrons emitted from the scraper and accelerated towards the anode. To re-capture these secondary electrons the scraper was biased with a positive 4.5 kV d.c. voltage. Beam current waveforms at various bias voltages are shown in Fig. 3. The scraper is recessed 0.8" from the diode exit to eliminate electric field effects due to the bias voltage. The beam phase-space and current density distributions for various scraper diameters (0.4", 0.45", and 0.5") were measured and found to be in agreement with EGUN simulations. An optimum scraper diameter of 0.45" was chosen leading to a current of 4.5 mA and a higher quality beam as shown in Fig. 4. Transverse emittance measurments along MBE-4 using the high quality 4.5 mA beam revealed a normalized transverse r.m.s. emittance of 0.03 mm-mrad, reasonably consistent with the 0.1 eV temperature of the source emitter. Furthermore, the 4.5 mA beam showed a significant reduction in the variations of the measured r.m.s. transverse emittance in the MBE-4 matching section as compared to the 10 mA beam (figure 5).

III. THE 10 mA IMPROVED SOURCE

The new 10mA source consists of a curved ion emitter, a modified Pierce electrode (graphite plate) and a new mechanical assembly. EGUN simulations of the new improved source show a uniform current density and a lower emittance for a beam emerging from the new diode and scraped with a 20 mm aperture to 10mA at the input to the MBE-4 matching section. Improved fabrication techniques were developed and used to produce the new curved emitters. The performance of the new emitters was evaluated in a test stand using a diode configuration with an anode-cathode spacing of 0.5" and voltages up to 20kV. Figure 6 shows a temperature profile taken on the emitter surface along the diameter and Faraday cup current density through crossed slits positioned 0.9" from the cathode grid. The current density measurements were found to be in agreement with EGUN simulations for the

*Work supported by the Office of Energy Research, Office of Basic Energy Sciences, U.S. Department of Energy. Contract DE-AC03-76SF00098.

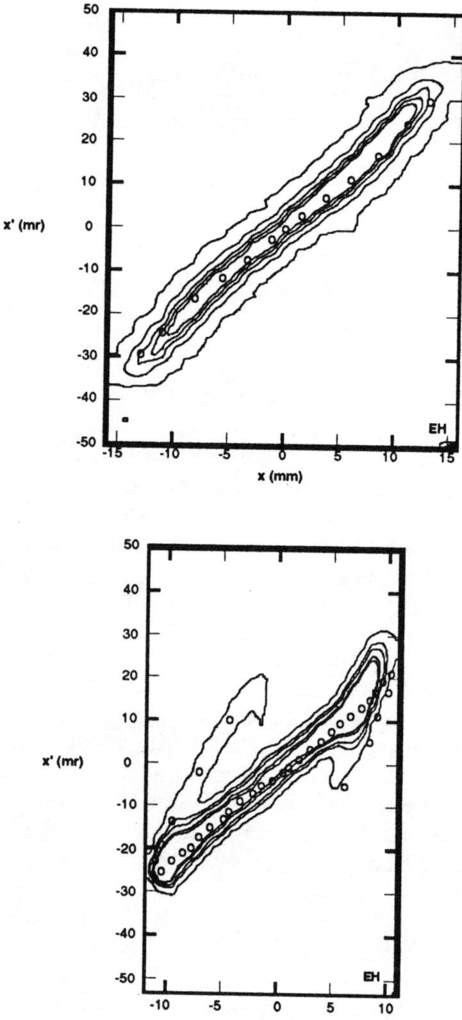

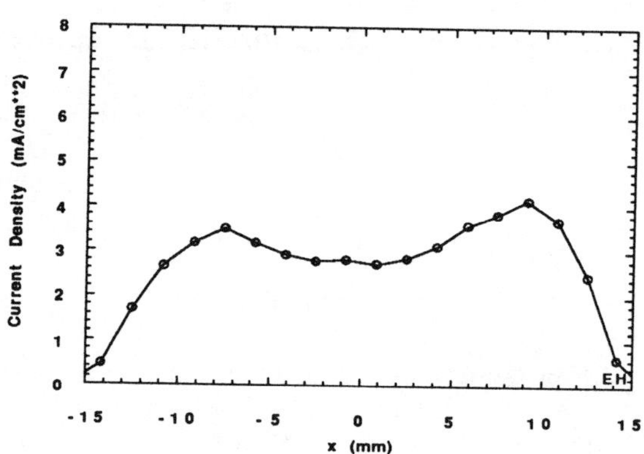

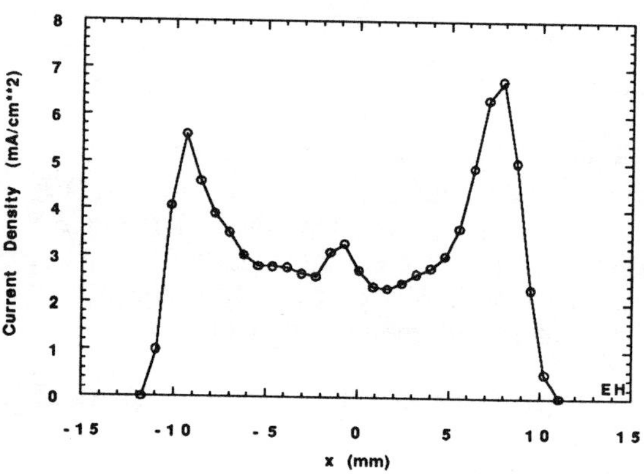

Fig. 1 Measured beam distribution in (x,x') phase space at diode exit. Comparison between experiment (-) and EGUN (o). Upper (lower) figure refers to the new (old) source.

Fig. 2 Measured beam current density distribution vs. radial position. Upper (lower) figure refers to the new (old) source.

same geometry , thus approving the emitter for further tests in the MBE-4 ion source. The new mechanical assembly, which stabilizes the diode geometry (especially when going from air to vacuum) and improves the emitter alignment, was installed in the MBE-4 ion source section. The new mechanical assembly uses four insulating posts connecting the graphite plate (anode) to the output aperture plate (cathode). The mechanical stabilty of the diode was surveyed using optical telescopes placed at the far end of MBE-4 and set to have a line of sight coinciding with the MBE-4 axis. To perform the survey we have replaced the left and right-hand sources with "targets" containing a point source of light. A displacement of less than 0.007" was measured when the system went from air to vacuum. The assembly was tested for high voltage breakdown and found to withstand high voltage pulses in excess of 220 kV and 10 micro-second duration applied between the cathode and anode.

The new curved source was placed at the right-hand beam

position and operated to evaluate the diode performance. Close agreement between the measured and the designed diode parameters, as predicted by previous EGUN simulations, was obtained when the source emitting surface was heated up to 1010 °C. The temperature was measured remotely through a window using a hot wire pyrometer. This temperature was found to be well above the diode current saturation temperature with the diode current determined by Child-Langmuir space-charge limit. The diode current, I_d, was measured for diode voltages, V_d, between ~100kV and ~200kV. The measured diode current was found to follow the well known Child-Langmuir law, $I_d = k\, V_d^{3/2}$, where k is the diode perveance, measured to be 2.3 x 10^{-4} μPervs.

The diode beam dynamics were measured using the aperture-slit technique described earlier. Figure 1 (upper) shows the beam phase-space profile. One can see that the optical

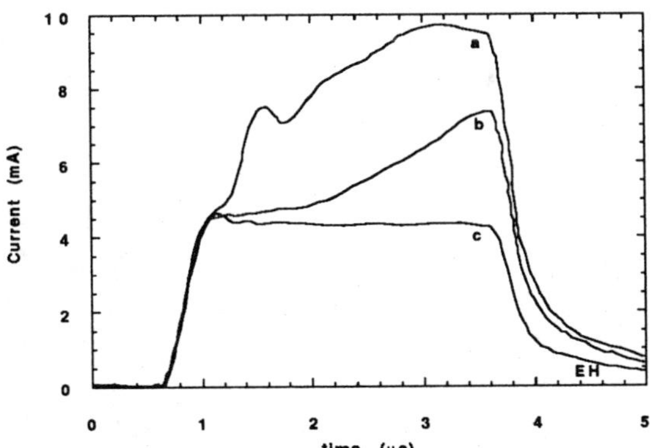

Fig. 3   Scraped beam current vs. time with a scraper bias of (a) 0.75kV, (b) 2.75 kV and (c) 4.75 kV.

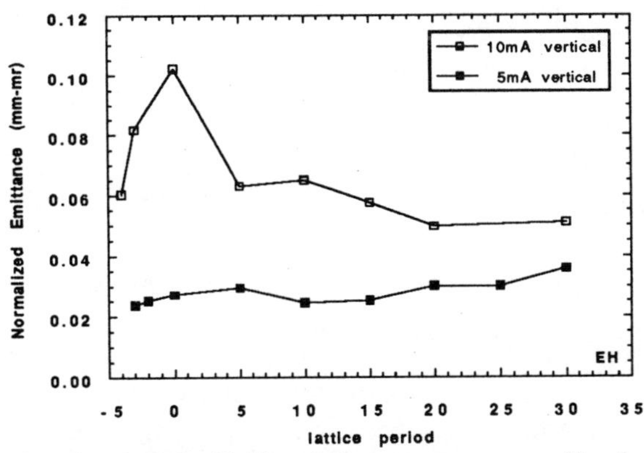

Fig. 5   5 mA and old 10 mA beam r.m.s. normalized transverse emittance vs. lattice period along MBE-4 . The matching section corresponds to lattice periods -4 through 0.

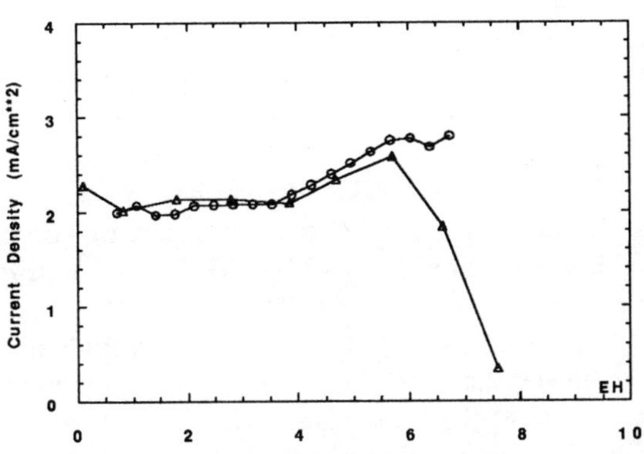

Fig. 4   5 mA beam current density at point of injection into first quadrupole of matching section. Comparison between experiment (Δ) and EGUN code (•).

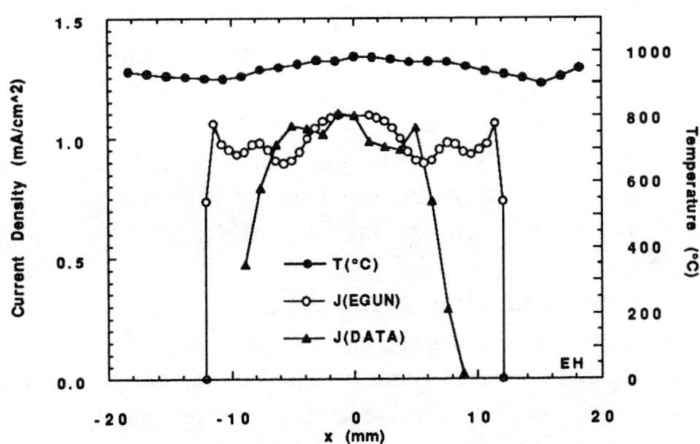

Fig. 6   Test stand measurements of the current density profile J(DATA) compared with EGUN simulation J(EGUN), and the temperature profile taken at the emitter surface.

aberrations in the diode are significantly reduced and that the previous filamentation in the profile is eliminated, thus leading to a more uniform current density profile as shown in Fig. 2 (upper). The full beam current of 16 mA measured using a Faraday cup at the exit of the MBE-4 matching section 1st quadrupole doublet and the beam divergence angle of 30 mrad. taken from the measured phase space profile shown in Fig. 1 (upper) were found to be in good agreement with the parameters predicted by the EGUN simulations. For applications in MBE-4 the beam is apertured to a diameter of 20 mm leading to a beam current of 10.6 mA consistent with the original nominal MBE-4 beam current. The normalized r.m.s. transverse emittance, ε_n, was measured using the double slit technique at the exit of the 1st MBE-4 quadrupole doublet and found to be 0.067 mm-mrad. As the intrinsic emittance of the beam is given by $\varepsilon_n = 2a\,(kT/m_i c^2)^{1/2}$, where m_i is the ion mass, a is the beam radius and T is the beam temperature, we obtain a temperature estimate of 1.4 eV. This is consistent with the source temperature and the adiabatic transverse

compression of the beam by the focusing elements. From the measurements of current and emittance we calculate the beam brightness, $I_b\,/\pi^2\varepsilon_n^2 = 2.4 \times 10^{11} A\,/(m\text{-rad})^2$.

ACKNOWLEDGEMENTS

We are indebted to T.J. Fessenden, L. Smith, and A. Faltens for helpful discussions. We thank D. Vanecek for mechanical design effort and W. Tiffany for technical support.

REFERENCES

[1] A.I . Warwick et al, "A four beam cesium injector for MBE-4", IEEE Trans. Nucl. Sci., NS-32, Oct. 1985, pp3196-3198.
[2] W.B. Herrmannsfeldt, "EGUN - An electron optics and gun design program", SLAC Report - 331, October 1988.
[3] H. Meuth et al., "Study of transverse emittance in MBE-4", LBL report, LBL-27230, December 1989.

Preliminary Design for a Thermionic R.F. - Gun*

G. D'Auria, J. Gonichon, T. Manfroi**
Sincrotrone Trieste
Padriciano 99, Trieste, Italy

Abstract

A high-gradient S-band RF-Gun with a thermionic LaB_6 cathode is now under study at Sincrotrone Trieste for future upgrading of the injection Linac. A geometry of the accelerating structure has been designed using the OSCAR2D code and a prototype has been constructed and measured. The transverse and longitudinal dynamics of the particles along the injector have been simulated with a modified version of the PARMELA code.

I. INTRODUCTION

The proper functioning of the injection system is of fundamental importance in order to have both high current and small emittances. Electron guns, however, have intrinsic limitations when combined with conventional R.F. systems (choppers-bunchers). In the last years special care has been devoted to R.F. gun development which promises to be a solution to these limitations with the production of very bright beams.

The system described here should be able to produce an electron beam for injection into a linac [1]; the beam characteristics should also fit the requirements for a FEL experiment, i.e. low emittance, low energy spread and high peak current. At injection into the Linac, the expected beam characteristics are summarized in table 1.

Table 1.

Average beam energy	1.1 MeV
Micropulse peak current	≥ 20 A
Transverse normalized emittance	$\leq 20\ \pi$ mm mrad
Relative energy spread	$\leq 20\%$

The injector will consist of an RF cavity working in the fundamental mode at 3 GHz provided with a thermionic LaB_6 cathode, an alpha-magnet, and four quadrupoles which focus the beam into the alpha-magnet and the Linac.

II. DESCRIPTION OF THE INJECTOR

Figure 1 shows a schematic diagram of the injector. The RF cavity is powered by a 2.6 MW magnetron (EEV. mod. M5193) operating with a 3 μsec macropulse length at a maximum repetition rate of 100 Hz.

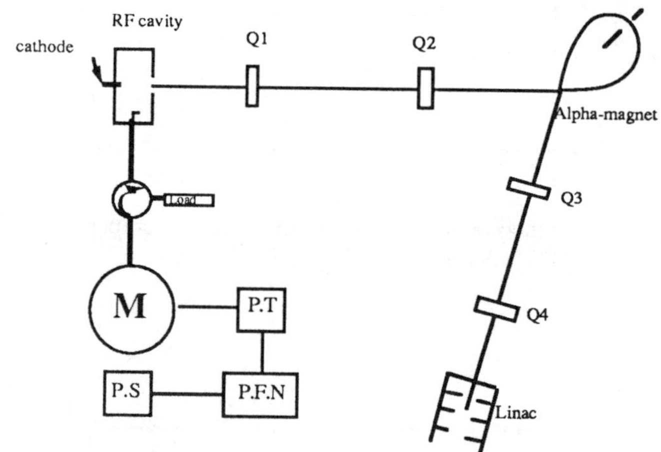

Figure 1. Injector layout.

A 5.2 MW line-type modulator supplies the negative H.V. pulse to the magnetron. The P.F.N. consists of 10 "lumped constant" L-C elements followed by a H.V. pulse transformer; the line is charged by a 20KV-0.1A D.C. power supply. A 3-port circulator is foreseen between the magnetron and the RF cavity. From the 2.6 MW delivered by the magnetron, only 1.8 MW can go to the beam due to the cavity losses and the feeding line attenuation. For an average beam kinetic energy of 1.1 MeV, this implies a maximum current of 1.63 A ($3.38 \cdot 10^9$ electrons per bunch for a 3 μsec pulse length). This current is produced by a 3 mm LaB_6 thermionic cathode heated to 1800-1900 K. The purpose of the alpha-magnet system [2] is to reduce the wide energy spread of the beam exiting the cavity and to compress the bunch. Electrons with different energies follow different alpha-shaped trajectories, allowing an energy range to be selected through the use of slits. Moreover, the more energetic electrons (emitted earlier) spend more time in the magnet than the less energetic ones (emitted later), resulting in a phase compression that can be adjusted by varying the magnetic field. Two quadrupoles are placed between the cavity and the alpha-magnet and two between the alpha-magnet and the Linac, allowing the matching of the beam into the alpha-magnet and the Linac, as well as the control of emittance.

III. CAVITY DESIGN AND MEASUREMENTS

The cavity design has been performed using the OSCAR2D code [3,4]. The geometry has been chosen in order to optimize the beam parameters at the cavity exit. The accelerating gradient along the beam axis has been maximized

---

* Work carried out in cooperation with the University of Trieste.
** Dipartimento di Fisica, Università di Trieste.

to accelerate the particles in as short a time as possible, reducing the space charge effects in the low-γ region. The cavity profile is shown in figure 2; figure 3 shows the behaviour of the electric field components along the z-axis for three values of r. The nose geometries are different: an optimum shape has been found for the one nearest the cathode to maximize the E-field in that region and to break the cavity symmetry. Furthermore, the behaviour of the radial electric field is linear [5] as it is shown in figure 4.

A prototype of the cavity has been realized and measured to verify the electric parameters. Table 2 shows the computed and measured values.

Table 2.

	Computed	Measured
Resonant frequency (MHz)	3016.73	3004.907
Q value	12000	9000
R_{SH} (MΩ)	2	1.54
$E_{max.sup}/E_{max.axis}$	1.6	-

IV. DYNAMIC SIMULATIONS

A modified version of PARMELA [6,7] was used to simulate the transverse and longitudinal motion of the particles all along the system. The beam represented by a collection of macro-particles, each located at three spatial and three momentum coordinates, is transformed through the system via numerical integration of the equations of motion, using the phase of the RF field as the independent variable.

At the cathode, the particles are emitted between 0 and 180° (in half a cycle of 3 GHz) according to the Richardson-Schottky law (most of the particles are emitted when $\phi = 90°$, i.e. when the electric field at the cathode is maximum). The distribution of the initial momenta is gaussian, as well as the one for the transverse spatial coordinates.

Just after leaving the cathode, the particles experience the E_z, E_r, and B_ϕ RF fields present in the cavity. About half the particles exit the cavity whereas the other return to hit the cathode. This back-bombardment has the effect of heating the cathode and impedes the maintainment of a constant current during the pulse duration. This problem could be solved by adding a transverse magnetic field near the cathode for a correct FEL operation. In the conditions of operation described in table 3, the average back-bombardment power was found to be 8.2 W (10 pps. for 3 μs pulse).

In the alpha-magnet, the equations of motion for each particle are solved for the position and the momenta in the alpha-magnet coordinates system, and then transformed to the injector coordinates [8]. The magnetic field, as well as the spacing between the two energy slits can be varied. The alpha-magnet further reduces the number of particles by one half, reducing in the same way the emittance and leaves roughly 20% of the total charge at the cathode going into the Linac.

A particle is considered lost when it touches the vacuum chamber wall or the energy slits in the alpha-magnet, or when its longitudinal velocity becomes lower than zero. The space charge forces are evaluated by a full 3 dimensions relativistic point by point calculation: the coulomb interaction between each macro-particles is calculated at every time step. Beam loading and wake field effects are not included in this simulation.

V. RESULTS

The system parameters used for the simulation are described in table 3.

Table 3.

0 cm	Cathode	R = 1.5 mm W = 0.5 eV
		T = 1900 °K
		Total charge = 0.5 nC
0-3.77 cm	Cavity	E_{zmax} = 63 MV/m
		$\overline{E}_z$ = 37 MV/m
		E_{zcat} = 30 MV/m
10.8-15.2 cm	Q1	G = 110 Gauss/cm
22.2-26.6 cm	Q2	G = -90 Gauss/cm
33.8-64.8 cm	Alpha-magnet	G = 140 Gauss/cm
71.8-76.2 cm	Q3	G = 25 Gauss/cm
83.2-87.6 cm	Q4	G = -50 Gauss/cm
93.8 cm	Linac	

Since the space charge calculations are very time consuming ($CPU_{time} \propto$ (number of particle)^2), Parmela has been installed on a very powerful workstation: Hewlett-Packard 9000/834, risc architecture, 17 Mips. This enable us to simulate a single bunch with 300 "macro-particles" taking a computation time of 50 mn. The simulation presented here has been made with a thousand of "macro-particles" (9 h).

In what follows, all the emittances are normalized and defined in the rms sense.

Just after the cavity, it can be seen in figures 5 and 6 that the bunch extends over roughly 200° of phase with an energy varying from 0 to 1.2 MeV. The transverse emittance is less than 10 π mm. mrad. Between the cavity and the alpha-magnet, the bunch becomes wider in phase because of the large energy spread. For the same reason, the transverse emittance growth is very important.

Just after the alpha magnet, the number of particles drops to 225 (23% of the total), and the transverse emittance becomes 9.1 π mm. mrad. (5.6 π mm. mrad) in the x (y) plane. The energy lies between 0.98 and 1.2 MeV, and the phase extension is about 25°.

At the end of the injector (figures 8 and 9), i.e. just before the Linac, we obtain a bunch of roughly 7° long (6.5 ps), with a relative energy spread of 19%, and a transverse normalized emittance of 9.2 (7.6) π mm. mrad. in the x (y) plane. The remaining charge is 112 pC, leading to a peak current of 17 A.

This bunch has been further simulated through the two accelerating sections of the ELETTRA Linac [1]; at 98 MeV the transverse normalized emittance of the beam is 9.7 π mm. mrad. and 11.6 π mm. mrad. in x and y plane respectively. The relative energy spread has been reduced to ± 0.25%. We find 84 pC in 5° leading to a peak current of 18 A.

This peak current is below the expected characteristics, but could be increased by upgrading the power source.

VI. REFERENCES

[1] C. Bourat et al., "The 100 MeV Preinjector for the Trieste Synchrotron", EPAC 1988.

[2] H.A. Enge, "Achromatic Magnetic Mirror for Ion Beams", Rev. Sci. Instrum. BF 23, No. 4, 385-389 (1963).

[3] R. Parodi, P. Fernandes, "OSCAR2D User's Guide", Genoa, November 1989.

[4] T. Manfroi, "Studi su sorgenti di elettroni a bassa emittanza", Degree Thesis, Univ. Trieste, April 1991.

[5] M.E. Jones et al., "Particle-in-cell simulations of the lasertron", IEEE Trans. on Nucl. Sc., Vol. NS-32, No. 5.

[6] User's manual from B.Mouton LAL/SERA 89-356, 1989.

[7] H. Liu, "Particle Generation for Simulation of RF Thermoionic Guns", LAL/ SERA 90-178, June 1990.

[8] H. Liu, "Description about How to Input an Alpha Magnet Card in PARMELA", private communication.

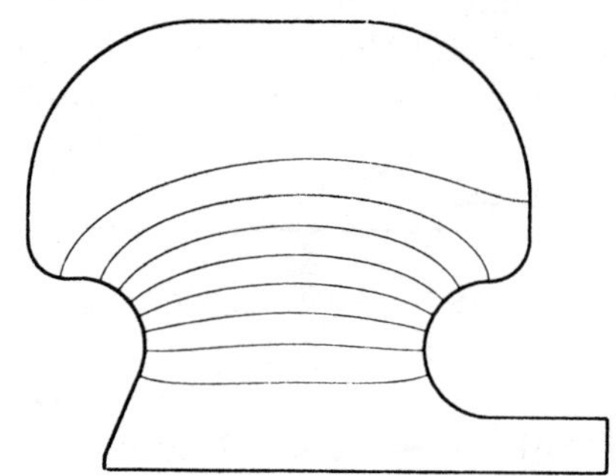

Figure 2. Cavity profile.

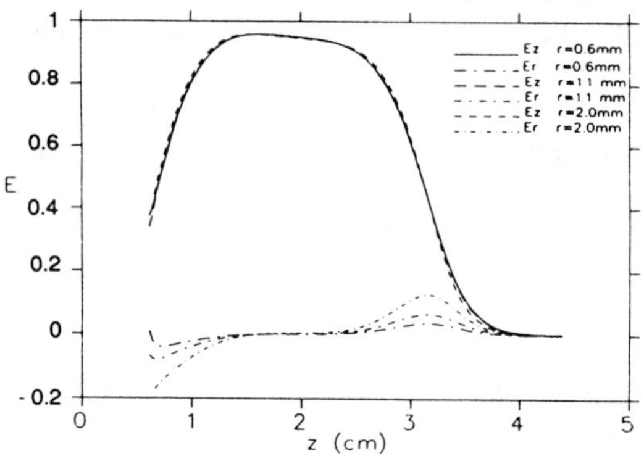

Figure 3. Relative amplitudes of E-field components.

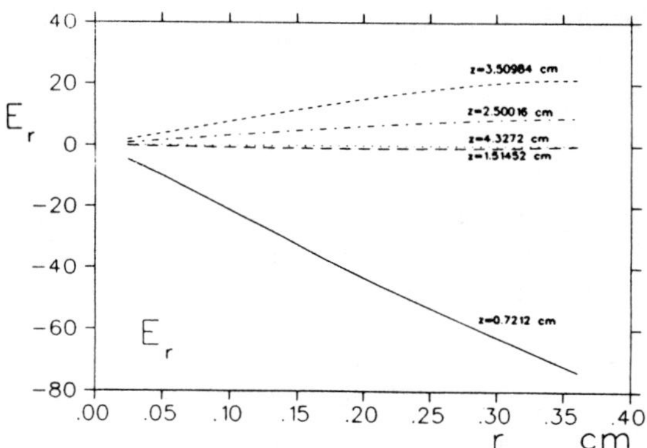

Figure 4. Radial electric field behaviour.

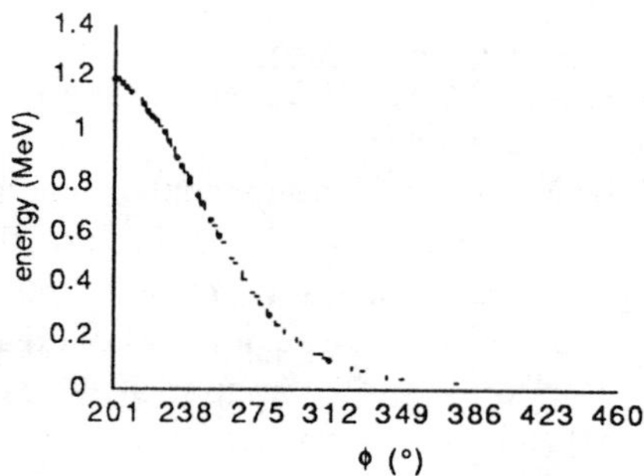

Figure 5. Longitudinal phase space after the cavity.

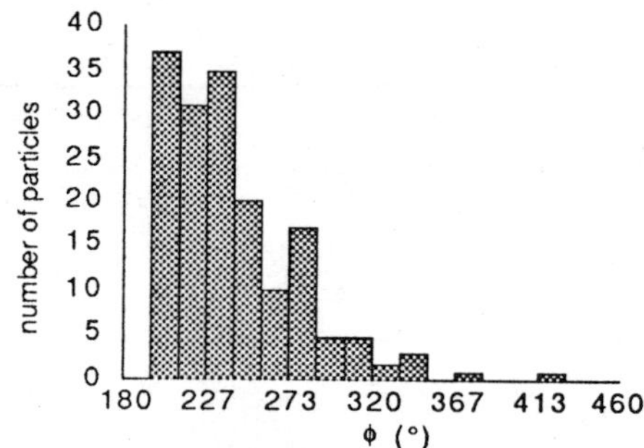

Figure 6. Phase distribution after the cavity.

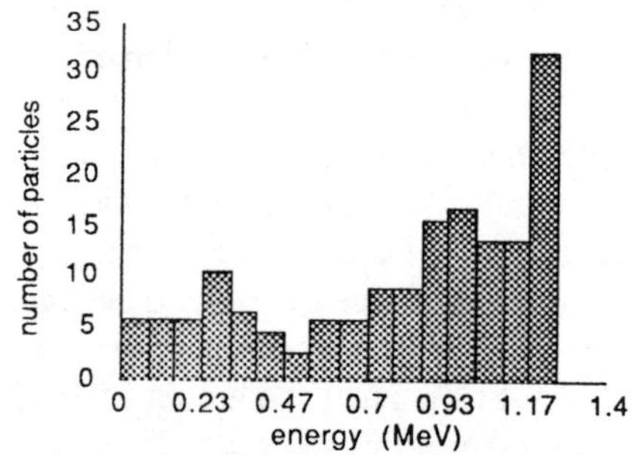

Figure 7. Energy distribution after the cavity.

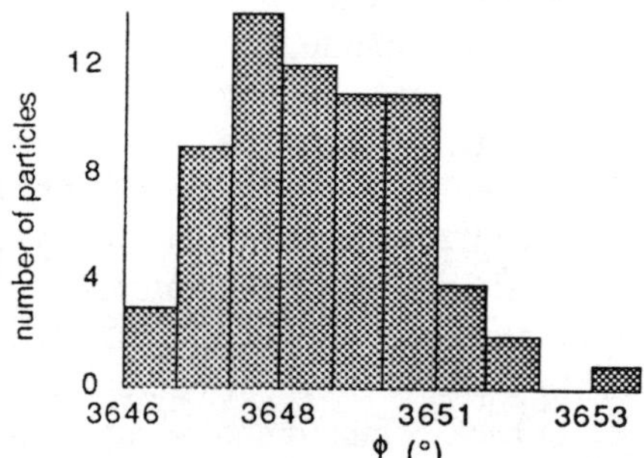

Figure 8. Phase distribution before the Linac.

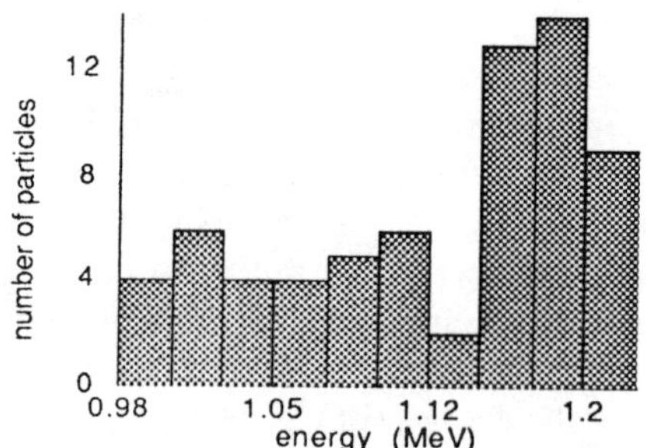

Figure 9. Energy distribution before the Linac.

Applications of Diamond Films to Photocathode Electron Guns and Accelerators

C. P. Beetz, B. Lincoln
ATM, Inc., Danbury, CT
and
K. Segall, D. Wall, M. Vasas, D. R. Winn
Dept. Physics, Fairfield University, Fairfield, CT
and
D. Doering and D. Carroll
Dept. Physics, Wesleyan University, Middletown, CT

Abstract
The unique thermal, chemical, electrical, surface and mechanical properties of synthetic diamond films are discussed for high current photocathode driven electron guns. The potential of diamond as a cold photocathode substrate and direct photocathode are discussed. Novel optic methods to create temporal and spatially profiled electron beams are presented.

I. INTRODUCTION

Compact laser driven electron sources of high brilliance, peak current, and cold temperatures that allow control over the spatial and temporal beam properties are of great interest to accelerator technology[1,2,3,4,5,6]. For example, a linac gun producing a bunch matched to the RF would operate without a buncher; accelerators are generally more efficient with the injected beam matched to the dynamic properties of the electromagnetic fields in the accelerator. Some accelerator techniques demand temporal and spatial properties difficult to achieve because of parasitic electrical properties. For example, in a wakefield accelerator, pulses may need to be separated by less than ~1 ns. Many of these issues can be addressed by an optically driven cathode.

We discuss 3 methods to extend photocathode electron gun technique: (1) a diamond photocathode substrate, both for heat-sinking and for rear illumination of vacuum photocathodes; (2) fiber-optic photocathode drivers to tailor the temporal and spatial properties of electron beams; (3) possibility of a robust, air-cycleable direct diamond film photocathode utilizing the negative electron affinity (NEA) properties of the diamond (111) surface.

II. DIAMOND TECHNIQUES FOR PHOTOCATHODES

A. Diamond Transparent Photocathode Substrate

Major advances have been made in high brilliance, kA/cm^2 electron beams with Cs_3Sb[7,8,9], Cs_3SbK_2[10] and other [11,12,13,14,15,16,17,18] photocathodes. A general schematic of a back illuminated laser driven photoemissive electron gun is shown in Figure 1. The primary deficit of photocathodes is the short photocathode life from thermal effects.[1,6,11] In order to reduce the thermal load and thermal shock on the photocathode, we propose a diamond film as a high thermal conductivity substrate for conventional photocathodes. The thermal conductivity of diamond films is 3-4 times higher than copper[19] at room temperature. Heat is extracted more efficiently from a diamond substrate than from any other material. The photoemission process from the photocathode also leads to thermionic emission because of optical absorption in the photocathode material. The resulting temperature rise can result in irreversible damage to the

photocathode. The temperature rise of the photocathode heated by either a continuous or a pulsed laser is given by:[6]

$$\Delta T \sim 2P'/\pi r k \qquad \text{(continuous illumination)}$$

$$\Delta T \sim W/\pi r^2/(2k\rho C_v\tau)^{1/2} \qquad \text{(pulsed illumination)}$$

where P' is the continuous absorbed power, r is the optical spot radius, k, ρ and C_v are the thermal conductivity, density and specific heat of the photocathode and substrate combination, W is the absorbed pulse energy and τ is the pulse duration. For typical metal substrates, k ~ 1 W/cm°K in the vicinity of room temperature, if r ~ 10^{-4} cm and P' ~ 100 mW, the temperature rise would be on the order of ~ 600°C. Pulsed operation with $W/\pi r^2$ ~ 0.1 J/cm^2 and ρC_v ~ 1 J/cm^3C would lead to a temperature rise of ~ 500°C. Cs_3Sb and Cs_3K_2Sb cannot be used as photoemitters at temperatures above 180°C. If we substitute a polycrystalline diamond film as the photocathode support, where k ~10-20 W/cm°K[19] then the temperature rise is ~x5-x7 lower, and the photocathode should continue to operate. The temperature rise of the photocathode is higher with other nonmetallic transparent substrates (sapphire, etc.). Diamond also has a relatively low dielectric constant ($\varepsilon/\varepsilon_o$=5.5), very high breakdown strength (10 MV/cm), and controllable ρ~10^{13}-10^{-3} Ω-cm (see below).

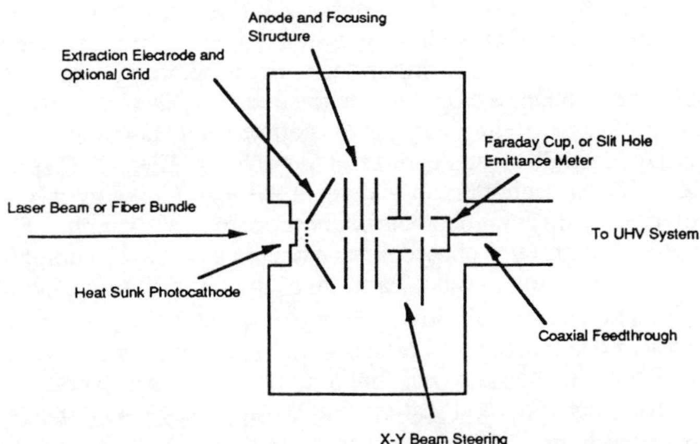

Fig. 1.Schematic of a laser driven photoemissive electron gun.

Moreover, the very low diffusion coefficients of foreign atoms into the diamond lattice, the chemical inertness of diamond surfaces, and the ability to be baked at high temperature may make the cesiation and photocathode processing easier than with competing metal or ceramic surfaces. Even so, cesiation is a complex process[20],[21] and

"

the lifetime of cesiated photocathodes is short, dependent on the intensity, temperature and pressure in the vacuum chamber[6]. Recent studies of Cs3Sb lifetime showed that the optimum operating temperature range is 280 - 290°K and that the lifetime decreased by a factor of >34 on increasing the temperature to 373°K[6]. The lifetime of the Cs3Sb also depends on pressure, decreasing by a factor of ~23 on going from 3×10^{-10} to 1.3×10^{-7} torr.[6] The use of a diamond film substrate for supporting the photocathode addresses the temperature and back-illumination issues only, and does not address the problems associated with (atmospheric) contamination. These observations point out the need for a more robust photocathode that can be cycled from UHV to atmosphere and still function after bakeout.

B. Diamond Film Photoemitter

A second approach to preparing a more robust photoemitter is to consider the diamond (111) surface which has been shown to have a negative electron affinity (NEA).[22],[23] The conduction band of diamond (111) is ~0.7 eV above the vacuum level, thus permitting electrons to escape from the diamond on excitation with band gap radiation (5.45 eV). This has the interesting prospect of making a photocathode that could be exposed to air at atmospheric pressure and not lose its photoemission properties. The (111) surface exhibits a (1x1) LEED pattern and appears to be terminated with hydrogen. This surface is stable up to ~ 1000°C before hydrogen desorption and subsequent relaxation occurs, thus permitting a simple photocathode design consisting of an epitaxially grown lightly B doped diamond layer on a single crystal diamond substrate. Figure 2 shows an SEM of a typical diamond film grown by us, used for diamond electro-optics. Note that polycrystalline diamond films have predominantly (111) crystallites. The major advantage is a more robust photocathode whose emission properties are potentially even less susceptible to degradation by temperature. Such a photocathode could be exposed to the air and subsequently baked in vacuum and retain its photoemission properties. During the course of this work, prototype photocathodes that we have investigated have been repeatedly exposed to air, and continue to function without processing, possibly a boon to electron gun practice.

Our preliminary investigation used diamond (111) oriented single crystal and textured polycrystalline CVD films as NEA photoemitters in a vacuum system. We know of one other report[21] on photoemission properties of diamond. The detectable onset of photoemission occurs at ~280-290 nm, 60-70 nm above the intrinsic band edge. By 260 nm, the quantum efficiencies are still low, $\sim 10^{-7}$-10^{-8}, but having risen by about x100, over 20-30 nm. Our light source cuts off below 260 nm, so the data was limited. However, we expect for typical photoemission scaling that the efficiency would rapidly increase below 220nm, to $\sim 10^{-3\text{-}4}$, where strong absorption occurs. The polycrystalline film had better emission properties. Research issues that need to be addressed are: (1) can the quantum efficiency improve in better synthetic polycrystalline samples, with activation processes, or with bias voltage, and (2) can the diamond film substrate conduct enough current to neutralize the photocathode in <1 ns, by conductivity changes through in-situ dopants, or a thin B-

doped film. Recent studies have shown that low bias voltages cause direct e⁻ emission from (111) surfaces into vacuum.[23]

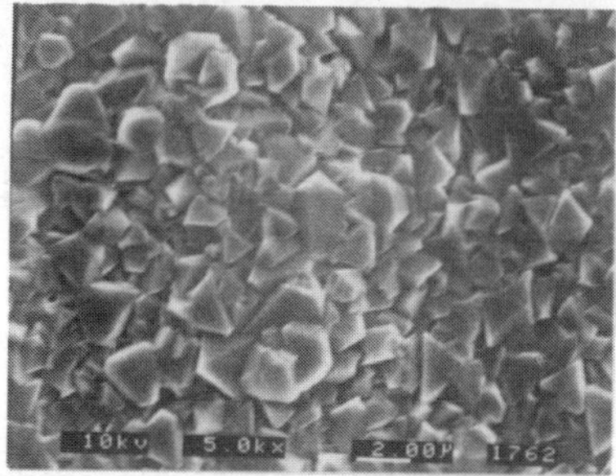

Fig. 2: SEM of a diamond film grown by us for diamond electro-optics. The crystallites are ~1-2 μ in size, ~50% (111).

III. FIBER OPTIC SPATIAL/TEMPORAL BEAMS

The second area of interest is to control the spatial and temporal properties of the photocathode-generated electron beam. A primary benefit of the diamond film is its transparency, ie the optical pulse can be injected from the back end of the gun, without intrusion into the downstream. We therefore propose back-coupling a dense fiber-optic array directly to the photocathode through the diamond film, driven by an array of laser(s). The coupling of the power of a laser to optical fibers can be up to ~20%, with kilowatt/mm² beams available from a single fiber. The spread in the spot size through a small fiber across a 0.5-3 mm thick film would be negligible and correctable. The vacuum standoff strength of the diamond film is good over several cm² of atmospheric pressure. (Note: small single-mode fibers with a small N.A. can interpenetrate the vacuum region for front illumination, with a very good HV standoff.)

The fiberoptic bundle would be fashioned to create the spatial and temporal emission current profile across the cathode, for example, to produce spatially hollow, smoothly shaped or multiple pulsed beams by adjusting the light intensity and temporal properties in the fiber array. For example, in Figure 3, by trimming the fibers into 2 bundles of 2 discrete lengths (delay times), a single short laser pulse injected into the bundle will produce 2 optical pulses, with their start separated by a time delay τ=ln/c.

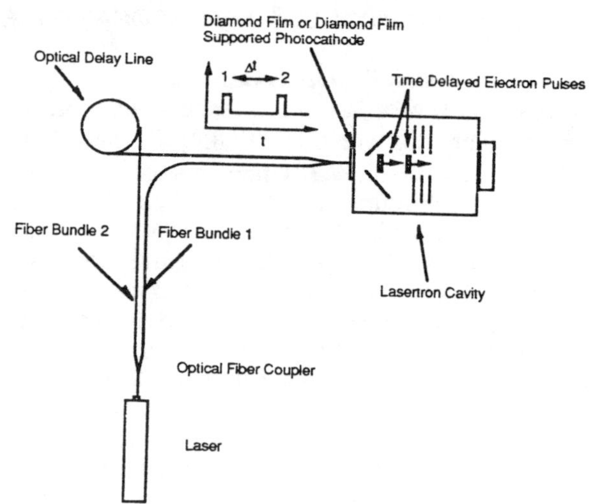

Figure 3. Fiber optic delay lines producing a double electron beam pulse.

A second example would be to arrange the fibers in a spatial pattern, such as a hollow beam. By appropriately adjusting the length, intensity and position of the fibers on the back of the photocathode, a beam of any property can be formed (granular on the beamlet size - the emittance properties of a beamlet are given partially by the photoemission angular dependence, which is $\sim\cos^2 t$, where t = angle to the normal of the surface).

A fiber-optic photocathode could have applications for FELs, microwave generators, direct drive & high frequency RF linacs, bunched beam-cavity & beam-beam interaction studies, and turbulent electron flow. The beamlets may provide a method of increasing the acceptance efficiency and reduce the electron beam power subtended by the grid in a gridded-gun. Narrow double pulses may allow studies of wakefield and beam breakup effects in conducting and dielectric channels. Variable optical delays producing multiple driving & probe beams is unique. Furthermore, short/multiple pulse electron sources can be used in several surface and materials analysis techniques.

IV. DIAMOND GROWTH AND CHARACTERIZATION

Recent advances in chemical vapor deposition (CVD) of diamond from dilute hydrocarbon in hydrogen gas mixtures have shown it is now possible to deposit polycrystalline diamond films over large areas (~8 cm dia.), with a predetermined impurity concentration via in-situ doping [24],[25], at a significantly lower cost than high pressure synthetic diamond. Polycrystalline films have demonstrated room temperature thermal conductivities as large as 10-20 W/cm°K[19], comparable to type Ia natural diamonds above room temperature. The electrical resistivity of diamond films can be varied from $\sim 10^{13}$ to 10^{-3} Ω-cm [26,27,28,29,30,31], depending on the dopant concentrations (both p- and n- type).

Our diamond films were grown using either hot filament-assisted or plasma-assisted CVD (polycrystalline films) or MBE (B-doped single crystal homoepitaxial films) using custom-built equipment. Boron is the shallowest (~360meV) known acceptor species dopant in diamond., and increases the film electrical conductivity, in order to make a good cathode connection (ohmic contact) and modify the performance of the photocathodes. The diamond films were characterized using Raman and photoluminescence spectroscopy, SEM and X-Ray diffraction, showing little sp2 carbon or hydrogen. Ohmic contacts were deposited using the method described in [32], using Ti and Au.

V. CONCLUSION

Because of its high thermal conductivity, optical transparency, controllable conductivity, breakdown strength, and negative electron affinity, diamond is an interesting material for applications in compact electron beam accelerator technology, possibly enhancing the potential of novel beam forming methods utilizing optical fiber driven photocathodes.

VI. REFERENCES

[1] D. Reid, "On the Horizon: New RF Power Sources for Linear Accelerators", LANL Rep LA-UR-88-3288, 1988.

[2] W. Willis, "Switched Power Linac", Laser Acceleration of Particles (Malibu, 1985), AIP Conf. Proc. 130, p. 242.

[3] R. B. Palmer, "The Microlasertron", SLAC publication 3890REV, December 1986.

[4] M. Yoshioka, INS-Rep.-726, December 1988.

[5] C. H. Lee et al.,IEEE Trans.Nuc.Sci.32, 3045, 1985.

[6] M. Boussoukaya, LAL-RT 98-03, 1989.

[7] C. Lee et al.,, Rev. Sci. Inst. 56, 560, 1985.

[8] P. Oettinger et al.,Proc.Part.Accel.Conf.,IEEE 87CH2387-9, 1705(1987)

[9] J. S. Fraser et al., "Photocathodes in Accelerator Applications", Proc. Particle Accelerator Conference, IEEE Cat. No. 87CH2387-9, 1987.

[10] R. L. Sheffield, "Photocathode Injectors", U.S. Particle Accelerator School, BNL, July 24, 1989.

[11] M. Boussoukaya et al., "Photoemission in Nanosecond and Picosecond Regimes",Proc.Part.Accel.Conf., IEEE No. 87CH2387-9, p. 325, 1987.

[12] Y. Fukushima et al.,"Lasertron.", NIM A238, 215(1985).

[13] T. J. Kauppila et al. "A Pulsed Electron Injector Using a Metal Photocathode",Proc.Part.Accel.Conf., IEEE 87CH2387-9, 273(1987).

[14] J. D. Saunders et al. , "Simple Laser Driven Metal Photocathodes Proc.Part.Accel.Conf., IEEE No. 87CH2387-9, p. 337, 1987.

[15] J. Fischer and T. Srinivasan-Rao, "UV Photoemission Studies", 4thWkshp on Pulsed Power Techniques for Future Accelerators, Erice, Italy, March 1988.

[16] T. Srinivasan-Rao et al., " Picosecond Photoemission ", BNL-43446, January 1989.

[17]L. H. Luthjens et al., " Feasibility of obtaining short electron beam pulses .", Rev.Sci.Inst.57, 2230, 1986.

[18] F. C. Tang et al.,"Operating experience with a GaAs .", Rev. Sci. Inst. 57, 3004, 1986.

[19] D. T. Morelli, C. P. Beetz, Jr. and T. A. Perry, "Thermal Conductivity of Synthetic Diamond Films", J. Appl. Phys. 64, 3063, 1988.

[20] A. Sommer, Photoemissive Materials, (1978) Wiley.

[21] H. Timan, "A Survey of Recent Advances in Theory and Practice of Vacuum Photoemitters," Adv. Elect. Electron Physics 63, 73 (1985).

[22] F. J. Himpsel et al., "Electronic surface properties and Schottky barriers diamond", J.Vac.Sci.T..17,1085, 1980.

[23] M. W. Geis, J. A. Gregory and B. B. Pate, "Capacitance-Voltage Measurements on Metal-SiO2-Diamond Structures Fabricated with (100)- and (111)- Oriented Substrates", IEEE Trans. Elect. Dev. 1990.

[24] R.F. Davis et al., Matl. Sci. and Eng. B1, 77 (1988).

[25] K. E. Spear, "Diamond - Ceramic Coating of the Future," J. Amer. Ceram. Soc. 72, 171 (1989).

[26] M.W. Geis et al. , 2nd Annual Diamond Technology Initiative Seminar, July (1987).

[27] M.W. Geis, 34th National Symposium, Amer. Vac. Soc., paper JS-Mo A1(1987).

[28] K. Nishimura, K. Kobashi, Y. Kawate and T. Horiuchi, Kobelco Technology Review No. 2, 49, (1987).

[29] N. Fujimori, T. Imai, and A. Doi, Proc. IPAT, (1985).

[30] J. Mort et al. Ext Absts Electrochemical Soc. Spring Mtg.Los Angeles, CA 89-1, 158,1989.

[31] M. Landstrass & K. Ravi, Appl.Phys.Let.55,975 (1989).

[32] K.L. Moazed et al., IEEE Elect.Dev. Lett. 9, 350 (1988).

Development of Laser Optics for the AWA Photocathode*

J. Norem, and W. Gai
Argonne National Laboratory, Argonne, IL 60439

I. INTRODUCTION

The photocathode electron source for the Argonne Wakefield Accelerator is being designed to produce a number of short (10 psec), 100 nC bunches per rf pulse.[1] Possible sources of fluctuations in the position and energy of these bunches are variations in the initial intensity, position and profile of the bunches which are amplified by wakefields. Since high pulse power lasers have historically produced large fluctuations in pulse to pulse energy and transverse beam profiles, we have been concerned about the stability of the RF photocathode as an injector for the linac. Unfortunately, there is no commercial devices will minimize those fluctuations. In order to minimize these problems, we have designed and begun to test systems which should reduce the initial fluctuations of the beam by stabilizing the laser pulse energy and beam profile.

The pulse to pulse energy fluctuation reduction device (noise eater) is a feedforward system, as shown schematically in Figure 1. The transmission of a polarized beam through the Pockels cell and analyser depends on the voltage applied on it. A fraction of the laser pulse incident on a vacuum photodiode will generate a voltage pulse on the Pockels cell. The amplitude of voltage will depend on the total number of the incident photons and bias voltage applied on the diode. If the transmission coefficient of the Pockels cell has negative slope vs applied voltage, it could stabilize the incoming laser pulses under conditions described below.

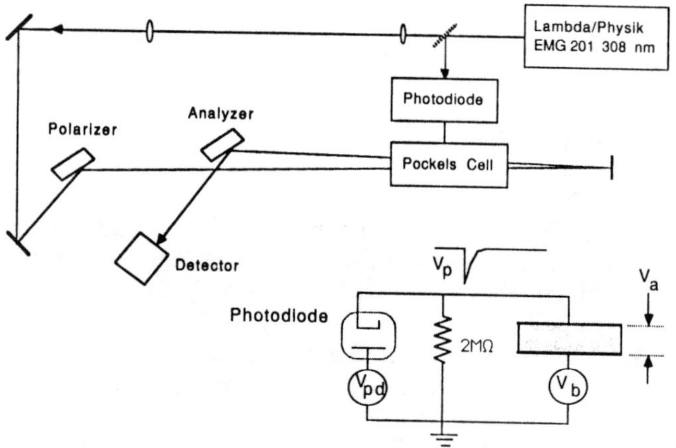

Figure 1. A schematic of the noise eater.

Fluctuations in the transverse profile of the laser will be magnified by the electron beam due to wake fields in the linac. This effect causes unstable electron beam positions in the wake field accelerator sections where further transverse amplification can occur. These fluctuations are

*Work supported by the US Department of Energy, Offices of High Energy and Nuclear Physics, including Contract W-31-109-ENG- 38.

stabilized by scattering the beam into small angles which is uncommon in classical optics.

II. PRINCIPLE OF THE NOISE EATER

The transmission coefficent of polarized light thru the Pockels cell and analyser vs applied voltage V_a is given by the expression[2]

$$T = T_a \cos^2 \left(\frac{\pi}{2} \frac{V_a}{V_{1/2}} \right) \tag{1}$$

where $V_a = V_b + V_p$ and $V_p = Q/C$ is induced voltage from photodiode with $Q = Q(I_{in}, V_{pd})$ the charge of total photo electrons, C is the capacitance, and $V_{1/2} = \lambda/2n_0^3 r_{63}$. Thus the transmission depends nonlinearly on the applied voltage as well as the wavelength, λ, refractive index, n_0, and electrooptic coefficient of the media, r_{63}. Single or double pass operation can be used, with two passes requiring half the applied voltage.

Attenuation of light in a medium is described by $dI(z)/dz = -[\alpha + \beta I(z)]I(z)$, where α and β are the one and two photon absorption coefficients.[3] If reflections are neglected, the transmission is the product of the linear and nonlinear terms, $T_a = T_l T_{nl}$. With short pulses and high powers, the aperture (and cost) of the Pockels cell must be large enough to minimize two photon absorption, which can become significant at power levels of $100 - 500$ MW/cm^2. Laser power density can reach damaging levels if the beam is inadvertently focused near the Pockels cell.

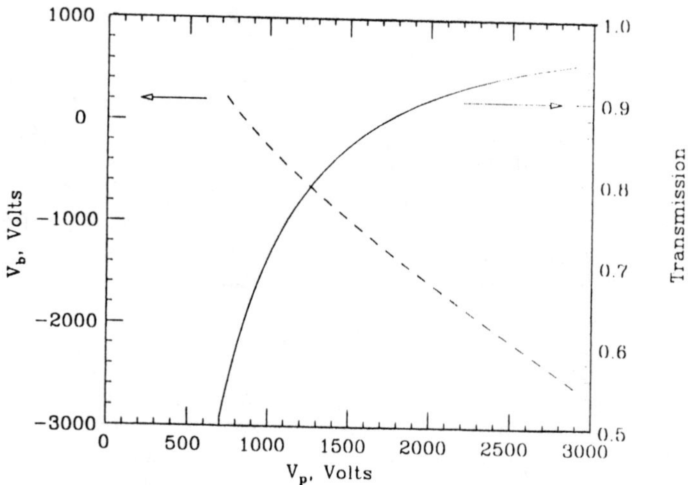

Fig 2 Transmission of the noise eater as a function of V_p and V_b neglecting attenuation effects for a single pass Pockels cell at 248 nm. A double pass device requires half the voltage.

For an incoming laser pulse with intensity I_i, the transmitted intensity $I_o = T I_i$, then the operating point will be where

$$\frac{\delta I_o}{I_o} = \frac{\delta T}{T} + \frac{\delta I_{in}}{I_{in}} = 0. \qquad (2)$$

For the purpose of noise eater under the ideal situations, one would like to have $\delta I_o = 0$ for $\delta I_{in} \neq 0$, which yields the condition for stabilization

$$\delta V_a = \frac{2V_{\lambda/2}}{\pi} \frac{\delta I_{in}}{I_{in}} \cot\left(\frac{\pi}{2}\frac{V_a}{V_{1/2}}\right). \qquad (3)$$

One can satisfy the above equation in many ways. If the bias voltage on the Pockels cell and the pulse voltage are free variables, it is possible to find solutions where the transmission of the system can be high, although at the expense of high V_p and V_b, see Figure 2. This also requires small capacitances high photodiode biases, V_{pd}.

III. EXPERIMENT

The test assembly is shown schematically in Figure 1. The Pockels cell used was a Model QX1020, with a $V_{\lambda/2} = 4.16$ kV at 633 nm, a capacitance of 7.3 pF, and a hard aperture of 9 mm, and was composed of 95% KD*P.[4] Plate polarizers used dichroic coatings on fused silica, tuned for 308 nm. Polarizers are being developed which can transmit 97%, reflect 99% and withstand high laser powers.[5] The vacuum photodiode used was a Hamamatsu R1193-02 with a Sb-Cs photocathode and a capacitance of about 2 pF. The short (15 cm) cable connecting the photodiode and Pockels cell contributed about 15 pF to the total capacitance of 25 pF. The laser used in the test was a Lambda Physik EMG201, operating at 308 nm, and had a pulse length of about 25 ns. Pulse energies were measured with a Molectron J25 probe and a silicon photodiode operated with no external bias, the pulse to pulse fluctuations of the laser varies but is about 5%.

The voltage induced by the vacuum photodiode on the Pockels cell is shown in Figure 3, for $V_{pd} = 500 - 2500$ V. At low incident pulse energies the induced voltage is independent of bias, however, for high energies space charge effects limit the photocurrent and the induced voltage is proportional to $V_{pd}^{3/2}$ due to Childs Law. The operating point during these tests was $V_p \sim 500$ V from 0.0002 mJ of laser light with $V_{pd} = 2500$ V. It is possible to work near the nonlinear regime of the vacuum photodiode, where the photon induced current is very close to space charge limited. Since the photodiode capacitance is small, it is possible to increase the voltage of this system by increasing the number of photodiodes.

A preliminary experiment was conducted to investigate the system as described above. The laser we used had a large and irregular emittance and a comparatively stable beam, and the laser intensity used was low, thus giving a large signal to noise ratio when the beam intensity was measured with detectors more suitable for high intensity

pulses. Nevertheless we have observed that transmission is sensitive to pockels cell voltages, and with $V_{pd} \sim 2100$ V, transmitted intensity first increases then stays constant then decreases with increases in the total beam power, demonstrating the basic operation. More precise measurements are under the way to determine the limitations of noise reduction device for pulsed power UV lasers.

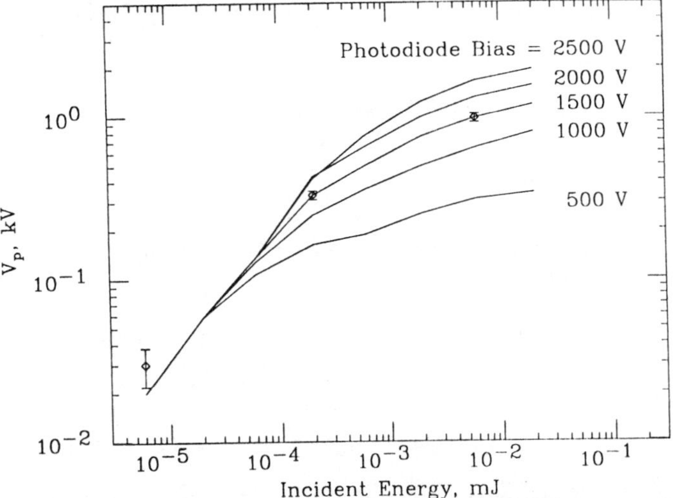

Fig 3 Voltage V_p as a function of light intensity, with $C = 25$ pf.

IV. SMALL ANGLE SCATTERERS

We are planning the use of components which will diffuse the light at small angles. In order to insure that the distribution of the light at the photocathode is as uniform as possible. Since most diffusers are composed of surfaces which have sharp discontinuities and small angle structures, most commercial diffusers scatter into large angles.[6] We have produced surfaces which scatter into small angles by 1) fire polishing and 2) etching rough ground surfaces. Fig . shows a microphotograph of a fused quartz surface produced using 1200 abrasive, and etching with HF for roughly 50 hrs. This surface scatters UV with $\sigma_\theta \sim 1°$.

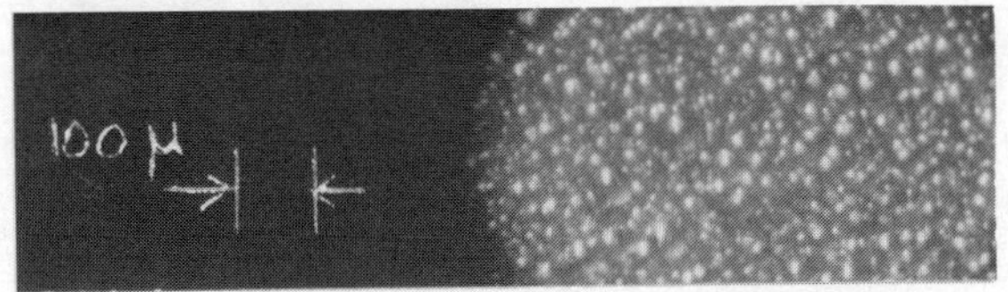

Fig. 4. Microphotograph of the etched surface.

V. SUMMARY

We have designed and constructed a noise eater to reduce the fluctuations of a UV laser. Preliminary experimental data demonstrate that the system works, though with large signal to noise due to problems in the test assembly that should not occur in the final device. Further experimental study should show the limits of this system.

We have also investigated the scattering effects of a variety
of different surfaces.

VI. ACKNOWLEDGEMENT

We would like to thank M. Pellin and C. Young of ANL-
CHM for their assistance and the use of their laser and Bill
Partlo of the University of California - Berkeley for helpful
discussions.

VII. REFERENCES

1. M. Rosing et al this conference

2. W. Koechner, Solid State Laser Engineering, Springer
Verlag, New York, (1976)

3. M. J. Weber CRC Handbook of Laser Science and
Technology, Vol III: Optical Materials, CRC Press, Inc.,
Florida, (1986)

4. Cleveland Crystals, Highland Heights, Ohio, 44143

5. T. Kardos, Broomer Research Inc, Islip, New York
11751

6. J. M. Elson and J. M. Bennett, J. Opt. Soc. Am.
69, (1979), 31

Flat-Beam Rf Photocathode Sources for Linear Collider Applications

J.B. Rosenzweig

UCLA Dept. of Physics, 405 Hilgard Ave., Los Angeles, CA 90024

Abstract

Laser driven rf photocathodes represent a recent advance in high-brightness electron beam sources. We investigate here a variation on these devices, that obtained by using a ribbon laser pulse to illuminate the cathode, yielding a flat beam ($\sigma_x \gg \sigma_y$) which has asymmetric emittances at the cathode proportional to the beam size each transverse dimension. The flat-beam geometry mitigates space charge forces which lead to intensity dependent transverse and longitudinal emittance growth, thus limiting the beam brightness. The fundamental limit on achievable emittance and brightness is set by the transverse momentum distribution and peak current density of the photoelectrons (photon energy and cathode material dependent effects) and appears to allow, taking into account space charge and rf effects, normalized emittances $\epsilon_x < 5 \times 10^{-5}$ m-rad and $\epsilon_y < 10^{-6}$ m-rad, with $Q = 5$ nC and $\sigma_z = 1$ mm. These source emittances are adequate for superconducting linear collider applications, and could preclude the use of a damping ring for the electrons in these schemes.

Introduction

The rapid development of rf photocathodes as high-brightness, low emittance electron sources has been spurred on by their potential for use as injectors for linacs which drive free-electron lasers (FELs) and linear colliders[1]. For linear collider applications it is natural to consider sources which give asymmetric ($\epsilon_x \gg \epsilon_y$) emittances, as this asymmetry is necessary for the final focus, which must produce flat beams ($\sigma_x^* \gg \sigma_y^*$). Collisions of flat beams is an almost universal feature of current linear collider designs, as beam flatness can be used to diminish the effects of the coherent beam-beam focusing, or disruption[2]. The strength of the beam-beam interaction can cause the colliding beam particles to radiate photons (beamstrahlung), which induces an undesirable spread in collision energies. Coherent production of e^+e^- pairs by beamstrahlung photons in the intense electromagnetic fields of the oncoming beam is another source of beam-beam related problems[2].

At the lowest energies, as the electron beam leaves the cathode, it is the self-fields of the beam (space-charge fields) which are problematic, as they can give rise to non-uniform defocusing of the beam, which produces transverse emittance growth. The same scheme can be applied for mitigating the strength of the beam's fields here as is applied at the final focus – flattening the beam profile while keeping the cross-sectional area of the beam constant. This not only reduces the effects of space-charge, but also naturally produces an asymmetric emittance.

A cursory analysis of the prospects for this scheme are presented here. It appears that the demands for the electron emittances of a superconducting linear collider[3], which does not utilize such small beams at collision as a normal-conducting machine, can be met using a flat-beam rf photocathode, thus eliminating the need for an electron damping ring. This device can also be considered for supplying the electron beam emittances demanded by the CLIC design[4], which are not too different than those of a superconducting machine. In addition, a recently proposed far infrared FEL scheme[5], which uses a wave-guide with a few mm height to eliminate slippage between the beam and optical pulse, requires very small beam heights (a few mm) at low energy (< 20 MeV), due to the aperture restriction of the wave-guide. Asymmetric emittances from an rf photocathode source may allow the small beam height combined with high peak current needed for this application.

Emittances from Photocathodes

The inherent spread in transverse energy of photoelectrons emitted from the cathode is a complicated function of the photon energy, the applied electric fields, and the electronic structure and work function of the cathode material. It is best to take a characteristic energy, typically the difference between the photon energy and the work function, as the transverse temperature of the beam at emission. We take this temperature to be approximately $kT_\perp = 0.5$ eV for the purposes of this paper. The normalized rms horizontal and vertical thermal emittances of the beam at the cathode are simply[6]

$$\epsilon_{n(x,y)}^{th} = \sqrt{\frac{kT_\perp}{mc^2}}\sigma_{x,y} \simeq \sigma_{x,y}(\text{mm}) \times 10^{-6} \text{ m}\cdot\text{rad}.$$

Thus, to take the example of a superconducting linear collider design, to achieve emittances of $\sigma_{y,x} \leq 1,50$ mm-mrad respectively (with a number of particles in a bunch $N_b = 3 \times 10^{10}$), one would illuminate the cathode with a ribbon laser pulse of dimensions less than or equal to $\sigma_{y,x} = 1,50$ mm. Since σ_x is not small compared to typical L or S-band cavity dimensions, it would be better to make σ_x smaller by a factor of at least two. The peak current density derived from the cathode is probably not limited by physics of the laser-surface interaction, but by the longitudinal space charge effects near the cathode. Since short pulse beams can be considered to be infinitesimally thin at very low energy ($\sigma_z \simeq v_z\sigma_t \ll \sigma_{x,y}$), the decelerating electric force at the trailing end of the beam pulse is approximately, including the contribution from image charges

$$eE_z \simeq 4\pi e^2 \Sigma_b = \frac{2N_b r_e m_e c^2}{\sigma_x \sigma_y}$$

where N_b is the number of electrons per pulse. If this field is as large as the rf accelerating field eE_{rf}, then one has a completely space-charge limited flow, analogous to the Child-Langmuir limit for gaps. In practice, the decelerating force must be kept much smaller than the applied accelerating field. We take the practical limit here to be a factor 10 times smaller than given above, as even at longitudinal space-charge fields much lower than the limit the bunch may lengthen significantly due to differential acceleration of the front and back of the beam, and we are interested in preserving short pulses. Thus the peak surface density allowed is

$$\frac{N_b}{\sigma_x \sigma_y} \leq \frac{eE_{rf}}{5 r_e m_e c^2}$$

and for $eE_{rf} = 80$ MeV/m, $N_b \simeq 1.1 \times 10^{10}\sigma_x\sigma_y(\text{mm}^2)$. For the present example we have $N_b \leq 5.5 \times 10^{11}$, and we can afford to make our initial beam sizes, and thus emittances, smaller by large factors in each plane and still run at the desired N_b; the constraint due to longitudinal space charge can be restated as $\sigma_x\sigma_y \geq 2.7$ mm^2.

Transverse Emittance Growth

The blowup of transverse emittance by phase dependent rf focusing and by transverse space-charge forces has been analyzed by K.J. Kim for round beams[7]. The emittance increase due to rf effects in a flat beam can be written as a simple extension of these results

$$\epsilon_{n(x,y)}^{rf} = \frac{eE_{rf}}{\sqrt{2}m_e c^2}(\Delta\phi)^2\sigma_{x,y}^2,$$

where $\Delta\phi = 2\pi\sigma_z/\lambda_{rf}$. It is apparent from this expression that the larger of the two emittances will be affected by

this limit first. For our example, with a pulse length of 1 mm, peak accelerating field of 80 MeV/m and a 1.3 GHz rf gun, the emittance is rf limited when $\sigma_x \geq 12$ mm ($\epsilon_x \geq 12$ mm-mrad), which is smaller than our maximum beam size due to thermal effects. The rf focusing effects can be ameliorated by running at a lower rf frequency, or by shortening the bunch. In addition, it has been proposed by Serafini et al. to remove the phase space correlations by use of an independent rf cavity after the gun[8]. This scheme may allow total cancellation of the rf derived emittance growth.

Calculation of space-charge driven emittance growth is properly done by by self-consistent simulations. This is a very difficult and computationally time-consuming in three-dimensions, and for now we shall only present calculations based on extrapolation of Kim's two-dimensional round beam result to flat, three-dimensional beams. If we require that an expression for the emittance growth reduce to Kim's result in the round beam limit, and have the correct scaling of geometrical factors if the beam is long in its own rest frame ($\gamma\sigma_z \gg \sigma_{x,y}$), then we find

$$\epsilon_{n(x,y)}^{sc} \simeq \frac{\pi I}{4I_A}\frac{mc^2}{eE_{rf}\sin(\phi_0)}\frac{1}{3\sqrt{\frac{\sigma_x\sigma_y}{\sigma_z^2}}+5}\frac{2\sigma_{x,y}}{\sigma_x+\sigma_y},$$

where I is the peak beam current, $I_A \simeq 1.7 \times 10^4$ A is the Alfven current, and ϕ_0 is the initial phase of the bunch on the rf wave, which in practice is often near 70 degrees. For our superconducting linear collider example, we now take beam sizes smaller than the thermally limited maximum by a factor of four, $\sigma_x = 12.5$ mm – approximately the size where the rf contribution asserts itself – and $\sigma_y = 0.25$ mm (note the constraint $\sigma_x\sigma_y \geq 2.7$ mm is satisfied). The horizontal and vertical emittances due to space charge are $\epsilon_{nx}^{sc} = 23$ mm-mrad and $\epsilon_{ny}^{sc} = 0.46$ mm-mrad, respectively. The total emittances for this example are now $\epsilon_{nx} \leq \sqrt{(\epsilon_{nx}^{th})^2 + (\epsilon_{nx}^{rf})^2 + (\epsilon_{nx}^{sc})^2} = 29$ mm-mrad, and $\epsilon_{ny} \leq \sqrt{(\epsilon_{ny}^{th})^2 + (\epsilon_{ny}^{rf})^2 + (\epsilon_{ny}^{sc})^2} = 0.52$ mm-mrad, both slightly over half the design tolerances. The $\leq$ sign is used to indicate that the space-charge and rf derived emittance growth is not independent – correlations should make the total emittance smaller than the sum of squares.

Some Practical Considerations

The model we have used thus far ignores many dynamical aspects of the problem which can only be calculated accurately by computer simulation, a task will be undertaken in the future to assess more completely the physics of this device. For example, should be noted that our model for space charge driven emittance growth is based on the assumption that the beam size does not grow significantly after leaving the cathode – this growth will aggravate the space-charge driven emittance blowup. This assumption may not be justified when a thermal beam is quite small, as the beam's 'depth of focus' is approximately proportional to the beam size at the cathode. Thus it may be necessary to provide transverse focusing in such a device.

The most common scheme for focusing low energy electron beams is by use of a magnetic solenoid. This is probably unacceptable for our purposes because of the mixing between horizontal and vertical motion due to the angular momentum induced at the entrance to the solenoid. The rotational velocity given to the electrons is caused by a transverse fringe-field which is purely radial. On the other hand, if the transverse is controlled by shaping a flux return yoke to give only a vertical component to the fringe field, then, in the approximation that the fringe field effects act as an impulse, the rotational motion is about the point $(x = x_0, y = 0)$, with radius y_0. In other words, vertical focusing is provided in this scheme, while the horizontal motion can be considered to be 'frozen'. Thus it is possible to provide focusing in using a solenoidal magnetic field without introducing unnecessary correlation of the two transverse phase planes.

Electrons which are created off-axis also traverse longer path lengths after encountering a transverse focusing element. It is possible to eliminate the pulse lengthening associated with this effect by shaping the laser pulse to liberate the off-axis electrons at slightly earlier times[9]. Since the beam in our scheme is wide in the horizontal dimension, it may be necessary to employ this technique to preserve pico-second pulse lengths.

Conclusions

Further theoretical and computational modelling of rf photocathodes in the flat-beam limit must be done to improve the level of understanding of this scheme. In addition, it is necessary to undertake experimental development of these devices. Tests which are scaled to smaller beam sizes and a shorter rf wavelength can be done at the UCLA compact linac source[10]. Also, it should be noted that the parameters given for the superconducting linear collider source is almost identical to the rf photocathode presently under development at for the Argonne Wake-field Accelerator injector[11]. Further tests at that facility would be valuable for judging the suitability of this electron source for linear colliders.

References

[1] R. Sheffield, "Progress in Photinjectors for Linacs", Proceedings of the 1990 Linear Accelerator Conference, 269 (Los Alamos, 1991), and references within.

[2] P. Chen and V. Telnov, *Phys. Rev. Lett.* **63**, 1796 (1989); P. Chen and K. Yokoya, *Phys. Rev. D* **38**, 987 (1988); and "Disruption effects from the interaction of flat e^+e^- beams" submitted to *Phys. Rev. D*.

[3] For an introduction to the literature on superconducting linear colliders, see the Proceedings of the 1st International TESLA Workshop, ed. by H. Padamsee (Cornell, 1990), in particular the article by U. Amaldi and references within; see also J. Rosenzweig "Conceptual Design of a Recirculating Superconducting Linear Collider Top Factory", these proceedings.

[4] G. Guignard, "Status of CERN Linear Collider Studies", Proceedings of the 1990 Linear Accelerator Conference, 8 (Los Alamos, 1991).

[5] S.K. Ride, R.H. Pantell and J. Feinstein, "Reducing Slip in a Far-Infrared Free Electron Laser Using a Parallel Plate Waveguide," *Appl. Phys. Lett.* **57**, 1283 (1990).

[6] M. Reiser, "Physics of High Brightness Sources" in *Advanced Accelerator Concepts*, Ed. C. Joshi, AIP Conf. Proc. **193**, 311 (1989).

[7] K.J. Kim, "Rf and Space-Charge Effects in Laser-Driven Rf Electron Guns, *Nucl. Instr. Meth. A* **275**, 201 (1989).

[8] L. Serafini, *et al*, to be published in the Proceedings of EPAC '90.

[9] J. Norem and C. Ho, private communication.

[10] S. Hartman, *et al.*, "The UCLA Compact Linac for FEL and Plasma Wake-field Experiments," these proceedings.

[11] J. Simpson, "Wake-field Accelerators", Proceedings of the 1990 Linear Accelerator Conference, 805 (Los Alamos, 1991).

Electron Beam Generation from a Superemissive Cathode*

T-Y Hsu, R-L Liou, G. Kirkman-Amemiya**, and M.A. Gundersen

Department of Electrical Engineering - Electrophysics

University of Southern California

Los Angeles, CA 90089-0484

Abstract

An experimental study of electron beams produced by a superemissive cathode in the Back-Lighted Thyratron (BLT) and the pseudospark is presented. This work is motivated by experiments demonstrating very high current densites (≥ 10 kA/cm^2 over an area of 1 cm^2) from the pseudospark and BLT cathode. This high-density current is produced by field-enhanced thermionic emission from the ion beam-heated surface of a molybdenum cathode. This work reports the use of this cathode as a beam source, and is to be distinguished from previous work reporting hollow cathode-produced electron beams. An electron beam of more than 260 A peak current has been produced with 15 kV applied voltage. An efficiency of $\approx 10\%$ is estimated. These experimental results encourage further investigation of the super-emissive cathode as an intense electron beam source for applications including accelerator technology.

I. INTRODUCTION

Pulsed electron beams have been produced by the pseudospark structure, and an intense electron beam of 1 kA peak current, 10^5 A/cm^2 current density, and 10-100 ns duration has been reported[1-3]. These reports of pseudospark-produced electron beams concentrated on the energetic, transient phase beams generated during and briefly before voltage breakdown, and related to hollow cathode (HC) emission.[4] Here we investigate the electron beams generated by the back-lighted thyratron (BLT) during the superemissive cathode (SEC) phase which follows the HC phase.[5-7] The BLT has an electrode structure similar to the pseudospark switch. The superemissive cathode in the BLT is self-heated, very robust, and produces extremely high, uniform current. In the low pressure glow discharge mode, typical of the BLT and the pseudospark, high current density ≥ 10 kA/cm^2 over an area of 1 cm^2 has been observed.[8-10] The heating of cathode surface by ion bombardment during the current build up is responsible for the observed high current density during the SEC phase. During the conduction phase of the BLT, the voltage across the gap is of few hundred volts. Most of this

voltage is across the cathode fall region, which is a thin layer of few microns thickness on the cathode surface. The electrons generated by the superemissive cathode are accelerated across the cathode fall region and some of them pass through the anode center hole. Modeling by Bauer and Gundersen suggests that the electrons can be extracted from the BLT.[11,12] We report here experiments to investigate the electron beams generated by this superemissive cathode.

II. EXPERIMENT

A. Setup

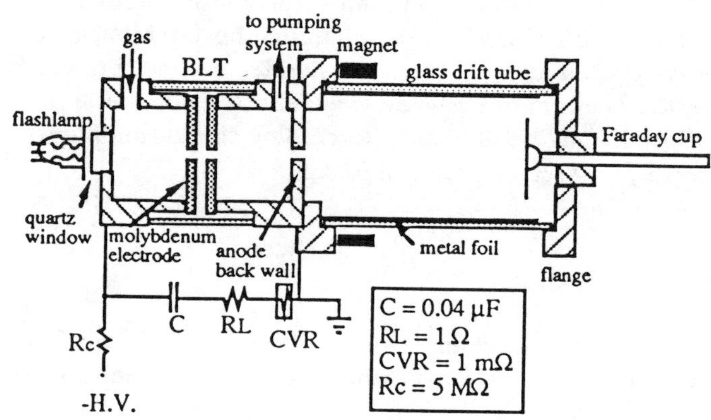

Figure 1. The experiment setup for BLT-produced electron beam diagnostics.

Figure 1 shows the experiment setup. The BLT has molybdenum electrodes with center holes of 5 mm diameter separated by a gap of 5 mm. The o-ring sealed BLT is electrically connected in parallel to a charging capacitor of 0.04 μF and a one-ohm load resistor. The cathode side is charged to a negative high potential through a 5 MΩ resistor unit and the anode side is at the ground potential. A one-miliohm current viewing resistor (CVR) is inserted for monitoring the circuit current pulse. The BLT was triggered by a UV flashlamp at the back of the cathode operating at one Hertz. There is a 5 mm diameter center hole on the back wall of the BLT anode bulk body for extracting the electron beam. The anode back wall is then connected to the diagnostic region through a glass tube which serves as the drift region for the electron beam. The beam current is monitored from the diagnostic port by a Faraday cup which is attached to the other end of the glass tube through a flange. The o-ring seal between the Faraday cup and the flange enables the anode-cup distance to be adjusted over a 10 cm range. For measurements, permanent magnets are

____

*This work was supported by the Army Research Office, and the Strategic Defense Initiative through the Office of Naval Research.

**G. Kirkman-Amemiya is now with Integrated Applied Physics, Inc.

placed transverse the drift tube to deflect the electron beam. A grounded metal layer is attached to the inside of the tube wall to collect the deflected electrons. The diagnostic port is sometimes replaced by a CRT screen with a decaying time of few microseconds. The minimum energy for electrons to excite the phosphor serves as a threshold to screen low energy electrons. A fiber bundle is used to couple the luminescence light onto a photodiode head. The photodiode signal is used to monitor the temporal behavior of beam electrons with energies above the threshold energy. Luminescence shows the beam shape and has been used to estimate the beam energy when combining with the use of transverse magnetic field and to measure beam emittance. The signals from the CVR and the Faraday cup are displayed by a fast digitizing oscilloscope which is outside the screen room surrounding the experiment setup.

B. Results and Discussion

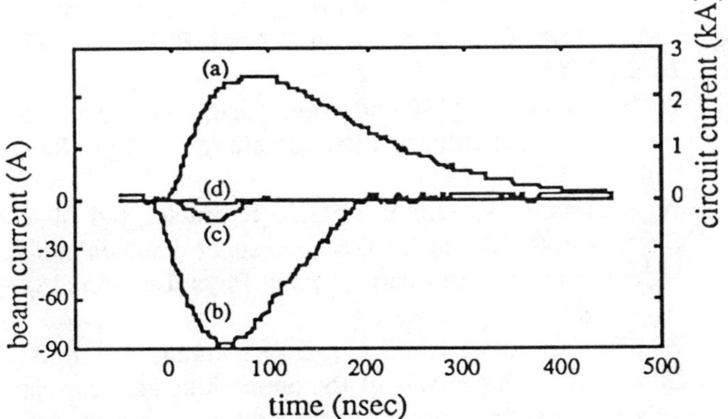

Figure 2. (a) The main discharge current and the beam currents at 7 cm downstream when (b) no magnet, (c) one pair of magnets, and (d) two pair of magnets are applied.

With argon gas at 55 mTorr, and 20 kV applied voltage, an electron beam of 90 A peak current, 120 ns duration was measured 7 cm downstream (figure 2b). The Faraday cup signal extended well into the falling portion of the CVR signal (main circuit current, figure 2a) indicating a beam generated by the superemissive cathode. These electrons are injected from the cathode fall region of the BLT during the high-current conduction phase. The penetration of these electrons into the bulk plasma and the hollow anode back space is possible and has been predicted by Bauer and Gundersen.[11,12] Two components of beam current were observed. The first component is a low current (<5 A), high electron energy, transient phase pulse of ~20 ns which begins before and shortly after the main discharge. The second component is a high current (>70 A), low electron energy, conducting phase pulse of >80 ns which is associated with the superemissive discharge. The two components are hard to distinguish in Figure 2 because of the second component of much higher current superimposing on the first component. 80 and 160

Gauss transverse magnetic fields were applied outside the drift tube to deflect the beam and a much smaller signal was obtained from the Faraday cup, as shown in figure 2c and 2d. The reduced signal indicates that the high energy electrons are created before and during the voltage breakdown. A simple calculation shows that most of the beam electrons have energy of few keV or less. Experiments also show that the beam current increases with applied voltage and with decreasing gas pressure. This suggests that the voltage holdoff capability of the BLT, thus the SEC-produced beam current, can be scaled by a multiple-gap structure.[13] The increase of beam current with decrease gas pressure is further studied by a differential pumping scheme.

The differential pumping effect on the electron beam has been investigated by changing the position of the gas pumping outlet of the system to the diagnostic port. The differential pumping effect comes from the center holes of the electrodes and the anode back wall. At drift tube gas pressure of 10 mTorr and 20 kV applied voltage, an electron beam current of 260 A peak current, 200 ns duration was measured 9 cm downstream (as compared to beam current of 105 A peak current, 120 ns duration without differential pumping). The large increase of beam current and duration with differential pumping is explained by the reduction in the numbers of collisions of beam electrons with neutral gas and plasma in the drift region.[11,12] Application of transverse magnetic field has dimmed the light coming out from the drift tube significantly, indicating bending of the beam electrons into the tube wall before they ionize the neutral gas. In principle the electron beam diagnostics in vacuum should be easily implemented with several stages of differential pumping. A two-stage differential pumping scheme is underway.

With a CRT screen 23 cm downstream, a luminescence light ~3.5 cm in diameter is seen on the screen center. The luminescence light is displaced and distorted into a more triangular shape when the transverse magnetic field is applied. This distortion of beam shape may result because the magnetic field is not uniform across the drift tube and the electron energies are decreasing outward from beam center. Figure 3 shows the detailed structure of the luminescence generated from the phosphor. The luminescence actually begins ~120 ns before the main discharge occurs indicating a beam with runaway electrons of energies comparable to applied voltage.[3] This portion of beam did not show in figure 2 because of its small current. Trace 3(c) represents the luminescence generated from the discharge of the self-capacitance of the BLT (no external capacitor). The electron beam corresponding to this is closely related to the initiation of hollow cathode discharge (HCD). Trace 3(b) represents the luminescence of the phosphor when external capacitor of 16 nF and load resistor of one ohm are connected. The interval between the two peaks on trace 3(b) is the electron beam from the HCD. The emittance of electron beam generated by the HCD is measured by a pepper pot mask and the phosphor screen. The normalized emittance is estimated to be about 20 π mm-mrad. The detailed experiment and the characteristics of the HCD-produced electron beam will be discussed in a separate paper. The second peak on trace 3(b) is evidence of the electron beam in the beginning of the superemissive discharge which also temporally coincides with the rising part of the discharge current. The voltage across the gap between the electrodes is dropping to the cathode fall voltage during the rise of the discharge current. Electrons with energies of the cathode fall are not likely to excite the phosphor and produce luminescence. Collisions experienced by the beam electrons propagating in the drift tube further reduce the beam energy. These two factors may explain why the luminescence trace did not follow the discharge current trace after the current maximum. The electron energies are estimated from the transverse magnetic field strength, the length of the interaction region, the distance between the screen and the interaction region, and the displacement of luminescence light on the screen. The electron energies at beam center and edge are estimated to be ~10 keV and ~600 eV at applied voltage of 15 kV. Because of the fringe effect of the magnets and the ununiformity of the magnetic field, the estimation is accurate only to an order of magnitude.

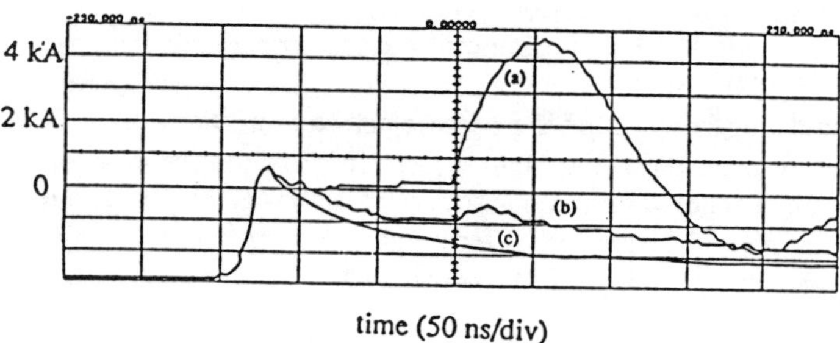

Figure 3. The temporal behavior of (a) the discharge current, and the luminescence from phosphor (b) with and (c) without an external capacitor connected.

III. CONCLUSIONS

Experimental evidence for electron beam production by the superemissive cathode in the BLT is reported. An electron beam of more than 260 A peak current has been extracted. The beam current can be scaled by adjusting the external circuit, using a multiple-gap BLT structure, and changing the gas pressure. With the multiple-gap BLT structure, the scaling of voltage holdoff capability from 10^4 volts to 10^5 or 10^6 volts is possible. The electron beam extraction into vacuum can be achieved by differential pumping. The electrode shape also is a factor affecting the beam generation and will be studied in the future work. These results show that the superemissive cathode in the BLT and pseudospark is potentially an important source for producing high current, high current density electron beams. The simple trigger, compact, and robust nature of the BLT and the data indicate that the superemissive cathode is a beam source that deserves further study for applications such as accelerators and high power klystrons.

IV. REFERENCES

[1] W. Benker, J. Christiansen, K. Frank, H. Gundel, W. Hartmann, T. Redel, and M. Stetter, "Generation of intense pulsed electron beams by the pseudospark discharge," IEEE Tran. Plasma Sci., 17 (5), 754 (1989).

[2] E. Boggasch, T. A. Fine and M. J. Rhee, "Measurement of Pseudospark Produced Electron Beam," Bull. Amer. Phys. Soc., 33, 1951 (1988).

[3] P. Choi, H. H. Chuaqui, M. Favre, E. S. Wyndham, "An Observation of Energetic Electron Beams in Low Pressure Linear Discharges," IEEE Trans. Plas. Sci. PS-15, 428 (1987).

[4] H. Pak and M. Kushner, "Multi-beam-bulk model for electron transport during communication in an optically triggered pseudospark thyratron," Appl. Phys. Lett., 57, 1619 (1990).

[5] W. Hartmann and M. Gundersen, "Origin of anomalous emission in superdense glow discharge," Phys. Rev. Lett., 60, 2371 (1988).

[6] W. Hartmann, V. Dominic, G.F. Kirkman, and M.A. Gundersen, "Evidence for large area super-emission into a high current glow discharge," App. Phys. Lett. 53 (18), 1699 (1988).

[7] W. Hartmann, V. Dominic, G. Kirkman, and M.A. Gundersen, "An analysis of the anomalous high current cathode emission in pseudo-spark and BLT switches," J. Appl. Phys., 65 (11), 4388 (1989).

[8] G. Kirkman and M.A. Gundersen, "Low pressure, light initiated, glow discharge switch for high power application," Appl. Phys. Lett. 49 (9), 494 (1986).

[9] K. Frank, E. Boggasch, J. Christiansen, A. Goertler, W. Hartmann, C Kozlik, G. Kirkman, C. G. Braun, V. Dominic, M.A. Gundersen, H. Riege and G. Mechtersheimer, "High power pseudospark and BLT switches," IEEE Trans. Plasma Science, vol. 16 (2), 317 (1988).

[10] G. Kirkman, W. Hartmann, and M.A. Gundersen, "A flashlamp triggered high power thyratron type switch with remarkable plasma characteristics," App. Phys. Lett. 52, 613 (1988).

[11] H. Bauer and M. A. Gundersen, "Penetration and equilibration of injected electrons into a high-current hydrogen pseudospark-type plasma," J. Appl. Phys., 68 (2), 512 (1990).

[12] H. Bauer and M. A. Gundersen, "High current plasma based electron source," Appl. Phys. Lett., 57 (5), 434 (1990).

[13] T-Y Hsu, G. Kirkman-Amemiya, and M.A. Gundersen, "Multiple-gap back-lighted thyratrons for high power applications," IEEE Trans. Elec. Devices, 38 (4), 717 (1991).

QUANTUM YIELD MEASUREMENTS OF PHOTOCATHODES ILLUMINATED BY PULSED ULTRAVIOLET LASER RADIATION

A. T. Young, P. Chen, W. B. Kunkel, K. N. Leung, C. Y. Li
Lawrence Berkeley Laboratory
Berkeley, CA 94720

and

J. M. Watson
Superconducting Super Collider Laboratory
Dallas, TX 75237

Abstract

The electron quantum yields from polycrystalline lanthanum hexaboride and barium irradiated by near ultraviolet laser excitation have been determined. These measurements show that the quantum yields from these materials are dependent on the processing and previous history of the photocathode material. For lanthanum hexaboride, a yield of 7×10^{-6} with 337 nm irradiation has been achieved. For barium, a yield of 1×10^{-6} has been measured with excitation at 308 nm. These results are discussed and future plans are outlined.

I. INTRODUCTION

The use of cathodes as a source of electrons has become the subject of much recent research. Photocathodes have many characteristics which meet special requirements. Short pulselength sources of electrons have applications in many fields, including free-electron lasers, high energy accelerators, microwave generation, and ion source cathodes and starter cathodes. With photoemission, free electrons are produced on a picosecond (or less) timescale, and the production of the electrons follows the temporal characteristics of the exciting light. Combining the good temporal characteristics of photoemission with the short pulses available from lasers leads to the possibility of well-defined picosecond-long pulses from photocathodes. Other applications of photocathodes take advantage of their "clean" emission characteristics. Unlike thermionic sources which may deposit cathode material throughout the discharge chamber, a photocathode does not intrinsically degrade and liberate any materials which could contaminate a chamber or a process.

Unfortunately, the application of photocathodes is not always possible. One problem is the low quantum efficiency of most materials, requiring the use of high optical powers to get usable currents. Allied with this problem is the large work function of some materials,

necessitating the use of radiation in the far ultraviolet. Finally, many of the most efficient materials used as photocathodes require very stringent environmental conditions, such as vacuum levels on the order of 10^{-8} -10^{-9} torr, in order to function. These problems have prevented the widespread use of photocathodes.

This paper describes preliminary work on two candidate photocathode materials, lanthanum hexaboride, LaB_6, and barium. These materials have low workfunctions and may be more "rugged" than many of their more conventional counterparts. Quantum yields for these materials are measured using near UV pulsed laser excitation, and the effects of preparation and treatment are studied.

II. EXPERIMENTAL APPARATUS

Several different experimental configurations are employed in the quantum yield measurements. A generic schematic is shown in Figure 1. A vacuum chamber/ion source chamber houses the photocathode holder. The cathode is at ground potential. The collection anode, a faraday cup biased at up to +2.2 kV is also located within

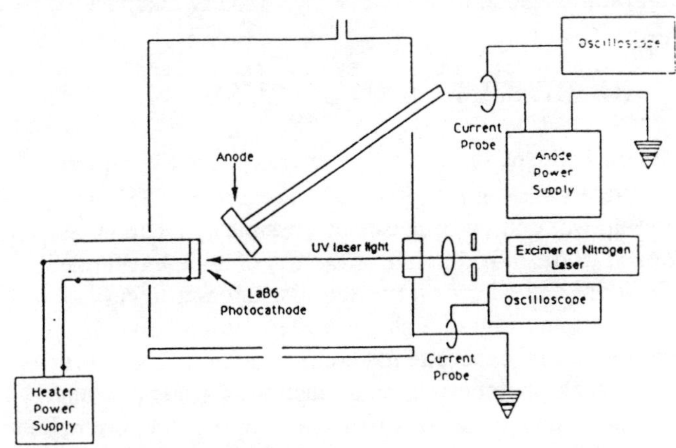

Fig. 1   Schematic diagram of quantum yield measurement apparatus shown here with a LaB_6 photocathode.

the chamber, approximately 5 cm from the photocathode surface. Laser radiation from a XeCl excimer laser operating at 308 nm or a nitrogen laser at 337 nm is used to stimulate photoemission. These lasers have pulsewidths of 30 ns and 10 ns, respectively. The laser beams are apertured using adjustable irises and directed into the chamber through quartz windows. The photo-induced current was detected using current transformers connected to either the anode/anode bias supply line or the cathode/ground line. This technique has a time resolution of 10 ns. This allows us to observe the pulselength of the photoelectrons and ensure we observe only photoemission current, rather than thermal emission current, which would last much longer than the laser pulse.

The disc-shaped polycrystalline LaB_6 photocathode is 19 mm in diameter. The LaB_6 is placed in a cathode holder that allowed an electric current to pass from the center of the disk to its outer edge[1]. By passing DC currents through the cathode, the LaB_6 could be heated resistively. The effects of this treatment on the quantum yield are discussed below.

The barium photocathode is prepared by melting, in an induction furnace and in an inert atmosphere, pieces of barium rod. The molten barium is contained in a shallow well in a copper disc 25 mm in diameter. The barium layer is about 4 mm thick. After cooling, the barium assembly is stored under paraffin oil. Just prior to installation in the experimental chamber, the surface of the barium is polished to remove any contamination. The photocathode is then rinsed with isopropyl alcohol and installed in the vacuum chamber, which is immediately pumped down.

The barium photocathode experiments are performed in a stainless steel high vacuum chamber equiped with a cryopump. This system could attain a vacuum of $\approx 5 \times 10^{-8}$ torr. The LaB_6 experiments are performed in a magnetic multicusp ion source. The base pressure of the chamber is 3×10^{-6}. Both the LaB_6 cathode and the anode are in the central, zero magnetic field volume of the source chamber.

III. RESULTS AND DISCUSSION

The experiments with LaB_6 were performed in an ion source chamber using nitrogen laser excitation at 337 nm. This set-up was chosen because one possible application, as mentioned in the introduction, is the use of a photocathode as a "starter" cathode for a pulsed discharge utilizing an RF drive system. Figure 2 shows some representative data obtained with this experimental configuration. This figure plots the peak photocurrent as a function of energy in the laser pulse. As can be seen from the figure, the current scales linearly with laser energy. This is indicative of photoemission. From this data, a quantum yield of 7×10^{-6} is calculated. This value lies in the range of yields

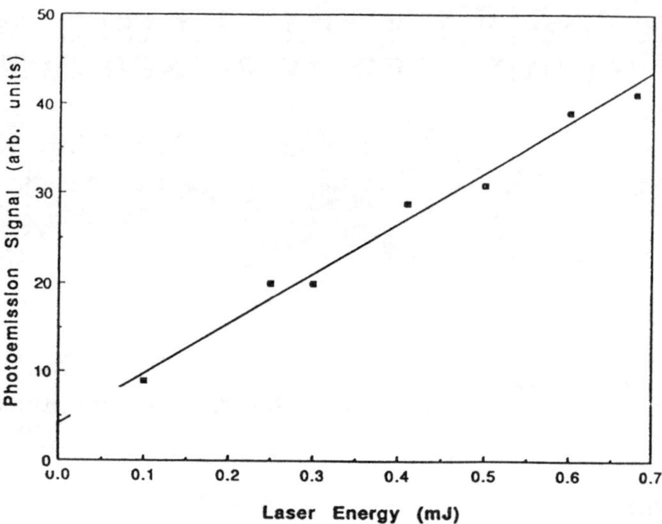

Fig. 2 LaB_6 photocurrent vs. N_2 laser energy. Linearity indicates photoemission is occuring.

previously reported[2,3], but is somewhat smaller than the value obtained by Lafferty[4].

To obtain this value, the LaB_6 was resistively heated. This heat treatment raises the temperature of the photocathode to approximately 600 K for 5 minutes. Prior to heating, no photocurrent could be observed, corresponding to a quantum yield of $< 10^{-8}$. Enhanced photoemission would last for hours. However if the heating was maintained too long, the photocurrent would

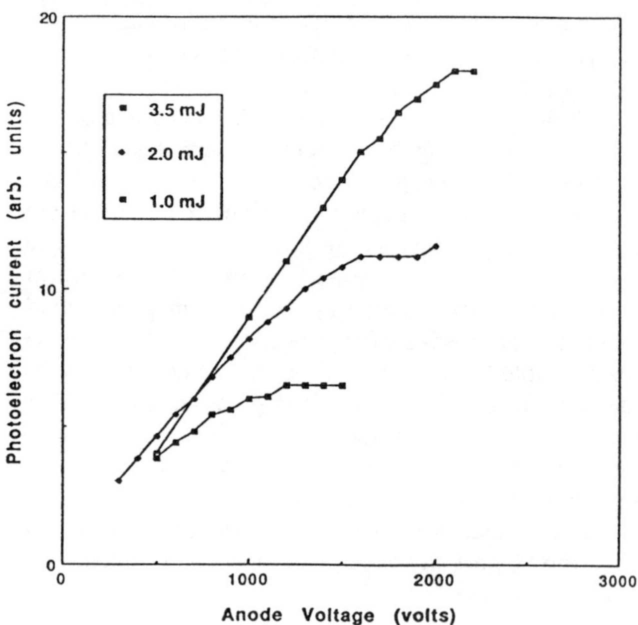

Fig. 3   Barium photocurrent vs. anode extraction voltage for 3 different laser energies. Variation of current with anode voltage indicates that the measurement is space-charge limited.

begin to decrease and would eventually disappear. It is possible that the initial heat treatment cleans away contamination that interferes with photoemission. However, too much heat may disrupt the crystal structure of the LaB_6 or alter the doping density of the lanthanum in the boron matrix. Lanthanum reportedly resides in interstices of a boron cage[4], and the photoelectric effect (both work function and quantum yield) may depend on the surface density or bonding of the lanthanum. Further experiments are being pursued to elucidate the mechanism of this effect.

Data for the barium photocathode was obtained using XeCl laser irradiation at 308 nm. Figure 3 shows the photocurrent collected at the anode as a function of the anode voltage for three different laser pulse energies. As can be seen, the current initially exhibits a dependence on the voltage before reaching an asymptotic value, something not predicted by photoemission theory. In addition, the voltage needed to reach this asymptotic value increases with increasing laser energy. These effects can be explained by assuming that at lower anode voltages or higher laser energies, the photocurrent collected is space charge limited. Increasing the extraction voltage or decreasing the photocurrent (by reducing the laser pulse energy) reduces the voltage necessary to collect all the photoelectrons. The three different pulse energies give approximately the same quantum efficiency, in this case 3×10^{-7}.

A quantum yield of 1×10^{-6} has been obtained for barium. However, the quantum efficiency of the barium is also dependent on the treatment history of the barium. This is exemplified in Figure 4, which shows the quantum yield as a function of laser irradiation time. An untreated barium surface shows little photoactivity. However, after irradiation of the surface with 30 mJ/cm^2 of 308 nm light for 5 minutes, the quantum yield improves markedly. For reference, the laser fluence while measuring the quantum yield is limited to 6 mJ/cm^2 due to the space charge considerations mentioned above. Increasing the irradiation time to 30 minutes results in an increase of the quantum yield by more than a factor of 10 over the original value. The cause of this increase is not known. However, the area that is subjected to the high power irradiation is whiter, which more closely approaches the color of pure barium. It is possible that the intense laser irradiation causes either thermal desorption of contaminants from the surface, or causes plasma formation which also cleans the surface. More experiments are underway to determine the cause of the increased quantum yields.

SUMMARY

Preliminary experiments have been performed to measure the quantum yield of LaB_6 and barium under UV laser radiation. These experiments indicate that modest levels of photoemission can be obtained with UV laser excitation under environmental conditions which "poison" more efficient photocathode materials. Preparation and handling of the photocathodes has been shown to be crucial in obtaining improved performance. Experiments are continuing to determine procedures to further improve the quantum yield. Experiments will also be performed on single crystal LaB_6 as well as on other alkali earth metals.

ACKNOWLEDGEMENTS

This work was supported by the Wright-Patterson Air Force Aero Propulsion and Power Laboratory, the SSC Laboratory, and the U.S. DOE under contract No. DE-AC03-76SF00098.

REFERENCES

1. P. Purgalis, K. N. Leung, S. B. Wilde, and M. D. Williams, "LaB_6 Cathode Development for Electron Beam Generation and for Steady-State Ion Source Operation" Bull. A. Phys. Soc. 33 2091 (1988).

2. A. T. Young, B. D'Etat, G. C. Stutzin, K. N. Leung, and W. B. Kunkel, "Nanosecond-length Electron Pulses from a Laser-Excited Cathode" Rev. Sci. Instrum. 61 650 (1990).

3. P. E. Oettinger, "Brightness Measurements of Electron Beams Photoemitted from Multicrystalline LaB_6 and the Effects of Environmental Pressure", in "Microwave and Particle Beam Diagnostics" Soc. Photo-Opt. Inst. Eng, (SPIE) Proceedings 1061 461 (1989).

4. J. M. Lafferty "Boride Cathodes", J. Appl. Phys. 22 299 (1951).

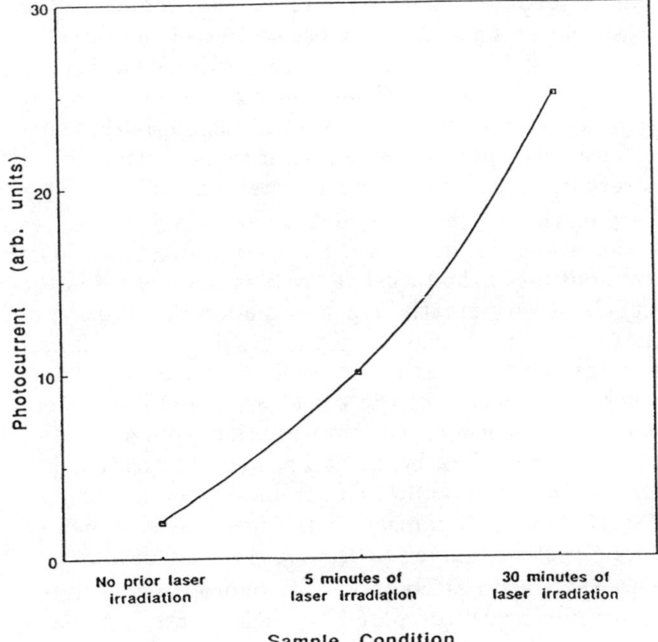

Fig. 4   Barium photocurrent as a function of laser treatment. Laser is 2 mJ/cm^2 while measuring the quantum yield and 30 mJ/cm^2 between measurements.

3D Numerical Thermal Stress Analysis of the High Power Target for the SLC Positron Source[*]

Eric M. Reuter and John A. Hodgson

Stanford Linear Accelerator Center Stanford University, Stanford, CA 94309 USA

Abstract

The volumetrically nonuniform power deposition of the incident 33 GeV electron beam in the SLC Positron Source Target is hypothesized to be the most likely cause of target failure. The resultant pulsed temperature distributions are known to generate complicated stress fields with no known closed-form analytical solution. 3D finite element analyses of these temperature distributions and associated thermal stress fields in the new High Power Target are described here. Operational guidelines based on the results of these analyses combined with assumptions made about the fatigue characteristics of the exotic target material are proposed.

I. BACKGROUND

The results of empirical beam tests conducted on various Positron Source target materials in the early stages of the SLAC Linear Collider (SLC) conceptual design resulted in several target failures.[1] Subsequent inspection of these failed targets indicated that the primary failure mechanism was catastrophic melting of the target material. However, in most of the failures, the estimated local power deposition in the target bodies (from the test beams) was not high enough to cause melting in homogeneous material. Therefore, the failures were hypothesized to be primarily caused by fatigue fracture initiated by the cyclic thermal stressing imposed by the nonuniform shower energy deposition of the incident 33 GeV electron beam. Cracks were thought to have propagated through the material, impeding the heat conduction paths and causing local melting.[2]

The tests provided a good rough idea of target limits, but were carried out over a short period of time and with too few samples to be statistically significant. Numerical calculations of the stresses in the target were deemed necessary and once the stress fields were calculated, some criteria had to be applied to estimate how these could be correlated to target life expectancy. This paper is a summary of the numerical study carried out to better estimate the limits on the new High Power Target (HPT) used in the SLC.

II. TARGET DESCRIPTION

The first Positron Target was stationary, and proved to be very reliable during the initial SLC operating cycle but was inherently limited in the amount of beam power that it could dissipate reliably. In order to reach the ultimate luminosity design goals of the SLC, it had to be replaced with a more robust version that could handle more than triple the power.

Initial analytical models of the target indicated that higher power levels could be reached if the new target were moved cyclically in the beam. The cycling concept chosen, known as "trolling," is shown (greatly simplified) in Fig. 1.

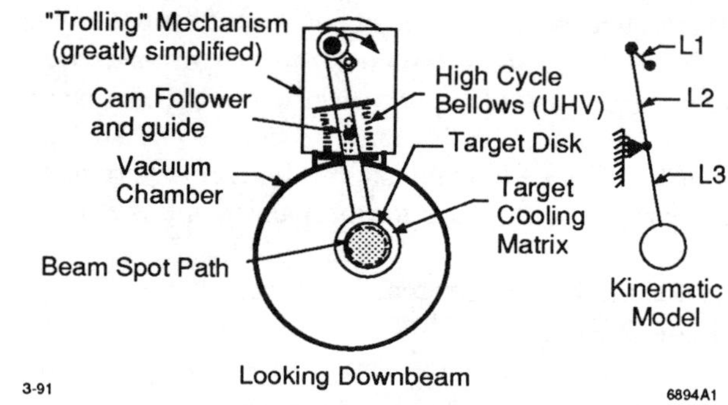

Figure 1. "Trolling" target mechanism.

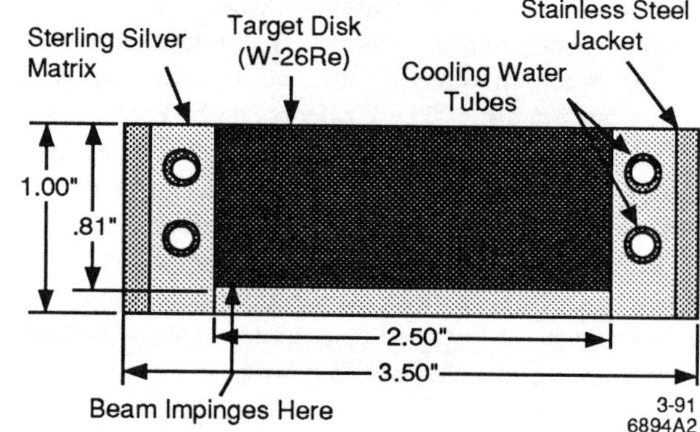

Figure 2. Target disk cross section.

The HPT itself is a 2 1/2 in dia. disk of W-26 Re, arc cast and forged, six radiation lengths long, copper plated and cast in sterling silver. A cross section of the target is shown in Fig. 2. The cooling lines are 0.25 in OD 304 stainless steel (0.020 in wall). They carry $\approx$ 3 gpm of cooling water in a counterflow direction at a differential pressure of $\approx$ 80 psi. The inlet water temperature is controlled (for RF structures in the same water system) at 43°C.

Varying the lengths of the linkage arms in Fig. 1 causes large deviations in the beam path on the target. The positron output of the target (as well as the stresses in the target) could vary greatly depending upon the position of the beam relative to the target edge (due to the difference in material properties between silver and W-26 Re). Therefore, the lengths of the target arms and crank were chosen to maintain a maximum deviation from a circle of about ± 0.032 in. The beam at its nominal position impinges on the target $\approx$ 0.125 in. from the outside edge of the target disk. A summary of the mechanical design of the target system is given in [3].

The SLC beam is composed of bunches of electrons which are far apart compared to their length. As each bunch impinges on the target, it generates a hot "spot." The average beam spot-to-spot spacing was set initially at 3 mm. The drive control system was designed to provide this spacing regardless of repetition (rep.) rate. This spacing corresponded approximately to the radius of the beam spot at the exit plane and was chosen (in advance

* Work supported by Department of Energy contract DE–AC03–76SF00515.

of detailed analysis) so that the successive pulses would have minimal influence on each other. If the pulses were moved closer together, component wear would be slightly reduced (due to slower cycle speed). However, a significant increase in both the material temperature and the associated stresses would be developed. In this preliminary analysis, the "worst case" beam-to-target-edge distance was chosen (i.e. with the beam furthest out on the target disk).

III. Analysis

Three basic steps were necessary to complete the analysis.

1. Power Deposition: The first step was to use SLAC's Monte Carlo Electromagnetic Gamma Shower code (EGS) to determine what the power deposition in the target would be (and in what spatial distribution) for a beam with the maximum SLC design parameters of 7×10^{10} electrons per pulse, 120 pulses per second, and a one σ incident beam spot radius of 0.6 mm. [4–5] The 7×10^{10} intensity criterion, along with a 0.8 mm spot size, was used as a goal for the project initially but recent studies and beam experiments have confirmed that a 15% increase in positron yield (for a given intensity) could be obtained if the spot size incident on the target were decreased to 0.6 mm. This tighter bunch was expected to cause a pronounced increase in the single pulse stresses and therefore a reduction in the allowable intensity.

The energy deposition output by EGS is in fractional energy per unit volume per pulse. For all practical purposes, the temperature rise per pulse is instantaneous and is found by scaling the EGS data by a constant ($121.5 \, \text{cm}^3 \, ^\circ\text{C}$). This results in very high temperature gradients being imposed on the target material during every beam pulse.

2. Model Generation and Solution: The second step was to model the geometry and physical properties of the target numerically. ANSYS, a commercial multi-purpose finite element code, was used for this task.

During operation, the beam pulses impinge sequentially on the target as they trace around the disk periphery. Therefore, the entire target had to be modeled as a single meshed entity and then solved iteratively. However, it was only necessary to calculate power deposition data in EGS for a local region cylindrically symmetric about the beam center. In order to map EGS output to thermal model input, temperature-rise values scaled from EGS output were input to a small dummy cylindrical thermal model and transferred, via a 3D interpolater in ANSYS, to a region near the periphery of the full 3D thermal model of the target.

The full model included all the components shown in Fig. 2. The mesh was fine near the edge of the W-Re disk in one area, and was made gradually coarser around the periphery. The initial pulse temperature-rise data was input in the finely meshed area, with the rest of the target set to 0° in order to facilitate future superposition. Convection boundary conditions were imposed on the inside surface of the cooling tubes, with a 0° bulk water temperature, again to facilitate superposition. The temperature distribution was then allowed to relax for 1/120 s, using ANSYS's time-

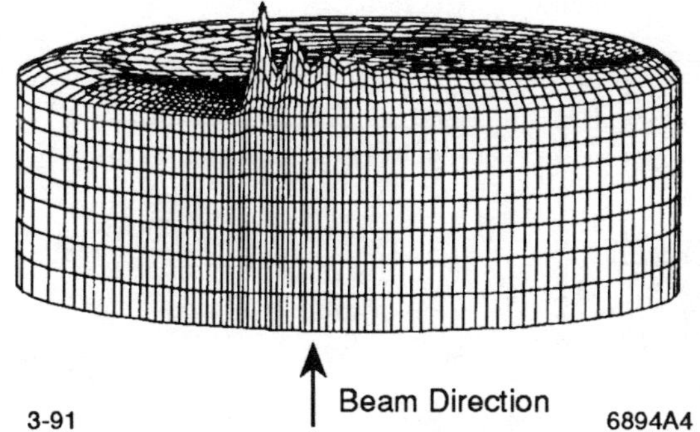

Figure 3. Raised contour temperature map of High Power Target in "steady state," just after a pulse.

step optimization routine. This took 19 cpu-hrs on a 7-MIP Sun 4 workstation.

A solution file was written at this point and the relaxation resumed. Using time-step optimization, the second 1/120 s of relaxation only took 5.5 cpu-hrs. This was repeated several times and the results used, in conjuction with the temperature interpolater, to create a superposition of temperatures representing ten "pulses," each 1/120 s older than its neighbor.

A seed time step for the new maximum gradient was determined and the ten-pulse superposition allowed to relax for 10/120 s. This took 31 cpu-hrs. A solution file was written and the relaxation resumed for another 10/120 s. By repeating this process several times, superpositions of more and more pulses could be generated until finally a superposition of 9193 pulses was created.

At this point a check of the total heat flowing out of the target model found that 7490 W was exiting, while 7950 W was being deposited. Since the iterative transient analysis only approximately conserves energy, and further iterations resulted in no significant change in the heat conducted through the cooling tubes, it was concluded that the "steady state" of cycling through the same sets of temperatures had been reached. The temperature distribution just after the 9193rd pulse is shown as a raised contour map in Fig. 3. Only the W-Re disk is shown in the figure. The silver cooling matrix and cooling tubes are deleted for clarity. The maximum temperature reached is only 720°C despite the fact that a single pulse creates a peak temperature rise of 530°C, indicating that the "piling up" of overlapping pulses has been reduced to a relatively small effect by the trolling action of the moving target.

To find the thermal stresses due to this temperature distribution a new mesh was generated, coarser than the thermal model everywhere but near the hottest spot, where the stresses were expected to be greatest. Since the silver surrounding the W-Re has a much lower modulus of elasticity, it was left out of this model. The interpolater was used to transfer temperatures from the thermal model to the structural model, and it was solved for stresses.

To accurately solve for the stresses perpendicular to the hot surface, a sub-model of the region around the hottest spot was created with a finer mesh in that direction. Temperatures from the thermal analysis were interpolated to its entirety and displacements from the larger structural model were interpolated to its cut surfaces. The plot in

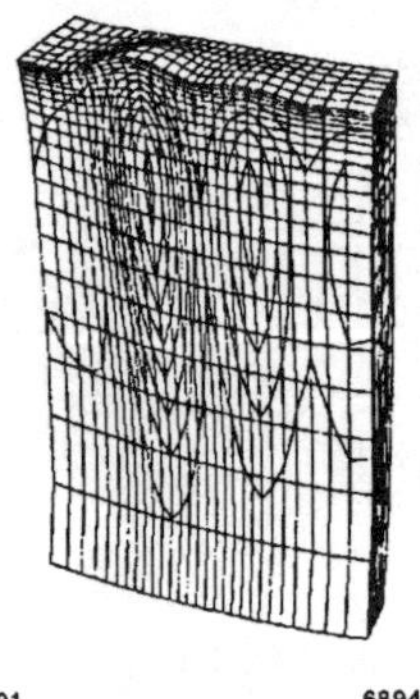

Figure 4. Axial stress contours in hottest region of High Power Target in "steady state," just after a pulse.

4-91   6894A5

Fig. 4 shows stress contours for the stress in the direction of the beam that resulted from solving the submodel. In this view the submodel is cut-away through the center of the pulses in order to show the most severe stress profiles.

The stress shown in Fig. 4 reaches a maximum of 180 ksi compressive. Where the stress is maximum the temperature is 650°C and so the yield strength of the W-Re is about 150 ksi. However, because the material is "captured" by compressive stresses in all directions, this situation is not the most critical. Distortion strain energy, or "Von Mises stress" is a better measure of the proximity to failure, and its value at that place is only 87 ksi. It is actually highest on the surface between two pulses, where a tensile radial stress of 94 ksi combines constructively with a compressive hoop stress of 63 ksi for a Von Mises stress of 140 ksi. At that place the temperature is 370°C, making the yield strength about 200 ksi.

3. Evaluation of Results: W-26 Re was chosen as a target material because of its excellent physics properties and extremely high strength and ductility at the anticipated target operating temperatures. Unfortunately, due to the relatively limited market for the material, only the most basic engineering properties were available.

Although the Von Mises stresses calculated and summarized above were below the actual yield strength of the material at operating temperature in all areas, the failure mechanism was hypothesized to be thermal stress fatigue fracture. Commonly accepted failure analysis principles state that the stress history of the body (alternating and mean stress values at the operating temperature) should be correlated in an S/N or similar statistical summary of material test data to estimate the safety of the design. However, no such data was available nor was it practical to have the material tested. It was necessary, therefore, to recommend operational limits based on advice from others who have worked with the material under similar conditions before, and empirical operational experience with the previous target. The criterion which was determined applicable was that the Von Mises stresses in all areas must be below 0.5 times the yield strength at the operating temperature (a compromise between suggestions from Dieter Walz (0.3) at SLAC and Prof. Tom Divine (0.8) at UC Berkeley).

III. Conclusion

Based on this criterion, the maximum allowable intensity for a .6 mm spot size is 5.3×10^{10} electrons per pulse. This is lower than the "break even" point (6.1×10^{10}) where the 0.6 mm beam yields the same number of positrons per incident electron as a 0.8 mm beam at 7×10^{10}.

The following is a list of some of the assumptions made and some practical considerations not quantified in the analysis.

• Beam Spot Shape and Path. The beam spot has been assumed to be round and its path circular.

• Surface Defects. The number and position of voids and other defects ultimately will have some effect on the useful lives of the targets. During the fabrication of the targets, the downstream faces are micro-polished to minimize these defects.

• Ductile-to-Brittle Transition Temperature. W-26 Re exhibits a relatively abrupt ductile-to-brittle transition at a temperature of about 90°C. Although the actual yield strength of the material does not change appreciably through this range, the ability of the material to elongate (i.e., absorb energy) does. At temperatures below this transition, the target may be much more susceptible to thermal shock and crack propagation.

• Thermal Shock Focussing. The beam energy deposition in the target occurs virtually instantaneously, causing a compression wave to be generated at the beam center and propagate outward. Some studies have shown that damage (spallation, cracking) may occur if these waves are symmetrically reflected back off an interfacing surface so that they constructively add at the beam axis. [6] However, no noticeable damage was seen on the previous target, which was centered on the beam (but ran at one third the intensity), and the beam does not hit the center of the new trolling target.

• Casting Pre-stress Effects. As the assembly cools after casting, the silver matrix shrinks considerably more than the target disk. However, sterling silver exhibits considerable plasticity and the elastic modulus is low. Therefore, the stresses imposed on the disk from the casting are probably negligible.

Operational experience with the new target has been very successful over the past year. Although the SLC has been unable to deliver beams of design intensity yet, the target has operated reliably at intensities $\approx 60\%$ of design.

IV. References

[1] Stan Ecklund "Positron Target Materials Tests" SLAC Collider Note 128 (Oct. 1981)

[2] Hobey DeStaebler "Temperature Calculations for the Positron Target," SLAC Collider Note 21, (Mar 1980).

[3] Eric Reuter "Mechanical Design and Development of a High Power Target System for the SLC Position Source," IEEE Particle Accelerator Conference Proceedings Record," (1991).

[4] W.R. Nelson, H. Hirayama and D.W.O. Rogers, "The EGS4 Code System," SLAC-265 (1985).

[5] S. Ecklund and W.R. Nelson, "Energy Deposition and Thermal Heating in Materials Due to Low Emittance Electron Beams," SLC Single Pass Collider Memo CN-135 (1981).

[6] J.G. Kelly "Deep Craters Produced by Shock Focussing in Relativistic Electron Beam Diodes." J. Appl. Phys. Vol. 40, No. 10, pp. 1415, (Oct. 1975).

Mechanical Design and Development of a High Power Target System for the SLC Positron Source*

Eric Reuter, Dean Mansour, Tom Porter, Werner Sax, Anthony Szumillo

Stanford Linear Accelerator Center Stanford University, Stanford, CA, 94309 USA

Abstract

In order to bring the SLC Positron Source luminosity up to design specifications, the previous (stationary) positron target had to be replaced with a version which could reliably dissipate the higher power levels and cyclic pulsed thermal stresses of the high intensity 33GeV electron beam. In addition to this basic requirement, the new target system had to meet SLAC's specifications for Ultra High Vacuum, be remotely controllable, "radiation hard", and designed in such a way that it could be removed and replaced quickly and easily with minimum personnnel exposure to radiation. It was also desirable to integrate the target and collection components into a compact, easily manufacturable, and easily maintainable module.

This paper briefly summarizes the mechanical design and development of the new modular target system, its associated controls and software, alignment, and the quick removal system. Operational experience gained with the new system over the first running cycle is also summarized.

I. MODULE DESCRIPTION

1. Target Drive System: Due to the extreme congestion of the Positron production and bunching area, the entire Module system had to be designed *around* the target and its drive system. An extensive amount of effort was initially spent roughly analyzing various target cycle concepts and geometries, calculating the response of these targets to beam induced thermal loads, and then estimating the effect each would have on other system component designs. Kinematic variables for the most promising models were optimized to make the system fit within the spatial constraints of the Positron Vault area.

The actuation system finally chosen, known as "trolling", is shown, greatly simplified, in Figure 1. This concept provided a compact and relatively simple way of providing the specified spacing between beam pulses required to reduce the risk of thermal stress fatigue fracture in the target [1].

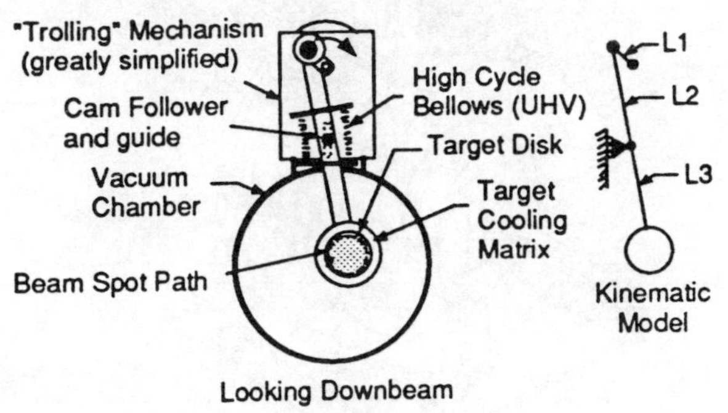

Figure 1. Trolling Target Model

The target itself is a 2-1/2" diameter, 0.812" thick disk of arc cast and forged W-26Re, ground to final diameter before casting. The disk is copper plated (to aid in metal bonding), cast in sterling silver with its cooling tubes, and post-machined to size. Diffusion bonding and shrink fitting were also considered as manufacturing techniques and both methods were prototyped. However, casting was chosen for production due to its superior thermal and mechanical interface.

The target arm assembly is suspended and supported in the vacuum chamber by the target housing mounted outside the chamber. A specially designed "high cycle" bellows is used to feed the target into the vacuum chamber and provide a flexible pivot. Figure 2 shows the target drive system as it appears when removed from the Module.

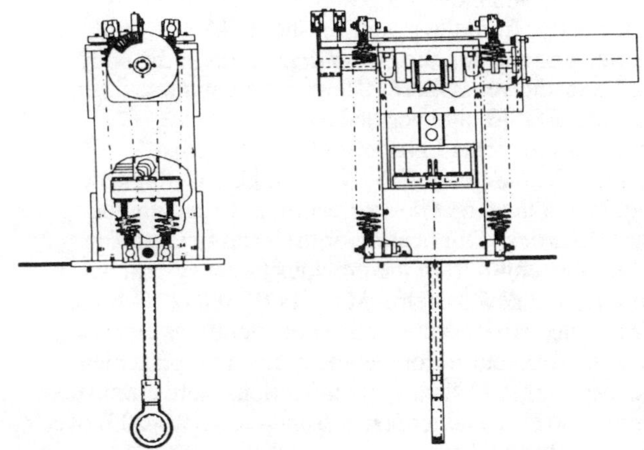

Figure 2 Trolling Target Drive Mechanism

Kinematic, environmental, and life cycle requirements were distributed to various bellows manufacturers throughout the conceptual design of the drive system for feedback.

The bellows chosen is a hydroformed, proprietrary design called a "nested convolute", engineered to withstand one full year's running at 100% duty cycle (~60,000,000 cycles) [2]. It is constructed of 4 hydroformed and heat treated bellows segments of Inconel 718 butt welded end-to-end using electron beam techniques. Special Conflat™ flanges are welded on to complete the assembly and make it interchangeable.

Welded bellows were suggested, but were ruled out due to their inherent stress concentrations, quality control problems, and susceptibility to fatigue failure.

The first two of the special bellows built were tested at an accelerated speed immediately after production. One surpassed 60,000,000 cycles and the other 40,000,000 cycles with no signs of damage or leakage.

All the components in the target drive system, in addition to the bellows, were designed for "infinite life" and high radiation resistance. Dynamic models of the system were constructed to help size components and predict system deflections during operation.

*Work supported by Department of Energy contract DE-AC03-76SF00515.

The springs which were installed to back-up the vacuum load imposed on the assembly due to the large I.D. of the bellows were custom designed with conical ends and matching end retainers to reduce stress concentrations and facilitate easy assembly and maintenance.

With the exception of the insulation on the control wires, the entire Module system is designed utilizing non-organic or radiation resistant components. All insulators and cooling water stand-offs are ceramic or Macor (machinable glass ceramic).

The complicated kinematics of the drive system produce a path of the beam on the target which is not perfectly circular. A more detailed discussion of the beam trace details and operational ramifications is given in [1].

2. Flux Concentrator: The original design of the Spiral Flux Concentrator was extremely simple and effective and had proven itself quite well over the past two years of running. The basic design was left intact in the new target system with some minor modifications to the supports and vacuum interface.

The Flux Concentrator is designed to run at 10kV and 16kA, with a 5μs pulse width at 120 pps. The peak field generated is 58kG and the RMS power deposited in the flux concentrator body (from the pulser) is about 3kW. At full beam rep. rate and current, the power deposition in the body due to the beam shower is 4kW. These deposition rates were estimated using EGS[3] and verified by extrapolating operational measurements from lower beam currents.

3. Tapered Field Solenoid (TFS): A high DC magnetic field is required at the target downstream face to capture the low energy positrons. Duplication of the axial field profile and 10kG field strength from the previous target system was a design goal for the new system. Many POISSON (2D finite difference magnetostatic/electrostatic code) iterations were run to optimize the flux return iron geometry and coil placement around the new target system. Field inflections were eliminated and the peak field at the target face was increased by ~20% over the old design. Practical requirements such as manufacturability and ease of assembly and maintenance were concurrently designed in as well.

Magnetic measurements conducted on the first Module verified the field profile to be within 2% of design. At a maximum current of 750A, the TFS generates a 13kG DC field at the target exit face.

4. Vacuum Chamber: The vacuum chamber used to house the flux concentrator and target is contained within the Tapered Field Solenoid. It is a custom design and was configured to maintain structural rigidity, provide the maximum pumping speed to the target and flux concentrator area, accomodate the pole tip section of the Tapered Field Solenoid flux return iron, mate up to the existing accelerator section, and allow the target and flux concentrator to be removed independently.

The downstream vacuum flange of the chamber was designed to be used with a remotely actuated quick-disconnect system. The disconnect utilizes an air driven motor to open and close a rigid hinged clamp by turning a single screw. This hinged clamp compresses a custom aluminum seal between the Module flange and the mating flange of the accelerator structure downstream. Only 10 ft-lbs. of torque is required to make the seal Helium leak-tight. A single convolution bellows between the chamber and the flange provides enough compliance to take up any slight angular misalignments of the two mating flanges while still being rigid enough to support the flange without vertical restraint. A picture of the completed Module is shown in Figure 3. A 3D isometric wireframe of the Module as designed is shown in Figure 4.

Figure 3
High Power Target Module

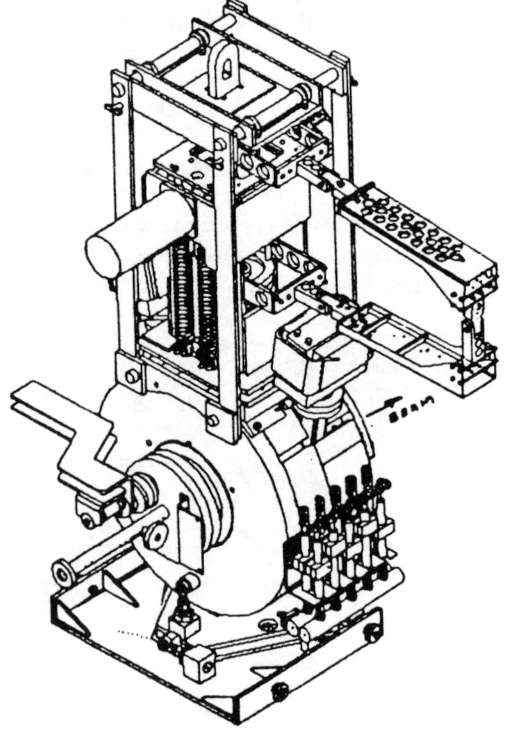

Figure 4
High Power Target Module 3D Model

II. TARGET CONTROL & INTERLOCK SYSTEM

The monitoring and control system hardware for the target drive was also largely custom designed due to the high radiation levels and RF noise in the Positron Vault. Two aluminum toothed wheels, one with 60 radial slots and fiducial with 1 slot are installed on the target drive crank opposite the motor drive in a shielded housing. Four sets of ferrite pickups (2 per wheel) are placed symmetrically around the outer rim of the disk. These are used to generate the incremental and fiducial pulses as the target crank is rotated. These pulses provide the basis for the entire control and interlock system. They are used in conjunction with a local ion chamber and toroids, a custom series of CAMAC modules, the Main Control Center VAX 8800, a local microVAX, and custom SLAC SCP software to control and interlock the drive system. Yield and status information on the incoming electron beam and outgoing positron beam are also provided by the system.

III. ALIGNMENT

The precise alignment of all the Module components to each other and to the beam is critical to optimize positron yield and prevent premature target failure. Since final alignment could not be completed until the system was fully assembled, the components each had to have their own adjustment system and fiducialization step or had to be designed to be self-aligning. Tolerances for the total system component-to-component alignment were held to +/- 0.010". The maximum total system alignement tolerance is +/- 0.020".

The target and Flux Concentrator are independently pre-aligned on the SLAC Leitz Coordinate Measuring Machine (CMM) The coordinates of the theoretical center of the downstream target face (as well as its planar orientation) are transferred to tooling balls mounted on the side of the target drive housing for easy viewing.

The Tapered Field Solenoid is optically fiducialized to the vacuum chamber contained within it. All the components are then assembled into the vacuum chamber and aligned using optical methods. The target position within the chamber is then nominally adjusted (using an external alignment adjustment system) to the T.F.S. pole piece and to the upstream face of the flux concentrator. This relative positioning must be maintained to withing 0.010" to optimize Positron capture.

The entire assembly is then placed on a test stand identical to the mounting system in the Positron Vault and aligned to theoretical beam center. The Module can then be installed and run with no on-line alignment. The initial installation and alignment check of the first production Module in the Vault verified the remote alignment techniques.

IV. REMOTE REMOVAL

The Module system was designed so that in the event of a component failure, the option to remove the entire Module and replace it quickly could be exercised rather than attempting to perform maintenance on the "hot" system in place.

Before removal, the Module instrumentation, power and water connections must be disconnected. All are designed to be manually removed in less than 10 seconds each. Most of the connects utilize standard components. However, the high current connections for the Tapered Field Solenoid are a custom fork design and will handle up to 1000 Amps (DC). They use Multilam™ contact bands to provide a reliable, repeatable contact.

The Module is retracted away from the accelerator section (and out of the vacuum quick disconnect clamp) and then translated out into the aisle by a 2 stage translation mechanism which is mounted permanently on the accelerator girder. This operation (including activating the vacuum quick-disconnect) can be done automatically or manually utilizing a remote control panel located away from the target but still in the Vault area. A remote crane (radio controlled) is hooked onto the Module and it is lifted and trollied along the Vault wall to another translation stage which moves the Module out into a specially designed penetration. A jib crane, mounted at the klystron gallery surface level (30 ft above) lifts the Module to the surface where it can be removed to repair or taken to radioactive storage.

V. OPERATIONAL EXPERIENCE

The first production prototype High Power Target Module was successfully run at full power on all systems and has logged over 20 million cycles in the beam over the past year with virtually no maintenance or beam down-time. Three spares have been built and are ready for beam. Initial problems encountered during start-up included a main bearing failure in the target drive (cause never determined) and a failure of the bellows on the osculating arm which feeds cooling water to the target from above the Module.

VI. AKNOWLEDGEMENTS

The author wishes to thank everyone at SLAC and elsewhere who contributed to the implementation and success of the project. Special thanks are extended to Bob Gardner and the staff of Gardner Bellows for their exceptional bellows design and to the individual members of the High Power Target Design Team whose teamwork and comradery were critical to the successful and timely completion of the project. The success of this design effort hopefully will provide an example of the results which can be achieved when concurrent engineering and design principles are embraced.

VII. REFERENCES

[1] Eric Reuter "3D Numerical Thermal Stress Analysis of the High Power Target for the SLC Positron Source" IEEE Particle Accelerator Conference Proceedings Record (1991)

[2] Bob Gardner, Gardner Bellows Corp. Chatsworth, CA, USA

[3] W.R. Nelson, H. Hirayama and D.W.O. Rogers, "The EGS4 Code System," SLAC-265 (1985).

Channeling Crystals for Positron Production

Franz-Josef Decker

Stanford Linear Accelerator Center, Stanford, California 94309

Abstract

Particles traversing at small angles along a single crystal axis experience a collective scattering force of many crystal atoms. The enormous fields can trap the particles along an axis or plane, called channeling. High energy electrons are attracted by the positive nuclei and therefore produce strongly enhanced so called coherent bremsstrahlung and pair production. These effects could be used in a positron production target: A single tungsten crystal is oriented to the incident electron beam within 1 mrad. At 28 GeV/c the effective radiation length is with 0.9 mm about one quarter of the amorphous material. So the target length can be shorter, which yields a higher conversion coefficient and a lower emittance of the positron beam. This makes single crystals very interesting for positron production targets.

1   Introduction

In linear e$^+$,e$^-$-colliders the positrons have to be produced for every cycle, if no recirculation scheme is considered. At the positron source of the Stanford Linear Collider, SLC [1] a 30 GeV electron beam of up to $5 \cdot 10^{10}$ particles strikes a tungsten target producing about 60 positrons. Four of these are captured with a pulsed high magnetic field [2], accelerated to 210 MeV and brought back to the damping ring. After the damping ring the overall yield (= number of positrons/number of electrons on the target) is about 0.7, below the desired value of 1.0 or more for tuning margins.

The main desired features of a positron source can be characterized by the following five points: It should produce many positrons (1) with a low emittance (2) and also polarization (3) would be preferable. The target should stand the average and peak power (4) with the high mechanical stresses, while a capture region should provide very high guiding magnetic fields (5).

Other ideas, like using a helical undulator for VLEPP [3] or back-scattered Compton photons from a laser [4] address mainly the issue of polarization. Here the possibility of channeling crystals (see Fig. 1) as a positron source is investigated, (compare also [5], where a crystal is considered only for producing photons for a normal target).

Compared with a normal amorphous target, a crystal behaves differently: More positrons (1, compare numbers with the five desired points from above) are produced in a shorter radiation length (2, less scattering provides smaller emittance) and polarization (3, probably linear [6]) has been observed. The energy transfer of a particle to the crystal is less since it reacts with a whole string of around 1000 atoms. So the target will receive less power (4), while

---

*Work supported by the Department of Energy contract DE-AC03-76SF00515.

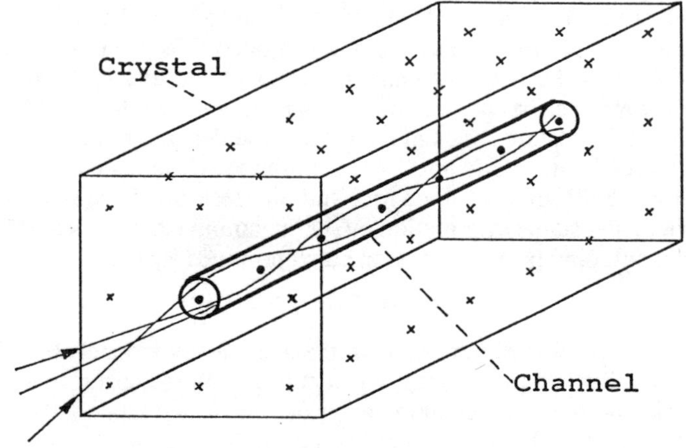

Figure 1: Principle of Channeling Crystal.

If a particle beam hits a crystal under a small angle to one of its axes, a particle reacts with many atoms collectively and can therefore be trapped in a small region around the axis (or plane), called channel. The large number of participating atoms (around 1000) generate enormous macroscopic fields of the order of $E = 10\,\mathrm{TV/m}$ (Tera-Volt) or $B = 30\,\mathrm{kT}$ (kilo-Tesla).

on the other hand a crystal (but out of tungsten) might be more sensitive. The enormous effective fields which guide the particles in channels are more than three orders of magnitude higher than the normally used fields (5) in accelerator physics. The main characteristics are summarized in the next section and afterwards the detailed target considerations are discussed.

2   Channeling Characteristics

The phenomenon of the channeling of particles in crystals is a wide field of physics; an overview and more introductions are given in [7] and references therein. The different behaviors with the energy of the particles, the characteristic angles of the beam to the crystal, the different kinds of particles (e$^+$, e$^-$, γ) and corresponding theoretical approaches are described.

2.1   Different Particle Energies

At low energies only a few transverse energy levels are quantum mechanically allowed in the potential valley of height U_0 between crystal axes. At higher energies the particles get heavier and more levels are possible, till it goes to a quasi-classical continuum.

	Particle Energy	γ-Emission
a:	low $\approx 10\,\mathrm{keV}$	Characteristic line spectrum
b:	med. $\approx 10\,\mathrm{MeV}$	” (more lines)
c:	high $\approx 10\,\mathrm{GeV}$	Classical synchrotron radiation

2.2 Important Angles at High Energies

Different characteristic angles influence the behavior of the beam at 25 GeV (compare Fig. 2):

1. Characteristic angle: $\theta_0 = \dfrac{U_0}{m_0 c^2} \approx 0.5\text{–}1.0\,\mathrm{mrad}$.

2. Lindhard angle (maximum angle being kept in channel): $\psi_L = \sqrt{\dfrac{2U_0}{\gamma m_0 c^2}} \approx 100\text{–}200\,\mu\mathrm{rad}$.

3. Gamma or pair production angle: $1/\gamma \approx 20\,\mu\mathrm{rad}$.

4. Beam divergence: $\sigma' = \sqrt{\varepsilon/\beta} \approx 5\,\mu\mathrm{rad}$, (at $\gamma\varepsilon = 5\cdot 10^{-5}\,\mathrm{m\,rad}$, $\beta = 30\,\mathrm{m}$).

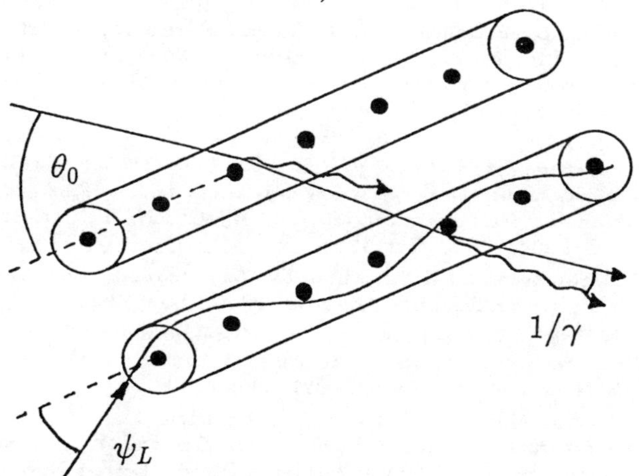

Figure 2: Angles in a Channeling Crystal.

Depending on energy different angles characterize the beam behavior in a crystal.

2.3 e^+, e^- or γ–Beams with $\psi \leq \psi_L$

In amorphous material the three kinds of particles in a particle shower behave similarly. In aligned crystals however there are big differences: Electrons are channeled near the atoms causing lots of radiation, positrons are between atoms and photons are not channeled, but create lots of pairs below an angle of θ_0.

2.4 Theoretical Approaches

Depending on the energies, angles and kinds of particles different theories have been developed [8, 9]. The behavior of a beam hitting a crystal at large angles to an axis or plane corresponds to an amorphous target and the normal Bremsstrahlung (BS) or for photons a pair production (PP) with a normal Bethe-Heitler rate (BH) occurs.

At smaller angles (around θ_0) the interference between the crystal axes or planes enhances the γ-production by coherent Bremsstrahlung (CBS) or CPP for photons. At even smaller angles below ψ_L (see section 2.2) channeling radiation (CR) occurs. It is different for electrons and positrons since electrons are trapped near the atoms while positrons are in the space between atoms. Although photons are not channeled, the corresponding channeling pair production (ChPP) [10] produce channeled pairs since $1/\gamma < \psi_L$.

For different crystal material the pair production enhancement over BH is twice at 120 GeV for Si, 50 GeV for Ge and 15 GeV for W [9], so it is only important at very high energies. Since the particles of high energies traverses along a large number of atoms, the fields can be averaged, leading to a constant field approximation (CFA). Photons create a pair in the field of the string rather than a single atom. The table below gives a rough comparison of the different theories with electrons in accelerator fields. A more detailed scenario of an electromagnetic shower is given in the next section.

Pairs	Photons	Accelerator
PP (BH)	BS	Bremsstrahlung
CPP	CBS	wiggler
ChPP	CR	undulator
CFA	CFA	synchrotron light

3 Target Consideration

A positron production target should produce many positron in a small angle. The different channeling consequences are briefly summarized. As material a tungsten crystal in <111> axis orientation is considered. At high energies an enhancement for pair production in W is at 15 GeV twice BH and at 30 GeV about 3.2 times BH. The shorter radiation length (normal is 3.5 mm) of 0.9 mm at 28 GeV leads to a shorter target length with less scattering (lower emittance) and less losses and therefore a higher yield $= N(e^+)/N(e^-)$. Channeled positrons scatter less than electrons resulting in more yield and a smaller angle. Polarization has been observed.

Single tungsten crystals can be made up to 10 cm long by 5 mm diameter with a small mosaic spread of about 120 μrad [11]. It is assumed that a similar temperature and mechanical behavior of the amorphous tungsten metal will be found with single tungsten crystals.

3.1 Detailed Shower Evolution

The following scenario of the e^+, e^-, γ-shower of a 30 GeV e^--beam in an oriented <111> tungsten crystal is expected: At the higher e^--energies (30 GeV) about three times as much photons have been measured after 1 mm length [14]. At small angles ($< \psi_L$) the higher energy photons (> 15 GeV) may contribute to an enhanced pair production by ChPP (hard shower [9]). At lower energies there is still CBS and more photons N_γ than charged particles N_e are present, $N_\gamma/N_e \approx 11$ [8]. These photons lead via BH to a normal pair production (only γ-production is enhanced = soft shower).

At lower energies the opening angle becomes as big as the Lindhardt angle and θ_0 (compare section 2.2):

$$2/\gamma = \psi_L = \sqrt{\frac{2U_0}{\gamma m_0 c^2}} = \frac{U_0}{m_0 c^2} = \theta_0 = 0.8 \,\mathrm{mrad} \ (\mathrm{for\,W}).$$

This energy of about $1.3\,\mathrm{GeV}$ occurs after about three effective radiation lengths. At this point in the shower the enhancement of the crystal decreases and a second, normal target could be considered from then on [5].

But also another effect of the crystal can be used: Low energy channeled positrons scatter about ten times less with the crystal [15] and will leave it with a small emittance (see below). Additionally there are transmissions from free states in transverse motion to bound states [15] increasing the number of channeled positrons. They oscillate like in a helical undulator and lower their transverse energy states via characteristic channeling radiation. This damps their effective transverse emittance. It is interesting to look at the normalized emittance $\gamma\varepsilon$ of the channeled particles. With $\varepsilon = \sigma^2/\beta = \beta\sigma'^2$ and assuming a low betatron function of $\beta = 0.1\,\mathrm{m}$ and $\sigma' = \psi_L$, it follows

$$\gamma\varepsilon = \gamma\beta\frac{2U_0}{\gamma m_0 c^2} = 2\beta\theta_0 = 16 \cdot 10^{-5}\,\mathrm{mrad}.$$

The angular part is independent of energy, while the spatial part

$$\gamma\varepsilon = \gamma\frac{\sigma^2}{\beta} = 16 \cdot 10^{-5}\,\mathrm{mrad} \quad (\mathrm{for}\ \gamma = 16,\ \sigma = 1\,\mathrm{mm})$$

decreases with energy, if the particles could be kept in the same channel. At very low energies, where $1/\gamma > \psi_L$, only a part of the positrons will be channeled immediately.

3.2 Beam Power and Polarization

Beam Power on Crystal The critical topic is the damage of the crystal by heat or stress and any lattice defects by radiation, which needs futher experimental investigations. Fluxes of up to 10^{18} particles causes no noticeable radiation damage [12]. Since one particle reacts with a string of many atoms less power is transferred to the target atoms. The vibration of single atoms round the string causes dechanneling of the beam particles. If the target is cooled e.g. by liquid nitrogen even less energy is transferred and more pairs are produced [16]. The observed Cu-crystals near the SLC target may lead to the hypothesis that an oriented crystal might be the most stable configuration of atoms in an environment of directed radiation.

Polarization Different hints and discussion of polarization have been reported [6, 16, 17]. A planar channeled particle will radiate linear polarized photons. The desired circular polarization would be achieved, if the particles could be trapped around an axis in an oriented helical motion. Helix-like crystals or crystals deformation of about $1\,\mu m$ wavelength might support such a motion. Or a normal crystal is oriented in such a way that the beam direction, the crystal axis and one crystal plane build a left-right asymmetry leading to different amounts of polarized photons.

4 Outlook and Conclusion

The enormous potentials in and of single crystals have been already and will be used more frequently in accelerator physics. With bending crystals SSC-energies could be kept in a PEP-like tunnel [18] and a TeV linear collider could be built within $10\,\mathrm{m}$ [13]. By starting small projects like a positron target, the confidence and experience in this quite new technology will grow. The feasibility and advantages of single crystals as a positron source have been discussed.

References

[1] Stan Ecklund, *The Stanford Linear Collider Positron Target*, SLAC-Pub-4437 Oct. 1987 or Workshop of Intense Positron Beams, Idaho, June 1987.

[2] A.V. Kulikov, S.D. Ecklund, E.M. Reuter, *SLC Positron Flux Concentrator*, PAC San Francisco, May 1991.

[3] T.A. Vsevolozhskaya et al., *Helical Undulator for Conversion System of the VLEPP Project*, Preprint 86-129-INP, Novossibirsk.

[4] J.E. Spencer, *High Brightness Sources for Colliders*, PAC, San Francisco, May 1991.

[5] R. Chebab, F. Couchot, A.R. Nyaieh, F. Richard, X. Artru *Study of a Positron Source Generated by Photons from Ultrarelativistic Channeled Particles*, LAL-RT 89-01 or PAC, Chicago, March 1989.

[6] R. Medenwaldt, S.P. Møller, A.H. Sørensen et al., *Investigations of the coherent hard photon yields from (50–300) GeV/c electrons/positrons in the strong crystalline fields of diamond, silicon and germanium crystals (and W)*, CERN/SPSC 90-31, SPSC/P234 Add. 3, Oct. 1990.

[7] A.H. Sørensen and E. Uggerhøj, *The Channeling of Electrons and Positrons*, Scientific American, 96pp., June 1989.

[8] V.N. Baier, V.M. Katkov and V.M. Strakhovenko, *Cascade Processes in the Fields of the Axes of Aligned Single Crystals*, Nucl. Instrum. Methods B27, 360pp., 1987.

[9] R. Medenwaldt, S.P. Møller, S. Tang-Petersen et al., *Detailed Investigations of Shower Formation in Ge- and W-Crystals traversed by 40–287 GeV/c Electrons*, Physics Letters B, Vol. 227, number 3,4 Aug. 1989.

[10] A.H. Sørensen, E. Uggerhøf, J. Bak and S.P. Møller, *New Analysis of the Two Competing Processes: Channeling- and Coherent Pair Production and a Comparison with Experimental Data*, CERN-EP/84-149, Nov. 1984.

[11] Edited by R.A. Carrigan, Jr. and J.A. Ellison, *Relativistic Channeling*, Workshop Maratea, NATO ASI Series: B, Physics, Vol. 165, Plenum Press, 1987, A. Seeger, *Growing Large Highly Perfect Single Crystals and its Limitations*, 423pp.

[12] Same book, S.I. Baker, *Radiation Damage effects in Channeling Applications*, 391pp.

[13] Same book, P. Chen, *Channeled Particle Acceleration by Plasma Waves in Metals*, 517pp.

[14] V.N. Baier, V.A. Baskov, V.B. Tanenko, *Radiation by 28-GeV Electrons in a Thick Tungsten Crystal*, JETP Lett., Vol. 49, No. 10, May 1989.

[15] A.H. Sørensen and E. Uggerhøf, *Channeling and Channeling Radiation*, Nature, Vol. 325 22 Jan. 1987.

[16] V.N. Baier, V.M. Katkov and V.M. Strakhovenko, *Theory of Pair Creation in Aligned Single Crystals*, Nucl. Instrum. Methods B16, 5-21, 1986.

[17] U. Timm, *Coherent Bremsstrahlung of Electrons in Crystals*, Fortschritte der Physik **17**, 765-808, 1969.

[18] *Bending Beams by Crystals*, CERN Courier, 5pp. May 1990.

SLC POSITRON SOURCE PULSED FLUX CONCENTRATOR

A.V. Kulikov, S.D. Ecklund, and E.M. Reuter[a]

Stanford Linear Accelerator Center, Stanford University, Stanford, CA 94309*

Abstract

SLC positron beams produced by a high energy electron beam, impinging on a high Z target, have initially small transverse size but large divergence, a situation ill matched to the following S-band accelerator. The flux concentrator is an adiabatic matching device placed between the target and this accelerator, which trades divergence versus size. It produces a magnetic field with a sharp rise over less than 5 mm to its peak value, and then falling off adiabatically over 10 cm. It is a 12 turn, 10 cm long copper coil with a cylindrical outside radius of 4 cm and a conical inside radius growing from 3.5 mm to 2.6 cm. The 0.2 mm gaps between the individual windings were manufactured by electric discharge machining out of one copper block. Excitation current and water cooling is provided by a hollow rectangular copper conductor brazed to the outside of the coil (also 12 turns). The pulsed magnetic field has a maximum strength of 58 kG at 16 kA. At the terminals, the coil has an inductance of 0.8 µH. Current shape is a half sinusoidal wave with a bottom width of 5 µs, and the system operates at a repetition rate of 120 Hz. The coil has only one supporting ceramic insulator at the low voltage front end. The flux concentrator has improved the positron yield approximately 2 times and had no failure in operation during several years.

I. INTRODUCTION

The design and operation of the SLC positron source has been described.[1] Recent improvements include the installation of a moving target.[2] The overall performance of the source is discussed in reference.[3]

Integral to the design of the source is the use of a large bandwidth phase-space transformer as in SLAC's original positron source.[4] Such a device utilizes a longitudinal magnetic field whose strength decreases slowly with longitudinal distance from the target.[5] It can be shown that this transforms the phase-space radii by the square root of the magnetic field ratio, *i.e.*, Ba^2 is a constant proportional to beam emittance, where B is the magnetic field and a the beam radius. The final field value, away from the target, must be large enough to keep the beam inside the accelerator aperture. The initial value then needs to be high to achieve a useful transformation of the phase-space aspect ratio.

II. DESIGN PRINCIPLES

The specifications for the SLC positron system require a high positron yield[6], achievable only if the peak longitudinal field at the target is at least 7 T. Since such a high field is not practical in the high radiation environment of the target with a simple tapered field solenoid (TFS), a 5.8 T (peak) field produced by a pulsed flux concentrator (FC) has been added to a 1.2 T TFS DC field as shown in Fig. 1. Beginning with an existing concept for a pulsed device[7], a vacuum insulated, physically robust, mechanically simple, highly reliable, pulsed FC has been designed and will be described below.

The field concentration is achieved by a pulsed coil having an internal conical shape. A sharp rise of the magnetic field to its peak value over less than 5 mm distance is achieved by limiting the smallest radius of the cone to 3.5 mm and locating it 2.5 mm from the moving target exit-surface. The strength of the longitudinal field along the axis of the cone is to first approximation simply inversely proportional to the cross-sectional area as indicated above. The dimensions of the cone are: minimum radius, 3.5 mm; maximum radius, 2.6 cm; length of cone along its axis, 10 cm. Twelve turns were chosen for the coil as a compromise between efficiency[8] and the maximum required current. More turns would require less peak current, but because of increased flux leakage would decrease the efficiency, requiring more power from the modulator. It should be noted that the flux leakage determines the exact shape of the longitudinal field. The loss of efficiency for a give number of turns can be limited by reducing the width of the gap between turns. Since 2 skin depths are present at each gap, the practical limit for the gap is about 1 skin depth. In this case, the skin depth is 0.2 mm.

III. FABRICATION TECHNIQUE

The FC is machined from a single block of OFHC Cu with internal cone and external groves for a rectangular water-cooled conductor as shown in Fig. 2 (a). The conductor is next brazed into the groves, after which a spiral gap of 0.2 mm is cut through by electric discharge machining (EDM) to form the 12-turn coil as shown in Fig. 2(b). No insulator between the turns is necessary--a significant achievement in this design. However, if there is a discharge between 2 turns, the resulting force from adjacent turns will push the 2 shorted turns together. This is a problem only if the Cu is annealed, such as is common when cleaning for vacuum. Thus, to insure the coil will spring back to its original position following a discharge, it must be work-hardened. Work-hardening is easily accomplished by mechanically collapsing then reopening the gaps a number of times.

The coils require only a single supporting insulator, chosen here to be at the low voltage end of the coils and also to be a vacuum-clean ceramic. The mechanical stability of the coils is ensured by choosing a short pulse width and a sufficiently large mass for the coils. On the other hand, to avoid excessive voltage requirements, the pulse width should not be too small. In this case, a half-sinusoidal wave with bottom width of 5 µs was chosen. (The positron pulse is <<1 ns long.)

Because this is a mechanically stiff structure with a complex shape, many mechanical resonances can be excited during operation. For this design, the highest frequence resonance observed was 49 Hz. Thus to avoid exciting any resonance, the coil repetition rate is restricted to be ≥60 Hz.

The assembly of the FC in the source vacuum system is shown in Fig. 3. The supporting insulator attaches the FC to the front end magnet yoke. The current and water cooling leads are carried through the vacuum wall with no water joints exposed

* Work supported by Department of Energy, contract
DE-AC03-76SF00515

[a] Present address: Siemens Medical Laboratory, Inc.,
Concord, CA 94520

to the vacuum system. The modulator design is discussed elsewhere.[9]

IV. PERFORMANCE

ETRANS[10], a computer simulation code that traces particles in electric and solenoidal magnetic fields, has been used with the measured magnetic fields to compute the effect of the FC on the positron phase space. The results for the transverse phase space are shown in Fig. 4(a) and (b). The longitudinal phase space of the positrons exiting the FC are shown in Fig. 4(c). Note that particles with <1 MeV/c are reflected by the rising magnetic field. Particles with small energy and relatively large transverse momentum are delayed because of the spiral trajectories in the strong field of the FC.

The performance of the FC can be judged by the effect on the yield. The first positron intensity monitor is after capture and acceleration to about 120 MeV. The yield here is improved by a factor of 2 as shown in Fig. 5.

The FC has been in operation in the SLC positron source for 3 years without a failure.

V. REFERENCES

[1] J. E. Clendenin, *Proc. of the 1989 Particle Accelerator Conf.*, Chicago (March 20-23, 1989) 1107; F. Bulos, *et al.*, *IEEE Trans. on Nucl. Sci.*, **NS-32** (1985) 1832.

[2] E. Reuter, *et al.*, "Mechanical Design and Development of a High Power Target System for the SLC Positron Source," this conference.

[3] R. Pitthan, *et al.*, "SLC Positron Source--Simulation and Performance," this conference.

[4] H. Brechna, *et al.*, in R. B. Neal, ed., *The Stanford Two-Mile Accelerator*, Benjamin (1968), section 16.

[5] R. H. Helm, SLAC-4 (1962).

[6] The positron yield at a give location is defined as the ratio of the number of positrons at that location to the number of electrons impinging on the conversion target.

[7] M. N. Wilson and K. D. Srivastava, *RSI* **36** (1965) 1096.

[8] The efficiency is defined as the ratio of the stored energy in the useful magnetic field to the total stored energy.

[9] A. Kulikov, *et al.*, "SLC Positron Source Flux Concentrator Modulator," this conference.

[10] The program ETRANS was written by H. Lynch

(SLAC).

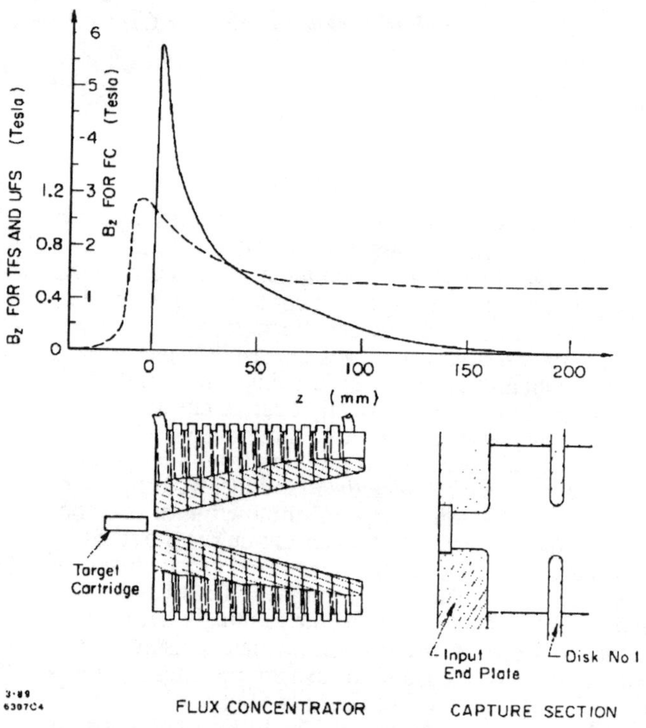

Figure 1. The positron source adiabatic system. The devices shown in cross section at bottom are to scale. The computed solenoidal fields and measured FC pulsed field are shown above with the same z-scale as for the devices.

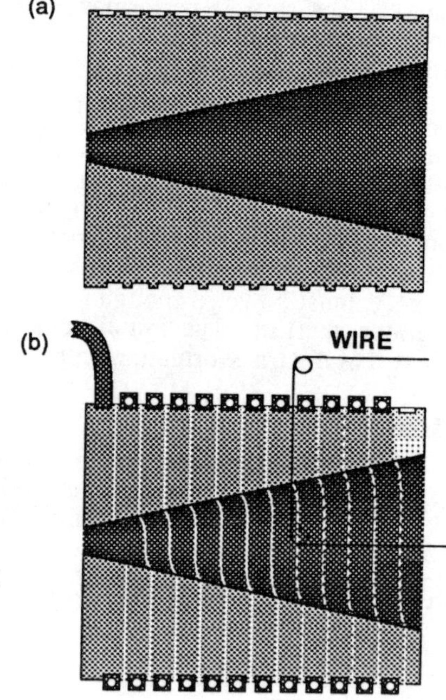

Figure 2. FC body cross section: (a) Showing internal cone and grooves fro rectangular conductor; (b) Showing EDM

wire cuts after conductor is brazed into grooves.

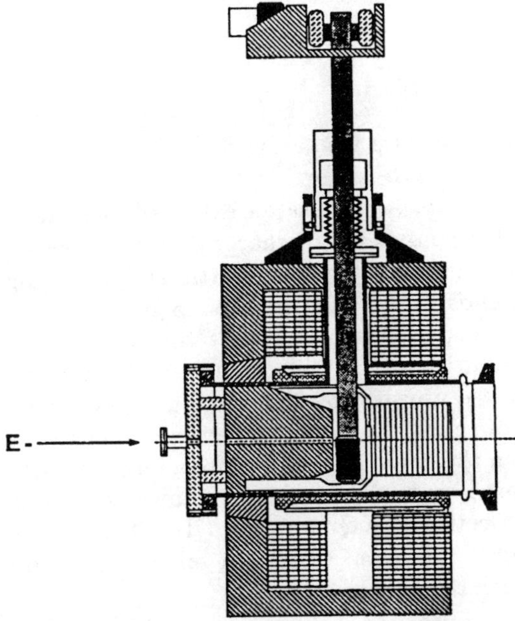

Figure 3. Complete target module assembly showing trolling target driven from above, TFS surrounding both the target (solid black) and FC, and the FC itself supported by the TFS yoke (hashed) on the upstream end.

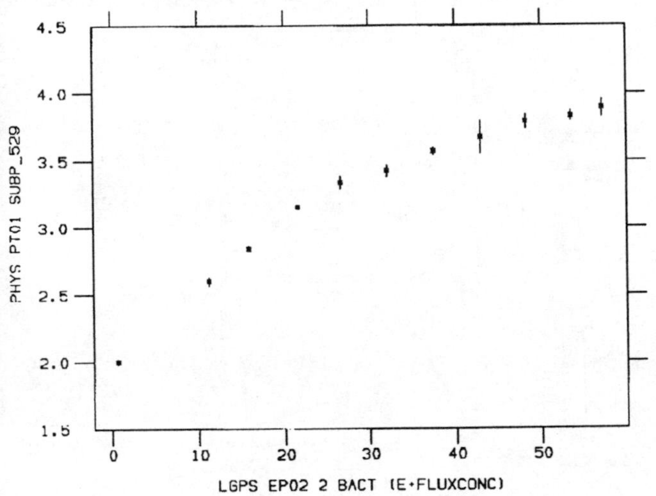

Figure 5. Positron yield at first intensity monitor (at 120 MeV location) as function of peak FC field in kG.

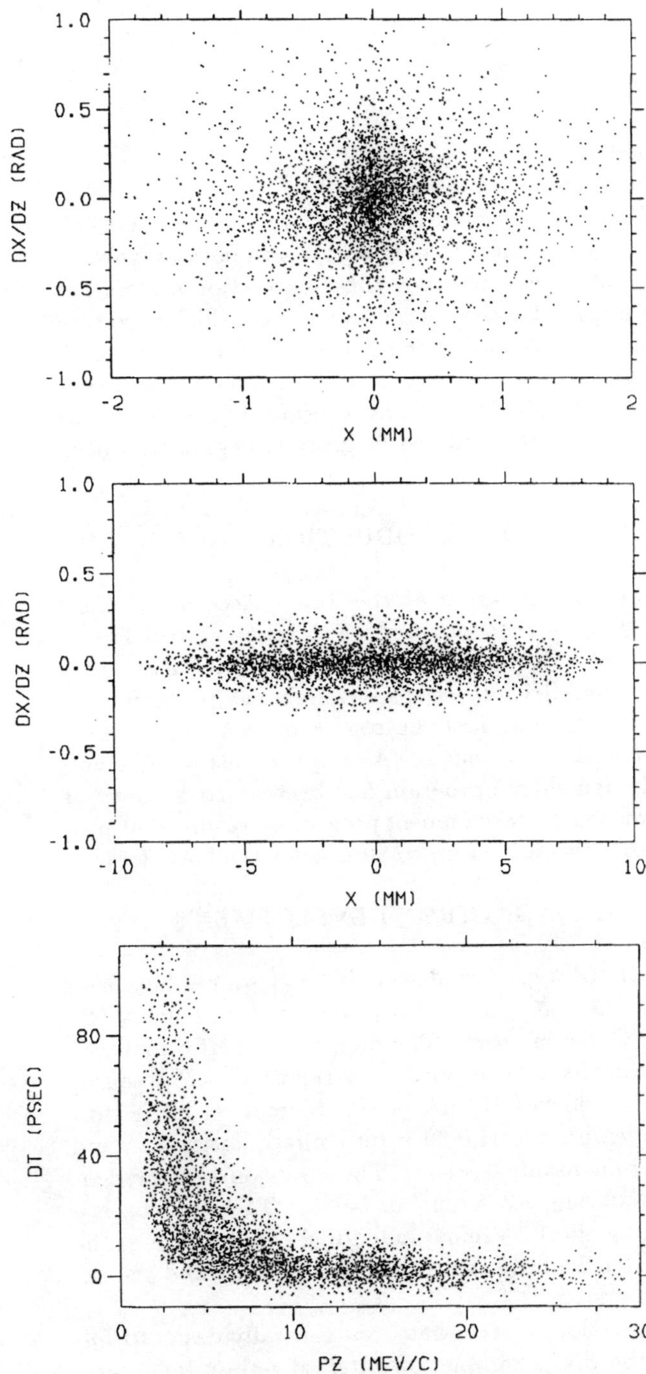

Figure 4. Phase space transformations. (a) EGS simulated phase space at target exit; (b) Transverse phase space at exit of FC (z = 140 mm) as simulated by ETRANS; (c) Longitudinal phase space at exit of FC, also ETRANS.

Progress in H⁻ Ion Source Development at TAC*

J. Culver, K. Antes, F.R. Huson, A. Larsson, C.R. Meitzler, and L. Xiu
Texas Accelerator Center†
4802 Research Forest Dr.
The Woodlands, TX 77381

Abstract

A program has been started at the Texas Accelerator Center to develop H⁻ ion sources for its 500 keV RFQ program. The group is working on both magnetron and volume ion sources. A tilted extraction system has been designed for the magnetron source which corrected a 25 mrad angular offset in the extracted beam. An LBL type volume H⁻ ion source has been designed and constructed. To date, the plasma generator of the source has been tested, with extracted beam tests beginning during Summer 1991.

I. INTRODUCTION

The linac program at the Texas Accelerator Center (TAC) has been centered on developing an RFQ as a preinjector for the SSC. To support the RFQ program, a parallel ion source development program had been started. The first ion source built at TAC was a magnetron H⁻ ion source. Within the past year, a volume H⁻ ion source program has been started. Both of these ion source development programs are aimed at providing H⁻ currents of up to 50 mA for the TAC RFQ.

II. MAGNETRON DEVELOPMENT

A BNL[1] type magnetron H⁻ ion source has been built at TAC as a part of a project to construct a 500 keV RFQ accelerator[2]. The design and performance of this source have been previously reported[3]. The source regularly achieves 10 mA of H⁻ current within a normalized emittance of 0.29 π mm–mrad, horizontal, and 0.36 π mm–mrad, vertical. The emission current density is routinely 1.3 A/cm² or better. There has been a continuing effort to refine individual components of the system, particularly the cesium transfer line and the extraction system.

The cesium system has been generally designed following the BNL example: an external cesium boiler and vacuum insulated transfer line entering the back of the source body. Several modifications are currently being developed for the system. First, the cesium boiler is connected to the valve by means of an EVAC[4] CF fitting instead of the Mini-conflat which is typically used. These fittings use a clamping chain and conflat type flange to make the seal, and will simplify the process of making a connection inside a glove bag. The second development is to remove the axial heater from the transfer line and place it in direct thermal contact with the outside of the

transfer line. A vacuum insulating jacket surrounds the transfer line back to the elbow.

Magnetron ion sources use an externally generated magnetic field to confine the discharge plasma to the gap between the anode and cathode. Unfortunately, the magnetic field extends far beyond the plasma generator into the extraction gap. This field causes the beam to deflect upwards during extraction. A 30 keV beam of H⁻ ions was deflected upwards by 25 mrad during extraction across a 4.5 mm gap[3]. Usually this deflection causes few problems when the beam is injected into a large aperture beam transport line; however, when it is injected into the TAC Helical Electrostatic Quadrupole (HESQ) lens[5], the deflection causes the beam to impact on the electrodes along the length of the lens.

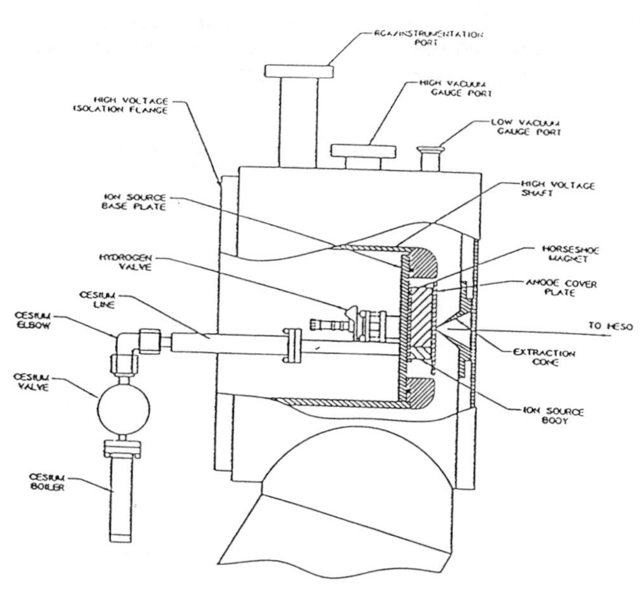

Figure 1. Original TAC magnetron ion source geometry from Ref 6.

A new extraction system was designed and constructed which corrected the trajectory of the ion beam. Figure 1 shows the old extraction geometry. In the modified system, the high voltage isolation flange, which holds the re-entrant high voltage shaft and source mounting flange, and the extraction cone were tilted upwards by

mrad while still maintaining the same extraction gap
geometry. The emittance of the beam was remeasured
after installation. It showed that the beam was now traveling in a direction parallel to the center of the center of
the beam line.

III. VOLUME SOURCE DEVELOPMENT

Volume ion sources are becoming increasingly wide
spread as H^- ion sources for accelerators. The lower
ion temperature can result in significantly smaller emittances than those obtained from magnetron ion sources,
making them attractive for low emittance accelerator systems. Within the past year the ion source group at TAC
has initiated a two phase volume ion source development
program aimed towards developing a volume H^- source
for the TAC 500 keV RFQ. Phase 1 is to build a LBL
type multicusp volume H^- ion source[7] to gain basic experience with volume sources. Phase 2, starting during
Summer 1991, will be to develop a compact toroidal volume source[8] to supply 10 mA of H^- to the 500 keV RFQ.
The toroidal geometry was chosen for the final source
geometry since it is capable of producing an electron-to-
H^- density on the order of $2-5$[9], thereby simplifying a
pulsed extractor system by reducing the capacitance and
switching currents needed to drive the electrodes.

The Phase 1 TAC volume ion source is a small LBL
type source. The entire source is constructed modularly
to provide an easy means of testing differentc source configurations. The plasma generator is a copper cylinder
with an inside diameter of 7.6 cm and a total length of
11.5 cm. Twelve 7.6 cm long longitudinal rows of Sm-Co
magnets form the plasma confinement field. The magnetic field at the surface of the copper has 1.0 kG. The
back flange of the ion source is magnetically insulated
with line cusps. A tungsten filament and PFN are used
to drive the discharge. Two filter magnets are used to
provide the magnetic filter. The separation between the
magnets is 3.8 cm resulting in a magnetic field of 300 G
in the center, and a surface field of 1 kG. Water cooling is
used to protect the filter and longitudinal cusp magnets.

Initial testing of the plasma generator is underway
at this time. Figure 2 shows the I–V characteristic of the
measured at a pressure of 27 microns. The discharge
power was varied from 0.7–29 kW. H^- yield was not
determined. The filament was heated by a DC current
which was not interupted during the plasma pulse. The
characteristic curve appears to quite linear over the entire range. The pressure was varied on either side of 27
microns, and we had observed a substantial decrease in
arc current at all voltages when compared to the 27 micron data.

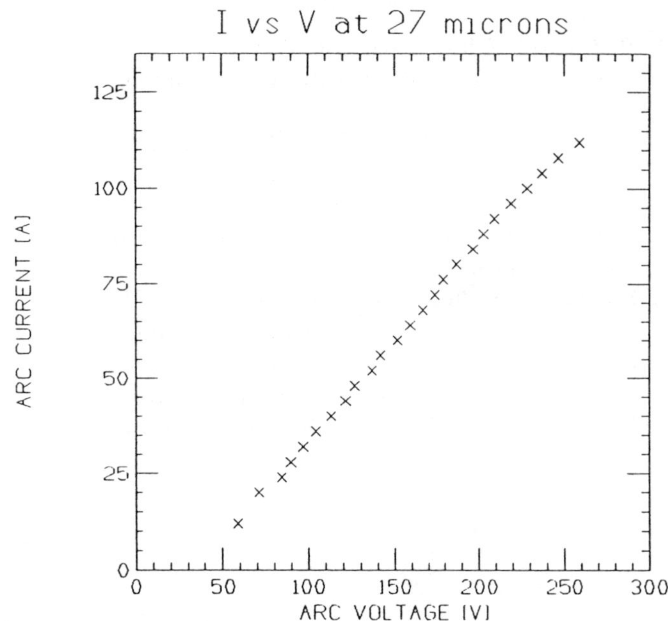

Figure 2. Characteristic curve of the plasma at a filling
pressure of 27 microns.

IV. SUMMARY

The ion source development program at TAC has
been actively involved in attempting to develop several
ion sources. The magnetron program has reached its design goals and is undergoing engineering refinement. The
volume source development at TAC is still in its beginning stages and is devoted to pursuing the toroidal volume source technology first developed at Brookhaven.

V. ACKNOWLEDGEMENTS

The authors would like to thank Larry Crowe and
John Colvin for their efforts as technicians on this
project. We would also like to thank James Burnett for
his help manufacturing the source components.

VI. REFERENCES

* This work supported by the Houston Advanced Research Center, Sam Houston State University, and
the SSC Laboratory.

† The Texas Accelerator Center at the Houston Ad-
vanced Research Center is a consortium of Rice Uni-
versity, Texas A&M University, the University of
Houston, the University of Texas, Prairie View A&M
University, Sam Houston State University, and the
Baylor College of Medicine MR Center.

1. J.G. Alessi, Proc. of the Fourth Int. Symp. on the
Production and Neutralization of Negative Ions and
Beams, AIP Conf. Proc. 158 (1986), 419.

2. R. Kazimi, F.R. Huson, and W.W. MacKay, "Progress of the 473 MHz Four-rod Type RFQ," these Proceedings.

3. C.R. Meitzler, P. Datte, F.R. Huson, and P. Tompkins, "Progress on the TAC ion source and LEBT," Proc. of the 1990 Linear Accelerator Conference, Los Alamos National Laboratory LA-12004-C Confernce (1990), p. 710.

4. EVAC, CH-9470 Buchs, Switzerland.

5. C.R. Meitzler, K. Antes, P. Datte, F.R. Huson, and L. Xiu, "Transport Properties of a Discrete Helical Electrostatic Quadrupole," these Proceedings.

6. P.A. Tompkins, "The TAC H^- Ion Source and Low Energy Beam Transport," Texas A&M University, College Station, Texas, Ph.D. Dissertation, 1990.

7. S.R. Walther, K.N. Leung, and W.B. Kunkel, "H^- production in a small multicusp ion source with the addition of barium," Appl. Phys. Lett. $\underline{54}$ (1989) 210.

8. K.Prelec and J.G. Alessi, "Performance of a Volume H^- Ion Source with a Toroidal Discharge Chamber," Rev. Sci. Instrum. $\underline{61}$ (1990) 415.

9. J.G. Alessi and K. Prelec, "Status of the BNL Toroidal Volume H^- Source," Proceedings of the 1990 Linear Accelerator Conference, Los Alamos National Laboratory, LA-12004-C, (1990), 671

Conference Author Index